Quantum Mechanics

Quantum Mechanics

P. K. Ghosh

Alpha Science International Ltd.
Oxford, U.K.

Quantum Mechanics

908 pgs. | 72 figs. | 5 tbls.

P. K. Ghosh *(Retd.)*
Department of Physics
Visva Bharati
Santiniketan

Copyright © 2014

ALPHA SCIENCE INTERNATIONAL LTD.
7200 The Quorum, Oxford Business Park North
Garsington Road, Oxford OX4 2JZ, U.K.

www.alphasci.com

ISBN 978-1-84265-842-0

Printed in India

To
Swapna, Sudeshna, Rishi and Niharika

Preface

Quantum mechanics plays the central role in our current understanding of essentially all physical phenomena. The present volume is a textbook of quantum mechanics designed for advanced undergraduate and graduate students of physics, mathematics and chemistry. It provides a pedagogical introduction to the formal apparatus of quantum mechanics, application to various physical problems, and the recent developments in the interpretation of quantum mechanics. Although this book contains considerably more material than can be accomodated in a standard two-semester course, it is hoped that the book will be suitable for a standard course of nonrelativistic quantum mechanics, omitting several sections in accord with the instructor's judgment and preference.

After an introductory Chapter on the mathematical concepts—linear operators, Hilbert space, generalized functions—we develop in Chapter 2 the formal framework of quantum mechanics. Here we introduce the basic postulates, different routes for quantization, uncertainty relations and complimentarity. Chapter 3 is devoted to wave mechanics. The Schrödinger equation and its solution in simple problems are discussed. Chapter 4 to 8 consider the theory of angular momentum and spin. Detailed discussions of rotation matrices, CG coefficients, 3j and 6j symbols and tensor operators are provided. Symmetry and invariance principles are considered in Chapters 9 and 10.

Discrete symmetry operations are studied in Chapter 10. Chapter 11 deals with the application of quantum mechanics to the standard material, e.g., the harmonic oscillator, the hydrogen, atom, Landau Levels, etc. This chapter contains a brief section on supersymmetric quantum mechanics. Approximation methods are often the only feasible tools to tackle interesting physical problems. These methods like the stationary state perturbation theories and nonperturbative methods are considered with illustrative examples in Chapters 12 and 13. In Chapter 14, we discuss the approximation methods for dealing with time-dependent phenomena. Chapter 15 is concerned with the interaction of radiation with matter. The symmetrization postulate is introduced in Chapter 16 and the permutation group is treated in detail. In Chapter 17, we study the occupation number representation and apply it to many-particle systems of bosons and fermions. Chapters 18 to 21 provide a detailed exposition of quantum scattering theory. The concluding Chapter addresses some of the conceptual issues of quantum theory and outlines the recent developments in the interpretation of quantum mechanics. The present volume contains discussion of several important topics. Some of these are:

   (i) Rigged Hilbert Space,

  (ii) Generalized functions,

 (iii) Path integrals,

 (iv) Deformation quantization,

  (v) Coherent states,

 (vi) Geometric phase,

(vii) Supersymmetric quantum mechanics,

(viii) The Aharonov-Bohm effect,

 (ix) Large orders of perturbation theory,

  (x) The Hanbury Brown and Twiss effect,

 (xi) Regge poles,

(xii) Quantum entanglement,

(xiii) Bell inequalities,

(xiv) Decoherence.

This work draws on many books and numerous journal articles. These sources are indicated in the bibliography. Certain books and articles are included in the bibliography for the benefit of the reader who requires detailed information about a specific topic/or a subject of interest. The author has relied mainly on the books of Ballentine, Baym, Bethe and Salpeter, Condon and Shortley, Dirac, Esposito, Marmo and Sudarshan, Galindo and Pascual, Gottfried, Gottfried and Yan, Holstein, Jordan, Landau and Lifshitz, Messiah, Pauli and Sakurai. The quotation at the top of Chapter 22 is taken from the book "I Won't Let You Go". Selected Poems of Rabindranath Tagore, Translated by K.K. Dyson (UBS Pub. & Dist. Ltd., New Delhi, 1992).

I would like to thank my teacher, Prof. R.R. Lewis, who not only passed his notes on quantum mechanics to the author, but also gave invaluable advice during the course of this work. The author has used his own notes on quantum mechanics taken during a course delivered by Prof. K.M. Case. In writing some topics, e.g., angular momentum, permutation symmetry, Regge poles, unpublished Lecture notes of Prof. K.T. Hecht and Prof. G.L. Kane have been used.

I would like to thank Prof. R. Dutt for reading some Chapters and suggesting improvements. Late Prof. H. P. Roy had been most helpful. I would like to express my gratitude to Prof. S.N. Ghosh for his constant support and encouragement. I am deeply indebted to Dr. A. Chakraborty for providing constant advice and support.

P. K. Ghosh

Contents

CHAPTER 1

Mathematical Preliminaries

The mathematical structure of quantum mechanics is built on the concepts of linear spaces, linear operators etc. Certain mathematical notions and results of functional analysis are essential for the development of the mathematical language of quantum mechanics. In this chapter we collect some of the basic concepts and results that are needed for our study of quantum mechanics. No attempt is made at mathematical rigor, generality or completeness. We omit the proofs of the theorems. Our study of quantum mechanics begins in Chapter 2.

This chapter is based on the works of Jordan (1969), Gel'fand and Shilov (1964), Gel'fand and Vilenkin (1964), Reed and Simon (1980), A. Böhm (1978), Gieres (2000), Esposito, Marmo and Sudarshan (2004) and de la Madrid (2005). For extensive treatments, the reader is referred to these works.

1.1 LINEAR SPACES AND LINEAR FUNCTIONALS

Definition A linear, or vector, space V is a set of elements, called vectors, with an operation of addition, which for each pair of vectors ψ and ϕ specifies a vector $\psi + \phi$, and an operation of scalar multiplication, which for each vector ψ and a number a specifies a vector $a\,\psi$ such that

   (i) $\psi + \phi = \phi + \psi$,

   (ii) $\psi + (\phi + \chi) = (\psi + \phi) + \chi$,

   (iii) a unique zero (or, null) vector 0 has the property that $\psi + 0 = \psi$,

   (iv) $a\,(\psi + \phi) = a\psi + a\phi$,

   (v) $(a + b)\,\psi = a\psi + b\psi$,

   (vi) $a(b\psi) = (ab)\psi$,

(vii) $1, \psi = \psi,$

(viii) $0.\psi = 0,$

for any vectors ψ, ϕ, and χ, the numbers a and b. In (viii), 0 on the left is the number zero, and 0 on the right is the zero vector. The numbers are called scalars. If the scalars belong to the field of complex (real) numbers the vector space is called complex (real) linear vector space. The element $(-1)\psi$ is denoted as $-\psi$. Also,

$$\phi + (-\phi) = (1 + (-1))\phi = 0\phi = 0.$$

The vectors $\vec{a}$, $\vec{b}$, $\vec{c}$, ... in the three-dimensional space $\boldsymbol{R}^3$ fulfill the above relations.

Examples of linear spaces

(i) n-dimensional Euclidean space – The set of all n-tuples of numbers with addition of two vectors $\psi = (x_1, ..., x_n)$ and $\phi = (y_1, y_2, ..., y_n)$ defined by $\psi + \phi = (x_1 + y_1, x_2 + y_2, ..., x_n + y_n)$ and multiplication of the vector ψ by a scalar a defined by $a\psi = (ax_1, ax_2, ..., ax_n)$.

(ii) The set of all infinite sequences of numbers $(x_1, x_2, ..., x_i, ...)$ such that $\sum\limits_{i=1}^{\infty} |x_i|^2$ is finite with addition and scalar multiplication defined componentwise. This space is called l^2.

(iii) The set of all continuous functions of a real variable x with addition of two vectors ψ and ϕ defined by $(\psi + \phi)(x) = \psi(x) + \phi(x)$ and multiplication of a vector ψ by a scalar a defined by $(a\psi)(x) = a\psi(x)$.

(iv) The set of all functions ψ of a real variable x for which the Lebesgue integral $\int |\psi(x)|^2$ is finite with addition and scalar multiplication defined pointwise. This space is referred to as L^2.

Each example can be a real or complex vector space, if the numbers or functions are real or complex respectively.

Definition

A linear manifold in a vector space is a subset of vectors which is a linear space.

Definition

A set of vectors ψ_1, ψ_2, ..., ψ_n is linearly independent if $\sum\limits_{i=1}^{n} a_i\psi_i = 0$ can hold only for $a_i = 0$ for all i. If this condition does not hold, the set of vectors is said to be linearly dependent. A linear space is n-dimensional if it contains n linearly independent vectors, but not $n + 1$. An infinite set of vectors is linearly independent if every finite subset is linearly independent; otherwise it is linearly dependent. A linear space is infinite dimensional if it contains n linearly independent vectors for every positive integer n. A maximal set of linearly independent vectors is called a basis for a linear space. A set of vectors ψ_1, ψ_2, ..., ψ_i is a basis for a linear space if it is a linearly independent set and spans the space.

These concepts are given in the following.

Theorem 1.1 A linear space is n-dimensional iff it has a basis of n vectors.

The Euclidean space of example (i) is n-dimensional. It has a basis of n vectors $\psi_1 = (1, 0, ..., 0)$, $\psi_2 = (0.1, 0, ..., 0)$, ..., $\psi_n = (0, 0, ..., 1)$. All other examples are infinite-dimensional.

A vector space is too general and structureless to be of much physical interest. We equip it by defining an inner product.

Definitions

The inner product or scalar product of two vectors ψ and ϕ in a linear space V is a scalar (ψ, ϕ) with the properties that for any vectors ψ, ϕ, χ and scalar a:

(i) $(\psi, \phi + \chi) = (\psi, \phi) + (\psi, \chi)$,

(ii) $(\psi, a\phi) = a(\psi, \phi)$,

(iii) $(\psi, \phi) = (\phi, \psi)^*$,

(iv) $(\psi, \psi) \geq 0$, the equality holding only

if $$\psi = 0$$

where * is the operation of complex conjugation. The norm or length of a vector ψ is the non-negative number $\|\psi\| = (\psi, \psi)^{1/2}$. The norm has the properties

(i) $\|\psi\| \geq 0$, equality holding only if $\psi = 0$,

(ii) $\|a\psi\| = a \|\psi\|$,

(iii) $\|\psi + \phi\| \leq \|\psi\| + \|\phi\|$ (triangle inequality),

for any vectors ψ, ϕ and scalar a.

Two vectors are orthogonal if their inner product is zero.

A vector space V on which an inner product is defined is called an inner product space.

A set of vectors $\{\psi_i\}$ is said to be orthonormal if the vectors are pairwise orthogonal and of unit norm, i.e., $(\psi_i, \psi_j) = \delta_{ij}$, where $\delta_{ij} = \begin{cases} 0, & i \neq j \\ 1, & i = j \end{cases}$ is the Kronecker delta. n-dimensional Euclidean space has an inner product defined by

$(\psi, \phi) = \sum\limits_{i=1}^{n} x_i^* y_i$ for vectors $\psi = (x_1, ..., x_n)$ and $\phi = (y_1, ..., y_n)$. The inner product for the space l^2 of infinite sequences may be defined by $(\psi, \phi) = \sum\limits_{i=1}^{\infty} x_i^* y_i$, for $\psi = (x_1, ..., x_i ...)$ and $\phi = (y_1, ..., y_i, ...)$. For the L^2 space of square-integrable functions an inner product is defined by $(\psi, \phi) = \int \psi^*(x)\phi(x)\, dx$.

Theorem 1.2 For any two vectors ψ and ϕ in an inner product space V, the Schwarz inequality holds

$$|(\psi, \phi)|^2 \le (\psi, \psi)(\phi, \phi). \tag{1}$$

Equality holds only if ψ is proportional to ϕ.

A sequence of complex numbers z_n converges if it is a Cauchy sequence – that is, if $|z_j - z_i| \to 0$ as $i, j \to \infty$. A sequence of vectors ψ_n is called a Cauchy sequence if $\|y_j - y_i\| \to 0$ as $i, j \to 0$. Every sequence of vectors which converges to a limit is a Cauchy sequence, and conversely, every Cauchy sequence converges. A space is said to be complete if every Cauchy sequence of vectors converges to a limit vector in the space. Every finite-dimensional space is complete.

A complete inner product space is called a Hilbert space. An inner product space (without the completeness property) is also called a pre-Hilbert space.

A Hilbert space is separable if it has an orthonormal basis which has a countable (finite or infinite) number of vectors. Then every orthonormal basis is countable. Every finite-dimensional space is a separable Hilbert space.

The space l^2 of square summable infinite sequences is an example of infinite-dimensional Hilbert space. The space l^2 is a separable Hilbert space. Another realization of the Hilbert space is the space L^2 of square-integrable functions. The space L^2 is a separable Hilbert space.

Two Hilbert spaces $\mathcal{H}_1$ and $\mathcal{H}_2$ are said to be isomorphic if there exist invertible maps U from $\mathcal{H}_1$ onto $\mathcal{H}_2$ such that

$$(U_x, U_y)_{\mathcal{H}_2} = (x, y)_{\mathcal{H}_1}, \ \forall\ x, y \in \mathcal{H}_1. \tag{2}$$

An isomorphism of a Hilbert space $\mathcal{H}$ onto itself is called an automorphism of a Hilbert space. The set of all automorphisms of $\mathcal{H}$ is a group, called the automorphism group of $\mathcal{H}$. A ray in a linear space is a set of vectors that differ only by a phase. A unit ray is the set of all vectors $\{\lambda\psi\}$, where $\|\psi\| = 1$, $\lambda = e^{i\alpha}$, $0 \le \alpha \le 2\pi$. In quantum mechanical description of a system, we consider a linear space $\mathcal{H}$ (a Hilbert space) whose unit rays are in one-to-one correspondence with the states (pure states) of the system.

Linear scalar-valued functions of vectors are called linear functionals. A linear functional F assigns a scalar $F(\phi)$ to each vector ϕ such that

$$F(a\phi + b\psi) = aF(\phi) + bF(\psi), \tag{3}$$

for any vectors ϕ and ψ, and any scalars a and b.

The set of linear functionals forms a linear space V' with the sum of two functionals

$$(F_1 + F_2)(\phi) = F_1(\phi) + F_2(\phi), \tag{4}$$

and the scalar multiple of a functional F and a scalar a

$$F(a\phi) = aF(\phi). \tag{5}$$

The space V' of linear functionals on a vector space V is called the algebraic dual space of V.

Theorem 1.3   (Riesz' theorem)   To every continuous linear functional F on a separable Hilbert space $\mathcal{H}$ there corresponds a unique vector $\psi_F \in \mathcal{H}$ such that

$$F(\phi) = (\psi_F, \phi). \tag{6}$$

for all $\phi \in \mathcal{H}$.

We thus have a one-to-one correspondence between continuous linear functionals and vectors.

The set of complex $n \times n$ matrices is a linear space. Linear functionals are the traces of the product of a matrix with some other matrix.

1.2   THE DIRAC NOTATION

Dirac introduced bra-ket formalism in his celebrated book on quantum mechanics. In Dirac's notation a vector ψ in V is denoted as $|\psi\rangle$ and is called a ket vector or simply a ket. The linear functional F_ψ defined by $F_\psi(\phi) = (\psi, \phi)$ is denoted by $\langle\psi|$ and $F_\psi(\phi)$ by $\langle\psi|\phi\rangle$. The linear functionals in the dual space (V') are called bra vectors and are denoted by $\langle\psi|$. We can consider $\langle\psi|\phi\rangle$ as the notation for the inner product of the vectors $|\psi\rangle$ and $|\phi\rangle$.

Thus, a ket $|\psi\rangle$ is a vector in a Hilbert space $\mathcal{H}$. With this vector we associate a linear functional $F_\psi \equiv \langle\psi|$ called a bra which is an element of the dual Hilbert space $\mathcal{H}^*$ and is defined by the scalar product ("bracket") $F_\psi(\phi) = \langle\psi|\phi\rangle$. By Riesz' theorem there is a one-to-one correspondence between $|\psi\rangle \in \mathcal{H}$ and $\langle\psi| \in \mathcal{H}^*$. We note that if $\langle\psi| \leftrightarrow |\psi\rangle$,

then

$$\alpha^*\langle\psi| + \beta^*\langle\psi| \leftrightarrow \alpha|\psi\rangle + \beta|\psi\rangle, \tag{7}$$

where α, β are arbitrary complex numbers. So there is an anti-linear correspondence between bras and kets. We have

$$\langle\psi|\phi\rangle = \langle\phi|\psi\rangle^*. \tag{8}$$

Dirac's elegant bra-ket formalism has proved to be of great power in formal calculations although it is awkward in certain cases and there are difficulties in finding mathematical justification for many calculations within the Hilbert space {see Gieres (2000)}.

In Section 1.5 we shall consider more general spaces (the rigged Hilbert space). As part of bra-ket formalism, Dirac introduced the delta function, a formal entity which has proved to be of great use in calculations. The delta function is, however, not a function in the classical sense. The theory of distributions (generalized functions) is sketched in Section 1.4.

We shall use the Dirac notation in Chapter 2 and other chapters whenever this is convenient and has become by now standard. In many places, however, we use the wave mechanics formalism.

1.3   LINEAR OPERATORS

Two key concepts in quantum mechanics are the linear superposition principle and the probability interpretation. These suggest that the space of the states in quantum mechanics be a linear one equipped with an inner product to obtain probability amplitudes. So we shall assume that a separable Hilbert space is the mathematical setting of quantum mechanics. In quantum mechanics physical observables are represented by linear operators on a Hilbert space.

An operator on a vector space maps vectors onto vectors. If A is an operator and ψ is a vector, then $\phi = A\psi$ is another vector. A function A, $A : V \to V$, that maps each vector ψ in a linear space V into a vector $\phi \in V$,

$$A\psi = \phi, \tag{9}$$

is called a linear operator if for all ψ, ϕ, $\in V$ and $a \in C$ it fulfills the conditions

$$A(\psi + \phi) = A\psi + A\phi, \tag{9a}$$

$$A(a\psi) = aA\psi. \tag{9b}$$

It is called anti linear if it satisfies

$$A(a\psi) = \alpha^* A\psi.$$

In Dirac's notation, one says that the operator A transforms the ket $|\psi\rangle$ into the ket $|\phi\rangle$:

$$A|\psi\rangle = |\phi\rangle. \tag{10}$$

Operators also act on bras:

$$\langle\psi|A = \langle\chi'|. \tag{11}$$

The simplest linear operator is the identity or unit operator I which leaves every vector unchanged; it multiplies every vector by 1:

$$I\psi = \psi \quad \text{for every } \psi \in V. \tag{12}$$

So every scalar defines an operator. The zero operator is defined by

$$0\psi = 0 \quad \text{for every } \psi \in V. \tag{13}$$

where the 0 on the left is the zero operator, while 0 on the right is the zero vector.

We have a one-to-one correspondence between linear operators on an n-dimensional vector space and $n \times n$ matrices.

On a function space a linear operator can be defined, for example, by

$$(A\psi)(x) = \frac{d\psi(x)}{dx}, \tag{14}$$

or,
$$(A\psi)(x) = \int a(x, y) \, \psi \, (y) \, dy. \tag{14a}$$

Definitions A linear operator A is continuous if $A\psi_n \to A\psi$ for any sequence of vectors ψ_n which converges to a limit vector ψ. A linear operator is bounded if there is a positive number a such that $\|A\psi\| \le a\|\psi\|$ for every vector ψ; the smallest number a with this property is called the norm of A and is denoted by $\|A\|$.

Theorem 1.4 A linear operator is continuous iff it is bounded.

Every operator on a finite-dimensional space is bounded. A bounded linear operator on an infinite-dimensional space can be represented by an infinite matrix.

Definition A linear operator U is unitary if its inverse U^{-1} exists and if $\|U\psi\| = \|\psi\|$ for every vector ψ. Also, $\|U\| = 1$.

Theorem 1.5 If U is unitary, then $(U\psi, U\phi) = (\psi, \phi)$ or in the Dirac notation, $\langle U\psi | U\phi \rangle = \langle \psi | \phi \rangle$ for every $|\psi\rangle$, $|\phi\rangle$.

The adjoint $A^\dagger$ of a bounded, linear operator A is defined in the Dirac notation,

$$\langle \phi | A^\dagger | \psi \rangle^* = \langle \psi | A | \phi \rangle \text{ for all } |\phi\rangle \text{ and } |\psi\rangle. \tag{15}$$

Theorem 1.6 Let A be a bounded linear operator. Then $A^\dagger$ is a bounded linear operator and $\|A^\dagger\| = \|A\|$.

A bounded linear operator A is called self adjoint or hermitian if $A^\dagger = A$. A hermitian operator A satisfies

$$\langle \phi | A | \psi \rangle = \langle \psi | A | \phi \rangle^*, \tag{16}$$

for all vectors ψ and ϕ.

Theorem 1.7 A linear operator U is unitary iff $U^\dagger U = I = UU^\dagger$.

Theorem 1.8 Let U be a unitary operator. Then

$$\langle U\psi | U\phi \rangle = \langle \psi | \phi \rangle, \tag{17}$$

for any vector $|\psi\rangle$ and $|\phi\rangle$.

Theorem 1.9   A bounded linear operator E is a projection operator iff $E^2 = E = E^\dagger$. The subspace onto which E projects is the set of vectors $E\psi$ for all vectors ψ.

Theorem 1.10   Let E_1 and E_2 be two different projection operators, where E_1 projects onto a subspace $\mathcal{M}_1$ and E_2 projects onto another subspace $\mathcal{M}_2$.

(i) If $E_1 E_2 = E_2 E_1$, then $E_1 E_2$ is a projection operator projecting onto the subspace which is the intersection of $\mathcal{M}_1$ and $\mathcal{M}_2$.

(ii) E_1 and E_2 are orthogonal if $\mathcal{M}_1$ and $\mathcal{M}_2$ are orthogonal. Then

$$E_1 E_2 = E_2 E_1 = 0. \tag{18}$$

Furthermore, $E_1 + E_2$ is the projector onto the subspace $\mathcal{M}_1 \oplus \mathcal{M}_2$.

In the Dirac notation, an operator called the outer product $|\psi\rangle\langle\phi|$ may be defined. We have

$$(|\psi\rangle\langle\phi|)\,|\chi\rangle = |\psi\rangle(\langle\phi|\chi\rangle). \tag{19}$$

Also,

$$(|\psi\rangle\langle\phi|)^\dagger = |\phi\rangle\langle\psi|. \tag{20}$$

Consider the one-dimensional subspace spanned by a vector ϕ with $\|\phi\| = 1$. Then $|\phi\rangle\langle\phi|$ is the projector onto the subspace spanned by a vector of length one.

Let A be a linear operator. If $|\psi\rangle$ is a nonzero vector and a is a scalar such that

$$A|\psi\rangle = a|\psi\rangle, \tag{21}$$

then we say that $|\psi\rangle$ is an eigenvector and a an eigenvalue of the operator A.

Theorem 1.11   Let T be a linear operator with an inverse T^{-1}. Then A and TAT^{-1} have the same eigenvalues.

Theorem 1.12   Every eigenvalue of a hermitian operator is real.

Theorem 1.13   Every eigenvalue of a unitary operator is a complex number of absolute value one.

Theorem 1.14   Two eigenvectors of a hermitian or unitary operator are orthogonal if they correspond to different eigenvalues.

The phenomenon of a single eigenvalue representing more than one eigenvectors is called degeneracy. This results from the occurrence of repeated roots of the characteristic polynomial. An eigenvalue is called f-fold degenerate if f eigenvalues are equal.

One can replace a nonorthogonal but linearly independent set of degenerate eigenvectors by a suitably chosen linear combination that is orthogonal. We shall always assume an orthonormal set of eigenvectors as a basis.

Theorem 1.15   For a finite dimensional space the eigenvectors of a hermitian or unitary operator span the whole space.

Hence for each hermitian or unitary operator A on a finite dimensional space there is an orthonormal basis of eigenvectors. If we express A in terms of its eigenvalues and eigenvectors, it is a diagonal matrix. A hermitian or unitary operator on an infinite-dimensional space may or may not possess a complete set of eigenvectors. The completeness depends on the nature of the operator and the vector space.

An operator is properly defined by the specification of an operating prescription and of a domain of definition for the operation. Two operators which act in the same way must be considered as different if they are defined on different subspaces of a Hilbert space. In infinite-dimensional Hilbert space, there exists subtle difference between a hermitian and a self-adjoint operator.

Definition   An operator on the Hilbert space $\mathcal{H}$ is a linear map

$$A : D(A) \to \mathcal{H}$$
$$\psi \to A\psi$$

where $D(A)$ represents a dense linear subspace of $\mathcal{H}$. $D(A)$ is called the domain of A and $R(A) \equiv AD(A)$ is its range.

Let B be another operator with domain $D(B)$. If

$$A\phi = B\phi \text{ for all } \phi \in D(A) = D(B)$$

then
$$A = B$$

The graph $\Gamma(A)$ of A is the set of pairs

$$\{\langle \phi, A\phi \rangle | \phi \in D(A)\}.$$

The operator A is said to be closed if the subspace $\Gamma(A)$ is closed

A is said to be closable if $\overline{\Gamma(A)}$, the closure of $\Gamma(A)$, is a graph:

$$\overline{\Gamma(A)} = \Gamma(\overline{A}).$$

$\overline{A}$ is called the closure of A and is its minimal closed extension.

If A_1 and A_2 are operators on $\mathcal{H}$ and if the graph of A_2, $\Gamma(A_2)$, includes the graph of A_1, $\Gamma(A_1)$, then we say that A_2 is an extension of A_1 and we write

$$A_1 \subset A_2 \text{ if } \Gamma(A_1) \subset \Gamma(A_2).$$

Given A, and $\lambda \in \mathbb{C}$, we say that λ belongs to the resolvent set $\rho(A)$ if $\overline{R(\lambda I - A)}$ $= \mathcal{H}$, $(\lambda I - A)^{-1} \equiv R_\lambda(A)$ exists and is bounded. $R_\lambda(A)$ is called the resolvent operator for A in λ. The set $\rho(A)$ is open. Its complement $\mathbb{C} - \rho(A) \equiv \sigma(A)$ is closed and is called the spectrum of A.

Definition   The spectrum $\sigma(A)$ is the disjoint union of the point spectrum $\sigma_p(A)$, continuous spectrum $\sigma_c(A)$, and residual spectrum $\sigma_r(A)$ which consist of all points which have the following respective properties:

$\sigma_p(A)$ of the points $\lambda \in \sigma(A)$ at which $R_\lambda(A)$ does not exist; [Then $(A - \lambda I)\phi = 0$ has a nontrivial solution ϕ; σ is called an eigenvalue of A and ϕ an eigenvector] $\sigma_c(A)$ of all $\lambda \in \sigma(A)$ for which $R_\lambda(A)$ exists, is unbounded and defined on a dense subset of $\mathcal{H}$;

$\sigma_r(A)$ of all $\lambda \in \sigma(A)$ at which $R_\lambda(A)$ exists, but is not defined in a dense subset of $\mathcal{H}$.

Theorem 1.16 (Hellinger-Toeplitz) Let A be a linear operator defined on the entire Hilbert space $\mathcal{H}$. If

$$(Ax, y) = (x, Ay), \ \forall x, y \in \mathcal{H} \tag{22}$$

then A is bounded, i.e. for some $C \geq 0$, $\|Ax\| \leq C\|x\|$, $\forall x \in \mathcal{H}$.

An unbounded operator B can only be defined on a subspace $D(B)$ called the domain of B, of the Hilbert space $\mathcal{H}$.

Definitions Let A be a densely defined linear operator on a Hilbert space $\mathcal{H}$. The domain $D(A^\dagger)$ is given by

$$D(A^\dagger) = \{\phi \in \mathcal{H} | \exists \tilde{\phi} \in \mathcal{H} \quad \text{such that}$$
$$\langle \phi, A\psi \rangle = \langle \tilde{\phi}, \psi \rangle \quad \text{for all } \psi \in D(A)\}.$$

For $\phi \in D(A^\dagger)$, one defines

$$A^\dagger \phi = \tilde{\phi}, \tag{23}$$

i.e.,
$$\langle \phi, A\psi \rangle = \langle A^\dagger \phi, \psi \rangle \quad \text{for all } \psi \in D(A) \tag{23a}$$

The operator $A^\dagger$ is called the adjoint of A.

A densely defined operator A on a Hilbert space $\mathcal{H}$ is called symmetric or hermitian if, $A \subset A^\dagger$, that is, if $D(A) \subset D(A^\dagger)$ and $A\phi = A^\dagger \phi$, $\forall \phi \in D(A)$. This implies that A is symmetric iff

$$\langle A\phi, \psi \rangle = \langle \phi, A\psi \rangle \ \forall \phi, \psi \in D(A) \tag{24}$$

In Theorem 1.16, the operator is symmetric.

An operator A on $\mathcal{H}$ is self-adjoint iff the operators A and $A^\dagger$ coincide ($A = A^\dagger$) i.e.,

$$D(A) = D(A^\dagger) \text{ and } A^\dagger \phi = A\phi \text{ for all } \phi \in D(A). \tag{25}$$

A symmetric operator A is called essentially self-adjoint if its closure $\overline{A}$ is self-adjoint i.e., $\overline{A} = A^\dagger$.

For a self-adjoint operator A, the residual spectrum $\sigma_r(A)$ is empty.

We now discuss the spectral decomposition of an operator with discrete and continuous spectrum.

It is convenient to use the Stieltjes integrals:

$$\int_a^b g(x)d\rho(x) = \lim_{n \to \infty} \sum_{k=1}^n g(x_k)[\rho(x_k) - \rho(x_k - 1)] \tag{26}$$

and $a < x_0 < x_1 < ... < x_n \leq b$ divides the range of integration into smaller and smaller invervals such that every interval $(x_k - x_{k-1}) \to 0$ as $n \to \infty$. The nondecreasing function $\rho(x)$ is called the measure. For $\rho(x) = x$ this reduces to the Riemann measure. If the measure is discontinuous then the integral contributes only at those points.

Definition   A family of projection operators E_x depending on a real parameter x is a spectral family if it satisfies the following requirements:

(i) If $x \leq y$ $E_x \leq E_y$ or $E_x E_y = E_x = E_y E_x$

(ii) $E_{x+\epsilon} y \xrightarrow[\substack{\epsilon > 0 \\ \epsilon \to 0}]{} E_x \psi$ for any vector ψ and any x,

(iii) $E_x \psi \to 0$ as $x \to -\infty$ and $E_x \psi \to \psi$ as $x \to \infty$ for any vector ψ.

Let A be a self-adjoint operator in a Hilbert space $\mathcal{H}$. We now state the fundamental theorem in spectral decomposition theory.

Theorem 1.17   Every self-adjoint operator A in $\mathcal{H}$ has the representation

$$A = \int_{-\infty}^{\infty} x \, dE_x, \tag{27}$$

where E_x is a unique spectral family such that

$$(\phi, A\psi) = \int_{-\infty}^{\infty} x \, d\,(\phi, E_x \psi) \tag{28}$$

for all vectors ψ and ϕ.

Equation (27) is called the spectral decomposition or spectral resolution of A. In Eq. (27), neither the domain of integral nor the operator-function is bounded; ψ is in $D(A)$, the domain of the operator A.

If A has only continuous spectrum, then all eigenvectors are outside the Hilbert space.

Given a self-adjoint operator A in a Hilbert space $\mathcal{H}$, we know that the spectrum $\sigma(A) = \sigma_p (A) \cup \sigma_c (A)$. If we remove the pure point (p.p.) part the remaining spectrum is the continuous spectrum which can be further decomposed into the absolutely continuous (a.c.) and singular continuous (s.c.) parts, according to the Lebesgue decomposition of the spectral measure.

Another useful decomposition of the spectrum of A is $\sigma(A) = \sigma_{disc} (A) \cup \sigma_{ess} (A)$. σ_{dis} is called the discrete spectrum and $\sigma_{ess}(A) \equiv \sigma (A) - \sigma_{disc} (A)$ is called the essential spectrum.

In quantum physics, observables are represented by self-adjoint operators on a Hilbert space. We note that there exists a difference between the properties of a hermitian operator and a self-adjoint operator acting on infinite-dimensional Hilbert space, the condition on the domains being crucial. This difference is important in quantum theory.

We now discuss the spectral properties of the Schrödinger operators. The Hamiltonian H acts on the wave functions ψ in the Hilbert space $\mathcal{H}$. The spectrum $\sigma(H)$ of the self-adjoint operator H coincides with the closure of the set of values E for which the eigenvalue equation (the time-independent Schrödinger equation) $H\psi = E\psi$ possesses a solution that is polynomially bounded. So in general, there are three different types of energy spectrum-point, absolutely continuous, and singular continuous. The spectral measure of the point spectrum is a set of δ-functions defined on a countable number of points $\{E_i\}$ (eigenvalues).

The a.c. spectrum has the spectral measure $d\mu = n(E)dE$ and has a smooth density of states $n(E)$. The a.c. spectrum is a closed set with non-empty interior. In the s.c. spectrum the number of states whose energies are below a fixed energy E is continuously increasing but non-differentiable at any E (the Cantor function). The density of states is not well-defined in this case. But the integrated density of states is well-defined in both the a.c. and s.c. spectra.

Wave functions are classified into three type – localized, extended, and critical.

The localized state is defined as the square integrable wave function $\displaystyle\int_{|r| < \infty} |\psi(\vec{r})|^2 \, d\vec{r}$

= constant. The values for which $\psi_E(i)$ is a square integrable function are called the eigenvalues of H and the closure of the set of eigenvalues is the point sepectrum. The extended state is defined as the wave function with asymptotically uniform amplitude $\displaystyle\int_{|r| < L} |\psi(\vec{r})|^2 \, d\vec{r} \sim L^d$, where d is the spatial dimension. The power-law

type behavior $\psi(\vec{r}) \sim |\vec{r}|^{-m} \left(m \leq \dfrac{d}{2} \right)$ of the wave function is typical of the critical state. The extended and critical wave functions correspond to a.c. and s.c. energy spectra, respectively.

The three types of the energy spectra with the three corresponding types of wave functions appear in the solutions to the Harper model depending on the strength of the incommensurate potential.

1.4 GENERALIZED FUNCTIONS

Dirac introduced the delta function while investigating the scattering problem in quantum mechanics. This singular function $\delta(x - x_0)$ is defined as "equal to zero everywhere except at x_0 where it is infinite, and its integral is one". According to the classical definition of a function and an integral, these conditions are inconsistent.

In a specific problem in physics, the delta function and other singular functions occur usually in an integrand where it is multiplied by a sufficiently well-behaved function. For a sufficiently well-behaved function $\phi(x)$, we have

$$\int_{-\infty}^{\infty} \delta(x - x_0) \, \phi(x) \, dx = \phi(x_0).$$

Schwartz gave a precise meaning to a singular function as a functional which associates with every "sufficiently good" function some well-defined number. For example, for the delta function the number corresponding to each "sufficiently good" function $\phi(x)$ is $\phi(x_0)$. This led to the development of a new branch of functional analysis called the theory of generalized functions or distributions (Gel'fand and coworkers).

(a) Definitions and Simple Properties of Generalized Functions

A test function $\phi(x) = \phi(x_1, \dots, x_n)$ is an infinitely differentiable function that vanishes outside a bounded region in R^n (has bounded support).

The set of all test functions is the test function space K. K is a linear space.

A distribution, or generalized function is a continuous linear functional on K. A generalized function includes both "singular" and ordinary functions.

A regular generalized function is a functional on K which can be written as

$$(f, \phi) = \int f(x) \, \phi(x) \, dx, \tag{29}$$

where $f(x)$ is a locally summable function.

All other generalized functions are called singular.

The regular generalized function f defined by

$$(f, \phi) = c \int \phi(x) \, dx = \int c \, \phi(x) \, dx, \tag{30}$$

is called the constant c. The unit generalized function is

$$(1, \phi) = \int \phi(x) \, dx. \tag{30a}$$

Although the delta function is not a function in the classical sense, we call the functional defined below as the delta function and denote it $\delta(x)$. We write

$$(\delta(x), \phi(x)) = \phi(0), \tag{31}$$

and also the functional $\delta(x - x_0)$ is written as

$$(\delta(x - x_0), \phi(x)) = \phi(x_0). \tag{31a}$$

A sequence $f_1, f_2, \dots, f_n, \dots$ of generalized functions is defined to converge to the generalized function f if for every $\phi(x) \in K$,

$$\lim_{n \to \infty} (f_n, \phi) = (f, \phi). \tag{32}$$

Depending on the space of test functions, different distributions can be defined.

We define complex generalized functions as continuous, linear functionals taking on complex values, in general, on this test function space. We denote the test function space by K and the generalized function space by K'.

A regular functional f is associated with a complex, locally summable function $f(x)$ according to

$$(f, \phi) = \int \overline{f(x)}\, \phi(x)dx, \tag{33}$$

where the bar denotes the complex conjugate. The generalized function f has the complex conjugate generalized function $\bar{f}$ according to

$$(\bar{f}, \phi) = \overline{(f, \bar{\phi})}. \tag{34}$$

The linear space S of rapidly decreasing functions consists of infinitely differentiable functions $f(x)$ which, together with their derivatives approach zero more rapidly than any power of $\dfrac{1}{|x|}$ as $|x| \to \infty$ (e.g., e^{-x^2}) such that

$$|x^k\, \phi^{(q)}(x)| \le C_{kq}, \tag{35}$$

for $\qquad\qquad k, q = 0, 1, 2, \ldots$

Every generalized function f has a derivative with respect to x_j, denoted by $\dfrac{\partial f}{\partial x_j}$ as

$$\left(\frac{\partial f}{\partial x_j}, \phi\right) = \left(f, -\frac{\partial \phi}{\partial x_j}\right). \tag{36}$$

Example 1.1 Let us consider the step function

$$\theta(x) = \begin{cases} 0, & \text{for} \quad x < 0 \\ 1, & \text{for} \quad x > 0. \end{cases} \tag{37}$$

The functional corresponding to this function is also denoted by $\theta(x)$. From Eq. (36), we have

$$\begin{aligned}
(\theta'(x), \phi(x)) &= (\theta(x), -\phi'(x)) \\
&= -\int_0^\infty \phi'(x)\, dx \\
&= \phi(0),
\end{aligned}$$

for any $\phi(x) \in K$.

Then $\qquad\qquad\qquad \phi'(x) = \delta(x).$ $\qquad\qquad$ (38)

Also, $\qquad\qquad \theta'(x - a) = \delta(x - a).$ $\qquad\qquad$ (39)

The derivative of the generalized function $\ln |x|$ is

$$\frac{d \ln |x|}{dx} = \frac{1}{x}. \tag{40}$$

The functional $\dfrac{1}{x}$ is not regular. It corresponds to the ordinary function $\dfrac{1}{x}$ for $x \ne 0$. The derivative of the generalized function $\ln (x + i0)$ is

$$\frac{d}{dx} \ln (x + i0) = \frac{1}{x} - i\pi\delta(x) \tag{41}$$

where
$$\frac{d}{dx} \ln |x| = P\,\frac{1}{x}. \tag{41a}$$

This is the Cauchy principal value of the integral. Equation (40) is often written as

$$\left(\frac{1}{x},\, \phi(x)\right) = \lim_{\epsilon \to 0} \int_{|x| > \epsilon} \frac{\phi(x)}{x}\, dx.$$

We write

$$\frac{1}{x \pm i0} = P\,\frac{1}{x} \mp i\pi\delta(x). \tag{41b}$$

so that

$$\frac{1}{x - i0} - \frac{1}{x + i0} = 2\pi i\delta(x) \tag{41c}$$

We consider the derivative of the δ-function. We have

$$(\delta'(x - a),\, \phi) = (\delta(x - a),\, -\phi') = -\phi'(a)$$

and in general

$$(\delta^{(k)}(x - a),\, \phi) = (-1)^k\, \phi^{(k)}(a), \quad k = 1, 2, \ldots$$

We write

$$\int \delta^{(k)}(x - a)\, \phi(x)\, dx = (-1)^k\, \phi^{(k)}(a). \tag{42}$$

One can construct in many ways a sequence of regular functions which converge to the δ-function. We list some these sequences below.

(i) $\dfrac{1}{\pi} \lim\limits_{\epsilon \to 0} \dfrac{\epsilon}{x^2 + \epsilon^2} = \delta(x),$

(ii) $\lim\limits_{t \to 0} \dfrac{1}{2\sqrt{\pi t}}\, e^{-x^2/4t} = \delta(x), \quad (t > 0)$

(iii) $\dfrac{1}{\pi} \lim\limits_{v \to \infty} \dfrac{\sin vx}{x} = \delta(x). \quad (0 < v < \infty)$

(iv) $\dfrac{1}{\pi} \lim\limits_{n \to \infty} \dfrac{1 - \cos nx}{nx^2} = \delta(x) \tag{43}$

From (i), we get

$$-\frac{2}{\pi} \lim_{\epsilon \to 0} \frac{\epsilon x}{(x^2 + \epsilon^2)} = \delta(x)$$

Since $\dfrac{1}{\pi} \dfrac{\sin vx}{x}$ can be written as the integral of $\dfrac{1}{2\pi}\, e^{i\xi x}$ over ξ from $-v$ to v, then in the sense of convergence in K',

$$\lim_{v \to \infty} \int_{-v}^{v} e^{i\xi x} \, d\xi = 2\pi\delta(x).\tag{44}$$

We note that the left hand side of Eq. (44) is the Fourier transform of unity. By differentiating the Eq. (44), we get

$$\lim_{v \to \infty} \int_{-v}^{v} i\xi e^{i\xi x} \, d\xi = 2\pi\delta'(x),\tag{45a}$$

$$\lim_{v \to \infty} \int_{-v}^{v} (i\xi)^2 \, e^{i\xi x} \, d\xi = 2\pi\delta''(x),\tag{45b}$$

and other relations of higher orders.

We list below some properties of the delta function. All these equalities can be proved in the theory of generalized functions.

$$\delta(x) = \delta(-x),$$

$$\delta(ax) = \frac{1}{|a|} \, \delta(x),$$

$$\delta(f(x)) = \sum_{n} \frac{1}{|f'(x_n)|} \, \delta(x - x_n), \qquad \begin{cases} f(x_n) = 0, \\[4pt] f'(x_n) \neq 0. \end{cases}$$

$$x\,\delta(x) = 0$$

$$f(x)\,\delta(x - a) = f(a)$$

The kth order derivative of $\delta(x)$ is defined by Eq. (42).

$$\delta^{(k)}(x) = (-1)^k \, \delta^{(k)}(-x),$$

$$x^{k+1} \, \delta^{(k)}(x) = 0,$$

$$x\delta'(x) = -\delta(x),\tag{46}$$

$$x^2 \, \delta'(x) = 0.$$

Let us consider the Laplacian in three-space applied to the regular functional corresponding to the function $\frac{1}{r}$, where $r^2 = x_1^2 + x_2^2 + x_3^2$. We write

$$\nabla^2 \frac{1}{r} = -4\pi\delta(x).\tag{47}$$

For $n > 2$,

$$\nabla^2 \frac{1}{r^{n-2}} = -(n-2)\Omega_n \, \delta(x),\tag{48}$$

where Ω_n is the hypersurface area of the unit sphere in n-space. Also

$$\nabla^2 \ln \frac{1}{r} = -2\pi\delta(x).\tag{49}$$

(b) Fourier Transforms of Generalized Functions

Let $\phi(x) \in K$. Its Fourier transform is

$$\psi(\sigma) = \int_{-\infty}^{\infty} \phi(x)\, e^{i\sigma x}\, dx. \tag{50}$$

The Fourier transform of $\phi(x)$ is also denoted by $\tilde{\phi}(x)$ or by $F[\phi(x)]$.

$\phi(\sigma)$ can be defined also for complex values of its argument $s = \sigma + i\tau$ by

$$\psi(\sigma + i\tau) = \int_{-\infty}^{\infty} \phi(x) e^{isx}\, dx. \tag{51}$$

The Fourier transform $\psi(s)$ of any $\phi(x) \in K$ which vanishes for $|x| \geq a$ is an entire analytic function of its argument s.

We define the space Z of slowly increasing functions, namely, of all entire functions $\psi(s)$ satisfying

$$|s|^q\, |\psi(s)| \leq C_q\, e^{a|\tau|} \quad (q = 0, 1, 2, \ldots) \tag{52}$$

The Fourier transform gives a one-to-one mapping between K and Z. To every linear operator in K there corresponds a dual operator in Z. We now discuss the case of several variables.

The Fourier transform of a function

$$\phi(x) \equiv \phi(x_1, \ldots, x_n) \quad \text{in } K \text{ is defined by}$$

$$\psi(\sigma) = \psi(\sigma_1, \ldots, \sigma_n)$$

$$= \int_{-\infty}^{\infty} \ldots \int_{-\infty}^{\infty} \phi(x_1, \ldots, x_n) \exp\left[i\,(x_1\sigma_1 + \ldots + x_n\sigma_n)\right] dx_1 \ldots dx_n \tag{53}$$

or, briefly, by

$$\psi(\sigma) = \int_{R^n} \phi(x)\, e^{i(x,\,\sigma)}\, dx, \tag{53a}$$

where (x, σ) denotes $x_1\sigma_1 + \ldots + x_n\sigma_n$, and $dx = dx_1 \ldots dx_n$.

We continue ψ to complex values of its argument

$$s = (s_1, \ldots, s_n) = (\sigma_1 + i\tau_1, \ldots, \sigma_n + i\tau_n)$$

$$\psi(s) = \int_{C^n} \phi(x) e^{i(x,\,s)}\, dx. \tag{53b}$$

$\psi(s)$, defined in C^n, is an entire analytic function of the complex variables $s_1, \ldots, s_n$.

The set of all $\psi(s)$ functions of this form is called the space Z of slowly increasing functions.

Let $f(x)$ be an absolutely integrable function whose Fourier transform is $g(\sigma)$. For any $\phi(x) \in K$ and its Fourier transform $\psi(\sigma)$ we have *Parseval's* relation

$$(f, \phi) = \int_{-\infty}^{\infty} \overline{f(x)}\, \phi(x)\, dx$$

$$= \frac{1}{2\pi} \int_{-\infty}^{\infty} \overline{g(\sigma)}\, \psi(\sigma)\, d\sigma$$

$$= \frac{1}{2\pi} (g, \psi), \tag{54}$$

or,
$$(g, \psi) = 2\pi (f, \phi). \tag{54a}$$

The Fourier transform of a generalized function f in K' is the functional $g = F[f]$ (also $\tilde{f}$) defined on Z by

$$(g, \psi) = 2\pi (f, \phi), \tag{55}$$

where $\psi(\sigma)$ is the Fourier transform of $\phi(x)$.

The set of all generalized functions on Z is denoted Z'.

The inverse operator F^{-1} defined on Z' maps g into f again:

$$F^{-1}[F[f]] = f, \quad F[F^{-1}[g]] = g,$$

$$(F^{-1}[g], \phi) = \frac{1}{2\pi} (g, F[\phi]). \tag{56}$$

If $P(t)$ is any polynomial with constant coefficients, then

$$P\left(\frac{d}{ds}\right) F[f] = F[P(ix)f], \tag{57}$$

$$F\left[P\left(\frac{d}{dx}\right)f\right] = P(-is)F[f]. \tag{58}$$

Also,

$$FF[f(x)] = 2\pi f(-x). \tag{59}$$

As an example, we want to find $F[\delta]$.

We have

$$(\tilde{\delta}, \tilde{\phi}) = 2\pi (\delta, \phi) = 2\pi\phi(0) = \int_{-\infty}^{\infty} \psi(\sigma) \, d\sigma = (1, \psi)$$

whence
$$F[\delta] \equiv \tilde{\delta} = 1, \quad F^{-1}[1] = \delta. \tag{60}$$

Again, we have

$$(\tilde{1}, \tilde{\phi}) = 2\pi (1, \phi) = 2\pi \int_{-\infty}^{\infty} \phi(x) \, dx$$

$$= 2\pi \int_{-\infty}^{\infty} \phi(x) \, e^{-ix, 0} \, dx$$

$$= 2\pi\psi(0)$$

$$= 2\pi (\delta, \psi)$$

whence

$$F[1] = \tilde{1} = 2\pi\delta, \quad F^{-1}[\delta] = \frac{1}{2\pi}. \tag{61}$$

Also,

$$F\left[P\left(\frac{d}{dx}\right), \delta(x)\right] = P(-is). \tag{62}$$

$$F[e^{bx}] = \sum_{n=0}^{\infty} \frac{b^n}{n!} \tilde{x}^n = 2\pi \sum_{0}^{\infty} \frac{b^n}{n!} \left(-i \frac{d}{ds}\right)^n \delta(s)$$

$$= 2\pi\delta\,(s - ib) \tag{63}$$

$$F\,[\sin bx] = F\left[\frac{e^{ibx} - e^{-ibx}}{2i}\right] = i\pi\,[\delta(s - b) - \delta(s + b)] \tag{64}$$

$$F\,[\cos bx] = F\left[\frac{e^{ibx} + e^{-ibx}}{2}\right] = \pi\,[\delta(s - b) + \delta(s + b)] \tag{65}$$

$$F\,[\sinh bx] = F\left[\frac{e^{ibx} - e^{-ibx}}{2}\right] = \pi\,[\delta(s - ib) - \delta(s + ib)] \tag{66}$$

$$F\,[\cosh bx] = F\left[\frac{e^{ibx} + e^{-ibx}}{2}\right] = \pi\,[\delta(s - ib) + \delta(s + ib)] \tag{67}$$

The Fourier transform of a functional f acting on K of test function $\phi(x) = \phi(x_1, ..., x_n)$ is defined as the functional g on Z of functions $\psi(s) = \psi(s_1, ..., s_n)$ according to

$$(g, \psi) = (2\pi)^n\,(f, \phi), \tag{68}$$

where ψ is the Fourier transform of $\phi(x)$. The linear continuous functional g is denoted as $\tilde{f}$ or $F[f]$.

If P is a polynomial with constant coefficients in n variables, then

$$P\left(\frac{\partial}{\partial s_1}, ..., \frac{\partial}{\partial s_n}\right) F[f] = F\,[P(ix_1, ..., ix_n)f] \tag{69}$$

$$P\left[P\left(\frac{\partial}{\partial x_1}, ..., \frac{\partial}{\partial x_n}\right)f\right] = P\,(-is_1, ..., -is_n)f \tag{70}$$

The Fourier transform of the direct product is given by

$$F\,[f \times g] = F[f] \times F[g]. \tag{71}$$

We have

$$F[\delta(x_1, ..., x_n)] = 1, \tag{72}$$

$$\tilde{1} = (2\pi)^n\,\delta(s). \tag{73}$$

1.5 THE RIGGED HILBERT SPACE

The Hilbert space formulation of quantum mechanics cannot accommodate continuous eigenvalues. When a continuous spectrum is present, the rigged Hilbert space (RHS) provides the proper mathematical setting for quantum mechanics. Gel'fand and collaborators introduced the RHS in the 1960s by combining von Neumann's Hilbert space with Schwartz' theory of distributions. In quantum mechanics, the RHS is the Hilbert space equipped with distribution theory. The RHS describes in a proper, mathematical manner the bra-ket formalism of Dirac. Observables with discrete spectrum and a finite number of eigenvectors (e.g., spin) do not require the use of the RHS. In general, unbounded operators with continuous spectra require the RHS. The RHS has turned out to be of great importance in scattering theory, the resonance spectrum, spectral decomposition of chaotic maps, etc.

The triplet of spaces

$$\Phi \subset \mathcal{H}' = \mathcal{H}^{\times} \subset \Phi^{\times} \tag{74}$$

is called a Gel'fand triplet on the rigged Hilbert space. Here, $\mathcal{H}$ is a Hilbert space, and $\mathcal{H}^{\times}$ is its conjugate space. Φ is a dense subspace of $\mathcal{H}$, and $\Phi^{\times}$ its conjugate space. The space Φ is a nuclear space and consists of all vectors of the form $\omega = \sum_{n} u_n \, \phi_n$ where the coefficients satisfy the infinite set of conditions $\sum_{n} |u_n|^2 \, n^m < \infty$, $m = 0, 1, 2, ...$, $\Phi^{\times}$ is the space of antilinear functionals over Φ. Mathematically speaking, Φ is the space of test functions and $\Phi^{\times}$ is the space of distributions. Associated with this RHS, there is another RHS.

$$\Phi \subset \mathcal{H} \subset \Phi' \tag{75}$$

where Φ', the dual space of Φ, is the space of linear functionals over Φ. $\Phi^{\times}$ is called the antidual space of Φ.

Unbounded operators are not defined on the whole of $\mathcal{H}$ but only on dense subspaces of $\mathcal{H}$ that are not invariant under the action of the observables. Then expectation values, uncertainties and commutation relations are not well defined on the whole of $\mathcal{H}$. The space Φ is the maximal invariant subspace of $\mathcal{H}$ on which such quantities are well defined. The bras and kets associated with the elements in the continuous spectrum of an observable belong, respectively, to Φ' and Φ^{x} rather than to $\mathcal{H}$.

In quantum mechanics, when the position, momentum and energy operators Q, P, H are unbounded, we seek a subspace Φ of $\mathcal{H}$ on which these quantities can be calculated yielding finite values.

Following de la Madrid (2005), we consider an exactly soluble example in quantum mechanics. We consider the one dimensional motion of a spinless particle of mass m in a rectangular barrier potential

$$V(x) = \begin{cases} 0, & -\infty < x < a \\ V_0, & a < x < b \\ 0, & b < x < \infty \end{cases} \tag{76}$$

In the position representation the position observable Q in one dimension (on the real line) is given by

$$Qf(x) = xf(x), \quad \text{for all } x \in \mathcal{R} \tag{77}$$

the momentum operator P by

$$Pf(x) = \frac{\hbar}{i} \frac{d}{dx} f(x), \tag{78}$$

and the energy operator H is realized by

$$Hf(x) = \left(-\frac{\hbar^2}{2m} \frac{d^2}{dx^2} + V(x) \right) f(x). \tag{79}$$

The Hilbert space on which these differential operators act is $\mathcal{H} = L^2$. The three operators are unbounded and can be defined only on subdomains of L^2. The Q, P, H operators have the spectra $(-\infty, \infty)$, $(-\infty, \infty)$, $(0, \infty)$ respectively. The spectrum of H is doubly degenerate. The eigenfunctions of the observables do not belong to L^2, that is, they are not square integrable.

We construct the rigged Hilbert spaces of Eqs. (74) and (75).

The subspace Φ is taken as the intersection of the domains of all the powers of Q, P and H:

$$\Phi = \bigcap_{n, m = 0}^{\infty} D(A^n B^m). \tag{80}$$

$$A, B = Q, P, H$$

Φ is the maximal invariant subspace of the algebra generated by Q, P and H.

We write

$$\Phi = \{ \phi \in L^2 | \phi \in C^\infty (\mathcal{R}), \phi^{(n)} (a) = \phi^{(n)} (b) = 0, \quad n = 0, 1, \ldots$$

$$P^n Q^m H^l \phi(x) \in L^2, n, m, l = 0, 1, \ldots \} \tag{81}$$

where $C^\infty (\mathcal{R})$ is the set of infinite differentiable functions and $\phi^{(n)}$ denotes the nth derivative of ϕ. The elements of Φ satisfy the following estimates

$$\|\phi\|_{n, m, l} \equiv \sqrt{\int_{-\infty}^{\infty} dx \, |P^n Q^m H^l \phi(x)|^2} \langle \infty,$$

$$n, m, l = 0, 1, \ldots \tag{82}$$

The estimates induce a topology on Φ, that is, they induce a meaning of convergence of sequences as follows. A sequence $\{\phi_\alpha\}$ Φ converges to ϕ when $\{\phi_\alpha\}$ converges to ϕ with respect to all the estimates Eq. (82):

$$\phi_\alpha \xrightarrow[\alpha \to \infty]{\tau_\Phi} \phi \text{ if } \|\phi_\alpha - \phi\|_{n, m, l} \xrightarrow[\alpha \to \infty]{} 0, n, m, l = 0, 1, \ldots \tag{83}$$

We see that Φ is very similar to the Schwartz space $S(\mathcal{R})$, of rapidly decreasing functions, the differences being that the derivatives of the elements of Φ vanish at $x = a$, b and that Φ is invariant under P, Q and H. Here we consider a domain that is invariant under the action of the operator. A function $f : \mathcal{R} \to \mathcal{C}$ belongs to $S(\mathcal{R})$ if it is differentiable an infinite number of times and if it and all of its derivatives decrease more rapidly at infinity than the inverse of any polynomial.

Thus, we write

$$\Phi \equiv S(\mathcal{R} - \{a, b\}). \tag{84}$$

Φ is invariant under the action of the operators,

$$AS\,(\mathcal{R} - \{a, b\}) \subset S(\mathcal{R} - \{a, b\}), \quad A = P, Q, H \tag{85}$$

The expectation values

$$(\phi, A^n\, \phi), \phi \in S(\mathcal{R} - \{a, b\}),$$

$$A = P, Q, H, \quad n = 0, 1, \ldots \tag{86}$$

are finite, and the commutation relations are well-defined.

The space $\Phi^\times$ is the collection of τ_Φ-continuous antilinear functionals over Φ.

The RHS of the system is

$$\Phi \subset \mathcal{H} \subset \Phi^\times, \tag{87}$$

which we denote in the position representation by

$$S(\mathcal{R} - \{a, b\}) \subset L^2 \subset S^\times\,(\mathcal{R} - \{a, b\}) \tag{88}$$

The eigenkets $|p\rangle$, $|x\rangle$ and $|E^\pm\rangle$ of P, Q and H belong to $\Phi^\times \equiv S^\times\,(\mathcal{R} - \{a, b\})$. The Dirac kets are distributions that belong to the space $\Phi^\times$.

The Dirac bras are distributions that belong to the space of Φ', the space of linear functionals over Φ. The bras $|p\rangle$, $|x\rangle$ and $\langle^\pm E|$ of P, Q and H belong to $S'\,(\mathcal{R} - \{a, b\})$.

The example discussed above shows that when the spectrum of an observable has a continuous part the RHS is the natural setting for quantum mechanics. Dirac's bra-ket formalism is fully implemented by the RHS.

CHAPTER 2

The Basic Concepts

The quantum mechanical descriptions of myriad systems in nature have been enormously successful. A vast range of systems – from electrons, nuclei, atoms, molecules, etc. to the interior of stars – can be properly understood in quantum mechanics. Quantum mechanics is a fundamental theory that applies to all physical systems regardless of size. For a vast range of macroscopic systems, however, a description by classical physics is adequate. Recent technological advances using transistors, lasers, superconductors, magnetic resonance, etc. are based on the underlying quantum principles. During the last eighty decades, all experiments performed have testified to the essential correctness of quantum mechanics. Indeed, quantum mechanics is the basis of most of present-day physics and chemistry.

Based on the early works in quantum theory by Planck, Einstein, Bohr, de Broglie, Sommerfeld, and others, quantum mechanics was formulated during 1925 – 1927 by Heisenberg, Pauli, Schrödinger, Born, Dirac, Jordan and others.

The fascinating world of quantum mechanics has brought about profound change in our thinking about the description of natural phenomena, especially on atomic scale. Quantum mechanics is based on concepts radically different from those of classical physics. In this chapter we introduce the basic concepts of quantum mechanics. Certain mathematical topics needed for the development of the mathematical structure of quantum mechanics have been discussed in the previous chapter.

2.1 STATES, OBSERVABLES, AND OPERATORS: THE BASIC POSTULATES

The description of a physical system requires a separable, complex Hilbert space $\mathcal{H}$. At any instant of time the state of the physical system is represented by a ket $|\Psi(t)\rangle$ in $\mathcal{H}$. More precisely, we say that the state is characterized by a ray in the corresponding Hilbert space. Usually, we take a representative of the ray by normalizing the state to unity. A phase factor of modulus one is unspecified.

We summarize all this in the basic postulate I.

Postulate I

The complete description of the state of a physical system S is provided by the elements of a separable, complex Hilbert space $\mathcal{H}$ associated with the system. At any instant of time t, a pure state of S is represented by a unit ray $|\psi(t)\rangle_R$ in $\mathcal{H}$. An element of the ray is called a state vector or ket. Every linear combination of such state vectors represents a possible physical state of S.

When S is in a pure state, it can be described by one ket. Otherwise, the state is a mixture or, is in a mixed state. The superposition principle is included in the postulate.

The physically meaningful attributes of the system, such as position, momentum, energy, etc. are measurable by a suitable device. These entities are called observables. The next postulate tells us that in quantum mechanics the observables are associated with operators.

Postulate II

An observable of a physical system is represented by a linear self-adjoint operator which acts in the Hilbert space associated with the system.

Postulate III introduces probabilities into quantum mechanics.

Postulate III

If a physical system is in a pure state $|\psi\rangle$, then a measurement corresponding to an observable A yields one of the eigenvalues a_i of A with the probability

$$P_\psi(a_i) = |\langle a_i|\psi\rangle|^2, \tag{1}$$

where all the kets are normalized, and

$$P_\psi(a_i) \geq 0, \quad \sum_i P(a_i) = 1. \tag{2}$$

Here, we use the same symbol for the observable and the corresponding self-adjoint operator. As a result of the measurement, the state of the system $|\psi\rangle$ is "thrown into" one of the eigenstates $|a_i\rangle$ of A. If the state is an eigenket $|a_i\rangle$, then a measurement will certainly yield the value a_i. Thus, the only measurable values of an observable are the eigenvalues of the corresponding operator.

Let us consider a set of identical replicas of a physical system in a pure state $|\psi\rangle$. We now measure an observable of all elements of this ensemble and obtain one of the possible eigenvalues, each one with a certain probability. The weighted mean value of all measured values is the average or mean value of the observable in that state.

$$\langle A \rangle_\psi = \sum_i a_i\, P_\psi\,(a_i)$$

$$= \sum_i a_i\, |\langle a_i|\psi\rangle|^2$$

$$= \sum_i a_i\, \langle a_i|\psi\rangle\, \langle\psi|a_i\rangle$$

$$= \sum_i a_i\, \langle\psi|a_i\rangle\, \langle a_i|\psi\rangle$$

$$= \sum_i a_i\, \langle\psi|A|a_i\rangle\, \langle a_i|\psi\rangle$$

$$= \langle\psi|A\left(\sum_i |a_i\rangle\, \langle a_i|\right)|\psi\rangle$$

$$= \langle\psi|A|\psi\rangle, \tag{3}$$

where we have used the closure relation

$$\sum_i |a_i\rangle\langle a_i| = 1. \tag{4}$$

$\langle A \rangle_\psi$ is called the "expectation value" of the observable in the state $|\psi\rangle$.

Let $|\phi\rangle$ be an arbitrary state of the system. Then the probability that the system in the state $|\psi\rangle$ will be measured to be in the state $|\phi\rangle$ is

$$P_\psi\,(\phi) = |\langle\phi|\psi\rangle|^2. \tag{5}$$

The scalar product between the states is called a "probability amplitude".

The previous postulates on the states and observables describe the kinematics of quantum mechanics. The next postulate gives the dynamics — it describes the time evolution of the state vectors.

Postulate IV

In the time interval between two consecutive measurements, pure states of a physical system continue to be pure, and there exists in every unit ray $|\Psi(t)\rangle_R$ some representative state vector $|\Psi(t)\rangle$ whose time evolution is given by the Schrödinger equation

$$i\hbar\,\frac{\partial}{\partial t}\,|\Psi(t)\rangle = H(t)||\Psi(t)\rangle, \tag{6}$$

where $H(t)$, called the Hamiltonian, is an observable of the system.

2.2   MEASUREMENTS AND OBSERVABLES: COMMUTABILITY AND COMPATIBILITY

Let us consider a physical system. When we make an observation we measure some dynamical variable of the system. The result of the measurement is always a real number. We consider the eigenvalue equation

$$A|a'\rangle = a'|a'\rangle \tag{7}$$

where $|a'\rangle$ is an eigenket of operator A corresponding to a dynamical variable of the system with the eigenvalue a'. If the physical system is in one of the eigenstates $|a'\rangle$ say of the observable, then a measurement of the observable A will certainly give as a result the eigenvalue a', which is real Conversely, if the system is in a state that the measurement of an observable A is certain to yield a particular value a', then the state of the system is an eigenstate $|a'\rangle$ of A, which belongs to the eigenvalue a'.

In general, before a measurement of the observable A is performed, the state of the system is assumed to be a linear combination of the eigenstates of A, i.e., an arbitrary state

$$|\alpha\rangle = \sum_{a'} c_{a'} |a'\rangle = \sum_{a'} |a'\rangle\langle a'|\alpha\rangle. \tag{8}$$

When a measurement is made, the system in the state $|\alpha\rangle$ jumps into one of the eigenstates, $|a'\rangle$ say, of the observable A, and the result of the measurement is the eigenvalue a', this eigenstate $|a'\rangle$ belongs to:

$$|\alpha\rangle \xrightarrow{\quad a \text{ measurement} \quad} |a'\rangle \tag{9}$$

Thus, a measurement in general causes a change of the state of the system. Only when the system is in one of the eigenstates, say $|a'\rangle$, of the observable A, then

$$|\alpha'\rangle \xrightarrow{\quad a \text{ measurement} \quad} |a'\rangle \tag{9a}$$

and the result of a measurement is the eigenvalue a' of A corresponding to $|a'\rangle$. Thus, we say that the result of a measurement of a real dynamical variable of a system is one of its eigenvalues. After the first measurement has been made then successive measurements performed immediately afterward on the same observable will yield the same result.

If $|\alpha\rangle$ given by Eq. (8) is the state ket of a physical system before the measurement, then we do not know in advance to which one of the various eigenkets $|a'\rangle$ the system will be thrown into as a result of the measurement. We postulate that we can predict the probability for jumping into a particular eigenket. The probability for going into $|a'\rangle$ is

$$P_{|a'\rangle} = |\langle a'|\alpha\rangle|^2, \tag{10}$$

where the state ket $|\alpha\rangle$ is normalized, i.e., $\langle\alpha|\alpha\rangle = 1$. In considering the probable value of a measurement, we perform a series of measurements on an ensemble, which is a collection of similarly prepared systems. The ensemble is a conceptual set of identical systems, all in the same state $|\alpha\rangle$. Such an ensemble is called a pure ensemble.

Let us consider a measurement device which selects only one of the eigenkets of A, say $|a'\rangle$ and rejects all other eigenkets. This selective measurement or filtration is obtained thus

$$\Lambda_{a'}|\alpha\rangle = |a'\rangle\langle a'|\alpha\rangle, \tag{11}$$

where $\Lambda_{a'}$ is the projection operator. Schwinger (2001) has given a detailed discussion of measurement algebra.

Given two observables A and B, a state may be simultaneously an eigenstate of A and B.

Theorem 2.1 If $|a_i\rangle$ ($i = 1, 2, \ldots$) are a complete set of orthonormal vectors in a linear vector space which are simultaneously eigenstate of two observables A and B, then the observables commute.

Let
$$A|a_i\rangle = \alpha_i|a_i\rangle,$$
$$B|a_i\rangle = \beta_i|a_i\rangle, \tag{12}$$

where α_i and β_i are eigenvalues of A and B respectively. Then

$$AB|a_i\rangle = A\beta_i|a_i\rangle = \alpha_i\beta_i|a_i\rangle,$$
$$BA|a_i\rangle = B\alpha_i|a_i\rangle = \alpha_i\beta_i|a_i\rangle,$$

or,
$$(AB - BA)|a_i\rangle = 0,$$

or,
$$(AB - BA) \sum_i c_i\, |a_i\rangle = 0. \tag{13}$$

Since any arbitrary ket $|\alpha\rangle$ can be written as $\sum_i c_i\, |a_i\rangle$, we get for any ket $|\alpha\rangle$

$$(AB - BA)|\alpha\rangle = 0. \tag{13a}$$

We thus obtain

$$AB - BA = 0,$$

or,
$$[A, B] = 0, \tag{14}$$

where $[A, B] \equiv AB - BA$ is the "commutator" of A and B. Observables A and B are said to be "compatible" if the corresponding operators commute, i.e., if $[A, B] = 0$. When $[A, B] \neq 0$, they are called "incompatible" observables

Let A and B be two compatible observables. Then we consider an eigenket of B, $|b_i\rangle$ say, belonging to the eigenvalue β_i, and expand it in terms of eigenkets of A as

$$|b_i\rangle = \int |a_i b_i c\rangle \, da_i + \sum_r |a_r b_i d\rangle, \tag{15}$$

where $|a_i b_i c\rangle$ and $|a_r b_i d\rangle$ are eigenkets of A and the labels c and d are put in to distinguish them when the eigenvalues a_i and a_r are equal. The label b_i is put in to indicate that these come from the expansion of a ket $|b_i\rangle$.

Each of these eigenkets of A is also an eigenket of B belonging to the eigenvalue b_i. We have

$$0 = (b - b_i)|b_i\rangle$$
$$= \int (b - b_i)|a_i b_i c\rangle \, da_i + \sum_r (b - b_i)|a_r b_i d\rangle \tag{16}$$

and

$$A(b - b_i)|a_r b_i d\rangle = (b - b_i)A|a_r b_i d\rangle$$
$$= (b - b_i)a_r|a_r b_i d\rangle$$
$$= a_r(b - b_i)|a_r b_i d\rangle.$$

Thus the ket $(b - b_i)|a_r b_i d\rangle$ is an eigenket of A belonging to the eigenvalue a_r. Also, the ket $(b - b_i)|a_i b_i c\rangle$ is an eigenket of A belonging to the eigenvalue a_i. Eq. (16) gives

$$(b - b_i)|a_i b_i c\rangle = 0, \quad (b - b_i)|a_r b_i d\rangle = 0,$$

so that all the eigenkets on the RHS of Eq. (16) are eigenkets of B and A. The simultaneous eigenkets of A and B form a complete set. The simultaneous eigenkets of A and B are denoted as $|a_i b_i\rangle$ or $|a_r b_i\rangle$.

The converse to the above Theorem 2.1 is

Theorem 2.2 If A and B are two observables such that their simultaneous eigenstates form a complete set, then A and B commute.

Proof If $|a_i b_i\rangle$ is a simultaneous eigenket belonging to the eigenvalues a_i and b_i, then

$$(AB - BA)|a_i b_i\rangle = (a_i b_i - b_i a_i)|a_i b_i\rangle = 0. \tag{17}$$

But any arbitrary ket $|P\rangle$ can be expanded in terms of the complete set of simultaneous eigenkets $|a_i b_i\rangle$. Hence

$$(AB - BA)|P\rangle = 0$$

or, $$AB - BA = 0. \tag{18}$$

If two observables are compatible, then there are simultaneous eigenstates. For incompatible observables, in general, a simultaneous eigenstate is not possible.

We can generalize the idea of simultaneous eigenstates to more than two mutually compatible observables. Let A, B, C, ... be compatible observables, i.e.,

$$[A, B] = [B, C] = [C, A] = ... = 0. \tag{19}$$

A set of observables is called (following Dirac) a complete set of commuting observables (c.s.c.o.) if all the operators corresponding to these observables are compatible (i.e., they commute with each other) and the basis formed by the simultaneous eigenvectors is unique (up to phases). For a linear harmonic oscillator, the c.s.c.o. consists of one operator, either the Hamiltonian operator or the position operator or some other operator. The eigenvalues of a c.s.c.o. are called quantum numbers.

We now consider measurements of two compatible observables A and B. We measure A first and obtain the result a_i. Next we measure B and obtain the result b_j. Then we measure A again. This measurement gives a_i with certainty, i.e., the second measurement of B does not destroy the information obtained in the first measurement of A. This is clear if the eigenvalues of A are non-degenerate. In case of degeneracy, after the first measurement of A which gives the result a_i, the system is thrown into a linear combination $\sum_{n=1}^{m} C_{a_i}^{(n)} \left| a_i, b_j^{(n)} \right\rangle$, where m is the degree of degeneracy and the kets $\left| a_i\, b_j^{(n)} \right\rangle$ all have the same eigenvalue a_i corresponding to A. The second B measurement selects one term of the linear combination, $\left| a_i\, b_j^{(k)} \right\rangle$ say. But the third measurement still yields the result a_i. Thus, irrespective of the degeneracy, the A and B measurements do not interfere if A and B are compatible.

Two observables are called compatible if the measurement of one does not interfere with the state that has been prepared by the measurement of the other operator; the operators corresponding to compatible observables commute. We can consider two or more commuting observables as a single observable, the result of a measurement of which consists of two or more numbers.

Observables represented by non-commuting operators are incompatible. Such observables do not have a complete set of simultaneous eigenkets. If two observables do not commute, then a simultaneous eigenket is not impossible, but is exceptional. Let us consider an s-state ($l = 0$). Even though L_x and L_y do not commute, this state is a simultaneous eigenstate of L_x and L_y (with eigenvalue zero for both). The subspace is one-dimensional.

For a given physical system, the question of what constitutes a c.s.c.o. depends not only on the system but also on the experimental facilities. For a free electron, the three components of linear momentum $\vec{P}$ are compatible and may form a c.s.c.o. The discovery of electron spin changed the situation. The subspace for an electron is two-dimensional. A c.s.c.o. consists of the three components of $\vec{P}$ and a component of the spin operator $\vec{S}$, i.e. $\{P_x, P_y, P_z, S_z\}$. The introduction of a new quantum number (the spin quantum number) enlarges the set of commuting observables.

2.3 THE CLASSICAL CONNECTION

There is a deep and subtle relationship between quantum mechanics and classical physics. The significance of the interface of quantum mechanics with classical physics is clarified by the following three principles:

(a) The Correspondence Principle.

(b) The Complementarity Principle.

(c) The Uncertainty Relations.

(a) The Correspondence Principle

We know that classical mechanics and classical electromagnetic theory describe very accurately most properties of macroscopic systems. Bohr provided the fundamental insight that there was a continuity between the two domains – the macro and the micro-world.

The correspondence principle, introduced during 1918-1923 by Bohr, states that the laws of quantum physics reduce in the limit of large quantum numbers to classical theory. This principle served as a general guide to the development of a great deal of atomic spectra and in the formulation of quantum mechanics. Bohr's profound ideas on the nature of the atom and the use of the correspondence principle helped in the passage from the earlier quantum theory to quantum mechanics.

An important feature of the correspondence principle is the correlation between the classical frequencies of motion and the spectroscopic frequencies. In the asymptotic limit of states with large quantum numbers, the quantum frequencies and the classical frequencies approach equality. It is postulated that the intensity of emission is determined by the asymptotic correlation with the corresponding Fourier components of the electric moment. The polarization of the radiation is postulated to asymptotically equal that of the classical radiation. The correspondence principle provides the basis of the treatment of selection rules governing "allowed" and "forbidden" transitions. It is postulated that a selection rule holds over the entire range of the concerned quantum number.

(b)   The Complementarity Principle

Classical experiments on interference and diffraction have established the wave nature of light. A series of experiments, mainly the photoelectric effect and the Compton effect, however, can be explained by assuming that the electromagnetic radiation consists of an assembly of massless particles (photons), each of energy E and momentum $\vec{p}$ given by

$$E = \hbar\omega$$

$$\vec{p} = \hbar\vec{k} \tag{20}$$

$$(\hbar = 1.05 \times 10^{-27} \text{ erg. sec.})$$

where $\vec{k}$ is the wave vector and ω the angular frequency. The dual aspect of light – particle attributes and wave attributes – is clearly seen in Eq. (20). We get a glimpse of wave-particle dualism in Einstein's famous 1905 paper on light (Einstein, (1905)). In a remarkable paper, in 1909, Einstein further developed his ideas on the light quanta using Planck's radiation law, Einstein calculated the mean square of the energy fluctuation in the frequency interval $(\nu, \nu + d\nu)$ of the radiation in the volume V as

$$\overline{\epsilon^2} = \hbar\nu E + \frac{c^3}{8\pi\nu^2\, d\nu} \frac{E^2}{V} \tag{21}$$

where E is the mean energy of the radiation in volume V (Einstein, 1909). The second term in Eq. (21) corresponds to the fluctuations according to the classical wave theory as due to the interferences among the partial waves. The first term, however, corresponds to the fluctuations in the number of light-quanta (photons). Thus, Eq. (21) reflects the wave particle duality of radiation. Einstein also considered the Brownian motion of a mirror that perfectly reflects the radiation in the frequency interval $(\nu, \nu + d\nu)$, and is transparent at all other frequencies. Einstein obtained the following expression for the mean square fluctuation $\overline{\Delta^2}$ of the light pressure during the time interval τ on a mirror of surface area f

$$\frac{\overline{\Delta^2}}{\tau} = \frac{1}{c}\left(h\nu\rho + \frac{c^3}{8\pi\nu^2}\rho^2\right) d\nu \cdot f \tag{22}$$

where ρ is the density of radiative energy in the interval $(\nu, \nu + d\nu)$. The second term in Eq. (22) follows from the wave theory. The first term in this formula can be explained in the corpuscular theory of light – quanta of the energy $h\nu$ and momentum $d\nu/c$. (Pauli (1959); El'yashevich (1977)).

In 1923 de Broglie extended the wave-particle duality for radiation to material particles. To a particle of momentum $\vec{p}$ a wave is associated with the wavelength given by

$$\lambda = \frac{h}{|\vec{p}|},\qquad(23)$$

or,
$$\vec{p} = \hbar\vec{k}.\qquad(24)$$

de Broglie's prediction was brilliantly confirmed by diffraction experiments with electrons by Davisson and Germer (1927) and G.P. Thomson (1927).

In order to understand the implications of the uncertainty principle and the wave-particle dualism Bohr (1927, 1928) introduced the point of view termed "complementarity". According to Bohr the complementarity principle provides a deep insight into the general epistemological situation underlying quantum mechanics.

The following points emerge from Bohr's analysis.

(i)   "However far the phenomena transcend the scope of classical physical explanation, the account of all evidence must be expressed in classical terms".

(ii) At the microscopic level, it is impossible to make "any sharp separation between the behavior of atomic objects and the interaction with the measuring instruments which serve to define the conditions under which the phenomena appear".

(iii) "Evidence obtained under different experimental conditions can not be comprehended within a single picture, but must be regarded as *complementary* in the sense that only the totality of the phenomena exhausts the possible information about the objects".

(The above quotations are taken from N. Bohr "Discussion with Einstein on Epistemological Problems in Atomic Physics" in Albert Einstein: Philosopher – Scientist (1949)).

According to Bohr's principle of complementarity, the description of a phenomenon in microscopic physics requires two or more complementary elements. If the use of a classical concept excludes that of another, we call both concepts complementary to each other. To determine the position and the corresponding momentum coordinates of a particle, we need to set up mutually exclusive experimental arrangements. Thus the position and momentum of a particle are complementary.

The complementary principle may be stated as follows (Messiah (1961)):

The description of the physical properties of microscopic objects in classical language requires pairs of complementary concepts; the accuracy in one element cannot be improved without a corresponding loss in the accuracy of the other.

The wave property and the particle property are two complementary aspects which are mutually exclusive, i.e., the two classical pictures hold under different arrangements. Observation of one aspect prevents the observation of the other. In the micro world, both pictures are present, but when one is observed the other is excluded.

(c)   The Uncertainty Relations I

Let us consider a system in a state characterized by a normalized ket $|\psi\rangle$. Let A and B be any two observables of the system. Let the self adjoint operators A and B on a Hilbert space $\mathcal{H}$, with domains $D(A)$ and $D(B)$ respectively, satisfy the commutation relation

$$[A, B] = iC. \tag{25}$$

We note that

$$(iC)^\dagger = -iC^\dagger = (AB)^\dagger - (BA)^\dagger$$

$$= B^\dagger A^\dagger - A\dagger B^\dagger$$

$$= BA - AB$$

$$= -(AB - BA)$$

$$= -iC$$

The factor i is inserted to make $C^\dagger = C$. Let us consider

$$(\Delta A)^2 (\Delta B)^2 = \langle\psi|(A - \langle A\rangle)^2|\psi\rangle\langle\psi|(B - \langle B\rangle)^2|\psi\rangle. \tag{26}$$

where $\langle A\rangle = \langle\psi|A|\psi\rangle$ and $\langle B\rangle = \langle\psi|B|\psi\rangle$. $(\Delta A)^2$ is called the dispersion or mean square deviation of A. Clearly, the dispersion vanishes when the state is an eigenstate of A. Now,

$$(\Delta A)^2_\psi = \left(\left(A - \langle A\rangle_\psi\right)^2\right)_\psi$$

$$= (\psi, \Delta A^2\,\psi)$$

$$= (\Delta A\psi, \Delta A\psi), \quad \text{using the hermiticity of } A.$$

So
$$(\Delta A)_\psi = \|\Delta A\psi\|$$

and similarly,
$$(\Delta B)_\psi = \|\Delta B\psi\|$$

Also,
$$[A, B] = [\Delta A, \Delta B].$$

If we apply the Schwarz inequality

$$|(\alpha, \beta)|^2 \le \|\alpha\|^2\,\|\beta\|^2 \tag{27}$$

to the vectors

$$|\alpha\rangle = \Delta A|\psi\rangle \tag{28}$$

$$|\beta\rangle = \Delta B|\psi\rangle, \tag{29}$$

for $\psi \in \mathcal{H}$, we get

$$\|\Delta A\psi\| \, \|\Delta A\psi\| \geq \|\langle \Delta A\psi, \Delta B\psi\rangle\|$$

$$= |(\psi, AB\psi)|, \tag{30}$$

where the last equality follows because ΔA is self-adjoint. Now, we consider the identity

$$AB = \frac{AB + BA}{2} + \frac{AB - BA}{2}$$

$$= \frac{1}{2}\{A, B\} + \frac{1}{2}[A, B] \tag{31}$$

where $\{A, B\} = AB + BA$ is the anticommutator of the two operators. Hence

$$(\Delta A)^2_\psi \, (\Delta B)^2_\psi \geq \left| \langle\psi| \frac{1}{2}\{A, B\} + \frac{1}{2}[A, B] |\psi\rangle \right|^2. \tag{32}$$

Since ΔA and ΔB are self-adjoint, the expectation value of $\{A, B\}$ is real and that of $[A, B]$ is pure imaginary. For any two real numbers a and b we have $|a + ib|^2 = |a|^2 + |b|^2$, and hence

$$(\Delta A)^2_\psi \, (\Delta B)^2_\psi \geq \frac{1}{4} \langle\{A, B\}\rangle^2_\psi + \frac{1}{4} \langle C\rangle^2_\psi. \tag{33}$$

This is the generalized uncertainly relation. Equation (33) leads to the weaker condition

$$(\Delta A)_\psi \, (\Delta B)_\psi \geq \frac{1}{2} \langle C\rangle_\psi \tag{34}$$

which is the most commonly quoted form. The relation Eq. (34) holds if $A\psi \in D(B)$ and $B\psi \in D(A)$.

An uncertainty relation is state dependent and does not refer to simultaneous measurement of the observables. $(\Delta A)_\psi$ refers to the dispersion or statistical spread in the measured values of the observable A obtained by making repeated identical and independent experiment on a set of large numbers of identically prepared systems, all in the same state described by the vector $|\psi\rangle$. If the system is in an eigenstate of A then $(\Delta A)_\psi = 0$, i.e., the state $|\psi\rangle$ is nondispersive in A. The uncertainty relation tells us that it is not possible to measure observable A and B in a system with arbitrarily small dispersions, unless $[A, B] = 0$. For $[A, B] \neq 0$, nondispersive states do not normally exist. The uncertainty relation imposes the fundamental limitation that if one prepares the system in a state $|\psi\rangle$, then the statistical dispersions in A and B must satisfy the inequality. This discussion pertains to pure state.

The above inequality becomes an equality only if

(a) there exists $\lambda \in \mathbb{C}$ such that $\lambda \Delta A |\psi\rangle = \Delta B |\psi\rangle$,

and (b) $\{\Delta A, \Delta B\}_\psi |\psi\rangle = 0$. $\tag{35}$

From these two conditions, we get

$$\lambda(\Delta A)_\psi^2 + \frac{1}{\lambda}(\Delta B)_\psi^2 = 0.$$

Using Eq. (25) on $|\psi\rangle$, we obtain

$$\lambda(\Delta A)_\psi^2 - \frac{1}{\lambda}(\Delta B)_\psi^2 = i\langle C\rangle_\psi.$$

Adding, we get for an equality when A and B have minimal dispersion

$$\lambda = \frac{i}{2}\frac{\langle C\rangle_\psi}{(\Delta A)_\psi^2} = \frac{[A, B]_\psi}{2(\Delta A)_\psi^2}. \tag{36}$$

We return to the uncertainty relations involving position-momentum and time-energy later.

2.4   THE DENSITY MATRIX

A physical system in a pure state is described by a single ket $|\psi\rangle$. It often happens, however, that the information about the system is less than complete. In the study of polarized beams, spin orientations, angular correlations and many other phenomena, we deal with a system with states of "less than maximum information". Here we merely say that the system has certain probabilities p_α, p_β, ... for being in the states represented by the kets $|\alpha\rangle$, $|\beta\rangle$, ... respectively. The state of the system cannot be described by a definite state ket, but by a statisfical mixture of vectors.

It is very convenient here to introduce the density (or statistical) operator formalism pioneered by von Neumann (1927) and Dirac (1929, 1930) (see Fano, (1957)).

We consider a single statistical ensemble of quantum mechanical systems prepared by identical procedures. We perform a measurement of an observable A on the system. The mean value $\langle A\rangle$ of the results of measurement has a probability p_m of being $\langle A\rangle_m = \langle m|A|m\rangle/\langle m|m\rangle$. When a large number of measurements are performed, the statistical average of these measurements is given by the ensemble average

$$\overline{\langle A\rangle} = \Sigma\, p_m\, \langle m|A|m\rangle, \tag{37}$$

assuming the vectors to be normalized to unity. The bar indicates the ensemble average.

Here, the ensemble averaging means that the expectation values taken with respect to definite states are further averaged with weight factors corresponding to the probability of being in a state, $|m\rangle$, when chosen randomly from the ensemble.

We now define the operator

$$\rho = \sum_{m} |m\rangle \, p_m \, \langle m|, \tag{38}$$

where the vectors $|m\rangle$ are normalized to unity (but not necessarily orthogonal) and the statistical weights p_m satisfy

$$p_m \geq 0 \quad \text{for all } m,$$

and

$$\sum_{m} p_m = 1. \tag{39}$$

The operator ρ is called the density operator (or, satistical (state) operator) or the density matrix. It contains all the necessary information about the ensemble.

The density matrix expresses the result of taking quantum-mechanical expectation values and ensemble averages in the same operation. Thus, there is a double averaging process – one due to the quantum mechanical probabilistic aspect and another due to the statistical aspect of the ensemble.

Let us calculate

$$Tr\rho A = \sum_{m} p_m \, Tr \, (|m\rangle\langle m|A).$$

where $Tr \, X$ indicates the trace of a matrix X.

Since $p_m = |m\rangle\langle m|$ is a projector of trace unity,

$$Tr \, p_m \, A = Tr \, p_m^2 \, A = Tr \, p_m \, A \, p_m$$

$$= Tr \, |m\rangle\langle m|A|m\rangle\langle m|$$

$$= \langle m|A|m\rangle \, Tr \, p_m$$

$$= \langle m|A|m\rangle.$$

Hence,

$$Tr \, \rho \, A = \sum_{m} p_m \, \langle m|A|m\rangle = \overline{\langle A\rangle}, \tag{40}$$

using Eq. (38)

Thus, for any operator Q, the ensemble average is

$$\overline{\langle Q\rangle} = Tr \, (\rho Q). \tag{41}$$

We may regard the density matrix as defined by Eq. (41) rather than by Eq. (38).

Since the unit operator I has the mean value 1 we get

$$Tr \, \rho = \sum_{n} \rho_{nn} = 1. \tag{42}$$

For any function of the observable Q, we have

$$\overline{\langle f(Q)\rangle} = Tr \, \rho \, f \, (Q). \tag{43}$$

The average $\overline{\langle Q\rangle}$ can be calculated in any suitable basis, because the trace is independent of representations.

The density matrix has the following properties.

(i) The condition that $\langle Q \rangle$ is real for every hermitian operator Q requires that ρ is hermitian, namely

$$\rho = \rho^{\dagger}. \tag{44}$$

(ii)
$$Tr\,\rho = 1. \tag{45}$$

(iii) ρ is non-negative:

$$\langle u|\rho|u \rangle \geq 0 \quad \forall |u\rangle. \tag{46}$$

Proof
$$\langle u|\rho|u \rangle \geq \sum_{m} p_m\,|\langle u|m\rangle|^2 \geq 0.$$

(iv) Every diagonal element of ρ in any matrix representation is non-negative:

$$\rho_{nn} \geq 0 \tag{47}$$

So, the eigenvalues of ρ are all non-negative.

(v) The eigenvalues of ρ all lie between 0 and 1.

(vi) For a pure ensemble,

$$\rho^2 = \rho. \tag{48}$$

Proof
$$\rho = \sum p_m|m\rangle\langle m|$$

$$\rho^2 = \sum_{m,\,n} p_m\,p_n|m\rangle\langle m|n\rangle\langle n|$$

$$= \sum_{m,\,n} p_m\,p_n\,\delta_{mn}|m\rangle\langle n|$$

$$= \sum_{n} p_n^2\,|n\rangle\langle n.$$

For a pure ensemble, all the eigenvalues p_n are zero except one which is 1, since $\sum_{n} p_n = 1$. Hence $\rho^2 = \rho$.

Thus for a pure ensemble, the density operator is the projector

$$\rho_\chi = |\chi\rangle\langle\chi|, \tag{49}$$

and
$$\rho_\chi^2 = \rho_\chi. \tag{50}$$

Conversely, if a density operator is a projector it represents a pure ensemble.

If ρ is a projector, it projects onto a one-dimensional subspace.

Let us consider a density matrix of N rows and columns characterizing a system with N independent pure states. In general, it takes $N^2 - 1$ separate measurements to identify the state of the system. Pure states of a system with $Tr\,\rho^2 = 1$ give the maximum of information.

(vii) A density operator can be written in several different ways as a linear combination of projectors in the form Eq. (38). A necessary and sufficient condition for ρ to be a projector, i.e., representing a pure state, is that all ket vectors $|m\rangle$ be identical to within a phase.

(viii) The trace of the quantum purity operator ρ^2 cannot exceed unity

$$Tr\,\rho^2 \leq 1 \tag{51}$$

where the equality holds only for a pure ensemble, i.e., only if ρ is a projection operator.

To prove it, we note that the hermitian matrix ρ can be diagonalized by a unitary transformation T.

$$\rho_j\,\delta_{jj'} = \sum_{nn'} T_{jn}\,\rho_{nn'}\,T_{n'j'}^{-1} \tag{52}$$

Since $0 \leq \rho_n \leq 1$, we have $\rho_n^2 \leq \rho_n$.

Now,
$$Tr\,(\rho^2) = \sum_j \rho_j^2 \leq \sum_j \rho_j = 1, \tag{53}$$

using Eq. (42).

Thus for a general state

$$Tr\,\rho^2 \leq 1.$$

The equality holds only if $\rho_n^2 = \rho_n$, $\forall n$, i.e., only if ρ is a projection operator.

(ix) Let ρ be the density operator in the representation $|\phi\rangle$. Then in the representation $|\chi\rangle = S|\phi\rangle$, where S is unitary, the density operator ρ' is given by

$$\rho' = S^{-1}\,\rho S. \tag{54}$$

Proof The expectation value of any observable is independent of the representation.

$$\overline{\langle A\rangle} = Tr\,(\rho A)$$
$$= Tr\,(\rho' S^{-1} A S),$$

where under the change in representation (change of basis)

$$A \rightarrow S^{-1} A S$$

Since the trace is unchanged by a cyclic permutation of its argument, the above condition is satisfied if $\rho' = S^{-1}\,\rho S$.

(x) The time dependence of ρ is given by

$$i\hbar\,\frac{\partial\rho}{\partial t} = [H,\,\rho(t)] \tag{55}$$

which is the quantum mechanical version of the Liouville equation of classical statistical mechanics. Equation (55) called the von Neumann equation describes the time evolution of the density matrix in the Schrödinger picture.

To prove this, consider the Schrödinger equation

$$i\hbar \frac{\partial}{\partial t} |\psi_i\rangle = H|\psi_i\rangle \quad \text{and} \quad -i\hbar \frac{\partial}{\partial t} \langle\psi_i| = \langle\psi_i|H,$$

where H is the Hamiltonian.

Then

$$i\hbar \frac{\partial}{\partial t} \rho = i\hbar \sum_i p_i \left(\frac{\partial}{\partial t} |\psi_i\rangle\langle\psi_i| + |\psi_i\rangle \frac{\partial}{\partial t} \langle\psi_i| \right)$$

$$= \sum_i p_i \left(H|\psi_i\rangle\langle\psi_i| - |\psi_i\rangle\langle\psi_i|H \right)$$

$$= [H, \rho]. \tag{56}$$

Let us write the time evolution of state vector in the form

$$|\psi(t)\rangle = U(t, t_0)| \psi(t_0)\rangle, \tag{57}$$

where the unitary operator $U(t, t_0)$ satisfies the equation

$$i\hbar \frac{d}{dt} U(t, t_0) = HU(t, t_0) \tag{58}$$

with the initial condition

$$U(t_0, t_0) = 1 \tag{59}$$

This yields

$$\rho(t) = U(t_0, t_0) \, \rho \, (t_0)U^\dagger (t, t_0). \tag{60}$$

(xi) $Tr \, \rho^2$ is time independent .

Proof
$$Tr \, \rho^2 (t) = TrU(t, t_0) \, \rho \, (t_0)U^\dagger (t, t_0)$$

$$\times \, U(t, t_0) \, \rho \, (t_0)U^\dagger (t, t_0)$$

$$= Tr \, \rho^2 (t_0)$$

by the cyclic invariance of the trace.

Hence, a pure (mixed) state remains pure (mixed).

In the scheme where the Hamiltonian is diagonal,

$$\rho_{nn'}(t) = \rho_{nn'}(0) \exp [i\hbar^{-1} (E_{n'} - E_n)t]. \tag{61}$$

The density or state operator ρ has been introduced in the conventional way. The initial state of a physical system can be described by an operator ρ which is self-adjoint, non-negative and of finite trace (= 1), and the expectation value of an observable A of the system is $\overline{\langle A \rangle} = \text{Tr}\{\rho A\}$. Gleason (1957) introduced the state operator through an existence theorem. Gleason's theorem asserts the existence of the operator if a probability is to exist for every property. The method relies on certain notions. Projection operator E is self-adjoint $E = E^*$ and idemptotent $E = E^2$ and has 0 and 1 as the eigenvalues. The projectors correspond to binary-valued observables and can be understood as propositions about properties of a system. A yes–no experiment which serves to select the values of a measurable physical quantity is called a proposition of the system. Propositions are represented by projectors or, equivalently, by closed linear subspaces of a separable Hilbert space Two propositions are compatible iff their corresponding projectors commute. We consider a proposition of the type $A \in \Delta$ which implies that the value of A is in a real set Δ. For a proposition P with a corresponding projector E, we associate on the measurement of A probability $p_{A,\xi}(P, E)$ in the state ξ. The probabilities satisfy

$$\left. \begin{array}{c} 0 \leq p(P; E) \leq 1, \quad \text{for all } P, \\ p(0) = 0 \\ p(1) = 1 \end{array} \right\} \tag{62}$$

Clearly, probabilities are expectation values for projection operators.

Gleason has proved the following theorem (see Jordan (1969)).

Theorem 2.3 Consider operators on a separable Hilbert space of dimension larger than two. If for each projection operator E there is a non-negative real number $\langle E \rangle$ such that $\langle 1 \rangle = 1$ and $\left\langle \sum_k E_k \right\rangle = \sum_k \langle E_k \rangle$ for every set of mutually orthogonal projection operators E_k, then there is a unique density matrix W such that

$$\langle E \rangle = TrWE \tag{63}$$

for every projection operator E.

The operator W is a self-adjoint, non-negative operator which satisfies the relations

$$W > 0, \quad Tr\, W = 1, \quad \text{and} \quad W^2 \leq W. \tag{64}$$

The expectation value of an observable A is

$$\langle A \rangle = Tr(WE) \tag{65}$$

Gleason's theorem asserts that a state operator or density operator W exists satisfying Eq. (64) due to the existence of probabilities. This theorem emphasizes the importance of the density operator in quantum mechanics and also puts strong constraints on any modification of the standard formulation of quantum mechanics.

The state represented by W is pure iff $W^2 = W$. In that case, W is a projector of rank 1.

Let us consider states in thermal equilibrium. The state of a quantum system in thermal equilibrium at a temperature T is represented by the incoherent superposition of eigenstates E_n with statistical weights proportional to the Boltzmann factor e^{-E_n/k_BT}. In order that the sum of weights of all eigenstates equal one, we write the weight or the probability of being a state of energy E_n as $\exp(-E_n/k_BT)$ divided by the normalization factor ("sum over states")

$$Z(T) = \sum_n e^{-E_n/k_BT}. \tag{66}$$

The canonical ensemble is defined by the density matrix

$$\rho_{mn} = \delta_{mn}\, e^{-\beta E}/Z(T), \tag{67}$$

where $\beta = \dfrac{1}{k_BT}$ with k_B, the Boltzmann constant.

Note that this is the fractional population for an energy eigenstate with eigenvalue E_n $(m = n)$.

The density operator in the basis of energy eigenstates is

$$\rho = \frac{e^{-\beta H}}{Z(T)} = \frac{e^{-\beta H}}{Tr(e^{-\beta H})}. \tag{68}$$

The partition function can be written in the form

$$Z(T) = Tr\,(e^{-\beta H}). \tag{69}$$

The ensemble average of any observable A in the canonical ensemble is

$$\overline{\langle A \rangle} = \frac{Tr\,(Ae^{-\beta H})}{Z(T)}. \tag{70}$$

The connection with thermodynamics is given by (see Huang (1963)) the following expressions

the Helmholtz free energy

$$A = -\beta^{-1}\ln Z(T), \tag{71}$$

the entropy

$$S = k_B\left(\ln Z - \beta\,\frac{\partial}{\partial\beta}\ln Z\right), \tag{72}$$

and the internal energy

$$U \equiv \langle H \rangle = -\frac{\partial}{\partial\beta}\ln Z. \tag{73}$$

Many examples of pure and mixed states and applications of the density matrix will be discussed in other sections.

2.5 UNITARY OPERATORS AND TRANSFORMATION THEORY

We consider two incompatible observables A and B. The ket space can be spanned either by a complete, orthonormal basis set $\{|a_i\rangle\}$ or by a complete, orthonormal basis set $\{|b_i\rangle\}$. Let U be a linear operator that transforms one set of basis kets into the other one:

$$|b_k\rangle = U_{ki}\, |a_k\rangle. \tag{74}$$

Now,
$$|b_k\rangle = \sum_i |a_i\rangle\langle a_i|b_k\rangle,$$

using the completeness and orthonormality of $|a_i\rangle$.

Then
$$U_{ki} = \langle a_i|b_k\rangle. \tag{75}$$

is the matrix representation of the U operation.

The adjoint of U is
$$U_{ij}^\dagger = \langle b_j|a_i\rangle. \tag{76}$$

From Eqs. (75) and (76), we get
$$\begin{aligned}
(UU^\dagger)_{jk} = \sum_i U_{ji}\, U_{ik}^\dagger &= \sum_i \langle a_i|b_j\rangle\langle b_k|a_i\rangle \\
&= \sum_i \langle b_k|a_i\rangle\langle a_i|b_j\rangle \\
&= \delta_{jk},
\end{aligned} \tag{77}$$

using $\langle b_k|b_j\rangle = \delta_{jk}$.

Thus,
$$UU^\dagger = 1. \tag{78}$$

In an analogous manner, we get
$$U^\dagger U = 1. \tag{79}$$

Therefore, U is a unitary operator and the change of basis or the change of representation is effected by a unitary operator.

The inverse of a unitary transformation is also a unitary transformation, since, if U is unitary, U^{-1} is also unitary. Further, if two unitary transformations are applied in succession then the result will be unitary transformation. Let U_1 and U_2 be unitary operators. Then

$$U_1 U_2 (U_1 U_2)^\dagger = U_1 U_2\, U_2^\dagger U_1^\dagger = 1. \tag{80}$$

The set of all the unitary operators on the vector space forms a group.

Unitary operators preserve the inner product of the vectors on which they act. Let

$$U|\alpha\rangle = U|\alpha'\rangle, \quad U|\beta\rangle = U|\beta'\rangle, \tag{81}$$

Then

$$\langle\alpha'|\beta'\rangle = \langle\alpha|U^\dagger U|\beta\rangle = \langle\alpha|\beta\rangle. \tag{82}$$

The probability overlap for $|\alpha\rangle$ and $|\alpha'\rangle$ is the same for the transformed states, i.e.,

$$|\langle\alpha'|\beta'\rangle|^2 = |\langle\alpha|\beta\rangle|^2. \tag{83}$$

We note that besides the solution Eq. (81) of (83), there is also another solution of Eq. (83) (antiunitary).

We shall now investigate the action of U on operators.

We consider an arbitrary ket $|\alpha\rangle$ and an operator A such that

$$A|\alpha\rangle = |\beta\rangle. \tag{84}$$

We apply the unitary transformation U:

$$|\alpha'\rangle = U|\alpha\rangle, \quad |\beta'\rangle = U|\beta\rangle. \tag{85}$$

Writing

$$A'|\alpha'\rangle = |\beta'\rangle, \tag{86}$$

we get

$$A'U|\alpha\rangle = U|\beta\rangle = UA|\alpha\rangle, \tag{87}$$

so that

$$A'U = UA. \tag{88}$$

From Eq. (88) it follows that

$$A' = UAU^\dagger, \quad A = U^\dagger A'U, \tag{89}$$

since U is unitary. $A' = UAU^{-1}$ is called a unitary transform of A. Equation (89) takes the form of a similarity transformation.

If A is a hermitian operator, so also is A'.

$$A'^\dagger = (UAU^\dagger)^\dagger = UA^\dagger U^\dagger = UAU^\dagger = A'. \tag{90}$$

A unitary transformation leaves invariant any algebraic equation between linear operators. In particular, commutation relations remain unchanged under a unitary transformation. Let $[A, B] = C$. Then

$$[A', B'] = UAU^{-1}UBU^{-1} - UBU^{-1}UAU^{-1}$$

$$= U[A, B]U^{-1}$$

$$= C.$$

If the operator is invariant under the transformation, i.e., if $A' = A$, then $[U, A] = 0$. Also,

$$Tr\, UAU^\dagger = TrA \quad \text{and} \quad \det UAU^\dagger = \det A \qquad (91)$$

Next, we consider the eigenvalue equation for A

$$A|a_i\rangle = a_i|a_i\rangle, \qquad (92)$$

where we consider two basis sets $\{|a_i\rangle\}$ and $\{|b_i\rangle\}$ connected by a unitary transformation Eq. (74). We rewrite Eq. (92) as

$$AU^\dagger U|a_i\rangle = a_i U^\dagger U|a_i\rangle,$$

or,

$$UAU^\dagger|a_i\rangle = a_i UU^\dagger|a_i\rangle$$

so that

$$A'|a_i\rangle = a_i|b_i\rangle. \qquad (93)$$

Thus, the eigenvalues of A are the same as those of $A' = UAU^{-1}$. A and $UAU^\dagger$ are called unitary equivalent observables. Equation (93) expresses the theorem: unitary equivalent observables have the same spectra.

Unitary transformations in quantum mechanics play the same role as the canonical transformations in classical mechanics.

We get an infinitesimal unitary transformation by taking U to be very close to unity and evolving continuously from unity. We put

$$U = 1 + i \in G, \qquad (94)$$

where $\in$ is infinitesimal. Then to first order in $\in$,

$$1 = U^\dagger U = (1 - i \in G^\dagger)(1 + i \in G^\dagger)$$
$$= 1 + i \in (G - G^\dagger)$$

so that

$$G - G^\dagger = 0, \quad \text{i.e.,} \quad G = G^\dagger. \qquad (95)$$

G is called the generator of the infinitesimal unitary transformation. The transformed operator is

$$A' = UAU^\dagger = (1 + i \in G)A(1 - i \in G)$$
$$= A + i \in (GA - AG) + O(\in^2)$$
$$= A + i \in [G, A], \qquad (96)$$

to first order in $\in$ so that if A commutes with G, then A remains unchanged by the transformation.

2.6 POSITION AND MOMENTUM OPERATORS

Every linear operator in a finite dimensional vector space is bounded. However, there are many examples of unbounded operators in an infinite-dimensional space. In quantum mechanics, there occur many unbounded operators. For example, the position and momentum operators are unbounded. These observables have continuous eigenvalue spectra, i.e., their eigenvalues range from $-\infty$ to ∞.

The position operator Q is defined in a subset D of the space $L^2(-\infty, +\infty)$ of all Lebesgue square-integrable functions $\psi(x)$ by

$$(Q\psi)(x) = x\psi(x) \tag{97}$$

The subset

$$D_Q = \{\psi(x)\} \in \mathcal{H}(R) : \int_{-\infty}^{\infty} x^2 \, |\psi(x)|^2 \, dx \, \langle\infty\} \tag{98}$$

is dense in $\mathcal{H}$.

The operator Q is symmetrical. Q is also self-adjoint for this subset D_Q.

The position of a particle is represented by the c.s.c.o. $\vec{Q} = (Q_1, Q_2, Q_3)$ defined by $\left(\vec{Q}\psi\right)(\vec{x}) = \vec{x}\,\psi(\vec{x})$ for a vector ψ which is a function of $\vec{x} = (x_1, x_2, x_3)$. Every bounded operator which commutes with Q_1, Q_2, Q_3 is a function of Q_1, Q_2, Q_3. Each of the operators Q_1, Q_2, Q_3 has a purely continuous spectrum which consists of all real numbers. The operators $\vec{Q}$ have no eigenvectors, but we can use delta functions as analogs of eigenvectors (see later).

We next consider the operator P, defined in $\mathcal{H} = L^2(R)$ on the subset D_P which consists of all absolutely continuously functions $\psi(x)$ which are differentiable and for which $\dfrac{d\psi}{dx} \in L^2(-\infty, +\infty)$ by

$$(P\psi)(x) = -i\,\frac{d\psi}{dx}. \tag{99}$$

The domain D_P is dense and this operator P is also self-adjoint.

The momentum (divided by $\hbar$) of a particle can be represented by the complete set of commuting operators $\vec{P} = (P_1, P_2, P_3)$ and is given by $\left(\vec{P}\psi\right)(\vec{x}) = -i\,\vec{\nabla}\,\psi(\vec{x})$.

The operators $\vec{P}$ have no eigenvectors but they have eigenfunctions which can be used as analogs of eigenvectors.

We consider a space translation

$$\vec{x} \to \vec{x}' = \vec{x} + \vec{r}$$

and we let
$$\psi \to \psi' = e^{-\vec{r}\cdot\vec{P}}\psi,$$

where
$$\psi'\,(\vec{x}) = \psi\,(\vec{x})$$

Thus,
$$\left(e^{-i\,\vec{r}\cdot\vec{P}}\psi\right)(\vec{x}) = \psi\,(\vec{x}-\vec{r}) = e^{-\vec{r}\cdot\vec{\nabla}}\,\psi\,(\vec{x})$$

and
$$\left(\vec{P}\psi\right)(\vec{x}) = -\,i\,\vec{\nabla}\,\psi\,(\vec{x}). \tag{100}$$

We see that the momentum (divided by $\hbar$) of the particle is a generator of space-translation.

2.7   CANONICAL QUANTIZATION

There are several schemes to obtain the quantum conditions which allow us to replace the classical mechanical description of a system by a quantum mechanical one. The original method of obtaining quantum conditions using the canonical coordinates and momenta is called the canonical quantization of a dynamical system.

(a)   The Canonical Commutation Relations

We consider a system of N-particles with $3N$ canonical coordinates q_r and corresponding momenta p_r $(r = 1, 2, ..., 3N)$. These operators are postulated to satisfy the following canonical (or fundamental) commutation relations

$$[q_r, q_s] = 0 \quad [p_r, p_s] = 0$$

$$[q_r, q_s] = i\hbar\delta_{rs}I. \tag{101}$$

These relations indicate where the lack of commutability among the canonical coordinates and momenta lie. These relations give the essential feature of quantum mechanics. These relations are essentially determined by assuming that the structure of physical space is homogeneous (see Isham (1995)). We now examine the commutation relations for position and momentum operators. If $Q_1, Q_2, ..., Q_n$ and $P_1, P_2, ... P_n$ are a set of $2n$ self-adjoint operators satisfying canonical commutation relations (c.c.r.), we write, symbolically,

$$[Q_j, Q_k] = [P_j, P_k] = 0 \quad [Q_j, P_k] = i\hbar\delta_{jk}\, I, \tag{102}$$

where Q_k, P_k are conjugate pairs. Since we are dealing with unbounded operators, $Q_j P_k - P_k Q_j$ is defined only on a dense domain which is a subspace of $\mathcal{H}$.

A solution to the problem Eq. (102) is the Schrödinger representation.

$$(Q_k\psi)\,(x_1, ..., x_n) = x_k\psi\,(x_1, ..., x_n)$$

$$(P_k\psi)\,(x_1, ..., x_n) = -\,i\hbar\,\frac{\partial}{\partial x_k}\,\psi\,(x_1, ..., x_n) \tag{103}$$

in which $\mathcal{H} = L^2\,(R^n)$.

A better representation is the Weyl relation which is an exponentiated version of the c.c.r. This relation involves only bounded operators defined on the entire Hilbert space.

We bring the c.c.r. into the Weyl form by using the Baker-Campbell Hausdorff (BCH) formula (Hassani, (1999)). Before proceeding farther, we first give certain important rules of commutator algebra and then establish the BCH theorem. Let A, B, C be three arbitrary linear operators. We have

$$[A, B] = -[B, A] \tag{104}$$

$$[\alpha A, \beta B] = \alpha\beta[A, B],$$

where $\alpha, \beta \in \mathbb{C}$ (or R) $\tag{104a}$

$$[A, B + C] = [A, B] + [A, C] \tag{105}$$

$$[A + B, C] = [A, C] + [B, C] \tag{105a}$$

$$[AB, C] = A[B, C] + [A, C]B \tag{106}$$

$$[A, BC] = [A, B]C + B[A, C] \tag{106a}$$

$$[A, [B, C]] + [B, [C, A]] + [C, [A, B]] = 0. \tag{107}$$

Equation (107) is known as the Jacobi identity. By repeated application of Eq. (106), one has

$$[A, B^n] = \sum_{s=0}^{n-1} B^s [A, B]B^{n-s-1}. \tag{108}$$

Let A, B be two linear operators. If $[A, [A, B]] = 0 = [B, [A, B]]$, then the BCH formula holds:

$$e^{tA} \, e^{tB} \, e^{-(t^2/2) \, [A, B]} = e^{t(A + B)}, \tag{109}$$

where $t \in R$

Iff $[A, B] = 0$, then

$$e^{tA} \, e^{tB} = e^{t(A + B)}. \tag{109a}$$

If $t = 1$, then Eq. (109) gives

$$e^A \, e^B \, e^{-\frac{1}{2} [A, B]} = e^{A + B}. \tag{110}$$

Putting $A = H(t)$ and $B = \Delta t \dfrac{dH}{dt}$ in Eq. (110)

we get

$$e^{H(t + \Delta t)} = e^{H(t) + \Delta t \frac{dH}{dt}}$$

$$= e^{H(t)} \, e^{\Delta t \frac{dH}{dt}} \, e^{-\frac{1}{2}\left[H(t), \, \Delta t \frac{dH}{dt}\right]}$$

assuming that both $H(t)$ and $\dfrac{dH}{dt}$ commute with $\left[H, \dfrac{dH}{dt}\right]$.

For infinitesimal Δt, we get

$$e^{H(t + \Delta t)} = e^{H(t)} \left(1 + \Delta t \,\frac{dH}{dt}\right)\left(1 - \frac{1}{2}\,\Delta t \left[H(t), \frac{dH}{dt}\right]\right)$$

$$= e^{H(t)} \left(1 + \Delta t \,\frac{dH}{dt} - \frac{1}{2}\,\Delta t \left[H(t), \frac{dH}{dt}\right]\right).$$

So

$$\frac{d}{dt}\, e^{H(t)} = e^{H}\,\frac{dH}{dt} - \frac{1}{2}\, e^{H}\left[H, \frac{dH}{dt}\right].$$

We may also write

$$e^{H(t + \Delta t)} = e^{\left(H(t) + \Delta t \,\frac{dH}{dt}\right)}$$

$$= e^{\left(\Delta t \,\frac{dH}{dt} + H(t)\right)}$$

$$= e^{\left(\Delta t \,\frac{dH}{dt}\right)} e^{H(t)}\, e^{-\frac{1}{2}\left[\Delta t \,\frac{dH}{dt},\, H(t)\right]},$$

whence

$$\frac{d}{dt}\, e^{H(t)} = \frac{dH}{dt}\, e^{H} + \frac{1}{2}\, e^{H}\left[H, \frac{dH}{dt}\right].$$

Thus,

$$\frac{d}{dt}\, e^{H(t)} = \frac{1}{2}\left(\frac{dH}{dt}\, e^{H} + e^{H}\,\frac{dH}{dt}\right)$$

$$= \frac{1}{2}\left\{\frac{dH}{dt},\, e^{H}\right\}.$$

If $\left[H, \dfrac{dH}{dt}\right] = 0$ then

$$\frac{d}{dt}\, e^{H(t)} = e^{H}\,\frac{dH}{dt} = \frac{dH}{dt}\, e^{H}.$$

We consider a family of operators $F(t)$, dependent on a real variable t:

$$F(t) = e^{tA}\, B\, e^{-tA},$$

where A and B are independent of t. Then

$$\frac{dF}{dt} = Ae^{tA}\, B\, e^{-tA} - e^{tA}\, B\, e^{-tA}\, A$$

$$= [A, F(t)]$$

and

$$\frac{d}{dt}\,[A, F(t)] = \left[A, \frac{dF}{dt}\right].$$

Hence,

$$\frac{d^2F}{dt} = \frac{d}{dt}\left(\frac{dF}{dt}\right)$$

$$= \frac{d}{dt}\,[A,\,F(t)]$$

$$= [A,\,[A,\,F(t)]],$$

$$\frac{d^3F}{dt} = [A,\,[A,\,[A,\,F(t)]]],$$

. . . .

We write

$$\frac{d^nF}{dt} = A^n\,[F(t)],$$

where $\qquad A^n\,[F(t)] = [A,\,A^{n-1}\,[F(t)]],\quad$ with

$$A^0\,[F(t)] \equiv F(t).$$

By Taylor expansion about $t = 0$,

$$F(t) = \sum_{n=0}^{\infty} \frac{t^n}{n!}\left.\frac{d^nF}{dt^n}\right|_{t=0} = \sum_{n=0}^{\infty} \frac{t^n}{n!}\,A^n\,[F(0)]$$

$$= \sum_{n=0}^{\infty} \frac{t^n}{n!}\,A^n\,[B]$$

i.e.,

$$e^{tA}\,Be^{-tA} = \left(\sum_{n=0}^{\infty} \frac{t^n}{n!}\,A^n\right)[B] = e^{tA}\,[B]$$

For $t = 1$, we get

$$e^A\,Be^{-A} = B + [A,\,B] + \frac{1}{2!}\,[A,\,[A,\,B]]$$

$$+ \frac{1}{3!}\,[A,\,[A,\,[A,\,B]]] + \dots \tag{111}$$

If A commutes with $[A,\,B]$, then we obtain

$$e^{tA}\,Be^{-tA} = B + t[A,\,B]. \tag{112}$$

If A and B both commute with $[A,\,B]$, then

$$[A,\,B^2] = B[A,\,B] + [A,\,B]B$$

$$= [A,\,B]2B.$$

Now, $\qquad\qquad [A,\,B^k] = [A,\,B]kB^{k-1}\quad$ for $k = 0, 1, 2,\dots$

We assume that $[A, B^k] = [A, B]kB^{k-1}$ is true for a positive integer k. Then

$$[A, B^{k+1}] = [A, BB^k]$$

$$= B[A, B^k] + [A, B]B^k$$

$$= B[A, B]kB^{k-1} + [A, B]B^k$$

$$= [A, B](k + 1)B^k,$$

so that it holds for any k.

For a function $f(B)$ given as a power series, i.e.,

$$f(B) = \sum_{n=0}^{\infty} a^n B^n, \quad \text{we get}$$

$$[A, f(B)] = \sum_{n=0}^{\infty} a_n [A, B^n]$$

$$= [A, B] \sum_{n=0}^{\infty} a_n \frac{dB^n}{dB}$$

$$= [A, B] \frac{df(B)}{dB}. \tag{113}$$

Let

$$g(t) = e^{tA} e^{tB}.$$

Then

$$\frac{dg}{dt} = Ae^{tA} e^{tB} + e^{tA} Be^{tB}$$

$$= [A + e^{tA} Be^{-tA}]g(t)$$

$$= (A + B + [A, B])g(t), \quad \text{using Eq. (112).}$$

Since $(A + B)$ commutes with $[A, B]$, the solution is

$$g(t) = e^{\left(t(A + B) + t^2/2 \, [A, B]\right)} g'(0)$$

or,

$$g(t) = e^{t(A + B) + \frac{t^2}{2} [A, B]},$$

since $g(0) = I$.

This is the BCH formula Eq. (109).

Another interesting result follows. From Eq. (111), we write for any two linear operators A and B whose commutator is a c number

$$e^A e^B e^{-A} = e^B + [A, e^B] + \frac{1}{2!} [A, [A, e^B]]$$

$$+ \frac{1}{3!} [A, [A, [A, e^B]]] + \dots$$

If A and B both commute with $[A, B]$, then using Eq. (113), we get

$$[A, e^B] = [A, B] e^B$$

and

$$[A, [A, e^B]] = [A, [A, B]e^B]$$
$$= [A, B] ([A, e^B])$$
$$= [A, B]^2 e^B.$$

Thus,

$$e^A e^B e^{-A} = e^B + e^B [A, B] + e^B \frac{[A, B]^2}{2!} + \dots$$
$$= e^B e^{[A, B]},$$

or,

$$e^A e^B = e^B e^A e^{[A, B]}$$
$$= e^{[A, B]} e^B e^A \tag{114}$$

The general form of BCH theorem (without any condition on the commutator of A and B) is

$$e^A e^B = \left(\exp A + B + \frac{1}{2} [A, B] + \frac{1}{12} [A, [A, B]] - [B, [A, B]] \right) + \dots \tag{115}$$

where the ellipsis refers to higher order multiple commutators.

We now return to the c.c.r. Setting $A = - i\alpha Q/\hbar$ and $B = - i\beta P/\hbar$, where α, β are real numbers, we obtain from Eq. (110),

$$U_\alpha V_\beta = e^{- i\alpha Q/\hbar} e^{- i\beta P/\hbar}$$
$$= e^{- i(\alpha Q + \beta P)/\hbar} e^{- i\alpha\beta/2\hbar} [Q, P]$$

and

$$V_\beta U_\alpha \equiv e^{- i\beta P/\hbar} e^{- i\alpha Q/\hbar}$$
$$= e^{- i(\alpha Q + \beta P)/\hbar} e^{i\alpha\beta/2\hbar} [Q, P]$$

so that we get the Weyl form of the commutation relations

$$U_\alpha V_\beta = e^{- i\alpha\beta/\hbar} V_\beta U_\alpha, \tag{116}$$

where $U_\alpha = e^{- i\alpha Q/\hbar}$ and $V_\beta = e^{- i\beta P/\hbar}$.

Now, the commutation relation is expressed in terms of unitary, one-parameter groups U_α and V_β. The Weyl relation involves only bounded operators defined on the entire Hilbert space. The infinitesimal form of the Weyl relations imply the canonical commutation relations. The converse is, in general, not true.

(b)   The Uncertainty Relations II

The uncertainty relations set a fundamental constraint on pairs of incompatible observables. In the commonly quoted form Eq. (34) of the uncertainty relations, we see that the commutator, the signature of quantum mechanics, is manifest here.

The position and momentum operators q_i and p_j satisfy the c.c.r. Eq. (101). The uncertainty relation Eq. (33) leads to the Heisenberg uncertainty relation.

$$(\Delta q_i)^2_\psi \, (\Delta p_j)^2_\psi \geq \frac{\hbar^2}{4} \, \delta_{ij} \tag{117}$$

This is not canonically invariant. To obtain canonical invariance, we consider the necessary and sufficient condition for positive-definiteness (from Eq. (31) and Eq. (32))

$$\det \begin{vmatrix} (\Delta q)^2_\psi & \Delta_\psi\,(qp) \\ \Delta_\psi\,(qp) & (\Delta p)^2_\psi \end{vmatrix} \geq 0, \tag{118}$$

or, using the c.c.r. for q, p and since (Δq), (Δp) satisfy the same commutation relations, we get the canonically invariant stronger Robertson–Schrödinger uncertainty relation (Sudarshan et al (1995), Esposito et al (2004))

$$(\Delta q)^2_\psi \, (\Delta p)^2_\psi - \Delta^2_\psi\,(qp) \geq \frac{\hbar^2}{4}, \tag{119}$$

where

$$\Delta_\psi(qp) = \left\langle \frac{qp + pq}{2} \right\rangle_\psi - \langle q \rangle_\psi \, \langle p \rangle_\psi.$$

From Eq. (119), we get the weaker Heisenberg uncertainty relation

$$(\Delta q)^2_\psi \, (\Delta p)^2_\psi \geq \frac{\hbar^2}{4}. \tag{120}$$

Santhanam (2000) obtained uncertainty relations among higher-order moments of q and p:

$$(\Delta q)^{2n} \, (\Delta p)^{2n} \geq \left(\frac{\hbar}{2} \right)^{2n} ((2n-1)!!)^2 \tag{121}$$

where

$$(2n-1)!! = 1 \times 3 \times 5 \times \dots \times (2n-1).$$

For uncertainty relations for mixed states, see Luo (2005).

We now discuss uncertainty relations for an angle variable and its conjugate angular momentum. It is necessary to take into account the periodicity. As pointed out by Carruthers and Nieto (1968), various pitfalls associated with the periodicity problem are avoided by introducing periodic variables to describe the phase variable.

Let ϕ be the azimuthal angle about the z-axis. Defining ϕ to be continuous ($-\infty$ to $+\infty$), the z-component of orbital angular momentum can be represented as a differential operator

$$L_z = - i\hbar \frac{\partial}{\partial \phi}. \tag{122}$$

The commutator of these two conjugate variables is

$$[\phi, L_z] = i\hbar. \tag{123}$$

This angle variable is not periodic and L_z is not hermitian. Then one cannot conclude that Eq. (123) implies the uncertainty relation.

$$\Delta L_z \, \Delta \phi \geq \frac{\hbar}{2}. \tag{124}$$

Although qualitatively, the result Eq. (124) is true, we must formulate the problem carefully. We use a pair of continuous periodic variables, say $\cos \phi$ and $\sin \phi$, instead of ϕ. The commutation relations

$$[\sin \phi, L_z] = i\hbar \cos \phi$$

$$[\cos \phi, L_z] = - i\hbar \sin \phi \tag{125}$$

lead to the uncertainty relations

$$(\Delta L_z)^2 \, (\Delta \sin \phi)^2 \geq \frac{\hbar^2}{4} \, \langle \cos \phi \rangle^2$$

$$(\Delta L_z)^2 \, (\Delta \cos \phi)^2 \geq \frac{\hbar^2}{4} \, \langle \sin \phi \rangle^2, \tag{126}$$

since L_z is now hermitian. A symmetrical result is

$$(\Delta L_z)^2 \, \frac{[(\Delta \cos \phi)^2 + (\Delta \sin \phi)^2]}{\langle \sin \phi \rangle^2 + \langle \cos \phi \rangle^2} \geq \frac{\hbar^2}{4}. \tag{127}$$

These relations reduce to Eq. (124) whenever the distribution of ϕ in the wave function is sufficiently localized.

(c) Wave Functions

Historically, Schrödinger formulated quantum mechanics from a different standpoint. His eponymous equation involves differential and multiplication operators acting on the wave function ψ, which has the interpretation of a probability amplitude. Schrödinger's wave mechanics follows from canonical quantization by considering the coordinate representation of all operators and kets.

We consider a system with $|\psi\rangle$ an arbitrary vector in the Hilbert space associated with the system. We choose a complete, orthonormal set of basis vectors $\{|u_i\rangle\}$. The ket $|\psi\rangle$ representing a state of the system can be expanded in terms of the basis set (assumed discrete)

$$|\psi\rangle = \Sigma \, |u_i\rangle\langle u_i|\psi\rangle.$$

In coordinate representation, the basis set is the set of eigenvectors $\{|\vec{x}\rangle\}$ of the position operator. Since this set is continuous, the expansion coefficients define a function of a continuous variable – the wave function $\psi(\vec{x}) = \langle\vec{x}|\psi\rangle$. The inner product is known as the wave function $\psi(\vec{x})$ for the state $|\psi\rangle$. The set of wave functions forms a (coordinate) representation of the vector space. The wave function is interpreted as a probability amplitude. For an operator O on the function space,

$$O\psi(\vec{x}) = \langle\vec{x}|O|\psi\rangle,$$

where
$$\psi(\vec{x}) = \langle\vec{x}|\psi\rangle. \tag{128}$$

2.8 QUANTUM DYNAMICS

Until now we have considered only the kinematics of a quantum system. The kinematical aspect pertains to the properties which can be measured at an instant of time. In this section, we consider the dynamical aspect of quantum systems. The time evolution of the states will now be discussed.

(a) Time Evolution and the Schrödinger Equation

Let us consider a system whose state ket at time t is $|\psi; t\rangle$. Then at a later time t', the state ket is $|\psi; t'\rangle (t' > t)$. The two kets are related by a linear, time – translation operator

$$|\psi; t'\rangle = U(t', t)|\psi; t\rangle. \tag{129}$$

The linearity of the operator follows from the superposition principle. During time evolution of the state ket, the norm dose not change (probability conservation). According to this, the time evolution operator U is a unitary operator.

$$U^{\dagger}(t', t)\, U(t', t) = I. \tag{130}$$

Thus the time evolution of a state is effected by a unitary transformation parametrized by a continuous parameter t. We note that time is a parameter in quantum mechanics. For an isolated system, U can only depend on time difference

$$|\psi; t'\rangle = U(t' - t)|\psi; t\rangle. \tag{129a}$$

Moreover, U statisfies the closure (composition) property

$$U(t_2, t_0) = U(t_2, t_1)\, U(t_1, t_0),\ (t_2 > t_1 > t_0), \tag{131}$$

since two successive time translations is also a time translation. Also, from Eq. (129),

$$U(t_0, t_0) = 1.$$

To obtain the form of U, we consider an infinitesimal time-evolution operator. For $t' = t + \delta t$, $U(t', t)$ evolves continuously from the identity and we write

$$U(t + \delta t, t) = 1 - \frac{i}{\hbar} H\delta t, \tag{132}$$

where $\hbar$ is introduced in order that the self-adjoint operator H has the dimension of energy. H is the Hamiltonian of the system. For finite time difference (which can be reached by repeated application of the infinitesimal ones),

$$U(t', t) = \lim_{N \to \infty} \left(1 - \frac{i}{\hbar} \delta t H\right)^{N}$$

$$= e^{-\frac{i}{\hbar}(t' - t)H}. \tag{133}$$

The unitary time-evolution operators U form a continuous one-parameter group and the Hamiltonian H is called the generator of time translations.

Using Eq. (131) with $t_1 = t$, $t_2 = t + dt$, we have

$$U(t + dt, t_0) = U(t + dt, t)U(t, t_0)$$

$$= \left(1 - \frac{i}{\hbar} Hdt\right) U(t, t_0)$$

or, $\qquad U(t + dt, t_0) - U(t, t_0) = -\frac{i}{\hbar} HdtU(t, t_0)$

so that

$$i\hbar \frac{\partial}{\partial t} U(t, t_0) = HU(t, t_0) \tag{134}$$

This is the Schrödinger equation for the time-evolution operator.

Multiplying both sides of Eq. (134) by $|\psi; t_0\rangle$ on the right, we have

$$i\hbar \frac{\partial}{\partial t} U(t, t_0)|\psi; t_0\rangle = HU(t, t_0)|\psi; t_0\rangle.$$

From Eq. (133),

$$i\hbar \frac{\partial}{\partial t} e^{-\frac{i}{\hbar}(t - t_0)H}|\psi; t_0\rangle$$

$$= He^{-\frac{i}{\hbar}(t - t_0)H}|\psi; t_0\rangle$$

$$= H|\psi; t\rangle,$$

i.e., $\qquad i\hbar \frac{\partial}{\partial t}|\psi; t\rangle = H|\psi; t\rangle. \tag{135}$

This is the Schrödinger equation for the state ket for a system developing with time.

If the Hamiltonian operator H is independent of t, then the solution to Eq. (134) is given by

$$U(t', t) = e^{-\frac{i}{\hbar}(t' - t)H}. \tag{136}$$

This follows from Eq. (132).

If H is time-dependent, then the differential Eq. (134) with the boundary condition $U(t_0, t_0) = 1$ leads to the integral equation.

$$U(t, t_0) = 1 - \frac{i}{\hbar} \int_{t_0}^{t} dt' H(t') U(t', t_0). \tag{137}$$

By iteration, we obtain the solution as the time-ordered infinite series (Dyson series)

$$U(t, t_0) = 1 + \sum_{n=1}^{\infty} \left(-\frac{i}{\hbar}\right)^n \int_{t_0}^{t} dt_1 \int_{t_0}^{t} dt_2 \ldots$$

$$\int_{t_0}^{t^{n-1}} dt_n \, H(t_1) \, H(t_2) \ldots H(t_n),$$

$$(t_1 \geq t_2 \geq \ldots \geq t_n). \tag{138}$$

If H is time-independent, then Eq. (138), becomes Eq. (137). If H is time-dependent but H's at different times commute, then the upper limit can be extended to t $\left(\text{with a factor } \dfrac{1}{n!}\right)$ giving the solution in this case as

$$U(t, t_0) = \exp\left(-\frac{i}{\hbar} \int_{t_0}^{t} dt' H(t')\right). \tag{139}$$

(b) Stationary States, Expectation Values

We consider the Schrödinger equation (135). The eigenkets of H, called energy eigenkets, are of great importance. These kets satisfy the eigenvalue equation

$$(H - E)|\psi_E; t\rangle = 0. \tag{140}$$

If a system is initially in a state

$$|\psi; 0\rangle = \sum_{E'} |\psi_{E'}; t\rangle\langle\psi_{E'}; t|\psi; 0\rangle, \tag{141}$$

then the state at time t is

$$|\psi; t\rangle = \sum_{E'} |\psi_{E'}; t\rangle\langle\psi_{E'}; t|\psi; 0\rangle \times \exp\left(- iE't/\hbar\right). \tag{142}$$

If the initial state happens to be an energy eigenstate, then

$$|\psi_E; t\rangle = |\psi_E; 0\rangle \exp\left(- iEt/\hbar\right). \tag{143}$$

Energy eigenstates are called stationary states because such states do not change in time, except for a phase factor. If the system is initially in one of the simultaneous eigenkets of the c.s.c.o. H, A, B, C, ..., then at any later time, the system will remain in this eigenstate.

Also, the matrix elements of any time-independent observable O between stationary states have only a phase factor

$$\langle \psi_E; t|O|\psi_{E'}; t \rangle = e^{\frac{i}{\hbar}(E - E')t} \langle \psi_E; 0|O|\psi_{E'}; 0 \rangle. \tag{144}$$

If we substitute the energy eigenket Eq. (143), into the Schrödinger equation (135), we get the eigenvalue equation

$$(H - E)|\psi_E; t\rangle = 0. \tag{145}$$

Equation (145) is known as the time-independent Schrödinger equation. From Eq. (145), we can show that there is no dispersion in energy in a stationary state. The precise value of the energy is the eigenvalue of the H operator.

We now study the time evolution of the expectation value of an observable. Let $A(t)$ denote an observable of a system. We calculate the expectation value $\langle A \rangle$ of A in a state $\psi(t)$ of the system:

$$\langle A \rangle \equiv \langle \psi(t)|A(t)|\psi(t)\rangle, \tag{146}$$

so that

$$\frac{d}{dt}\langle A \rangle = \left[\frac{d}{dt}\langle \psi(t)|\right] A(t)|\psi(t)\rangle$$

$$+ \langle \psi(t)|A(t)\left[\frac{d}{dt}|\psi(t)\rangle\right]$$

$$+ \left\langle \psi(t)\,\frac{dA(t)}{dt}\middle|\psi(t)\right\rangle. \tag{147}$$

From the Schrödinger Eqs. (135) and (147), we have

$$i\hbar\frac{d}{dt}\langle A \rangle = \langle \psi(t)|[A(t), H(t)]|\psi(t)\rangle$$

$$+ i\hbar\langle \psi(t)|\left|\frac{dA(t)}{dt}\right||\psi(t)\rangle. \tag{148}$$

In conservative systems (H time-independent) if $|\psi\rangle$ is an energy eigenket (stationary), then

$$\frac{d}{dt}\langle \psi(t)|A(t)|\psi(t)\rangle = \langle \psi(t)\left|\frac{dA(t)}{dt}\right|\psi(t)\rangle. \tag{149}$$

If A has no explicit time dependence, then its expectation value in a stationary state is constant in time.

In general, the expectation value of an observable is a constant of motion if it satisfies

$$i\hbar\frac{d}{dt}\langle A \rangle + \langle [A(t), H(t)]\rangle = 0. \tag{150}$$

(c) The Schrödinger, the Heisenberg, and the Interaction Pictures

The above formulation of time development considers the time evolution of the state vectors while the operators are time independent. This description of quantum dynamics is known as the Schrödinger picture. Another alternative but equivalent description where observables, rather than state vectors vary with time is known as the Heisenberg picture.

(i) The Schrödinger Picture

The state of the system at time t is given by $|\psi_S(t)\rangle$. The time evolution of the system is described by the time dependence of the state vectors governed by the Schrödinger equation

$$i\hbar \frac{d}{dt}|\psi_S(t)\rangle = H|\psi_S(t)\rangle. \tag{151}$$

We assume that the Hamiltonian H is independent of time. We further assume that the operators, O_S, are time independent. Equation (151) has the formal solution

$$|\psi_S(t)\rangle = e^{-\frac{i}{\hbar}Ht}|\psi_S(0)\rangle. \tag{152}$$

The expectation value of an operator O_S changes with time as

$$i\hbar \frac{d}{dt}\langle O_S\rangle = \langle \psi_S(t)|[O_S, H]|\psi_S(t)\rangle. \tag{153}$$

This description in which the state vectors depend on time, while the operators are time independent is known as the Schrödinger picture.

(ii) The Heisenberg Picture

In this picture, the state vectors are fixed in time along with time-dependent operators in such a way that all expectation values are identical with the expectation values calculated in the Schrödinger picture.

We choose

$$|\psi_H\rangle = |\psi_S(0)\rangle = e^{\frac{i}{\hbar}Ht}|\psi_S(t)\rangle. \tag{154}$$

From Eqs. (152) and (154), we get the expectation value

$$\langle O_S\rangle \equiv \langle \psi_S(t)|O_S|\psi_S(t)\rangle = \left\langle \psi_S(0)\left|e^{\frac{i}{\hbar}Ht} O_s\, e^{-\frac{i}{\hbar}Ht}\right|\psi_S(0)\right\rangle$$

$$= \langle \psi_H|O_H|\psi_H\rangle$$

$$= \langle O_H\rangle, \tag{155}$$

where we have defined a time-dependent operator in the Heisenberg picture

$$O_H(t) = e^{\frac{i}{\hbar}Ht} O_S\, e^{-\frac{i}{\hbar}Ht}, \tag{156}$$

so that the expectation values of the operators remain the same in the two descriptions. Differentiating the relation Eq. (156) with respect to time, we get the equation of motion

$$\frac{dO_H}{dt} = \frac{\partial O_H}{dt} + \frac{1}{i\hbar} [OH, H] \tag{157}$$

where we put

$$\frac{\partial O_H}{dt} = e^{\frac{i}{\hbar} Ht} \frac{\partial O_S}{dt} e^{-\frac{i}{\hbar} Ht}. \tag{158}$$

This Eq. (157) describe the time development of operators in the Heisenberg picture.

(iii)   The Interaction Picture

Between these two pictures there is a third extremely useful one in which part of the time evolution appears as a change in the state vectors while another part as a change in the observables.

We assume that the Hamiltonian for the system is a sum of two terms $H = H_0 + V$. We define

$$|\psi_I (t)\rangle = e^{-\frac{i}{\hbar} H_0 t} |\psi_S (t)\rangle. \tag{159}$$

Then

$$\langle O \rangle = \langle \psi_S (t)| O_S |\psi_S (t)\rangle$$

$$= \left\langle \psi_S (t)\left| e^{-\frac{i}{\hbar} H_0 t} e^{\frac{i}{\hbar} H_0 t} O_S e^{-\frac{i}{\hbar} H_0 t} e^{\frac{i}{\hbar} H_0 t}\right|\psi_S (t)\right\rangle$$

$$= \left\langle \psi_I (t)\left| e^{\frac{i}{\hbar} H_0 t} O_S e^{-\frac{i}{\hbar} H_0 t}\right|\psi_S (t)\right\rangle. \tag{160}$$

Since the expectation value of an operator is physically observable, it must be independent of the picture. So we define the operator $O_I(t)$ in the interaction picture as

$$O_I(t) = e^{\frac{i}{\hbar} H_0 t} O_S e^{-\frac{i}{\hbar} H_0 t} \tag{161}$$

giving

$$\frac{dO_I(t)}{dt} = \frac{\partial O_I(t)}{\partial t} + \frac{1}{i\hbar} [O_I, H_0]. \tag{162}$$

Differentiating Eq. (159) will respect to the time, we get

$$\frac{\partial}{\partial t}|\psi_I (t)\rangle = \frac{i}{\hbar} H_0 e^{\frac{i}{\hbar} H_0 t}|\psi_S (t)\rangle$$

$$+ e^{\frac{i}{\hbar} H_0 t} \frac{\partial}{\partial t}|\psi_S (t)\rangle,$$

and using the Schrödinger equation

$$i\hbar \frac{\partial}{\partial t}\bigg| \psi_S(t)\rangle = (H_0 + V)|\psi_S(t)\rangle, \tag{163}$$

we obtain

$$i\hbar \frac{\partial}{\partial t}\bigg| \psi_I(t)\rangle = V(t)|\psi_I(t)\rangle. \tag{164}$$

$|\psi_I(t)\rangle$ satisfies a Schrödinger equation driven only by V.

In this picture, called the Dirac or the interaction picture, the state vectors evolve due to the perturbative or the interaction part of the Hamiltonian, and the operators due to the free part H_0.

We give below a table comparing the three pictures. The three pictures coincide at $t = 0$. Note that $H = H_0 + V$.

Table 2.1   The three pictures

	Schrödinger picture	Heisenberg picture	Interaction picture
State vectors	$|\psi_S(t)\rangle$	$|\psi_H\rangle = e^{\frac{i}{\hbar}Ht}|\psi_S(t)\rangle$	$|\psi_I(t)\rangle = e^{\frac{i}{\hbar}H_0 t}|\psi_S(t)\rangle$
	$i\hbar \frac{\partial}{\partial t}|\psi_S(t)\rangle = H|\psi_S(t)\rangle$	$i\hbar \frac{\partial}{\partial t}|\psi_H\rangle = 0$	$i\hbar \frac{\partial}{\partial t}|\psi_I(t)\rangle = V(t)|\psi_I(t)\rangle$
Observables	O_S	$O_H(t) = e^{\frac{i}{\hbar}Ht}\, O_S\, e^{-\frac{i}{\hbar}Ht}$	$O_I(t) = e^{\frac{i}{\hbar}H_0 t}\, O_S\, e^{-\frac{i}{\hbar}H_0 t}$
	$i\hbar \frac{d}{dt}O_S = 0$	$\frac{dO_H}{dt} = \frac{\partial O_H}{\partial t} + \frac{1}{i\hbar}[O_H, H]$	$\frac{dO_I}{dt} = \frac{\partial O_I}{\partial t} + \frac{1}{i\hbar}[O_I, H_0]$

(d)   The Energy – Time Uncertainty Relations

In this section we study the time-energy uncertainty relation

$$\Delta E \cdot \Delta t \geq \frac{\hbar}{2}. \tag{165}$$

This relation has to be used with caution. The interpretation of an energy-time uncertainty relation and even the existence of such a relation has for long remained problematic. (Peres (1993); Hilgevoord. (1996), (2002)).

In quantum theory, the time variable is a parameter, not an operator corresponding to an observable. Time is not determined by measuring a time variable, but by observing the rate of change (time variation) of some other dynamical variable. This acts as a clock.

Let us consider a conservative system with an observable A that does not depend explicitly on time. For an arbitrary state $|\psi(t)\rangle$, the mean value of A in the state $|\psi\rangle$ evolves, according to Eq. (157):

$$\frac{d}{dt}\langle A\rangle_\psi = \frac{1}{i\hbar}\langle\psi|[A, H]|\psi\rangle. \tag{166}$$

The mean quadratic deviation for the observation of A in the state $|\psi\rangle$ is

$$\Delta A = \left[\langle A^2\rangle_\psi - \langle A\rangle_\psi^2\right]^{1/2}. \tag{167}$$

The expectation value of the energy, $\langle\psi|H|\psi\rangle$, has the uncertainty

$$\Delta E = \left[\langle H^2\rangle_\psi - \langle H\rangle_\psi^2\right]^{1/2}. \tag{168}$$

Then, using the generalized uncertainty relation we obtain

$$(\Delta E)(\Delta A) \geq \frac{1}{2}\langle\psi|[A, H]|\psi\rangle. \tag{169}$$

Combining Eqs. (166) and (169), we obtain the inequality

$$(\Delta E)(\Delta A) \geq \frac{\hbar}{2}\left|\frac{d}{dt}\langle A\rangle_\psi\right|. \tag{170}$$

This is the Mandelstam-Tamm inequality (1945).

Now, we denote by Δt the time interval during which the mean value of A changes by ΔA. We put

$$\Delta t = \frac{(\Delta A)}{\left|\dfrac{d\langle A\rangle_\psi}{dt}\right|} = \tau_\psi(A), \tag{171}$$

which, on insertion into Eq. (170), gives an energy-time uncertainty relation of the form

$$\tau_\psi(A) \cdot \Delta t \geq \frac{\hbar}{2}. \tag{172}$$

Here, $\tau_\psi(A)$ is not an uncertainty or statistical spread in a time variable. It is a time characteristic of the evolution of the statistical distribution of A in the state $|\psi\rangle$. It is the time required for the center $\langle A\rangle$ of this distribution to be displaced by an amount $\langle\Delta A\rangle_\psi$.

If A is a constant of motion, then $\dfrac{d}{dt}\langle A\rangle_\psi = 0$, and one has to assume that

$\tau_\psi(A) = \infty$, since, in general, $(\Delta A)_\psi \neq 0$.

If the system is in an eigenstate of A, then $(\Delta A)_\psi = 0$, one should take $\tau_\psi(A) = \infty$.

If we consider the set of all observables $\{A\}$ which do not depend explicitly on time, we can define, at a given instant of time t, a set of characteristic times of evolution $\{\tau_\psi(A)\}$:

$$\tau_\psi \equiv \inf_{A \in \{A\}} \{\tau_\psi(A)\}. \tag{173}$$

$\tau_\psi(t)$ may be considered a time interval characteristic for the evolution of the system, independently of the observable that is under consideration, the inequality Eq. (172) is written as

$$\tau_\psi \cdot \Delta_\psi H \geq \frac{\hbar}{2}. \tag{174}$$

This is the time-energy uncertainty relation which states that for a system in an arbitrary state $|\psi(t)\rangle$, the uncertainty in the energy and the time interval characteristic of the evolution of the system should satisfy Eq. (174) at all times. Note that if $|\psi(t)\rangle$ is an energy eigenstate, i.e., a stationary state, then $\Delta_\psi H = 0$, we require τ_ψ to be ∞.

An important application of the time-energy uncertainty relation is the well known connection of life-time and width for radioactive systems (e.g., radioactive nucleus, excited state of an atom, unstable elementary particle). Such an unstable system decays with a mean lifetime τ and does not have a fixed energy, but an energy band $E \in [E_0 - \Delta E, E_0 + \Delta E]$. As an example, consider an excited atom which decays by emitting radiation. The spectral line has a certain line width. The mean life-time τ is the characteristic time for the unstable system. If $\Gamma \equiv 2\Delta E$ is the width of the energy band of the unstable system, then

$$\tau\Gamma \approx \hbar. \tag{175}$$

Following Uffink (1993), we discuss the Mandelstam-Tamm inequality in detail and obtain some interesting results. Let $|\psi(0)\rangle$ be the (pure) state at $t = 0$. We choose A in Eq. (170) to be the projector on the state $|\psi(t)\rangle$, i.e., $A = |\psi(t)\rangle\langle\psi(t)|$. Then $(\Delta A)^2 \equiv \langle A^2\rangle - \langle A\rangle^2 = \langle A\rangle (1 - \langle A\rangle)$, with $\langle A\rangle = |\langle\psi(t)|\psi(0)\rangle|^2$. Inserting $|\langle\psi(0)|\psi(t)\rangle|^2 = \cos^2\phi(t)$ in Eq. (170), we get

$$\left|\frac{d\phi}{dt}\right| \leq \frac{\Delta E}{\hbar},$$

which, on integration, gives

$$|\phi(t)| \leq \frac{\Delta E \cdot t}{\hbar}$$

or,

$$|\langle\psi(0)|\psi(t)\rangle|^2 \geq \cos^2\left(\frac{\Delta E \cdot t}{\hbar}\right) \tag{176}$$

for
$$0 \leq t \leq \frac{\pi\hbar}{2\Delta E}.$$

This inequality was given by Mandelstam and Tamm.

If we consider now that $|\psi(t)\rangle$ is orthogonal to $|\psi(0)\rangle$, we see from Eq. (176) that the system needs a time longer than $\dfrac{\pi\hbar}{2\Delta E}$ to reach an orthogonal state.

The survival or nondecay probability $P(t)$ of the initial state $|\psi(0)\rangle$ is the probability of finding the system in the initial or original state after a time t has passed:

$$P(t) = |\langle\psi(t)|\psi(0)\rangle|^2. \tag{177}$$

In terms of P, the inequality Eq. (176) becomes

$$E \cdot t \geq \hbar \, \cos^{-1} \sqrt{P}. \tag{176a}$$

$P(t)$ provides a definition of the lifetime of a state. The (shortest) time in which P becomes $\dfrac{1}{2}$ is the half-life $\tau_{\frac{1}{2}}$ of the state $|\psi(0)\rangle$, i.e., $\tau_{\frac{1}{2}}$ is the smallest time at which the probability of finding the system in its initial state has decreased to $\dfrac{1}{2}$. From Eq. (176a), we get the inequality

$$\Delta E \cdot \tau_{\frac{1}{2}} \geq \frac{\pi \hbar}{4}. \tag{178}$$

We define the "average lifetime" as

$$\tau_{av} = \int |\langle \psi(0)|\psi(t)\rangle|^2 \, dt. \tag{179}$$

By integrating the inequality Eq. (176), we get

$$\Delta E \cdot \tau_{av} \geq \frac{\pi}{4}. \tag{180}$$

There are cases where the Mandelstam-Tamm inequality is not applicable. An example is the Breit-Wigner state where

$$|\langle E|\psi(0)\rangle|^2 = \frac{\gamma}{\pi} \frac{1}{(E - E_0)^2 + \gamma^2}$$

$$|\langle \psi(0)|\psi(t)\rangle|^2 = e^{-(\gamma/2 + iE_0)t/\hbar} \tag{181}$$

Here, ΔE is infinite, so that the inequality gives a trivial bound only for $t = 0$. It is therefore desirable to have an inequality that does not rely on the standard deviation. Let $|\psi(0)\rangle$ be any state and we have to find a definition for the width of the energy distribution $|\langle E|\psi(0)\rangle|^2$. We define $W_\alpha(E)$ be the size of the smallest energy interval W such that

$$\int_W |\langle E|\psi(0)\rangle|^2 \, dE = \alpha. \tag{182}$$

$W_\alpha(E)$ is always finite and $W_\alpha(E)$ gives a reasonable measure of the uncertainty in energy if α is less than but close to one. Let τ_β be the minimal time it takes for $|\psi(0)\rangle$ to evolve to a state $|\psi(\tau)\rangle$ (survival amplitude):

$$|\langle \psi(0)|\psi(\tau)\rangle| = \beta. \tag{183}$$

Let P_W be the projector on the energy interval W and P_{W^c} be the projector on its complement. We write the state $|\psi(t)\rangle$ as

$$|\psi(t)\rangle = P_W |\psi(t)\rangle + P_{W^c} |\psi(t)\rangle$$

$$= \sqrt{\alpha} \, |\psi_W(t)\rangle + \sqrt{1 - \alpha} \, |\psi_{W^c}(t)\rangle, \tag{184}$$

where the vectors $|\psi_W(t)\rangle$ and $|\psi_W c(t)\rangle$ are orthogonal and normalized to unity. Since $P_W P_W c = 0$, we get the inner product of Eq. (184) with $\langle\psi(0)|$:

$$\langle\psi(0)|\psi(t)\rangle = \alpha \, \langle\psi_W(0)|\psi_W(t)\rangle + (1 - \alpha)$$
$$\times \langle\psi_W c(0)|\psi_W c(t)\rangle. \tag{185}$$

We get the inequality

$$|\langle\psi(0)|\psi(t)\rangle| + (1 - \alpha) \, |\langle\psi_W c(0)|\psi_W c(t)\rangle|$$
$$\geq \alpha \, |\langle\psi_W(0)|\psi_W(t)\rangle|. \tag{185}$$

For $t = \tau_\beta$, the first term has the value β. The absolute value of the second term is bounded by one. So

$$\frac{\beta + 1 - \alpha}{\alpha} \geq |\langle\psi_W(0)|\psi_W(\tau_\beta)\rangle|. \tag{186}$$

We apply now the inequality Eq. (176a) to $|\psi_W(0)\rangle$:

$$\Delta_W E \times \tau_\beta \geq \hbar \cos^{-1} (|\langle\psi_W(0)|\psi_W(\tau_\beta)\rangle|), \tag{187}$$

where $\Delta_W E$ is the standard deviation for the energy distribution of the state ψ_W. And, by construction, the width of this state is W_α. Therefore, $\Delta_W E \leq \dfrac{W_\alpha(E)}{2}$. The inequality Eq. (186) now reads

$$\tau_\beta \, W_\alpha(E) \geq 2\,\hbar \cos^{-1}\left(\frac{\beta + 1 - \alpha}{\alpha}\right), \tag{188}$$

for
$$\beta \leq 2\alpha - 1.$$

This relation is valid for all states in terms of reasonable measures of the uncertainty in energy even if the standard deviation of the energy distribution is infinite.

For the Breit-Wigner state.

$$W_\alpha(E) = 2\gamma \tan\left(\frac{\alpha\pi}{2}\right), \tag{189}$$
$$\tau_\beta = -2\,\hbar\,\gamma^{-1} \log \beta.$$

For $\beta = \sqrt{\dfrac{1}{2}}$ $\left(\tau_{\sqrt{\frac{1}{2}}} \text{ is the half-life of the state}\right)$

and $\alpha = 0.9$, from Eq. (188), we get

$$\tau_{\sqrt{\frac{1}{2}}} \, W_{0.9} \geq 0.9\hbar. \tag{190}$$

2.9 THE PATH INTEGRAL

The path integral approach to quantum mechanics was developed by Feynman following up on hints by Dirac. His space-time approach provides a deep insight into quantum dynamics. Due to computational complexity the path integral formalism is not a convenient tool for solving problems in nonrelativistic quantum mechanics. Path integration provides an excellent method for quantizing quantum fields and has become a powerful tool in quantum field theory, statistical mechanics, and numerical computation.

(a)   The Transition Amplitude

In the path integral formulation of quantum mechanics, the basic object of interest is the transition amplitude. Feynman brilliantly developed the path integral representation of the finite transition amplitude and a detailed exposition of his ideas are given in the book by Feynman and Hibbs (1965). We shall, however, present the path integral from the canonical structure of quantum mechanics.

For simplicity, let us discuss a system with one degree of freedom, say, the motion of a one-dimensional particle in a potential. We assume that the Hamiltonian H has no explicit time dependence. Let the Hamiltonian be

$$H(P, Q) = \frac{p^2}{2m} + V(Q). \tag{191}$$

The instantaneous eigenstates of the operators $Q(t)$ and $P(t)$ in the Heisenberg picture are

$$|q, t\rangle = e^{iHt/\hbar}|q\rangle, \tag{192}$$

$$|p, t\rangle = e^{iHt/\hbar}|p\rangle,$$

we note that

$$Q(t)|q, t\rangle = q|q, t\rangle, \tag{193}$$

$$P(t)|p, t\rangle = p|p, t\rangle,$$

and these states are complete, as

$$\int_{-\infty}^{\infty} dq|q, t\rangle\langle q, t| = e^{iHt/\hbar} \int_{-\infty}^{\infty} dq|q\rangle\langle q| \, e^{-iHt/\hbar}$$

$$= 1.$$

We suppose that the particle is in an eigenstate of position $|q_a\rangle$ at time t_a. The amplitude for it to be at q_b at a later time t_b is

$$Z(q_b, t_b; q_a, t_a) \equiv \langle q_b, t_b|q_a, t_a\rangle$$

$$= \langle q_b|U(t_b, t_a)|q_a\rangle, \tag{194}$$

where $U(t_b, t_a) = e^{-iH(t_b - t_a)/\hbar}$ is the time evolution operator.

The transition amplitude $Z(q_b, t_b; q_a, t_a)$ is the amplitude of a particle which is localized at q_a at time t_a to be at the position q_b at time t_b $(t_b > t_a)$. Once we know the amplitude, we can use it to determine the time evolution of a state.

$$\psi(q'', t) \equiv \langle q''|\psi(t)\rangle$$

$$= \int_{-\infty}^{\infty} dq'\, Z(q'', t, q', t_0)\, \langle q'|\psi(t_0)\rangle$$

$$= \int_{-\infty}^{\infty} dq'\, Z(q'', t, q', t_0)\, \psi(q', t_0) \qquad (195)$$

The kernel is known as the propagator in quantum mechanics, since we can use it to determine how an arbitrary state (wave function) propagates in time.

We define the (causal) propagator (that moves into the future) as the function

$$K(q', t'; q, t) = \langle q'|U(t', t)|q\rangle\theta(t' - t), \qquad (196)$$

where $\theta(x)$ is the unit step (or Heaviside) function

$$\theta(x) = \begin{cases} 1, & x > 0 \\ 0, & x < 0 \end{cases} \qquad (197)$$

$$\frac{d}{dx}\,\theta(x) = \delta(x).$$

As $t \to t'$, $U(t', t) \to 1$, and we have

$$\lim_{t \to t'} K(q', t'; q, t) = \delta(q' - q). \qquad (198)$$

The propagator is a probability amplitude for finding a particle at q' at time t' if it was originally at q at time t.

The propagator K is often called the (retarded) Green function for the time-dependent Schrödinger equation. The Green function $G(x, x', t, t')$ for the one-dimensional form of Schrödinger's equation satisfies

$$\left(H\left(-\,i\hbar\,\frac{d}{dx}, x\right) - i\hbar\,\frac{d}{dt}\right) G(x, x', t, t')$$

$$= i\hbar\,\delta(x - x')\delta(t - t'). \qquad (199)$$

In this case, the Green function is

$$G(x, x', t, t') = \theta(t - t')Z(x', t'; x, t,)$$

$$= K(x', t'; x, t,). \qquad (200)$$

Now,

$$i\hbar\,\frac{d}{dt}\,G(x, x', t, t') = -\,i\hbar\,\delta(t' - t)Z(x', t'; x, t,)$$

$$+ \theta(t' - t)\langle x|He^{iH(t' - t)/\hbar}|x'\rangle,$$

and since $Z(x', t; x, t,) = \langle x|x'\rangle = \theta(x - x')$, relation Eq. (199) is obtained.

If the Hamiltonian operator (no explicit time dependence) has a purely discrete spectrum with eigenvectors ϕ_n, then by virtue of the resolution of the identity

$$\sum_{n=0}^{\infty} |\phi_n\rangle\langle\phi_n| = 1, \tag{201}$$

we write the Green function as

$$G(x', t'; x, t) = \theta(t' - t) \sum_{n=0}^{\infty} \phi_n^*(x)\phi_n(x')$$

$$\times \, e^{-iH(t' - t)/\hbar} \tag{202}$$

As an illustrative example, let us evaluate the propagator for a free particle moving one-dimensionally by operator techniques. The Hamiltonian is

$$H(P, Q) = \frac{P^2}{2m}. \tag{203}$$

Inserting a complete set of momentum states, the transition amplitude is

$$\begin{aligned}
Z(q_b, t_b; q_a, t_a) &= \langle q_b | e^{-iH(t_b - t_a)/\hbar} | q_a \rangle \\
&= \int_{-\infty}^{\infty} dp \, e^{-i\frac{P^2}{2m\hbar}(t_b - t_a)} \langle q_b | p \rangle \langle p | q_a \rangle \\
&= \frac{1}{2\pi\hbar} \int_{-\infty}^{\infty} dp \, e^{i\frac{p}{\hbar}(q_b - q_a) - i\frac{p^2}{2m\hbar}(t_b - t_a)}.
\end{aligned} \tag{204}$$

Here we have used

$$\langle q | p \rangle = \frac{1}{\sqrt{2\pi\hbar}} \, e^{\frac{ipq}{\hbar}}. \tag{205}$$

This is a Gaussian integral, and the result is

$$Z(q_b, t_b; q_a, t_a) = \left[\frac{m}{2\pi i\hbar \, (t_b - t_a)} \right]^{1/2} e^{\frac{i}{\hbar}\frac{m(q_b - q_a)^2}{2(t_b - t_a)}}. \tag{206}$$

This is the free particle propagator.

(b) The Feynman Path Integral

The path integral representation of the propagator or the transition amplitude for a finite period of time is obtained by first considering the amplitudes for infinitesimal time intervals and then building up the finite-time propagator out of a large number of infinitesimal time ones in a limiting sense.

The time interval $t_b - t_a$ is first partitioned into N equal segments, with

$$t_0 = t_a \qquad \frac{t_b - t_a}{N} = \epsilon$$

$$t_n = t_a + n\epsilon. \tag{207}$$

Using the composition property for the time-evolution operators

$$U(t_2, t_3)U(t_3, t_1) = U(t_2, t_1), \tag{208}$$

and inserting a complete set of positon states sequentially at each t_n,

we obtain

$$K(q_b, t_b; q_a, t_a) = \int_{-\infty}^{\infty} dq_{N-1} \dots dq_1$$
$$\times \langle q_b, t_b; q_{N-1}, t_{N-1} \rangle \langle q_{N-1}, t_{N-1}; q_{N-2}, t_{N-2} \rangle \dots$$
$$\times \langle q_1, t_1; q_a, t_a \rangle. \tag{209}$$

Thus the propagator is reduced to a product of N propagators, each for short time intervals. The jth such propagator is

$$\langle q_{j+1}, t_{j+1}; q_j, t_j \rangle = \langle q_{j+1} | e^{-i \epsilon H(P, Q)/\hbar} | q_j \rangle. \tag{210}$$

In order to evaluate the matrix element of the exponentiated Hamiltonian, we need to expand the exponential in a power series. Then we adopt a convention for the ordering of the operators. After the expansion, we use coordinate ordering, i.e., all the P operators are always to the left of the Q operators. At this stage, it is tempting to replace $e^{-iH\epsilon/\hbar}$ by $1 - iH\epsilon/\hbar$. This would, however, be inconsistent with the unitarity property.

We write

$$e^{-i \epsilon H(P, Q)/\hbar} = \left\{ e^{-i \epsilon H(P, Q)/\hbar} \right\} + O(\epsilon^2), \tag{211}$$

where $\{....\}$ indicates coordinate ordering. Thus, if the infinitesimal evolution operator is approximated by the coordinate-ordered evolution operator, the error is of order ϵ^2 times an operator which may be expressed in terms of multiple commutators of P and Q. To proceed further, we use the Trotter formula

$$e^{t(A+B)} = \lim_{n \to \infty} \left(e^{\frac{t}{n}A} e^{\frac{t}{n}B} \right)^n, \tag{212}$$

$$t > 0$$

for any two self-adjoint operators that are bounded below. The time t may be continued analytically to imaginary values and the formula used for the evolution operator. For $\epsilon \to 0$, the $O(\epsilon^2)$ term is ignorable. If we assume that the Hamiltonian has the Cartesian structure

$$H(P, Q) = T(P) + V(Q), \tag{213}$$

where T, the kinetic energy, is a function of P only, and V, the potential energy, is a function of Q only, then the Trotter formula permits the approximation as $\epsilon \to 0$:

$$e^{-iH\epsilon/\hbar} = e^{-iT\epsilon/\hbar} e^{-iV\epsilon/\hbar}. \tag{214}$$

Next, we analyze the jth transition amplitude Eq. (210), for $\epsilon = 0$,

$$\langle q_{j+1} | e^{-i \epsilon H(P, Q)/\hbar} | q_j \rangle = \int_{-\infty}^{\infty} dp_j \, \langle q_{j+1} | p_j \rangle \langle p_j | e^{-i \epsilon H(P, Q)/\hbar} | q_j \rangle$$

with insertion of the resolution of identity in the momentum basis

$$= \int_{-\infty}^{\infty} dp_j \, e^{-i \epsilon H(P, Q)/\hbar} \langle q_{j+1} | p_j \rangle \langle p_j | q_j \rangle. \tag{215}$$

Using $\langle q_j | p_j \rangle = \dfrac{1}{\sqrt{2\pi\hbar}} \, e^{\frac{i}{\hbar} p_j q_j}$, we have

$$\langle q_{j+1} | p_j \rangle \langle p_j | q_j \rangle = \frac{1}{2\pi\hbar} \, e^{\frac{i}{\hbar} p_j (q_{j+1} - q_j)}. \tag{216}$$

Since

$$\lim_{\epsilon \to 0} \frac{1}{\epsilon} (q_{j+1} - q_j) = \frac{dq}{dt} \equiv \dot{q}_j, \tag{217}$$

we write the infinitesimal transition element as

$$(q_{j+1}, t_{j+1}; q_j, t_j) = \frac{1}{2\pi\hbar} \int_{-\infty}^{\infty} dp \; e^{\frac{i}{\hbar} \epsilon \, (p_j \dot{q}_j - H(p_j, q_j))}$$

$$= \frac{1}{2\pi\hbar} \int_{-\infty}^{\infty} dp_j \; e^{\frac{i}{\hbar} \epsilon \, L(q_j, \dot{q}_j)}, \tag{218}$$

where $L(q_j, \dot{q}_j)$ is the Lagrangian density.

The finite transition amplitude Z can be written as the product of N infinitesimal elements, giving

$$\langle q_b, t_b; q_a, t_a \rangle = \int_{-\infty}^{\infty} \frac{dp_0}{2\pi\hbar} \cdots \frac{dp_{N-1}}{2\pi\hbar} \, dq_1 \cdots dq_{N-1} \, e^{\frac{i}{\hbar} \sum_{j=0}^{N-1} \epsilon \, L} \tag{219}$$

where we assume $q_0 = q_a$ and $q_N = q_b$.

The Riemann sum in the exponent may be indicated symbolically

$$\lim_{\epsilon \to 0} \sum_{j=0}^{N-1} \epsilon \, L \to \int_{t_a}^{t_b} dt L$$

$$= S[\dot{q}(t), q(t), t_a, t_b], \tag{220}$$

where the functional S is the classical action. The path integral measure is written symbolically as

$$\lim_{N \to \infty} \frac{dp_0}{2\pi\hbar} \cdots \frac{dp_{N-1}}{2\pi\hbar} \, dq_1 \cdots dq_{N-1} \equiv DpDq. \tag{221}$$

The measure spans the phase space available to the particle.

Mathematical subtleties associated with this formalism arise from the fact that the measure is ill-defined and the integrand an oscillatory function A well behaved entity in an Euclidean theory may be obtained after a Wick rotation to imaginary time corresponding to a change from the Schrödinger equation to the heat or diffusion equation. Wiener considered the path integral representation for Brownian motion in a mathematical context. It has been shown that the path integral has a well-defined measure in the Euclidean region.

The Feynman path integral for the propagator may be written as a functional integral

$$K(q_b, t_b; q_a, t_a) = \int_{q_a}^{q_b} DpDq \, \exp\left[\frac{i}{\hbar} \int_{t_a}^{t_b} (p\dot{q} - H(p, q)) \, dt\right]. \tag{222}$$

Expression (222) is the fundamental form for the path integral representation of the propagator.

When the Hamiltonian has the form

$$H = \frac{p^2}{2m} + V(q) \tag{223}$$

then

$$K(q_b, t_b; q_a, t_a) = \lim_{N \to \infty} \int \prod_{n=1}^{N-1} dq_n \prod_{n=1}^{N-1} \frac{dp_n}{2\pi\hbar}$$

$$\times \exp \frac{i}{\hbar} \left[\sum_{n=1}^{N-1} p_n \, (q_{n+1} - q_n) \right.$$

$$\left. - \epsilon \, \frac{p_n^2}{2m} - \epsilon \, V(q_n) \right]. \tag{224}$$

Now,

$$\int_{-\infty}^{\infty} \frac{dp_j}{2\pi\hbar} \exp \frac{i}{\hbar} \left[p_j \, (q_{j+1} - q_j) - \epsilon \, \frac{p_j^2}{2m} - \epsilon \, V(q_j) \right]$$

$$= \sqrt{\frac{m}{2\pi i\hbar\epsilon}} \exp \frac{i}{\hbar} \epsilon \left[\frac{1}{2} m \left(\frac{q_{j+1} - q_j}{\epsilon} \right)^2 - V(qj) \right]$$

$$= \sqrt{\frac{m}{2\pi i\hbar\epsilon}} \exp \left[\frac{i}{\hbar} \epsilon \left(\frac{1}{2} m\dot{q}_j^2 - V(q_j) \right) \right], \tag{225}$$

so

$$K(q_b, t_b; q_a, t_a)$$

$$= \lim_{N \to \infty} \left(\frac{m}{2\pi i\hbar\epsilon} \right)^{N/2} \int \prod_{n=1}^{N-1} dq_n \, \exp \left[\frac{i}{\hbar} \epsilon \left(\frac{1}{2} m\dot{q}_n^2 - V(q_n) \right) \right]$$

$$= \int_{q_a}^{q_b} \overline{D}q \, \exp \left[\frac{i}{\hbar} \int_{t_a}^{t_b} dt \left(\frac{1}{2} m\dot{q}^2 - V(q) \right) \right]$$

$$= \int_{q_a}^{q_b} \overline{D}q \, \exp \left[\frac{i}{\hbar} \int_{t_a}^{t_b} L(\dot{q}, q) dt \right]$$

$$= \int_{q_a}^{q_b} \overline{D}q \, \exp \left[\frac{i}{\hbar} S[q(t)] \right]. \tag{226}$$

Here, the factors arising from the momentum integrations are absorbed in the measure

$$\overline{D}q \equiv \lim_{N \to \infty} \left(\frac{m}{2\pi i\hbar\epsilon} \right)^{N/2} dq_1 \, ... \, dq_{N-1}, \tag{227}$$

the classical action is

$$S[q(t)] = \int_{t_a}^{t_b} dt L \, [q(t)], \tag{228}$$

and the Lagrangian is

$$L[q(t)] = \frac{1}{2} m \left(\frac{dq}{dt}\right)^2 - V(q(t)). \tag{229}$$

This form of the Feynman path integral will be used in the next section. The generalization to many particles in three dimensions is straight-forward in the absence of velocity dependent forces.

(c)   The Sum Over Histories

The basic quantity in the Feynman path integral is the propagator. Here we divide a finite time interval into infinitesimal ones, and evaluate the transition amplitude for each one, and finally chain these together to get the result for the finite time interval. We consider the transition amplitude for a particle going from an initial space-time point (q_a, t_a) to the final space-time point (q_b, t_b). The time interval $t_b - t_a$ is divided into N slices of width ϵ, i.e., $t_b - t_a = N\epsilon$. At each time t_i we select a point q_i. To visualize this pictorially we consider the space-time plane (Fig. 2.1). The intial and final space-time points (q_a, t_a) and (q_b, t_b) are fixed.

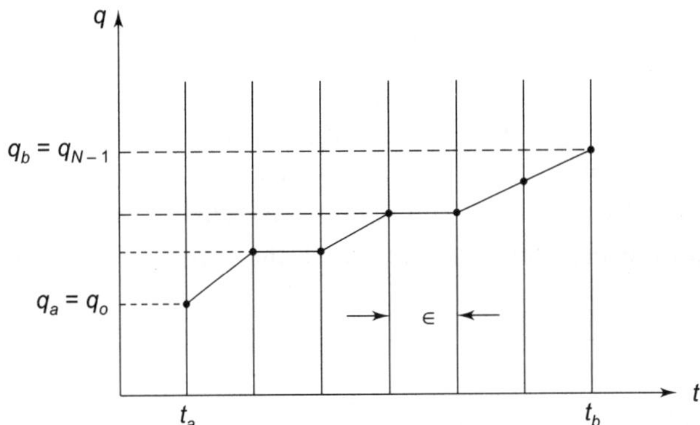

Fig. 2.1   A path between (q_a, t_a) and (q_b, t_b) in space-time plane

The set of points $\{q_0, \ldots, q_{N-1}\}$ defines a path or trajectory. In the limit $N \to \infty$, $\epsilon \to 0$ such that $N\epsilon = t_b - t_a$ we denote this trajectory as $q(t)$ with the initial point $q(t_a) = q_a$ and final point $q(t_b) = q_b$. But this does not imply continuity or differentiability. The trajectory is specified by a set of N points $q(t_i)$ indexed by the discrete times t_i. This represents the path integral as a multiple Riemann integral over this set of coordinates. In the limit, the path integral can be viewed as a sum over all piecewise continuous paths starting at position q_a at time t_a and ending at q_b at time t_b, with each path receiving a weight factor given by the exponential $\frac{i}{\hbar}$ times the classical action along that path. Thus, every piecewise continuous path represents a possible "history" of the particle's motion and the path integral representation of the propagator is a (weighted) sum over all possible histories. In classical physics there is only one single path connecting the two end points, the path which extremizes the action. In Feynman's formulation of quantum mechanics, all paths contribute to the amplitude – they are all on the same footing. Here, all paths contribute equally in amplitude but differently in phase, the phase of each path being proportional to the classical action. This is called Feynman's principle of the democratic equality of all histories.

The superposition principle may be written in terms of path integrals as

$$\int_{q_a t_a}^{q_b t_b} D[q(t)]\, e^{\frac{i}{\hbar}\int_{t_a}^{t_b} dt' L[q(t')]} = \int dq \int_{qt}^{q_b t_b} D[q(t)]\, e^{\frac{i}{\hbar}\int_{t}^{t_b} dt' L[q(t')]}$$

$$\times \int_{q_a t_a}^{qt} D[q(t)]\, e^{\frac{i}{\hbar}\int_{t_a}^{t} dt' L[q(t')]}. \tag{230}$$

Quantum mechanical interference arises directly from the sums over paths.

In path integral formalism all paths connecting the space time end points are included in the summation, whereas in classical mechanics there is a well defined tranjectory. The path integral prescription for the propagator is

$$\sum_{\text{Paths}} e^{iS/\hbar}. \tag{231}$$

Classical physics results in the limit as $\hbar \to 0$ (the stationary phase approximation) and in this limit small changes in the path $q(t)$ produce a great change in phase (ΔS very large compared to $\hbar$). Hence little or no contribution to the summation except for paths for which the action is stationary, i.e.,

$$\left.\frac{\delta S[q(t)]}{\delta q}\right|_{q(t) = \bar{q}(t)} = 0. \tag{232}$$

To find such a path, we write

$$0 = S[\bar{q}(t) + \delta q(t)] - S[\bar{q}(t)]$$

$$= \int_0^t dt' \left[\frac{1}{2} m(\dot{\bar{q}}(t') + \delta\dot{q}(t'))^2 - \frac{1}{2} m(\dot{\bar{q}}(t'))^2\right.$$

$$\left. - V(\bar{q}(t') + \delta q(t')) + V(\bar{q}(t'))\right]$$

$$= \int_0^t dt' \{m\dot{\bar{q}}(t')\, \delta\dot{q}(t') - V'(\bar{q}(t'))\, \delta q(t') + O((\delta q)^2). \tag{233}$$

Integrating by parts, we get

$$0 = -\int_0^t dt' \,(m\ddot{\bar{q}}(t') + V'(\bar{q}(t')))\, \delta q(t'), \tag{234}$$

since the end points of the path are fixed, i.e., $\delta q(0) = \delta q(t) = 0$. Thus, the path which satisfies the stationary phase condition for arbitrary $\delta q(t)$ must obey the classical mechanics equation of motion for a particle:

$$m\ddot{\overline{q}}(t) + V'(\overline{q}) = 0. \tag{235}$$

In the limit $h \to 0$, the classical trajectory $q(t) = q_{cl}(t)$ is the only path which contributes to the integral.

The main contribution to the propagator comes from paths near the classical path, and since they all have approximately the same action as the classical trajectory, we may write as a first approximation

$$K(q_b, t_b; q_a, t_a) \sim e^{\frac{i}{\hbar} \int_{t_a}^{t_b} dt L \left(q_{cl}, \frac{dq_{cl}}{dt}, t \right)}, \tag{236}$$

Equation (236) is known as the "classical approximation". The dominant contribution to the propagator comes from paths in a "strip" about the classical path joining q_a to q_b. In this strip the action varies slowly so that $\Delta S < \pi\hbar$. The stationary phase approximation will generate the semi-classical expansion of the evolution operator in powers of $\hbar$.

(d) Schrödinger's Wave Equation

We have previously shown that the propagator K is the unique solution to the Schrödinger equation. The path integral representation of the propagator must satisfy the Schrödinger equation. Here we show that the path integral representation of the propagator indeed satisfies Schrödinger's time-dependent wave equation.

Now,

$$K(q_N, t_N; q_1, t_1) = \int dq_{N-1} \, \langle q_N, t_N | q_{N-1}, t_{N-1} \rangle \langle q_{N-1}, t_{N-1} | q_1, t_1 \rangle$$

$$= \int_{-\infty}^{\infty} dq_{N-1} \left(\frac{m}{2\pi i \hbar \epsilon} \right)^{1/2} e^{\left[\frac{im}{2\hbar} (q_N - q_{N-1})^2 / \epsilon - \frac{iV\epsilon}{\hbar} \right]}$$

$$\times \langle q_{N-1}, t_{N-1} | q_1, t_1 \rangle. \tag{237}$$

Putting

$$\eta = q_N - q_{N-1},$$

and letting $q_N \to q$ and $t_N \to t + \epsilon$, we get

$$\langle q, t + \epsilon | q_1, t_1 \rangle = \left(\frac{m}{2\pi i \hbar \epsilon} \right)^{1/2} \int_{-\infty}^{\infty} d\eta \, e^{\left(\frac{im\eta^2}{2\hbar\epsilon} - \frac{iV\epsilon}{\hbar} \right)}$$

$$\times \langle q - \eta, t | q_1, t_1 \rangle. \tag{238}$$

Expanding $\langle q - \eta, t | q_1, t_1 \rangle$ in powers of η, and $\langle q, t + \epsilon | q_1, t_1 \rangle$ and $e^{-\frac{iV\epsilon}{\hbar}}$ in powers of ϵ, we obtain

$$\langle q, t | q_1, t_1 \rangle + \epsilon \, \frac{\partial}{\partial t} \langle q, t | q_1, t_1 \rangle$$

$$= \left(\frac{m}{2\pi i\hbar\epsilon}\right)^{1/2} \int_{-\infty}^{\infty} d\eta\, e^{\frac{im\eta^2}{2\hbar\epsilon}} \left(1 - \frac{iV\epsilon}{\hbar} + \cdots\right)$$

$$\times \left[\langle q,\, t|q_1,\, t_1\rangle + \left(\frac{\eta^2}{2}\right) \frac{\partial^2}{\partial q^2} \langle q,\, t|q_1,\, t_1\rangle + \cdots\right]. \tag{239}$$

Collecting terms to order ϵ, we get

$$\epsilon\,\frac{\partial}{\partial t} \langle q,\, t|q_1,\, t_1\rangle = \left(\frac{m}{2\pi i\hbar\epsilon}\right)^{1/2} (2\pi)^{1/2} \left(\frac{i\hbar\epsilon}{m}\right)^{3/2}$$

$$\times \frac{1}{2} \frac{\partial^2}{\partial q^2} \langle q,\, t|q_1,\, t_1\rangle$$

$$- \left(\frac{i}{\hbar}\right) \epsilon\, V\langle q,\, t|q_1,\, t_1\rangle, \tag{240}$$

where we have used

$$\int_{-\infty}^{\infty} d\eta\, \eta^2\, e^{\frac{im\eta^2}{2\hbar\epsilon}} = (2\pi)^{1/2} \left(\frac{i\hbar\epsilon}{m}\right)^{3/2}.$$

We see that $\langle q,\, t|q_1,\, t_1\rangle$ satisfies Schrödinger's time-dependent wave equation

$$i\hbar\,\frac{\partial}{\partial t} \langle q,\, t|q_1,\, t_1\rangle = -\frac{\hbar^2}{2m} \frac{\partial^2}{\partial q^2} \langle q,\, t|q_1,\, t_1\rangle + V\langle q,\, t|q_1,\, t_1\rangle. \tag{241}$$

Feynman's space-time approach to quantum mechanics in terms of path integral representation of propagators is completely equivalent to Schrödinger's wave mechanics.

(e) Imaginary Time and Statistical Mechanics

There is a deep connexion between the path integral representation of the transition amplitude ("sum over paths") and the quantum mechanical partition function ("sum over states").

The partition function for a single particle is

$$Z_\beta = \sum_n e^{-\beta E_n}$$

$$= \sum_n \langle E_n|e^{-\beta H}|E_n\rangle$$

$$= Tr\, e^{-\beta H}. \tag{242}$$

Here, we work in the canonical ensemble and H is assumed time independent. H is assumed to possess a complete set of eigenstates $|E_n\rangle$. The sum in Eq. (242) runs over the complete set of states and $\beta = 1/k_B T$.

Now, the transition amplitude

$$Z_{ab} = (q_b,\, t_b;\, q_a,\, t_a)$$

$$= \left\langle q_b\middle|e^{-\frac{iH(t_b - t_a)}{\hbar}}\middle|q_a\right\rangle, \tag{243}$$

becomes with a Wick rotation $t_b - t_a \rightarrow i\beta\hbar$ and $q_b = q_a$,

$$\overline{Z}_{aa} = \langle q_a | e^{-\beta H} | q_a \rangle. \tag{244}$$

Integrating, we obtain

$$\int dq_a \overline{Z}_{aa} = \int dq_a \langle q_a | e^{-\beta H} | q_a \rangle$$

$$= \sum_n \langle E_n | e^{-\beta H} | E_n \rangle$$

$$= Z_\beta. \tag{245}$$

Using the path integral representation of $\overline{Z}_{aa}$, given by Eq. (228), we perform analytic continuation to imaginary time. This is known as Wick rotation and may be considered as a rotation of the integration contour in the complex t-plane. The transformation

$$t = -i\tau \tag{246}$$

is introduced so that

$$\frac{dq}{d\tau} = \frac{dt}{d\tau} \cdot \frac{dq}{dt} = -i\frac{dq}{dt}.$$

We get, with $t_a = 0$

$$Z_\beta = \int dq_a \int_{q_a}^{q_a} DpDq \; e^{\left\{ -\int_0^\beta d\tau[H(p,\, q) - ip\dot{q}] \right\}}. \tag{247}$$

In this case, the action, which is called the Euclidean action, becomes

$$\frac{i}{\hbar} \int_{t_1}^{t_2} dt \left[\frac{1}{2} m \left(\frac{dq}{dt} \right)^2 - V(q(t)) \right] = -\frac{1}{\hbar} \int_{\tau_1}^{\tau_2} d\tau \left[\frac{1}{2} m \left(\frac{dq}{d\tau} \right)^2 - V(q(\tau)) \right]. \tag{248}$$

The partition function is

$$Z_\beta = \int dq \int_{q(0)\,=\,q}^{q(\beta\hbar)\,=\,q} D[q(\tau)] \; e^{-\frac{1}{\hbar} \int_0^{\beta\hbar} d\tau \left(\frac{m}{2} \frac{dq(\tau)}{d\tau} \right)^2 + V(q(\tau))}$$

$$= \int_{q(\beta\hbar)\,=\,(o)} \overline{D}[q(\tau)] \; e^{-\frac{1}{\hbar} \int_0^{\beta\hbar} d\tau \left(\left(\frac{m}{2} \frac{dq(\tau)}{d\tau} \right)^2 + V(q(\tau)) \right)}. \tag{249}$$

The partition function is a sum over all periodic trajectories of period $\beta\hbar$. From Eq. (248), we see that the kinetic energy changes sign, and the Lagrangian becomes $L = -\frac{1}{2} m \left(\frac{dq}{d\tau} \right)^2 - V(q(\tau))$ in imaginary time, which has the form of the negative of the classical energy. In classically forbidden regimes, this provides the interpretation of a particle moving in an inverted potential. The Wick-rotated path integral Eq. (249) is well-behaved as all the contributions are real and positive.

The free energy, F_β, can be expressed in terms of the path integral as

$$F_\beta \equiv -\frac{1}{\beta} \ln Z_\beta = -\frac{1}{\beta} \ln \int_{q_a}^{q_a} DpDq \; e^{-\frac{1}{\hbar} \int_0^{\beta\hbar} d\tau[H(p,\, q) - ip\dot{q}]}. \tag{250}$$

From Eq. (242), we see that in the limit $\beta \rightarrow \infty$ (low temperature), the summation in the partition function is dominated by the ground state term, E_g. Then

$$\lim_{\beta \to \infty} e^{\beta E_g} Z_\beta = 1 \tag{251}$$

so that

$$E_g = \lim_{\beta \to \infty} \frac{1}{\beta} \ln Z_\beta. \tag{252}$$

The ground state energy can be obtained from the Euclidean form of the path integral.

(f) Gaussian Integrals

In this section we present several useful Gaussian integrals for reference.

We begin with the well-known integral

$$I_0(\alpha) = \int_{-\infty}^{\infty} dx\, e^{-\alpha x^2} = \sqrt{\frac{\pi}{\alpha}}, \tag{253}$$

where α is real and positive.

By differentiating with respect to α, we can get all the integrals of the form

$$I_{2n}(\alpha) = \int_{-\infty}^{\infty} x^{2n} e^{-\alpha x^2}\, dx$$

$$= \frac{1 \cdot 3 \cdot 5 \dots (2n-1)}{2^n \cdot \alpha^n} \times \sqrt{\frac{\pi}{\alpha}}. \tag{254}$$

Next,

$$I_0(\alpha, \beta) = \int_{-\infty}^{\infty} dx\, e^{-\alpha x^2 \pm \beta x}$$

$$= e^{\beta^2/4\alpha} \int_{-\infty}^{\infty} dx\, e^{-\alpha \left(\frac{x \pm \beta}{2\alpha}\right)^2},$$

by completing the square on the exponent

$$= e^{\beta^2/4\alpha} \sqrt{\frac{\pi}{\alpha}}. \tag{255}$$

The Gaussian integral may be evaluated for imaginary α by analytic continuation. The result is

$$I(\alpha) \equiv \int_{-\infty}^{\infty} dx\, e^{-i\alpha x^2} = \sqrt{\frac{\pi}{i\alpha}}. \tag{256}$$

Integrals of this type are also known as Fresnel integrals.

Let us consider a multiple integral over n real variables of the form

$$I(a) = \int_{-\infty}^{\infty} \dots \int dx_1 \dots dx_n \exp\left(-x_i a_{ij} x_j\right), \tag{257}$$

where a_{ij} are the elements of matrix A and summation over repeated Latin indices is understood. If $I(a)$ is to be a real number, the A must be a real, symmetric matrix.

To evaluate Eq. (257) we change variables to reduce it to diagonal form: $x_i = U_{ij}y_j$, where U is a unitary matrix. The new variables are y_i and

$$a_{ij}y_j = \lambda_i y_i, \tag{258}$$

where λ_i is the ith eigenvalue of A.

In terms of the new variable, the integral is

$$I(a) = \int_{-\infty}^{\infty} \ldots \int dy_1 \ldots dy_n \, J(y) \times \exp(-\lambda_i y_i^2), \tag{259}$$

where $J(y)$ is the Jacobian of the transformation.

For a unitary transformation, the Jacobian is unity. Hence

$$I(a) = \pi^{n/2} \prod_{i=1}^{n} (\lambda_i)^{-\frac{1}{2}}$$

$$= \pi^{n/2} (\det A)^{-\frac{1}{2}}$$

$$= (\pi^n \det A^{-1})^{\frac{1}{2}}. \tag{260}$$

By analytic continuation, we obtain

$$\int_{-\infty}^{\infty} dx_1 \ldots dx_n \exp(-ix_i a_{ij} x_j) = [(-i\pi)^n (\det A^{-1}]^{\frac{1}{2}} \tag{261}$$

This is valid if the eigenvalues of A are nonzero and positive.

From Eqs. (255) and (256) we have

$$I(\alpha, \beta) = \int_{-\infty}^{\infty} dx e^{-i\alpha x^2 \pm i\beta x}$$

$$= \sqrt{\frac{\pi}{i\alpha}} \exp\left(\frac{i\beta^2}{4\alpha}\right). \tag{262}$$

(g)   Illustrative Examples

In this section we workout the path integral evaluation of propagators in two cases.

(i)   Free-particle Propagator

Here
$$L = \frac{1}{2} m\dot{q}^2.$$

The propagator is given by

$$K(q_b, t_b; q_a, t_a)$$

$$= \lim_{\substack{N \to \infty \\ \epsilon \to 0 \\ N\epsilon \to t_b - t_a}} \int_{q_a}^{q_b} \left(\frac{m}{2\pi i\hbar\epsilon}\right)^{N/2} dq_, \ldots, dq_{N-1} \exp\left(\frac{i}{\hbar} \sum_{k=0}^{N-1} \frac{m}{\epsilon} (q_{k+1} - q_k)^2\right). \tag{263}$$

Let us start by doing the q_1 integration which includes the q_0 and q_2 factors. Considering this part of the integration, we get

$$\int_{-\infty}^{\infty} dq_1 \exp\left(\frac{i}{\hbar}\frac{m}{2\epsilon}\left[(q_2 - q_1)^2 + (q_1 - q_a)^2\right]\right)$$

$$= \int_{-\infty}^{\infty} dq_1 \exp\left(\frac{i}{\hbar}\left[\frac{m}{4\epsilon}(q_2 - q_a)^2 + \frac{m}{\epsilon}\left(q_1 - \frac{1}{2}q_2 - \frac{1}{2}q_a\right)^2\right]\right)$$

$$= \sqrt{\frac{i\pi\hbar\epsilon}{m}}\,\exp\left(\frac{i}{\hbar}\frac{m}{4\epsilon}(q_2 - q_a)^2\right). \tag{264}$$

Next, we do the integration over q_2. Considering the part of the integrand involving q_2 and combining it with the result above we get

$$\sqrt{\frac{i\pi\hbar\epsilon}{m}}\int_{-\infty}^{\infty} dq_2 \exp\left(\frac{i}{\hbar}\left[\frac{m}{4\epsilon}(q_2 - q_a)^2 + \frac{m}{2\epsilon}(q_3 - q_2)^2\right]\right)$$

$$= \sqrt{\frac{4}{3}\left(\frac{i\pi\hbar\epsilon}{m}\right)^2}\,\exp\left(\frac{i}{\hbar}\frac{m}{2.3\epsilon}(q_3 - q_2)^2\right). \tag{265}$$

After the $N - 1$ integrations, we get, putting $q_N = q_b$,

$K(q_b, t_b; q_a, t_a)$

$$= \lim_{N \to \infty}\left(\frac{m}{2\pi i\hbar\epsilon}\right)^{N/2}\prod_{j=1}^{N-1}\sqrt{\frac{2\pi i\hbar\epsilon j}{(j+1)m}}\,\exp\left[\frac{i}{\hbar}\frac{m(q_b - q_a)^2}{2N\epsilon}\right]$$

$$= \lim_{N \to \infty}\left(\frac{m}{2\pi i\hbar N\epsilon}\right)^{1/2}\exp\left[\frac{i}{\hbar}\frac{m(q_b - q_a)^2}{2N\epsilon}\right] \tag{266}$$

Since $N\epsilon = t_b - t_a$, the $\substack{N \to \infty \\ \epsilon \to 0}$ limit gives the result

$K(q_b, t_b; q_a, t_a)$

$$= \left(\frac{m}{2\pi i\hbar(t_b - t_a)}\right)^{1/2}\exp\left[\frac{i}{\hbar}\frac{m(q_b - q_a)^2}{2(t_b - t_a)}\right], \tag{267}$$

which is identical to that obtained earlier.

(ii)   The Harmonic Oscillator

Here, we consider the one-dimensional oscillator.

Let us consider the general quadratic Lagrangian density of the form

$$L(q, \dot{q}, t) = a(t)q^2 + b(t)\dot{q}^2 + c(t)q\dot{q} + d(t)q$$

$$+ e(t)\dot{q} + f(t) \tag{268}$$

Then the Feynman propagator will have the general form

$$K(q_b, t_b; q_a, t_a) = F(t_b, t_a)\exp\left(\frac{i}{\hbar}S_{cl}\right), \tag{269}$$

where $S_{cl} = \int_{t_a}^{t_b} L(q_{cl}, \dot{q}_{cl}, t)\,dt$ is the action evaluated along the classical path. The prefactor depends only on t_a and t_b. If the coefficients a, b, and c are constants, then the prefactor depends only on the elapsed time $(t_b - t_a)$.

The above quadratic Lagrangian includes the case (i) of a free particle, and the case of a simple harmonic oscillator for which the Lagrangian density is

$$L = \frac{1}{2} m\dot{q}^2 - \frac{1}{2} m\omega^2 q^2. \tag{270}$$

The corresponding equation of motion is

$$\frac{d^2q}{dt^2} + \omega^2 q = 0. \tag{271}$$

The general solution is

$$q_{cl}(t) = A \sin (\omega t + \phi), \tag{272}$$

where A, ϕ are constants.

The boundary conditions are

$$q_b = A \sin (\omega t_b + \phi)$$

and
$$q_a = A \sin (\omega t_a + \phi). \tag{273}$$

Then

$$\cot \phi = \frac{q_b}{q_a \sin \omega T} - \cot \omega T$$

$$A = \frac{q_a}{\sin \phi}, \tag{274}$$

where $T = t_b - t_a$ is the elapsed time. The classical action i.e., the value of the action for the classical trajectory is

$$S_{cl} = \int_{t_a}^{t_b} dt \left[\frac{1}{2} m\dot{q}_{cl}^2 - \frac{1}{2} m\omega^2 q_{cl}^2 \right]$$

$$= \frac{m\omega}{2 \sin \omega T} \left[(q_a^2 + q_b^2) \cos \omega T - 2q_a q_b \right]. \tag{275}$$

This is singular at $\omega T = n\pi$.

We proceed to calculate the prefactor[*].

We introduce new variables of integration as

$$q_j = q_{cl}(t_j) + x_j. \tag{276}$$

Now, $x(t)$ vanishes at the times t_a and t_b. So we expand

$$x(t) = \sum_{n=1}^{N-1} a_n \sin \frac{n\pi}{T} (t - t_a), \tag{277}$$

where a_n are arbitrary constants. The set a_n constitutes the new variables of integration, and

$$x_j = \sum_{n=1}^{N-1} a_n \sin \frac{n\pi}{T} (t_j - t_a). \tag{278}$$

---

[*] The details are given in Swanson (1992).

It can be shown that the Jacobian of the transformation J is independent of a_n and the path integral satisfies

$$\left(\frac{m}{2\pi i\hbar T}\right)^{1/2} = J \int \left(\frac{Nm}{2\pi i\hbar T}\right)^{N/2} da_1 \dots da_{N-1}$$

$$\times \exp\left(\frac{i}{\hbar} \prod_{n=1}^{N-1} m \, \frac{a_n^2 \, \pi^2 \, n^2}{4T}\right),$$

where $\epsilon = T/N$ has been used

$$= \left(\frac{m}{2\pi i\hbar T}\right)^{1/2} J \prod_{n=1}^{N-1} \left(\frac{2\pi i\hbar T}{m\pi^2}\right)^{1/2} \cdot \frac{1}{n}. \tag{279}$$

Then

$$J(a_1, \dots, a_{N-1}) = N^{-N/2} \, \pi^{N-1}(N-1)! \tag{280}$$

To determine the amplitude $F(T)$, the action is

$$\frac{i}{\hbar} \int_{t_a}^{t_b} dt \left[\frac{1}{2} m\dot{x}^2 - \frac{1}{2} m\omega^2 x^2\right] = \frac{i}{\hbar} \sum_{n=1}^{N-1} \frac{1}{2} m \left(\frac{n^2\pi^2}{T} - \omega^2 T\right) a_n^2. \tag{281}$$

Then

$$F(T) = \left(\frac{mN}{2\pi i\hbar T}\right)^{N/2} J(a_1, \dots, a_{N-1})$$

$$e^{\frac{i}{\hbar} \sum_{n=1}^{N-1} \frac{m}{2} \left(\frac{n^2\pi^2}{T} - \omega^2 T\right) a_n^2}. \tag{282}$$

We note that if $\omega T = n\pi$, the coefficient of a_n^2 vanishes so that the nth mode is undefined. To remedy the defect, we make the analytic continuation using the Wick rotation $T \to iT$. Then

$$\frac{i}{\hbar} \sum_{n=1}^{N-1} \frac{m}{2} \left(\frac{n^2\pi^2}{T} - \omega^2 T\right) a_n^2 \to -\frac{1}{\hbar} \sum_{n=1}^{N-1} \frac{m}{2} \left(\frac{n^2\pi^2}{T} - \omega^2 T\right) a_n^2. \tag{283}$$

This is defined for all modes.

Performing the Gausian integrals, we get the Wick-rotated prefactor as

$$F(T) = \left(\frac{m}{2\pi\hbar T}\right)^{1/2} \prod_{n=1}^{N-1} \left(1 + \frac{\omega^2 T^2}{n^2\pi^2}\right)^{-1/2}. \tag{284}$$

In the limit $N \to \infty$ the Wick-rotated perfactor is

$$F(T) = \left(\frac{m\omega}{2\pi\hbar \, \sinh \omega T}\right)^{1/2}, \tag{285}$$

using the identity

$$\prod_{n=1}^{\infty} \left(1 + \frac{\omega^2 T^2}{n^2\pi^2}\right)^{-1/2} = \left(\frac{\omega T}{\sinh \omega T}\right)^{1/2}. \tag{286}$$

We analytically continue this to real time by the inverse Wick rotation $T \to iT$. We obtain the prefactor as

$$F(T) = \left(\frac{m\omega}{2\pi\hbar \ \sinh \omega T}\right)^{1/2}, \tag{287}$$

using $\sinh \omega T = i \sin \omega T$.

From Eqs. (275) and (287), we get the final result

$$K\,(q_b,\,t_b;\,q_a,\,t_a) = \left(\frac{m\omega}{2\pi\hbar \ \sinh \omega T}\right)^{1/2} \times$$

$$\times \exp\left(\frac{im\omega}{2\hbar \ \sinh \omega T}\,(q_a^2 + q_b^2)\,\cos\,\omega T - 2q_a q_b]\right), \tag{288}$$

where $T = t_b - t_a$.

2.10 DEFORMATION QUANTIZATION

There are three different routes to quantization. The first is the standard one – the operator formalism – in which the observables are represented by operators in Hilbert space. The second one is the path integral approach to quantum mechanics. The third one is deformation quantization or the phase space formulation of quantum mechanics.

The phase space formalism grew from the pioneering works of Wigner (1932) on the quasi-probability distribution function and Weyl (1927) on the correspondence between operators and c-number phase space functions. The basic formulation was completed by Groenewold (1946) and Moyal (1949). Important contributions were made by Takabayasi (1954), Baker (1958), Fairlie (1964), Bayen et al (1978) and others. This formalism can be interpreted as a deformation of classical mechanics and hence called deformation quantization.

In the last quarter century, this new way of doing quantum mechanics has been found to be of importance in quantum transport processes, quantum optics, nuclear physics, condensed matter, quantum chaos and decoherence. The mathematical structure of this formulation is of relevance in non-commutative quantum mechanics.

(a) The Wigner Distribution

For a single particle in one dimension, the Wigner distribution function is defined as

$$P_W(q, p) = (2\pi\hbar)^{-1} \int_{-\infty}^{\infty} dy \left\langle q - \frac{1}{2}\,y \,\middle|\, \rho \,\middle|\, q + \frac{1}{2}\,y \right\rangle e^{ipy/\hbar} \tag{289}$$

where ρ is the state operator or density matrix. For a pure state

$$P_W(q, p) = (2\pi\hbar)^{-1} \int_{-\infty}^{\infty} dy \ \psi^*\left(q + \frac{y}{2}\right) \psi\left(q - \frac{y}{2}\right) e^{ipy/\hbar}, \tag{290}$$

where $\psi(q) = \langle q | \psi \rangle$ is the wave function of the state.

For N-particles in three dimensions, all variables are considered as $3N$-dimensional vectors and $(2\pi\hbar)^{-1}$ must be replaced by $(2\pi\hbar)^{-3N}$.

Using the momentum representation, we write

$$P_W(q, p) = (2\pi\hbar)^{-1} \int_{-\infty}^{\infty} ds \left\langle p - \frac{1}{2}s \left| \rho \right| p + \frac{1}{2}s \right\rangle e^{iqs/\hbar}. \tag{291}$$

The Wigner function (WF) has several properties.

(i) It is normalized:

$$\int dq dp P_W(q, p) = Tr\rho = 1. \tag{292}$$

(ii) $P_W(q, p)$ is real.

(iii) The integral over momentum gives the probability distribution of position

$$\int_{-\infty}^{\infty} P_W(q, p)\, dp = |\psi(q)|^2 = \langle q|\rho|q\rangle. \tag{293}$$

The integral over position gives the probability distribution of momentum

$$\int_{-\infty}^{\infty} P_W(q, p)\, dq = \langle p|\rho|p\rangle. \tag{294}$$

The WF is a distribution function in phase space.

p – and q – projections lead to marginal probability densities (positive semidefinite). The WF is not a probability distribution as P_W is not non-negative for arbitary states. The WF can and usually does become negative in some areas of phase space. The only pure state WF which is non-negative is the Gaussian. The WF of the ground state of the harmonic oscillator is positive definite (see Section 11.2).

The Wigner distribution function is analytic in p and q and is square-integrable. It is not non-negative for almost all wave functions. The WF is not a probability distribution, but a quasi-probability distribution. Nevertheless, it has been found useful in different areas of physics.

Let us consider pure state $\rho = |\psi\rangle\langle\psi|$ and $\rho' = |\phi\rangle\langle\phi|$. If $P_W(q, p)$ and $P'_W(q, p)$ are the distributions corresponding to these states respectively, then

$$|\langle\phi|\psi\rangle|^2 = (2\pi\hbar) \iint dq dp P_W(q, p)\, P'_W(q, p). \tag{295}$$

If $\rho = \rho'$, then

$$\iint dq dp [P_W(q, p)]^2 = (2\pi\hbar)^{-1}, \tag{296}$$

for a pure state.

If we choose the two states vectors $|\phi\rangle$ and $|\psi\rangle$ orthogonal, then

$$\iint dq dp P_W(q, p)\, P'_W(q, p) = 0, \tag{297}$$

which implies that $P_W(q, p)$ cannot be everywhere positive. This conclusion is rather general.

Also,

$$(2\pi\hbar)^{-1} \iint dq\,dp\,A(q, p)\, B(q, p) = Tr\,(\hat{A}\hat{B}), \qquad (298)$$

where $A(q, p)$ is the classical function corresponding to the quantum operator $\hat{A}$, and is given, according to Wigner's prescription, by

$$A(q, p) = \int dz\, e^{ipz/\hbar} \left\langle q - \frac{1}{2}\, z \middle| \hat{A} \middle| q + \frac{1}{2}\, z \right\rangle. \qquad (299)$$

A similar relation exists between $B(q, p)$ and $\hat{B}$.

Next we consider the time evolution of the WF. The time dependence of P_W may be written as the sum of two parts

$$\frac{\partial P_W}{\partial t} = \frac{\partial_K P_W}{\partial t} + \frac{\partial_V P_W}{\partial t}, \qquad (300)$$

the first part resulting from the kinetic energy and the second term from the potential.

We know that the time dependence of the state operator ρ is given by

$$\frac{d\rho}{dt} = \frac{i}{\hbar}\, [\rho, H], \qquad (301)$$

where the Hamiltonian $H = \dfrac{p^2}{2m} + V$ in this case.

Equation (301) may be written as

$$\frac{d\rho}{dt} = \frac{\partial_K \rho}{\partial t} + \frac{\partial_V \rho}{\partial t}, \qquad (302)$$

where

$$\frac{\partial_K \rho}{\partial t} = \frac{i}{2m\hbar}\, (\rho p^2 - p^2 \rho), \qquad (303)$$

and

$$\frac{\partial_V \rho}{\partial t} = \frac{i}{\hbar}\, (\rho V - V\rho). \qquad (304)$$

Now,

$$\frac{\partial K}{\partial t}\, \langle p|\rho|p'\rangle = \frac{i}{2m\hbar}\, \langle p|\rho|p'\rangle\, (p'^2 - p^2)$$

$$= \frac{i}{2m\hbar}\, \langle p|\rho|p'\rangle\, (p' + p)\, (p' - p) \qquad (305)$$

Using Eq. (291), we get

$$\frac{\partial_K P_W\, (q, p, t)}{\partial t} = -\frac{p}{m}\, \frac{\partial P_W\, (q, p, t)}{\partial q}. \qquad (306)$$

Using Eq. (304) in the position representation, and with Eq. (289), we get

$$\frac{\partial_V}{\partial t} P_W (q, p, t) = \frac{i}{\hbar} \cdot \frac{1}{2\pi\hbar} \int_{-\infty}^{\infty} dy \left\langle q - \frac{1}{2} y \left| \rho \right| q + \frac{1}{2} y \right\rangle$$

$$\times \left[V \left(q + \frac{y}{2} \right) - V \left(q - \frac{y}{2} \right) \right] \times e^{ipy/\hbar}. \quad (307)$$

Assuming that V can be expanded in a Taylor series, we write

$$V \left(q + \frac{y}{2} \right) - V \left(q - \frac{y}{2} \right) = \sum_\lambda \frac{2}{\lambda!} \left(\frac{1}{2} y \right)^\lambda V^{(\lambda)}(q), \quad (308)$$

where λ is restricted to all odd positive integers, and $V^{(\lambda)}(q) = \dfrac{d^\lambda V}{dq^\lambda}$. We obtain

$$\frac{\partial_V}{\partial t} P_W (q, p, t) = \sum_{\lambda \ \text{odd}} \frac{1}{\lambda!} \left(-\frac{1}{2} i\hbar \right)^{\lambda - 1} \frac{d^\lambda V(q)}{dq^\lambda} \frac{\partial^\lambda}{\partial p^\lambda} P_W (q, p, t) \quad (309)$$

The sum of Eqs. (306) and (309) gives the time dependence of the WF.

The lowest term of Eq. (309) occurs when $\lambda = 1$ and then the expression

for $\dfrac{\partial}{\partial t} P_W (q, p, t)$, the sum of Eqs. (306) and (309), reproduces the classical

Liouville equation. The $\hbar^2$ terms give the quantum correction if this is very small. In certain cases, the correction terms do not vanish in the limit $\hbar \to 0$.

The use of a smearing function (smoothing P_W by a filter of size larger than $\hbar$) was first proposed by Husimi (1940). It leads to a distribution function which is non-negative for all p and q. Besides the Husimi distribution, other distribution functions have been proposed.

(b)   The Star Product and Quantization

Weyl proposed the association of a quantum mechanical operator to every function of q and p. The Weyl correspondence maps invertible c-number phase-space functions $f (q, p)$ (called classical kernels) to operators $\hat{F}$ with a specific ordering prescription. We write $p \to \hat{p}$, $q \to \hat{q}$, and in general

$$f \xrightarrow{\ \theta\ } \hat{F}$$

$$\hat{F}(\hat{p}, \hat{q}) = \frac{1}{(2\pi)^2} \int d\tau d\sigma dq dp f(q, p)$$

$$\times \exp \left(i\tau (\hat{p} - p) + i\sigma (\hat{q} - q) \right). \quad (310)$$

The Weyl ordering prescription requires that an arbitrary operator considered as a power series in $\hat{p}$ and $\hat{q}$ be ordered in a completely symmetrized manner. A classical term $q^n p^m$ becomes

$$q^n p^m \to \frac{1}{2^n} \sum_{r=0}^{n} \binom{n}{r} \hat{q}^{n-r} \hat{p}^m \hat{q}^r \quad (311)$$

as can be seen by considering the $\sigma^n \tau^m$ coefficient in the expansion of $(\sigma \hat{q} + \tau \hat{p})^{m+n}$. As an example,

$$6p^2 q^2 \rightarrow \hat{p}^2\,\hat{q}^2 + \hat{q}^2\,\hat{p}^2 + \hat{p}\hat{q}\hat{p}\hat{q} + \hat{p}\hat{q}^2\hat{p} + \hat{q}\hat{p}\hat{q}\hat{p} + \hat{q}\hat{p}^2\hat{q}.$$

We note that

$$Tr\hat{F} = \int dq dp f. \tag{312}$$

Conversely, the c-number phase space kernels $f(q, p)$ of Weyl-ordered operators $\hat{F}(\hat{q}, \hat{p})$ are specified by $\hat{p} \rightarrow p$, $\hat{q} \rightarrow q$, and

$$f(q, p) = \frac{1}{(2\pi)^2} \int d\tau d\sigma e^{i(\tau p + \sigma q)}\, Tr^{(-i(\tau \hat{p} + \sigma \hat{q})\hat{F})}$$

$$= \frac{1}{2\pi\hbar} \int dy e^{-iyp/\hbar} \left(q + \frac{\hbar}{2} y \middle| \hat{F}(\hat{q}, \hat{p}) \middle| q - \frac{\hbar}{2} y \right). \tag{313}$$

The expectation value of the measurement of the operator $\hat{F}$, if carried out on a system in the state $|\psi\rangle$, is equal to the expectation value of the classical function $f(q, p)$ to which $\hat{F}$ corresponds *a'* la Weyl

$$\langle \psi | \hat{F} | \psi \rangle = \int dq \int dp P_W(q, p)\, f(q, p). \tag{314}$$

For any density matrix $\hat{\rho}$

$$Tr(\hat{\rho}\hat{F}) = \int dq \int dp P_W(q, p)\, f(q, p). \tag{315}$$

θ is called the Weyl map. It takes a c-number phase space function $f(q, p)$ to an operator $\hat{F}(\hat{q}, \hat{p}) \equiv \theta(f)$ by the Weyl correspondence. Such an operator may be called a Weyl operator. If we multiply two Weyl operators we get another Weyl operator.

Groenewold showed that

$$\theta(f)\, \theta(g) = \theta(f * g), \tag{316}$$

where the Groenewold-Moyal $*$ product (star product) is

$$* \equiv \exp i\hbar/2 \left\{ (\overleftarrow{\partial}_q \overrightarrow{\partial}_p - \overleftarrow{\partial}_p \overrightarrow{\partial}_q) \right\} \tag{317}$$

with $\partial_q \equiv \dfrac{\partial}{\partial q}$, etc. and the arrows indicate the directions in which the derivatives act. The $*$product of two functions f and g is

$$f(q, p) * g(q, p) = f(q, p) \exp \left\{ \frac{i\hbar}{2} (\overleftarrow{\partial}_q \overrightarrow{\partial}_p - \overleftarrow{\partial}_p \overrightarrow{\partial}_q) \right\} g(q, p). \tag{318}$$

This may be evaluated by shift operators (Bopp operators)

$$f(q, p) * g(q, p) = f\left(q + \frac{i\hbar}{2} \overrightarrow{\partial}_p, p - \frac{i\hbar}{2} \overrightarrow{\partial}_q \right) g(q, p). \tag{319}$$

The $*$product is non-commutative, but associative:

$$f * g \neq g * f, \tag{320}$$

$$(f * g) * h = f * (g * h). \tag{321}$$

Also,

$$(f * g)* = (f)^* * (g)*. \tag{322}$$

The equivalent Fourier representation of the *product is

$$f * g = \frac{1}{\hbar^2 \pi^2} \int dp' dp'' dq' dq'' f(q', p') g(q'', p'')$$

$$\times \exp\left(-\frac{2i}{\hbar}\left(p(q' - q'') + p'(q'' - q) + p''(q - q')\right)\right). \tag{323}$$

Another integral representation is

$$f * g = \frac{1}{\hbar^2 \pi^2} \int du \, dv \, dw \, dz f(q + u, p + v) g(q + w, p + z)$$

$$\times \exp\left(-\frac{2i}{\hbar}(uz - vw)\right). \tag{324}$$

*multiplication of c-number phase space functions is in complete isomorphism to Hilbert-space operator multiplication

$$\hat{F}\hat{G} = \frac{1}{(2\pi)^2} \int d\tau \, d\sigma \, dq \, dp \, (f * g) \exp\left(i\tau(\hat{p} - p)\right.$$

$$\left. + i\sigma(\hat{q} - q)\right). \tag{325}$$

The cyclic phase-space property of trace is

$$\int dq \, dp \, f * g = \int dq \, dp \, g * f = (f * g - g * f). \tag{326}$$

The *commutator, or Moyal bracket (Mb), is defined as

$$[f, g]_* \equiv \{f, g\}_{Mb} = (f * g - g * f). \tag{327}$$

This is the essentially unique one-parameter ($\hbar$) associative deformation of the Poisson bracket (Pb). The leading term in the *commutator expressed as a power series expansion in $\hbar$ is the Poisson bracket

$$(f * g - g * f) = i\hbar\{f, g\}_{Pb} + \ldots. \tag{328}$$

The correspondence principle is

$$\lim_{\hbar \to 0} \frac{1}{i\hbar} [f, g]_* = \{f, g\}_{Pb}. \tag{329}$$

In this language, the equation of motion of a system with Hamiltonian $H(q, p)$ is

$$\frac{\partial}{\partial t} f(q, p) = \frac{1}{i\hbar} [f(q, p), H(p, q)]_*. \tag{330}$$

The dynamical evolution of the WF is given by such an equation. Stationary WFs obey the following equation

$$H(p, q) * P_W(p, q) = H\left(q + \frac{i\hbar}{2}\,\overrightarrow{\partial}_p,\, p - \frac{i\hbar}{2}\,\overrightarrow{\partial}_q\right) P_W(q, p)$$

$$= P_W(q, p) * H(q, p)$$

$$= EP_W(q, p), \tag{331}$$

where E is the energy eigenvalue of $\hat{H}|\psi\rangle = E|\psi\rangle$.

Deformation quantization is an alternative route to the problem of quantizing a classical system. Although it contains rather strange features like a quasi-probability distribution function in phase space and a deformed law of multiplication (*product), it is a logically complete and self-standing formulation of quantum mechanics. The star product deforms the commutative classical algebra of observables into the non-commutative quantum algebra of observables. One starts with the algebra of observables $\mathcal{A}$ of the classical problem. With a phase space M this gives the algebra of functions $C(M)$. The *commutator then provides a deformation of this algebra to a family of algebras $\mathcal{A}_\hbar$ depending on a parameter $\hbar$ which reduces to $C(M)$ in the limit $\hbar \to 0$, and the leading term in its expansion is the Poisson bracket. This clarifies the connection between classical and quantum mechanical behavior of systems.

A state is characterized by its energy E. The states are described by distributions on phase space that are projectors. The state of a system with energy E is denoted by $\pi_E\,(q, p)$. We have

$$\frac{1}{2\pi\hbar} \int dq dp \pi_E\,(q, p) = 1, \tag{332}$$

and

$$(\pi_E * \pi_{E'})(q, p) = \delta_{EE'}\,\pi_E\,(q, p). \tag{333}$$

The eigenvalue equation for the Hamiltonian $H(q, p)$ is

$$(H * \pi_E)(q, p) = E\pi_E\,(q, p), \tag{334}$$

where the spectral decomposition of H is

$$H(q, p) = \sum_E E\pi_E\,(q, p). \tag{335}$$

The summation may indicate integration if the spectrum is continuous.

Also,

$$E = \frac{1}{2\pi\hbar} \int dq dp (H * \pi_E)(q, p)$$

$$= \frac{1}{2\pi\hbar} \int dq dp H(q, p)\,\pi_E(q, p) \tag{336}$$

Equation (334) corresponds to the time-independent Schrödinger equation an is called the *genvalue equation.

The time evolution of the system is expressed by

$$i\hbar \, \frac{d}{dt} \, \mathrm{Exp}\,(Ht) = H * \mathrm{Exp}\,(Ht). \tag{337}$$

The solution to this differential equation is the star exponential

$$\mathrm{Exp}\,(Ht) = \sum_{n=0}^{\infty} \frac{1}{n!} \left(\frac{-it}{\hbar} \right)^n (H*)^n, \tag{338}$$

where

$$(H*)^n = H* \, H* \, ... \, H* \quad (n \text{ factors}). \tag{339}$$

The time-evolution function may be expressed through the Fourier-Dirichlet expansion

$$\mathrm{Exp}\,(Ht) = \sum_E \pi_E \, e^{-\frac{i}{\hbar} Et}. \tag{340}$$

This formulation can be generalized to more than one degree of freedom in a straight forward manner.

(c) Non-commutative Quantum Mechanics

Recent developments in string theory have aroused growing interest in non-commutative spaces. Spatial non-commutativity can be realized by the coordinate operators satisfying the commutation relation

$$[\hat{x}_\mu, \hat{x}_\upsilon] = i\theta_{\mu\upsilon}, \tag{341}$$

where θ is an antisymmetric (constant) tensor of dimension (length)^2. By taking the position coordinates to be noncommuting, a large class of "non-commutative field theories" have been studied (Douglas and Nekrasov (2001)). Apart from the current activity in field theory, there is considerable interest in the phenomenological implication of non-commutative spaces. Quantum mechanics on non-commutative space has been extensively studied in recent years.

Non-commutativity of space, as postulated in Eq. (341), implies an uncertainty relation between position measurements. This leads to a nonlocal theory. Also, the non-commutativity of space-time leads to the violation of the Lorentz symmetry. Possible tests of such symmetry violations are being actively considered. Non-commutative time implies nonlocality in time and breakdown of unitarity.

To develop quantum mechanics on non-commutative space, we specify the phase space. We consider a quantum particle whose coordinates satisfy the deformed Heisenberg algebra

$$[x_i, x_j] = i\theta_{ij},$$

$$[x_i, p_j] = i\hbar\delta_{ij}, \tag{342}$$

$$[p_i, p_j] = 0.$$

The parameter θ is called the non-commutativity parameter. If $\theta = 0$, we go back to the commutative case. Time is taken as commutative so that unitarity is assured. The standard formulation of classical mechanics and canonical quantization follow. In non-commutative space the ordinary product is replaced by the star product. Usually, one considers non-commutative associative algebras which are deformed with respect to the parameter θ (see earlier discussion). The Hilbert space is taken to be the same as the Hilbert space of the corresponding commutative space. This formulation is consistent with the rules of standard quantum mechanics with an uncertainty relation between the non-commuting coordinate variables. Spatial non-commutativity violates rotational symmetry. Some recent articles on non-commutative quantum mechanics are listed in the bibliography.

PROBLEMS

2.1 Let $A(x)$ be an operation which depends on a continuous variable x. The derivative of A is defined as

$$\frac{dA(x)}{dx} \equiv A'(x) = \lim_{\epsilon \to 0} \frac{A(x + \epsilon) - A(x)}{\epsilon}.$$

Show that

(i) $\dfrac{d(e^{xA})}{dx} = Ae^{xA}$,

(ii) $\dfrac{d}{dx}(A^{-1}) = -A^{-1} A' A^{-1}$,

if A has an inverse.

2.2 Show that

$$[A, e^{-\beta H}] = e^{-\beta H} \int_0^\beta e^{\lambda H} [A, H] e^{-\lambda H} \, d\lambda.$$

for any two operators A and H.

2.3 The Hamiltonian of a two-state system is given by

$$H = a(|1><2| + |2><1|),$$

where a is a real number having the dimension of energy.

(i) Is H a projection operator? What about $a^{-2} H$?

(ii) Find the energy levels and the normalized eigenstates.

2.4 Let A be a hermitian operator with non-degenerate eigenvalues a_i, and corresponding eigenkets $|a_i\rangle$. Show that

(i) $\prod_i (A - a_i)$ is a null operator.

(ii) $\displaystyle\prod_{i \neq 1} \frac{A - a_i}{a_1 - a_i}$ is a projection operator which projects $|a_1\rangle$ only.

2.5 Let the position x be an operator in the Schrödinger picture. Find the corresponding operator x_H in the Heisenberg picture for (i) a free particle, (ii) a linear harmonic oscillator whose Hamiltonian is

$$H = \frac{p^2}{2m} + \frac{1}{2} m\omega^2 x^2.$$

2.6 Calculate the density matrix $\rho(x, x'; \beta)$ for a one-dimensional free particle. Use the definition given in Eq. (60).

2.7 Derive the Green function for a one-dimensional free particle.

2.8 Consider a particle of mass m moving in one dimension subject to a constant gravitational acceleration g. The Lagrangian is $\mathcal{L}(x, \dot{x}) = \dfrac{1}{2} m\dot{x}^2 - mgx$.

Calculate the classical action and obtain the propagator for the linear potential. (See problem 3.5).

2.9 A linear harmonic oscillator is acted on by an external time-varying force $f(t)$. The Lagrangian of the system is

$$\mathcal{L}(x, \dot{x}, t) = \frac{1}{2} m\dot{x}^2 - \frac{1}{2} m\omega^2 x^2 + xf(t).$$

(a) Obtain the classical solution $x_{cl}(t)$ with the boundary conditions $x(t_1) = x_1$; $x(t_2) = x_2$.

(b) Find the action $S[x_{cl}(t)]$.

(c) Calculate the propagator.

2.10 Consider the general quadratic Lagrangian

$$\mathcal{L}(x, \dot{x}, t) = a(t)x^2 + b(t)\,\dot{x}^2 + c(t)x\dot{x} + d(t)x$$
$$+ e(t)\dot{x} + f(t).$$

(a) Show that the Feynman propagator can be represented as

$$K(x_2, t_2; x_1, t_1) = A(t_2, t_1) \exp\left[\frac{i}{\hbar}\right] \int_{t_1}^{t_2} \mathcal{L}(x_{cl}, \dot{x}_{cl}, t)$$

where x_{cl} is the classical path and A is an arbitrary function.

(b) If a, b and e are constants, then show that $A = A(t_2 - t_1)$.

CHAPTER 3

The Schrödinger Equation and Its Solution

Thus far, our discussion of quantum mechanics has concentrated on the states and observables of a system in an abstract Hilbert space, though occasionally we have referred to specific functional realizations of the Hilbert space.

In the present chapter we begin our consideration of wave mechanics.

In the Schrödinger wave function approach, we employ a continuous Hilbert space, i.e., a function space. We then choose a specific representation — the coordinate or the momentum representation — in which the results of measurements of position or momentum of a point particle are considered. Here we discuss the basic features of wave mechanics in terms of wave functions in a specific representation.

In this chapter, we introduce the wave function associated with a physical system and its probabilistic interpretation, the Schrödinger equation of motion, the wave packets, and also the solutions of the wave equation for different simple systems. For making the presentation fairly complete we discuss some of the concepts which have been introduced in Chapter 2 though from the wave mechanics viewpoint. Other important systems, e.g., the harmonic oscillator, the hydrogen atom are considered in the chapter on applications (Chapter 11).

3.1 THE SCHRÖDINGER EQUATION AND THE PROBABILITY INTERPRETATION OF THE WAVE FUNCTION

Let us consider a physical system and let A, B, C .... be the c.s.c.o. The set of their eigenvectors $\{|a, b, c, ...\rangle\}$ forms an orthonormal basis. Given a ket $|\psi\rangle$ in the Hilbert space associated with the system, we write a function

$$\psi(a, b, c, ...) = \langle a, b, c, ... |\psi\rangle. \tag{1}$$

This function is called the wave function in the representation of the set of observables. ψ characterizes a pure state of the system equivalent to the ket $|\psi\rangle$. A wave function is thus the representative of a ket expressed as a function of the observables $a, b, c \ldots$. As an example, the position or coordinate representation is obtained by choosing the eigenvectors $\{|\vec{x}\rangle\}$ of the position operator. The wave function representing a pure state $|\psi\rangle$ in the position representation is written as

$$\psi(\vec{x}, t) = \langle \vec{x}|\psi(t)\rangle. \tag{2}$$

We now proceed to study the equation of motion for a pure state of a system. When one performs an observation on a system, the state of the system changes in an unpredictable way. In between observations, however, causality applies, and the state of the system is governed by an equation of motion which makes the state at one time determine the state at a later time.

Let us consider a (spinless) particle of mass m moving in a force-free region between measurements. We begin with the Einstein-de Broglie relation

$$\vec{p} = \hbar\vec{k}; \quad E = \hbar\omega, \tag{3}$$

which attributes wave-like properties to the particle. Eq. (3) relates the momentum $\vec{p}$ and the energy E of the particle to the wave vector $\vec{k}$ and the frequency ω, respectively.

We make the following basic assumptions. In every state of the system and indeed for a free particle there exists at each instant of time t, a probability $W(\vec{x}, t)d^3x$ that the particle is found in the interval $(\vec{x}, \vec{x} + d\vec{x})$. $W(\vec{x}, t)$ is called the position or coordinate probability density. We further assume that for every state of a system and indeed for a free particle, there exists at every time t a probability $W(\vec{p}, t)d^3p$ of finding the momentum of the particle in the interval $(\vec{p}, \vec{p} + d\vec{p})$. $W(\vec{p}, t)$ is called the momentum probability density.

We suppose that a wave function $\psi(\vec{x}, t)$ exists and it satisfies a linear, homogeneous equation so that the "matter waves" possess the property of superposition.

The probability densities for the coordinate and momentum $W(\vec{x}, t)$ and $W(\vec{p}, t)$ must have the following properties.

(i) positive definite functions of time

$$W(\vec{x}, t) \geq 0; \quad W(\vec{p}, t) \geq 0; \tag{4}$$

(ii) conservation of probability

$$\frac{d}{dt} \int d^3 x W(\vec{x}, t) = \frac{d}{dt} \int d^3 p W(\vec{p}, t) = 0 \tag{5}$$

(iii) normalizability

this provides the existence of the integrals in Eq. (5).

We may write

$$\int d^3 x\, W(\vec{x}, t) = 1; \quad \int d^3 p\, W(\vec{p}, t) = 1 \tag{6}$$

(iv) probability distribution, e.g., $\int W(\vec{x}, t)d^3x$ must be a world scalar.

In nonrelativistic case this means it must be invariant under space rotations, time reversal, and Galilei transformation.

We also assume that the differential equation for ψ must be of the first order with respect to time. Thus, the specification of ψ at a given instant of time determines its evolution until the next measurement. Once ψ is given, then the state of the system is uniquely determined so that all observable quantities can be computed.

We must also ensure that the predictions of this theory must agree with those of classical mechanics in the domain of small de Broglie wavelength where the latter is valid.

We also assume that ψ is a scalar. Multicomponent wave functions are required for particles with spin.

The most general wave function $\psi(\vec{x}, t)$ is a superposition of monochromatic plane waves $e^{i(\vec{k}\cdot\vec{x} - \omega t)}$ and is written as

$$\psi(\vec{x}, t) = \frac{1}{(2\pi)^{3/2}} \int A(\vec{k})\, e^{i(\vec{k}\cdot\vec{x} - \omega t)}\, d^3k. \tag{7}$$

$$= \frac{1}{(2\pi)^{3/2}} \int A(\vec{p})\, e^{i(\vec{p}\cdot\vec{x} - Et)/\hbar}\, d^3p, \tag{8}$$

using Eq. (1).

The functions $A(\vec{k})$ and $A(\vec{p})$ differ by a numerical factor such that $|A(\vec{k})|^2\, d^3k = |A(\vec{p})|\, d^3p$. Now, with the appropriate choice of the energy scale, the energy of the particle is related to its momentum by

$$E = \frac{\vec{p}^{\,2}}{2m}. \tag{9}$$

We see that $\psi(\vec{x}, t)$ as given in Eqs. (7) with (9) is the general solution of

$$i\hbar\, \frac{\partial}{\partial t}\, \psi(\vec{x}, t) = -\frac{\hbar^2}{2m}\, \nabla^2\, \psi(\vec{x}, t). \tag{10}$$

This is the Schrödinger equation for a free particle. It satisfies the requirements discussed earlier. $\psi^*(\vec{x}, t)$ satisfies

$$i\hbar\, \frac{\partial}{\partial t}\, \psi^*(\vec{x}, t) = \frac{\hbar^2}{2m}\, \nabla^2\, \psi^*(\vec{x}, t). \tag{10a}$$

Formally Eq. (10) results from Eq. (9) if we introduce the differential operators acting on ψ according to

$$E \rightarrow i\hbar \frac{\partial}{\partial t}; \quad p_i \rightarrow \frac{\hbar}{i} \frac{\partial}{\partial x_i}. \tag{11}$$

The wave function $\psi(\vec{x}, t)$ acquires physical significance only when it is related to the probability $W(\vec{x}, t)$.

The simplest ansatz for W satisfying the conditions (i) to (iv) is that $W(\vec{x}, t)$ is a definite quadratic form constructed from the values of the function ψ and ψ^*. Neither $\int \psi^2 \, d^3x$ nor $\int \psi^{*2} \, d^3x$ nor any linear combination of them is constant in time, but we have

$$\int \psi\psi^* d^3x = \text{constant in time.} \tag{12}$$

Absorbing any constants in ψ we see that the most general form for the position probability distribution $W(\vec{x}, t)$ is

$$W(\vec{x}, t) = \left| \psi(\vec{x}, t) \right|^2. \tag{13}$$

Now, we consider the Schrödinger wave equation (10) and its complex conjugate Eq. (10a). We multiply Eq. (10) by $\psi^*(\vec{x}, t)$ and Eq. (10a) by $\psi(\vec{x}, t)$ and subtracting the resulting equations one from the other, we get

$$i\hbar \left(\psi^* \frac{\partial \psi}{\partial t} + \psi \frac{\partial \psi^*}{\partial t} \right) = \frac{\hbar^2}{2m} \left[(\nabla^2 \psi^*)\psi - \psi^* (\nabla^2 \psi) \right]. \tag{14}$$

We write this in the form of a continuity equation

$$\frac{\partial}{\partial t} (\psi^* \psi) + \vec{\nabla} \cdot \vec{j} = 0, \tag{15}$$

with the current

$$\vec{j}(\vec{x}, t) = \frac{\hbar}{2mi} \left[\psi^* \vec{\nabla} \psi - \psi \vec{\nabla} \psi^*) \right]. \tag{16}$$

Equation (15) is a local conservation law. Applying Gauss' theorem to Eq. (15), we get the global conservation property

$$\frac{d}{dt} \int_{R^3} \psi^* \psi \, d^3x = 0, \tag{17}$$

if ψ vanishes in a sufficiently rapid way as $|\vec{x}| \rightarrow \infty$ and its first derivatives are bounded. Eq. (17) shows that $\int (\psi^* \psi) \, d^3x$ is independent of time. Rescaling over all space the wave function, we write

$$\int_{R^3} d^3x \, \psi^* \psi = 1. \tag{18}$$

This justifies our interpretation of $W(\vec{x}, t) = \left| \psi(\vec{x}, t) \right|^2$ as the position probability distribution. Equation (18) tells us that ψ has the dimension of $(\text{length})^{-3/2}$. The wave functions belong to the Hilbert space of square-integrable functions on R^3.

Next we consider $W(\vec{p}, t)$, the momentum space probability distribution. Setting

$$\phi(\vec{p}, t) = A(\vec{p})\, e^{-\frac{i}{\hbar} Et} \tag{19}$$

with $E = E(\vec{p})$ in Eq. (8), we have

$$\psi(\vec{x}, t) = \frac{1}{(2\pi\hbar)^{3/2}} \int \phi(\vec{p}, t)\, e^{\frac{i}{\hbar}\vec{p},\vec{x}}\, d^3p, \tag{20}$$

and also for its complex conjugate

$$\psi^*(\vec{x}, t) = \frac{1}{(2\pi\hbar)^{3/2}} \int \phi^*(\vec{p}, t)\, e^{-\frac{i}{\hbar}\vec{p},\vec{x}}\, d^3p. \tag{20a}$$

The spatial Fourier transform of the wave function is

$$\phi(\vec{p}, t) = \frac{1}{(2\pi\hbar)^{3/2}} \int \psi(\vec{x}, t)\, e^{-\frac{i}{\hbar}\vec{p}\cdot\vec{x}}\, d^3x, \tag{21}$$

and hence

$$A(\vec{p}) = \frac{1}{(2\pi\hbar)^{3/2}} \int \psi(\vec{x}, t)\, e^{-\frac{i}{\hbar}(\vec{p},\vec{x} - Et)}\, d^3x. \tag{21a}$$

Parseval's relation holds

$$\int \psi^*\psi\, d^3x = \int \phi^*\phi\, d^3p = \int A^*A\, d^3p. \tag{22}$$

The probability density $W(\vec{p}, t)$ in momentum space is itself (not merely its integral) constant in time for a free particle, since its momentum is constant. This probability density is given by

$$W(\vec{p}, t) = |\phi(p, t)|^2, \tag{23}$$

or, in terms of $\psi(\vec{x}, t)$

$$W(\vec{p}, t) = \frac{1}{(2\pi\hbar)^3} \int d^3x\, d^3x'\; \psi(\vec{x}, t)\, \psi^*(\vec{x}', t)$$

$$\times\, e^{-\frac{i}{\hbar}\vec{p}\cdot(\vec{x} - \vec{x}')}. \tag{23a}$$

The statistical description of any state of a free particle is now complete. A state is described by a wave packet $\psi(\vec{x}, t)$ of the form Eq. (8) from which the wave packet in momentum space $\phi(\vec{p}, t)$ follows. All observable quantities are determined by ψ. $\psi(\vec{x}, t)$ and $\phi(\vec{p}, t)$, called the "probability amplitudes", are not directly observable with regard to their phases; this holds only for the probability densities $W(\vec{x}, t)$ and $W(\vec{p}, t)$.

On going from the x-integration to the p-integration by a Fourier transform, the position operator becomes a first order differential operator:

$$x \rightarrow i\hbar\, \frac{\partial}{\partial p}, \tag{24}$$

the momentum operator acts in a multiplicative way:

$$p \rightarrow p. \tag{25}$$

This is the momentum representation. In this representation the Schrödinger equation for a free particle is

$$i\hbar \frac{\partial \psi}{\partial t} = \frac{p^2}{2m}\, \psi. \tag{26}$$

Born (1926) pointed out in his treatment of the collision process the necessity for a statistical interpretation of the wave function. If $\psi(\vec{x}, t)$ is the wave function, then Born asserted that $\left|\psi(\vec{x}, t)\right|^2 d^3x$ should be regarded as the probability of observing a particle at time t in the interval $(\vec{x}, \vec{x} + d\vec{x})$. Born's introduction of the probability in quantum mechanics reveals an essential feature of quantum physics. To quote Born, :.... I myself am inclined to give up determinism in the world of atoms. But this is a philosophical question for which physical arguments alone are not decisive (See Omnes (1999)), Thus began the era of interpretation of quantum mechanics.

The Schrödinger equation does not satisfy the principle of relativity. The form of the Schrödinger equation is invariant under Galilei transformation. Let us consider two inertial frames S and S' with coordinates (x, t) and (x', t'), respectively.

For simplicity, we consider only one spatial dimension.

The coordinates are related by

$$x' = x - vt, \quad t' = t. \tag{27}$$

In S the wave function is $\psi(\vec{x}, t)$ and it satisfies

$$\left(i\hbar \frac{\partial}{\partial t} + \frac{\hbar^2}{2m} \frac{\partial^2}{\partial x^2}\right) \psi(x, t) = 0. \tag{28}$$

In S', the wave function is $\psi'(x', t')$ and it solves

$$\left(i\hbar \frac{\partial}{\partial t'} + \frac{\hbar^2}{2m} \frac{\partial^2}{\partial x'^2}\right) \psi'(x', t') = 0. \tag{29}$$

The probability density at a space-time point must be the same in S and S' (the Jacobian of the transformation is 1)

$$|\psi'(x', t')|^2 = |\psi(x, t)|^2. \tag{30}$$

We have

$$\psi' = e^{-\frac{i}{\hbar} f} \psi, \tag{31}$$

where f is a real function of the coordinates. The differential operators transform as

$$\frac{\partial}{\partial t'} = \frac{\partial}{\partial t} + v\,\frac{\partial}{\partial x},$$

$$\frac{\partial}{\partial x'} = \frac{\partial}{\partial x}. \tag{32}$$

Substitution of Eqs. (31) into (29) gives

$$\left(i\hbar\,\frac{\partial}{\partial t} + \frac{\hbar^2}{2m}\,\frac{\partial^2}{\partial x^2}\right)\psi + e^{\frac{i}{\hbar}f}\left\{-\frac{\partial f}{\partial t}\,\psi - v\,\frac{\partial f}{\partial x}\,\psi + i\hbar v\,\frac{\partial \psi}{\partial x} + \frac{\hbar^2}{2m}\right.$$

$$\left.\left[\frac{2i}{\hbar}\,\frac{\partial f}{\partial x}\,\frac{\partial \psi}{\partial x} + \frac{i}{\hbar}\,\frac{\partial^2 f}{\partial x^2}\,\psi - \frac{1}{\hbar^2}\left(\frac{\partial f}{\partial x}\right)^2\psi\right]\right\} = 0 \tag{33}$$

All the extra terms vanish if

$$f = \frac{1}{2}\,mv^2 t - mvx. \tag{34}$$

Then in the frame S', $\psi'(x', t')$ satisfies Eq. (29), if $\psi(x, t)$ satisfies Eq. (28) in the frame S. The wave function transforms as under Galilei transformation

$$\psi'(x', t') = e^{\frac{i}{\hbar}\left(mxv - \frac{1}{2}mv^2 t\right)}\,\psi(x, t). \tag{35}$$

We note that the Lagrangian $\mathcal{L} = \dfrac{1}{2}\,m\dot{x}^2$ transforms as

$$\mathcal{L}' = \frac{1}{2}\,m(\dot{x}')^2$$

$$= \frac{1}{2}\,m\dot{x}^2 + \frac{1}{2}\,mv^2 - m\dot{x}v,$$

since $\dot{x}' = \dot{x} - v$.

$$= \mathcal{L} - \frac{d}{dt}\left(mxv - \frac{1}{2}\,mv^2 t\right)$$

$$= \mathcal{L} + \frac{df}{dt}. \tag{36}$$

The phase of a plane wave transforms as

$$\frac{i}{\hbar}\,(p'x' - E't') = \frac{i}{\hbar}\,[(px - Et) + f]. \tag{37}$$

Also, the current transforms as

$$\vec{j}' = \vec{j} - \vec{v}\,\psi^*\psi, \tag{38}$$

where $$\vec{v} = (v, 0, 0).$$

The transformation Eq. (31) is known as a gauge transformation.

The initial value problem of wave mechanics consists in finding the state $\psi(\vec{x}, t)$ at all times t, given the state $\psi(\vec{x})$ at an initial time t' such that

$$\psi(\vec{x}, t) = \psi(\vec{x}, t') \quad (t' = t) \tag{39}$$

The dynamics is given by the Schrödinger equation which is of the first order in time.

We now construct $\psi(\vec{x}, t)$ from $\psi'(\vec{x}', t')$ $(t' \neq t)$ by means of the Green function

$$G_S(\vec{x} - \vec{x}'; t - t') = (2\pi)^{-3} \int e^{i\vec{k}.(\vec{x} - \vec{x}') - i(\hbar\vec{k}^2/2m)(t - t')} d^3k, \tag{40}$$

where G_S satisfies the equation

$$\left(i\hbar \frac{\partial}{\partial t} + \frac{\hbar^2}{2m} \nabla^2 \right) G_S(\vec{x} - \vec{x}'; t - t') = 0, \quad (t \neq t') \tag{41}$$

and the initial condition is

$$G_S(\vec{x} - \vec{x}'; 0) = \delta(\vec{x} - \vec{x}'). \tag{42}$$

If $\psi'(x', t')$ is given,

$$\psi(\vec{x}, t) = \int G_S(\vec{x} - \vec{x}'; t - t') \, \psi(\vec{x}', t') \, d^3x' \tag{43}$$

is a solution of the Schrödinger equation (41) and satisfies the initial condition Eq. (39).

We write

$$G_S = \left[\frac{m}{2\pi\hbar \, (t - t')} \right]^{3/2} e^{-\frac{3}{4} i\pi} \, e^{\frac{im \, (\vec{x} - \vec{x}')^2}{2\hbar(t - t')}}. \tag{44}$$

From the concepts so far developed we can compute a number of simple results which can be directly compared with experiment. For example, we can calculate the average values of any function of $\vec{x}$ or $\vec{p}$. We have

$$\bar{x}_l \equiv \langle x_l \rangle = \int \psi^* x_l \, \psi \, d^3x;$$
$$\bar{p}_l \equiv \langle \bar{p}_l \rangle = \int \phi^* p_l \, \phi \, d^3p. \tag{45}$$

The corresponding mean square deviations are

$$\overline{(\Delta x_l)^2} \equiv \langle (\Delta x_l)^2 \rangle = \int \psi^* \, (x_l - \bar{x}_l)^2 \, \psi \, d^3x;$$
$$\overline{(\Delta p_l)^2} \equiv \langle (\Delta p_l)^2 \rangle = \int \phi^* \, (p_l - \bar{p}_l)^2 \, \phi \, d^3p. \tag{46}$$

Let $F(x_l)$ be any rational integral function of x_l and $G(p_l)$ any rational function p_l. Then

$$\langle F(x_l) \rangle = \int \psi^* \, F(x_l) \, \psi \, d^3x$$
$$= \int \phi^* \, F\left(i\hbar \frac{\partial}{\partial p_l} \right) d^3p,$$
$$\langle G(x_l) \rangle = \int \psi^* \, G\left(\frac{\hbar}{i} \frac{\partial}{\partial x_l} \right) \psi \, d^3x$$
$$= \int \phi^* \, G(p_l) \, \phi \, d^3p. \tag{47}$$

We consider a material particle acted on by force. Evidently, we require that in the limit of no force we get back to the case of a free particle and also contain the laws of classical particle mechanics as a limiting case.

We assume that the probability densities are still well defined concepts satisfying the properties as in the force-free case. We retain the concept of $\psi(\vec{x}, t)$ as the wave function associated with the particle. This assumption follows since the measurement of the position can take place in such a short time that the presence of forces has no role to play. We assume that the wave equation in this case is linear and of first order in $\dfrac{\partial}{\partial t}$.

We set

$$i\hbar \frac{\partial}{\partial t}\, \psi(\vec{x}, t) = H\psi(\vec{x}, t), \tag{48}$$

where in order to retain the superposition principle H is considered a linear operator. Now,

$$i\hbar \frac{\partial}{\partial t} \int \psi^{*}\psi\, d^3x = -\int [(H\psi)^{*}\psi - \psi^{*}(H\psi)]\, d^3x, \tag{49}$$

which must vanish for all ψ. So

$$\int [(H\psi)^{*}\psi - \psi^{*}(H\psi)]\, d^3x = 0 \tag{50}$$

On account of the linearity of H, it follows from Eq. (50) for any two solutions ψ and ψ_2

$$\int [(H\psi_1)^{*}\psi_2 - \psi_1^{*}(H\psi_2)]\, d^3x = 0. \tag{51}$$

The operator H is thus hermitian. In nonrelativistic wave mechanics the only way to determine the operator H for a particular system is by comparing the behavior of the general solution of Eq. (48) with the corresponding limiting cases in classical mechanics. In the free particle case, H is obtained from the classical Hamiltonian $\dfrac{\vec{p}^{2}}{2m}$ by the correspondence $p_i \rightarrow \dfrac{\hbar}{i} \dfrac{\partial}{\partial x_i}$.

As a simple example, we consider a particle in an external force field derived from a scalar potential $V(\vec{x})$. We write the classical Hamiltonian as

$$H(\vec{p} \cdot \vec{x}) = \sum_{i} \frac{p^{2}i}{2m} + V(\vec{x}). \tag{52}$$

With the substitution $p_i \rightarrow \dfrac{\hbar}{i} \dfrac{\partial}{\partial x_i}$, we write the Schrödinger equation for a material particle in a potential as

$$\frac{i}{\hbar} \frac{\partial}{\partial t}\, \psi(\vec{x}, t) = -\frac{\hbar^{2}}{2m} \nabla^{2}\, \psi(\vec{x}, t) + V(\vec{x})\, \psi(\vec{x}, t). \tag{53}$$

The Schrödinger equation just obtained corresponds formally to the heat equation. The imaginary coefficient in the equation describes oscillations. This equation is invariant under the transformation $t \rightarrow -t$, $\psi \rightarrow \psi^*$ whereby the probability distribution $\psi^*\psi$ remains constant.

From Eq. (53) the continuity equation (15) follows with the expression (16) for the current.

Now,

$$\langle x_k \rangle = \int \psi^* x_k \, \psi \, d^3x;$$

$$\langle p_k \rangle = \int \phi^* p_k \, \phi \, d^3p$$

$$= \int \psi^* \left(\frac{\hbar}{i} \frac{\partial \psi}{\partial x_k} \right) d^3x. \tag{54}$$

So

$$\frac{d}{dt} \langle x_k \rangle = \int x_k \frac{\partial \psi^* \psi}{\partial t} \, d^3x,$$

since x_k does not depend on time explicitly

$$= -\int x_k \frac{\partial}{\partial x_i} j_i \, d^3x,$$

since $-\,\text{div.}\, \vec{j} = \frac{\partial}{\partial t}(\psi^*\psi)$, from Eq. (15)

$$= -\int j_i \frac{\partial x_k}{\partial x_i} \, d^3x$$

$$= \int j_k \, d^3x$$

$$= \frac{1}{2m} \int \left\{ \psi^* \frac{\hbar}{i} \frac{\partial \psi}{\partial x_k} - \psi \frac{\hbar}{i} \frac{\partial \psi^*}{\partial x_k} \right\} d^3x,$$

using Eq. (16)

$$= \frac{1}{m} \int \psi^* \frac{\hbar}{i} \frac{\partial}{\partial x_k} \psi \, d^3x$$

$$= \frac{\langle p_k \rangle}{m}. \tag{55}$$

Next we calculate $\dfrac{d}{dt} \langle p_k \rangle$. For the free particle case it vanishes but it is not so in the present case.

$$\frac{d}{dt} \langle p_k \rangle = \frac{d}{dt} m \int j_k \, d^3x$$

$$= m \int \frac{\partial j_k}{\partial t} \, d^3x. \tag{56}$$

But,

$$m \frac{\partial j_k}{\partial t} = \frac{i\hbar}{2} \left\{ \frac{\partial \psi}{\partial t} \frac{\partial \psi^*}{\partial x_k} + \psi \frac{\partial^2 \psi^*}{\partial x_k \partial t} - \frac{\partial \psi^*}{\partial t} \frac{\partial \psi}{\partial x_k} - \psi^* \frac{\partial^2 \psi}{\partial t \partial x_k} \right\}$$

$$= \frac{1}{2} \left\{ \frac{\partial \psi^*}{\partial x_k} H\psi - \psi \frac{\partial}{\partial x_k} (H^*\psi^*) + \frac{\partial \psi}{\partial x_k} H^*\psi^* - \psi^* \frac{\partial}{\partial x_k} (H\psi) \right\}$$

$$= -\frac{\hbar^2}{4m} \left\{ \frac{\partial \psi^*}{\partial x_k} \nabla^2\psi - \psi \frac{\partial}{\partial x_k} (\nabla^2\psi^*) + \frac{\partial \psi}{\partial x_k} \nabla^2\psi^* - \psi^* \frac{\partial}{\partial x_k} (\nabla^2\psi) \right\}$$

$$+ \frac{1}{2} \left\{ \frac{\partial \psi^*}{\partial x_k} V\psi - \psi \frac{\partial}{\partial x_k} (V\psi^*) + \frac{\partial \psi}{\partial x_k} V\psi^* - \psi^* \frac{\partial}{\partial x_k} (V\psi) \right\}.$$

using Eq. (53).

The second bracket on the right is

$$-\frac{\partial V}{\partial x_k} \psi^*\psi.$$

Using $u\nabla^2 v - v\nabla^2 u = \sum_i \frac{\partial}{\partial x_i} \left(u \frac{\partial v}{\partial x_i} - v \frac{\partial u}{\partial x_i} \right)$ for arbitrary function u, v, we get

for the first bracket

$$-\frac{\hbar^2}{4m} \frac{\partial}{\partial x_i} \left\{ \frac{\partial \psi}{\partial x_k} \frac{\partial \psi^*}{\partial x_i} + \frac{\partial \psi}{\partial x_i} \frac{\partial \psi^*}{\partial x_k} - \psi \frac{\partial^2 \psi^*}{\partial x_i \partial x_k} - \psi^* \frac{\partial^2 \psi}{\partial x_i \partial x_k} \right\}.$$

Introducing the force

$$K_l = -\frac{\partial V}{\partial x_l} = -\frac{\partial H}{\partial x_l}, \tag{57}$$

and the stress tensor

$$T_{kl} = \frac{\hbar^2}{4m} \left\{ \frac{\partial \psi}{\partial x_k} \frac{\partial \psi^*}{\partial x_l} + \frac{\partial \psi^*}{\partial x_k} \frac{\partial \psi}{\partial x_l} - \psi^* \frac{\partial^2 \psi}{\partial x_k \partial x_l} - \psi \frac{\partial^2 \psi^*}{\partial x_k \partial x_l} \right\}$$

$$= T_{lk}, \tag{58}$$

we have

$$m \frac{\partial j_k}{\partial t} = K_k \psi^*\psi - \sum_l \frac{\partial}{\partial x_l} T_{kl}. \tag{59}$$

So

$$\frac{\partial}{\partial t} \langle p_k \rangle = m \frac{d^2\langle x_k \rangle}{dt^2}$$

$$= m \int \frac{\partial j_k}{\partial t} d^3x$$

$$= \int K_k \, \psi^* \psi \, d^3x$$

$$= \langle K_k \rangle$$

$$= - \left\langle \left(\frac{\partial V}{\partial x_k} \right) \right\rangle$$

$$= - \left\langle \left(\frac{\partial H}{\partial x_k} \right) \right\rangle. \tag{60}$$

The time derivative of the mean value of p_k is equal to the mean value of the force taken over the wave packet. The latter is in general different from the value of the force at the midpoint of the wave packet. Only when the force varies slowly within the packet we get a packet which behaves as a classical particle whose trajectory satisfies the equation of motion

$$m \frac{d^2 x_k}{dt^2} = - \frac{\partial V}{\partial x_k}. \tag{61}$$

The result, expressed in Eqs. (55) and (60) is known as the Ehrenfest theorem. To obtain a criterion for the validity of such an approximation, we expand F $(\langle \vec{x} \rangle)$ at the position $\langle \vec{x} \rangle$ (Pauli (1980)),

$$F_i \, (\vec{x}) = F_i(\langle \vec{x} \rangle) + (x_j - \langle x_j \rangle) \, F_{i,j}(\langle \vec{x} \rangle)$$

$$+ \frac{1}{2} \, (x_j - \langle x_j \rangle) \, (x_l - \langle x_l \rangle) \, F_{i,jl}(\langle \vec{x} \rangle) + ...,$$

where $F_{i,j} = \dfrac{\partial F_i}{\partial x_j}$ and sum over repeated indices is indicated.

So

$$\langle F_i \, (\vec{x}) \rangle = F_i(\langle \vec{x} \rangle) + \frac{1}{2} \, \langle (x_j - \langle x_j \rangle)$$

$$\times (x_l - \langle x_l \rangle) \rangle \, F_{i,jl} + ..., \tag{62}$$

since $\langle (x_j - \langle x_j \rangle) \rangle = 0$.

Therefore, the approximation that $\langle x_i \rangle$ and $\langle x_k \rangle$ move according to laws of classical mechanics is valid only when the second and higher derivatives of the force vanish, so the potential $V(\vec{x}, t)$ depends at the most, quadratically on $\vec{x}$. This occurs, for example, in the case of a free particle, a charged particle in an electric field or a harmonic oscillator.

We now discuss another example. We consider a particle of mass m and charge e in an electromagnetic field given by

$$\vec{E}(\vec{x}, t) = - \frac{1}{c} \frac{\partial}{\partial t} A(\vec{x}, t) - \vec{\nabla} \, \bar{\phi} \, (\vec{x}, t),$$

$$\vec{B}(\vec{x}, t) = \vec{\nabla} \times \vec{A}(\vec{x}, t) \tag{63}$$

where $\vec{A}(\vec{x}, t)$ is the vector potential and $\bar{\phi}(\vec{x}, t)$ is the scalar potential.

In the presence of a scalar, non-electromagnetic potential $V(\vec{x})$, the Newtonian equation of motion of the particle is

$$m \frac{d\vec{v}}{dt} = e\vec{E} + \frac{e}{c} \vec{v} \times \vec{B} - \vec{\nabla} V$$

$$= -e\vec{\nabla}\bar{\phi} - \frac{e}{c} \frac{\partial \vec{A}}{\partial t} + \frac{e}{c} \vec{v} \times \text{curl } \vec{A} - \vec{\nabla} V, \quad (64)$$

using Eq. (63).

Equation (64) may be written as

$$m \frac{dv_i}{dt} = -e \frac{\partial \phi}{\partial x_i} - \frac{e}{c} \frac{\partial A_i}{\partial t} + \frac{e}{c} \left(v_j \frac{\partial A_i}{\partial x_i} - v_j \frac{\partial A_i}{\partial x_j} \right) - \frac{\partial V}{\partial x_i},$$

$$= -\frac{e}{c} \left(\frac{\partial A_i}{\partial x_i} + \frac{\partial A_i}{\partial x_j} \frac{dx_j}{dt} \right) \quad i, j = 1, 2, 3$$

$$+ \frac{\partial}{\partial x_i} \left(\frac{e}{c} A_j v_j - e\bar{\phi} - V \right),$$

where A_i is a component of $\vec{A}$ evaluated at the position of the particle, or,

$$\frac{d}{dt} \left(mv_i + \frac{e}{c} A_i \right) = \frac{\partial}{\partial x_i} \left(\frac{e}{c} A_j v_j - e\bar{\phi} - V \right). \quad (65)$$

This can be easily put in the Lagrangian form. We put

$$\frac{\partial \mathcal{L}}{\partial v_i} = mv_i + \frac{e}{c} A_i$$

or,

$$\mathcal{L} = \frac{e}{c} A_j v_j - e\bar{\phi} - V + f(v_i)$$

so that

$$mv_i + \frac{e}{c} A_i = \frac{e}{c} A_i + f'(v_i)$$

So

$$\frac{df}{dv_i} = mv_i \quad \text{or} \quad f = \frac{1}{2} mv_i^2.$$

Hence,

$$\mathcal{L} = \frac{1}{2} mv_i^2 + \frac{e}{c} A_i v_i - e\bar{\phi} - V, \quad (66)$$

or,

$$\mathcal{L} = \frac{m}{2} \vec{\dot{x}}^2 + \frac{e}{c} \vec{A} \cdot \vec{\dot{x}} - e\bar{\phi} - V. \quad (66a)$$

Defining

$$p_j = \frac{\partial \mathcal{L}}{\partial v_j} = mv_i + \frac{e}{c} A_j, \tag{67}$$

we write the classical Hamiltonian as

$$H = \sum_i p_i v_i - \mathcal{L}$$

$$= \sum_i \frac{1}{2m} \left(p_i - \frac{e}{c} A_i \right)^2 + e\bar{\phi} + V. \tag{68}$$

$\vec{p}$ is called the canonical momentum and $\vec{\pi} = \vec{p} - \frac{e}{c} \vec{A}$ is called the kinetic momentum. The classical Hamiltonian may be written as

$$H = \frac{\vec{\pi}^2 (\vec{x}, t)}{2m} + e\bar{\phi} (\vec{x}, t) + V(\vec{x}). \tag{68a}$$

We make the transition to quantum mechanics by the correspondence $p_j \rightarrow \frac{\hbar}{i} \frac{\partial}{\partial x_j}$ and obtain the Schrödinger equation for a charged particle in an electromagnetic field

$$i\hbar \frac{\partial}{\partial t} \psi(\vec{x}, t) = \frac{1}{2m} \left[\left(\frac{\hbar}{i} \frac{\partial}{\partial \vec{x}} - \frac{e}{c} \vec{A} \right)^2 + e\bar{\phi} + V \right] \psi(\vec{x}, t). \tag{69}$$

Expanding the squared term on the right, we obtain, as $\vec{\nabla}$ and $\vec{A}$ do not commute,

$$i\hbar \frac{\partial \psi}{\partial t} = \frac{1}{2m} \left\{ -\hbar^2 \nabla^2 - \frac{e\hbar}{ic} \vec{\nabla} \cdot \vec{A} - \frac{e\hbar}{ic} \vec{A} \cdot \vec{\nabla} \right.$$

$$\left. + \frac{e^2}{c^2} \vec{A}^2 \right\} \psi + (e\Phi + V)\psi. \tag{70}$$

Using the operator equation

$$\vec{\nabla} \cdot \vec{A} = \vec{A} \cdot \vec{\nabla} + (\vec{\nabla} \cdot \vec{A}), \tag{71}$$

we write the Schrödinger equation in the coordinate representation as

$$i\hbar \frac{\partial \psi}{\partial t} = \frac{1}{2m} \left\{ -\hbar^2 \nabla^2 - \frac{e\hbar}{ic} \vec{\nabla} \cdot \vec{A} - \frac{2e\hbar}{ic} \vec{A} \cdot \vec{\nabla} \right.$$

$$\left. + \frac{e^2}{c^2} \vec{A}^2 \right\} \psi + (e\Phi + V)\psi. \tag{72}$$

In the Coulomb gauge

$$\vec{\nabla} \cdot \vec{A} = 0, \tag{73}$$

we have

$$i\hbar \frac{\partial \psi}{\partial t} = \frac{1}{2m} \left\{ -\hbar^2 \nabla^2 - \frac{2e\hbar}{ic} \vec{A} \cdot \vec{\nabla} \right.$$

$$\left. + \frac{e^2}{c^2} \vec{A}^2 \right\} \psi + (e\Phi + V)\psi. \tag{72a}$$

We again have a continuity equation

$$\frac{\partial}{\partial t}(\psi^* \psi) + div\,\vec{j} = 0, \tag{74}$$

which also shows that H is hermitian. The current in this case is

$$\vec{j} = \frac{1}{2m} \left\{ \psi^* \left(\frac{\hbar}{i} \vec{\nabla} - \frac{e}{c} \vec{A} \right) \psi - \psi \left(\frac{\hbar}{i} \vec{\nabla} + \frac{e}{c} \vec{A} \right) \psi^* \right\}$$

$$= \frac{1}{2m} \left\{ \psi^* (m\vec{v}) \psi + \psi (m\vec{v})^* \psi^* \right\}$$

$$= \frac{1}{2m} \left\{ \psi^* (m\vec{v}) \psi + \left(\psi^* (m\vec{v}) \psi \right)^* \right\}$$

Since $m\vec{v}$ is hermitian, the total current is

$$\int \vec{j} \, d^3x = \frac{1}{2m} \int 2\psi^* (m\vec{v})\psi \, d^3x$$

$$= \langle \vec{v} \rangle$$

$$= \frac{d}{dt} \langle \vec{x} \rangle. \tag{75}$$

We have

$$\langle p_k \rangle = \int \phi^* \, p_k \, \phi \, d^3p$$

$$= \int \psi^* \frac{\hbar}{i} \frac{\partial \psi}{\partial x_k} \, d^3x,$$

and

$$\frac{d}{dt} \langle x_k \rangle = \frac{d}{dt} \int \psi^* x_k \, \psi \, d^3x$$

$$= \int j_k \, d^3x,$$

so

$$\frac{d\langle x_k \rangle}{dt} = \frac{1}{m} \left(\langle p_k \rangle - \frac{e}{c} \langle A_k \rangle \right). \tag{76}$$

Further we find

$$m \frac{\partial j_k}{\partial t} = -\sum_i \frac{\partial T_{kl}}{\partial x_l} + \psi^* \left(-\frac{\partial V'}{\partial x_k} - \frac{e}{c} \frac{\partial A_k}{\partial t} \right) \psi + \frac{e}{c} \sum_l B_{kl} j_l, \tag{77}$$

where
$$B_{kl} = \frac{\partial A_l}{\partial x_k} - \frac{\partial A_k}{\partial x_l},$$

$$V' = e\Phi + V,$$

and

$$
\begin{aligned}
T_{kl} &= \frac{\hbar^2}{4m}\Bigg[-\psi^*\left(\frac{\partial}{\partial x_l} - \frac{ie}{\hbar c}A_l\right)\left(\frac{\partial \psi}{\partial x_k} - \frac{ie}{\hbar c}A_k\psi\right) \\
&\quad -\psi\left(\frac{\partial}{\partial x_l} - \frac{ie}{\hbar c}A_l\right)\left(\frac{\partial \psi^*}{\partial x_k} + \frac{ie}{\hbar c}A_k\psi^*\right) \\
&\quad +\left(\frac{\partial \psi}{\partial x_k} - \frac{ie}{\hbar c}A_k\psi\right)\left(\frac{\partial \psi^*}{\partial x_l} + \frac{ie}{\hbar c}A_l\psi^*\right) \\
&\quad +\left(\frac{\partial \psi^*}{\partial x_k} + \frac{ie}{\hbar c}A_k\psi^*\right)\left(\frac{\partial \psi}{\partial x_l} + \frac{ie}{\hbar c}A_l\psi\right)\Bigg] \\
&= \frac{\hbar^2}{4m}\Bigg\{\Bigg[-\psi^*\frac{\partial^2 \psi}{\partial x_l \partial x_k} - \psi\frac{\partial^2 \psi^*}{\partial x_l \partial x_k} + \frac{\partial \psi}{\partial x_l}\frac{\partial \psi^*}{\partial x_k} + \frac{\partial \psi^*}{\partial x_l}\frac{\partial \psi}{\partial x_k}\Bigg] \\
&\quad + \frac{2ie}{\hbar c}\Bigg[A_k\left(\psi^*\frac{\partial \psi}{\partial x_l} - \psi\frac{\partial \psi^*}{\partial x_l}\right) + A_l\left(\psi^*\frac{\partial \psi}{\partial x_k} - \psi\frac{\partial \psi^*}{\partial x_k}\right)\Bigg] \\
&\quad + \frac{4e^2}{\hbar^2 c^2}A_k A_l \psi^* \psi\Bigg\},
\end{aligned}
\tag{78}
$$

so that
$$T_{kl} = T_{lk}.$$

Setting

$$\langle K_k\rangle = \int\left\{-\psi^*\left(-\frac{\partial V'}{\partial x_k} - \frac{e}{c}\frac{\partial A_k}{\partial t}\right)\psi + \frac{e}{c}\sum_l B_{kl}j_l\right\} d^3x, \tag{79}$$

we obtain from Eq. (77), the analogue of the equation of motion

$$m\frac{d^2}{dt^2}\langle x_k\rangle = \langle K_k\rangle. \tag{80}$$

In classical electromagnetic theory, the electric and magnetic field strengths remain unchanged when the gauge transformation is made on the potentials

$$\vec{A} \to \vec{A}' = \vec{A} + \vec{\nabla}\chi$$

$$\Phi \to \Phi' = \Phi - \frac{1}{c}\frac{\partial \chi}{\partial t}, \tag{81}$$

where $\chi(\vec{x}, t)$ is an arbitrary scalar function. Now we must see if the Schrödinger equation

$$i\hbar \frac{\partial}{\partial t}\psi(\vec{x}, t) = \left\{ \frac{1}{2m}\left(\frac{\hbar}{i}\vec{\nabla} - \frac{e}{c}\vec{A}(\vec{x}, t)\right) \right. $$
$$\left. - e\Phi(\vec{x}, t)\right\}\psi(\vec{x}, t) \tag{82}$$

is invariant under the gauge transformation. After transformation, the Schrödinger equation is

$$i\hbar \frac{\partial \psi'}{\partial t} = \left\{ \frac{1}{2m}\left(\frac{\hbar}{i}\vec{\nabla} - \frac{e}{c}\vec{A}'\right)^2 - e\Phi'\right\}\psi'. \tag{83}$$

If we put

$$\psi'(\vec{x}, t) = \psi(\vec{x}, t)\, e^{\frac{ie}{\hbar c}\chi(\vec{x}, t)}, \tag{84}$$

then

$$i\hbar \frac{\partial \psi'}{\partial t} = -\left\{ \frac{e}{c}\frac{\partial \chi}{\partial t}\psi + i\hbar \frac{\partial \psi}{\partial t}\right\} e^{\frac{ie}{\hbar c}\chi}$$

$$\frac{\hbar}{i}\vec{\nabla}\psi' = \left\{ \frac{e}{c}\vec{\nabla}\chi + \frac{\hbar}{i}\vec{\nabla}\right\}\psi\, e^{\frac{ie}{\hbar c}\chi}$$

so

$$i\hbar \frac{\partial \psi'}{\partial t} = H'\psi'$$

becomes

$$-\frac{e}{c}\frac{\partial \chi}{\partial t}\psi + i\hbar \frac{\partial \psi}{\partial t} = \left\{ \frac{1}{2m}\left(\frac{\hbar}{i}\vec{\nabla} - \frac{e}{c}\vec{A}\right)^2 + e\Phi - \frac{e}{c}\frac{\partial \chi}{\partial t}\right\}\psi,$$

or,

$$i\hbar \frac{\partial \psi}{\partial t} = H\psi.$$

The Schrödinger equation is invariant under the gauge transformation

$$\vec{A} \to \vec{A}' = \vec{A} + \vec{\nabla}\chi,$$

$$\Phi \to \Phi' = \Phi - \frac{1}{c}\frac{\partial \chi}{\partial t}, \tag{85}$$

$$\psi \to \psi' = \psi e^{(ie/\hbar c)\chi}.$$

The gauge transformation introduces a space and time dependent phase into the complex wave function. Its squared modulus, $|\psi(\vec{x}, t)|^2$, which is interpreted as a probability density, does not change. It is easily verified that $\langle x_k \rangle$, $\frac{d}{dt}\langle x_1 \rangle$ are invariant quantities. The current $\vec{j}$ and the stress tensor T_{kl} also are gauge invariant quantities. The kinetic momentum $\vec{\pi}$ is gauge-unvariant, although the canonical momentum is not. The group of substitutions, Eq. (85), called the gauge transformation, is an Abelian group, the gauge group.

In classical physics, only the field strengths have invariant physical meanings. The potentials are auxiliary constructs used for the convenience of calculating the fields. The vector potential (its line integral) has in quantum mechanics, a measurable physical significance (see § 11.7(b)).

3.2   MANY PARTICLE INTERACTIONS

We now consider a system of many particles which interact with each other and with any external fields that may be present. The basic assumptions for the description of an N-particle system are as follows.

(i) At an instant of time t there exists the probability density $W(\vec{x}_1, \vec{x}_2, ..., \vec{x}_N; t)$ such that $W \prod_i d\vec{x}_i$ is the probability for finding simultaneously the coordinates of the first particle in the interval $(\vec{x}_1 + d\vec{x}_1)$, ..., those of the jth particle in the interval $(\vec{x}_j + d\vec{x}_j)$, ..., those of the Nth particle in the interval $(\vec{x}_N, + d\vec{x}_N)$. Here, we have assumed the distinguishability of particles.

(ii) We assume the existence of a function

$$\psi(\vec{x}_1, ..., \vec{x}_N; t)$$

such that

$$W(\vec{x}_1, ..., \vec{x}_N; t) = \psi^* \psi.$$

(iii) The function ψ satisfies an equation

$$i\hbar \frac{\partial \psi}{\partial t} = H\psi,$$

where H is a linear operator.

In order to satisfy the condition

$$\frac{d}{dt} \int \psi^* \psi \, dx = 0,$$

where $dx = \prod_{i=1}^{N} d^3 x_i$, H must be a hermitian operator.

(iv) The operator H is constructed from the classical Hamiltonian by the correspondence $\vec{p}_j \rightarrow \dfrac{\hbar}{i} \dfrac{\partial}{\partial \vec{x}_j}$.

As a generalization of the equation, we assume that the momentum space wave function is

$$\phi(\vec{p}_1, ..., \vec{p}_N; t) = \frac{1}{(2\pi\hbar)^{3N/2}} \int \psi(\vec{x}_1, ..., \vec{x}_N; t) \, e^{-\frac{i}{\hbar} \Sigma_j \vec{p}_j \cdot \vec{x}_j} \, dx \qquad (86)$$

so that

$$\psi(\vec{x}_1, ..., \vec{x}_N; t) = \left(\frac{1}{(2\pi\hbar)}\right)^{3N/2} \int \phi(\vec{p}_1, ..., \vec{p}_N; t)\, e^{\frac{i}{\hbar}\sum_j \vec{p}_j \cdot \vec{x}_j}\, dp \tag{87}$$

where $\quad dp = \prod_{i=1}^{N} d^3 p_i.$

Also,
$$\int \phi^* \phi\, dp = \int \psi^* \psi\, dx, \tag{88}$$

whence
$$\frac{d}{dt} \int \phi^* \phi\, dp = 0. \tag{89}$$

The momentum space probability density is

$$W(\vec{p}_1, ..., \vec{p}_N; t) = \left|\phi(\vec{p}_1, ..., \vec{p}_N; t)\right|^2. \tag{90}$$

Let us consider a system of non-interacting particles in an external force field. Then the Hamiltonian operator can be written as a sum of independent, individual operators

$$H = \prod_{i=1}^{N} H_i \tag{91}$$

We assume that $\psi_i(\vec{x}_i, t)$ are arbitrary solutions of the wave equations

$$i\hbar\,\frac{\partial \psi_i}{\partial t} = H_i \psi_i \tag{92}$$

of the isolated subsystems. Then

$$\psi(\vec{x}_1, ..., \vec{x}_N; t) = \prod_{i=1}^{N} \psi_i(\vec{x}_i, t) \tag{93}$$

is a solution of the equation

$$i\hbar\,\frac{\partial \psi}{\partial t} = H\psi = \left[\sum_{i=1}^{N} H_i\right] \prod_{i=1}^{N} \psi_j. \tag{94}$$

Let H_0 be the Hamiltonian for a set of N non-interacting particles subject to external forces. For interacting particles, if the interaction can be described from a potential, which depends only on their position coordinates, then we write

$$i\hbar\,\frac{\partial \psi}{\partial t} = H\psi = (H_0 + V)\psi. \tag{95}$$

As a simple example, we consider a system of N charged particles (having charges e_i and masses m_i) which interact through Coulomb forces, in an applied electromagnetic field described by the potentials $\vec{A}$ and Φ. The Schrödinger equation is

$$i\hbar\,\frac{\partial}{\partial t}\,\psi(\vec{x}_1, ..., \vec{x}_N; t) = \sum_{i=1}^{N}\left[\frac{1}{2m_j}\left(\frac{\hbar}{i}\frac{\partial}{\partial \vec{x}_j} - \frac{e_j}{c}\vec{A}(\vec{x}_j, t)\right)^2\right.$$

$$\left. + e_j\,\Phi(\vec{x}_j, t)] + \frac{1}{2}\sum_{i \neq j}\frac{e_i e_j}{|\vec{x}_i - \vec{x}_j|}\right] \cdot \psi(\vec{x}_1, ..., \vec{x}_N; t). \tag{96}$$

The continuity equation may be written as

$$\frac{d}{dt}\, W(\vec{x}_1, \ldots, \vec{x}_N;\, t) + \sum_{k=1}^{N} \frac{\partial}{\partial \vec{x}_k} \cdot \vec{j}_k(\vec{x}_1, \ldots, \vec{x}_N;\, t) = 0, \qquad (97)$$

where

$$\vec{j}_k(\vec{x}_1, \ldots, \vec{x}_N;\, t) = Re\,\psi^* \vec{v}_k \psi, \qquad (98)$$

with

$$\vec{v}_k = \frac{1}{m_k}\left(\vec{p}_k - \frac{e}{c}\vec{A}_k\right), \qquad (99)$$

and

$$\vec{A}_k = \vec{A}(\vec{x}_k,\, t).$$

The time-dependent Schrödinger equation for the N-particle system is

$$i\hbar \frac{\partial}{\partial t}\, \psi(\vec{x}_1, \ldots, \vec{x}_N;\, t) = H\psi(\vec{x}_1, \ldots, \vec{x}_N;\, t) \qquad (100)$$

We run into difficulties if we try to transform a Hamiltonian in curvilinear coordinates into the quantum Hamiltonian operator — we run into the problem of which order to take the non-commuting factors. The transcription $p_i \to \dfrac{\hbar}{i}\dfrac{\partial}{\partial x_i}$ is valid only in rectangular coordinates. We will, therefore, use the Cartesian coordinates in making the transition to quantum mechanics by the usual methods of substitution and make the coordinate transformation if necessary, afterward. Pauli (1980) has given a discussion of this by formulating the wave equation in any arbitrary curvilinear coordinates.

Quantum mechanics of a system described by the Schrödinger wave equation is termed as wave mechanics. The introduction of the wave function ψ satisfying a linear equation and for which therefore, the principle of superposition holds has proved to be extremely fruitful in the study of the properties of atomic systems. Each solution of the wave equation corresponds to a state of motion of the system. It is important to note that the wave function $\psi(\vec{x}, t)$ is purely symbolic and abstract that is quite different from the wave functions of the classical theory (elastic waves, electromagnetic waves etc). The superposition principle of quantum mechanics is of an essentially different nature from any occurring in the classical theory. The quantum superposition principle incorporates uncertainty in the results of observation. The correct interpretation of ψ is as a statistical state function from which a statistical distribution of values obtained in measurements for each dynamical variable may be calculated.

3.3 COORDINATE AND MOMENTUM REPRESENTATIONS

We consider a single, spinless, material particle. Experiments suggest that the set of the three position observables $\{Q_1,\, Q_2,\, Q_3\}$ constitutes a c.s.c.o. The simultaneous eigenvectors $\{|\vec{x}\rangle\}$

$$Q_i \, |\vec{x}\rangle = x_i \, |\vec{x}\rangle, \tag{101}$$

form an orthonormal basis. Wave mechanics is usually formulated in the coordinate or position representation in which this basis set $\{|\vec{x}\rangle\}$ is chosen. A pure state $|\psi(t)\rangle$ in this basis is a function of a continuous variable, called the wave function and is written as

$$\psi(\vec{x}, t) = \langle \vec{x} | \psi(t)\rangle. \tag{102}$$

The position eigenkets satisfy the orthonormality and completeness conditions

$$\langle \vec{x} | \vec{x}'\rangle = \delta(\vec{x} - \vec{x}')$$
$$\int d^3x \, |\vec{x}\rangle\langle\vec{x}| = I. \tag{103}$$

In the Schrödinger picture in this representation the rate of change with time of $\psi(\vec{x}, t)$ is given by the Schrödinger equation.

The action of the position operators Q_i in function space is simply to multiply by the position coordinate,

$$(Q_i\psi)\,(\vec{x}, t) = x_i\psi(\vec{x}, t). \tag{104}$$

The translation operator

$$V_\beta = e^{-i\vec{\beta}\cdot\vec{P}/\hbar}$$

gives

$$V_\beta|\vec{x}\rangle = \left| \vec{x} + \vec{\beta}\, \right\rangle, \tag{105}$$

where the relative phases have been appropriately chosen. In terms of wave function, Eq. (105) is written as

$$(V_\beta\psi)\,(\vec{x}, t) = \psi\!\left(\vec{x} - \vec{\beta}, t\right). \tag{106}$$

For infinitesimal $\vec{\beta}$, Eq. (106) reduces to

$$\psi(\vec{x}, t) - \frac{i}{\hbar}\,\vec{\beta}\cdot\left(\vec{P}\psi\right)(\vec{x}, t) + O(\beta^2)$$
$$= \psi(\vec{x}, t) - \vec{\beta}\cdot\left(\vec{\nabla}\psi\right)(\vec{x}, t) + O(\beta^2), \tag{107}$$

whence

$$\left(\vec{P}\psi\right)(\vec{x}, t) = \left\langle \vec{x} \left| \vec{P} \right| \psi(t)\right\rangle$$
$$= -\,i\hbar\,\vec{\nabla}\psi(\vec{x}, t). \tag{108}$$

Therefore, the expression for the momentum operator $\vec{P}$ in coordinate representation is

$$\vec{P} = -\,i\hbar\,\vec{\nabla}, \quad P_i = -\,i\hbar\,\frac{\partial}{\partial x_i}, \tag{109}$$

The operators Q_i, P_i ($i = 1, 2, 3$) satisfy canonical commutation relations. On the basis of the discussion above, we can consider the wave mechanics of N-particle systems.

Instead of the set $\{Q_i\}$($i = 1, 2, 3$), the set $\{P_j\}$($j = 1, 2, 3$) also constitute a c.s.c.o. for the single spinless, particle. The simultaneous eigenvectors

$$P_i \,|\vec{p}\,\rangle = p_i \,|\vec{p}\,\rangle \tag{110}$$

form an orthonormal basis, satisfying

$$\langle p|\vec{p}'\rangle = \delta(\vec{p} - \vec{p}')$$
$$\int d^3 \,|\vec{p}\,\rangle\langle \vec{p}\,| = I. \tag{111}$$

The wave function in this basis, called the momentum representation, is

$$\tilde{\psi}(\vec{p}, t) = \langle \vec{x}|\psi(t)\rangle \tag{112}$$

This procedure leads to the same Hilbert space with the state vectors as the wave functions $\tilde{\psi}$. The momentum operators P_i have the representation

$$(P_i\tilde{\psi})\,(\vec{p}, t) = p_i\tilde{\psi}(\vec{p}, t). \tag{113}$$

($i = 1, 2, 3$).

The almost complete symmetry under the interchange of position and momentum observables shows that

$$(Q_k\tilde{\psi})\,(\vec{p}, t) = i\hbar \,\frac{\partial \tilde{\psi}(\vec{p}, t)}{\partial p_k}. \tag{114}$$

($k = 1, 2, 3$).

We now establish a connection between the two representations. We write the inner product $\langle \vec{x}|\vec{p}\rangle$ as

$$-i\hbar \,\vec{\nabla} \,\langle \vec{x}|\vec{p}\rangle = \langle \vec{x}|\vec{P}|\vec{p}\rangle$$
$$= \vec{p} \,\langle x|\vec{p}\rangle, \tag{115}$$

whose solution is

$$\langle x|\vec{p}\rangle = C(\vec{p}) \,e^{i\vec{p} \,\cdot\, \vec{x}/\hbar}. \tag{116}$$

The normalization factor $C(\vec{p})$ is determined from the first equation in Eq. (111):

$$\delta(\vec{p} - \vec{p}') = \langle \vec{p}|\vec{p}'\rangle$$
$$= \int \langle \vec{p}|\vec{x}\rangle \langle \vec{x}|\vec{p}'\rangle \,d^3x$$
$$= |C|^2 \int e^{\frac{i}{\hbar}(\vec{p} - \vec{p}')\,\cdot\,\vec{x}} \,d^3x$$
$$= |C|^2 \,(2\pi\hbar)^3 \,\delta(\vec{p} - \vec{p}'),$$

whence

$$C(\vec{p}) = \frac{1}{(2\pi\hbar)^{3/2}}.$$

Thus,

$$\langle \vec{x}|\vec{p} \rangle = \frac{1}{(2\pi\hbar)^{3/2}}\, e^{\frac{i}{\hbar}\vec{p}\cdot\vec{x}}. \tag{117}$$

Given Eq. (117), we see that $\psi(\vec{x})$ and $\tilde{\psi}(\vec{p})$ are Fourier transforms of each other:

$$\psi(\vec{x}) = \frac{1}{(2\pi\hbar)^{3/2}} \int d^3p\, e^{\frac{i}{\hbar}\vec{p}\cdot\vec{x}}\, \tilde{\psi}(\vec{p})$$

$$\tilde{\psi}(\vec{p}) = \frac{1}{(2\pi\hbar)^{3/2}} \int d^3x\, e^{-\frac{i}{\hbar}\vec{p}\cdot\vec{x}}\, \psi(\vec{x}). \tag{118}$$

3.4 STATIONARY STATES AND THE EIGENVALUE PROBLEM

Given a Hamiltonian which does not depend explicitly on time, we seek from the solutions of the Schrödinger wave equation

$$i\hbar\, \frac{\partial \psi}{\partial t} = H\psi \tag{119}$$

those for which the probability density. $\psi^*\psi$ and the current density $\vec{j}$ are

constant in time. In order that $\psi^*\psi$ and $\left(\psi^* \dfrac{\partial \psi'}{\partial x_k} - \psi\, \dfrac{\partial \psi^*}{\partial x_k}\right)$ be independent of

time the wave function ψ must have the form

$$\psi(\vec{x}, t) = u(\vec{x})\, e^{-if(t)}, \tag{120}$$

where u depends only on the coordinates of configuration space but not on time and $f(t)$ is a function of time only. Substitution of Eq. (120) into Eq. (119) leads to

$$\hbar\, \frac{df}{dt}\, u = H(u)$$

and this holds only if $\dfrac{df}{dt}$ is independent of t. We, therefore, write Eq. (120) as

$$\psi_E = u e^{-\frac{i}{\hbar}Et}. \tag{121}$$

On substitution of Eq. (121) into Eq. (119) we obtain

$$Hu = Eu \tag{122}$$

which is the time-independent Schrödinger equation. This is a homogeneous, linear differential equation containing a parameter. We have an eigenvalue problem. When the system is in a state given by Eq. (121), it is said to be in a stationary state of energy E. Such energy eigenstates are of special interest.

In order that the eigenvalue problem be well defined so as to obtain solutions of the problem, we must specify certain regularity and boundary conditions to be satisfied by u.

Only those functions that maintain the "hermiticity" of H are permitted. We demand the stability requirement that H must have a lower bound. The condition of "quadratic integrability", i.e., the requirement that

$$\int u^* u\, d^3x \le K(K < \infty), \tag{123}$$

is imposed on the wave function. In many cases this requirement gives rise to discrete eigenvalues E_i. The requirement Eq. (123) has to be weakened for continuous eigenvalues. For any hermitian operator A, $u_1 A u_2$ must be single-valued. The condition is slightly weaker than requiring that the wave function be single-valued. Actually, multiple-valuedness and singularity of a function go together.

Let E_i be an eigenvalue, and ψ_i the corresponding eigenfunction. Then

$$H\psi_i = E_i\psi_i, \tag{124}$$

and

$$\begin{aligned}
\langle H \rangle_i &= \int \psi_i^* H\psi_i\, d\vec{x} \\
&= E_i \int \psi_i^* \psi\, d^3x = E_i
\end{aligned} \tag{125}$$

where the wave function is normalized. The eigenvalue is the expectation value of the energy in that state. Also

$$\langle H^2 \rangle_i = E_i^2, \tag{126}$$

so that

$$\langle \Delta H \rangle_i = \langle H^2 \rangle_i - \langle H \rangle_i^2 = 0. \tag{127}$$

We conclude that the wave function ψ_i corresponds to a precise value E_i in that state. Note that

$$\langle f(H) \rangle_i = f(E_i) \tag{128}$$

for a function of the energy.

Let us consider an eigenstate for which $p_x = p_0$, a constant. Then

$$\frac{\hbar}{i}\frac{\partial}{\partial x} u = p_0 u, \tag{129}$$

the solution of which is a plane wave

$$u = Ce^{\frac{i}{\hbar} p_0 x}. \tag{130}$$

We consider a free particle.

Then

$$H\psi \equiv -\frac{\hbar^2}{2m}\,\vec{\nabla}^2\,\psi = E\psi. \tag{131}$$

From Eq. (130) we see that for a free particle with fixed energy (and so fixed momentum p_0) the solution of Eq. (131) is a monochromatic plane wave

$$\psi = Ce^{ip_0 x/\hbar}. \tag{132}$$

Clearly all eigenvalues greater than zero are permissible for the energy. $\psi^*\psi$ is not integrable.

Let us consider a particle of mass m in a time independent potential $V(\vec{x})$. The time independent Schrödinger equation is

$$H\psi(\vec{x}) \equiv \left[-\frac{\hbar^2}{2m}\,\vec{\nabla} + V(\vec{x})\right]\psi(\vec{x})$$

$$= E\psi(\vec{x}). \tag{133}$$

The boundaries to be considered for this second order differntial equation are the "point at infinity" and the points or domains where the potential energy becomes infinite. Such domains are called "singular domains" of the differential equation.

We require that a wave function and its gradient be continuous, finite, and single-valued in the neighbourhood of every nonsingular point.

We have thus far encountered two classes of wave functions. Wave functions which satisfy the continuity conditions and are bounded and quadratically integrable belong to a type or class of wave functions. For the other type or class of wave functions we consider the example of a free particle. The eigenfunctions are $\psi_E(\vec{x}) = e^{ik \cdot \vec{x}}$, and the corresponding eigenvalue is $\dfrac{\hbar^2 k^2}{2m}$. For a finite integration volume Ω, $\int \psi_E^* \, \psi_E \, d^3x \propto \Omega$ and so the limit of this quantity as $\Omega \to \infty$ does not exist.

The difficulty is that the state described by the wave function $e^{ik \cdot \vec{x}}$ is not localized, and not normalizable. Here we have completely specified p_i and hence x_i is completely indeterminate. Such wave functions which are not quadratically integrable but satisfy the continuity condition are admitted. They can be compounded by integration to build up wavepacket solutions of the Schrödinger equation. The probability of the particle being in a given finite volume is zero. Thus only the relative probabilities of the particle being in different finite volumes will be physically significant. We require that the wave function be bounded or quadratically integrable, depending on whichever is the weaker condition.

We note that at finite singularities (singularities in finite plane) the condition of quadratic integrability is the weaker condition, for example, $\psi = \dfrac{1}{(\sqrt{x})^{1/3}}$ is integrable but not bounded. However, for convergence of integrals over infinite domains, boundedness is the weaker condition. For example, a function can be bounded but not quadratically integrable — a constant. We can obtain integrable solutions by superposition of bounded solutions, e.g., a wave packet. Physically, we only work with wave packets, not quantum states.

Let us consider the case of particle of mass m moving in one dimension in a potential $V(x)$. We assume that $V(x)$ has a single minimum (V_{min}) placed at the origin for convenience, a pole at $x = -\infty$, and a finite asymptotic limit V_+ at $x = +\infty$ (Fig. 3.1)

We consider three cases corresponding to different ranges of the energy E.

Case I $E\langle V_{min} = 0$;

Case II $0\langle E\langle V_+$;

Case III $V_+\langle E$.

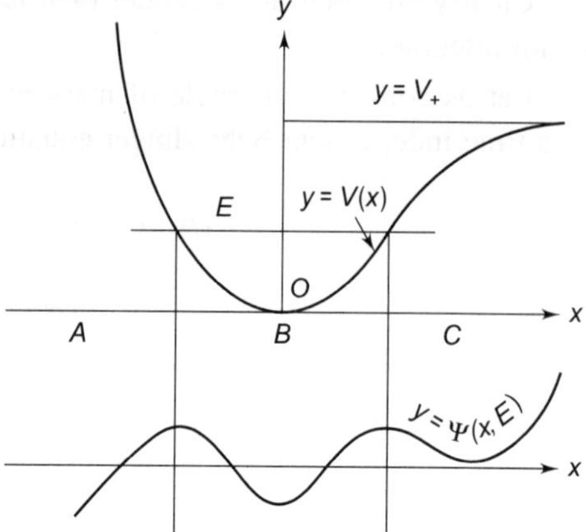

Fig. 3.1 Potential energy and integral curves for an anharmonic oscillator

Case I is classically impossible. Here $V(x) - E$ is always positive. There is no acceptable solution to the wave equation.

Case II gives oscillatory solution in classical theory. It gives a series of discrete quadratically integrable eigenfunctions. In Case III the classical motion is aperiodic. The eigenfunctions are not quadratically integrable.

Integral curves (corresponding to real solutions) are concave (convex) to the x-axis when $\dfrac{1}{\psi}\dfrac{d^2\psi}{dx^2}$ is negative (positive). In region B, where $E > V(x)$, the curves are oscillatory, crossing and recrossing the x-axis. In regions A and C the integral curves are convex to x-axis. In $V_+ > E$, then every quadratically integrable eigenfunction must vanish at $x = +\infty$, and conversely.

In Case III ($E > V_+$) the integral curve cannot vanish at $x = \infty$. For large values of x, ψ has the asymptotic form corresponding to a free particle. So for $E > V_+$, the eigenfunctions are not quadratically integrable but remain finite as $x \to \infty$. In this case, we have the continuous range of energy eigenvalues and the eigenfunctions are of the second class or type.

The potential energy and integral curves for the one-dimensional anharmonic oscillator are shown in Fig. 3.1. This discussion follows Kemble (1937) where further details of the conditions to be imposed on wave functions are given.

If $V(x)$ is infinite at $x = \pm \infty$, then there is only a discrete spectrum with an infinite range of values. As an example we mention the case of a simple harmonic oscillator.

In summary we see that the eigenfunction must be bounded at infinity. Then either (i) the eigenfunction is quadratically integrable, giving the discrete energy spectrum or, (ii) the eigenfunction is not quadratically integrable, giving a continuous spectrum. The eigenfunctions corresponding to the discrete part of the spectrum represent the localized bound states. Thus, the eigenvalue problem in wave mechanics gives the quantized energy levels. The energy eigenfunctions in the continuum describe a collision phenomenon. The scattering states belong to the continuous part of the spectrum. Similar conclusion can obtained for a particle moving in three dimensions. Instead of giving a general discussion of the spectrum of H (see Section 1.3 and Galindo and Pascual (1990)), we shall work out several problems to illustrate various ideas about the energy spectrum.

3.5 THE FREE PARTICLE

We consider a system consisting of a free particle, or a particle not acted on by any force. To begin, we consider the one-dimensional motion of the particle. The time-independent Schrödinger equation is

$$-\frac{\hbar^2}{2m}\frac{d^2\psi(x)}{dx^2} = E\psi(x), \tag{134}$$

or
$$\left(\frac{d^2}{dx^2} + k^2\right)\psi(x) = 0, \tag{134a}$$

with $k^2 = \dfrac{2mE}{\hbar^2}$. The general solution is

$$\psi_k(x) = A_+ e^{ikx} + A_- e^{-ikx}, \tag{135}$$

where A_+, A_- are two arbitrary constants. The eigenvalue equation for the momentum operator p_x is

$$-i\hbar\frac{\partial}{\partial t}\psi_p(x) = p_x\psi_p(x). \tag{136}$$

Here $p_x \rightarrow -i\hbar\dfrac{\partial}{\partial t}$ and $p_x = \hbar k_x$ is the momentum of the particle with magnitude $p = |p_x| = \hbar k$. We see that the energy eigenfunctions of the free-particle are also momentum eigenfunctions $\sim Ce^{\pm \frac{i}{\hbar} p_x x}$. The eigenfunctions are bounded, but not quadratically integrable. The spectrum of the operator p_x is continuous. Since all non-negative values of E are allowed the energy spectrum is continuous ($E = 0$ to $E = \infty$). Each energy eigenvalue is doubly degenerate. A stationary state of the energy $E > 0$ for a free particle may be written as

$$\psi(x,\, t) = A_+ e^{i(kx - \omega t)} + A_- e^{i(kx + \omega t)}, \tag{137}$$

where $\omega = \dfrac{E}{\hbar}$. It represents two plane waves traveling in opposite directions with phase velocity $v_{ph} = \dfrac{\omega}{k} = \dfrac{\hbar k}{2m}$. The particle velocity $v = \dfrac{p}{m} = \dfrac{\hbar k}{m}$ is different from v_{ph}. This justifies the usage of wave function for ψ. The position probability density corresponding to either solution is

$$P_\pm = |\psi_\pm(x,\, t)|^2 = |A_\pm|^2. \tag{138}$$

which is constant (of t, x). Hence, a particle moving with well-defined momentum p_x along the x-axis cannot be localized on this axis. The probability current density for either plane wave is

$$j_\pm = \frac{\hbar k}{m} |A_\pm|^2 = v\, |A_\pm|^2 = vP_\pm. \tag{139}$$

We can construct physical solutions by constructing wave packets (see § 3.6).

Born suggested a method to "normalize" these wave functions. We enclose the particle in a box (in the one-dimensional case a box of length L) at the walls of which we impose boundary conditions on the wave function. We require that the wave function satisfies periodic boundary condition:

$$\psi(x) = \psi(x + L), \tag{140}$$

or,

$$A e^{ikx} = A e^{ik(x + L)}. \tag{140a}$$

This gives

$$e^{ikL} = 1 = e^{2\pi in}, \quad n = 0,\, \pm 1,\, \pm 2,\, \ldots$$

Hence k_x is restricted to the values

$$k_x = \frac{2\pi}{L}\, n. \tag{141}$$

The energy spectrum becomes discrete

$$E_n = \frac{\hbar^2 k_x^2}{2m} = \frac{2\pi^2 \hbar^2}{mL^2}\, n^2. \tag{142}$$

The energy difference between two adjacent levels is

$$\Delta E = \frac{2n^2\hbar^2 \cdot (2n + 1)}{mL^2},$$ (143)

which goes to o as $L \to \infty$, i.e., the spectrum then becomes continuous.

To find the normalization constant, we write for two different eigenstates

$$\delta_{nm} = N^2 \int_{-L/2}^{L/2} e^{i(k_n - k_m)x}\, dx$$

$$= N^2 \cdot 2\,\frac{\sin (k_n - k_m)\frac{L}{2}}{(k_n - k_m)}$$

$$= N^2 \cdot L \cdot \delta_{nm},$$ (144)

where we have assumed that the particle lies in the interval $\frac{-L}{2} \leq x \leq \frac{L}{2}$.

The normalization constant is $\frac{1}{\sqrt{L}}$ and the orthonormalized wave function is

$$\psi_n(x) = \frac{1}{\sqrt{L}}\, e^{ik_n x} = \frac{1}{\sqrt{L}}\, e^{\frac{i2\pi n}{L}x}.$$ (145)

One can also use the delta function normalization as follows. We have the momentum eigenfunctions

$$\psi_{k_x}(x) = \frac{1}{\sqrt{2\pi}}\, e^{ik_x x},$$ (146)

satisfying the orthonormality relation

$$\int_{-\infty}^{\infty} \psi_{k_x'}^{*}(x)\, \psi_{k_x}(x)\, dx = \delta(p_x - p_x')$$ (147)

and the closure relation

$$\int_{-\infty}^{\infty} \psi_{k_x'}^{*}(x')\, \psi_{k_x}(x)\, dk_x = \delta(x - x').$$ (148)

We may also write the momentum eigenfunctions as

$$\psi_{p_x}(x) = \frac{1}{\sqrt{2\pi\hbar}}\, e^{\frac{i}{\hbar}p_x x},$$ (149)

satisfying

$$\int_{-\infty}^{\infty} \psi_{p_x'}^{*}(x)\, \psi_{p_x}(x)\, dx = \delta(p_x - p_x').$$ (150)

We now give hints about normalization of wave functions.

(i) Discrete Case Let $u_n(x)$ be the un-normalized, but quadratically integrable eigenfunction. Then the normalized wave function can be written as

$$v_n(x) = \frac{u_n(x)}{\sqrt{\int |u_n(x)|^2\, dx}}.\tag{151}$$

(ii) Continuum case   Here we can use the "box normalization" concept or the delta function normalization.

We normalize in the λ scale if we set

$$\int u_\lambda^*(q)\, u_\lambda(q)\, dq = \delta(\lambda - \lambda'),\tag{152}$$

Given a monotonic function $\mu(x)$ with $\dfrac{d\mu}{dx} > 0$, we show that the wave function normalized in μ scale is equal to the function normalized in λ scale multiplied by $\sqrt{\dfrac{d\lambda}{d\mu}}$.

For normalization of a wave function in a specific coordinate, we write

$$\int \rho(x) u_{\lambda'}^*(x)\, u_\lambda(x)\, dx = \delta(\lambda - \lambda'),\tag{153}$$

where $\rho(x)\, dx$ is the differential line element in this coordinate. Now we take the first terms in the asymptotic expansion (x large) of these functions and compare with

$$\delta(x - x') = \frac{1}{2\pi} \int_{-\infty}^{\infty} e^{-ik(x - x')}\, dk,$$

$$\delta(k - k') = \frac{1}{2\pi} \int_{-\infty}^{\infty} e^{-ix(k - k')}\, dx,$$

in two scales and accordingly adjust constants. We therefore normalize by considering the asymptotic behavior of a wave function.

We mention the general expansion theorem. Let $f(x)$ be an arbitrary twice-differentiable function with certain conditions. Then $f(x)$ has an absolutely and uniformly convergent expansion of the form

$$f(x) = \sum_n a_n u_n(x) + \int_{\substack{\text{over} \\ \text{continuum}}} a(\lambda) u_\lambda(x)\, d\lambda.\tag{154}$$

Here the discrete eigenfunctions are normalized in the usual way and

$$a_n = \int u_n^*(x)\, f(x)\rho(x)\, dx.\tag{155}$$

$u_\lambda(x)$ is a continuum type eigenfunction and λ denotes a continuous-spectrum eigenvalue so that

$$a(\lambda) = \int u_\lambda^*(x)\, f(x)\rho(x)\, dx.\tag{156}$$

We note that the wave functions for continuous and discrete eigenvalues are orthogonal to one another.

We generalize to the case of the motion of a free particle in three dimensions. The time independent Schrödinger equation for a free particle of mass m is

$$H\psi(\vec{r}) = E\psi(\vec{r}), \tag{157}$$

or,

$$-\frac{\hbar^2}{2m}\nabla^2\psi(\vec{r}) = E\psi(\vec{r}) \tag{157a}$$

or,

$$\left(\frac{\partial^2}{\partial x^2} + \frac{\partial^2}{\partial y^2} + \frac{\partial^2}{\partial z^2} + k^2\right)\psi(\vec{x}) = 0, \tag{157b}$$

where

$$k^2 = \frac{2m}{\hbar^2}E. \tag{158}$$

Here, we work in the coordinate representation. The solutions are of the form $\psi_{\vec{k}} \sim e^{i\vec{k}\cdot\vec{r}}$ Imposing the periodic boundary conditions on the wave function, we have

$$\psi(x + L, y, z) = \psi(x, y, z),$$
$$\psi(x, y + L, z) = \psi(x, y, z),$$
$$\psi(x, y, z + L) = \psi(x, y, z), \tag{159}$$

The general solution for a particle with momentum of eigenvalue $\hbar k$ and energy $E = \dfrac{\hbar^2 k^2}{2m}$ is

$$\psi_{\vec{k}}(\vec{r}) = \frac{1}{L^{3/2}}e^{i(\vec{k}\cdot\vec{r})}, \tag{160}$$

where

$$k_x \equiv k_1 = \frac{2\pi}{L}n_1,$$
$$k_y \equiv k_2 = \frac{2\pi}{L}n_2,$$
$$k_z \equiv k_3 = \frac{2\pi}{L}n_3, \tag{161}$$

with

$$n_i(i = 1, 2, 3) = 0, \pm 1, \pm 2, \ldots$$

The orthonormalization condition is

$$\int \psi^*_{n_1'n_2'n_3'}\,\psi_{n_1 n_2 n_3}\,d^3x = \delta_{n_1 n_1'}\,\delta_{n_2 n_2'}\,\delta_{n_3 n_3'}. \tag{162}$$

Using the delta-function normalization we write

$$\psi_{\vec{k}}(\vec{x}) = \frac{1}{(2\pi)^{3/2}}e^{i(\vec{k}\cdot\vec{r})} \tag{163}$$

which satisfy the following orthogonality and completeness relations

$$\int d^3r \; \psi_{\vec{k}}^*(\vec{r}) \; \psi_{\vec{k}'}(\vec{r}) = \delta(\vec{k} - \vec{k}')$$

$$\int d^3k \; \psi_{\vec{k}}^*(\vec{r}) \; \psi_{\vec{k}'}(\vec{r}) = \delta(\vec{r} - \vec{r}\,'). \tag{164}$$

The time-dependent wave function is

$$\psi(\vec{r}, t) = \frac{1}{(2\pi)^{3/2}} \; e^{i(\vec{k}\,\cdot\,\vec{r} - \omega t)}. \tag{165}$$

This function of x, y, z, t describes plane waves in space-time.

Here $$\rho(\vec{r}, t) = (2\pi)^{-3}$$

and

$$\vec{j}(\vec{r}, t) = (2\pi)^{-3} \frac{\hbar \vec{k}}{m} = (2\pi)^{-3} \; \vec{v}, \tag{166}$$

where $$\vec{v} = \frac{\vec{p}}{m} = \frac{\hbar \vec{k}}{m}.$$

Rotational invariance leads to another c.s.c.o. ($E = p^2$, $\vec{L}^2$ and L_z) where $\vec{L}$ is the angular momentum of the particle. We separate the variables of Eq. (157a) in spherical polar coordinates and obtain another set of solutions.

3.6 THE MOTION OF WAVE PACKETS

We have seen that to describe a particle sufficiently localized we need to associate a "wave packet" with it. A wave packet describes a localized wave disturbance and is a function of space-time whose value is very small outside a spatial region, and inside this domain of width Δq is approximately periodic with a definite frequency. They are called "wave packets" as a Fourier analysis will show that they are superpositions of infinite plane monochromatic waves involving a narrow range of wave lengths and directions of wave normals. The amplitudes of all the Fourier components are small, except those in the neighborhood of the definite frequency. The Fourier components whose amplitudes are not small fill up a frequency band whose width is of order $1/\Delta q$. The coordinates and momenta of the particle which is represented in wave mechanics as a wave packet have their accuracy limited by the uncertainty relation. The evolution of the wave packet in time is described by Schrödinger wave equation.

Let us begin by considering the one-dimensional case. We consider a wave packet, or a wave train composed of a small range of wave numbers

$$\psi(x, t) = \frac{1}{\sqrt{2\pi}} \int_{-\infty}^{\infty} A(k) e^{i(kx - \omega t)} \; dk. \tag{167}$$

We assume that the amplitude A peaks at k_0 and is appreciably different from zero only in a small range Δk about k_0. Given the dispersion law $\omega = \omega(k)$ we can expand the exponent in a power series about k_0:

$$\omega t - kx = \omega_0 t - k_0 x + (k - k_0)\left[\frac{d\omega}{dk}\bigg|_{k_0} t - x\right]$$

$$+ \frac{1}{2}(k - k_0)^2 \frac{d^2\omega(k)}{d\omega^2}\bigg|_{k_0} t + \dots \tag{168}$$

where $\omega(k_0) = \omega_0$.

If the higher order terms can be neglected, we can write

$$\psi(x,\,t) = \frac{1}{\sqrt{2\pi}}\int_{-\infty}^{\infty} A(k)\,e^{-\,i(\omega_0 t - k_0 x + (k - k_0)(v_g t - x) + \dots)dk}.$$

$$= \frac{1}{\sqrt{2\pi}}\,e^{-\,i(\omega_0 - k_0 v_g)t}\int_{-\infty}^{\infty} e^{ik(x - v_g t) + \dots}\,A(k)\,dk.$$

Thus we write

$$\psi(x,\,t) = \frac{1}{\sqrt{2\pi}}\,e^{-\,i(\omega_0 - k_0 v_g)t}\int_{-\infty}^{\infty} A(k)e^{ik(x - v_g t) + \dots}\,dk$$

$$= \frac{1}{\sqrt{2\pi}}\,e^{-\,i(\omega_0 - k_0 v_g)t}\,\psi(x - v_g t). \tag{169}$$

The integral $\int_{-\infty}^{\infty} A(k)e^{ik(x - v_g t)}\,dk$ can be approximated by

$$A(k_0)\int_{k_0 - \frac{\Delta k}{2}}^{k_0 + \frac{\Delta k}{2}} e^{-\,i(\omega_0 - k_0 v_g)t} = A(k_0)\Delta k\,\frac{\sin \xi}{\xi}$$

where $\xi = \dfrac{\Delta k}{2}(x - v_g t)$.

The discarded terms in the exponent describe the spreading of the packet.

It follows from Eq. (169) that the wave packet $\psi(x,\,t)$ can be approximated by a plane wave contained in an envelope described by the modulating function. The mean location of the wave packet moves with the group velocity.

$$v_g = \frac{d\omega(k)}{dk}\bigg|_{k\,=\,0} = \frac{dE}{dp}. \tag{170}$$

This velocity is, in general, different from the phase velocity

$$v_{ph} = \frac{\omega(k)}{k}\bigg|_{k\,=\,0} = \frac{E}{p}, \tag{171}$$

which is the velocity of propagation for the phase of a single plane wave. The group velocity is the velocity of propagation of the group of waves which make up the wave packet. For a nonrelativistic particle of mass m, $E = \dfrac{p^2}{2m}$. The group velocity is $v_g = \dfrac{p}{m} = v_{\text{classical}}$, and $v_{ph} = \dfrac{p}{2m} = \dfrac{1}{2} v_{\text{classical}}$.

We consider a wave packet

$$\psi(x, t) = \frac{1}{\sqrt{2\pi}} \int_{-\infty}^{\infty} A(k) e^{i(kx - \omega t)} \, dk. \tag{172}$$

Here we want to look at the wave packet at a fixed time which we take to be $t = 0$. We shall consider the behavior of a wave packet in time later. So we write

$$\psi(x) = \frac{1}{\sqrt{2\pi}} \int A(k) \, \exp(ikx) \, dk,$$

$$A(k) = \frac{1}{\sqrt{2\pi}} \int \psi(k) \, \exp(-ikx) \, dx. \tag{173}$$

The symmetry is obvious – all equations are invariant under the following substitutions:

$$
\begin{array}{ccccc}
\psi & A & x & k & i \\
A & \psi & k & x & -i
\end{array}
\tag{174}
$$

Parseval's relation holds

$$\int |\psi(x)|^2 \, dx = \int |A(k)|^2 \, dk. \tag{175}$$

As the wave packet becomes more localized the number of different wavelengths contained in the Fourier spectrum increases. So we expect a reciprocal relationship of the form $\Delta x_i \Delta k_i \rangle$ a constant.

We define the variance

$$(\Delta x)^2 = \left\langle (x - \bar{x})^2 \right\rangle = \overline{x^2} - \bar{x}^2$$

$$(\Delta k)^2 = \left\langle (k - \bar{k})^2 \right\rangle = \overline{k^2} - \bar{k}^2. \tag{176}$$

For the sake of simplicity, let

$$\bar{x} = 0 \quad \text{and} \quad \bar{k} = 0,$$

which can be achieved by a coordinate translation. Now,

$$\overline{x^2} = \int A^*(k) \left[\left(-\frac{\partial^2}{\partial k^2} \right) A(k) \right] dk$$

$$= + \int \frac{\partial A^*}{\partial k} \cdot \frac{\partial A}{\partial k} \, dk \tag{177}$$

$$\overline{k^2} = \int \psi^*(x) \left[\left(-\frac{\partial^2}{\partial x^2} \right) \psi(x) \right] dx$$

$$= + \int \frac{\partial \psi^*}{\partial x} \cdot \frac{\partial \psi}{\partial x}\, dx. \tag{178}$$

To get a lower bound for the quantity $\overline{x^2} \cdot \overline{k^2}$, we consider an expression which is non-negative:

$$D \equiv \left| \frac{x}{2\overline{x^2}} \psi + \frac{\partial \psi}{\partial x} \right|^2 \geq 0. \tag{179}$$

Expanding this,

$$D = \frac{x^2}{4(\overline{x^2})^2} \psi^* \psi + \frac{x}{2\overline{x^2}} \left(\frac{\partial \psi^*}{\partial x} \psi + \psi^* \frac{\partial \psi}{\partial x} \right) + \frac{\partial \psi^*}{\partial x} \cdot \frac{\partial \psi}{\partial x}$$

$$= \frac{1}{4} \left(\frac{x}{\overline{x^2}} \right)^2 \psi^* \psi + \frac{1}{2} \cdot \frac{\partial}{\partial x} \left(\frac{x}{\overline{x^2}} \psi^* \psi \right) - \frac{1}{2} \cdot \frac{1}{\overline{x^2}} \cdot \psi^* \psi$$

$$+ \frac{\partial \psi^*}{\partial x} \cdot \frac{\partial \psi}{\partial x}$$

$$= \frac{1}{4} \cdot \frac{1}{\overline{x^2}} \left(x^2 - 2\overline{x^2} \right) \psi^* \psi + \frac{1}{2} \cdot \frac{\partial}{\partial x} \left(\frac{x}{\overline{x^2}} \psi^* \psi \right)$$

$$+ \frac{\partial \psi^*}{\partial x} \cdot \frac{\partial \psi}{\partial x}$$

Integrating over all space and discarding boundary terms at $|x| \to \infty$, we get

$$\int D(x)\, dx = - \frac{1}{4\overline{x^2}} + \overline{k^2} \geq 0.$$

Thus,

$$\overline{x^2} \cdot \overline{k_x^2} \geq \frac{1}{4}$$

or, more generally,

$$\overline{(\Delta k_x)^2}\; \overline{(\Delta x)^2} \geq \frac{1}{4}. \tag{180}$$

We note that to obtain a lower bound for $(\Delta x)^2\, (\Delta k)^2$ we need to look only for a lower bound on

$$\overline{x^2}\, \overline{k^2} = \left\{ (\Delta x)^2 + \overline{x}^2 \right\} \left\{ (\Delta k)^2 + \overline{k}^2 \right\},$$

in general. This lower bound is attained for $\overline{x} = 0$, $\overline{k} = 0$.

With

$$\Delta k_x \equiv + \sqrt{\overline{(\Delta k_x)^2}},$$

$$\Delta x \equiv + \sqrt{\overline{(\Delta x)^2}},$$

we obtain

$$\Delta p_x \equiv \sqrt{\overline{(\Delta p_x)^2}},$$

$$\Delta k_x \cdot \Delta x \geq \frac{1}{2}, \tag{181}$$

$$\Delta p_x \cdot \Delta x \geq \frac{\hbar}{2}. \tag{181a}$$

The uncertainty relation Eqs. (181) or (181a) follows on a purely wave-kinematical law. In quantum mechanics, the uncertainty relation is a universal constraint applicable to massive particles and photons. The uncertainty relation is a consequence of the incompatibility of the pair of observables. The de Broglie relation establishes a relationship between wave number and momentum. A classical electromagnetic wave with a fixed wave number k can have arbitrary amplitude, and therefore arbitrary momentum. The uncertainty relation reconciles the wave-particle duality.

The equality sign in Eq. (181) holds — the lower bound is attained — only when $D = 0$, or

$$\frac{\partial \psi}{\partial x} = -\frac{x}{2\overline{x^2}}\,\psi.$$

The solution of this differential equation is

$$\psi(x) = Ce^{\frac{-x^2}{4\overline{x^2}}}. \tag{182}$$

This is a gaussian distribution which when normalized, is

$$\psi(x) = \left(\frac{1}{2\pi\overline{x^2}}\right)^{\frac{1}{4}} e^{\frac{-x^2}{4\overline{x^2}}}, \tag{183}$$

with

$$\overline{x^2}\,\overline{p_x^2} = \frac{1}{4}\,\hbar^2, \quad \overline{x} = 0, \overline{p_x} = 0.$$

For the general case when $\overline{x}$ and $\overline{p_x}$ are not zero, we have

$$\psi(x) = \left(\frac{1}{(2\pi(\Delta x)^2)}\right)^{\frac{1}{4}} e^{-(x-\overline{x})^2/4(\Delta x)^2 + \frac{i}{\hbar}\overline{p_x}x}, \tag{184}$$

with

$$(\Delta x)^2\,(\Delta p_x)^2 = \frac{1}{4}\,\hbar^2.$$

Thus a gaussian wave packet is called a minimum uncertainty wave packet. We now go to momentum space. We have

$$A(k) = \frac{1}{\sqrt{2\pi}} \cdot \left(\frac{2a}{\pi}\right)^{\frac{1}{4}} \int \exp\left(-ax^2 - ikx\right) dx, \quad \text{from Eq. (173)}$$

where
$$a = \frac{1}{4x^2}$$

$$= \frac{1}{\sqrt{2\pi}} \cdot \left(\frac{2a}{\pi}\right)^{\frac{1}{4}} \int \exp\left[-a\left(x + \frac{ik}{2a}\right)^2\right] dx \cdot \exp\left(-\frac{k^2}{4a}\right)$$

$$= \frac{1}{\sqrt{2\pi}} \cdot \left(\frac{2a}{\pi}\right)^{\frac{1}{4}} \cdot \sqrt{\frac{\pi}{a}} \cdot \exp\left(-\frac{k^2}{4a}\right).$$

So

$$A(k) = \left(\frac{1}{\sqrt{2\pi a}}\right)^{\frac{1}{4}} \exp\left(-\frac{k^2}{4a}\right), \tag{185}$$

which is also a gaussian distribution.

We consider the behavior of a wave packet in time. We have

$$\psi(x, t) = \frac{1}{\sqrt{2\pi}} \int A(k) \exp\left[i\left(kx - \frac{\hbar k^2}{2m} t\right)\right] dk$$

$$= \frac{1}{\sqrt{2\pi}} \int A(k, t)\, e^{ikx}\, dk, \tag{186}$$

where

$$A(k, t) = A(k) \exp\left[-i\frac{\hbar k^2}{2m} t\right]$$

$$= \frac{1}{\sqrt{2\pi}} \int \psi(x, t) \exp\left(-ikx\right) dx. \tag{187}$$

$$|A(k, t)|^2 = |A(k)|^2. \tag{188}$$

Introducing the momentum p in place of the wave number $k : p = \hbar k$. Now,

$$\phi(p, t) = \frac{1}{\sqrt{\hbar}} A(k, t)$$

$$\phi(p) = \frac{1}{\sqrt{\hbar}} A(k) \tag{189}$$

$$|\phi(p, t)|^2\, dp = |A(k, t)|^2\, dk. \tag{190}$$

We write the free particle wave packet as

$$\psi(x, t) = \frac{1}{\sqrt{2\pi\hbar}} \int dp\, e^{-\frac{i}{\hbar}\left(\frac{p^2}{2m} t - px\right)} \phi(p), \tag{191}$$

where $\phi(p)$ is fixed by initial conditions.

We put

$$\psi(x, 0) = \frac{1}{\sqrt{2\pi\hbar}} \int_{-\infty}^{\infty} dp\, \phi(p)\, e^{\frac{i}{\hbar} px} \equiv \delta(x)$$

where
$$\delta(x) = \frac{1}{2\pi\hbar} \int_{-\infty}^{\infty} e^{\frac{i}{\hbar} px}\, dp.$$

The packet is concentrated at the origin initially.

So
$$\phi(p) = \frac{1}{\sqrt{2\pi\hbar}}$$

Thus,

$$\psi(x, t) = \frac{1}{2\pi\hbar} \int_{-\infty}^{\infty} dp\, e^{\frac{-i}{\hbar}\left(\frac{p^2}{2m}t - px\right)}.$$

$$= \frac{1}{2\pi\hbar}\, e^{\frac{imx^2}{2\hbar t}} \int_{-\infty}^{\infty} e^{-\frac{it}{2m\hbar}\left(p - \frac{mx}{t}\right)^2} dx.$$

$$= \frac{1}{2\pi}\left(\frac{2m}{\hbar t}\right)^{\frac{1}{2}} e^{imx^2/2\hbar t} \int_{-\infty}^{\infty} e^{is^2}\, ds. \qquad (192)$$

with the change of variables

$$p - \frac{mx}{t} = \frac{(2m\hbar)^{1/2}}{t^{1/2}}\, s.$$

Let us now evaluate the integral.

$$I = \int_{-\infty}^{\infty} e^{is^2}\, ds = 2 \int_{0}^{\infty} e^{is^2}\, ds.$$

The integral is along the real axis of s. Let $s = \dfrac{y}{\sqrt{i}}$ and since the integrand is

analytic everywhere in the finite plane, we can bend the contour back to the real axis of y, to get

$$I = \frac{2}{\sqrt{i}} \int_{0}^{\infty} e^{-y^2}\, dy$$

$$= \frac{2}{\sqrt{i}} \cdot \frac{\sqrt{\pi}}{2} = \sqrt{\frac{\pi}{i}}.$$

Collecting all this, we get

$$\psi(x, t) = \left(\frac{m}{2\pi i\hbar t}\right)^{1/2} e^{\frac{imx^2}{2\hbar t}}, \qquad (193)$$

which is the solution evolving out of $\delta(x)$ at $t = 0$.

This is the free particle propagator in one dimension. We note that this was a packet initially concentrated at a point, which then spreads. This expression is very similar to the heat source function with an imaginary heat conductivity.

Since $\phi(p, t) = \dfrac{1}{\sqrt{2\pi\hbar}}\, e^{-\frac{i}{\hbar}\frac{p^2 t}{2m}}$, $W(p) = \dfrac{1}{2\pi\hbar}$, and all values of p are equally likely.

We now discuss examples of the motion of wave packets. First we consider the free particle packet in one dimension. So

$$\dot{\bar{p}}_x = 0, \quad \text{i.e.,} \quad \bar{p}_x = \text{constant.} \qquad (194)$$

We have
$$\dot{\bar{x}} = \frac{\bar{p}_x}{m} = \text{constant,} \qquad (195)$$

so that

$$\bar{x} = \bar{x}_0 + \frac{1}{m}\,\bar{p}_x t. \tag{196}$$

Also,

$$\frac{d}{dt}(\Delta p_x)^2 = \left\langle \frac{d}{dt}(p_x - \bar{p}_x)^2 \right\rangle = 0,$$

giving

$$\Delta p_x^2 = \text{constant}, \tag{197}$$

since

$$\dot{p}_x = 0 \quad \text{and} \quad \dot{\bar{p}}_x = 0.$$

We have

$$\frac{d}{dt}(\Delta x)^2 = \left\langle \frac{d}{dt}(x - \bar{x})^2 \right\rangle$$

$$= \frac{1}{m}\left\langle (p_x - \bar{p}_x)(x - \bar{x}) + (x - \bar{x})(p_x - \bar{p}_x) \right\rangle$$

$$= \left\langle (\dot{x} - \dot{\bar{x}})(x - \bar{x}) + (x - \bar{x})(\dot{x} - \dot{\bar{x}}) \right\rangle. \tag{198}$$

This is the expression for twice the correlation average for $\dot{x}$ and x (Furry (1963)).

Therefore,

$$\frac{d^2}{dt^2}(\Delta x)^2 = \frac{2}{m^2}\left\langle (p_x - \bar{p}_x)^2 \right\rangle$$

$$= \frac{2}{m^2}(\Delta p_x)^2, \tag{199}$$

whence the Mclaurin series for $(\Delta x)^2$, using Eqs. (198) and (199),

$$(\Delta x)^2 = (\Delta x)_0^2 + \left\langle (\dot{x} - \dot{\bar{x}})(x - \bar{x}) + (x - \bar{x})(\dot{x} - \dot{\bar{x}}) \right\rangle t$$

$$+ \frac{1}{m^2}(\Delta p_x)^2 t^2. \tag{200}$$

We note that the higher derivatives of (Δx) are all zero, using Eq. (197).

The variance in x of a moving wave packet always eventually increases quadratically with time. If the initial spread of the position wave packet is $(\Delta x)_0$, then a typical estimate of the spreading time may be made from the ratio $\dfrac{t\Delta p_x}{m(\Delta x)_0}$:

$$\frac{t\Delta p_x}{m(\Delta x)_0} \sim \frac{\hbar t}{m(\Delta x)_0^2}.$$

For a macroscopic particle with $m = 1g$ and $(\Delta x)_0 = 0.01$ cm, the uncertainty of position has increased by a factor of two in a time $\approx 3 \times 10^{23}$ s. On the other hand, for an electron with $m \approx 9 \times 10^{-28}$ g and $(\Delta x)_0 = 10^{-8}$ cm. the required time is only 3×10^{-16} s.

Next we choose a gaussian wave packet. So $A(k)$ is the gaussian distribution Eq. (185) giving

$$\psi(x,\,t) = \frac{1}{\sqrt{2\pi}} \left(\frac{1}{2\pi a}\right)^{\frac{1}{4}} \int \exp\left[-\frac{k^2}{4a}\left(1 + \frac{2i\hbar t}{m}\right) + ikx\right] dk. \tag{201}$$

We get

$$\psi(x,\,t) = \left(\frac{2a}{\pi}\right)^{\frac{1}{4}} \frac{1}{\sqrt{1 + 2i\hbar at/m}} \cdot \exp\left(-\alpha(t)x^2+, \right. \tag{202}$$

with

$$\alpha(t) = \frac{a}{1 + 2i\hbar at/m},$$

and

$$\int \exp\left[-\frac{k^2}{4\alpha} + ikx\right] dk = \int \exp\left[-\frac{1}{4\alpha}(k - 2ix\alpha)^2\right] dk$$

$$\times \exp\left(-\alpha x^2\right)$$

$$= 2\sqrt{\alpha\pi}\,\exp\left(-\alpha x^2\right).$$

For $t = 0$, we get back Eq. (183).

The position probability distribution is

$$|\psi(x,\,t)|^2 = \left(\frac{2a}{\pi}\right)^{\frac{1}{2}} \cdot \frac{1}{\sqrt{1 + (2a\hbar t/m)^2}}\,\exp\left[-\beta x^2\right], \tag{203}$$

with

$$\beta = \alpha + \alpha^* = \frac{2a}{1 + (2a\hbar t/m)^2}.$$

We know that the variance of a gaussian distribution

$$W(x) = |\psi(x)|^2 = \sqrt{\frac{\beta}{\pi}} \cdot \exp\left[-\beta x^2\right],$$

implies

$$\overline{x^2} = \frac{1}{2\beta}, \quad \text{or} \quad \beta = 2a.$$

Hence,

$$(\Delta x)^2 = \overline{x^2} = \frac{1}{4a}\left[1 + \left(\frac{2a\hbar t}{m}\right)^2\right]. \tag{204}$$

Also,

$$(\Delta k)^2 = a, \text{ with } |(\Delta k)|^2 \text{ given by Eq. (185).}$$

Therefore,

$$(\Delta x)^2 = \frac{1}{4(\Delta k)^2} + \frac{\hbar^2(\Delta k)^2}{m^2}\,t^2. \tag{205}$$

The variance of a moving gaussian wave packet grows quadratically in time. As we have seen, this result is generally true.

Let us next consider a particle under the action of a constant force. For the x-component of the force

$$\dot{p}_x = f_x = -\frac{\partial V}{\partial x} = \text{constant},$$

so

$$\bar{f}_x = f_x = \text{constant}. \tag{206}$$

We have

$$\frac{d}{dt}(\Delta p_x)^2 = \left\langle \frac{d}{dt}(p_x - \bar{p}_x)^2 \right\rangle$$

$$= \left\langle (f_x - \bar{f}_x)(p_x - \bar{p}_x) + (p_x - \bar{p}_x)(f_x - \bar{f}_x) \right\rangle$$

$$= 0, \tag{207}$$

$$\frac{d}{dt}(\Delta x)^2 = \left\langle \frac{d}{dt}(x - \bar{x})^2 \right\rangle$$

$$= \frac{1}{m} \left\langle (p_x - \bar{p}_x)(x - \bar{x}) + (x - \bar{x})(p_x - \bar{p}_x) \right\rangle, \tag{208}$$

$$\frac{d^2}{dt^2}(\Delta x)^2 = \frac{2}{m^2} \left\langle (p_x - \bar{p}_x)^2 \right\rangle$$

$$= \frac{2}{m^2}(\Delta p_x)^2, \tag{209}$$

and all higher derivatives are zero.

These are the same as in the free particle case. So the Eq. (200) holds in this case also. The generalization of our discussion of one-dimensional wave packets to three dimensions is fairly straight forward. We write the wave packet

$$\psi(\bar{x}, t) = \frac{1}{(2\pi\hbar)^{3/2}} \int e^{\frac{i}{\hbar}[\bar{p}\cdot\bar{r} - E(p)t]} \phi(\vec{p})d\vec{p} \tag{210}$$

Writing
$$\psi(\vec{r}) = \psi(\vec{r}, t = 0),$$

we have

$$\psi(\vec{r}) = \frac{1}{(2\pi\hbar)^{3/2}} \int e^{\frac{i}{\hbar}\bar{p}\cdot\bar{r}} \phi(\vec{p})d\vec{p}, \tag{211}$$

and

$$\phi(\vec{p}) = \frac{1}{(2\pi\hbar)^{3/2}} \int e^{-\frac{i}{\hbar}\bar{p}\cdot\bar{r}} \psi(\vec{r})d\vec{r}, \tag{212}$$

as Fourier transforms of each other.

We may introduce the wave function $\psi(\vec{r}, t)$ and $\phi(\vec{p}, t)$ as Fourier transforms of each other.

$$\psi(\vec{r}, t) = \frac{1}{(2\pi\hbar)^{3/2}} \int e^{\frac{i}{\hbar}\vec{p} \cdot \vec{r}} \, \phi(\vec{p}, t)d\vec{p}, \tag{213}$$

$$\phi(\vec{p}, t) = \frac{1}{(2\pi\hbar)^{3/2}} \int e^{-\frac{i}{\hbar}\vec{p} \cdot \vec{r}} \, \psi(\vec{r}, t)d\vec{r}, \tag{213a}$$

The center of a three-dimensional free particle wave packet

$$\vec{\nabla}_{\vec{p}}\left(\vec{p} \cdot \vec{r} - E(p)t\right)\Big|_{\vec{p} = \vec{p}_0} = 0$$

moves uniformly according to $\vec{r} = \vec{v}_g t$,

where $\vec{v}_g = \vec{\nabla}_{\vec{p}} \, E(p)\Big|_{\vec{p} = \vec{p}_0}$ is the group velocity of the packet.

A wave packet corresponding to classical motion in the Coulomb potential is discussed by Nauenberg (2000).

3.7 THE CLASSICAL LIMIT

Classical mechanics has been enormously successful in describing a vast range of natural phenomena. Quantum mechanics must therefore reduce to classical mechanics under appropriate conditions. The relation between the wave equation and classical mechanics has been discussed in connection with the motion of wave packets in the previous section. A wave packet gives a probability distribution that corresponds in an appropriate limit to an ensemble of classical trajectories. Of course, this still does not mean a complete limiting transition to classical mechanics. The correspondence principle have played an important role in the establishment of quantum theory.

Quantum mechanics must make a limiting transition to classical mechanics in the limit $\hbar \to 0$. The validity of classical mechanics in a wide-ranging domain is a reflection of the fact that $\hbar$ is a very small number on the ordinary scale. However, we must note that taking the limit $\hbar \to 0$ is quite subtle as classical concepts cannot be considered as limiting cases of quantum concepts. The limiting transition of quantum mechanics to classical mechanics is analogous to the transition of wave optics to geometrical optics.

In the Schrödinger wave equation

$$-\frac{\hbar^2}{2m} \nabla^2\psi(\bar{x}, t) + V\psi(\bar{x}, t) = i\hbar \frac{\partial}{\partial t} \psi(\bar{x}, t) \tag{214}$$

we make the substitution for the time dependent wave function

$$\psi(\bar{x}, t) = A(\bar{x}, t) \, e^{iS(\bar{x}, t)/\hbar}, \tag{215}$$

where A and S are real functions of $\bar{x}$, and t which are slowly varying with their arguments. The wave function is now in the form of waves with A and S determining the amplitude and phase respectively. We get

$$- A \frac{\partial S}{\partial t} = \frac{1}{2m} \left\{ (\vec{\nabla} S)^2 A - \hbar^2 \nabla^2 A \right\} + VA, \qquad (216)$$

$$\frac{\partial A}{\partial t} = - \frac{1}{2m} \left\{ A \nabla^2 S + 2 (\vec{\nabla} A \cdot \vec{\nabla} S) \right\}. \qquad (216a)$$

The second equation is the continuity equation. The probability density $P = \psi^* \psi = A^2$ and the current density $\vec{j} = A^2 \dfrac{\vec{\nabla} S}{m}$ so that Eq. (216a) may be written as

$$\frac{\partial}{\partial t} P + \vec{\nabla} \cdot \left(\frac{P \vec{\nabla} S}{m} \right) = 0. \qquad (217)$$

Let us now suppose that $\hbar$ can, in some sense be regarded as small and let us neglect terms involving $\hbar$ in Eq. (216).

We then get

$$\frac{\partial S}{\partial t} + \frac{1}{2m} (\vec{\nabla} S)^2 + V = 0, \qquad (218)$$

This is a first order partial differential equation which the phase function S has to satisfy Eq. (218) has the form of the Hamilton-Jacobi equation in classical mechanics.

We introduce a velocity field

$$\vec{v}(\vec{x}, t) = \frac{\vec{j}}{P} = \frac{\vec{\nabla} S}{m}, \qquad (219)$$

and the momentum

$$\vec{p} = \vec{\nabla} S. \qquad (219a)$$

In presence of a vector potential $\vec{A}$, we have

$$\vec{v} = \frac{1}{m} \left(\vec{\nabla} S - \frac{e}{c} \vec{A} \right) \qquad (219b)$$

Taking the gradient of the left hand side of Eq. (218), we have

$$\left(\frac{\partial}{\partial t} + (\vec{v} \cdot \vec{\nabla}) \right) m \vec{v} + \vec{\nabla} V = 0, \qquad (220)$$

using Eq. (219). In this approximation, a partcle following the flow of a classical fluid defined by $\vec{v}$ obeys the equation of motion

$$m \frac{d \vec{v}}{dt} = - \vec{\nabla} V. \qquad (221)$$

We note that one must not take such an analogy too literally.

Let us expand S in ascending powers of $\hbar/i$:

$$S = S_0 + \left(\frac{\hbar}{i}\right) S_1 + \left(\frac{\hbar}{i}\right)^2 S_2 + \ldots \tag{222}$$

We substitute this into Eq. (216) and equate the coefficients of $\hbar^n$ ($n = 0, 1, 2, \ldots$). The $\hbar$-independent terms give Eq. (218)

$$\frac{\partial S_0}{\partial t} + \frac{1}{2m}\left(\vec{\nabla} S_0\right) + V = 0, \tag{218a}$$

Further,

$$-\frac{\partial S_1}{\partial t} = \frac{1}{2m}\left(2\vec{\nabla} S_0 \cdot \vec{\nabla} S_1 + \nabla^2 S_0\right). \tag{223}$$

$S_0(\vec{x}, t)$ is called Hamilton's principal function in classical mechanics. The solution of Eq. (218a) S_0 is a real function in regions where the classical trajectories lie. According to Eq. (223), S is also real in this region. In this approximation the wave packets behave like an ensemble of classical trajectories. The non-classical aspects (e.g. the spreading of a wave packet) are described by the terms S_2, S_3, ... in Eq. (222). We mention here one crucial point - the Hamilton-Jacobi equation is non-linear in S.

The classical limit for stationary states is of great importance. For a stationary state with a time-independent Hamiltonian,

$$\psi = e^{\frac{i}{\hbar} Et}\, u, \text{ we set}$$

$$S(\vec{x}, t) = \bar{S}(\vec{x}) - Et, \quad u = e^{\frac{i}{\hbar}\bar{S}} \tag{224}$$

where $\bar{S}$, Hamilton's characteristic function, and u are independent of t. Also,

$$\frac{\partial S}{\partial t} = -E, \quad \frac{\partial A}{\partial t} = 0.$$

We obtain

$$\frac{1}{2m}\left(\vec{\nabla}\bar{S}\right)^2 + V = E + \frac{\hbar^2}{2m}\frac{\nabla^2 A}{A},$$

$$\vec{\nabla} \cdot \left(A^2 \vec{\nabla}\bar{S}\right) = 0. \tag{225}$$

In the classical limit ($\hbar \to 0$), we have

$$\frac{1}{2m}\left(\vec{\nabla}\bar{S}_0\right)^2 + V = E,$$

$$\vec{\nabla} \cdot \left(A^2 \vec{\nabla}\bar{S}_0\right) = 0. \tag{225a}$$

where $\bar{S}_0$ is the approximate value of $\bar{S}$. As time goes, a surface $S = \text{constant}$

advances as a wave front. The velocity vector $\vec{v} = \dfrac{\vec{\nabla}S}{m}$. Since $\vec{\nabla}S$ traces the trajectory normal to the wave front, the particle trajectory is tangential to the velocity vector. In this sense the particle trajectory is like a ray in geometrical optics. Hence the classical approximation to wave mechanics is equivalent to the geometrical optics approximation to wave optics.

We shall now consider the one-dimensional case where it is easier to determine the condition of validity of the classical approximation. From Eq. (225a),

$$\overline{S}_0(x) = \pm \int^x dx' \sqrt{2m\,[E - V(x')]}. \tag{226}$$

The local (or, reduced) deBroglie wavelength $\lambdabar(\lambda/2\pi)$ is given by

$$\lambdabar = \frac{\hbar}{\sqrt{2m\,[E - V(x)]}}, \tag{227}$$

whence

$$\frac{d\overline{S}_0}{dx} = \pm \frac{\hbar}{\lambdabar(x)}. \tag{228}$$

The semi-classical wave function is the approximate wave function given by

$$\psi(x,\,t) \sim \exp\left(\pm i \int^x \frac{dx'}{\lambdabar(x')}\right) e^{-\frac{i}{\hbar} Et}. \tag{229}$$

The first equation in Eq. (225a) can be written as

$$(\vec{\nabla}S)^2 = \frac{\hbar^2}{\lambdabar^2}\left(1 + \lambdabar^2\,\frac{\nabla^2 A}{A}\right). \tag{230}$$

In the classical approximation,

$$\lambdabar^2\,\frac{\nabla^2 A}{A} \ll 1, \tag{231}$$

over all space. Then Eq. (230) may be replaced by the approximate equation

$$\left(\vec{\nabla}\overline{S}_0\right)^2 = \frac{\hbar^2}{\lambdabar^2}, \tag{232}$$

which is the equation of the wave fronts in geometrical optics.

As a general rule, the classical approximation is justified when

$$\left|\vec{\nabla}\lambdabar\right| \ll 1, \tag{233}$$

that is, the wavelength associated with a particle should change only slightly within a wavelength.

3.8 ILLUSTRATIVE SOLUTIONS OF THE SCHRÖDINGER EQUATION

In this section we shall consider a few simple one-dimensional quantum mechanical problems which will acquaint us with some of the methods used in dealing with the Schrödinger equation. We shall study some interesting quantum mechanical phenomena in course of our discussion.

We consider a one-dimensional stationary Schrödinger equation for a particle of mass m with no internal degree of freedom in a certain potential $V(x)$:

$$\left[-\frac{\hbar^2}{2m}\frac{d^2}{dx^2} + V(x)\right]\psi(x) = E\psi(x), \tag{234}$$

where E is the energy of the stationary state.

$V(x)$ must satisfy certain conditions for the eigenvalue problem to be well-defined so that the solutions are acceptable. The examples discussed below will illustrate many of the general properties of the energy spectra.

(a) The Potential Step, Reflection and Transmission Coefficients

We consider a particle moving in a potential $V(x)$ (Fig. 3.2):

$$V(x) = V_0\theta(x);$$

$$\theta(x) = \begin{cases} 1, x > 0 \\ 0, x < 0 \end{cases} \tag{235}$$

where $V_0 \geq 0$.

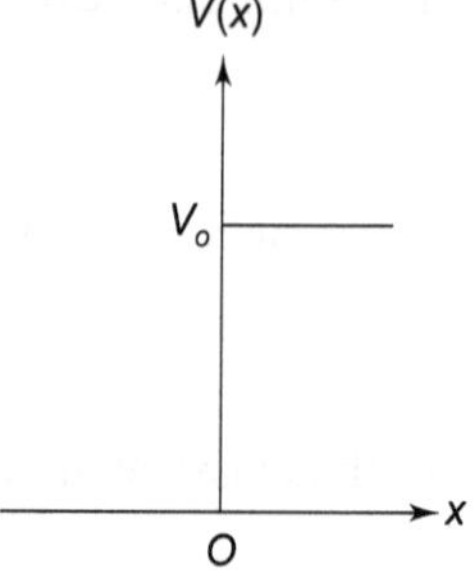

Fig. 3.2 The step potential

There is no acceptable solution for $E < 0$.

So we consider first the case $0 < E < V_0$. Classically, a particle of this energy incident from the left will move freely until totally reflected at the potential step. The motion of the particle is reversed at the step.

The Schrödinger equation becomes

$$\frac{d^2\psi(x)}{dx^2} + k^2\psi(x) = 0,$$

$$k^2 = \frac{2mE}{\hbar^2}, \quad x < 0 \tag{236}$$

and

$$\frac{d^2\psi(x)}{dx^2} - \kappa^2\psi(x) = 0,$$

$$\kappa^2 = \frac{2m}{\hbar^2}(V_0 - E), \quad x > 0. \tag{236a}$$

The general solution of Eq. (236) is

$$\psi(x) = Ae^{ikx} + Be^{-ikx}, \quad x < 0 \tag{237}$$

where A and B are arbitrary constants.

The general solution of Eq. (236a) is

$$\psi(x) = Ce^{\kappa x} + De^{-\kappa x}, \quad x > 0 \tag{238}$$

where C and D are arbitrary constants.

Since the wave function ψ must remain finite as $x \rightarrow \infty$, the constant $C = 0$.

Therefore, the complete wave function is

$$\psi(x) = \begin{cases} Ae^{ikx} + Be^{-ikx}, & x < 0 \\ De^{-\kappa x}, & x > 0 \end{cases}. \tag{239}$$

Since one of the two linearly independent solutions is excluded, the energy eigenvalues for $E < V_0$ are nondegenerate.

To be an acceptable solution of the Schrödinger equation, the wave function and its first derivative must be finite and continuous for all values of x.

By matching the wave function and its derivative at the discontinuity of the potential at $x = 0$, we obtain

$$A + B = D,$$

$$ik(A - B) = -\kappa D, \tag{241}$$

whence

$$\frac{B}{A} = \frac{ik + \kappa}{ik - \kappa} = e^{i\alpha} \quad \text{with } \alpha \text{ real} \tag{242}$$

$$\frac{D}{A} = \frac{2ik}{ik - \kappa} = 1 + e^{i\alpha}.$$

Eq. (239) gives

$$\psi(x) = \begin{cases} 2Ae^{i\alpha/2} \cos\left(kx - \dfrac{\alpha}{2}\right), & x < 0 \\ 2Ae^{i\alpha/2} \cos \alpha/2\, e^{-\kappa x}, & x > 0 \end{cases}, \tag{243}$$

The wave function (no degeneracy) is real except for a multiplicative complex constant. The classical turning point is a point of inflection of the wave function. The oscillatory and the exponential parts of ψ can be joined smoothly at $x = 0$ for all values of E between 0 and V_0. The spectrum is continuous.

The functions e^{ikx} and e^{-ikx} in Eq. (239) when multiplied by $e^{-\frac{i}{\hbar}Et}$ represent in the region $x < 0$ plane waves incident from the left with an amplitude A and a reflected wave propagating toward left with amplitude B. According to Eq. (241), $|B|^2 = |A|^2$. The reflection coefficient R is defined as the ratio of the intensity of the reflected wave to that of the incident wave. Then

$$R = \frac{|B|^2}{|A|^2}, \tag{242a}$$

which is unity in this case. Hence there is total reflection, as in classical mechanics. Using the eigenfunctions in Eq. (242) we can construct a wave packet describing the motion of a particle incident from the left. The wave packet would move classically and would be reflected at the wall, giving a vanishing probability of finding the particle in the region $x > 0$ after the wave packet has receded.

The current density in the region $x < 0$ calculated from Eq. (239) is

$$j \equiv \frac{\hbar}{2im} \left(\psi^* \frac{d\psi}{dx} - \frac{d\psi^*}{dx} \psi \right)$$

$$= \frac{\hbar k}{m} \left(|A|^2 - |B|^2 \right) \tag{244}$$

j vanishes in the region $x < 0$. This is seen from Eq. (242). There is no net current and hence no net momentum anywhere.

The solution for $x > 0$ is $De^{-\kappa x}$. But $D = \dfrac{2ik}{ik - \kappa} A$ does not vanish. The position probability density

$$P(x) = |D|^2 \exp\left(-2\kappa x\right), \quad x > 0 \tag{245}$$

decreases very rapidly in increasing x.

There is a finite probability of finding the particle in this classically forbidden region. This penetration of waves in the forbidden region is a characteristic quantum phenomenon. There is no flux in this region.

For an infinitely high potential barrier ($V_0 \to \infty$ or $\kappa \to \infty$), $\psi(x) \to 0$ in the region under the barrier, regardless of the value of D. From Eq. (241), we see that

$$\lim_{\kappa \to \infty} \frac{B}{A} = -1,$$

$$\lim_{\kappa \to \infty} \frac{D}{A} = 0 \tag{246}$$

or, $A + B = 0$ and $D = 0$ as $V_0 \to \infty$.

Hence,

$$\psi(x) = \begin{cases} A\left(e^{ikx} - e^{-ikx}\right), & x < 0 \\ 0, & x > 0 \end{cases} \tag{247}$$

The wave function vanishes at $x = 0$ where the potential makes an infinite jump whereas its derivative jumps discontinuously from a finite value $2ikA$ to zero. We note that the condition of smooth joining which requires that the wave function and its slope be continuous holds for the potential to be either continuous or with finite jumps.

We next consider the potential step where the particle energy is above the step: $E > V_0$. Classically, such a particle passes the potential step with altered velocity but with no change of direction. The particle can be incident either from the left or from the right.

The Schrödinger equation in the two regions is

$$\frac{d^2\psi(x)}{dx^2} + k^2\psi(x) = 0, \quad x < 0$$

$$\frac{d^2\psi(x)}{dx^2} + q^2\psi(x) = 0, \quad x > 0 \tag{248}$$

where $k^2 = \dfrac{2mE}{\hbar^2}$ and $q^2 = \dfrac{2m(E - V_0)}{\hbar^2}$.

The general solutions are

$$\psi(x) = \begin{cases} Ae^{ikx} + Be^{-ikx} & (x < 0) \\ Ce^{iqx} + De^{-iqx} & (x > 0) \end{cases} \tag{249}$$

We assume that the particle is incident from the left. In region $x < 0$, the wave function is a combination of a plane wave, whose amplitude can be put equal to unity, incident from the left and a reflected wave. In region $x > 0$, the wave function is a transmitted wave. So we write

$$\psi(x) = \begin{cases} e^{ikx} + re^{-ikx}, & x < 0 \\ te^{iqx}, & (x > 0) \end{cases} \tag{250}$$

By the condition of smooth joining at $x = 0$, i.e., by the continuity of ψ

and $\dfrac{d\psi}{dx}$ at $x = 0$, we obtain

$$1 + r = t; \quad ik(1 - r) = iqt$$

or,

$$r = \frac{k - q}{k + q}, \quad t = \frac{2k}{k + q}. \tag{251}$$

The energy spectrum is continuous.

The current density is constant and is

$$j = \begin{cases} \dfrac{\hbar k}{m}\,(1 - |r|^2), & (x < 0) \\ \dfrac{\hbar q}{m}\,|t|^2 \end{cases} \tag{252}$$

The flux on the left can be split into two parts

$$j_0 = j_{in} - j_{ref}, \quad x < 0 \tag{253}$$

and the flux on the right is the transmitted flux

$$j = j_{tr}, \quad x > 0 \tag{253a}$$

From the conservation law, the flux on the left is equal to that on the right, i.e.,

$$\frac{\hbar k}{m}\left(1 - |r|^2\right) = \frac{\hbar q}{m}\,|t|^2, \tag{254}$$

or,

$$j_{in} = j_{ref} + j_{tr}. \tag{255}$$

Let us now evaluate the reflection coefficient R and transmission coefficient T:

$$R = \left|\frac{j_{ref}}{j_{in}}\right| = |r|^2 = \frac{(k - q)^2}{(k + q)^2} = \frac{(1 - K)^2}{(1 + K)^2},$$

$$T = \left|\frac{j_{tr}}{j_{in}}\right| = \frac{q}{k}\,|t|^2 = \frac{4kq}{(k + q)^2} = \frac{4K}{(1 + K)^2}, \tag{256}$$

where $K = \dfrac{q}{k} = \sqrt{1 - V_0/E}$. Clearly,

$$R + T = 1. \tag{257}$$

r is called the reflection amplitude and t is called the transmission amplitude.

Classically, in case $E > V_0$, the particle passes over the potential i.e., there is no reflection. Quantum mechanically, a certain fraction of the incident particles are reflected $(R \neq 0)$

For $E \gg V_0$, that is, for $q \to k$ from below, we have $K \to 1$ and hence $R \to 0$, $T \to 1$. This means that at very high energies, the presence of the potential step makes negligibly small effect on the motion. On the other hand, for $E \to V_0$, $K \to 1$ so that $R \to 1$ and $T \to 0$.

(b) The Infinite Square Well

Let us consider the one-dimensional potential

$$V(x) = \begin{cases} 0, & -a \leq x \leq a, \\ \infty, & |x| > a, \end{cases} \tag{258}$$

We consider the one dimensional motion of a particle of mass m constrained to move in a region of width $2a$ bounded by perfectly rigid, impenetrable walls at $x \pm a$.

Classically, the particle remains confined within the well executing a (periodic) back and forth motion as a result of repeated reflections at the walls. As long as $E > 0$, the particle can move with any energy. Quantum mechanically, the energy spectrum is discrete.

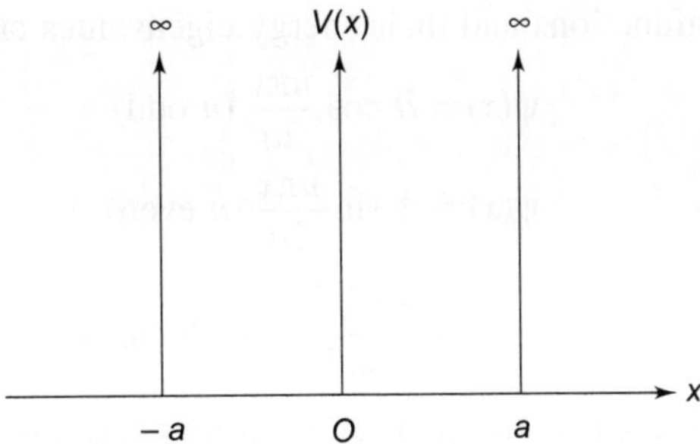

Fig. 3.3   The infinite square well potential

The eigenvalue problem for |x| < a is

$$-\frac{\hbar^2}{2m}\frac{d^2\psi(x)}{dx^2} = E\psi(x). \tag{259}$$

The wave function must vanish in the region outside the well, as the probability of finding the particle outside is zero. The wave function must vanish at the walls since it must be continuous

$$\psi(x) = 0 \quad \text{at } x = \pm\, a. \tag{260}$$

At these points where the potential makes infinite jumps, the slope $\dfrac{d\psi(x)}{dx}$ is discontinuous at $x = \pm\, a$.

We write the general solution as

$$\psi(x) = A\,\sin kx + B\,\cos kx,$$

$$k = +\left(\frac{2mE}{\hbar^2}\right)^{\frac{1}{2}}. \tag{261}$$

The boundary conditions at $x = \pm\, a$ give

$$A\,\sin ka + B\,\cos ka = 0,$$

$$-A\,\sin ka + B\,\cos ka = 0,$$

whence

$$A\,\sin ka = 0 \quad B\,\cos ka = 0.$$

If both A and B are zero, then we get a trivial solution $\psi = 0$ everywhere. For a given value of k or E, both sin ka and cos ka cannot be zero. So there are two classes of solution:

(i) $A = 0$ and cos $ka = 0$ so that $ka = \dfrac{n\pi}{2}$ $(n = 1, 3, 5, \ldots)$

(ii) $B = 0$ and sin $ka = 0$ giving $ka = \dfrac{n\pi}{2}$ $(n = 2, 4, 6, \ldots)$

Hence the eigenfunctions and their energy eigenvalues are

$$\psi(x) = B \cos \frac{n\pi x}{2a} \quad (n \text{ odd})$$

$$\psi(x) = A \sin \frac{n\pi x}{2a} \quad (n \text{ even})$$

$$E = \frac{\pi^2 \hbar^2 n^2}{8ma^2} \text{ in both cases.} \tag{262}$$

It is evident that $n = 0$ gives the trivial solution. The eigenfunctions in each case can easily be normalized.

There is an infinite sequence of discrete energy levels corresponding to bound states for all positive integer values of the quantum number n. The energy spectrum is non-degenerate as there is just one eigenfunction for each level. The number of nodes of the nth eigenfunction that are within the well is $n - 1$. The real eigenfunctions $\psi_n(x)$ and $\psi_m(x)$ corresponding to different eigenvalues E_n and E_m respectively are orthogonal.

The eigenfunctions corresponding to odd quantum numbers $n = 1, 3, 5, \ldots$ are symmetric, $\psi_n(-x) = \psi_n(x)$ and are even functions of x, while the eigenfunctions corresponding to $n = 2, 4, 6, \ldots$ are antisymmetric, $\psi_n(-x) = -\psi_n(x)$ and are odd functions of x.

The lowest energy occurs when $n = 1$, it is

$$E_1 = \frac{\hbar^2 \pi^2}{8ma^2}.$$

The energy of nth excited state is

$$E_n = n^2 E_1.$$

The general solution of the time-dependent Schrödinger equation is

$$\Psi(x, t) = \sum_{n=1}^{\infty} \psi_n(x) \, e^{-\frac{i}{\hbar} E_n t}$$

$$= \sum_{n=1}^{\infty} \psi_n(x) \, e^{-\frac{i}{\hbar} n^2 E_1 t} \tag{263}$$

We have seen that the eigenfunctions can be classified as even (symmetric) functions of x and odd (anti-symmetric) functions of x. Under the parity operation, i.e., the operation of reflection about the origin, $x \rightarrow -x$, the Hamiltonian H is invariant if the potential is symmetric about $x = 0$, $V(x) = V(-x)$. The eigenfunctions of the one-dimensional Schrödinger equation can always be chosen to have definite parity (even or odd) when the potential is symmetric.

(c)   The Finite Square Well

Let us consider a particle of mass m moving under the action of a square (or rectangular) well potential

$$V(x) = \begin{cases} 0, & |x| > a \\ -V_0, & |x| < a \end{cases} \qquad (264)$$

where V_0 (> 0) is the depth of the well and a is its range.

The eigenvalue equation is

$$-\frac{\hbar^2}{2m}\frac{d^2u}{dx^2} + V(x)u = Eu \qquad (265)$$

Putting $k^2 = \dfrac{2mE}{\hbar^2}$, $\alpha^2 = \dfrac{2mV_0}{\hbar^2}$, we have

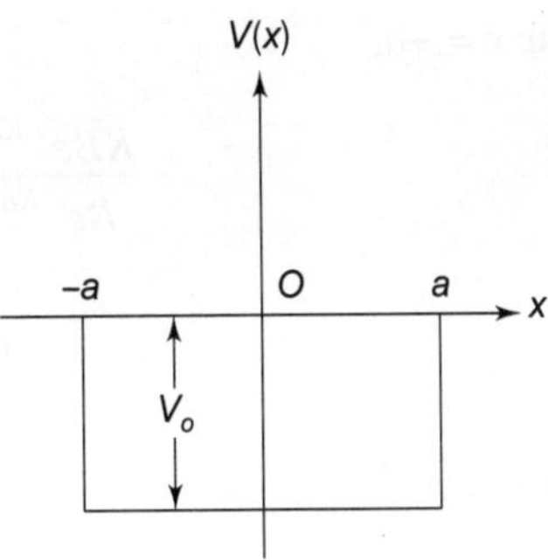

Fig. 3.4　The finite square well potential

$$\frac{d^2u}{dx^2} + (k^2 + \alpha^2)u = 0, \quad \text{inside the well}$$

$$\frac{d^2u}{dx^2} + k^2u = 0, \quad \text{outside} \qquad (266)$$

The boundary conditions are boundedness of u at infinity, and u, $\dfrac{du}{dx}$ continuous everywhere (even where V is discontinuous).

When $E > 0$, the particle is not confined and the scattering solutions will be discussed later. When $E < 0$, the motion of the particle is confined and the discrete solutions describing bound states are considered first. The energy E cannot be lower than the absolute minimum of the potential. We must have $-V_0 \le E < 0$ for bound states. Let us prove first that for $E < -V_0$, there are no possible eigenfunctions.

We have

$$|x| > a, \quad \frac{d^2u}{dx^2} + k^2u = 0, \quad k^2 = -K^2 \quad \text{with } K \text{ real} \qquad (267)$$

since E negative,

$$|x| < a, \quad \frac{d^2u}{dx^2} - (K^2 + \alpha^2)u = 0, \quad \text{with } K^2 > \alpha^2.$$

The acceptable solutions are

$$u(x) = \begin{cases} Ae^{+Kx}, & x < -a \\ Be^{-Kx}, & x > a \\ Ce^{\sqrt{K^2-\alpha^2}\,x} + De^{-\sqrt{K^2-\alpha^2}\,x}, & -a < x < +a \end{cases} \qquad (268)$$

Since u', u are both continuous at $x = \pm a$, so is the logarithmic derivative $\dfrac{u'}{u}$. Therefore,

$$\frac{u'}{u}\bigg|_{a-0}^{a+0} = 0; \quad \frac{u'}{u}\bigg|_{-a-0}^{-a+0} = 0 \qquad (269)$$

or, at $x = +a$,

$$-\frac{KBe^{-Ka}}{Be^{-Ka}} = \sqrt{K^2 - \alpha^2} \cdot \frac{Ce^{+\sqrt{K^2 - \alpha^2}\,a} - De^{-\sqrt{K^2 - \alpha^2}\,a}}{Ce^{+\sqrt{K^2 - \alpha^2}\,a} + De^{-\sqrt{K^2 - \alpha^2}\,a}}$$

$$-K = \sqrt{K^2 - \alpha^2} \cdot \frac{Cq^2 - D}{Cq^2 + D}$$

with $q = e^{+\sqrt{K^2 - \alpha^2}\,a}$

and at $x = -a$,

$$K \frac{Ae^{-Ka}}{Ae^{-Ka}} = \sqrt{K^2 - \alpha^2} \cdot \frac{Ce^{-\sqrt{K^2 - \alpha^2}\,a} - De^{+\sqrt{K^2 - \alpha^2}\,a}}{Ce^{-\sqrt{K^2 - \alpha^2}\,a} + De^{+\sqrt{K^2 - \alpha^2}\,a}}$$

$$K = \sqrt{K^2 - \alpha^2} \cdot \frac{C - Dq^2}{C + Dq^2}.$$

Therefore, our conditions to determine C, D are

$$C\left(K - \sqrt{K^2 - \alpha^2}\right) - D\left(K + \sqrt{K^2 - \alpha^2}\right)q^2 = 0,$$

$$C\left(K + \sqrt{K^2 - \alpha^2}\right)q^2 - D\left(K - \sqrt{K^2 - \alpha^2}\right) = 0,$$

for which the determinant is

$$\Delta = +\left(K + \sqrt{K^2 - \alpha^2}\right)^2 q^4 - \left(K - \sqrt{K^2 - \alpha^2}\right)^2$$

$$= 0 \quad \text{for a nontrivial solution}$$

or,

$$q^4 = \left(\frac{K - \sqrt{K^2 - \alpha^2}}{K + \sqrt{K^2 - \alpha^2}}\right)^2. \tag{270}$$

However, K is a real number and q is a number greater than one, since its exponent is positive. So, Eq. (270) can never be satisfied – one side is always greater than one and the other side always less than one. Therefore, there are no nontrivial eigenfunctions in this case.

For $-V_0 \leq E < 0$, there are discrete levels. We have

$$\frac{d^2u}{dx^2} - K^2 u = 0, \quad |x| > a \quad \text{with } K \text{ real}$$

$$\frac{d^2u}{dx^2} + (\alpha^2 - K^2)u = 0, \quad |x| < a \quad \text{with } \alpha^2 > K^2. \tag{271}$$

The potential $V(x)$ is symmetric about $x = 0$, $V(-x) = V(x)$. So the eigenfunctions of H have definite parity – they are even or odd functions of x.

The even solutions of Eq. (271) are

$$x < -a \quad u(x) = Ae^{Kx}$$

$$x > a \quad u(x) = Ae^{-Kx}$$

and for $|x| < a$, we have in general

$$u(x) = B \cos \sqrt{\alpha^2 - K^2}\, x \tag{272}$$

We just satisfy conditions at one boundary, since the problem is symmetric. We have

$$\left.\frac{u'}{u}\right|_{a-0} = \left.\frac{u'}{u}\right|_{a+0} = -K\frac{Ae^{-Ka}}{Ae^{-Ka}}$$

$$= -\sqrt{\alpha^2 - K^2}\,\frac{B \sin \sqrt{\alpha^2 - K^2}\,a}{B \cos \sqrt{\alpha^2 - K^2}\,a}$$

and so

$$\tan \sqrt{\alpha^2 - K^2}\,a = \frac{K}{\sqrt{\alpha^2 - K^2}}. \tag{273}$$

The odd solutions of Eq. (271) are

$$u(x) = Ae^{+kx} \quad \text{for } x < a$$

$$u(x) = Ae^{-kx} \quad x > a$$

and for $|x| < a$,

$$u(x) = B \sin \sqrt{\alpha^2 - K^2}\,x. \tag{274}$$

The continuity of $\dfrac{u'}{u}$ at $x = a$ yields

$$\left.\frac{u'}{u}\right|_{a+0} = -K$$

$$= \left.\frac{u'}{u}\right|_{a-0}$$

$$= \sqrt{\alpha^2 - K^2}\,\frac{B \cos \sqrt{\alpha^2 - K^2}\,a}{B \sin \sqrt{\alpha^2 - K^2}\,a}$$

or,
$$\tan \sqrt{\alpha^2 - K^2}\,a = -\frac{\sqrt{\alpha^2 - K^2}}{K}. \tag{275}$$

The eigenvalues of H are completely determined by the conditions

$$\tan \sqrt{\alpha^2 - K^2}\,a = \begin{cases} \dfrac{Ka}{\sqrt{\alpha^2 - K^2}} & \text{for even solutions} \\[2ex] -\dfrac{\sqrt{\alpha^2 - K^2}\,a}{Ka} & \text{for odd solutions} \end{cases} \tag{276}$$

The bound state energies are obtained by solving these transcendental equations, either numerically or graphically. If we put

$\xi = \sqrt{\alpha^2 - K^2}\,a$, then $Ka = \sqrt{(\alpha a)^2 - \xi^2}$ and Eq. (276) is written as

$$\tan \xi = \begin{cases} \dfrac{\sqrt{\alpha^2 a^2 - \xi^2}}{\xi} = R_1\,(\xi), & \text{say} \\[2em] -\dfrac{\xi}{\sqrt{\alpha^2 a^2 - \xi^2}} = R_2\,(\xi), & \text{say} \end{cases} \qquad (277)$$

Therefore,

there is at least one even solution, for any V_0;

if $\alpha a < \dfrac{\pi}{2}$, there will be only one solution, an even one;

if $\dfrac{\pi}{2} < \alpha a < \pi$, there will be one even and one odd solutions;

if $\pi < \alpha a < \dfrac{3\pi}{2}$, there will be two even and one odd solutions; etc.

Note that the wave functions are not zero outside the well, where kinetic energy is negative. To measure the particle in this position

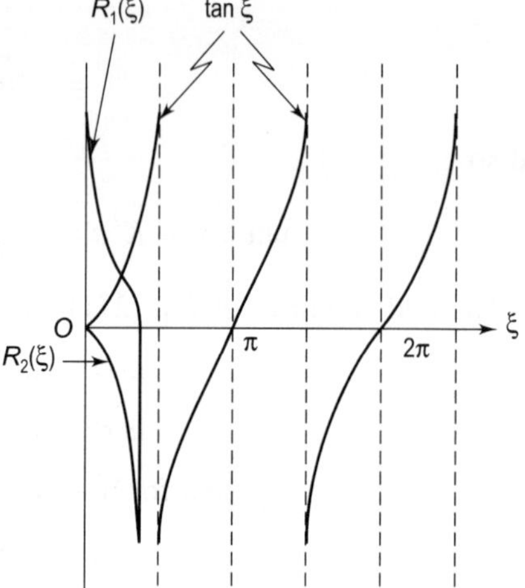

Fig. 3.5 Graphical determination of the energy levels in a square well (schematic diagram)

we must use photons with energy large with respect to this "negative k.e." The photon wave train must fall off faster than e^{-kx} in order to observe the particle.

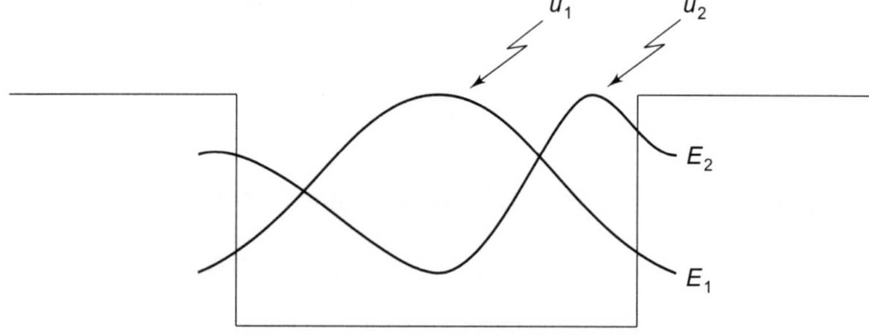

Fig. 3.6 The energy eigenfunctions in a square well: even ($E = E_1$); odd ($E = E_2$)

Hence, $p \gg \hbar K$, and so the particle recoil energy $E_{rec} \gg \dfrac{\hbar^2 K^2}{2m}$ which is greater than E_2.

It remains for us to consider the stationary states for positive energies $E > 0$. These are the scattering states as the motion of the particle is unconfined.

With
$$k^2 = \frac{2mE}{\hbar^2},$$

and
$$q^2 = \frac{2m(E + V_0)}{\hbar^2},$$

we write the solutions of the eigenvalue equation

$$u(x) = \begin{cases} e^{ikx} + re^{-ikx}, & x < -a \\ Ae^{iqx} + Be^{-iqx}, & -a < x < a \\ te^{ikx}, & a < x \end{cases} \tag{278}$$

Here there is an incoming flux $\dfrac{\hbar k}{m}$ from the left, and a reflected flux $\dfrac{\hbar k}{m} |r|^2$ and a transmitted flux $\dfrac{\hbar k}{m} |t|^2$ to the right. Inside the well, waves are going in both directions. Flux conservation demands

$$\frac{\hbar k}{m} (1 - |r|^2) = \frac{\hbar q}{m} (|A|^2 - |B|^2) = \frac{\hbar k}{m} |t|^2. \tag{279}$$

Continuity of the wave functions and derivatives at $x \pm a$ give

$$e^{-ika} + re^{ika} = Ae^{-iqa} + Be^{iqa},$$

$$ik(e^{-ika} - re^{ika}) = iq(Ae^{-iqa} - Be^{iqa}),$$

$$Ae^{iqa} + Be^{-iqa} = te^{ika},$$

$$iq(Ae^{iqa} - Be^{-iqa}) = ikte^{ika}, \tag{280}$$

hence

$$r = ie^{-2ika} \frac{(q^2 - k^2) \sin 2qa}{2kq \cos 2qa - i(q^2 + k^2) \sin 2qa}, \tag{281}$$

$$t = e^{-2ika} \frac{2kq}{2kq \cos 2qa - i(q^2 + k^2) \sin 2qa},$$

The reflection coefficient $R = |r|^2$ and the transmission coefficient $T = |t|^2$ are given by

$$R = \left[1 + \frac{4k^2 q^2}{(k^2 - q^2)^2 \sin^2(2qa)} \right]^{-1}$$

$$= \left[1 + \frac{4\left(\dfrac{E}{V_0}\right)\left(1 + \left(\dfrac{E}{V_0}\right)\right)}{\sin^2(2qa)} \right]^{-1},$$

$$T = \left[1 + \frac{(k^2 - q^2)^2 \sin^2(2qa)}{4k^2q^2} \right]^{-1}$$

$$= \left[1 + \frac{\sin^2(2qa)}{4\left(\dfrac{E}{V_0}\right)\left(1 + \dfrac{E}{V_0}\right)} \right]^{-1}. \tag{282}$$

Clearly, $\qquad\qquad R + T = 1.$

When $E = 0$, the transmission is zero ($T = 0$).

When $2qa = n\pi$ ($n = 1, 2, 3, ...$), T reaches its maximum value ($T = 1$) and its minimum value occurs when $2qa = \dfrac{(2n + 1)\pi}{2}$. If $E \gg V_0$, there is practically no reflection, since $q^2 - k^2 \ll 2kq$, and T almost becomes one.

The maximum value of unity ($T = 1$) occurs when $\sin 2qa = 0$, or $2qa = n\pi$, that is, for the energies given by

$$E_R = \frac{\hbar^2 q^2}{2m} - V_0 = n^2\frac{\pi^2\hbar^2}{8ma^2} - V_0$$

there is no reflection. The maxima of T are called resonances. A low energy particle scattered by a strong potential well shows a series of sharp maxima, the transmission coefficient being small almost everywhere except at these points. Note that the resonance condition may be written as $\lambda = \dfrac{2\pi}{q} = \dfrac{4a}{n}$, which describes a Fabry-Perot interferometer.

(d)   The Potential Barrier and The "Tunnel" Effect

We now study the important case of the motion of a particle of mass m in a potential $V(x)$

$$V(x) = \begin{cases} 0, & x < -a \\ V_0\ (> 0), & -a < x < a \\ 0, & a < x \end{cases}. \tag{283}$$

This rectangular (or, square) barrier potential supports no bound states ($E < 0$). The problem is a one-dimensional scattering problem in which a particle with energy $E > 0$ interacts with a potential on a given interval and is free on both sides of the potential. In the present case, we consider two cases — $E > V_0$ and $E < V_0$. In case $E < V_0$, classically a particle arriving at the barrier will be totally reflected back and there will be no penetration of the barrier. In case $E > V_0$, classically a particle will emerge on the other side so that there is no reflection but total transmission takes place.

We consider the case $E < V_0$. Assuming that the particle is incident on the potential barrier from the left, we write the solutions of the Schrödinger equation for $E < V_0$ in the three regions as

$$u(x) = \begin{cases} e^{ikx} + re^{-ikx}, & x < -a \\ Ae^{Kx} + Be^{-Kx}, & -a < x < a, \\ te^{ikx}, & a < x \end{cases} \tag{284}$$

where

$$k^2 = \frac{2mE}{\hbar^2}, \quad K^2 = \frac{2m(V_0 - E)}{\hbar^2}.$$

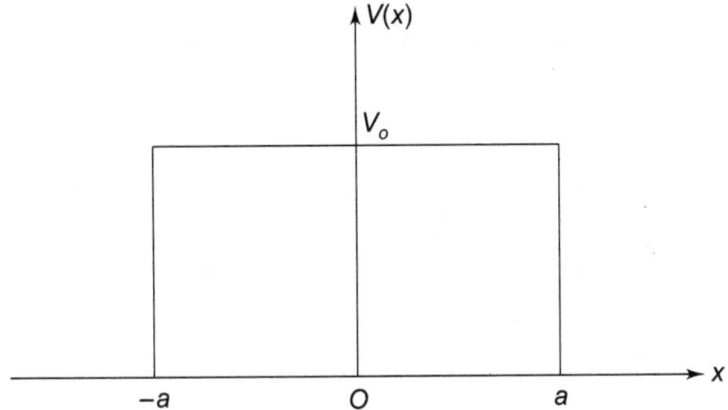

Fig. 3.7 A potential barrier

With the substitution $q \to iK$ in Eq. (281), we get the result

$$r = -ie^{-2ika} \frac{(k^2 + K^2) \sinh 2Ka}{2kK \cosh 2Ka - i(k^2 - K^2) \sinh 2Ka},$$

$$t = e^{-2ika} \frac{2kK}{2kK \cosh 2Ka - i(k^2 - K^2) \sinh 2Ka} \tag{285}$$

so that the reflection and transmission coefficients are

$$R = \frac{(k^2 + K^2)^2 \sinh^2 2Ka}{(k^2 + K^2)^2 \sinh^2 2Ka + (2kK)^2}$$

$$= \left[1 + \frac{4E(V_0 - E)}{V_0^2 \sinh^2(2Ka)} \right]^{-1},$$

$$T = \frac{(2kK)^2}{(k^2 + K^2)^2 \sinh^2 2Ka + (2kK)^2}$$

$$= \left[1 + \frac{V_0^2 \sinh^2(2Ka)}{4E(V_0 - E)} \right]^{-1}. \tag{286}$$

Note that $R + T = 1$.

We see that there is a finite probability for the transmission of the particle to the region $x > a$ (on the other side of the barrier) even when the energy of the particle is lower than the height of the barrier. This purely quantum mechanical phenomenon is known as "barrier penetration" or the "tunnel effect". Tunneling effect provides the explanation of the emission of α-particles from radioactive nuclei. We shall study α-decay of nuclei and cold emission of electrons from metals in Chapter 13 in connection with the WKB approximation where we learn to calculate approximately the transmission through a general potential barrier.

From Eq. (286), we see that $T \to 0$ as $E \to O$, and as $E \to V_0$ from below,

$$T \to \left(1 + \frac{2mV_0a^2}{\hbar^2}\right)^{-1}.$$ In this region $0 < E \le V_0$, T is a monotonically increasing function of E.

When $Ka \gg 1$ in the case of a very high and wide potential barrier,

$$\sinh 2Ka \simeq \frac{1}{2}\, e^{2Ka} \gg 1,$$

$$T = \left(\frac{4kK}{k^2 + K^2}\right)^2 e^{-4Ka}$$

$$= \frac{16E(V_0 - E)}{V_0^2}\, \exp\left\{-4\,\sqrt{2m(V_0 - E)}\,\frac{a}{\hbar}\right\}. \tag{287}$$

This is very small and is a very sensitive function of the width of the barrier and the amount by which the barrier height V_0 exceeds the energy E. We may write Eq. (287) as

$$T = \exp\left\{-4\,\sqrt{2m(V_0 - E)}\,\frac{a}{\hbar} + \log\left(\frac{16E(V_0 - E)}{V_0}\right)\right\} \tag{287a}$$

Under most circumstances the logarithmic term can be neglected, and we have

$$T = \exp\left\{-4\,\sqrt{2m(V_0 - E)}\,\frac{a}{\hbar}\right\}. \tag{288}$$

Next, we consider the case $E > V_0$. Here the wave function will be oscillatory in all three regions.

In the region $-a < x < a$,

$$u(x) = Ae^{ik'x} + Be^{-ik'x}, \tag{289}$$

where $k'^2 = 2m\,(E - V_0)/\hbar^2$.

The reflection and transmission coefficients are obtained by substituting $K \to ik'$ in Eq. (286)

$$R = \left[1 + \frac{(2kk')^2}{(k^2 - k'^2)^2 \sin^2(2k'a)}\right]^{-1}$$

$$= \left[1 + \frac{4E(E - V_0)}{V_0^2 \sin^2(2k'a)}\right]^{-1},$$

$$T = \left[1 + \frac{(k^2 - k'^2)^2 \sin^2(2k'a)}{(2kk')^2}\right]^{-1}$$

$$= \left[1 + \frac{V_0^2 \sin^2(2k'a)}{4E(E - V_0)}\right]^{-1}. \tag{290}$$

Again, $\qquad\qquad R + T = 1.$

We see that in general T is less than 1 which implies that even though $E > V_0$, there is no total transmission. If $E >> V_0$, $T = 1$, $R = 0$. So there is total transmission at very high energies and weak barriers. Total transmission ($T = 1$) occurs when $\sin(2k'a) = 0$ or $2k'a = n\pi$, $n = 1, 2, 3, \ldots$. Note the similarity with the resonance phenomena discussed earlier.

(e) The Delta Function Potential

Let us consider a particle of mass m moving in an attractive one-dimensional δ-function potential

$$V(x) = - g\delta(x), \tag{291}$$

where g is a real, positive constant. g has the dimension of energy × length. Although unphysical, this zero-range potential serves as a useful toy model, and is known as a one-dimensional hydrogen atom or a delta-hydrogen atom. The simple quantum mechanical problem in this case admits both bound and continuum solutions.

The Schrödinger equation for the δ-function potential well is

$$-\frac{\hbar^2}{2m}\frac{d^2u(x)}{dx^2} - g\delta(x)\, u(x) = Eu(x). \tag{292}$$

We consider first the case of bound state ($E < 0$). The equation to be solved is

$$\frac{d^2u(x)}{dx^2} - K^2 u(x) = -\frac{2m}{\hbar^2}\, g\delta(x)\, u(x), \tag{293}$$

where $K^2 = \dfrac{2m|E|}{\hbar^2}$.

Since $\delta(x)$ vanishes everywhere except at $x = 0$, the solution must satisfy $\dfrac{d^2 u(x)}{dx^2} - K^2 u(x) = 0$ everywhere (except at $x = 0$). Assuming $u(x)$ vanishes at $x \to \pm \infty$, we have

$$u(x) = \begin{cases} Ae^{Kx}, & x < 0 \\ Be^{-Kx}, & x > 0 \end{cases} \tag{294}$$

Now, $u(x)$ is continuous at $x = 0$, and so $A = B$, and

$$u(x) = \begin{cases} Ae^{Kx}, & x \le 0 \\ Ae^{-Kx}, & x \ge 0 \end{cases}. \tag{295}$$

We integrate the Schrödinger equation (292) from $-\epsilon$ to $+\epsilon$, and then take the limit $\epsilon \to 0$:

$$\lim_{\epsilon \to 0} \left(\frac{du}{dx}\bigg|_{+\epsilon} - \frac{du}{dx}\bigg|_{-\epsilon} \right) + \frac{2m}{\hbar^2} g u(0) = 0. \tag{296}$$

Using Eq. (295), we get

$$-2KA + \frac{2m}{\hbar^2} gA = 0,$$

or,
$$K = \frac{mg}{\hbar^2}. \tag{297}$$

Since $K = \sqrt{\dfrac{2m|E|}{\hbar^2}}$ we have $\dfrac{mg}{\hbar^2} = \sqrt{\dfrac{2m|E|}{\hbar^2}}$, the allowed energy is

$$E = -\frac{mg^2}{2\hbar^2}. \tag{298}$$

The energy is negative, and there is only one bound state. We normalize $u(x)$:

$$1 = \int_{-\infty}^{\infty} |u(x)|^2 \, dx = 2|A|^2 \int_0^{\infty} e^{-Kx} \, dx,$$

hence $A = \sqrt{K}$. The normalized wave function is

$$u(x) = \sqrt{\frac{mg}{\hbar^2}} \, e^{-mg|x|/\hbar^2}. \tag{299}$$

To obtain the continuum solutions for positive energy ($E > 0$), we assume that a particle is incident from the left. The solutions may be written as

$$u(x) = \begin{cases} e^{ikx} + re^{-ikx}, & x < 0 \\ te^{ikx}, & x > 0 \end{cases} \tag{300}$$

where we have chosen the normalization so that the coefficient of the first term is one. The continuity of $u(x)$ at $x = 0$ gives

$$1 + r = t \tag{301}$$

and

$$\left.\frac{du}{dx}\right|_+ - \left.\frac{du}{dx}\right|_- = -\frac{2mg}{\hbar^2}\, u(0)$$

or,

$$ik(t - 1 + r) = -\frac{2mg}{\hbar^2}\,(1 + r) \tag{301a}$$

whence

$$r = \frac{\dfrac{i\alpha}{k}}{1 - \dfrac{i\alpha}{k}},$$

$$t = \frac{1}{1 - \dfrac{i\alpha}{k}} \tag{302}$$

with

$$\alpha = \frac{mg}{\hbar^2}.$$

The reflection and transmission coefficients are

$$R = |r|^2 = \frac{\dfrac{\alpha^2}{k^2}}{1 + \dfrac{\alpha^2}{k^2}} = \frac{\alpha^2}{k^2 + \alpha^2} = \frac{1}{1 + \dfrac{2E\hbar^2}{mg^2}},$$

$$T = \frac{1}{1 + \dfrac{\alpha^2}{k^2}} = \frac{k^2}{k^2 + \alpha^2} = \frac{1}{1 + \dfrac{mg^2}{2E\hbar^2}} \tag{303}$$

so that

$$R + T = 1.$$

We note that if we consider a delta potential barrier obtained by changing the sign of g in the previous case, R and T remain the same. Of course, there is no bound state.

We now write the continuum eigenfunctions corresponding to a particle incident from the left

$$\psi_k(x) = \begin{cases} \dfrac{1}{\sqrt{2\pi}}\left[e^{ikx} - \dfrac{1}{1 + \dfrac{i\hbar^2 k}{mg}}\, e^{-ikx}\right] & \text{for } x < 0 \\[4ex] \dfrac{1}{\sqrt{2\pi}}\, \dfrac{i\hbar^2 k/mg}{1 + \dfrac{i\hbar^2 k}{mg}}\, e^{ikx} & \text{for } x > 0 \end{cases} \tag{304}$$

with $k > 0$.

These eigenfunctions are δ-function normalized:

$$\langle \psi_k | \psi_{k'} \rangle = \delta(k - k'). \tag{305}$$

The transmission amplitude $t(E)$ has a pole at $i\,\dfrac{\alpha}{k} = 1$, or $\dfrac{img}{k\hbar^2} = 1$, or,

$p\hbar = img$. Since p is imaginary, this pole occurs for negative energy. At the value $k = \dfrac{img}{\hbar^2}$, there is no incident wave but only an outgoing wave. This imaginary value gives a physically admissible solution. At large distance the amplitude decreases to zero and the solution corresponds to a bound state which occurs at a negative energy $E = \dfrac{\hbar^2 k^2}{2m} = -\dfrac{mg^2}{2\hbar^2}$.

Alternatively, we can write the continuum eigenfunctions as even-parity and odd-parity standing waves. For one dimensional scattering we define the polar variables (Liprin (1973))

$$r = |x|$$
$$\theta = \begin{cases} 0 & \text{if } x > 0 \\ \pi & \text{if } x < 0 \end{cases} \tag{306}$$

so that the solution Eq. (300) can be written as

$$u(x) = e^{ikx} + \xi(\theta)e^{ikr} \tag{307}$$

where

$$\xi(0) = t - 1,$$
$$\xi(\pi) = r. \tag{308}$$

We define the one-dimensional scattering amplitude as

$$f(\theta) = \frac{\xi(\theta)}{ik}. \tag{309}$$

The odd-parity solution vanishes at the origin and is written as

$$u_k^{\text{odd}} = \sin\,(kx \pm \delta_1) \tag{310}$$

the $\pm$ sign correspond to $x > 0$, $x < 0$, and $\delta_1 = 0$ for a δ-potential.

The even-parity solution is

$$u_k^{\text{even}} = \cos\,(kx \pm \delta_0), \tag{310a}$$

the $\pm$ sign correspond to $x > 0$, $x < 0$.

But

$$\left(\frac{du_k^{\text{even}}}{dx}\bigg|_{0+} - \frac{du_k^{\text{even}}}{dx}\bigg|_{0-} \right) = -2k \sin \delta_0$$
$$= \frac{-2mg}{\hbar^2} \cos \delta_0, \tag{311}$$

whence the phase shift for the even-parity solution for the δ-function potential is

$$\tan \delta_0 = \frac{mg}{\hbar^2 k}. \tag{312}$$

The phase shifts depend upon the potential and these states differ from the free-particle parity eigenstates by the phase shifts.

The S-matrix relates the amplitudes of the outgoing and incoming waves. The S-matrix elements are

$$e^{2i\delta_0} = \frac{2k + i2mg/\hbar^2}{2k - i2mg/\hbar^2}$$

and

$$e^{2i\delta_0} = 1. \tag{313}$$

The scattering amplitude is

$$f(\theta) = \frac{1}{2ik}\left(e^{2i\delta_0} - 1\right)$$

$$= \frac{2mg/\hbar^2}{k[2k - i2mg/\hbar^2]}, \tag{314}$$

and

$$\xi(\theta) = \left[\frac{2mg/\hbar^2}{2k - i\dfrac{2mg}{\hbar^2}}\right]. \tag{315}$$

This gives the solution to the scattering problem.

3.9 THE TRANSFER MATRIX

In this section we develop a transfer matrix formalism for dealing with one dimensional quantum mechanical problems. The scattering properties of a potential barrier or well is characterized by a transfer matrix which relates the flux of one side to that on the other. The transfer matrix is closely related to the scattering or S matrix which relates the incoming flux to the outgoing flux. In the general theory of collisions S-matrix plays a central role. For one dimensional problems the transfer matrix is, however, more convenient to use owing to its multiplicative property.

In exploring one-dimensional scattering problems, the transfer matrix formalism has proved to be of great importance. A potential barrier (or well) is represented by a transfer matrix belonging to one of the four homomorphic, noncompact groups $SU(1, 1)$, $SO(2, 1)$, $Sp(2, R)$, and $SL(2, R)$. (see Peres (1983)). This approach has found applications in the study of electron transmission and electron states in disordered media, quasiperiodic and aperiodic media, tunneling in superlattices, quantum transport, etc.

Several methods may be used to introduce the transfer matrix to represent a potential barrier of finite range. In different applications one of these may turn out to be more convenient.

We introduce the transfer matrix which connects the wave function and its derivative at two ends of a potential. We consider the motion of a particle of mass m with energy E in an arbitrary shaped single-valued, real potential $V(x)$, defined on the interval $x_n < x < x_{n+}$. Such a potential is called localized.

The eigenvalue equation is

$$\left(-\frac{\hbar^2}{2m}\frac{d^2}{dx^2} + V(x)\right)\psi(x) = E\psi(x). \tag{316}$$

The general solution is

$$\psi(x) = a_1\phi_1(x) + a_2\phi_2(x), \tag{317}$$

where ϕ and ϕ_2 are two linearly independent solutions. Also,

$$\psi'(x) = a_1\phi'_1(x) + a_2\phi'_2(x) \tag{318}$$

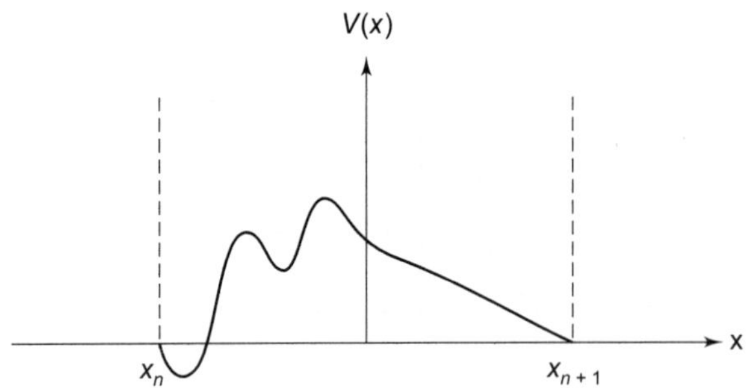

Fig. 3.8 An arbitrary shaped potential on the interval $x_n < x < x_{n+1}$

Now, we may write

$$\Psi_{n+1} = T\Psi_n, \tag{319}$$

where T is the transfer matrix and Ψ_n is given as a two-component column vector

$$\Psi_n = \begin{pmatrix} \psi(x_n) \\ \psi'(x_n) \end{pmatrix}. \tag{320}$$

T is a 2×2 matrix having the following elements

$$T_{11} = \frac{1}{W}\begin{vmatrix} \phi_1(x_{n+1}) & \phi'_1(x_n) \\ \phi_2(x_{n+1}) & \phi'_2(x_n) \end{vmatrix},$$

$$T_{12} = -\frac{1}{W}\begin{vmatrix} \phi_1(x_{n+1}) & \phi_1(x_n) \\ \phi_2(x_{n+1}) & \phi_2(x_n) \end{vmatrix},$$

$$T_{21} = \frac{1}{W} \begin{vmatrix} \phi_1'(x_{n+1}) & \phi_1'(x_n) \\ \phi_2'(x_{n+1}) & \phi_2'(x_n) \end{vmatrix},$$

$$T_{22} = -\frac{1}{W} \begin{vmatrix} \phi_1'(x_{n+1}) & \phi_1(x_n) \\ \phi_2'(x_{n+1}) & \phi_2(x_n) \end{vmatrix}, \tag{321}$$

where

$$W = \begin{vmatrix} \phi_1 & \phi_2 \\ \phi_1' & \phi_2' \end{vmatrix} \tag{322}$$

is the Wronskian of the two independent solutions.

For convenience, we may choose ϕ_1 and ϕ_2 at x_n as a real basis set

$$\phi_1(x_n) = 0, \quad \phi_1'(x_n) = 1$$

$$\phi_2(x_n) = 0, \quad \phi_2'(x_n) = 1$$

The determinant of T is unity.

The transfer matrix has the property that the state of the system at x_n represented by Ψ_n is transferred to the next state Ψ_{n+1} through the matrix T.

Next we introduce an alternative form of the transfer matrix. We consider a localized potential $V(x)$, defined in the region (x_n, x_{n+1}). The solution of the eigenvalue equation is

$$\psi(x) = \begin{cases} Ae^{ikx} + Be^{-ikx}, & x \le x_n \\ Ce^{ikx} + De^{-ikx}, & x \ge x_{n+1} \end{cases} \tag{323}$$

We relate the coefficients C, D to A, B by means of a 2×2 matrix R with complex elements

$$v = Ru, \tag{324}$$

where $v = \begin{pmatrix} C \\ D \end{pmatrix}$, $u = \begin{pmatrix} A \\ B \end{pmatrix}$ are the column vectors. The matrix R has several interesting properties. (Erdös and Herndon (1982)).

(i) R is independent of A and B (or, C and D).

(ii) The matrix elements satisfy

$$R_{22} = R_{11}^* \tag{325}$$

$$R_{21} = R_{12}^* \tag{326}$$

and

$$\det R = |R_{11}|^2 - |R_{12}|^2 = 1 \tag{327}$$

Clearly, R is completely determined by three real parameters which depend only on the potential V and the energy E (or, k).

We invoke the time-reversal invariance and flux conservation. T-invariance gives that $\psi^*(x)$ is also another solution, given that $\psi(x)$ is a solution. We get

$$u' = \sigma_x u^*,$$
$$v' = \sigma_x v^*, \tag{328}$$

where the primed quantity denotes the column vector for the time-reversed solution, $\psi^*(x)$ and $\sigma_x = \begin{pmatrix} 0 & 1 \\ 1 & 0 \end{pmatrix}$. Eq. (324) yields

$$v' = Ru, \tag{329}$$

since R is T-invariant. Complex conjugation of Eq. (324) yields

$$v^* = R^* u^*. \tag{330}$$

From these equations,

$$R^* u^* = \sigma_x^{-1} R\sigma_x u^*$$

or,
$$R^* = \sigma_x^{-1} R\sigma_x. \tag{331}$$

Eq. (325) immediately follows.

The probability current density is

$$j = \frac{\hbar}{2mi}\left(\psi^* \frac{d\psi}{dx} - \frac{d\psi^*}{dx} \psi\right). \tag{332}$$

Flux conservation gives

$$u^\dagger \sigma_z u = v^\dagger \sigma_z v = u^\dagger R^\dagger \sigma_z Ru, \tag{333}$$

using Eq. (324) with $\sigma_z = \begin{pmatrix} 0 & 1 \\ 1 & 0 \end{pmatrix}$. Hence

$$R^\dagger \sigma_z R = \sigma_z. \tag{334}$$

This equation gives

$$|R_{11}|^2 - |R_{21}|^2 = 1,$$
$$|R_{12}|^2 - |R_{22}|^2 = -1,$$
$$R_{11}R_{12}^* = R_{21}R_{22}^*,$$

whence

$$\det R = 1. \tag{335}$$

Therefore, we may write the general form of the transfer matrix as

$$M = \begin{pmatrix} \alpha & \beta \\ \gamma & \delta \end{pmatrix}, \tag{336}$$

with $\gamma = \beta^*$, $\delta = \alpha^*$, and $|\alpha|^2 - |\beta|^2 = 1$. \hfill (337)

If the potential is reflection symmetric, i.e., $V(-x) = V(x)$, then β is imaginary.

These properties follow from the time reversal invariance and the conservation of probability. The matrix M is pseudounitary and unimodular, i.e.,

$$M^{\dagger}\sigma_z M = \sigma_z, \quad \det M = 1. \tag{338}$$

The potential barrier is represented by a transfer matrix M belonging to the two-dimensional representation of the group $SU(1, 1)$.

We may denote the amplitudes of the transmitted and reflected waves by t and r, respectively (Fig. 3.9). For an incident beam of unit amplitude, we see that we can write M in the form

$$M = \begin{pmatrix} \dfrac{1}{t} & \dfrac{r}{t} \\ \dfrac{r^*}{t^*} & \dfrac{1}{t^*} \end{pmatrix}. \tag{339}$$

Another useful representation of the transfer matrix is as follows. We write

$$M = \begin{pmatrix} \alpha & \beta \\ \beta^* & \alpha^* \end{pmatrix}$$

with $\quad |\alpha|^2 - |\beta|^2 = 1.$

$$= \begin{pmatrix} \sqrt{1+\rho}\, e^{-i(\mu+\nu)} & \sqrt{\rho}\, e^{-i(\mu-\nu)} \\ \sqrt{\rho}\, e^{i(\mu-\nu)} & \sqrt{1+\rho}\, e^{i(\mu+\nu)} \end{pmatrix}$$

$$= \begin{pmatrix} e^{-i\mu} & 0 \\ 0 & e^{i\mu} \end{pmatrix} \begin{pmatrix} \sqrt{1+\rho} & \sqrt{\rho} \\ \sqrt{\rho} & \sqrt{1+\rho} \end{pmatrix} \begin{pmatrix} e^{-i\nu} & 0 \\ 0 & e^{i\nu} \end{pmatrix}, \tag{336a}$$

where the real parameters ρ, μ and ν are called Bargmann parameters and

$$-\pi \leq \mu, \nu < \pi$$

$$0 \leq \rho < \infty$$

The advantage of this parametrization is that the parameter ρ is the "dimensionless" resistance of a system.

The importance of using the transfer matrix is its multiplicative property. If we consider the one dimensional scattering by a succession of N scatterers, then the total transfer matrix for the system can be written as

$$M = \prod_{i=1}^{N} M_i, \tag{340}$$

where M_i is the transfer matrix of the ith scatterer.

S-matrix expresses the outgoing amplitudes in terms of the incoming amplitudes:

$$\begin{pmatrix} B \\ C \end{pmatrix} = S \begin{pmatrix} A \\ D \end{pmatrix}. \tag{341}$$

S-matrix may be written in terms of the reflection amplitude r and the transmission amplitude t as

$$S = \begin{pmatrix} r & t \\ t & r \end{pmatrix}. \tag{342}$$

The scattering properties of a scatterer is completely described by the unitary S-matrix. We note that for elastic scattering

$$|r|^2 + |t|^2 = 1. \tag{343}$$

Of course, S and M are related.

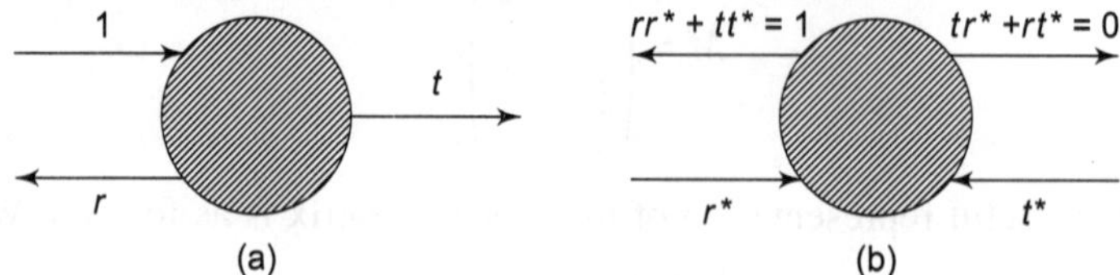

Fig. 3.9   (a) Definition of r, t; (b) Time-reversed state of (a)

We consider the transmission through a set of N localized potentials. Each potential is represented by a transfer matrix. In defining r, t we have taken the center of the potential as the origin. Here, between two successive potentials there is a region of constant potential which we take as zero potential region ($V = 0$). The wave function at any point between the cells can be expressed as

$$Ae^{-ikx} + Be^{-ikx}.$$

Now, it is convenient to take as the origin the point midway between the scattering centers. Clearly, there is an extra path length a (Fig. 3.10)

We can relate the coefficients of the wave function on the right and left of the ith potential

$$\begin{pmatrix} A_{i+1} \\ B_{i+1} \end{pmatrix} = M_i \begin{pmatrix} A_i \\ B_i \end{pmatrix}, \tag{344}$$

where M_i is the transfer matrix of the ith potential. The transfer matrix of the set of N potentials is given by

$$M^{(N)} = M_N \cdots M_2 M_1, \tag{345}$$

where M_i is the transfer matrix of the ith potential. If all the potentials are identical, then

$$M^{(N)} = M^N. \tag{346}$$

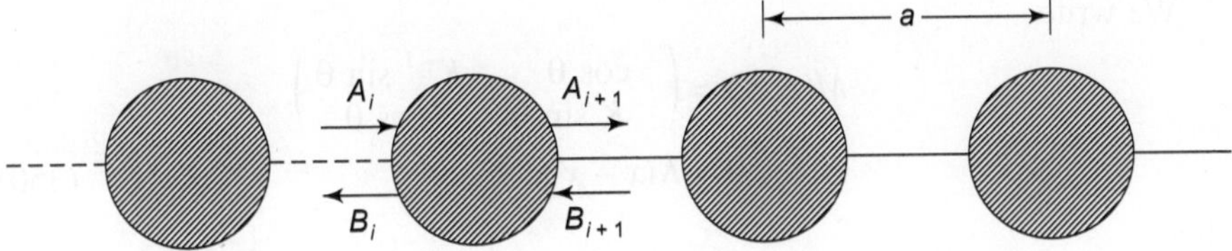

Fig. 3.10   Wave propagation along a linear chain of scatterers

Here, the transfer matrix M_i is the previous expression of the transfer matrix multiplied by a translation matrix $\begin{pmatrix} e^{ika} & 0 \\ 0 & e^{-ika} \end{pmatrix}$. Some authors incorporate this into the expression of the transfer matrix, while others use it separately. It can be introduced into the transfer matrix by the transformation $r \to re^{ika}$, $t \to te^{ika}$ (Heine (1970)).

Notice that the redefined transfer matrix is unimodular and satisfies all the properties as the earlier one.

Let us construct the transfer matrix for a single δ-function potential.

$$V(x) = g\delta(x). \tag{347}$$

Using the results from Eq. (302) we get from Eq. (339), the transfer matrix

$$M = \begin{pmatrix} 1 + \dfrac{img}{\hbar^2 k} & -\dfrac{img}{\hbar^2 k} \\[2ex] \dfrac{img}{\hbar^2 k} & 1 - \dfrac{img}{\hbar^2 k} \end{pmatrix} \tag{348}$$

The scattering matrix in this case may be written as

$$S = \begin{vmatrix} -\dfrac{\dfrac{img}{\hbar^2 k}}{1 + \dfrac{img}{\hbar^2 k}} & \dfrac{1}{1 + \dfrac{img}{\hbar^2 k}} \\[4ex] \dfrac{1}{1 + \dfrac{img}{\hbar^2 k}} & -\dfrac{\dfrac{img}{\hbar^2 k}}{1 + \dfrac{img}{\hbar^2 k}} \end{vmatrix}. \tag{349}$$

A simple pole of the S-matrix corresponds to a bound state. Here, $\det S = 0$ gives $\dfrac{m^2 g^2}{\hbar^2 k^2} = -1$ whence $E_b = -\dfrac{mg^2}{2\hbar^2}$. Clearly, a delta potential well of strength g supports only one bound state at the energy E_b.

Next we construct the transfer matrix according to the representation given in Eqs. (320) – (323). Let us construct the transfer matrix for a constant potential V_0.

We write

$$M(x, x') = \begin{pmatrix} \cos\theta & K^{-1}\sin\theta \\ -K\sin\theta & \cos\theta \end{pmatrix},$$
$$\theta = K(x - x'), \tag{350}$$

where

$$K = \begin{cases} \left[\dfrac{2m}{\hbar^2}(E - V_0)\right]^{\frac{1}{2}}, & E > V_0, \\[4mm] i\left[\dfrac{2m}{\hbar^2}(V_0 - E)\right]^{\frac{1}{2}}, & E < V_0 \end{cases} \tag{351}$$

For a δ-function potential of strength g the transfer matrix $M(x, x')$, which transfers from x' to x, is given by

$$M(x, x') = \begin{pmatrix} \cos(k\Delta) & k^{-1}\sin(k\Delta) \\ -k\sin(k\Delta) & \cos(k\Delta) \\ +g_0\cos(k\Delta) & +g_0 k^{-1}\sin(k\Delta) \end{pmatrix}, \tag{352}$$

where $\Delta = x - x'$, $k^2 = \dfrac{2m}{\hbar^2}E$ and $g_0 = \dfrac{2m}{\hbar^2}g$.

These transfer matrices belong to $SL(2, R)$.

As an application of the transfer matrix approach, we consider the transmission properties of a one-dimensional system of N non-overlapping equispaced identical potentials (symmetric) $v(x)$:

$$V(x) = \sum_{n=1}^{N} v(x - na). \tag{353}$$

Scattering by each potential may be described by a transfer matrix. With use of unitarity and time reversal invariance (we are not considering magnetic interactions), the transfer matrix associated with a potential is written as

$$M = \begin{pmatrix} \dfrac{1}{t} & -\dfrac{r}{t} \\[3mm] \dfrac{r}{t} & \dfrac{1}{t^*} \end{pmatrix}, \tag{354}$$

where the transmission and reflection amplitudes t and r respectively are functions of k with $E = \dfrac{\hbar^2}{2m}k^2$, the incident energy. M is a pseudounitary, unimodular matrix.

For our purpose, we use the following form of the transfer matrix (where we have used the same symbol for the matrix)

$$M = \begin{pmatrix} \omega & -\sigma \\ \sigma & \omega^* \end{pmatrix}, \tag{355}$$

where $\sigma = \dfrac{\rho}{\tau}$ and $\omega = \dfrac{1}{\tau}$ with $\rho = re^{ika}$ and $\tau = te^{ika}$.

The combined transfer matrix for two successive potentials separated by a distance a is obtained by multiplying two transfer matrices together. This gives, σ_2, the total σ, and ω_2, the total ω, as follows

$$\sigma_2 = \sigma(\omega + \omega^*),$$

or,

$$\left(\frac{r}{t}\right)_2 = \left(\frac{r}{t}\right)_1 \left(\frac{e^{-ika}}{t_1} + \frac{e^{ika}}{t_1^*}\right), \tag{356}$$

and

$$\omega_2 = (\omega^2 - \sigma^2)$$

or,

$$t_2^{-1} = t_1^{-2}\, e^{ika} + \left|\frac{r}{t}\right|_1^2 e^{ika}. \tag{356a}$$

Such recursion relations are important in the study of a hierarchical, or self-similar Schrödinger equation in one dimension.

Continuing with this manner, we see that the scattering by the series of N potentials can be described by a total transfer matrix which is given by

$$M_N = M^N, \tag{357}$$

where M^N is the Nth power of M.

Now, the problem is reduced to finding the Nth power of a 2×2 unimodular matrix. The answer is provided by the Cayley-Hamilton identity which states that for a matrix A belonging to $SL\,(2,\,C)$, the nth power of A is given by

$$A^n = U^{(2)}_{n-1}\,(X)A - U^{(2)}_{n-2}\,(X)I, \tag{358}$$

where I is the unit 2×2 matrix, $U^{(2)}_n\,(X)$ is the Chebyshev polynomial of the second kind of order n, and $X = \dfrac{1}{2}\,Tr(A) \cdot U^{(2)}_n\,(X)$ satisfies the recurrence relation $U^{(2)}_{n+1}\,(X) - 2XU^{(2)}_n\,(X) - U^{(2)}_{n-1}\,(X) = 0$ and $U^{(2)}_0 = 1,\ U^{(2)}_1 = 2X,\ U^{(2)}_2 = 4X^2 - 1$, etc. Now,

$$M^N = \begin{pmatrix} U^{(2)}_N - \omega^* U^{(2)}_{N-1}\,(X) & -\sigma U^{(2)}_{N-1}\,(X) \\[2mm] \sigma U^{(2)}_{N-1}\,(X) & U^{(2)}_N - \omega U^{(2)}_{N-1}\,(X) \end{pmatrix} \tag{359}$$

with

$$X = \frac{1}{2}\,(\omega + \omega^*) = Re\,\omega.$$

r and t are $\dfrac{\pi}{2}$ out of phase, so that we may write

$$t = |t|e^{i\theta},$$

$$r = |r|e^{i\theta}, \tag{360}$$

where θ is the phase shift. Thus,

$$X = |t|^{-1} \cos\,(ka + \theta). \tag{361}$$

From Eq. (359),

$$\omega_N = U_N^{(2)}(X) - \omega^* U_{N-1}^{(2)}(X)$$
$$= U_N^{(1)}(X) + iYU_{N-1}^{(2)}(X), \tag{362}$$

and

$$\sigma_N = \sigma U_{N-1}^{(2)}(X), \tag{363}$$

where $U_N^{(1)}(X)$ is the Chebyshev polynomial of the first kind of order n and $Y = Im\omega \cdot U_n^{(1)}(X)$ and $U_n^{(2)}(X)$ are related as $U_n^{(1)}(X) = U_n^{(2)}(X) - XU_{n-1}^{(2)}(X)$.

Also, $U_0^{(1)} = 1$, $U_1^{(1)} = X$, $U_2^{(1)} = 2X^2 - 1$, etc.

$U_n^{(1)}(X)$ and $U_n^{(2)}(X)$ can be expressed in terms of trigonometric functions:

$$U_n^{(1)}(X) = \cos(n\cos^{-1}X),$$
$$U_n^{(2)}(X) = \frac{\sin[(n+1)\cos^{-1}X]}{\sin(\cos^{-1}X)}. \tag{364}$$

The wave vector χ is related to the energy in the following way

$$\cos\chi a = X = |t|^{-1}\cos(ka + \delta). \tag{365}$$

For N large we note that the Bloch condition (see next section) gives a constant phase factor $e^{-i\chi a}$ in each cell. This may be looked upon as the requirement for periodic boundary condition.

Therefore, the transmission amplitude t_N and the reflection amplitude r_N for N symmetric equispaced (spacing a) identical potentials are given by

$$t_N = \frac{e^{-i(N+1)ka}}{\cos N\chi a - i|t|^{-1}\sin(ka + \delta)\dfrac{\sin N\chi a}{\sin \chi a}}$$

$$r_N = \left(\frac{r}{t}\right)t_N \frac{\sin N\chi a}{\sin \chi a} \tag{366}$$

using periodic boundary condition for large N.

The transmission coefficient

$$T_N = |t_N|^2$$

and the reflection coefficient $R_N = |r_N|^2$ are

$$T_N = \frac{1}{1 + \left|\dfrac{r}{t}\right|^2 \dfrac{\sin^2 N\chi a}{\sin^2 \chi a}}, $$

$$R_N = T_N \left|\frac{r}{t}\right|^2 \frac{\sin^2 N\chi a}{\sin^2 \chi a}. \tag{367}$$

Of course, $\qquad\qquad R_N + T_N = 1.$

Laandauer (1970) has shown that for a one-dimensional conductor the resistance (in unit of $2\pi\hbar/e^2$) is given by

$$\mathcal{R} = \frac{R}{T}. \tag{368}$$

The resistance of the one-dimensional chain is

$$\mathcal{R}_N = \frac{R_N}{T_N} = |\sigma|^2 \left[U^{(2)}_{N-1}(X) \right]^2$$

or,
$$\mathcal{R}_N = T_N^{-1} - 1 = \left|\frac{r}{t}\right|^2 \frac{\sin^2 N\chi a}{\sin^2 \chi a}.$$

$$= \mathcal{R}_1 \frac{\sin^2 N\chi a}{\sin^2 \chi a}, \tag{369}$$

where $\mathcal{R}_1$ is the resistance of the elementary cell.

In terms of the phase shifts corresponding to the "even parity" and "odd parity" solutions, δ_0 and δ_1, respectively, the reflection and transmission amplitudes may be expressed as follows

$$r = i\, e^{i(\delta_0 + \delta_1)} \sin (\delta_0 - \delta_1),$$

$$t = e^{i(\delta_0 + \delta_1)} \cos (\delta_0 - \delta_1). \tag{370}$$

From Eqs. (366) and (370), we obtain

$$\tan (\delta_0^{(N)} - \delta_1^{(N)}) = \tan (\delta_0 - \delta_1)\, U^{(2)}_{N-1}(X),$$

$$e^{-i(\delta_0^{(N)} - \delta_1^{(N)} + (N+1)ka)} \sec (\delta_0^{(N)} - \delta_1^{(N)})$$

$$= U^{(1)}_N(X) + iYU^{(2)}_{N-1}(X) \tag{371}$$

where now

$$X = \cos (\delta_0 + \delta_1 + ka) \cos (\delta_0 - \delta_1),$$

$$Y = \sin (\delta_0 + \delta_1 + ka) \cos (\delta_0 - \delta_1). \tag{372}$$

For the calculation of the phase shifts for scattering by any number of potential bumps, Eq. (371) are very useful.

To illustrate the formalism, we consider the simple case of a one dimensional lattice of N identical delta potentials, each of strength g, the lattice constant being a. Then for a single delta potential

$$t^{-1} = 1 + \frac{img}{\hbar^2 k} \quad \text{and} \quad \frac{r}{t} = -\frac{img}{\hbar^2 k},$$

and from Eq. (370), we get the well-known results $\tan \delta_0 = -\dfrac{mg}{\hbar^2 k}$ and $\delta_1 = 0$.

For two identical delta potentials, we get

$$\tan \delta_0^{(2)} = \frac{\cos ka + 1}{\sin ka - \dfrac{2\hbar^2 k}{mg}}, \tag{373}$$

$$\tan \delta_1^{(2)} = \frac{\cos ka - 1}{\sin ka + \dfrac{2\hbar^2 k}{mg}}. \tag{373a}$$

From Eq. (369), the resistance of the lattice of N identical delta potentials is

$$\mathcal{R}_N = \mathcal{R}_1 \frac{\sin^2 \chi L}{\sin^2 \chi a}, \tag{374}$$

where $\mathcal{R}_1 = \dfrac{m^2 g^2}{\hbar^4 k^2}$, $L = Na$ is the length of the lattice and $\cos \chi a = \cos ka - \dfrac{mg}{\hbar^2 k}$ $\times \sin ka$ with χ the quasimomentum. In the allowed zone ($|\cos \chi a| < 1$) of a periodic potential, the typical oscillatory behavior of T_N is easily seen from the expression for T_N. $\mathcal{R}_N$ does not increase monotonically with N and in the allowed region, $\mathcal{R}_N \to 0$ as $N \to \infty$. In the forbidden or energy gap region ($|\cos \chi a| > 1$), the resistance becomes very large (infinite) since in this case $U_N^{(2)}$ diverges as N grows. Transitions occur at $\chi a = m\pi$ ($m = 0, \pm 1, \pm 2, \dots$).

Using the properties of Chebyshev polynomials, we obtain the following interesting relations

$$\sigma_{N+1} - (2 \cos \chi a)\, \sigma_N + \sigma_{N-1} = 0, \tag{375}$$

$$\lim_{N \to \infty} \frac{\sigma_{N+1}}{\sigma_N} = \Phi, \tag{376}$$

$$\lim_{N \to \infty} \frac{\mathcal{R}_{N+1}}{\mathcal{R}_N} = \Phi^2, \tag{377}$$

where $\Phi \equiv \lim\limits_{N \to \infty} \dfrac{U_N^{(2)}(X)}{U_{N-1}^{(2)}(X)} = X + \sqrt{X^2 - 1}$, with $|X| > 1$.

3.10 THE PERIODIC POTENTIAL

We consider the motion of a particle in a periodic potential. Such potentials are produced in three dimensions by a "crystal" having a lattice structure. The nature of the eigenvalue spectrum in such a periodic potential is of great importance in the theory of conduction and insulation in solids. In a metal the positive ions form a lattice exhibiting spatial periodicity. Conduction electrons move through the periodic potential supported by the ions. We shall limit ourselves to one dimension. We consider a potential that is periodic in space

$$V(x) = V(x + a), \tag{378}$$

where a, is a fixed distance, is called the lattice constant. The Hamiltonian for an electron in this potential is

$$H = \frac{p^2}{2m} + V(x). \tag{379}$$

Let T be the translation operator

$$Tf(x) = f(x + a). \tag{380}$$

Since T commutes with H

$$[T, H] = 0, \tag{381}$$

we can choose the energy eigenfunctions to be simultaneously eigenfunctions of T and H. See Section 10.1, where we consider lattice translation as a discrete symmetry operation. An eigenfunction of T has the property

$$T\psi(x) = \lambda\psi(x),$$

or, $$\psi(x + a) = \lambda\psi(x) \tag{382}$$

λ can be expressed as e^{iKa} where K is constant (independent of x). Soon we shall see that K is real. Eq. (382) can be written as

$$\psi(x + a) = e^{iKa}\psi(x). \tag{382a}$$

If we write

$$\psi(x) = e^{iKa}u_K(x), \tag{383}$$

then it follows from Eq. (382a),

$$u_K(x + a) = u_K(x). \tag{384}$$

$u_K(x)$ is a periodic function of x with period a.

Bloch's theorem (or, Floquet's theorem) states that for a periodic potential Eq. (378), the energy eigenstates can be written as

$$\psi(x) = e^{iKa}u_K(x), \tag{385}$$

or, $$\psi(x + na) = e^{inKa}\psi(x), \quad n = 0, \pm 1, \pm 2, \ldots \tag{385a}$$

where $u_K(x)$ is periodic with the periodicity of the lattice. A wave function of this form is called a Bloch wave function which is a plane wave modulated by a periodic function with the same period as that of the lattice. Notice that the energy eigenfunctions Eq. (385) are also simultaneous eigenfunctions of the translation operator T. All e^{iKa} are eigenvalues of T. Since $e^{2\pi in} = 1$, the values of K in the interval $-\frac{\pi}{a} \leq K \leq \frac{\pi}{a}$ give all the eigenvalues of T. K is real for physically acceptable solutions. Bloch wave functions are extended wave functions.

The quantity $\hbar K$ is called the quasi-momentum of the particle. For a free particle ($V = 0$), this quantity represents the momentum of the particle. In the general case, the Bloch functions are not eigenfunctions of the momentum operator.

A real crystal is, of course, finite, extending from $x = 0$ to $L = Na$, N large. To avoid the edge effects, we impose periodic (or, cyclic) boundary conditions on the wave functions:

$$\psi(0) = \psi(L) = \psi(Na), \tag{386}$$

Bloch condition gives

$$u_K(0) = e^{iKL}u_K(L)$$
$$= e^{iKNa}u_K(Na)$$

whence

$$e^{iKNa} = 1,$$

since

$$u_K(Na) = u_K(0).$$

Therefore, $NKa = 2\pi n$ and

$$K = \frac{2\pi n}{Na} \quad (n = 0, \pm 1, \pm 2, \ldots). \tag{387}$$

Clearly, we need to solve the eigenvalue equation within a single cell. Between $-\dfrac{\pi}{a}$ and $\dfrac{\pi}{a}$, K has N independent values

$$K = \begin{cases} (2n - N)\dfrac{\pi}{L}, & N \text{ even} \\[2ex] (2n - N - 1)\dfrac{\pi}{L}, & N \text{ odd} \end{cases},$$

$$n = 1, 2, \ldots, N. \tag{388}$$

We now consider the eigenvalue equation

$$\left[-\frac{\hbar^2}{2m}\frac{d^2}{dx^2} + V(x) \right] \psi(x) = E\psi(x), \tag{389}$$

where $V(x)$ is a periodic potential.

There are no bound states. The energy spectrum is purely continuous and has a band structure, i.e., intervals of allowed energies with alternate intervals of forbidden energy (gap regions). The allowed bands are labeled by the band index $n = 1, 2, \ldots$ according to increasing energy.

The work of Kronig and Penney (1931) initiated the study of energy levels in a one-dimensional periodic lattice. The simple and exactly solvable Kronig-Penney model of a particle moving in a linear array of rectangular or delta potentials demonstrates many of the basic features of the quantum theory of periodic systems. These include the effect of Bloch's theorem on the eigenstates, the occurrence of van Hove singularities in the density of states, and the concept of effective masses of electrons and holes.

This model continues to attract attention in recent studies on the effects of impurities, electric and magnetic fields and other features in periodic structures. In the study of disordered or quasiperiodic systems, the Kronig-Penney model is often used for getting some insight into localization and other associated features.

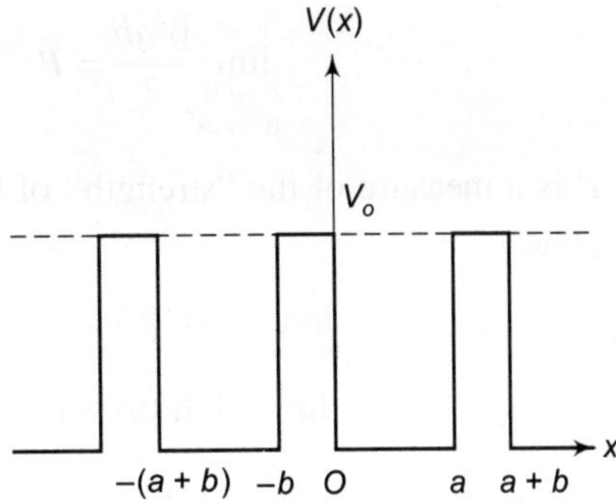

Fig. 3.11 A periodic potential with rectangular sections (the Kronig-Penney potential)

We consider the motion of electrons in an infinite array of rectangular potential structures in one dimension. In the Kronig-Penney model, the potential field (see Fig. 3.11) is a linear array of rectangular barriers, each of height V_0 and width b. In the region $0 \leq x \leq a$, the potential is zero, and in the region, $-b \leq x \leq 0$, it is V_0.

The solution of the Schrödinger equation is of the form

$$\psi_1(x) = Ae^{ikx} + Be^{-ikx},$$

$$k^2 = \frac{2mE}{\hbar^2} \tag{390}$$

in the region of zero potential, and

$$\psi_2(x) = Ce^{\beta x} + De^{-\beta x},$$

$$\beta^2 = \frac{2m(V_0 - E)}{\hbar^2} \tag{390a}$$

in the region of the potential.

According to Bloch's theorem, the solutions must be of the form

$$\psi(x) = e^{iKx}u_K(x), \tag{391}$$

where $\hbar K$ is the quasimomentum and $u_K(x)$ is a function periodic in x with the same period as the lattice, i.e., $(a + b)$. To find $u_K(x)$ we can confine ourselves to one cell of the lattice.

We obtain the dispersion relation

$$\cos K(a + b) = \frac{\beta^2 - k^2}{2k\beta} \sinh \beta b \sin ka$$

$$+ b \cosh b \cos ka. \tag{392}$$

This is a transcendental equation relating the electronic energy E to K.

The problem may be simplified by representing the potential field by a Dirac comb, a periodic chain of delta function potentials. We, therefore, pass to the limit where $b \to 0$, $V_0 \to \infty$ in such a way that $V_0 b$ remains finite, or $\beta^2 b$ stays finite. We put

$$\lim_{\substack{b \to 0 \\ \beta \to \infty}} \frac{\beta^2 ab}{2} = P. \tag{393}$$

P is a measure of the "strength" of the delta potentials.

Now,

$$\lim_{\beta b \to 0} \cosh \beta b = 1,$$

$$\lim_{\beta b \to 0} \sinh \beta b = \beta b.$$

So, in this limit, Eq. (392) becomes

$$\cos Ka = \cos ka + \lim_{\beta b \to 0} \frac{\beta}{2k} \cdot \beta b \cdot \sin ka$$

$$= \cos ka + P \frac{\sin ka}{ka}$$

$$= F(ka). \tag{394}$$

This equation is the basic equation describing the energy eigenvalues. This equation may be satisfied for real values of K whenever $F(ka)$ lies between the values -1 and $+1$. Thus only these values of ka are allowed for which $F(ka)$ falls within this range. In Fig. 3.12 we have plotted $F(ka)$ as a function of ka for $P = \dfrac{3\pi}{2}$. It is seen that there are alternate allowed and forbidden energy regions but that the gaps become smaller and smaller as the energy incrases. Within a given band almost any energy is allowed, since $Ka = \dfrac{2\pi n}{N}$, with N a very large number and n any integer.

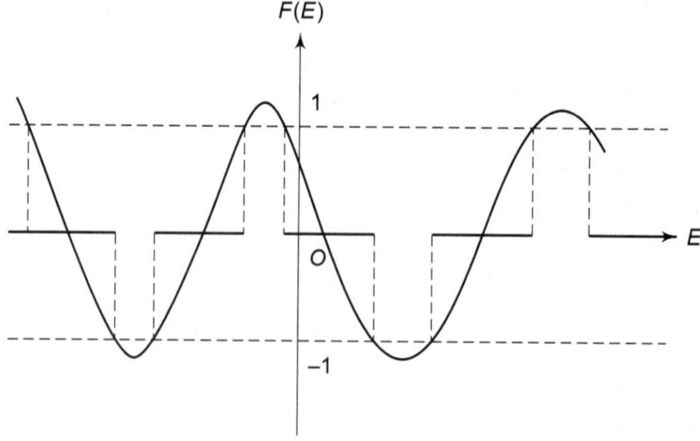

Fig. 3.12   Plot of $F(E)$ versus E. The heavy lines on the abscissa show the allowed values of E, separated by forbidden energy gaps

The boundaries of the allowed energies of ka correspond to the values $\frac{n\pi}{a}$ for K (n integer). In the limit $P \to 0$, Eq. (394) gives $ka = Ka$ whence $E = \frac{\hbar^2 K^2}{2m}$, which is the free particle solution. In the limit $P \to \infty$, the energy levels are independent of K and we can have solutions of Eq. (394) only if $\sin ka = 0$. Then $ka = n\pi$ ($n = \pm 1, \pm 2, ...$) so that $E = \frac{n^2 \pi^2 \hbar^2}{2ma^2}$. The discrete energy spectrum is the same as that of a particle in a box of length a.

The width of an energy band increases with energy. Energy bands become narrower and narrower with potential strength P and in the limit $P \to \infty$ the energy spectrum becomes discrete. On the other hand, for $P \to 0$ it changes into the free particle continuum. The band width increases as the lattice spacing decreases.

We suppose that the potential consists of a periodic array of delta potentials

$$V(x) = g \sum_{n=-\infty}^{\infty} \delta(x - na). \tag{395}$$

In this case, a traditional analysis yields

$$\cos Ka = \cos ka + \left(\frac{mga}{\hbar^2}\right) \frac{\sin ka}{ka}. \tag{396}$$

The wave function (up to normalization) is

$$\psi(x) = \sin kx + e^{-iKa} \sin k(a - x). \tag{397}$$

Hence, the probability distribution

$$\psi^*\psi = \sin^2 kx + \sin^2[k(a - x)]$$

$$+ 2\sin kx \sin [k(a - x)] \cos Ka \tag{398}$$

is periodic.

We now want to derive the band structure from the scattering properties using the transfer matrix formalism. We begin by determining TrM, where M is given in Eq. (355). M is a complex, unimodular 2×2 matrix so that $\det M = 1$. The eigenvalues are determined from the secular equation

or, $$\lambda^2 - (TrM)\lambda + 1 = 0$$

$$\lambda^2 - 2Re\left(\frac{1}{\tau}\right) \lambda + 1 = 0. \tag{399}$$

In the periodic Kronig-Penney structure, the eigenvalues must be of the form

$$\lambda_1 = e^{iKa}, \quad \lambda_1 = e^{-iKa}, \tag{400}$$

where $\hbar K$ is the crystal momentum. Therefore,

$$2 \cos Ka = TrM(E), \tag{401}$$

which is the dispersion relation. We see that the allowed solutions for a real K must satisfy the condition

$$-1 \le \frac{1}{2} TrM \le 1. \tag{402}$$

which gives rise to the band structure.

Periodic boundary conditions on the wave function impose a further constraint:

$$K = \left(\frac{2\pi}{Na}\right) \times \text{integer}. \tag{403}$$

From Eq. (355), we get the dispersion relation

$$\cos Ka = \frac{1}{2}(\omega + \omega^*) = \frac{1}{2}\left(\frac{1}{\tau} + \frac{1}{\tau^*}\right)$$

$$= \frac{1}{2}\left(\frac{e^{-ika}}{t} + \frac{e^{ika}}{t^*}\right)$$

$$= \cos Ka + \left(\frac{mga}{\hbar^2}\right)\frac{\sin ka}{ka},$$

since

$$t^{-1} = 1 + \frac{img}{\hbar^2 k}.$$

PROBLEMS

3.1 Establish equation (44).

3.2 A wave packet is represented at the time $t = 0$ by

$$\psi(x) = Ne^{\frac{i}{\hbar}p_0 x}\, e^{-\frac{|x|}{2\Delta x}}.$$

 (a) Find the normalization constant N.

 (b) Obtain $\tilde{\psi}(p_x)$, the momentum space wave function.

 (c) Calculate Δp_x, and show that $\Delta x\, \Delta p_x \ge \hbar$.

3.3 Given a wave function for a free particle of mass m at the time $t = 0$

$$\psi(x) = \begin{cases} 1, & |x| < \dfrac{1}{2} \\[2mm] 0, & |x| > \dfrac{1}{2} \end{cases}$$

how will it propagate in time, i.e., what will be the form of the wave function $\psi(x, t)$? Find $W(x, t) = \psi^*(x, t)\, \psi(x, t)$ and plot $W(x, 0)$ and $W\left(x, \dfrac{m}{\pi\hbar}\right)$ versus x.

3.4 Consider a one-dimensional system. The oscillation theorem states that if the discrete energies are placed in order of increasing magnitude $E_1 < E_2 < E_3 < ...$, then the corresponding eigenfunctions $\psi_1(x)$, $\psi_2(x)$, $\psi_3(x)$, ... will occur in increasing order of the number of their zeros, the nth eigenfunction $\psi_n(x)$ having $(n-1)$ zeros. Show that, between two consecutive zeros of $\psi_n(x)$, $\psi_{n+1}(x)$ will have at least one zero.

3.5 A particle of mass m is moving in the homogeneous gravitational field over the earth's surface, which is assumed to reflect the particle elastically. The gravitational potential is $V(x) = mgx$, where x is the height over the earth's surface. The stationary Schrödinger equation is

$$\left[\frac{d^2}{dx^2} + \frac{2m}{\hbar^2}(E - mgx) \right] u(x) = 0,$$

with $x \geq 0$. The boundary condition is elastic reflection at $x = 0$:

$$u(0) = 0.$$

Also, the condition

$$\lim_{x \to \infty} u(x) = 0$$

ensures normalizability.

Introduce the dimensionless parameter λ, and the parameter l with dimension of length by

$$\frac{2mE}{\hbar^2} = \frac{\lambda}{l^2}; \quad \frac{2m^2g}{\hbar^2} = \frac{1}{l^3},$$

and the variable

$$\xi = \frac{x}{l} - \lambda.$$

The Schrödinger equation becomes

$$\left(\frac{d^3}{d\xi^2} - \xi \right) u = 0$$

with the boundary conditions $u(-\lambda) = 0$,

$$\lim_{\xi \to \infty} u(\xi) = 0.$$

This differential equation is solved by Bessel functions of order 1/3. Determine the energy levels and the wave functions.

3.6 Consider the first Pöschl-Teller potential

$$V(x) = \frac{1}{2} V_0 \left\{ \frac{\mu(\mu - 1)}{\sin^2 \alpha x} + \frac{\lambda(\lambda - 1)}{\cos^2 \alpha x} \right\}$$

where $V_0 = \dfrac{\hbar^2\alpha^2}{m}$, α is a positive constant, and μ, λ are constants greater than one. The potential is periodic with period $\dfrac{\pi}{2\alpha}$; it is infinite at $x = 0$ and $\dfrac{\pi}{2\alpha}$. Solve the Schrödinger equation in the interval $0 \le x \le \dfrac{\pi}{2\alpha}$ with boundary conditions $u = 0$ at $x = 0$, $\dfrac{\pi}{2\alpha}$, and obtain the energy spectrum.

3.7 The modified Pöschl-Teller potential is

$$V(x) = -\frac{\hbar^2\alpha^2}{2m}\frac{\lambda(\lambda - 1)}{\cosh^2\alpha x},$$

with $\lambda > 1$.

   (a) For positive energies, find the reflection and transmission coefficients.

   (b) Find the eigenvalues for negative energies.

3.8 A particle of mass m is confined to the double potential well — two attractive square wells in one dimension each of width a, well depth $-V_0$, and separated by a distance $2x_0$. The origin of the coordinate system is placed midway between the two wells.

   (a) Obtain the transcendental equations for the allowed energies of the bound states.

   (b) Find the two lowest energies in the high barrier limit.

   (c) Write the time-dependent wave function for the ground and the first excited state energies.

   (d) What is the probability of finding a particle at any time and at any distance?

   (e) What is the frequency of tunneling?

3.9 A particle of mass m is moving in a potential

$$V(x) = \begin{cases} \infty, & x \le 0 \\ -g\delta(x - a), & x > 0 \end{cases}$$

with $g > 0$. Find a transcendental equation for the existence of bound-state energies, and discuss the dependence of bound states on a.

3.10 A particle of mass m moves in an attractive double-delta potential

$$V(x) = -V_0\left[\delta(x - a) + \delta(x + a)\right],$$

where $V_0 > 0$. Note that the potential is symmetric so one can restrict to even and odd solutions.

(a) Consider the case of negative energies. Find the wave functions of the bound states and obtain the eigenvalue equations. Show that

$$E_{\text{even}} < E_{\text{odd}}.$$

(b) Calculate the transmission coeffeicient for $V_0 < 0$.

3.11 Discuss the continuity equation in the presence of a complex-valued potential. For a particle of mass m in a potential $V = - V_0 (1 + i\xi)$ with V_0, ξ positive constants, solve the Schrödinger equation. Interpret the solutions if $\xi << 1$.

3.12 This problem concerns the Kronig-Penney model discussed in Section 3.10.

(a) Consider the motion of an electron of mass m in an infinite array of rectangular potentials in one dimension (see Fig. 3.11). Carry out the traditional analysis and obtain the dispersion relation given in Eq. (392).

(b) Assume that the potential consists of a periodic array of delta potentials. Carry out the traditional analysis and obtain the dispersion relation given in Eq. (396).

(c) The Kronig-Penney dispersion relation is

$$\frac{P}{ka} \sin ka + \cos ka = \cos Ka,$$

where $P = \dfrac{mga}{\hbar^2}$.

The density of states in one dimension is given by

$$n(E) = \frac{2}{\pi} \frac{dK}{dE}.$$

Plot n as a function of E, choosing values of g and a. Notice the van Hove singularities at the zone boundaries.

(d) The effective mass m_{eff} of electrons at the Brillonin zone boundaries $K = \dfrac{n\pi}{a}$ ($n = 0, 1, 2, ...$) is

$$\frac{1}{m_{\text{eff}}} = \frac{1}{\hbar^2} \frac{d^2E}{dk^2} \left(\frac{n\pi}{a}\right).$$

Obtain an expression for m_{eff}.

3.13 Consider a one-dimensional semi-infinite Kronig-Penney lattice of strength V. The surface is terminated with a potential barrier of strength V_0. Use the transfer matrix formalism to find the energy expression for Tamm-like surface states.

CHAPTER 4

Angular Momentum

In this chapter and the following four chapters we deal with the theory of angular momentum and related topics. Angular momentum theory plays a central role in quantum mechanics. Angular momentum considerations are important for atomic and molecular physics, nuclear physics and particle physics.

4.1 COMMUTATION RELATIONS

In classical mechanics the angular momentum of a particle about a point O is defined as

$$\vec{L} = \vec{r} \times \vec{p}, \tag{1}$$

where $\vec{r}$ is the position vector of the particle with respect to O and $\vec{p}$ is its linear momentum. We define the angular momentum operator in quantum mechanics by this same formula in terms of the position and momentum operators. In components

$$
\begin{aligned}
L_x &= yp_z - zp_y \\
L_y &= zp_x - xp_z \\
L_z &= xp_y - yp_x
\end{aligned} \tag{2}
$$

The angular momentum operator is self – adjoint

$$
\begin{aligned}
L_z^\dagger &= (xp_y)^\dagger - (yp_x)^\dagger \\
&= p_y^\dagger x^\dagger - p_x^\dagger y^\dagger \\
&= p_y x - p_x y \\
&= xp_y - yp_x \\
&= L_z
\end{aligned} \tag{3}
$$

$$L_y^\dagger = L_y$$

$$L_x^\dagger = L_x$$

Next, we evaluate the commutation relations satisfied by the angular momentum operators, with themselves, with the position operator and with the momentum operator. To evaluate these commutators we use the following two identities

$$[A, BC] = [A, B]C + B[A, C]$$

$$[AB, C] = A[B, C] + [A, C]B. \tag{4}$$

We get

$$[L_i, r_j] = i\hbar \in_{ijk} r_k \tag{5}$$

$$[L_i, p_j] = i\hbar \in_{ijk} p_k$$

and the angular momentum commutation rule

$$[L_i, L_j] = i\hbar \in_{ijk} L_k \tag{6}$$

where the sum over the repeated intex $k = 1, 2, 3$ is understood. Here $\in_{ijk}$, the totally anti-symmetric unit tensor, is defined by

$$\in_{ijk} = \begin{cases} +1, & \text{if } ijk \text{ is an even permutation of 1, 2, 3} \\ -1, & \text{if } ijk \text{ is an odd permutation of 1, 2, 3} \\ 0, & \text{if any two indices are equal} \end{cases} \tag{7}$$

In vector form, the angular momentum commutation relation is

$$\vec{L} \times \vec{L} = i\hbar\vec{L}. \tag{8}$$

We may express the operators of angular momentum in the differential operator form. Using $p_x = i\hbar \dfrac{\partial}{\partial x}$ etc. we get

$$L_x = -i\hbar \left(y \frac{\partial}{\partial z} - z \frac{\partial}{\partial y} \right)$$

$$Ly = -i\hbar \left(z \frac{\partial}{\partial x} - x \frac{\partial}{\partial y} \right) \tag{9}$$

$$Lz = -i\hbar \left(x \frac{\partial}{\partial y} - y \frac{\partial}{\partial x} \right).$$

These equations may be written in terms of the spherical polar coordinates (r, ϑ, ϕ) as

$$L_x = i\hbar \left(\sin \phi \frac{\partial}{\partial \vartheta} + \cot \vartheta \cos \phi \frac{\partial}{\partial \phi} \right).$$

$$L_y = i\hbar \left(- \cos \phi \, \frac{\partial}{\partial \vartheta} + \cot \vartheta \, \sin \phi \, \frac{\partial}{\partial \phi} \right) \tag{10}$$

$$L_z = -i\hbar \, \frac{\partial}{\partial \phi}$$

where

$$z = r \cos \vartheta$$

$$y = r \sin \vartheta \, \sin \phi \tag{11}$$

$$x = r \sin \vartheta \, \cos \phi.$$

Thus the angular momentum operators involve only the angular coordinates and do not depend at all on the *radial* coordinate r. From Eq. (10), we see that the angular momentum operators are proportional to the operators of infinitesimal rotations (see later).

The operator describing the square of the angular momentum

$$\vec{L}^2 = L_x^2 + L_y^2 + L_z^2 \tag{12}$$

commutes, according to Eq. (6), with each component of $\vec{L}$:

$$[\vec{L}^2, \vec{L}] = 0. \tag{13}$$

It is given in terms of the spherical differential operators by

$$\vec{L}^2 = - \hbar^2 \left[\frac{1}{\sin \vartheta} \frac{\partial}{\partial \vartheta} \left(\sin \vartheta \, \frac{\partial}{\partial \vartheta} \right) + \frac{1}{\sin^2 \vartheta} \frac{\partial^2}{\partial \phi^2} \right]. \tag{14}$$

We note that the Laplacian operator ∇^2, using Eq. (14) takes the following form in spherical polar coordinates

$$\nabla^2 = \frac{1}{r^2} \frac{\partial}{\partial r} \left(r^2 \frac{\partial}{\partial r} \right) - \frac{\vec{L}^2/\hbar^2}{r^2} \tag{15}$$

where $\vec{L}^2$ is the differential operator in Eq. (14).

An important consequence of the quantization of angular momentum of a system is that the components no longer commute. Therefore, the uncertainty principle does not allow us to measure simultaneously the values of all three (or even two) components of angular momentum. We know that the angular momentum vector of a classical system in a central field is fixed in space and its three components are constants of motion. But only one of them can actually be used for the description of the motion since their Poisson brackets do not vanish. In quantum mechanics the corresponding commutators do not vanish and therefore the angular momentum vector precesses around some direction in space. In an eigenstate of L_z, the angular momentum vector precesses around the z-axis such that its projection on this axis remains constant and its projection on the x and y-axis average out to zero.

4.2 EIGENVALUES AND EIGENFUNCTIONS OF THE ANGULAR MOMENTUM OPERATORS

We now find the spectra of the operators L_z and $\vec{L}^2$ and evaluate their eigenfunctions. From Eq. (13) we know that $\vec{L}^2$ commutes with each L_k, i.e.

$$[\vec{L}^2, L_k] = 0. \quad [k = 1, 2, 3] \tag{16}$$

Because L_x, L_y and L_z do not commute with each other, we can choose only one of them to be the observable to be diagonalized simultaneously with $\vec{L}^2$.

By convention, we choose this component to be L_z. Thus, there exixts a complete set of common eigenvectors of $\vec{L}^2$ and L_z.

For any state $|\Psi\rangle$

$$\langle\Psi|\,\vec{L}^2|\,\Psi\rangle = \sum_{k=1}^{3} \langle\Psi|\,L_k^2|\,\Psi\rangle = \sum_{k} (\langle\Psi|\,L_k)\,(L_k|\,\Psi\rangle) \geq 0$$

since the inner product of any vector with itself is nonnegative. Thus, if $|\Psi\rangle$ is an eigenket of $\vec{L}^2$ with eigenvalues λ, then

$$\langle\Psi\,|\vec{L}^2|\,\Psi\rangle = \lambda\,\langle\Psi|\,\Psi\rangle \geq 0$$

Therefore, we write the eigenvalues of $\vec{L}^2$ as $\hbar^2 l\,(l+1)$ with $l \geq 0$. The common and convenient way of writing the eigenvalues in this way is that $2l$ shall turn out to be an integer. We also write $\hbar_m$ to denote an eigenvalue of L_z and the eigenstates of $\vec{L}^2$ and L_z we denote as $|l, m\rangle$.

To determine the allowed values for l and m, we introduce two non hermitian operators

$$L_+ = L_x + iL_y \tag{17}$$

$$L_- = L_x - iL_y$$

These have the differential operator representations

$$L_\pm = -i\hbar e^{\pm i\phi}\left(\pm i\,\frac{\partial}{\partial\vartheta} - \cot\vartheta\,\frac{\partial}{\partial\phi}\right). \tag{18}$$

They satisfy the commutation relations

$$[L_z, L_\pm] = \pm\,\hbar L_\pm \tag{19}$$

and
$$[L_+, L_-] = 2\hbar L_z$$

Notice that

$$[\vec{L}^2, L_\pm] = 0$$

and
$$L_+\,L_- = \vec{L}^2 - L_z^2 + \hbar L_z$$

and
$$L_- L_+ = \vec{L}^2 - L_z^2 - \hbar L_z$$
$$L_-^\dagger = L_+.$$

We consider how L_z acts on $L_\pm |l, m)$

$$L_z \left(L_\pm |l, m\rangle \right) = \left([L_z, L_\pm] + L_\pm L_z \right) |l, m\rangle \tag{20}$$
$$= (m \pm 1)\hbar \left(L_\pm |l, m\rangle \right).$$

Thus $L_+ |l\ m)$ is an eigenvector of L_z, belonging to the eigenvalue $(m + 1)\hbar$, while $L_- |l, m)$ is an eigenvector of L_z belonging to the eigenvalue $(m - 1)\hbar$. since $L_\pm$ raises (lowers) L_z eigenvalues by one unit of $\hbar$, $L_\pm$ are called the raising (lowering) operators or ladder operators.

Now, $\langle l, m| L_+ L_-| l, m\rangle \geq 0$, since $(L_-)^\dagger = L_+$, Therefore,

$$\langle l, m| L_+ L_-| l, m\rangle = \langle l, m| \vec{L}^2 - L_z^2 + \hbar L_z| l, m\rangle$$
$$= \hbar^2 \{l (l + 1) - m^2 + m\} \langle l, m| l, m\}$$

Hence,

$$l (l + 1) - m^2 + m \geq 0$$

or,
$$\left(l + \frac{1}{2}\right)^2 \geq \left(m - \frac{1}{2}\right)^2 \tag{21}$$

From $\langle l, m |L_- L_+| l, m\rangle \geq 0$, we get

$$\left(l + \frac{1}{2}\right)^2 \geq \left(m + \frac{1}{2}\right)^2 \tag{22}$$

Eqs. (21) and (22) imply that

$$- l \leq m \leq l \tag{23}$$

Even though $L_\pm$ changes the eigenvalue of L_z by one unit of $\hbar$, it does not change the eigenvalue of $\vec{L}^2$, since $\vec{L}^2$ and L_z commute.

We have shown that

$$L_+ |l, m\rangle = c| l, m + 1\rangle \tag{24}$$

where c is a numerical factor that may depend on l and m. The value of $|c|$ is determined by calculating the norm of Eq. (24):

$$\left(\langle l, m| L_-\right) \left(L_+ |l, m\rangle\right) = |c|^2.$$

Using
$$L_- L_+ = \vec{L}^2 - L_z^2 - \hbar L_z, \text{ we get}$$
$$|c|^2 = \hbar^2 [l (l + 1) - m^2 - m]$$
$$= \hbar^2 [l (l + 1) - m (m + 1)].$$

The phase of c is arbitrary. We choose the phase of c to be real positive. We have

$$L_+ |l, m\rangle = \hbar \, [l \, (l + 1) - m \, (m + 1)]^{1/2} | \, l, m + 1\rangle$$

$$= \hbar \, [(l + m + 1) \, (l - m)]^{1/2} | \, l, m + 1\rangle. \qquad (25)$$

By a similar argument,

$$L_- | \, l, m\rangle = \hbar \, [l \, (l + 1) - m \, (m - 1)]^{1/2} | \, l, m - 1\rangle$$

$$= \hbar \, [(l - m + 1) \, (l + m)]^{1/2} | \, l, m - 1\rangle. \qquad (26)$$

Equation (23) shows that m is bounded both above and below. Therefore, for some m not exceeding l, $[l \, (l + 1) - m \, (m + 1)]^{1/2}$ must vanish so that the series of eigenstates terminates with $m \leq l$. This occurs for $m = l$. In order for the series to terminate the m values must be smaller than l by integers only and the maximum value of m is l. From Eq. (26) we see that for the lowest m value, $[l \, (l + 1) - m \, (m - 1)]^{1/2} = 0$ for $m = -l$. All possible values of m are greater than $-l$ by integers only and the minimum value of m is $-l$. m can differ from both l and $-l$ by integer only if $2l$ is an integer (or zero).

To summarize:

(a) The eigenvalues of $\vec{L}^2$ have the form $l \, (l + 1)$, where $l = 0, \dfrac{1}{2}, 1, \dfrac{3}{2}, 2, \, ...;$

(b) The eigenvalues of L_z are $-l, -l + 1, \, ..., \, l - 1, l.$

$$(27)$$

Clearly, for each l there are $2l + 1$ different possible m values.

We have the following picture of the angular momentum vector. In an $|l, m|$ state, we choose to measure the component along the z-axis obtaining a value $m\hbar$ for the z-component while it has the length $\hbar \, \sqrt{l \, (l + 1)}$. The transverse component $\sqrt{L_x^2 + L_y^2} = \sqrt{\vec{L}^2 - L_z^2}$ has length $\hbar \, \sqrt{l \, (l + 1) - m^2}$, but the direction of the transverse component cannot be fixed becasue L_x and L_y do not commute with L_z. We have for the minimum uncertainties of L_x and L_y

$$\overline{(\Delta L_x)^2} \cdot \overline{(\Delta L_y)^2} \geq \frac{1}{4} \, \overline{L_z^2} = \frac{m^2 \hbar^2}{4}. \qquad (28)$$

We have

$$\overline{(\Delta L_x)^2} = \overline{(L_x - \overline{L_x})^2} = \overline{L_x^2} - \overline{(L_x)^2} = \overline{L_x^2}$$

and a similar expression for L_y. The expectation value of the square of the angular momentum is given by

$$\vec{L}^2 = \hbar^2 l \, (l + 1) = \overline{L_x^2} + \overline{L_y^2} + \overline{L_z^2}$$

$$= \overline{(\Delta L_x)^2} + \overline{(\Delta L_y)^2} + m^2 \hbar^2$$

Hence,

$$\overline{(\Delta L_x)^2} + \overline{(\Delta L_y)^2} = \hbar^2 (l^2 + l - m^2). \tag{29}$$

The minimum fluctuation in the measurements of L_x and L_y clearly occurs for $|m| = l$, ie when the angular momentum vector points as nearly as possible along the z-axis. The angular momentum vector moves in an unobservable way about the z-axis, keeping the angle between the vector and the axis fixed. When the total angular momentum is zero, the components L_x and L_y are sharply defined. The results Eq. (27) obtained in this section giving the eigenvalue spectrum of the angular momentum operator have been determined using only the commutation rules Eq. (5) and the requirement that the norm of any wave function is nonnegative. The explicit form of the operators was not used in reaching the conclusions. Consequently any angular momentum operator satisfying the above commutation relations possesses the spectrum given by Eq. (27). We define the vector operator $\vec{J}$ an angular momentum operator if its components are hermitian operators satisfying the commutation relations

$$[J_i, J_j] = i\hbar \in_{ijk} J_k \tag{30}$$

or, in vector form

$$\vec{J} \times \vec{J} = i\hbar \, \vec{J}. \tag{31}$$

We choose as basis of our representation the simultaneous normalized eigenvectors of the commuting operators $\vec{J} \times \vec{J}$ and J_z. The eigenkets are labeled $|j, m\rangle$, where j and m are given by

$$J^2 |\vec{j}, m\rangle = \hbar^2 j (j + 1) |j, m\rangle$$

and

$$J_z |j, m\rangle = \hbar m |j, m\rangle. \tag{32}$$

The possible values of j are $0, \frac{1}{2}, 1, \frac{3}{2}, 2, \ldots$ and there are $2j + 1$ value of m allowed for a particular j, namely $m = -j, -j + 1, \ldots, j - 1, j$.

A particular case of an angular momentum operator is the orbital angular momentum operator $\vec{L}$ expressed in terms of the coordinate and momentum operators. The orbital angular momentum of a system has always integer eigenvalues. The half-integer eigenvalues also occur in nature. They are incorporated in the theory by introducing a non-classical internal degree of freedom – spin. For a given type of particle, the spin angular momentum eigenvalues can be integer of half-integer. To show that the orbital angular momentum eigenvalues are restricted to only the integer values, we follow Ballentine (1998). We seek the eigenvalues of the orbital angular momentum operator.

$$L_x = xp_y - yp_x$$

where $[x, y] = [p_x, p_y] = 0$, and $[x_i, p_j] = i\hbar\,\delta_{ij}$. We use a system of units in which x_i and p_j are dimensionless and $\hbar = 1$. We introduce

$$x_1 = \frac{x + p_y}{\sqrt{2}}, \quad p_1 = \frac{p_x - y}{\sqrt{2}}$$

$$x_2 = \frac{x - p_y}{\sqrt{2}}, \quad p_2 = \frac{p_x + y}{\sqrt{2}}$$

It is easy to show that

$$[x_1, x_2] = [p_1, p_2] = 0, \quad [x_i, p_j] = i\delta_{ij}.$$

Then

$$L_z = \frac{1}{2}\,(p_1^2 + x_1^2) - \frac{1}{2}\,(p_2^2 + x_2^2),$$

is expressed as the difference of the two independent harmonic oscillator Hamiltonians, each has mass $m = 1$ and angular frequency $\omega = 1$. Since the two terms commute, the eigenvalue of L_z is equal to

$$\left(n_1 + \frac{1}{2}\right) - \left(n_2 + \frac{1}{2}\right) = n_1 - n_2$$

where n_1 and n_2 are nonnegative integers. Therefore the orbital angular momentum eigenvalues must be integer multiples of $\hbar$ (see Gottfried (1966) for a treatment of this point).

We now discuss explicit matrix representations of the angular momentum operators. Assuming $|j, m\rangle$ to be normalized, we have

$$\langle j', m'|\vec{J}^2| j, m\rangle = j\,(j + 1)\,\hbar^2\delta_{j'j}\,\delta_{m'm}. \tag{33}$$

$$\langle j', m'|J_z| j, m\rangle = m\hbar\delta_{j'j}\,\delta_{m'm}. \tag{34}$$

Our choice of a representation in which $\vec{J}^2$ and J_z are diagonal has led to discrete sequences of values for j and m. The infinite matrices on the Hilbert space are expressed as an infinite set of finite (hermitian) matrices, each of which is characterized by a particular value of j and has $2j + 1$ rows and columns.

The matrix elements of $J_\pm = J_x \pm iJ_y$ are

$$\langle j', m'|\,J_\pm| j, m\rangle = \sqrt{(j \mp m)\,(j \pm m + 1)}\,\hbar\delta_{j'j}\,\delta_{m'm\pm 1}. \tag{35}$$

For $j = 0$, $\vec{J}^2$ and the components of $\vec{J}$ are all represented by null matrices of unit rank: (0). The matrices of $\vec{j} = \frac{1}{2}$ are the spin $\frac{1}{2}$ representation; they are namely,

$$j = \frac{1}{2};$$

$$\left\langle \frac{1}{2}m' | J_x | \frac{1}{2}m \right\rangle \qquad \left\langle \frac{1}{2}m' | J_y | \frac{1}{2}m \right\rangle \qquad \left\langle \frac{1}{2}m' | J_z | \frac{1}{2}m \right\rangle$$

m' \ m	$\frac{1}{2}$	$-\frac{1}{2}$
$\frac{1}{2}$	0	$\frac{\hbar}{2}$
$-\frac{1}{2}$	$\frac{\hbar}{2}$	0

m' \ m	$\frac{1}{2}$	$-\frac{1}{2}$
$\frac{1}{2}$	0	$-i\frac{\hbar}{2}$
$-\frac{1}{2}$	$i\frac{\hbar}{2}$	0

m' \ m	$\frac{1}{2}$	$-\frac{1}{2}$
$\frac{1}{2}$	$\frac{\hbar}{2}$	0
$-\frac{1}{2}$	0	$-\frac{\hbar}{2}$

$$(36)$$

These are the Pauli spin matrices and are usually written as $\left(\frac{\hbar}{2}\right)\sigma_x$, or, $\left(\frac{\hbar}{2}\right)\sigma_y$ and $\left(\frac{\hbar}{2}\right)\sigma_z$, where

$$\sigma_x = \begin{pmatrix} 0 & 1 \\ 1 & 0 \end{pmatrix}; \quad \sigma_y = \begin{pmatrix} 0 & -i \\ i & 0 \end{pmatrix}; \quad \sigma_z = \begin{pmatrix} 1 & 0 \\ 0 & -1 \end{pmatrix}. \qquad (37)$$

(see Chapter 6).

Also,

$$\left\langle \frac{1}{2}m' | \vec{J}^2 | \frac{1}{2}m \right\rangle \qquad \left\langle \frac{1}{2}m' | J_+ | \frac{1}{2}m \right\rangle$$

m' \ m	$\frac{1}{2}$	$-\frac{1}{2}$
$\frac{1}{2}$	$\frac{3}{4}\hbar^2$	0
$-\frac{1}{2}$	0	$\frac{3}{4}\hbar^2$

m' \ m	$\frac{1}{2}$	$-\frac{1}{2}$
$\frac{1}{2}$	0	$\hbar$
$-\frac{1}{2}$	0	0

$$(38)$$

For $j = 1$:

$$\langle 1\,m' | J_x | 1\,m \rangle \qquad \langle 1\,m' | J_y | 1\,m \rangle$$

m' \ m	$+1$	0	-1
$+1$	0	$\frac{\hbar}{\sqrt{2}}$	0
0	$\frac{\hbar}{\sqrt{2}}$	0	$\frac{\hbar}{\sqrt{2}}$
-1	0	$\frac{\hbar}{\sqrt{2}}$	0

m' \ m	$+1$	0	-1
$+1$	0	$-\frac{\hbar}{\sqrt{2}}$	0
0	$\frac{\hbar}{\sqrt{2}}$	0	$-\frac{\hbar}{\sqrt{2}}$
-1	0	$\frac{\hbar}{\sqrt{2}}$	0

$$\langle 1\,m'|J_z|1\,m\rangle$$

$m'\ \backslash\ m$	1	0	-1
+1	$\hbar$	0	0
0	0	0	0
-1	0	0	$-\hbar$

$$\langle 1\,m'|\vec{J}^{\,2}|1\,m\rangle$$

$m'\ \backslash\ m$	1	0	-1
+1	$2\hbar^2$	0	0
0	0	$2\hbar^2$	0
-1	0	0	$2\hbar^2$

$$\langle 1\,m'|J_+|1\,m\rangle$$

$m'\ \backslash\ m$	1	0	-1
+1	0	$\sqrt{2}\,\hbar$	0
0	0	0	$\sqrt{2}\,\hbar$
-1	0	0	0

$$\tag{39}$$

For $j = 3/2$,

$$\left\langle \tfrac{3}{2}\,m'|J_x|\tfrac{3}{2}\,m\right\rangle$$

$m'\ \backslash\ m$	$\tfrac{3}{2}$	$\tfrac{1}{2}$	$-\tfrac{1}{2}$	$-\tfrac{3}{2}$
$\tfrac{3}{2}$	0	$\dfrac{\sqrt{3}}{2}\hbar$	0	0
$\tfrac{1}{2}$	$\dfrac{\sqrt{3}}{2}\hbar$	0	$\hbar$	0
$-\tfrac{1}{2}$	0	$\hbar$	0	$\dfrac{\sqrt{3}}{2}\hbar$
$-\tfrac{3}{2}$	0	0	$\dfrac{\sqrt{3}}{2}\hbar$	0

$$\left\langle \tfrac{3}{2}\,m'|J_y|\tfrac{3}{2}\,m\right\rangle$$

$m'\ \backslash\ m$	$\tfrac{3}{2}$	$\tfrac{1}{2}$	$-\tfrac{1}{2}$	$-\tfrac{3}{2}$
$\tfrac{3}{2}$	0	$-i\dfrac{\sqrt{3}}{2}\hbar$	0	0
$\tfrac{1}{2}$	$i\dfrac{\sqrt{3}}{2}\hbar$	0	$i\hbar$	0
$-\tfrac{1}{2}$	0	$i\hbar$	0	$-i\dfrac{\sqrt{3}}{2}\hbar$
$-\tfrac{3}{2}$	0	0	$i\dfrac{\sqrt{3}}{2}\hbar$	0

$$\left\langle \tfrac{3}{2}\,m'|J_z|\tfrac{3}{2}\,m\right\rangle$$

$m'\ \backslash\ m$	$\tfrac{3}{2}$	$\tfrac{1}{2}$	$-\tfrac{1}{2}$	$-\tfrac{3}{2}$
$\tfrac{3}{2}$	$\dfrac{3}{2}\hbar$	0	0	0
$\tfrac{1}{2}$	0	$\dfrac{1}{2}\hbar$	0	0
$-\tfrac{1}{2}$	0	0	$-\dfrac{1}{2}\hbar$	0
$-\tfrac{3}{2}$	0	0	0	$-\dfrac{3}{2}\hbar$

$$\left\langle \tfrac{3}{2}\,m'|\vec{J}^{\,2}|\tfrac{3}{2}\,m\right\rangle$$

$m'\ \backslash\ m$	$\tfrac{3}{2}$	$\tfrac{1}{2}$	$-\tfrac{1}{2}$	$-\tfrac{3}{2}$
$\tfrac{3}{2}$	$\dfrac{15}{4}\hbar^2$	0	0	0
$\tfrac{1}{2}$	0	$\dfrac{15}{4}\hbar^2$	0	0
$-\tfrac{1}{2}$	0	0	$\dfrac{15}{4}\hbar^2$	0
$-\tfrac{3}{2}$	0	0	0	$\dfrac{15}{4}\hbar^2$

$$\left\langle \frac{3}{2}\, m' | J_+ | \frac{3}{2}\, m \right\rangle$$

m' \ m	$\frac{3}{2}$	$\frac{1}{2}$	$-\frac{1}{2}$	$-\frac{3}{2}$
$\frac{3}{2}$	0	$\sqrt{3}\hbar$	0	0
$\frac{1}{2}$	0	0	$\sqrt{4}\hbar$	0
$-\frac{1}{2}$	0	0	0	$\sqrt{3}\,\hbar$
$-\frac{3}{2}$	0	0	0	0

$$\tag{40}$$

Notice that the matrix for J_- is the transpose of the matrix J_+, since

$$J_- = (J_+)^\dagger.$$

Next, we discuss the eigenfunctions of orbital angular momentum. The eigenfunctions of $\vec{L}^2$ and L_z play a very important role in atomic and nuclear physics. In fact, in any problem involving central field these functions are extremely useful. We have to construct the simultaneous eigenfunctions of the two eigenvalue equations

$$L_z\, |l,\, m\rangle = m\hbar|\, l,\, m\rangle \tag{41}$$

and

$$\vec{L}^2\, |l,\, m\rangle = l\,(l+1)\hbar^2\, |l,\, m\rangle. \tag{42}$$

Here we have not made any assumption about l and m. Using Eq. (9) in Eq.(41), we obtain

$$-i\hbar\, \frac{\partial}{\partial \phi}\, \langle \hat{n}\, |l,\, m\rangle = m\hbar\, |l,\, m\rangle, \tag{43}$$

where we have defined a direction eigenket $|\hat{n}\rangle = |\vartheta,\, \phi\rangle$. Equation (43) shows that ϕ – dependence is of the form $e^{im\phi}$ where $m = 0,\, \pm1,\, +2,\, \dots.$ We now use the notation $(\vartheta,\, \phi\, |l,\, m\rangle = Y_{lm}(\vartheta,\, \phi)$ to refer to the angular part of the wave function of the state $|l,\, m\rangle$. Thus, $\langle r,\, \vartheta,\, \phi\, |l,\, m\rangle = f(r)\, Y_{lm}(\vartheta,\, \phi)$ where $f(r)$ is an arbitrary function of r. Instead of working from the partial differential equation satisfied by Y_{lm} obtained from Eq. (42), we find the function at the top of the chain using the relation $L_+\, Y_{l,\, l} = 0$ and then use Eq. (26) to go down m values. From Eq. (18)

$$-i\hbar e^{i\phi}\left[i\, \frac{\partial}{\partial \vartheta} - \cot \vartheta\, \frac{\partial}{\partial \phi} \right] \langle \vartheta,\, \phi|\, l, l\rangle = 0. \tag{44}$$

Since the ϕ – dependence must be $e^{il\phi}$, the solution of the partial differential equation is

$$\langle \vartheta, \phi | \, l, l \rangle = Y_{l,l}(\vartheta, \phi) = c_l e^{il\phi} \sin^l \vartheta \tag{45}$$

where c_l is a normalization constant. The normalization condition is

$$\int_0^\pi \sin \vartheta d\vartheta \int_0^{2\pi} d\phi \, Y^*_{l'm'}(\vartheta, \phi) \, Y_{lm}(\vartheta, \phi) = \delta_{l'l}\delta_{m'm}. \tag{46}$$

Eqs. (45) and (46) give

$$(Y_{ll}, Y_{ll}) = |c_l|^2 4\pi \, \frac{(2^l \times l!)^2}{(2l + 1)}.$$

By convention the phase of c_l is chosen so that

$$c_l = (-1)^l \, \frac{1}{2^l \cdot l!} \, \sqrt{\frac{(2l + 1)!}{4\pi}}. \tag{47}$$

To find the other $Y_{lm}(\vartheta, \phi)$, we repeatedly use Eqs. (26) and (18). The result is $(m \geq 0)$

$$Y_{lm}(\vartheta, \phi) = \frac{(-1)^{l+m}}{2^l \cdot l!} \, \sqrt{\frac{(2l + 1)\,(l + m)!}{4\pi\,(l - m)!}} \, e^{im\phi} \, (\sin \theta)^m \times \left(\frac{d}{d(\cos \vartheta)}\right)^{l+m} (\sin \vartheta)^{2l} \tag{48}$$

For $m = 0$, we get

$$Y_{l0}(\vartheta, \phi) = \sqrt{\frac{2l + 1}{4\pi}} \, P_l(\cos \vartheta) \tag{49}$$

where P_l is the Legendre polynomial.

For negative m, one uses

$$Y_{l,-m}(\vartheta, \phi) = (-1)^m \, Y^*_{lm}(\vartheta, \phi). \tag{50}$$

The functions $Y_{lm}(\vartheta, \phi)$ are defined for l (and hence m) integers. The $Y_{lm}(\vartheta, \phi)$ are called the spherical harmonics. In terms of the associated Legendre functions

$$P_l^{|m|}(\cos \theta) = \frac{(-1)^{m+l}}{2^l \cdot l!} \cdot \frac{(l + |m|)!}{(l - |m|)!} \, \sin^{-|m|} \vartheta \left(\frac{d}{d\cos \vartheta}\right)^{l-|m|} \sin^{2l} \vartheta, \tag{51}$$

the spherical harmonics are expressed as follows

$$Y_{lm}(\vartheta, \phi) = (-1)^m \, \sqrt{\frac{(2l + 1)}{4\pi} \cdot \frac{(l - m)!}{(l + m)!}} \, e^{im\phi} \, P_l^m(\cos \theta) \; (m \geq 0). \tag{52}$$

The first few Y_{lm} are

$$Y_{00}(\vartheta, \phi) = \frac{1}{\sqrt{4\pi}}, \qquad\qquad Y_{1,\pm 1}(\vartheta, \phi) = \mp\sqrt{\frac{3}{8\pi}}\, e^{\pm i\phi}$$

$$Y_{10}(\vartheta, \phi) = \sqrt{\frac{3}{4\pi}}\,\cos\vartheta, \qquad\qquad Y_{20}(\vartheta, \phi) = \sqrt{\frac{5}{16\pi}}\,(3\cos^2\vartheta - 1)$$

$$Y_{2,\pm 1}(\vartheta, \phi) = \mp\sqrt{\frac{15}{8\pi}}\,\sin\vartheta\,\cos\vartheta\, e^{\pm i\phi}, \quad Y_{2,\pm 2}(\vartheta, \phi) = \sqrt{\frac{15}{32\pi}}\,\sin^2\vartheta\, e^{\pm 2i\phi}. \tag{53}$$

Under the parity operation $\vec{r} \to -\vec{r}$, i.e., $r \to r$, $\vartheta \to \pi - \vartheta$, $\phi \to \pi + \phi$,

$$Y_{lm}(\vartheta, \phi) \to (-1)^l\, Y_{lm}(\vartheta, \phi). \tag{54}$$

The even l states have even parity ($+1$) and the odd l states have odd parity (-1). $|Y_{lm}(\vartheta, \phi)|$ is a complete set of single valued functions on the unit sphere:

$$\sum_{l=0}^{\infty} \sum_{m=-l}^{l} Y_{lm}(\vartheta, \phi)\, Y_{lm}^{*}(\vartheta', \phi') = \frac{\delta(\vartheta - \vartheta')\,\delta(\phi - \phi')}{2\pi}. \tag{55}$$

The addition theorem for spherical harmonics is

$$P_l(\cos\omega) = \frac{4\pi}{2l+1} \sum_{m=-l}^{l} Y_{lm}^{*}(\vartheta_1, \phi_1)\, Y_{lm}(\vartheta_2, \phi_2), \tag{56}$$

where ω is the angle between the directions (ϑ_1, ϕ_1) and (ϑ_2, ϕ_2). Orthonormality is expressed as

$$\int_{0}^{2\pi} \int_{0}^{\pi} Y_{lm}^{*}(\vartheta, \phi)\, Y_{l'm'}(\vartheta, \phi)\, \sin\vartheta\, d\vartheta\, d\phi = \delta_{ll'}\delta_{mm'}. \tag{57}$$

PROBLEMS

4.1 Calculate $<lm|L_x^2|lm>$ and $<lm|L_x L_y|lm>$.

4.2 An electron is moving in the Coulomb field of a nucleus of charge Ze. Show that the Hamiltonian for the electron may be written

$$H = \frac{1}{2m}\,\vec{p}_r^2 + \frac{\vec{L}^2}{2mr^2} - \frac{Ze^2}{r},$$

where $\vec{p}_r = \frac{1}{r}\,(\vec{r}\cdot\vec{p} - i\hbar)$ is the conjugate momentum operator and $\vec{L}$ is the angular momentum operator.

4.3 The spherical harmonics $Y_{lm}(\theta, \phi)$ are simultaneous eigenfunctions of $\vec{L}^2$ and L_z operators, as given in Eqs. (35a) and (35b). Assume $Y_{lm}(\theta, \phi) = \Phi(\phi) \times P_{lm}(\theta)$.

   (a) Solve the $\vec{L}^2$ equation by the method of separation of variables.

   (b) Determine the normalized function $\Phi(\phi)$.

4.4 Consider the representation where J_x is diagonal. Obtain the matrices for J_y, J_z, and J^2.

4.5 Consider a particle of mass m moving in a potential $V(r)$. For an eigenstate of the energy and the angular momentum, show that

$$\hbar^2 l\,(l+1) <\frac{1}{r^3}> = m <\frac{\partial V}{\partial r}>.$$

CHAPTER 5

Rotations

The angular momentum operator is closely related to rotations in space. Rotations form a symmetry group which plays a central role in many fields of physics. In this chapter we discuss this intimate connection of the theory of the group of rotations with the concept of angular momentum.

5.1 ROTATIONS

Three parameters are required to describe an arbitrary rotation in three dimensions. We write R_ϑ (λ, μ, ν) for the rotation through angle ϑ with the axis of rotation specified by a unit vector $\hat{n}$ whose components are given by the direction cosines λ, μ, ν. The angle ϑ is positive for the rotation sense forming a right handed system with the vector direction. Since

$$R_\vartheta (\lambda, \mu, \nu) = R_{2\pi-\vartheta} (\lambda, \mu, \nu), \tag{1}$$

we take $0 \le \vartheta \le \pi$. The result of carrying out a rotation $R_\beta^{(2)}$ through an angle β about the axis 2 after the rotation $R_\alpha^{(1)}$ through an angle α about the axis 1 is the same as that of another rotation $R_\gamma^{(3)}$ through a certain angle γ about an axis 3 in a certain direction. Thus the composition of rotations is expressed as

$$R_\beta^{(2)} R_\alpha^{(1)} = R_\gamma^{(3)}. \tag{2}$$

Given the directions of the axes 1, 2 and the angles α, β, it is possible geometrically to find the direction 3 and the angle γ.

The set of rotations $R_\vartheta(\lambda, \mu, \nu)$ with the product defined by Eq. (2) forms a group. If R and R' are conjugate

$$R' = QRQ^{-1}, \tag{3}$$

where Q is the rotation which sends the rotation axis of R into that of R', both R and R' having the same angle of rotation. We notice that in the rotation group, rotations with an equal angle belong to one and the same class, regardless of the axis.

5.2 THE EULER ANGLES

A convenient way to characterize the most general rotation in three dimensions is by the Euler angles (Fig. 5.1). Any rotation can be regarded as a result of the following three successive rotations.

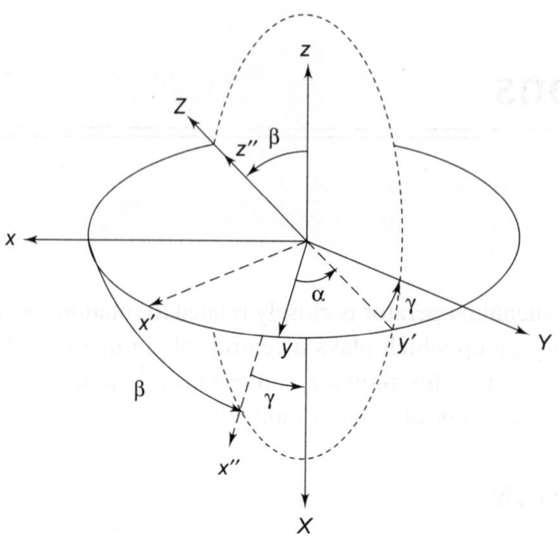

Fig. 5.1 The Euler angles

(a) Rotation $R_\alpha(z)$ through an angle α about the z-axis ($0 \le \alpha \le 2\pi$). This rotation carries the coordinate axes x, y and z into x', y' and $z' = z$ (see Fig. 5.1).

(b) Rotation $R_\beta(y')$ through an angle β about the y'-axis ($0 \le \beta \le \pi$). This carries x', y' and z' into x'', $y'' = y'$ and z''. According to Eq. (3)

$$R_\beta(y') = R_\alpha(z)\, R_\beta(y)\, R_\alpha(z)^{-1}.$$

(c) Rotation $R_\gamma(z'')$ through an angle γ about the z''-axis ($0 \le \gamma \le 2\pi$). The axes x'', y'', and z'' are carried into X, Y, and $Z = z''$. Since

$$R_\gamma(z'') = R_\alpha(z)\, R_\gamma(z)\, R_\alpha(z)^{-1} = R_\gamma(z),$$

we obtain

$$R(\alpha,\, \beta,\, \gamma) = R_\gamma(z'')\, R_\beta(y')\, R_\alpha(z)$$

$$= R_\alpha(z)\, R_\beta(y)\, R_\gamma(z). \tag{4}$$

In Eq. (4) the notation is expressed in the original fixed coordinate system xyz. The coordinates (x, y, z) and (X, Y, Z) of a point fixed in space are related to each other by the following expression

$$(X,\, Y,\, Z) = (x,\, y,\, z)\,\hat{R} \tag{5}$$

where $\hat{R}$ is the matrix given by

$$R = \begin{pmatrix} \cos\alpha & -\sin\alpha & 0 \\ \sin\alpha & \cos\alpha & 0 \\ 0 & 0 & 1 \end{pmatrix} \begin{pmatrix} \cos\beta & 0 & \sin\beta \\ 0 & 1 & 0 \\ -\sin\beta & 0 & \cos\beta \end{pmatrix} \begin{pmatrix} \cos\gamma & -\sin\gamma & 0 \\ \sin\gamma & \cos\gamma & 0 \\ 0 & 0 & 1 \end{pmatrix}. \tag{6}$$

Then matrix $\hat{R}$ is a 3×3 orthogonal matrix with the value of its determinant equal to unity, because only proper rotations (not involving inversions) are considered here. Such matrices are called special orthogonal matrices. The set of $n \times n$ special, orthogonal matrices form a group $SO(n)$ with respect to ordinary matrix multiplication. Every proper rotation can be put in one-to-one correspondence with a 3×3 special orthogonal matrix. Thus the rotation group is isomorphic to $SO(3)$.

5.3 ACTIVE AND PASSIVE ROTATIONS

Rotations may be viewed in either of two approaches, In "active rotations", the physical system is rotated with respect to a fixed coordinate system. In "passive rotations" the physical system remains fixed but the coordinate system is rotated. We usually use the active point of view, but both approaches are valid. We discuss below the relation between the two (see Fig. 5.2) Under an active rotation the system is rotated by a positive angle α so that a point in the system goes from (x, y) to (x', y'). The active rotation matrix $R_{ij}^{(\alpha)}$ connects the coordinates of the two points

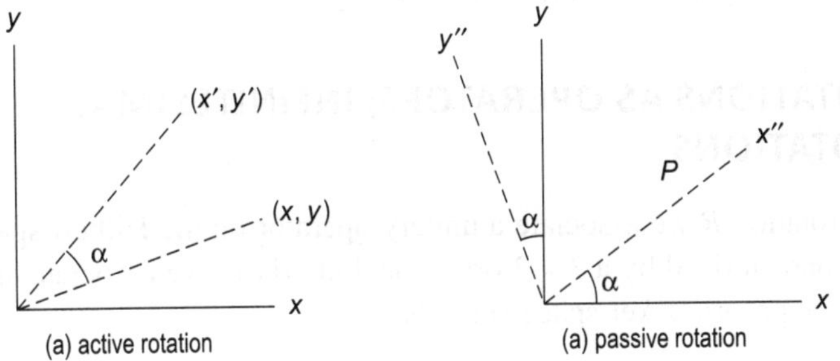

Fig. 5.2 Active and Passive rotations

$$x_i' = R_{ij}^{(a)} x, \tag{7}$$

where

$$R_{ij}^{(a)}(\alpha, 0, 0) = \begin{pmatrix} \cos\alpha & -\sin\alpha & 0 \\ \sin\alpha & \cos\alpha & 0 \\ 0 & 0 & 1 \end{pmatrix} \tag{8}$$

In general, the rotation matrix is a function of three Euler angles. If the object of the rotation is a scalar field $\Psi(\vec{x})$, then the rotated function is given by $\Psi'(\vec{x})' = R(\alpha)\,(\alpha,\,\beta,\,\gamma)\,\Psi(\vec{x})$. The value of the new function Ψ' at the rotated point $\vec{x}' = (x',\,y',\,z')$ is equal to the value of the old function Ψ at the original point $\vec{x} = (x,\,y,\,z)$. Thus $\Psi'\,(\vec{x}') = \Psi\,(x')$, i.e.

$$R^{(a)}\,(\alpha,\,\beta,\,\gamma)\,\Psi(\vec{x}) = \Psi\,([R^{(a)}\,(\alpha,\,\beta,\,\gamma)]^{-1}\,\vec{x}) \tag{9}$$

In a passive rotation, a point on the physical system has two sets of coordinates, the old coordinates $\vec{x} = (x,\,y,\,z)$ and the new coordinates $\vec{x}'' = (x'',\,y'',\,z'')$. The two sets of coordinates are connected by the passive rotation matrix.

$$x_i'' = R_{ij}^{(p)}\,x_j \tag{10}$$

With reference to the Fig. 5.2(b)

$$R_{ij}^{(p)}\,(\alpha,\,0,\,0) = \begin{pmatrix} \cos\alpha & \sin\alpha & 0 \\ -\sin\alpha & \cos\alpha & 0 \\ 0 & 0 & 1 \end{pmatrix}. \tag{11}$$

Since the value of the state functions is the same at a point, we get

$$\Psi''\,(\vec{x}'') = \Psi(\vec{x}) = \Psi\,([R^{(p)}]^{-1}\,\vec{x}) \tag{12}$$

Although the two Eqs. (8) and (12) have the same form the active and passive rotation matrices are inverse of each other. This is true in general

$$R^{(p)}(\alpha,\,\beta,\,\gamma) = [R^{(a)}(\alpha,\,\beta,\,\gamma)]^{-1} = R^{(a)}\,(-\gamma,\,-\beta,\,-\alpha). \tag{13}$$

The geometric interpretation of Eq. (13) is clear. This discussion is based on Bouten (1969) as given in Ballentine (1998).

5.4　ROTATIONS AS OPERATORS; INFINITESIMAL ROTATIONS

With each rotation R we associate a unitary operator on the Hilbert space. Given a rotation characterized by a 3×3 orthogonal matrix R, we associate an operator $D(R)$ in the appropriate ket space such that

$$|\alpha\rangle_R = D(R)|\,\alpha\rangle \tag{14}$$

where $|\alpha\rangle_R$ and $|\alpha\rangle$ are the state kets of the rotated and the original system respectively. The operator $D(R)$ has the following properties

$$D(R_2)\,D(R_1) = D(R_2 R_1) \tag{15}$$

$$D(R) = D^\dagger(R^{-1}) \tag{16}$$

The set of objects $\{D(R)\}$ forms a group.

We first consider a subgroup – the rotations about an axis, the z-axis, say. These rotations are defined by one parameter – an angle α, say. The unitary operators that correspond to these rotations are described by

$$D_z(\alpha_1)\, D_z(\alpha_2) = D_z(\alpha_1 + \alpha_2)$$

$$D_z(\alpha) = D_z^\dagger(-\alpha) \tag{17}$$

This subgroup is Abelian.

To construct the rotation operator we first study its properties under an infinitesimal rotation. From Eq. (6), we see that the infinitesimal forms of $R_z(\epsilon)$, (ϵ) and $R_y(\epsilon)$ are given by

$$R_z(\epsilon) = \begin{pmatrix} 1 - \dfrac{\epsilon^2}{2} & -\epsilon & 0 \\ -\epsilon & 1 - \dfrac{\epsilon^2}{2} & 0 \\ 0 & 0 & 1 \end{pmatrix}$$

$$R_y(\epsilon) = \begin{pmatrix} 1 - \dfrac{\epsilon^2}{2} & 0 & \epsilon \\ 0 & 1 & 0 \\ -\epsilon & 0 & 1 - \dfrac{\epsilon^2}{2} \end{pmatrix} \tag{18}$$

$$R_x(\epsilon) = \begin{pmatrix} 1 & 0 & 0 \\ 0 & 1 - \dfrac{\epsilon^2}{2} & -\epsilon \\ 0 & -\epsilon & 1 - \dfrac{\epsilon^2}{2} \end{pmatrix}$$

neglecting terms of order ϵ^3 and higher. From Eq. (17) we obtain

$$R_x(\epsilon)\, R_y(\epsilon) = R_y(\epsilon)\, R_x(\epsilon) \tag{19}$$

i.e., infinitesimal rotations, about different axes commute if terms of order ϵ^2 and higher are ignored. If, however, terms upto order ϵ^2 are retained,

$$R_x(\epsilon)\, R_y(\epsilon) = -R_y(\epsilon)\, R_x(\epsilon)$$

$$= \begin{pmatrix} 0 & -\epsilon^2 & 0 \\ \epsilon^2 & 0 & 0 \\ 0 & 0 & 0 \end{pmatrix} \tag{20}$$

$$= R_z(\epsilon^2) - 1 = R_z(\epsilon^2) - R(0)$$

where $R(0)$ stands for rotation through zero angle about any axis.

We now determine the operator of an infinitesimal rotation. An infinitesimal rotation is characterized by a vector $\overrightarrow{\delta\phi}$, the length of which is equal to the angle $\delta\phi$ over which we rotate, and the direction of which is along the axis of rotation. A radius vector $\vec{r}$ becomes $\vec{r}'$ after an infinitesimal rotation $\overrightarrow{\delta\phi}$ where

$$\vec{r}' = \vec{r} + \overrightarrow{\delta\phi} \times \vec{r}. \tag{21}$$

The value of the rotated state vector for the rotated radius vector is equal to the value of the original state vector at the original radius vector. Keeping only the first order terms the corresponding change in the state vector is

$$\Psi\,(\vec{r} - (\overrightarrow{\delta\phi} \times \vec{r})\,) = \{1 - (\overrightarrow{\delta\phi} \times \vec{r})\cdot\vec{\nabla}\}\,\Psi(\vec{r})$$

$$= \left\{1 - \frac{i}{h}\,(\overrightarrow{\delta\phi} \times \vec{r})\cdot\vec{p}\right\}\Psi(\vec{r}) \tag{22}$$

$$= \left(1 - \frac{i}{h}\,\overrightarrow{\delta\phi}\cdot\vec{L}\right)\Psi(\vec{r}).$$

Thus, for a rotation by an infinitesimal angle $d\phi$ about an axis defined by a unit vector $\hat{n}$, the infinitesimal rotation operator is

$$D\,(\hat{n},\,d\phi) = 1 - \frac{i}{h}\,(\vec{L}\cdot\hat{n})\,d\phi. \tag{23}$$

Since $\vec{L}$ is hermitian, the infinitesimal rotation operator is unitary and becomes the identity operator in the limit $d\phi \to 0$.

A finite rotation about an axis can be obtained by compounding successively infinitesimal rotations about the same axis. For a rotation through a finite angle ϕ about the z-axis.

$$D_z(\phi) = \lim_{N\to\infty}\left[1 - \frac{i}{h}\,L_z\left(\frac{\phi}{N}\right)\right]^N$$

$$= \mathrm{Exp}\left(-\frac{iL_z\phi}{h}\right) = 1 - \frac{iL_z\phi}{h} - \frac{L_z^2\phi^2}{2h^2} + \dots \tag{24}$$

The generator of the transformation of a rotation is determined by the angular momentum component along that axis

$$I_n = -\frac{i}{h}\,(\vec{L}\cdot\hat{n}) \tag{25}$$

The connection between the angular momentum components and the infinitesimal rotation operators is given below.

The operators of the infinitesimal rotations about the x, y and z-axes can be expressed by the following 3×3 real, anti-symmetric matrices

$$I_x = \left.\frac{\partial R_a^{(x)}}{\partial\alpha}\right|_{\alpha=0} = \begin{pmatrix} 0 & 0 & 0 \\ 0 & 0 & -1 \\ 0 & 1 & 0 \end{pmatrix}$$

and

$$I_y = \left.\frac{\partial R_\alpha^{(y)}}{\partial \alpha}\right|_{\alpha=0} = \begin{pmatrix} 0 & 0 & 1 \\ 0 & 0 & 0 \\ -1 & 0 & 0 \end{pmatrix}$$

$$I_z = \left.\frac{\partial R_\alpha^{(z)}}{\partial \alpha}\right|_{\alpha=0} = \begin{pmatrix} 0 & -1 & 0 \\ 1 & 0 & 0 \\ 0 & 0 & 0 \end{pmatrix},$$

these matrices satisfy the commutation relations

$$[I_x, I_y] = I_z \tag{27}$$

and cyclic permutations.

Since $I_x = -\dfrac{i}{\hbar} L_x$, etc., the angular momentum commutation rules are

$$[L_x, L_y] = i\hbar L_z, \text{ cyclic.} \tag{28}$$

Using Eq. (20) in the basic commutation relations for rotations, we obtain

$$\left(1 - \frac{iL_x\epsilon}{\hbar} - \frac{L_x^2\epsilon^2}{2\hbar^2}\right)\left(1 - \frac{iL_y\epsilon}{\hbar} - \frac{L_y^2\epsilon^2}{2\hbar^2}\right)$$

$$-\left(1 - \frac{iL_y\epsilon}{\hbar} - \frac{L_y^2\epsilon^2}{2\hbar^2}\right)\left(1 - \frac{iL_x\epsilon}{\hbar} - \frac{L_x^2\epsilon^2}{2\hbar^2}\right) = 1 - \frac{iL_z\epsilon^2}{\hbar} - 1.$$

Equating terms of order ϵ^2 on both sides, we get $[Lx, Ly] = ihLz$ and similarly for the other relations.

We now define the angular momentum operator $\vec{J}$ such that the unitary operator $D(R)$ for an infinitesimal rotation has the form.

$$D(\hat{n}, d\phi) = 1 - i/\hbar \, \vec{J} \cdot \hat{n} \tag{29}$$

and hence for a finite rotation α about the z-axis

$$D_z(\alpha) = e^{-iJ_z\alpha/\hbar} \tag{30}$$

Equation (30) means that J_k is the generator of rotation about the kth-axis. Also, this leads to the fundamental commutation relations of angular momentum

$$[J_i, J_j] = i\hbar \, \epsilon_{ijk} J_k. \tag{31}$$

Notice that the I's operate in three dimensions, while J's operate in Hilbert space

We now write the general rotation operator parametrized by the Euler angles. Corresponding to the product of orthogonal matrices Eq. (4), we can write a product of rotation operators in the ket space as

$$D\ (\alpha,\ \beta,\ \gamma) = D_z(\alpha)\ D_y(\beta)\ D_z(\gamma) \tag{32}$$

Writing $\qquad\qquad\qquad R \equiv (\alpha,\ \beta,\ \gamma)$, we get

$$D(R) = e^{-iJ_z\alpha/\hbar}\ e^{-iJ_y\beta/\hbar}\ e^{-iJ_z\gamma/\hbar}|\alpha\rangle. \tag{33}$$

5.5   ROTATION MATRICES

We now study the matrix elements of $D(R)$. Since $[D(R),\ \vec{J}^2] = 0$ the matrix of D in this representation is diagonal in j. Clearly, rotations cannot change the j-value. Let $|jm\rangle$ be a simultaneous eigenket of $\vec{J}^2$ and J_z.

The matrix representation of the rotation operator in the basis formed by the angular momentum eigenvectors gives the rotation matrices

$$D^{(j)}_{m'm}\ (R) \equiv D^{(j)}_{m'm'}\ (\alpha,\ \beta,\ \gamma)$$

$$= \langle jm'|\ e^{-i\alpha J_z/\hbar}\ e^{-i\beta J_y/\hbar}\ e^{-i\gamma J_z/\hbar}|\ jm\rangle$$

$$= e^{-i\alpha m'/\hbar}\ d^{(j)}_{m'm}\ (\beta)\ e^{-i\gamma m/\hbar} \tag{34}$$

where

$$d^{(j)}_{m'm}\ (\beta) = \langle jm'|\ e^{-i\beta J_y/\hbar}|\ jm\rangle \tag{35}$$

These matrix elements are functions of the Euler angles and are called the Wigner functions, the generalized spherical functions or D-functions.

The matrix $D^{(j)}$ has demension $(2j + 1)$, and is called the $(2j + 1) -$ dimensional irreducible representation of the rotation group. This means that the matrix which correspond to an arbitrary rotation operator in ket space not necessarily of a single j-value can be brought to diagonal blocks of dimension $(2j + 1)$. But it is not possible to reduce the matrices $D^{(j)}$ to blocks of dimension smaller than $(2j + 1)$ by a change of basis. It is clear that this follows from that fact that J_y does not commute with J_z.

The rotation matrices (for a definite j) is a representation since from Eq. (14), we see that the product of two matrices corresponding to rotations R_1, and R_2 is the matrix corresponding to the rotation R_1R_2:

$$\sum_{m'} D^{(j)}_{m''m'}(R_1)\ D^{(j)}_{m'm}\ (R_2) = D^{(j)}_{m''m}\ (R_1R_2). \tag{36}$$

The inverse exists. From Eq. (15), we see that the matrix corresponding to R and the matrix corresponding to R^{-1} are related by

$$D^{(j)}_{mm'}(R) = D^{(j)}_{m'm}(R^{-1})^*. \tag{37}$$

The identity element is the $(2j + 1) \times (2j + 1)$ identity matrix corresponding to no rotation:

$$D^{(j)}_{m'm}(0) = \delta_{m'm} \tag{38}$$

$$\sum_{m''} D^{(j)}_{m'm''}(R)\, D^{(j)}_{m''m}(R^{-1}) = \delta_{m'm}. \tag{39}$$

The inverse matrix is obtained by reversing the Euler angles as follows:

$$D(\alpha, \beta, \gamma) = D(-\gamma, -\beta, -\alpha). \tag{40}$$

Combining Eqs. (37) and (39), we write the unitarity relations as

$$\sum_{m''} D^{(R)}_{m'm''}\, D^{(j)}_{mm''}(R)^*$$

$$= \sum_{m''} D^{(j)}_{m''m'}(R)^*\, D^{(j)}_{m''m}(R)$$

$$= \delta_{m'm}. \tag{41}$$

For the case of $j = \dfrac{1}{2}$, it is easy to evaluate $d^{(j)}_{m'm}(\beta)$. Replacing J_y by $\dfrac{1}{2}\hbar\sigma_y$,

$$\exp\left(-\frac{i}{\hbar}\beta J_y\right) = \exp\left(-\frac{i\beta\sigma_y}{2}\right).$$ Using the properties of the Pauli spin matrices,

$$\exp\left(-\frac{i\beta}{2}\sigma_y\right) = \sum_{n=0}^{\infty} \frac{(-i\beta/2)^n}{n!}\sigma_y^n$$

$$= \sum_{n\,\text{even}} \frac{1}{n!}\left(-\frac{i\beta}{2}\right)^n + \sigma_y \sum_{n\,\text{odd}} \frac{1}{n!}\left(-\frac{i\beta}{2}\right)^n$$

$$= I\cos(\beta/2) - i\sigma_y\sin(\beta/2) \tag{42}$$

Thus,

$$d^{\left(\frac{1}{2}\right)}(\beta) = \begin{pmatrix} \cos(\beta/2) & -\sin(\beta/2) \\ \sin(\beta/2) & \cos(\beta/2) \end{pmatrix}. \tag{43}$$

We obtain the unitary, unimodular 2×2 matrix

$$D^{\left(\frac{1}{2}\right)}(\alpha, \beta, \gamma) = \begin{pmatrix} e^{-i(\alpha+\gamma)/2}\cos(\beta/2) & -e^{-i(\alpha-\gamma)/2}\sin(\beta/2) \\ e^{i(\alpha-\gamma)/2}\sin(\beta/2) & e^{i(\alpha+\gamma)/2}\cos(\beta/2) \end{pmatrix} \tag{44}$$

This is the $j = \frac{1}{2}$ irreducible representation of the rotation operator. The matrix is periodic in β with period 4π, but it changes sign when 2π is added to β. Note that for a rotation of 2π about any axis, say the z-axis,

$$D^{\left(\frac{1}{2}\right)} (2\pi, 0.0) = - I. \tag{45}$$

The representation $D^{\left(\frac{1}{2}\right)}$ is double valued. In fact, $D^{(j)}$ is a double-valued representation for any half-integral j. The representation is single-valued under rotation by 2π whenever j is an integer (see discussion later).

5.6   THE SPHERICAL HARMONICS

We have introduced the spherical harmonics in (Section 4.2) as the eigenfunctions of orbital angular momentum. These are the coordinate representation of the angular momentum eigenvectors $|lm>$ for integer values of l and m. Under rotation, these transform as

$$|lm\rangle_R = \sum_{m'} |lm'\rangle \, \langle lm'|D(R)|lm\rangle$$

$$= \sum_{m'} |lm' > D^{(l)}_{m'm} (R). \tag{46}$$

Clearly, Eq. (46) implies the relation between the spherical harmonics and the rotation matrices. It is convenient to consider the equation for the inverse of the transformation given in Eq. (46). We write

$$R^{-1}(\alpha, \beta, \gamma) \, Y_{lm} (\theta, \phi) = Y_{lm} (\theta', \phi') \tag{47}$$

$$= \sum_{m'} Y_{lm'} (\theta, \phi) \, [D^{(l)}_{mm'} (\alpha, \beta, \gamma)]^{*}$$

where the unitarity of D has been used. Here, $R(\alpha, \beta, \gamma)$ takes a vector in the direction $\hat{n}$ described by (θ, ϕ) into the direction $\hat{n}'$ described by (θ', ϕ') and $\langle \hat{n}/lm \rangle = Y_{lm} (\theta, \varphi)$.

Let us evaluate
$$\mathcal{I} = \sum_{m} Y_{lm}{}^{*}(\theta'_1, \varphi'_1) \, Y_{lm} (\theta'_2, \varphi'_2) \tag{48}$$

Substituting for $Y_{lm} (\theta'_1, \varphi'_1)$ from Eq. (47), we get

$$\sum_{m} Y^{*}_{lm} (\theta'_1, \varphi'_1) \, Y_{lm} (\theta'_2, \varphi'_2)$$

$$= \sum_{m} \sum_{\mu} \sum_{\nu} Y^{*}_{l\mu} (\theta_1, \varphi_1) \, Y_{l\nu} (\theta_2, \varphi_2) \, D^{(l)}_{\mu m} (\alpha, \beta, \gamma) \, D^{(l)^{*}}_{m\nu} (\alpha, \beta, \gamma)$$

$$= \sum_{m} \sum_{\mu} \sum_{\nu} Y^{*}_{l\mu} (\theta_1, \varphi_1) \, Y_{l\nu} (\theta_2, \varphi_2) \, D^{(l)}_{\mu m} (\alpha, \beta, \gamma) \, D^{(l)^{-1}}_{m\nu} (\alpha, \beta, \gamma)$$

$$= \sum_{\mu} \sum_{v} \delta_{\mu v} \, Y^*_{l\mu} \, (\theta_1, \, \varphi_1) \, Y_{lv} \, (\theta_2, \, \varphi_2) \text{ performing the sum over } m \text{ using the fact}$$

$$\text{that } \sum_m D^{(l)}_{\mu m} \, D^{(l)^{-1}}_{mv} = \delta\mu v.$$

$$= \sum_{\mu} Y^*_{l\mu} \, (\theta_1, \, \varphi_1) \, Y_{l\mu} \, (\theta_2, \, \varphi_2).$$

Hence, $\mathcal{I}$ is an invariant under rotations. Therefore, to evaluate $\mathcal{I}$, we can use any coordinate system $x'y'z'$ we wish. We choose $\theta'_1 = 0$, $\varphi'_1 = 0$ (see figure 5.3) Now, with respect to the primed axis we can write

$$\mathcal{I} = \sum_m Y^*_{lm} \, (0, \, \varphi'_1) \, Y_{lm} \, (\theta_{12}, \, 0)$$

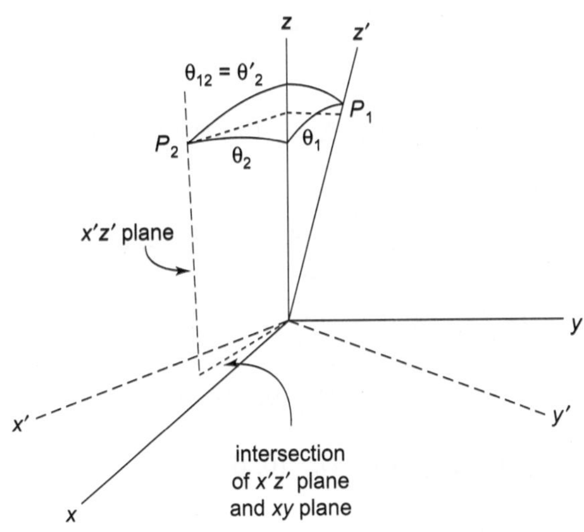

Fig. 5.3 The coordinate systems and angles

But, since the spherical harmonics Y_{lm} are proportional to $\sin^m\theta$ which is zero for $\theta = 0$, except in the case $m = 0$, the sum collapses to a single term.

We get

$$\mathcal{I} = Y_{lo} \, (0, \, \varphi'_1) \, Y_{l0} \, (\theta_{12}, \, 0)$$

But from Eq. (4.42)

$$Y_{lo} \, (\theta, \, \varphi) = \sqrt{\frac{2l + 1}{4\pi}} \, P_l \, (\cos \theta),$$

and $P_l(1) = 1$. Therefore, we have evaluated the invariant:

$$\mathcal{I} = \sqrt{\frac{2l + 1}{4\pi}} \, Y_{l0} \, (\theta_{12}, \, 0) = \frac{2l + 1}{4\pi} \, P_l \, (\cos \theta_{12}) \qquad (49)$$

We note that Eq. (49) can be written as

$$P_l(\cos\theta_{12}) = \frac{4\pi}{2l+1} \sum_m Y^*_{lm}(\theta_1, \phi_1)\, Y_{lm}(\theta_2, \phi_2), \qquad (49)$$

where θ_{12} is the angle between the directions (θ_1, ϕ_1) and (θ_2, ϕ_2) i.e., $\cos\theta_{12} = \cos\theta_1 \cos\theta_2 + \sin\theta_1 \sin\theta_2 \cos(\phi_1 - \phi_2)$, Eq. (49) is known as the addition theorem for spherical harmonics.

Writing the spherical harmonic with respect to P_2,

$$Y_{l0}(\theta_{12}, 0) = Y_{l0}(\theta'_2, 0) = \sqrt{\frac{4\pi}{2l+1}}\, \mathcal{I}$$

$$= \sqrt{\frac{4\pi}{2l+1}} \sum_m Y^*_{lm}(\theta_1^{=\beta}, \varphi_1^{=\alpha})\, Y_{lm}(\theta_2, \varphi_2).$$

But, $\qquad Y_{l0}(\theta'_2, 0) = \sum_m Y_{\ell m}(\theta_2, \varphi_2)\, D^{(\ell)*}_{m0}(\alpha, \beta, 0)$

Comparing, we obtain

$$D^{(l)}_{m0}(\alpha, \beta, 0) = \sqrt{\frac{4\pi}{2l+1}}\, Y_{lm}(\beta, \alpha) \qquad (50)$$

Notice that $\qquad D^{(l)*}_{m0}(\alpha, \beta, 0) = [D^{(l)}_{0m}(\alpha, \beta, 0)]^{-1}$

$$= D^{(l)}_{0m}(0, -\beta, -\alpha)$$

or, $\qquad D^{(l)*}_{m0}(-\alpha, -\beta, 0) = D^{(l)}_{0m}(0, \beta, \alpha) \qquad (51)$

From Eq. (50), we get with $\alpha \to -\alpha$, $\beta \to -\beta$

$$D^{(l)*}_{m0}(-\alpha, -\beta, 0) = D^{(l)}_{0m}(0, \beta, \alpha) = \sqrt{\frac{4\pi}{2l+1}}\, Y^*_{lm}(-\beta, -\alpha)$$

$$= \sqrt{\frac{4\pi}{2l+1}}\, (-1)^m\, Y_{l,-m}(-\beta, -\alpha),$$

using $\qquad Y_{lm}(\theta, \phi)^* = (-1)^m\, Y_{l,-m}(\theta, \phi).$

or, $\qquad D^{(l)}_{0m}(0, \beta, \alpha) = \sqrt{\frac{4\pi}{2l+1}}\, (-1)^m\, f_{lm}(-\beta)\, e^{-im(-\alpha)}$

where $f_{lm}(-\beta)$ is some function of $-\beta$ only.

For spherical harmonics with $m = 0$, we have just a Legendre polynomial which depends only on $\cos\beta$. So $f_{l0}(\beta) = f_{l0}(-\beta)$. For $m \neq 0$, we can generate spherical harmonics by operating with $L_\pm$ operators on f_{l0}. Writing $L_\pm$ in terms of

β and α, we see that $L_{\pm}(\beta, \alpha)$ is the same as $L_{\pm}(-\beta, \alpha)$. Hence $f_{lm}(-\beta) = f_{lm}(\beta)$. Changing dummy α to γ, we obtain

$$D_{0m}^{(l)}(0, \beta, \gamma) = (-1)^m \sqrt{\frac{4\pi}{2l+1}}\; Y_{lm}(\beta, \gamma). \tag{52}$$

In a similar way it can be shown that

$$D_{m0}^{(\ell)}(\alpha, \beta, 0) = \sqrt{\frac{4\pi}{2l+1}}\; Y_{lm}(\beta, \alpha) \tag{52a}$$

In particular,

$$D_{00}^{(l)}(0, \beta, 0) = P_l(\cos\beta).$$

Equations (52) and (52a) enable us to call the matrix elements of the matrix of a finite rotation the generalized spherical functions of jth order.

In many physical problems one has to evaluate the integrals of products of D-functions. To simplify the notation, we introduce a shortened notation for the set of the three Euler angles: $\omega \equiv (\alpha, \beta, \gamma)$. We determine integrations over all rotations by

$$\int \ldots d\omega = \int_0^{2\pi} d\alpha \int_0^{\pi} \sin\beta d\beta \int_0^{2\pi} d\gamma. \tag{53}$$

We now evaluate

$\int D_{mk}^{(l)}(\omega)\, d\omega$. Substituting from the definition of $d_{mk}^{(l)}$, we obtain

$$\int D_{mk}^{(l)}(\omega)\, d\omega = \int_0^{2\pi} e^{-im\alpha}\, d\alpha \int_0^{2\pi} e^{-ik\gamma}\, d\gamma \int_0^{\pi} d_{mk}^{(l)}(\beta)\, \sin\beta d\beta$$

$$= 4\pi^2\, \delta_{m0}\delta_{k0} \int_0^{\pi} d\beta\, \sin\beta\, d_{mk}^{(l)}(\beta), \quad \text{since} \int_0^{2\pi} e^{-ip\alpha}\, d\alpha = 2\pi\delta_{p0}.$$

$$= 4\pi^2\, \delta_{m0}\delta_{k0} \int_0^{\pi} d\beta\, \sin\beta\, d_{00}^{(l)}(\beta).$$

But,

$$d_{00}^{(l)} = D_{00}^{(l)}(0, \beta, 0) = \sqrt{\frac{4\pi}{2l+1}}\; Y_{l0}(\beta, 0) = P_l(\cos\beta).$$

$$\therefore \quad \int_0^{\pi} d\beta\, \sin\beta\, d_{00}^{(l)}(\beta) = \int_{-1}^{+1} dx\, P_0(x)\, P_l(x), \quad \text{since } P_0(x) = 1$$

$$= \delta_{l0}$$

$$\therefore \quad \int D_{mk}^{(l)}(\omega)\, d\omega = 8\pi^2\, \delta_{l0}\, \delta_{m0}\, \delta_{k0}. \tag{54}$$

A few important integrals involving D-functions will be evaluated in the next section. We discuss explicit formulas for $d^{(j)}_{m'm}$ in § 5.7. $D^{(j)}_{m'm}(R)$ functions are very important in the theory of molecular spectra and nuclear shell theory.

5.7 GROUP THEORETICAL CONSIDERATIONS

In this section, we discuss the group theoretical underpinnings of some of the concepts introduced in the theory of angular momentum.

The set of real $n \times n$ orthogonal matrices is the orthogonal group $O(n)$. In $O(3)$, there are 3 independent parameters. We now study the best known and most useful non-abelian continuous group – SO(3), the group of rotations in three dimensions (S stands for special, O stands for orthogonal, 3 for three dimensions).

The SO(3) group consists of all continuous linear transformations in three-dimensional Euclidean space which leave the length of coordinate vectors invariant. Let $\vec{x}$ be an arbitrary vector defined with respect to a Cartesian coordinate frame specified by the orthonormal vectors $\hat{e}_i$ $i = 1, 2, 3$. $\vec{x} = \hat{e}_i x^i$.

Then $\vec{x} \rightarrow \vec{x}'$ under rotation R such that $x'^i = R^i_j x^j$ (55)

where R^i_j are elements of 3×3 rotational matrix. The requirement that $|\vec{x}| = |\vec{x}'|$, i.e. $x_i x^i = x'_i x'^i$ gives

$$RR^T = R^T R = E, \tag{56}$$

for all rotational matrices. Real matrices satisfying this condition have determinants equal ± 1. In this section we consider only proper rotations which can be reached continuously from the identity transformation (zero angle of rotation) and for this case the determinant is $+1$. We see that all (proper) rotation matrices have determinant $+1$. Therefore, the rotation matrices satisfy the condition. det $R = +1$ (57)

Matrices satisfying Eq. (56) but with determinant equal to -1 correspond to rotations combined with discrete spatial reflection transformations

$$I = \begin{pmatrix} -1 & 0 & 0 \\ 0 & -1 & 0 \\ 0 & 0 & -1 \end{pmatrix}.$$

The set of real $n \times n$ proper orthogonal matrices is the special orthogonal group SO(n). These matrices depend on $n(n-1)/2$ real parameters.

If we perform rotation R_1, followed by rotation R_2 the effect is equivalent to a single rotation R_3:

$$R_3 = R_2 R_1, \tag{58}$$

where matrix multiplication on the RHS is understood. It is easy to see that the product of two SO(3) matrices is an SO(3) matrix, that the identity matrix is an SO(3) matrix, and that each SO(3) matrix has an inverse. Therefore, the (proper) rotation matrices from a group – the SO(3) group.

If we include the operation I, namely, inversion with respect to the origin, we obtain the full rotation group or the rotation-reflection group consisting of all proper and improper rotations. This is the group O(3) = SO(3) $\times$ C_i with C_i = {E, I}. O(3) is the group of all 3×3 real orthogonal matrices whose determinants can be equal to either $+1$ or -1.

A general SO(3) group element depends on three continuous group parameters. It is possible to choose these parameters in an infinite number of ways. The two commonly used parametrizations of rotations are described below.

A rotation in three dimension can be denoted as $R(\psi, \hat{n})$ where the unit vector $\hat{n}$ specifies the direction of the axis of rotation and ψ is the angle of rotation about the axis. The unit vector $\hat{n}$ is determined by two angles–say the polar and the azimuthal angles (θ, φ) of its direction. Hence R is characterized by three parameters (ψ, θ, φ) where $0 \leq \psi \leq \pi$, $0 \leq \theta \leq \pi$, and $0 \leq \phi \leq 2\pi$. There is a redundancy in this parametrization

$$R(\pi, -\hat{n}) = R(\pi, \hat{n}). \tag{59}$$

We associate with each rotation a three-dimensional vector $\vec{k} = \psi\hat{n}$. The tips of these vectors fill a three-dimensional sphere of radius π. There exists a one-to-one correspondence between the points in the sphere and the rotations for a given fixed point, except for the points on the surface (corresponding to rotations of π) for which opposite points (i.e., on opposite ends of a diameter) on the surface refer to the same rotation, because of the redundancy. A sphere with this added feture is compact (i.e., closed and bounded) and doubly connected. The latter means that this group manifold allows two distinct classes of closed curves: those that can be continuously deformed into a point, and those that must wrap around the sphere once. Thus the rotation group is doubly connected.

A rotation can also be described by the Euler angles (α, β, γ) (see Fig. 5.1)

Rotations about any fixed axis in the direction $\hat{n}$ form a subgroup of SO(3). Each subgroup is isomorphic to SO(2). Associated with each of these subgroups there is generator J_n.

All elements of the subgroup may be written as

$$R(\psi, \hat{n}) = e^{-i\psi J_n}. \tag{60}$$

They form a one parameter subgroup of SO(3). Each J_n is a 3×3 matrix and under rotations J_n behaves as a vector in the direction of $\hat{n}$ Using infinitesimal angles of rotation, we can write

$$(J_k)^l_m = -i \in_{klm}. \tag{61}$$

$\{J_1, J_2, J_3\}$ from a basis for the generators of all the one parameter abelian subgroups of SO(3). In terms of the generators the Euler angle representation of a rotation can be written as

$$R(\alpha, \beta, \gamma) = e^{-i\alpha J_3}\, e^{-i\beta J_2}\, e^{-i\gamma J_3} \tag{62}$$

The three basis generators $\{J_k\}$ ($k = 1, 2, 3$) satisfy the Lie algebra

$$[J_k, J_l] = i\in_{klm} J^m \tag{63}$$

Combined with Eq. (62), the commutation relations, Eq. (63), determine the most important properties (local) of the group structure and group representations Each component J_i commutes with the operator

$$\vec{J}^2 = \vec{J} \cdot \vec{J} = (J_1)^2 + (J_2)^2 + (J_3)^2, \ i.e.$$

$$[J_i, J^2] = 0 \quad (i = 1, 2, 3). \tag{64}$$

$\vec{J}^2$ is the Casimir operator.

We have discussed the irreducible representations in Section 5.5. In this section, we summarize the results. The IR's of the Lie algebra of SO(3) are each characterized by an angular momentum eigenvalue j from the set of positive integers and half-integers. The orthonormal basis vectors and the matrix elements have been studied. For j a positive integer, the representations are all single-valued and for j a half-odd integer, the representations are all double valued. The existence of the double-valued representations is due to double-connectedness of the group manifold. The generators $\{J_i\}$ are the components of the vector operator $\vec{J}$ (the angular momentum-operator measured in units of $\hbar$). All fermions (with half-odd integer spin) are described by wave functions corresponding to the double-valued representations of SO(3). All bosons (with integer spin) are described by the wave functions corresponding to the single-valued representations of SO(3).

We now study the non-abelian continuous group SU(2). The unimodular unitary group SU(2) – the group of two dimensional unitary matrices with unit determinant-has the rotation group in three dimensions (SO(3)) as a homomorphic (in fact, 2 to 1) image. SU(2) has the same Lie algebra as SO(3). The group SU(2) is a three-parameter, semi-simple compact Lie group. All IR's for the Lie algebra are single-valued representations of SU(2), in contrast to the case of SO(3). In addition to the vector representations (for which the phase is one) of SO(3) (the integral angular momenta), the ray representations (for which the phase is different from one) of SO(3) (the half-integral angular momenta) play an important role in physics. In order to include these, we go to the "universal covering group" SU(2) in SO(3). This means that there exists a larger group ("covering group") SU(2) in which the ray representations of the original group SO(3) now appear as vector representations of the covering group.

The group of SU(n) is the set of complex $n \times n$ unitary and unimodular matrices:

$$UU^\dagger = U^\dagger U = 1, \ \det U = 1 \tag{65}$$

The matrices of $SU(n)$ depend upon $n^2 - 1$ real parameters.

An arbitrary 2×2 unitary matrix U can be written in the form.

$$U = e^{i\lambda} \begin{pmatrix} \cos\theta\, e^{i\xi} & -\sin\theta\, e^{i\eta} \\ \sin\theta\, e^{-i\eta} & \cos\theta\, e^{-i\xi} \end{pmatrix} \tag{66}$$

where $0 \le \theta \le \pi$, $0 \le \lambda < \pi$, and $0 \le \eta, \xi \le 2\pi$. An arbitrary 2×2 SU(2) matrix A can be parametrized in terms of three real parameters (θ, η, ζ), as in Eq. (66) without the overall phase factor. The general SU(2) matrix can be put in the form of $D^{(\frac{1}{2})}(\alpha, \beta, \gamma)$ Eq. (44) with

$$\theta = \beta/2$$

$$\zeta = (-\alpha - \gamma)/2$$

$$\eta = (-\alpha + \gamma)/2.$$

The ranges of the new variables then becomes

$$0 \le \beta \le \pi$$

$$0 \le \alpha < 2\pi$$

$$0 \le \gamma < 4\pi.$$

Notice that range of γ is twice that of the physical Euler angle γ, because the SU(2) matrices form a double-valued representation of SO(3).

The SU(2) matrix can also be written as

$$A = \begin{pmatrix} r_0 - ir_3 & -r_2 - ir_1 \\ r_2 - ir_1 & r_0 + ir_3 \end{pmatrix} \tag{67}$$

subject to the condition

$$\det A = r_0^2 + r_1^2 + r_2^2 + r_3^2 = 1,$$

where r_i are real numbers. If we regard r_i ($i = 0, 1, 2, 3$) as the Cartesian coordinates in a four-dimensional Euclidean space, the group parameter space is the surface of a sphere of radius one. Thus, there is a one-to-one correspondence between the elements of SU(2) and the surface points of the sphere. Two diametrically opposite points correspond to A and $-A$. The group manifold is compact and simply connected.

Let us write a 2×2 hermitian traceless matrix

$$X = \sigma_i x^i,$$

where σ_i ($i = 1, 2, 3$) are the Pauli matrices and $\vec{x} = (x^1, x^2, x^3)$ is a coordinate vector. Now,

$$-X = - \begin{vmatrix} x^3 & x^1 - ix^2 \\ x^1 + ix^2 & -x^3 \end{vmatrix} = |\vec{x}|^2.$$

Let A be an arbitrary SU(2) matrix, which induces a linear transformation on X:

$$X \to X' = AXA^{-1}.$$

X' is a hermitian traceless 2×2 matrix, and we can write

$$X' = \sigma_i x'^i \text{ where}$$

$\vec{x}' = (x'^1, x'^2, x'^3)$. Since $\det X = \det X'$, we see that $|\vec{x}'^2| = |\vec{x}^2|$ Thus, the SU(2) transformation induces an SO(3) transformation in the three-dimensional Euclidean space. The mapping $A \in$ SU(2) to $R \in$ SO(3) is two-to-one, as the two SU(2) matrices $\pm A$ correspond to the same rotation.

In Eq. (67), we may consider r_1, r_2, r_3 as the independent variables with $r_0 = [1 - (r_1^2 + r_2^2 + r_3^2)]^{1/2}$.

The identity element E corresponds to $r_1 = r_2 = r_3 = 0$.

We now consider an infinitesimal transformation about the identity. Then $r_k \to dr^k$ ($k = 1, 2, 3$) and $r_0 \to 1+$ (second order terms in dr^k). Hence we may write Eq. (67) as

$$A = E - i\, dr^k\, \sigma_k. \tag{68}$$

Thus, $\{\sigma_k\}$ is a basis for the Lie algebra of SU(2). σ_i satisfy the following commutation relations

$$[\sigma_k, \sigma_l] = 2i \in^{klm} \sigma_m. \tag{69}$$

Comparing Eqs. (69) with (63) we see that SU(2) and SO(3) have the same Lie algebra with the correspondence

$$J_k \to \sigma_k/2. \tag{70}$$

We have constructed all the IR's of this Lie algebra. As SU(2) is a simply connected group, all the IR's of its Lie algebra are single-valued IR's of the group.

We now obtain the differential equation satisfied by $d^{(j)}(\beta)$. Since

$$R(\alpha, \beta, \gamma) = e^{-i\alpha J_3}\, e^{-i\beta J_2}\, e^{-i\gamma J_3},$$

we get

$$i \frac{\partial}{\partial \alpha} R(\alpha, \beta, \gamma) = J_3 R(\alpha, \beta, \gamma) = R\,[R^{-1} J_3 R],$$

$$i \frac{\partial}{\partial \beta} R(\alpha, \beta, \gamma) = R\,[e^{i\gamma J_3} J_2\, e^{-i\gamma J_3}],$$

$$i \frac{\partial}{\partial \gamma} R(\alpha, \beta, \gamma) = R J_3.$$

Using the generators J_3, $J_\pm$, we may write

$$R^{-1} J_3 R = -\sin\beta \left(J_+ e^{i\gamma/2} + J_- e^{-i\gamma/2}\right) + J_3 \cos\beta$$

$$e^{i\gamma J_3} J_2\, e^{-i\gamma J_3} = i\,[-J_+ e^{i\gamma/2} + J_- e^{-i\gamma/2}].$$

Therefore,

$$e^{-i\gamma} \left[-\frac{\partial}{\partial\beta} - \frac{i}{\sin\beta}\left(\frac{\partial}{\partial\alpha} - \cos\beta\, \frac{\partial}{\partial\gamma}\right)\right] R = R J_+$$

$$e^{i\gamma} \left[\frac{\partial}{\partial\beta} - \frac{i}{\sin\beta}\left(\frac{\partial}{\partial\alpha} - \cos\beta\, \frac{\partial}{\partial\gamma}\right)\right] R = R J_-$$

$$i \frac{\partial}{\partial\gamma} R = R J_3.$$

Taking matrix elements between $|jm\rangle$ and $\langle jm'|$, we obtain

$$\left[-\frac{d}{d\beta} - \frac{1}{\sin\beta}(m' - m\cos\beta)\right] d^{(j)}_{m'm}$$

$$= d^{(j)}_{m', m+1}(\beta)\,[j(j+1) - m(m+1)]^{1/2},$$

$$\left[\frac{d}{d\beta} - \frac{1}{\sin\beta}(m' - m\cos\beta)\right] d^{(j)}_{m'm}$$

$$= d^{(j)}_{m', m-1}(\beta)\,[j(j+1) - m(m-1)]^{1/2}$$

Now,

$$R J^2 = R \lfloor J_+ J_- + J_3^2 - J_3 \rfloor$$

$$= \left\{ e^{-i\gamma} \left[-\frac{\partial}{\partial\beta} - \frac{i}{\sin\beta}\left(\frac{\partial}{\partial\alpha} - \cos\beta\, \frac{\partial}{\partial\gamma}\right)\right] e^{i\gamma} \left[\frac{\partial}{\partial\beta} - \frac{i}{\sin\beta}\left(\frac{\partial}{\partial\alpha} - \cos\beta\, \frac{\partial}{\partial\gamma}\right)\right] \right.$$

$$\left. - \frac{\partial^2}{\partial\gamma^2} - i\frac{\partial}{\partial\gamma} \right\} R.$$

Taking matrix elements between the states $|jm>$ and $<jm'|$, we obtain the differential equation for $d^{(j)}(\beta)$

$$\left\{\frac{1}{\sin\beta}\frac{d}{d\beta}\sin\beta\frac{d}{d\beta}-\frac{1}{\sin^2\beta}(m^2+m'^2-2mm'\cos\beta)+j(j+1)\right\}$$

$$\times d^{(j)}_{m'm}(\beta)=0 \tag{71}$$

Using our phase convention

$$J_{\pm}|jm\rangle=[j(j+1)-m(m\pm1)]^{1/2}|jm\pm1\rangle$$

and the requirement that $d^{(j)}$ is a unitary matrix, the solution is

$$d^{(j)}_{m'm}(\beta)=\frac{(-1)^{j-m'}}{(m+m')!}\sqrt{\frac{(j+m)!\,(j+m')!}{(j-m)!\,(j-m')!}}\left(\sin\frac{1}{2}\beta\right)^{2j}\left(\cot\frac{1}{2}\beta\right)^{m+m'}$$

$$\times\,{}_2F_1\left(m-j,\,m'-j;\,m+m'+1;\,-\cot^2\frac{1}{2}\beta\right), \tag{72}$$

when $m\geq m'$; for other values of m and m' we use

$$d^{(i)}_{m'm}(\beta)=(-1)^{m'-m}\,d^{(j)}_{mm'}(\beta). \tag{73}$$

In terms of the Jacobi polynomial, we write

$$d^{(j)}_{m'm}(\beta)=\left[\frac{(j+m')!\,(j-m')!}{(j+m)!\,(j-m)!}\right]^{\frac{1}{2}}(\cos\beta/2)^{m'+m}\,(\sin\beta/2)^{m'-m}$$

$$\times\,P^{(m'-m,\,m'+m)}_{j-m'}(\cos\beta) \tag{74}$$

$$=\left[\frac{(j+m')!\,(j-m')!}{(j+m)!\,(j-m)!}\right]^{\frac{1}{2}}\sum_\sigma\binom{j+m}{j-m'-\sigma}\binom{j-m}{\sigma}(-1)^{j-m'-\sigma}$$

$$\times\,(\cos\beta/2)^{2\sigma+m'+m}\,(\sin\beta/2)^{2j-2\sigma-m'-m} \tag{75}$$

using the series expression

$$P^{(\alpha,\beta)}_n(x)=2^{-n}\sum_{v=0}^{n}\binom{n+\alpha}{v}\binom{n+\beta}{n-v}(x-1)^{n-v}(x+1)^v \tag{76}$$

From Eqs. (75) and (76) we get, for the case $j=1$, the matrix $d^{(1)}(\beta)$:

m' \ m	$+1$	0	-1
$+1$	$\frac{1}{2}(1+\cos\beta)$	$\frac{1}{\sqrt{2}}\sin\beta$	$\frac{1}{2}(1-\cos\beta)$
0	$-\frac{1}{\sqrt{2}}\sin\beta$	$\cos\beta$	$\frac{1}{\sqrt{2}}\sin\beta$
-1	$\frac{1}{2}(1-\cos\beta)$	$-\frac{1}{\sqrt{2}}\sin\beta$	$\frac{1}{2}(1+\cos\beta)$

$$\tag{77}$$

The rotation matrices satisfy several symmetry relations.

From Eq. (76), $\qquad d^{(j)}_{m'm}(\beta) = d^{(j)}_{-m,-m'}(\beta)$.

Since $\qquad d^{(j)}(\beta)\, d^{(j)}(-\beta) = 1$ and $d^{(j)}$ is an orthogonal matrix, we have

$$d^{(j)}_{mm'}(\beta) = d^{(j)}_{m'm}(-\beta),$$

Putting $-\beta$ in place of β in Eq. (76), we get

$$d^{(j)}_{m'm}(\beta) = d^{(j)}_{m'm}(-\beta),$$

Combining the above two relations, we have

$$d^{(j)}_{mm'}(\beta) = (-1)^{m'-m}\, d^{(j)}_{-m',-m}(-\beta).$$

and so

$$D^{(j)}_{mm'}(R)^{*} = (-1)^{m-m'}\, D^{(j)}_{-m,-m'}(R). \tag{78}$$

Also, $\qquad d^{(j)}_{mm'}(\pi) = (-1)^{j+m}\, \delta_{m,-m'}$

and $\qquad d^{(j)}_{m,-m'}(\pi - \beta) = (-1)^{j+m'}\, d^{(j)}_{m,-m'}(\beta).$

5.8 THE ROTATION OF A RIGID BODY

Let us consider the rotational motion of a rigid body, the atoms being rigidly fixed. The rotational motion of such a solid body (a rotor or "top") is a useful model of the rotational energy levels of a polyatomic molecule or a nucleus.

Let ξ, η, ζ be a system of coordinates with axes along the principal axes of a rigid body. Let $\vec{a}$, $\vec{b}$, $\vec{c}$ be mutually orthogonal unit vectors along the ξ, η, ζ axes. These vectors characterize the position of the rigid body in space. These three commuting vectors are the dynamical variables.

The quantum-mechanical Hamiltonian is obtained from the classical expression of the energy as

$$H = \frac{1}{2}\,\hbar^2 \left(\frac{J_a^2}{I_a} + \frac{J_b^2}{I_b} + \frac{J_c^2}{I_c} \right), \tag{79}$$

where I_a, I_b, and I_c are the principal moments of inertia and J_a, J_b, and J_c are the angular momentum operators for components along the principal axes.

The commutation relations of the angular momentum operators for components along space-fixed axes are

$$[J_x, J_y] = i\hbar\, J_z. \text{ (cyclic)} \tag{80}$$

We write $\vec{J} = J_a\vec{a} + J_b\vec{b} + J_c\vec{c} = \sum J_i\hat{e}_i$ $(i = x, y, z)$ (81)

Now,

$$[J_a, J_b] = [J_i a_i, J_j b_j] = J_i\,[a_i, J_j]\,b_j + [J_i, J_j]\,a_i b_j + J_j\,[j_i, b_j]\,a_i. \quad (82)$$

Here, we have used the identity

$$[AB, CD] = A\,[B, C]\,D + AC\,[B, D] + [A, C]\,DB + C\,[A, D]\,B$$

and $[a_i, b_j] = 0$.

But, $\qquad\qquad\qquad [J_i, a_j] = i\hbar \in_{ijk} a_k$

Hence,

$$[J_a, J_b] \equiv (\vec{J}\cdot\vec{a})\,(\vec{J}\cdot\vec{b}) - (\vec{J}\cdot\vec{b})\,(\vec{J}\cdot\vec{a})$$

Since $\qquad\qquad\qquad \vec{a}\times\vec{b} = \vec{c}$, we have

$$[J_a, J_b] = -i\hbar J_c \;(\text{cyclic}). \quad (82)$$

Next, we find the energy eigenvalues of a rotating top. There are special cases of a top, depending on the symmetries of the moment of inertia tensor.

If all three principal moments of inertia of the body are equal $(I_a = I_b = I_c = I)$ the body is called a spherical top. The Hamiltonian Eq. (79) takes the form

$$H = \frac{\hbar^2\vec{J}^2}{2I} \quad (83)$$

and its eigenvalues are

$$E = \frac{\hbar^2 J(J+1)}{2I}. \quad (84)$$

To consider the degeneracy of these energy levels, we consider a

c.s.c.o.$-\vec{J}^2 = \vec{J}\cdot\vec{J} = J_x^2 + J_y^2 + J_z^2 = J_a^2 + J_b^2 + J_c^2;\ J_z;\ J_c.$

$I_c = \vec{c}\cdot\vec{J}$ is a psendoscalar operator, it commutes with all the components of $\vec{J} : [\vec{J}, J_c] = 0$ etc. So each of these energy levels is $(2J + 1)$ fold degenerate with respect to the value of J_c, $\hbar K$ is the eigenvalue of the psendoscalar J_c, ranging from $-J$ to $+J$. There is also $(2J + 1)$ fold degeneracy corresponding to the eigenvalues of J_z (fixed coordinate system). The total degree of degeneracy of a spherical top is $(2J + 1)^2$.

On quantization, the spherical top shows a high degree of degeneracy characteristic of the four dimensional rotation group. If only two of the moments of inertia are equal, but distinct from the third $(I_a = I_b \neq I_c)$ the body is called a symmetric top. The Hamiltonian takes the form

$$H = \frac{\hbar^2}{2I_a}(J_a^2 + J_b^2) + \frac{\hbar^2}{2I_c}J_c^2 = \frac{1}{2}\hbar^2\frac{\vec{J}^2}{I_a} + \frac{1}{2}\hbar^2\left(\frac{1}{I_c} - \frac{1}{I_a}\right)J_c^2. \tag{85}$$

The energy eigenvalues of a state with given values of J, K are

$$E_{jk} = \frac{\hbar^2}{2I_a}J(J+1) + \frac{\hbar^2}{2}\left(\frac{1}{I_c} - \frac{1}{I_a}\right)K^2. \tag{86}$$

The three quantum numbers characterizing the stationary states of a symmetric top are J. M, and K. The energy is independent of M. There is degeneracy with respect to $\pm K$. The total degree of degeneracy is $2(2J+1)$.

The wavefunctions of the stationary states of a symmetric top with respect to the body-fixed axes and to the space-fixed axes are related by the rotation $R(\alpha, \beta, \gamma)$. We denote the states in the frame in which the body is instantaneously at rest by $|JK>_B$ where B stands for body-fixed. The amplitude for finding $|JK>_B$ in the space-fixed state $|JM>$ is the wave function

$$\psi_{JMK} = {}_B<JK|JM>. \tag{87}$$

$$|JK>_B = D^{(J)}_{MK}(\alpha, \beta, \gamma)|JM>, \tag{88}$$

$$\psi = <JM|D|JK>.$$

Since

The wave functions are

$$\psi_{JMK}(\alpha, \beta, \gamma) = \sqrt{2J+1}\, D^{(J)}_{MK}(\alpha, \beta, \gamma)^* \tag{89}$$

The normalization is as follows

$$\int |\psi_{JMK}|^2 \sin\beta\, d\alpha\, d\beta\, d\gamma = 1. \tag{90}$$

Here, we consider only integral values of J. When all three moments of inertia are different ($I_a \neq I_b \neq I_c$), the body is called an "asymmetrical top". It is not possible to obtain a general closed form expression for the eigenvalues for an asymmetrical top. Surprisingly, the degree of degeneracy in its energy levels is $(2J+1)$, characteristic of spherical symmetry. The reader should consult Landau and Lifshitz (1977).

5.9   SCHWINGER OPERATOR METHOD

According to Schwinger, the theory of angular momentum can be considered from the algebra of two independent harmonic oscillators. This elegant method, given by Schwinger in his report "On Angular Momentum" [AEC Report NYO-3071 (1952)] and reprinted in Biedenharn and van Dam (1965)] derives from Wigner (1959).

Let us consider two independent harmonic oscillators. One oscillator is described by the annihilation and creation operators a_+ and $a_+^\dagger$, while the other is described by the operators a_- and $a_-^\dagger$. The particle number operators are

$$N_+ = a_+^\dagger a_+, \ N_- = a_-^\dagger a_-, \ N = N_+ + N_- = a_+^\dagger a_+ + a_-^\dagger a_-. \tag{91}$$

For each oscillator, the usual commutation relations are satisfied:

$$[a_+, a_+^\dagger] = 1 \qquad [a_-, a_-^\dagger] = 1$$

$$[N_+, a_+] = -a_+ \qquad [N_-, a_-] = -a_-$$

$$[N_+, a_+^\dagger] = a_+^\dagger \qquad [N_-, a_-^\dagger] = a_-^\dagger. \tag{92}$$

Since the two oscillators are independent, the "+" operators commute with the "–" operators

$$[a_+, a_-^\dagger] = [a_-, a_+^\dagger] = 0, \text{ etc.} \tag{93}$$

N_+ and N_- commute, and so we can have simultaneous eigenkets of N_+ and N_- with eigenvalues n_+ and n_- respectively. We have

$$N_+ \, |n_+, n_-\rangle = n_+|n_+, n-\rangle,$$

$$N_- \, |n_+, n\rangle = n_-|n_+, n_-\rangle. \tag{94}$$

Further, we have

$$a_+^\dagger \, |n_+, n_-\rangle = \sqrt{n_+ + 1} \, |n_+ + 1, n_-\rangle,$$

$$a_-^\dagger \, |n_+, n_-\rangle = \sqrt{n_- + 1} \, |n_+, n_- + 1\rangle,$$

$$a_+|n_+, n_-\rangle = \sqrt{n_+} \, |n_+ - 1, n_-\rangle,$$

$$a_-|n_+, n_-\rangle = \sqrt{n_-} \, |n_+, n_- - 1\rangle. \tag{95}$$

Let us denote by $|0, 0\rangle$ the vacuum ket in the Fock space.

$$a_+ \, |0, 0\rangle = 0,$$

$$a_- \, |0, 0\rangle = 0. \tag{96}$$

We write the normalized state with prescribed occupation numbers n_+ and n_- (see Chapter 17):

$$|n_+, n_-\rangle = \frac{(a_+^\dagger)^{n_+} (a_-^\dagger)^{n_-}}{\sqrt{n_+! \, n_-!}} \, |0, 0\rangle. \tag{97}$$

We define three operators as

$$J_i = (a_+^\dagger, a_-^\dagger) \frac{\hbar \sigma_i}{2} \begin{pmatrix} a_+ \\ a_- \end{pmatrix}, \quad i = 1, 2, 3 \tag{98}$$

or,

$$J_1 = \frac{\hbar}{2} \left(a_+^\dagger a_- + a_-^\dagger a_+ \right)$$

$$J_2 = \frac{\hbar}{2i} \left(a_+^\dagger a_- - a_-^\dagger a_+ \right)$$

$$J_3 = \frac{\hbar}{2} \left(a_+^\dagger a_+ - a_-^\dagger a_- \right). \tag{99}$$

These operators satisfy the angular momentum commutation relations

$$\lfloor J_i, J_j \rfloor = i\hbar \in_{ijk} J_k.$$

We find that

$$\vec{J}^2 = \sum_i J_i^2 = \hbar^2 \frac{N}{2} \left(\frac{N}{2} + 1 \right), \tag{100}$$

where $N = N_+ + N_- = a_+^\dagger a_+ + a_-^\dagger a_-$ is the operator for the total particle number. The eigenvalues of $\vec{J}^2$ are of the form $j\,(j+1)$ where $j = \frac{n}{2}$ take the possible values $0, \frac{1}{2}, 1, \frac{3}{2}, 2, \dots$ Also,

$$J_3\, |n_+, n_-\rangle = \frac{1}{2}\,(n_+ - n_-)|\, n_+, n_-\rangle \tag{101}$$

$$\vec{J}^2\, |n_+, n_-\rangle = \frac{n}{2} \left(\frac{n}{2} + 1 \right) \Big| n_+, n_- \rangle, \tag{102}$$

where $n = n_+ + n_-$. So the state $|n_+, n_-\rangle$ is an eigenstate of $\vec{J}^2$ and J_3 with eigenvalues

$$j\,(j+1),\ j = \frac{1}{2}\,(n_+ + n_-),\ m = \frac{n_+ - n_-}{2}. \tag{103}$$

Clearly, there are $n_+ + n_- + 1 = 2j + 1$ possible values of m (with $n_+ + n_-$ fixed); these are $j, j-1, \dots, -j+1, -j$.

From Eq. (52), the operators $J_\pm = J_1 \pm iJ_2$ are

$$J_+ = \hbar a_+^\dagger a_-,\ J_- = \hbar a_-^\dagger a_+ \tag{104}$$

satisfying the usual commutation relations

$$[J_3, J_\pm] = \pm\,\hbar J_\pm,$$

$$[J_+, J_-] = 2\hbar J_3.$$

Furthermore,

$$J_+\, |n_+, n_-\rangle = \hbar a_+^\dagger a_- |n_+, n_-\rangle = \hbar \sqrt{n_-\,(n_+ + 1)}\ |n_+ + 1, n_- - 1\rangle$$

$$J_-\, |n_+, n_-\rangle = \hbar a_-^\dagger a_+ |n_+, n_-\rangle = \hbar \sqrt{n_+\,(n_- + 1)}\ |n_+ - 1, n_- + 1\rangle \tag{105}$$

The structure of the algebra of two independent harmonic oscillators suggests the following physical interpretation.

We associate each + quantum unit of the oscillator with plus one-half angular momentum in the z-direction $\left(m = \dfrac{1}{2}\right)$ and each − quantum unit of the oscillator with minus one-half unit of angular momentum in the z-direction $\left(m = -\dfrac{1}{2}\right)$. We may imagine one spin $\left(\dfrac{1}{2}\right)$ particle with spin up (down) with each quantum unit of the + (−) type of oscillator. The total number of units $\left(\dfrac{1}{2}\right)n_+ + \left(\dfrac{1}{2}\right)n_-$ is the total j value, while the net angular momentum in the z-direction (m value) is $\left(\dfrac{1}{2}\right)n_+ - \left(\dfrac{1}{2}\right)n_-$, the eigenvalues n_+ and n_- give the number of spins up and spins down, respectively. The operator J_+ increases the m value by one keeping j fixed. It reduces the number of − quanta by one and increases the number of + quanta by one, thereby increasing the z-component of angular momentum by $\hbar$. Similarly, J_- increases n_- by one and reduces n_+ by one, thereby decreasing the z-component of angular momentum by $\hbar$.

We now label the state $|n_+, n_-\rangle$ by j, m values. Notice that $n_+ = j + m$ and $n_- = j - m$ so that we write Eq. (51) as the eigenket

$$|j, m\rangle = \frac{(a_+^\dagger)^{j+m} (a_-^\dagger)^{j-m}}{\sqrt{(j + m)! \, (j - m)!}} \, |0\rangle, \tag{106}$$

where $|0\rangle$ denotes the vacuum ket, denoted previously as $|0, 0\rangle$. Using Eq. (59), it is easy to show that

$$J_\pm |j, m\rangle = \sqrt{j\,(j + 1) - m\,(m \pm 1)} \, \hbar |j, m \pm 1\rangle. \tag{107}$$

Let us now evaluate the rotation matrix elements. We consider $D(R)|\,j, m\rangle$. As we know, the α, γ dependence is in simple exponential factors, so we consider rotation about the y-axis, i.e.,

$$D(R) \equiv D(\alpha, \beta, \gamma)|_{\alpha=\gamma=0} = e^{-\frac{i}{\hbar} J_y \beta}. \tag{108}$$

We now have

$$D(R)|\,j, m\rangle = \frac{[D(R)\, a_+^\dagger \, D^{-1}(R)]^{j+m} \, [D(R)\, a_-^\dagger \, D^{-1}(R)]^{j-m}}{\sqrt{(j + m)! \, (j - m)!}}$$

$\times \, D(R)|0\rangle$, using the explicit form for $|j, m\rangle$ given in Eq. (59)

Clearly, $\qquad\qquad D(R)|0\rangle = |0\rangle$ and

$$DR)\, a_{\pm}^{\dagger} D^{-1}(R) = e^{-\frac{i}{\hbar} J_y \beta}\, a_{\pm}^{\dagger}\, e^{-\frac{i}{\hbar} J_y \beta}.$$

We have

$$\frac{\partial}{\partial \beta}\left(e^{-\frac{i}{\hbar} J_y \beta}\, a_{\pm}^{\dagger}\, e^{\frac{i}{\hbar} J_y \beta}\right) = \frac{1}{i\hbar}\, e^{-\frac{i}{\hbar} J_y \beta}\, [J_y, a_{\pm}^{\dagger}]\, e^{\frac{i}{\hbar} J_y \beta}.$$

But,

$$[J_y, a_{\pm}^{\dagger}] = \frac{1}{2i}\, [a_{+}^{\dagger} a_{-} - a_{-}^{\dagger} a_{+},\, a_{\pm}^{\dagger}]$$

$$= \mp \frac{1}{2i}\, a_{\mp}^{\dagger}.$$

Thus,

$$\frac{\partial}{\partial \beta}\left(e^{-\frac{i}{\hbar} J_y \beta}\, a_{\pm}^{\dagger}\, e^{\frac{i}{\hbar} J_y \beta}\right) = \pm \frac{1}{2\hbar}\left(e^{-\frac{i}{\hbar} J_y \beta}\, a_{\pm}^{\dagger}\, e^{\frac{i}{\hbar} J_y \beta}\right)$$

The solutions of these two equations are

$$e^{-\frac{i}{\hbar} J_y \beta}\, a_{+}^{\dagger}\, e^{\frac{i}{\hbar} J_y \beta} = a_{+}^{\dagger} \cos\left(\frac{\beta}{2}\right) + a_{-}^{\dagger} \sin\frac{\beta}{2},$$

$$e^{-\frac{i}{\hbar} J_y \beta}\, a_{-}^{\dagger}\, e^{\frac{i}{\hbar} J_y \beta} = a_{-}^{\dagger} \cos\left(\frac{\beta}{2}\right) - a_{+}^{\dagger} \sin\left(\frac{\beta}{2}\right). \tag{109}$$

Putting Eqs. (60) and (61) into (59), we obtain

$$D(\alpha = 0,\, \beta,\, \gamma = 0) | j, m \rangle = \sum_{k} \sum_{l} \frac{(j+m)!\,(j-m)!}{(j+m-k)!\,(j-m-l)!\,l!}$$

$$\times \frac{\left\{a_{+}^{\dagger} \cos\left(\frac{\beta}{2}\right)\right\}^{j+m-k} \left\{a_{-}^{\dagger} \sin\left(\frac{\beta}{2}\right)\right\}^{k}}{\sqrt{(j+m)!\,(j-m)!}}$$

$$\times \left\{- a_{+}^{\dagger} \sin\left(\frac{\beta}{2}\right)\right\}^{j-m-l} \left\{a_{-}^{\dagger} \cos\left(\frac{\beta}{2}\right)\right\}^{l} |0\rangle. \tag{110}$$

The LHS of Eq. (63) may be written as

$$\sum_{m'} | j, m' \rangle\, d^{(j)}_{m'm}(\beta)$$

$$= \sum_{m'} d^{(j)}_{m'm}(\beta)\, \frac{(a_{+}^{\dagger})^{j+m'}\,(a_{-}^{\dagger})^{j-m'}}{\sqrt{(j+m)!\,(j-m')!}}\, |0\rangle.$$

Equating the coefficients of equal powers of $a_{+}^{\dagger}$, one obtains Wigner's formula for $d^{(j)}_{m'm}(\beta)$:

$$d_{m'm}^{(j)}(\beta) = \sum_k (-1)^{k-m+m'} \frac{\sqrt{(j+m)!\,(j-m)!\,(j+m')!\,(j-m')!}}{(j+m-k)!\,k!\,(j-k-m')!\,(k-m+m')!}$$

$$\times \left(\cos\frac{\beta}{2}\right)^{2j-2k+m-m'} \left(\sin\frac{\beta}{2}\right)^{2k-m+m'}, \tag{111}$$

where the sum over k is taken such that the arguments of factorials in the denominator are not negative.

In the Schwinger operator method the representations $d^{(j)}$ are generated by symmetrized tensorial product of the fundamental representation $d^{\frac{1}{2}}$ [SU (2)] by Fock method. One considers the elementary spin $\frac{1}{2}$ particles which can exist in two possible states in second quantization. The spin creation and annihilation operators satisfy boson commutation relations.

The isospin operators satisfy all the properties of the angular momentum operators. So an isospin state $|I, I_3\rangle$ can be described in the Schwinger scheme as a state of two oscillators with the numbers of quanta $n_+ + I + I_3$ and $n_- = I - I_3$.

The Schwinger oscillator method has recently been extended in "deformed algebra"

PROBLEMS

5.1 Find the transformation rule of the spherical harmonics Y_{11}, Y_{10}, Y_{1-1} under a rotation of the coordinate system by the Euler angles α, β, γ.

5.2 Consider a particle with spin 1. Find the matrix transforming the components of the spin under a rotation of the coordinate system.

5.3 Obtain explicit expressions for the reduced rotation matrices $d_{mm'}^{(j)}$, for $j = 1, \frac{3}{2}, 2$.

5.4 Show that

$$\langle jm|\,J_+^{(m-m')}\,|jm'\rangle = \left[\frac{(j+m)!\,(j-m')!}{(j+m')!\,(j-m)!}\right]^{1/2}.$$

5.5 Establish the Schrödinger equation for a symmetric rotor.

5.6 (a) Calculate the matrix elements $\langle JK'\,|H|\,JK\rangle$ for an asymmetrical top.

(b) Consider the states with $J = 1$. Find the energy eigenvalues.

5.7 Use Schwinger's method to derive Wigner's formula for $d_{mm'}^{(j)}$, (β), (Eq. (75)).

CHAPTER 6

Spin

The intrinsic angular momentum of a particle is called its spin angular momentum or, spin. This is distinct from the angular momentum due to its motion in space called the orbital angular momentum. The particle may be an elementary particle or a composite system behaving in some sense as an elementary one. For a particle with spin, we denote the rest-frame angular momentum (spin) by the vector operator $\vec{s}$. Every elementary particle carries a spin s (in unit of $\hbar$) : pi-meson s = 0; electron, mu-meson, neutron, proton $s = \dfrac{1}{2}$; photon, W, Z s = 1, etc. s is fixed and immutable. This is in contrast to the orbital angular momentum quantum number which can take only integer value, and can change if the system is perturbed.

Uhlenbeck and Goudsmit (1925) introduced the notion of an intrinsic angular momentum (spin) of $\dfrac{1}{2}\hbar$ with an electron in addition to the orbital angular momentum and a magnetic moment of one Bohr magneton, $\mu_B = \dfrac{|e|\hbar}{2m_e c}$.

In this chapter, we shall first develop the mathematical description of a particle with spin. We then consider the Stern-Gerlach experiment. We discuss the spin precession in Section 6.3. The Pauli equation is studied in Section 6.4. In the Section 6.5, the density matrix description of a particle with spin is considered. Finally, in Section 6.6, the helicity formalism is introduced.

6.1 THE DESCRIPTION OF A SPIN-S PARTICLE

A particle with spin is characterized by three dynamical variables, in addition to its position and momentum. These are called the components of spin.

The total angular momentum of a particle is

$$\hbar\vec{J} = \hbar\vec{L} + \hbar\vec{s}. \tag{1}$$

$\hbar\vec{s}$ is the spin angular momentum, and $\hbar\vec{L}$ is the orbital angular momentum. $\vec{L}$ and $\vec{s}$ commute. In an isolated system with rotational symmetry, the Hamiltonian H commutes with $\vec{J}$, but not with $\vec{L}$ or $\vec{s}$ separately.

Corresponding to the spin variables, there are the observables s_i which satisfy the usual commutation relations of angular momentum components

$$[s_i, s_j] = i\hbar \sum_k \in_{ijk} s_k. \tag{2}$$

The eigenvalues of the square of the spin are

$$\vec{s}^{\,2} = \hbar^2 \, s(s+1), \tag{3}$$

for a given value s such that $2s$ is a non-negative integer. s can therefore take an integer (including zero) or a half odd integer value: $s = 0, \frac{1}{2}, 1, \frac{3}{2}, \dots$. For given s, the component s_z can take the values $s, s-1, \dots, -s$, i.e. $2s+1$ components. So the wave function of a spin-s particle has $2s+1$ components. Clearly, the operators $(\vec{s}^{\,2}, s_z)$ form a complete set of compatible spin observables. Although s is fixed for a particle, the orientation of $\vec{s}$ can change.

For a spin-s particle in motion a c.s.c.o. consists of four operators, e.g., $\vec{p}$, s_z or $\vec{r}$, s_z. We do not include s^2 since its eigenvalue is fixed by the nature of the particle. Other c.s.c.o. are $(H, \vec{L}^2, \vec{L}z, s_z)$, and $H, \vec{L}^2, \vec{J}^2, J_z$.

The Hilbert space $\mathcal{H}$ for a spin-s particle can be constructed from the direct product of the infinite-dimensional $\varepsilon^{\text{space}}$ on which the observables $\vec{r}$ and $\vec{p}$ act and a $(2s+1)$ – dimensional complex vector space $\varepsilon^{\text{spin}}$ on which the observables s_i act:

$$\mathcal{H} = \varepsilon^{\text{space}} \times \varepsilon^{\text{spin}}, \tag{4}$$

We can take the eigenvectors $|m_s>$ of s_z to span the spin space. For the observables $\vec{r}$, s_z we write the basis as

$$|\vec{r}, m_s> = |\vec{r}> \otimes |m_s>; \tag{5}$$

for the observables $\vec{p}$, s_z, we write

$$|\vec{p}, m_s> = |\vec{p}> \otimes |m_s>; \tag{5a}$$

and for (H, L^2, L_z, s_z),

$$|Elm_l\, m_s> = |Elm_l> \otimes |m_s>. \tag{5b}$$

For simplicity, we omit the subscript on m_s.

The total wave function depends on both position (or, momentum) and spin:

$$\psi_m(\vec{r}) = <\vec{r}, m|\psi>. \tag{6}$$

The number of components of $\psi_m(\vec{r})$ is equal to the number of possible values of m.

$|m>$ is the eigenket in

$$s_z|m> = \hbar m|m>.\tag{7}$$

In the space-spin representation, the Schrödinger equation involves differential operators which act on $\vec{r}$ and matrices which act on the spin index.

The mathematical theory of spin is the general theory of angular momentum which has been discussed in the previous two chapters. The addition of angular momenta is considered in the next chapter.

Let us look at some examples.

(a) spin zero:

This is the trival one-dimensional representation. There is only one state $|0>$.

(b) spin $\dfrac{1}{2}\hbar$.

The spin operators act on a two-dimensional vector space. The eigenvalues of s_z are $\pm\dfrac{\hbar}{2}$. A complete set of eigenvectors are $\left|\dfrac{1}{2}>, \left|-\dfrac{1}{2}>\right.\right.$ such that

$$s_z\left|\frac{1}{2}> = \frac{\hbar}{2}\left|\frac{1}{2}>,\right.\right.$$

$$s_z\left|-\frac{1}{2}> = -\frac{\hbar}{2}\left|-\frac{1}{2}>,\right.\right.\tag{8}$$

we denote the basis states as $|+>$ and $|->$. These are orthonormalized as

$$\left.\begin{array}{l}<+|+> = <-|-> = 1\\<+|-> = <-|+> = 0\end{array}\right\}.\tag{9}$$

The matrix elements of s_z are

$$<+|s_z|+> = \frac{\hbar}{2}$$

$$<-|s_z|+> = 0$$

$$<-|s_z|-> = \frac{\hbar}{2}.\tag{10}$$

In a two-dimensional matrix representation, we write

$$|+> = \begin{pmatrix}1\\0\end{pmatrix} \quad |-> = \begin{pmatrix}0\\1\end{pmatrix}.\tag{11}$$

A state vector in this space can be written as a linear combination of these two basis vectors

$$|\psi> = C_+|+> + C_-|->. \tag{12}$$

The representation of the state vector is

$$\psi = \begin{pmatrix} \psi_+ \\ \psi_- \end{pmatrix}. \tag{13}$$

The basis vectors in the space-spin space are

$$|\vec{r}, +> = |\vec{r}> \otimes \begin{pmatrix} 1 \\ 0 \end{pmatrix} = \begin{pmatrix} |\vec{r}> \\ 0 \end{pmatrix},$$

$$|\vec{r}, -> = |\vec{r}> \otimes \begin{pmatrix} 0 \\ 1 \end{pmatrix} = \begin{pmatrix} 0 \\ |\vec{r}> \end{pmatrix}. \tag{14}$$

We define

$$< \vec{r}, |\psi_+ > = \psi_+(\vec{r}),$$

$$< \vec{r}, |\psi_- > = \psi_-(\vec{r}). \tag{15}$$

For an electron $\left(\text{spin } \dfrac{1}{2} \right)$, we denote the spinor wave function as

$$\psi(\vec{r}) = \begin{pmatrix} (\psi_+(\vec{r}) \\ \psi_-(\vec{r}) \end{pmatrix}. \tag{16}$$

A general wave function of the electron is a linear combination of the two states $\begin{pmatrix} \psi_+(\vec{r}) \\ 0 \end{pmatrix}$ and $\begin{pmatrix} 0 \\ \psi_-(\vec{r}) \end{pmatrix}$ with definite spin along the z-direction.

We define the ladder or shift operators for m as

$$s_+ = s_x + is_y,$$

$$s_- = s_x - is_y. \tag{17}$$

We have $\qquad s_+ |+> = 0, \ s_- |+> = \hbar|->,$

$$s_+ |-> = \hbar|+>, \ s_- |-> = 0. \tag{18}$$

The matrix representations of the spin operators are

$$s_x = \frac{\hbar}{2} \begin{pmatrix} 0 & 1 \\ 1 & 0 \end{pmatrix} \qquad s_y = \frac{\hbar}{2} \begin{pmatrix} 0 & -i \\ 1 & 0 \end{pmatrix}$$

$$s_z = \frac{\hbar}{2} \begin{pmatrix} 1 & 0 \\ 0 & -1 \end{pmatrix} \qquad \vec{s}^2 = \frac{3\hbar^2}{4} \begin{pmatrix} 1 & 0 \\ 0 & 1 \end{pmatrix}$$

$$s_+ = \hbar \begin{pmatrix} 0 & 1 \\ 0 & 0 \end{pmatrix} \qquad s_- = \hbar \begin{pmatrix} 0 & 0 \\ 1 & 0 \end{pmatrix}. \tag{19}$$

We see that we can define the spin operator as

$$\vec{s} = \frac{\hbar}{2}\,\vec{\sigma}, \tag{20}$$

where the three-components of $\vec{\sigma}$ are called the Pauli spin matrices

$$\sigma_x = \begin{pmatrix} 0 & 1 \\ 1 & 0 \end{pmatrix} \quad \sigma_y = \begin{pmatrix} 0 & -i \\ i & 0 \end{pmatrix}$$

$$\sigma_z = \begin{pmatrix} 1 & 0 \\ 0 & -1 \end{pmatrix}. \tag{21}$$

Also,
$$\sigma_+ = \sigma_x + i\sigma_y = \begin{pmatrix} 0 & 1 \\ 0 & 0 \end{pmatrix}$$

$$\sigma_- = \sigma_x - i\sigma_y = \begin{pmatrix} 0 & 0 \\ 1 & 0 \end{pmatrix}. \tag{22}$$

We have

$$\sigma_3|\pm\rangle = \pm|\pm\rangle, \ \sigma_\pm|\mp\rangle = |\pm\rangle, \ \sigma_1|\pm\rangle = |\mp\rangle$$

$$\sigma_2|\pm\rangle = \pm i|\mp\rangle, \tag{23}$$

here, we use the notation $\sigma_x = \sigma_1$, $\sigma_y = \sigma_2$, $\sigma_z = \sigma_3$,

We collect here the properties of the Pauli matrices.

(i) The Pauli matrices satisfy the angular momentum commutation rule

$$[\sigma_i, \sigma_j] = 2i \in_{ijk} \sigma_k. \tag{24}$$

(ii) Since the only eigenvalues are $\pm\frac{1}{2}\hbar$, the Cayley-Hamilton theorem gives $s_i^2 = \frac{\hbar^2}{4}$. Hence

$$\sigma_i^2 = I, \tag{24a}$$

where I is the two-dimensional unit matrix.

(iii) Now,

$$0 = s_+^2 = (s_1 + is_2)^2 = s_1^2 - s_2^2 + i\{s_1, s_2\}$$

$$= i\{s_1, s_2\}.$$

So distinct Pauli matrices anti-commute

$$\{\sigma_i, \sigma_j\} = 0 \quad (i \neq j) \tag{24b}$$

i.e., $$\sigma_i\sigma_j = -\sigma_j\sigma_i \quad (i \neq j).$$

(iv) From Eqs. (24) and (24b),

$$\sigma_i\sigma_j = i \in_{ijk}\cdot\sigma_k \quad (i \neq j). \tag{24c}$$

Combining, we obtain

$$\sigma_i \sigma_j = i \in_{ijk} \sigma_k + \delta_{ij} I. \tag{24d}$$

From Eqs. (24a) and (24c), we have

$$\sigma_x \sigma_y \sigma_z = iI \tag{24e}$$

(v) For $i \neq j$, we see $Tr(\sigma_i \sigma_j) = Tr(\sigma_j \sigma_i) = - Tr(\sigma_j \sigma_i)$,

or, $\qquad\qquad Tr(\sigma_i \sigma_j) = 0 \quad (i \neq j).$

Using Eq. (24c),

$$Tr(\sigma_i \sigma_j) = i \in_{ijk} Tr(\sigma_k) = 0$$

whence the Pauli matrices are traceless:

$$Tr\,\sigma_i = 0. \tag{24f}$$

Since the eigenvalues are equal (in magnitude) and opposite,

$$det\,\sigma_i = -1. \tag{24g}$$

(vi) If $\vec{A}$ and $\vec{B}$ are two vectors that commute with $\vec{\sigma}$, but not necessarily with each other, then

$$(\sigma \cdot \vec{A})\,(\sigma \cdot \vec{B}) = (\sigma_i A_i)\,(\sigma_j B_j),\ \text{(a sum over repeated indices is indicated)}$$

$$= A_i B_j\,(\sigma_i \sigma_j)$$

$$= A_i B_j \left(\frac{1}{2}\,\{\sigma_i, \sigma_j\} + \frac{1}{2}\,[\sigma_i, \sigma_j] \right)$$

$$= A_i B_j\,(\delta_{ij} I + i \in_{ijk} \sigma_k)$$

$$= A_i B_i I + i \in_{ijk} \sigma_k A_j B_i$$

$$= (\vec{A} \cdot \vec{B}) + i\vec{\sigma} \cdot (\vec{A} \times \vec{B}).$$

Thus, we have the identity

$$(\vec{\sigma} \cdot \vec{A})\,(\vec{\sigma} \cdot \vec{B}) = \vec{A} \cdot \vec{B} + i\sigma \cdot (\vec{A} \times \vec{B}), \tag{24h}$$

where $\vec{A}, \vec{B}$ are two vectors which commute with σ, but not necessarily with each other.

If the components of $\vec{A}$ commute among themselves, then

$$(\vec{\sigma} \cdot \vec{A})^2 = A^2.$$

If $\hat{n}$ is a unit vector, then

$$(\hat{n} \cdot \vec{\sigma})^2 = I,$$

And if we choose $\hat{n}$ to be along a fixed axis, then

$$(\vec{\sigma})^2 = 3.$$

(vii) The Pauli matrices σ_i and the two-dimensional unit matrix I, together form a complete set of 2×2 matrices.

Therefore, an arbitrary 2×2 matrix M can always be written as

$$M = \Sigma m_\mu \sigma_\mu. \tag{24i}$$

$$\mu = 0, 1, 2, 3 \text{ where } m_\mu = \frac{1}{2} Tr\sigma_\mu M.$$

and $\sigma_0 = I$, σ_i $(i = 1, 2, 3)$ are the Pauli matrices.

Clearly, $\qquad Tr\,(\sigma_\mu \sigma_\nu) = 2\delta_{\mu\nu}$, $\mu, \nu = 0, 1, 2, 3$

Also, the four matrices σ_μ are linearly independent, that is,

$$\sum_{\mu=0}^{3} C_\mu \sigma_\mu = 0$$

implies that for all μ, $\qquad C_\mu = 0.$

To prove this, we note that

$$\sigma_\nu \sum_{\mu=0}^{3} C_\mu \sigma_\mu = 0,$$

or, $\qquad Tr \sum_{\mu=0}^{3} C_\mu\,(\sigma_\nu \sigma_\mu) = 0,$

or, $\qquad \sum_{\mu=0}^{3} C_\mu \cdot 2\delta_{\nu\mu} = 0,$

whence $\qquad 2C_\nu = 0,$

or, $\qquad C_\mu = 0$ for all μ.

The four matrices σ_μ form a set of linearly independent operators in this space. This set is a basis for any operator in this space.

The "spin representation" of rotations has been discussed in the previous chapter. We have constructed the spin $\frac{1}{2}$ rotation matrix and discussed the SU(2) group in this context. We recall a peculiar feature. Since a 2π rotation is represented by -1, a state $|\psi>$ that transforms under rotation as a half-integral representation is rotated by 2π, the rotated state is $-|\psi>$. Of course, the expectation value of any observable is unchanged. This representation of any half integral spin is double-valued.

(c) Spin $\hbar$

The state space is three-dimensional. We have

$$s_x = \frac{\hbar}{\sqrt{2}} \begin{pmatrix} 0 & 1 & 0 \\ 1 & 0 & 1 \\ 0 & 1 & 0 \end{pmatrix} \qquad s_y = \frac{\hbar}{\sqrt{2}} \begin{pmatrix} 0 & -i & 0 \\ i & 0 & -i \\ 0 & i & 0 \end{pmatrix}$$

$$s_+ = \hbar \begin{pmatrix} 0 & \sqrt{2} & 0 \\ 0 & 0 & \sqrt{2} \\ 0 & 0 & 0 \end{pmatrix} \qquad s_- = \hbar \begin{pmatrix} 0 & 0 & 0 \\ \sqrt{2} & 0 & 0 \\ 0 & \sqrt{2} & 0 \end{pmatrix} \tag{25}$$

$$s_z = \hbar \begin{pmatrix} 1 & 0 & 0 \\ 0 & 0 & 0 \\ 0 & 0 & -1 \end{pmatrix}.$$

The square of the spin angular momentum $s^2 = s_x^2 + s_y^2 + s_z^2$ has the eigenvalue $2\hbar^2$. A basis set is the set of eigenvectors $|m\rangle$ of s_z with eigenvalues $m\hbar$, where $m = 1, 0, -1$. The set is

$$\begin{pmatrix} 1 \\ 0 \\ 0 \end{pmatrix}, \begin{pmatrix} 0 \\ 1 \\ 0 \end{pmatrix}, \begin{pmatrix} 0 \\ 0 \\ 1 \end{pmatrix}. \tag{26}$$

6.2 THE STERN-GERLACH EXPERIMENT

The existence of the spin angular momentum of a particle was demonstrated in 1922 by Stern and Gerlach (SG). The SG experiment ranks as a landmark experiment that firmly established the quantum nature of the spin.

The SG experiment is now considered as a paradigm of the measurement process. A two-state system of this type can serve as a model of state preparation and measurement. The SG experiment is often cited for its conceptual simplicity and clarity.

We now consider the experiment itself. A schematic diagram is shown in Fig. 6.1. A collimated beam of paramagnetic silver atoms (Ag^{47}) is produced by heating silver in an oven and allowing the beam to go through a collimator. The beam is then subjected to an inhomogeneous magnetic field produced by the poles of a magnet, one of which has a very sharp tip, the other one being flat. After passing through the magnetic field, the beam is allowed to fall on a cool plate. The density of the deposit is proportional to the intensity of the beam and the time during which the beam falls on the plate.

A silver atom is composed of 47 electrons and a nucleus. 46 electrons form a spherically symmetric charge distribution with no net angular momentum. The 47th electron occupies a $5s$ orbital. Ignoring the nuclear spin, the atom as a whole has the angular momentum $l = 0$. For a field in the z-direction, we would

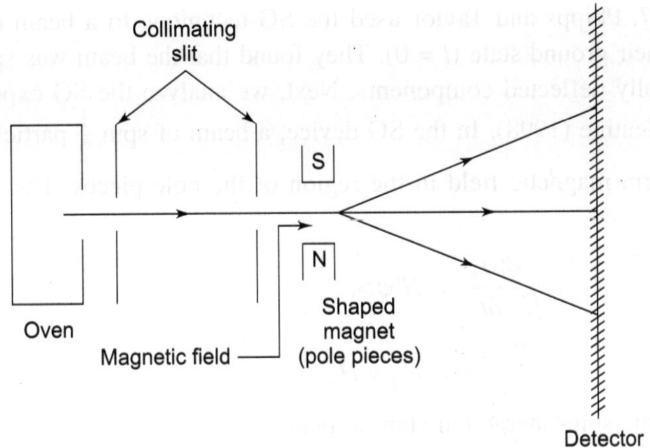

Fig. 6.1 The Stern-Gerlach experiment

then expect classically to see a continuous distribution on the screen about the undeflected direction ($z = 0$). In case of a $5s$ electron, no splitting should occur, and there would be one spot on the screen. If the electron from the fifth shell were in a $5p$ state, it is expected that there would be three spots. In any case, if the atom had an orbital angular momentum l, there would be an odd ($2l + 1$) discrete spots. In the experiment itself, the beam was split into two distinct components. To explain, it was postulated that the electron, in addition to the orbital angular momentum, possesses an intrinsic spin angular momentum which can assume only the values $+ \hbar/2$ and $- \hbar/2$ about an arbitrarily chosen direction.

Now the atom as a whole has an angular momentum due to the spin of the 47th electron. Because the nucleus is very heavy ($\sim 2 \times 10^5$ times) compared with the electron, the atom as a whole possesses a magnetic moment $\vec{\mu}_s = g_s \dfrac{e}{2m_e c} \vec{S}$ ($g_s \approx 2$ for electron). The force on the atom is

$$\vec{F} = \vec{\nabla} (\mu \cdot \vec{B}), \text{ the interaction energy is } - \vec{\mu} \cdot \vec{B}$$

$$\approx \mu_z \frac{\partial B_z}{\partial z} \hat{z}, \tag{27}$$

ignoring the components of $\vec{B}$ in directions other than the z-direction. The atom experiences a upward or a downward force depending on the signature of μ_z. The beam is then split into two components. The SG device measures μ_z (or S_z). The SG experiment provides the evidence of the existence of a spin degree of freedom of an electron.

This also provides a practical method for measuring the magnetic moments of atoms (and nuclei). It is useful for neutral atoms or molecules. For charged particles, e.g. electrons, the deflections produced by the Lorentz force will mask the spin-dependent deflection.

In 1927, Phipps and Taylor used the SG technique to a beam of hydrogen atoms in their ground state ($l = 0$). They found that the beam was split into two symmetrically deflected components. Next, we analyze the SG experiment. We follow Ballentine (1998). In the SG device, a beam of spin $\frac{1}{2}$ particles traverses a nonuniform magnetic field in the region of the pole pieces. The Schrödinger equation is

$$i\hbar \, \frac{\partial |\psi>}{\partial t} = H|\psi>, \tag{28}$$

where
$$H = H_0 + H_{int}.$$

H_{int} is the spin-interaction Hamiltonian $-\vec{\mu} \cdot \vec{B}$.

The incident beam is in the y-direction and the magnetic field lies in the xz-direction. We choose a reference frame comoving with the beam in the y-direction. In this frame, a particle experiences a magnetic field which is nonzero only during the time when the particle is within the pole pieces. The interaction Hamiltonian is

$$H_{int} \, (t) = 0, \, (t < 0)$$

$$= - \, c \, (B_0 + \gamma z) \, \sigma_z, \, (0 \le t \le T)$$

$$= 0. \, (t > T) \tag{29}$$

The magnetic field must satisfy $\vec{\nabla} \cdot \vec{B} = 0$. So the x-component of $\vec{B}$ is nonvanishing. Platt (1992) has shown that if $B_0 \gg B_x$, then the force on the particle due to B_x vanishes.

We suppose that for $t \le 0$ the state is

$$|\psi_0 > = \alpha|+> + \beta|->, \tag{30}$$

where
$$|\alpha|^2 + |\beta|^2 = 1$$

Due to the action of H_{int}, the state evolves according to

$$|\psi> = \alpha e^{-\frac{i}{\hbar}E_+ t} \, | + > + \beta e^{-\frac{i}{\hbar}E_- t}|->$$

$$(0 \le t \le T) \tag{31}$$

Here

$$E_\pm = \mp c \, (B_0 + \gamma z). \tag{32}$$

For $t \ge T$, the state vector will be

$$|\psi> = \alpha e^{\frac{i}{\hbar}c \, (B_0 + \gamma z) T}|+> + \beta e^{-\frac{i}{\hbar}c \, (B_0 + \gamma z) T}|-> \tag{33}$$

The initial wavepacket, after traversing the magnetic field ($t > T$), emerges as two distinct, localized wavepackets depending on the spin quantum number. The two terms carry momentum

$$p_{z_\pm} = \pm c\gamma T, \tag{34}$$

corresponding to $\qquad \sigma_z = \pm 1.$

Thus, the beam is split into two macroscopically distinct trajectories.

6.3   SPIN PRECESSION IN A MAGNETIC FIELD

A spin s particle in a magnetic field has interesting dynamical properties.

We consider a spin $\frac{1}{2}$ particle, say an electron, in a static magnetic field. The Hamiltonian, neglecting all orbital motion, is

$$H = -\vec{\mu}\cdot\vec{B} = \frac{eg}{2m_e c}\,\vec{S}\cdot\vec{B}(t) = \frac{eg\hbar}{4m_e c}\,\sigma\cdot\vec{B}(t). \tag{35}$$

The equation of motion of the spin in the Heisenberg picture is

$$i\hbar\,\frac{dS_i}{dt} = [S_i(t), H(t)] = i\,\frac{eg\hbar}{2m_e c}\,\epsilon_{ijk}\,S_k B_j(t) \tag{36}$$

or, $\qquad$
$$\frac{d\vec{S}(t)}{dt} = \frac{eg}{2m_e}\,\vec{S}(t)\times\vec{B}(t) = \vec{M}(t)\times\vec{B}(t), \tag{37}$$

where $\vec{M} = -\dfrac{eg}{2m_e c}\,\vec{S}$ is the intrinsic magnetic moment of the electron. Therefore, the time rate of change of the spin is the torque exerted on the magnetic moment of the electron by the magnetic field. The electron, being negatively charged, will precess in a positive sense about the field (Fig. 6.2).

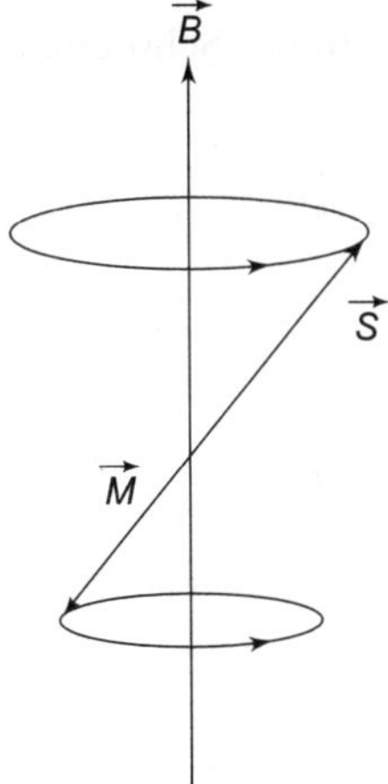

Fig. 6.2   The precession of the spin and magnetic moment about a constant magnetic field

For a static magnetic field pointing in the z-direction, $\vec{B} = B_0\hat{z}$, the Eq. (37) becomes

$$\frac{dS_z}{dt} = 0, \quad \frac{dS_x(t)}{dt} = \frac{egB_0}{2m_ec} S_y(t), \quad \frac{dS_y(t)}{dt} = -\frac{egB_0}{2m_ec} S_x(t). \tag{38}$$

The solution is

$$S_x(t) = S_x(0) \cos \omega_0 t + S_y(0) \sin \omega_0 t,$$

$$S_y(t) = -S_x(0) \sin \omega_0 t + S_y(0) \cos \omega_0 t, \tag{39}$$

$$S_z(t) = -S_z(0),$$

where

$$\omega_0 = \frac{egB_0}{2m_ec}.$$

We suppose, for example, that initially, the spin is in an eigenstate of S_x with eigenvalue $+ \hbar/2$. Then,

$$\langle S_x(0) \rangle = + \frac{\hbar}{2}, \ \langle S_y(0) \rangle = \langle S_z(0) \rangle = 0. \tag{40}$$

Equation (39) then yields the expectation values

$$\langle S_x(t) \rangle = \frac{\hbar}{2} \cos \omega_0 t,$$

$$\langle S_y(t) \rangle = -\frac{\hbar}{2} \sin \omega_0 t,$$

$$\langle S_z(t) \rangle = 0. \tag{41}$$

This set of equations for the spin precession has been employed in the case of muon.

Let us look at the state ket. In the Schrödinger picture, the state ket evolves in time according to

$$|\psi(t)> = e^{-\frac{i}{\hbar}Ht}|\psi(0)>$$

$$= e^{-\frac{i}{\hbar}S_z\omega_0 t}|\psi(0)>. \tag{42}$$

6.4   THE PAULI EQUATION

The Hamiltonian for the motion of an electron in an external electromagnetic field is

$$H_{em} = \frac{1}{2m_e}\left(\vec{p} - \frac{e}{c}\vec{A}\right)^2 + e\Phi, \tag{43}$$

where $\vec{A}$, Φ are the vector and scalar potentials.

There is an additional potential energy due to the interaction $-\vec{M}_s \cdot \vec{B}$ between the magnetic field $\vec{B}$ and an intrinsic magnet moment $\vec{M}_s$ of the electron due to spin.

The intrinsic magnetic moment is

$$\vec{M}_s = g\left(\frac{-e}{2m_e c}\right)\vec{S} = 2\left(\frac{-e}{2m_e c}\right)\vec{B} = -\mu_B \vec{S}, \tag{44}$$

where

$$\mu_B = \frac{e\hbar}{2m_e c} \tag{45}$$

is the Bohr magneton, and the spin gyromagnetic ratio $g = 2$.

The spin-interaction Hamiltonian is

$$H_{sp} = -\vec{M}_s \cdot \vec{B}$$

$$= \mu_B \vec{\sigma} \cdot \vec{B}. \tag{46}$$

Therefore, the Hamiltonian of an electron with spin $\left(\frac{1}{2}\hbar\right)$ in an electromagnetic field is

$$H = H_{em} + H_{sp}$$

$$= \frac{1}{2m}\left(\vec{p} - \frac{e}{c}\vec{A}\right)^2 + e\Phi + \mu_B \vec{\sigma} \cdot \vec{B}. \tag{47}$$

The Schrödinger equation for a particle with spin $\frac{1}{2}$, known as the Pauli equation, is

$$i\hbar \frac{\partial}{\partial t}\Psi = H\Psi, \tag{48}$$

where

$$\Psi = \begin{pmatrix} \chi_1 \\ \chi_2 \end{pmatrix}$$

are the spinor wave functions.

The Pauli equation

$$i\hbar \frac{\partial}{\partial t}\begin{pmatrix} \chi_1 \\ \chi_2 \end{pmatrix}$$

$$= \left[\left\{\frac{1}{2m}\left(\frac{\hbar}{i}\vec{\nabla} - \frac{e}{c}\vec{A}(\vec{r}, t)\right)^2 + e\Phi(\vec{r}, t)\right\}I + \mu_B \vec{\sigma} \cdot \vec{B}\right]\begin{pmatrix} \chi_1 \\ \chi_2 \end{pmatrix} \tag{49}$$

is a system of two coupled differential equations for χ_1 and χ_2.

We note that only the term $\mu_B \vec{\sigma} \cdot \vec{B}$ depends on the spin operators. With the operators x, y, z, and S^2, S_z as fundamental operators, we seek a solution of the

Pauli equation (49) by the method of the separation of variables with the wave function as the product of a "coordinate part" and a "spin part". The coordinate part satisfies the Schrödinger equation with the Hamiltonian H_{em}, while the time evolution of the spin part of the wave function satisfies

$$i\hbar \frac{\partial \chi}{\partial t} = \frac{2\mu}{\hbar} \vec{B} \cdot \vec{S} \chi. \tag{50}$$

Equation (50) describes the time evoluation of spinor wave function of a spin-s particle

$$\chi(\vec{S}, t) = \begin{pmatrix} \chi_1 \; (\vec{S}, \, t) \\ \chi_2 \; (\vec{S}, \, t) \end{pmatrix}. \tag{51}$$

Here μ denotes the intrinsic magnetic moment of the particle.

Following Landau and Lifshitz (1977), we consider an electrically neutral spin $\frac{1}{2}$ particle having a magnetic moment μ in a magnetic field $\vec{B}$ which is uniform but varies with time, the only nonvanishing component is $\vec{B}_z = B(t)$. In this case, the spin part equations are decoupled.

$$\left. \begin{aligned} i\hbar \, \frac{\partial \chi_1}{\partial t} &= - \frac{\mu B}{\hbar} \, \chi_1 \\[2mm] i\hbar \, \frac{\partial \chi_2}{\partial t} &= - \frac{\mu B}{\hbar} \, \chi_2 \end{aligned} \right\} \tag{52}$$

which can be easily solved.

The Pauli equation has been used to solve several interesting physical problems. We note the Dirac equation for the electron reduces to the Pauli equation in the non-relativistic limit.

6.5   THE DENSITY MATRIX FOR A SPIN-S SYSTEM

As examples of the general method, we now consider the density matrix for a spin-s system.

(i) Orientation of Spin $\frac{1}{2}$ particles

The spin orientation in this case is represented by a density matrix with two rows and columns, corresponding to two pure states with opposite spin orientation (e.g. up and down). Since the unit 2×2 matrix I and the three Pauli spin matrices σ_x, σ_y, and σ_z form a complete set of 2×2 matrices, the general state in this state space can be described by a density matrix of the following form.

$$\rho = \frac{1}{2}(I + \vec{P}\cdot\vec{\sigma}), \tag{53}$$

where $\vec{P}$ is a real, numerical vector. The factor $\frac{1}{2}$ has been chosen as $Tr\rho = 1$.

The vector $\vec{P}$ is called the polarization vector. The degree and direction of spin orientation are given by the magnitude and direction of $\vec{P}$:

$$\vec{P} = \langle\vec{\sigma}\rangle = Tr\rho\,\vec{\sigma}. \tag{54}$$

Now, $$Tr\rho^2 = \frac{1}{2}(1 + P^2).$$

But, $$Tr\rho^2 \leq 1, \text{ and so}$$

$$0 \leq |\vec{P}| \leq 1. \tag{55}$$

A pure state with definite spin orientation has $P = 1$, while a state with random orientation has $P = 0$. For particles with magnetic moment μ in a magnetic field B, we have

$$\vec{P} = \frac{\mu\vec{B}}{\frac{1}{2}k_B T} \text{ under conditions of paramagnetic polarization.}$$

The knowledge of $\vec{P}$ is sufficient to determine the density matrix. Explicitly, the Eq. (53) is

$$\rho = \frac{1}{2}\begin{pmatrix} 1 + P_z & P_x - iP_y \\ P_x + iP_y & 1 - P_z \end{pmatrix}. \tag{56}$$

Let us diagonalize ρ. We obtain

$$\rho = \frac{1}{2}\begin{pmatrix} 1 + P & 0 \\ 0 & 1 - P \end{pmatrix} \tag{57}$$

where $$P = \pm |\vec{P}|.$$

In this representation in which ρ is diagonal, $P_x = P_y = 0$, and $P = P_z$. Also from Eq. (53), we see that $\vec{\sigma}\cdot\vec{P}$ is also diagonal. We write

$$\vec{\sigma}\cdot\vec{P}|+> = P|+>,$$
$$\vec{\sigma}\cdot\vec{P}|-> = -P|->, \tag{58}$$

From Eq. (57), the eigenvalues of ρ are $\frac{1}{2}(1 + |\vec{P}|)$ and $\frac{1}{2}(1 - |\vec{P}|)$.

Let us consider a beam of spin $\frac{1}{2}$ particle with N_+ the number of pure spin-up states and N_- the number of pure spin-down states. Then

$$\frac{1}{2}(1 + P) = \frac{N_+}{N_+ + N_-},$$

$$\frac{1}{2}(1 - P) = \frac{N_-}{N_+ + N_-}, \qquad (59)$$

Hence, $$P = \frac{N_+ - N_-}{N_+ + N_-}. \qquad (60)$$

An electron of magnetic moment μ_B is placed in a magnetic field $\vec{B} = B\hat{z}$. The density matrix in the canonical ensemble is

$$\rho = \frac{e^{-\beta H_s}}{Tr\,(e^{-\beta H_s})}, \qquad (61)$$

where $$H_s = -\mu_B\,(\vec{\sigma}\cdot\vec{B})$$

$$= -\mu_B\,B\sigma_z \text{ is the}$$

spin-interaction Hamiltonian.

Therefore, $$\rho = \frac{1}{e^{\mu_B B/k_B T} + e^{-\mu_B B/k_B T}}\begin{pmatrix} e^{\mu_B B/k_B T} & 0 \\ 0 & e^{-\mu_B B/k_B T} \end{pmatrix}. \qquad (62)$$

Then the expectation value of σ_z is given by

$$\langle\sigma_z\rangle = Tr\,(\rho\sigma_z) = \frac{e^{\mu_B B/k_B T} - e^{-\mu_B B/k_B T}}{e^{\mu_B B/k_B T} + e^{-\mu_B B/k_B T}}$$

$$= \tanh\,(\mu_B B/k_B T). \qquad (63)$$

The representation Eq. (56) leads to a simple treatment of the precession spin in a magnetic field. The spin interaction Hamiltonian is $H_s = -g_s\left(\frac{1}{2}\hbar\right)\vec{\sigma}\cdot\vec{B}$. The rate of change of the density matrix is

$$\frac{\partial\rho}{\partial t} = i\hbar^{-1}\,(H_s\rho - \rho H_s)$$

$$\frac{1}{2}\frac{\partial}{\partial t}\vec{p}\cdot\vec{\sigma} = i\frac{1}{4}g_s\,(\vec{B}\cdot\vec{\sigma}\vec{P}\cdot\vec{\sigma} - \vec{P}.\vec{\sigma}\vec{B}.\vec{\sigma})$$

$$= -\frac{1}{2}g_s\vec{B}\times\vec{P}\cdot\vec{\sigma}$$

whence $$\frac{\partial\vec{P}}{\partial t} = -g_s\vec{B}\times\vec{P}, \qquad (64)$$

or,
$$\frac{\partial}{\partial t}\langle \vec{\sigma}\rangle = g_s \langle \vec{\sigma}\rangle \times \vec{B}. \tag{64a}$$

This is just the classical equation for the precession.

(ii) Orientation of Spin-1 particles

The state spin space is now three-dimensional. The density matrix ρ depends on nine parameters. In the $s = \frac{1}{2}$ case, all operators in the state space can be constructed as linear combinations of the unit 2×2 matrix and the three angular momentum 2×2 matrices conveniently denoted by the Pauli matrices. To describe the state of a spin-1 system we require the unit 3×3 matrix, the angular momentum operators $S_i = (i = 1, 2, 3)$, and another set of five matrices. We write the density matrix

$$\rho = \frac{1}{2}(I + \vec{P}\cdot\vec{S} + a_{ij}T_{ij}), \tag{65}$$

where
$$T_{ij} = \frac{1}{2}(S_iS_j + S_jS_i) - \frac{2}{3}\delta_{ij}$$

and a_{ij} is a set of five real numbers.

The polarization vector provides only three parameters, and so, in contrast to the case $S = \frac{1}{2}$, cannot determine the state uniquely.

(iii) Orientation of Spin-s particles

The density matrices has $2s + 1$ rows and columns. A complete set of matrices in this $2s + 1$ dimensional space must have $(2s + 1)^2$ linearly independent members. For $s = 1/2$, this set includes the unit matrix, and the three angular momentum operators. For $s > 1/2$, we can use the representation to include, besides the unit matrix and terms depending on dipole polarization, additional terms depending on multipole polarization. For spin s, the density matrix representation consists of the unit matrix, the three S_i, the five components of T_{ij}, and higher multipole polarizations upto irreducible tensors of rank $2s$.

(iv) Photon States

Photon states have the eigenvalues ± 1 in the helicity basis. For one-photon states with no definite polarization, we represent the density matrix as

$$\rho = \frac{1}{2}(I + \vec{\xi}\cdot\vec{\tau}) \tag{66}$$

where $\vec{\xi}$ is a real vector and τ_i are a set of three Pauli matrices. This representation has proved to be convenient and it is of the same form as that of a spinning particle with $s = 1/2$. We use a different notation for the Pauli matrices since here there is no connexion with rotations. ξ are the Stokes' parameters. Since $Tr\rho^2 \leq 1$, we have

$$0 \leq \vec{\xi}^2 \leq 1.$$

The polarization of a one-photon state can be represented by a real vector $\vec{\xi}$ which lie on or inside a unit sphere (Poincaré sphere). For a pure state, the vector $\vec{\xi}$ lies on the surface of the sphere.

6.6 THE HELICITY FORMALISM

The helicity λ of a particle with spin $\vec{s}\hbar$ is defined as the projection of the spin in the direction of its linear momentum $\vec{p}$. We define a (pseudoscalar) operator

$$\Lambda = \frac{\vec{J} \cdot \vec{p}}{|\vec{p}|} = \frac{\vec{s} \cdot \vec{p}}{|\vec{p}|} \tag{67}$$

The helicity is defined as the eigenvalue of this operator, the eigenvalues being (in units of $\hbar$)

$$\lambda = -s, -s + 1, ..., s \tag{68}$$

Clearly, Λ is invariant under rotations, i.e. Λ commutes with $\vec{J}$, though not with either $\vec{l}$ or $\vec{s}$. For a particle of spin s, there are $(2s + 1)$ spin-states for each $\vec{p}$. The associated orthonormal basis of one particle states $\{|p, \lambda\}$ provides an extremely useful description of spinning particle. This basis has several advantages for particles with spin over the one labeled by $\vec{p}$ and m, where $m\hbar$ is the eigenvalue of $s_z(s_3)$.

We now consider the basis $\{|\vec{p}, \lambda >\}$ relative to the basis $\{|\vec{p}, m >\}$. Let us consider states where the momentum is along the 3-axis (z-axis); then for these states the helicity coincides with s_3. We write

$$|\vec{p}, \lambda> = |\vec{p}> \otimes | s_3 = \lambda>, \text{ where } \vec{p} = p\hat{e}_3.$$

We consider a state $|p\hat{n}, \lambda>$, where the momentum is of the same magnitude but in the direction $\hat{n}$ (θ, ϕ), where θ is the polar angle measured from 3-axis, ϕ the azimuthal angle. Since the helicity is a scalar operator under rotations, one can obtain a state $|p\hat{n}, \lambda>$ from the state $|p\hat{e}_3, \lambda>$ by applying a specific rotation. There are an infinite number of such rotations all of which differ in phase. We parametrize a rotation in terms of Euler angles:

$$R(\phi, \theta, \psi) = e^{-i\phi J_3} e^{-i\theta J_2} e^{-i\psi J_3}. \tag{69}$$

We choose the rotation $R(\phi, \theta, -\phi)$ and write

$$|p\hat{n}, \lambda> = R(\phi, \theta, -\phi)|p\hat{e}_3, \lambda>, \tag{70}$$

where $\hat{n} = (\theta, \phi)$

The state $|p\hat{e}_3, \lambda>$ is part of the basis $\{\vec{p}, s_3\}$ when $\vec{p} = pe_3$, $\Lambda = s_3$. The relation between the two bases is

$$|\vec{p}, \lambda> = \sum_{m'} D^{(s)}_{\lambda m'} (\phi, \theta - \phi)| \vec{p}, m'>). \tag{71}$$

We have thus an orthonormal basis for the one particle helicity states satisfying

$$< \vec{p}', \lambda'|\vec{p}, \lambda> = \delta (\vec{p}' - \vec{p}) \, \delta_{\lambda'\lambda}. \tag{72}$$

There are other basis in common use. We consider simultaneous eigenstates of the commuting operators $\vec{p}^2$, $\vec{j}^2$, j_3, Λ:

$$\left.\begin{aligned}
\vec{p}^2|p, j, m, \lambda> &= p^2|p, j, m, \lambda> \\
\vec{j}^2|p, j, m, \lambda> &= \hbar^2 j \, (j + 1)|p, j, m, \lambda> \\
j_3|p, j, m, \lambda> &= \hbar m|p, j, m, \lambda> \\
\Lambda|p, j, m, \lambda> &= \hbar\lambda|p, j, m, \lambda>
\end{aligned}\right\}. \tag{73}$$

Also, we have

$$1 = \sum_{j,m,\lambda} \int_0^\infty \frac{p^2 dp}{(2\pi\hbar)^3} \, |p, j, m, \lambda><p, j, m, \lambda|.$$

$$< p', j', m', \lambda'|p, j, m, \lambda> = (2\pi\hbar)^3 \, \frac{\delta(p' - p)}{p'p} \, \delta_{j'j}\delta_{m'm}\delta_{\lambda'\lambda}. \tag{74}$$

We obtain

$$|\vec{p}, \lambda> = \sum_{j,m} \sqrt{\frac{2j + 1}{4\pi}} \, D^{(j)}_{\lambda m} (\phi, \theta, -\phi)|p, j, m, \lambda>. \tag{75}$$

Inverting this relation, we get

$$|\vec{p}, j, m, \lambda> = \sqrt{\frac{2j + 1}{4\pi}} \int d\Omega \, (D^{(j)}_{\lambda m} (\phi, \theta, - \phi)^*|\vec{p}, \lambda>. \tag{76}$$

So $|p, j, m, \lambda>$ is a complete basis.

Equation (75) is a generalization of the expansion of a plane wave in terms of spherical waves to the case of a particle of spin. The one particle helicity state which is a simultaneous free-particle eigenstate of energy, and angular momentum and helicity is given by Eq. (76).

Two particle states are important for both bound state and scattering problems. If one or both of the particles have nonzero spin, then the helicity formalism provides an elegant characterization of the quantum states.

One can easily construct a two particle basis by the product

$$|\vec{p}_1, \lambda_1 > \otimes |\vec{p}_2, \lambda_2 >.$$

For collisional problems, we consider the two particle states in the center of mass (CM) frame in which $\vec{p}_1 = -\vec{p}_2$. In the CM frame the state in which particle 1 moves up the three-axis with momentum $\vec{p}$ and particle 2 moves down the three-axis with momentum $-\vec{p}$ may be written as

$$|\vec{p}, \lambda_1, \lambda_2 > = |\vec{p} > \otimes |s_3^{(1)} = \left|\lambda_1 > \otimes | s_3^{(2)} \right\rangle \right| = -\lambda_2 >, \tag{77}$$

as before, we write

$$|\vec{p}, \lambda_1, \lambda_2 > = R_{12}(\phi, \theta, -\phi)|p\hat{e}_3, \lambda_1, \lambda_2 >$$

$$\vec{p} = p\hat{e}_3 \tag{78}$$

where R_{12} is a two particle rotation operator generated by the total angular momentum.

We have now an orthonormal basis for the relative motion. The normalization is

$$<\vec{p}', \lambda_1', \lambda_2'|\vec{p}, \lambda_1\lambda_2 > = \delta(\vec{p}' - \vec{p})\, \delta_{\lambda_1'\lambda_1}\, \delta_{\lambda_2'\lambda_2}. \tag{79}$$

The helicity formalism introduced by Jacob and Wick (1959) has several advantages. The relationship of the plane-wave states and the angular momentum states is direct, and free of the various coupling schemes like L-S, j-j, etc. The behavior of the helicity states under discrete symmetry transformations is direct. This formalism is applicable to both massless and massive particles. In the relativistic domain the formalism is very advantageous. The helicity states are very useful in the context of general collisional problems.

PROBLEMS

6.1 (a) The matrices S_x, S_y, S_z for a spin $s = 1$ operator are given in (25). Show that their squares commute. Construct their common eigenvector.

(b) In the basis formed by the eigenvectors in (a) find the $s = 1$ spin matrices.

6.2 Consider a neutral particle of spin $\frac{1}{2}$ in a magnetic field with the only component $B_z = B(t)$. Determine the spin wave function.

6.3 Repeat the above problem for a uniform magnetic field constant in absolute magnitude but whose direction rotates with angular velocity ω about the z-axis at an angle θ to it.

6.4 The Hamiltonian of a particle of spin 1 is $H = AS_x^2 + BS_y^2 + CS_z^2$, where A, B, C are constants. Find the energy levels and the wave functions.

CHAPTER 7

The Addition of Angular Momenta

The total angular momentum $\vec{L}$ of a classical mechanical system composed of two parts each having angular momentum $\vec{L}_1$ and $\vec{L}_2$ is simply $\vec{L} = \vec{L}_1 + \vec{L}_2$. The analogous quantum mechanical problem deals with systems in which there are two or more angular momentum vector operators, e.g. orbital angular momentum of a particle and its spin angular momentum or, two particles with individual angular momentum $\vec{j}_1$ (with component m_1 along some axis) and $\vec{j}_2$ (with component m_2 along some axis). The problem then is to construct the eigenkets and determine the eigenvalues pertaining to the total angular momentum. This problem is of considerable complexity but Wigner, Racah and others have developed a formal theory of addition of angular momenta. The theory is of great importance in many areas of physics, e.g., atomic spectroscopy, nuclear shell theory and particle collisions.

7.1 TWO SIMPLE EXAMPLES

Before studying the general theory, we consider two simple examples. In the first, we consider a spin $\frac{1}{2}$ particle. The space of physical states for the particle is the direct-product space of the infinite dimensional ket space spanned by the position eigenkets $\{|\vec{x}>\}$ and the two dimensional spin space spanned by $|+>$ and $|->$. The base ket in the product space is

$$|\vec{x}, \pm\rangle = |\vec{x}\rangle \otimes |\pm\rangle. \tag{1}$$

The rotation operator is $\exp(-i\vec{J}.\hat{n}\phi/\hbar)$,

where
$$\vec{j} = \vec{l}* + \vec{s}. \tag{2}$$

We define the operator $\vec{j}$ in the product space by

$$\vec{j} = \vec{l} \otimes I + I \otimes \vec{s}. \tag{3}$$

where the I in $\vec{l} \otimes I$ stands for the identity operator in the spin space and the I in $I \otimes \vec{s}$ stands for the identity operator in the space spanned by the position eigenkets. Since $\vec{l}$ and $\vec{s}$ commute, we write

$$D(R) = D^{(orb)}(R) \otimes D^{(spin)}(R) = \exp\left(\frac{-i\vec{l}\cdot\hat{n}\phi}{\hbar}\right) \otimes \exp\left(\frac{-i\vec{s}\cdot\hat{n}\phi}{\hbar}\right). \tag{4}$$

The wave function for a particle with spin is written as

$$\langle \vec{x}, \pm | \alpha \rangle = \psi_{\pm}(\vec{x}). \tag{5}$$

The two components ψ_+ and ψ_- are usually written in a column matrix form:

$$\begin{pmatrix} \psi_+(\vec{x}) \\ \psi_-(\vec{x}) \end{pmatrix}. \tag{6}$$

A state ket of a particle with spin can be expanded in terms of simultaneous eigenkets of $\vec{l}^2$, $\vec{s}^2$, l_z and s_z or in terms of simultaneous eigenkets of $\vec{j}^2$, j_z, $\vec{l}^2$ and $\vec{s}^2$. We will study the relationship of the two descriptions.

As a second example, we consider two spin $\frac{1}{2}$ particles (say, two electrons) with the orbital degree of freedom suppressed. The total spin operator $\vec{s}$ is in the direct product space of the spin space of particle 1 with spin operator $\vec{s}_1$ and the spin space of partice 2 with spin operator $\vec{s}_2$. The total spin angular momentum operator

$$\vec{s} = \vec{s}_1 \otimes I + I \otimes \vec{s}_2, \tag{7}$$

where the I in the first (second) term stands for the identity operator in the spin space of particle 2(1).

We write Eq. (7) in an abbreviated notation:

$$\vec{s} = \vec{s}_1 + \vec{s}_2. \tag{8}$$

Because $\vec{s}_1$ and $\vec{s}_2$ refer to different particles

$$[\vec{s}_1, \vec{s}_2] = 0. \tag{9}$$

$\vec{s}_1$ and $\vec{s}_2$ individually obey the angular momentum commutation relations and since $\vec{s}_1$ and $\vec{s}_2$ commute, $\vec{s}$ satisfies the angular momentum commutation relations

$$[s_i, s_j] = i\hbar \in_{ijk} s_k. \tag{10}$$

There are two linearly independent orientations for each spin. Therefore, there are four linearly independent states for the two spins.

The eigenvalues are as follows:

$$\vec{s}^2 = (\vec{s}_1 + \vec{s}_2)^2 : s\,(s+1)\hbar^2$$

$$s_z = s_{1z} + s_{2z} : m\hbar \qquad (11)$$

$$s_{1z} \qquad\qquad : m_1\hbar$$

$$s_{2z} \qquad\qquad : m_2\hbar.$$

We can expand an arbitrary ket corresponding to a spin state of the two particles in terms of either the eigenkets of $\vec{s}^2$ and s_z or the eigenkets of s_{1z} and s_{2z}. In the (m_1, m_2) representation, we write the four base kets as

$$|++>, |+->, |-+>, \text{ and } |-->|, \qquad (12)$$

where the first symbol refers to the eigenvalue of s_{1z}, and the second symbol to the eigenvalue of s_{2z}. In the (s, m) representation, the spin states $|s, m\rangle$ are labeled by the eigenvalues of $\vec{s}^2$ and s_z, where $\vec{s}^2\,|s, m> = s\,(s+1)|s, m>$.

In this set there are four base kets. The basic problem is to find the connexion between the two representations. We now try to construct the $|s, m>-$ states in terms of the states Eq. (12).

We have

$$s_z|++> = \left(\frac{1}{2} + \frac{1}{2}\right)|++> = |++>,$$

$$s_z|+-> = \left(\frac{1}{2} - \frac{1}{2}\right)|+-> = 0, \qquad (13)$$

$$s_z|-+> = \left(-\frac{1}{2} + \frac{1}{2}\right)|-+> = 0,$$

$$s_z|--> = \left(-\frac{1}{2} - \frac{1}{2}\right)|--> = -|-->.$$

The eigenvalues of s_z are $0, \pm 1$. There are a triplet of states with $s_z = +1, 0, -1$ corresponding to total spin one and a singlet state with $s_z = 0$ corresponding to total spin zero.

The state $|++>$ is a spin one state, i.e.,

$$\vec{s}^2\,|++> = s\,(s+1)|++> = 2|++>.$$

The state $|-->$ is also a spin one state.

By applying the ladder operator $s_- \equiv s_{1-} + s_{2-}$

$$= (s_{1x} - is_{1y}) + (s_{2x} - is_{2y}) \text{ on } |s, m>, \text{ we get}$$

$$s_-|s, m> = \sqrt{s\,(s+1) - m\,(m-1)}|\,s, m-1>. \qquad (14)$$

Then

$$s_- |s = 1, m = 1 > = (s_{1_-} + s_{2_-})|++>$$

i.e.

$$\sqrt{1\,(1 + 1) - 1\,(1 - 1)}\,|s = 1, m = 0 >$$

$$= \sqrt{\frac{1}{2}\left(\frac{1}{2} + 1\right) - \frac{1}{2}\left(\frac{1}{2} - 1\right)}\,|-+> + \sqrt{\frac{1}{2}\left(\frac{1}{2} + 1\right) - \frac{1}{2}\left(\frac{1}{2} - 1\right)}\,|+->$$

or,

$$\sqrt{2}\,|s = 1, m = 0\rangle = |-+\rangle + |+-\rangle$$

i.e.

$$|s = 1, m = 0> = \frac{1}{\sqrt{2}}\,(|-+>+|+->).$$

The state $|s = 0, m = 0>$ must be orthogonal to the above state and therefore,

$$|s = 0, m = 0> = \frac{1}{\sqrt{2}}\,(|+->-|-+\rangle).$$

The relationship between the sets of base kets is given below

$$|s = 1, m = 1> = |++>,$$

$$|s = 1, m = 0> = \frac{1}{\sqrt{2}}\,(|+->+|-+>),$$

$$|s = 1, m = -1> = |-->,$$

$$|s = 0, m = 0 > = \frac{1}{\sqrt{2}}\,(|+->-|-+>). \tag{15}$$

The three states with $s = 1$, i.e. $|s = 1, m = \pm 1, 0>$ are called spin-triplet states, while the state $s = 0$, i.e. $|s = 0, m = 0$ is called a spin-singlet state. The coefficients on the LHS of Eq. (15) are examples of Clebsch – Gordan coefficients. These are the elements of transformation matrix connecting the (m_1, m_2) basis to the (s, m) basis.

We can also write the $|m_1, m_2 >$ states in terms of the $|s, m >$ states:

$$|++> = |1, 1>,$$

$$|+-> = \frac{1}{\sqrt{2}}\,(|1, 0> + |0, 0>),$$

$$|-+> = \frac{1}{\sqrt{2}}\,(|1, 0> - |0, 0>),$$

$$|--> = |1, -1>. \tag{16}$$

The $|s, m>$ states are eigenfunctions of the operator $\vec{s}_1 \cdot \vec{s}_2$. Since $\vec{s}_1 \vec{s}_2 = \frac{1}{2}$

$$[(\vec{s}_1 + \vec{s}_2)^2 - \vec{s}_1^2 - \vec{s}_2^2] = \frac{\vec{s}^2}{2} - \frac{3}{4}\left(\text{note: } \vec{s}_1^2 = \vec{s}_2^2 = \frac{1}{2}\left(\frac{1}{2} + 1\right) = \frac{3}{4}\right).$$

Thus,

$$\vec{s}_1 \cdot \vec{s}_2 \, |s = 1, m> = \frac{1}{4} \, |s = 1, m>$$

and

$$\vec{s}_1 \cdot \vec{s}_2 \, |s = 0, m = 0> = \frac{-3}{4} |s = 0, m = 0>. \qquad (17)$$

Thus, the operator $\vec{P}_1 = \frac{3}{4} + \vec{s}_1 \cdot \vec{s}_2$ is a projection operator for the triplet state

and $\vec{P}_2 = 1 - P_1 = \frac{1}{4} - \vec{s}_1 \cdot \vec{s}_2$ is a projection operator for the singlet state.

7.2   ADDITION OF TWO ANGULAR MOMENTA

We consider the general problem of adding two angular momenta. Let $\vec{J}_1$ and $\vec{J}_2$ be two independent angular momentum operator in two subspaces. The space of physical states of the entire system is the direct-product space of the two spaces. Clearly,

$$\lfloor \vec{J}_1, \vec{J}_2 \rfloor = 0. \qquad (18)$$

The components J_{1i} and J_{2i} satisfy the commutation relations

$$\lfloor J_{1i}, J_{1j} \rfloor = i\hbar \in_{ijk} J_{1k}, \qquad (19)$$

$$\lfloor J_{2i}, J_{2j} \rfloor = i\hbar \in_{ijk} J_{2k}. \qquad (19a)$$

We define the total angular momentum operator in the direct-product space by

$$\vec{J} = \vec{J}_1 \otimes I + I \otimes \vec{J}_2, \qquad (20)$$

which we write in an abbreviated notation

$$\vec{J} = \vec{J}_1 + \vec{J}_2. \qquad (20a)$$

It follows from Eqs. (19) and (20a), that J_i fulfil the commutation relations

$$\lfloor J_i, J_j \rfloor = i\hbar \in_{ijk} J_k. \qquad (21)$$

The basis system of the direct-product space consists of eigenkets of the complete set of commuting operators:

$$\vec{J}_1^2, J_{1z}, \vec{J}_2^2, J_{2z}$$

plus whatever other hermitian operators are needed to give a complete specification (we suppress these), denoted by $|j_1 m_1 j_2 m_2>$, with the eigenvalues

$$\vec{J}_1^2 \, |j_1 m_1 j_2 m_2> = j_1 \, (j_1 + 1) \, \hbar^2 \, |j_1 m_1 j_2 m_2 >,$$

$$J_{1z} \,|j_1 m_1 \, j_2 m_2> = m_1 \hbar \,|j_1 m_1 \, j_2 m_2 >, \tag{22}$$

$$\vec{J}_2^{\,2} \,|j_1 m_1 \, j_2 m_2> = j_2 \, (j_2 + 1)\, \hbar^2 |j_1 m_1 \, j_2 m_2 >,$$

$$J_{2z} \,|j_1 m_1 \, j_2 m_2> = m_2 \hbar \,|j_1 m_1 \, j_2 m_2 >.$$

But we may choose an alternate basis consisting of simultaneous eigenkets of the commuting operators

$$\vec{J}_1^{\,2},\ \vec{J}_2^{\,2},\ \vec{J}^{\,2},\ \vec{J}_z.$$

We denote this basis as $|j_1 \, \hat{j}_2 \, jm>$:

$$\vec{J}_1^{\,2} \,|j_1 j_2 \, jm> = j_1 \, (j_1 + 1)\hbar^2 |j_1 j_2 \, jm >,$$

$$\vec{J}_2^{\,2} \,|j_1 j_2 \, jm> = j_2 \, (j_2 + 1)\hbar^2 \,|j_1 j_2 \, jm >,$$

$$\vec{J}^{\,2} \,|j_1 j_2 \, jm> = j \, (j + 1)\hbar^2 |\, j_1 j_2 \, jm >,$$

$$\vec{J}_z \,|j_1 j_2 \, jm> = m\hbar \,|j_1 \, j_2 \, jm >. \tag{23}$$

We shall be concerned with the translation of one scheme to the other.

The ket space of j_1 and j_2 is spanned by kets $|j_1 m_1 \, j_2 m_2 >$, i.e.

$$\sum_{m_1} \sum_{m_2} |j_1 m_1 \, j_2 m_2 > < j_1 m_1 \, j_2 m_2| = I_{j_1 j_2}$$

where $I_{j_1 j_2}$ is the identity operator in the ket space. The basis defined by Eq. (23) also spans the space:

$$\sum_{j} \sum_{m} |j_1 \, j_2 \, jm > < j_1 \, j_2 \, jm| = I_{j_1 j_2}.$$

We may write the kets $|j_1 \, j_2 \, jm >$ in terms of the kets $|j_1 m_1 \, j_2 m_2 >$ as

$$|j_1 \, j_2 \, jm\rangle = \sum_{m_1} \sum_{m_2} |j_1 m_1 \, j_2 m_2 > < j_1 m_1 \, j_2 m_2 | j_1 m j_2 \, jm\rangle. \tag{24}$$

The coefficients $< j_1 m_1 \, j_2 m_2 | j_1 j_2 \, jm >$ are called Clebsch-Gordan (CG) or vector addition coefficients.

We may write Eq. (24) as

$$|j_1 j_2 jm\rangle = \sum_{m_1, m_2} |j_1 m_1 \, j_2 m_2 > U_{m_1 m_2 jm} \tag{24a}$$

with
$$U_{m_1 m_2 jm} = \langle j_1 m_1 \, j_2 m_2 \,|\, jm\rangle,$$

where we have shortened the notation by dropping j_1, j_2 in the last part of the coefficient. The inverse relation is

$$|j_1 m_1 \, j_2 m_2 = \sum_{jm} |j_1 j_2 \, jm > U^{-1}_{m_1 m_2 jm} \tag{25}$$

with
$$U^{-1}_{m_1 m_2 jm} = <jm\,|j_1 m_1\, j_2 m_2>.$$

The total degeneracy of this sum is

$$\sum_{j'=j_1-j_2}^{j_1+j_2} (2j' + 1) = (2j_1 + 1)\,(2j_2 + 1).$$

There are $2j_1 + 1$ different values of m_1 possible for each value of j_1. For each of these values, there are $2j_2 + 1$ values of m_2 possible for each value of j_2. So the degeneracy is just the product above. The same degeneracy must occur in the other scheme–one can verify it algebraically (see Gottfried (1966)).

Thus, the spectrum of $\vec{J}^2$ is $j = j_1 + j_2, j_1 + j_2 - 1, ..., |j_1 - j_2|$ and there is only one sequence of $(2j + 1)$ kets to each value of j. The possible j obeys the triangular rule

$$|j_1 - j_2| \le j \le j_1 + j_2 \tag{26}$$

which is equivalent to the condition on the possible lengths of the third side of a triangle, the other two sides being of length j_1 and j_2. Equation (26) has a classical interpretation in terms of the vector model of angular momentum addition. The numbers j_1, j_2, and j occur symmetrically in the triangle relation Eq. (26). The inequality Eq. (26) is often written in the form $\Delta\,(j_1\, j_2\, j)$.

U is a transformation matrix which is square and has $(2j_1 + 1)\,(2j_2 + 1)$ rows and the same number of columns. The elements of this matrix are the CG coefficients. Also, U^{-1} exists. But this is a transformation between the eigenkets of two sets of commuting hermitian operators. U, is therefore, a unitary matrix. Since the coefficients are taken to be real, so the matrix is orthogonal (we write o for U). So the inverse coefficient $<j_1 j_2\, jm|\, j_1 m_1 j_2 m_2 >$ is the same as $< j_1 m_1\, j_2 m_2|\, jm>$. Since O is an orthogonal matrix, we have

$$\sum_i O_{i\alpha}\, O_{i\beta} = \delta_{\alpha\beta}$$

$$\sum_i O_{\alpha i}\, O_{\beta i} = \delta_{\alpha\beta}$$

where α, β, i stand for the two labels that are needed : $m_1 m_2$ for rows and jm for columns. In conventional form, we write the orthogonality condition on the CG coefficients:

$$\sum_{m_1}\sum_{m_2} <j_1 m_1\, j_2 m_2\,|jm><j_1 m_1\, j_2 m_2\,|j'm'> = \delta_{jj'}\,\delta_{mm'} \tag{27}$$

$$\sum_{j}\sum_{m} <j_1 m_1\, j_2 m_2\,|jm><j_1 m'_1\, j_2 m'_2|jm> = \delta_{m_1 m'_1}\,\delta_{m_2 m'_2} \tag{28}$$

For $j = j'$, $m = m'$, Eq. (27) reduces to

$$\sum_{m_1} \sum_{m_2} <j_1 m_1 \, j_2 m_2| \, jm >^2 = 1,$$

$$m_1 + m_2 = m$$

which is the normalization condition for $| \, jm >$.

7.3 THE CLEBSCH-GORDAN COEFFICIENTS

There are several methods to obtain explicit formulas for the CG coefficients. Wigner (1931) obtained an explicit formula for the vector coupling coefficients using group theoretical methods. Later, Racah (1942), Schwinger (1952) and others gave a number of derivations for the general formula. The explicit expressions for the CG coefficients are quite complicated. We first construct CG coefficients via recursion relations.

We operate and form $J_\pm \, |j_1 \, j_2 \, jm >$ which we can write the other way in terms of the uncoupled system:

$$J_\pm \, |j_1 j_2 \, jm \rangle = (J_{1\pm} + J_{2\pm}) \sum_{m_1} \sum_{m_2} |j_1 m_1 \, j_2 m_2 \rangle \, \langle j_1 m_1 j_2 m_2| \, jm \rangle \tag{29}$$

Using Eqs. (24) and (25), we get

$$\sqrt{(j \mp m) \, (j \pm m + 1)} \, | \, j_1 \, j_2 \, (m \pm 1) \rangle$$

$$= \sum_{m_1'} \sum_{m_2'} \left(\sqrt{(j_1 \mp m_1') \, (j_1 \pm m_1' + 1)} |j_1 j_2 \, (m_1' \pm 1) \, m_2' \rangle \right.$$

$$\left. + \sqrt{(j_2 \mp m_2') \, (j_1 \pm m_2' + 1)} \, |j_1 j_2 \, m_1' \, (m_2' \pm 1) \rangle \right)$$

$$\times \langle j_1 j_2 \, m_1' \, m_2'| jm \rangle. \tag{30}$$

Next we multiply by $< j_1 m_1 \, j_2 m_2|$ on the left and use orthonormality. On the LHS we get non-vanishing contributions only with $m_1 = m_1' \pm 1$, $m_2 = m_2'$ for the first term and $m_1 = m_1'$, $m_2 = m_2' \pm 1$ for the second one. We get the following recursion relations

$$\sqrt{(j \mp m) \, (j \pm m + 1)} \, < j_1 m_1 \, j_2 m_2| \, j(m \pm 1) >$$

$$= \sqrt{(j_1 \mp m_1 + 1) \, (j_1 \pm m_1)} \, < j_1 \, (m_1 \mp 1) \, j_2 m_2| \, jm >$$

$$+ \sqrt{(j_2 \mp m_2 + 1) \, (j_2 \pm m_2)} \, < j_1 m_1 j_2 \, (m_2 \mp 1)| \, jm >. \tag{31}$$

Here are two recursion relations – (a) the J_+ relation (upper sign) is a formula for higher m coefficient in terms of lower m ones and (b) the J_- relation (lower sign) giving lower m coefficient in terms of higher m ones. These relations are not very simple to use since they are three-term formulas. These become two term recursion formulas when $m = +j$ in (a) and $m = -j$ in (b). The coefficients of the LHS of these relations in both cases must become zero and by inspection this is seen to be so.

We can use the recursion relations by first calculating these coefficients with say $m = j$ and arbitrary m_1, m_2. Once these are known, we can calculate any other term where m is lower than j by repeated application of (b). Recursion formula (a) with $m = j$ becomes

$$\frac{<j_1 (m_1 - 1) j_2 m_2 | jj >}{<j_1 m_1 j_2 (m_2 - 1)| jj >} = -\sqrt{\frac{(j_2 + m_2)(j_2 - m_2 + 1)}{(j_1 + m_1)(j_1 - m_1 + 1)}}.$$

Now, we can use this formula repeatedly to go along one step at a time to go through all possible m_1, m_2 values. Let us start with $m_1 = j_1$. Then, since $m_1 + m_2 = m$ we may write

$$(m_1 - 1) + m_2 = j$$

or,
$$m_2 = j - m_1 + 1 = j - j_1 + 1$$

The two-term recursion formula in this case is

$m_1 = j_1$:

$$\frac{<j_1 (j_1 - 1) j_2 (j - j_1 + 1)| jj >}{<j_1 j_1 j_2 (j - j_1)| jj >} = \sqrt{\frac{(j_2 + j - j_1 + 1)(j_1 + j_2 - j)}{2 j_1 \cdot 1}}$$

Now we write it again but with $m_1 = j_1 - 1$

$m_1 = j_1 - 1$:

$$\frac{<j_1 (j_1 - 2) j_2 (j - j_1 + 2)| jj >}{<j_1 (j_1 - 1) j_2 (j - j_1 + 1)| jj >} = -\sqrt{\frac{(j_2 + j - j_1 + 2)(j_1 + j_2 - j - 1)}{(2 j_1 - 1) \cdot 2}}$$

$\cdot \quad \cdot$

$m_1 = $ arbitrary:

$$\frac{<j_1 m_1 j_2 (j - m_1)| jj >}{<j_1 (m_1 + 1) j_2 (j - m_1 + 1)| jj >} = -\frac{\sqrt{(j_2 + j - m_1)(j_2 - j + m + 1)}}{(j_1 + m_1 + 1)(j_1 - m_1)}$$

Multiplying all of the above together, we get

$$\frac{<j_1 m_1 j_2 (j - m_1)| jj >}{<j_1 j_1 j_2 (j - j_1)| jj >}$$

$$= (-1)^{j_1 - m_1} \sqrt{\frac{(j_1 + m_1)! (j_2 + j - m)! (j_1 + j_2 - j)!}{(2 j_1)! (j_1 - m_1)! (j_2 + j - j_1)! (j_2 - j + m_1)!}}.$$

This gives one specific class of coefficients-namely, the ones with $m = j$. If we know all the coefficients with $m = j$, we can use the recursion formula (b) to generate the coefficient with $m = j - 1$, ...: they will all be expressed in terms of the first coefficient $< j_1 j_1 j_2 (j - j_1)| jj >$, but by use of the orthogonality relation a numerical value can be assigned to this coefficient and hence to all the remaining of them. We thus have an explicit method of calculating the CG coefficients. Beyond the first few j_1 and j_2 the calculation becomes quite cumbersome. As the CG coefficients are extensively tabulated, in actual practice one looks up the coefficients in books of tables. The coefficients are well determined only up to a phase factor which is arbitrary. The standard convention (due to Condon and Shortley (1935)) for the CG coefficients is

$$\langle j_1 j_1 j_2 (j - j_1)| jj \rangle \text{ real and positive.} \tag{32}$$

Let us consider as an example the case $j_1 = j_2 = \dfrac{1}{2}$. When j and m take up their maximum value $j = m = 1$, we see, from (24), that the sum contains only one term

$$\left|\frac{1}{2}\,\frac{1}{2}\,11\right\rangle = \left|\frac{1}{2}\,\frac{1}{2}\,\frac{1}{2}\,\frac{1}{2}\right\rangle \left\langle\frac{1}{2}\,\frac{1}{2}\,\frac{1}{2}\,\frac{1}{2}\right|\left|11\right\rangle. \tag{33}$$

Since the vectors have unit norm and using the phase convention (32), $\left\langle\frac{1}{2}\,\frac{1}{2}\,\frac{1}{2}\,\frac{1}{2}\right|\left|11\right\rangle = 1$. From Eq. (31) recursion relation (b)

we get $\qquad \sqrt{2}\,\left\langle\frac{1}{2}\,\frac{1}{2}\,\frac{1}{2}\,\frac{1}{2}\right|\left|10\right\rangle = 0$ i.e. $\left\langle\frac{1}{2}\,\frac{1}{2}\,\frac{1}{2}\,\frac{1}{2}\right|\left|10\right\rangle = 0$

Proceeding in this manner we obtain the following table

Table 7.1   Values of $\left\langle\frac{1}{2}\,m_1\,\frac{1}{2}\,m_2\right|\left|jm\right\rangle$

	Triplet			*Singlet*
$j, m =$	1, 1	1, 0	1, −1	0, 0
$m_1,\,m_2 = \dfrac{1}{2},\,\dfrac{1}{2}$	1	0	0	0
$\dfrac{1}{2},\,-\dfrac{1}{2}$	0	$\sqrt{\dfrac{1}{2}}$	0	$\sqrt{\dfrac{1}{2}}$
$-\dfrac{1}{2},\,\dfrac{1}{2}$	0	$\sqrt{\dfrac{1}{2}}$	0	$-\sqrt{\dfrac{1}{2}}$
$-\dfrac{1}{2},\,-\dfrac{1}{2}$	0	0	1	0

Notice that the fourth column is obtained by requiring that the singlet state $|0, 0\rangle$ be orthogonal to the three triplet states.

Let us consider another example $j_1 = \dfrac{1}{2}$, $j_2 = l$ (spin of an electron added to its orbital angular momentum). Since there is only one state with $m = l + \dfrac{1}{2}$,

$$\left|\frac{1}{2}\,\frac{1}{2}\,ll\right\rangle = \left|\frac{1}{2},\,l,\,l + \frac{1}{2},\,l + \frac{1}{2}\right\rangle$$

and so

$$\left\langle \frac{1}{2}\,\frac{1}{2}\,ll\,\middle|\left(l+\frac{1}{2}\right)\left(l+\frac{1}{2}\right)\right\rangle = 1.$$

From the recursion relation (b) in Eq. (31), we get

$$\left\langle \frac{1}{2}\,\frac{1}{2}\,l\left(m-\frac{1}{2}\right)\middle|\left(l+\frac{1}{2}\right)m\right\rangle$$

$$= \sqrt{\frac{l+m+\dfrac{1}{2}}{l+m+\dfrac{3}{2}}}\ \left\langle \frac{1}{2}\,\frac{1}{2}\,l\left(m+\frac{1}{2}\right)\middle|\left(l+\frac{1}{2}\right)(m+1)\right\rangle.$$

By repeated application of the recursion relation, we get

$$\left\langle \frac{1}{2}\,\frac{1}{2}\,l\left(m-\frac{1}{2}\right)\middle|\left(l+\frac{1}{2}\right)m\right\rangle = \sqrt{\frac{l+m+\dfrac{1}{2}}{2l+1}}.$$

In general

$$\left\langle \frac{1}{2}\,\pm\frac{1}{2}\,l\left(m-\frac{1}{2}\right)\middle|\left(l+\frac{1}{2}\right)m\right\rangle = \sqrt{\frac{l\pm m+\dfrac{1}{2}}{2l+1}}. \tag{34}$$

and

$$\left\langle \frac{1}{2}\,\pm\frac{1}{2}\,l\left(m-\frac{1}{2}\right)\middle|\left(l-\frac{1}{2}\right)m\right\rangle = \pm\sqrt{\frac{l\mp m+\dfrac{1}{2}}{2l+1}}. \tag{34a}$$

With the help of these CG coefficients, we write the spin-orbital eigenfunctions (in the standard matrix representation) as

$$y_l^{l+\frac{1}{2},m} = \frac{1}{\sqrt{2l+1}}\begin{pmatrix} \sqrt{\left(l+m+\dfrac{1}{2}\right)}\ Y_l^{m-\frac{1}{2}}(\theta,\phi) \\[2ex] \sqrt{\left(l-m+\dfrac{1}{2}\right)}\ Y_l^{m+\frac{1}{2}}(\theta,\phi) \end{pmatrix} \tag{35}$$

$$y_l^{l-\frac{1}{2},m} = \frac{1}{\sqrt{2l+1}}\begin{pmatrix} -\sqrt{\left(l-m+\dfrac{1}{2}\right)}\ Y_l^{m-\frac{1}{2}}(\theta,\phi) \\[2ex] \sqrt{\left(l+m+\dfrac{1}{2}\right)}\ Y_l^{m+\frac{1}{2}}(\theta,\phi) \end{pmatrix}. \tag{35a}$$

Since these are eigenfunctions of $\vec{L}^2$, $\vec{S}^2$, $\vec{J}^2$, and J_z, they are eigenfunctions of the spin-orbit interaction operator $\vec{L} \cdot \vec{S} = \frac{1}{2} (\vec{J}^2 - \vec{L}^2 - \vec{S}^2)$. The eigenvalues are

$$\frac{1}{2} \hbar^2 \left[j\,(j+1) - l\,(l+1) - \frac{3}{4} \right] = \frac{1}{2} \hbar^2 l \text{ for } j = l + \frac{1}{2} \tag{36}$$

$$= \frac{1}{2} \hbar^2 (l+1), \text{ for } j = l - \frac{1}{2}. \tag{36a}$$

An explicit formula for the vector coupling coefficient is given below (see Edmonds (1957):

$$\langle j_1 m_1 j_2 m_2 | jm \rangle = \delta_{m_1 + m_2, m} \times$$

$$\left[\frac{(2j+1)\,(j_1 + j_2 - j)!\,(j_1 - j_2 + j)!\,(-j_1 + j_2 + j)!}{(j_1 + j_2 + j + 1)} \right]^{\frac{1}{2}}$$

$$\times \left[(j_1 + m_1)!\,(j_1 - m_1)!\,(j_2 + m_2)!\,(j_2 - m_2)!\,(j + m)!\,(j - m)! \right]^{1/2} \times$$

$$\sum_z (-1)^z \times$$

$$\frac{1}{z!\,(j_1 + j_2 - j - z)!\,(j_1 - m_1 - z)!\,(j_2 + m_2 - z)!\,(j - j_2 + m_1 + z)!\,(j - j_1 - m_2 + z)!} \tag{37}$$

We now briefly discuss a method due to Goldberger for computing the vector coupling coefficients (see Goldberger and Watson (1964)). In this method, a set of projection operators is constructed as follows. Let $\Sigma = \vec{J}_1 \cdot \vec{J}_2$. For given j_1, j_2, the eigenvalues of Σ corresponding to any resultant j are

$$\Sigma_j = \frac{[j\,(j+1) - j_1\,(j_1 + 1) - j_2\,(j_2 + 1)]\hbar^2}{2}, \tag{38}$$

$(|j_1 - j_2| \leq j \leq j_1 + j_2)$, and these are all different.

We construct the operator

$$P_j = \prod_{i \neq j} (\Sigma - \Sigma_i) \Big/ \prod_{i \neq j} (\Sigma_j - \Sigma_i), \tag{39}$$

where i is one of the possible j values and the value $i = j$ is omitted from the sum. P_j's are hermitian and $P_j^2 = P_j$. Thus, P_j is a projection operator which gives unity when the vector on which it operates contains the eigenvalue j and gives zero otherwise: $P_j |j_1 j_2 j'm\rangle = \delta_{jj'} |j_1 j_2 j'm\rangle$. The diagonal element of P_j is

$$\langle j_1 m_1 j_2 m - m_1 \,|P_j|\, j_1 m_1 j_2 m - m_1 \rangle$$

$$= \langle j_1 j_2 m_1\, m - m_2 |\, jm \rangle^2. \tag{40}$$

From the diagonal elements of P_j we can obtain the absolute value of the CG coefficients. The relative phases can be determined by considering the off-diagonal elements

$$\langle j_1 m_1' j_2 m - m_1' \,|P_j|\, j_1 m_1 j_2 m - m_1 \rangle$$

$$= \langle j_1 j_2\, m_1'\, m - m_1' |\, jm \rangle \times$$

$$\times \langle j_1 j_2 m_1\, m - m_1 |\, jm \rangle. \tag{41}$$

This method of computing the CG coefficients is useful when one of the j's is small. When both j_1 and j_2 are large $\left(\geq \dfrac{1}{2}\right)$ the method becomes complicated.

We now consider the symmetry properties of the CG coefficients. When we deal with the addition of two angular momenta, say $\vec{J}_1$, and $\vec{J}_2$, we must pay attention to the order in which they are coupled. Although in absolute value $\langle j_1 m_1 j_2 m_2 |\, j_3 m_3 \rangle$ is equal to $\langle j_2 m_2 j_1 m_1 |\, j_3 m_3 \rangle$, we must also be concerned with the phase. The standard phase convention (CS), Eq. (32), states that when the m positions on the extreme left and the extreme right are at their maximum possible values, then the coefficient must be real and positive, i.e.

$$\langle j_1 j_1 j_2\, (m_2 = j_3 - j_1) |\, j_3 j_3 \rangle > 0.$$

But then what about $\langle j_2\, (j_3 - j_1)\, j_1 j_1 |\, j_3 j_3 \rangle$? In this case, on the extreme left the m value is not at its maximum possible value (in fact, it is at its minimum value) so we cannot apply the CS phase convention directly. $j_3 - j_1$ is less than j_2 by $j_2 - (j_3 - j)$ steps so we can step the last formula up to the previous one by use of the recursion formula making $j_2 + j_1 - j_3$ steps and thereby getting a factor $(-1)^{j_1 + j_2 - j_3}$. So we obtain the symmetry property (I)

$$\langle j_1 m_1 j_2 m_2 |\, j_3 m_3 \rangle = (-1)^{j_1 + j_2 - j_3} \langle j_2 m_2 j_1 m_1 |\, j_3 m_3 \rangle \tag{42}$$

In this case, we have interchanged $1 \leftrightarrow 2$. Let us now interchange $1 \leftrightarrow 3$. Instead of writing $\vec{J}_1 + \vec{J}_2 = \vec{J}_3$ we now write $\vec{J}_3 - \vec{J}_2 = \vec{J}_1$. We see $j_1 \leftrightarrow j_3$, $m_1 \leftrightarrow m_3$ and the relation $m_1 + m_2 = m_3 \leftrightarrow m_3 + m_2 = m_1$. Then $m_2 \leftrightarrow -m_2$. We expect $\langle j_3 m_3 j_2 - m_2 |\, j_1 m_1 \rangle$ to be related to $\langle j_1 m_1 j_2 m_2 |\, j_3 m_3 \rangle$. To find this relation we interchange $1 \leftrightarrow 3$ in recursion relation (a) getting

$$\sqrt{(j_3 + m_3)\,(j_3 - m_3 + 1)}\ \langle j_3\,(m_3 - 1)\, j_2 - m_2 |\, j_1 m_1 \rangle$$

$$= \sqrt{(j_1 - m_1)\,(j_1 + m_1 + 1)}\ \langle j_3 m_3 j_2 - m_2 |\, j_1\,(m_1 + 1) \rangle$$

$$- \sqrt{(j_2 - m_2)(j_2 + m_2 + 1)} \; \langle j_3 m_3 j_2 - (m_2 + 1) | j_1 m_1 \rangle.$$

We write $\langle j_3 m_3 j_2 - m_2 | j_1 m_1 \rangle = K \langle j_1 m_1 j_2 m_2 | j_3 m_3 \rangle$ where K may depend on j's but is independent of m. The LHS of the above equation (relabeled recursion equation (a)) is replaced by $K \langle j_1 m_1 j_2 m_2 | j_3 (m_3 - 1) \rangle$ and the two terms on the LHS by $K \langle j_1 (m_1 + 1) j_2 m_2 | j_3 m_3 \rangle$ and $K \langle j_1 m_1 j_2 (m_2 + 1) | j_3 m_3 \rangle$ respectively. This would be exactly (b) if signs were different. Summing over all m,

$$\sum_{m_1 (m_2)} \sum_{m_3} |\langle j_3 m_3 j_2 - m_2 | j_1 m_1 \rangle|^2 = |K|^2 \sum_{m_1 (m_2)} \sum_{m_3} |\langle j_1 m_1 j_2 m_2 | j_3 m_3 \rangle|^2.$$

On LHS the first sum is for fixed m_1 over all m_3 (and therefore m_2) and we are left with a sum over m_1 alone. But by the orthogonality relation, we see that the summation yields $1 + 1 + \ldots$ to $(2j_1 + 1)$ terms (values of m_1). On the LHS we get the summation to yield $(2j_3 + 1)$ (values of m_3). Thus

$$(2j_1 + 1) = |K|^2 (2j_3 + 1)$$

or,

$$|K| = \sqrt{\frac{2j_1 + 1}{2j_3 + 1}}.$$

Now, we write $\langle j_3 m_3 j_2 - m_2 | j_1 m_1 \rangle = \sqrt{\dfrac{2j_1 + 1}{2j_3 + 1}} \; (-1)^{m_2} C \langle j_1 m_1 j_2 m_2 | j_3 m_3 \rangle.$

For $m_1 = j_1$, and $m_3 = j_3$, both CG coefficients must be real and positive. We require

$$(-1)^{m_2} C = 1$$

giving

$$C = (-1)^{-m_2} = (-1)^{j_1 - j_3}$$

We obtain the symmetry property (II)

$$\langle j_3 m_3 j_2 - m_2 | j_1 m_1 \rangle = (-1)^{j_1 - j_2 + m_3} \sqrt{\frac{2j_1 + 1}{2j_3 + 1}} \; \langle j_1 m_1 j_2 m_2 | j_3 m_3 \rangle. \tag{43}$$

Combining symmetry property (I) and (II) we get another symmetry property (III).

$$\langle j_1 - m_1 j_2 - m_2 | j_3 - m_3 \rangle = (-1)^{j_1 + j_2 - j_3} \langle j_1 m_1 j_2 m_2 | j_3 m_3 \rangle. \tag{44}$$

From Eq. (43),

$$\langle j m j - m | 0 0 \rangle = \frac{(-1)^{j-m}}{\sqrt{2j + 1}} \; \langle 0 0 j m | j m \rangle = (-1)^{j-m} \cdot \frac{1}{\sqrt{2j + 1}},$$

and from Eq. (44)

$$\langle l_1 0 l_2 0 | l_3 0 \rangle = (-1)^{l_1 + l_2 - l_3} \langle l_1 0 l_2 0 | l_3 0 \rangle.$$

i.e., $\qquad \langle l_1 0\, l_2 0 | l_3 0 \rangle = 0$ if $l_1 + l_2 - l_3$ is odd.

Symmetry property (III) can be obtained using the concept of time reversal, to be discussed in Section 10.3. Time reversal is effected by an antiunitary and antilinear operator T which may be written as

$$T = UK \qquad (45)$$

where U is unitary and K the complex conjugation operator. For the time reversed state the signs of all linear and angular momenta are reversed, while T leaves the coordinate vector unchanged.

$$T \vec{p}\, T^{-1} = -\vec{p}$$
$$T \vec{J}\, T^{-1} = -\vec{J}$$
$$T \vec{r}\, T^{-1} = \vec{r} \qquad (46)$$

For a spin-zero particle, we can choose $U = 1$, and $T = K$. For a single spin $\frac{1}{2}$ particle, we can write $U = -i\sigma_y$ where

$$\sigma_y = \begin{pmatrix} 0 & -i \\ i & 0 \end{pmatrix}, \text{ and } T = -i\sigma_y\, K.$$

From Eq. (45), we see $\quad TJ_+ = -J_-\, T$

$$TJ_- = -J_+\, T. \qquad (47)$$

Now, let us find the result of $T| jm\rangle$. Since $J_z| jm\rangle = m| jm\rangle$ $(\hbar = 1)$, $TJ_z| jm\rangle = mT| jm\rangle$, or, using Eq. (46), $-J_z T| jm\rangle = mT| jm\rangle$

i.e., $\qquad J_z T| jm\rangle = -mT| jm\rangle. \qquad (48)$

Thus, $T| jm\rangle$ is an eigenvector of J_z with the eigenvalue $-m$. So $T| jm\rangle = C_{jm}| j - m\rangle$ where C is a phase. We choose the phase such that

$$T| jj\rangle = | j - j\rangle, \qquad (49)$$

i.e., $C_{jj} = 1$. Let us operate on Eq. (49) with J_+^{j-m} to get

$$J_+^{j-m}\, T| jj\rangle = J_+^{j-m}| j - j\rangle.$$

But, since $J_+| jm\rangle = \sqrt{(j - m)(j + m + 1)}| j (m + 1)\rangle$ and $TJ_+ = -J_-\, T$, we get from above

$$(-1)^{j-m}\, TJ_-^{j-m}| jj\rangle = \sqrt{\frac{2j!\,(j - m)!}{(j + m)!}}| j - m\rangle.$$

Using $\qquad J_-| jm\rangle = \sqrt{(j + m)(j + m - 1)}| j (m - 1)\rangle$

we get

$$(-1)^{j-m}\, T|\,jm\rangle = |\,j-m\rangle,$$

and so
$$T|\,jm\rangle = (-1)^{j-m}|\,j-m\rangle. \tag{50}$$

Since

$$|\,jm\rangle = \sum_{m_1(m_2)} \langle j_1 m_1 j_2 m_2|\,jm\rangle|\,j_1 m_1\rangle|\,j_2 m_2\rangle, \text{ we get}$$

$$T|\,jm\rangle = \sum_{m_1(m_2)} \langle j_1 m_1 j_2 m_2|\,jm\rangle\, T|\,j_1 m_1\rangle|\,j_2 m_2\rangle$$

Using Eq. (50)

$$(-1)^{j-m}|\,j-m\rangle = \sum_{m_1(m_2)} \langle j_1 m_1 j_2 m_2|\,jm\rangle\, (-1)^{j_1-m_1}\,(-1)^{j_2-m_2}\times|\,j_1-m_1\rangle|\,j_2-m_2\rangle.$$

We relabel $m_1,\ m_2,\ m \to -m_1,\ -m_2,\ -m$ and so

$$|\,jm\rangle = \sum_{m_1(m_2)} \langle j_1-m_1 j_2-m_2|\,j-m\rangle|\,j_1 m_1\rangle|\,j_2 m_2\rangle\,(-1)^{j_1+j_2-j}.$$

Comparing the two expressions for $|\,jm\rangle$ we get

$\langle j_1-m_1 j_2-m_2|\,j-m\rangle\,(-1)^{j_1+j_2-j} = |\,j_1 m_1 j_2 m_2|\,jm\rangle$ which is just the symmetry property (III).

We have obtained 3 basic symmetry properties which can be combined to get $2 \times 3! = 12$ symmetry properties altogether. This means that we can rearrange these coefficients in 12 ways and still get the same number (up to a factor ± 1). To make these symmetries more apparent we can introduce a new coefficient called the 3j-symbol. The high degree of symmetry of these coefficients is obtained by the coupling of three angular momenta $\vec{J}_1$, $\vec{J}_2$, and $\vec{J}_3$ to a resultant zero. The 3j-symbol of Wigner (1951) is defined by

$$\begin{pmatrix} j_1 & j_2 & j_3 \\ m_1 & m_2 & m_3 \end{pmatrix} = (-1)^{j_1-j_2-m_3}\,(2j_3+1)^{-\frac{1}{2}}\, \langle j_1 m_1 j_2 m_2|\,j_3-m_3\rangle. \tag{51}$$

The above 12 symmetry properties of the CG coefficients imply the following symmetry properties of the 3j-symbols. The values of the 3j-symbol are unchanged under an even permutation of the columns of the symbol. Under an odd permutation it is multiplied by $(-1)^{j_1+j_2+j_3}$. The symbol changes sign by the same factor under the replacement $m_1, m_2, m_3 \to -m_1, -m_2, -m_3$. There are $3! = 6$ column permutations, and in addition we may change the signs of the m's. This gives 12 symmetry properties. The invariance of the 3j-symbols under these permutations leads to a symmetry group of order 72 (Regge (1958)):

(3! Permutations of rows) × (3! Permutations of columns) × (a reflection about the diagonal of the 3 × 3 square symbol). (See Esposito, Marmo, and Sudarshan (2004)).

The orthogonality properties are not so convenient. These are given below

$$\sum_{m_1 m_2} \begin{pmatrix} j_1 & j_2 & j_3 \\ m_1 & m_2 & m_3 \end{pmatrix} \begin{pmatrix} j_1 & j_2 & j_3' \\ m_1 & m_2 & m_3' \end{pmatrix} = \delta\,(j_1 j_2 j_3)\, \frac{1}{2j_3 + 1}\, \delta_{j_3 j_3'}\, \delta_{m_3 m_3'}$$

$$\sum_{m_1 m_2 m_3} \begin{pmatrix} j_1 & j_2 & j_3 \\ m_1 & m_2 & m_3 \end{pmatrix} \begin{pmatrix} j_1 & j_2 & j_3 \\ m_1 & m_2 & m_3 \end{pmatrix} = \delta\,(j_1 j_2 j_3) \tag{52}$$

$$\sum_{j_3 m_3} (2j_3 + 1) \begin{pmatrix} j_1 & j_2 & j_3 \\ m_1 & m_2 & m_3 \end{pmatrix} \begin{pmatrix} j_1 & j_2 & j_3 \\ m_1' & m_2' & m_3 \end{pmatrix} = \delta\,(j_1 j_2 j_3)\, \delta_{m_1 m_1'}\, \delta_{m_2 m_2'}$$

where $\delta\,(j_1 j_2 j_3) = 1$ if j_1, j_2, j_3 satisfy the triangular condition, and is zero otherwise.

We now discuss the angular momentum addition, Eq. (20a) from the point of view of rotation matrices. We have

$$D(R) = (e^{-i\alpha J_{1z}}\, e^{-i\beta J_{1y}}\, e^{-i\gamma J_{1z}})\, (e^{-i\alpha J_{2z}}\, e^{-i\beta J_{2y}}\, e^{-i\gamma J_{2z}})$$

$$= D(R)_1\, D(R)_2 \tag{53}$$

$$\therefore \qquad \langle j_1 j_2\, jm' |\, D(R)|\, j_1 j_2\, jm \rangle$$

$$= \sum_{\substack{m_1, m_1' \\ m_2, m_2'}} \langle jm' |\, j_1 m_1\, j_2 m_2 \rangle\, \langle j_1 m_1 | D(R)_1 |\, j_1 m_1' \rangle \times$$

$$\times \langle j_2 m_2 |\, D(R)_2 |\, j_2 m_2' \rangle\, \langle j_1 m_1'\, j_2 m_2' |\, jm \rangle,$$

or, $$D^{(j)}_{m'm}(R) = \sum_{\substack{m_1, m_1' \\ m_2, m_2'}} \langle jm' |\, j_1 m_1\, j_2 m_2 | D^{(j_1)}_{m_1 m_1'}\, D^{(j_2)}_{m_2 m_2'} \cdot \langle j_1 m_1'\, j_2 m_2' |\, jm \rangle. \tag{54}$$

The inverse relation is

$$D^{(j_1)}_{m_1 m_1'}\, D^{(j_2)}_{m_2 m_2'} = \sum_{jmm'} \langle j_1 m_1 |\, j_2 m_2 |\, jm' \rangle\, D^{(j)}_{m'm}(R)\, \langle jm |\, j_1 m_1'\, j_2 m_2' \rangle. \tag{55}$$

In group theory, we write this as the Clebsch-Gordan decomposition of the Kronecker (or direct) product

$$D^{(j_1)} \otimes D^{(j_2)} = D^{(j_1 + j_2)} \oplus D^{(j_1 j_2 - 1)} \oplus \ldots \oplus D^{(|j_1 - j_2|)} \tag{56}$$

Using Eq. (56), we can construct all the matrices $D^{(j)}$ from $D^{(1/2)}$ (not a practical way!).

From Eq. (56), we obtain

$$D^{(j_1)}_{m_1 m_1'}(R)\, D^{(j_2)}_{m_2 m_2'}(R) = \sum_{jmm'} \langle j_1 m_1\, j_2 m_2 |\, jm \rangle \times$$

$$\times \langle j_1 m_1' j_2 m_2' | jm' \rangle D_{mm'}^{(j)}(R) \qquad (57)$$

where the sum over j is from $|j_1 - j_2|$ to $j_1 + j_2$. Equation (57) is known as the Clebsch-Gordan series.

The orthogonality relation for the representation matrices is

$$\int D_{m_1 m_1'}^{(j_1)}(\omega)^* \, D_{m_2 m_2'}^{(j_2)}(\omega) \, d\omega$$

$$= \int (-1)^{m_1' - m_1} D_{-m_1, -m_2'}^{(j_1)}(\omega) \, D_{m_2 m_2'}^{(j_2)}(\omega) \, d\omega, \text{ using } (5.78)$$

$$= \int \left\{ \sum_{j=|j_1-j_2|}^{j_1+j_2} (-1)^{m_1' - m_1} \langle j_1 - m_1 j_2 - m_2 | jm \rangle D_{mm'}^{(j)}(\omega) \times \langle jm' | j_1 - m_1' j_2 m_2' \rangle \right\} d\omega$$

$$\text{from the properties of CG coeffs.}$$
$$m = -m_1 + m_2$$
$$m' = -m_1' + m_2'$$

$$= \frac{8\pi^2}{2j_1 + 1} \, \delta_{j_1 j_2} \, \delta_{m_1 m_2} \, \delta_{m_1' m_2'}. \qquad (58)$$

Next, we evaluate the integral

$$\mathcal{I} = \int D_{m_3 m_3'}^{(j_3)}(\omega)^* \, D_{m_1 m_1'}^{(j_1)}(\omega) \, D_{m_2 m_2'}^{(j_2)}(\omega) \, d\omega.$$

This integral is extremely useful and is needed in evaluating matrix elements of operators in the theory of atomic and nuclear spectra. We have

$$\mathcal{I} = \frac{8\pi^2}{2j_3 + 1} \, \langle j_1 m_1 j_2 m_2 | j_3 m_3 \rangle \, \langle j_1 m_1' j_2 m_2' | j_3 m_3' \rangle. \qquad (59)$$

A useful special case: $m_1' = m_2' = m_3' = 0$.

Since $\qquad D_{mo}^{(l)} = \sqrt{\frac{4\pi}{2l+1}} \, Y_{lm}(\beta, \alpha)$, we have

$$\frac{2\pi \cdot (4\pi)^{3/2}}{\sqrt{(2l_1 + 1)(2l_2 + 1)(2l_3 + 1)}} \int_0^{2\pi} d\alpha \int_0^{\pi} d\beta \sin \beta Y_{l_3 m_3}^* \, Y_{l_1 m_1} \, Y_{l_2 m_2}$$

$$= \frac{8\pi^2}{2l_3 + 1} \, \langle l_1 m_1 \, l_2 m_2 | l_3 m_3 \rangle \times \langle l_1 0 l_2 0 | l_3 0 \rangle,$$

or,
$$\int_0^{2\pi} d\alpha \int_0^{\pi} d\beta \sin \beta Y_{l_3 m_3}^*(\beta, \alpha) \, Y_{l_1 m_1}(\beta, \alpha) \, Y_{l_2 m_2}(\beta, \gamma)$$

$$= \sqrt{\frac{(2l_1 + 1)(2l_2 + 1)}{4\pi (2l_3 + 1)}} \, \langle l_1 m_1 \, l_2 m_2 | l_3 m_3 \rangle \times \langle l_1 0 l_2 0 | l_3 0 \rangle \qquad (60)$$

This is Gaunt's formula.

There is a closed expression for $\langle l_1\, 0 l_2\, 0 | l_3\, 0 \rangle$:

$$\langle l_1\, 0 l_2\, 0 | l_3\, 0 \rangle = 0 \text{ for } l_1 + l_2 + l_3 \text{ odd}$$

$$\langle l_1\, 0 l_2\, 0 | l_3\, 0 \rangle = (-1)^{l_2+L} \times \frac{L! \sqrt{2 l_2 + 1}}{(L - l_2)!\, (L - l_3)!\, (L - l_1)!} \times$$

$$\frac{(l_2 + l_3 - l_1)!\, (l_3 + l_1 - l_2)!\, (l_1 + l_2 - l_3)!}{(l_1 + l_2 + l_3 + 1)!} , \qquad (61)$$

if $l_1 + l_2 + l_3$ is even, with $2L = l_1 + l_2 + l_3$.

7.4 THE RECOUPLING OF THREE ANGULAR MOMENTA

We now consider three mutually commuting operators $\vec{J}_1$, $\vec{J}_2$, $\vec{J}_3$ with corresponding eigenfunctions $|j_1 m_1\rangle$, $|j_2 m_2\rangle$, and $|j_3 m_3\rangle$. The operator

$$\vec{J} = \vec{J}_1 + \vec{J}_2 + \vec{J}_3 \qquad (62)$$

is called the total angular momentum operator. From the functions $|j_1 m_1\rangle$, $|j_2 m_2\rangle$, and $|j_3 m_3\rangle$ we can construct eigenfunctions of the operators $\vec{J}^2$ and $\vec{J}_z$ corresponding to the eigenvalues

$$\vec{J}^2 = \hbar^2\, j\, (j + 1) \text{ and } J_z = \hbar m \qquad (63)$$

The addition of three angular momenta can be carried out successively:

$$\vec{J}_1 + \vec{J}_2 = \vec{J}_{12}, \ \vec{J}_{12} + \vec{J}_3 = \vec{J}. \qquad (64)$$

This corresponds to the reduction of the product representation $D^{(j_1)} \times D^{(j_2)} \times D^{(j_3)}$ is two steps:

$$D^{(j_1)} \times D^{(j_2)} = \sum_{j_{12}} D^{(j_{12})}; \ D^{(j_{12})} \times D^{(j_3)} = \sum_{j} = D^{(j)} \qquad (65)$$

We denote the state vector in this case as $|[j_1 j_2)\, J_{12}\, j_3]\, JM\rangle$. But this is just one way of adding three angular momenta. The addition may be carried out in another order:

$$\vec{J}_2 + \vec{J}_3 = \vec{J}_{23}, \ \vec{J}_1 + \vec{J}_{23} = \vec{J}. \qquad (66)$$

This gives another basis which spans the product space. The basis function corresponding to this coupling process is denoted as $|[j_1\, (j_2 j_3)\, J_{23}]\, JM\rangle$. We note that we may couple $\vec{J}_1$ and $\vec{J}_3$ to give a resultant $\vec{J}_{13}$, and then couple this to $\vec{J}_2$ to give $\vec{J}$. The order of coupling determines the phase of the resulting state vector.

The two sets of bases from Eqs. (64) and (66) are both eigenfunctions of $\vec{J}^2$ and J_z, and therefore, are related to each other by a unitary transformation:

$$|[(j_1 j_2) J_{12} j_3] JM\rangle = \sum_{j_{23}} |[j_1 (j_2 j_3) J_{23}] JM\rangle \times$$

$$\times \langle \{j_1 (j_2 j_3) J_{23}\} JM | \{(j_1 j_2) J_{12} j_3\} JM\rangle. \quad (67)$$

The transformation coefficients are independent of m. We also denote the above coefficient as.

$$U^{-1}_{J_{23} J_{12}} (j_1\, j_2\, Jj_3).$$

The inverse relation is

$$|[j_1 (j_2 j_3) J_{23}] JM\rangle$$

$$= \sum_{j_{12}} |[(j_1 j_2) J_{12} j_3] JM\rangle \times$$

$$\times \langle \{(j_1 j_2) J_{12} j_3\} JM | \{j_1 (j_2 j_3) J_{23}\} JM\rangle \quad (68)$$

We denote the coefficient as $U_{J_{12} J_{23}} (j_1\, j_2\, Jj_3)$. [also denoted as

$$U|j_1\, j_2\, Jj_3; J_{12}\, J_{23}\rangle$$

We note that we always put 1, 2, 3 in the same canonical order. Next, we introduce a pictorial notation (due to J.B. French (1966), Hecht (unpub.)) We denote

$$|(j_1 j_2) JM\rangle \equiv$$

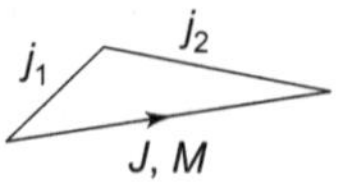

We must be careful of the order in which we couple the angular momenta. Whether we couple 1 to 2 or 2 to 1 makes a difference of a phase. From symmetry property (I) Eq. (42), we write

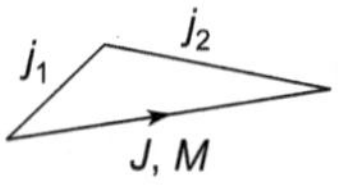

$$= (-1)^{j_1 + j_2 - J}$$

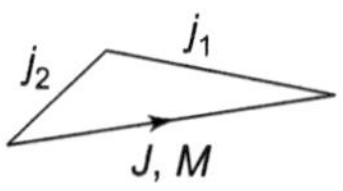

By extension, we denote the LHS Eq. of (67) as

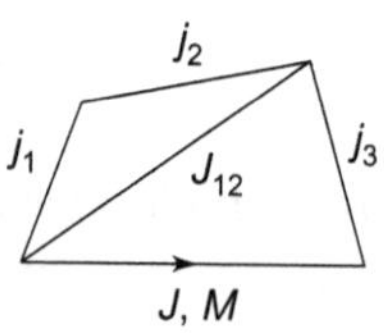

And the LHS of Eq. (68) as

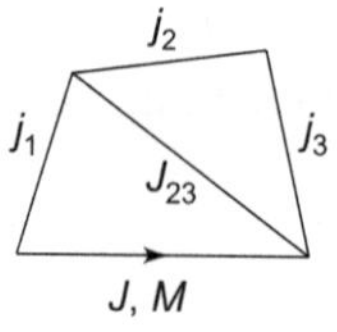

We can write Eq. (67) in terms of the individual two angular momenta couplings used.

$$\vcenter{\hbox{[diagram]}} \;\; = \sum_{m_1 m_2 (m_3)} \langle j_1 m_1\, j_2 m_2|\, J_{12} m_{12}\rangle \,\langle J_{12} m_{12}\, j_3 m_3|\, JM\rangle$$

$$\times\, |j_1 m_1\rangle |j_2 m_2\rangle |j_3 m_3\rangle.$$

Since the transformation is unitary, we have

$$U^{-1}_{J_{23} J_{12}} = U^{*}_{J_{12} J_{23}}.$$

But the complex conjugation is not necessary since U is real, orthogonal. The orthogonality relations are

$$\sum_{J_{12}} U_{J_{12} J_{23}}\, U_{J_{12} J'_{23}} = \delta_{J_{23} J'_{23}}$$

$$\sum_{J_{23}} U_{J_{12} J_{23}}\, U_{J'_{12} J_{23}} = \delta_{J_{12} J'_{12}} \tag{69}$$

It can be shown that the $U's$ are independent of m.

Next we express U in terms of the C.G. coefficients. By definition,

$$|\{j_1\,(j_2\, j_3)\, J_{23}\}\, JM\rangle = \sum \langle j_1 m_1\, J_{23} m_{23}|\, JM\rangle \,\langle j_2 m_2\, j_3 m_3|\, J_{23} m_{23}\rangle$$

$$\times\, |j_1 m_1\rangle|\, j_2 m_2\rangle|\, j_3 m_3\rangle.$$

The LHS may be written as

$$\sum \langle\,\rangle \langle\,\rangle \sum_{J'_{12}} \langle j_1 m_1\, j_2 m_2|\, J'_{12} m_{12}\rangle |(j_1 j_2)\, J'_{12} m_{12}\rangle|\, j_3 m_3\rangle$$

$$= \sum \langle\,\rangle\, (\,) \sum_{J'_{12}} \langle j_1 m_1\, j_2 m_2|\, J'_{12} m_{12}{}' \sum_{J'} \langle J'_{12} m_{12}\, j_3 m_3|J'M\rangle|\, \{(j_1 j_2)\, J'_{12} j_3\}|\, J'M\rangle$$

$$\times\, \langle\{(j_1 j_2)\, J'_{12} j_3\}\, JM\rangle, \text{ where } \langle\,\rangle\langle\,\rangle = \langle j_1 m_1\, J_{23} m_{23}|\, JM\rangle$$

$$\times\, \langle j_2 m_2\, j_3 m_3|\, J_{23} M_{23}\rangle.$$

If we multiply on the left with specific J_{12} and J and integrate over all the variables of the space we get zero in the sum unless $J'_{12} = J_{12}$ and $J' = J$ where we get 1.

$$\therefore\;\; U\,(j_{1\, j2}\, Jj_3|\, J_{12}\, J_{23}) = \sum_{m_1 m_2 (m_3)} \langle j_1 m_1\, J_{23} m_{23}|\, JM\rangle \times$$

$$\langle j_2 m_2\, j_3 m_3|\, J_{23} m_{23}\rangle \,\langle j_1 m_1\, j_2 m_2|\, J_{12} m_{12}\rangle\langle J_{12} m_{12}\, j_3 m_3|\, JM\rangle \tag{70}$$

with fixed $M = m_1 + m_2 + m_3$.

In pictorial language,

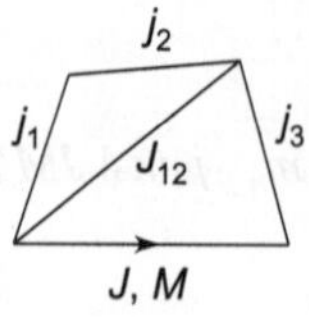

$$= \sum_{J_{23}} U(j_1\, j_2\, Jj_3;\, J_{12}\, J_{23})$$

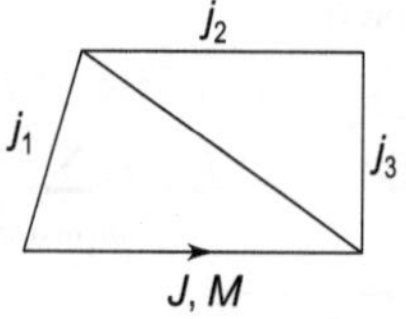

We note that the four couplings in the four C.G. coefficients are just the four couplings that come into the four triangles above.

Eq. (70) provides us a method of calculating the *U's* by multiplying four C.G. coefficients.

Other useful relations are

$$\langle j_1 m_1\, J_{23} m_{23} |\, JM \rangle\, U\,(j_1 j_2\, J j_3;\, J_{12}\, J_{23})$$

$$= \sum_{m_2} \langle j_1 m_1\, j_2 m_2 |\, J_{12} m_{12} \rangle\, \langle J_{12} m_{12} j_3 m_3 |\, JM \rangle\, \langle j_2 m_2\, j_3 m_3 |\, J_{23} m_{23} \rangle \tag{71}$$

and $\langle j_1 m_1\, j_2 m_2 |\, J_{23} m_{12} \rangle\, \langle J_{12} m_{12}\, j_3 m_3 |\, JM \rangle$

$$\sum_{J_{23}} U\,(j_1 j_2\, J j_3;\, J_{12}\, J_{23})\, \langle j_2 m_2\, j_3 m_3 |\, J_{23} m_{23} \rangle\, \langle j_1 m_1\, J_{23} m_{23} |\, JM \rangle. \tag{72}$$

We define the Racah coefficient (or *W* coefficient) as follows:

$$U\,(j_1 j_2\, J j_3;\, J_{12}\, J_{23}) = \sqrt{(2J_{12} + 1)\,(2J_{23} + 1)} \times W\,(j_1 j_2\, J j_3;\, J_{12}\, J_{23}) \tag{73}$$

The 6j-symbol is defined as follows:

$$U\,(j_1 j_2\, J j_3;\, J_{12}\, J_{23}) = (-1)^{j_1 + j_2 + j_3 + J} \sqrt{(2J_{12} + 1)\,(2J_{23} + 1)} \times \begin{Bmatrix} j_1 & j_2 & J_{12} \\ j_3 & J & J_{23} \end{Bmatrix} \tag{74}$$

The 6j-symbol $\begin{Bmatrix} j_1 & j_2 & j_3 \\ j_4 & j_5 & j_6 \end{Bmatrix}$ has the following symmetry properties. It is invariant under any permutation of the columns. It is also invariant against interchange of the upper and lower arguments in each of any two columns. In fact there are 24 operations generated by permutations of these types which leave a 6j-symbol invariant, and these form a group isomorphic with the symmetry group of a regular tetrahedron.

Because of the unitary nature of the recoupling transformations, the 6j-symbols have the orthogonality property:

$$\sum_{j} (2j + 1)\,(2j'' + 1) \begin{Bmatrix} j_1 & j_2 & j' \\ j_3 & j_4 & j \end{Bmatrix} \begin{Bmatrix} j_3 & j_2 & j \\ j_1 & j_4 & j'' \end{Bmatrix} = \delta_{j'j''}. \tag{75}$$

The 6j-symbol is of importance whenever recoupling of angular momenta is involved. Even when the recoupling of 4 or more angular momenta is done, the coefficients can be expressed in terms of 6j-symbols.

7.5 FOUR ANGULAR MOMENTA

Let us consider four independent systems with angular momenta $\vec{j}_1$, $\vec{j}_2$, $\vec{j}_3$ and $\vec{j}_4$. We construct a state with a given total angular momentum $\vec{J}$. We can couple $\vec{j}_1$ and $\vec{j}_2$ to $\vec{J}_{12}$, then couple $\vec{j}_3$ and $\vec{j}_4$ to $\vec{J}_{34}$ and finally couple $\vec{J}_{12}$ and $\vec{J}_{34}$ to $\vec{J}$. There are many coupling schemes. For example, we can first couple $\vec{j}_1$ and $\vec{j}_3$ to $\vec{J}_{13}$, $\vec{j}_2$ and $\vec{j}_4$ to $\vec{J}_{24}$, and then $\vec{J}_{13}$ and $\vec{J}_{24}$ to $\vec{J}$.

Let us, for example, relate the following two schemes:

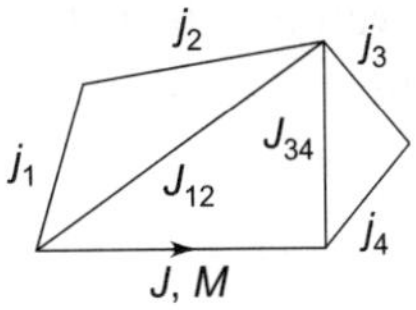

$$\cdots = \sum_{J_{234}} U(j_1 j_2\, JJ_{34};\, J_{12} J_{234}) \times \cdots \qquad (76)$$

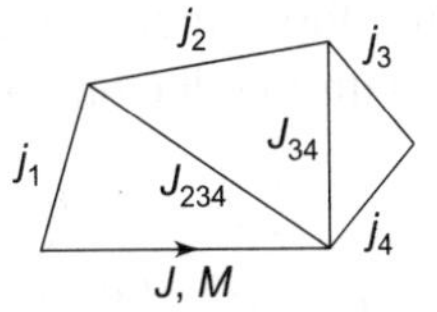

But this is just 3 angular momenta coupling in which J_{34} carries j_3 and j_4. So we can express the coupling in terms of U function, or 6j-symbol, But

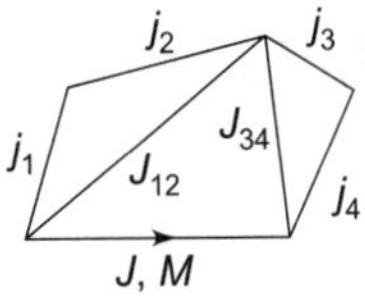

$$\cdots = \sum_{J_{13}, J_{24}} (\text{coeff.}) \times \cdots \qquad (77)$$

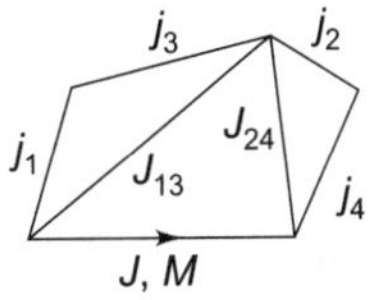

where the coefficient is

$$U = \left\langle [(j_1 j_3)\, J_{13}\, (j_2 j_4)\, J_{24}]\, J \,|[(j_1 j_2)\, J_{12}\, (j_3 j_4)\, J_{34}]\, J \right\rangle \qquad (78)$$

The transformation matrix is unitary and does not depend on M.

Interchanging the order of j_3 and j_4 on the LHS of Eq. (76),

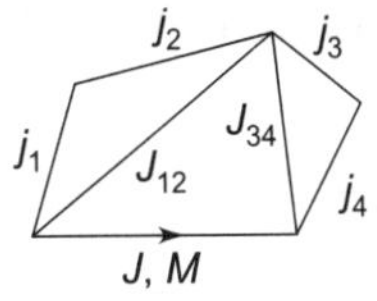

$$\cdots = \sum_{J_{234}} U()\, (-1)^{j_3 + j_4 - J_{34}} \times \cdots \qquad (79)$$

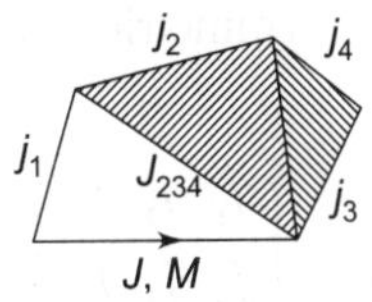

where

$$U(\ ) = U(j_1 j_2\, JJ_{34};\, J_{12} J_{234})$$ and using the symmetry property (I).

We now wish to put j_2 and j_4 together to get the LHS of Eq. (77). We do this by considering the shaded triangle alone and using 3 angular momenta coupling U functions to get the LHS of Eq. (79) as

$$\sum_{J_{234}} U\,() \,(-1)^{j_3+j_4-J_{34}} \sum_{J_{24}} U\,(j_2 j_4 J_{234} j_3;\, J_{24}\, J_{34}) \times j_1$$

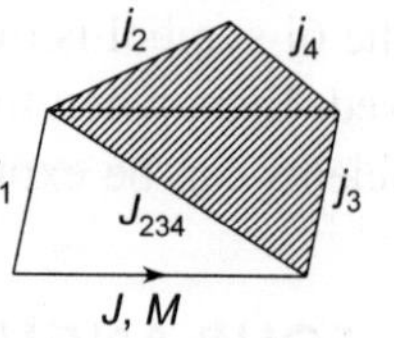

Now, we can interchange the order of the coupling J_{24} and j_3. Thus, the above expression may be written as

$$\sum_{J_{234}} U\,() \,(-1)^{j_3+j_4-J_{34}} \sum_{J_{24}} U\,() \,(-1)^{j_{24}+j_3-J_{234}} \times j_1$$

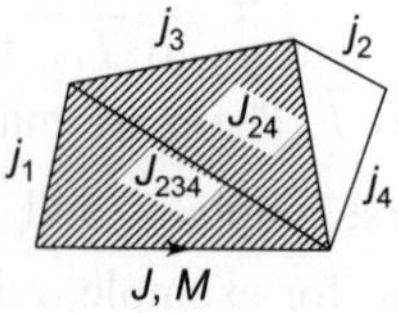

We use three angular momenta coupling scheme to convert the shaded region into one with the diagonal running the other way. Therefore,

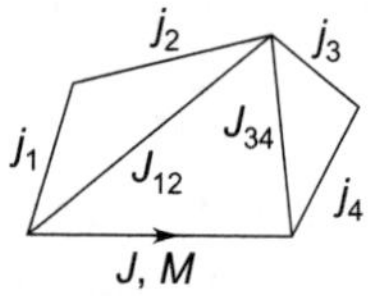

$$= \sum_{J_{234}} \sum_{J_{24}} U\,() \,(-1)^{j_3+j_4-J_{34}+J_{24}+j_3-J_{23}}\, U\,() \times$$

$$\sum_{J_{13}} U\,(j_1 j_3 JJ_{24};\, J_{13}\, J_{234}) \times j_1$$

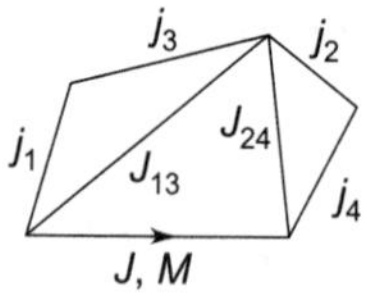

We write this result as

$$\langle [(j_1 j_3)\, J_{13}\, (j_2 j_4)\, J_{24}]\, J\| (j_1 j_2)\, J_{12}\, (j_3 j_4)\, J_{34}]\, J\rangle = X$$

$$= \sum_{J_{234}} \sum_{J_{24}} \sum_{J_{13}} (-1)^{2j_3+j_4+J_{24}-J_{34}-J_{234}} \,\{ U\,(j_1 j_2 \, JJ_{34};\, J_{12}\, J_{234})$$

$$\times U\,(j_2 j_4 J_{234} j_3;\, J_{24}\, J_{34})\, U\,(j_1 j_3 \, JJ_{24};\, J_{13}\, J_{234})\}. \tag{80}$$

A more symmetric way of writing this expression is to introduce the 9j-symbol:

$$X \equiv \begin{Bmatrix} j_1 & j_2 & J_{12} \\ j_3 & j_4 & J_{34} \\ J_{13} & J_{24} & J \end{Bmatrix} \sqrt{(2J_{12}+1)(2J_{34}+1)(2J_{13}+1)(2J_{24}+1)} \tag{81}$$

The 9j-symbol can be expressed in terms of 6j-symbols as

$$\begin{Bmatrix} j_1 & j_2 & J_{12} \\ j_3 & j_4 & J_{34} \\ J_{13} & J_{24} & J \end{Bmatrix} = \sum_{J_{234}} (-1)^{2J_{234}} \,(2J_{234}+1) \begin{Bmatrix} j_1 & j_2 & J_{12} \\ J_{34} & J & J_{234} \end{Bmatrix} \times$$

$$\begin{Bmatrix} j_3 & j_4 & J_{34} \\ J_2 & J_{234} & J_{24} \end{Bmatrix} \begin{Bmatrix} j_1 & j_3 & J_{13} \\ J_{24} & J & J_{234} \end{Bmatrix} \tag{82}$$

or, in terms of CG coefficients,

$$\begin{Bmatrix} j_1 & j_2 & J_{12} \\ j_3 & j_4 & J_{34} \\ J_{13} & J_{24} & J \end{Bmatrix} \sqrt{(2J_{12} + 1)\,(2J_{34} + 1)\,(2J_{13} + 1)\,(2J_{24} + 1)}$$

$$= \sum_{m_1, m_2 (m_3)} \langle j_1 m_1 \, j_2 m_2 | J_{12} m_{12} \rangle \langle j_3 m_3 \, j_4 m_4 | J_{34} m_{34} \rangle \times$$

$$\langle J_{13} m_{13} \, J_{24} m_{24} | JM \rangle \langle j_1 m_1 \, j_3 m_3 | J_{13} m_{13} \rangle \times$$

$$\langle j_2 m_2 \, j_4 m_4 | J_{24} m_{24} \rangle \langle J_{12} m_{12} \, J_{34} m_{34} | JM \rangle. \tag{83}$$

Notice that the CG coefficients correspond to the three rows and the three columns of the 9j-symbol. The 9j-symbols are the elements of the transformation matrix between two schemes for the coupling of four angular momenta to a given total angular momentum. A 9j-symbol is an invariant function built by contracting completely the product of six Wigner coefficients.

The 9j-symbols satisfy some symmetry properties and orthogonality. A 9j-symbol vanishes identically unless each row and each column satisfies the triangular condition. An even permutation of the rows or columns leaves the symbol unchanged. An odd permutation of the rows or columns of a 9j-symbol multiplies it by a factor $(-1)^s$ where s is the sum of all arguments. The symbol is invariant under reflections about either diagonal.

The orthogonality relation is

$$\sum_{J_{13} J_{24}} (2J_{13} + 1)\,(2J_{24} + 1) \begin{Bmatrix} j_1 & j_2 & J_{12} \\ j_3 & j_4 & J_{34} \\ J_{13} & J_{24} & J \end{Bmatrix} \begin{Bmatrix} j_1 & j_2 & J'_{12} \\ j_3 & j_4 & J'_{34} \\ J_{13} & J_{24} & J \end{Bmatrix}$$

$$= \frac{\delta_{J_{12} J'_{12}} \, \delta_{J_{34} J'_{34}}}{(2J_{12} + 1)\,(2J_{34} + 1)}. \tag{84}$$

The multiplicative property of the transformations gives the sum rule

$$\sum_{J_{13} J_{24}} (-1)^{2j_2 + J_{23} + J_{24} - J_{34}} (2J_{13} + 1)\,(2J_{24} + 1) \times$$

$$\begin{Bmatrix} j_1 & j_2 & J_{12} \\ j_3 & j_4 & J_{34} \\ J_{13} & J_{24} & J \end{Bmatrix} \begin{Bmatrix} j_1 & j_3 & J_{13} \\ j_4 & j_2 & J_{24} \\ J_{14} & J_{23} & J \end{Bmatrix} = \begin{Bmatrix} j_1 & j_2 & J_{12} \\ j_4 & j_3 & J_{34} \\ J_{14} & J_{23} & J \end{Bmatrix}. \tag{85}$$

A 9j-symbol with one argument zero reduces to a 6j-symbol times a factor. For example,

$$\begin{Bmatrix} j_1 & 0 & j_1 \\ j_3 & j_4 & J_{34} \\ J_{13} & j_4 & J \end{Bmatrix} = \frac{(-1)^{J+j_3+j_1+j_4}}{\sqrt{(2j_1+1)(2j_4+1)}} \begin{Bmatrix} J_{34} & J & j_1 \\ J_{13} & j_3 & j_4 \end{Bmatrix}. \quad (86)$$

The recoupling of four angular momenta is involved in the transition from LS to jj coupling. The coefficients are very important in the evaluation of matrix elements of tensor products of tensor operators, and in the computation of the coefficients of fractional parentage.

The addition of n angular momenta is achieved by introducing $(n-2)$ intermediate angular momenta. The transformation coefficients relating two different coupling schemes involves $(n+1) + 2(n-2) = 3(n-1)$ j's. Thus the coefficients involved in the addition of n angular momenta are called $3(n-1)$ j-symbols. For $n = 3$ we have the 6j-symbols, $n = 4$, the 9j-symbols, for $n = 5$, the 12j-symbols, etc.

These $3nj$-symbols and coefficients separate the geometrical aspects of various problems from their physical contents. The calculations are greatly simplified by use of their symmetry and orthogonality properties and sum rules.

Next, we discuss the transformation between LS and jj couplings. We consider two interacting particles in a central field. We couple first $\vec{l}_1$ and $\vec{l}_2$ to $\vec{L}$, $\vec{s}_1$ and $\vec{s}_2$ to $\vec{S}$, and then $\vec{S}$ and $\vec{L}$ to $\vec{J}$.

Since $s_1 = \dfrac{1}{2}$, $s_2 = \dfrac{1}{2}$, S may take the values 0 or 1. Pictorially, we have the

LS coupling scheme as

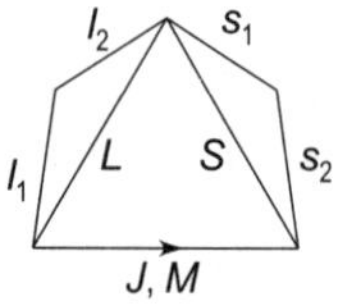

In another coupling scheme for the four angular momenta $\vec{l}_1$, $\vec{l}_2$, $\vec{s}_1$, $\vec{s}_2$ to a total $\vec{J}$, we couple $\vec{s}_1$ and $\vec{l}_1$ to $\vec{J}_1$, $\vec{s}_2$ and $\vec{l}_2$ to $\vec{J}_2$, and then $\vec{J}_1$ and $\vec{J}_2$ to $\vec{J}$.

Pictorially, we represent the jj-coupling scheme as

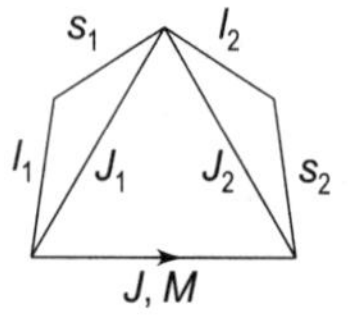

We can express a wave function in the LS-coupling scheme in terms of wave functions in the jj-coupling scheme. The LS-jj-coupling transformation coefficient is given by

$$\langle [(l_1 l_2) L (s_1 s_2) S] J | [(l_1 s_1) j_1 (l_2 s_2) j_2] J \rangle$$

$$= [(2L + 1)(2S + 1)(2j_1 + 1)(2j_2 + 1)]^{1/2} \begin{Bmatrix} l_1 & l_2 & L \\ s_1 & s_2 & S \\ j_1 & j_2 & J \end{Bmatrix} \tag{87}$$

If $S = 0$, the LHS reduces to

$$(-1)^{l_1 + j_2 + J + \frac{1}{2}} \left[\frac{(2j_1 + 1)(2j_2 + 1)}{2} \right]^{\frac{1}{2}} \begin{Bmatrix} J & j_1 & j_2 \\ \frac{1}{2} & l_2 & l_1 \end{Bmatrix}, \tag{88}$$

$$\text{using Eq. (86)}$$

For $S = 1$, we evaluate the 9j-symbol

$$3 \begin{Bmatrix} J & L & l \\ \frac{1}{2} & \frac{1}{2} & \lambda \end{Bmatrix} \begin{Bmatrix} l_1 & l_2 & L \\ \frac{1}{2} & \frac{1}{2} & 1 \\ j_1 & j_2 & J \end{Bmatrix} + \frac{(-1)^{l_1 + j_2 - \lambda}}{2(2J + 1)} \begin{Bmatrix} J & l_2 & l_1 \\ \frac{1}{2} & j_1 & j_2 \end{Bmatrix} \delta_{LJ}$$

$$= (-1)^{2\lambda} \begin{Bmatrix} j_1 & j_2 & J \\ \frac{1}{2} & \lambda & l_2 \end{Bmatrix} \begin{Bmatrix} l_2 & l_1 & L \\ \frac{1}{2} & \lambda & j_1 \end{Bmatrix} \tag{89}$$

where

$$\lambda = \begin{cases} L + \dfrac{1}{2}, & \text{if } L = J \\[2mm] \dfrac{L + J}{2}, & \text{if } L \neq J \end{cases}.$$

For example,

$$\left\langle \left[(l_1 l_2) L \left(\tfrac{1}{2} \tfrac{1}{2} \right) 1 \right] J = L + 1 \left| \left[\left(l_1 \tfrac{1}{2} \right) l_1 + \tfrac{1}{2} \left(l_2 \tfrac{1}{2} \right) l_2 + \tfrac{1}{2} \right] J = L + 1 \right. \right\rangle$$

$$= \left[\frac{(l_1 + l_2 + J + 1)(l_1 + l_2 + J + 2)(l_1 - l_2 + J)(-l_1 + l_2 + J)}{2J(2J + 1)(2l_1 + 1)(2l_2 + 1)} \right]. \tag{90}$$

--- **PROBLEMS** ---

7.1 Use the identity

$$(2j_1 + 1)(2j_2 + 1) = \sum_{i=j}^{j_1 + j_2} (2j + 1)$$

to prove that , in the addition of two angular momentum operators, $j_{\min} = |j_1 - j_2|$.

7.2 Use lowering and raising operators to compute all Clebsch-Gordan coefficients for the addition of two angular momentum operators (a) $j_1 = 1$, $j_2 = 1/2$; (b) $j_1 = 1$, $j_2 = 1$.

7.3 (a) Let $\vec{S} = \vec{S}_1 + \vec{S}_2 + \vec{S}_3$ be the total spin of three spin $\frac{1}{2}$ particles. What are the possible values of the total spin? Derive the corresponding Clebsch-Gordan coefficients.

(b) Consider a system of three distinguishable particles of spin $\frac{1}{2}$ particles.

The Hamiltonian of the system is $H = -\dfrac{a}{\hbar^2}(\vec{S}_1 + \vec{S}_2)\cdot\vec{S}_3$, where a is a constant. Determine the energy eigenvalues and their degeneracies.

7.4 Let $\vec{J} = \vec{J}_1 + \vec{J}_2 + \vec{J}_3$ be the total angular momentum of three spin 1 particles.

    (a) Find the possible values of the total angular momentum J, where $J(J+1)$ is the eigenvalue of $\vec{J}^2$, How many angular momentum eigenstates are there corresponding to each value of J?

    (b) Construct the state $J = 0$ explicityly.

7.5 Use the recursion relation to obtain the general formula for the 3j-symbol $\begin{pmatrix} j_1 & j_2 & j_3 \\ 0 & 0 & 0 \end{pmatrix}$.

CHAPTER 8

Tensor Operators

In this chapter "tensor operators" are introduced and the product of tensors are discussed and the Wigner-Eckart theorem is derived. Applications of the WE theorem are pointed out.

8.1 IRREDUCIBLE TENSOR OPERATORS

We define a tensor by its transformation properties under rotations in three dimensional space. A vector is defined as a set of three quantities which, under rotations, transforms as

$$x_i' = \sum_j O_{ij} x_j, \tag{1}$$

where O_{ij} are the elements of a real orthogonal matrix with determinant +1. A second rank tensor is a set of $3^2 = 9$ quantities which transforms as

$$T_{ij}' = \sum_{k,l} O_{ik} O_{jl} T_{kl}. \tag{2}$$

In general, a set of 3^n numbers is called a tensor of rank n, if these numbers $A_{ijk....}$ undergo the following linear transformation under rotation :

$$A_{ijk....}' = \sum_{l,m,n...} O_{il} O_{jm} O_{kn} ... A_{lmn...} \tag{3}$$

The subscripts run over the values 1, 2, or 3. $A_{lmn....}$ has n subscripts. Such a tensor is known as a Cartesian tensor.

As a simple example, a Cartesian tensor of rank 2 can be formed by taking all products of the components of two vectors $\vec{x}$ and $\vec{y}$. A typical tensor component is thus

$$T_{ik} = x_i y_k \ (i, \, k = 1, \, 2, \, 3).$$

Every tensor T_{ik} of rank 2 can be split into a symmetric tensor and an antisymmetric tensor

$$T_{ik} = S_{ik} + A_{ik}$$

with

$$S_{ik} = \frac{1}{2} \, (T_{ik} + T_{ki}) = S_{ki},$$

$$A_{ik} = -A_{ki} = \frac{1}{2} \, (T_{ik} - T_{ki}).$$

The components of the symmetric part transform among themselves under rotations and the same holds for the components of the antisymmetric part. Can the symmetric and antisymmetric tensor be further reduced? It is not possible to reduce the antisymmetric part. But the symmetric part is reducible; we impose the condition that the trace of the symmetric part vanishes. No further reduction is possible. Thus a tensor of rank 2 can be decomposed into a sum of three irreducible tensors as

$$T_{ik} = S_{ik} + A_{ik} + \mathfrak{S}_{ik}, \tag{4}$$

with

$$\mathfrak{S}_{ik} = \frac{1}{3} \, [Tr(T_{ik})] \, \delta_{ik} = \frac{\delta_{ik}}{3} \sum_{\alpha=1}^{3} T_{\alpha\alpha},$$

$$A_{ik} = \frac{1}{2} \, (T_{ik} - T_{ki}),$$

$$S_{ik} = \frac{1}{2} \, (T_{ik} + T_{ki}) - \frac{\delta_{ik}}{3} \sum_{\alpha=1}^{3} T_{\alpha\alpha}.$$

In Eq. (4), the last term is a scalar and thus an invariant under rotation. The second term is an antisymmetric tensor with three independent components which transforms as a vector under rotation. The first term is a 3×3 traceless symmetric tensor with five independent components which transform among themselves under rotation. The single component of the trace, the three components of the antisymmetric tensor and five components of the symmetric tensor are completely equivalent to the nine components of the original tensor T_{ik}. The advantage of the reduction of a Cartesian tensor into irreducible tensors is that under rotations each one of them transforms separately.

We can generalize the procedure described above to tensors of higher rank. We see that for the second rank tensor the transformation matrix A in $T'_{ij} = \sum A_{ij;nm} T_{nm}$ can be reduced to the form

$$A \rightarrow \begin{pmatrix} D^{(0)} & & 0 \\ & D^{(1)} & \\ 0 & & D^{(2)} \end{pmatrix}$$

In summary, a general second rank tensor transforms under rotation as the $D^{(1 \times 1)}$ representation; it is reducible. The trace of the tensor is invariant under SO(3) – it transforms as $D^{(0)}$; the three independent components of the antisymmetric part behave like a vector under rotation – they transform as $D^{(1)}$; the five independent components of the traceless symmetric part of the tensor transform into each other; they transform as $D^{(2)}$.

We now define an irreducible tensor operator of rank k as a set of $(2k + 1)$ operators $T_q^{(k)}$ with $q = k, k - 1, k - 2, ..., -k$ which transforms under a rotation R according to

$$T_q'^{(k)} = R(\alpha, \beta, \gamma) \, T_q^{(k)} \, R^{-1}(\alpha, \beta, \gamma). \tag{5}$$

We assume that the operators are such that they transform under rotations as angular momentum eigenfunctions. Thus

$$T_q'^{(k)} = \sum_{q'} T_{q'}^{(k)} D_{q'q}^{(k)}(\alpha, \beta, \gamma). \tag{6}$$

Such an operator is called a spherical tensor operator of rank k. Equation (6) means that $T_q^{(k)}$ transforms like the qth partner of the representation $D^{(k)}$ under rotation. These operators have the same transformation properties under rotations as the angular momentum functions. Operators such as the three irreducible ones with 5, 3, 1 components must be combined further among themselves so that they posses the proper behavior acting like the 5, 3, 1 components of an angular momentum function with $l = 2, 1, 0$.

An irreducible tensor is defined in Eq. (5) in terms of its properties under finite rotations. We may characterize it by its transformation under infinitesimal rotations. We write Eqs. (5) and (6) as

$$R T_q^{(k)} R^{-1} = \sum_{q'} T_{q'}^{(k)} \langle kq'|R|kq \rangle.$$

If we consider a rotation through an infinitesimal angle ϵ about an axis in the direction $\hat{n}$ the rotation operator is $R = \mathbf{1} - i \epsilon \, \hat{n} \cdot \vec{J}/h$, to the first order in ϵ. The above equation becomes

$$\left[\hat{n} \cdot \vec{J}, T_q^{(k)} \right] = \sum_{q'} T_{q'}^{(k)} \langle kq'|\hat{n} \cdot \vec{J}|kq \rangle.$$

Choosing $\hat{n}$ in the z, x, y-directions and using the raising (lowering) operators $J_\pm$ we get

$$[J_z, T_q^{(k)}] = \hbar q T_q^{(k)}$$

$$[J_\pm, T_q^{(k)}] = \hbar\{(k \mp q)\,(k \pm q + 1)\}^{1/2}\,T_{q\pm1}^{(k)} \tag{7}$$

or, in a compact form

$$[J_m, T_q^{(k)}] = \sqrt{k(k+1)}\,\langle kq|m|kq + m\rangle\,T_{q+m}^{(k)} \tag{7a}$$

$$(m = -1, 0, +1).$$

Irreducible tensor operators may also be defined through these commutation relations.

Equation (7) shows that a component $T_q^{(k)}$ of an irreducible tensor of degree k can be considered as a wave function and is an eigenfunction of J_z with the eigenvalue q. Since $\vec{J}^{(2)}T_q^{(k)} = k(k+1)T_q^{(k)}$ each component is simultaneously an eigenfunction of $\vec{J}^2$ with the eigenvalue $k(k+1)$.

As given by Racah, this definition reads:

A tensor operator of rank k is a set of $(2k + 1)$ operators (in spherical components) $T_q^{(k)}(q = k, k-1, ..., -k)$ that satisfy the following commutation relations with respect to the total angular momentum operator of the system:

$$[J_0, T_q^{(k)}] = q T_q^{(k)}$$

$$[J_\pm, T_q^{(k)}] = \sqrt{(k \mp q)\,(k \pm q + 1)}\,T_{q\pm1}^{(k)}. \tag{8}$$

This is the infinitesimal form of the definition; Eq. (6) is the integral form of the definition.

The hermitian adjoint $T^\dagger$ of a tensor operator T is defined by

$$T_q^{\dagger(k)} = (-1)^q\,(T_{-q}^{(k)})^\dagger \tag{8a}$$

$T^\dagger$ transforms like T under rotations; self-adjoint tensor $T^\dagger = T$ can exist only for integer k.

Equation (8) is very useful to determine if a particular operator is a tensor operator or not.

The simplest example of such an operator is a scalar operator – a spherical tensor of rank zero $(k = 0)$. The Hamiltonian H of a spherically symmetric system is a scalar operator.

Another example is a set of three operators $V_i(i = 1, 2, 3)$ satisfying the commutation relation with angular momentum

$$[V_i, J_j] = i\hbar\epsilon_{ijk}\,V_k\,(i, j, k = 1, 2, 3) \tag{9}$$

Such a set of operators is called a vector operator. The position vector $\vec{r}$ and the linear momentum $\vec{p}$ of a particle are vector operators. The angular momentum $\vec{j}$ of a system is a vector operator.

A spherical tensor of rank one ($k = 1$) is a vector opertor: for example, the coordinate vector of a particle. But we must put these in standard form, i.e., rather than using the (Cartesian) components V_i we use the spherical components ("q components") $V_{q=0}$, $V_{q=\pm 1}$ defined by

$$T_1^{(1)} \equiv V_1 = \frac{V_1 + iV_2}{\sqrt{2}}$$

$$T_0^{(1)} \equiv V_0 = V_3$$

$$T_{-1}^{(1)} \equiv V_{-1} = -\frac{V_1 - iV_2}{\sqrt{2}} \tag{10}$$

If $\vec{V} = \vec{r}$, then

$$r_1 = -\frac{x + iy}{\sqrt{2}} = -\frac{r \sin\theta}{\sqrt{2}} e^{i\varphi}, \text{ in polar coordinates}$$

$$r_0 = z = r \cos\theta$$

$$r_{-1} = \frac{x - iy}{\sqrt{2}} = \frac{r \sin\theta}{\sqrt{2}} e^{-i\phi}. \tag{11}$$

We see that

$$r_q = \sqrt{\frac{4\pi}{3}} \, rY_{1q}(\theta, \phi), \tag{12}$$

where $Y_{1q}(\theta, \varphi)$ is the qth component of the spherical harmonics for $l = 1$.

The quantity

$\mathcal{Y}_{1,q}(x, y, z) = rY_{1q}(\theta, \varphi)$ is a homogeneous polynomial in x, y, z. Thus

$$\mathcal{Y}_{1,1} = -\sqrt{\frac{3}{8\pi}} \, (x + iy)$$

$$\mathcal{Y}_{1,0} = \sqrt{\frac{3}{8\pi}} \, z$$

$$\mathcal{Y}_{1,-1} = \sqrt{\frac{3}{8\pi}} \, (x - iy).$$

We define the normalized harmonic polynomial or solid harmonic as

$$\mathcal{Y}_{lm}(\vec{r}) = r^l Y_{lm}(\theta, \varphi). \tag{13}$$

$\mathcal{Y}_{lm}$ is a homogeneous polynomial of degree l in x, y, z.

In terms of the spherical components of a vector $\vec{V}$, we write

$$\mathcal{Y}_{1q}(\vec{V}) = \sqrt{\frac{3}{4\pi}}\, V_q \ (q = \pm 1, 0). \tag{14}$$

The set of $(2l + 1)$ polynomials $\mathcal{Y}_{lm}(\vec{V})$ is an irreducible tensor operator of order l. For example, for $l = 2$.

$$\mathcal{Y}_{2,\pm 2} = \sqrt{\frac{15}{8\pi}}\, V_{\pm 1}^2$$

$$\mathcal{Y}_{2,\pm 1} = \sqrt{\frac{15}{16\pi}}\, (V_0 V_{\pm 1} + V_{\pm 1} V_0) \tag{15}$$

$$\mathcal{Y}_{2,0} = \sqrt{\frac{5}{16\pi}}\, (2V_0^2 + V_1 V_{-1} + V_{-1} V_1).$$

The five independent components of the standard irreducible tensor of rank 2 are given below in terms of the Cartesian components of the symmetric tensor S_{ij}:

$$T_0^{(2)} = 2S_{33} - S_{11} - S_{22}$$

$$T_{\pm 1}^{(2)} = \mp \sqrt{6}\, (S_{13} \pm iS_{23})$$

$$T_{\pm 2}^{(2)} = \sqrt{\frac{3}{2}}\, (S_{11} - S_{22} \pm 2iS_{12}). \tag{16}$$

Irreducible tensors obtained by reduction of a general tensor of order k have $(2k + 1)$ components which is always an odd number. So the spherical tensors are of integral rank (implied by their definition in terms of the spherical harmonics). However, there are irreducible tensors of half-integral rank as allowed by the general definition. Such tensors with even number of components represent wavefunctions of particles with half-integral spin.

8.2 PRODUCT OF TENSOR OPERATORS

The product of two irreducible tensor operators is usually not an irreducible tensor. However, we can construct irreducible tensor operators from such a product as follows. Let $T_{q_1}^{(k_1)}$ and $T_{q_2}^{(k_2)}$ be irreducible tensor operators of rank k_1 and k_2, respectively. Then

$$T_{(q)}^{(k)} = \sum_{q_1} \sum_{q_2} T_{q_1}^{(k_1)}\, T_{q_2}^{(k_2)}\, \langle k_1 q_1 k_2 q_2 | kq \rangle \tag{17}$$

is an irreducible tensor of rank k. To prove this, we write

$$D^{\dagger}(R)T_q^{(k)}D(R) = \sum_{q_1}\sum_{q_2}(D^{\dagger}(R)\,T_{q_1}^{(k_1)}D(R))(D^{\dagger})\,(R)T_{q_2}^{(k_2)}D(R) \times \langle k_1 q_1 k_2 q_2 | k q \rangle$$

$$= \sum_{q_1}\sum_{q_2}\sum_{q_1'}\sum_{q_2'} T_{q_1}^{(k_1)}D_{q_1'q_1}^{(k_1)}(R^{-1})T_{q_2'}^{(k_2)}D_{q_2'q_2}^{(k_2)}(R^{-1}) \times \langle k_1 q_1 k_2 q_2 | k q \rangle$$

$$= \sum_{k''}\sum_{q_1}\sum_{q_2}\sum_{q_1'}\sum_{q_2'}\sum_{q'} D_{q''q'}^{(k'')}(R^{-1}) \times T_{q_1'}^{(k_1)}T_{q_2'}^{(k_2)}\langle k_1 q_1 k_2 q_2 | k q \rangle \langle k_1 q_1' k_2 q_2' | k'' q' \rangle$$

$$\times \langle k_1 q_1 k_2 q_2 | k'' q' \rangle, \text{ using the CG series formula.}$$

$$= \sum_{k''}\sum_{q_1'}\sum_{q_2'}\sum_{q'} D_{q'q''}^{(k'')}(R^{-1}) \times T_{q_1'}^{(k_1)}T_{q_1'}^{(k_1)}T_{q_2'}^{(k_2)}\,\delta_{kk'}\,\delta_{qq''} \times \langle k_1 q_1' k_2 q_2' | k'' q' \rangle,$$

$$\text{using the orthogonality of CG coefficients.}$$

$$= \sum_{q'}\left(\sum_{q_1'}\sum_{q_2'}\langle k_1 q_1' k_2 q_2' | k q'\rangle\, T_{q_1'}^{(k_1)}\,T_{q_2'}^{(k_2)}\right)D_{q'q}^{(k_2)}(R^{-1})$$

$$= \sum_{q'} T_{q'}^{(k)}\,D_{q'q}^{(k)}(R^{-1})$$

$$= \sum_{q'} D_{qq'}^{(k)*}(R)T_q^{(k)}$$

which is the correct transformation law.

Note that we can also invert Eq. (17):

$$T_{q_1}^{(k_1)}T_{q_2}^{(k_2)} = \sum\sum T_q^{(k)}\,\langle k_1 q_1 k_2 q_2 | k q\rangle. \tag{18}$$

Thus, we can construct tensor operators of higher or lower ranks by multiplying two tensor operators.

The tensor $T^{(k)}$ is called the tensor product of degree k of $T^{(k_1)}$ and $T^{(k_2)}$ and is denoted as

$$T^{(k)} = \left[T^{(k_1)} \times T^{(k_2)}\right]^{(k)}. \tag{19}$$

The construction of a state with a definite total angular momentum J from two states with well-defined angular momenta is equivalent to the reduction of the tensor product to irreducible parts and chosing the one which transforms like a tensor of degree J.

We can build a scalar operator from two tensor operators of the same rank:

$$T_0^{(0)} = \sum_q U_q^{(k)}V_{-q}^{(k)}\,\langle k q k - q | 0 0\rangle$$

$$= (2k + 1)^{-1/2} \sum_q (-1)^{k-q} U_q^{(k)} V_{-q}^{(k)}. \qquad (20)$$

Let us consider the product of two vector operators. We have discussed in the preceding section the reduction of a second rank tensor (built from products of components of vectors) into a linear combination of irreducible tensors. Evaluating Eq. (17) for the product of two vectors $\vec{V}$ and $\vec{U}$ we obtain

$$T_0^{(0)} = -\frac{(\vec{V} \cdot \vec{U})}{3} = \frac{(V_1 U_{-1} + V_{-1} U_1 - V_0 U_0)}{3} \quad \text{(scalar)}$$

$$T_q^{(1)} = \frac{(\vec{V} \times \vec{U})_q}{i\sqrt{2}} \quad \text{(vector)} \qquad (21)$$

$$\left.\begin{aligned}
T_{\pm 2}^{(2)} &= V_{\pm 1} U_{\pm 1} \\[2mm]
T_{\pm 1}^{(2)} &= \frac{V_{\pm 1} U_0 + V_0 U_{\pm 1}}{\sqrt{2}} \\[2mm]
T_{\pm 0}^{(2)} &= \frac{V_1 U_{-1} + 2 V_0 U_0 + V_{-1} U_1}{\sqrt{6}}
\end{aligned}\right\} \text{second-rank tensor}$$

For example,
the orbital angular momentum operator $\vec{l}$ is proportional to the tensor product of degree 1 of the two tensors of degree 1, $\vec{r}$ and $\vec{p}$:

$$\vec{l} = -i\sqrt{2} \, [\vec{r} \times \vec{p}]^{(1)} \qquad (22)$$

The Legendre polynomial of order l is proportional to the tensor product of degree zero of two spherical harmonics of degree l:

$$P_l(\cos \omega_{12}) = \frac{4\pi(-1)^l}{\sqrt{2l+1}} \, [Y_l(\theta_1, \varphi_1) \times Y_l(\theta_2, \varphi_2)]$$

$$= \frac{4\pi}{2l+1} \sum_m (-1)^m Y_{l,-m}(\theta_1, \varphi_1) Y_{l,m}(\theta_2, \varphi_2)$$

$$= \frac{4\pi}{2l+1} \sum_m Y_{lm}^*(\theta_1, \varphi_1) Y_{lm}(\theta_2, \varphi_2), \qquad (23)$$

where ω_{12} is the angle between the directions (θ_1, φ_1) and (θ_2, φ_2). Thus it is the addition theorem for spherical harmonics.

In nuclear physics, we are often concerned with a scalar operator built from the product of two second degree tensors— one in coordinate space and another in spin space. Thus

$$\left[[\vec{r}_1 \times \vec{r}_2]^{(2)} \times [\vec{\sigma}_1 \times \vec{\sigma}_2]^{(2)}\right]_0^{(0)}$$

$$= \frac{r_{12}^2}{\sqrt{5}} \left[\frac{(\vec{\sigma}_1 \cdot \vec{r}_{12})\,(\vec{\sigma}_2 \cdot \vec{r}_{12})}{r_{12}^2} - \frac{1}{3}\,(\vec{\sigma}_1 \cdot \vec{\sigma}_2)\right], \tag{24}$$

where $\vec{r}_{12} = \vec{r}_1 - \vec{r}_2$ is the vector joining the nucleons 1 and 2, and $\vec{\sigma}_1$ and $\vec{\sigma}_2$ are their respective spin operators. This is usually called the "tensor interaction" (tensor part of the nucleon-nucleon interaction) – though, of course, it is a scalar.

8.3 MATRIX ELEMENTS OF TENSOR OPERATORS AND THE WIGNER-ECKART THEOREM

The evaluation of matrix elements of irreducible tensor operators between two states of given angular momenta can be considerably simplified by applying the concept of tensor products.

The matrix elements of tensor operators have an important property which is expressed by the Wigner-Eckart theorem:

Let $T_q^{(k)}$ be a tensor operator.

The matrix element of $T_q^{(k)}$ between the angular momentum eigenstates may be written as

$$\langle \alpha', j'm' | T_q^{(k)} | \alpha, jm \rangle = \langle jmkq | j'm' \rangle \times \frac{\langle \alpha' j' \| T^{(k)} \| \alpha j \rangle}{\sqrt{2j'+1}}, \tag{25}$$

where $\langle jmkq | j'm' \rangle$ is Clebsch-Gordan coefficient and the symbol $\langle \alpha' j' \| T^{(k)} \| \alpha j \rangle$ denotes a quantity that depends on α, j', k, α, j and the nature of the tensor operator but is independent of m', m and q. The quantity $\langle \alpha' j' \| T^{(k)} \| \alpha j \rangle$ is called the reduced or double-bar matrix element of the tensor operator (Note: The normalization and phase of the "reduced matrix element" vary with various authors). α denotes all the other quantum numbers required to specify the states.

To prove the theorem, we write

$$\langle \alpha', j'm' | T_q^{(k)} | \alpha, jm \rangle = \langle \alpha', j'm' | R^\dagger R T_q^{(k)}\, R^\dagger R | \alpha, jm \rangle$$

$$= \sum_{\mu'} \sum_{\sigma} \sum_{\mu} \{D_{\mu'm'}^{(j')}(R)\}\, D_{\sigma q}^{(k)}(R)\, D_{\mu m}^{(j)}(R)$$

$$\times \langle (\alpha', j'm' | T_\sigma^{(k)} | \alpha, j\mu \rangle, \text{ using (5), (6) and } R^{+*} = R^{-1}.$$

The LHS is independent of R, so we integrate this expression over R, and using Eq. (7.59), we obtain

$$\langle \alpha', j'\mu' | T_q^{(k)} | \alpha, jm \rangle = \frac{\langle jmkq | j'm' \rangle}{(2j'+1)} \times \sum_{\mu'} \sum_{\sigma} \sum_{\mu} \langle \alpha', j'\mu' | T_\sigma^{(k)} | \alpha, j\mu \rangle \times \langle j\mu k\sigma | j'\mu' \rangle.$$

Conventionally, the RHS of the above expression is written as

$$\langle jmkq | j'm' \rangle \, \frac{\langle \alpha' j \| T^{(k)} \| \alpha j \rangle}{\sqrt{2j'+1}}.$$

The Wigner-Eckart (WE) theorem is a basic theorem in the algebra of irreducible tensor operators. In fact, it is one of the most important theorems in quantum mechanics. The WE theorem expresses the matrix element of a component of an irreducible tensor operator as the product of two factors. The first factor is a CG coefficient which takes care of the geometrical properties of the tensor and the states considered. We note that m, m' and q can be prescribed only when the orientation of the coordinate frame is specified whereas α, α', j, j' and k do not depend on the orientation. The CG coefficients are known numbers and are calculated from the group representation. The second factor includes all the specific physical information contained in the tensor and the system. The reduced matrix element is independent of m, m' and q. This factorization is fundamental in the theory of tensor operators and results in great simplification in the evaluation of matrix elements of tensor operators. The generality of Eq. (25) is to be noted. The result holds if the eigenstates are of definite angular momenta and $T^{(k)}$ is an irreducible tensor operator. These can be functions or operators of a single partile or of many particles. The states may refer to scalar fields, spinor fields, vector fields, etc. In all cases, the WE theorem holds.

From the requirement that the CG coefficient be non-vanishing we have two selection rules:

$$\langle \alpha, j'm' | T_q^{(k)} | \alpha, jm' \rangle = 0$$

unless
$$m' = q + m \tag{26}$$

and
$$|j - k| \le j' \le j + k. \tag{27}$$

These two selection rules are of great importance in atomic and nuclear spectroscopy.

Another important consequence of the WE theorem is that we can obtain the branching ratios (B):

$$B = \frac{\langle j'm' | O_\lambda^{(k)} | jm \rangle}{\langle j'n' | O_\mu^{(k)} | jn \rangle} = \frac{\langle jmk\lambda | j'm' \rangle}{\langle jnk\mu | j'n' \rangle},$$

where the RHS involves only known CG coefficients and O is the specific operator.

We now consider simple examples and calculation of certain reduced matrix elements.

Example 8.1 Scalar operator $T_0^{(0)} = S$.

The matrix element of a scalar operator is

$$\langle \alpha', j', m' | S | \alpha, jm \rangle = \langle jm00 | j'm' \rangle \, \frac{\langle \alpha' j' \| S \| \alpha j \rangle}{\sqrt{2j' + 1}}$$

$$= \delta_{jj'} \delta_{mm'} \, \frac{\langle \alpha' j' \| S \| \alpha j \rangle}{\sqrt{2j + 1}}$$

Clearly, a scalar operator cannot change j, m values

$$\langle \alpha' j' \| S \| \alpha j \rangle = \langle \alpha, j'm' | S | \alpha, jm \rangle \sqrt{2j + 1}. \tag{28}$$

Example 8.2 The angular momentum operator $\vec{J}$ or $J_q (q = -1, 0, 1)$ is a tensor of rank 1.

Thus,

$$\langle \alpha', J'M' | J_q^{(1)} | \alpha, JM \rangle = \langle JM1q | j'M' \times \frac{\langle \alpha', J' \| J^{(1)} \| \alpha, J \rangle}{\sqrt{2J' + 1}} \tag{29}$$

But, we know

$$\langle \alpha', J'M' | J_0 | \alpha, JM \rangle = \langle JM10 | J'M' \rangle \, \frac{\langle \alpha' J' \| J^{(1)} \| \alpha J \rangle}{\sqrt{2J' + 1}}$$

$$= \delta_{JJ'} \delta_{MM'} \, \frac{M}{\sqrt{J(J + 1)}} \, \frac{\langle \alpha' J' \| J^{(1)} \| \alpha J \rangle}{\sqrt{2J' + 1}}.$$

Clearly, the LHS is $hM\delta_{\alpha'\alpha}\delta_{J'J}\delta_{M'M}$
Thus,

$$\langle \alpha' J' \| \vec{J} \| \alpha J \rangle = \delta_{J'J}\delta_{\alpha'\alpha} h \sqrt{J(J + 1)(2J + 1)} \tag{30}$$

For the particular cases of the orbital angular momentum and spin of an electron, we have

$$\langle l' \| \vec{L} \| \alpha J \rangle = \hbar \sqrt{l(l + 1)\,(2l + 1)} \, \delta_{l'l},$$

$$\langle s' \| \vec{S} \| s \rangle = \hbar \sqrt{\frac{3}{2}} \, \delta_{s's}.$$

Example 8.3   Spherical harmonics $Y_q^k(\theta, \varphi)$ as a tensor operator.

$$\langle \alpha', l'm'|Y_q^k| \alpha, lm\rangle = \langle lmkq|l'm'\rangle \frac{\langle l'\|Y^{(k)}\|l\rangle}{\sqrt{2l'+1}}$$

But,

$$\langle \alpha', l'm'|Y_q^k| \alpha, lm\rangle = \sqrt{\frac{(2l+1)(2k+1)}{4\pi(2l'+1)}} \langle lmkq|l'm'\rangle \times \langle l0k0|l'0\rangle.$$

Using the WE theorem, the LHS of the above expression is

$$\langle lmkq|l'm'\rangle \frac{\langle \alpha'\|Y^k\|\alpha l\rangle}{\sqrt{2l'+1}}$$

Therefore,

$$\langle \alpha'l'\|Y^k\|\alpha l\rangle = \sqrt{\frac{(2l+1)(2k+1)}{4\pi}} \langle l0k0|l'\rangle. \tag{31}$$

Example 8.4   We now evaluate the electric quadrupole moment in a quantum state of definite angular momentum. The quadrupole moment is defined as

$$Q(\rho) = e\langle \psi(\vec{r})|(3z^2 - r^2)| \psi(\vec{r})\rangle, \tag{32}$$

where the tensor operator $Q^{(2)}$ is

$$Q_0^{(2)} = (3z^2 - r^2) \tag{33}$$

Conventionally, the quadrupole moment is

$$Q = \langle jj|Q_0^{(2)}|jj\rangle$$

$$= \langle jj20|jj\rangle \frac{\langle j\|Q^{(2)}\|j\rangle}{\sqrt{2j+1}}. \tag{34}$$

We evaluate

$$\langle jm|Q_0^{(2)}|jm\rangle = \langle jm20|jm\rangle \frac{\langle j\|Q^{(2)}\|j\rangle}{\sqrt{2j+1}}$$

$$= Q\,\frac{3m^2 - j(j+1)}{j(2j-1)}. \tag{35}$$

Example 8.5   Vector operator V_q

$$V_q|jm\rangle = \sum_{j'm'} |j'm'\rangle \langle j'm'|V_q|jm\rangle$$

$$= \sum_{j'm'} |j'm'\rangle \langle jm q|j'm'\rangle \langle j'\|V\|j\rangle$$

From the selection rules Eqs. (26) and (27), we see that the only non-vanishing terms in the above expression are in which $j' = j + 1, j, j - 1$ and $m' = m + q$. Thus,

$$V_q|jm\rangle = |j - 1m + q\rangle \langle jm1q|j - 1m + q\rangle \langle j - 1\|V\|j\rangle$$

$$+ |jm + q\rangle \langle jm1q|jm + q\rangle \langle j\|V\|j\rangle$$

$$+ |j + 1m + q\rangle \langle jm1q|j + 1m + q\rangle \langle j + 1\|V\|j\rangle. \qquad (36)$$

Thus, we can specify a vector operator by three reduced matrix elements. It can be shown that only two such functions are required to specify a vector operator.

For $j = j'$, the result is an interesting one.

We have

$$\langle \alpha', jm|\vec{J} \cdot \vec{V}|\alpha, jm\rangle = \langle \alpha', jm|(J_0 V_0 - J_{+1} V_{-1} - J_{-1} V_{+1})|\alpha, jm\rangle,$$

Note: $J_{\pm 1} = \mp \dfrac{1}{\sqrt{2}} J_\pm; J_0 = J_z$ $\qquad$ since $\vec{J} \cdot \vec{V}$ is a scalar operator, the matrix is diagonal in angular momentum indices

$$= m\hbar\langle \alpha', jm|V_0|\alpha, jm\rangle$$

$$+ \frac{\hbar}{\sqrt{2}} \sqrt{(j + m)(j - m + 1)} \langle \alpha', jm + 1|V_{-1}|\alpha, jm\rangle$$

$$- \frac{\hbar}{\sqrt{2}} \sqrt{(j - m)(j + m + 1)} \langle \alpha', jm + 1|V_{+1}|\alpha, jm\rangle$$

$$= c_j\langle \alpha'j\|\vec{V}\|\alpha j\rangle,$$

where c_j is a function of j.

$\therefore$ Putting $\vec{V} = \vec{J}$ and $\alpha' = \alpha$, we have

$$\langle \alpha, jm|\vec{J}^2|\alpha, jm\rangle = c_j\langle \alpha j\|\vec{J}\|\alpha j\rangle$$

Further, using the WE theorem,

$$\frac{\langle \alpha', jm'|V_q|\alpha, jm\rangle}{\langle \alpha, jm'|V_q|\alpha, jm\rangle} = \frac{\langle \alpha'j\|\vec{V}\|\alpha j\rangle}{\langle \alpha j\|\vec{J}\|\alpha j\rangle}$$

$$= \frac{\langle \alpha', jm|\vec{J} \cdot \vec{V}|\alpha, jm\rangle}{\langle \alpha, jm|\vec{J}^2|\alpha, jm\rangle}$$

$$= \frac{\langle \alpha', jm|\vec{J} \cdot \vec{V}|\alpha, jm\rangle}{j(j+1)\,h^2}.$$

Thus,

$$\langle \alpha', jm'|V_q^{(1)}|\alpha, jm\rangle = \frac{\langle \alpha', jm|\vec{J} \cdot \vec{V}|\alpha, jm\rangle}{j(j+1)h^2}\, \langle jm'|J_q^{(1)}| jm\rangle. \tag{37}$$

This expression Eq. (37) is called the "projection theorem".

The magnetic moment operator for an atom may be written as

$$\vec{\mu} = \frac{-e}{2m_e c}\,(g_L \vec{L} + g_s \vec{S}) \tag{38}$$

where $g_L \approx 1$, $g_S \approx 2$. Using Eq. (38), we get

$$\langle \alpha, jm'|\vec{\mu}|\alpha, jm\rangle = \frac{-e}{2m_e c}\, g_{eff}\, \langle jm'|\vec{J}|jm\rangle$$

whence

$$g_{eff} = \frac{\langle \alpha, jm|\vec{\mu} \cdot \vec{J}|\alpha, jm\rangle}{\hbar^2\, j(j+1)}$$

$$= \frac{\langle \alpha, jm|(g_L \vec{L} \cdot \vec{J} + g_S \vec{S} \cdot \vec{J})|\alpha, jm\rangle}{\hbar^2 j(j+1)}$$

$$= \frac{\langle \alpha, jm|\{(g_L + g_S)\,\vec{J} \cdot \vec{J} + (g_L - g_S)\,(\vec{L} \cdot \vec{L} - \vec{S} \cdot \vec{S})\}|\alpha, jm\rangle}{2\hbar^2 j\,(j+1)}$$

since

$$\vec{L} \cdot \vec{J} = \frac{1}{2}\,(\vec{J} \cdot \vec{J} + \vec{L} \cdot \vec{L} - \vec{S} \cdot \vec{S})$$

$$\vec{S} \cdot \vec{J} = \frac{1}{2}\,(\vec{J} \cdot \vec{J} + \vec{S} \cdot \vec{S} - \vec{L} \cdot \vec{L}).$$

In the $L - S$ coupling scheme, the atomic state vector is an eigenvector of $\vec{L} \cdot \vec{L}$ and $\vec{S} \cdot \vec{S}$ with eigenvalues $\hbar^2 L(L+1)$ and $\hbar^2 S(S+1)$ respectively.

For $g_L = 1$ and $g_S = 2$,

$$g_{eff} = \left[1 + \frac{J(J+1) - L(L+1) + S(S+1)}{2J(J+1)}\right]. \tag{39}$$

Equation (39) is called the Landé factor.

Next, we discuss the matrix elements of the tensor product of two tensor operators. Let

$$T_q^{(k)} = \left[U^{(k_1)} \times V^{(k_2)} \right]_q^{(k)}$$

$$= \sum_{q_1 q_2} \langle k_1 q_1 k_2 q_2 | kq \rangle U_{q_1}^{(k_1)}(1) \, V_{q_2}^{(k_2)}(2) \tag{40}$$

where $U^{(k_1)}$ and $V^{(k_2)}$ are tensor operators of rank k_1 and k_2 and operate in different subspaces, say 1 may indicate spatial coordinates and 2 spin coordinates of a particle, or 1 may be the coordinates of particle 1 and 2 the coordinates of particle 2. We want to calculate the matrix element of T in a scheme in which the angular momentum of the total system is built from the coupling of the angular momenta of systems 1 and 2. The eigenkets are eigenfunctions of $\vec{j}_1^2$, $\vec{j}_2^2$, $J^2 = (\vec{j}_1 + \vec{j}_2)^2$, J_z, where $\vec{j}_1$ and $\vec{j}_2$ are the angular momenta of the first and second parts of the system. We want to derive relations between the reduced matrix elements of tensor product $T_q^{(k)}$ and those of $U_{q_1}^{(k_1)}$ and $V_{q_2}^{(k_2)}$.

The matrix element of $T_q^{(k)}$ is

$$\langle \alpha' j_1' j_2' J'M' | T_q^{(k)} | \alpha j_1 j_2 JM \rangle,$$

$$= \langle JMkq | J'M' \rangle \frac{\langle \alpha' j_1' j_2' J' \| T^{(k)} \| \alpha j_1 j_2 J \rangle}{\sqrt{2J'+1}},$$

where α are the additional quantum numbers needed to specify the state of the system. We want to obtain relations between the reduced matrix elements of tensor products $T_q^{(k)}$ and those of $U_{q_1}^{(k_1)}$ and $V_{q_2}^{(k_2)}$, i.e., we want to relate $\langle \alpha' j_1 j_2 J' \| T^{(k)} \| \alpha j_1 j_2 J \rangle$ to $\langle \beta' j_1' \| U^{(k_1)} \| \beta j_1 \rangle$ and $\langle \gamma j_2' \| V^{(k_2)} \| \gamma j_2 \rangle$, where $\beta(\gamma)$ denote those quantum numbers among α which apply to 1(2).

We have

$$\langle \alpha' j_1' j_2' J'M' | T_q^{(k)} | \alpha j_1 j_2 JM \rangle$$

$$= \sum_{q_1 (q_2)} \langle k_1 q_1 k_2 q_2 | kq \rangle \, \langle \alpha' j_1' j_2' J'M' | U_{q_1}^{(k_1)}(1) \, V_{q_2}^{(k_2)}(2) \, | \alpha \, j_1 j_2 JM \rangle$$

$$= \sum_{\substack{q_1, q_2 \\ m_1, m_2 \\ m_1', m_2'}} \langle k_1 q_1 k_2 q_2 | kq \rangle \, \langle j_1 m_1 j_2 m_2 | JM \rangle \, \langle j_1' m_1' j_2' m_2' | J'M' \rangle$$

$$\times \left[\langle \alpha' j_1' m_1' j_2' m_2' \left| U_{q_1}^{(k_1)} V_{q_2}^{(k_2)} \right| \alpha j_1 m_1 j_2 m_2 \rangle \right]$$

$$\times \sum_{\alpha''} |\alpha''j_1''m_1''j_2''m_2''\rangle \langle\alpha''j_1''m_1''j_2''m_2''|.$$

Since $U(V)$ depends only on the variables 1(2), the expression in square brackets simplifies to

$$\sum_{\alpha''} \langle\alpha'j_1'm_1'|U_{q_1}^{(k_1)}|\alpha''j_1m_1\rangle \langle j_2m_2|j_2m_2\rangle \times \langle\alpha''j_2'm_2'|V_{q_2}^{(k_1)}|\alpha j_2m_2\rangle \langle j_1m_1|j_1m_1\rangle.$$

The additional quantum numbers α depend on 1 and 2, but it is likely that we may choose operators such that the dependence is separable, i.e., $\alpha = \alpha_1(1) \times \alpha_2(2)$. This is a possible source of trouble but generally can be avoided by proper choice of operators associated with α to ensure the separation. Assuming that this is possible, we write the above expression as

$$\langle\alpha_1'j_1'm_1'|U_{q_1}^{(k_1)}|\alpha_1 j_1m_1\rangle \langle\alpha_2'j_2'm_2'|V_{q_2}^{(k_2)}|\alpha_2 j_2m_2\rangle$$

(if the above-mentioned separation is not possible then we must carry the summation over α'' as given in the expression)

Next, we apply the WE theorem to each of the above operators in their own subspaces. Then the above expression becomes

$$\langle j_1m_1k_1q_1|j_1'm_1'\rangle \langle j_2m_2k_2q_2|j_2'm_2'\rangle \frac{\langle\alpha_1'j_1'\|U^{(k)}\|\alpha_1 j_1\rangle}{\sqrt{2j_1'+1}} \times \frac{\langle\alpha_2'j_2'\|V^{(k_2)}\|\alpha_2 j_2\rangle}{\sqrt{2j_2'+1}}.$$

Now, we obtain

$$\langle\alpha'j_1'j_2'J'M'|T_q^{(k)}|\alpha j_1 j_2 JM\rangle = \langle JMkq|J'M'\rangle \times \frac{\langle\alpha'j_1'j_2'J'\|T^{(k)}\|\alpha j_1 j_2 J\rangle}{\sqrt{2J'+1}}$$

$$\times \langle k_1q_1k_2q_2|kq\rangle \langle j_1m_1j_2m_2|JM\rangle = \langle j_1' m_1'j_2'm_2'|J'M'\rangle$$

$$\times \langle j_1m_1k_1q_1|j_1'm_1'\rangle\langle j_2m_2k_2q_2|j_2'm_2'\rangle$$

$$\times \frac{\langle\alpha_1'j_1'\|U^{(k_1)}\|\alpha_1 j_1\rangle}{\sqrt{2j_1'+1}} \times \frac{\langle\alpha_2'j_2'\|V^{(k_2)}\|\alpha_2 j_2\rangle}{\sqrt{2j_2'+1}}.$$

Multiplying both sides by $\langle JMkq|J'M'\rangle$ and summing over $M(q)$ (fixed M') and using 9j-symbol, we get

$$\langle\alpha'j_1'j_2'J'\|T^{(k)}\|\alpha j_1 j_2 J\rangle$$

$$= \sqrt{(2J'+1)(2J+1)(2k+1)} \begin{Bmatrix} j_1 & k_1 & j_1' \\ j_2 & k_2 & j_2' \\ J & k & J' \end{Bmatrix} \times \langle\alpha_1'j_1'\|U^{(k_1)}\|\alpha_1 j_1\rangle$$

$$\times \langle\alpha_2'j_2'\|V^{(k_2)}\|\alpha_2 j_2\rangle. \tag{41}$$

The reduced matrix element of the tensor product of $U^{(k_1)}$ and $V^{(k_2)}$ is proportional to the product of their reduced matrix elements. The proportionality factor in general depends on all angular momenta and degrees of the tensors involved.

We mention two important special cases:

(i) $k = 0$, $k_1 = k_2$; $T^{(k)}$ is a scalar built from the contraction of 2 tensors. In this case

$$\langle \alpha' j_1'' j_2'' J' \| U^{(k_1)} V^{(k_1)} \| \alpha j_1 j_2 J \rangle$$

$$= (-)^{J + j_1 + j_2''} \begin{Bmatrix} j_1 & j_2 & J \\ j_2'' & j_1'' & k_1 \end{Bmatrix} \langle \alpha_1' j_1'' \| U^{(k_1)} \| \alpha_1 j_1 \rangle \langle \alpha_2' j_2'' \| V^{(k_1)} \| \alpha_2 j_2 \rangle. \tag{42}$$

(ii) $k_1 = k$ and $k_2 = 0$. Thus, now, if we have $T^{(k)}$ (1) – there is no "2" dependence – then

$$\langle \alpha' j_1'' j_2'' J' \| T^{(k)}(1) \| \alpha j_1 j_2 J \rangle = \sqrt{(2J + 1)(2J' + 1)}\, (-1)^{J + J_2 + k + j_1'}$$

$$\times \begin{Bmatrix} J & j_1 & J_2 \\ J_1' & J' & k \end{Bmatrix} \langle \alpha_1' j_1'' \| T^{(k_1)}(1) \| \alpha_1 j_1 \rangle. \tag{43}$$

The derivation of an expression for the reduced matrix elements of the tensor product of two tensors operating on the same system is left as an exercise for the reader.

PROBLEMS

8.1 Express $\vec{r}$, $\vec{p}$ and $\vec{l} = \vec{r} \times \vec{p}$ as irreducible tensor operators.

8.2 The "tensor interaction" between two nucleons with spin operators $\vec{\sigma}_1$ and $\vec{\sigma}_2$ is written as

$$S_{12} = J(r_{12}) \left[\frac{(\vec{\sigma}_1 \cdot \vec{r}_{12})(\vec{\sigma}_2 \cdot \vec{r}_{12})}{r_{12}^2} - \vec{\sigma}_1 \cdot \vec{\sigma}_2 \right]$$

where $\vec{r}_{12}$ is the position vector joining the nucleons 1 and 2. Show that the operator S_{12} may be written as the scalar product

$$S_{12} = (\vec{S}^{(2)} \cdot \vec{L}^{(2)}),$$

where $\vec{S}^{(2)}$ is the irreducible tensor operator of rank 2 formed from $\vec{\sigma}_1$ and $\vec{\sigma}_2$ and $\vec{L}^{(2)}$ is a product of the scalar $J(r_{12})$ and the irreducible tensor operator of rank 2 formed from the unit vector $\vec{r}_{12}/r_{12}$.

8.3 Use the Wigner-Eckart theorem to obtain an explicit expression for the matrix elements of the form

$$<j_1, j_2, J - 1, M \,|j_{1z}|\, j_1, j_2, J, M>$$

where j_1 and j_2 have been coupled to form J and $J - 1$ in the initial and final states, respectively.

CHAPTER 9

Symmetry Transformations

The applications of symmetry consideration and invariance principles have led for a deep understanding of nature. In general, a symmetry principle or a conservation law is a consequence of an invariance property under a certain transformation group. In this and the following chapter we shall discuss these symmetry principles and their consequences.

9.1 INTRODUCTION

Symmetries, both exact and approximate, play a fundamental role in physics. Symmetry principles have served very useful purposes Symmetries imply conservation laws through Noether's theorem. Although very important in classical physics, symmetry considerations are of the greatest relevance in quantum physics. Approximate symmetries and the various asymmetries recently observed have turned out to be extremely meaningful.

There are four main groups of symmetries (exact or broken) in quantum physics :

  (i) Permutation symmetry

 (ii) Continuous space-time symmetries such as translation, rotation, etc.

(iii) Discrete symmetries, such as space inversion, time reversal etc.

(iv) Unitary symmetries:

    $U(1)$–symmetries which lead to conservation of charge, baryon number, etc.

    $SU(2)$–isopin symmetry

    $SU(3)$–color symmetry

    $SU(n)$–flavor symmetry

Among these, the first two groups and some unitary symmetries (those leading to conservation of charge, and $SU(3)$ color perhaps) are believed to be "exact" while the others are "broken".

The concept of symmetry is associated with the basic idea that it is impossible to observe or measure certain quantities–these are called "non-observables". As examples of such non-observables, we mention absolute position, absolute time, etc. The assumption of a non-observable and the underlying invariance principle lead to a conservation law or a selection rule. See Lee (1988) for an excellent discussion of these non-observables, the associated symmetry transformations and the conserved quantities.

Since non-observables imply symmetry, the detection of asymmetries of a physical law in a system must imply observables. A conservation law is established by means of selection rules in various processes or by directly measuring the conserved quantities.

The fundamental interactions in nature are classified into four groups (1) Strong interaction, (2) Electromagnetic interaction (3) Weak interaction, and (4) Gravitational interaction. The characteristics of these interactions are summarized in the table below.

Table 9.1   Fundamental Interactions

Interaction	*Quantum of the mediating field*	*Dimensionless coupling constant*	*Ranges (m)*
Strong	Color gluon	~ 1	$\leq 10 - 15$
Electromagnetic	Photon	$\alpha = \dfrac{e^2}{4\pi\hbar c} \simeq \dfrac{1}{137}$	∞
Weak	Intermediate bosons $W^{\pm}, Z^{0}$	$(Mc/\hbar)^2 G_F/\hbar c \sim 10^{-5}$	10^{-18}
Gravitational	Graviton	$GM^2/\hbar c \sim 10^{-39}$	∞

[Here α is the fine structure constant, G_F is the Fermi constant for β-decay, M = nucleon mass and G is gravitational constant].

The next table summarizes the validity of invariance principles in the three types of interactions.

Table 9.2   The Validity of Invariance Principles

Symmetry operations or conserved quantity	*Interaction*		
	Strong	*Electromagnetic*	*Weak*
Electric charge	Yes	Yes	Yes
Baryon number	Yes	Yes	Yes
Lepton number	Yes	Yes	Yes
Parity (P)	Yes	Yes	No
Charge conjugation (C)	Yes	Yes	No
Time reversal (T)	Yes	Yes	No
CPT	Yes	Yes	Yes
Isospin (I)	Yes	No	No $\left(\Delta I = 1 \text{ or } \dfrac{1}{2}\right)$

Contd...

Contd...

Strangeness (S)	Yes	Yes	No
			($\Delta S = 1, 0$)
Charm ($\mathscr{C}$)	Yes	Yes	No
			($\Delta \mathscr{C} = 1, 0$)

9.2 SYMMETRY TRANSFORMATIONS AND WIGNER'S THEOREM

In quantum mechanics the state of a system is described by a (normalized) vector $|\psi\rangle$ (or, density matrix ρ) of a Hilbert space $\mathscr{H}$. A more precise statement is that one assigns a "ray" to a state i.e., the (unit) vector assigned is defined up to an arbitrary multiplicative constant λ ($|\lambda| = 1$).

It frequently occurs that a system is invariant under certain symmetries. A symmetry transformation leaves the system unchanged i.e. it takes the system into itself. The symmetry operation permits different but equivalent descriptions of the system.

Let the same physical system be described in two different ways in the same (coherent subspace) Hilbert space $\mathscr{H}$ by rays $\underline{\overline{\psi}}_1, \underline{\overline{\phi}}_1, \ldots$ and by rays $\underline{\overline{\psi}}_2, \underline{\overline{\phi}}_2, \ldots$. For a symmetry operation the two descriptions must be equivalent. The transition probabilities must be the same, i.e., there is a norm preserving mapping T between the rays $\underline{\overline{\psi}}_1$, and $\underline{\overline{\psi}}_2$. Mathematically, we study the map $\mathscr{H} \to \mathscr{H}$ between the (unit) vectors $\overline{\psi}_1, \overline{\phi}_1, \ldots$ from the rays $\underline{\overline{\psi}}_1 \, \underline{\overline{\phi}}_1, \ldots$ and the (unit) vectors $\overline{\psi}_2, \overline{\phi}_1, \ldots$ from the rays $\underline{\overline{\psi}}_2, \underline{\overline{\phi}}_2, \ldots$. A theorem due to Wigner, which plays a central role in the theory of symmetry transformations, states that a transformation of the rays of a Hilbert space which preserves the inner product of rays can be regarded as the result of either a unitary or an antiunitary transformation of the vectors of the space. Therefore, symmetries are represented by either unitary or antiunitary operators which map the Hilbert space onto itself and conserves the transition probabilities.

Theorem 9.1 (Wigner) Let $\underline{\overline{\psi}}_2 = \hat{T} \underline{\overline{\psi}}_1$ be a mapping of the rays of a Hilbert space $\mathscr{H}$ which conserves the inner product of rays. Then there exists a mapping $\psi_2 = T\underline{\overline{\psi}}_1$ of all vectors of $\mathscr{H}$ such that $T\underline{\overline{\psi}}$ belongs to the ray $\hat{T}\,\underline{\overline{\psi}}$ if $\psi \in \underline{\overline{\psi}}$ and (a) $T(\psi + \phi) = T\psi + T\phi$, (b) $T(\lambda\psi) = \chi(\lambda) T(\psi)$, (c) $(T\psi, T\phi) = \chi\,[(\psi, \phi)]$, where either $\chi(\lambda) = \lambda$ (unitary case) or $\chi(\lambda) = \overline{\lambda}$ (antiunitary case) for all λ. Wigner's theorem is proved in his book, and in Bargmann ((1964)). See also Wick (1966), and Gottfried (1966). Here $\overline{a}$ denotes complex conjugate of a.

We follow the proof given in Barut & Raczka (1986).

Let $\psi_1, \phi_1, \ldots$ and $\psi_2, \phi_2, \ldots$ be two sets of orthonormal basis vectors from the set $\overline{\psi}_1, \overline{\phi}_1, \ldots$ and the set $\overline{\psi}_2, \overline{\phi}_2, \ldots$ of rays, respectively of the Hilbert space $\mathcal{H}$. We are given a transformation which conserves the inner product of the rays. The problem is to construct the corresponding transformation T on the vectors by suitable choices of the phases. T conserves the absolute value of the scalar products of the corresponding unit vectors. We want to prove that T can be so defined that either it is a unitary or an antiunitary operator.

We single out the unit vector ψ_1 and choose ψ_2 such that $\psi_2 = T\psi_1$. (1) Once this is done, all other phases are uniquely determined. Thus T is determined up to an overall phase factor. Let us next consider the vector $\psi_1 + \phi_1$, with ϕ_1 orthogonal to ψ_1. A representative vector of the corresponding ray in the second description is $a\psi_2 + b\phi_2$. Then

$$T(\psi_1 + \phi_1) = c\,(a\psi_2 + b\phi_2) = \psi_2 + b'\phi \tag{2}$$

Clearly, $c = 1/a$. We put $cb = b' = b/a$. We define $T\phi_1$ by $T(\psi_1 + \phi_1) - \psi_2$ or by $b'\phi_2$. Hence

$$T(\psi_1 + \phi_1) = T\psi_1 + T\phi_1. \tag{3}$$

Similarly, for a general $f_1 = \vec{a}_\psi\psi + \vec{a}_\phi\phi + \ldots$ we choose a representative $f_2 = \hat{a}_\psi\psi_2 + \hat{a}_\phi\phi_2 + \ldots$ to write $Tf_1 = cf_2$ with $c\hat{a}_\psi = a_\psi$, so that

$$T(a_\psi\psi_1 + a_\phi\phi_1 + \ldots) = a_\psi\psi_2 + c\hat{a}_\phi\phi_2 + \ldots$$

$$= a_\psi T\psi_1 + a'_\phi T\phi_1 + \ldots \tag{4}$$

The absolute values of the scalar products

$$|(\psi_1 + \phi_1, \phi_1)| = |a_\psi + a_\phi|, \tag{5}$$

and
$$|(T\psi_1 + T\phi_1, Tf_1)| = |a_\psi + a'_\phi|. \tag{6}$$

Since T conserves the absolute values of the scalar products, these two Eqs. (5) and (6) are equal. Thus with $|a'_\phi| = |a_\phi|$ we can calculate a'_ϕ in terms of a_ϕ and a_ψ. We get two solutions

$$a'_\phi = a_\phi \tag{7}$$

and
$$a'_\phi = \overline{a}_\phi\,\frac{a_\psi}{\overline{a}_\psi} \tag{8}$$

For the first solution Eq. (7) T is linear and unitary. For the second solution (8) we find

$$Tf_1 = \frac{a_\psi}{\overline{a}_\psi}\,[\overline{a}_\phi T\psi_1 + \overline{a}_\phi T\phi_1 + \ldots]. \tag{9}$$

Since an overall phase is unimportant and by a new normalization (which does not change our choice $T\psi_1 = \psi_2$) we get an antiunitary operator. Q.E.D.

For a given symmetry transformation, one or the other case occurs. For the rays of a two dimensional vector space, Wigner's theorem follows from the geometrical proposition that every transformation of the surface of a sphere which preserves the arc-distances between points is either a rotation (about the center of the sphere) or a pseudorotation (rotation accompanied by a reflection in a plane through the center).

The symmetry transformation of a state vector $|\psi\rangle$ is implemented by an operator U:

$$|\psi\rangle \rightarrow |\psi'\rangle = U\,|\psi\rangle \tag{10}$$

If U is unitary $\qquad UU^\dagger = U^\dagger U = I$. Clearly.

$\langle\phi'|\,\psi'\rangle = \langle\phi|\,U^\dagger U|\,\psi\rangle = \langle\phi|\,\psi\rangle$. If U is antilinear, then $Uc\,|\psi\rangle = c^*U|\,\psi\rangle$. If U is antiunitary, then $\langle\phi'|\,\psi'\rangle = \langle\phi|\,\psi\rangle^*$ (see the following section for antiunitary operators).

Only linear operators can represent continuous transformations, like translation, rotation, etc. If the group of symmetry transformations is continuous and connected, then the representation by U is unitary.

The transformation of state vectors is accompanied by simultaneous transformation of the operators for observables ($\Omega \rightarrow \Omega'$) Since the transformation leaves the system unchanged, we must have $\Omega'|\,\psi'_n\rangle = a_n|\,\psi'_n\rangle$ if we have $\Omega\,|\psi_n\rangle = a_n|\,\psi_n\rangle$. Using Eq. (10), we get $\Omega'U|\,\psi_n\rangle = a_nU|\,\psi_n\rangle$, or $U^{-1}\,\Omega'\,U|\,\psi'_n\rangle = a_n|\,\psi_n\rangle$. Hence the transformation of operators for observables.

$$\Omega \rightarrow \Omega' = U\Omega U^{-1}. \tag{11}$$

The transformation described by Eqs. (10) and (11) leaves the system unchanged. Infinitesimal unitary transformations are of particular importance. Such a transformation is written as

$$U = 1 + \frac{i}{\hbar}\,\epsilon\,G. \tag{12}$$

where, the real parameter ϵ is infinitesimal and G is hermitian. G is called the generator of the operator. To first order in ϵ,

$$U^{-1} = 1 - \frac{i\epsilon}{\hbar}\,G$$

and also $\qquad U^* = 1 - \frac{i\epsilon}{\hbar}\,G.$

Hence, $\qquad U^{-1} = U^*. \tag{13}$

Under Eq. (12), an operator Ω changes as follows:

$$\Omega \to \Omega' = \left(1 + \frac{i\epsilon}{\hbar}\, G\right) \Omega \left(1 - \frac{i\epsilon}{\hbar}\, G\right)$$

$$= \Omega + \frac{i\epsilon}{\hbar}\, [G, \Omega], \text{ to terms first order in } \epsilon.$$

The change is

$$\delta\Omega \equiv \Omega' - \Omega = \frac{i\epsilon}{\hbar}\, [G, \Omega]. \tag{14}$$

If the unitary transformation leaves the Hamiltonian H invariant, then (11) gives

$$UH - HU = 0,$$

i.e.,
$$[U, H] = 0. \tag{15}$$

Hence,

$$\frac{d}{dt}\, \langle U \rangle = \frac{1}{i\hbar}\, \langle [U, H] \rangle = 0. \tag{16}$$

The unitary operator U is a constant of the transformation which leaves H unchanged for the infinitesimal transformation Eq. (12),

$$[G, H] = 0 \tag{17}$$

so that G is a constant of the motion. Since $U(\epsilon)$ is a function of G, it follows that H commutes with $U(\epsilon)$ for all ϵ. Thus, invariance under an infinitesimal unitary transformation implies invariance under a finite continuous unitary transformation $U = e^{\frac{i}{\hbar}\epsilon G}$. For example, rotational invariance of the Hamiltonian leads to the conservation of angular momentum. As a consequence of Eq. (15), we see that if $|\psi\rangle$ is a solution of the equation of motion

$$i\hbar\, \frac{d}{dt}\bigg|\, \psi\rangle = H|\,\psi\rangle, \text{ then so is } |\psi'\rangle = U|\,\psi\rangle.$$

Let S be a symmetry operator so that

$$[H, S] = 0, \tag{18}$$

and $|\alpha\rangle$ an eigenket with energy eigenvalue E_α:

$$H|\alpha\rangle = E_\alpha |\alpha\rangle. \tag{19}$$

From Eqs. (18) and (19), $S|\alpha\rangle$ is an eigenstate of H with the same eigenvalue E_α:

$$H\,(S|\alpha\rangle) = SH|\alpha\rangle = E_\alpha\,(S|\alpha\rangle). \tag{20}$$

There are two possibilities.

(a) The eigenvalue is non degenerate if $|\chi\rangle = S|\alpha\rangle$ differ from $|\alpha\rangle$ only by a normalization factor

$$|\chi\rangle = c\,|\alpha\rangle \tag{21}$$

with $|c|^2 = 1$, due to the normalization condition.

(b) If $|\chi\rangle$ is linearly independent of $|\alpha\rangle$, the eigenvalue is degenerate. If the energy level is f-fold degenerate, then there are f linearly independent eigenfunctions $|\psi_i\rangle$ ($i = 1, 2, \ldots f$) with the particular eigenvalue. Then $|\chi\rangle$ can be expressed as a linear combination of $|\psi_i\rangle$'s. Incase S is characterized by a continuous parameter λ, all states $S_\lambda|\alpha\rangle$ have the same energy E_α.

9.3 ANTIUNITARY OPERATORS

A one-to-one mapping A of a linear space V over the complex numbers $\mathbb{C}$ is called a *semilinear* operator if

$$A\,(|\psi_1\rangle + |\psi_2\rangle) = A|\psi_1\rangle + A|\psi_2\rangle \text{ for all } \psi_1, \psi_2 \in V \tag{22}$$

$$A|\psi\rangle = 0 \Rightarrow |\psi\rangle = 0$$

and for all $c \in \mathbb{C}$

$$Ac\,|\psi\rangle = c'A|\psi\rangle, \text{ where } c' = c \text{ or } c' = c^* \tag{22a}$$

A semilinear operator is called *linear* if $c' = c$.

A semilinear operator is called *antilinear* if $c' = c^*$, i.e., if

$$Ac|\psi\rangle = c^* A|\psi\rangle. \tag{22b}$$

The product of an antilinear operator A with another operator B is

$$(AB)|\psi\rangle = A\,(B|\psi\rangle) \quad \text{for all } |\psi\rangle \in V \tag{23}$$

It is easy to show that if B is linear, then AB is antilinear; if B is antilinear, then AB is a linear operator. The sum of two antilinear operators can be defined in the usual way. The inverse of A, if it exists, is defined in the usual way. A^{-1} is antilinear. Note that

$$(cA)^{-1} = c^{*-1} A^{-1}. \tag{24}$$

The adjoint of A, denoted by $A^\dagger$, is defined as

$$\langle A\alpha|\beta\rangle = \langle A^\dagger\beta|\alpha\rangle = \langle\alpha|\,A^\dagger\beta\rangle^*, \ |\alpha\rangle, |\beta\rangle \in V \tag{25}$$

If L is a linear operator, and A_1, A_2 are antilinear operators, then

$$(A_1 A_2)^\dagger = A_2^\dagger A_1^\dagger, \ (AL)^\dagger = L^\dagger A^\dagger,$$

$$(A_1 LA_2)^\dagger = A_2^\dagger L^\dagger A_1^\dagger. \tag{26}$$

An *antiunitary* operator A is an antilinear operator whose inverse A^{-1} exists and which satisfies

$$A^\dagger = A^{-1} \tag{27}$$

For an antiunitary operator A if $A|\psi_1\rangle = |\psi_1'\rangle$ and $A|\psi_2\rangle = |\psi_2'\rangle$, then

$$\langle\psi_2'|\psi_1'\rangle = \langle\psi_2|\psi_1\rangle^* = \langle\psi_1|\psi_2\rangle. \tag{28}$$

Then it is easy to show that an antilinear operator A is antiunitary if A^{-1} exists and it is isometric

$$\|A|\psi\rangle\| = \||\psi\rangle\|. \tag{29}$$

The action of an antilinear operator A to the right on a ket is given by Eqs. (22) and (22a)

$$A\left(c_1|\psi_1\rangle + c_2|\psi_2\rangle\right) = c_1^* A|\psi_1\rangle + c_2^* A|\psi_2\rangle. \tag{30}$$

Now, we have to define the action of A to the left on a bra. Let $\langle\phi_1|$ $(\langle\phi_1'|)$ be the dual to $|\psi_1\rangle$ $(|\psi_1'\rangle)$. Then the dual to $A\left(c_1|\psi_1\rangle + c_2|\psi\rangle\right)$ is $(\langle\phi_1'|c_1 + \langle\phi_2'|c_2)$. Hence we define

$$\langle\phi_1|A = \langle\phi_1'|, \ \langle\phi_2|A = \langle\phi_2'| \tag{31}$$

$$\left(\langle\phi_1| c_1^* + \langle\phi_2| c_2^*\right) A = \langle\phi_1'| c_1 + \langle\phi_2'| c_2. \tag{32}$$

Then

$$\langle\phi| (A|\psi\rangle) = [(\langle\phi|A)| \psi\rangle], \tag{33}$$

In this case, the expression $\langle\phi|A|\psi\rangle$ is undefined as $(\langle\phi|A)|\psi\rangle \neq \langle\phi|(A|\psi\rangle)$. For linear operators, we can unambiguously define $\langle\phi |L| \psi\rangle$. Hence we adopt the convention used by Messiah (1961), Gottfried (1966), and others. (See Ballentine (1998) for another convention). We now write an antiunitary operator as

$$\Theta = UK \tag{34}$$

where U and K are defined as follows. Let us expand $|\alpha\rangle$ in terms of base kets. $|a'\rangle$. Then

$$|\psi\rangle = \sum_{a'} |a'\rangle \langle a'|\alpha\rangle. \tag{35}$$

We now define a complex-conjugation operator K which takes the complex conjugate of all the expansion coefficients of the arbitrary ket in terms of the particular basis. The action of K on the ket is

$$|\alpha\rangle \xrightarrow{K} |\tilde{\alpha}\rangle = \sum_{a'} \langle a'|\alpha\rangle^* UK|a'\rangle$$

$$= \sum_{a'} \langle a'|\alpha\rangle|a'\rangle. \tag{36}$$

Next we consider the action of Θ on the ket

$$|\alpha\rangle \xrightarrow{\Theta} |\tilde{\alpha}\rangle = \sum_{a'} \langle a'|\alpha\rangle^* UK |a'\rangle.$$

$$= \sum_{a'} \langle a'|\alpha\rangle^* U|a'\rangle. \tag{37}$$

Also,

$$|\beta\rangle \xrightarrow{\Theta} |\tilde{\beta}\rangle = \sum_{a'} \langle a'|\beta\rangle^* U|a'\rangle$$

so that

$$\langle \tilde{\beta}| = \sum_{a'} \langle a'|\beta\rangle \langle a'| U^{\dagger}.$$

Therefore,

$$\langle \tilde{\beta}|\tilde{\alpha}\rangle = \sum_{a'',a'} \langle a''|\beta\rangle \langle a''|U^{\dagger}U|a'\rangle \langle \alpha|a'\rangle$$

$$= \langle \alpha|\beta\rangle \tag{38}$$

$$= \langle \beta|\alpha\rangle, \text{ in agreement with Eq. (28)}$$

provided

$$U^{\dagger}U = I. \tag{39}$$

Thus the antiunitary operator is the product of complex conjugation and a unitary operator.

It is to be emphasized that a particular form of U depends on the representation used.

9.4 SYMMETRY GROUPS

The set G of all summetry transformations operations R constitutes a group and is called a symmetry group of the Hamiltonian (i.e., the physical system). For two symmetry operations R and $S \in G$, we have

$$\text{RHR}^{-1} = H \text{ and SHS}^{-1} = H \tag{40}$$

Then

$$(RS)\, H\, (RS)^{-1} = RSHS^{-1}R^{-1} = RHR^{-1} = H \tag{41}$$

so

$$RS \in G.$$

All these operators satisfy the associative law. The identity operation is the unit element of G. From Eq. (40), $(R^{-1})\, H\, (R^{-1}) = H$, and so $R^{-1} \in G$. Hence G is a group. Corresponding to a symmetry property, a set of symmetry transformations froms a group.

Let us consider a group of symmetry transformations $G\ (g_0,\ g_1,\ ...,\ g_i,\ ...)$. We have already seen that to each symmetry transformation g_i, we associate a unitary or antiunitary operator U_{g_i} in the Hilbert space of the system. Each U_g is determined only upto a phase, so that if $g_i g_j = g_k$, then

$$U_{g_i g_j} = e^{ic_{ij}} U_{g_i} U_{g_j}. \tag{42}$$

The set of operators U forms a (ray) representation of the group G in Hilbert space, upto a phase factor. In case $c_{ij} = 0$ for all i, j then the representation is exact, and there is isomorphism between G and the group of operators U in Hilbert space.

In general, this is not possible, but we can restrict c_{ij} to 0 or π so that apart from the sign the operator U_{g_i} corresponding to g_i represents G uniquely. Let G be a continuous connected Lie group. For any infinitesimal transformation there is a unitary operator in Hilbert space. This may be written as

$$U\,(\alpha_1, \alpha_2, \ldots, \alpha_n) = 1 + \frac{i}{\hbar} \sum_{j=1}^{n} \alpha_j \mathcal{G}_j, \tag{43}$$

where $\mathcal{G}$'s are hermitian and $\alpha_1, \ldots, \alpha_1$ are n-parameters. A finite transformation can be built from the infinitesimal transformations. For n-parameters, an arbitrary group element is given by

$$U\,(\alpha_1, \alpha_2, \ldots, \alpha_n) = e^{-i/\hbar \sum_{j=1}^{n} \alpha_j \mathcal{G}_j.} \tag{44}$$

We have seen that as a consequence of the existence of the symmetry group, all generators $\mathcal{G}_j$ must commute with the Hamiltonian and the corresponding physical observables are conserved.

Let us now enumerate some examples of the symmetries in physics.

(1) Continuous space-time symmetries-

    (i) Translations in space ($\vec{x} \to \vec{x} + \vec{a}$, where $\vec{a}$ is a constant 3-vector). All isolated systems exhibit this symmetry. It leads to the conservation of linear momentum.

    (ii) Translations in time ($t \to t + a_0$, where a_0 is a content). All isolated systems exhibit this symmetry. It leads to the conservation of energy.

    (iii) Rotations in three-dimensional space (Rotation group) ($x^i \to x'^i = R^i_j\, x^j$, where $i, j = 1, 2, 3$; R is an orthogonal 3×3 matrix; x^i are the components of a 3-vector). All isolated systems exhibit this symmetry. It leads to the conservation of angular momentum

    (iv) Lorentz transformations or Galilei transformations (for nonrelativistic theories) (Lorentz or Galilei group). Lorentz transformation is

$$\begin{pmatrix} t \\ \vec{x} \end{pmatrix} \to \begin{pmatrix} t' \\ \vec{x}' \end{pmatrix} = \alpha \begin{pmatrix} t \\ \vec{x} \end{pmatrix}$$

where α is a 4×4 matrix. This symmetry is the combined space-time symmetry.

(2) Discrete space-time symmetries

    (i) Space inversion or parity transformation $(\vec{x} \rightarrow -\vec{x})$

    (ii) Time reversal

    (iii) Discrete translations on a lattice (Translation Group) – An electron experiences a periodic potential in a crystal. The Hamiltonian is invariant with respect to all lattice translations

    (iv) Discrete rotational symmetry of a lattice (Point group) – A given lattice is invariant under the symmetry operations of three-dimensional rotations and reflections. There are 32 crystallographic point groups which combined with discrete translations give 230 space groups. These are the basic symmetry groups of crystalline solids.

    Besides the above geometrical symmetry groups there are nongeometrical symmetries.

(3) Permutation symmetry (Symmetric or permutation group)

(4) Gauge transformations

(5) Internal symmetries – Examples are isospin, SU(3), charge-conjugation, etc.

In this book we discuss only some of these symmetry groups in detail. (See Fonda and Ghirardi (1970); Greiner (1989)).

In the table below we list some symmetry transformation groups with the associate conserved generators.

Table 9.3 Symmetries and generators

Symmetry transformations	*Conserved generators or physical implications*
(1) Spatial translations	Linear momentum
(2) Rotations	Angular momentum
(3) Time translations	Energy
(4) Space inversion	Parity
(5) Time reversal	Antiunitary
(6) Permutation symmetry of identical particles	Statistics (symmetry nature of wave functions)
(7) Gauge transformations	Charge Baryon number Lepton number

We now show that the invariance of the Hamiltonian under a symmetry group G implies that the time-development operator $U\,(t, t_0)$ is invariant under the group. The time evolution operator satisfies the integral equation

$$U(t, t_0) = 1 - \frac{i}{\hbar} \int_{t_0}^{t} H(t') \, U(t', t_0) \, dt'. \qquad (45)$$

Multiplying this equation from the left by $U(\alpha)$ and from the right by $U^{\dagger}(\alpha)$, and assuming $[U(\alpha), H\,(t')] = 0$, $t_0 \leq t', \leq t$, we obtain

$$U(\alpha)\, U(t, t_0)\, U^\dagger(\alpha) = 1 - \frac{i}{\hbar} \int_{t_0}^{t} H(t')\, U(\alpha)\, U(t', t_0)\, U^\dagger(\alpha)\, dt'. \tag{46}$$

Hence, $U(t, t_0)$ and $U(\alpha)\, U(t, t_0)\, U^\dagger(\alpha)$ satisfy the same integral equation and so they are equal. Thus,

$$[U(\alpha),\, U(t, t_0)] = 0. \tag{47}$$

The converse is also true. Hence the invariance of the Hamiltonian and dynamical invariance are equivalent.

We discuss the connection of symmetry with degeneracy. Let us consider that the eigenvalue E_n is f_n-fold degenerate. Then we may choose a set of f_n orthonormal eigenfunctions belonging to E_n. Then

$$H|\,\psi_k^{(n)}\rangle = E_n|\,\psi_K^{(n)}\rangle, \quad K = 1, 2, \ldots, f_n. \tag{48}$$

A symmetry operation R belonging to the symmetry group G satisfies $RH = HR$ so that

$$HR|\,\psi_K^{(a)}\rangle = E_n R|\,\psi_K^{(n)}\rangle \tag{49}$$

Thus $R|\,\psi_K^{(n)}\rangle$ must be an eigenfunction of H having the same energy eigenvalue. Then $R|\,\psi_K^{(n)}\rangle$ is expressible as a linear combination of the $\{\psi_n^K\}$. The f_n degenerate functions form the basis vectors in f_n-dimensional vector space. This space is an invariant subspace of the entire Hilbert space of eigenfunctions of H. We write

$$R|\,\psi_\upsilon^{(n)}\rangle = \sum_{K=1}^{f_n} |\psi_K^{(n)}\rangle\, \Gamma^{(n)}\,(R)_{K\upsilon.} \tag{50}$$

These f_n-dimensional matrices $\Gamma(R)$ form a representation of the group G and the set $\{\psi_n^K\}$ forms a basis for that representation. To show that these matrices form a representation of G, we consider two successive operations: $SR|\,\psi_\upsilon\rangle = S \sum_K |\psi_K\rangle\, \Gamma(R)_{K\upsilon}$, using Eq. (50) and suppressing the superscript n

$$= \sum_{K,\mu} |\psi_\mu\rangle\, \Gamma(S)_{\mu K}\, \Gamma(R)_{Kv}$$

$$= \sum_{\mu} |\psi_\mu\rangle\, [\Gamma(S)\, \Gamma(R)]_{\mu v}$$

But,

$$SR|\,\psi_v\rangle = \sum_{\mu} |\psi_\mu\rangle\, \Gamma(SR)_{\mu v}$$

Therefore,

$$\Gamma(SR) = \Gamma(S)\, \Gamma(R). \tag{51}$$

Hence the matrices indeed form a representation of the group G. The set of f_n degenerate eigenfunctions $\{\psi_K^n\}$ of energy E_n forms a basis for that representation. The representation Γ is unitary. Let us evaluate $\langle R\psi_K | R\psi_\upsilon \rangle$

$$= \sum_{\lambda,\mu} \langle \psi_\lambda | \psi_\mu \rangle \, \Gamma^*_{\lambda K}(R) \, \Gamma(R)_{\mu v}$$

$$= \sum_{\lambda} \Gamma(R)^*_{\lambda K} \, \Gamma(R)_{\lambda v} \text{ using the orthonormality of the eigenfunctions.}$$

$$= \sum_{\lambda} \Gamma(R)^\dagger_{K\lambda} \, \Gamma(R)_{\lambda v},$$

$$= [\Gamma(R)^\dagger \, \Gamma(R)]_{Kv}$$

From unitary of the symmetry operation R, we get

$$\langle R\psi_K | R\psi_v \rangle = \langle \psi_K | \psi_v \rangle = \delta_{Kv}.$$

Then

$$\Gamma(R)^\dagger \, \Gamma(R) = I. \tag{52}$$

This representation Γ is irreducible. If it were reducible, one may regroup the functions ψ_K into at least two sets closed under symmetry operations. Then, in general, the above two sets of functions have different eigenvalues which contradicts our assumption that the ψ's are degenerate to the same eigenvalue. So the representation is irreducible. It is, of course, possible that the two eigenvalues coincide by accident ("accidental degeneracy"), not due to the stated symmetries. Thus, the orthonormal eigenfunctions belonging to a degenerate energy level of a Hamiltonian form a basis for an irreducible representation of the symmetry group. The degeneracy f is equal to the dimension of the representation.

Normal degeneracy occurs if all the eigenfunctions of an energy eigenvalue span an irreducible invariant subspace. The degeneracy is known as accidental if the subspace is reducible. A set of vectors $\{\phi_i\}$ is said to span an invariant subspace V_S under a given set of operations $\{R_j\}$ if $R_j \phi_i$ (for all i, j) is also in V_S. If V_S can be divided into smaller subspaces also invariant under $\{R_j\}$, then the original subspace V_S is said to be "reducible", otherwise, V_S is said to be irreducible.

Let us consider an electron (assumed spinless) in a central field. In this case, the bound state eigenfunctions can be separated in spherical coordinates:

$$\psi_{nlm}(\vec{r}) = R_{nl}(r) \, Y_{lm}(\theta, \phi). \tag{53}$$

$R_{nl}(r)$ depends on the potential $V(r)$. The rotation group is the symmetry group and under rotation R we get

$$RY_{lm} = \sum_{m'=-l}^{l} Y_{lm'} \, \Gamma_{m'm}^{(l)} (R).$$

(54)

The matrices $\Gamma^{(l)}(R)$ form an *IR* of the symmetry group and the spherical harmonics Y_{lm} are the basis functions for the representation. The dimension of the $\Gamma^{(l)}$ is $(2l + 1)$ which is the degeneracy $(m = l, l - 1, ..., -l)$. For the special form of $V(r) = \dfrac{1}{r}$, e.g., for the case of an electron in a Coulomb potential (H atom), the solution of the Schrödinger equation that in addition to the quantum numbers l and m, the states are characterized by a quantum member n (the principal quantum number). The energy levels depend on n only so that all states with the same n but different l belong to the same energy. The degree of degeneracy is $\displaystyle\sum_{l=0}^{n-1} (2l + 1) = n^2$ This degeneracy of states in a Coulomb potential with the same n but different values of l is termed "accidental degeneracy". Notice that the set of all eigenfunctions ψ_{nlm} (for a fixed value of n) forms an invariant subspace for the group of rotations, but it is not irreducible. But the $(2l + 1)$ states of ψ_{nlm} (for fixed values of n and l) span an irreducible subspace.

The problem of perturbation theory (see Chapter 12) is considerably simplified by taking advantage of the symmetries present. Let us consider a system with Hamiltonian H_0 which is invariant under its symmetry group G_0. Then

$$H_0 \phi^{(a)} = E^{(0)} \phi^{(a)}.$$

(55)

In general, the energy level E^0 is degenerate and the states $\psi_i^{(a)}$ ($i = 1, 2, ... n$) span an n-dimentional representation of the symmetry group G_0. Now the system is subjected to a perturbation H_1 which is invariant under the symmetry group G_1. The states of the system are now described by the equation

$$(H_0 + H_1) \psi^{(a)} = E\psi^{(a)}$$

(56)

and are classified according to the representations of the symmetry group G of this equation. G is the intersection of G_0 and G_1, i.e., the elements of G are common to both G_0 and G_1. We distinguish between two cases: (a) The group G_1 is either larger than G_0 or coincides with it. In this case $H = H_0 + H_1$ and H_0 are both invariant under G_0. (b) H_1 has a lower symmetry than H_0. The group G_1 is a subgroup of G_0 so that H is invariant under G_1. For (a) if the representation is irreducible, i.e., the degeneracy of $E^{(0)}$ is normal, then the states of the perturbed system are classified according to the *IR*'s of G_0. The level $E^{(0)}$ will not split (still n-fold degenerate) but shifted by $\langle \phi_i^{(a)} | H_1 | \phi_i^{(a)} \rangle$ after the perturbation is switched on. In case (*a*), if the representation spanned by $\phi_r^{(a)'}$ s is reducible then a level $E^{(0)}$ belonging to a *RR* containing *rIR*'s of dimensions $d_1, ..., d_r$ respectively can be split at most into r levels $E_1, E_2, ..., E_r$ and the level E_r will have a degeneracy d_r after the perturbation is imposed.

We now turn to case (b) in which the perturbation has a lower symmetry than H_0. The total Hamiltonian H has a symmetry group G_1 which is a subgroup of G_0. From representation $\Gamma(G_0)$ of the group G_0, we can obtain a representation of the subgroup G_1 by selecting from among the matrices $\Gamma(G_0)$ those which correspond to elements in G_1. Even if $\Gamma(G_0)$ is an *IR* of G_0 the representation of G_1 obtained in the manner above $[\Gamma'(G_1)]$ may be reducible. Therefore, even though the eigenfunctions belonging to a level $E^{(0)}$ form the basis of an *IR* of G_0, this representation may be reducible for the subgroup G_1. Then the perturbation H_1 will split the level. Using the character tables, one can determine the splitting of the degenerate level $E^{(0)}$ by the application of the perturbation.

In the calculation of transition probabilities between the states of a system, it is necessary to calculate matrix elements of different operators. Such calculations are considerably simplified by taking advantage of the symmetries present. In many cases, some of these matrix elements vanish due to symmetry. Then the corresponding transitions are forbidden. These give rise to the "selection rules" which we now discuss.

Let us consider the matrix element of the irreducible tensor operator $T^{(\alpha)}$:

$$\langle \psi_m^{(\gamma)} |\, T_j^{(\alpha)} |\, \psi_l^{(\beta)} \rangle. \tag{57}$$

The functions $T_j^{(\alpha)} | \psi_l^{(\beta)} \rangle$ transform as basis functions of the product representation $G^{(\alpha)} \times G^{(\beta)}$. Then $T_j^{(\alpha)} |\, \psi_l^{(\beta)} \rangle$ belong to irreducible components that can be obtained by decomposition of $\Gamma^{(\alpha)} \times \Gamma^{(\beta)}$. By the basis function orthogonality theorem $\langle \psi_m^{(\alpha)} |\, \psi_l^{(\beta)} \rangle = c\delta_{\alpha\beta}\delta_{ml}$, where c is a constant (independent of m and l) the matrix element is zero unless $\Gamma^{(\gamma)}$ is contrained in $\Gamma^{(\alpha)} \times \Gamma^{(\beta)}$ i.e. the representation $\Gamma^{(\gamma)}$ is found in the decomposition of $\Gamma^{(\alpha)} \times \Gamma^{(\beta)}$. We may also say that the matrix element must vanish unless $\Gamma^{(\gamma)*} \times \Gamma^{(\alpha)} \times \Gamma^{(\beta)}$ includes Γ_1, the identical representation.

Next, we consider the diagonal matrix elements

$$\langle \psi_m |\, T_j^{(\alpha)} |\, \psi_m \rangle \tag{58}$$

where we suppress the superscript specifying the *IR* to which the ψ's belong. We consider real functions ψ_i (neglecting electron spin). For a real hermitian operator $T^{(\alpha)}$, the matrix element Eq. (58) does not vanish if the symmetric product representation $[\Gamma \times \Gamma]$ contains $\Gamma^{(\alpha)}$. For an imaginary hermitian operator $T^{(\alpha)}$ the matrix element is nonvanishing if the antisymmetric product representation $\{\Gamma \times \Gamma\}$ contain $\Gamma(\alpha)$. We shall illustrate applications of the selection rules in appropriate places.

9.5 SPATIAL TRANSLATIONS

We assume that space is homogeneous. This statement means that the properties of a closed system are invariant under a parallel displacement of the system.

Space translation invariance implies that two frames of reference differing by a uniform translation

$$\vec{r} \to \vec{r}\,' = \vec{r} + \vec{a} \tag{59}$$

are equivalent in describing the properties of the system. Let us consider a quantum mechanical system whose state at a time t is given by the ket $|\alpha\rangle$ or the wave function $\psi_\alpha (\vec{r}, t)$. For a spatial translation by $\vec{a}$, $|\alpha\rangle \to |\alpha'\rangle$ or $\psi_\alpha \to \psi_{\alpha'}$. Clearly

$$\psi_{\alpha'} (\vec{r} + \vec{a}, t) = \psi_\alpha (\vec{r}, t) \tag{60}$$

In this operation, we have shifted the wave function keeping the reference frame the same. This operational point of view is called the "active" interpretation of the symmetry operation (a shift in this case). In the active point of view a new state is generated in the original reference frame. In the "passive" interpretation, the wave function remains fixed but the reference frame is shifted (by $-\vec{a}$) with respect to the original frame (see § 5.3).

We write (suppressing t)

$$\psi_{\alpha'} (\vec{r}) = U_{\vec{r}} (\vec{a}) \, \psi_\alpha (\vec{r}) \tag{61}$$

or,
$$\psi_{\alpha'} (\vec{r} - \vec{a}) = U_{\vec{r}} (\vec{a}) \, \psi_\alpha (\vec{r}) \tag{62}$$

where U is the spatial translation operator (unitary).

Spatial translations form a continuous Abelian group which has three real parameters, corresponding to the displacements in the directions of the coordinate axes. Let us first discuss the continuous translation group in one dimension. An arbitrary element is denoted by g_{a_x} corresponding to a translation by a_x. Then the composition law is

$$g_{a_x} \, g_{a'_x} = g_{a''_x}$$

i.e., in terms of the parameters

$$a''_x = a_x + a'_x.$$

The unit element is $g_0 = E$ and $g_{a_x}^{-1} = g_{-a_x}$.

These properties specify the one-dimensional group.

For an infinitesimal displacement denoted by d_{a_x}, we have

$$g da_x = E - \frac{i}{\hbar} \, da_x P_x, \tag{63}$$

where P_x is the generator of the translation.

Since
$$g_{a_x} + d_{a_x} = g_{a_x} + d_{a_x} \frac{dg_{a_x}}{dx} \tag{64}$$

and

$$g_{a_x} + d_{a_x} = g_{da_x} g_{a_x}, \tag{65}$$

substituting Eq. (63) in Eq. (65), and comparing with Eq. (64), we get

$$\frac{dg_{a_x}}{dx} = -\frac{i}{\hbar} P_x g_{a_x}. \tag{66}$$

Using the boundary condition $g_0 = E$, the solution of the differential equation is

$$g_{a_x} = e^{-\frac{i}{\hbar} P_x}. \tag{67}$$

For unitary representations, the generator P_x corresponds to a hermitian operator with real eigenvalues (say, p_x). Therefore, for infinitesimal displacements,

$$U(a_x)\, \psi(\vec{r}) = \psi\,(x - \delta_{a_x},\, y,\, z) \tag{68}$$

$$= \psi\,(x,\, y,\, z) - \delta_{a_x} \left(\frac{\partial \psi}{\partial x}\right) \text{ to terms of order } \delta_{a_x}.$$

Hence, from Eqs. (67) and (68) the hermitian operator P_x may be written as

$$P_x = -i\hbar\, \frac{\partial}{\partial x}. \tag{69}$$

Thus, the operator representing a finite displacement by a_x in the x-direction is expressed in terms of the linear momentum operator P_x. For translations along the y- and z-directions, the corresponding operators are P_y and P_z respectively. P_x, P_y, P_z commute with each other. In general, the translation operator may be written as

$$U\vec{r}(\vec{a}) = \exp\left(-\frac{i}{\hbar}\, \vec{a} \cdot \vec{P}\right) \tag{70}$$

where $\vec{P}$ is the linear momentum operator and in coordinate representation $\vec{P} = -i\hbar\, \vec{\nabla}$. Similar considerations apply to the case where several particles are involved.

From the general discussion above, we see that linear momentum is a constant of motion as a result of translational invariance.

9.6 TIME TRANSLATION

We assume that time is homogeneous. This means tha a closed system is invariant under time translation, i.e., the description of a system is independent of the choice of the origin with respect to which time is measured. Time translations form a continuous one-parameter Abelian group. In physical time, time displacement is given by

$$t \rightarrow t' = t + \tau. \tag{71}$$

The state of a system as observed by O at his time is given by $\underline{\overline{\Psi}}_O(t)$ (suppressing $\vec{r}$) while O' observes the state at his time t' so that

$$\underline{\overline{\Psi}}_{O'}(t') = \underline{\overline{\Psi}}_O(t)$$

or,
$$\underline{\overline{\Psi}}_{O'}(t + \tau) = \underline{\overline{\Psi}}_O(t) \tag{72}$$

The mapping between the time-displaced and the initial state is given by a linear, unitary (time-evolution) operator $U_t(\tau)$:

$$\underline{\overline{\Psi}}_{O'}(t) = U_t(\tau)\, \underline{\overline{\Psi}}_O(t) = \underline{\overline{\Psi}}_O(t - \tau). \tag{73}$$

The Taylor expansion of $\underline{\overline{\Psi}}_O(t - \tau)$ gives

$$U_t(\tau)\, \underline{\overline{\Psi}}_O(t) = \underline{\overline{\Psi}}_O(t) + \frac{(-\tau)}{1!} \frac{d}{dt} \underline{\overline{\Psi}}_O(t) + \frac{(-\tau)^2}{2!} \frac{d^2}{dt^2} \underline{\overline{\Psi}}_O(t) + \dots \tag{74}$$

Therefore, the representation of an arbitrary time translation τ is given by

$$U_t(\tau) = e^{-\tau \frac{\partial}{\partial t}} = e^{\frac{i}{\hbar}\tau E} = e^{\frac{i}{\hbar}\tau H} \tag{75}$$

where the energy operator $E = i\hbar \dfrac{\partial}{\partial t} = H$. Thus, the generator of time translations is the Hamiltonian of the system.

Let us consider the Schrödinger equation for $\underline{\overline{\Psi}}_O(t)$:

$$i\hbar\, \frac{\partial \underline{\overline{\Psi}}_O(t)}{\partial t} = H \underline{\overline{\Psi}}_O(t)$$

$$\tag{76}$$

Then

$$i\hbar\, \frac{\partial \underline{\overline{\Psi}}_{O'}(t)}{\partial t} = i\hbar\, \frac{\partial}{\partial t}\, U_t(\tau)\, \underline{\overline{\Psi}}_O(t)$$

$$= i\hbar\, U_t(\tau)\, \frac{\partial \underline{\overline{\Psi}}_O(t)}{\partial t}$$

$$= U_t(\tau)\, H \underline{\overline{\Psi}}_O(t)$$

$$= H \underline{\overline{\Psi}}_{O'}(t)$$

$$= H U_t(\tau)\, \underline{\overline{\Psi}}_O(t) \tag{77}$$

Thus,
$$[H,\, U_t(\tau)] = 0. \tag{78}$$

Then H is time independent. The assumption that the time-displaced state satisfies the Schrödinger equation implies Eq. (78), and vice versa. We obtain

$$\frac{dH}{dt} = 0 \tag{79}$$

which means conservation of energy. Conservation of energy is a consequence of the temporal translation invariance, i.e., the assumed homogeneity of time.

9.7 SPACE ROTATIONS

We assume that space is isotropic i.e., all directions are equivalent. This implies that a closed system is invariant under rotations in space.

Rotations about arbitrary axes form a Lie group (noncommutative). We have already studied rotations in detail in Chapters 4 and 5. If the Hamiltonian of the system is invariant under rotations, then

$$[D\,(R),\, H] = 0. \tag{80}$$

This implies conservation of angular momentum

$$[\vec{J},\, H] = 0. \tag{81}$$

Also, $\qquad\qquad [\vec{J}^2,\, H] = 0. \tag{82}$

Hence we can form simultaneous eigenkets of H, $\vec{J}^2$, and J_z. Since H commutes with $J_{\pm}$ operators for the eigenvalues of J_z, there is $(2j + 1)$ fold degeneracy of the energy levels of an atom, a nucleus or a particle.

9.8 GALILEI TRANSFORMATIONS AND GALILEI GROUP

Let us consider space-time transformations including rotations, displacements connecting uniformly moving frames of reference. In general, such transformations are Lorentz transformations, but for velocities small compared to the speed of light, these are Galilei transformations. Galilei transformations may be written as

$$t \rightarrow t' = t + \tau,$$

$$\vec{r} \rightarrow r' = R\vec{r} + \vec{v}t + \vec{a}, \tag{83}$$

where R is a rotation (a 3×3 orthogonal matrix), $\vec{a}$ is the spatial translation, τ the temporal translation and $\vec{v}$ is the (constant) relative velocity of the two inertial systems. For Galilei transformation, the time interval between any two events is invariant and the spatial distance between simultaneous events are invariant. Pure Galilei transformations are the transformations

$$t \rightarrow t' = t$$

$$\vec{r} \rightarrow \vec{r}' = \vec{r} + \vec{v}t \tag{84}$$

The set of all such transformations described in Eq. (83) form the (proper) Galilei group. Since the rotations R are described by 3 parameters, $\vec{v}$ by 3, $\vec{a}$ by 3 and τ by one, the Galilei group G is a 10 parameter connected noncompact Lie group. The Galilei group is a subgroup of the general linear group in 5 dimensions.

The Galilei group has as a subgroup the Euclidean group of spatial translations and rotation.

We denote the generic element of the Galilei group G by

$$g = (\tau, \vec{a}, \vec{v}, R) \tag{85}$$

The composition law is

$$(\tau', \vec{a}', \vec{v}', R') \cdot |(\tau, \vec{a}, \vec{v}, R)|$$

$$= (\tau' + \tau, \vec{a}' + R'\vec{a} + \tau\vec{v}', \vec{v}' + R'\vec{v}, R'R). \tag{86}$$

This can be represented by the multiplication of the 5×5 matrices

$$\begin{pmatrix} R & \vec{v} & \vec{a} \\ 0 & 1 & \tau \\ 0 & 0 & 1 \end{pmatrix}. \tag{87}$$

The identity for the group is

$$e = (0, \vec{0}, \vec{0}, 1) \tag{88}$$

and the inverse of the generic element is

$$(\tau, \vec{a}, \vec{v}, R)^{-1} = (-\tau, R^{-1}(\vec{a} - \vec{v}\tau), -R^{-1}\vec{v}, R^{-1}). \tag{89}$$

Among subgroups, we mention

$\mathcal{T} = \{(\tau, 0, 0, 1)\}$, the subgroup of time translations.

$\mathcal{S} = \{(0, \vec{a}, 0, 1)\}$, the subgroup of space translations.

$\mathcal{V} = \{(0, 0, \vec{v}, 1)\}$, the subgroup of pure Galilei transformations.

$\mathcal{R} = \{(0, 0, 0, R)\}$ the subgroup of rotations. $\tag{90}$

There is a 9 parameter invariant subgroup consisting of space translations, pure Galilei transformations, and rotations.

Corresponding to a Galilei transformation g there is a unitary operator $U(g)$ which acts on the state vectors and operators for observables thus :

$$|\psi\rangle \rightarrow |\psi'\rangle = U(g)\,\psi\rangle. \tag{91}$$

$$A \rightarrow A = U(g)\,AU^{-1}(g). \tag{92}$$

The action of g on a state $|\vec{r}, t\rangle$ is

$$|\vec{r}, t\rangle \xrightarrow{\;g\;} e^{i\omega_g/\hbar}\,|R\vec{r} + \vec{v}t + \vec{a}; t + \tau\rangle, \tag{93}$$

where ω_g is a phase

Corresponding to a general Galilei transformation, the unitary operator may be written as

$$U(g) = \Pi \, e^{is_\mu G_\mu}, \tag{94}$$

where G_μ ($\mu = 1, 2, \ldots, 10$) are the ten hermitian generators and s_μ denote ten parameters. For infinitesimal transformations ($s_\mu << 0$)

$$U = I + i \sum_{\mu=1}^{10} s_\mu G_\mu. \tag{95}$$

The ten generators are

H generating the time-translation subgroup $\mathcal{T}$;

P_i ($i = 1, 2, 3$) generating space-translation $\mathcal{S}_i$ along the axis i;

K_i ($i = 1, 2, 3$) generating pure Galilei transformations $\mathcal{V}_i$ along the axis i;

J_i ($i = 1, 2, 3$) generating rotations $\mathcal{R}_i$ about the axis i.

In the standard representation,

$$H = \frac{\partial}{\partial t}$$

$$\vec{P} = \vec{\nabla}_r \tag{96}$$

$$\vec{J} = \vec{r} \times \vec{\nabla}_r$$

$$\vec{K} = t\vec{\nabla}r$$

and the commutation relations defining the Lie algebra of the Galilei group are

(a) $[J_i, J_j] = i\in_{ijk} J_k,$

(b) $[J_i, K_j] = i\in_{ijk} K_k,$

(c) $[J_i, \mathrm{P}j] = i\in_{ijk} P_k$

(d) $[J_i, H] = 0,$

(e) $[K_i, K_j] = 0,$ (97)

(f) $[K_i, P_j] = 0,$

(g) $[K_i, H] = iP_i,$

(h) $[P_i, P_j] = 0,$

(i) $[P_i, H] = 0,$ $(i, j, k = 1, 2, 3)$

where $\in_{ijk}$ is the completely anti-symmetric tensor of order three. For details, consult Levy–Leblond (1971).

We have described $\vec{P}, \vec{J}, \vec{K}$ and H as the generators of Galilei transformations. These can be identified with dynamical variables. For example, $\vec{P}, \vec{J}$ and H denote the momentum, angular momentum and the energy of the system. In standard notation,

$$H = i\hbar \frac{\partial}{\partial t}$$

$$P = -i h \vec{\nabla}_r$$

$$\vec{J} = \vec{r} \times \vec{P} \tag{98}$$

so that the unitary operator for spatial displacement is $e^{-i\vec{a}\cdot\vec{P}/\hbar}$, for rotation $e^{-i\theta\hat{n}\cdot\vec{J}/\hbar}$ and for time translation $e^{-iHt/\hbar}$. The commutation relations are suitably modified.

The full Galilei group contains the proper Galilei transformations and the discrete transformations of space reflection and time reversal.

9.9 ISOSPIN

So far we have considered symmetry transformations of space-time. In this section we turn to one of the "internal symmetries".

The concept of isopin (or isobaric spin, isotopic spin, i-spin) is of great significance in nuclear physics and in the theory of elementary particles. Basically, the proton and the neutron are similar particles, although having different charges ($Q_p = 1$, $Q_n = 0$ in units of e). Their masses are nearly equal ($m_n = 939.573$, $m_p = 938.279$ in MeV; the small difference is attributed to electromagnetic interaction), they each have spin $\frac{1}{2}$, and they interact strongly with about the same strength.

For example, the ground states of the mirror pairs $^{7}_{3}Li$ and $^{7}_{4}Be$ have the same binding energies, the small difference $\sim 1.5\,MeV$ can be attributed to the extra Coulomb energy of $^{7}_{4}Be$. The excited states show similarity. $^{7}_{4}Be.$ has an extra pp bond and $^{7}_{3}Li$ an extra nn bond but with the same np bonds. Thus, the strong nn and pp interactions are equal (the hypothesis of charge symmetry of nuclear forces). A similar consideration of the nuclear energy levels of (B^{12}, C^{12}, N^{12}) or (C^{14}, N^{14}, O^{14}) implies that the nn, pp and np forces are equal in the same spatial-spin state (the hypothesis of charge independence of nuclear forces). Further, this is supported by the approximate equivalence of pp (after subtracting the Coulomb repulsion effect) nn, and np scattering lengths.

We thus treat the neutron and proton as two different charge states of a single particle, the nucleon. There exists a symmetry operator called "isospin" and we ascribe a quantum number I (isospin) with a value, $I = \frac{1}{2}$ and two states with $I_3 = \pm \frac{1}{2}$. We adopt the convention that $I_3 = +\frac{1}{2}$ for the proton and $I_3 = -\frac{1}{2}$ for the neutron. The charge is given by $Q/e = \frac{1}{2} + I_3$.

The wave function of a nucleon depends on its space and spin coordinates and also must indicate whether it is a proton or a neutron. We denote the states of a nucleon by $|p\rangle$ and $|n\rangle$ (suppressing the space-spin state). Thus,

$$I_3|p\rangle = \frac{1}{2}\Big|p\rangle, \tag{99}$$

$$I_3|n\rangle = -\frac{1}{2}\Big|n\rangle, \tag{100}$$

and the general nucleon state may be written as

$$|N\rangle = a|p\rangle + b|n\rangle \tag{101}$$

where a and b are arbitrary coefficients.

We write the matrix representation as

$$|p\rangle = \begin{pmatrix} 1 \\ 0 \end{pmatrix} \quad \text{and} \quad |n\rangle = \begin{pmatrix} 0 \\ 1 \end{pmatrix} \tag{102}$$

so that $|N\rangle = \begin{pmatrix} a \\ b \end{pmatrix}$. By analogy with the angular momentum operators for a spin $\frac{1}{2}$ particle, we introduce the isospin operators.

$$I_1 = \frac{1}{2}\begin{pmatrix} 0 & 1 \\ 1 & 0 \end{pmatrix}, \quad I_2 = \frac{1}{2}\begin{pmatrix} 0 & -i \\ i & 0 \end{pmatrix}, \quad I_3 = \frac{1}{2}\begin{pmatrix} 1 & 0 \\ 0 & -1 \end{pmatrix} \tag{103}$$

$$= \frac{1}{2}\tau_1, \qquad = \frac{1}{2}\tau_2, \qquad = \frac{1}{2}\tau_3,$$

where τ_1, τ_2, τ_3 are the Pauli spin matrices. Any operator acting on the nucleon wave function $|N\rangle$ can be represented by a linear combination of τ_1, τ_2, τ_3 and the unit 2×2 matrix.

We can visualize isospin as a vector $\vec{I}$ in a three-dimensional "isospin space" with the components I_1, I_2, I_3. Both p and n are eigenstates of $I^2 = I_1^2 + I_2^2 + I_3^2$ with the eigenvalue $I(I + 1) = \frac{1}{2}\left(\frac{1}{2} + 1\right) = \frac{3}{4}$. The eigenvalue of I_3 is $+\frac{1}{2}$ for p and $-\frac{1}{2}$ for n.

We introduce the operators

$$\tau_\pm = \frac{1}{\sqrt{2}}\,(\tau_1 \pm i\tau_2) \tag{104}$$

Then
$$\tau_+|p\rangle = 0, \quad \tau_+|n\rangle = \sqrt{2}|p\rangle \tag{105}$$

$$\tau_-|p\rangle = \sqrt{2}|n\rangle, \quad \tau_-|n\rangle = 0$$

I_i operators are analogous to the spin operators $S_i = \frac{1}{2}\,\sigma_i \cdot I_i$ satisfy the angular momentum commutation relations

$$[I_i, I_j] = i\,\epsilon_{ijk}\,I_k \quad (i, j, k = 1, 2, 3). \tag{106}$$

The electric charge operator Q of the nucleon (in units of e) is given by

$$Q = I_3 + \frac{1}{2}\cdot 1 = \frac{1}{2}\,(1 + \tau_3) = \begin{pmatrix} 1 & 0 \\ 0 & 0 \end{pmatrix}, \tag{107}$$

so that

$$Q|p\rangle = |p\rangle,\; Q|n\rangle = 0, \tag{108}$$

The nucleon wave function can now be written as a function of space-spin coordinates x and an isospin (charge) coordinate η. The dichotomic variable has only two possible values ± 1 with $+1$ for the proton state and -1 for the neutron state. We may write the wave function for a nucleon as

$$\overline{\Psi}\,(x, \eta) = \psi(x)|N(\eta)\rangle. \tag{109}$$

Let us consider a two-nucleon system. The isospin states of this system are denoted as $\chi\,(I, I_3)$:

$$\chi\,(1, 1) = |p_1\rangle|p_2\rangle \tag{110}$$

$$\chi(1, 0) = \frac{1}{\sqrt{2}}\,[|p_1 n_2\rangle + |n_1 p_2\rangle] \tag{110a}$$

$$\chi(1,-1) = |n_1\rangle|n_2\rangle \tag{110b}$$

$$\chi(0, 0) = \frac{1}{\sqrt{2}}\,[|p_1 n_2\rangle - |n_1 p_2\rangle]. \tag{110c}$$

The first three states belong to $I = 1$ (isotriplet) and are symmetric under $1 \to 2$, and the last one, $I = 0$, is an isosinglet, which is antisymmetric.

The two nucleon wave function $\overline{\Psi}$ must be antisymmetric under the exchange of space, spin and charge coordinates. For the diproton and dineutron systems, we see that the charge part is symmetric under the exchange $\eta_1 \leftrightarrow n_2$. So the space-spin part $(-1)^{s+1}$ of the wavefunction must be anti-symmetrical. The deuteron (bound proton-neutron system) has spin 1 so that the spin part is symmetric under the interchange of the two nucleons. In the ground state of the deuteron, the two nucleons are in $L = 0$ state (with a small admixture of $L = 2$ state). The space part $(-1)^L$ of the ground state of the deuteron is symmetric. Hence the isospin part χ must be anti-symmetric so that the total wave function is antisymmetric under the particle exchange. Thus the ground state of the deuteron is an isosinglet. Note that all states of the two nucleon system with $I = 1$ are unstable. For a system of A nucleons, the charge operator Q (in units of e) may be written as

$$I_3 + \frac{1}{2}A. \tag{111}$$

The three observed pions π^+, π^0 and π have very similar properties with nearly equal masses aside from their different charges. Consequently, the three pions are treated as three charge states of a single particle – the pion – which is assigned $I = 1$. The three pions are an isotriplet with the eigenvalues of I_3 as +1, 0, –1 with the charge given by $\frac{Q}{e} = I_3$:

$$I_3|\pi^+\rangle = |\pi^+\rangle,$$

$$I_3|\pi^0\rangle = 0.$$

$$I_3|\pi^-\rangle = -|\pi^-\rangle. \tag{112}$$

By introducing the baryon number B, we can write the general formula for Q (in units of e) as

$$Q = I_3 + \frac{B}{2}. \tag{113}$$

Notice that since the pion has zero spin, the total wave function of pions must be symmetric under the interchange of space, spin and isospin coordinates of any two particles.

For strange particles like kaons, Σ or Λ-hyperon, Eq. (113) is generalized to the following formula

$$Q = I_3 + \frac{1}{2}Y \tag{114}$$

where $Y = S + B$ is the hypercharge, S being the strangeness quantum number. Equation (114) is known as the Gell-Mann-Nishijima formula.

The concept of isospin is extremely useful because it is conserved in strong interactions:

$$[H_{st}, \vec{I}] = 0. \tag{115}$$

Consequently, an energy eigenstate of a system can be specified by I and I_3 and states with the same I but different I_3 are degenerate states. In a transition, I and I_3 remain unchanged. Electromagnetic interactions are manifestly charge dependent.

Let us consider

$$
\begin{array}{ccccc}
 & d\ + & d\ \rightarrow & {}^4He\ + & \pi^0 \\
I & 0 & 0 & 0 & 1 \\
I_3 & 0 & 0 & 0 & 0
\end{array}
\tag{116}
$$

This reaction is forbidden as a strong interaction, but can proceed as an electromagnetic interaction (isospin changing by one unit) with a small cross section. Experimentally the selection rule is at least 99.6% "good".

Next consider

$$\begin{array}{ccc} n + p & \to & \pi^0 + d \\ I \quad 0 \text{ or } 1 & & 1 \quad 0 \end{array} \tag{117}$$

$$\begin{array}{ccc} p + p & \to & \pi^+ + d. \\ I \quad 1 & & 1 \quad 0 \end{array} \tag{117a}$$

In both cases, the final state has isospin 1. If I is a good quantum number, then the initial state in both reactions must have isospin I.

From Eqs. (110), (110a), (110b) we see that the p-p system is in the $I = 1$ state all the time while the n-p system is in the $I = 1$ state only half the time. Then

$$\frac{\sigma \, (\pi^\circ + d)}{\sigma(\pi^+ + d)} = \frac{1}{2}. \tag{118}$$

Experimental evidence confirms the ratio $1:2$ to an accuracy of ~10%.

We now turn to pion-nucleon scattering:

$$\pi + N \to \pi' + N' \tag{119}$$

where π and π' denote one of the three different charge states of pion and N' denotes a neutron or a proton or an excited state of the nucleon.

Since $I_\pi = 1$, and $I_N = \frac{1}{2}$, a system of a pion and a nucleon can be either in $I_{tot} = \frac{1}{2}$ or $I_{tot} = \frac{3}{2}$.

If the isospin is conserved in strong interactions, then the 6 pion-nucleon scattering processes can be described in terms of the two isospin amplitudes.

Let us now construct states of a pion and a nucleon that are eigenstates of $\vec{I}^2$. We denote the possible states of a pion and a nucleon by $|I_3(\pi), I_3(N)\rangle$ and the eigenstate of total $\vec{I}^2$ and total I_3 by $|I, I_3\rangle$. Now,

$$|I, I_3\rangle = \sum_{I_3(\pi), I_3(N)} |I_3(\pi), I_3(N)\rangle \left\langle 1, \frac{1}{2}, I_3(\pi), I_3(N) \middle| 1, \frac{1}{2}, I, I_3 \right\rangle, \tag{120}$$

where $\left\langle 1, \frac{1}{2}, I_3(\pi), I_3(N) \middle| 1, \frac{1}{2}, I, I_3 \right\rangle$ is a CG coefficient. Since the isospin algebra is isomorphic to the angular momentum algebra we use the results from the sections on the angular momentum. The CG coefficients for the coupling of $I^{(1)} = 1, I^{(2)} = \frac{1}{2}$ and for completeness, the coupling of $I^{(1)} = \frac{1}{2}, I^{(2)} = \frac{1}{2}$ are listed in the Table below (see Chapter 7 and Table 7.1). Here the possible I values are $\frac{3}{2}$ and $\frac{1}{2}$.

$$\left\langle 1, \frac{1}{2}, I_3 \mp \frac{1}{2}, \pm \frac{1}{2} \,\middle|\, 1, \frac{1}{2}, I = \frac{3}{2}, I_3 \right\rangle = \sqrt{\frac{1}{2} \pm I_3/3}.$$

$$\left\langle 1, \frac{1}{2}, I_3 \mp \frac{1}{2}, \pm \frac{1}{2} \,\middle|\, 1, \frac{1}{2}, I = \frac{1}{2}, I_3 \right\rangle = \mp \sqrt{\frac{1}{2} \mp I_3/3}.$$

Table 9.4 Clebsch-Gordan coefficients

$\frac{1}{2} \otimes 1$	$I = \frac{1}{2}$		$I = \frac{3}{2}$			
	$I_3 = \frac{1}{2}$	$I_3 = -\frac{1}{2}$	$I_3 = \frac{3}{2}$	$I_3 = \frac{1}{2}$	$I_3 = -\frac{1}{2}$	$I_3 = \frac{-3}{2}$
$I_3^{(1)} = \frac{1}{2}$	$\sqrt{\frac{1}{3}}$	$\sqrt{\frac{2}{3}}$	1	$\sqrt{\frac{2}{3}}$	$\sqrt{\frac{1}{3}}$	0
$I_3^{(1)} = -\frac{1}{2}$	$-\sqrt{\frac{2}{3}}$	$-\sqrt{\frac{1}{3}}$	0	$\sqrt{\frac{1}{3}}$	$\sqrt{\frac{2}{3}}$	1

$\frac{1}{2} \otimes \frac{1}{2}$	$I = 0$	$I = 1$		
	$I_3 = 0$	$I_3 = 0$	$I_3 = 1$	$I_3 = -1$
$I_3^{(1)} = \frac{1}{2}$	$\sqrt{\frac{1}{2}}$	1	$\sqrt{\frac{1}{2}}$	0
$I_3^{(1)} = -\frac{1}{2}$	$-\sqrt{\frac{1}{2}}$	0	$\sqrt{\frac{1}{2}}$	1

Therefore, the $|I, I_3\rangle$ states are given as follows:

$$\left| \frac{3}{2}, \frac{3}{2} \right\rangle = \left| \pi^+ p \right\rangle$$

$$\left| \frac{3}{2}, \frac{1}{2} \right\rangle = \sqrt{\frac{1}{3}} \left| \pi^+ n \right\rangle + \sqrt{\frac{2}{3}} \left| \pi^0 p \right\rangle$$

$$\left| \frac{3}{2}, -\frac{1}{2} \right\rangle = \sqrt{\frac{1}{3}} \left| \pi^- p \right\rangle + \sqrt{\frac{2}{3}} \left| \pi^0 n \right\rangle$$

$$\left| \frac{3}{2}, -\frac{3}{2} \right\rangle = \left| \pi^- n \right\rangle$$

$$\left|\frac{1}{2}, \frac{1}{2}\right\rangle = \sqrt{\frac{2}{3}}\left|\pi^+ n\right\rangle - \sqrt{\frac{1}{3}}\left|\pi^0 p\right\rangle$$

$$\left|\frac{1}{2}, -\frac{1}{2}\right\rangle = \sqrt{\frac{1}{3}}\left|\pi^0 n\right\rangle - \sqrt{\frac{2}{3}}\left|\pi^- p\right\rangle. \tag{121}$$

From Eq. (121), we can easily express the initial pion nuclear states in terms of eigenstates of I^2 as

$$\left|\pi^+ p\right\rangle = \left|\frac{3}{2}, \frac{3}{2}\right\rangle$$

$$\left|\pi^0 p\right\rangle = \sqrt{\frac{2}{3}}\left|\frac{3}{2}, \frac{1}{2}\right\rangle - \sqrt{\frac{1}{3}}\left|\frac{1}{2}, \frac{1}{2}\right\rangle$$

$$\left|\pi^- p\right\rangle = \sqrt{\frac{1}{3}}\left|\frac{3}{2}, -\frac{1}{2}\right\rangle - \sqrt{\frac{2}{3}}\left|\frac{1}{2}, -\frac{1}{2}\right\rangle$$

$$\left|\pi^+ n\right\rangle = \sqrt{\frac{1}{3}}\left|\frac{3}{2}, \frac{1}{2}\right\rangle + \sqrt{\frac{2}{3}}\left|\frac{1}{2}, \frac{1}{2}\right\rangle$$

$$\left|\pi^0 n\right\rangle = \sqrt{\frac{2}{3}}\left|\frac{3}{2}, -\frac{1}{2}\right\rangle + \sqrt{\frac{1}{3}}\left|\frac{1}{2}, -\frac{1}{2}\right\rangle$$

$$\left|\pi^- n\right\rangle = \left|\frac{3}{2}, -\frac{3}{2}\right\rangle. \tag{122}$$

Of the ten possible reactions described by Eq. (119) six are elastic scattering processes and the rest are charge exchange processes. The following two elastic scattering processes

$$\pi^+ p \rightarrow \pi^+ + p$$

and $$\pi^- + n \rightarrow \pi^- + n \tag{123}$$

have $I_3 = \pm\frac{3}{2}$ and so these are expressed by a pure $I = \frac{3}{2}$ amplitude. Clearly the two processes will have identical cross sections at a given energy.

Let us next consider the three easily observed processes

$$\pi^+ + p \rightarrow \pi^+ + p \text{ (elastic stattering)} \tag{124}$$

$$\pi^- + p \rightarrow \pi^- + p \text{ (elastic scattering)} \tag{124a}$$

and $$\pi^- + p \rightarrow \pi^0 + n \text{ (charge exchange)} \tag{124b}$$

The differential cross section for any scattering process is proportial to the square of the transition matrix element connecting the initial and final states. All possible processes are described by the two independent matrix elements

$$M_1 = \left\langle \psi_f \left(\tfrac{1}{2}, I_3\right) \middle| \mathcal{J} \middle| \psi_i \left(\tfrac{1}{2}, I_3\right) \right\rangle, \tag{125}$$

$$M_3 = \left\langle \psi_f \left(\tfrac{3}{2}, I_3\right) \middle| \mathcal{J} \middle| \psi_i \left(\tfrac{3}{2}, I_3\right) \right\rangle. \tag{126}$$

Thus there are three real parameters for any reaction, since a common real phase cancels in taking the $|...|^2$.

The $\pi^+ p$ elastic scattering process Eq. (124) can go only through the channel $I = \dfrac{3}{2}$. Therefore,

$$\left(\frac{d\sigma}{d\Omega}\right)_{\pi^+ p} = K |M_3|^2, \tag{127}$$

where K is a constant.

For the $\pi^- p$ elastic scattering process Eqs. (124a) we may write from (122),

$$\psi_i = \psi_f = \sqrt{\tfrac{1}{3}} \left| \tfrac{3}{2}, -\tfrac{1}{2} \right\rangle - \sqrt{\tfrac{2}{3}} \left| \tfrac{1}{2}, -\tfrac{1}{2} \right\rangle, \tag{128}$$

so that

$$\left(\frac{d\sigma}{d\Omega}\right)_{\pi^- p \, \text{elastic}} = K \left| \tfrac{1}{3} M_3 + \tfrac{2}{3} M_1 \right|^2. \tag{129}$$

For the charge exchange reaction Eq. (124b),

$$\psi_f = \sqrt{\tfrac{2}{3}} \left| \tfrac{3}{2}, -\tfrac{1}{2} \right\rangle + \sqrt{\tfrac{1}{3}} \left| \tfrac{1}{2}, -\tfrac{1}{2} \right\rangle \tag{130}$$

and ψ_i as given in Eq. (128), so that

$$\left(\frac{d\sigma}{d\Omega}\right)_{ch.\,ex} = K \left| \frac{\sqrt{2}}{3} M_3 - \sqrt{\tfrac{2}{3}} M_1 \right|^2. \tag{131}$$

Therefore, the differential cross section ratios are

$$\left(\frac{d\sigma}{d\Omega}\right)_{\pi^+ p} : \left(\frac{d\sigma}{d\Omega}\right)_{\pi^- p \, \text{elastic}} : \left(\frac{ds}{d\Omega}\right)_{ch.\,ex}$$

$$= |M_3|^2 : \tfrac{1}{9} |M_3 + 2M_1|^2 : \tfrac{2}{9} |M_3 - M_1|^2 \tag{132}$$

If, under experimental conditions,

$\quad M_3 \gg M_1 \quad$ the ratios are $\qquad 9:1:2$

$\quad M_3 \ll M_1 \quad$ the ratios are $\qquad 0:2:1$

$$M_3 \approx M_1 \quad \text{the ratios are} \qquad 1:1:0$$

$$M_3 \approx -M_1 \quad \text{the ratios are} \qquad 9:1:8$$

Several experiments have been made of the differential and total cross sections in pion-nucleon scattering. At a pion kinetic energy (Lab). ~200 MeV there is a strong peak for both π^+ and π^-. The ratio $\sigma_{\pi^+ p}/\sigma_{\pi^- p}$ is close to 3, indicating that the $I = \dfrac{3}{2}$ amplitude dominates in this region. This bump is known as a resonance (a short lived intermediate state N^*)

$$\pi + N \rightarrow N^* \rightarrow \pi + N \tag{133}$$

This resonance is denoted as Δ (1236), 1236 MeV being the invariant pion-nucleon mass, the width at half height is 120 MeV, spin-parity assignment is $J^p = \dfrac{3^+}{2}$. This resonance is called the 3-3 resonance. At higher inergies, more resonances are observed.

The isospin symmetry group is of great importance in nuclear physics and elementary particle physics. Although the isospin concept has been introduced by analogy with the formalism of angular momentum, we prefer to use the continuous group SU(2) for the description of isospin. We take the isospin symmetry group as SU(2). Therefore, the strong (hadronic) interactions are invariant under the isospin group SU(2). This implies isospin conservation in strong interactions. The hadronic mass multiplets e.g., the doublet $\begin{pmatrix} p \\ n \end{pmatrix}$, belong to the IR's of the isospin symmetry group.

The group SU(2) is the group of unitary, unimodular transformations in two dimensions. This non-abelian continuous group is the group of two-dimensional unitary matrices with unit determinant i.e.,

$$UU^\dagger = U^\dagger U = I, \quad U^\dagger = U^{-1}$$

$$\text{Det } (U) = 1. \tag{134}$$

This group is locally equivalent to SO(3), so it has the same Lie algebra as the latter. In fact, the SU(2) matrices form a double-valued representation of SO(3). We may write

$$U = \begin{pmatrix} a & -b^* \\ b & a^* \end{pmatrix} \tag{135}$$

with $aa^* + bb^* = 1$.

An arbitrary SU(2) matrix A can be parametrized in terms of three real parameters θ, η, ζ as

$$A = \begin{pmatrix} \cos\theta\, e^{i\zeta} & -\sin\theta\, e^{i\eta} \\ -\sin\theta\, e^{-i\eta} & \cos\theta\, e^{-i\zeta} \end{pmatrix}, \tag{136}$$

where $0 \le \theta \le \pi$, $0 \le \eta, \zeta \le 2\pi$. For details, see § 5.7.

Corresponding to an infinitesimal transformation, the unitary operator U is given by

$$U = 1 - i\vec{\in}\cdot\vec{I} = 1 - i\sum_{j=1}^{3} \in_j I_j, \tag{137}$$

where $\in_1$, $\in_2$ and $\in_2$ are infinitesimal real parameters and I_1, I_2 and I_3 are the generators. These are hermitian and traceless, i.e.,

$$I^{\dagger}_j = I_j$$

$$TrI_j = 0 \qquad (j = 1,2,3) \tag{138}$$

The generators satisfy the Lie algebra

$$[I_i, I_j] = i\in ijk\, I_k \tag{139}$$

SU(2) is compact and simply connected group of rank 1. The only Casimir operator is $\vec{I}^2$ with eigenvalues $I(I + 1)$. The states within an IR can be specified uniquely by I_3 which is conserved.

9.10 SUPERSELECTION RULES

There are often limitations on the validity of the superposition principle. It may not always be possible to realize a pure state out of the superposition of certain states. For example, one cannot form a pure state by the linear superposition of a pure state with positively charged particle and another with negatively charged particle, or a pure state consisting of fermions and a boson. This implies that such a linearly superposed state is not a physically realizable pure state (superselection rule). The existence of superelection rules depends on the measurability of the relative phase of the superposition. In all such cases we can decompose the Hilbert space $\mathcal{H}$ of the system into a direct sum of subspaces in each of which the superposition principle holds. These subspaces are called coherent subspaces. We know that the common phase of state vectors is non-measurable. Now if there are several states vectors belonging to different subspaces such that there are no matrix elements between any two vectors, then not only the common phase but also the phase of each vector is non-measurable. Then we say that a superselection rule operates to forbid us to compare the phases of the different state vectors. As examples of superselection rules for gauge groups, we mention the conservation laws of electric charge, baryon number, lepton number.

Next, we consider rotational invariance and superselection rule. The rotation operator for the rotation 2π along the direction $\hat{n}$ is $R_n(2\pi) = e^{-2\pi i \hat{n}.\vec{J}/\hbar}$ so that

$$R_n(2\pi)| J,M\rangle = (-1)^{2J}|J,M\rangle. \tag{140}$$

Thus, we get an extra relative phase of $(-1)^{2J}$ in the linear combination of two states. Under the action of $R(2\pi)$ any observable O is invariant, i.e.

$$[R(2\pi), O] = 0. \tag{141}$$

This gives rise to a superselection rule which states that if rotational invariance holds, then states corresponding to a linear superposition of states with integer J and those with half-odd integer J-values are not physically realizable as pure states. The vector space is decomposed into the sum of two subspaces – one subspace with integer J values and another with half – odd integer J-values – by the operator $R(2\pi)$ The matrix element of any physical observable vanishes between states with integer J and states with half-odd integer J-values states. There is a superselection rule for states with different permutation symmetry. For the superselection rule and intrinsic parity, see Section 10.2.

9.11 DYNAMICAL SYMMETRIES

The relationship between symmetry and degeneracy has been discussed earlier. In certain quantum systems there are more degeneracy present than required by the geometric symmetry group (e.g., rotational symmetry). In such cases we say that there is an accidental degeneracy present, in addition to the degeneracy due to the symmetry group. The accidental degeneracy occurs due to the existence of a larger symmetry group than the one considered. The accidental degeneracy is thus associated with a "hidden symmetry" not of geometrical origin. Such symmetries are called "dynamical symmetries", since the symmetry is established only for the inteaction and the groups are called "dynamical groups". We shall examine two examples, the hydrogen atom and the three-dimentional isotropic harmonic oscillator.

The accidental degeneracy of the hydrogen atom (see Section 11.5) is closely connected with the existence of a second vector constant of motion in the corresponding classical problem. So we first study the classical Kepler problem where the Hamiltonian is in relative coordinates

$$H = \frac{\vec{p}^2}{2\mu} - \frac{K}{r} \tag{142}$$

where μ is the reduced mass and K is a constant. The bound orbits are ellipses; these do not precess. Due to the rotational symmetry of H, the orbital angular momentum $\vec{L} = \vec{r} \times \vec{p}$ is a constant of motion and is perpendicular to the plane of the orbit. There is another constant vector $\vec{A}$ which points from focus F along the semi major axis of the orbit to P or to A (see Figure 9.1). This vector $\vec{A}$ is called the Runge-Lenz vector.

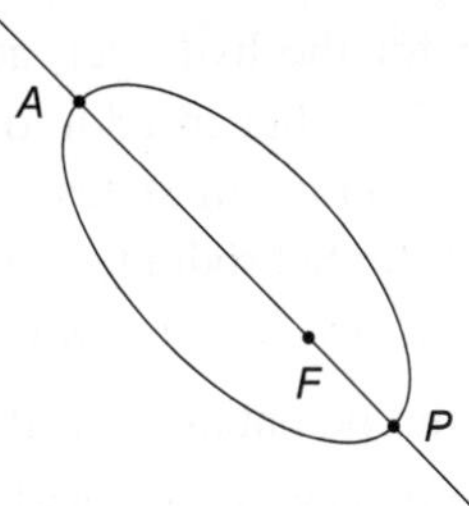

Fig. 9.1   Classical Kepler orbit with F focus, P perihelion and A aphelion.

Explicitly,
$$\vec{A} = \frac{1}{\mu}\vec{p} \times \vec{L} - \frac{K\vec{r}}{r}. \tag{143}$$

We note that

$$\vec{A} \cdot \vec{L} = \frac{1}{\mu}\,\vec{p} \times \vec{L} \cdot \vec{L} - \frac{K}{r}\,\vec{r} \cdot \vec{L} = 0, \tag{144}$$

because $\vec{r}$ is perpendicular to $\vec{L}$ and the triple product vanishes.

Also,

$$\frac{d}{dt}\vec{A} = \frac{1}{\mu}\,\dot{\vec{p}} \times \vec{L} - \frac{K}{r}\,\dot{\vec{r}} + \frac{K\vec{r}(\vec{r}\cdot\vec{p})}{\mu r^3}, \text{ since } \dot{\vec{L}} = 0, \text{ and } \vec{p} = \mu\,\dot{\vec{r}}.$$

Using

$$\vec{L} = \vec{r} \times \vec{p} \text{ and } \dot{\vec{p}} = -\frac{K\vec{r}}{r^3}, \text{ we see that } \dot{\vec{A}} = 0. \tag{145}$$

Therefore, $\vec{A}$ is a vector constant of motion.

For the quantum mechanical problem, we write the Hamiltonian of the hydrogen atom as

$$H = \frac{\vec{p}^2}{2\mu} - \frac{K}{r}, \text{ where } K = Ze^2. \tag{146}$$

H is invariant under $SO(3) \approx SU(2)/Z_2$, the rotation group in three-dimensional Euclidean space.

This orbital angular momentum $\vec{L}$ commutes with H and is a constant of motion and this symmetry leads to the $(2l + 1)$ fold degeneracy.

We define the Runge-Lenz hermitian operator

$$\vec{A}' = \frac{1}{2\mu}\,(\vec{p} \times \vec{L} - \vec{L} \times \vec{p}) - K\frac{\vec{r}}{r}. \tag{147}$$

The Hamiltonian commutes with the Runge-Lenz operator:

$$[\vec{A}', H] = 0, \tag{148}$$

and therefore $\vec{A}'$ is a constant of the motion. The existence of another constant of the motion $\vec{A}'$ explains the n^2 degeneracy of the hydrogen atom.

The Schrödinger equation for the hydrogen atom is separable in spherical polar coordinates as a central field, but in addition it is separable in parabolic coordinates. The accidental degeneracy of states with the same n but different l is related to the separability of the Schrödinger equation in different coordinate systems. The accidental degeneracy in this case is due to the nature of the $\frac{1}{r}$ potential. Any deviation of the potential from the $\frac{1}{r}$ nature causes classically the major axis of the eilipse to precess slowly and hence the orbit is not closed. This deviation removes the degeneracy.

As classically,

$$\vec{L} \cdot \vec{A}' = \vec{A}' \cdot \vec{L} = 0. \tag{149}$$

Also,

$$\vec{A}'^2 = K^2 + \frac{2H(\vec{L}^2 + \hbar^2)}{\mu}. \tag{150}$$

The 15 commutation relations for the 6 generators $\vec{L}'$ and $\vec{A}'$ are

$$[L_i, L_j] = i\hbar \in_{ijk} L_k, \tag{151}$$

$$[A_i', L_j] = i\hbar \in_{ijk} A_k', \tag{152}$$

$$[A_i', A_j'] = i\hbar \frac{-2H}{\mu} \in_{ijk} L_k. \tag{153}$$

We absorb the factor $-2H/\mu$ into $\vec{A}'$ and define a new vector

$$\vec{A} = \sqrt{\frac{-\mu}{2H}}\, \vec{A}'. \tag{154}$$

Here, we restrict ourselves to a Hilbert subspace corresponding to bound states of the hydrogen atom. $\vec{A}$ is hermitian acting on the eigenstates of H with negative eigenvalues. Combining Eqs. (152), (153) and (154), we get

$$[A_i, A_j] = i\hbar \in_{ijk} L_k, \tag{155}$$

$$[A_i, L_j] = i\hbar \in_{ijk} A_k. \tag{156}$$

We write

$$J_{ij} = \in_{ijk} L_k \qquad \text{for } i, j, k = 1, 2, 3 \tag{157}$$

$$J_{jl} = -J_{lj} = A_j \qquad \text{for } l = 0$$

Then the operators $J_{\rho\sigma}$ are the six infinitesimal generators of the group SO(4), the proper rotation group in 4 dimensions. The four-dimensional orthogonal group SO(4) is the group of all transformations in the four-dimensional Euclidean space that leaves invariant the quadratic form $x_0^2 + x_1^2 + x_2^2 + x_3^2$. Its subgroup SO(4) contains only transformation with determinant +1. The group SO(4) describes a dynamical symmetry of the hydrogen atom. From Eqs. (150) and (154), we get

$$H = -\frac{\mu K^2}{2(\vec{A}^2 + \vec{L}^2 + \hbar^2)}. \tag{158}$$

We define the linear combinations

$$\vec{M} = \frac{1}{2}(\vec{L} + \vec{A}), \tag{159}$$

$$\vec{N} = \frac{1}{2}(\vec{L} - \vec{A}). \tag{160}$$

which satisfy the commutation relations

$$[M_i, M_j] = i\hbar \in_{ijk} M_k, \tag{161}$$

$$[N_i, N_j] = i\hbar \in_{ijk} N_k, \tag{162}$$

$$[M_i, N_j] = 0. \tag{163}$$

Therefore, $\vec{M}$ and $\vec{N}$ behave like angular momenta which commute with each other. We see that each $\vec{M}$ and $\vec{N}$ constitutes a SO(3) or SU(2) algebra and $m(m + 1)\hbar^2$ and $n(n + 1)\hbar^2$ are eigenvalues of $\vec{M}^2$ and $\vec{N}^2$ respectively, with $m, n = 0, \frac{1}{2}, 1 \dots$ Since the two algebras commute, SO(4) is locally isomorphic to $SU(2) \otimes SU(2)$. Then there are two Casimir operators $\vec{M}^2$ and $\vec{N}^2$ or any combination of these, e.g.,

$$\vec{M}^2 + \vec{N}^2 = \frac{1}{2}(\vec{L}^2 + \vec{A}^2), \tag{164}$$

$$\vec{M}^2 - \vec{N}^2 = \frac{1}{2}(\vec{L} \cdot \vec{A} + \vec{A} \cdot \vec{L}) = \vec{L} \cdot \vec{A}. \tag{165}$$

In terms of $\vec{M}^2$ and $\vec{N}^2$, Eq. (158) may be written as

$$H = -\frac{\mu K^2}{2(2\vec{M}^2 + 2\vec{N}^2 + \hbar^2)}, \tag{166}$$

Since $\vec{L} \cdot \vec{A} = 0$, $\vec{M}^2 = \vec{N}^2$, so that $m = n$ and the energy eigenvalue of the bound states of the hydrogen atom is

$$E = -\frac{\mu K^2}{2\hbar^2(4m(m + 1) + 1)}$$

$$= - \frac{\mu K^2}{2\hbar^2 \, (2m + 1)^2}$$

$$= - \frac{\mu K^2}{2\hbar^2 n^2} \tag{167}$$

where $n = 2m + 1$ can take the values 1, 2, 3, ... and is called the principal quantum number. For a fixed value of $m = n$, M_z and N_z can have $(2m + 1)^2$ independent eigenvalues, so that there are $(2m + 1)^2 = n^2$ different states with the same energy E_n. Thus, the extra symmetry due to the second constant of motion leads to the explanation of the accidental degeneracy of the hydrogen atom.

It is to be noted that SO(4) symmetry is obtained for the bound states ($E < 0$) of the hydrogen atom. For continuum states ($E > 0$) the dynamical group is the restricted homogeneous Lorentz group SO(3,1).

It is possible to find a larger group, containing SO(4) as a subgroup, which has, an IR that is the direct sum of the IR's belonging to the eigenvalues of H. This group, the noncompact SO(4,1) is not a symmetry group. Such a group is called a spectrum generating group.

Next we discuss the degeneracy of the energy levels of the harmonic oscillator which is due to the symmetry of the Hamiltonian under a dynamical group. In fact for the n-dimensional harmonic oscillator the symmetry group turns out to be SU(n). We consider here the case of the three-dimensional harmonic oscillator (see § 11.2 (c)).

The harmonic oscillator Hamiltonian is

$$H_0 = \sum_{i=1}^{3} \left(\frac{1}{2} \mu^{-1} p_i^2 + \frac{1}{2} k r_i^2 \right)$$

$$= \frac{1}{2} \mu^{-1} \sum_{i=1}^{3} (p_1^2 + \mu^2 \omega^2 r_i^2) \tag{168}$$

where
$$\omega = (k\mu^{-1})^{1/2}.$$

It is useful to introduce the lowering and raising operators

$$a_j = \frac{1}{(2\mu\hbar\omega)^{1/2}} (p_j - i\omega\mu r_j), \tag{169}$$

$$(j = 1,2,3)$$

$$a_j^\dagger = \frac{1}{(2\mu\hbar\omega)^{1/2}} (p_j + i\omega\mu r_j). \tag{170}$$

These satisfy the boson commutation relations

$$[a_i , a_j], = \left[a_i^\dagger, a_j^\dagger \right] = 0, \tag{171}$$

$$\left[a_i, a_j^\dagger \right] = \delta_{ij}. \tag{172}$$

In terms of these operators, the Hamiltonian is

$$H_0 = \hbar\omega \sum_i \left(a_i^\dagger a_i + \frac{3}{2} \right)$$

$$= \frac{1}{2} \hbar\omega \sum_i \{a_i^\dagger, a_i\}. \tag{173}$$

Also,

$$[H_0, a_i] = - \hbar\omega a_i, \tag{174}$$

$$[H_0, a_i^\dagger] = \hbar\omega a_i^\dagger \tag{175}$$

From Eq. (173), we obtain the eigenvalues of H

$$E_n = \hbar\omega \left(n + \frac{3}{2} \right), \; n = 0,1,2, \ldots. \tag{176}$$

using the properties of the occupation number operator $\sum_i a_i^\dagger a_i$.

We define the nine shift operators

$$A_{ij} = \frac{1}{2} \{a_i, a_j^\dagger\}. \tag{177}$$

These operators just transfer an oscillator quantum $\hbar\omega$ from the j-direction to the i-direction. These satisfy the commutation relations

$$[A_{ij}, A_{kl}] = \delta_{jk} A_{il} - \delta_{il} A_{kj}. \tag{178}$$

In terms of these operators, the Hamiltonian is

$$H_0 = \frac{1}{2} \hbar\omega \sum_i A_{ii} \tag{179}$$

and using Eq. (178), we obtain

$$[H_0, A_{ij}] = 0. \tag{180}$$

The algebra can be decomposed in SU(3) $\times$ U(1), where U(1) is generated by $\sum_i A_{ii}$.

To see this explicitly, we define the operators a_μ:

$$a_{+1} = - \frac{1}{\sqrt{2}} (a_1 - ia_2)$$

$$a_0 = a_3 \tag{181}$$

$$a_{-1} = \frac{1}{\sqrt{2}} (a_1 + ia_2)$$

and also the operators $\bar{a}_\mu$:

$$\bar{a}_{+1} = - \frac{1}{\sqrt{2}} (a_1^\dagger - ia_2^\dagger)$$

$$\bar{a}_0 = a_0^\dagger$$

$$\bar{a}_{-1} = \frac{1}{\sqrt{2}} \left(a_1^\dagger + i a_2^\dagger \right).$$ (182)

Each form an irreducible tensor belonging to the three-dimensional representation of SU(2).

The commutation relations are

$$[a_\mu, \bar{a}_{\mu'}] = (-1)^\mu \, \delta_{\mu,-\mu'}$$ (183)

and all other commutators vanish. If

$$b_{\mu\mu'} = \bar{a}_{-\mu^a - \mu'}$$ (184)

with commutation relations

$$[b_{\mu_1\mu_2}, b_{\mu_3\mu_4}] = (-)^{\mu_2} \, \delta_{\mu_3 - \mu_2} \, b_{\mu_1\mu_4}$$
$$- (-)^{\mu_1} \, \delta_{\mu_4 - \mu_1} \, b_{\mu_3\mu_2}.$$ (185)

Putting

$$I_+ = b_{1,1} \qquad\qquad I_- = b_{-1,-1}$$
$$U_+ = -b_{-1,0} \qquad\qquad U_- = b_{0,1}$$
$$V_+ = b_{1,0} \qquad\qquad V_- = -b_{0,-1}$$

$$I_3 = \frac{1}{2} \left(b_{-1,1} - b_{1,-1} \right) = \frac{1}{2} L_3$$

$$Y = -\frac{1}{3} \left(2b_{0,0} + b_{1,-1} + b_{-1,1} \right)$$

$$= -\frac{1}{3} \left(3b_{0,0} - \frac{H}{\hbar\omega} + \frac{3}{2} \right)$$

we get the algebra of SU(3). $I_\pm$, I_0 are called the quasispin operators.

Thus, the three-dimensional harmonic oscillator has as a symmetry group SU(3). The only possible IRs of SU(3) are the $(n, 0)$. The two Casimir operators of SU(3) are related so that the energy levels are labeled by one index. For the $(n, 0)$ representation, the eigenvalue of C_1 is

$$C_1 = \frac{n}{3} \left(n + 3 \right),$$ (186)

and since

$$\left(\frac{H_0}{\hbar\omega} \right)^2 = \frac{3}{2} \left(2C_1 + \frac{3}{2} \right),$$ (187)

we get for the $(n, 0)$ representation, the eigenvalues of H_0 are

$$E_n = \hbar\omega \left(n + \frac{3}{2} \right), \quad (n, 0) \text{ representation}$$ (188)

n being the number of quantum. The dimensionality of the $(n, \ 0)$ representation $\frac{1}{2}(n + 1).(n + 2)$ is the degeneracy of the level E_n. The parity of the states is positive (n even) or negative (n odd). The possible values of the angular momentum L are 0, 2, 4, ... N (n even) or 1, 2, 3, ..., n (odd n).

A different classification of the states is more convenient. The eight generators are now the three components of $\vec{L}$ and the five components of the quadrupole moment tensor. We define the irreducible tensor belonging to the $D^{(j=1)}$ representation of SU(2):

$$B_\lambda^m = -\sqrt{2}\,(-)^m \sum_\mu \ C(11\lambda;\,\mu,\,m-\mu)\,b_{\mu,m-\mu}, \tag{189}$$

where C is the CG coefficient for SU(2). For $\lambda = 0$

$$B_0^0 = \sqrt{\frac{2}{3}} \sum_\mu (-)^\mu\, b_{\mu,-\mu} = \sqrt{\frac{2}{3}}\left(\frac{H}{\hbar\omega} - \frac{3}{2}\right) \tag{190}$$

and for $\lambda = 1, 2$ we get the set of eight operators

$$B_1^0 = L_0 = (b_{-1,1} - b_{1,-1})$$

$$B_1^{\pm 1} = L_{\pm 1} = \pm\,(b_{\pm 1,0} - b_{0,+1})$$

$$B_2^0 = \frac{1}{\sqrt{3}}\,(2b_{0,0} + b_{1,-1} + b_{-1,1}) = \sqrt{3}\,Y,$$

$$B_2^{\pm 1} = (b_{\pm 1,0} + b_{0,\pm 1})$$

$$B_2^{\pm 2} = -\sqrt{2}\,b_{\pm 1,\pm 1}. \tag{191}$$

The tensor of rank one $B_1^m = L_m$ ($m = 0, \pm 1$) is the angular momentum and the tensor of rank two $B_2^m = Q_m$ ($m = 0, \pm 1, \pm 2$) is the quadrupole tensor.

It can be shown that the Casimir operator C_1 is related to $\vec{L}$ and Q by

$$C_1 = \frac{1}{4}\,(\vec{L}^2 + Q^2), \tag{192}$$

where
$$\vec{L}^2 = \sum_m (-)^m\, B_1^m\, B_1^{-m} \quad \text{and} \quad Q^2 = \sum_m (-)^m\, B_2^m\, B_2^{-m}.$$

Since

$$Q^2 = \frac{4}{3}\left(\frac{H}{\hbar\omega}\right)^2 - \vec{L}^2 - 3, \tag{193}$$

we obtain

$$C_1 = \frac{1}{3}\left(\frac{H_0}{\hbar\omega}\right)^2 - \frac{3}{4}. \tag{194}$$

In this case, there are two spectrum-generating groups – the noncompact Lie groups Sp (3, R) and SU (3, 1). For details, consult Fonda and Ghirardi (1970).

There is a higher order symmetry group which accounts for the accidental or nongeometric degeneracy. In quantum mechanics, the oscillator and the Coulomb problems are closely related and both systems exhibit accidental degeneracy and have O(4) for Kepler and SU(3) for three-dimensional oscillator potentials as the symmetry group respectively. The uniqueness of the oscillator and Coulomb potentials in exhibiting accidental degeneracy is in correspondence with the fact that in classical mechanics these are the only spherically symmetric potentials that yield bounded closed orbits. This follows from Bertrand's theorem (1873) which states that in problems where the kinetic energy is the sum of the squares of velocities and the potential is spherically symmetric the only potentials with bounded closed orbits are those for $V(r) \propto - 1/r$ and $V(r) \propto r^2$. We remark that the separability in several coordinates systems of the Hamilton-Jacobi equation is related to the existence of non space-filling orbits. This is in correspondence with the separability of the Schrödinger equation.

PROBLEMS

9.1 Consider the scale transformation $\vec{x} \to \vec{x}' = e^{\alpha}\,\vec{x}$, where α is a parameter. The unitary operator is $e^{-i\alpha D}$, where D is the dilation generator. Find $[D, \vec{P}]$.

9.2 Let α be a vector in L^2. $F\psi$ is the Fourier transform of ψ:

$$(F\psi)\,(\vec{k}) = (2\pi)^{-3/2} \int e^{-i\vec{k}\cdot\vec{x}}\ \psi(\vec{x})\,d\vec{x},$$

and $\psi(\vec{x}) = (2\pi)^{-3/2} \int (F\psi)\,(\vec{k})\ e^{\vec{k}\cdot\vec{x}}\,d\vec{x} \cdot F$ is a unitary operator on L^2.

   (a) Show that $F\vec{Q} = -\,\vec{P}F$ and $F\vec{L} = \vec{L}F$, where $\vec{L} = \vec{Q} \times \vec{P}$

   (b) If ψ is an eigenvector of $\vec{L}^2$ and L_3, then show that $F\psi$ is an eigenvector of $\vec{L}^2$ and L_3 with the same eigenvalues.

9.3 Consider the Galilei transformation $U\vec{V}U^{-1} = \vec{V} - \vec{v}I$, $U\vec{Q}U^{-1} = Q - \vec{v}t\,I$. $U\,(\vec{v}, t) = e^{i\vec{v}\cdot\vec{G}_t}$ is the corresponding unitary operator.

   (a) Find $\vec{G}_t$.

   (b) Show that $\vec{G}_t$ satisfies the Galilei-group commutation relations with $\vec{P}, \vec{J}$, and H.

9.4 This problem concerns the Runge-Lenz vector.

   (a) Show that the Runge-Lenz vector $\vec{A}$ commutes with the Hamiltonian of the hydrogen atom.

(b) Prove Eqs. (155) and (156) for the commutators of the components of $\vec{A}$ and $\vec{L}$.

(c) Show that $\vec{M}$ and $\vec{N}$ behave like angular momenta which commute with each other, i.e., prove the commutation relations Eqs. (161), (162) and (163).

CHAPTER 10

Discrete Symmetries

In the preceding chapter we considered continuous space-time symmetry transformations (except isospin) which are generated as a continuous unfolding of infinitesimal transformations. Such symmetry transformations are represented by linear operators. In this chapter we consider other kinds of symmetry transformations that cannot be generated from infinitesimal transformations. These are called discrete symmetry transformation and are represented by either a linear or by an antilinear operator as proved by Wigners theorem.

The three discrete symmetry transformations-space inversion, time reversal and charge (particle-antiparticle) conjugation are related to the non-measurability of right-left, absolute sign of t, and particle-antiparticle. Recently, some of these non-observables have become actually measurable leading to the violation of the symmetry principles in certain interactions. In this chapter, we discuss only two (space inversion and time reversal) of these three symmetry operations.

10.1 LATTICE TRANSLATION

In this section we consider lattice translation as a discrete symmetry transformation.

The motion of an electron inside a crystal is governed by the Schrödinger equation

$$H\psi(\vec{r}) \equiv \left[-\frac{\hbar^2}{2m} \nabla^2 + V(\vec{r}) \right] \psi(\vec{r})$$

$$= E\psi(\vec{r}). \tag{1}$$

If the crystal potential is periodic

$$V(\vec{r} + \vec{l}) = V(\vec{r}), \tag{2}$$

then so is the Hamiltonian H. The symmetry properties of the Hamiltonian provide information about the nature of the eigenfunctions and eigenvalues of Eq. (1). We discuss the consequences of the translational symmetry of the Hamiltonian. The periodic nature of a crystal means that the lattice is invariant under translations by an integer multiple of the lattice period. The lattice or primitive translations can be written as

$$\vec{R}_n = n_1\vec{a}_1 + n_2\vec{a}_2 + n_3\vec{a}_3, \tag{3}$$

where $\vec{a}_i$ $(i = 1, 2, 3)$ are the primitive or basis vectors and n_i are integers. To say that the Hamiltonian has translational symmetry means that the Hamiltonian is invariant under primitive translations.

Let T_l be an operator associated with a primitive translation R_l such that

$$T_l f(\vec{r}) = f(\vec{r} + \vec{R}_l). \tag{4}$$

Since the Hamiltonian H is invariant to all T_l, we get

$$[H, T_l] = 0. \tag{5}$$

From Eq. (1), we have for any solution $\psi(\vec{r})$

$$T_l(H\psi_n) = T_l(E_n\psi_n)$$

or,
$$H(T_l\psi_n) = E_n(T_l\psi_n). \tag{6}$$

Thus, all $T_l\psi_n$ are simultaneous eigenfunctions with ψ_n to the same eigenvalue E_n. If E_n is non-degenerate, then $T_l\psi_n$ is equal to ψ_n, apart from some factor. Since both eigenfunctions ψ_n and $T_l\psi_n$ are normalizable, this factor can only be a phase factor, i.e.,

$$T_l\psi_n = C_l\psi_n$$

with
$$|C_l|^2 = 1. \tag{7}$$

Clearly, C_l are the eigenvalues of T_l and can be written as $e^{i\alpha_l}$. As $\vec{R}_l + \vec{R}_m = \vec{R}_p$, $T_lT_m = T_p$ and $e^{i(\alpha_l + \alpha_m)} = e^{i\alpha_p}$,

so we write

$$C_l = e^{i\vec{k}.\vec{R}_l}, \tag{8}$$

where $\vec{k}$ is a vector common to all α. If E_n is degenerate, then we have to take appropriate linear combinations of the degenerate solutions.

Thus, ψ is an eigenfunction of T_l with the eigenvalue $e^{i\vec{k}.\vec{R}_l}$ and so in this case we characterize $\psi(\vec{r})$ as $\psi_{\vec{k}}(\vec{r})$. Obviously, the eigenvalues E_n also depend on $\vec{k}$:

$$E_n = E_n(\vec{k}). \tag{9}$$

The non-degenerate solutions and appropriately chosen linear combinations of the degenerate solutions of the Schrödinger equation (1) are also eigenfunctions $\psi_{\vec{k}}(\vec{k}, \vec{r})$ of the lattice translation operator T_l with the eigenvalue $e^{i\vec{k}.\vec{R}_l}$:

$$T_l\psi_n(\vec{k}, \vec{r}) = e^{i\vec{k}.\vec{R}_l}\,\psi_n(\vec{k}, \vec{r}) \tag{10}$$

or,

$$\psi_n(\vec{k}, \vec{r} + \vec{R}_l) = e^{i\vec{k}.\vec{R}_l}\,\psi_n(\vec{k}, \vec{r}). \tag{11}$$

If we put

$$\psi_n(\vec{k}, \vec{r}) = e^{i\vec{k}.\vec{r}}\,u_n(\vec{k}, \vec{r}) \tag{12}$$

in Eq. (11), we get

$$u_n(\vec{k}, \vec{r} + \vec{R}_l) = u_n(\vec{k}, \vec{r}). \tag{13}$$

So $u_n(\vec{k}, \vec{r})$ has the periodicity of the lattice.

We arrive at the usual form of Bloch's theorem. The eigenstates ψ of the one-electron Hamiltonian $H = -\hbar^2\nabla^2/2m + V(\vec{r})$, where $V(\vec{r} + \vec{R}_l) = V(\vec{r})$ for all primitive translations $\vec{R}_l$, can be written in the form of a plane wave times a function with the periodicity of the lattice

$$\psi_n(\vec{k}, \vec{r}) = e^{i\vec{k}.\vec{r}}\,u_n(\vec{k}, \vec{r}) \tag{14}$$

where $u_n(\vec{k}, \vec{r})$ satisfies Eq. (13). An alternative form of Bloch's theorem is given in Eq. (11). In one dimension, this theorem is frequently called Floquet's theorem.

All symmetry operations which leave an ideal, infinite crystal invariant belong to its space group. A general space-group operation can be described by an orthogonal transformation of the form

$$\vec{r} \rightarrow r' = R\vec{r} - t \equiv \{R, t\}, \tag{15}$$

where R is a rotation (proper or improper) and t is a translation, not necessarily a primitive one. The space group G consists of all operations $\{R, t\}$ which leave a given lattice invariant. Primitive translations are denoted by $\left\{E|\vec{R}_n\right\}$ where E is unit operation. All the elements of G of the form $\left\{E|\vec{R}_n\right\}$ constitute the translation group $\mathcal{J}$. $\mathcal{J}$ is an abelian invariant subgroup of G.

Now,

$$\left\{E|\vec{R}_n\right\}\vec{r} = \vec{r} - \vec{R}_n, \tag{16}$$

and the group of operators $T_{\left\{E|\vec{R}_n\right\}}$, isomorphic with $\mathcal{J}$, is defined as

$$T_{\left\{E|\vec{R}_n\right\}}f(\vec{r}) = f\left(\left\{E|\vec{R}_n\right\}^{-1}\vec{r}\right)$$
$$= f(\vec{r} + \vec{R}_n). \tag{17}$$

Since

$$\left\{E|\vec{R}_n\right\} = \left\{E|\vec{R}_n^{(1)}\right\}\left\{E|\vec{R}_n^{(2)}\right\}\left\{E|\vec{R}_n^{(3)}\right\}, \tag{18}$$

where $R_n^{(i)} = n_i \vec{a}_i$ $(i = 1, 2, 3)$ and each one commutes with the other two. $\mathcal{J}$ is a direct product of three commuting subgroups. In an infinite crystal there are infinite elements in the (discontinuous) translation group, i.e., a discrete group of translations along three non-coplanar directions. If we assume that the crystal has a finite volume V_g with periodic boundary conditions, then the translation group becomes finite. We shall work in the following with periodic boundary conditions. The number of different $\vec{R}_l$ is then equal to the number of lattice points in V_g. Now, two primitive translations $\vec{R}_l$ and $\vec{R}_l + N_i \vec{a}_i$ are identical. Then

$$\psi(\vec{k}, \vec{r} + N_i \vec{a}_i) = \psi_n(\vec{k}, \vec{r}) \tag{19}$$

or using Eq. (12),

$$e^{ik \cdot (\vec{r} + N_i \vec{a}_i)} u_n(\vec{k}, \vec{r}) = e^{i\vec{k} \cdot \vec{r}} u_n(\vec{k}, \vec{r})$$

or,
$$e^{iN_i \vec{k}.\vec{a}_i} = 1. \tag{20}$$

We now define a reciprocal lattice. This is a point lattice described by the three basis vectors $\vec{b}_1, \vec{b}_2, \vec{b}_3$ as

$$\vec{K}_m = m_1 \vec{b}_1 + m_2 \vec{b}_2 + m_3 \vec{b}_3, \tag{21}$$

where m_i are integers. Further,

$$\vec{a}_i.\vec{b}_j = 2\pi \delta_{ij}, \quad (i, j = 1, 2, 3) \tag{22}$$

and
$$\vec{b}_i = 2\pi \frac{\vec{a}_j \times \vec{a}_k}{\vec{a}_i \cdot (\vec{a}_j \times \vec{a}_k)},$$

$$\vec{a}_i = 2\pi \frac{\vec{b}_j \times \vec{b}_k}{\vec{b}_i \cdot (\vec{b}_j \times \vec{b}_k)}. \tag{23}$$

The direct lattice and the reciprocal lattice are reciprocal to each other. We now write $\vec{k}$ as a vector in the reciprocal lattice, $\vec{k} = \sum_{i=1}^{3} K_i \vec{b}_i$, then Eq. (20) gives

$$e^{2\pi i (N_1 K_1 + N_2 K_2 + N_3 K_3)} = 1 \tag{24}$$

so that

$$K_i = \frac{n_i}{N_i}, \quad n_i = 1, 2, \dots N_i \quad (i = 1, 2, 3). \tag{25}$$

There are $N = N_1 N_2 N_3$ different sets of $\{K_1 K_2 K_3\}$, and so there are N different, allowed $\vec{k}$ vectors. Two vectors $\vec{k}$ and $\vec{k}' = \vec{k} + \vec{K}_m$ give equivalent representations. To find a complete set of nonequivalent representations, we find the set of $\vec{k}$ vectors such that the length of $\vec{k}$ is less than or equal to the length of any vector $\vec{k} + \vec{K}_m$. The set of $\vec{k}$ vectors lie in the first Brillouin zone.

The group of lattice translation operators is finite and of order N. Since two translations commute, the group is abelian. Since $T_m^{-1} T_l T_m = T_l$ for any T_m, every element forms a class. There are N classes, and so there are N IR's. If we write $\vec{R}_l = l_1\vec{a}_1 + l_2\vec{a}_2 + l_3\vec{a}_3$, then $T_l = T^{l_1}_{\vec{a}_1} T^{l_2}_{\vec{a}_2} T^{l_3}_{\vec{a}_3}$. The translation group is a direct product of the three commuting subgroups, each in a direction $\vec{a}_i$. Each group has N_i elements $T^{l_i}_{\vec{a}_i}$ where $l_i = 0, 1, ..., N_i - 1$. Because of the periodic boundary conditions, $\left(T^{l_i}_{\vec{a}_i}\right)^{N_i} = E$. Thus $D\left(T^{l_i}_{\vec{a}_i}\right)^{N_i} = 1$, since the IR's are one-dimensional. So

$$D\left(T^{l_i}_{\vec{a}_i}\right)^{N_i} = \left[e^{(2\pi i n_i/N_i)}\right]^{l_i}, \quad n = 1, ..., N_i \tag{26}$$

and

$$D(T_l) = e^{2\pi i \left(\frac{n_1 l_1}{N_1} + \frac{n_2 l_2}{N_2} + \frac{n_3 l_3}{N_3}\right)}.$$

Therefore,

$$D(T_l) = e^{i\vec{k}.\vec{R}_l}, \tag{27}$$

where $\vec{k}$, can take $N = N_1 N_2 N_3$ different values. The IR's of the lattice translation group are the Bloch factors:

$$\Gamma^{(\vec{k})}\left(\left\{E | \vec{R}_l\right\}\right) = e^{i\vec{k}.\vec{R}_l}. \tag{28}$$

10.2 SPACE INVERSION

In classical mechanics, each particle in a system is described by space-time coordinates $\vec{r}$ and t. By space-inversion invariance in classical mechanics, we mean that the dynamical laws are invariant under the transformation

$$\vec{r} \rightarrow \vec{r}' = -\vec{r}'; \quad t \rightarrow t' = t. \tag{29}$$

In the passive interpretation, space inversion changes a right-handed coordinate system to a left handed one. Operationally, space inversion is equivalent to reflection with respect to a plane followed by a continuous rotation of π about an axis normal to the mirror plane.

In three-dimensional Euclidean space, the space reflection transformation I_s is defined as

$$I_s\hat{e}_i = -\hat{e}_i \quad (i = 1, 2, 3) \tag{30}$$

$$I_s\vec{r} = -\vec{r} \quad \text{for all } \vec{r}$$

where $\hat{e}_i$ is coordinate unit vector and $\vec{r}$ is any coordinate vector. For a system invariant under space inversion, the symmetry group. ("parity group") is a finite abelian group with two elements only, the unit element E and the reflection I_s represented by the matrix $-E$:

$$I_S = \begin{pmatrix} -1 & 0 & 0 \\ 0 & -1 & 0 \\ 0 & 0 & -1 \end{pmatrix}. \tag{31}$$

The space-inversion transformation commutes with all three-dimensional rotations and their generaters:

$$I_S\vec{R}I_S^{-1} = \vec{R} \quad \text{for all } \vec{R} \in \text{SO(3)} \tag{32}$$

$$[I_S, J_i] = 0 \quad \text{for } i = 1, 2, 3$$

The space inversion preserves the length of all three-dimensional vectors like the rotations. The three-dimensional rotation and reflection groups taken together form the three-dimensional orthogonal group O(3). The elements of this group are divided into two classes — $\left\{\vec{R}; \vec{R} \in \text{SO(3)}\right\}$ and $\left\{I_S\vec{R}; \vec{R} \in \text{SO(3)}\right\}$. The determinant of the transformation matix is +1 in the first case and −1 in the second case.

The basis vectors of the IR's of O(3) can be chosen as simultaneous eigenvectors of J_3 and I_s:

$$J_3|m\rangle = m|m\rangle \tag{33}$$

$$I_S|m\rangle = \eta|m\rangle$$

with $\eta = \pm 1$. The unitary IR's are denoted as (j, η), where $j = 0, \dfrac{1}{2}, 1, \dfrac{3}{2}, \dots$

In quantum mechanics, let P be the operator in the Hilbert space corresponding to space inversion. The parity operator reverses the position operator $\vec{r}$ and the momentum operator $\vec{p}$:

$$P\vec{r}\,P^{-1} = -\vec{r}, \tag{34}$$

$$P\vec{p}\,P^{-1} = -\vec{p}. \tag{35}$$

Vectors like $\vec{r}, \vec{p}$ that change sign under reflection are called polar vectors. Since the orbital angular momentum $\vec{L} = \vec{r} \times \vec{p}$, we have

$$P\vec{L}\,P^{-1} = \vec{L}. \tag{36}$$

A vector like $\vec{L}$ which is invariant under space inversion is called an axial or a pseudo vector. Since rotation and space inversion commute, any angular momentum operator $\vec{J}$ commutes with P.

$$[P, \vec{J}] = 0. \tag{37}$$

This relation applies to spin angular momentum

$$P\vec{s}\,P^{-1} = \vec{s}. \tag{38}$$

The helicity operator

$$\Lambda = \frac{\vec{s} \cdot \vec{p}}{|\vec{p}|}.$$

satisfies

$$P \Lambda P^{-1} = -\Lambda. \tag{39}$$

This is an example of a pseudoscalar quantity.

A state vector $|\psi\rangle$ is related to the space-inverted state vector $|\psi'\rangle$ as

$$|\psi'\rangle = P|\psi\rangle. \tag{40}$$

In the state $|\psi\rangle$ the positions and momenta of all the particles are inverted with respect to those for the $|\psi\rangle$.

Since the commutation relations

$$[x_i, p_j] = i\hbar\delta_{ij} \quad \text{and} \quad [J_i, J_j] = i\hbar\epsilon_{ijk} J_k,$$

$(i, j, k = 1, 2, 3)$ are preserved under space inversion, P is unitary operator and $P = P^{-1}$, i.e., $P^2 = I$. Two successive space inversions produce no change at all so that $|\psi\rangle$ and $P^2|\psi\rangle$ must be the same state (to a phase factor). It is customary to take the phase factor unity, so that P is not only unitary, but also hermitian:

$$P = P^{-1} = P^\dagger. \tag{41}$$

All eigenvalues of P are either $+1$ or -1. Let $\psi(\vec{r})$ be the wave function of a spinless particle whose state ket is $|\alpha\rangle$:

$$\psi(\vec{r}) = \langle \vec{r}|\alpha\rangle. \tag{42}$$

The wave function of the space-inverted state is

$$\psi(-\vec{r}) = \langle -\vec{r}|\alpha\rangle = \langle \vec{r}|P|\alpha\rangle. \tag{43}$$

If $|\alpha\rangle$ is an eigenket of parity, then

$$P|\alpha\rangle = \pm |\alpha\rangle \tag{44}$$

so that

$$\langle \vec{r}|P|\alpha\rangle = \pm \langle \vec{r}|\alpha\rangle. \tag{45}$$

But,

$$\langle \vec{r}|P|\alpha\rangle = \langle -\vec{r}|\alpha\rangle. \tag{46}$$

So the state $|\alpha\rangle$ is even (odd) under parity transformation if the corresponding wave function satisfies

$$\psi(-\vec{r}) = \pm \psi(\vec{r}) \quad \begin{cases} \text{even} \\ \text{odd} \end{cases}. \tag{47}$$

Space inversion is a symmetry of the system if

$$[H, P] = 0. \tag{48}$$

Then H and P are simultaneously diagonalizable. Energy eigenfunctions can be chosen to be eigenfunctions of parity that is to have a definite parity.

Next, we consider the behavior of an eigenket of orbital angular momentum under space inversion. Its wave function is

$$\langle \vec{r} | \alpha, l, m \rangle = R_\alpha (r) Y_l^m (\theta, \varphi). \tag{49}$$

Under space inversion $\vec{r} \rightarrow - \vec{r}$ which in spherical polar coordinates implies

$$r \rightarrow r$$

$$\theta \rightarrow \pi - \theta$$

$$\phi \rightarrow \pi + \phi \tag{50}$$

the spherical harmonics undergoes the transformation

$$Y_l^m (\theta, \phi) \rightarrow Y_l^m (\pi - \theta, \pi + \phi)$$

$$= (- 1)^l Y_l^m (\theta, \phi) \tag{51}$$

Thus,

$$P | \alpha, l, m \rangle = (- 1)^l | \alpha, l, m \rangle. \tag{52}$$

$(- 1)^l$ is called the parity of the state with orbital angular momentum l. Thus,

$s, d, g, ...$ waves have even parity

$p, f, h, ...$ waves have odd parity

Since spin is unchanged by parity, a two component Pauli spinor transforms as

$$u \rightarrow u' = \eta u, \tag{53}$$

where $$|\eta|^2 = 1.$$

If H is invariant under P, then the S-matrix is also invariant, i.e., if $PHP^{-1} = P$, then

$$PSP^{-1} = S \tag{54}$$

See Sakurai (1964) for a detailed discussion. For parity considerations on scattering, see Section 19.4 in our discussion of scattering.

When particles are created or annihilated we have to assign an intrinsic parity to each particle.

The intrinsic parity of a particle is +1 or −1 if its internal wave function does not or does change sign under parity. The concept of intrinsic parity is meaningful if the interactions are invariant under space inversion. The concept of intrinsic parity can be defined for a system of particles. Formally if a particle A is represented by $|A, \vec{p}\rangle$, the intrinsic parity η_P^A of the particle is defined by

$$P|A, \vec{p}\rangle = \eta_P^A|A, -\vec{p}\rangle, \tag{55}$$

where $\eta_P^A = \pm 1$.

A photon is represented by a vector potential $\vec{A}$ of the form $\vec{A}(\vec{x}) \propto \vec{\epsilon}\, f(\vec{x})$, where $\vec{\epsilon}$ is the polarization vector and f a scalar function. Since $\vec{x}$ and $\vec{p}$ change sign under P, we have

$$\vec{A}(\vec{x}) \xrightarrow{\ P\ } -\vec{A}(-\vec{x})$$

$$\Phi(\vec{x}) \xrightarrow{\ P\ } \Phi(-\vec{x}) \tag{56}$$

if the interaction is to be invariant under P (Φ is the scalar potential). From (52), we see that the electric and magnetic field vectors transform as

$$\vec{E}(\vec{x}) = -\vec{\nabla}\Phi + \frac{1}{c}\frac{\partial \vec{A}}{\partial t} \xrightarrow{\ P\ } -\vec{E}(-\vec{x}) \tag{57}$$

$$\vec{B}(\vec{x}) = \vec{\nabla} \times \vec{A} \xrightarrow{\ P\ } \vec{B}(-\vec{x}).$$

Therefore, under parity $\vec{\epsilon} \rightarrow -\vec{\epsilon}$. The behavior of the polarization vector $\vec{\epsilon}$ characterizes the intrinsic parity of the field. Thus the intrinsic parity of a photon is odd.

Parity is a multiplicative quantum number. The intrinsic parity of a system of particles is the product of the intrinsic parities of the particles times the parity of the relative orbital angular momentum. As an example, let us consider a reaction.

$$A + B \rightarrow C + D. \tag{58}$$

If $\eta_A^P, \eta_B^P, \eta_C^P, \eta_D^P$ are the intrinsic parities of A, B, C, D respectively, then

$$\eta_{\text{initial}}^P = \eta_A^P\eta_B^P\,(-1)^{L_i} \tag{59}$$

$$\eta_{\text{final}}^P = \eta_C^P\eta_D^P\,(-1)^{L_f} \tag{59a}$$

where L_i and L_f are the total relative orbital angular momenta of the initial and final state, respectively.

Conservation of parity in the reaction Eq. (55) implies

$$\eta_{\text{initial}}^P = \eta_{\text{final}}^P \tag{60}$$

or, $$\eta_A^P\eta_B^P\,(-1)^{L_i} = \eta_C^P\eta_D^P\,(-1)^{L_f}. \tag{60a}$$

The intrinsic parity of the charged pion has been determined from observations of the capture of slow π^- in deuterium. The reactions are

$$\pi^- + d \to n + n$$

$$\to n + n + \gamma. \tag{61}$$

The spin of the pion is zero and the deuteron is predominantly in a $^3S_1$ state, i.e., for the deuteron $J^P = 1^+$. The π^- is captured from an atomic S state of deuterium, as proved by studies of mesonic X-rays and by the direct experimental determinantion of the π^- cascade time. So initially there is a spin zero particle in an S state of a spin 1 nucleus. The total angular momentum of the initial state $J_{in} = 1$. Angular momentum conservation requires $J_{fi} = 1$ so that the only possible final states are $^3S_1$, $^3P_1$, $^1P_1$, $^3D_1$. Since we have two identical fermions in the final state, antisymmetrization implies that a triplet (singlet) spin state must combine with an odd (even) orbital state. So the only possibility is the $^3P_1$ state of two neutrons, which has $(-1)^L = -1$. Therefore the initial state must have odd parity, i.e., π^- has odd intrinsic parity. The intrinsic parity of π^+ must be odd. For neutral pion, the existence of the decay $\pi^0 \to 2\gamma$ shows that the intrinsic parity of π^0 is -1.

The pion has zero spin and odd intrinsic parity, that is, $J^P = 0^-$. Such mesons are called pseudoscalar particles. Particles with $J^P = 0^+$ are called as scalar, $J^P = 1^-$ as vector and $J^P = 1^+$ as pseudo (axial) vector.

By convention, the intrinsic parity of a proton is $+1$. A neutron is assigned the same parity $+1$. Note that the Dirac theory predicts opposite parity for fermion and antifermion. For bosons, however, particles and antiparticles have the same intrinsic parity.

We now discuss some consequences of parity conservation.

(i) For a discussion of parity in scattering and reaction processes, see § 19.4.

(ii) Parity selection rule –

Let $\mathcal{E}$ be an even and O an odd observable under space inversion. So

$$P \, \mathcal{E} \, P^{-1} = \mathcal{E} \tag{62}$$

$$P O P^{-1} = - O \tag{62a}$$

Then between two parity eigenstates $|\alpha\rangle$, $|\beta\rangle$

$$\langle \beta | \mathcal{E} | \alpha \rangle = \langle \beta | P^{-1} (P \mathcal{E} P^{-1}) P | \alpha \rangle$$

$$= \eta_\beta \eta_\alpha \, \langle \beta | \mathcal{E} | \alpha \rangle \tag{63}$$

and

$$\langle \beta | O | \alpha \rangle = \langle \beta | P^{-1} (P O P^{-1}) P | \alpha \rangle$$

$$= - \eta_\beta \eta_\alpha \, \langle \beta | O | \alpha \rangle \tag{63a}$$

where η_α, η_β are the parity eigenvalues. From Eqs. (59) and (59a), we see that an even-parity observable has nonvanishing matrix elements between states of the same parity, while an odd parity observable has nonvanishing matrix elements between states of opposite parity. This important selection rule has many applications, some of which are discussed below.

(iii) Static electric and magnetic moments – In classical electromagnetic theory, the electric moments arise from the distribution of charge of the source, the magnetic moments from the distribution of currents and magnetizations. This quantum mechanical 2^l static moments are expressed is the expectation values of the corresponding operators obtained from the classical moments. From, rotational invariance the 2^l-pole can only occur for a system with angular momentum j if $2j \geq l$.

The charge density $\rho(\vec{r})$ of a quantum mechanical system of N-particles with charges e_i ($i = 1, 2, ..., N$) is given by

$$\rho(\vec{r}) = \sum_{i=1}^{N} e_i \, \delta(\vec{r} - \vec{x}_i), \tag{64}$$

where $\vec{x}_i$ is the position operator for the ith particle. The electric multipole moments are defined as

$$Q_{lm} = \int r^l \, Y_{lm}(\hat{r}) \, \rho(\vec{r}) \, d^3r. \tag{65}$$

Q_{lm} is a tensor operator of rank l. Since $P \, \rho(\vec{r}) \, P^{-1} = \rho(\vec{r})$ and $Y_{lm}(-\hat{r}) = (-1)^l \times Y_{lm}(\hat{r})$, we have

$$PQ_{lm}P^{-1} = (-1)^l \, Q_{lm}. \tag{66}$$

From the parity selection rule, we see that all electric multipole moments with odd values of l vanish, if the state in question is nondegenerate. As an example, the electric dipole moment operator $O_E^{(1)} \equiv \vec{d} = \Sigma e_i \vec{r}_i$ has odd parity ($l = 1$).

Therefore, if a system possesses a space-invariant Hamiltonian, and if the state is nondegenerate, then the system cannot have permanent electric dipole moment in that state.

If, however, the state is a linear combination of even and odd-parity components, then the system can have a non-zero electric dipole moment in that state. See the discussion on the linear Stark effect in Section 12.7.

The magnetostatic multipole operator is defined as

$$M_{lm} = \int r^l \, Y_{lm}(\hat{r}) \, \rho(\vec{r}) \, d^3r, \tag{67}$$

where the magnetic charge density is

$$\rho_m(\vec{r}) = -\vec{\nabla} \cdot \vec{M}(\vec{r}) \tag{68}$$

and $\vec{M}(\vec{r})$ is the magnetic moment density. The moment density is given by

$$\vec{j}(\vec{r}) = c\vec{\nabla} \cdot \vec{M}(\vec{r}). \tag{69}$$

M_{lm} is a tensor operator of rank l with the property

$$PM_{lm}P^{-1} = -(-1)^l M_{lm} = (-1)^{l+1} M_{lm}. \tag{70}$$

A system can only have odd magnetic multipole elements in a state of definite parity.

Clearly, $M_0 = 0$ and the magnetic dipole moment M_{lm} (usually denoted as $\vec{\mu}$) is important in most cases. We write

$$\vec{\mu} = \frac{1}{2c} \int \vec{r} \times \vec{j}(\vec{r}) \, d^3 r. \tag{71}$$

For a nucleus or other nonrelativistic system the magnetic dipole operator is expressed as

$$O_M^{(1)} \equiv \vec{\mu} = \Sigma(g_{iL} \vec{L}_{iL} + g_{is} \vec{S}_i), \tag{72}$$

where L_i is the orbital angular momentum operator (in units of $\hbar$) for the ith particle, S_i is the spin angular momentum, and g_{iL} and g_{iS} are the corresponding gyromagnetic ratios.

(iv)   Transition electromagnetic moments -

The interaction of an atom or a nucleus with electromagnetic radiation leads to transitions with the emission or absorption of photons and take place between bound states of the system or between bound and continuum states. Space-time symmetries put restrictions on the transition electromagnetic moments.

The radiation field is expanded in multipole fields in terms of which the selection rules are expressed. It is convenient to write the fields as superpositions of photon wave functions. A photon wave function with the quantization axis along the beam direction is expressed as

$$\vec{A}(\vec{r}, t) = \sum_{l=0}^{\infty} f_l(r, \vec{e}) \sum_{J=l-1}^{l+1} C_{l1}(J, \pm 1; 0, \pm 1)\vec{Y}_{J, l, 1}^{\pm 1} \tag{73}$$

where the vector spherical harmonics (see § 15.7)

$$\vec{Y}_{J, l, 1(\theta, P)}^{M} = \sum_{m=-l}^{l} C_{l1}(J, M; m, m)Y_{lm}(\theta, \phi)\chi_{m'} \tag{74}$$

are vector functions of the angles and are egenfunctions of $J = \vec{L} + \vec{S}$, the total angular momentum operator of the photon. $\vec{L}$ is the orbital, angular momentum and $\vec{S}$ is the photon spin. The spin eigenfunctions are expressed as

$$\chi_{\substack{+1 \\ -1}} = \mp \frac{1}{\sqrt{2}} \, (\hat{e}_x \pm i e_y).$$

For each eigenvalue J, there are three different vector spherical harmonics corresponding to $l = J - 1, J, J + 1$. In the case $J = 0$, we get only one vector spherical harmonics. The vector functions $\vec{Y}^M_{J,\,l,\,1}$ have parity $(-1)^l$.

It is useful to classify the electromagnetic waves according to their various J and l components, as given below.

J	l		Parity	Name
0	Impossible			
1	Dipole	$\begin{cases} J \pm 1 = 0, 2 \\ \quad J = 1 \end{cases}$	$\begin{matrix} - \\ + \end{matrix}$	$\begin{matrix} E1 \\ M1 \end{matrix}$
2.	Quadrupole	$\begin{cases} J \pm 1 = 1, 3 \\ \quad J = 2 \end{cases}$	$\begin{matrix} + \\ - \end{matrix}$	$\begin{matrix} E2 \\ M2 \end{matrix}$

The parity of the electric 2^J pole is $-(-1)^{J+1} = (-1)^J$ and the parity of the magnetic 2^J pole is $-(-1)^J = (-1)^{J+1}$, the odd intrinsic parity of the photon contributing the extra minus signs.

One quantum of multipole radiation l, m carries with it an angular momentum l with z component m (both measured in units of $\hbar$). If J, M and $J'M'$ are the angular momenta and their z components of the intitial and final states, respectively, then angular momentum conservation gives that multipole radiation l, m can be emitted only if

$$|J - J'| \leq |J + J'|, \tag{75}$$

$$M - M' = m. \tag{75a}$$

Since there is no multipole radiation with $l = 0$, Eq. (71) implies that radiative transitions between two states with angular momentum $J = J' = 0$ are absolutely forbidden. If the total Hamiltonian of the system (including its interaction with the electromagnetic field) is invariant under space inversion, then the initial and final states have definite parities η and η'. Parity conservation then allows those multipoles which satisfy

$$\eta_l = \eta\eta'. \tag{76}$$

Combined with the parity selection rules, Eq. (73) leads to the allowed electromagnetic transitions, given below

Parities	**Allowed multipoles**	
	Electric	Magnetic
$\eta' = \eta$	even l	odd l
$\eta' = -\eta$	odd l	even l

If the wavelength of the emitted radiation is much larger than the size of the radiating system, then only electric dipole emission is important. Since $E1$ has odd parity, parity conservation leads to Laporte's rule: electric dipole transitions are allowed only between states of opposite parity.

10.3 TIME REVERSAL

We now study another discrete symmetry operator, called the time reversal. We introduce the time reversal transformation

$$T : t \rightarrow t' = -t \tag{77}$$

This is an improper space-time symmetry. The transformation reverses the velocity. The transformation T is often referred to as "motion reversal". The effect of this transformation is to reverse the linear momentum and angular momentum with the position remaining unchanged. We have

$$T \vec{x} \, T^{-1} = \vec{x}, \tag{78}$$

$$T \overline{O} \, T^{-1} = -\vec{p}, \tag{79}$$

$$T \vec{J} \, T^{-1} = -\vec{J}, \tag{80}$$

and
$$T \wedge T^{-1} = \wedge, \tag{81}$$

where $\wedge$ is the helicity.

The novel property of the time inversion operator is seen from the commutation relation

$$[x_i, p_j] = i\hbar \delta_{ij}. \tag{82}$$

This is preserved provided that $Ti T^{-1} = -i$ Therefore, T is an antilinear operator. The fundamental angular momentum commutation relation

$$[J_i, J_j] = i\in_{ijk} J_k, \tag{83}$$

is preserved provided T is antilinear. The time evolution of a system is given by the nonrelativistic time-dependent Schrödinger equation, a diffusion-like equation:

$$i\hbar \, \frac{\partial \psi(t)}{\partial t} = H\psi(t). \tag{84}$$

$\psi(-t)$ does not satisfy Eq. (84). But if $\psi(t)$ is a solution then $\psi'(t) = \psi^*(-t)$ is also a solution, if the Hamiltonian is real. This solution represents a wave packet whose density distribution describes the time reversed motion by comparison with the density distribution of the wave packet ψ. In the time reversed state, the linear and the angular momenta have the opposite sign, while positions remain unchanged.

The time reversal transformation is written in the Schrödinger picture,

$$|\psi'(t')\rangle = T|\psi(t)\rangle \tag{85}$$

and

$$Q' = TQT^{-1} \tag{86}$$

for Q any hermitian operator, or, in the Heisenberg picture,

$$|\psi'\rangle = T|\psi\rangle \tag{87}$$

and

$$Q'(t') = TQ(t)T^{-1}. \tag{88}$$

We discuss the Wigner time reversal in quantum mechanics. Following Wigner (1932), we write T as

$$T = U_T K, \tag{89}$$

where K is the operator of complex conjugation and U_T is a linear transformation. For any complex number z, $KzK^{-1} = z^*$ and $K^{-1} = K$. We note that the commutation relations are invariant under any linear similarity transformation.

K is antilinear, i.e.

$$K(c_1|\psi_1\rangle + c_2|\psi_2\rangle) = c_1^*|\psi_1\rangle^* + c_2^*|\psi_2\rangle^*. \tag{90}$$

The operator T is antilinear:

$$T(a_1|\psi_1\rangle + a_2|\psi_2\rangle) = a_1^* T|\psi_1\rangle + a_2^* T|\psi_2\rangle. \tag{91}$$

Under T, the norm of ψ is preserved. i.e.,

$$(T\psi, T\psi) = (\psi, \psi).$$

We have

$$\langle T\phi(t)|T\psi(t)\rangle = \langle \phi(-t)|\psi(-t)\rangle^* \tag{92}$$

$$= \langle \psi(-t)|\phi(-t)\rangle \tag{93}$$

U_T is unitary:
$$U_T^\dagger U_T = U_T U_T^\dagger = I.$$

Equation (92) is called the antiunitary condition.

For all time – reversible systems a T operator with the properties Eqs. (91) and (92) exists. Such an operator is called antilinear-unitary, or simply, antiunitary. Since it is antiunitary, it cannot lead to a conservation law. Now, we consider the matrix element $\langle \phi|O|\psi \rangle$.

We have

$$\langle T\phi|O_T|T\psi\rangle = \langle \phi(-t)|O|\psi(-t)\rangle \tag{94}$$

where

$$O_T = TOT^{-1} = U_T O^* U_T^{-1}. \tag{95}$$

T is an involutional operator. We write

$$T^2 = \eta I \tag{96}$$

where $\eta = \pm 1$ as Wigner showed.

The proof is as follows. We know

$$T^2 = U_T K U_T K = U_T U_T^*.$$

Since U_T is unitary

$$U_T^{-1} = U_T^{\dagger} = \tilde{U}_T^*.$$

or, $\qquad\qquad\qquad U_T^* = \tilde{U}_T^{-1}.$

So $\qquad\qquad\qquad T^2 = U_T \tilde{U}_T^{-1} = \eta I,$

and $\qquad\qquad\qquad U_T = \eta \tilde{U}_T$

or, $\qquad\qquad\qquad \tilde{U}_T = \eta U_T$

whence $\qquad\qquad\qquad U_T = \eta(\eta U_T)$

Hence, $\qquad\qquad\qquad \eta^2 = 1.$

We may write

$$T^2 = \pm I \tag{97}$$

or, $\qquad\qquad\qquad T^{-1} = \pm T. \tag{98}$

The sign of T^2 depends on the properties of U.

The operator U_T is basis dependent. The form of the T operator depends on spin.

Let us consider a spinless particle without internal structure. In the coordinate representation $\psi(\vec{x}, t)$ is a solution of the Schrödinger time-dependent equation then so is $\psi^*(\vec{x}, -t)$ provided that the Hamiltonian is T-invariant. For spin 0 and in the coordinate representation, time reversal operator $T = K$. $\hfill (99)$

Momentum eigenstates are not invariant under time reversal: $T|p\rangle = |-\vec{p}\rangle$.

So the momentum space wave function $\phi(\vec{p}, t) \equiv \left\langle \vec{p} \left| e^{-\frac{i}{\hbar}Ht} \right| \psi \right\rangle$ has the following behavior under time reversal if the Hamiltonian is time reversal invariant:

$$\phi(\vec{p}, t) \xrightarrow{\ T\ } \phi^*(-\vec{p}, -t). \tag{100}$$

The operator T must reverse the spin operator $\vec{S}$:

$$T\vec{S}T^{-1} = -\vec{S}. \tag{101}$$

For a particle having spin $\frac{1}{2}$, the Pauli spin matrices σ satisfy

$$T\vec{\sigma}T^{-1} = -\vec{\sigma}. \tag{102}$$

In the standard representation σ_x and σ_z are real matrices whereas σ_y is purely imaginary. The operation K yields

$$K\sigma_x K^{-1} = \sigma_x,$$
$$K\sigma_y K^{-1} = -\sigma_y, \tag{103}$$
$$K\sigma_z K^{-1} = \sigma_z,$$

From Eq. (102), we obtain

$$U_T \sigma_x U_T^{-1} = \sigma_x,$$
$$U_T \sigma_y U_T^{-1} = \sigma_y, \tag{104}$$
$$U_T \sigma_z U_T^{-1} = -\sigma_z,$$

We know that σ_y commutes with σ_y and anticommutes with σ_x and σ_z. Since σ_y is unitary, we write

$$T = \eta \sigma_y K, \quad |\eta| = 1. \tag{105}$$

Clearly,

$$T \begin{pmatrix} 1 \\ 0 \end{pmatrix} = i\eta \begin{pmatrix} 0 \\ 1 \end{pmatrix},$$
$$T \begin{pmatrix} 0 \\ 1 \end{pmatrix} = -i\eta \begin{pmatrix} 1 \\ 0 \end{pmatrix}. \tag{106}$$

i.e., in the spin basis $\left| \pm \dfrac{1}{2} \right\rangle$ with the standard phase conventions,

$$T \left| \frac{1}{2} \right\rangle = ie^{i\phi} \left| -\frac{1}{2} \right\rangle,$$
$$T \left| -\frac{1}{2} \right\rangle = -ie^{i\phi} \left| \frac{1}{2} \right\rangle. \tag{107}$$

For a single spin $\dfrac{1}{2}$ particle, let $|\vec{p}, m_s\rangle$ and $|\vec{x}, m_s\rangle$ be the momentum and coordinate representation eigenkets with $m_s = \pm \dfrac{1}{2}$. Then

$$T |\vec{p}, m_s\rangle = e^{i\phi} i^{2m_s} |-\vec{p}, -m_s\rangle, \tag{108}$$
$$T |\vec{x}, m_s\rangle = e^{i\phi} i^{2m_s} |\vec{x}, -m_s\rangle. \tag{109}$$

For a spin 0 particle,

$$T^2 = K^2 = I, \tag{110}$$

and for a spin $\dfrac{1}{2}$ particle,

$$\begin{aligned}
T^2 &= e^{i\phi} \sigma_y K e^{i\phi} \sigma_y K \\
&= e^{i\phi} \sigma_y e^{-i\phi} \sigma_y^* K \\
&= -\sigma_y^2 \\
&= -I.
\end{aligned} \tag{111}$$

Hence,

$$T^2 = I \quad \text{for a spin 0 particle}$$

$$T^2 = -I \quad \text{for a spin } \frac{1}{2} \text{ particle.}$$

For system of N spin 0 particles, $T^2 = I$ for any N. For a system of N spin $\frac{1}{2}$ particles,

$$T = \prod_{i=1}^{N} \sigma_y^{(i)} \times K \tag{112}$$

aside from an arbitrary phase. Therefore

$$T^2 = (-1)^N, \tag{113}$$

We discuss the time reversal operation in the angular momentum formalism (see Section 7.3). The definition of the orbital angular momentum $\vec{L}$ in terms of $\vec{r}$ and $\vec{p}$ shows that L_x, L_y, L_z must be replaced by $-L_x$, $-L_y$, $-L_z$ respectively and the orbital angular momentum eigenfunctions are replaced by their complex conjugation. In general, the total angular momentum $\vec{J}$ behaves under T as follows

$$T\vec{J}T^{-1} = -\vec{J}. \tag{114}$$

Therefore,

$$TJ_z T^{-1} = -J_z,$$

$$TJ_\pm T^{-1} = -J_\pm. \tag{115}$$

Now, let

$$T|\alpha, j, m> = c_m|\alpha, j, -m>, \quad |c_m| = 1 \tag{116}$$

where α refers to quantum numbers other than angular momentum. Now,

$$TJ_+|\alpha, jm\rangle = c_{mj1}[j(j+1) - m(m+1)]^{1/2} |\alpha, j, -m-1> \tag{117}$$

and

$$J_- T|\alpha, jm\rangle = c_m [j(j+1) - m(m+1)]^{1/2} |\alpha, j, -m-1> \tag{118}$$

Thus,

$$c_m = -c_{m+1},$$

So

$$c_m = (-1)^{j-m} e^{i\phi}. \tag{119}$$

Then

$$T^2|\alpha, j, m\rangle = (-1)^{j-m} e^{i\phi} T|\alpha, j, -m>$$

$$= (-1)^{2j} |\alpha, j, m>. \tag{120}$$

This is independent of the phase convention. The eigenvalue of T^2 is $+1$ for states of integral spin, and -1 for states of half integral spin.

We write

$$T|\alpha, j, m> = (-1)^{j-m} |\alpha, j, -m>. \tag{121}$$

This function transforms in the convential way with the rotation matrix $D^{(j)}_{mm'}$.

Let us return to the Eqs. (92) and (93)

$$\langle\phi|\psi\rangle^* = \langle\psi|\phi\rangle = \langle T\phi|T\psi\rangle.$$

This suggests an alternative approach to the time inversion operation (Schwinger (1951)). In the Schwinger procedure the T operation requires interchange of initial and final states, i.e., bras and kets are transformed into each other under time reversal: $|\psi> \leftrightarrow <\psi|$. Instead of the complex conjugation, transposition is used for transformation under the T operation. Hence T is an antiunitary operator. The commutation relation $[x_i, p_j] = i\hbar\delta_i$ is preserved since $[x_i, p_j] = [(-p_j)^T, x_i^T]$ without changing $i \rightarrow -i$. Notice that for hermitian operators, complex conjugation and transposition amount to the same thing.

We now discuss some consequences of T invariance.

Let us consider the energy eigenvalue equation

$$H|\psi\rangle = E|\psi\rangle \tag{122}$$

where the Hamiltonian H is time-reversal invariant

$$THT^{-1} = H.$$

Then $HT|\psi\rangle = TH|\psi\rangle = ET|\psi\rangle$, and so both $|\psi\rangle$ and $T|\psi\rangle$ are eigenstates of H with the same eigenvalue E. If $|\psi\rangle$ and $T|\psi\rangle$ are linearly dependent, they describe the same state. In this case $T|\psi\rangle = c|\psi'\rangle$, $|c| = 1$. Then $T^2|\psi\rangle = Tc|\psi\rangle = c^*T|\psi\rangle = c^*c|\psi\rangle = |\psi\rangle$. The nondegenerate case occurs for states with $T^2 = I$. But for states for which $T^2 = -I$, each eigenvalue of H is at least two-fold degenerate. For such states the state $|\psi'\rangle$ is orthogonal to $|\psi\rangle$:

$$\langle T\psi|\psi\rangle = \langle TT\psi|T\psi\rangle^* = -\langle\psi|T\psi\rangle^* = -\langle T\psi|\psi\rangle = 0 \tag{123}$$

So $T|\psi\rangle$ and $|\psi\rangle$ are independent eigenstates.

This type of degeneracy is known as Kramers' degeneracy (1930). We have given Wigner's proof (1932) of Kramers' theorem. For a system with an odd number of spin $\frac{1}{2}$ particles "(odd system"} each energy state will be at least two-fold degenerate if the Hamiltonian is T invariant.

If an odd system is subjected to an external electric field $\vec{E}$, every stationary state of such a system is at least doubly degenerate, if the internal interactions are

T invariant. A static magnetic field $\vec{B}$ changes sign under T, and will in general split this degeneracy.

Let us consider the rotational levels of a diatomic molecule. The quantum number representing the total electron angular momentum is $\Omega = |\wedge + \Sigma|$ where $\wedge$ is the projection of the electron orbital angular momentum on the molecular axis, and Σ is the projection of the total electron spin. Every level except that with $\wedge = 0$ of the molecule in the total angular momentum state $J = 0$ is doubly degenerate, since the projection of the orbital angular momentum on the axis may point either way. The degeneracy is split if $J \neq 0$, the molecule has a two-fold degeneracy if the molecule contains an odd number of electrons. In this case it is assumed that all the interactions are T invariant and hyperfine structure effects are neglected. An analogous situation occurs for odd – A deformed nuclei.

An important example of the Kramers' degeneracy occurs in solid state physics in the treatment of paramagnetic ions in crystals.

It was shown in Section 10.2 that the static electric dipole moment (EDM) in a nondegenerate eigenstate must be zero if the Hamiltonian is invariant under parity. Thus a measurement on the absence of such EDM for elementary particles gives also an upper limit of F, the probability amplitude for parity mixing. Smith, Purcell and Ramsay (1957) have measured the EDM of the neutron by a resonance-beam apparatus. The EDM was found to be

$$D(n) = - (0.1 \pm 2.4) \times 10^{-20} \ e \ \text{cm} \tag{124}$$

where e is the electronic change. Assuming the size of the neutron to be $\sim 10^{-13}$ cm, this gives a very severe limit on F:

$$|F|^2 < 3 \times 10^{-10}. \tag{125}$$

It is known that the weak interactions are not invariant under parity.

However we can show that if T invariance holds then EDM must still be zero even though parity may not be conserved. If parity is not conserved, then we write

$$\psi = \psi_p + F\psi_{-p}, \tag{126}$$

where ψ_p and ψ_{-p} have opposite parities. But if time reversal is invariant then ψ_p and ψ_{-p} will be $\dfrac{\pi}{2}$ out of phase and hence cannot contribute to EDM which is a real quantity.

Let us consider a particle with spin $\vec{J}$. Then

$$T|\alpha, j, m> = e^{i\phi} \ |\alpha, j, - m>, \tag{116$'$}$$

assuming T invariance.

The EDM is
$$\vec{D} = \Sigma \, e_i \, \vec{r}_i. \tag{127}$$

The average value of $\vec{D}$ must be proportion to $\vec{J}$:
$$<\alpha, j, m|\vec{D}|\alpha, j, m> = k <\alpha, j, m|\vec{J}|\alpha, j, m> \tag{128}$$

where k is real numerical constant. If we take the complex conjugation of both sides in Eq. (128) and replace 1 by $U_T^\dagger U_T$, we get

$$<\alpha, j, m|U_T^\dagger U_T \vec{D} \, U_T^\dagger U_T |\alpha, j, m>^*$$
$$= k <\alpha, j, m|U_T^\dagger U_T \vec{J} \, U_T^\dagger U_T |\alpha, j, m>^*$$

From Eqs. (78) and (80), (128)

we get
$$<\alpha, j, -m|\vec{D}|\alpha, j, -m> = -k <\alpha, j, -m|\vec{J}|\alpha, j, -m> \tag{129}$$

Comparison of Eqs. (128) and (129) shows that
$$<\alpha, j, m|\vec{D}|\alpha, j, m> = 0 \tag{130}$$

Thus the EDM of the state must vanish under time reversal invariance.

A system having rotational invariance cannot have a static electric dipole moment in the absence of an accidental degeneracy if the dynamics are time reversal invariant.

Static odd electric and even magnetic moments are forbidden by P and or by T invariance (Henley (1969)). Both P and T must be violated for such moments to occur.

Dress, Miller and Ramsey (1973) have measured the EDM of the neutron:
$$D(n) = (3.2 \pm 7.5) \times 10^{-23} \, e \text{ cm} \tag{131}$$

Stein et al (1969) measured the EDM of neutral Cs atom with spin half:
$$D(Cs) < 3 \times 10^{-21} \, e \text{ cm} \tag{132}$$

This gives
$$D(e^-) < 2.5 \times 10^{-23} \, e \text{ cm} \tag{133}$$

Some recent bounds on the EDM of various particles are given below.
$$D(n) = -(0.3 \pm 0.5) \times 10^{-25} \, e \text{ cm}$$
$$D(e^-) = (-2.7 \pm 8.3) \times 10^{-27} \, e \text{ cm}$$
$$|D(v)| \lesssim 2 \times 10^{-27} \, e \text{ cm}$$
$$|D(p)| < 2 \times 10^{-21} \, e \text{ cm}.$$

The standard model predictions of the EDM are many orders of magnitude below these upper limits, for example, $D(n) \sim 10^{-33} \, e$ cm; zero for the charged leptons (Marshak (1993)).

Recently, several experiments are under way to measure the EDM of elementary particles. These are expected to have important impact on particle physics. An EDM of an elementary particle implies that T symmetry is violated. This implies a violation of CP symmetry by the CPT theorem. Landau (1957) first pointed out that EDM must be zero if CP is invariant. Although no EDM has been found, the limits set have had decisive influence on particle theories. The discovery of an EDM will give insights into physics "beyond the standard model" of elementary particles.

PROBLEMS

10.1 Prove the following theorem. If the negative part of the spectrum of the Hamiltonian H is bounded and the positive part is unbounded, the parity transformation operator P is linear and the time-reversal operator T is antilinear.

10.2 (a) Show that the momentum inversion operator $U\psi_\alpha(\vec{p}) = \psi_\alpha(-\vec{p})$ is unitary.

    (b) For a spin 1/2 particle, show that

$$e^{-i\pi s_y/\hbar} = -i\sigma_y.$$

    (c) Show that $T^2 = \pm 1$.

CHAPTER 11

Applications

We consider some applications of the methods and techniques discussed in previous chapters to solve problems of interest. In Section 11.1 the rigid rotator is considered. We discuss the harmonic oscillator and the coherent states in Section 11.2 and 11.3 respectively. In Sections 11.4 and 11.5 the motion of a particle in a central potential and the hydrogen atom are studied. The decays of K-mesons (kaons) provide us with examples of the superposition principle in quantum mechanics. This is discussed in Section 11.6. We consider the theory of the motion of a charged particle in a magnetic field in Section 11.7. The Aharonov-Bohm effect is considered in this section. Finally, we discuss supersymmetric quantum mechanics in Section 11.8.

11.1 THE RIGID ROTATOR

A system of two particles of masses m_1 and m_2, constrained to be at a distance R apart, is termed a rigid rotator. It serves as a model of a diatomic molecule in respect of its rotational motion.

As in classical mechanics, here we can separate the motion of the center of mass of the system. We shall ignore the motion of the center of mass. The moment of inertia for this system about its center of mass is $I = \mu R^2$, where $\mu = \dfrac{m_1 m_2}{m_1 + m_2}$ is the reduced mass. The rotator may now be treated as a one particle problem. We consider the motion of a particle of mass μ which is free except that it is constrained to move on a sphere of radius R. The energy is purely kinetic:

$$E = \frac{\vec{L}^2}{2I},\tag{1}$$

where $\vec{L}$ is the angular momentum about the axis of rotation passing through the center of mass.

The Hamiltonian is

$$H = \frac{\vec{L}^2}{2I}.$$ (2)

The time-independent Schrödinger equation for this system is

$$\frac{\vec{L}^2}{2I}\,\psi(\theta, \phi) = E\psi(\theta, \phi).$$ (3)

Since the eigenfunctions of $\vec{L}^2$ are the spherical harmonics $Y_{lm}(\theta, \phi)$ and its eigenvalues are $l(l + 1)\hbar^2$ with $l = 0, 1, 2, \ldots$, the eigenfunctions of the rigid rotator Hamiltonian are $Y_{lm}(\theta, \phi)$ and the energy eigenvalues are

$$E_l = \frac{\hbar^2}{2I}\, l(l + 1), \quad l = 0, 1, 2, \ldots$$ (4)

The energy levels do not depend on m, the azimuthal quantum number ($m = 0, \pm 1, \pm 2, \ldots, \pm l$). Each energy level E_l (except for $l = 0$) is $(2l + 1)$ fold degenerate, since for a given l, there are $(2l + 1)$ independent wave functions Y_{lm} with $m = -l, -l + 1, \ldots, l$. This is due to the fact that H commutes with $\vec{L}$.

The normalized wave functions ψ_{lm} satisfying

$$\int_0^\pi \int_{-\pi}^\pi \psi_{lm}^* \, \psi_{lm} \sin\theta \, d\theta \, d\phi = 1$$ (5)

may be written as

$$\psi_{lm} = N_{lm} Y_{lm},$$ (6)

where

$$N_{lm} = (-1)^{(m + |m|)/2} \left[\frac{2l + 1}{4\pi} \frac{(l - |m|)!}{(l + |m|)!} \right]^{1/2}.$$ (7)

The wave functions form a complete, ortho-normal set:

$$\int_0^\pi \int_{-\pi}^\pi \psi_{l'm'}^* \, \psi_{lm} \sin\theta \, d\theta \, d\phi = \delta_{l'l} \, \delta_{m'm}.$$ (8)

A diatomic molecule may be considered to be a rigid dumb-bell where the nuclei of the two atoms are held a fixed distance R apart. In this approximation, the Hamiltonian of a diatomic molecule with respect to rotational states is that of a rigid rotator. Transitions between rotational states with the selection rule $\Delta l = \pm 1$ give rise to a band spectrum. For diatomic molecules with typical values: $R \sim 1$ to 2Å, $\mu \sim 1$ to 20 amu, and $l \sim 1$ to 10, the photon energies in transitions between rotational states are $\sim 10^{-3}$ to 10^{-4} eV, corresponding to wavelength $\lambda \sim 1$ mm to 1 cm. Thus, pure rotational spectra lie in the far infrared and microwave regions.

11.2 THE HARMONIC OSCILLATOR

The study of the harmonic oscillator is of great importance and relevance in quantum mechanics. Many complicated potentials can be approximated by the harmonic oscillator potential in the vicinity of their equilibrium positions. It has applications in molecular vibrational spectra and crystalline vibrations. In the quantum theory of radiation, the electromagnetic field may be viewed as an infinite set of uncoupled simple harmonic oscillators. It provides the basis of the second quantization formalism for many-particle systems and quantum fields. Newton in "Principia" pointed out that a large number of physical systems may be considered as a collection of oscillators. From a pedagogical point of view, the harmonic oscillator illustrates the general principles and methods of quantum theory.

(a) The Linear Harmonic Oscillator

We shall now study the one-dimensional motion of a particle of mass m in a parabolic potential

$$V(x) = \frac{s}{2}\,x^2,\tag{9}$$

with $H = \dfrac{p^2}{2m} + \dfrac{s}{2}\,x^2$ where s is a constant.

The eigenvalue equation $H|\psi\rangle = E|\psi\rangle$, in the coordinate representation is

$$\frac{d^2\psi(x)}{dx^2} + \frac{2m}{\hbar^2}\left(E - \frac{s}{2}\,x^2\right)\psi(x) = 0.\tag{10}$$

We put

$$y = x\left(\frac{ms}{\hbar^2}\right)^{1/4}$$

$$\alpha = \frac{2E}{\hbar}\left(\frac{m}{s}\right)^{1/2} = \frac{2E}{\hbar\omega}\tag{11}$$

with $\omega = \left(\dfrac{s}{m}\right)^{1/2}$, the frequency.

If we put $\psi(x) = u(y)$, Eq. (10) becomes

$$\frac{d^2u}{dy^2} + (\alpha - y^2)u = 0\tag{12}$$

Now clearly, for $E < 0$, there are no eigenvalues, and for $E > 0$, the spectrum will be discrete with an infinite set of values in the interval $(0, \infty)$, since for any E, $V > E$ at large distances.

(i) *Sommerfeld Polynomial Method*

We discuss the quantization of the oscillator in coordinate basis.

This differential equation (12) has an irregular singularity at infinity. The asymptotic behavior of $u(y)$ cnn be inferred from the equation by neglecting α compared with y^2 for large y. The equation then becomes $\dfrac{d^2u}{dy^2} \approx y^2u.$ The functions $u \sim e^{\pm \frac{1}{2} y^2}$ satisfy this equation. The boundary condition on the wave function $\psi(x) = 0$ at $x = \pm \infty$ is equivalent to the constraints $u(y)$ bounded as $y \to \pm \infty$. The solution $\sim e^{+\frac{1}{2} y^2}$ is unacceptable, as it diverges to be outside the Hilbert space and the rigged Hilbert space. The acceptable solution is $\sim e^{-\frac{1}{2} y^2}$ for large y. We, therefore, seek the solution of Eq. (12) for all y in the form

$$u(y) = h(y)\, e^{\frac{-y^2}{2}}, \tag{13}$$

where the functions $h(y)$ must not affect the asymptotic behavior of u, Substitution of Eq. (13) in Eq. (12) yield

$$h'' - 2yh' + (\alpha - 1)\, h = 0, \tag{14}$$

where the primes denote differentiation with respect to y.

A simple solution, for $\alpha = 1$, $h = $ constant. This is a solution without nodes: this solution corresponds to the lowest eigenfunction with eigenvalue $E_0 = \dfrac{\hbar\omega}{2}$. So the lowest energy is not zero — this is very important.

Since the solutions are regular except at infinity, we can therefore expand around the origin in a power series in y. The eigenfunctions must have definite parity (even or odd) as $V(x) = V(-x)$.

Let
$$h(y) = \sum_{n=0}^{\infty} c_n y^n, \tag{15}$$

where for the even solutions $c_0 = 1$, $c_1 = 0$ and for the odd solution $c_0 = 0$, $c_1 = 1$. Substitution of Eq. (15) into Eq. (14) gives the recursion relation

$$c_{n+2} = \frac{-\alpha + 1 + 2n}{(n+1)(n+2)}\, c_n \quad (n \geq 0). \tag{16}$$

Now either the series Eq. (15) terminates, or not. If it does not terminate, the asymptotic ratio of the coefficients is

$$\frac{c_{n+2}}{c_n} \xrightarrow[n \to \infty]{} \frac{2}{n}. \tag{17}$$

This is the same as in the series $y^p e^{y^2}$ with p having any finite positive value. This behavior of $h(y) \sim e^{+y^2}$ for large y gives $u(y) \sim e^{y^2/2}$ for large y as the solution which we want to discard. Therefore the series must terminate. If

$$\alpha = 2n + 1 \tag{18}$$

for n, a nonnegative integer, $h(y)$ will be a polynomial of degree n and $u(y) \to 0$ at large y. The eigenvalue condition Eqs. (18) with (11) yields the energy spectrum of a linear harmonic oscillator (for both even and odd eigenfunctions)

$$E_n = \left(n + \frac{1}{2}\right)\hbar\omega, \quad n = 0, 1, 2, \ldots \tag{19}$$

The energy spectrum is an infinite sequence of equispaced, discrete energy levels.

The eigenfunctions are

$$u_n(x) = A_n h_n(y)\, e^{\frac{-y^2}{2}}, \tag{20}$$

where A_n is the normalization constant, with $h_j(y) = \sum_j c_j y^j$, $c_{j+2} = \dfrac{(2j+1-\alpha)}{(j+2)(j+1)} c_j$.

This gives us an explicit method of calculating the coefficients, and thus the eigenfunctions.

We consider the generating function

$$Z(t, y) = e^{-t^2 + 2ty} = \sum_{n=0}^{\infty} h_n(y)\, \frac{t^n}{n!}. \tag{21}$$

Then

$$\frac{\partial Z}{\partial t} = (2y - 2t)Z = \sum_0^{\infty} h_{n+1}\, \frac{t^n}{n!}, \tag{22}$$

$$\frac{\partial Z}{\partial y} = 2tZ = \sum h'_n\, \frac{t^n}{n!}. \tag{23}$$

Adding Eqs. (22) and (23), we get

$$2yZ = \sum_{n=0}^{\infty} 2y h_n\,(y)\, \frac{t^n}{n!}$$

$$= \sum_{n=0}^{\infty} (h_{n+1}(y) + h'_n\,(y))\, \frac{t^n}{n!},$$

or,
$$h_{n+1}(y) + h'_n\,(y) = 2y h_n\,(y). \tag{24}$$

From Eq. (22).

$$\sum_0^{\infty} 2y h_n\, \frac{t^n}{n!} - \sum 2h_n\, \frac{t^{n+1}}{n!} = \sum h_{n+1}\, \frac{t^n}{n!},$$

or,
$$2y h_n - 2n h'_{n-1} = h_{n+1}. \tag{24a}$$

From Eq. (23),

$$\sum 2h_n\, \frac{t^{n+1}}{n!} = \sum h'_n\, \frac{t^n}{n!},$$

or,
$$h'_n\,(y) = 2n h_n - 1. \tag{24b}$$

Adding Eqs. (24a) and (24b), we get Eq. (24).

We will now find the differential equation satisfied by the coefficients $h_n(y)$.

Differentiating Eq. (24) we obtain

$$h'_{n+1} + h''_n = 2h_n + 2y h'_n,$$

and from Eq. (24b),

$$h'_{n+1} = 2(n+1)h_n$$

so that

$$h''_n(y) - 2yh'_n(y) + 2nh_n(y) = 0, \tag{25}$$

which is the same equation as we had Eq. (14).

Therefore, the $h_x(y)$ are those which we want as eigenfunctions. These polynomials $h_x(y)$ are called Hermite polynomials.

The second-order linear differential equation Eq. (25) is not self-adjoint. To put Eq. (25) into self-adjoint form we multiply by e^{-y^2}. So the orthogonality integral is

$$\int_{-\infty}^{\infty} h_m(y)h_n(y)e^{-y^2}\, dy = 0$$

$$m \neq n \tag{26}$$

We may define the Hermite functions as

$$v_n(y) = e^{-y^2/2}\, h_n(y), \tag{27}$$

satisfying the self-adjoint differential equation

$$v''_n(y) + (2n + 1 - y^2)v_n(y) = 0. \tag{28}$$

The solutions v_n are orthogonal for $-\infty < y < \infty$ with a unit weight function.

Now,

$$h_n(y) = \frac{\partial^n}{\partial t^n} Z \bigg|_{t=0} = \frac{\partial^n}{\partial t^n} e^{-(t-y)^2 + y^2} \bigg|_{t=0}$$

$$= e^{y^2} (-1)^n \frac{\partial^n}{\partial y^n} e^{-(t-y)^2} \bigg|_{t=0}$$

so that

$$h_n(y) = e^{y^2} (-1)^n \frac{d^n}{dy^n} \left(e^{-y^2}\right). \tag{29}$$

This is Rodrigues' representation of $h_n(y)$. Let us now find the normalized linear harmonic oscillator eigenfunctions. We have

$$\psi_n(x) = A_n\, e^{-y^2/2}\, h_n(y), \tag{30}$$

where

$$x = y \left(\frac{ms}{\hbar^2}\right)^{-1/4}.$$

We require

$$\int_{-\infty}^{\infty} |\psi_n(x)|^2\, dx = 1. \tag{31}$$

We obtain

$$|A_n|^2 \left(\frac{ms}{\hbar^2}\right)^{-1/4} \int_{-\infty}^{\infty} dy \, e^{-y^2} h_n^2(y)| = 1.$$

Let

$$I = \int_{-\infty}^{\infty} h_n^2(y) \, e^{-y^2} \, dy$$

$$= \int_{-\infty}^{\infty} e^{-y^2} \left(\frac{d^n}{dy^n} e^{-y^2}\right)^2 dy$$

$$= \int_{-\infty}^{\infty} (-1)^n h_n(y) \frac{d^n}{dy^n} e^{-y^2} \, dy$$

$$= (-1)^n \int_{-\infty}^{\infty} dy \frac{d^n}{dy^n} [(-1)^n h_n(y)] e^{-y^2},$$

integrating by parts n times

But $h_n(y)$ is a polynormial of degree n, and so we get the coefficient of the highest term.

$$I = \int_{-\infty}^{\infty} n! \, \{\text{coeff of } y^n \text{ in } h_n(y)\} e^{-y^2} \, dy.$$

From the differentiated form of the generating function, we see that this coefficient is

$$(-1)^n \cdot 2^n \cdot (-1)^n = 2^n.$$

So

$$I = \int_{-\infty}^{\infty} n! \, 2^n \, e^{-y^2} \, dy$$

$$= n! \, 2^n \, \sqrt{\pi}.$$

Therefore,

$$|A_n| = \left(\frac{4ms}{h^2}\right)^{1/8} (2^n \, n!)^{-1/2},$$

and the normalized eigenfunctions are

$$\psi_n(x) = \left[\left(\frac{m\omega}{\pi\hbar}\right)^{1/2} \frac{1}{2^n \, n!}\right]^{1/2} h_n\left(\sqrt{\frac{m\omega}{\hbar}} \, x\right) e^{-\frac{mw}{2\hbar} x^2}, \quad (32)$$

where $h_n(z)$ is the Hermite polynomial of degree n.

The first few Hermite polynomials are

$$h_0(z) = 1,$$

$$h_1(z) = 2z,$$

$$h_2(z) = 4z^2 - 2$$

$$h_3(z) = 8z^3 - 12z,$$

$$h_4(z) = 16z^4 - 48z^2 + 12. \tag{33}$$

The wave functions have a definite parity — even (odd) when n is even (odd). Clearly, the expectation values of x and p vanish:

$$\langle x \rangle = x_{nn} = 0 \tag{34}$$

$$\langle p \rangle = p_{nn} = 0$$

The ground-state wave function $\psi_0(x)$ is a gaussian.

To obtain the wave functions in the momentum representation, we note that the Hamiltonian $H = \dfrac{p^2}{2m} + \dfrac{s}{2}\, x^2$ and the commutation rules are invariant under the transformation $x \to \dfrac{p}{m\omega}$, $p \to -\, m\omega x$. So we write

$$\tilde{\psi}_n(p) = \frac{a_n}{\sqrt{m\omega}}\; \psi_n\,(p/m\omega), \tag{35}$$

where

$$a_n = (-\,1)^n.$$

Next, we discuss the comparison of the behaviors of the classical and quantal oscillators. For small quantum numbers ($n \sim 1$), the difference in the behaviors is noticeable. The agreement between the classical and the quantam mechanical probability densities improves as n increases. Let us calculate the classical probability density of $P_{cl}(x)$ of a simple harmonic oscillator with frequency ω. Let the particle be at the origin at $t = 0$ with speed $x_0\omega$. The displacement of the particle at time t is $x = x_0 \sin \omega t$, and speed $v = x\, \omega \cos \omega t$, and the energy is $E = \dfrac{m\omega^2 x_0^2}{2}$. x_0 is the amplitude. The classical motion takes place between the turning points at $\pm\, x_0 = \pm\, (2E/m\omega^2)^{1/2}$. The probability $P_{cl}(x)\, dx$ is the probability of finding the particle in the interval $(x, x + dx)$ at any time. It is equal to the fraction of the time spent by the particle in that interval to the total time. If $T = \dfrac{2\pi}{\omega}$ is the period of oscillation, then

$$P_{cl}(x)\, dx = \frac{dt}{T} = \frac{\omega dt}{2\pi}$$

But,

$$dt = \frac{dx}{x} = \frac{dx}{\omega\,\sqrt{x_0^2 - x^2}}$$

So that

$$P_{cl}(x)\, dx = \frac{dx}{2\pi\,\sqrt{x_0^2 - x^2}}\,.$$

The normalized classical probability density, over the interval, $-x_0 < x < x_0$, is

$$P_{cl}(x) = \frac{1}{\pi \sqrt{x_0^2 - x^2}}, \tag{36}$$

where

$$\int_{-x_0}^{+x_0} P_{cl}(x)\, dx = 1$$

For x small, $|\psi_n|^2$ for a quantal oscillator is very different from P_{cl} of a classical oscillator. The ground state ($n = 0$) position probability density has its maximum at $x = x_0$, although, classically the particle is most likely to be found at or near its turning points. As n increases, the agreement improves, in accordance with Bohr's correspondence principle.

Using the virial theorem, (Section 13.4), for the quantum mechanical oscillator in the state ψ_n,

$$\langle T \rangle_n = \langle V \rangle_n = \left(n + \frac{1}{2} \right) \frac{\hbar\omega}{2} = \frac{E_n}{2}, \tag{37}$$

and so

$$\langle p^2 \rangle_n = \left(n + \frac{1}{2} \right) m\hbar\omega = mE_n,$$

$$\langle x^2 \rangle_n = \left(n + \frac{1}{2} \right) \frac{\hbar}{m\omega} = \frac{E_n}{m\omega^2} \tag{38}$$

These agree with the classical expressions. For the harmonic oscillator, the agreement is true for all n (not only for large n). Since $\langle x \rangle_n = \langle p \rangle_n = 0$, we obtain

$$(\Delta x)_n \cdot (\Delta p)_n = \left(n + \frac{1}{2} \right) \hbar. \tag{39}$$

The ground state is, therefore, a minimum uncertainty state. Note that ψ_0 is just a gaussian distribution: The Hamiltonian H is very symmetric in p, x — quadratic in both. This is the reason for the minimum uncertainty in the lowest state of the oscillator.

We consider the time-dependent behavior of a harmonic oscillator. The Green function must here satisfy

$$i\hbar \frac{\partial K}{\partial t} = \left[-\frac{\hbar^2}{2m} \frac{\partial^2}{\partial x^2} + \frac{s}{2} x^2 \right] K, \tag{40}$$

with

$$\lim_{t \to t_0} K(t, t_0) = \delta (x - x_0).$$

Then

$$\psi(x, t) = \int_{-\infty}^{\infty} K(x, t; x_0, t_0)\, \psi(x_0, t_0)\, dx_0 \tag{41}$$

We may write

$$K(x, t; x_0, t_0) = \sum_i \psi_i(x, t)\, \psi_i^*(x_0, t_0), \tag{42}$$

with

$$\psi_n(x, t) = u_n(x)\, e^{-\frac{i}{\hbar} E_n t}$$

$$= u_n(x)\, e^{-in t \omega}\, e^{-i\omega t/2}$$

$$u_n(x) = \left(\frac{4ms}{\hbar^2}\right)^{\frac{1}{8}} \cdot \left(\frac{1}{2^n \cdot n!}\right)^{\frac{1}{2}} h_n(y) e^{-y^2/2},$$

$$x = y \left(\frac{ms}{\hbar^2}\right)^{\frac{-1}{4}}.$$

Therefore,

$$K(x, t; x_0, t_0) = \left(\frac{4ms}{\hbar^2}\right)^{\frac{1}{4}} e^{-\frac{(y^2 + y_0^2)}{2}}\, e^{-\frac{i\omega}{2}(t - t_0)}$$

$$\times \sum_{n=0}^{\infty} h_n(y) h_n(y_0)\, \frac{1}{2^n \cdot n!}\, e^{-in\omega(t - t_0)}. \tag{43}$$

We have

$$\sum_n h_n(y) h_n(y_0)\, \frac{T^n}{2^n \cdot n!}\quad \text{with } T = e^{-iw(t - t_0)}$$

$$= e^{y^2 + y_0^2} \sum_{n=0}^{\infty} \frac{d^n}{dy^n} e^{-y^2} \cdot \frac{d^n}{dy_0^n} e^{-y_0^2}\, \frac{1}{n!} \left(\frac{T}{2}\right)^n$$

$$= e^{y^2 + y_0^2} \sum_{n=0}^{\infty} \left\{ \left(\frac{TD_y}{2}\right)^n \cdot \frac{1}{n!}\, \frac{d^n}{dy_0^n} e^{-y_0^2} \right\} e^{-y^2},$$

$$D_y \equiv \frac{d}{dy}$$

which is just the Taylor series expansion

$$\sum \{...\} = e^{y^2 + y_0^2} \left[e^{-\left(y_0 + \frac{TD_y}{2}\right)^2} \right] e^{-y^2}.$$

Now, to interpret this operator, let us use the Fourier integral theorem. We put

$$e^{-y^2} = \frac{1}{\sqrt{2\pi}} \int_{-\infty}^{+\infty} e^{i\omega y}\, \phi(\omega)\, d\omega.$$

Then

$$\phi(\omega) = \frac{1}{\sqrt{2\pi}} \int_{-\infty}^{+\infty} e^{-y^2}\, e^{-i\omega y}\, dy$$

$$= \frac{e^{-\omega^2/4}}{\sqrt{2}}.$$

But now, operating on the function e^{-y^2} with the differential operator [...] is the same as multiplying its transform ϕ.

Now,

$$e^{-\left(y_0 + \frac{TD_y}{2}\right)^2} e^{-y^2} = \frac{1}{\sqrt{2\pi}} \int_{-\infty}^{+\infty} e^{-\left(y_0 + \frac{i\omega T}{2}\right)^2} e^{i\omega y}\, \phi(\omega)\, d\omega$$

$$= \frac{1}{\sqrt{4\pi}} \int_{-\infty}^{+\infty} e^{-\left(y_0 + \frac{i\omega T}{2}\right)^2} e^{i\omega y}\, e^{-\omega^2/4}\, d\omega.$$

$$= \frac{1}{\sqrt{4\pi}} \int_{-\infty}^{+\infty} \exp - \left[y_0^2 + i\omega(y_0 T - y) + \frac{\omega^2}{4}(1 - T^2) \right] d\omega$$

$$= \frac{1}{\sqrt{4\pi}}\, e^{-y_0^2}\, e^{-\frac{(y - Ty_0)^2}{1 - T^2}} \int_{-\infty}^{+\infty} e^{-\frac{1}{2}\sqrt{1 - T^2}\, y + \frac{i(y_0 T - y)}{\sqrt{1 - T^2}}}\, dy.$$

$$= \frac{1}{\sqrt{4\pi}}\, e^{-y_0^2}\, e^{-\frac{(y - Ty_0)^2}{1 - T^2}} \cdot 2\sqrt{\frac{\pi}{1 - T^2}}.$$

$$= \frac{e^{-y_0^2 - \frac{(y - Ty_0)^2}{1 - T^2}}}{\sqrt{1 - T^2}}.$$

Note our use of two different ω-one as constant of oscillation, and as dummy integration variable.

So

$$K = \left(\frac{4ms}{\hbar^2}\right)^{\frac{1}{4}} e^{\frac{(y^2 + y_0^2)}{2}}\, e^{-\frac{i\omega}{2}(t - t_0)}\, e^{-\left(y_0^2 + \frac{(y - Ty_0)^2}{1 - T^2}\right)} \times \frac{1}{\sqrt{1 - T^2}}.$$

But,

$$\frac{1}{\sqrt{1 - T^2}} = \frac{1}{\sqrt{T}}\,(2i\,\sin\,\omega(t - t_0))^{-1/2}.$$

Therefore,

$$K = \left(\frac{4ms}{\hbar^2}\right)^{\frac{1}{4}} e^{\frac{(y^2 + y_0^2)}{2}} \left(2i\,\sin\,\omega(t - t_0)^{-1/2}\right) \frac{i}{e^{2\,\sin\,\omega(t - t_0)}} \times$$

$$\times \left[\frac{y_0^2 + y^2}{T} - 2yy_0\right] \tag{44}$$

For $t_0 = 0$,

$$K_{osc}(x, t; x_0, 0) = \left[\frac{2\pi i\hbar\,\sin\,\omega t}{m\omega}\right]^{-1/2} \exp\left(i\,\frac{m\omega}{2\hbar\,\sin\,\omega t}\,\{(x^2 + x_0^2)\,\cos\,\omega t - 2xx_0\}\right). \tag{45}$$

The multiplicative factor can be written as

$$[2i\,\sin\,(\omega t)]^{-1/2} = e^{-i\omega/2}\,(1 - e^{-2i\omega t})^{-1/2}.$$

So K_{osc} is periodic with a period which is twice the classical oscillation period.

Also, we note

$$K_{osc}\left(x, t + \frac{n\pi}{\omega}; x_0, 0\right) = (-i)^n \, K_{osc}\left(x, t; (-1)^n x_0, 0\right)$$

with the help of this relation, we can study the time evolution of K_{osc} over the zeros of $\sin(\omega t)$.

For $t_0 \neq 0$, simply replace t by $(t - t_0)$.

Expression Eq. (45) gives the propagator for the simple harmonic oscillator. This is also called the fundamental solution of harmonic oscillator (Pauli (1973)).

For $\omega \to 0$, we get back the free particle case

$$K \xrightarrow[\omega \to 0]{} \left[\frac{2\pi i\hbar t}{m}\right]^{-\frac{1}{2}} e^{\frac{im}{2\hbar t}(x - x_0)^2}. \tag{46}$$

The reason that we could carry out this solution is that $e^{-\frac{i}{\hbar}E_n t} \approx \left(e^{-\frac{i}{\hbar}E_0 t}\right)^n$ and so leads to a particularly simple series.

The lowest eigenfunction is a gaussian function:

$$u_0(x) = \left(\frac{m\omega}{\pi\hbar}\right)^{1/4} e^{-\frac{m\omega x^2}{2\hbar}}, \tag{47}$$

and so the position probability density in that state is

$$P_0(x) = |u_0(x)|^2 = \left(\frac{m\omega}{\pi\hbar}\right)^{1/4} e^{-\frac{m\omega x^2}{\hbar}} \tag{48}$$

Consider now a "displaced oscillator" such that at $t = 0$,

$$u_0(x, 0) = \left(\frac{m\omega}{\pi\hbar}\right)^{1/4} e^{-\frac{m\omega x^2}{2\hbar}(x_0 - \bar{x}_0)^2}$$

$$= u_0 \, (x_0 - \bar{x}_0). \tag{49}$$

so that

$$P_0(x_0) = \left(\frac{m\omega}{\pi\hbar}\right)^{1/2} e^{-\frac{m\omega}{\hbar}(x_0 - \bar{x}_0)^2}. \tag{50}$$

This state is a non-stationary state and so it will change in time. We can find the time behavior of the wave packet using the principal solution. This is just an example of the expansion theorem – the state is describable as a superposition of harmonic oscillators of frequency ω, and all energies E_n.

Then the wave packet is

$$u(x, t) = \left(\frac{m\omega}{\pi\hbar}\right)^{1/4} \left[\frac{2\pi i\hbar \, \sin \omega t}{m\omega}\right]^{-1/2} \int_{-\infty}^{\infty} dx_0 \; e^{-\frac{m\omega}{\hbar}(x_0 - \bar{x})^2}$$

$$\times \; \exp\left(\frac{im\omega}{e^{2\pi} \sin \omega t} \left\{(\bar{x}^2 - x^2) \cos \omega t - 2\bar{x}x\right\}\right) \qquad (51)$$

and so

$$u(x, t) = \left(\frac{m\omega}{\pi\hbar}\right)^{1/4} e^{-\frac{i\omega t}{2}} \, e^{-\frac{m\omega}{2\hbar}\left\{\bar{x}^2 - i \cot \omega t - ie^{-i\omega t} \sin \omega t \left(\bar{x} - \frac{ix}{\sin \omega t}\right)^2\right\}} \qquad (52)$$

Then

$$P(x, t) = u^* u$$

$$= \left(\frac{m\omega}{\pi\hbar}\right)^{1/4} e^{-\frac{m\omega}{\hbar}\{x - \bar{x} \cos \omega t\}^2}, \qquad (53)$$

which describes a bell-shaped curve about $\bar{x}(t) = \bar{x} \cos \omega t$.

Physically, this represents the classical behavior of an oscillator – the pulse moves harmonically on the average. A wave packet will come back to its shape after a time $2\pi/\omega$.

(ii) *Operator or Algebraic Method*

Let us consider a simple harmonic oscillator of mass m and frequency ω. The classical Hamiltonian is

$$H = \frac{1}{2m} (p^2 + m^2 \omega^2 q^2). \qquad (54)$$

We assume the same Hamiltonian in quantum mechanics. This Hamiltonian together with the commutation relation $[q, p] = i\hbar$, where q, and p are self-adjoint operators, defines the system completely. A clever method, due to Dirac, enables us to consider the quantization of the oscillator in the energy basis.

In order to factorize H we introduce two dimensionless, non-hermitian operators

$$a = \frac{1}{\sqrt{2m\hbar\omega}} (m\omega x + ip),$$

$$a^\dagger = \frac{1}{\sqrt{2m\hbar\omega}} (m\omega x - ip), \qquad (55)$$

Using the canonical commutation relations, we obtain

$$[a, a^\dagger] = I. \qquad (56)$$

Also, $\qquad\qquad [a, I] = [a^\dagger, I] = 0 \qquad (57)$

The operators a, $a^\dagger$, and I are generators of a Lie algebra.

We define the (level) number operator

$$N = a^\dagger a. \qquad (58)$$

N is formally a self-adjoint operator. It gives the number of quanta of excitation of the oscillator.

Now, the Hamiltonian H can be written as

$$H = \frac{1}{2}\,\hbar\omega\,(aa^\dagger + a^\dagger a)$$

$$= \hbar\omega\left(aa^\dagger - \frac{1}{2}\,I\right)$$

$$= \hbar\omega\left(a^\dagger a + \frac{1}{2}\,I\right) = \hbar\omega\left(N + \frac{1}{2}\right). \tag{59}$$

Clearly, N commutes with H, and its eigenstates are also eigenstates of H. Hence, the problem of finding the spectrum of H is now reduced to finding the spectrum of N.

From Eq. (56), we obtain

$$[N, a] = -a, \tag{60}$$

$$[N, a^\dagger] = a^\dagger. \tag{61}$$

The operators a, $a^\dagger$, I, and N are also generators of a Lie algebra. The corresponding Lie group is the Heisenberg-Weyl group (H_4).

In terms of a, $a^\dagger$, we can express the operators x and p:

$$x = \sqrt{\frac{\hbar}{2m\omega}}\,(a^\dagger + a), \tag{62}$$

$$p = i\,\sqrt{\frac{m\hbar\omega}{2}}\,(a^\dagger - a). \tag{63}$$

From Eq. (59), we get (in units of $\hbar\omega$)

$$aa^\dagger = H + \frac{1}{2}\,I, \tag{64}$$

$$a^\dagger a = H - \frac{1}{2}\,I, \tag{65}$$

Thus, the operators a and $a^\dagger$ provide "two-way" factorization.

By Eq. (58), for any arbitrary ket $|u\rangle$,

$$\langle u|N|u\rangle = \langle u|a^\dagger a|u\rangle = \langle au|au\rangle = \|a|u\rangle\|^2, \tag{66}$$

so that $\qquad\qquad (Nu, u) \geq 0.$

N is a positive definite operator.

Since $H = N + \dfrac{1}{2}$ (in units of the $\hbar\omega$), the energy spectrum has a lower bound $\dfrac{1}{2}$.

Let us assume that there is a normalizable (energy) eigenket of N unique except for a phase factor, with eigenvalue n:

$$(N - n)|n\rangle = 0. \tag{67}$$

Here $n \geq 0$.

N has no negative eigenvalues. Using Eqs. (60) and (61), we obtain from Eq. (67),

$$Na|n\rangle = (n - 1)a|n\rangle, \tag{68}$$

$$Na^{\dagger}|n\rangle = (n + 1)a^{\dagger}|n\rangle. \tag{69}$$

Hence, $a|n\rangle$ is an eigenvector of N with eigenvalue $n - 1$, if $a|n\rangle$ is in L^2 and $a|n\rangle$ does not vanish. N has only isolated eigenvalues. Repeated applications of the operator a on $|n\rangle$ will generate a sequence of eigenkets with eigenvalues $n - 1, n - 2, \ldots$ since $n \geq 0$, and since $a|0\rangle = 0$, this sequence must terminate at $n = 0$, if n is any positive integer. If we start with a noninteger n, the series will not terminate, leading to eigenkets with negative values of n. But since $n \geq 0$, n has to be a nonnegative integer.

From Eq. (69), we see that $a^{\dagger}|n\rangle$ is an eigenstate of N with eigenvalue $(n + 1)$. By repeated application of the operator $a^{\dagger}$ on $|n\rangle$, we can generate an unlimited sequence of eigenkets $|n + 1\rangle, |n - 2\rangle, \ldots$ with eigenvalues $n + 1, n + 2, \ldots$ The sequence begins with $n = 0$. Therefore, the spectrum of N consists of non-negative integers.

The state $a^{\dagger}|n\rangle$ is proportional to $|n + 1\rangle$. So we write

$$a^{\dagger}|n\rangle = c_n|n + 1\rangle.$$

Then

$$\begin{aligned}
|c_n|^2 &= |c_n|^2 \langle n + 1|n + 1\rangle \\
&= \langle n|aa^{\dagger}|n\rangle \\
&= \langle n|a^{\dagger}a + 1|n\rangle \\
&= (n + 1). \tag{70}
\end{aligned}$$

Hence $|c_n| = \sqrt{n + 1}$. The phases of c_n are arbitrary. We choose the phase such that c_n is real and positive. Then

$$a^{\dagger}|n\rangle = \sqrt{n + 1}|n + 1\rangle.$$

From Eq. (68), $a|n\rangle \propto |n - 1\rangle$. The norm of $a|n\rangle$ is $\langle n|aa^{\dagger}|n\rangle = n$, since $\langle n|n\rangle = 1$. With appropriate choice of phase, we have

$$a|n\rangle = \sqrt{n}|n - 1\rangle. \tag{71}$$

The nonzero matrix elements of $a^{\dagger}$ and a are

$$\langle n|a^{\dagger}|n'\rangle = \sqrt{n}\ \delta_{n,\, n' + 1},$$

$$\langle n|a|n'\rangle = \sqrt{n + 1}\ \delta_{n,\, n' - 1}, \tag{72}$$

The spectrum of N, the number operator, is $n = 0, 1, 2, \ldots$; the energy spectrum of the harmonic oscillator is

$$E_n = \left(n + \frac{1}{2}\right) \hbar\omega, \quad n = 0, 1, 2, \ldots \tag{73}$$

Because the minimum value of n is zero, the ground state of the harmonic oscillator has the energy

$$E_0 = \frac{1}{2} \hbar\omega. \tag{74}$$

Let $|n\rangle$ be the energy eigenstates:

$$H|n\rangle = E_n|n\rangle. \tag{75}$$

Then

$$Ha|n\rangle = aH|n\rangle + [H, a]|n\rangle$$
$$= (E_n - \hbar\omega)a|n\rangle, \tag{76}$$

and

$$Ha^\dagger|n\rangle = a^\dagger H|n\rangle + \hbar w a^\dagger|n\rangle$$
$$= (E_n + \hbar\omega)a^\dagger|n\rangle. \tag{77}$$

If $|n\rangle$ is an eigenstate of H with energy E_n, i.e., $H|n\rangle = E_n|n\rangle$, then $a^\dagger|n\rangle$ is an eigenstate of H with energy $E_n + \hbar\omega$, and $a|n\rangle$ is an eigenstate of H with energy $E_n - \hbar\omega$. So the actions of the operators and $a^\dagger$ on $|n\rangle$ generate ladders of energy eigenstates with energies increasing or decreasing respectively by units of $\hbar\omega$. The operator a is called a lowering operator or the destruction or the annihilation operator. The operator $a^\dagger$ is called a raising operator or the creation operator. They are also known as the ladder operators. We now repeatedly apply the creation operator $a^\dagger$ to the ground state $|0\rangle$. From Eq. (70), we obtain

$$|1\rangle = a^\dagger|0\rangle,$$

$$|2\rangle = \frac{a^\dagger}{\sqrt{2}} |1\rangle = \left(\frac{(a^\dagger)^2}{\sqrt{2}}\right) |0\rangle,$$

$$|3\rangle = \frac{3}{\sqrt{3}} |2\rangle = \left(\frac{(a^\dagger)^3}{\sqrt{3!}}\right) |0\rangle, \tag{78}$$

$$\cdots\cdots\cdots\cdots\cdots\cdots$$

$$|n\rangle = \left(\frac{(a^\dagger)^n}{\sqrt{n!}}\right) |0\rangle.$$

We have a set of orthonormal vectors.

$$|0\rangle, |1\rangle, |2\rangle, \ldots, |n\rangle, \tag{79}$$

corresponding, respectively, to the following eigenvalues of N:

$$0, 1, 2, ..., n, ... \tag{80}$$

The states $|0\rangle$, $|1\rangle$, $|2\rangle$, ..., $|n\rangle$, ... are simultaneous eigenstates of N and H. The set $\{|n\rangle\}$ forms an orthonormal and complete basis of a certain representation (the $\{N\}$ representation):

$$\langle n'|n\rangle = \delta_{n', n}$$

$$\sum_{n=0}^{+\infty} |n\rangle\langle n| = 1 \tag{81}$$

We now give the representative matrices of the operators in this representation. N and H are represented by infinite diagonal matrices:

$$N = \begin{pmatrix} 0 & 0 & 0 & \cdot & \cdots \\ 0 & 1 & 0 & & \\ 0 & 0 & 2 & 0 & \cdots \\ \vdots & 0 & 0 & 3 & \cdots \\ & & \cdot & \cdot & \cdot \end{pmatrix}$$

$$H = \frac{\hbar\omega}{2} \begin{pmatrix} 1 & 0 & 0 & \cdot & \cdots \\ 0 & 3 & 0 & \cdot & \cdots \\ 0 & 0 & 5 & 0 & \cdots \\ \vdots & \vdots & \vdots & \vdots & \vdots \end{pmatrix} \tag{82}$$

The operators a, $a^\dagger$, x, p are not diagonal.

a is the real matrix

$$a = \begin{pmatrix} 0 & \sqrt{1} & 0 & 0 & \cdots \\ 0 & 0 & \sqrt{2} & 0 & \cdots \\ 0 & 0 & 0 & \sqrt{3} & \cdots \\ \vdots & \vdots & \vdots & \vdots & \vdots \end{pmatrix},$$

and $a^\dagger$ is its transpose

$$a^\dagger = \begin{pmatrix} 0 & 0 & 0 & 0 & \cdots \\ \sqrt{1} & 0 & 0 & 0 & \cdots \\ 0 & \sqrt{2} & 0 & 0 & \cdots \\ 0 & 0 & \sqrt{3} & 0 & \cdots \\ \cdot & \cdot & \cdot & \cdot & \cdot \end{pmatrix} \tag{83}$$

From Eqs. (62) and (63), we form the matrices representing x, p in the $\{N\}$ representation:

$$x = \sqrt{\frac{\hbar}{2m\omega}} \begin{pmatrix} 0 & \sqrt{1} & 0 & 0 & \cdots \\ \sqrt{1} & 0 & \sqrt{2} & 0 & \cdots \\ 0 & \sqrt{2} & 0 & \sqrt{3} & \cdots \\ 0 & 0 & \sqrt{3} & 0 & \cdots \\ \cdot & \cdot & \cdot & \cdot & \cdot \end{pmatrix},$$

$$p = i \sqrt{\frac{m\hbar\omega}{2}} \begin{pmatrix} 0 & -\sqrt{1} & 0 & 0 & \cdots \\ \sqrt{1} & 0 & -\sqrt{2} & 0 & \cdots \\ 0 & \sqrt{2} & 0 & -\sqrt{3} & \cdots \\ 0 & 0 & \sqrt{3} & 0 & \cdots \\ & & \cdot & & \end{pmatrix} \qquad (84)$$

The matrix elements of the x and p operators are

$$\langle n'|x|n \rangle = \sqrt{\frac{\hbar}{2m\omega}} \left(\sqrt{n}\ \delta_{n',\,n-1} + \sqrt{n+1}\ \delta_{n',\,n+1} \right) \qquad (85)$$

$$\langle n'|p|n \rangle = i \sqrt{\frac{m\hbar\omega}{2}} \left(-\sqrt{n}\ \delta_{n',\,n-1} + \sqrt{n+1}\ \delta_{n',\,n+1} \right) \qquad (85a)$$

So

$$\langle n|x|n \rangle = \langle n|p|n \rangle = 0. \qquad (86)$$

Next, we consider the expectation values of x^2 and p^2. Since

$$x^2 = \left(\frac{\hbar}{2m\omega} \right) (a^2 + a^{\dagger 2} + a^\dagger a + a a^\dagger)$$

$$= \left(\frac{\hbar}{2m\omega} \right) (a^2 + a^{\dagger 2} + 2a^\dagger a + 1)$$

$$p^2 = - \left(\frac{m\hbar\omega}{2} \right) (a^2 + a^{\dagger 2} - a^\dagger a - a a^\dagger)$$

$$= - \left(\frac{m\hbar\omega}{2} \right) (a^2 + a^{\dagger 2} - 2a^\dagger a - 1),$$

and the expectation values of a^2, and $a^{\dagger 2}$ are both zero, we get

$$\langle x^2 \rangle_n = \frac{\hbar}{2m\omega} \langle 2a^\dagger a + 1 \rangle_n$$

$$= \frac{\hbar}{2m\omega} (2n + 1) = \frac{E_n}{m\omega^2}$$

$$\langle P^2 \rangle_n = \frac{m\hbar\omega}{2} \langle 2a^\dagger a + 1 \rangle_n$$

$$= \frac{m\hbar\omega}{2} (2n + 1) = mE_n. \qquad (87)$$

The expectation values of the kinetic and the potential energies, are respectively

$$\langle T \rangle_n = \left\langle \frac{p^2}{2m} \right\rangle = \frac{E_n}{2} = \frac{\langle H \rangle_n}{2},$$

$$\langle V \rangle_n = \left\langle \frac{m\omega^2}{2} x^2 \right\rangle_n = \frac{E_n}{2} = \frac{\langle H \rangle_n}{2}, \qquad (88)$$

in accordance with the virial theorem.

We have

$$\langle (\Delta x)^2 \rangle_n = \frac{E_n}{m\omega^2}$$

$$\langle (\Delta p)^2 \rangle_n = mE_n$$

so that

$$\Delta x \Delta p = \left(n + \frac{1}{2} \right) \hbar. \tag{89}$$

Hence the ground state of the harmonic oscillator attains the minimum concertainty product. We note that the ground state wave function has a gaussian shape.

Let us find the energy eigenfunctions in position space. Because $a|0\rangle = 0$, we have

$$\langle x'|a|0 \rangle = (2m\hbar\omega)^{-1/2} \langle x'|m\omega x' + ip|0 \rangle = 0$$

$$= (2m\hbar\omega)^{-1/2} \left(m\omega x' + \hbar \frac{\partial}{\partial x'} \right) \langle x'|0 \rangle = 0.$$

Since $\psi_n(x') = \langle x'|n \rangle$, we can consider the above equation as a differential equation for the ground state wave function $\langle x'|0 \rangle$:

$$\left(\frac{\partial}{\partial x'} + \frac{m\omega}{\hbar} x' \right) \psi_0(x') = 0,$$

or, using the dimensionless variable

$$\xi = \sqrt{\frac{m\omega}{\hbar}} \, x',$$

we have

$$\left(\frac{\partial}{\partial \xi} + \xi \right) \psi_0(\xi) = 0.$$

The solution, normalized to

$$\sqrt{\frac{\hbar}{m\omega}} \int_{-\infty}^{\infty} d\xi |\psi_0(\xi)|^2 = 0,$$

is

$$\psi_0(x') = \left(\frac{m\omega}{\pi\hbar} \right)^{1/4} e^{-\frac{1}{2}\xi^2}. \tag{90}$$

By repeated application of the operator $a^\dagger$ on the ground state wave function $\psi_0(x)$ (dropping the prime on x) we can find the wave function of any excited state. Using Eq. (78), the wave function of the nth excited state $\psi_n(x)$ is

$$|n\rangle = \frac{1}{\sqrt{n!}} (a^\dagger)^n |0\rangle$$

and so

$$\langle x|n \rangle = (n!)^{-\frac{1}{2}} (2m\hbar\omega)^{-\frac{1}{2}n} \langle x|(m\omega x - ip)^n|0\rangle$$

$$= [n! \, (2m\hbar\omega)^n]^{-\frac{1}{2}} \left(m\omega x - \hbar \frac{\partial}{\partial x}\right)^n \psi_0(x)$$

or,
$$\psi_n(x) = (2^n \cdot n!)^{-\frac{1}{2}} \left(\xi - \frac{d}{d\xi}\right)^n \psi_0(x). \tag{91}$$

using the operator identity

$$\xi - \frac{d}{d\xi} = -e^{\xi/2} \frac{d}{d\xi} e^{-\xi^2/2},$$

and Eq. (29), we obtain

$$\psi_n(x) = (2^n \cdot n!)^{-\frac{1}{2}} \left(\frac{m\omega}{\pi\hbar}\right)^{1/4} h_n\left(\sqrt{\frac{m\omega}{\hbar}}x\right) e^{-\frac{m\omega x^2}{2\hbar}} \tag{92}$$

where $h_n(z)$ is the Hermite polynomial of degree n in z.

So far everything we have done is supposed to hold at an instant of time, say at $t = 0$. We now consider the time development of the harmonic oscillator. We work in the Heisenberg picture in which the operators, e.g., x, p, a, and $a^\dagger$ are all time-dependent, and the eigenkets are time-independent.

The expression for $H = \hbar\omega\left(a^\dagger a + \frac{1}{2}\right)$ yields the commutation rules

$$[a(t), H] = \hbar\omega a(t);$$
$$[a^\dagger(t), H] = -\hbar\omega a^\dagger(t). \tag{93}$$

The Heisenberg equations of motion for p and x are

$$i\hbar \frac{d}{dt} p(t) = [p(t), H] = -\hbar m\omega^2 x(t)$$

$$i\hbar \frac{d}{dt} x(t) = [x(t), H] = \frac{i\hbar}{m} p(t) \tag{94}$$

or,
$$\frac{d}{dt} p(t) = -m\omega^2 x(t),$$

$$\frac{d}{dt} x(t) = \frac{p(t)}{m} \tag{94a}$$

whence

$$\frac{d^2 x(t)}{dt} + \omega^2 x(t) = 0 \tag{94b}$$

Introducing the non-hermitian operator

$$a(t) = (2\hbar m\omega)^{-1/2} (p(t) - im\omega x(t)), \tag{95}$$

we obtain from Eq. (94),

$$\left(\frac{d}{dt} + i\omega\right) a(t) = 0, \tag{96}$$

and its conjugate equation

$$\left(\frac{d}{dt} - i\omega\right) a^\dagger(t) = 0. \tag{96a}$$

The solutions are

$$a(t) = a(0)\, e^{-i\omega t}$$

and

$$a^\dagger(t) = a^\dagger(0)\, e^{i\omega t} \tag{97}$$

Solving for x and p, we get

$$p(t) = \left(\frac{m\hbar\omega}{2}\right)^{1/2} [a(t) + a^\dagger(t)]$$

$$= p(0)\cos\omega t - m\omega q(0)\sin\omega t$$

$$x(t) = \left(\frac{\hbar}{2m\omega}\right)^{1/2} [a(t) - a^\dagger(t)]$$

$$= q(0)\cos\omega t + \frac{1}{m\omega} p(0)\sin\omega t. \tag{98}$$

These expressions are the same as for the classical harmonic oscillator. The mean values follow the classical laws of motion:

$$\langle x\rangle_t = \langle x\rangle_0 \cos\omega t + \frac{1}{m\omega}\langle p\rangle_0 \sin\omega t,$$

$$\langle p\rangle_t = \langle p\rangle_0 \cos\omega t - m\omega\langle x\rangle_0 \sin\omega t. \tag{99}$$

Therefore, for a wave packet, the mean values of the position and momentum operators follow the classical trajectories. However, as we have seen the form of the wave packet changes with time. There exists a class of nonstationary states (coherent states) that does not change form in time. Such states are discussed in the next section.

(iii) *Wigner Distributions*

Since the ground state of the simple harmonic oscillator is gaussian, the Wigner distribution of the ground state is a positive definite distribution in phase space and therefore can be considered as a probability distribution in phase space. The Wigner distributions of the excited states all have zeros (due to the nodes in the higher states) and go negative. The nth eigenstate of the linear harmonic oscillator may be written as

$$u_n(x) = \left(\frac{\alpha^2}{4}\right)^{1/4} \left(\frac{1}{2^n \cdot n!}\right)^{\frac{1}{2}} e^{-\alpha^2 x^2/2}\, h_n(\alpha x), \tag{100}$$

where

$$\alpha = \sqrt{\frac{m\omega}{\hbar}}.$$

The Wigner distribution function is

$$P_W = \frac{1}{\pi\hbar} \int_{-\infty}^{\infty} dy \, \psi^*(x + y) \, \psi(x - y) \, e^{2ipy/\hbar}. \tag{101}$$

Substituting the expression for u_n in Eq. (101), we have

$$P_W(x, p) = \frac{1}{\pi\hbar} \frac{\alpha}{\sqrt{\pi}} \frac{1}{2^n \cdot n!} e^{-\alpha^2 x^2}$$

$$\times \int dy e^{2ipy/\hbar} \, e^{-\alpha^2 y^2} \, h_n(\alpha(x + y)) h_n(\alpha(x - y)). \tag{102}$$

In terms of a new variable

$$z = \alpha(y - ip/\alpha^2\hbar), \tag{103}$$

we write

$$P_W(x, p) = \frac{1}{\sqrt{\pi}} \frac{1}{\pi\hbar} \frac{1}{2^n \cdot n!} e^{-\alpha x^2} e^{\beta^2}$$

$$\times \int dz e^{-z^2} \, h_n(\alpha x + z + \beta) h_n(\alpha x - z - \beta), \tag{104}$$

where

$$\beta = \frac{ip}{\alpha\hbar}.$$

Because $h_n(-x) = (-1)^n h_n(x)$,

we get

$$P_W(x, p) = \frac{1}{\sqrt{\pi}} \frac{1}{\pi\hbar} \frac{(-1)^n}{2^n \cdot n!} e^{-\alpha x^2} e^{\beta^2}$$

$$\times \int dz e^{-z^2} \, h_n(\alpha x + z + \beta) h_n(z + \beta - \alpha x)$$

Now, the integral

$$\int dz e^{-z^2} \, h_n(z + \beta + \alpha x) \, h_n(z + \beta - \alpha x)$$

$$= 2^n \sqrt{\pi} \, n! \, L_n \left(2(\alpha^2 x^2 - \beta^2)\right), \tag{105}$$

where L_n is the nth Laguerre polynomial (Gradshteyn and Ryzhik (1980)).

But

$$\alpha^2 x^2 - \beta^2 = \frac{2}{\hbar\omega} \frac{p^2}{2m} + \frac{1}{2} m\omega^2 x^2$$

$$= \frac{2}{\hbar\omega} H(x, p), \tag{106}$$

where $H(x, p)$ is the Hamiltonian.

So the Wigner distribution for the harmonic oscillator is

$$P_W(x, p) = \frac{1}{\pi\hbar} (-1)^n E^{-2H/\hbar\omega} L_n (4H/\hbar\omega). \tag{107}$$

The first few Laguerre polynomials are

$$L_0(x) = 1, \quad L_1(x) = 1 - x, \quad L_2(x) = 1 - 2x + x^2, \ ...,$$

So, for the ground state of the oscillator $P_W > 0$, while for the excited states P_W can assume negative values. For a canonical ensemble of harmonic oscillators at a temperators $T \equiv (k_B \beta)^{-1}$ (here, $\beta = (k_B T)^{-1}$)

$$P_W(x, p) = \frac{1}{\pi \hbar} \tanh\left(\frac{\hbar \omega \beta}{2}\right)$$

$$\times \exp\left[-\frac{2}{\hbar \omega} \tanh\left(\frac{\hbar \omega \beta}{2}\right) H(x, p)\right] \quad (108)$$

So the thermal Wigner distribution is always positive for all temperature.

(b)   The Two-Dimensional Oscillator

The Hamiltonian for an isotropic harmonic oscillator in two dimensions is

$$H_{iso} = \frac{p_1^2}{2\mu} + \frac{p_2^2}{2\mu} + \frac{\mu}{2}\omega^2(q_1^2 + q_2^2). \quad (109)$$

The Schrödinger equation is

$$\left(-\frac{\hbar^2}{2\mu}\frac{d^2}{dq_1^2} + \frac{\mu}{2}\omega^2 q_1^2\right)u + \left(-\frac{\hbar^2}{2\mu}\frac{d^2}{dq_2^2} + \frac{\mu}{2}\omega^2 q_2^2\right)u = Eu,$$

or,
$$\frac{d^2u}{dx_1^2} + \frac{d^2u}{dx_2^2} + (\lambda - x_1^2 - x_2^2)u = 0 \quad (110)$$

where

$$\lambda = \frac{2E}{\hbar \omega}, \quad x_i = \sqrt{\frac{\mu \omega}{\hbar}}\,q_i \quad (i = 1, 2)$$

Equation (110) is the sum of two equations in x_1, and x_2. The solution of Eq. (110) is

$$u = h_{n_1}(x_1)\,h_{n_2}(x_2), \quad (111)$$

and the eigenvalues are

$$\lambda = \lambda_1 + \lambda_2 = 2n_1 + 1 + 2n_2 + 1$$

$$= 2(n_1 + n_2 + 1)$$

$$= 2(n + 1), \quad (112)$$

where $n = n_1 + n_2$, with $n_1, n_2 = 0, 1, 2, 3, \ ....$

In this case different states (except for $n = 0$ when $n_1 = n_2 = 0$) are associated with the same eigenvalue (degeneracy). In this case the degeneracy is $(n + 1)$ fold.

The degeneracy can be removed if we consider an anisotropic harmonic oscillator with the Hamiltonian.

$$H_{ani} = \frac{p_1^2}{2\mu} + \frac{p_2^2}{2\mu} + \frac{\mu}{2}\,(\omega_1^2 q_1^2 + \omega_2^2 q_2^2). \tag{113}$$

The equation is still separable and the eigenfunction is

$$u = h_{n_1}\left(\sqrt{\frac{\mu\omega_1}{\hbar}}\,q_1\right) h_{n_2}\left(\sqrt{\frac{\mu\omega_2}{\hbar}}\,q_2\right), \tag{114}$$

with

$$E = \hbar\left\{\omega_1\left(n_1 + \frac{1}{2}\right) + \omega_2\left(n_2 + \frac{1}{2}\right)\right\}. \tag{115}$$

The degeneracy is removed when ω_1/ω_2 is irrational. Note that if $\dfrac{\omega_1}{\omega_2} \to 1$, we go to the case of an isotropic oscillator.

A degenerate system is separable in various coordinate systems.

The isotropic harmonic oscillator is also separable in plane polar coordinates:

$$x_1 = \rho\cos\phi, \quad x_2 = \rho\sin\phi$$
$$dx_1 dx_2 = \rho\, d\rho\, d\phi. \tag{116}$$

Following Pauli (1973), we write
the Schrödinger equation as

$$\frac{\partial^2 u}{\partial\rho^2} + \frac{1}{\rho}\frac{\partial u}{\partial\rho} + \frac{1}{\rho^2}\frac{\partial^2 u}{\partial\phi^2} + (\lambda - \rho^2)u = 0. \tag{117}$$

We put

$$u(\rho, \phi) = v_m(\rho)e^{im\phi}, \tag{118}$$

where m is a positive or negative integer and obtain the separated differential equation for $v_m(\rho)$:

$$\frac{\partial^2 v_m}{\partial\rho^2} + \frac{1}{\rho}\frac{\partial v_m}{\partial\rho} - \frac{m^2}{\rho^2}v_m + (\lambda - \rho^2)v_m = 0. \tag{119}$$

Using the coordinate transformation

$$\rho^2 = x, \quad \rho = \sqrt{x}, \quad \rho\frac{\partial}{\partial\rho} = 2x\frac{\partial}{\partial x}, \tag{120}$$

Eq. (119) becomes

$$x\frac{d^2 v_m}{dx^2} + \frac{dv_m}{dx} - \frac{m^2}{4x}v_m + \frac{\lambda - x}{4}v_m = 0 \tag{121}$$

with the substitution

$$v_m = x^{|m|/2}\,e^{-x/2}\,y, \tag{122}$$

Eq. (121) becomes

$$y'' + \left(\frac{m+1}{x} - 1\right)y' + \frac{1}{2x}\left(\frac{\lambda}{2} - m - 1\right)y = 0, \tag{123}$$

where
$$\frac{dy}{dx} = y',$$

or,
$$xy'' + (m + 1 - x)\,y' + ky = 0, \tag{124}$$

with
$$k = \frac{1}{2}\left(\frac{\lambda}{2} - m - 1\right).$$

From Eq. (112), $\lambda = 2(n + 1)$, and so $n = 2k + m$.

Here, we consider m and k to be non-negative integers.

Rodrigues' formula for Lagurre polynomials is
$$L_k(x) = \frac{e^x}{k!}\frac{d^k}{dx^k}\,(x^k e^{-x}). \quad (k\ \text{integer}) \tag{125}$$

In series form
$$L_k(x) = \sum_{m=0}^{k}(-1)^m\,\frac{k!}{(k-m)!\,m!\,m!}\,x^m. \tag{126}$$

Laguerre's differential equation is
$$xL_k''(x) + (1 - x)L_k'(x) + kL_k(x) = 0. \tag{127}$$

Equation (127) is the same as Eq. (124) with $m = 0$.

The generating function for the Laguerre polynomials is
$$g(x, z) = \frac{e^{-\frac{xz}{(1-z)}}}{1 - z}$$

$$= \sum_{k=0}^{\infty} L_k(x)z^k, \quad |z| < 1. \tag{128}$$

We have
$$L_k(x) = \frac{e^x}{2\pi i}\oint_c \frac{s^k e^{-s}}{(s - x)^{k+1}}\,ds. \tag{129}$$

The associated Lagurre polynomials are defined by
$$L_k^m = \frac{d^m}{dx^m}\,L_k. \tag{130}$$

Differentiating Eq. (127) m times, we get
$$x\,(L_k^m(x))'' + (m + 1 - x)(L_k^m(x))' + (k - m)L_k^m(x) = 0. \tag{131}$$

We have
$$L_{k+m}^m(x) = (-1)^m\,\frac{m[(k + m)!]^2}{k!}$$

$$\times \frac{1}{2\pi i}\oint_c e^s\,(s - x)^k\,s^{-(k+m+1)}\,ds. \tag{132}$$

The generating function is given by

$$\sum_{k=0}^{\infty} \frac{L_{k+m}^{m}(x)}{(k+m)!} z^{k} = (-1)^{m} \frac{\exp[-xz/(1-z)]}{(1-z)^{m+1}}. \tag{133}$$

If we replace k by $k+m$ in Eq. (131) we get Eq. (124). So $L_{k+m}^{m}(x)$ is a solution of Eq. (124) with $m \geq 0$. The solution of Eq. (121) (and of the Eq. (119)) is

$$v_{k,m} = \text{const.} \ x^{m/2} \ e^{-x/2} \ L_{k+m}^{m}(x). \tag{134}$$

We find the normalization constant. Let

$$\int_{0}^{\infty} v_{k,m} \, v_{k',m} \, dx = \int_{0}^{\infty} x^{m} \, e^{-x} \, L_{k+m}^{m}(x) \, L_{k'+m}^{m}(x) \, dx$$

$$= N_{km} \, \delta_{kk'}. \tag{135}$$

From Eq. (133), we get

$$\sum_{k'=0}^{\infty} \frac{L_{k'+m}^{m}(x)}{(k'+m)} \, t^{k'} = (-1)^{m} \frac{e^{-\frac{xt}{1-t}}}{(1-t)^{m+1}}. \tag{136}$$

Now,

$$\sum_{k=0}^{\infty} \sum_{k'=0}^{\infty} \frac{z^{k} t^{k'}}{(k+m)!(k'+m)!} \int_{0}^{\infty} x^{m} \, e^{-x} \, L_{k+m}^{m}(x) \, L_{k'+m}^{m}(x) \, dx$$

$$= \frac{1}{(1-z)^{m+1}} \frac{1}{(1-t)^{m+1}} \int_{0}^{\infty} x^{m} \exp\left[-x\,(1-zt)/(1-t)(1-z)\right] dx$$

$$= \frac{1}{(1-zt)^{m+1}} \int_{0}^{\infty} y^{m} \, e^{-y} \, dy = \frac{m!}{(1-zt)^{m+1}},$$

where $y = x \dfrac{1-zt}{(1-z)(1-t)}$ and $\int_{0}^{\infty} y^{m} \, e^{-y} \, dy = m!$

$$= m! \sum_{k=0}^{\infty} \binom{-m-1}{k} (-1)^{k} \, (zt)^{k}$$

$$= \sum_{k=0}^{\infty} \frac{(k+m)!}{k!} \, (zt)^{k} \tag{137}$$

By comparing coefficients in Eq. (137), we get

$$N_{k,m} = \frac{[(k+m)!]}{k!}. \tag{138}$$

The coefficients of terms of the form $z^{k} \, t^{k'}$ $(k \neq k')$ on the right-hand side are zero. The orthogonality follows.

For non-integral values of α,

$$\binom{-\alpha}{n} (-1)^{n} = \frac{\alpha(\alpha+1) \, ... \, (\alpha+n-1)}{n!}$$

$$= \frac{\Gamma(\alpha + n)}{\Gamma(\alpha)n!}.$$

The allowed eigenvalues for an orthonormal solution are $k = 0, 1, 2, \ldots$ The asymptotic solution of the plane harmonic oscillator is (large x)

$$v_{k,m} \sim e^{i\pi k} \frac{m!}{\Gamma(m + 1 + k)} x^{\frac{m}{2} + k} e^{-x/2}. \tag{139}$$

The wave function is

$$u(\rho, \phi) = \frac{1}{\sqrt{2\pi}} e^{im\phi} v_{k,m}(\rho), \tag{140}$$

where $n = 2k + m$, $\rho^2 = x$. The wave functions are complete and orthonormal. We write

$$u(\rho, \phi) = \frac{1}{\sqrt{2\pi}} e^{im\phi} P_{n,m}(\rho), \tag{141}$$

where P's are the normalized radial wave functions

$$P_{n,m}(\rho) = (-1)^{\frac{n - |m|}{2}} \sqrt{2 \frac{\left(\dfrac{n - |m|}{2}\right)!}{\left(\dfrac{n + |m|}{2}\right)!}} \, \rho^{|m|} \, e^{-\frac{1}{2}\rho^2} L_{\frac{n - |m|}{2}}^{|m|}(\rho^2)$$

$$= P_{n,-m}. \tag{142}$$

Note that

$$\int_0^\infty d\rho \, P_{n,m}(\rho) P_{n',m}(\rho) = \delta_{n'n}.$$

We write,
$$n = |m| + 2n_\rho, \quad k = n_\rho = 0, 1, \ldots \tag{143}$$

where n_ρ is called the radial quantum number. Equation (119) in which $\lambda = 2n + 1$ is now written as

$$\left[\frac{d^2}{d\rho^2} + \frac{1}{\rho}\frac{d}{d\rho} - \frac{m^2}{\rho^2} + 2|m| + 4n_\rho + 2 - \rho^2 \right] P(\rho) = 0, \tag{144}$$

where $|m|$ is arbitrary.

For future reference, we write the differential equation (144) with the first derivative removed.

The following identity

$$\rho^{-a} \frac{d^2}{d\rho^2} \rho^a = \left(\rho^{-a} \frac{d}{d\rho} \rho^a \right)^2 = \left(\frac{d}{d\rho} + \frac{a}{\rho} \right)^2$$

$$= \frac{d^2}{d\rho^2} + \frac{2a}{\rho}\frac{d}{d\rho} + \frac{a^2 - a}{\rho^2},$$

or,

$$\rho^a \left(\frac{d^2}{d\rho^2} + \frac{2a}{\rho}\frac{d}{d\rho} + \frac{a^2 - a}{\rho^2} \right)\rho^{-a} = \frac{d^2}{d\rho^2}, \tag{145}$$

is useful.

With the substitution

$$P(\rho) = \frac{1}{\sqrt{\rho}}\, \chi\,(\rho), \tag{146}$$

in Eq. (144) and using Eq. (145) with $a = \dfrac{1}{2}$, we obtain

$$\left[\frac{d^2}{d\rho^2} - \frac{m^2 - \dfrac{1}{4}}{\rho^2} + 2|m| + 4n_\rho + 2 - \rho^2 \right]\chi(\rho) = 0. \tag{147}$$

The radial wave functions

$$\chi^{(\rho)}_{|m|,n_\rho} = \sqrt{2\,\frac{n_\rho!}{(n_\rho + |m|)!}}\; \rho^{|m| + \frac{1}{2}}\, L^{(|m|)}_{n_\rho}\,(\rho^2)\, e^{-\frac{1}{2}\rho^2}, \tag{148}$$

satisfy,

$$\int_0^\infty d\rho\, \chi^{(\rho)}_{|m|,n_\rho}\, \chi^{(\rho)}_{|m|,n'_\rho} = \delta_{n_\rho,n'_\rho}. \tag{149}$$

(c)   The Harmonic Oscillator in Three Dimensions

We consider the motion of a particle of mass m in a potential

$$V(\vec{r}) = \frac{1}{2}\,(s_1 x^2 + s_2 y^2 + s_3 z^2), \tag{150}$$

The Schrödinger equation for this anisotropic harmonic oscillator in three dimensions is separable in Cartesian coordinates. The three separated equations are each of the form of a Schrödinger equation of a linear harmonic oscillator, e.g.

$$\frac{d^2 X(x)}{dx^2} + \frac{2\mu}{\hbar^2}\left(E_x - \frac{1}{2}\,s_1 x^2 \right) X\,(x) = 0, \tag{151}$$

and similar equations for $Y(y)$ and $Z(z)$, where $\psi(x,\, y,\, z) = X(x)Y(y)Z(z)$. The spectrum is discrete and the energy levels are

$$E_{n_x n_y n_z} = \left(n_x + \frac{1}{2} \right)\hbar\omega_1 + \left(n_y + \frac{1}{2} \right)\hbar\omega_2 + \left(n_z + \frac{1}{2} \right)\hbar\omega_3, \tag{152}$$

where $n_x,\, n_y,\, n_z$ are 0, 1, 2, …. and

$$\omega_i = \sqrt{\frac{s_i}{m}} \quad (i = 1,\, 2,\, 3).$$

In general, if $\omega_1,\, \omega_2,\, \omega_3$ are all different, then the energy levels are non-degenerate.

The normalized eigenfunctions are

$$\psi_{n_x n_y n_z}(x, y, z) = N_{n_x} N_{n_y} N_{n_z} \, e^{-\frac{1}{2}\left(\alpha_1 x^2 + \alpha_2 y^2 + \alpha_3 z^2\right)}$$

$$\times \, h_{n_x}(\alpha_1 x) \, h_{n_y}(\alpha_2 y) \, h_{n_z}(\alpha_3 z). \tag{153}$$

where

$$\alpha_i = \left(\frac{\mu s_i}{\hbar^2}\right)^{1/4};$$

$$N_{n_j} = \left(\frac{\alpha_i}{\sqrt{\pi} \cdot 2^n \cdot n_j!}\right) \quad (j = x, y, z)$$

and h_{n_j} are the Hermite polynomials.

We consider the case of an isotropic oscillator for which the potential

$$V(q^2) = \frac{1}{2} \, s\vec{q}^{\,2} \tag{154}$$

is central (see § 11.4).

We can solve the problem in Cartesian coordinates by setting $s_1 = s_2 = s_3 = s$. The energy levels are $E_n = \left(n + \frac{3}{2}\right)\hbar\omega$, where $n = n_x + n_y + n_z$ and $\omega = \sqrt{\frac{s}{\mu}}$. Here, $n = 0, 1, 2, \ldots.$ The ground state is $E_0 = \frac{3}{2}\hbar\omega$. Except the ground state all the energy levels are degenerate – the nth level is $(n + 1)(n + 2)$ fold degenerate.

The Schrödinger equation is

$$\left[-\frac{\hbar^2}{2\mu} \vec{\nabla}^2 + \frac{1}{2}\mu\omega^2 \vec{q}^{\,2}\right]\psi(\vec{r}) = E\psi(\vec{r}), \tag{155}$$

where $V(q) = \frac{1}{2} s\vec{q}^{\,2} = \frac{1}{2} m\omega^2 q^2$ with $\omega = \sqrt{\frac{s}{\mu}}$.

We introduce spherical polar coordinates (ρ, θ, ϕ). $\vec{L}^2$ and L_z both commute with the Hamiltonian H. We recall $\vec{\nabla}^2(\rho^l Y_{lm}) = 0$,

$$\text{and} \qquad \left(\frac{\partial^2}{\partial\rho^2} + \frac{2}{\rho}\frac{\partial}{\partial\rho}\right)\rho^l Y_{lm} = l(l+1)\frac{\rho^l Y_{lm}}{\rho^2},$$

and so

$$(\vec{L}^2)Y_{lm} = l(l+1)Y_{lm},$$

$$L_z Y_{lm} = mY_{lm}.$$

We write

$$\psi(\rho, \theta, \phi) = \frac{u(\rho)}{\rho} \, Y_{lm}(\theta, \phi), \tag{156}$$

and from Eq. (155), we get the reduced radial equation

$$\left[\frac{d}{d\rho^2} - \frac{l(l+1)}{\rho^2} + \frac{2\mu E}{\hbar^2} - \frac{\mu^2\omega^2}{\hbar^2}\rho^2\right] u(\rho) = 0, \tag{157}$$

or, using the dimensionless eigenvalue.

$$\lambda = \frac{2E}{\hbar\omega} \quad \text{and putting} \quad \frac{\mu\omega}{\hbar} = 1,$$

we get

$$\left[\frac{d^2}{d\rho^2} - \frac{l(l+1)}{\rho^2} + \lambda - \rho^2\right] u(\rho) = 0. \tag{158}$$

This Eq. (158) has the same form as that of the two-dimensional oscillator (Eq. (147)). We note the following correspondence (see Schwinger (2001))

two-dimensional oscillator   three-dimensional oscillator

$$m^2 - \frac{1}{4} \qquad\qquad\qquad l(l+1)$$

$$2|m| + 4n_\rho + 2 \qquad\qquad\qquad \lambda$$

So

$$|m| = l + \frac{1}{2},$$

and

$$\frac{2E}{\hbar\omega} = 2\left(l + \frac{1}{2}\right) + 4n_\rho + 2$$

$$= 2(l + 2n_\rho) + 3.$$

The energy levels are

$$E_n = \left(n + \frac{3}{2}\right)\hbar\omega, \tag{159}$$

with

$$n = 2n_\rho + l.$$

The radial quantum number $n_\rho = 0, 1, 2, \ldots$ and so $n = 0, 1, 2, \ldots$ The ground state ($n = 0$, with $n_\rho = l = 0$) is unique. The nth level is $\frac{1}{2}(n+1)(n+2)$ fold degenerate. The multiplicity is

n even

$$g(n) = \sum_{l=0,2,\ldots}^{n} \sum_{m=-l}^{l} m = \sum_{l=0,2,\ldots}^{n}(2l+1),$$

since for each l, m takes $(2l + 1)$ values, $-l, -l + 1, \ldots. l.$

$$= \frac{n+2}{2}(n+1).$$

n odd

$$g(n) = \sum_{l=1,3,\ldots}^{n}(2l+1) = \frac{n+1}{2}(n+2).$$

The normalized wave functions are

$$\psi_{n_\rho l m}(\vec{q}) = \frac{1}{\rho}\, u_{n_\rho l}(\rho)\, Y_{lm}(\theta,\phi) \tag{160}$$

where

$$u_{n_\rho l}(\rho) = (-1)^{n_\rho} \sqrt{2\,\frac{n_\rho!}{\left(n_\rho + l + \frac{1}{2}\right)}} \cdot$$

$$\times\, \rho^{l+1}\, L_{n_\rho}^{\left(l+\frac{1}{2}\right)}(\rho)\, e^{\frac{-\rho^2}{2}}, \tag{161}$$

using the correspondence mentioned above.

The spherical harmonics are orthonormalized, and

$$\int_0^\infty d\rho\, u_{n_{\rho'}}(\rho)\, u_{n'_\rho l}(\rho) = \delta_{n_\rho n'_\rho}. \tag{162}$$

We can also use the operator formalism of the creation, destruction and number operators for the study of isotropic harmonic oscillators in several dimensions. The Hamiltonian of an isotropic harmonic oscillator in p dimensions is

$$H = \sum_{i=1}^{p} H_i = \sum_{i=1}^{p} \frac{1}{2m}\,(p_i^2 + m^2\omega^2 q_i^2). \tag{163}$$

The Hilbert space $\mathcal{H}$ of this system is the tensor product of the individual spaces

$$\mathcal{H} = \mathcal{H}_1 \otimes \mathcal{H}_2 \otimes ... \otimes \mathcal{H}_p. \tag{164}$$

We now consider $\{N\}$ representation. Let

$$a_i = \frac{1}{\sqrt{2m\hbar\omega}}\,(m\omega q + ip)$$

$$a_i^\dagger |n_i\rangle = \frac{1}{\sqrt{2m\hbar\omega}}\,(m\omega q - ip) \tag{165}$$

which satisfy the commutation relations

$$[a_i, a_j] = [a_i^\dagger, a_j^\dagger] = 0,$$

$$[a_i, a_j^\dagger] = \delta_{ij} \quad (i, j = 1,2, ..., p). \tag{166}$$

Let $|n_i\rangle$ (i fixed, $n_i = 0, 1, 2, ..., \infty$) be the eigenvectors of H_i in $\mathcal{H}_i$. These form a complete, orthonormal set in $\mathcal{H}_i$. $|n_i\rangle$ satisfies the relations

$$a_i^\dagger |n_i\rangle = (n_i + 1)^{1/2}\, |n_i + 1\rangle,$$

$$a_i |n_i\rangle = n_i^{1/2}\, |n_i - 1\rangle,$$

$$a_i |0\rangle = 0. \tag{167}$$

In terms of the number operator

$$N_i = a_i^\dagger a_i, \tag{168}$$

we write

$$H = (H_1 + \dots + H_p)$$

$$\sum_{i=1}^{p} \left(N_i + \frac{1}{2}\right) \hbar\omega. \tag{169}$$

The vectors

$$|n_1 n_2 \dots n_p\rangle = |n_1\rangle|n_2\rangle\dots|n_p\rangle \tag{170}$$

from a complete orthonormal set in $\mathcal{H}$.

Now,

$$a_i |\dots n_i \dots\rangle = \sqrt{n_i}|\dots n_i - 1 \dots\rangle$$

$$a_i^\dagger |\dots n_i \dots\rangle = \sqrt{n_i + 1}\,|\dots n_i + 1 \dots\rangle \tag{171}$$

So $|\dots n_i \dots\rangle$ is an eigenket of N with eigenvalue $n_i.|\dots n_i \dots\rangle$ is also an eigenket of H with the eigenvalue

$$E_n = \sum_{i=1}^{p} \left(n_i + \frac{1}{2}\right) \hbar\omega$$

$$= \left(n + \frac{1}{2} p\right) \hbar\omega. \tag{172}$$

where

$$n = n_1 + n_2 + \dots + n_p.$$

The eigenvalue $\left(n + \dfrac{1}{2} p\right) \hbar\omega$ is $C_{n+p-1}^{n} = \dfrac{(n + p - 1)!}{n!\,(p - 1)!}$ fold degenerate.

In one dimension, the energy levels are non-degenerate. For a two-dimensional oscillator, $(p = 2)$, the multiplicity is

$$g(n) = (n + 1),$$

and in three dimensions $(p = 3)$,

$$g(n) = \frac{1}{2} (n + 1)(n + 2).$$

The operators H_i or N_i $(i = 1, \dots, p)$ constitute a c.s.c.o. for the oscillator. For the two-dimensional oscillator, N_1 and N_2 form a c.s.c.o. Alternatively, $N = N_1 + N_2$ and the angular momentum operator L form a c.s.c.o. An alternative c.s.c.o. provides the angular momentum basis.

In the $\{q\} = \{q_1 q_2 \dots q_p\}$ representation, the wave function is

$$\psi_{n_1 n_2 \dots n_p}(\vec{q}) = \langle q_1|n_1\rangle\langle q_2|n_2\rangle\dots\langle q_p|n_p\rangle.$$

$$= \psi_{n_1}(q_1)\psi_{n_2}(q_2)\dots\psi_{n_p}(q_p). \tag{173}$$

11.3 COHERENT STATES

In this section we want to consider a special class of sates known as coherent states. These are an overcomplete and non-orthogonal system of Hilbert-space vectors.

Way back in 1926, Schrödinger discovered the concept of these states in connection with his study of the classical states of a quantum harmonic oscillator. The modern study of this important field was initiated by Glauber (1963) and Sudarshan (1963). Coherent states are of great importance in the quantum mechanical description of coherent light sources. These have been used in investigations of various phenomena in many different fields. (Zhang et al. (1990)).

(a) Definitions and Properties

Let $D(\alpha)$ be the unitary displacement operator for a complex number α:

$$D(\alpha) = e^{(\alpha a^\dagger - \alpha^* a)} \tag{174}$$

The coherent states $|\alpha\rangle$ can be defined as the displaced states of the vacuum

$$|\alpha\rangle = D(\alpha)|0\rangle, \tag{175}$$

where $|0\rangle$ denotes the vacuum state of the harmonic oscillator.

Using the BCH theorem, we may write

$$\begin{aligned}
e^{\alpha a^\dagger - \alpha^* a} &= e^{\left(\frac{1}{2}\alpha\alpha^*\right)} e^{-\alpha^* a} e^{\alpha a^\dagger} \\
&= e^{-\frac{1}{2}\alpha\alpha^*} e^{\alpha a^\dagger} e^{-\alpha^* a}.
\end{aligned} \tag{176}$$

$D(\alpha)$ has the following properties

$$\begin{aligned}
D^{-1}(\alpha) a D(\alpha) &= a + \alpha, \\
D^{-1}(\alpha) a^\dagger D(\alpha) &= a^\dagger + \alpha^*.
\end{aligned} \tag{177}$$

We expand the coherent states in terms of particle number states $|n\rangle$:

$$\begin{aligned}
|\alpha\rangle &= D(\alpha)|0\rangle \\
&= e^{-\frac{1}{2}|\alpha|^2} \sum_{n=0}^{\infty} \frac{\alpha^n}{n!} (a^\dagger)^n |0\rangle \\
&= e^{-\frac{1}{2}|\alpha|^2} \sum_{n=0}^{\infty} \frac{\alpha^n}{(n!)^{1/2}} |n\rangle.
\end{aligned} \tag{178}$$

The probability of finding n excitations (or, n photons) is a Poisson distribution

$$P_n = |\langle n|\alpha\rangle|^2 = e^{-|\alpha|^n} \frac{|\alpha|^{2n}}{n!} \tag{178a}$$

The distribution is centered at $n \simeq |\alpha|^2$ with width $|\alpha|$. Therefore, in a coherent state the occupation number population has a Poissonian distribution.

The coherent states $|\alpha\rangle$ may also be defined as the eigenstates of the harmonic oscillator annihilation operators with eigenvalue α:

$$a|\alpha\rangle = \alpha|\alpha\rangle. \tag{179}$$

From Eq. (178), we get

$$a|\alpha\rangle = e^{-\frac{1}{2}|\alpha|^2} \sum_{n=0}^{\infty} \frac{\alpha^n}{\sqrt{n!}} \frac{aa^{\dagger n}}{\sqrt{n!}} |0\rangle$$

$$= e^{-\frac{1}{2}|\alpha|^2} \sum_{1}^{\infty} \frac{\alpha^n n}{\sqrt{n!}} \frac{a^{\dagger n-1}}{\sqrt{n!}} |0\rangle$$

$$= e^{-\frac{1}{2}|\alpha|^2} \sum_{1}^{\infty} \frac{\alpha^n \sqrt{n}}{\sqrt{n!}} |n-1\rangle. \text{ using } [a, (a^\dagger)^n] = n(a^\dagger)^{n-1},$$

$$= \alpha|\alpha\rangle,$$

where we have used a dummy label $n' = n - 1$ which goes from 0 to ∞.

We have labeled the coherent state by α since apart from a normalization factor, it depends only on α and not α^*. We denote the coherent state bra as $\langle\alpha|$, it obeys

$$\langle\alpha|a^\dagger = \langle\alpha|\alpha^*. \tag{180}$$

There are two important Hilbert space properties.

(i)   Non-Orthogonality

The inner product

$$\langle\beta|\alpha\rangle = e^{-\frac{1}{2}|\alpha|^2 + \beta^*\alpha - \frac{1}{2}|\beta|^2} \tag{181}$$

This is a nowhere vanishing, continuous function of α, β. Also

$$|\langle\beta|\alpha\rangle|^2 = e^{(-|\beta - \alpha|^2)}. \tag{181a}$$

So the coherent states $|\alpha\rangle$ are not orthogonal but normalized.

(ii)   Over-completeness

The resolution of identity in terms of coherent states is not unique. A useful resolution is

$$I = \pi^{-1} \int d^2\alpha |\alpha\rangle\langle\alpha|, \tag{182}$$

where

$$d^2\alpha \equiv d(Re\alpha)d(Im\alpha) = \frac{1}{2}\, d\alpha d\alpha^*.$$

Now,

$$\pi^{-1} \int d^2\alpha |\alpha\rangle\langle\alpha| = \frac{1}{\pi} \sum_{n,m=0}^{\infty} \frac{|m\rangle\langle n|}{\sqrt{m!n!}} \int_0^\infty dr\, e^{-r^2}\, r^{m+n+1}$$

$$\times \int_0^{2\pi} e^{i(m-n)\phi}\, d\phi,$$

using Eq. (178) and putting $\alpha = re^{i\phi}$ so that $d^2\alpha = r\,dr\,d\phi$,

$$= 2 \sum_n \frac{|n\rangle\langle n|}{n!} \int_0^\infty dr\, e^{-r^2}\, r^{2n+1}$$

where the angular integral is $2\pi\delta_{mn}$. Since the radial integral is $n!/2$, and the number states are complete $\sum_n |n\rangle\langle n| = 1$, the Eq. (182) follows.

The coherent states are over-complete, since they are labeled by a continuous index in a Hilbert space with a countable basis. They form an over-complete basis; it is not linearly independent. This means that an arbitrary vector can be expanded in terms of this basis which has more states than are necessary to do so.

From Eq. (182), we see that the trace of any operator A is

$$TrA = \frac{1}{\pi} \int d^2\alpha \langle \alpha|A|\alpha\rangle. \tag{183}$$

A normal ordered function in which all the destruction operators are to the right and creation operators to the left can be easily evaluated in a coherent state:

$$\langle \alpha|f(a^\dagger)g(a)|\alpha\rangle = f(\alpha^*)g(\alpha). \tag{183a}$$

A third definition is the coherent states $|\alpha\rangle$ are quantum states with a minimum-uncertainty relation

$$(\Delta p)^2\, (\Delta q)^2 = \left(\frac{\hbar}{2}\right)^2 \tag{184}$$

This definition is not unique. For the field coherent states $\Delta p = \Delta q = \dfrac{\hbar}{2}$.

Let us prepare a coherent state at time $t = 0$: $|\alpha, 0\rangle = |\alpha\rangle$. The time evolution is

$$|\alpha(t)\rangle = e^{-Ht/\hbar}|\alpha\rangle$$
$$= e^{-\frac{|\alpha|^2}{2}} \sum_n \frac{\alpha^n}{\sqrt{n!}}\, e^{-i\omega t\left(n+\frac{1}{2}\right)} |n\rangle$$
$$= e^{-\frac{i\omega t}{2}} |\alpha e^{-i\omega t}\rangle. \tag{185}$$

Therefore, a coherent state remains a coherent state (with a new label) under time evolution. For a Hamiltonian which is linear in the operators of H_4, the coherence of these states is maintained under evolution.

In terms of a and $a^\dagger$, we write the position and momentum operators

$$x = x_0\, (a + a^\dagger),$$
$$p = -\, im\omega x_0\, (a - a^\dagger) \tag{186}$$

where
$$x_0 = \sqrt{\frac{\hbar}{2m\omega}}.$$

From Eqs. (179) and (180), we see
$$\langle \alpha|a|\alpha\rangle = \alpha, \quad \langle \alpha|a^\dagger|\alpha\rangle = \alpha^*,$$

so that the expectation values with the coherent states are
$$\langle \alpha|x|\alpha\rangle = x_0\,\langle \alpha|a + a^\dagger|\alpha\rangle$$
$$= x_0\,(\alpha + \alpha^*)$$
$$= 2x_0\,Re\alpha, \tag{187}$$
$$\langle \alpha|p|\alpha\rangle = 2m\omega x_0\,Im\alpha.$$

We obtain
$$(\Delta x)_\alpha^2 = \langle \alpha|x^2|\alpha\rangle - \langle \alpha|x|\alpha\rangle^2$$
$$= x_0^2$$
$$= \frac{\hbar}{2m\omega}, \tag{188}$$
$$(\Delta p)_\alpha^2 = \langle \alpha|p^2|\alpha\rangle - \langle \alpha|p|\alpha\rangle^2$$
$$= m^2\omega^2 x_0^2 \tag{189}$$
$$= \frac{m\hbar\omega}{2}.$$

Hence for all coherent states
$$(\Delta x)^2\,(\Delta p)^2 = \frac{\hbar^2}{4}. \tag{190}$$

Note that
$$\frac{1}{2m}\,(\Delta p)^2 + \frac{1}{2}\,m\omega^2\,(\Delta x)^2 = \frac{1}{2}\,\hbar\omega.$$

The coherent states are minimum-uncertainty states which follow the classical motion of an oscillator:
$$\langle x(t)\rangle \sim Re\alpha(t) = Re(q(0) + ip(0))e^{i\omega t}$$
$$\langle p(t)\rangle \sim Im\alpha(t) = Im(q(0) + ip(0))e^{i\omega t} \tag{191}$$

Now,
$$|\alpha\rangle = D(\alpha)|0\rangle$$
$$= e^{-\frac{1}{2}|\alpha|^2}\,e^{\alpha a^\dagger}\,|0\rangle$$
$$= e^{-\frac{1}{2}(|\alpha|^2 - \alpha^2)}\,e^{\left(-\frac{i}{\hbar}\sqrt{\frac{2\hbar}{m\omega}}\,\alpha p\right)}\,|0\rangle, \tag{192}$$

putting $\alpha a^\dagger = \alpha(a^\dagger - a) + \alpha a$ and using the BCH theorem and Eq. (186).

The quantity on the right operating on the wave function $\psi_0(x)$, a gaussian, produces a translation by the amount $\alpha \sqrt{\dfrac{2\hbar}{m\omega}}$.

So the wave function of the coherent state is

$$\psi_\alpha(x) = \langle x | \alpha \rangle$$

$$= \left(\frac{m\omega}{\pi\hbar}\right)^{1/4} e^{-\frac{1}{2}(|\alpha|^2 - \alpha^2)} \times e^{\left(-\frac{1}{2}\left(\sqrt{\frac{m\omega}{\hbar}}\, x - \sqrt{2}\alpha\right)^2\right)}$$

$$= \left(\frac{m\omega}{\pi\hbar}\right)^{1/4} e^{\frac{-|\alpha|^2}{2}} e^{-\left(\frac{m\omega}{2\hbar}\right)x^2} e^{\sqrt{\frac{2m\omega}{\hbar}}\,\alpha x}. \tag{193}$$

The wave packet is

$$\psi_\alpha(x,\,t) = \left(\frac{m\omega}{\pi\hbar}\right)^{1/4} e^{\frac{-i\omega t}{2}} \exp\left(-\frac{1}{2}\left(|\alpha|^2 - \alpha^2 e^{-2i\omega t}\right)\right)$$

$$\times \exp\left(-\frac{1}{2}\left(\sqrt{\frac{m\omega}{\hbar}}\, x - \sqrt{2}\,\alpha e^{-i\omega t}\right)^2\right). \tag{194}$$

The position probability density is

$$|\psi_\alpha(x,\,t)|^2 = \sqrt{\frac{m\omega}{\pi\hbar}}\; e^{-\frac{m\omega}{\hbar}(x - \langle \alpha(t)|x|\alpha(t)\rangle)^2} \tag{195}$$

So the probability density remains invariant in time relative to its center. The wave packet oscillates about this position without change of shape with frequency ω.

Let us consider the propagator or the transition element between an initial and final coherent state.

$$Z_{ab} = \langle \alpha_b,\, t_b;\, \alpha_a,\, t_a \rangle. \tag{196}$$

In terms of the time-evolution operator $U(t_b,\, t_a)$, we write

$$Z_{ab} = \langle \alpha_b | U(t_b,\, t_a) | \alpha_a \rangle$$

$$= \langle \alpha_b | U(t_b,\, 0) U(t_a,\, 0)^{-1} | \alpha_a \rangle, \tag{197}$$

as the coherent state at time t is given by

$$|\alpha(t)\rangle \equiv |\alpha,\, t\rangle = U(t,\, 0)^{-1} |\alpha\rangle$$

$$= e^{\frac{i}{\hbar}Ht} |\alpha\rangle, \tag{198}$$

where the coherent state $|\alpha\rangle$ is the eigenstate of a, the destruction operator, at time $t = 0$. We can express the coherent-state propagator in terms of a path integral. We proceed as in Section 2.9, so that the time interval is sliced into $N - 1$ sub-intervals of duration ϵ, and N resolutions of identity are inserted at the respective times. We obtain

$$Z_{ab} = \int \mathcal{D}\alpha \prod_{j=1}^{N-1} \langle \alpha_{j+1},\, t_{j+1};\, \alpha_j,\, t_j \rangle, \tag{199}$$

where the measure is

$$\mathcal{D}\alpha = \prod_{j=1}^{N} \frac{1}{2\pi i} \, d\alpha_j^* \, d\alpha_j. \tag{200}$$

Here we have set $\lambda_N = \lambda_b$, $\lambda_0 = \lambda_a$ and have assumed that H is normally ordered. Now, consider the individual infinitesimal elements.

$$\langle \alpha_{j+1}, t_{j+1} \, \alpha_j, t_j \rangle = \langle \alpha_{j+1} | T \exp\left(-\frac{i}{\hbar} \int_{t_j}^{t_{j+1}} d\tau \, H(\tau) | \alpha_j \right)$$

$$= \left\langle \alpha_{j+1} \left| 1 - \frac{i}{\hbar} \int_{t_j}^{t_{j+1}} d\tau \, H(a^\dagger, a; \tau) \right| \alpha_j \right\rangle$$

where T is the Dyson time-ordering operator,

$$= \langle \alpha_{j+1} | \alpha_j \rangle \left[1 - \frac{i}{\hbar} \in H\!\left(\alpha_{j+1}^*, \alpha_j; t_j \right) \right]$$

$$= \exp\left[-\frac{1}{2} |\alpha_{j+1}|^2 + |\alpha_j|^2 + \alpha_{j+1}^* \, \alpha_j \right.$$

$$\left. -\frac{i}{\hbar} \in H(\alpha_{j+1}^*, \alpha_j; t_j) \right].$$

So

$$Z_{ab} = \int \mathcal{D}\alpha \, \exp\left[\sum_{j=0}^{N-1} \left(-\frac{1}{2} |\alpha_{j+1}|^2 + |\alpha_j|^2 + \alpha_{j+}^* \, \alpha_j \right.\right.$$

$$\left.\left. -\frac{i}{\hbar} \in H(\alpha_{j+1}^*, \alpha_j; t_j) \right].$$

We note that

$$\left[\sum_{j=0}^{N-1} \left(-\frac{1}{2} |\alpha_{j+1}|^2 + |\alpha_j|^2 + \alpha_{j+}^* \, \alpha_j - \frac{i}{\hbar} \in H(\alpha_{j+1}^*, \alpha_j; t_j) \right) \right]$$

$$= \Sigma \left[-\frac{1}{2} \alpha_{j+1}^* \left(\frac{\alpha_{j+1} - \alpha_j}{\in} \right) \in + \frac{1}{2} \alpha_j \left(\frac{\alpha_{j+1}^* - \alpha_j^*}{\in} \right) \in \right.$$

$$\left. -\frac{i}{\hbar} \in H(\alpha_{j+1}^*, \alpha_j; t_j) \right]$$

$$\rightarrow \int_{t_a}^{t_b} d\tau \left[\frac{1}{2} (\alpha\dot{\alpha}^* - \dot{\alpha}^*\alpha) - \frac{i}{\hbar} H(\alpha^*, \alpha; \tau) \right]$$

as $N \to \infty$, $\in \to 0$.

So in the continuum limit,

$$Z_{ab} = \int \mathcal{D}[\alpha(\tau)] \, \exp\left[\int_{t_a}^{t_b} d\tau \left[\frac{1}{2} (\alpha\dot{\alpha}^* - \dot{\alpha}^*\alpha) - \frac{i}{\hbar} H(\alpha^*, \alpha; \tau) \right] \right]. \tag{201}$$

Equation (201) can be rewritten as

$$Z_{ab} = \int \mathcal{D}[\alpha(\tau)] \exp\left[\frac{\alpha_b^*\alpha_b + \alpha_a^*\alpha_a}{2} + \frac{i}{\hbar}\int_{t_a}^{t_b}\left\{\frac{i\hbar}{2}\left(\alpha^*\frac{d\alpha}{dt} - \frac{d\alpha^*}{dt}\alpha\right) - H(\alpha^*,\alpha)\right\}dt\right].$$

(202)

Here we have taken the average of two equivalent expressions to obtain a symmetric formal expression in continuum limit. By doing an integration by parts, we obtain an asymmetric form of the result

$$Z_{ab} = \int D[\alpha(t)]\, \exp\left[\alpha_b^*\alpha_a + \frac{i}{\hbar}\left[\int_{t_a}^{t_b}\left[i\hbar\alpha^*\frac{d\alpha}{dt} - H(\alpha^*,\alpha)\,dt\right]\right]\right].$$

(202a)

We note that the Lagrangian operator is $i\hbar\dfrac{\partial}{\partial t} - H$. Consult Shanker (1994).

Field coherent states have been found to be extremely useful in quantum optics. The electromagnetic field can be considered to be an infinite set of independent harmonic oscillators. We have discussed. boson coherent states. The reader should consult the books and articles mentioned in the bibliography for spin coherent states, fermion coherent states and general coherent states.

(b)   Squeezed States

Squeezed states of the light field have a lower level of quantum noise in one of its field quadratures than coherent light from a laser. In a squeezed state one arranges to deform or squeeze the uncertainty circle to an ellipse thereby reducing the uncertainty in one of the quadratures with a concomitant increase in uncertainty in a quadrature not involved in transporting information. Squeezed states of the electromagnetic field were first experimentally realized by Slusher et al. (1985) using four-wave mixing in sodium atoms.

Squezed states are important in optical communication as they represent a nonclassical light field.

Let a and $a^\dagger$ be the annihilation and creation operators respectively. We define a pure (single-photon) coherent state

$$|\alpha\rangle = D(\alpha)|0\rangle,$$

(203)

and a pure squeezed state

$$|\beta\rangle = S(\beta)|0\rangle.$$

(204)

The unitary displacement operator is

$$D(\alpha) = \exp(\alpha a^\dagger - \alpha^* a).$$

(205)

A general quadratic generator of squeezed states is given by the unitary squeeze operator

$$S(\beta) = \exp\left(\frac{1}{2}\beta^* a^2 - \frac{1}{2}\beta a^{\dagger 2}\right),$$

(206)

β is the complex squeeze parameter

$$\beta = se^{i\theta} \quad (0 \leq s < \infty, \ 0 \leq \theta \leq 2\pi). \tag{207}$$

A general squeezed state is defined as

$$|\alpha, \beta\rangle = D(\alpha)S(\beta)|0\rangle. \tag{208}$$

If $\beta = 0$, Eq. (208) reduces to the coherent state $|\alpha\rangle = D(\alpha)|0\rangle$.

The squeeze operator transforms the annihilation and creation operators as

$$a \rightarrow S^{-1}(\beta)aS(\beta) = a \cosh s - a^{\dagger}e^{i\theta} \sinh s,$$

$$a^{\dagger} \rightarrow S^{-1}(\beta)a^{\dagger}S(\beta) = a^{\dagger} \cosh s - ae^{-i\theta} \sinh s. \tag{209}$$

It is convenient to express a and $a^{\dagger}$ by two quadrature operators

$$X = \frac{a + a^{\dagger}}{2}, \tag{210}$$

$$Y = \frac{a - a^{\dagger}}{2i}. \tag{210a}$$

These hermitian operators describing the two quadratures satisfy

$$[X, Y] = \frac{i}{2}, \tag{211}$$

and

$$\Delta X \Delta Y \geq \frac{1}{4}. \tag{212}$$

The squeezed state can remain minimum uncertainty state if the uncertainty in one quadrature is compressed with a concomitant stretching of the other one. The modified variances are written as

$$(\Delta X)^2 = \frac{1}{4} e^{-2s},$$

$$(\Delta Y)^2 = \frac{1}{4} e^{2s}, \tag{213}$$

The squeezed coherent state has also been defined as the eigenstates of

$$b = \mu a + \upsilon a^{\dagger}, \tag{214}$$

with

$$|\mu|^2 - |\upsilon|^2 = 1.$$

The operator b is called the Bogolibov-Valatin transformation of a, $a^{\dagger}$.

11.4 MOTION IN A CENTRAL FIELD OF FORCE

We now study the problem of the motion of a particle in a central field of force. The simplest model of an atom consists of each electron moving independently in a central field of force, namely that of a massive positively charged, fixed nucleus, together with some kind of average of the forces due to the other electrons.

Let us consider the motion of an electron of mass μ in a central potential, that is a potential $V(r)$ that only depends on the magnitude $r = (x^2 + y^2 + z^2)$ of the position vector x, y, z are the Cartesi an coordinates of the particle. Neglecting electron spin and relativity effects, the Hamiltonian is

$$H = \frac{1}{2\mu} \left(p_x^2 + p_y^2 + p_z^2 \right) + V, \tag{215}$$

where V, the potential energy, is a central potential.

We introduce the dynamical variable

$$p_r = r^{-1} \left(\vec{r} \cdot \vec{p} \right)$$
$$= r^{-1} \left(x p_x + y p_y + z p_z \right). \tag{216}$$

and the orbital angular momentum of the particle about the origin

$$\vec{L} = \vec{r} \cdot \vec{p}. \tag{217}$$

We get

$$H = \frac{1}{2\mu} \left(p_r^2 + \frac{\vec{L}^2}{r^2} \right) + V(r). \tag{218}$$

The radial momentum operator is hermitian

$$p_r = \frac{\hbar}{i} \left(\frac{\partial}{\partial r} + \frac{1}{r} \right) \tag{219}$$

and satisfies the commutation relation

$$[r, p_r] = i\hbar. \tag{220}$$

$V(r)$ is spherically symmetric. We use the spherical polar coordinates in which the Schrödinger equation is

$$\left\{ \left[-\frac{\hbar^2}{2\mu} \frac{1}{r^2} \frac{\partial}{\partial r} \left(r^2 \frac{\partial}{\partial r} \right) + \frac{\vec{L}^2}{2\mu r^2} \right] + V(r) \right\} \psi(\vec{r}) = E\psi(\vec{r}). \tag{221}$$

Since H commutes with L_x, L_y, L_z and $\vec{L}^2$ we can look for simultaneous eigenstates of H, $\vec{L}^2$ and L_z.

The spherical harmonics $Y_{lm}(\theta, \phi)$ are simultaneous eigenfunctions of $\vec{L}^2$ and L_z. We write

$$\psi_{lm}(\vec{r}) = R_l(r) Y_{lm}(\theta, \phi), \tag{222}$$

in Eq. (221), and the radial wave function satisfies

$$\left[-\frac{\hbar^2}{2\mu} \left(\frac{d^2}{dr^2} + \frac{2}{r} \frac{d}{dr} \right) + \frac{l(l+1)\hbar^2}{2\mu r^2} + V(r) \right] R_l = ER_l(r). \tag{223}$$

Let us introduce the reduced radial wave function

$$u_l(r) = rR_l(r). \tag{224}$$

$u_l(r)$ satisfies the reduced radial equation

$$\left[-\frac{\hbar^2}{2\mu}\frac{d^2}{dr^2} + \frac{l(l+1)\hbar^2}{2\mu r^2} + V(r) \right] u_l(r) = Eu_l(r). \tag{225}$$

The Eq. (225) is formally that of a one-dimensional motion under the effective potential energy.

$$V_{eff}(r) = V(r) + \frac{\hbar^2}{2\mu}\frac{l(l+1)}{r^2}, \tag{226}$$

the second term being the repulsive centrifugal barrier term.

For a particular value of l, the states are distinguished by means of a total quantum number n which has the value $l + 1$ for the lowest state with $n - l - 1$ being the number of nodes of the radial eigenfunction not counting the origin as a node. It is customery to denote a letter code for the values of l.

Values of l 0 1 2 3 4 5 ……..

Designation   s p d f g h ……….

The energy levels are labeled by the quantum number n and l. The quantum numbers nl specify the configuration of the electron.

So far our discussion has been for a fixed center of force. We now consider the two particle system in which the electron moves under the influence of the field due to a nucleus of mass M. As in classical mechanics, we can separate the motion of the center of mass of the system and the motion of the electron relative to the nucleus. From now on we shall always consider only the relative motion of the two particles with appropriate coordinates, energy, etc.

11.5 THE HYDROGEN ATOM

In this section we begin our study of one of the most completely treated fields of application of quantum mechanics by considering the simplest atomic system – the hydrogen atom. The hydrogen atom consists of a proton and an electron interacting by an attractive Coulomb potential. Other one electron systems (hydrogenic atoms) e.g., He^+, Li^{++}, deuterium, tritium, etc. will also be considered here. Hydrogen-like atoms provide an excellent way of testing quantum mechanics in a simple system.

The model consists of an atomic nucleus of charge Ze and a single spinless electron interacting by the Coulomb potential

$$V(r) = -\frac{Ze^2}{r}, \tag{227}$$

where r is the distance between the two particles. Since the potential is a central one we use the results of the preceding section.

(a) Spherical Polar Coordinates

The reduced radial equation is

$$\frac{d^2 u_l(r)}{dr^2} + \left[\frac{2\mu}{\hbar} E + \frac{2\mu Z e^2}{\hbar^2 r} - \frac{l(l+1)}{r^2} \right] u_l(r) = 0, \tag{228}$$

where μ is the reduced mass. We also have the conditions

$$\int_0^\infty |u(r)|^2 \, dr < \infty, \tag{229}$$

and
$$u(0) = 0. \tag{229a}$$

We consider the discrete spectra for which $E < 0$. First we study the asymptotic behavior. For small values of r, the centrifugal term dominates and the second order differential equation becomes

$$\left[-\frac{d^2}{dr^2} + \frac{l(l+1)}{r^2} \right] u_l(r) = 0,$$

which has the general solution

$$u_l(r) = A r^{l+1} + B r^{-l},$$

with A, B constants. Since $u(0) = 0$, the second term is discarded. So in the limit $r \to 0$, $u_l(r) \sim r^{l+1}$ and we can write, in general,

$$u_l(r) = r^{l+1} (a_0 + a_1 r + ...).$$

For very large values of r, V_{eff} is negligible and the equation becomes

$$\left[\frac{d^2}{dr^2} + \frac{2\mu E}{\hbar^2} \right] u_l(r) = 0.$$

For bound states, the solutions are $\sim e^{\pm Kr}$ with $K = \frac{1}{\hbar} \sqrt{2\mu(-E)}$. Only the minus sign is acceptable. So for $r \to \infty$, $u_l(r) \sim e^{-Kr}$.

We now introduce the dimensionless variable $\rho = Kr$.

The differential equation (228) is

$$\left[\frac{d^2}{d\rho^2} - \frac{l(l+1)}{\rho^2} + \frac{\rho_0}{\rho} - 1 \right] u_l(\rho) = 0, \tag{230}$$

where
$$\rho_0 = \frac{2\mu Z e^2}{K \hbar^2} = \sqrt{\frac{2\mu}{|E|}} \cdot \frac{Z e^2}{\hbar}.$$

And so
$$\frac{V}{|E|} = -\frac{\rho_0}{\rho}.$$

We now substitute for $u_l(\rho)$, taking into account the asymptotic solutions, into Eq. (230)

$$u(\rho) = \rho^{l+1} e^{-\rho} v(\rho). \tag{231}$$

We obtain a differential equation in v:

$$\frac{d^2v}{d\rho^2} + 2\left(\frac{l+1}{\rho} - 1\right)\frac{dv}{d\rho} + \frac{\rho_0 - 2(l+1)}{\rho}v = 0. \tag{232}$$

Let us expand v in a power series,

$$v(\rho) = \sum_{k=0}^{\infty} a_k\rho^k, \quad a_0 \neq 0. \tag{233}$$

Substituting Eq. (233) in Eq. (232), we obtain

$$\sum_{k=0}^{\infty} a_k \left[k(k-1)\rho^{k-1} + 2(l+1)k\rho^{k-1} - 2k\rho^k\right.$$

$$\left. + (\rho_0 - 2(l+1)\rho^k\right] = 0. \tag{234}$$

The coefficient of each power of ρ must vanish in the above equation, and thus for ρ^k

$$[(k+1)k + 2(l+1)(k+1)]a_{k+1} + [-2k$$

$$+ (\rho_0 - 2(l+1))]a_k = 0.$$

This leads to a recursion relation

$$a_{k+1} = \frac{2(k+l+1) - \rho_0}{(k+1)(k+2l+2)}a_k. \tag{235}$$

In the limit of large values of k, the ratio of successive coefficients is

$$\frac{a_{k+1}}{a_k} \sim \frac{2}{k} \quad \text{for} \quad k \to \infty. \tag{236}$$

For large k the terms resemble the exponential function $e^{2\rho}$ so $v(\rho)$ behaves like $e^{2\rho}$ and $u(\rho)$ would increase as $e^{\rho} = e^{Kr}$ for large r. The series must therefore terminate to ensure that the eigenfunction be bounded at infinity. Let us assume that $v(\rho)$ is a polynomial of order n. Then Eq. (235) gives

$$\rho_0 = 2(n_r + l + 1). \tag{237}$$

n_r, the radial quantum number, is the degree of the polynomial and must be 0, 1, 2,

The energy eigenvalues of the bound states of the hydrogenic atoms are given by

$$\rho_0 = 2n,$$

where

$$n = n_r + l + 1$$

or,

$$E_n = -\frac{\mu Z^2 e^4}{2n^2\hbar^2}, \quad n = 1, 2, 3, \ldots \tag{238}$$

$n = n_r + l + 1$ is the principal quantum number and takes on the values 1, 2, 3, ..., as n_r, l can take positive integer or zero values. Here, there is "accidental" degeneracy – greater than the "normal" degeneracy $2n + 1$. For given n, $l = 0, 1, 2, ..., n - 1$ and for each l there are $2l + 1$ distinct values of $m(- l \leq m \leq l)$.

So the total degeneracy is

$$g_n = \sum_{l = 0}^{n - 1} (2l + 1) = n \cdot \left(\frac{1 + 2n - 1}{2} \right) = n^2. \tag{239}$$

$$\underset{\substack{\text{no. of} \\ \text{terms}}}{} \quad \underset{\substack{\text{average of} \\ \text{terms}}}{}$$

Equation (238) is the Balmer formula for the discrete energy levels of hydrogen ($Z = 1$) and other one-electron systems. There are an infinite number of the discrete energy levels. The energy eigenvalues depend only on n and are degenerate with respect to l and m. The degeneracy with respect to m is the normal one for any central potential. The l-degeneracy is, however, characteristic of the Coulomb potential.

We shall now deal with the radial eigenfunctions of the discrete spectrum. From Eq. (232) with $\rho_0 = 2n$, we obtain

$$\left[\rho \frac{d^2}{d\rho^2} + (2l + 2 - 2\rho) \frac{d}{d\rho} + 2(n - l - 1) \right] v = 0$$

or, with $z = 2\rho$,

$$z \frac{d^2v(z)}{dz^2} + (2l + 2 - z) \frac{dv(z)}{dz} + (n - l - 1)v(z) = 0. \tag{240}$$

This equation can be identified with Kummer's equation

$$z \frac{d^2v(z)}{dz^2} + (\gamma - z)\frac{dv(z)}{dz} - \alpha v(z) = 0 \tag{241}$$

with $\gamma = 2l + 2$, and $\alpha = l + 1 - n$.

The solution of Eq. (241), regular at the origin, is the confluent hypergeometric function

$$_1F_1(\alpha, \gamma; z) = 1 + \frac{\alpha z}{\gamma 1!} + \frac{\alpha(\alpha + 1)z^2}{\gamma(\gamma + 1)2!} + ...$$

$$= \frac{\Gamma(\gamma)}{\Gamma(\alpha)} \sum_{k = 0}^{\infty} \frac{\Gamma(\alpha + k)}{\Gamma(k + 1)\Gamma(\gamma + k)} z^k, \tag{242}$$

within a multiplicative constant.

An argument similar to that used earlier will show that $e^{-z/2} \, _1F_1(l + 1 - n, 2l + 2; z)$ will be infinite unless the power series for $_1F_1$ terminates (becomes a polynomial). To obtain acceptable solutions, we must have the condition that α be a negative integer when the series terminates. This implies $l + 1 - n = - n_r$ ($n_r = 0, 1, 2, ...$) or $n = n_r + l + 1$. The confluent hypergeometric function $_1F_1(- n_r, 2l + 2; \rho)$ reduces to a polynormial of degree n_r.

The radial eigenfunctions may be written as

$$R \sim e^{-\rho/2}\, \rho^l\ {}_1F_1(-(n-l-1),\, 2l+2;\, \rho).$$

The radial wave functions are often expressed in terms of the associated Laguerre functions (See Eq. (130)).

We note that $L_q^P(\rho)$ satisfies the differential equation.

$$\left[\rho\, \frac{d^2}{d\rho^2} + (p+1-\rho)\, \frac{d}{d\rho} + (q-p)\right] L_q^P(\rho) = 0.$$

So the acceptable solution of $v(\rho)$ is given by the associated Laguerre polynomial.

$L_{n+l}^{2l+1}(\rho)$ which is a polynomial of degree

$$n+l-(2l+1) = n-l-1 = n_r.$$

We write the hydrogenic radial function as

$$R_{nl}(r) = N_{nl}\, e^{-\rho/2}\, \rho^l\, L_{n+l}^{2l+1}(\rho), \tag{243}$$

where N_{nl} are the normalization constants and

$$\rho = 2r\, \frac{\sqrt{-2\mu E}}{\hbar}$$

$$= \frac{2Z}{n}\, \frac{\mu e^2}{\hbar^2}\, r$$

$$= \frac{2Z}{na_\mu}\, r, \tag{244}$$

with $a_\mu = a_0\, \dfrac{m_e}{\mu}$ and $a_0 = \dfrac{\hbar^2}{m_e e^2}$ is the first Bohr radius.

The radial functions are normalized by the condition.

$$\int_0^\infty r^2 |R_{nl}(r)|^2\, dr = 1. \tag{245}$$

Using the normalization of the associated Laguerre polynomials (see Bethe and Salpeter (1957))

$$\int_0^\infty e^{-\rho}\, \rho^{2l}\, \left[L_{n+l}^{2l+1}(\rho)\right]^2 \rho^2\, d\rho$$

$$= \frac{2n[(n+l)!]^3}{(n-l-1)!}, \tag{246}$$

we get

$$N_{nl} = -\left(\frac{2Z}{na_\mu}\right)^{3/2} \left(\frac{(n-l-1)!}{2n[(n+l)!]^3}\right)^{1/2}. \tag{247}$$

The normalized radial functions are

$$R_{nl}(r) = - \left\{ \left(\frac{2Z}{na_\mu} \right)^3 \left(\frac{(n-l-1)!}{2n[(n+l)!]^3} \right)^{1/2} \right\}$$

$$\times \, e^{-\rho/2} \, \rho^l \, L_{n+l}^{2l+1}(\rho), \tag{248}$$

with $\rho = \dfrac{2Z}{na_\mu} r, \quad a_\mu = \dfrac{\hbar^2}{\mu e^2}.$

We write explicit expressions for the first few radial eigenfunctions

$n = 1, l = 0$ (K shell, s orbital)
$$R_{10}(r) = 2 \left(\frac{Z}{a_\mu} \right)^{3/2} e^{-Zr/a_\mu},$$

$n = 2, l = 0$ (L shell, s orbital)
$$R_{20}(r) = 2 \left(\frac{Z}{2a_\mu} \right)^{3/2} (1 - Zr/2a_\mu) \, e^{-Zr/2a_\mu},$$

$n = 2, l = 1$ (L shell, p orbital)
$$R_{21}(r) = \frac{1}{\sqrt{3}} \left(\frac{Z}{2a_\mu} \right)^{3/2} (Zr/a_\mu) \, e^{-Zr/2a_\mu},$$

$n = 3, l = 0$ (M shell, s orbital)
$$R_{30}(r) = 2 \left(\frac{Z}{3a_\mu} \right)^{3/2} (1 - 2Zr/3a_\mu + 2Z^2 r^2/27a_\mu^2) e^{-Zr/3a_\mu},$$

$n = 3, l = 1$ (M shell, p orbital)
$$R_{31}(r) = \frac{4\sqrt{2}}{9} \left(\frac{Z}{3a_\mu} \right)^{3/2} (1 - Zr/6a_\mu) \cdot (Zr/a_\mu) \, e^{-Zr/3a_\mu},$$

$n = 3, l = 2$ (M shell, d orbital)
$$R_{32}(r) = \frac{4}{27\sqrt{10}} \left(\frac{Z}{3a_\mu} \right)^{3/2} (Zr/a_\mu)^2 \, e^{-Zr/3a_\mu}, \tag{249}$$

The states with $l = 0$ are finite at the origin, while the states with $l \geq 1$ vanish there.

The complete eigenfunctions of the discrete spectrum for a one-electron atom are

$$\psi_{nlm}(r, \theta, \phi) = R_{nl}(r) \, Y_{lm}(\theta, \phi), \tag{250}$$

where the radial functions are given by Eq. (248) and the spherical harmonics provide the angular part. The orthonormality condition holds

$$\int d^3r \, \psi_{nlm}^* \, \psi_{n'l'm'} = \delta_{nn'} \, d_{ll'} \, d_{mm'} \tag{251}$$

We note that the parity operator P acts on the hydrogenic wave function ψ as follows

$$P\psi_{nlm}(r, \theta, \phi) = (-1)^l \, \psi_{nlm}(r, \theta, \phi). \tag{252}$$

We can calculate the expectation of mean value or the distance r between the electron and the nucleus, raised to various powers.

As an example, we calculate

$$\langle r \rangle_{lmn} = \int \psi^*_{nlm}(\vec{r}) \, r \, \psi_{nlm}(\vec{r}) \, d\vec{r}$$

$$= \int_0^\infty r^3 \, |R_{nl}(r)|^2 \, d\vec{r}$$

$$= a_\mu \frac{1}{2Z} [3n^2 - l(l+1)], \quad \text{using Eq. (248)}$$

The average values of r^k where k is a positive or a negative integer are given below.

k	$\langle r^k \rangle_{nlm}$ (in units a_μ^k)
1	$\dfrac{1}{2Z}[3n^2 - l(l+1)]$
2	$\dfrac{n^2}{2Z^2}[5n^2 + 1 - 3l(l+1)]$
3	$\dfrac{n^2}{8Z^3}[35n^2(n^2-1) - 30n^2(l+2)(l-1) + 3(l+2)(l+1)l(l-1)]$
-1	$\dfrac{Z}{n^2}$
-2	$\dfrac{Z^2}{n^3\left(l+\dfrac{1}{2}\right)}$
-3	$\dfrac{Z^3}{n^3(l+1)\left(l+\dfrac{1}{2}\right)l}$

$$(253)$$

One may use Kramers' recursion relation to obtain the averages

$$\frac{k+1}{n^2} \langle r^k \rangle_{nlm} - (2k+1)a_\mu \langle r^{k-1} \rangle_{nlm}$$

$$+ \frac{k}{4}[(2l+1)^2 - k^2]a_\mu^2 \langle r^{k-2} \rangle_{nlm} = 0. \qquad (254)$$

(b) Parabolic Coordinates

Schrödinger's equation for the hydrogen atom can also be separated and solved in parabolic coordinates. The wave equation for an electron moving in a central potential has evident spherical symmetry and can always be separated in spherical polar coordinates. In case the central field is Coulombic, then a separation can also be carried out in parabolic coordinates. This alternative is connected with higher symmetry and is an example of a general result – namely, when a problem has accidental degeneracy, more than the normal one ($2l+1$), it implies and is implied by the fact that the problem is separable in other orthogonal coordinate systems. The separation of the wave equation in parabolic coordinates is especially useful in all perturbation problems in which there is a privileged direction in space as distinguished by some external force, e.g., Stark effect, photoelectric effect, Compton effect, and collision of electrons.

The parabolic coordinates ξ, η, ϕ are defined by the relations

$$x = \sqrt{\xi\eta}\ \cos\phi, \quad \xi = r + z,$$

$$y = \sqrt{\xi\eta}\ \sin\phi, \quad \eta = r - z,$$

$$\phi = \tan^{-1}\frac{y}{x},$$

$$z = \frac{1}{2}(\xi - \eta), \quad r = \frac{1}{2}(\xi + \eta), \tag{255}$$

where $\xi \geq 0$, $0 < \eta < \infty$, $0 \leq \phi \leq 2\pi$,

The surfaces $\xi = $ const. and $\eta = $ const. are paraboids of revolution about the z-axis with the nucleus at the origin $(x = y = z = 0)$ as focus. The line and volume elements are

$$ds^2 = g_1^2\ d\xi^2 + g_2^2\ d\eta^2 + g_3^2\ d\phi^2$$

$$= \frac{\xi + \eta}{4\xi}\ d\xi^2 + \frac{\xi + \eta}{4\eta}\ d\eta^2 + \xi\eta d\phi^2,$$

$$d\tau = g_1 g_2 g_3 d\xi d\eta d\phi$$

$$= \frac{1}{4}(\xi + \eta)d\xi d\eta d\phi \tag{256}$$

where $\quad g_1 = \frac{1}{2}\sqrt{\frac{\xi + \eta}{\xi}}, \quad g_2 = \frac{1}{2}\sqrt{\frac{\xi + \eta}{\eta}}, \quad g_3 = \sqrt{\xi\eta},$

The Laplacian operator is

$$\vec{\nabla}^2 \equiv \frac{1}{g_1 g_2 g_3}\left\{\frac{\partial}{\partial\xi}\frac{g_2 g_3}{g_1}\frac{\partial}{\partial\xi} + \frac{\partial}{\partial\eta}\frac{g_3 g_1}{g_2}\frac{\partial}{\partial\eta}\right.$$

$$\left. + \frac{\partial}{\partial\phi}\frac{g_1 g_2}{g_3}\frac{\partial}{\partial\phi}\right\}$$

$$= \frac{4}{\xi + \eta}\left\{\frac{\partial}{\partial\xi}\left(\xi\frac{\partial}{\partial\xi}\right) + \frac{\partial}{\partial\eta}\left(\eta\frac{\partial}{\partial\eta}\right)\right.$$

$$\left. + \frac{\xi + \eta}{4\xi\eta}\frac{\partial^2}{\partial\phi^2}\right\}. \tag{257}$$

With the Coulomb potential

$$V = -\frac{Ze^2}{r} = -\frac{2Ze^2}{\xi + \eta}.$$

the Schrödinger equation for a hydrogenic atom in parabolic coordinates is

$$\frac{\partial}{\partial \xi}\left(\xi \frac{\partial \psi}{\partial \xi}\right) + \frac{\partial}{\partial \eta}\left(\eta \frac{\partial \psi}{\partial \eta}\right) + \frac{\xi + \eta}{4\xi\eta}\frac{\partial^2 \psi}{\partial \phi^2}$$

$$+ \left(\frac{\mu E}{\hbar^2} \cdot \frac{\xi + \eta}{2} + \frac{\mu Z e^2}{\hbar^2}\right)\psi = 0. \tag{258}$$

This equation is readily separable:

$$\psi = f_1(\xi)\, f_2(\eta)\, \Phi(\phi). \tag{259}$$

We know

$$\Phi(\phi) = \frac{1}{\sqrt{2\pi}}\, e^{im\phi}, \tag{260}$$

where m is the magnetic quantum number f_1 and f_2 satisfy the equations

$$\frac{d}{d\xi}\left(\xi \frac{df_1}{d\xi}\right) + \left(\frac{\mu E}{2\hbar^2}\,\xi - \frac{m^2}{4\xi}\right)f_1 + \frac{1}{2}\left(\frac{Z}{a_\mu} + \lambda\right)f_1 = 0 \tag{261}$$

$$\frac{d}{d\eta}\left(\eta \frac{df_2}{d\eta}\right) + \left(\frac{\mu E}{2\hbar^2}\,\eta - \frac{m^2}{4\eta}\right)f_2 + \frac{1}{2}\left(\frac{Z}{a_\mu} - \lambda\right)f_2 = 0 \tag{262}$$

where the parameter λ is the arbitrary separation constant. These two equations have the same form of Eq. (121) for the plane harmonic oscillator.

Let $\qquad A = \dfrac{\mu E}{2\hbar^2}, \quad B = \dfrac{1}{4}\left(\dfrac{Z}{a_\mu} + \lambda\right), \quad C = -\dfrac{m^2}{4}$

For discrete spectrum, $\qquad A < 0.$

Equation (261) is now

$$\frac{d^2 f_1}{d\xi^2} + \frac{1}{\xi}\frac{df_1}{d\xi} + \left(A + \frac{2B}{\xi} + \frac{C}{\xi^2}\right)f_1 = 0. \tag{263}$$

We proceed to solve this equation by the polynomial method. We see that f behaves as $e^{-\sqrt{-A}\,\xi}$ asymptotically for large ξ and as ξ^{m+2} for small ξ. So we put $\rho = 2\sqrt{-A}\,\xi$, which is a real variable for discrete states and write

$$f = e^{-\rho/2}\, v(\rho).$$

Substituting this in Eq. (263), we obtain

$$\frac{d^2 v}{d\rho^2} + \left(\frac{1}{\rho} - 1\right)\frac{dv}{d\rho} + \left\{\left(\frac{B}{\sqrt{-A}} - \frac{1}{2}\right)\frac{1}{\rho} + \frac{C}{\rho^2}\right\}v = 0. \tag{264}$$

The indicial equation shows that at the origin $v \sim \rho^{\frac{\pm|m|}{2}}$ and so we require a solution of the form $\rho^{+|m|/2}$ at the origin and bounded at infinity. Putting $v = \rho^{|m|/2} \times \sum_{n=0}^{\infty} a_n \rho^n$, we obtain the recursion formula

$$a_{n+1} = \frac{\left(n + \dfrac{|m|}{2} + \dfrac{1}{2} - \dfrac{B}{\sqrt{-A}}\right)}{(n+1)^2 + (n+1)|m|}\, a_n.$$

For large n, $a_n \approx \dfrac{1}{n}\, a_{n-1}$, which is unacceptable as the function resembles e^{+x} at infinity. The power series must terminate, giving

$$n_\xi + \frac{|m|}{2} = -\frac{1}{2} + \frac{B}{\sqrt{-A}}$$

$$= -\frac{1}{2} + \frac{1}{\sqrt{-A}} \cdot \frac{1}{4}\left(\frac{Z}{a_\mu} + \lambda\right),$$

where n_ξ is the degree of the polynomial. The corresponding η equation yields

$$n_\eta + \frac{|m|}{2} = -\frac{1}{2} + \frac{1}{\sqrt{-A}} \cdot \frac{1}{4}\left(\frac{Z}{a_\mu} - \lambda\right).$$

Adding,

$$n_\xi + n_\eta + |m| + 1 = \frac{1}{2\sqrt{-A}}\frac{Z}{a_\mu}. \tag{265}$$

Putting the principal quantum number

$$n = n_\xi + n_\eta + |m| + 1, \tag{266}$$

in Eq. (265), we get the energy eigenvalues

$$E_n = \frac{-\hbar^2 Z^2}{2\mu a_\mu^2 n^2} = -\frac{\mu Z^2 e^4}{2\hbar^2 n^2}, \quad \text{as before.} \tag{267}$$

We have

$$\lambda = \frac{n_\xi - n_\eta}{n} \cdot \frac{Z}{a_\mu}.$$

n_ξ and n_η are both non-negative integers, and m is the magnetic quantum number.

Now that we have the energy levels, let us evaluate the eigenfunctions. If we put $v = \rho^{\frac{|m|}{2}} w$, then w satisfies the equation

$$\rho_\xi w'' + (|m| + 1 - \rho_\xi)w' + n_\xi w = 0,$$

and similarly for η,

$$\rho_n w'' + (|m| + 1 - \rho_n)w' + n_n w = 0,$$

where

$$\rho_\xi = \frac{Z}{na_\mu}\, \xi, \quad \text{and} \quad \rho_n = \frac{Z}{na_\mu}\, \eta.$$

So

$$w = L_{n_\xi + |m|}^{|m|}(\rho_\xi) \quad \text{or} \quad L_{n_\eta + |m|}^{|m|}(\rho_\eta)$$

The eigenfunctions are

$$\psi(\xi, \eta, \phi) = N e^{-\frac{1}{2}(\rho_\xi + \rho_\eta)} (\rho_\xi \, \rho_\eta)^{|m|/2}$$

$$\times L_{n_\eta + |m|}^{|m|} (\rho_\eta) \cdot \frac{e^{im\phi}}{\sqrt{2\pi}}$$

$$= N \frac{e^{im\phi}}{\sqrt{2\pi}} e^{-\frac{Z}{2na_\mu}(\xi + \eta)} \left(\frac{Z}{na_\mu}\right)^{|m|} (\xi\eta)^{\frac{|m|}{2}}.$$

$$\times L_{n_\xi + |m|}^{|m|} \left(\frac{Z\xi}{na_\mu}\right) L_{n_\eta + |m|}^{|m|} \left(\frac{Z\eta}{na_\mu}\right), \tag{268}$$

where N is a normalization constant. What is the degeneracy of the nth eigenvalue? If m is fixed, n_ξ can assume the n-m values 0, 1, ..., $n - m - 1$. But we have n states for $m = 0$ and twice $\frac{1}{2} n(n - 1)$ states for $m \neq 0$. The multiplicity is

$$g(n) = n + 2 \cdot \frac{1}{2} n(n - 1) = n^2, \quad \text{as before.}$$

(c) Schwinger's Method

Let us consider the differential Eq. (147) for the energy eigenstates of the plane isotropic oscillator

$$\left[\frac{d^2}{d\rho^2} - \frac{m^2 - \frac{1}{4}}{\rho^2} + 2|m| + 4n_\rho + 2 - \rho^2 \right] \chi(\rho) = 0. \tag{269}$$

Substituting

$$\rho^2 = 2\lambda r \ (\lambda > 0), \tag{270}$$

we obtain

$$\left[\frac{d^2}{dr^2} + \frac{1}{2r} \frac{d}{dr} - \frac{m^2 - \frac{1}{4}}{4r^2} + \frac{\lambda}{r}(|m| + 2n_\rho + 1) - \lambda^2 \right] \chi(\rho) = 0. \tag{271}$$

Writing
$$\chi(\rho) = C(\lambda r)^{-1/4} u(r), \tag{272}$$

we get

$$\left[\frac{d^2}{dr^2} - \frac{m^2 - 1}{4r^2} + \frac{\lambda(|m| + 2n_r + 1)}{r} - \lambda^2 \right] u(r) = 0, \tag{273}$$

where we have put n_r for n_ρ.

The reduced radial equation for a hydrogenic atom is Eq. 228.

$$\left[\frac{d^2}{dr^2} - \frac{l(l + 1)}{r^2} + \frac{2\mu Ze^2}{\hbar^2 r} + \frac{2\mu E}{\hbar^2} \right] u(r) = 0 \tag{274}$$

Comparing the Eqs. (273) and (274), we have the correspondence

2D oscillator	**3D Coulomb**

$$\frac{m^2 - 1}{4} \qquad\qquad l(l + 1)$$

$$\lambda(|m| + 2n_r + 1) \qquad\qquad \frac{2\mu Z e^2}{\hbar^2}$$

$$\lambda^2 \qquad\qquad \frac{2\mu E}{\hbar^2}(-E).$$

So

$$m^2 \to 4l(l + 1) + 1 = (2l + 1)^2,$$

or,
$$|m| \to 2l + 1,$$

then
$$\lambda(2l + 2n_r + 2) \to \frac{2\mu E}{\hbar^2} Z e^2$$

or,
$$\lambda \to \frac{Z}{na_\mu},$$

where
$$n = n_r + l + 1.$$

Therefore,
$$\lambda^2 = \frac{Z^2}{n^2 a_\mu^2} = \frac{2\mu E}{\hbar^2}(-E)$$

whence
$$E_n = -\frac{Z^2 e^2}{2n^2 a_\mu}, \quad n = 1, 2, 3, \dots. \tag{275}$$

The wave functions of the hydrogenic atom can be determined as follows. Now,

$$1 = \int_0^\infty d\rho \, [u(\rho)]^2$$

$$= C^2 \int_0^\infty \frac{1}{\sqrt{2}} \, dr \, \frac{1}{r} [u(r)]^2,$$

so that

$$C = 2^{\frac{1}{4}} \left(\int_0^\infty dr \, \frac{1}{r} [u(r)]^2 \right)^{-\frac{1}{2}}$$

$$= 2^{\frac{1}{4}} \left\langle \frac{1}{r} \right\rangle^{-\frac{1}{2}} \tag{276}$$

But,
$$\left\langle \frac{1}{r} \right\rangle = \frac{Z}{n^2 a_\mu}. \tag{277}$$

The hydrogenic wave functions are found by using the oscillator wave functions from Eq. (142):

$$u_{n,\,l}\,(r) = (-\,1)^{n\,-\,l\,-\,1}\,\sqrt{\frac{Z}{n^2 a_\mu}\,\frac{(n-l-1)!}{(n+l)!}}\,\left(\frac{2Zr}{na_\mu}\right)^{l+1}$$

$$\times\, e^{-\frac{Zr}{na_\mu}}\, L_{n-l-1}^{(2l+1)}\left(\frac{2Zr}{na_\mu}\right).$$

or,

$$R_{n,\,l}\,(r) = (-\,1)^{n\,-\,l\,-\,1}\,\left(\frac{Z}{a_\mu}\right)^{\frac{3}{2}}\frac{2}{n^2}\left(\frac{(n-l-1)!}{(n+l)!}\right)^{\frac{1}{2}}$$

$$\times\, e^{-\,\rho/2}\times \rho^l\, L_{n-l-1}^{(2l+1)}\,(\rho), \tag{278}$$

where $$\rho = \frac{2Z}{na_\mu}\,r.$$

(d)   Wave Functions in Momentum Space

The wave function in momentum space, $\phi(\vec{p})$ is defined as the Fourier transform of the ordinary wave function $\psi(\vec{r})$ in position space (see § 3.1).

Explicitly,

$$\phi(\vec{p}) = (2\pi\hbar)^{-\,3/2}\int e^{-\,i\vec{p}\,\cdot\,\vec{r}/\hbar}\,\psi(r)\,d^3r,$$

$$\psi(\vec{r}) = (2\pi\hbar)^{-\,3/2}\int e^{+\,i\vec{p}\,\cdot\,\vec{r}/\hbar}\,\phi(\vec{p})\,d^3p. \tag{279}$$

If $\psi(\vec{r})$ is normalized to unity, then $\phi(\vec{p})$ is also normalized to unity:

$$\int d^3p\,\,|\phi(\vec{p})|^2 = 1. \tag{280}$$

For some simple potential problems (bound state) one can solve the Schrödinger equation in position space and then evaluate the Fourier transform of the wave function in position space.

A better method is to rewrite the Schrödinger equation as an equation involving $\phi(\vec{p})$ directly by replacing $\vec{r}$ by the operator $i\hbar\vec{\nabla}_{\vec{p}}$. The Schrödinger equation then assumes the form of a differential equation in momentum space. For example, the isotropic oscillator in three dimensions with the Hamiltonian

$$H = \frac{\vec{p}^{\,2}}{2m} + \frac{1}{2}\,m\omega^2\vec{r}^{\,2},\ \text{where}\ \vec{r}^{\,2} = x^2 + y^2 + z^2,\ \text{yields the equation.}$$

$$\left(\frac{\vec{p}^{\,2}}{2m} - \frac{1}{2}\,m\omega^2\hbar^2\vec{\nabla}_{\vec{p}}^{\,2}\right)\phi(\vec{p}) = E\phi(\vec{p}),$$

or,

$$\left\{\vec{\nabla}_{\vec{p}}^{\,2} + \left(\frac{2\gamma}{\hbar\omega}\,E - \gamma^2 p^2\right)\right\}\phi(\vec{p}) = 0,$$

where $\gamma = \dfrac{1}{m\hbar\omega}$. Since this equation is formally identical with the equation in Section 11.2(c), one can obtain the wave functions in a similar manner. Since one usually deals with potentials which depend on $\vec{r}$, but not on $\vec{p}$, such an approach usually is not easy. Hylleraas (1932), however, used this method to obtain the momentum space wave functions for the discrete spectrum of hydrogen.

A more convenient method due to Fock (1935) involves rewriting the Schrödinger equation in the form of an integral equation in momentum space.

Let us consider a particle in a potential $V(\vec{r})$. The time-independent Schrödinger equation in position space is

$$\left[\frac{\vec{p}^{2}}{2m} + V(\vec{r})\right] \psi(\vec{r}) = E\psi(\vec{r}). \tag{281}$$

Multiplying both sides by $(2\pi\hbar)^{3/2} e^{-i\vec{p}\cdot\vec{r}/\hbar}$ and integrating over all space, we find

$$\frac{\vec{p}^{2}}{2m} \phi(\vec{p}) + (2\pi\hbar)^{-3/2} \int e^{-i\vec{p}\cdot\vec{r}/\hbar} V(\vec{r})\psi(\vec{r})\, d^{3}r = E\phi(\vec{p}). \tag{282}$$

Using Eq. (279) and introducing

$$V'(\vec{p} - \vec{p}') = \frac{1}{(2\pi\hbar)^{3}} \int e^{-i(\vec{p}-\vec{p}')\cdot\vec{r}/\hbar} V(\vec{r})d^{3}r \tag{283}$$

we get

$$\left(\frac{\vec{p}^{2}}{2m} - E\right) \phi(\vec{p}) = - \int d^{3}p' V'(\vec{p} - \vec{p}')\, \phi(\vec{p}'). \tag{284}$$

For a state of negative energy (bound state), we obtain an integral equation

$$(\vec{p}^{2} + \beta^{2})\, \phi(\vec{p}) = - 2m \int V'(\vec{p} - \vec{p}')\, \phi(\vec{p}')\, d^{3}p', \tag{285}$$

where $2mE = -\beta^{2}$. The kernel of the integral equation, $V(\vec{p} - \vec{p}')$, is a function of a single vector variable $\vec{q} = (\vec{p}' - \vec{p})$

If the potential is central, $V(r)$, then the momentum space potential $V'(\vec{p})$ is a function of the absolute value of $\vec{p}$ only and is real.

In this case, the wave equation (285) is separable in spherical polar coordinates. If (p, θ, ϕ) are the polar coordinates of the momentum $\vec{p}$, then the solutions exist of the form

$$\phi(\vec{p}) = F_{l}(p)\, Y_{lm}(\theta, \phi). \tag{286}$$

We expand $V'(|\vec{p} - \vec{p}'|)$ in Legendre polynomials

$$V'(\vec{p} - \vec{p}') = \sum_{l=0}^{\infty} W_{l}\,(p.p')P_{l}(x), \tag{287}$$

where

$$x = \frac{\vec{p}.\vec{p}'}{pp'}$$

and

$$W_{l}\,(p.p') = \frac{2l + 1}{2} \int_{-1}^{+1} P_{l}(x)\, V'(|\vec{p} - \vec{p}'|)\, dx.$$

We obtain a one-dimensional integral equation for $F_l(p)$ of the form

$$(p^2 + \beta^2)\, F_l(p) = -\,2m \int_0^\infty p'^2\, K_l(p, p')\, F_l(p')\, dp', \tag{288}$$

where

$$K_l(p, p') = 4\pi \int_{-1}^{+1} V'\left(\sqrt{p^2 + p'^2 - 2pp'x}\right) P_l(x)\, dx, \tag{289}$$

is a symmetric kernel in p and p'.

The Coulomb potential is $V(r) = -\dfrac{Ze^2}{r}$ so that

$$V'(p) = -\frac{Ze^2}{2\pi^2\hbar}\frac{1}{p^2}$$

The three-dimensional integral equation then is

$$(p^2 + \beta^2)\,\phi(\vec{p}) = \frac{mZe^2}{\hbar\pi^2} \int d^3p'\, \frac{\phi(\vec{p'})}{|\vec{p} - \vec{p'}|^2}. \tag{290}$$

The kernel is

$$\begin{aligned}
K_l(p, p') &= -\frac{2Z}{\pi} \cdot \frac{1}{2pp'} \int_{-1}^{+1} \frac{1}{(p^2 + p'^2)/2pp' - x}\, P_l(x)dx \\[2mm]
&= -\frac{Z}{\pi}\frac{1}{pp'}\, Q_l\!\left(\frac{p^2 + p'^2}{2pp'}\right),
\end{aligned} \tag{291}$$

where Q_l is a Legendre function of the second kind, given by

$$Q_l(z) = \frac{1}{2} \int_{-1}^{+1} \frac{1}{z - x}\, P_l(x)dx. \tag{292}$$

The one-dimensional integral equation is then

$$(p^2 + \beta^2)\, F_l(p) = \left(\frac{2mZe^2}{\pi\hbar p}\right) \int_0^\infty p'Q_l\!\left(\frac{p^2 + p'^2}{2pp'}\right) F_l(p')dp'. \tag{293}$$

Equation (290) is converted into a symmetric integal equation which is invariant under O(4).

Fock recognized the kernel as the Jacobian determinant for a stereographic projection from the surface of a four-dimensional sphere to three dimensions. Here the integral equation is the same as the integral equation for spherical harmonics in four-dimensional space. Fock suggested that the accidental or Coulomb degeneracy is related to the symmetry properties of four-dimensional space. The proper rotation group in four dimensions (SO(4)) describes a dynamical symmetry for the bound states of the hydrogen atom.

Fock solved this equation and obtained the energy eigenvalues for the discrete spectrum of hydrogen. The spectrum E_n are of course identical with that obtained earlier.

The normalized radial momentum space functions for hydrogen are given (in a.u.) by

$$F_{nl}(p) = \left[\frac{2}{\pi}\frac{(n-l-1!)}{(n+l)!}\right]^{\frac{1}{2}} n^2 2^{2(l+1)}\, l!\, \frac{n^l p^l}{(n^2p^2+1)^{l+2}}$$

$$\times\, C^{l+1}_{n-l-1}\left(\frac{n^2p^2-1}{n^2p^2+1}\right), \tag{294}$$

where $C_N^{\nu}(x)$ is the Gegenbauer function defined by the following relation

$$(1-2hx+\hbar^2)^{-\nu} = \sum_{N=0}^{\infty} C_N^{\nu}(x)h^N, \quad |h| < 1. \tag{295}$$

The explicit expressions for C_N^{ν} for a few values of N are

$$C_0^{\nu}(x) = 1, \quad C_1^{\nu}(x) = 2\nu x,$$

$$C_2^{\nu}(x) = 2\nu(\nu+1)x^2 - \nu. \tag{296}$$

The first few radial wave functions $F_{nl}(p)$ are

$$F_{10}(p) = 4\sqrt{\frac{2}{\pi}}\,\frac{1}{(p^2+1)^2},$$

$$F_{20}(p) = \frac{32}{\sqrt{\pi}}\,\frac{4p^2-1}{(p^2+1)^3},$$

$$F_{21}(p) = \frac{128}{\sqrt{3\pi}}\,\frac{p}{(p^2+1)^3}, \tag{297}$$

We now describe the method due to Hylleraas (1932). The problem of the hydrogen atom in momentum space is done here with the usual operator calculus yielding a differential equation (an advantage!) which is then solved by the polynomial method. The Hamiltonian is

$$H \equiv \frac{\vec{p}^{\,2}}{2m} - \frac{e^2}{r} = E. \tag{298}$$

All that is necessary to "rationalize" and get rid of the operator $\frac{1}{r}$. This can be done easily by expressing the wave equation only in terms of r^2 and functions of p.

Hylleraas derives the expression

$$gf = \left\{ e^{\frac{\hbar}{i}[\vec{\nabla}_{\vec{r}}\,\cdot\,\vec{\nabla}_{\vec{p}}]} \right\} fg, \tag{299}$$

where g is an arbitrary function of $p_x,\,p_y,\,p_z$ and f an arbitrary function of x, y, z. The operator is interpreted in terms of its power series expansion. Here $\vec{\nabla}_{\vec{r}}$

operates only on f and $\vec{\nabla}_{\vec{p}}$ only on g. This gives an expression for the commutator of a pair of functions

$$gf = fg + \frac{\hbar}{i}\,\nabla f \cdot \nabla g + \frac{1}{2}\left(\frac{\hbar}{i}\right)^2 \left[\frac{\partial^2 f}{\partial x^2}\frac{\partial^2 g}{\partial p_x^2}\right.$$

$$\left. + \frac{\partial^2 f}{\partial y^2}\frac{\partial^2 g}{\partial p_y^2} + \frac{\partial^2 f}{\partial z^2}\frac{\partial^2 g}{\partial p_z^2}\right] + \cdots \tag{300}$$

For the present application, however, one can easily derive all the necessary commutation relations from the fundamental formula

$$pf = fp + \frac{\hbar}{i}\,\nabla f \tag{301}$$

Writing Eq. (298) as

$$(\vec{p}^2 - 2mE)r = 2me^2$$

and squaring, we get

$$(\vec{p}^2 - 2mE)r(\vec{p}^2 - 2mE)r = 4m^2 e^4. \tag{302}$$

We now apply the operator formula to pull r out of the middle to the right in Eq. (302). Now,

$$r(p^2 - 2mE) = (p^2 - 2mE)r + 2i\hbar(\vec{p}\cdot\vec{r})\frac{1}{r} + (i\hbar)^2\frac{2}{r}.$$

Multiplying this relation on the right by r and on the left by $(p^2 - 2mE)$, Eq. (302) becomes

$$(p^2 - 2mE)^2 r^2 + (p^2 - 2mE)\left[2i\hbar(\vec{p}\cdot\vec{r}) + 2(i\hbar)^2\right] = 4m^2 e^4.$$

By substituting $x_i = i\hbar\,\dfrac{\partial}{\partial p_i}$, we obtain Schrödinger's equation for the hydrogen atom in the form

$$(p^2 - 2mE)^2\,\nabla_p^2\psi + 2(p^2 - 2mE)\left[\vec{p}\cdot\vec{\nabla}_{\vec{p}} + 1\right]\psi + \frac{8m^2 e^4}{\hbar}\,\psi = 0, \tag{303}$$

which can now be solved by the Sommerfeld method.

If we introduce the spherical polar coordinate p, θ, ϕ, then

$$\nabla_p^2 = \frac{\partial^2}{\partial p^2} + \frac{2}{p}\frac{\partial}{\partial p} - \frac{\vec{L}^2}{p^2\hbar^2},$$

$$\vec{p}\cdot\vec{\nabla}_p = p\hat{p}\cdot\left[\hat{p}\frac{\partial}{\partial p} + \hat{\theta}\frac{1}{p}\frac{\partial}{\partial\theta} + \hat{\phi}\frac{1}{p\sin\theta}\frac{\partial}{\partial\phi}\right].$$

We split off the angular part as before and the radial equation becomes

$$\left\{ (p^2 - 2mE)^2 \left[\frac{d^2}{dp^2} + \frac{2}{p}\frac{d}{dp} - \frac{l(l+1)}{p^2} \right] + 2(p^2 - 2mE) \right.$$

$$\left. \times \left[p\frac{d}{dp} + 1 \right] + \frac{8m^2 e^4}{\hbar^2} \right\} P(p) = 0.$$

and with $p = \sqrt{-2mE}\,\xi$ and $a = \dfrac{-me^4}{2\hbar^2 E}$, it becomes

$$\left\{ (\xi^2 + 1)^2 \left[\frac{d^2}{d\xi^2} + \frac{2}{\xi}\frac{d}{d\xi} - \frac{l(l+1)}{\xi^2} \right] + (\xi^2 + 1)\left[2\xi\frac{d}{d\xi} + 2 \right] + 4a \right\} P(\xi) = 0. \quad (304)$$

If we put $P(\xi) = \xi^l v$, and then put $\eta = \xi^2$, we obtain

$$\left[(\eta^2 + 1)^2\,\eta\,\frac{d^2}{d\eta^2} + \left(l + \frac{3}{2} \right)\frac{d}{d\eta} \right] v + (\eta + 1)\left[\eta\frac{d}{d\eta} + \frac{l+1}{2} \right] v + av = 0.$$

$$(305)$$

Now, clearly we cannot expect to find a polynomial solution for v about $\xi = 0$, since we would obtain the wrong behavior at infinity when we put $P = \xi^l v$. We can, however, use a solution about $\eta = -1$ which diverges there, but converge at infinity, at zero. Thus, we try a solution of the form

$$v = \sum_{k=0}^{\infty} \frac{a_k}{(\eta + 1)^{k + \lambda}}.$$

This will certainly behave well at infinity for positive k, λ and can also be made to converge at $\eta = 0$. Putting this ansatz into (305) we get a recursion relation

$$\left\{ (k + \lambda)\left(k + \lambda - l - \frac{3}{2} \right) + \frac{l+1}{2} \right\} a_k$$

$$= a_{k-1}\left\{ (k + \lambda - 1)^2 - a \right\}. \quad (306)$$

The solution must have only positive terms in k for acceptable behavior at infinity and so we want a_{-1} to vanish. This gives us

$$\left\{ \lambda\left(\lambda - l - \frac{3}{2} \right) + \frac{l+1}{2} \right\} = 0$$

whence

$$\lambda = l + 1;\; \frac{1}{2}.$$

Thus the first term will be either $\dfrac{1}{(\eta + 1)^{1/2}}$ or $\dfrac{1}{(\eta + 1)^{l+1}}$ corresponding

to behavior at infinity of terms like $\dfrac{1}{\xi}$, $\dfrac{1}{\xi^{2l+2}}$ of which only the second is sufficiently strongly diminishing. Therefore, we want the solutions v starting out like $\dfrac{1}{(\eta + 1)^{l+1}}$ – poles of order $l + 1$ and higher at $\eta = -1$, outside the range of interest. At $\eta = 0$, $v = \sum\limits_{k=0}^{\infty} a_k$. For large k, we see that $a_k \approx a_{k-1}$ and so the series will diverge. If we require that the highest term be $\dfrac{1}{(\eta + 1)^{\lambda + n_r}}$ then $a_{n_r} + 1 = 0$ and also all higher terms. The series terminates if $\{(k + \lambda - 1)^2 - a\} = 0$ with $k = n_r + 1$. The condition gives $a = (n_r + l + 1)^2 = n^2$, where $n = n_r + l + 1$. Substituting the value of a, the energy eigenvalues are found

$$E_n = -\frac{me^4}{2\hbar^2 n^2}. \quad n = 1, 2, 3, \ldots$$

11.6   KAON PHYSICS

The K-meson (kaon) and the π-meson (pion) are pseudoscalar bosons having zero spin and odd intrinsic parity. In nature there are four different kinds of kaons (K^+, K^-, K^0, $\overline{K}^0$) and three different kinds of pions ($\pi^\pm$, π^0). The kaons are a rich source of highly interesting phenomena. The study of the production and decay of the kaons provides the testing ground for the linear superposition principle of quantum mechanics, and the conservation laws.

The two neutral kaons are denoted as K^0 and $\overline{K}^0$, its antiparticle. These mesons are produced in strong interaction processes, e.g.,

$$\pi + p \rightarrow \Lambda^0 + K^0,$$

while the neutral kaons decay by weak interaction. In the Gell-Mann-Nishijima classification scheme

$$\frac{Q}{e} = I_3 + \frac{Y}{2}, \tag{307}$$

where $Y = B + S$ is the hypercharge, B the baryon number, S the strangeness quantum number. K^0 is assigned $S = 1$, while $\overline{K}^0$ has $S = -1$.

$$B = 0 \left\{ \begin{array}{ccc} & S & I_3 \\ & & +\frac{1}{2} \quad -\frac{1}{2} \\ & +1 & K^+ \quad K^0 \\ & -1 & \overline{K}^0 \quad K^- \end{array} \right.$$

Clearly, the strangeness scheme implies that K^0 and its antiparticle $\overline{K}^0$ are distinct. We define a strangeness operator S by

$$S|K^0\rangle = |K^0\rangle$$

$$S|K^0\rangle = -|\overline{K}^0\rangle \tag{308}$$

so that in the $|K^0\rangle$, $|\overline{K}^0\rangle$ basis

$$S = \begin{pmatrix} 1 & 0 \\ 0 & -1 \end{pmatrix}. \tag{308a}$$

We introduce the unitary operator C (charge conjugation operator or the particle-antiparticle conjugation operator) which converts the wave function of a particle of momentum $\vec{p}$ into the wave function of the antiparticle with momentum $\vec{p}$ and vice versa. Let $|P, \vec{p}\rangle$ and $|\overline{P}, \vec{p}\rangle$ denote the particle and antiparticle states respectively. Then

$$C|P, \vec{p}\rangle = \eta_c |\overline{P}, \vec{p}\rangle,$$

$$C|\overline{P}, \vec{p}\rangle = \eta_c^* |P, \vec{p}\rangle, \tag{309}$$

where η_c is a constant phase factor. Since $C^2 = +1$, the eigenvalues of C are ± 1.

For the neutral kaons, we write

$$C|K^0\rangle = -|\overline{K}^0\rangle,$$

$$C|\overline{K}^0\rangle = -|K^0\rangle \tag{310}$$

and for the three pions, we have

$$C|\pi^+\rangle = -|\pi^-\rangle$$

$$C|\pi^-\rangle = -|\pi^+\rangle$$

$$C|\pi^0\rangle = |\pi^0\rangle \tag{311}$$

If P is the parity operator, then for the neutral kaons at rest

$$P|K^0\rangle = -|K^0\rangle,$$

$$C|\overline{K}^0\rangle = -|\overline{K}^0\rangle, \tag{312}$$

Strong interactions conserve I_3 and S. If H_{st} denotes the strong interaction Hamiltonian, then

$$[H_{st}, S] = 0,$$

$$[H_{st}, C] = 0.$$

In weak interactions, strangeness is not conserved. If H_{wk} denotes the weak interaction Hamiltonian, then

$$[H_{wk}, S] \neq 0,$$

and
$$[H_{wk}, C] \neq 0,$$

$$[H_{wk}, P] \neq 0.$$

Let us assume that CP invariance holds in weak decays. The Gell-Mann-Pais scheme (1955) originally assumed just C invariance. We define

$$|K_1^0\rangle = \frac{1}{\sqrt{2}}\,(|K^0\rangle + CP|K^0\rangle)$$

$$|K_2^0\rangle = \frac{1}{\sqrt{2}}\,(|K^0\rangle - CP|K^0\rangle) \tag{313}$$

with
$$CP|K^0\rangle = \eta|\overline{K}^0\rangle,$$

$$CP|\overline{K}^0\rangle = \eta'|\overline{K}^0\rangle, \tag{314}$$

where η, η' are arbitrary phase factors. We can adjust such that $\eta = \eta' = 1$. We have

$$|K_1^0\rangle = \frac{1}{\sqrt{2}}\,(|K^0\rangle + |\overline{K}^0\rangle),$$

$$|K^0\rangle = \frac{1}{\sqrt{2}}\,(|K_1^0\rangle + |K_2^0\rangle),$$

$$|K_2^0\rangle = \frac{1}{\sqrt{2}}\,(|K^0\rangle - |\overline{K}^0\rangle),$$

$$|\overline{K}^0\rangle = \frac{1}{\sqrt{2}}\,(|K_1^0\rangle - |K_2^0\rangle), \tag{315}$$

Clearly, $|K^0\rangle$ and $|\overline{K}^0\rangle$ are not eigenstates of CP, but are eigenstates of S with eigenvalues $+1$ and -1 respectively. $|K_1^0\rangle$ and $|K_2^0\rangle$ are not eigenstates of S, but are eigenstates of CP with eigenvalues $+1$ and -1 respectively.

Looking at the decay modes of neutral kaons, we see an interesting feature. They decay faster into two pions than into three pions. Assuming that CP is good, we consider the decay modes of neutral kaons. The first type of decay is into a 2π state ($\pi^+ + \pi^-$; $2\pi^0$). A state of 2π with a definite J in the center of mass (c.m.) system has $CP = +1$. By CP invariance, we see that only $K_1^0 \to 2\pi$. These processes occur in a lifetime $\sim 0.9 \times 10^{-10}$ s. The other type of decay is into a 3π state ($\pi^+ \pi^- \pi^0$, $3\pi^0$) or leptonic modes ($\pi^+ e^- \nu$ etc.). A state ($J = 0$) has $CP = -1$ (C gives $(+1)^3 = +1$ and P gives -1). For $\pi^+ \pi^- \pi^0$ configuration with the relative angular momentum $l = 0$ for the $\pi^+ \pi^-$ pair, and $L = 0$, where L is the orbital angular momentum of the π^0 relative to the c.m. system of $\pi^+ \pi^-$, CP is again odd. Since the 2π phase space factor is much larger than that for 3π, $\pi e\nu$, ..., it is expected that K_1^0 with the dominant decay mode of 2π should have a much shorter lifetime than K_2^0 which is forbidden to decay into 2π. In case the three pions are in a relation $l > 0$ state, both positive and negative eigenvalues of

CP can exist. But such decays are highly suppressed because of the centrifugal barrier. These processes occur in a lifetime of $\sim 0.6 \times 10^{-7}$ s. The decay modes of the short-lived neutral kaon ($K_S \equiv K_1^0$) and the long-lived neutral kaon ($K_L \equiv K_2^0$) are given below.

	Lifetime (s)	Main decay mode	Main decay (%)
K_S	$(0.8923 \pm 0.0022) \times 10^{-10}$	$\pi^+ \pi^-$	70
		$\pi^0 \pi^0$	30
K_L	$(5.183 \pm 0.040) \times 10^{-8}$	$3\pi^0$	22
		$\pi^- + \pi^- + \pi^0$	12
		$\pi^\pm + \mu^\pm + \nu_\mu$	27
		$\pi^\pm + e^\mp + \nu_e$	39

Let us now see how the neutral kaon beam changes with time. We suppose that initially we produce a neutral K beam in the $|K^0\rangle$ state. The non-relativistic wave function at time $t = 0$ is

$$|\psi(0)\rangle = |K^0\rangle = \frac{1}{\sqrt{2}} \left(|K_S\rangle + |K_L\rangle \right). \tag{316}$$

At a later time t,

$$|\psi(t)\rangle = \left\{ \frac{1}{\sqrt{2}} \, e^{-\frac{t}{2\tau_S} - im_S t} \Big| K_S \Big\rangle \right.$$

$$\left. + \, e^{-\frac{t}{\tau_L} - m_L t} \Big| K_L \Big\rangle \right\}, \tag{317}$$

when m_S and m_L denote the masses and τ_S and τ_L the lifetimes of K_S and K_L respectively. The probability of observing K^0 at time t is

$$\left| \langle K^0 | \psi(t) \rangle \right|^2 = \frac{1}{4} \left\{ e^{\frac{-t}{\tau_S}} + e^{\frac{-t}{\tau_L}} + 2e^{\frac{-t(\tau_S^{-1} + \tau_L^{-1})}{2}} \cos\Delta mt \right\}, \tag{318}$$

where $\Delta m = |m_S - m_L|$. Similarly for $\overline{K}^0$

$$\left| \langle \overline{K}^0 | \psi(t) \rangle \right|^2 = \frac{1}{4} \left\{ e^{\frac{-t}{\tau_S}} + e^{\frac{-t}{\tau_L}} - 2e^{\frac{-t(\tau_S^{-1} + \tau_L^{-1})}{2}} \cos\Delta mt \right\}, \tag{318a}$$

Therefore, the K^0 and $\overline{K}^0$ intensities oscillate with the frequency Δm.

By measuring $\overline{K}^0$ interaction events (i.e. the hyperon yield) as a function of position from the K^0 source, one can find Δm. The extremely small mass difference between K_L and K_S has been measured very accurately

$$m_L - m_S = (3.521 \pm 0.015) \times 10^{-6} \text{ eV}.$$

or, a fractional mass difference

$$\frac{\Delta m}{m} = 0.7 \times 10^{-14}$$

where $\qquad\qquad\qquad m_k \approx 497.7$ MeV

Another interesting effect was pointed out Pais and Piccioni (1955). Let us start by producing a pure K^0 beam by associated production in pion-nucleon collisions. If the K^0 beam is allowed to travel (in vacuo) for a time t where then all K_S component decays and we have a pure beam of K_L. Then let this K_L beam be allowed to pass through matter (a Pb slab) and interact. The neutral kaon particles that interact in the Pb plate are either K^0 or $\overline{K}^0$. Because of different strong interactions, $\overline{K}^0$ is absorbed more strongly (e.g. $\overline{K}^0 + p \rightarrow \pi^\dagger + \Lambda$). Under idealized conditions all $\overline{K}^0$ are absorbed and a beam of K^0 emerges and the whole process repeats. This is called the regeneration effect since the short-lived K_S component reappears. This has been confirmed experimentally.

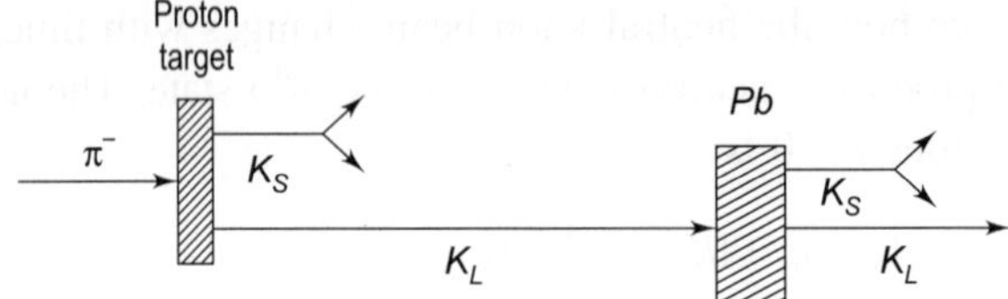

Fig. 11.1 Schematic drawing of regeneration experiment

The two-state system of the neutral K mesons K^0 and $\overline{K}^0$ mix with one another to form the short-lived and long-lived kaons.

This system provides an illustration of two basic principles of quantum mechanics: (i) the linear superposition principle, (ii) the interpretation of probabilities as squares of amplitudes. This system also enables us to test the validity of various conservation principles.

Following the discovery of the downfall of parity (and also C) in weak interactions in 1957, it was believed that weak interactions were invariant under the combined CP operations. Thus far we have assumed CP invariance, and provided the description of the CP eigenstates of neutral kaons, K_1^0 and K_2^0, given above. The experimental discovery of CP violation came in 1964 by Christenson, Cronin, Fitch and Turlay who demonstrated that the long-lived kaon (K_L) can decay into $2\pi(\pi^+ + \pi^-)$ with a branching ratio $\sim 10^{-3}$. The amplitude of CP violation can be characterized by the parameter.

$$\eta_{+-} \equiv \frac{\text{amplitude } (K_L^0 \rightarrow \pi^+ \pi^-)}{\text{amplitude } (K_S^0 \rightarrow \pi^+ \pi^-)}.$$

The value obtained is

$$\eta_{+-} \equiv (2.274 \pm 0.022) \times 10^{-3}.$$

This shows that the "observed" long-lived and short-lived kaons cannot be the CP eigenstates K_1^0 and K_2^0.

Assuming that *CP* is violated but with *CPT* invariance, we write

$$|K^0\rangle = e^{iS\theta}\, CPT|K^0\rangle, \tag{319}$$

where S is the strangeness quantum number and θ an arbitrary phase angle.

The amplitude of a coherent beam of K^0 and $\overline{K}^0$ may be written as

$$\psi(t) = a_1(t)|K^0\rangle + a_2(t)|\overline{K}^0\rangle, \tag{320}$$

or in a compact notation

$$\psi(t) = \begin{pmatrix} a_1(t) \\ a_2(t) \end{pmatrix}. \tag{320a}$$

Here t is the proper time.

In natural units, we have

$$i\frac{d}{dt}\psi(t) = \left(M - \frac{i}{2}\Gamma\right)\psi(t), \tag{321}$$

where the decay matrix Γ and the mass matrix M are 2×2 hermitian matrices.

The eigenstates of the mass matrix are

$$|S\rangle = \frac{1}{\sqrt{2(1 + |\epsilon|^2)}}\,[(1 + \epsilon)|K^0\rangle + (1 - \epsilon)|\overline{K}^0\rangle],$$

$$|S\rangle = \frac{1}{\sqrt{2(1 + |\epsilon|^2)}}\,[(1 + \epsilon)|K^0\rangle - (1 - \epsilon)|\overline{K}^0\rangle], \tag{322}$$

or, simply,

$$|S\rangle = \frac{1}{\sqrt{2(1 + |\epsilon|^2)}}\begin{pmatrix} 1 + \epsilon \\ 1 - \epsilon \end{pmatrix},$$

$$|L\rangle = \frac{1}{\sqrt{2(1 + |\epsilon|^2)}}\begin{pmatrix} 1 + \epsilon \\ -(1 - \epsilon) \end{pmatrix}, \tag{322a}$$

where is a complex number. Here S, L denote K_S and K_L. We note

$$\xi = \langle S|L\rangle = \frac{\epsilon + \epsilon^*}{(1 + |\epsilon|^2)}. \tag{323}$$

Since ϵ is a small number ($|\epsilon|\ 4.3 \times 10^{-3}$), neglecting $0(\epsilon^2)$ we may write the eigenstates approximately as

$$|S\rangle = \frac{1}{\sqrt{2}}\begin{pmatrix} 1 + \epsilon \\ 1 - \epsilon \end{pmatrix} \tag{324}$$

and

$$|L\rangle = \frac{1}{\sqrt{2}}\begin{pmatrix} 1 + \epsilon \\ -(1 - \epsilon) \end{pmatrix}. \tag{325}$$

Also,

$$\left(M - \frac{1}{2}i\Gamma\right)|j\rangle = \left(m_j - \frac{1}{2}i\gamma_j\right)|j\rangle, \tag{326}$$

where $|j\rangle$ is either $|S\rangle$ or $|L\rangle$. M_j $(j = L, S)$ are the masses and $\tau_j = \gamma_j^{-1}(j = L, S)$ are lifetimes

We have

$$m_j = \langle j|H|j\rangle + \sum_n P \, \frac{\langle j|H_{wk}|n\rangle + \langle n|H_{wk}|j\rangle}{m_k - m_n}$$

$$\gamma_j = 2\pi \sum_n \langle j|H_{wk}|n\rangle \, \langle n|H_{wk}|j\rangle \, \delta(m_k - m_n) \qquad (327)$$

Here we sum over all eigenstates $|n\rangle$ of $H_{st} + H_\gamma$, but $n \neq K^0$ or $\overline{K}^0$.

A detailed analysis of $K^0 - \overline{K}^0$ system can be carried out (see Lee (1988); Sachs (1988). Rosner and Slezak (2001) provide classical illustrations of *CP* violation of kaon decays.

11.7 MOTION OF A CHARGED PARTICLE IN A MAGNETIC FIELD

The motion of charged particles in an electromagnetic field is an important problem in quantum mechanics. In this section we first discuss the motion in a magnetic field leading to the Landau levels. We then discuss a startling effect, the Aharonov-Bohm effect, which is a purely quantum phenomenon.

(a) Equation of Motion and The Energy Levels

The Hamiltonian for a particle of mass m and charge q in a static magnetic field $\vec{B}$ is

$$H = \frac{1}{2m} \left(\vec{p} - \frac{q}{c} A(\vec{r}) \right)^2, \qquad (328)$$

where $\vec{A}$ is the vector potential. Here $\vec{p}$ is the canonical momentum and $\vec{\pi} = \vec{p} - \frac{q}{c}\vec{A}$

is the kinetic momentum. The velocity operator is

$$v_x = \frac{i}{\hbar} \, [H, x]$$

$$= \frac{1}{m} \left(p_x - \frac{q}{c} A_x \right)$$

or,
$$\vec{v} = \frac{1}{m} \left(\vec{p} - \frac{q}{c} A(\vec{r}) \right). \qquad (329)$$

Combining Eqs. (328) and (329), we have

$$H = \frac{1}{2} \, m|\vec{v}|^2. \qquad (330)$$

Since $[p_i, A_j] = \frac{\hbar}{i} \, \frac{\partial A_j}{\partial r_i}$, where $r_i = x, y, z$ for $i = 1, 2, 3$, we get

$$[v_i, v_j] = i\,\frac{\hbar q}{m^2 c}\,\epsilon_{ijk}\,B_k. \tag{331}$$

Clearly, this commutation relation is gauge-independent. The equation of motion is

$$m\,\frac{dv_\alpha}{dt} = \frac{i}{\hbar}\,m[H, v_\alpha] \quad (\alpha = x, y, z)$$

We evaluate

$$[H, v_\alpha] = \frac{m}{2} \sum_\beta [v_\beta^2, v_\alpha]$$

$$= \frac{m}{2} \sum_\beta v_\beta\{[v_\beta, v_\alpha] + [v_\beta, v_\alpha]v_\beta\}$$

$$= \frac{i\hbar q}{2mc} \sum_{\beta,\,\gamma} (v_\beta \epsilon_{\beta\alpha\gamma} B_\gamma + \epsilon_{\alpha\beta\gamma} B_\gamma v_\beta).$$

Therefore, the equation of motion is

$$m\dot{v} = \frac{q}{2c}\left(\vec{v} \times \vec{B} - \vec{B} \times \vec{v}\right). \tag{332}$$

We now consider the case of a uniform field $\vec{B} = B\hat{z}(B > 0)$. Here we treat only the orbital motion; the effect of spin will be discussed later.

From Eq. (331), we see that the only monzere commutator among the velocity components is $[v_x, v_y] = \dfrac{iq\hbar B}{m^2 c}$. So we write the Hamiltonian H as

$$H = \frac{1}{2}\,m(v_x^2 + v_y^2) + \frac{1}{2}\,mv_z^2. \tag{333}$$

Since v_z commutes with v_x, v_y, the momentum along the field is a constant of motion. So the motion along $\vec{B}$ is that of a free particle, i.e. unaffected by the field.

The Hamiltonian for the transverse motion is

$$H_\perp \equiv H_{xy} = \frac{1}{2}\,m(v_x^2 + v_y^2). \tag{334}$$

We have seen that v_x and v_y do not commute with each other or with H. So the Hamiltonian for motion in the xy plane (perpendicular to $\vec{B}$) is that of a harmonic oscillator. In terms of the operators

$$a = \sqrt{\frac{m}{2\hbar\omega_c}}\,(v_x + iv_y),$$

$$a^\dagger = \sqrt{\frac{m}{2\hbar\omega_c}}\,(v_x - iv_y), \tag{335}$$

where $\omega_c = qB/mc$ is the cyclotron frequency, which satisfy the commutation relation

$$[a, a^\dagger] = 1. \tag{336}$$

The Hamiltonian for motion in the plane perpendicular to $\vec{B}$ is

$$H_\perp = \hbar\omega_c \left(a^\dagger a + \frac{1}{2} \right) \tag{337}$$

The energy levels, known as the Landau levels, are given by

$$E_n = \hbar\omega_c \left(n + \frac{1}{2} \right) \quad n = 0, 1, 2, \ldots . \tag{338}$$

Now, we consider the classical case. The equation of motion is

$$m \frac{d^2\vec{r}}{dt^2} = \frac{q}{c} \, (\vec{v} \times \vec{B}) \tag{339}$$

with the solution

$$x = x_0 + \frac{mcv_\perp}{qB} \sin \left(\frac{qB}{mc} t + \alpha \right)$$

$$= x_0 + \frac{v_\perp}{\omega_c} \sin (\omega_c t + \alpha), \tag{340}$$

$$y = y_0 + \frac{v_\perp}{\omega_c} \cos (\omega_c t + \alpha)$$

$$z = z_0 + v_z t,$$

where $v_\perp$ is the projection of $\vec{v}$ onto the plane perpendicular to $\vec{B}$.

In the classical case, the particle moves uniformly along the z-axis and the projection onto the transverse plane is a circle with center at (x_0, y_0), of radius $R = \dfrac{v_\perp}{\omega_c} = \sqrt{\dfrac{2E}{m\omega_c^2}}$. The particle moves with the angular frequency ω_c. We note that the orbital positions and velocity in the transverse plane can be written as

$$x = x_0 + R \cos (\omega_c t + \alpha),$$

$$y = y_0 - R \sin (\omega_c t + \alpha),$$

and

$$v_x = - \omega_c R \sin (\omega_c t + \alpha),$$

$$v_y = - \omega_c R \cos (\omega_c t + \alpha), \tag{341}$$

In quantum theory the equation of motion for a uniform field is, from Eq. (332),

$$\frac{d}{dt} v_x = \omega_c v_y,$$

$$\frac{d}{dt} v_y = - \omega_c v_x. \tag{342}$$

The solution looks like the classical motion

$$x(t) = x_0 - \frac{1}{\omega_c} v_y(t),$$

$$y(t) = y_0 + \frac{1}{\omega_c} v_x(t). \tag{343}$$

The orbit center coordinates are, in this case, operators that do not commute:

$$[x_0, y_0] = -\frac{i\hbar c}{qB} = il_m^2, \tag{344}$$

where the magnetic length l_m is given by

$$l_m = \sqrt{\frac{\hbar c}{qB}} = \sqrt{\frac{\hbar}{m\omega_c}}. \tag{345}$$

If we fix the value of one of x_0 or y_0 then the other one is undetermined. Also, since

$$[H, x_0] = [H, y_0] = 0, \tag{346}$$

the orbit center coordinates are constants of motion.

Let us now solve the Schrödinger equation for a charged particle in a magnetic field $\vec{B} = B\hat{z}$ in rectangular coordinates. We use the Landau gauge $\vec{A} = (-By, 0, 0)$.

The Hamiltonian is

$$H = \frac{1}{2m}(p_y^2 + p_z^2) + \frac{1}{2m}\left(p_x + \frac{q}{c}By\right)^2$$

$$= \frac{1}{2m}\left[p_x^2 + p_y^2 + p_z^2 + \frac{2q}{c}Byp_x + \left(\frac{qB}{c}\right)^2 y^2\right]. \tag{347}$$

Clearly, H commutes with p_x and p_z. So we write

$$\psi(x, y, z) = e^{ip_x x/\hbar} e^{ip_z z/\hbar} \phi(y), \tag{348}$$

where $\phi(y)$ satisfies

$$-\frac{\hbar^2}{2m}\frac{d^2\phi(y)}{dy^2} + \left[\frac{1}{2}m\omega_c^2 (y - y_0)^2 - E'\right]\phi(y) = 0, \tag{349}$$

where E' is the energy of motion in the xy plane. The energy eigenvalues are

$$E_n = \hbar\omega_c\left(n + \frac{1}{2}\right) + \hbar^2 k_z^2/2m, \tag{350}$$

and the eigenfunction is

$$\psi(x, y, z) \sim e^{\frac{i}{\hbar}(p_x x + p_z z)} h_n(l_m^{-1}(y - y_0) \times e^{-\frac{1}{2}l_m^{-2}(y - y_0)^2} \tag{351}$$

The Landau levels for fixed n and k_z are infinitely degenerate.

We consider a description in terms of angular momentum and work in a particular gauge.

We choose the symmetric gauge satisfying the transverse gauge condition $\vec{\nabla} \cdot \vec{A}(\vec{r}) = 0$,

$$\vec{A}(\vec{r}) = \frac{1}{2}\,(\vec{B} \times \vec{r}), \tag{352}$$

or,

$$A_x = -\frac{1}{2}\,yB,$$

$$A_y = \frac{1}{2}\,xB, \tag{352a}$$

for a constant managic field in the z-direction ($\vec{B} = B\hat{z}$). Then

$$v_x = \frac{p_x}{m} + \frac{1}{2}\,\omega_c y,$$

$$v_y = \frac{p_y}{m} - \frac{1}{2}\,\omega_c x, \tag{353}$$

The angular momentum along $\vec{B}$ is

$$L_z = \frac{xp_y - yp_x}{\hbar}. \tag{354}$$

We confine our attention to the xy plane.

The Hamiltonian is then

$$H = \frac{1}{2m}\,(p_x^2 + p_y^2) + \frac{1}{8}\,m\omega_c^2\,(x^2 + y^2)$$

$$-\frac{1}{2}\,\hbar\omega_c L_z. \tag{355}$$

L_z is a constant of motion, i.e., it commutes with H, since H is invariant under rotation about the z-axis. We note that L_z is conserved in the transverse gauge, but not in general.

We see that the first two terms in the expression Eq. (355) represent an isotropic two-dimensional harmonic oscillator of frequency $\frac{1}{2}\,\omega_c$. The eigenstates are

$$\left| n_x, n_y \right> = \frac{1}{\sqrt{n_x!n_y!}}\,(a_x^\dagger)^{n_x}\,(a_y^\dagger)^{n_y}\left| 0, 0 \right>. \tag{356}$$

The energy eigenvalues are

$$E(n_x, n_y) = \frac{1}{2}\,\hbar\omega_c\,(n_x + n_y + 1) \tag{357}$$

where $n_x, n_y = 0, 1, \ldots$ But n_x and n_y both cannot be specified in a representation in which L_z is diagonal. The energy spectrum is

$$E_{n_\perp,\,\mu} = \frac{1}{2}\,\hbar\omega_c\,(n_\perp + 1) \tag{358}$$

where $n_\perp = n_x + n_y$. The degeneracy of a level is $n_\perp + 1$.

A convenient way is to define the operators

$$a^\dagger_\pm = \frac{1}{\sqrt{2}} \, (a^\dagger_x \pm i a^\dagger_y)$$

and

$$a_\pm = \frac{1}{\sqrt{2}} \, (a_x \mp i a_y). \tag{359}$$

The oscillator Hamiltonian is

$$H_{osc} = \frac{1}{2} \, \hbar \omega_c \, (a^\dagger_+ a_+ + a^\dagger_- a_- + 1), \tag{360}$$

a sum of the hamiltonians of two independent oscillators. The eigenstates are

$$\left| n_+, \, n_- \right> \sim (a^\dagger_+)^{n_+} (a^\dagger_-)^{n_-} \left| 0, \, 0 \right>. \tag{361}$$

Clearly, we may write

$$n_\perp = n_+ + n_-. \tag{362}$$

We can show that

$$L_z = a^\dagger_+ a_+ - a^\dagger_- a_-. \tag{363}$$

We note that $a^\dagger_+ (a^\dagger_-)$ raises (lowers) the eigenvalue of L_z by one unit, while $a^\dagger_+$ and $a^\dagger_-$ both raise the eigenvalue of the oscillator part of the Hamiltonian by $\frac{1}{2} \, \hbar \omega_c$.

All the eigenstates can be constructed from the ground state $n_+ = n_- = 0$. These states are linear combinations of the states in the basis $|n_x, \, n_y>$ with $n_x + n_y = n_+ + n_-$. The eigenstates on the basis $|n_+, \, n_->$ are also eigenstates of L_z, and hence of the Hamiltonian given in Eq. (355).

We now use the basis $|n_\perp, \, \mu>$, where $n_\perp$ is the eigenvalue of H_{osc} and μ the eigenvalue of L_z:

$$H_{osc} \left| n_\perp + \mu \right> = (n_\perp + 1) \, \frac{1}{2} \, \hbar \omega_c \left| n_\perp, \, \mu \right>,$$

$$L_z \left| n_\perp + \mu \right> = \frac{1}{2} \, \hbar \omega_c \, \mu \left| n_\perp, \, \mu \right>. \tag{364}$$

The operators $a^\dagger_+$ and $a^\dagger_-$ are given by

$$a^\dagger_+ \left| n_\perp, \, \mu \right> = \sqrt{n_\perp + 1} \left| n_\perp + 1, \, \mu + 1 \right>,$$

$$a^\dagger_- \left| n_\perp, \, \mu \right> = \sqrt{n_\perp} \left| n_\perp + 1, \, \mu - 1 \right>. \tag{365}$$

We can obtain an eigenstate of the transverse Hamiltonian H from the state $|0, \, 0>$:

$$(a^\dagger_+)^{n_+} (a^\dagger_-)^{n_-} |0, \, 0> \sim |n_+ + n_-, \, n_+ - n_->. \tag{366}$$

The energy spectrum is

$$E_{n_\perp, \mu} = \frac{1}{2} \, \hbar \omega_c \, (n_+ + n_- + 1 + (n_+ - n_-)). \tag{367}$$

For positive (negative) charged particle, the angular momentum spectrum is bounded from below (above). For the first case, we may write the Hamiltonian in the xy plane as

$$H = \frac{1}{2}\,\hbar\omega_c\,(2a^\dagger_+ a_+ + 1), \tag{368}$$

and the energy spectrum is

$$E_{n_\perp,\,\mu} = \frac{1}{2}\,\hbar\omega_c\,(2n_+ + 1)$$

$$= \hbar\omega_c\left(n_+ + \frac{1}{2}\right).$$

So the energy spectrum is characterized by a real, nonnegative integer n_+ which we now denote by n, and can have any nonnegative value for n_-:

$$E_{n,\,\mu} = \hbar\omega_c\left(n + \frac{1}{2}\right) \quad n = 0,\,1,\,\dots \tag{369}$$

The equally spaced energy levels for motion of a charged particle in a uniform magnetic field are known as the Landau levels. These levels are infinitely degenerate.

The Landau levels are of great importance in the study of quantum Hall effect (see Shankar (1994). Prange and Girvin (1987)).

We conclude this section by considering a particle of mass m, charge q with spin in a uniform magnetic field.

We assume that the managnetic field is in the z-direction, $\vec{B} = B\hat{z}$, and use the Landau gauge $\vec{A} = (-By,\,0,\,0)$. The total Hamiltonian is

$$H = \frac{1}{2m}\left(\vec{p} - \frac{q}{c}\vec{A}\right)^2 - \vec{\mu}\cdot\vec{B}, \tag{370}$$

where $\vec{\mu}$ is the intrinsic magnetic moment of the particle. We have

$$H = \frac{1}{2m}\left[p_x^2 + p_y^2 + p_z^2 + \frac{2qBp_x}{c}y + \left(\frac{qB}{c}\right)^2 y^2\right]$$

$$- \frac{\mu}{s}Bs_z, \tag{371}$$

where $\vec{s}$ is the spin operator. s_z is a constant of motion.

The space part of the wavefunction $\psi(x,\,y,\,z)$ satisfies the equation

$$\frac{1}{2m}\left[p_x^2 + p_y^2 + p_z^2 + \frac{2qBp_x}{c}y + \left(\frac{qB}{c}\right)^2 y^2\right]\psi$$

$$- \frac{\mu}{s}\sigma B\psi = E\psi. \tag{372}$$

The energy eigenvalues are

$$E_n = \left(n + \frac{1}{2}\right)\hbar\omega_c - \frac{\mu\sigma}{s}B + \frac{p_z^2}{2m}, \quad n = 0, 1, \ldots \tag{373}$$

For an electron, $\dfrac{\mu}{s} = -\dfrac{|e|\hbar}{mc}$, and $\omega_c = \dfrac{|e|B}{mc}$, $\tag{374}$

and so

$$E_n = \left(n + \frac{1}{2} + \sigma\right)\hbar\omega_c + \frac{p_z^2}{2m}. \tag{375}$$

The eigenfunction is $\psi(x, y, z)\chi_{\text{spin}}$.

(b)   The Aharonov-Bohm Effect

In classical physics, only the electric and magnetic fields have physical significance. The vector and scalar potentials are convenient mathematical constructs from which the fields may be calculated. The potentials can be redefined via a guage transformation without changing the fields, that is, without changing any physical laws. In quantum mechanics, the vector potential does have a significance of its own, as shown by Aharonov and Bohm (1959, 1961).

Let us consider a double slit diffraction set up shown schematically in Fig. 11.2. Charged particles (e.g., electrons) from a source at S are allowed to fall on a screen with two slits and are detected at a detecter screen at D. An electron can reach D by traversing one or another of the two paths. Along the classical path P_i ($i = 1, 2$), the phase of the electron wave function at D (relative to S) is

$$\phi_{P_1} = \exp\left\{\frac{i}{\hbar}\int_{t_1}^{t_2} dt\, \mathcal{L}\right\}$$

$$= \exp\left\{\frac{i}{\hbar}\int_{t_1}^{t_2} dt\, \frac{1}{2}mv^2\right\}$$

$$\sim \exp 2\pi i\left(\frac{d + d_1}{\lambda}\right),$$

$$\phi_{P_2} \sim \exp 2\pi i\left(\frac{d + d_2}{\lambda}\right). \tag{376}$$

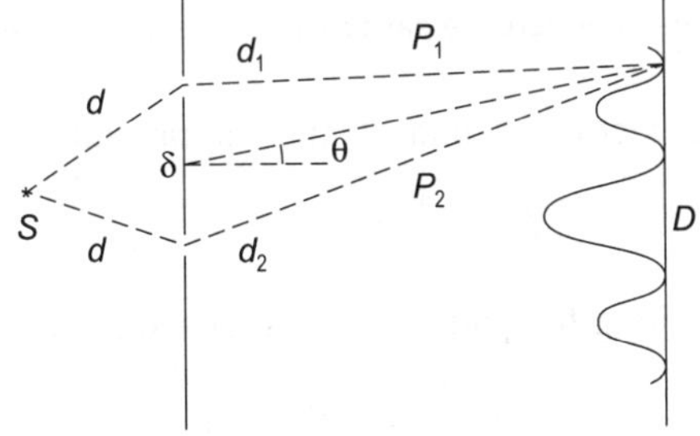

Fig. 11.2   Interference pattern in a two-slit diffraction experiment

The electron wave function has the de Broglie wavelength $\lambda = \dfrac{2\pi\hbar}{p}$, where the electron has the characteristic momentum p.

The phase difference is

$$\exp \frac{2\pi i}{\lambda}\,(d_2 - d_1). \tag{377}$$

The wave function at D is

$$\psi(D) = \psi_{P_1}(D) + \psi_{P_2}(D)$$

$$\sim \psi_0 \left(1 + \exp \frac{2\pi i}{\lambda}\,(d_2 - d_1)\right). \tag{378}$$

The intensity is given by

$$|\psi|^2 = |\psi|^2\, 4\,\cos^2\!\pi\frac{d_2 - d_1}{\lambda},$$

or,
$$I(\theta) = 4 I_0 \cos^2\!\pi\delta\,\frac{\sin\theta}{\lambda}. \tag{379}$$

Here, δ is the slit separation, θ the angle of deflection and I_0 the intensity if only a single slit were open.

Next we consider the situation in which an infinite solenoid is introduced in the experimental set-up (Fig. 11.3). The solenoid carries a current $\vec{j}$ and is surrounded by an impenetrable cylinder. There is no magnetic field outside the solenoid. However, the vector potential $\vec{A}$ is non-vanishing everywhere.

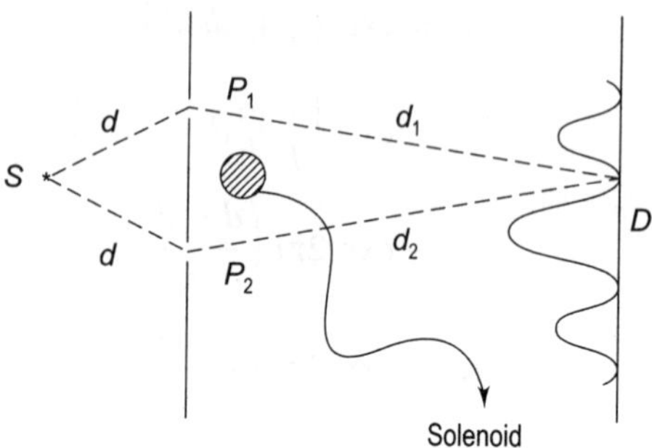

Fig. 11.3   Shift in the intereference pattern in the presence of an infinite solenoid

In presence of the vector potential $\vec{A}$, the Lagrangian for the electron is

$$\mathcal{L}_{\vec{A}} = \frac{1}{2}\,m\dot{r}^2 + \frac{e}{c}\,\vec{A}\cdot\dot{r}. \tag{380}$$

The path integral along the path P_i ($i = 1, 2$) acquires an extra factor

$$\exp\left\{\frac{ie}{\hbar c}\int_{P_i}\vec{A}\cdot\frac{d\vec{r}}{dt}\,dt\right\} = \exp\left\{\frac{ie}{\hbar c}\int_{P_i}\vec{A}\cdot d\vec{r}\right\}. \tag{381}$$

The line integral is evaluated along the path P_i. The difference between the line integrals along P_1 and P_2 is

$$\int_{P_2} \vec{A} \cdot d\vec{r} - \int_{P_1} \vec{A} \cdot d\vec{r} = \int_{S_{21}} (\vec{\nabla} \times \vec{A}) \cdot d\vec{S}$$

$$= \int_{S_{21}} \vec{B} \cdot d\vec{S}$$

$$= \Phi_B. \tag{382}$$

Φ_B is the magnetic flux crossing the surface S_{21} bounded by P_1 and P_2.

The interference pattern at a point D on the screen is (Holstein (1992))

$$I(\theta) = 4I_0 \cos^2 \left[\pi\delta \, \frac{\sin \theta}{\lambda} + \frac{e}{2\hbar c} \, \Phi_B \right]. \tag{383}$$

This shift in the interference pattern due to the presence of a non-vanishing vector potential was suggested by Aharonov and Bohm. The interference pattern at a point on the screen varies sinusoidally as a function of the magnetic flux Φ_B with a period given by a fundamental unit or quantum of magnetic flux.

$$\Phi_0 = \frac{2\pi\hbar c}{|e|}$$

$$= 4.135 \times 10^{-7} \text{ gauss cm}^2. \tag{384}$$

This was first observed by Chambers (1960) using a magnetic one-domain iron whisker. A beautiful experimental demonstration of the *AB* effect using electron holography was given by Tonomura et al. (1982).

The *AB* effect is a realization of $U(1)_{em}$ gauge principle in quantum theory. In the region outside the solenoid there is no electromagnetic field. But in this region the vector potential cannot be gauged away. The nonintegrable $U(1)$ phases of the wave function cause the shift in the interference pattern.

Another intriguing effect was pointed out by Aharonov and Casher (1984) in which a beam of uncharged magnetic dipoles is affected by an electric field. In the double-slit diffraction setup the infinite solenoid is replaced by an infinite line charge per unit length λ and the electrons are replaced by a polarized beam of neutrons (a neutral beam of magnetic dipoles). There is a shift of the interference pattern depending on λ. We note that this is an electric field $\vec{E} = \dfrac{\lambda}{2\pi r} \, \hat{r}$, but no magnetic field. Hagen (1990) showed that the *AB* and *AC* effects are dual. The *AC* effect is a consequence of the $SU(2)_{spin}$ gauge symmetry. The *AC* effect was demonstrated experimentally by Cimmino et al. (1989) using neutrons and by Sangster et al. (1993) using a beam of TlF molecules.

Summarizing, in classical physics, the electric and the magnetic fields are the physically relevant quantities, and there is no Lorentz force on a charged particle where the fields vanish. Here the electromagnetic potentials are only auxiliary quantities. In quantum theory, the electromagnetic potentials are themselves physically significant. These can affect the motion of charged particles even though the field vanishes at the location of the particle. But the observable consequence, i.e., the response is guage-invariant. Although the vector potential is significant, the flux Φ_B is a guage-invariant quantity. The AB effect is a topological effect, and the factor $e^{\frac{ie}{\hbar c}\Phi_B}$ is a topological phase. Berry pointed out that the AB effect is a particular consequence of a geometrical phase factor. There is another school which regards the AB effect as originating in the interaction of the magnetic field of the electron with the distant $\vec{B}$ field. It is maintained that the AB effect is the consequence of a non-local theory but of classical origin. (O'Raifeartaigh, Straumann, and Wipf (1991)).

11.8 SUPERSYMMETRIC QUANTUM MECHANICS

In field theory, supersymmetry generates transformations between bosons and fermions. The generators of supersymmetry transform the component fields whose intrinsic spins differ by $\frac{1}{2}\hbar$ and hence the corresponding multiplets contain in general both bosons and fermious. It was realized that this symmetry may provide a unified description of elementary particles. Supersymmetric field theories possess many attractive features which have aroused an enormous interest in such theories.

We now consider a simple example which exhibits some of the basic features of such a theory (Marshak (1993). We consider a pair of bosonic creation and annihilation operators a, $a^\dagger$ satisfying the commutation relations

$$[a, a^\dagger] = 1, \quad [a, a] = [a^\dagger, a^\dagger] = 0, \tag{385}$$

and a pair of fermionic creation and annihilation operators b, $b^\dagger$ statisfying the anti-commutation relations

$$\{b, b^\dagger\} = 1, \quad \{b, b\} = \{b^\dagger, b^\dagger\} = 0. \tag{386}$$

We consider a quantum system with the Hamiltonian

$$H = \hbar\omega_a\, a^\dagger a + \hbar\omega_b\, b^\dagger b. \tag{387}$$

We introduce a "supercharge" operator (fermionic)

$$Q = b^\dagger a + a^\dagger b. \tag{388}$$

Since
$$[Q, a^\dagger] = b^\dagger, \quad \{Q, b^\dagger\} = a^\dagger, \tag{389}$$

a bosonic state $a^\dagger |0\rangle$ is transformed by Q into a fermionic state $b^\dagger |0\rangle$.

The commutator of Q and H is

$$[Q, H] = \hbar\, (\omega_a - \omega_b)\, Q \tag{390}$$

which shows that H is supersymmetric only for equal energies (masses) for basons and fermions. The anticommutator

$$\{Q, Q^\dagger\} = \frac{2}{\hbar\omega} H \tag{391}$$

closes upon the Hamiltonian. Q can couple the bosonic and fermionic degrees of freedom, provided there is mass degeneracy and there are equal numbers of them. In this case, the energy eigenvalues $E = \hbar\omega \, (n_a + n_b)$, $n_a = 0, 1, 2, ..., n_b = 0, 1$, are doubly degenerate, except for the ground state $E = 0$, $n_a = n_b = 0$. Degeneracy is the manifestation of an underlying symmetry. The extra symmetry of this system is called supersymmetry and is manifested in the degeneracy of bosonic and fermionic states.

The application of supersymmetry to quantum mechanics was first elucidated by Witten (1981). The study of supersymmetric (SUSY) quantum mechanics has aroused considerable interest by providing deep insights into many aspects of quantum mechanics, e.g., the factorization method. SUSY quantum mechanics has been applied with considerable success to the studies of exactly soluble potentials and shape invariance, inverse scattering problems, and extension of the WKB method. There has been considerable interest in the application of SUSY quantum mechanics to the diverse fields of particle physics, atomic, nuclear, and solid state physics.

We give the defining algebra of SUSY quantum mechanics. According to Witten, SUSY quantum mechanics is characterized by the charge operators Q_i and the supersymmetric Hamiltonian H_{ss} satisfying the algebra

$$\{Q_i, Q_j\} = \delta_{ij} \, H_{ss}, \; i, j = 1, 2, ..., N \tag{392}$$

$$[Q_i, H_{ss}] = 0, \tag{392a}$$

where N is the number of generators of supersymmetry. For one dimensional quantum mechanics $N = Z$, and we consider a system with two charge operators Q_1 and Q_2. In terms of $Q = (Q_1 + iQ_2)/\sqrt{2}$ and $Q^\dagger(Q_1 - iQ_2)/\sqrt{2}$, the algebra is given by

$$H_{ss} = \frac{1}{\hbar} \{Q, Q^\dagger\}, \; Q^2 = 0 \text{ and } Q^{\dagger 2} = 0. \tag{393}$$

From these equations, we see that

$$[Q, H_{ss}] = [Q^\dagger, H_{ss}] = 0 \tag{394}$$

The charge operator Q is nilpotent and commutes with the Hamiltonian showing the invariance of the Hamiltonian under supersymmetry. We have

$$\{Q, Q\} = \{Q^\dagger, Q^\dagger\} = 0, \tag{395}$$

showing that Q and $Q^\dagger$ are fermionlike. The anticommutator of Q with $Q^\dagger$ closes the algebra. The algebra is a Graded Lie Algebra. The unitary operator for infinitesimal supersymmetry transformation is

$$U = 1 + \frac{i}{\hbar}\,(i\delta\lambda^* Q - iQ^\dagger\delta\lambda), \tag{396}$$

where $\delta\lambda$, $\delta\lambda^*$ are infinitesimal anticommuting c-numbers (Grassmann algebra).

For any state $|\psi\rangle$ in the domain of Q, we get

$$\langle\psi|H_{ss}|\psi\rangle = \frac{1}{\hbar}\,\langle\psi|QQ^\dagger|\psi\rangle + \frac{1}{\hbar}\,\langle\psi|Q^\dagger Q|\psi\rangle$$

$$= \frac{1}{\hbar}\,\|Q^\dagger\psi\|^2 + \frac{1}{\hbar}\,\|Q\psi\|^2 \geq 0. \tag{397}$$

Hence, the spectrum of H_{ss} is non-negative. Let us consider a simple realization of the algebra Eq. (393) by setting

$$Q = \begin{pmatrix} 0 & 0 \\ A & 0 \end{pmatrix} \text{ and } Q^\dagger = \begin{pmatrix} 0 & A^\dagger \\ 0 & 0 \end{pmatrix}, \tag{398}$$

where A is a linear differential operator and $A^\dagger$ is the adjoint. By construction, $Q^2 = Q^{\dagger 2} = 0$.

From Eq. (398), the supersymmetric Hamiltonian is

$$H_{ss} = \begin{pmatrix} A^\dagger A & 0 \\ 0 & AA^\dagger \end{pmatrix} = \begin{pmatrix} H_- & 0 \\ 0 & H_+ \end{pmatrix}, \tag{399}$$

where $\qquad\qquad H_- = A^\dagger A$ and $H_+ = AA^\dagger$,

Since $\qquad\quad Q\begin{pmatrix} \alpha \\ 0 \end{pmatrix} = \begin{pmatrix} 0 \\ A\alpha \end{pmatrix}$ and $Q^\dagger\begin{pmatrix} 0 \\ \beta \end{pmatrix} = \begin{pmatrix} A^\dagger\beta \\ 0 \end{pmatrix}, \tag{400}$

Q and $Q^\dagger$ induce transformations between the "bosonic" sector α and the "fermionic" sector β. H_- and H_+ are the diagonal elements of the supersymmetric Hamiltonian, and act on the bosonic and the fermionic sectors respectively of the two component basis states. H_+ and H_- are supersymmetric partners.

Let us write, in detail, the operators

$$A = \frac{\hbar}{\sqrt{2m}}\,\frac{d}{dx} + W(x), \tag{401}$$

$$A^\dagger = -\frac{\hbar}{\sqrt{2m}}\,\frac{d}{dx} + W(x),$$

where $W(x)$ is a real function. Then

$$H_+ = AA^\dagger = \frac{p^2}{2m} + V_+(x),$$

and $\qquad\qquad H_- = A^\dagger A = \frac{p^2}{2m} + V_-(x), \tag{402}$

with $\qquad\qquad V_\pm(x) = W^2 \pm \frac{\hbar}{\sqrt{2m}}\,\frac{dW}{dx}. \tag{403}$

We note $\qquad\qquad H_- = H_+ - \sqrt{\frac{2}{m}}\,\hbar\,\frac{dW}{dx}. \tag{404}$

$W(x)$ is referred to as the superpotential and $V_+(x)$ and $V_-(x)$ the supersymmetric partner potentials.

We assume that H_- has a discrete spectrum

$$H_- |\psi_N^-\rangle = E_N^- |\psi_N^-\rangle, \quad N = 0, 1, 2, \ldots \tag{405}$$

Since $\qquad\qquad H_- = A^\dagger A, \ E_N^- \geq 0.$

We suppose that

$$A|\psi_0^-\rangle = 0.$$

Then $\qquad\qquad A^\dagger A|\psi_0^-\rangle = 0, \tag{406}$

and $|\psi_0^-\rangle$ corresponds to the ground state with the eigenvalue $= E_0^- = 0$.

In coordinate representation, Eq. (406) is

$$\left[\frac{\hbar}{\sqrt{2m}} \frac{d}{dx} + W(x) \right] \psi_0^- = 0, \tag{407}$$

The solution relates the ground state wave function to the superpotential:

$$\psi_0^-(x) = C_0 \exp\left(-\frac{\sqrt{2m}}{\hbar} \int^x W(x) \, dx \right). \tag{408}$$

We may also write

$$W(x) = -\frac{\hbar}{\sqrt{2m}} \frac{d}{dx} \log \psi_0^-. \tag{408a}$$

Here we assume that $|\psi_0^-\rangle$ is normalizable and so Eq. (408) is square-integrable.

We write the eigenvalue equation for $H_+ \, H_+$:

$$H_+|\psi_N^+\rangle = E_N^+|\psi_N^+\rangle, \quad N = 0, 1, 2, \ldots \tag{409}$$

Then $\qquad\qquad H_+(A\psi_N^-) = AA^\dagger(A\psi_N^-)$

$$= A(H_- \psi_N^-)$$

$$= E_N^-(A\psi_N^-),$$

so that $A\psi_N^-(x)$ is an eigenfunction of H_+ with the eigenvalue E_N^-.

The partner wave functions are related thus

$$\psi_N^+(x) = \frac{1}{\sqrt{E_{N+1}^-}} A\psi_{N+1}^-(x). \tag{410}$$

We can similarly show, if ψ_N^+ is an eigenfunction of H_+ with eigenvalues E_N^+, then $A^\dagger \psi_N^+$ is an eigenfunction of H_- with the same eigenvalue. Here, we have assumed normalized kets and put a phase factor equal to unity.

Thus, we have the following relations

$$E_N^+ = E_{N+1}^-; \quad E_0^- = 0$$

$$\psi_N^+ = \frac{1}{\sqrt{E_{N+1}^-}} A\psi_{N+1}^-, \quad N = 0, 1, 2, \ldots$$

$$\psi_{N+1}^- = \frac{1}{\sqrt{E_n^+}} A^\dagger \psi_N^+$$

$$(411)$$

We conclude that H_+ and H_- have the same spectra except for the ground state. Only H_- has a normalized ground state with eigenvalue zero. A and $A^\dagger$ connect states of the same energy for two different potentials. We know

$$V_+(x) = V_-(x) + \sqrt{\frac{2}{m}}\, \hbar \frac{d}{dx} \log \psi_0^-(x). \tag{412}$$

If the solutions of a one dimensional potential $V_-(x)$ are known, then the solutions of a partner potential can be obtained.

The operators A and $A^\dagger$ not only connect the eigenfunctions of H_- and H_+ but also destroy and create a node.

The spectra of H_+ and H_- are schematically shown in Fig. 11.4.

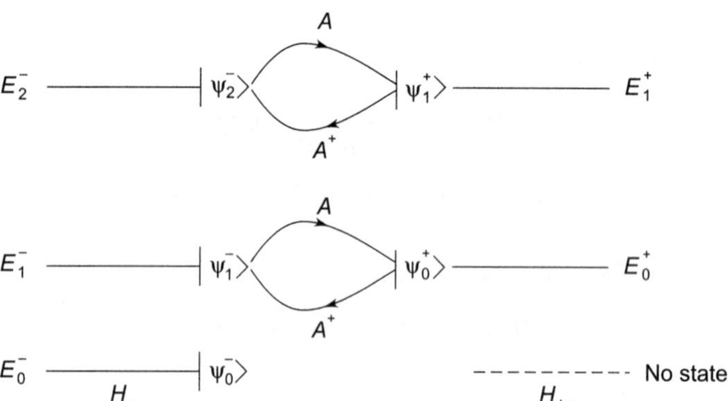

Fig. 11.4 Schematic illustration of the eigenvalues of $H_\pm$

We note that in addition to Eq. (406), two other possibilities arise. We suppose $A^\dagger|\psi_0^+\rangle = 0$. Then $E_0^+ = 0$, and all the relations are valid with the interchange of the labels $+$ and $-$. If there are no normalizable eigenstates of H_- or H_+ such that either $A^\dagger|\psi_0^-\rangle = 0$ or $A|\psi_0^+\rangle = 0$, then the eigenvalues of both H_- and H_+ begin at positive values, and every eigenvalue of H_- is an eigenvalue of H_+ and vice versa.

We discuss a few examples.

(i)   We consider the simple harmonic oscillator with the Hamiltonian

$$H_{\text{osc}} = \frac{p^2}{2m} + \frac{1}{2} m\omega^2 x^2. \tag{413}$$

which has the eigenvalue spectrum

$$E_n = \hbar\omega\left(n + \frac{1}{2}\right), \quad n = 0, 1, 2, \ldots. \tag{414}$$

The zero energy ground state wave function $\psi_0(x) = \psi_0^-(x)$ in a potential $V(x) = \frac{1}{2}m\omega^2 x^2 - \frac{\hbar\omega}{2}$ is

$$\psi_0(x) = \alpha \exp\left(-\frac{m\omega}{\hbar}\frac{x^2}{2}\right), \quad \alpha = \left(\frac{m\omega}{\pi\hbar}\right)^{1/4} \tag{415}$$

From Eq. (408a),

$$W(x) = -\frac{\hbar}{\sqrt{2m}}\frac{\psi_0'}{\psi_0} = \frac{\hbar}{\sqrt{2m}}\left(\frac{m\omega}{\hbar}x\right) = \sqrt{\frac{m}{2}}\,\omega x \tag{416}$$

Then
$$\left.\begin{aligned}
H_+ &= AA^\dagger = -\frac{\hbar^2}{2m}\frac{d^2}{dx^2} + \frac{1}{2}m\omega^2 x^2 + \frac{1}{2}\hbar\omega \\[2mm]
H_- &= AA^\dagger = -\frac{\hbar^2}{2m}\frac{d^2}{dx^2} + \frac{1}{2}m\omega^2 x^2 - \frac{1}{2}\hbar\omega,
\end{aligned}\right\} \tag{417}$$

and so the supersymmetric Hamiltonian is

$$H_{ss} = \begin{pmatrix} H_{\text{osc}} - \dfrac{1}{2}\hbar\omega & 0 \\[3mm] 0 & H_{\text{ose}} + \dfrac{1}{2}\hbar\omega \end{pmatrix} \tag{418}$$

and the partner potentials are

$$V_{\mp}(x) = \frac{1}{2}m\omega^2 x^2 \mp \frac{1}{2}\hbar\omega. \tag{419}$$

The ground state of H_- is

$$\psi_0^-(x) = \alpha\, e^{-\frac{m\omega}{\hbar}x^2/2} \quad \text{and } E_0^- = 0.$$

The lowest state of H_+ is

$$\psi_1^+(x) = \alpha\, e^{-\frac{m\omega}{\hbar}x^2/2}, \quad E_1^+ = \hbar\omega$$

The first excited state of H_- is obtained by applying $A^\dagger$ to $\psi_1^+(x) = \psi_0^-(x)$

$$\psi_1^-(x) = \frac{1}{\sqrt{\hbar\omega}}A^\dagger\psi_0^-(x) = \sqrt{\frac{2}{\sqrt{\pi}\left(\sqrt{\frac{\hbar}{m\omega}}\right)^3}}\, x\, e^{-\frac{m\omega}{\hbar}\frac{x^2}{2}} \tag{420}$$

and
$$E_1^- = \hbar\omega.$$

The procedure is then iterated. We get

$$\left.\begin{aligned}
E_n^- &= n\hbar\omega, & n = 0, 1, 2, \ldots \\
E_n^+ &= (n + 1)\hbar\omega, & n = 0, 1, 2, \ldots
\end{aligned}\right\} \tag{421}$$

(ii)   Let us consider the superpotential

$$W(x) = \frac{\hbar}{\sqrt{2m}}\ \tanh x. \tag{422}$$

The partner potentials are

$$\left. \begin{aligned} V_+ &= \frac{1}{2},\ \text{putting } \hbar = m = 1. \\[2ex] V_- &= \frac{1}{2} - \operatorname{sech}^2 x = \frac{1}{2}\left(1 - \frac{2}{\cosh^2 x}\right) \end{aligned} \right\} \tag{423}$$

The Hamiltonians are

$$H_- = \frac{p^2}{2} - \operatorname{sech}^2 x + \frac{1}{2}, \tag{424}$$

$$H_+ = \frac{p^2}{2} + \frac{1}{2} \tag{424a}$$

The potential $-\dfrac{1}{\cosh^2 x}$ has a constant potential $\dfrac{1}{2}$ as the supersymmetric partner potential. The Hamiltonian H_+ corresponds to a free particle.

The ground state of H_- is

$$\psi_0^-(x) = \frac{1}{\sqrt{2}}\,\frac{1}{\cosh x},\ \text{using Eq. (408)}$$

and

$$E_0^- = 0 \tag{425}$$

The eigenfunction and the energy eigenvalue of the lowest state of H_+ are

$$\psi_k^+ = e^{ikx}\ \text{and}\ E_1^+ = \frac{1}{2}\,(1 + k^2). \tag{426}$$

The remaining normalized eigenfunctions of H_- are

$$\psi_k^- = \frac{1}{\sqrt{E_1^+}}\,A^\dagger \psi_k^\dagger = \frac{(-ik + \tanh x)}{\sqrt{1 + k^2}}\,e^{ikx} \tag{427}$$

with the eigenvalues E_1^+.

The continuum states Eq. (427) posses no reflected wave, and the potential $-\dfrac{1}{\cosh^2 x}$ is called a reflectionless potential.

(iii)   The δ-function potential

We consider the superpotential

$$W(x) = \frac{\lambda\hbar}{2\sqrt{2m}\,a}\,(\theta(x) - \theta(-x)), \tag{428}$$

where $\theta(x)$ is the step function and $\lambda > 0$. Then

$$V_-(x) = \frac{\lambda^2\hbar^2}{8ma^2} - \frac{\lambda\hbar^2}{2ma}\,\delta(x),$$

$$V_+(x) = \frac{\lambda^2 \hbar^2}{8ma^2} + \frac{\lambda \hbar^2}{2ma}\, \delta(x), \tag{429}$$

are the partner potentials. Clearly, the supersymmetric partner potentials V_- and V_+ are respectively the attractive and the repulsive δ-potentials. $V_-(x)$ has a single zero-energy bound state with normalized wave function $\psi_0^-(x) = \sqrt{K}\, e^{-K|x|}$, where $K = \dfrac{\lambda}{2a}$.

Since $E_n^+ = E_{n+1}^-$, and $V_-(x)$ has only one (zero-energy) bound state, $V_+(x)$ has no bound states, which is obvious because $V_+(x)$ is a purely repulsive potential.

Usually the supersymmetric partner potentials $V_\pm(x)$ depend on a set of parameters ξ_0, and we write them as $V_\pm(x, \xi_0)$. $V_+(x, \xi_0)$ and $V_-(x, \xi_0)$ are said to be shape-invariant [Gendenshtein (1983)] if

$$V_+(x, \xi_0) = V_-(x, \xi_1) + R(\xi_1), \tag{430}$$

where the remainder $R(\xi_1)$ is independent of x, $R(\xi_1) \neq 0$, and there exists some function f such that

$$f(\xi_0) = \xi_1. \tag{431}$$

We see that the SUSY partner potentials $V_\pm$, apart from the additive x-independent term $R_1(\xi_1)$, have similar shapes when parametrized with, in general, different parameters. From Eq. (402), the shape invariance condition Eq. (430) can be written in terms of the superpotential W:

$$W^2(x, \xi_0) + \frac{\hbar}{\sqrt{2m}}\, W'(x, \xi_0) = W^2(x, \xi_0) - \frac{\hbar}{\sqrt{2m}}\, W'(x, \xi_1) + R(\xi_1), \tag{432}$$

where $$W'(x, \xi_0) = \frac{\partial W(x, \xi_0)}{\partial x}.$$

This condition Eq. (432) is equivalent to the factorization condition.

It has been shown that for the class of shape invariant partner potentials satisfying Eq. (430), the eigenvalues and eigenkets can be determined by a purely algebraie method.

Let us consider a Hamiltonian $H_-(\xi_0)$, depending on some parameter ξ_0,

$$H_-(\xi_0) = \frac{p^2}{2m} + V_-(x, \xi_0). \tag{433}$$

If the "shape invariance" condition holds for the partner potentials, then SUSY partner Hamiltonian is

$$H_+(\xi_0) = \frac{p^2}{2m} + V_+(x, \xi_0) = \frac{p^2}{2m} + V_-(x, \xi_1) + R(\xi_1)$$

$$= H_-(\xi_1) + R(\xi_1). \tag{434}$$

We construct a sequence of Hamiltonians for shape-invariant potentials $H^{(s)}$ ($s = 0, 1, 2, \ldots$), where $H^{(0)} = H_-(\xi_0)$ and

$$H^{(s)} = \frac{p^2}{2m} + V_-(x, \xi_s) + \sum_{k=1}^{s} R(\xi_k), \quad (s = 1, 2, \ldots) \tag{435}$$

Here $\xi_s = f(\xi_{s-1})$ and $H^{(0)} = H_-(\xi_0)$ and $H^{(1)} = H_+(\xi_0)$ are SUSY partners. Now,

$$H^{(s+1)} = \frac{p^2}{2m} + V_-(x, \xi_{s+1}) + R(\xi_{s+1}), \quad s = 0, 1, 2, \ldots \tag{436}$$

and using the shape invariance condition Eq. (430) in terms of ξ_s and ξ_{s+1},

$$V_+(x, \xi_s) = V_-(x, \xi_{s+1}) + R(\xi_{s+1}), \quad s = 0, 1, 2, \ldots \tag{437}$$

We obtain

$$H^{(s+1)} = \frac{p^2}{2m} + V_-(x, \xi_{s+1}) + \sum_{R=1}^{s+1} R(\xi_k)$$

$$= \frac{p^2}{2m} + V_+(x, \xi_s) + \sum_{R=1}^{s} R(\xi_k). \tag{438}$$

So $H^{(s)}$ and $H^{(s+1)}$ are SUSY partner

Hamiltonians. From Eq. (411), the eigenvalues are related as

$$E_N^{(s)} = E_{N-1}^{(s+1)} \quad (N = 1, 2, \ldots) \tag{439}$$

The ground state energy $E_0^{(N)}$ of $H^{(N)}$ is equal to the eigenvalue E_N^- of $H^{(0)}$. We have

$$E_N^- = \sum_{R=1}^{N} R(\xi_k); \quad E_0^- = 0. \tag{440}$$

From $|\psi_0^{(0)}(\xi_N)\rangle$ we can generate all the excited states of H_-. We have

$$|\psi_N^{(0)} \xi_0\rangle = C_N A^\dagger(x, \xi_0) A^\dagger(x, \xi_1) \ldots A^\dagger(x, \xi_{N-1})|\psi_0^{(0)}(\xi_N)\rangle,$$

$$N = 1, 2, \tag{441}$$

where

$$C_N = \prod_{k=1}^{N} \left(E_N^{(0)} - E_{k-1}^{(0)} \right)^{-1/2}. \tag{442}$$

As an example, we consider the linear harmonic oscillator for which the superpotential is $W(x) = \sqrt{\dfrac{m}{2}}\,\omega x$. The SUSY partner potentials are

$$V_\pm(x, \xi_0) = \frac{1}{2} m\omega^2 x^2 \pm \frac{1}{2} \hbar\omega. \tag{443}$$

Let $\xi_s = \omega$ ($s = 0, 1, 2, \ldots$). Clearly, $V_\pm(x, \xi_s)$

satisfy the shape invariance condition Eq. (437)

where $R(\xi_s) = \hbar\omega$. Using Eq. (440), we get

$$E_N^- = N\hbar\omega \quad (N = 0, 1, 2, \ldots) \tag{444}$$

Using $A = \sqrt{\hbar\omega}\, a$ and $A^\dagger = \sqrt{\hbar\omega}\, a^+$, where a, $a^\dagger$ are the harmonic oscillator annihilation and creation operators, we can calculate the eigenfunctions.

The bound state spectra for twelve types of shape-invariant potentials have been obtained using the concept of shape invariance (Dutt, Khare and Sukhatme (1988)); Barclay and Maxwell (1991)).

PROBLEMS

11.1 The time-indepedent Schrödinger equation for a free particle of mass m is

$$-\frac{\hbar^2}{2m} \vec{\nabla}^2 \psi_E(\vec{r}) = E\psi_E(\vec{r})$$

Using spherical polar coordinates, find free particle solutions of the form:

$$\psi_{Elm}(\vec{r}) \sim R_{El}(r)\, Y_{lm}(\theta, \phi).$$

Show that the orthonormal free particle wave functions are

$$\psi_{Elm}(\vec{r}) = \frac{2}{\pi} \left(\frac{k^2 dk}{dE} \right)^{1/2} j_l\,(kr)\, Y_{lm}(\theta, \phi).$$

11.2 Prove that for the nth stationary states of a one-dimensional harmonic oscillator

$$<(\Delta x)^2><(\Delta p)^2> = \left(\frac{n+1}{2} \right)^2 \hbar^2$$

11.3 Find the probability distribution of the various values of the momentum of an oscillator.

11.4 Write the *-genvalue equation for a harmonic oscillator and solve it to obtain *-gen Wigner functions.

11.5 A particle is moving in the "half" harmonic oscillator potential

$$V(x) = \begin{cases} + \infty, & x \le 0 \\ \dfrac{1}{2} m\omega^2 x^2, & x > 0 \end{cases}$$

Obtain the eigenenergies and the eigenfunctions.

11.6 The delta-shell potential is

$$V(r) = -\,V_0\, \delta(r - a).$$

For a particle of mass m moving in this potential, show that if

$$\frac{2m}{\hbar^2} V_0 a = 2l_0 + 1,$$

there is a bound state of angular momentum l_0 with energy zero. What is the condition that there is an $l = 0$ bound state?

11.7 Consider a quark-antiquark bound system, called quarkonium. The two particles interact by a potential $V(r) = kr$. Find the energy levels and the radial functions for $l = 0$ state.

11.8 Determine the energy levels of a particle moving in a central field with potential energy.

$$V(r) = \frac{A}{r^2} + Br^2,$$

where A, B are positive constants.

11.9 Find the energy levels of a particle moving in Kratzer's molecular potential

$$V(r) = \frac{A}{r^2} - \frac{B}{r}.$$

11.10 The vibrational spectra of a diatomic molecule are well described by a potential function proposed by Morse:

$$V(r) = D(e^{-2\alpha x} - 2e^{-\alpha x}); \quad x = \frac{r - r_0}{r_0},$$

where D and α are positive constants. Determine the vibrational energy levels. Consider the Schrödinger equation for $l = 0$ bound state.

11.11 Derive Kramers' relation for $<r^k> = <nl|r^k|nl>$ for the hydrogen atom:

$$\frac{(k + 1)}{n^2} <r^k> - (2k + 1)\, a_0 <r^{k-1}>$$

$$+ \frac{k}{4}\, [(2l + 1)^2 - k^2] \times a_0^2 <r^{k-2}> = 0,$$

where $a_0 = \dfrac{\hbar^2}{m_e e^2}$ is the Bohr radius.

11.12 Calculate the expectation value $<nl|r^k|nl>$ for the hydrogen atom for $k = 1, 2, -1, -2, -3$ and -4.

11.13 Determine the eigenvalues and the eigenfunctions for a two-dimensional hydrogen atom which is constrained to move in a plane. Use cylinderical coordinates and the method of the separation of variables.

11.14 A charged particle is moving in a constant magnetic field in the z-direction, $\vec{B} = B\hat{z}$. Take the vector potential in the Landau gauge.

$$\vec{A} = xB\hat{y}.$$

(In the text the symmetric gauge has been used).

Find the energy spectrum and the eigenfunctions.

11.15 Consider a particle of mass m and charge q moving in a uniform magnetic field $\vec{B}$. Solve the Schrödinger equation to find the energy levels and the wave functions. Use the cylindrical coordinates (p, ϕ, z) and take the vector potential as

$$A_\phi = \frac{B}{2}\, p, \ \hat{A}_\rho = A_z = 0.$$

11.16 A charged particle is moving in uniform crossed electric fields with the magnetic field $\vec{B}$ along the z-direction and the electric field $\vec{\varepsilon}$ along x-direction. Find the energy levels and the wave functions.

11.17 Consider the bound state Aharonov-Bohm effect for a simple system in which a particle of mass m and charge q is constrained to move on a circle of radius a in the xy plane. The center of the circle is chosen as the origin. A thin solenoid with a magnetic flux $\bar{\phi}$ is placed along the z-axis. Use cylindrical coordinates (p, ϕ, z) and take the vector potential as

$$A_\rho = 0, \quad A_\phi = \frac{\bar{\phi}}{2\pi\rho}, \quad A_z = 0$$

Find the energy eigenvalues.

11.18 A particle is moving in an infinite well potential (see Section 3.8). Find the supersymmetric partner potential of the infinite well and the wave function $\psi_n^{(+)}(x)$.

11.19 Consider the motion of a particle of mass m in a twin attractive delta-function potential;

$$V = -\frac{\lambda\hbar^2}{2ma}\,[\delta(x - a) + \delta(x + a)] \quad (\lambda > 0)$$

Discuss the supersymmetric partner potential.

11.20 Consider the bound states of the Coulomb potential. Solve the problem by means of SUSY quantum mechanics.

CHAPTER 12

Stationary State Perturbation Theory

In quantum machanics, as in classical mechanics, few problems can be solved exactly. For the vast majority of problems of physical interest we resort to some form of an approximation method. In this and the following chapter we develop several approximation methods for bound states and apply these techniques to solve interesting problems. Approximation methods dealing with scattering states are discussed later.

In this chapter we develop steady state perturbation theory which deals with finding the changes in discrete energy eigenvalues and eigenfunctions of a quantum mechanical system due to the application of a small disturbance. The Rayleigh-Schrödinger (RS) and Brillouin–Wigner (BW) perturbataion expansions are discussed and then applied to illustrative problems. In the next chapter we introduce nonperbative methods, e.g., the variation method and the WKB approximation. In chapter 14, we formulate approximation methods for time-dependent phenomena. We discuss the semiclassical theory of radiation in chapter 15.

12.1 THE RAYLEIGH–SCHRÖDINGER PERTURBATION EXPANSION–THE NONDEGENERATE CASE

We consider a time-independent Hamiltonian of a system. The purpose of the theory is the evaluation of the eigenvalues E_n and eigenvectors $|n\rangle$:

$$H |n\rangle = E_n| n\rangle. \tag{1}$$

We suppose that H can be split into two parts, namely,

$$H = H_0 + V, \tag{2}$$

where the spectrum $E_n^{(0)}$ and eigenkets $|n^{(0)}\rangle$ of H_0 are known:

$$H_0|\,n^{(0)}\rangle = E_n^{(0)}|n^{(0)}\rangle \tag{3}$$

H_0 is called the "unperturbed" or "free" Hamiltonian and $|n^{(0)}\rangle$ the "free" eigenvector. H is called the "exact" Hamiltonian and $|n>$ the exact eigenket. V is called the "perturbation" Hamiltonian. The set $\{|n^{(0)}\rangle\}$ is complete and orthonormal. We have to find the approximate eigenkets and eigenvalues for the problem

$$(H_0 + V)\,|n\rangle = E_n\,|n\rangle. \tag{4}$$

Customarily, we write Eq. (4) in the form

$$(H_0 + \lambda V)|n\rangle = E_n|n\rangle \tag{5}$$

to keep track of the order of the perturbation.

λ is a continuous real parameter ($0 \le \lambda \le 1$).

The $\lambda = 0$ case corresponds to the unperturbed problem, while $\lambda = 1$ case corresponds to the full problem. In the case where the approximation method works, we have

$$E_n^{(0)} = \lim_{\lambda \to 0} E_n \text{ and } |n^{(0)}\rangle = \lim_{\lambda \to 0} |\,n\rangle. \tag{6}$$

We assume that there exists a neighborhood Λ of $\lambda = 0$ such that for $\lambda \in \Lambda$, $H_0 + \lambda V$ has a single nondegenerate eigenvalue $E_n(\lambda)$ with eigenket $|n(\lambda)\rangle$. We assume analyticity of energy eigenvalues and eigenkets in a complex-plane in Λ.

We require for sufficiently small λ,

$$\langle n^{(0)}|\,n(\lambda)\rangle = \langle n^{(0)}|\,n^{(0)}\rangle = 1. \tag{7}$$

We now assume that for $\lambda \in \Lambda$, we can expand $|n(\lambda)\rangle$ and $E_n(\lambda)$ in powers of λ according to

$$E_n(\lambda) = \sum_{i=0}^{\infty} \lambda^i E_n^{(i)}, \tag{8}$$

$$|n(\lambda)\rangle = \sum_{i=0}^{\infty} \lambda^i |n^{(i)}(\lambda)\rangle. \tag{9}$$

The series Eq. (8) is called the Rayleigh-Schrödinger series.

From Eqs. (5), (8) and (9), we get

$$H_0 + \sum_{i=0}^{\infty} \lambda^i |n^{(i)}\rangle + V \sum_{i=0}^{\infty} \lambda^{i+1}|n^{(i)}\rangle = \sum_{i=0}^{\infty} \lambda^i \sum_{r=0}^{\infty} E_n^{(r)}|n^{(i-r)}\rangle. \tag{10}$$

Here, for simplicity we write $|n\rangle$ and E_n for $|n(\lambda)\rangle$ and $E_n(\lambda)$ respectively

Assuming that $|n\rangle$ and E_n are continuous functions of λ, we obtain by equating the coefficients of like powers of λ

Order

$$0: \qquad (H_0 - E_n^{(0)})|\, n^{(0)}\rangle = 0,$$

$$1: \qquad (H_0 - E_n^{(0)})|\, n^{(1)}\rangle + (V - E_n^{(1)})|\, n^{(0)}\rangle = 0,$$

$$2: \qquad (H_0 - E_n^{(0)})|\, n^{(2)}\rangle + (V - E_n^{(1)})|\, n^{(1)}\rangle = E_n^{(2)}|\, n^0\rangle,$$

$$\cdot$$
$$\cdot$$
$$\cdot$$

k

$$(H_0 - E_n^{(0)})|\, n^{(k)}\rangle + (V - E_n^{(1)})|\, n^{(k-1)}\rangle = \sum_{j=2}^{k} \cdot E_n^{(j)}|\, n^{(k-j)}\rangle,$$

i.e.

$$(H_0 - E_n^{(0)})|\, n^{(0)}\rangle = 0,$$

$$(H_0 - E_n^{(0)})|\, n^{(k)}\rangle = (E_n^{(1)} - V)|\, n^{(k-1)}\rangle + \sum_{j=2}^{k} \cdot E_n^{(j)}|\, n^{(k-j)}\rangle, \; k \geq 1. \tag{11}$$

Substituting Eq. (9) in the normalization condition we get

$$\langle n^{(0)}|\, n^{(i)}\rangle = 0, \quad i \geq 1 \tag{12}$$

The scalar product of the second equation in Eq. (11) with $|n^{(0)}\rangle$, together with Eq. (12) gives

$$E_n^{(i)} = \langle n^{(0)}\, |V|\, n^{(i-1)}\rangle. \tag{13}$$

The scalar product of Eq. (11) with $|m^{(0)}\rangle$ for $m \neq n$, together with $H_0|\, m^{(0)}\rangle = E_m^{(0)}|m^{(0)}\rangle$ gives

$$\langle m^{(0)}|\, n^{(0)}\rangle = \frac{1}{E_n^{(0)} - E_m^{(0)}}\, [\langle m^{(0)}|V|\, n^{(i-1)}\rangle$$

$$- \sum_{k=1}^{i-1} E_n^{(k)} \langle m^{(0)}|\, n^{(i-k)}\rangle]. \tag{14}$$

Using Eq. (12), we obtain

$$|n^{(i)}\rangle = \sum_{m \neq n} \frac{1}{E_n^{(0)} - E_m^{(0)}}\, [\langle m^{(0)}|V|\, n^{(i-1)}\rangle$$

$$- \sum_{k=1}^{i-1} E_n^{(k)} \langle m^{(0)}|\, n^{(i-k)}\rangle]|m^{(0)}\rangle. \tag{15}$$

For a compact description, we introduce the notation

$$\langle O \rangle = \langle n^{(0)} | O | n^{(0)} \rangle \tag{16}$$

for any operator O and the complementary projection operator

$$Q_n^{(0)} = 1 - P_n^{(0)}$$

$$= 1 - |n^{(0)}\rangle\langle n^{(0)}|$$

$$= \sum_{k \neq n} |k^{(0)}\rangle\langle k^{(0)}| \tag{16a}$$

and the propagator q given by

$$\frac{1}{q_n} = \frac{Q_n^{(0)}}{E_n^{(0)} - H_0} = \sum_{k \neq n} \frac{|k^{(0)}\rangle\langle k^{(0)}|}{E_n^{(0)} - E_k^{(0)}}. \tag{16b}$$

Then

$$|n^{(i)}\rangle = \frac{1}{q_n} [V|n^{(i-1)}\rangle - \sum_{k=1}^{i-1} E_n^{(k)}| n^{(i-k)}\rangle]. \tag{17}$$

From the Eqs. (13) and (17) one can iteratively determine

$$E_n^{(1)}, E_n^{(2)}, \dots \text{ and } |n^{(1)}\rangle, | n^{(2)}\rangle, \dots.$$

Explicitly, to lowest orders in perturbation we have (Dalgarno, (1961))

$$E_n^{(0)} = \langle H_0 \rangle,$$

$$E_n^{(1)} = \langle V \rangle,$$

$$E_n^{(2)} = \left\langle V \frac{1}{q_n} V \right\rangle,$$

$$E_n^{(3)} = \left\langle V \frac{1}{q_n} \bar{V}_n \frac{1}{q_n} V \right\rangle, \quad \bar{V}_n = V - E_n^{(1)}$$

$$E_n^{(4)} = \left\langle V \frac{1}{q_n} \left[\bar{V}_n \frac{1}{q_n} \bar{V}_n - \left\langle \bar{V}_n \frac{1}{q_n} \bar{V}_n \right\rangle \right] \frac{1}{q_n} V \right\rangle,$$

$$E_n^{(5)} = \left\langle V \frac{1}{q_n} [\bar{V}_n \frac{1}{q_n} \bar{V}_n \frac{1}{q_n} \bar{V}_n - \left\langle \bar{V}_n \frac{1}{q_n} \bar{V}_n \frac{1}{q_n} \bar{V}_n \right\rangle \right.$$

$$\left. - \bar{V}_n \frac{1}{q_n} \left\langle \bar{V}_n \frac{1}{q_n} \bar{V}_n \right\rangle - \left\langle \bar{V}_n \frac{1}{q_n} \bar{V}_n \right\rangle \frac{1}{q_n} \bar{V} \right] \frac{1}{q_n} V \right\rangle \dots$$

$$|n^{(0)}\rangle,$$

$$|n^{(1)}\rangle = \frac{1}{q_n} V| n^{(0)}\rangle,$$

$$= \sum_{m \neq n} \frac{\langle m^{(0)}| V |n^{(0)}\rangle}{E_n^{(0)} - E_m^{(0)}} \left| m^{(0)} \right\rangle,$$

$$|n^{(2)}\rangle = \sum_{k \neq n} \sum_{l \neq n} \frac{\langle k^{(0)}| V |l^{(0)}\rangle \langle l^{(0)}| V |n^{(0)}\rangle}{(E_n^{(0)} - E_k^{(0)})(E_k^{(0)} - E_l^{(0)})} \left| k^{(0)} \right\rangle$$

$$- \sum_{k \neq n} \frac{\langle V \rangle \langle k^{(0)}| V |n^{(0)}\rangle}{(E_n^{(0)} - E_k^{(0)})^2} \left| k^{(0)} \right\rangle, \tag{18}$$

It is to be noted that for the ground state energy the second-order energy correction is non positive. The first-order correction to the energy is the mean value of the perturbation averaged over the unperturbed eigenfunctions. If H_0 also contains continuous spectrum, then the summations above include integration over the continuum together with the sum over the discrete part.

According to Eq. (13), a knowledge of the wave function to order p-1 is sufficient to determine the *p*th order correction $E_n^{(p)}$ to the energy. Careful analyses by Silverman (1952), Dalgarno and Stewart (1956) and others have shown that to calculate the energy correction to order $2p+1$, it is sufficient to know the wave function only to the order *p*. As an example, let us consider $E_n^{(3)}$ Using Eq. (17), we have from Eq. (13)

$$E_n^{(3)} = \langle n^{(0)}| V |n^{(2)}\rangle$$

$$= \left\langle n^{(0)} \left| V \frac{1}{q_n} \right| (V - E_n^{(1)}) |n^{(1)}\rangle \right\rangle, \text{ using (17) and (18)}$$

$$= \langle n^{(1)}| \overline{V}_n |n^{(1)}\rangle. \tag{19}$$

This shows that the energy eigenvalue is determined to third-order by a wave function correction to first-order. Similarly, one obtains.

$$E_n^{(4)} = \langle n^{(1)} + | \overline{V}_n |n^{(2)}\rangle - \frac{1}{2} E_n^{(2)} \langle n^{(1)}|n^{(1)}\rangle, \tag{19a}$$

$$E_n^{(5)} = \langle n^{(2)}| \overline{V}_n |n^{(2)}\rangle - 2 E_n^{(2)} \langle n^{(1)}|n^{(2)}\rangle. \tag{19b}$$

Next, we consider the normalization of the perturbed ket. As we see from Eq. (12) that the perturbed ket is not normalized in the usual manner. We can renormalize the perturbed ket by defining

$$|\overline{n}(\lambda)\rangle = Z_n^{1/2}(\lambda)|n(\lambda)\rangle, \tag{20}$$

where Z_n is the renormalization constant with $\langle \overline{n}|\overline{n}\rangle = 1$.

The scalar product of Eq. (20) with $|n^{(0)}\rangle$ gives

$$Z_n^{1/2} = \langle n^{(0)}|\bar{n}\rangle. \tag{21}$$

Z_n measures the probability for the perturbed energy eigenstate to be observed in the corresponding unperturbed energy eigenstate. Since $\langle\bar{n}|\bar{n}\rangle = Z_n\langle n|n\rangle = 1$, we have

$$Z_n^{-1} = \langle n|n\rangle$$

$$= 1 + \lambda^2\langle n^{(1)}|n^{(1)}\rangle + O(\lambda^3), \text{ expending } |n\rangle = |n^{(0)}\rangle + \lambda|n^{(1)}\rangle + \lambda^2|n^{(2)}\rangle + \dots$$

$$= 1 + \lambda^2\sum_{k\neq n}\frac{|\langle k^{(0)}|V|n^{(0)}|^2}{(E_n^{(0)} - E_k^{(0)})^2} + O(\lambda^3) \tag{22}$$

Hence, to second order in λ,

$$Z_n = 1 - \lambda^2\sum_{k\neq n}\frac{|V_{kn}|^2}{(E_n^{(0)} - E_k^{(0)})^2}. \tag{23}$$

From Eqs. (18) and (23), we see that to order,

$$Z_n(\lambda) = \frac{\partial E_n(\lambda)}{\lambda E_n^{(0)}} = \frac{\partial}{\partial E_n^{(0)}}\left[E_n^{(0)} + \lambda\langle n^{(0)}|V|n^{(0)}\rangle + \lambda^2\sum_{k\neq n}\frac{|\langle k^{(0)}|V|n^{(0)}\rangle|^2}{E_n^{(0)} - E_k^{(0)}}\right]$$

$$= 1 - \lambda^2\sum_{k\neq n}\frac{|\langle k^{(0)}|V|n^{(0)}\rangle|^2}{(E_n^{(0)} - E_k^{(0)})^2}. \tag{24}$$

Actually, this result is true to all orders of perturbation.

The perturbation expansions Eqs. (8) and (9) frequently do not converge. It is difficult to give sufficient conditions for the convergence of the series. The expansions converge if $|\langle i^{(0)}|V|l^{(0)}\rangle| << |(E_i^{(0)} - E_l^{(0)})|$.

The *RS* perturbation theory is useful in many cases of physical interest when it is an asymptolic expansion.

A general, rigorous perturbation theory has been given by Kato (see Kato (1949, 1989) Reed and Simon (1978), Galindo and Pascual (1991)).

12.2 THE RAYLEIGH–SCHRÖDINGER PERTURBATION EXPANSION – THE DEGENERATE CASE

The perturbation method discussed above is valid for nondegenerate unperturbed states of the discrete part of the spectrum. It is essentially an expansion in quantities like $\langle k^{(0)}|V|n^{(0)}\rangle/(E_n^{(0)} - E_k^{(0)})$ and is rapidly convergent for $V_{kn} << (E_n^{(0)} - E_k^{(0)})$. The method fails in case the unperturbed energy eigenkets are degenerate, i.e., when there are states for which $V_{kn} \neq 0$ and $E_n^{(0)} = E_k^{(0)}$. We modify the discussion given in the previous section to the degenerate case.

Let us suppose that there is a k-fold degeneracy for the unperturbed Hamiltonian, i.e., there are k different eigenkets $|n_a^{(0)}\rangle$, $|n_b^{(0)}\rangle$, ..., $|n_k^{(0)}\rangle$ all with the same unperturbed energy $\in_n^{(0)}$. Hence

$$H_0 |n_i^{(0)}\rangle = \in_n^{(0)} |n_i^{(0)}\rangle, \ i = a, b, \ldots, k. \tag{25}$$

Clearly, if now $\langle n_a^{(0)} | V | n_b^{(0)} \rangle \neq 0$ for $a \neq b$, the perturbation theory given above fails. However, any linear combination of those degenerate states $|n_i^{(0)}\rangle$ is also an eigenstate with the same energy $\in_n^{(0)}$. So we choose a basis set of k orthogonal states

$$|n_\alpha^{(0)}\rangle = \sum_{i=a}^{k} c_{\alpha i} |n_i^{(0)}\rangle \tag{26}$$

in which

$$\langle n_\alpha^{(0)} | V | n_\beta^{(0)} \rangle = 0 \quad \text{if } \alpha \neq \beta \tag{27}$$

Now, we can use the perturbation theory as discussed, since there will be zero matrix elements corresponding to zero energy denominators.

Thus, the proper basis set of states is the one which diagonalizes V within each group of degenerate states.

To diagonalize V within the k-fold degenerate unperturbed eigenkets $|n_i^{(0)}\rangle$, $i = a, b, \ldots k$

we construct the $k \times k$ perturbation matrix

$$\begin{vmatrix} \langle n_a^{(0)} | V | n_a^{(0)} \rangle & \langle n_a^{(0)} | V | n_b^{(0)} \rangle & \langle n_a^{(0)} | V | n_c^{(0)} \rangle & \cdots \\ \langle n_b^{(0)} | V | n_a^{(0)} \rangle & \langle n_b^{(0)} | V | n_b^{(0)} \rangle & \langle n_b^{(0)} | V | n_c^{(0)} \rangle & \cdots \\ \cdots & \cdots & \cdots & \cdots \\ \cdots & \cdots & \cdots & \cdots \end{vmatrix} \tag{28}$$

and find the egenvector. The matrix Eq. (28) is a hermitian matrix and can be disagonalized by a unitary transformation. The basis states Eq. (26) give the matrix elements

$$V_{\alpha\beta} = \langle n_\alpha^{(0)} | V | n_\beta^{(0)} \rangle = \sum_{i,j} c_{i\alpha}^* V_{ij} c_{j\beta}. \tag{29}$$

We can always choose $c_{i\alpha}$ in such a way that Eq. (27) is satisfied. One obtains from Eqs. (27) and (29),

$$\sum_{i,j} c_{i\alpha}^* V_{ij} c_{j\beta} = E_{n\alpha}^{(1)} \delta_{\alpha\beta}, \tag{30}$$

where $E_{n\alpha}^{(1)}$ is the eigenvalue.

Using unitarity conditions $[\sum_i c_{i\alpha}^* c_{i\beta} = \delta_{\alpha\beta}, \sum_\alpha c_{i\alpha} c_{j\alpha}^* = \delta_{ij}]$, and multiplying by $c_{i\alpha}$, we get the eigenvalue equation

$$\sum_j V_{ij} c_{j\beta} = E_{n\beta}^{(1)} c_{i\beta}. \tag{31}$$

This set of linear homogenous equations has a nontrivial solution only if the determinant of the coefficients vanishes:

$$\det \left(V_{ij} - E_n^{(1)} \delta_{ij} \right) = 0 \tag{32}$$

$$(i, j = a, b, c \ldots, k)$$

Thus, the basis set $|n_\alpha^{(0)}\rangle$ given by Eq. (26), satisfies

$$\langle n_\beta^{(1)}| V |n_\alpha^{(1)}\rangle = E_{n\alpha}^{(1)} \delta_{\alpha\beta}, \tag{33}$$

where the eigenvalue $E_{n\alpha}^{(1)}$ is the first-order energy shift of the state $|n_\alpha^{(0)}\rangle$. The k-values of $E_n^{(1)}$ give the first-order perturbed energies. Equation (32) gives k real roots $E_{na}^{(1)}$, $E_{nb}^{(1)}$, ..., $E_{nk}^{(1)}$. If all the roots are *distinct* the degeneracy is *completely removed* in first-order by the perturbation. This means that there will be k perturbed eigenkets all with different energies. Then the unpurturbed states $|n_a^{(1)}\rangle$, $|n_b^{(1)}\rangle$, ... $|n_k^{(1)}\rangle$, all with energy $\in_n^{(0)}$ split into k-states $|n_\alpha'\rangle$, $|n_\beta'\rangle$, ... which are given in first-order by

$$|n_\alpha'\rangle = |n_\alpha^{(0)}\rangle + \sum_m{}' \frac{|m^{(0)}\rangle \langle m^{(0)}| V |n_\alpha^{(0)}\rangle}{\in_n^{(0)} - \in_m^{(0)}}. \tag{34}$$

The prime on the sum means that the summation is over all states except the k-states $|n_\alpha^{(0)}\rangle$, $|n_\beta^{(0)}\rangle$, ...

The energy shift to second-order in perturbation is

$$E_{n\alpha} = \in_n^{(0)} + \lambda \langle n_\alpha^{(0)}| V | n_\alpha^{(0)}\rangle + \lambda^2 \sum_m{}' \frac{|\langle m^{(0)}| V |n_\alpha^{(0)}\rangle|^2}{\in_n^{(0)} - \in_m^{(0)}}, \tag{35}$$

where $\langle n_\alpha^{(0)}| V |n_\alpha^{(0)}\rangle = E_{n\alpha}^{(1)}$.

For higher orders, one proceeds just as for the non-degenerate case, except that the summations are replaced by summations with primes in Eq. (35) (see Dalgarno (1961)).

If some or all roots of Eq. (32) are equal then the degeneracy is only partially or not at all removed. The remaining degeneracy may or may not be removed in higher orders of perturbation.

Next, we discuss almost degenerate perturbation theory which is applicable when the unperturbed level is one of a group of levels closely spaced in energy. In this case, $|\langle n^{(0)}| V | m^{(0)}\rangle|$ is large compared with $|E_n^{(0)} - E_m^{(0)}|$ and the convergence is slow. Usually, it may be best to proceed as if the levels are degenerate, although in a particular case the non-degenerate theory may be preferable.

We consider the simple case when only two eigenstates $|n^{(0)}\rangle$ and $|m^{(0)}\rangle$ of the unperturbed Hamiltonian H_0 are very closely spaced in energy ($E_n^{(0)} \approx E_m^{(0)}$). We separate the perturbation V into two parts V_1 and V_2 where V_1 takes into account that part of V referring to $|n^{(0)}\rangle$ and $|m^{(0)}\rangle$ only and V_2 the rest of the perturbation.

We write
$$V = V_1 + V_2 \tag{36}$$
where

$$V_1 = |m^{(0)}\rangle\langle m^{(0)}| V | m^{(0)}|\rangle\langle m^{(0)}| + |n^{(0)}\rangle\langle n^{(0)}| V | n^{(0)}\rangle\langle n^{(0)}|$$
$$+ |m^{(0)}\rangle\langle m^{(0)}| V | n^{(0)}|\rangle\langle n^{(0)}| + |n^{(0)}\rangle\langle n^{(0)}| V | m^{(0)}\rangle\langle m^{(0)}| \tag{37}$$

First, we find the exact eigenfunctions and eigenvalues of the Hamiltonian $H_1 = H_0 + V_1$ and then treat V_2 as perturbation.

Now, any eigenstate $|i^{(0)}\rangle$ ($i \neq m, n$) of H_0 is also an eigenstate of H_1, since $V_1|i^{(0)}\rangle = 0$. But, $|n^{(0)}\rangle$ and $|m^{(0)}\rangle$ are not eigenstates of H_1. A linear combination $\alpha|n^{(0)}\rangle + \beta|m^{(0)}\rangle$ is an eigenstate of H_1:

$$H_1\left(\alpha|n^{(0)}\rangle + \beta|m^{(0)}\rangle\right) = E\left(\alpha|n^{(0)}\rangle + \beta|m^{(0)}\rangle\right) \tag{38}$$

Also,

$$\left.\begin{aligned}
H_1|n^{(0)}\rangle &= E_n^{(1)}|n^{(0)}\rangle + \langle m^{(0)}| V | n^{(0)}\rangle|m^{(0)}\rangle \\
H_1|m^{(0)}\rangle &= E_m^{(1)}|m^{(0)}\rangle + \langle n^{(0)}| V | m^{(0)}\rangle|n^{(0)}\rangle
\end{aligned}\right\} \tag{39}$$

where

$$E_n^{(1)} = E_n^{(0)} + \langle n^{(0)}| V | n^{(0)}\rangle, \text{ etc.}$$

From Eqs. (37) and (38), we get

$$\begin{pmatrix} E_n^{(1)} & \langle n^{(0)}| V | m^{(0)}\rangle \\ \langle m^{(0)}| V | n^{(0)}\rangle & E_m^{(1)} \end{pmatrix}\begin{pmatrix} \alpha \\ \beta \end{pmatrix} = E\begin{pmatrix} \alpha \\ \beta \end{pmatrix}. \tag{40}$$

The solutions are

$$\alpha = \langle n^{(0)}| V | m^{(0)}\rangle$$

$$\beta_{\pm} = \frac{E_m^{(1)} - E_n^{(1)}}{2} \pm \left[\left(\frac{E_m^{(1)} - E_n^{(1)}}{2} \right)^2 + |\langle n^{(0)}| V| m^{(0)}\rangle|^2 \right]^{1/2} \tag{41}$$

giving the eigenstates of H_1 as $\alpha| n^{(0)}\rangle + \beta| m^{(0)}\rangle$.

The eigenvalues of H_1 are

$$E_{\pm} = \frac{E_m^{(1)} - E_n^{(1)}}{2} \pm \left[\left(\frac{E_m^{(1)} - E_n^{(1)}}{2} \right)^2 + |\langle n^{(0)}| V| m^{(0)}\rangle|^2 \right]^{1/2}. \tag{42}$$

In the special case when $V_{nn} = V_{mm} = 0$ and the two unperturbed levels are very close; then

$$E_{\pm} = E^{(0)} \pm V_{mn},$$

i.e., due to the perturbation there is a repulsion of the two levels, one being pushed up by V_{mn} and the other one being pushed down by the same amount.

As an illustrative example, we consider the motion of an electron in a one dimensional lattice of length $L = Na$, where a is the lattice constant. The nonzero lattice potential $V(x)$ is periodic in the lattice, $V(x) = V(x + a)$. We assume that the periodic potential is weak. When the periodic potential is zero, the solutions to Schrödinger equation are plane waves. So the unperturbed wave functions are

$$\psi_k(x) = \frac{e^{ikx}}{\sqrt{L}}, \tag{43}$$

and the unperturbed energy are

$$E_k^{(0)} = \hbar^2 k^2/2m. \tag{43a}$$

Here the allowed values of k are integral multiples of $2\pi/L$, since we impose periodic boundary conditions $\psi(x) = \psi(x + L)$. Because V is periodic, we can expand it in the following Fourier series

$$V(x) = \sum_{-\infty} V_n e^{iKnx} \quad (n \text{ integer}) \tag{44}$$

where $K = \dfrac{2\pi}{a}$ is a reciprocal lattice vector.

The only non-vanishing matrix elements of V are

$$\langle k + nK| V| k\rangle = V_n. \tag{45}$$

If k is near a single Bragg plane $(\pm nK/2)$, then these energies are close. If k is not near $\pm nK/2$ (n = integer), the wave function to first-order is

$$\psi_k(x) = \frac{e^{ikx}}{\sqrt{L}} + \sum_n \frac{e^{i(k+nK)x}}{\sqrt{L}} \cdot \frac{V_n}{E_k^{(0)} - E_{k+nK}^{(0)}}, \tag{46}$$

and the energy to second order is

$$E_k = E_k^{(0)} + V_0 + \sum_n \frac{|V_n|^2}{E_k^{(0)} - E_{k+nK}^{(0)}}. \tag{47}$$

So in the case of no near degeneracy, the shift in energy is second-order in V_n. In the near degenerate case, we diagonalize the following matrix

$$\begin{pmatrix} E_k^{(0)} + V_0 & V_n^* \\ V_n & E_{k+nK}^{(0)} + V_0 \end{pmatrix}$$

and obtain the eigenvalues (using Eq. (42)),

$$E_{k_\pm} = V_0 + \frac{E_k^{(0)} + E_{k+nK}^{(0)}}{2} \pm \sqrt{\left(\frac{E_k^{(0)} + E_{k+nK}^{(0)}}{2}\right)^2 + |V_n|^2} \tag{48}$$

For $|E_k^{(0)} - E_{k+nK}^{(0)}| \gg |V_n|$, Eq. (48) reduces to the eigenvalues $E_{k+} = E_k^{(0)} + V_0$ and $E_{k-} = E_{k+nK}^{(0)} + V_0$.

The shift in energy in this case is linear in V. So the nearly degenerate free electron levels are most significantly shifted due to a weak periodic potential. When the wave vector k is on the Bragg plane then $E_{k\pm} = E_{nK/2}^{(0)} \pm |V_n|$, i.e. one level is raised by $|V_n|$ and the other level lowered by the same amount, giving rise to a band gap of magnitude $2|V_n|$ at the zone boundaries (see figure 12.1 below)

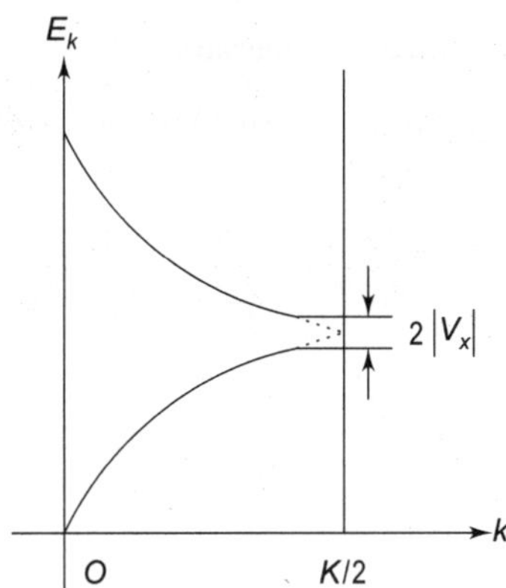

Fig. 12.1   Structure of the energy bands. Dotted lines-free electron

12.3   BRILLOUIN-WIGNER PERTURBATION THEORY

The *RS* perturbation theory yields complicated expressions for higher order terms beyond the second, the complexity increasing as the order increases. We now discuss a different form of the perturbation theory, due to Brillouin and Wigner, in which the expression to any arbitrary order is easy.

We consider the time-independent Schrödinger equation

$$(E_n - H_0)|n\rangle = \lambda V|n\rangle \tag{49}$$

and multiply on the left by $\langle m^{(0)}|$.

Thus, $\qquad (E_n - E_m^{(0)})\,\langle m^{(0)}|n\rangle = \lambda\langle m^{(0)}|V|n\rangle. \tag{49a}$

We choose the normalization $\langle n^{(0)}|n\rangle = 1$. If $|m^{(0)}\rangle = n^{(0)}\rangle$, this gives the perturbed energy eigenvalue

$$E_n = E_n^{(0)} + \lambda\,\langle n^{(0)}|V|n\rangle. \tag{50}$$

Expanding $|n\rangle$ in unperturbed states

$$|n\rangle = \sum_m |m^{(0)}\rangle\langle m^{(0)}|n\rangle$$

$$= |n^{(0)}\rangle\langle n^{(0)}|n\rangle + {\sum_m}'|m^{(0)}\rangle\langle m^{(0)}|n\rangle, \tag{51}$$

where the prime indicates that in the summation $m = n$ term is excluded, and substituting Eq. (49) into Eq. (51), we get

$$|n\rangle = |n^{(0)}\rangle + {\sum_m}'\,|m^{(0)}\rangle\,\frac{1}{E_n - E_m^{(0)}}\,\lambda\langle m^{(0)}|V|n\rangle. \tag{52}$$

Formally, we may write Eq. (52) as

$$|n\rangle = |n^{(0)}\rangle + (E_n - H_0)^{-1}\,Q_n V|n\rangle, \tag{53}$$

where the complementary projection operator

$$Q_n = 1 - P_n = 1 - |n^{(0)}\rangle\langle n^{(1)}|. \tag{54}$$

We solve Eq. (53) by iteration. Thus,

$$|n\rangle = |n^{(0)}\rangle + \frac{1}{E_n - H_0}\,Q_n V\left(|n^{(0)}\rangle + \frac{1}{E_n - H_0}\,Q_n V|n\rangle\right)$$

$$= |n^{(0)}\rangle + \frac{1}{E_n - H_0}\,Q_n V|n^{(0)}\rangle + \frac{1}{E_n - H_0}\,Q_n V\,\frac{1}{E_n - H_0}\,Q_n V|n^{(0)}\rangle + \ldots \tag{55}$$

This series is known as the Brillouin-Wigner perturbation expansion. Explicitly,

$$|n\rangle = |n^{(0)}\rangle + \lambda\,{\sum_m}'\,|m^{(0)}\rangle\,\frac{1}{E_n - E_m^{(0)}}\,\langle m^{(0)}|V|n^{(0)}\rangle$$

$$+ \lambda^2\,{\sum_{j,m}}'\,|j^{(0)}\rangle\,\frac{1}{E_n - E_j^{(0)}}\,\langle j^{(0)}|V|m^{(0)}\rangle\,\frac{1}{E_n - E_m^{(0)}}\,\langle m^{(0)}|V|n^{(0)}\rangle + \ldots \tag{55a}$$

Since the perturbed energy E_n appears in each denominator on the right hand side instead of the unperturbed energy $E_n^{(0)}$, this is an implicit equation for E_n, If we substitute Eq. (55) into the expression Eq. (50) for the energy eigenvalue then we obtain a series in powers of V. Using the spectral representation of $\dfrac{Q_n}{E_n - H_0}$ as

$$\frac{Q_n}{E_n - H_0} = \sum_{m \neq n} \frac{|m^{(0)}\rangle\langle m^{(0)}|}{E_n - E_m^{(0)}}$$

in this expression we obtain the perturbation expansion

$$E_n = E_n^{(0)} + \lambda\langle n^{(0)}|V|n^{(0)}\rangle + \lambda^2 \sum_m' \frac{|\langle m^{(0)}|V|n^{(0)}\rangle|^2}{E_n - E_m^{(0)}}$$

$$+ \lambda^3 \sum_m' \sum_{m'}' \frac{\langle n^{(0)}|V|m^{(0)}\rangle\langle m^{(0)}|V|m'^{(0)}\rangle\langle m'^{(0)}|V|n^{(0)}\rangle}{(E_n - E_m^{(0)})\,(E_n - E_{m'}^{(0)})} + \dots \qquad (56)$$

In a compact notation

$$E_n = E_n^{(0)} + \langle n^{(0)}|V|n^{(0)}\rangle + \langle n^{(0)}| + V \frac{Q_n}{E_n - H_0}\, V|n^{(0)}\rangle + \dots \qquad (56a)$$

This formula is similar to the *RS* series with E_n in stead of $E_n^{(0)}$. If we expand the terms such as $(E_n - E_m^{(0)})^{-1}$ on the right hand side of Eq. (56) in powers of λ, we recover the *RS* perturbation series. The kth order term in the B-W series is the same as the leading term of order k in the *RS* series with E_n instead of $E_n^{(0)}$. To all orders beyond the second the *RS* series contains many more terms, the extra terms to order k are the corrections to the energy denominators of the lower order terms $(k-1, k-2, \dots)$. The *RS* perturbataion theory is less convenient than the B-W series. If we have a good idea of the value of E_n, e.g. from a variational calculation, then the B-W perturbation series for $|n\rangle$ converges more rapidly than the *RS* series. The B-W series is usually superior in the case of nearly degenerate states.

The B-W perturbation theory is formally simple and is usually considered an improvement on the *RS* perturbation expansion in the sense that it avoids the complexity of the higher order formulae of the latter. The B-W method contains the exact energy of the perturbed system and yields an implicit equation for its determination. Hence the B-W method may be less convenient than the *RS* scheme. It is best suited for problems with one energy level at a time. The B-W scheme is quite unsuitable for extended systems.

We analyze the N-dependence of the terms in the B-W expansion, where N is the number of the particles of the system (in the limit $N \to \infty$, but the number density remaining constant). Let H_0 be the Hamiltonian for non-interacting particles while V the two-body interaction between the particles. V can either change the state of two particles or of none. The unperturbed ket is a determinant of plane waves.

The diagonal matrix elements of V are proportional to N, while the off diagonal terms are proportional to N^{-1}. The first two terms on the right in Eq. (56a) has the linear N-dependence while the third one is independent of N. In the limit of large N, the B-W series converges very slowly. The RS scheme is an expansion of the energy as a power series in the strength of the interaction so that this procedure has the property that each term has the correct linear N-dependence, due to the replacement of $\dfrac{1}{E_n - H_0}$ by $\dfrac{1}{E_n^{(0)} - H_0}$. This makes the RS scheme suitable for many-body problems.

12.4   THE ANHARMONIC OSCILLATOR AND LARGE ORDERS OF PERTURBATION THEORY

The anharmonic oscillator has become a testing ground for checking different approximate methods for computing energy levels in quantum mechanics and quantum field theory. Also, a study of anharmonic interaction has important applications in various branches of physics, e.g., condensed matter physics, atomic and molecular physics, etc. Indeed, numerous methods, including perturbation expansion, WKB method, functional integral method, Padé approximants, the Hill determinant method, supersymmetric quantum mechanics, the phase-integral method, the algebraic method, etc. have been used to investigate anharmonic potentials.

The standard method of finding the energy spectrum and eigenfunctions of an anharmonic potential is to apply the perturbation theory. However, the RS perturbation series diverges for most problems. The perturbation series in many quantum mechanical problems and in quantum field theory are asymptotic ones and hence, diverge for all values of the expansion parameter. The three perturbation series associated with three standard textbook problems in time-independent perturbation theory the quartic anharmonic oscillator, the Zeeman effect in atoms and the Stark effect in atoms are divergent.

First, let us consider a one-dimensional oscillator perturbed by a cubic and a quartic terms.

The Hamiltonian is

$$H = H_0 + V = \frac{p^2}{2m} + \frac{1}{2}kx^2 + \alpha x^3 + \beta x^4 \tag{57}$$

where $k = m\omega^2$. The spectrum of the unperturbed Hamiltonian $H_0 = \dfrac{1}{2}\dfrac{p^2}{2m} + \dfrac{1}{2}m\omega^2 x^2$ is simple and discrete and the energy eigenvalues are

$$E_n^{(0)} = \hbar\omega\left(n + \frac{1}{2}\right), \; n = 0, 1, 2, \ldots. \tag{58}$$

To calculate the matrix elements of x^3 and x^4, we use the following expression for the non-vanishing values of the matrix elements of x:

$$x_{n,n-1} = x_{n-1,n} = (n\hbar/2m\omega)^{1/2}. \tag{59}$$

Clearly, the diagonal elements $(x^3)_{nn}$ vanish, because the perturbation is odd and the states are of definite parity. The correction to the energy in the first-order due to the term αx^3 is zero. The correction in the second-order due to this term αx^3 is of the same order due to the term βx^4 in the first-order.

Now,
$$(x^4)_{nn} = \sum_k (x^2)_{nk}\,(x^2)_{kn}$$

$$= ((x^2)_{n,n-2})^2 + ((x^2)_{n,n})^2 + ((x^2)_{n,n+2})^2 \tag{60}$$

Using Eq. (59), $(x^2)_{n-2,n} = \left(\dfrac{\hbar}{2m\omega}\right)(n(n-1))^{1/2}$

$$(x^2)_{n+2,n} = \left(\frac{\hbar}{2m\omega}\right)[(n+2)(n+1)]^{1/2} \tag{61}$$

$$(x^2)_{n,n} = \left(\frac{\hbar}{2m\omega}\right)(2n+1)$$

Combining Eqs. (60) and (61) the diagonal matrix elements of x^4 are

$$(x^4)_{n,n} = \left(\frac{\hbar}{m\omega}\right)^2 \cdot \frac{1}{2}\,(3n^2 + 2n + 1). \tag{62}$$

The following matrix elements of x^3 are different from zero

$$(x^3)_{n,n-1} = (x^2)_{n,n}\,(x)_{n,n-1} + (x^2)_{n,n-2}\,(x)_{n-2,n-1}$$

$$= 3\left(\frac{\hbar}{m\omega}\right)^{3/2}\left(\frac{n}{2}\right)^{3/2},$$

$$(x^3)_{n,n-3} = (x^2)_{n,n-2}\,(x)_{n-2,n-3}$$

$$= \left(\frac{\hbar}{m\omega}\right)^{3/2}\left(\frac{n(n-1)(n-2)}{8}\right)^{1/2}$$

$$(x^3)_{n,n+1} = 3\left(\frac{\hbar}{m\omega}\right)^{3/2}\left(\frac{n+1}{2}\right)^{3/2}$$

$$(x^3)_{n,n+3} = \left(\frac{\hbar}{m\omega}\right)^{3/2}\left(\frac{(n+3)(n+2)(n+1)}{8}\right)^{1/2}. \tag{63}$$

Using the general formulas for $E_n^{(1)}$ and $E_n^{(2)}$, the energy levels of the anharmonic oscillator to the second order is given by

$$E_n = \hbar\omega\left(n + \frac{1}{2}\right) - \frac{15}{4}\frac{\alpha^2}{\hbar\omega}\left(\frac{\hbar}{m\omega}\right)^3\left(n^2 + n + \frac{11}{30}\right)$$

$$+ \frac{3}{2}\beta\left(\frac{\hbar}{m\omega}\right)^2\left(n^2 + n + \frac{1}{2}\right). \tag{64}$$

so that the energy of the perturbed ground state is

$$E_0 = \frac{1}{2}\hbar\omega + \frac{3}{4}\beta\frac{\hbar^2}{m^2\omega^2} - \frac{11}{8}\alpha^2\frac{\hbar^2}{m^3\omega^4} + \cdots \tag{65}$$

We now discuss the behavior of perturbation theory at large orders. The motivation for studying the behavior of perturbation theory at large orders is that a better understanding of the nature of the asymptotic behavior will lead to the development of methods to extract numerical information from the divergent series.

We present the general ideas basic to the analysis of the behavior of perturbation theory at large orders with little mathematical details. For details, consult Itzykson and Zuber (1985), Negele and Orland (1988), Le Guillou and Zinn-Justin (1970), Bender and Orszag (1978). To understand the nature of the divergence, we consider the integral

$$Z(g) = \frac{1}{\sqrt{2\pi}}\int_{-\infty}^{\infty} dx\, e^{-\left(x^2/2 + \frac{g}{4}x^4\right)} \tag{66}$$

which corresponds to the classical partition function of a particle in a quartic potential. $Z(g)$ has an essential singularity at $g = 0$. Negative values of g correspond to the case where the potential is not bounded from below. $Z(g)$ is analytic in the complex g-plane with a cut along the negative real axis. The perturbative expansion is easily obtained by expanding $e^{-g/4x^4}$.

$$Z(g) = \sum_{k=0}^{\infty} g^k Z_k \tag{67}$$

where

$$Z_k = \frac{(-)^k}{k!}\frac{1}{4^k}\int_{-\infty}^{\infty}\frac{dx}{\sqrt{2\pi}}\, e^{-\frac{x^2}{2}}x^{4k}$$

$$= \frac{(-1)^k\,(4k-1)!!}{k!4^k}$$

$$= \frac{(-1)^k\,(4k)!}{k!16^k(2k)!} \tag{68}$$

The perturbation series diverges since Z_k grows like $k!$

Using Stirling's formula

$$k! \approx \sqrt{2\pi}\, k^{k+\frac{1}{2}}\, e^{-k} \text{ for large } k,$$

$$Z_k \underset{k\to\infty}{\sim} \frac{(-1)^k}{\sqrt{k\pi}} \left(\frac{4k}{e}\right)^k \tag{69}$$

$$k \underset{\to}{\sim} \infty \;\; \frac{(-1)^k\, 4^k\, k!}{\pi\sqrt{2}\; k}. \tag{70}$$

The asymptotic behavior of Z_k is obtained by using Stirling's formula. We now use the method of steepest descent for large k to obtain the integral:

$$Z_k = \frac{(-1)^k}{k!\,4^k} \int_{-\infty}^{\infty} \frac{dx}{\sqrt{2\pi}}\, e^{-\frac{x^2}{2}}\, x^{4k}$$

$$= \frac{(-1)^k}{k!\,4^k}\, e^{2k\ln k}\, \sqrt{k} \int_{-\infty}^{\infty} \frac{dy}{\sqrt{2\pi}}\, e^{-k(y^2/2 - 2\ln y^2)},\; \sqrt{k}\,y = x.$$

Let

$$A = \frac{y^2}{2} - 2\ln y^2.$$

The stationary condition

$$\frac{dA}{dy} = 0$$

gives two stable saddle points $y_c = \pm 2$.

For large k,

$$Z_k \underset{k\to\infty}{\sim} 2\frac{(-1)^k}{k!\,4^k}\, e^{2k\ln k}\, \sqrt{k}\, e^{-k(2-2\ln 4)}\, \frac{1}{\sqrt{2\pi}}\, \sqrt{\frac{\pi}{k}},$$

$$\sim (-1)^k/\sqrt{\pi k}\; e^{k\ln 4k - k}, \tag{71}$$

in agreement with Eq. (69).

As in the example above, perturbation methods usually yield divergent series. There are several methods for summation of divergent series so that a finite expression can be assigned. If the coefficients in the series $\sum_{k=0}^{\infty} Z_k x^k$ grow faster than a power of k [$Z_k \sim k!$, *for example*], then it is useful to consider Borel summation. Let the Borel transform

$$B(x) = \sum_{k=0}^{\infty} \frac{Z_k x^k}{k!} \tag{72}$$

converge for sufficiently small x and the inverse transform

$$Z_B(x) = \int_0^{\infty} e^{-t}\, B(xt)\, dt \tag{73}$$

exist. If we expand the integral

$$Z_B(x) = \int_0^\infty e^{-t/x}\, B(t)\, dt/x \text{ for small } x \text{ by substituting Eq. (72) and then integrate}$$

term by term, then

$$Z_B(x) \sim \sum_{k=0}^\infty \frac{Z_k}{k!} \int_0^\infty B(t) e^{-t/x} t^k \frac{dt}{x}$$

$$= \sum_{k=0}^\infty Z_k x^k, \; x \to 0_+ \tag{74}$$

The series Eq. (74) is asymptotic to $Z_B(x)$ as $x \to 0_+$ since $Z_B(x)$ exists.

Although the asymptotic series diverges, we define the Borel sum of $\sum_{k=0}^\infty Z_k x^k$

as $Z_B(x)$. Also, the sum of $\sum_{k=0}^\infty Z_k$ is denoted as $Z_B(1)$.

As an example, the series $\sum_{k=0}^\infty (-1)^k k!$ diverges but $B(x) = \sum_{k=0}^\infty (-x)^k$ converges

for $|x| \le 1$ to $\dfrac{1}{1+x}$. The Borel sum of $\sum_{k=0}^\infty (-x)^k\, k!$ is $Z_B(x) = \int_0^\infty [e^{-t(1+x)}]\, dt$ and

the Borel sum of $\sum_{k=0}^\infty (-x)^k\, k!$ is $Z_B(1) = \int_0^\infty [e^{-t(1+t)}]\, dt$.

Another alternative summation method is the Padé summation. A power

series $\sum_{n=0}^\infty a_n x^n$ is replaced by a sequence of rational functions of the form

$$P_M^N(x) = \frac{\displaystyle\sum_{n=0}^N A_n x^n}{\displaystyle\sum_{n=0}^M B_n x^n} \tag{75}$$

where we choose $B_0 = 1$. The remaining $(M + N + 1)$ coefficients $A_0, A_1, \ldots$ $A_N, B_1, B_2, \ldots B_M$ are chosen so that the first $(M + N + 1)$ terms in the Taylor-expansion of $P_M^N(x)$ agree with the first $(M + N + 1)$ terms of power series $\sum_{n=0}^\infty a_n x^n$. The rational function $P_M^N(x)$ is called a $[N, M]$ Padé approximant.

If $f(x) = \sum_{n=0}^\infty a_n x^n$ then in many cases $P_M^N(x) \to f(x)$ as $N, M \to \infty$, even if

$\sum_{n=0}^\infty a_n x^n$ is a divergent series. Usually one considers only the convergence of the

Padé sequences $P_0^J\, P_1^{1+J},\, P_2^{2+J},\, \ldots$, having $N = M + J$ with J fixed and $M \to \infty$.

Let us consider the quartic anharmonic oscillator with the Hamiltonian

$$H = \frac{p^2}{2} + \frac{x^2}{2} + \frac{g}{4} x^4 \tag{76}$$

For $g > 0$, the energy eigenvalues $E^K(g)$, $K = 0, 1, 2 \, \dots \dots$ of H can be written as a *RS* perturbation expansion

$$E^K(g) = K + \frac{1}{2} + \sum_{n=1}^{\infty} A_n^K g^n. \tag{77}$$

For a negatively coupled anharmonic oscillator ($g = -\epsilon$, $\epsilon > 0$) the Hamiltonian is not bounded from below. The large-order behavior of perturbation theory is connected with the lifetimes of the unstable states of a negatively coupled anharmonic oscillator. The nth *RS* coefficient for the Kth energy level (A_n^K) and the lifetime of the Kth unstable state of negatively coupled anharmonic oscillator and decaying by barrier penetration is

$$A_n^K = \frac{1}{2\pi i} \int_{-\infty}^{0} D^K(x)\, x^{-n}\, dx \tag{78}$$

where

$$D^K(x) = \lim_{\epsilon \to 0} \left[F^K(x + i\epsilon) - F^K(x - i\epsilon) \right] \tag{79}$$

$$F^K(g) = \left[E^K(g) - K - \frac{1}{2} \right] g^{-1}. \tag{80}$$

Bender and Wu (1969, 1971, 1972) used WKB technique to calculate Im $E^K(\epsilon)$ which goes as the reciprocal of the lifetime. Using the WKB analysis, we find for a potential $V(x)$

$$\text{Im } E^K(\epsilon) \sim \exp\left(-2 \int_{0}^{x_0} dx\, \sqrt{2V(x)} \right) \tag{81}$$

where x_0 is a zero of the potential. For an anharmonic oscillator

$$\text{Im } E^K(\epsilon) \sim e^{-4/3\epsilon}. \tag{82}$$

A careful analysis by Bender and Wu showed that the lowest order WKB result gives the leading behavior of A_n^K for large n:

$$A_n^K = \frac{(-1)^{n+1}\, 12^K\, \sqrt{6}}{K! \pi^{3/2}}\, 3^n \Gamma\left(n + K + \frac{1}{2} \right) \left(1 + O\left(\frac{1}{n} \right) \right). \tag{83}$$

The ground state ($K = 0$) energy coefficient is

$$A_n^0 \sim (-1)^{n+1} \left(\frac{6}{\pi^3} \right)^{1/2} 3^n \Gamma\left(n + \frac{1}{2} \right) \left(1 + O\left(\frac{1}{n} \right) \right). \tag{84}$$

The results are in spectacular agreement with numerical fits to order 150.

The WKB method is not well suited for quantum field theory where an alternative method based on functional integral technique is adopted.

Graffi, Grecchi and Simon (1990) have proved that $E(g)$ for an anharmonic oscillator (x^{2m}) is Borel summable, i.e., may be represented by its Borel transform by a relation of the type Eq. (73).

The anharmonic oscillator is also expressible as a ϕ^4 field theory. The integral $Z(g)$ in Eq. (66) gives the number of Feynman diagrams contributing to the partition function (the vacuum amplitude) for the anharmonic oscillator as well as the ϕ^4 field theory. From Eq. (70), we see that the number of diagrams grows like $k!$ and hence the $k!$ behavior in perturbation theory at large orders.

As example of potentials with a discrete set of minima the double-well potential (see Section 13.8) has been studied. A system with the double-well potential $V(x) = \dfrac{1}{2} x^2 \times (1 - \sqrt{g}\, x)^2$ has two degenerate ground states. If we expand using perturbation theory around the minima, then we find that the degeneracy exists at all orders. Actually, the barrier penetration effects remove this degeneracy, through the difference in energies of the two states $\Delta E \sim e^{-c/g}$ (c = const) is exponentially small. It has been shown that the parturbation series in the case of degenerate minimum is not Borel-summable. The large order behavior of the ground state energy has been found to be

$$E_n^0 \sim -\frac{1}{\pi} 3^{n+1}\, n! \tag{85}$$

The large-order behavior of perturbation theory has suggested various summation methods and their convergence properties. The Zeeman effect, the Stark effect and the H_2^+ ion have been studied from the point of view of large-order behavior of perturbation theory.

12.5 HYDROGENIC ATOMS: FINE STRUCTURE AND OTHER CORRECTIONS

In this section we study the various corrections to the spectra of hydrogen-like atoms. The zeroth order Hamiltonian is given by

$$H_0 = \bar{p}^2/2\mu - \frac{Ze^2}{r} \tag{86}$$

where the first term represents the nonrelativistic kinetic energy of the atom, and the second term represents the electrostatic interaction between the electron and the nucleus, μ being the reduced mass. The expression is approximate. There are many correction terms to be added to the previous Hamiltonian. A proper treatment of the energy level of the hydrogen atom is given in quantum electrodynamics and in the Dirac theory of the electron (see Bjorken and Drell (1964), Itzykson and Zuber (1985)). The most important correction terms arise from relativistic effects and the electron spin. The leading corrections can be computed from the nonrelativistic theory. Since the relativistic corrections are very small (provided Z is not too large), we can conveniently make use of perturbation theory. The relativistic correction terms are: (1) the relativistic kinetic energy, (2) the spin-orbit interaction, and (3) the Darwin term. These three terms are discussed below. The relativistic energy

$$E = \sqrt{\vec{p}^2 c^2 + \mu^2 c^4}$$

is expanded in powers of $\dfrac{|\vec{p}|}{\mu c}$ giving

$$E = \mu c^2 + \frac{\vec{p}^2}{2\mu} - \frac{1}{2}\frac{1}{\mu c^2}\left(\frac{\vec{p}^2}{2\mu}\right)^2 + \ldots \tag{87}$$

The relativistic kinetic energy is

$$\frac{\vec{p}^2}{2\mu} - \frac{\vec{p}^4}{8\mu^3 c^2}.$$

Therefore, the term

$$H_1' = \frac{-\vec{p}^4}{8\mu^3 c^2} \tag{88}$$

represents the relativistic correction to the kinetic energy and is the first energy correction. To calculate the order of magnitude of this term, we estimate

$$\frac{H_1'}{H_0} \simeq \frac{\vec{p}^4/8\mu^3 c^2}{\vec{p}^2/2\mu} = \frac{\vec{p}^2}{4\mu^2 c^2} = \frac{1}{4}\left(\frac{v}{c}\right)^2 \simeq (Z\alpha)^2. \tag{89}$$

So H_1' is small for low Z. For hydrogen, $H_0 \simeq 10eV$, and hence $H_1' \simeq 10^{-3}eV$.

The electron moves with velocity $v = \vec{p}/m_e$ in the electric field $\vec{E} = -\vec{\nabla}\bar{\phi} = -\dfrac{d\bar{\phi}}{dr}\hat{r}$, where $\bar{\phi}(r)$ is the electrostatic potential of the nucleus. In the rest frame of the electron, there is a magnetic field

$$\vec{B} = -\frac{1}{c} v \times \vec{E}$$

$$= \frac{1}{c} v \times \hat{r} \frac{d\bar{\phi}}{dr}$$

$$= -\frac{1}{m_e c} \vec{L} \frac{1}{r} \frac{d\bar{\phi}(r)}{dr}, \tag{90}$$

where $\vec{L}$ is the orbital angular momentum of the electron. The intrinsic magnetic moment of the electron

$$\vec{\mu} = \frac{e}{m_e c}\,\vec{S} \tag{91}$$

when $\vec{S}$ is the spin of the electron, interacts with this field and gives the energy

$$U' \equiv -\,\vec{\mu}\,.\,\vec{B} = \frac{1}{m_e^2 c^2}\,\frac{1}{r}\,\frac{dV(r)}{dr}\,(\vec{L}\,.\,\vec{S}). \tag{92}$$

where $V(r) = e\bar{\phi}(r)$ is the Coulomb energy. This term represents the interaction of the spin of the electron with its own orbital angular momentum. Since the rest frame of the electron is not an inertial frame (it is an accelerated frame), there is an extra factor of $\dfrac{1}{2}$ in Eq. (92) arising from the "Thomas precession". (see Jackson (1975), Møller (1952)).

$\vec{E}$ and $\vec{B}$ are fields in the laboratory frame in which the electron has velocity $\vec{\upsilon}$. The correct expression for the magnetic interaction energy is

$$U = U' - \vec{S}\,.\,\vec{\omega}_T \tag{93}$$

where the internal state of the electron appears in the laboratory inertial frame (moving with velocity $\vec{\upsilon}$ with respect to the electron) to be precessing with angular velocity ω_T. We suppose that the particle accelerates to velocity $\vec{\upsilon} + \delta\vec{\upsilon}$, where $\delta\vec{\upsilon} = \vec{\upsilon}\delta t$. So we consider the product of two boosts of velocities $-\vec{\upsilon}$ and $\vec{\upsilon} + \delta\vec{\upsilon}$. This is a Lorentz transformation which is factorised into a boost of velocity

$$\Delta\vec{\upsilon} = \frac{1}{\sqrt{1-\beta^2}}\left[\delta\vec{\upsilon} + \left(\frac{\vec{\upsilon}}{\sqrt{1-\beta^2}} - \vec{\upsilon}\right)\frac{\vec{\upsilon}.\delta\vec{\upsilon}}{\vec{\upsilon}^2}\right],\ \beta = \upsilon/c \tag{94}$$

and a rotation

$$\Delta\vec{\omega} = \omega_T\delta t = \left(\frac{1}{\sqrt{1-\beta^2}} - 1\right)\frac{\delta\vec{\upsilon}\times\vec{\upsilon}}{\vec{\upsilon}^2} \simeq \frac{\delta\vec{\upsilon}\times\vec{\upsilon}}{2c^2} \tag{95}$$

in the small velocity approximation ($\vec{\upsilon} \ll c$).

So the internal state of the particle appears in the laboratory frame to be precessing with angular velocity

$$\omega_T = \frac{\Delta\vec{\omega}}{\delta t} = \frac{\vec{\dot{\upsilon}}\times\vec{\upsilon}}{2c^2} \tag{96}$$

This is the Thomas precession.

From Eqs. (92), (93) and (96), we obtain the correct expression for the spin-orbit coupling energy

$$U = \frac{1}{m_e^2 c^2} \frac{1}{r} \frac{dV(r)}{dr} (\vec{L} \cdot \vec{S}) - \vec{S} \cdot \frac{\vec{v} \times \dot{\vec{v}}}{2c^2}$$

$$= \left(\frac{1}{2 m_e^2 c^2} \right) \frac{1}{r} \frac{dV(r)}{dr} (\vec{L} \cdot \vec{S}). \tag{97}$$

Note that even a neutral spin $\frac{1}{2}$ particle moving in a potential V will experience a spin-orbit interaction $\dfrac{1}{2 m^2 c^2} \vec{S} \cdot (\vec{F} \times \vec{p})$, where $\vec{F} = -\vec{\nabla} V$.

For a particle with gyromagnatic ratio g and mass m, the expression is

$$\left(\frac{g-1}{2 m^2 c^2} \right) \frac{1}{r} \frac{dV(r)}{dr} (\vec{L} \cdot \vec{S}).$$

From the Dirac theory, $g = 2$ for an electron and one obtains the correct value Eq. (97).

Hence the correction term to the Hamiltonian due to the spin-orbit interaction is

$$H_2' = \frac{1}{2 (\mu c)^2} \frac{1}{r} \frac{dV(r)}{dr} (\vec{L} \cdot \vec{S}), \tag{98}$$

where $V(r) = -Ze^2/r$ and m_e has been replaced by μ.

Since $\vec{L}$ and $\vec{S}$ are of the order of $\hbar$, we estimate

$$\frac{H_2'}{H_0} \approx \frac{\dfrac{Zc^2}{2 m_e^2 c^2} \dfrac{\hbar^2}{a_0^3}}{Ze^2 * a_0} = (Z\alpha)^2 \tag{99}$$

The predicted splitting is about one part in 10^4. Let us estimate the magnitude of the magnetic field acting on the spin magnetic moment of an electron. Since $|U| \sim |\vec{\mu} \cdot \vec{B}|$, we have

$$B \sim \frac{|H_2'|}{\mu_B} \sim \frac{10^{-6} \, \text{erg}}{10^{-20} \, \text{erg} - \text{gauss}^{-1}} = 10^4 \, \text{gauss}$$

Although in the Dirac equation the interaction of the electron with the Coulomb field of the nucleus is local, the nonrelativistic approximation yields an equation in which this interaction is non local. The electron located at $\vec{r}$ now interacts with the field in a region $\left(|\vec{\delta r}| = \dfrac{\hbar}{m_e c} \right)$ whose size is of the order of its Compton wavelength.

The electron feels the average potential

$$\langle V(\vec{r} + \delta\vec{r})\rangle \approx V(\vec{r}) + \langle \delta\vec{r}\cdot\vec{\nabla}V\rangle + \frac{1}{2}\langle(\delta\vec{r}\cdot\vec{\nabla})(\delta\vec{r}\cdot\vec{\nabla})V(\vec{r})\rangle$$

$$= V(r) + \frac{1}{6}\langle(\delta\vec{r})^2\rangle\,\vec{\nabla}^2 V(r), \text{ assuming spherical symmetry.}$$

The correction term is

$$\sim \frac{1}{6}\left(\frac{\hbar}{m_e c}\right)^2 \vec{\nabla}^2 V(r), \text{ where } \langle(\delta r)^2\rangle \sim \left(\frac{\hbar}{m_e c}\right)^2.$$

For a Coulomb potential $V(r) = -Ze^2/r$ and using $\vec{\nabla}^2\left(\frac{1}{r}\right) = -4\pi\delta(\vec{r})$, we obtain the correction to the potential term of the same form, order of magnitude, and sign as the Darwin term from the Dirac theory of the electron.

Therefore, the correction term (the Darwin term) to the Hamiltonian is

$$H_3' = \frac{\pi\hbar^2 Z e^2}{2\mu^2 c^2}\,\delta(\vec{r}) \tag{100}$$

Due to the δ-function, a contribution of order

$$\frac{\pi\hbar^2 Z e^2}{2\mu^2 c^2}\,|\psi(0)|^2$$

$\sim \mu c^2 (Z\alpha)^4$, non-vanishing for $l = 0$

Since $H_0 \simeq \mu\,(Z\alpha c)^2$, we see that

$$\frac{H_3'}{H_0} \simeq (Z\alpha)^2. \tag{101}$$

Thus, all the three correction terms are about 10^4 times smaller than H_0 (provided Z is not too large), the nonrelativistic Hamiltonian.

We write the Hamiltonian of hydrogenic atoms as

$$H = H_0 + H' \tag{102}$$

with

$$H_0 = \frac{\hbar^2}{2\mu}\vec{\nabla}^2 - \frac{Ze^2}{r} \tag{103}$$

and

$$H' = H_1' + H_2' + H_3' \tag{104}$$

(relativistic correction to the kinetic energy)

$$H_1' = \frac{1}{2\mu c^2}\left(-\frac{\hbar^2}{2\mu}\vec{\nabla}^2\right)$$

(spin-orbit interaction)

$$H_2' = \frac{Ze^2}{2(\mu c)^2} \frac{1}{r^3} (\vec{L} \cdot \vec{S}) \tag{105}$$

(the Darwin term)

$$H_3' = \frac{\pi Ze^2 \hbar^2}{2(\mu c)^2} \delta(\vec{r}).$$

The fine structure of the energy levels of hydrogenic atoms is due to the above terms H_1', H_2', and H_3' all of which are much smaller than the non-relativistic energy. Since these relativistic corrections are very small (if Z is not very large) we use perturbation theory to calculate the energy corrections due to the three terms.

The unperturbed energy level $E_n^{(0)}$ is $2n^2$-fold degenerate. For H_0, we can use as an eigenstate basis either the set $\{|nlm_l m_s\rangle\}$ or $\{|nljm\rangle\}$. The two base sets are related by

$$|nljm\rangle = \sum_\sigma C(1/2\, j;\, m - \sigma,\, \sigma,\, m)|nlm - \sigma\sigma\rangle. \tag{106}$$

H_1' does not act on the spin and commutes with all the orbital angular momentum operators. So H_1' is diagonal in l, m_l and m_s. The energy correction $\Delta E_n^{(1)}$ due to H_1' is given in first-order perturbation theory by

$$\Delta E^{1(1)} = -\frac{1}{2\mu c^2} \left\langle nl \left| \left(H_0 + \frac{Ze^2}{r} \right) \right| nl \right\rangle$$

$$= -\frac{1}{2\mu c^2} \left[E_n^{(0)^2} + 2 E_n Z e^2 \left\langle \frac{1}{r} \right\rangle_{nl} + (Ze^2)^2 \left\langle \frac{1}{r^2} \right\rangle_{nl} \right]. \tag{107}$$

From Eq. (11.253)., we have

$$\left\langle \frac{1}{r} \right\rangle_{nl} = \frac{1}{a_\mu n^2}$$

$$\left\langle \frac{1}{r^2} \right\rangle_{nl} = \frac{1}{a_\mu^2 n^3 (l + 1/2)} \tag{108}$$

where $a_\mu = \hbar/\mu Z\alpha c$ is the modified Bohr radius and using these values we obtain

$$\Delta E_{nl}^{1(1)} = -E_n^{(0)} \frac{(Z\alpha)^2}{n^2} \left[\frac{3}{4} - \frac{n}{l + 1/2} \right] \tag{109}$$

$$a_0 = \frac{\hbar^2}{me^2} = \frac{\hbar}{m_e \alpha c} \text{ is the Bohr radius.}$$

Next, we consider the spin-orbit term $H'_2 = \xi(r)\, \vec{L}.\vec{S}.\vec{L}^2$ commutes with H'_2, but H'_2 does not commute with L_z, S_z. So we choose the representation in which $\vec{L}.\vec{S}$ is diagonal. The set $\lfloor \vec{L}^2, \vec{J}^2, J_z \rfloor$ of energy eigenkets are used as the base set instead of the set $\lfloor \vec{L}^2, L_z, S_z \rfloor$. Since $\vec{J} = \vec{L} + \vec{S}$, we have $\vec{L}.\vec{S} = \dfrac{1}{2}(\vec{J}^2 - \vec{L}^2 - \vec{S}^2)$ and so the diagonal elements are

$$\langle \vec{L}.\vec{S} \rangle_{lj} = \frac{1}{2}\left[(j(j+1) - l(l+1)) - \frac{3}{4} \right]$$

$$= \begin{cases} \dfrac{1}{2}\, l, & j = l + \dfrac{1}{2} \\[2mm] -\dfrac{1}{2}(l+1), & j = l - \dfrac{1}{2} \end{cases} \tag{110}$$

For $\qquad l = 0, \langle \vec{L}.\vec{S} \rangle = 0.$

Since the only J dependence of the spin-orbit energy is through the term $[J(J+1) - L(L+1) - S(S+1)]$, we evaluate the difference

$$\Delta E_{s.o}(J) - \Delta E_{s.o}(J-1)$$

$$= A\,(L.S)\,J,$$

where A depends only on L, S and the configuration.

This is Landé's interval rule which states that the separation of two J states belonging to the same L, S is proportional to the larger value of J. The splitting constant $A(LS)$ is different for different terms. If $A > 0$, the multiplet component with the smallest possible value $J = |L - S|$ has the lowest energy value. Such multiplets are called normal. If $A < 0$, the possible value $J = L + S$ has the lowest energy value. These multiplets are called inverted.

Therefore, the correction to the energy due to H'_2 is

$$\Delta E^{(2)}_{nlj} = \frac{Ze^2\hbar^2}{4\,(\mu c)^2}\left[j(j+1) - l(l+1) - \frac{3}{4} \right]\left\langle \frac{1}{r^3} \right\rangle_{nl}. \tag{111}$$

Using $\qquad \left\langle \dfrac{1}{r^3} \right\rangle_{nl} = \dfrac{1}{l(l+1/2)(l+1)n^3 a_\mu^3}$ from Eq. (11.253)

we obtain

$$\Delta E^{(2)}_{nlj} = -E_n^{(0)}\,\frac{(Z\alpha)^2}{n(2l+1)} \times \begin{cases} \dfrac{1}{l+1}, & j = l + \dfrac{1}{2} \\[2mm] -\dfrac{1}{l}, & j = l - \dfrac{1}{2} \end{cases} \quad l > 0. \tag{112}$$

Finally, the Darwin term gives a contribution

$$\Delta E_{nl}^{\prime(3)} = \frac{\pi Z e^2 \hbar^2}{2\mu^2 c^2} \langle nl | \delta(\vec{r}) | nl \rangle$$

$$= \frac{\pi Z e^2 \hbar^2}{2\mu^2 c^2} |\psi_{n00}^{(0)}|^2$$

$$= -E_n^{(0)} \frac{(Z\alpha)^2}{n}, \quad l = 0. \tag{113}$$

Combining the three relativistic corrections, the total energy shift to first-order of perturbation is for all l

$$\Delta E_{nj}^{\prime} = E_n^{(0)} \frac{(Z\alpha)^2}{n^2} \left(\frac{n}{j + 1/2} - \frac{3}{4} \right). \tag{114}$$

The energy levels of one-electron atoms are given by

$$E_{nj} = E_n^{(0)} \left[1 + \frac{(Z\alpha)^2}{n^2} \left(\frac{n}{j + \dfrac{1}{2}} - \frac{3}{4} \right) \right]. \tag{115}$$

This agrees to order $(Z\alpha)^2$ with the result obtained from the Dirac equation. In the nonrelativistic theory for each value of n, the principal quantum number, there are $2n^2$ linearly independent eigenstates, all having the same energy, the factor two arising from the spin. The relativistic corrections give rise to the fine structure splitting of the energy levels. In the Dirac theory, a nonrelativistic energy level E_n depending only on n is split into n components (different levels), one for each value of $j = \dfrac{1}{2}, \dfrac{3}{2}, \ldots, n - \dfrac{1}{2}$. This splitting is called the fine structure splitting and the set of n components form a fine structure multiplet. In the Dirac theory, the pair of levels with $l = j \pm 1/2$ have exactly the same energy. Let us consider the levels $n = 2$ for hydrogen ($Z = 1$). The energy difference of the level with $k = j + \dfrac{1}{2} = 1$ (both the $2S_{1/2}$ and $2P_{1/2}$ state) and with $k = 2(2P_{3/2})$ is $\alpha^2 E_n^{(0)}/2n = 1.33 \times 10^{-5} \, E_n^{(0)}$

$$= 0.365 \, \text{cm}^{-1}$$

$$= 1.10 \times 10^4 \, Mc/s. \tag{116}$$

For any Z and n the energy separation between the two extreme components of the fine structure multiplet (total fine structure width), i.e., between $k = 1$ and $k = n$ (between $j = n - \dfrac{1}{2}$ and $j = \dfrac{1}{2}$) is

$$\Delta E_n = E_n^{(0)} \frac{(Za)^2 (n - 1)}{n^2} \tag{117}$$

Let us consider a sodium atom. The electronic configuration for the ground state is $(1s)^2 (2s)^2 (2p)^6 (3s)$.

The eleventh electron can go to a $3p$ state which is split due to fine structure into $3p_{1/2}$ and $3p_{3/2}$. The transitions give rise to the doublet–NaD lines with $\lambda = 5896A$ and 5890 A.

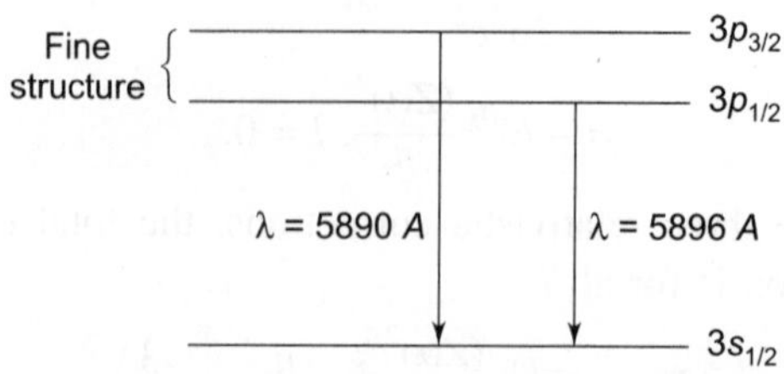

Fig. 12.2 Schematic diagram of the doublet sodium D lines

Next we discuss hyperfine splitting of atomic energy levels due to the interaction of electrons with multipole moments of the nucleus. The most important of these are the lowest-order moments e.g., the magnetic dipole and the electric quadrupole moments. We consider a hydrogenic atom with a nucleus of charge Ze ($Z\alpha \ll 1$) and a magnetic moment $\vec{M}$. In this case, the atom may be considered as nonrelativistic to lowest order, the ordinary fine structure as a perturbation and the hyperfine structure as an even smaller perturbation. The Hamiltonian may then be written as

$$H = H_0 + H'_M \tag{118}$$

where H_0 now includes the Coulomb interaction $-\dfrac{Ze^2}{r}$ and the fine structure (relativistic) corrections and H'_M is the perturbation due to the magnetic moment. The orbital angular momentum $\vec{L}$ and the spin $\vec{S}$ of the electron will interact with the magnetic moment of the nucleus. So we write

$$H'_M = H'_L + H'_s \tag{119}$$

where H'_L and H'_s are the interaction with the orbital and the spin angular momenta respectively. Now

$$H'_L = \frac{-i\hbar e}{mc}\,\vec{A}.\vec{\nabla} \tag{120}$$

where $\vec{A}(\vec{r})$ is the vector potential due to a dipole located at the origin

$$\vec{A}(\vec{r}) = -\,\vec{M} \times \vec{\nabla}\left(\frac{1}{r}\right) = (\vec{M} \times \vec{r})\,\frac{1}{r^3} \tag{121}$$

so that

$$H'_L = +\,\frac{e}{mc}\,\frac{1}{r^3}\,\vec{M}.\vec{L} \tag{122}$$

with
$$\vec{L} = \vec{r} \times \vec{p}$$

This term is the interaction of the nuclear magnetic moment with the magnetic field at the nucleus due to the rotation of the electronic charge.

Next, we calculate H_S'. The magnetic field $\vec{B}$ due to $\vec{A}$ is

$$\vec{B} = \vec{\nabla} \times \vec{A}$$

$$= - \left[\vec{M} \vec{\nabla}^2 \left(\frac{1}{r} \right) - \vec{\nabla} (\vec{M}. \vec{\nabla}) \frac{1}{r} \right]. \tag{123}$$

The interaction energy of the spin magnetic moment of the electron with the magnetic field $\vec{B}$ of the nucleus is

$$H_S' = - \vec{M}_e. \vec{B} = \frac{e}{mc} \vec{S}. \vec{B}. \tag{124}$$

The magnetic moment $\vec{M}$ of a nucleus is related to the nuclear spin $\vec{I}$ by the relation

$$\vec{M} = g \mu_N \vec{I} \tag{125}$$

where

$$\mu_N = \frac{|e| \hbar}{2 M_p^c} = \mu_B \frac{m}{M_p} \tag{126}$$

is the nuclear magneton. Here m is the mass of the electron, M_p the mass of the proton and $\mu_B = \dfrac{e\hbar}{2 m_e c}$ is the Bohr magneton. g is a dimensionless constant called the gyromagnetic ratio or g-factor and may be positive or negative (lies between $- 4.2$ to $+ 5.5$ for most nuclei). For the proton, $g_p \simeq 5.58$ and for the neutron $g_n \simeq -3.82$. Clearly, μ_N is about 1836 times smaller than μ_B.

From Eqs. (123), (124) and (125),

$$H_S' = \frac{e}{mc} g \mu_N \vec{S}. \left[- \vec{I} \vec{\nabla}^2 \left(\frac{1}{r} \right) + \vec{\nabla} (\vec{I}. \vec{\nabla}) \frac{1}{r} \right]. \tag{127}$$

Since $\vec{\nabla}^2 \left(\dfrac{1}{r} \right) = - 4\pi \delta (\vec{r})$, the first term on RHS in Eq. (127) vanishes for $r \neq 0$. Also, for $r \neq 0$,

$$\vec{S}. \vec{\nabla} (\vec{I}. \vec{\nabla}) \frac{1}{r} = - \frac{1}{r^3} \left[\vec{S}. \vec{I} - 3 \frac{(\vec{S}. \vec{r})(\vec{I}. \vec{r})}{r^2} \right]. \tag{128}$$

Hence,

$$H_S' = - \frac{e}{mc} g \mu_N \frac{1}{r^3} \left[\vec{S}. \vec{I} - 3 \frac{(\vec{S}. \vec{I})(\vec{I}. \vec{r})}{r^2} \right], \quad r \neq 0. \tag{129}$$

This is the dipote-dipole interaction term between the electronic and nuclear magnetic moments. Combining Eqs. (122), (125) and (129), we get from Eq. (119),

$$H'_M = + \frac{e}{mc}\, g\mu_N\, \frac{1}{r^3} \left[\vec{L}.\vec{I} - \vec{S}.\vec{I} + 3\,\frac{(\vec{S}.\vec{r})\,(\vec{I}.\vec{r})}{r^2} \right],\ r \neq 0. \tag{130}$$

Now, we discuss the case $r = 0$ for H'_S. This is relevant for s-states ($l = 0$). For $l = 0$ (spherically symmetric states),

$$(\vec{S}.\vec{I})\,(\vec{I}.\vec{\nabla})\,\frac{1}{r} = \frac{1}{3}\,(\vec{S}.\vec{I})\,\vec{\nabla}^2\left(\frac{1}{r}\right) = -\frac{4\pi}{3}\,\vec{S}.\vec{I}\,\delta(\vec{r})$$

so that Eq. (127) gives

$$H'_S = H_M\ (\text{since } H'_L = 0 \text{ for } l = 0 \text{ states})$$

$$= \frac{e}{mc}\, g\mu_N\, \frac{8\pi}{3}\, \delta(\vec{r})\, \vec{S}.\vec{I} \tag{131}$$

The term is known as the "Fermi contact term".

The order of magnitude of the two terms (the first term H'_L and H'_S term in Eq. (130)) is $\sim \dfrac{e^2\hbar}{m_e M_p c^2}\,\dfrac{1}{a_0^3}$ (a_0 is the Bohr radius).

These terms are about 1836 times smaller than the spin-orbit coupling term. The contact term is about 1836 times smaller than the Darwin term.

Next, we calculate the energy shifts due to these terms Eq. (130) and (131) in first order perturbation theory. The energy shifts are small compared to those due to the fine-structure Hamiltonian. Hence these terms constitute the hyperfine structure Hamiltonian.

To calculate the energy shift, we as in our discussion of the spin-orbit coupling diagonalize $\vec{F}$, the total angular momentum of the atom.

$$\vec{F} = \vec{I} + \vec{J}. \tag{132}$$

F_z and $\vec{F}^2$ with eigenvalues $M_F = -F, -F + 1, \ldots, + F$ and $F(F + 1)$ are constants of the motion.

The energy shift is (for the case $l \neq 0$)

$$\Delta E = \frac{e}{mc}\, g\mu_N \left\langle lsjFM_F \left| \frac{1}{r^3}\, \vec{G}.\vec{I} \right| lsjIFM_F \right\rangle,\quad l \neq 0 \tag{133}$$

where

$$\vec{G} = \vec{L} - \vec{S} + 3\,\frac{(\vec{S}.\vec{r})\,\vec{r}}{r^2}$$

$$= \mu_B\mu_N g\, \frac{l(l + 1)}{j(j + 1)}\, [F\,(F + 1) - I\,(I + 1) - j\,(j + 1)] \times \left\langle \frac{1}{r^3} \right\rangle$$

$$= \mu_B \mu_N g \, \frac{l(l+1)}{j(j+1)} \, [F\,(F+1) - I\,(I+1) - j\,(j+1)]$$

$$\times \frac{Z^3}{a_0^3 \left(\dfrac{m_e}{M}\right)^3} \left[n^3 l \left(l + \frac{1}{2} \right) (l+1) \right]^{-1}, \quad (l \neq 0) \tag{134}$$

where M is the reduced mass.

The expression Eq. (134) vanishes for an s-state $l = 0$. For an *s*-state, the term Eq. (131) gives a non-vanishing contribution. The first order energy shift due to this perturbation is

$$\Delta E_0 = \frac{e}{mc} \, g\mu_N \, \frac{8\pi}{3} \, \langle \delta(\vec{r}) \, \vec{S} \cdot \vec{I} \rangle \quad (l = 0)$$

$$= g\mu_B \, \mu_N \, \frac{8\pi}{3} \, [F\,(F+1) - I\,(I+1) - s\,(s+1)] \, \langle \delta(\vec{r}) \rangle,$$

since $\vec{L} = 0$, $\vec{F} = \vec{I} + \vec{S}$ and so $\vec{S} \cdot \vec{I} = \frac{1}{2} (\vec{F}^2 - \vec{I}^2 - \vec{S}^2)$

But $\langle \delta(\vec{r}) \rangle = |\psi_{n00}(0)|^2 = \dfrac{Z^3}{\pi a_0^3 \left(\dfrac{m_e}{M}\right)^3 n^3}$

where $\psi_{n00}(0)$ is the normalized radial wave function at the origin, and hence

$$\Delta E_0 = \frac{8}{3} \, g\mu_B\mu_N \, \frac{Z^3}{a_0^3 \left(\dfrac{m}{M}\right)^3 n^3} \times [F\,(F+1) - I\,(I+1) - s\,(s+1)] \tag{135}$$

For both $l = 0$ and $l \neq 0$ cases, we have (for *s*-state, $j = s$)

$$\Delta E = A \, [F\,(F+1) - I\,(I+1) - j\,(j+1)] \tag{136}$$

where

$$A = g\mu_B\mu_N \, \frac{1}{j\,(j+1)\,(l+1/2)} \cdot \frac{Z^3}{a_0^3 \left(\dfrac{m}{M}\right)^3 n^3} \tag{137}$$

A fine structure level with fixed values of l and j is thus split further into hyperfine components with the possible values of F being $(j + I)$, $(j + I - 1)$, ..., $|j - I|$. The multiplicity of such a level is the smaller of the two numbers $(2j + 1)$ and $(2I + 1)$. These components form a hyperfine structure multiplet. Since A does not depend on F, the energy difference between two neighbouring hyperfine levels is proportional to F:

$$\Delta E(F) - \Delta E\,(F - 1) = 2AF. \tag{138}$$

This is an example of an interval rule.

For the ground state ($1S_{1/2}$) of hydrogen (with a proton of spin $\frac{1}{2}$ as nucleus) is split into two hyperfine components ($F = 0$ and $F = 1$), the state $F = 0$ being the ground state.

The hyperfine splitting is

$$\Delta E_{h.f} = \frac{16}{3}\,\alpha^2\left(\frac{g_P\mu_N}{2\mu_B}\right) Ry \tag{139}$$

where $g_P = 5.5883$ is the g factor for a proton. The frequency $v = \Delta E/h$ of the (magnetic dipole) transition between these two hyperfine levels is $v = 1420.4$ MHz (corresponding to $\lambda = 21$ cm.). Due to the hyperfine interaction, each level of the hydrogen is split into doublets. There are various corrections to the above formula (139). A large part comes from the electron's anomalous magnetic moment g_S, which is slightly different from the value 2 predicted by the Dirac theory. The theoretical expression for the frequency v (for the hyperfine splitting of the hydrogen ground state) is

$$\frac{v}{cR_\infty} = \frac{16}{3}\,\alpha^2\left(\frac{M_p}{M_p + m}\right)^3\left(\frac{g_p\mu_N}{g_s\mu_B}\right)\frac{g_s}{2}$$

$$\times\left[\frac{g_s}{2} + \alpha^2\,(4 - \log 2) - 0.2 \times 10^{-5} + \delta\right], \tag{140}$$

where (Bethe and Salpeter (1957))

$$\frac{g_s}{2} = 1 + \frac{\alpha}{2\pi} - \frac{2.973}{\pi^2}\,\alpha^2$$

$$R_{0s} = \text{Rydberg for infinite mass in wave numbers}$$

$$\delta = \text{nuclear structure correction}$$

$$\alpha = \text{fine structure constant.}$$

The experimental value of v is

$$v = (142040575.1800 \pm 0.028)\ Hz \tag{141}$$

The 21 cm radiation is very important in radio astronomy. From its intensity, Doppler shift and line broadening, a great deal of information has been obtained about the distribution and motion of interstellar and intergalactic hydrogen atoms.

The first-order energy shift due to the electric quadrupole interaction (Bransden and Joachain (1983)) is

$$\Delta E = \frac{B}{4}\,\frac{\dfrac{3}{2}\,K(K + 1) - 2I(I + 1)\,j(j + 1)}{I(2I - 1)\,j(2j - 1)} \tag{142}$$

where

$$B = Q \left(\frac{\partial^2 V}{\partial z^2} \right) \quad \text{(quadrupole coupling constant)}$$

$$K = F (F + 1) - I (I + 1) - j (j + 1).$$

Here Q is the magnitude of the electric quadrupole moment and V is the electrostatic potential created by an electron at the nucleus. There is no quadrupole energy shift for s-states, since $\left(\frac{\partial^2 V}{\partial z^2} \right)$ vanishes for a spherically symmetric charge distribution.

The total hyperfine structure energy correction due to the magnetic dipole and electric quadrupole interactions is

$$\Delta E = AK + \frac{B}{4} \frac{\frac{3}{2} K(K + 1) - 2I(I + 1) j(j + 1)}{I(2I - 1) j(2j - 1)} \tag{143}$$

According to the Dirac theory, energy levels of one-electron atoms with the same n and j but different values of l are degenerate. The two $n = 2, j = \frac{1}{2}$ levels of opposite parity $2S_{1/2}$ and $2P_{1/2}$ of hydrogen atom are degenerate. In 1947, Lamb and Retherford showed that $2S_{1/2}$ level lies above the $2P_{1/2}$ level by about 1000 MH$_Z$. This "Lamb shift" breaking the degeneracy of energy levels with the same value of n and j but different values of l arises from the interaction of electrons with the fluctuations of the quantized electromagnetic field. A complete treatment of radiative corrections including Lamb shift requires the methods of quantum field theory (see Itzykson & Zuber (1985)). Following Welton (1948) we give below an intuitive description of the Lamb shift by considering the interaction of a bound electron (treated nonrelativistically) with the vacuum fluctuations of the electromagnetic field. Each mode upon quantization has a zero-point energy of $\frac{1}{2} \hbar\omega$ and the vacuum energy of the field is the sum of zero-point energies. Although the average field strengths are zero their mean-square values are nonvanishing and the interaction with the field leads to mean square fluctuation in the electron coordinate. So we estimate the mean square displacement of a bound electron in a hydrogen atom. This contributes an additional interaction energy due to the smearing out of the Coulomb potential as seen by the electron. This discussion is similar to our discussion of the Darwin term where the fluctuation in the position is due to zitterbewegung.

We treat the electron motion classically and nonrelativistically. Let $\vec{\delta r}$ be the displacement of the electron from its equilibrium position. Then its equation of motion is

$$m\delta\ddot{\vec{r}} = e\vec{E} \tag{144}$$

where $\vec{E}(t)$ is the fluctuating electric field. Here we have neglected the term on the RHS with the magnetic field, proportional to $\dfrac{|\delta\ddot{r}|}{c} \ll 1$. The nucleus, being heavier, is assumed to the at rest.

We consider the x-component and make a Fourier analysis. We write

$$\delta x(t) = \frac{1}{2\pi}\int_{-\infty}^{\infty} d\omega\, e^{-i\omega t}\, \delta x(\omega) \tag{145}$$

and since δx is real,

$$\delta x(\omega) = \delta x^*(-\omega). \tag{146}$$

Also,

$$E_x(t) = \frac{1}{2\pi}\int_{-\infty}^{\infty} i\omega e^{-i\omega t}\, E_x(\omega) \tag{147}$$

with

$$E_x(\omega) = E_x^*(-\omega). \tag{148}$$

From the equation of motion Eq. (144), we get

$$\delta x(\omega) = -\frac{e}{m\omega^2}\, E_x(\omega). \tag{149}$$

Next, we compute the average value of $(\delta x)^2$ over a long time interval T.

$$\langle(\delta x)^2\rangle = \frac{1}{T}\int_{T/2}^{-1/2} dt\, \delta x(t)\, \delta x^*(t)$$

$$= \frac{1}{T}\int_{T/2}^{-1/2} dt \int_{-\infty}^{\infty} \frac{d\omega}{2\pi}\, e^{-i\omega t}\, \delta x(\omega)$$

$$\times \int_{-\infty}^{\infty} \frac{d\omega'}{2\pi}\, e^{-i\omega' t}\, \delta x^*(\omega')$$

$$\simeq \frac{1}{T}\int_{-\infty}^{\infty} \frac{d\omega}{2\pi} \int_{-\alpha}^{\alpha} \frac{d\omega'}{2\pi}\, 2\pi\delta\,(\omega' - \omega)\, \delta x(\omega)\, \delta x^*(\omega')$$

$$= \frac{1}{T}\int_{-\infty}^{\infty} \frac{d\omega}{2\pi}\, \delta x(\omega)\, \delta x^*(\omega).$$

Using Eq. (149), we obtain

$$\langle(\delta x)^2\rangle = \frac{1}{T}\frac{e^2}{m^2}\int_{-\infty}^{\infty} \frac{d\omega}{2\pi}\,\frac{1}{\omega^4}\, E_x(\omega)\, E_x^*(\omega) \tag{150}$$

Now,

$$\langle E_x^2 \rangle = \frac{1}{3} \langle \vec{E}^2 \rangle = \frac{1}{T} \int_{-\infty}^{\infty} \frac{d\omega}{2\pi} E_x(\omega)\, E_x^*(\omega)$$

$$= \frac{2}{T} \int_0^{\infty} \frac{d\omega}{2\pi} E_x(\omega)\, E_x^*(\omega), \tag{151}$$

where $\langle \vec{E}^2 \rangle$ is the mean square field strength. The total vacuum energy of the electromagnetic field is the sum of zero-point energies

$$\frac{1}{2} \int d^3x (\vec{E}^2 + \vec{B}^2) = \sum_{\lambda=1}^{2} \sum_k \frac{1}{2} \hbar\omega \tag{152}$$

over all wave numbers k and transverse polarizations λ. In a large box of volume L, $\vec{k} = \frac{2\pi}{L} \vec{n}$ where n_x, n_y, n_z are integers, and

$$\frac{1}{2} \int d^3x (\vec{E}^2 + \vec{B}^2) = 2L^3 \int \frac{d^3k}{(2\pi)^3} \frac{\hbar\omega_k}{2}. \tag{153}$$

Since $\int d^3x \vec{E}^2 = \int d^3x \vec{B}^2$ and $\omega = |\vec{k}| c$ for free electromagnetic waves, the mean square field strength in free space is

$$\langle \vec{E}^2 \rangle = \int d\omega \langle \vec{E}_\omega^2 \rangle$$

$$= \frac{1}{L^3} \int d^3x \vec{E}^2$$

$$= \int \frac{d^3k}{(2\pi)^3} \hbar\omega_k$$

$$= \frac{4\pi\hbar}{2\pi^2 c^3} \int \omega^3 d\omega. \tag{154}$$

From Eqs. (151) and (154)

$$\frac{2}{T} \int_0^{\infty} \frac{d\omega}{2\pi} E_x(\omega)\, E_x^*(\omega) = \frac{\hbar}{6\pi^2 c^3} \int_0^{\infty} \omega^3\, d\omega$$

or,

$$\frac{1}{T} E_x(\omega)\, E_x^*(\omega) = \frac{\hbar}{6\pi c^3} \omega^3. \tag{155}$$

So we get

$$\langle (\delta\vec{r})^2 \rangle = 3 \langle (\delta x)^2 \rangle = \frac{2e^2\hbar}{\pi\, m^2 c^3} \int_0^{\infty} \frac{d\omega}{\omega}$$

$$= \frac{2}{\pi} \left(\frac{e^2}{\hbar c} \right) \left(\frac{\hbar}{mc} \right)^2 \int_0^{\infty} \frac{d\omega}{\omega}$$

$$= \frac{2\alpha}{\pi} \left(\frac{\hbar}{mc} \right)^2 \int_0^\infty \frac{d\omega}{\omega}. \tag{156}$$

The frequency integral diverges logarithmically at both limits. Since the electron is non-relativistically treated, the momentum acquired by it must not exceed mc, i.e. $\hbar k < mc$. Thus, there is a high frequency cut-off at distances $\sim$ the electron Compton wavelength $\hbar/mc$. This leads to an effective cut-off at

$$\omega_{max} \sim \frac{mc^2}{\hbar}. \tag{157}$$

The infrared cut-off is obtained from the condition that the frequency should not be less than the frequency corresponding to the binding energy of the electron. Thus,

$$\omega_{min} \sim \frac{|E|}{\hbar} = \frac{Z^2 e^4 m}{2\hbar^3 n^2} \tag{158}$$

We have

$$\langle (\delta \vec{r})^2 \rangle = \frac{2\alpha}{\pi} \left(\frac{\hbar}{mc} \right)^2 \ln \frac{\omega_{max}}{\omega_{min}}$$

$$= \frac{2\alpha}{\pi} \left(\frac{\hbar}{mc} \right)^2 \ln \frac{2n^2 \hbar^2 c^2}{Z^2 e^4}$$

$$= \frac{2\alpha}{\pi} \left(\frac{\hbar}{mc} \right)^2 \ln \left(\frac{2n^2}{(Z\alpha)^2} \right). \tag{159}$$

Since the electron coordinate fluctuates its interaction with the nucleus is affected. Due to the spreading of the electron, it sees a smeared out Coulomb potential. For the potential energy, we write

$$V(\vec{r}(t) + \delta \vec{r}(t)) = V(\vec{r}) + \delta \vec{r} \cdot \vec{\nabla} V(\vec{r}) + \frac{1}{2} (\delta \vec{r} \cdot \vec{\nabla})^2 V(\vec{r}) + \dots \tag{160}$$

where $\vec{r}(t)$ is the classical trajectory and

$$V(r) = - \frac{Ze^2}{r}$$

is the Coulomb potential. Since $\langle \delta \vec{r} \rangle = 0$, the second term on RHS of (160) vanishes, and the third term is

$$\frac{1}{2} \sum_{i,j} \delta r_j \, \delta r_j \frac{\partial^2 V}{\partial r_i \, \partial r_j}.$$

$$= \frac{1}{6} (\delta \vec{r})^2 \vec{\nabla}^2 V.$$

Then the additional interaction energy is

$$\delta V = \langle V(\vec{r} + \delta\vec{r}) - V(\vec{r}) \rangle$$

$$= \frac{1}{6} \langle (\delta\vec{r})^2 \rangle \, \vec{\nabla}^2 V.$$

$$= \frac{4\pi Z e^2}{6} \langle (\delta\vec{r})^2 \rangle \, \delta(\vec{r}), \tag{161}$$

where Poisson's law gives $\vec{\nabla}^2 V = - 4\pi Z e^2 \delta(\vec{r})$.

This term may be considered as a perturbation in the Hamiltonian. In first order this leads to an energy shift for the nth level.

$$\Delta E_n = \frac{4\pi Z e^2}{6} \langle (\delta\vec{r})^2 \rangle \int d^3r \, \psi_n^*(\vec{r}) \, \delta(\vec{r}) \, \psi_n(\vec{r}).$$

$$= \frac{2\pi Z e^2}{3} \langle (\delta r)^2 \rangle \, |\psi_{n0}(0)|^2 \tag{162}$$

where $\psi_{n0}(0)$ is the nonrelativistic wave function at the origin.

In this approximation, only s-waves are affected and the nth level is shifted by the amount

$$\Delta E_n(Lamb) = \frac{2\pi Z e^2}{3} \cdot \frac{Z^3}{\pi n^3 a^3} \cdot \frac{2\alpha}{\pi} \left(\frac{\hbar}{mc} \right)^2 \ln \frac{2n^2}{(Z\alpha)^2}$$

$$= \frac{4}{3\pi} \cdot \frac{Z^4 \alpha^5}{n^3} \, mc^2 \ln \left(\frac{2n^2}{(Z\alpha)^2} \right). \tag{163}$$

In 1947, Bethe made a non-relativistic calculation of Lamb shift, based on Kramers' idea of mass renormalization and obtained a result similar to Eq. (163). For the $2S_{1/2}$, state of hydrogen atom ($Z = 1$, $n = 2$, $l = 0$), the Lamb shift given by Eq. (163) is 1040 MHz raising it above the $2P_{1/2}$ level, in remarkable agreement with the experimental value. The beautiful, accurate and remarkable experiments of Lamb and coworkers stimulated the development of quantum field theory. A comparison of this term with the Darwin term shows a reduction factor $\sim \alpha \ln \alpha$. A quantum electrodynamic calculation gives

$$\Delta E_{n,l,j}(Lamb) = \frac{4\alpha}{3\pi} \frac{(Z\alpha)^4}{n^3} \, mc^2 \begin{cases} \ln \dfrac{m}{2\langle E_{n',0}\rangle} + \dfrac{19}{30} & \text{for } l = 0 \\[2em] \ln \dfrac{m(Z\alpha)^2}{2\langle E_{n,l}\rangle} + \dfrac{3}{8}\dfrac{C_{l,j}}{2l+1} & \text{for } l \neq 0 \end{cases} \tag{164}$$

where
$$C_{lj} = \begin{cases} \dfrac{1}{l+1}, & j = l + \dfrac{1}{2} \\ \dfrac{-1}{l}, & j = l - \dfrac{1}{2}, \end{cases} \quad l \geq 1$$

For atomic hydrogen, the $2S_{1/2} - 2P_{1/2}$ splitting is, from Eq. (164), 1052.1 MHz. Improved calculations give

    Erickson (1971) 1057.915 ± 0.010 MHz

    Mohr (1976) 1057.864 ± 0.014 MHz

Experimental results

Lamb et al (1953)

    1057.77 ± 0.10 MHz

Robiscoe & Shyn

    1057.90 ± 0.06 MHz

Lumdeen & Pipkin (1975)

    1057.893 ± 0.020 MHz

Andrews & Newton (1976)

    1057.862 ± 0.020 MHz

$n = 2$ levels of atomic hydrogen are shown in Fig. 12.3.

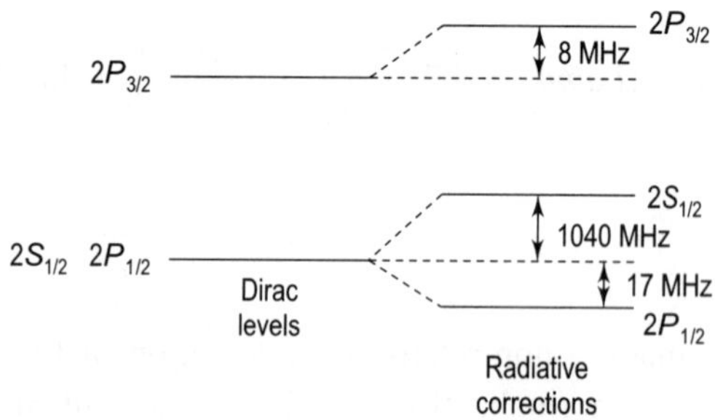

Fig. 12.3   Schematic diagram of n = 2 levels of atomic hydrogen

12.6   THE HYDROGEN ATOM IN A MAGNETIC FIELD

We now consider a hydrogen or hydrogenic (one electron) atom in a uniform (over atomic dimensions) magnetic field and study its effect on atomic states and energy levels.

The Hamiltonian of a hydrogenic atom in a uniform external magnetic field $\vec{B}$ is given by

$$H = H_0 + H' - \frac{e}{2mc}\,(\vec{p}\cdot\vec{A} + \vec{A}\cdot\vec{p}) + \frac{e^2\vec{A}^2}{2mc^2} \tag{165}$$

where $\vec{B} = \vec{\nabla} \times \vec{A}$ and H_0 and H' are given by Eq. (102). Equation (165) is obtained using the substitution

$$\vec{p} \to \vec{p} - \frac{e\vec{A}}{c}. \tag{166}$$

The uniform magnetic field $\vec{B}$ can be obtained from the vector potential

$$A = \frac{1}{2}(\vec{B} \times \vec{r}) \tag{167}$$

Note that this choice of $\vec{A}$ is not unique. If $\vec{B}$ points in the z-direction, then.

$$A_z = 0, \ A_x = -\frac{1}{2}By \text{ and } A_y = -\frac{1}{2}Bx, \text{ where } B = |\vec{B}|. \tag{168}$$

Since, $\lfloor \vec{p}, \vec{A} \rfloor$ and $i\hbar\vec{\nabla}\cdot\vec{A}$ and $\vec{\nabla}\cdot\vec{A} = 0$ for our choice of $\vec{A}$ in (168) we replace $\vec{p}\cdot\vec{A}$ by $\vec{A}\cdot\vec{p}$ in Eq. (165). Now,

$$\vec{A}\cdot\vec{p} = \frac{1}{2}(\vec{B} \times \vec{r})\cdot\vec{p} = \frac{1}{2}\vec{B}\cdot(\vec{r} \times \vec{p}) = \frac{1}{2}\vec{B}\cdot\vec{L},$$

where $\vec{L}$ is the orbital angular momentum operator. Therefore, Eq. (165) becomes

$$H = H_0 + H' - \frac{e}{2mc}\vec{B}\cdot\vec{L} + \frac{e^2}{8mc^2}(\vec{B} \times \vec{r})^2. \tag{169}$$

To this we add the spin magnetic moment interaction

$$-\vec{\mu}\cdot\vec{B} = \frac{-e}{mc}\vec{S}\cdot\vec{B}. \tag{170}$$

The quadratic form in $|\vec{B}|$ is important for very strong fields and large orbits. Also, the diamagnetic susceptibility can be obtained from the quadratic term in B.

So, keeping only the linear terms in B, the total Hamiltonian of a hydrogen like (one electron) atom in a constant uniform magnetic field is given by

$$H = H_0 + H' + H_M \tag{171}$$

where H_0, H' are given in Eq. (102), and

$$H_M = \frac{-e\,|\vec{B}|}{2mc}(L_z + 2S_z), \tag{172}$$

the magnetic field is being along the z-direction.

We now evaluate the change in the bound state spectrum of H_0 due to the perturbation $H' + H_M$ to first order in perturbation theory. H_M commutes with $\vec{L}^2$, L_z and S_z, but the fine structure energy does not. Therefore, one cannot diagonalize Eq. (171), and so we use degenerate perturbataion theory. But, we first consider two limiting cases $H_M \ll H'$ and $H_M \gg H'$, depending on the magnetic field strength.

If $B << 10^5\, G$ (10 T) then H_M is treated as a small perturbation and we use the eigenkets of $H_0 + H'$ as the base kets. The first-order energy shift is

$$\Delta E_M = \frac{-e\,|\vec{B}|}{2m_e c}\,\langle J_z + S_z\rangle_{nljm}$$

$$= \frac{-eB}{2m_e c}\,[m\hbar + \langle S_Z\rangle_{nljm}]. \tag{173}$$

The expectation value of S_z is obtained as follows.

$$\left\langle n, j = l \pm \tfrac{1}{2}, m, l \,|S_z|\, n, j = l \pm \tfrac{1}{2}, m, l\right\rangle$$

$$= \frac{\hbar}{2}\,(\alpha_\pm^2 - \beta_\pm^2)$$

$$= \frac{\hbar}{2}\cdot\frac{1}{(2l+1)}\left[\left(l \pm m + \tfrac{1}{2}\right) - \left(l \mp m + \tfrac{1}{2}\right)\right]$$

$$= \pm\frac{m\hbar}{(2l+1)}, \tag{174}$$

where $\left|j = l \pm \tfrac{1}{2}, m\right\rangle = \alpha_\pm\left|m_l = m - \tfrac{1}{2}, m_s = \tfrac{1}{2}\right\rangle + \beta_\pm\left|m_l = m + \tfrac{1}{2}, m_s = \tfrac{1}{2}\right\rangle$,

and

$$\alpha_\pm = \pm\sqrt{\frac{l \pm m + \tfrac{1}{2}}{2l+1}} = \pm\beta_\pm.$$

From Eqs. (173) and (174), we obtain the energy shift.

$$\Delta E_M = \mu_B\,Bm\left[1 \pm \frac{1}{2l+1}\right]$$

$$= \mu_B\,Bm\,\frac{2l+1\pm 1}{2l+1}\quad\left(j = l \pm \tfrac{1}{2}\right). \tag{175}$$

This splitting of levels in a magnetic field is known as the Zeeman effect. The magnitude of the splitting depends on l. In the anomalous Zeeman effect, the $j = l + \tfrac{1}{2}$ level splits into $2l + 2$ levels while the $j = l - \tfrac{1}{2}$ level splits into $2l$ levels. Electric dipole transitions ($\Delta m = 0, \pm 1$; $\Delta l = \pm 1$) between the $n = 2$ and $n = 1$ levels of hydrogen atom in a (weak) magnetic field are shown in Fig. 12.4.

The expectation value of S_z can also be obtained by using the projection theorem (see § 8.3).

We have

$$j\,(j+1)\,\hbar^2\,\langle lsjm\,|S_z|\,lsjm\rangle = \langle lsjm\,|(\vec{S}\cdot\vec{J})\,J_z|\,lsjm\rangle$$

$$= m\hbar\,\langle lsjm\,|\vec{S}\cdot\vec{J}|\,lsjm\rangle$$

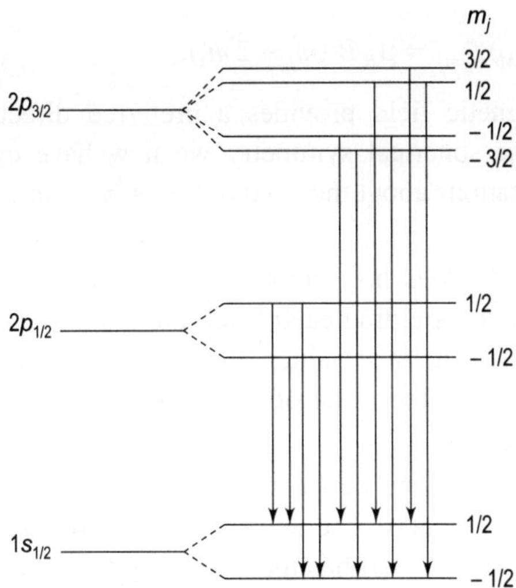

Fig. 12.4 Schematic diagram showing the electric dipole transitions between the $n = 2$ and $n = 1$ levels of atomic hydrogen in a weak magnetic field

Since $\vec{S} \cdot \vec{J} = (\vec{J}^2 + \vec{S}^2 - \vec{L}^2)/2$ the expectation value of S_z is

$$\langle S_z \rangle_{lsjm} = m\hbar \left[\frac{j(j+1) + s(s+1) - l(l+1)}{2j(j+1)} \right] \quad (176)$$

which agrees with Eq. (174) for $j = l \pm \dfrac{1}{2}$, $s = \dfrac{1}{2}$.

The energy shift due to the magnetic field is

$$\Delta E_m = g\mu_B Bm,$$

where
$$g = 1 + \frac{j(j+1) + s(s+1) - l(l+1)}{2j(j+1)} \quad (177)$$

is the Landé g factor.

The LSJ multiplet of an atom splits into $2J + 1$ magnetic sublevels ($M = 0$, ± 1, ± 2, ..., $\pm J$). The splitting is linear in B and symmetrical. When $S = 0$, $g = 1$; when $L = 0$, $g = 2$ and when $L = S$, $g = 3/2$. For some levels, e.g. $^4D_{1/2}$, $^5F_1$, $g = 0$. Thus, a magnetic field completely removes the degeneracy of levels with respect to M ($H' \ll H_M$).

We now consider the case where the magnetic field is strong enough so the Zeeman energy is large compared to the relativistic correction term. In this case, the main splitting of the LS multiplet is due to the maganetic field. The relativistic corrections can then be added by first-order perturbation theory. The base kets are the $\left| nls = \dfrac{1}{2} m_l m_s \right\rangle$ and the shift in energy is

$$\langle \Delta E_M \rangle_{m_l, m_s} = \mu_B B \, (m_l + 2 m_s). \tag{178}$$

The strong magnetic field provides a preferred direction (z-direction) in space. Instead of the spherical symmetry we now have cylindrical symmetry (invariance under rotations about the z-axis). So, only L_z and S_z are good quantum numbers.

The magnetic field does not remove the designeracy in l but removes the degeneracy in m_l and m_s, splitting each level into a number of components, each of which corresponds to definite values of $(m_l + 2 m_s)$. Some of the components are degenerate since the same value of $m_l + 2 m_s$ can be obtained from different combinations of m_l and m_s.

Electric dipole transitions between the levels n and n' is split into three components (Lorentz triplets). The selection rules are $\Delta m_s = 0$ and $\Delta m_l = 0$, ± 1. The π-component ($\Delta m_l = 0$) has the frequency $\nu_{n'n}$ while the σ-components ($\Delta m_l = \pm 1$) have the frequencies $\nu_{n'n}^{\sigma} = \nu_{n'n} \pm \nu_L$,

where $\nu_L = \dfrac{\mu_B B}{\hbar}$ is called the Larmor frequency. This splitting is called the normal Zeeman effect.

If the magnetic field strength is such the relativistic corrections are appreciable but small compared to the Zeeman energy, then these corrections can be treated in first order perturbation theory:

$$\langle H' \rangle_{nlm_L m_s'} = -\frac{mc^2 \, (Z\alpha)^4}{2n^4} \left(\frac{n}{l + \dfrac{1}{2}} - \frac{3}{4} \right) + \lambda_{nl} \times m_l m_s, \tag{179}$$

where

$$\lambda_{nl} = \frac{mc^2 \, (Z\alpha)^4}{2n^3 l \left(l + \dfrac{1}{2} \right)(l + 1)}, \quad l \neq 0.$$

The degeneracy in l is removed. The observed splitting is called the Paschen-Back effect.

After discussing the two limiting cases, we now consider the Zeeman effect in a magnetic field of arbitrary strength (the regime intermediate between the ordinary Zeeman and Paschen-Back effects). To determine the eigenvalues and eigenkets of H we use degenerate perturbation theory treating $H' + H_M$ as a perturbation.

We use the base kets $|njml\rangle$ in which H' is diagonal. H_M is cylindrically symmetric and has matrix elements between states of the same parity. $H' + H_M$ is diagonal in n, l, m. So the diagonal matrix elements are

$$\langle n, l_{\pm}, m, l | J_z + S_z | n, l_{\mp}, m, l \rangle = \hbar m \left(l \pm \frac{1}{2l_+} \right), \ l_{\pm} = l \pm \frac{1}{2} \tag{180}$$

and $\langle n,\, l_{\pm},\, m,\, l \| J_z + S_z | n,\, l_{\mp},\, m,\, l \rangle = -\dfrac{\hbar}{2l_+}\sqrt{l_+^2 - m^2}.$ (181)

The states $s_{1/2}$, $m = \pm\dfrac{1}{2}$ and $p_{3/2}$, $m = \pm\dfrac{3}{2}$ occur only in diagonal elements. There is no admixture due to H_M for these states and in general for $n,\, j = l_+$, $m = \pm\, l_+,\, l\rangle$. The energy shift is given by Eq. (175) and is

$$(\Delta E_M)_{n,\,j=l\pm\frac{1}{2},\,m=\pm\left(l+\frac{1}{2}\right),\,l} = \pm\,\mu_B B\,(l+1).$$ (182)

The submatrices for $n,\, l,\, m \neq \pm\left(l + \dfrac{1}{2}\right)$ are 2×2 matrices and give the characteristic equation

$$\begin{vmatrix} \Delta E_{n,\,l+\frac{1}{2}} + 2\mu_B Bm\,\dfrac{l+1}{2l+1} - \epsilon & -\dfrac{\mu_B B}{2l+1}\sqrt{l_+^2 - m^2} \\[2ex] -\dfrac{\mu_B B}{2l+1}\sqrt{l_+^2 - m^2} & \Delta E_{n,\,l-\frac{1}{2}} + 2\mu_B Bm\,\dfrac{l}{2l+1} - \epsilon \end{vmatrix} = 0,$$ (183)

where

$$\Delta E_{n,\,j} = \frac{mc^2(Z\alpha)^4}{2n^4}\left(\frac{3}{4} - \frac{n}{j+\frac{1}{2}}\right)$$

is the fine-structure energy shift due to the relativistic corrections H'. Let Δ be the fine structure splitting when $B = 0$:

$$\Delta = \Delta E_{n,\,l+\frac{1}{2}} - \Delta E_{n,\,l-\frac{1}{2}}$$

$$= \frac{mc^2(Z\alpha)^4}{2n^3 l(l+1)}.$$ (184)

From Eq. (183), the eigenvalues are

$$\epsilon_{\pm} = \Delta E_{n,\,l-\frac{1}{2}} + \frac{1}{2}\Delta + \mu_B Bm \pm \frac{1}{2}\sqrt{\Delta^2 + \frac{4}{2l+1}m\Delta\mu_B B + (\mu_B B)^2}.$$ (185)

For $\Delta \gg \mu_B B$, we get

$$\epsilon_{\pm} = \Delta E_{n,\,j}\,l \pm \frac{1}{2} + \mu_B Bm\left(1 \pm \frac{1}{2l+1}\right) \pm \frac{1}{4}\frac{\mu_B^2 B^2}{\Delta}\left(1 - \frac{4m^2}{(2l+1)^2}\right)\ldots$$ (186)

The terms linear in B are in agreement with those of Eq. (175). In the other limiting case, $\mu_B B \gg \Delta$ we get

$$\epsilon_{\pm} = \Delta E_{n,\,l\pm\frac{1}{2}} + \mu_B B\left\{\left(m \pm \frac{1}{2}\right) + \frac{\Delta}{2\mu_B B}\left(1 \pm \frac{2m}{2l+1}\right)\right.$$

$$\pm \left(\frac{\Delta}{2\mu_B B}\right)^2 \left(1 - \frac{4m^2}{(2l+1)^2}\right) \cdots \Biggr\}. \tag{187}$$

Substituting $m = m_l \pm \frac{1}{2}$, $m_s = \mp \frac{1}{2}$, one finds agreement with the results in the Paschen-Back limit. For details, consult Condon and Shortley (1951).

When the atom is placed in a constant magnetic field $\vec{B}$ then the interaction Eq. (172) is not invariant under arbitrary rotations, because there is a preferred direction (the direction of $\vec{B}$). But, there is invariance with respect to rotations about this axis (usually taken to be the z-direction), and so $H_0 + H_M$ will be invariant under this subgroup of the full rotation group. Since the two-dimensional rotation group (rotations in a plane) is Abelian, it has only one-dimensional representations. So an atomic level in a magnetic field will split into $(2j + 1)$ sublevels, each corresponding to a value of j_3 (j_3 ranging from $+j$ to $-j$ in steps of one). So the j_3-degeneracy is removed.

For strong magnetic field and large orbits or n values, effects of the term quadratic in B in the Hamiltonian become appreciable. In this case the term H' is neglected and since m_s is a constant of the motion (the electron spin commutes with the Hamiltonian) the spin can be ignored. Also, L_z commutes with H and so m_l is a constant of the motion. Hence the term containing L_z in H shifts each energy level by $\left(\dfrac{eB}{2mc}\right) \hbar m_l$. Thus for strong magnetic field and large n, we consider only the quadratic term in the Hamiltonian

$$H_B^{\text{II}} = \frac{e^2}{8mc^2} B^2 r^2 \sin^2\theta, \tag{188}$$

where we take $\vec{B}$ along the $\theta = 0$ axis of a polar coordinate system (r, θ, ϕ). The base kets are taken as $|nlm_l m_s\rangle$ and the matrix elements are diagonal in m_l, m_s, and parity, but not in n and l. Δn may have any value but $\Delta l = 0$, or ± 2. The energy shift in first order $\dfrac{e^2 B^2}{8\,mc^2} \langle r^2 \sin\theta\rangle_{nlm_l m_s}$ is given by

$$\Delta E_Q = \frac{e^2 B^2}{8mc^2}\left(\frac{a_0}{Z}\right)^2 \frac{n^2[5n^2 + 1 - 3l\,(l+1)]\,(l^2 + l - 1 + m_l^2)}{(2l-1)\,(2l+3)} \tag{189}$$

for a nuclear charge Z. Here a_0 is the first Bohr radius.

The diamagnetic susceptibility of the atom can be obtained from H_B^{II}. (Prob. 12.9).

In recent years there has been considerable interest in the study of the behavior of atoms in huge magnetic fields. Such huge magnetic fields have been observed on the surfaces of neutron stars ($\sim 5 \times 10^8\,T$) and on some white dwarf stars ($\sim 10^3$ to $10^4\,T$). Neutron stars endowed with superstrong magnetic fields

$B > 10^{10}\,T$ are called magnetars. Although in the laboratory atoms are studied at high field in the range 1-6T, high field effects may appear in solids even under laboratory conditions. The motion of an electron in a magnetic field can be studied in an effective mass approximation in which the effective mass m^* may be one or two order of magnitude smaller than m_e. In addition, the electron is considered to move in a medium of large dielectric constant. These conditions increase the magnetic interaction energy and decrease the Coulomb energy. In indium antimonide, a magnetic field of 2.4 T in laboratory conditions corresponds to an effective magnetic field of $3.6 \times 10^6\,T$. The properties of matter in intense magnetic fields ($10^5\,T - 10^{12}\,T$) are reviewed by Lai (2001).

For weak magnetic fields ($H_{Coul} \gg H_{spin} \gg H_B^I$) with H_B^{II}, H_{nucl} negligible, one has the usual linear (in B) Zeeman effect ($H_B^I \sim \mu_B\,\vec{B}.\vec{L}$). Then occurs the first intermediate regime ($H_{Coul} \gg H_{spin} \sim H_B^I$) and finally, the usual high-field Paschen-Back regime (upto 10T) ($H_{Coul} \gg H_B^I \gg H_{spin}$) with H_B^{II}, H_{nucl} negligible. In all these cases the Coulomb interaction energy H_{Coul} is the largest. For increasingly large values of B, H_B^{II} becomes significant and gives rise to the quadratic Zeeman effect. For $B > 10^2\,T$, $H_B^{II} > H_B^I$.

For extremely large fields, $H_B^{II} \gg H_{Coul}$, one reaches the Landau regime where one is dealing with the motion of free electron in a magnetic field (see Section 11.7). However, in between the Paschen-Back (with quadratic Zeeman effect corrections) and the Landau regimes there is the second intermediate regime where the magnetic interaction energies are comparable to the Coulomb energies. This interesting regime relevant to astrophysics and solid state physics cases mentioned above, is difficult to study.

In atomic units, the Hamiltonian of a hydrogen atom (without spin) in a magnetic field for the ground state is

$$H_B = -\frac{1}{2}\vec{\nabla}^2 + \frac{b^2}{8}(x^2 + y^2) - \frac{1}{(x^2 + y^2 + x^2)^{1/2}} \tag{190}$$

The direction of the field is the z-direction. We have defined a dimensionless magnetic field strength

$$b = \frac{B}{B_0};\ B_0 = \frac{m_e^2 e^3 c}{\hbar^3} = 2.35 \times 10^9\,G = 2.35 \times 10^5\,T \tag{191}$$

At a field strength B_0 the oscillator energy $\hbar\omega$ is equal to the Rydberg energy $R = m_e e^4/2\hbar^2 \simeq 13.6\,ev$, where is ω half the cyclotron frequency.

$$\omega = \frac{1}{2}\,\omega_c = \frac{eB}{2m_e c} \tag{192}$$

Note that the Rydberg in atomic units is one-half.

At $B = B_0$, the cyclotron radius R is equal to the first Bohr radius and for higher fields R is less than a_0. Thus for $b \gg 1$, the electron cyclotron energy $\hbar\omega$ is much larger than the typical Coulomb energy and the atoms will have an elongated shape (cigar-shape). The relativistic corrections are negligible for fields with $b < 10^4$. The effects of spin-orbit coupling can be neglected for fields with $bn^3 > 10^{-4}$, where n is the principal quantum number.

The basic difficulty in calculating the energy levels of hydrogen in a magnetic field is that the Schrödinger equation is not separable, due to the different symmetries (spherical or cylindrical) of the Coulomb and magnetic terms. The azimuthal quantum number m and the parity are good quantum numbers.

The Rayleigh-Schrödinger perturbation series for the ground state energy of the hydrogen atom in a magnetic field (with $H_B - H_{B=0}$ as perturbation) is

$$E_0(B) = \frac{1}{2}\left[-1 + \frac{1}{2}b^2 - \frac{53}{96}b^4 + \frac{5581}{2302}b^6 - \frac{21577397}{1105920}b^8 + \dots\right] \tag{193}$$

and is very likely divergent. However, the perturbation series is known to be Borel summable.

Let us now consider the Schrödinger equation with the Hamiltonian in Eq. (190) for calculating the energy levels of a hydrogen atom in a large magnetic field. In sufficiently large fields, the atoms will become cigear-shaped and the oscillations in the field direction are much slower than the cyclotron frequency. This allows us to make the adiabatic approximation, in which we take the motion along the field direction as occurring in the Coulomb potential averaged over the plane perpendicular to the field. In this approximation, the wave function is taken as

$$\underline{\Psi}(\vec{r}) \sim \underline{\bar{\phi}}(r_l)\,f(z) \tag{194}$$

where $\vec{r}_\perp = x^2 + y^2$ and $\bar{\phi}$ denotes Landau state. $f(z)$ is a function to be determined from the equation

$$\left[-\frac{1}{2}\frac{d^2}{dz^2} + V(z)\right]f(z) = E_{11}f(z) \tag{195}$$

where $V(z)$ is an effective potential obtained by averaging the Coulomb term over the Landau states and $E = E_\perp + E_{11}$. The binding energy ε is related to the energy E_0 of the ground state by

$$E_0 = \frac{b}{2} - \varepsilon. \tag{196}$$

Equation (195) has been solved in various ways (see Lai (2001), Le Guillou & Zinn-Justin (1985)). The following implicit equation has been obtained

$$\epsilon = \ln(b/\epsilon^2) - (\gamma + \ln 2) + \frac{\pi^2}{3\epsilon} - 4\left(\frac{2}{b}\right)^{1/2}$$

$$\times \ln\left(\frac{b}{2\epsilon^2}\right) + 2\sqrt{\pi}\left(\frac{2}{b}\right)^{1/2}\epsilon + 2\sqrt{\pi}\left(\frac{2}{b}\right)^{1/2}(-0.03648) + \ldots \qquad (197)$$

where $\quad \epsilon = \sqrt{2\varepsilon}$ and γ is Euler's constant.

We write the ground state energy in powers of the magnetic field

$$E_0 = \sum_{n=0}^{\infty} E_n g^n \qquad (198)$$

where $$g = \frac{b^2}{8}.$$

The large order behavior of the coefficients is given by

$$E_n = (-1)^{n+1}(4/\pi)^{5/2}(8/\pi^2)^n \,\Gamma\left(2n + \frac{3}{2}\right) \times [1 + O\,(1/n)] \text{ as } n \to \ldots \qquad (199)$$

Le Guillou and Zinn-Justin (1985) have calculated the ground state energy of the hydrogen atom in a magnetic field for values of the field upto about $10^9\,T$. The perturbative expansion has been summed by an order-dependent mapping method. They have obtained reasonably accurate values up to $10^9\,T$.

We note that due to the scaling relation

$$E\,(Z, B) = Z^2 E\,(1, {}^B\!/Z^2) \qquad (200)$$

the calculation of the energy of an hydrogen like atom with nuclear charge Z reduces to that of the hydrogen atom.

The hydrogen atom in a uniform magnetic field is a real and physical example of a simple nonintegrable system capable of exhibiting chaos. See Friedrich and Wintgen (1989) for a review.

12.7 THE STARK EFFECT

The splitting and shifting of atomic energy levels under the action of an external electric field is known as the Stark effect. The theory of the Stark effect in hydrogen was the first application of the perturbation theory in quantum mechanics.

In an atom placed in a uniform electric field $\vec{\varepsilon}$ directed say, along the z-axis, we have a system of electrons in an axially symmetric field. This field is invariant under translation, under rotation about z-axis and under reflection in planes parallel to the z-axis. The invariance group is the product of the translation group and the group of reflections in planes containing Oz.

In chapter 10, we have seen that atoms do not possess permanent electric dipole moment. But in an external electric field, an induced dipole moment proportional to the field strength is developed.

For an hydrogenic atom, the perturbation due to the electric field is

$$H' = -\vec{d} \cdot \vec{\varepsilon} = e\vec{\varepsilon} \cdot \vec{r}, \tag{201}$$

where $\vec{d} = -e\vec{r}$ is the dipole moment operator ($-e$ is the charge of the electron) and $\vec{r}$ is the radius vector of the electron relative to the nucleus. We choose the polar axis and $\vec{\varepsilon}$ in the direction of z-axis. Then

$$H' = e\varepsilon r \cos\theta \tag{202}$$

where $\varepsilon = |\vec{\varepsilon}|$ is the electric field strength. Perturbation theory can be used if the additional energy due to the applied field is small compared to the internal electric field of the atom $\left(\sim \dfrac{E^{(0)}}{ea_0} \approx 10^{10} \; Vm^{-1} \right)$.

We also suppose that ε is large enough for fine structure effects to be negligible. For usual field strengths $\sim 10^7 Vm^{-1}$, the assumptions are well satisfied.

From Eq. (202), we see that H' is a component of an irreducible tensor of the type $T_0^{(1)}$. According to the Wigner-Eckart theorem, the matrix element $\langle nlm |H'| n'l'm' \rangle$ vanishes unless $m = m'$ and $l, l', 1$ satisfy the triangle rule. Since $\cos\theta \rightarrow -\cos\theta$ under inversion ($\vec{r} \rightarrow \vec{r}$), the state vectors must have opposite parity. Thus for the matrix element to be non vanishing, we must have $l' = l \pm 1$ and $m' = m$.

Therefore, the first-order contribution to the energy vanishes. In complex atoms there is no first-order Stark effect (linear Stark effect). The hydrogen atom in the ground state ($n = 1$), which is a nondegenerate state of even parity, has no first-order Stark effect. For weak field the interaction energy is given by second-order perturbation theory and is quadratic in the field strength (quadratic Stark effect). However, hydrogenic atoms form an exception to this rule because for fixed n states of different l are degenerate. Since H' has non zero matrix elements for transitions between states of odd and even l, the perturbation H' will remove (partially) this l-degeneracy. First-order perturbation theory gives an effect which is linear in the field strength (for weak fields). We shall discuss later the linear Stark effect of the first excited state of hydrogen as an example of degenerate perturbation theory.

The second-order shift in ground state energy is given by

$$E_1^{(2)} = \sum_n{}' \frac{|\langle n| H'|0\rangle|^2}{E_1^{(0)} - E_n^{(0)}} = e^2\varepsilon^2 \sum_{nlm}{}' \frac{|\langle nlm |z| 100 \rangle|^2}{E_1^{(0)} - E_n^{(0)}} \tag{203}$$

where the sum is over all states except the particular state for which the perturbed energy is being calculated. The infinite sum involves a sum over the discrete bound states and an integral over the continuum of unbound states.

The matrix element is

$$\langle nlm|z|100\rangle = \int d^3r\, R_{nlm}(r)\, Y_{lm}^*(\theta,\phi)\, r\cos\theta R_{100}(r)\, Y_{00}(\theta,\phi). \tag{204}$$

The angular part is

$$\int d\Omega\, Y_{lm}^*(\theta,\phi)\,\frac{1}{\sqrt{3}}\,Y_{10}(\theta,\phi) = \frac{1}{\sqrt{3}}\,\delta_{l1}\delta_{m0}, \tag{205}$$

since

$$Y_{00} = \frac{1}{\sqrt{4\pi}},\quad \cos\theta = \frac{\sqrt{4\pi}}{3}\,Y_{10}.$$

The radial integral is

$$R = \int_0^\infty dr\, r^2 R_{nlm}(r)\, r R_{10m}(r)\,\delta_{10}\delta_{m0}$$

$$= \int dr\, r^3 R_{n10}(r)\, R_{100}(r). \tag{206}$$

This integral can be done (see Bethe & Salpeter (1957), p. 262) and the result is

$$|\langle n10|z|100\rangle|^2 = \frac{1}{3}\left(\frac{2^8 n^7 (n-1)^{2n-5}}{(n+1)^{2n+5}}\right)a_0^2 = f(n)\,a_0^2. \tag{207}$$

Therefore, the second-order shift in the ground state energy is

$$E_{100}^{(2)} = -e^2\varepsilon^2 a_0^2 \sum_{n=2}^{\infty} \frac{f(n)}{\frac{1}{2}\mu c^2\alpha^2\left(1-\dfrac{1}{n^2}\right)}$$

$$= -\frac{2e^2\varepsilon^2 a_0^2}{\mu c^2\alpha^2}\sum_{n=2}^{\infty}\frac{n^2 f(n)}{n^2-1}$$

$$= -2a_0^3\varepsilon^2 \sum_{n=2}^{\infty}\frac{n^2 f(n)}{n^2-1}$$

$$= -2.25\,\varepsilon^2 a_0^3, \tag{208}$$

where

$$\sum_{n=2}^{\infty}\frac{n^2 f(n)}{n^2-1} = 1.125 \tag{209}$$

For hydrogenic atoms, one replaces a_0 by a_0/Z.

By differentiating the expression Eq. (203) with respect to the electric field strength, we get an expression for the dipole moment.

$$d = -\frac{\partial E_{100}^{(2)}}{\partial\varepsilon} = \overline{\alpha}\varepsilon, \tag{210}$$

where

$$\bar{\alpha} = 2e^2 \sum_{nlm}{}' \frac{|\langle nlm|z|100\rangle|^2}{E_1^{(0)} - E_n^{(0)}} = 4.5\, a_0^3, \tag{211}$$

is called the dipole polarizability of the atom in the state Eq. (100). Since d is proportional to the electric field strength, the dipole moment is induced.

Instead of working directly with Eq. (203) we use an alternative method. We consider

$$(H_0 - E_0)\, |\underline{\psi}_{100}^{(1)}\rangle = (\langle 100|\, H'\,|100\rangle - H')|\,100\rangle) \tag{212}$$

This method is effective only when Eq. (212) is easier to solve than the full eigenvalue equation. As an example of this method, we consider the quadratic Stark effect. Let $H_0 = -\dfrac{\hbar^2}{2\mu}\, \vec{\nabla}^2 - \dfrac{e^2}{r}$, $E_0 = -\dfrac{e^2}{2a_0}$, and $\underline{\psi}^{(0)}(r) = (\pi a_0^3)^{-1/2}\, \exp\left(-\dfrac{r}{a_0}\right)$. Equation (212) is an inhomogeneous differential equation for $\underline{\psi}^{(1)}$:

$$\left(-\frac{\hbar^2}{2\mu}\, \vec{\nabla}^2 - \frac{e^2}{r} - E_0\right) \underline{\psi}^{(1)} = -\, e\varepsilon r \cos\theta\, \underline{\psi}^{(0)}. \tag{213}$$

We write

$$\underline{\psi}^{(1)}(\vec{r}) = f(r) \cos\theta \tag{214}$$

and the condition $\langle \underline{\psi}^{(0)}|\underline{\psi}^{(1)}\rangle = 0$ is automatically satisfied. Substituting Eqs. (214) in (213) we get (Schiff (1968)):

$$\frac{d^2 f}{dr^2} + \frac{2}{r}\frac{df}{dr} - \frac{2}{r^2}f + \frac{2}{a_0 r}\, f - \frac{1}{a_0^2}f = \frac{2\varepsilon r e^{-r/a_0}}{e a_0 (\pi a_0^3)^{1/2}}. \tag{215}$$

The solution of Eq. (215) is expected to be of the form

$$f(r) = p(r) e^{-r/a_0}. \tag{216}$$

The solution of Eq. (215) is

$$f(r) = -\, (\pi a_0^3)^{-1/2} \frac{\varepsilon}{e}\left(a_0 r + \frac{1}{2}r^2\right) e^{-r/a_0}. \tag{217}$$

Thus the wave function correct to first-order in ε is

$$\underline{\psi}_{100}^{(1)} = (\pi a_0^3)^{-1/2}\left[1 - \frac{\varepsilon}{e}\left(a_0^r + \frac{1}{2}r^2\right)\cos\theta\right] e^{-r/a_0}. \tag{218}$$

The second-order perturbed energy is

$$E^{(2)} = \langle \underline{\psi}^{(0)}|H'|\underline{\psi}_{100}^{(1)}\rangle$$

$$= e\varepsilon\, (\pi a_0^3)^{-1/2} \int d^3 r\, r \cos^2\theta f(r)\, e^{-r/a_0}$$

$$= -\frac{4\varepsilon^2}{3a_0^3} \int_0^\infty \left(a_0 r^4 + \frac{1}{2} r^5 \right) e^{-2r/a_0}\, dr$$

$$= -\frac{9}{4}\,\varepsilon^2\, a_0^3. \tag{219}$$

The rigorous evaluation of the infinite summation, e.g. for the second-order perturbation energy, in all higher order perturbation calculations is a difficult problem. The general method, due to Dalgarno and Lewis (1955) and applied by them for the calculation of dipole polarizabilities of atoms replaces the evaluation of the summation by the solution of an inhomogeneous differential equation which may be simpler. Let us consider

$$E^{(2)} = \sideset{}{'}\sum_n \frac{\langle 0|H'|n\rangle\langle n|H'|0\rangle}{E_0^{(0)} - E_n^{(0)}}, \tag{220}$$

and suppose that an operator F can be found which satisfies the equation.

$$\frac{\langle n|H'|0\rangle}{E_0^{(0)} - E_n^{(0)}} = \langle n|F|0\rangle \tag{221}$$

where n is any state other than the ground state.

Then

$$E^{(2)} = \sideset{}{'}\sum_n \langle 0|H'|n\rangle\langle n|F|0\rangle$$

$$= \langle 0|H'F|0\rangle - \langle 0|H'|0\rangle\langle 0|F|0\rangle. \tag{222}$$

From Eq. (222), we see that if we are able to find F, then the summation over states can be evaluated by the calculation of the integrals over the unperturbed groundstate wave functions.

We write Eq. (221) as

$$\langle n|H'|0\rangle = (E_0^{(0)} - E_n^{(0)})\,\langle n|F|0\rangle.$$

$$= \langle n|[F, H_0]|0\rangle.$$

This gives the operator equation for F

$$[F, H_0] = H' + C$$

where C is a constant.

Instead of this general equation, it is enough that F statisfies the following equation

$$[F, H_0]|0\rangle = H'|0\rangle + C|0\rangle \tag{223}$$

which gives

$$C = - \langle 0|H'|0\rangle$$

If we write $|1\rangle = F|0\rangle$, then Eq. (223) may be written as

$$(E_0^{(0)} - H_0)|1\rangle = H'|0\rangle - \langle 0|H'|0\rangle|0\rangle. \tag{224}$$

Clearly $|1\rangle$ is not unique since we can always add to it an arbitrary multiple of $|0\rangle$. This multiple is chosen so that $\langle 0|1\rangle = 0$. If the inhomogeneous differential Eq. (224) can be solved for $|1\rangle$, then the second-order perturbed energy Eq. (222) can be written as

$$E^{(2)} = \langle 0|H'|1\rangle \tag{225}$$

Also,

$$|\underline{\psi}^{(1)}\rangle = \sum_n{}' \frac{|n\rangle\langle n|H'|0\rangle}{E_0^{(0)} - E_n^{(0)}}$$

$$= \sum_n{}' |n\rangle\langle n|F|0\rangle$$

$$= F|0\rangle - |0\rangle\langle 0|F|0\rangle$$

$$= |1\rangle. \tag{226}$$

Even if the differential equation can be solved only approximately, the method may be much simpler than the infinite sum.

The first excited state ($n = 2$) of hydrogen is four-fold degenerate. The four degenerate $n = 2$ states are $|2S_0\rangle = |200\rangle$, $|2P_1\rangle = |211\rangle$, $|2P_0\rangle = |210\rangle$ and $|2P_{-1}\rangle = |21 - 1\rangle$. We now use degenerate perturbation theory to calculate the first-order (linear in ε) Stark effect for the $n = 2$ states of hydrogen atom. We have to diagonalize the 4×4 matrix of the perturbation H' in the four-dimensional subspace spanned by the four degenerate unperturbed basis vectors. The non-vanishing off – diagonal matrix elements of H' exist only for the states with the same quantum number m. Of the 16 matrix elements in the 4×4 matrix of H' between the four degenerate unperturbed states the only non-vanishing ones are the off-diagonal elements between the opposite parity states $|200\rangle$ and $|210\rangle$. These non-vanishing matrix elements are

$$-w = \langle 210|H'|200\rangle = \langle 200|H'|210\tilde{n}*$$

$$= e\varepsilon \langle 210|r \cos \theta|200\rangle$$

$$= e\varepsilon \int_0^\infty dr r^4 (2a_0)^{-3} e^{-r/a_0} \frac{1}{\sqrt{3}\, a_0} \left(2 - \frac{r}{a_0}\right)$$

$$\times \int d\Omega\, Y_{00}^* \left(\sqrt{\frac{4\pi}{3}}\, Y_{10}\right) Y_{10} = - 3eea_0. \tag{227}$$

The secular equation is

$$\begin{vmatrix} -E^{(1)} & -w & 0 & 0 \\ -w & -E^{(1)} & 0 & 0 \\ 0 & 0 & -E^{(1)} & 0 \\ 0 & 0 & 0 & -E^{(1)} \end{vmatrix} = 0. \tag{228}$$

The four roots of Eq. (228) are $0, 0, w, -w$ and the corresponding normalised eigenvectors are $(0,1,0,0)$, $(0,0,0,1)$, $\dfrac{1}{\sqrt{2}}(1,0,-1,0)$ and $\dfrac{1}{\sqrt{2}}(1,0,1,0)$. The four-fold degeneracy is partly (half) removed in the first-order. The $m = \pm 1$ states remain degenerate. The linear Stark splitting of a hydrogen atom in its first excited state is shown schematically below.

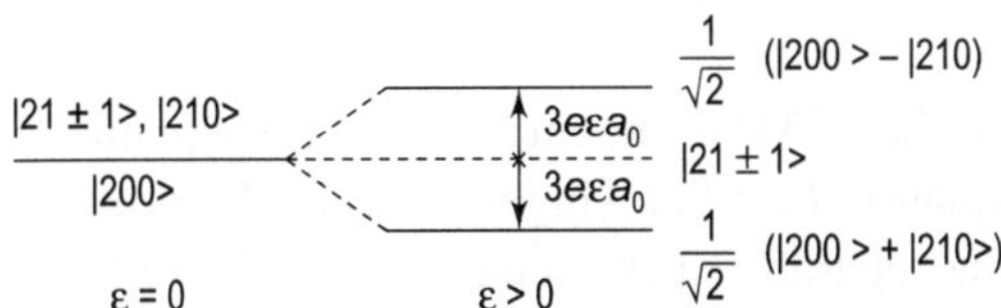

Fig. 12.5 The linear Stark spittings for atomic hydrogen in its first excited state ($n = 2$)

Thus, a hydrogen atom in the first excited state behaves as though it has a permanent electric dipole moment of magnitude $3ea_0$ orientable in three different ways: one state parallel to the electric field, one state antiparallel to the electric field (so its energy is raised by $3e\varepsilon a_0$), and two states with no components along the field.

The Schrödinger equation of the hydrogen atom can be separated in parabolic coordinates (see Section 11.5 (b)). This separability is maintained in the presence of an electric field. The potential energy of an electron in the field of the nucleus in the presence of an applied weak homogeneous electric field ε is

$$V = -\frac{Ze^2}{r} + e\varepsilon z, \tag{229}$$

which, in parabolic coordinates, may be written as

$$V = -\frac{2Ze^2}{\xi + \eta} + \frac{1}{2} e\varepsilon \, (\xi - \eta) = -\frac{1}{\xi + \eta}\left[2Ze^2 - \frac{1}{2} e\varepsilon \, (\xi^2 - \eta^2)\right]. \tag{230}$$

The Sahrödinger equation (in atomic units)

$$\left(\frac{1}{2}\vec{\nabla}^2 + E - V\right) u = 0 \tag{231}$$

in parabolic coordinates assumes the form

$$\frac{\partial}{\partial \xi}\left(\xi \, \frac{\partial u}{\partial \xi}\right) + \frac{\partial}{\partial \eta}\left(\eta \, \frac{\partial u}{\partial \eta}\right) + \left(\frac{1}{4\xi} + \frac{1}{4\eta}\right)\frac{\partial^2 u}{\partial \varphi^2}$$

$$+ \left[\frac{1}{2} E(\xi + \eta) + Z - \frac{1}{4} \varepsilon \left(\xi^2 - \eta^2 \right) \right] u = 0. \qquad (232)$$

This differential equation is separated by the following assumption

$$u(\xi_S, \eta, \varphi) = u_1(\xi) u_2(\eta) e^{im\varphi}$$

$$Z = Z_1 + Z_2 \qquad (233)$$

Now, u_1 and u_2 satisfy the differntial equations

$$\left. \begin{array}{l} \dfrac{d}{d\xi} \left(\xi \dfrac{du_1}{d\xi} \right) + \left(\dfrac{1}{2} E\xi + Z_1 - \dfrac{m^2}{4\xi} - \dfrac{1}{4} \varepsilon \xi^2 \right) u_1 = 0 \\[3ex] \dfrac{d}{d\eta} \left(\eta \dfrac{\partial u_2}{\partial \eta} \right) + \left(\dfrac{1}{2} E\eta + Z_2 - \dfrac{m^2}{4\eta} \dfrac{1}{4} \varepsilon \eta^2 \right) u_2 = 0 \end{array} \right\} . \qquad (234)$$

The ξ equation in Eq. (234) is known as the up-hill equation, and the η equation as the down-hill equation (see Schwinger (2001)). We now use the perturbation method in which the separation parameters Z_1 and Z_2 are the eigenvalues.

The unperturbed value of Z_1 (in the absence of ξ) is

$$Z_1^{(0)} = \varepsilon \left(n_1 + \frac{m+1}{2} \right) \qquad (235)$$

where $\varepsilon = (-2E)^{1/2}$, n_1 is the electric quantum number, and m the magnetic quantum number. The unperturbed eigenfunction is

$$u_1(\xi) = \frac{(n_1!)^{1/2}}{((n_1 + m)!)^{3/2}} \, e^{-\frac{1}{2}\varepsilon\xi} \, \xi^{\frac{1}{2}m} \, \varepsilon^{\frac{1}{2}(m+1)} \, L^m_{n_1+m}(\varepsilon \xi) \qquad (236)$$

which is normalized according to the condition $\int_0^\infty u_1^2(\xi) \, d\xi = 1$.

Now,

$$Z_1^{(1)} = \frac{1}{4} \varepsilon \int_0^\infty \xi^2 u_1^2 d\xi$$

$$= \frac{1}{4} \varepsilon \varepsilon^{-2} \left(6n_1^2 + 6n_1 m + m^2 + 6n_1 + 3m + 2 \right). \qquad (237)$$

To first-order,

$$Z_1 = Z_1^{(0)} + Z_1^{(1)}$$

$$= \varepsilon \left(n_1 + \frac{m+1}{2} \right) + \frac{1}{4} \frac{\varepsilon}{\varepsilon^2} \left(6n_1^2 + 6n_1 m + m^2 + 6n_1 + 3m + 2 \right), \qquad (238)$$

and similarly,

$$Z_2 = Z_2^{(0)} + Z_2^{(1)}$$

$$= \in \left(n_2 + \frac{m+1}{2} \right) - \frac{1}{4} \frac{\varepsilon}{\in^2} \left(6n_2^2 + 6n_2 m + m^2 + 6n_2 + 3m + 2 \right), \quad (239)$$

$$\therefore \qquad Z = Z_1 + Z_2$$

$$= \in n + \frac{3}{2} \frac{\varepsilon}{\in^2} (n_1 - n_2) n, \qquad (240)$$

where the principal quantum number

$$n = n_1 + n_2 + m + 1. \qquad (241)$$

From Eq. (240)

$$\in \; = \frac{Z}{n} - \frac{3}{2} \varepsilon \left(\frac{n}{Z} \right)^2 (n_1 - n_2)$$

and hence the energy is given by

$$E^{(1)} = \frac{-1}{2} \in^2 = -\frac{1}{2} \frac{Z^2}{n^2} + \frac{3}{2} \frac{\varepsilon n}{Z} (n_1 - n_2). \qquad (242)$$

This is the expression for the linear Stark effect. The quantum numbers n_1 and n_2 range from 0 to $n - 1$ so that $(n_1 - n_2)$ ranges from $(n - 1)$ to $-(n - 1)$ giving in all $(2n - 1)$ equally spaced values. For $n = 2$, there are $4(=n^2)$ states, and we get only $2.2 - 1 = 3$ states in presence of ε.

With increasing field strength the quadratic term in the electric field becomes important.

The expression for the energy to second-order is (Bethe and Salpeter (1957)).

$$E^{(2)} = -\frac{Z^2}{2n^2} + \frac{3}{2} \varepsilon \left(\frac{n}{Z} \right) (n_1 - n_2)$$

$$- \frac{1}{16} \varepsilon^2 \left(\frac{n}{Z} \right)^4 [17 \, n^2 - 3 (n_1 - n_2)^2 - 9m^2 + 19]. \text{ in a.u.} \qquad (243)$$

It should be noted that the electric field can remove an electron from the atom. From Eq. (229), we see that the atomic center is not the only location at which the potential is a minimum. The potential energy tends to $- \infty$ for large negative values of z. Thus the electron can tunnel troughs the potential barrier (whose width decreases as ε increases) between the two troughs and will be accelerated toward the anode leaving the ionized atom. So all energy levels are allowed and there is, strictly speaking, no bound states. The perturbation series for the energy in powers of ε diverges. For ordinary field strengths, however, the spectrum has a quasi-discrete structure. which is discussed by treating $- e\varepsilon z$ as a perturbation. Oppenheimer (and Lanczos) calculated the probability of ionization of a hydrogen atom in an electric field and discussed "cold emission" (the emission of electrons from a cold metal in a strong electric field). In weak field the probability of ionisation is extremely small. It increases exponentially with ε and is quite large in strong fields.

The perturbation series for the Stark effect in atoms is divergent. The Stark problem in hydrogen is Borel summable and the behavior of resonances has been studied using the perturbation theory at large orders. The perturbation series for the ground state for any real value of $\varepsilon \neq 0$ is

$$E(\varepsilon) = -\frac{1}{2} - \frac{9}{4}\,\varepsilon^2 - \frac{3555}{64}\,\varepsilon^4 - \frac{2512779}{512}\,\varepsilon^6 \ldots \qquad (244)$$

It has been found that in small field regime, the Oppenheimer formula for the width Γ of the ground state is obtained

$$\Gamma = -2\,\mathrm{Im}\,E(\varepsilon) = 4\varepsilon^{-1}\exp\left(-\frac{2}{3}\,\varepsilon^{-1}\right)[1 + O(\varepsilon)]. \qquad (245)$$

In large field one obtains for $\varepsilon \to \infty$

$$\arg E \sim -\frac{\pi}{3} + O\left((\ln\varepsilon)^{-1}\right),$$

$$|E| \sim \alpha\varepsilon^{2/3}(\ln\varepsilon)^{2/3} + \beta\varepsilon^{2/3}(\ln\ln\varepsilon)$$

$$\times (\ln\varepsilon)^{-1/3} + O\left(\varepsilon^{2/3}(\ln\varepsilon)^{-1/3}\right), \qquad (246)$$

where

$$\alpha = 2^{-5/3},$$

$$\beta = \frac{8}{3}\,\alpha.$$

PROBLEMS

12.1 A simple harmonic oscillator is acted on by a constant force F. Compute the second-order correction to the energy and compare with the exact answer.

12.2 Consider a particle of mass m moving in a one-dimensional infinite square well of length $2L$ with its center at the origin:

$$V(x) = \begin{cases} 0, & -L \leq x \leq L \\ \infty, & \text{otherwise} \end{cases}$$

There is a δ-function perturbation at the origin:

$$H' = \lambda\,E_0\,L\,\delta(x),$$

where λ is a small dimensionless parameter and E_0 is the unperturbed ground-state energy.

   (a) Calculate the first, second, and third-order perturbation corrections to the unperturbed energy levels and the first and second-order corrections to the wave functions.

   (b) Compare with the exact expressions.

12.3 The rotational motion of a diatomic molecule is described by the Hamiltonian

$$H_0 = \frac{\vec{L}^2}{2I},$$

where $\vec{L}$ is the angular momentum operator, and I is the moment of inertia about the axis of rotation. This molecule possesses a permanent electric dipole moment $\vec{D}$ and is placed in a uniform electric field $\vec{\varepsilon}$, that lies in the plane of rotation. The interaction energy is $-\vec{D}\cdot\vec{\varepsilon}$. Find the second-order correction to the energy levels. Comment on the degeneracy.

12.4 Consider a two-dimensional isotropic harmonic oscillator whose Hamiltonian is

$$H_0 = \frac{p_x^2 + p_y^2}{2m} + \frac{m\omega^2}{2}(x^2 + y^2).$$

We now apply a perturbation

$$V(x, y) = \lambda m\omega^2 xy$$

where λ is a real, dimensionless number much smaller than unity.

   (a) Find the zeroth order energy eigenket and the corresponding energy to first-order.

   (b) Solve the problem exactly and compare with the results obtained by perturbation theory in (a).

12.5 A one dimensional harmonic oscillator with mass m and charge q moves in a constant electric field. (a) Find the corrections to the energy levels and the eigenstates to the first non-vanishing order in perturbation. (b) Find the electric dipole moment. (c) Solve the problem exactly and compare with the results obtained in (a).

12.6 Discuss the linear Stark effect to the lowest non-vanishing order for the $n = 3$ level of the hydrogen atom.

12.7 Calculate the dielectric polarizability of the hydrogen atom in the ground state.

12.8 Calculate the splitting of a term with the total electron spin 1/2 of an atom placed in a strong magnetic field (Paschen-Back effect).

12.9 Compute the diamagnetic susceptibility of a neon atom ($Z = 10$). Use hydrogen-like wave functions with the following screening constants:

$$\sigma_{n,l} = \begin{matrix} 0.23 & n = 1, \; l = 0 \\ 3.26 & n = 2, \; l = 0 \\ 4.11 & n = 2, \; l = 1 \end{matrix}$$

The experimental value is $\chi_{Ne} = -5.6 \times 10^{-6}$ cm^3/mole.

12.10 Calculate the relative intensites of all possible $\pi \, (\Delta M = 0)$ and $\sigma \, (\Delta M = \pm 1)$ Zeeman components for the transition between a $^2S_{1/2}$ state and a $^2P_{3/2}$ state in a weak magnetic field. Assume linear polarization along x and z for σ and π transitions, respectively.

CHAPTER 13

Nonperturbative Methods

In many cases, the stationary perturbation method developed in Chapter 12 cannot be applied successfully as there exists no closely related problem which is exactly soluble. In this chapter we discuss nonperturbative methods for the approximate determination of the energy levels and wave functions of the discrete spectrum. Apart from the WKB method, the variational method is a general method which is very useful in finding approximate bound-state energies and wave functions of a time-independent Hamiltonian. The variational method has numerous applications in atomic and molecular physics, nuclear physics, and solid state physics. The widely used Hartree-Fock approximation is based on this method.

13.1   THE VARIATIONAL METHOD

The variational method (the Rayleigh-Ritz method) is based on the following two theorems.

Theorem 13.1   Let H be the Hamiltonian of a system with at least one bound state. The average value of the energy

$$E\left[\psi\right] = \frac{\langle|\psi|H|\psi\rangle}{\langle|\psi|\psi\rangle}.$$

(1)

Any state vector for which the functional $E[\psi]$ is stationary, is an eigenvector of the discrete spectrum of H, and conversely. The corresponding eigenvalue is the stationary value of the functional $E[\psi]$.

We note that the vectors $|\psi\rangle$ have a finite norm and the functional $E[\psi]$ is independent of the normalization and of the phase of $|\psi\rangle$.

The theorem states that the eigenfunctions of H in the Hilbert space are solutions of the variational equation

$$\delta E = 0.$$

(2)

Proof The variation of $E[\psi]$ is

$$\langle \psi | \psi \rangle \, \delta E = \delta(\langle \psi | H | \psi \rangle - E\delta \langle | \psi | \psi \rangle)$$

$$= \langle \delta\psi | (H - E) | \psi \rangle + \langle \psi | (H - E) | \delta\psi \rangle.$$

So long as $\langle \psi | \psi \rangle$ remains finite and non-null, the variational equation is equivalent to

$$\langle \delta\psi | (H - E) | \psi \rangle + \langle \psi | (H - E) | \delta\psi \rangle = 0. \tag{3}$$

Though the variations $| \delta\psi \rangle$ and $\langle \delta\psi |$ are not independent, they may be treated as such. To see this, we may replace $| \delta\psi \rangle$ by $i | \delta\psi \rangle$ in Eq. (3), so that we get

$$-i \, \langle \delta\psi | (H - E) | \psi \rangle + i \, \langle \psi | (H - E) | \delta\psi \rangle = 0. \tag{3a}$$

By taking linear combination of Eqs. (3) and (3a), we obtain two equivalent equations.

$$\langle \delta\psi | (H - E) | \psi \rangle = 0, \ \langle \psi | (H - E) | \delta\psi \rangle = 0. \tag{4}$$

These two equations are equivalent to Eq. (3) if we consider the variations $| \delta\psi \rangle$ and $\langle \delta\psi |$ as arbitrary and independent. Therefore, we have

$$(H - E) | \psi \rangle = 0, \ \langle \psi | (H - E) = 0.$$

i.e.,
$$(H - E[\psi]) | \psi \rangle = 0, \tag{5}$$

$$(H^{\dagger} - E^{*}[\psi]) | \psi \rangle = 0. \tag{5a}$$

Since $H = H^{\dagger}$, the relations Eqs. (5) and (5a) are identical. The variational equation (2) is therefore equivalent to the Eq. (5), i.e., any ket $| \psi_n \rangle$ for which the functional E is stationary is an eigenket of H belonging to the eigenvalue $E_n = E[\psi_n]$.

Conversely, if $| \psi_n \rangle$ is an eigenket of finite norm and E_n is the corresponding eigenvalue:

$$H | \psi_n \rangle = E_n | \psi_n \rangle,$$

then scalar multiplication on the left by $\langle \psi_n |$ gives $E_n = E[\psi_n]$.

Therefore, $| \psi_n \rangle$ satisfies the Eqs. (5) and (5a). Hence the functional $E[\psi_n]$ is stationary for $\psi = \psi_n$.

In this procedure, a small first order error in the wave function leads to a quadratic error in the energy. Let $| \psi \rangle = | \psi_n \rangle + | \epsilon \rangle$ where $| \epsilon \rangle$ is an infinitesimally small error vector. For simplicity, we assume $\langle \psi | \psi \rangle = \langle \psi_n | \psi_n \rangle = 1$. Then

$$E[\psi] = \langle \psi | H | \psi \rangle$$

$$= \langle \psi_n | H | \psi_n \rangle + \langle \epsilon | H | \psi_n \rangle + \langle \psi_n | H | \epsilon \rangle + \langle \epsilon | H | \epsilon \rangle$$

$$= E_n + E_n \left\{ \langle \in | \psi_n \rangle + \langle \psi_n | \in \rangle + O(\in^2) \right\}$$

Since $\quad \langle \psi | \psi \rangle = \langle \psi_n | \psi_n \rangle = 1$, we get

$$\left\{ \langle \in | \psi_n \rangle + \langle \psi_n | \in \rangle \right\} + \langle \in | \in \rangle = 0.$$

So

$$E[\psi] = E_n + O(\in^2). \tag{6}$$

The error in the approximate energy value will be second order in $\in$ from the corresponding energy eigenvalue if the wave function differs from an energy eigenstate by order $\in$.

Theorem 13.2   The average value of the energy in an arbitrary state is equal to or greater than the energy of the ground state.

$$E[\psi] \geq E_0 \tag{7}$$

Proof   We assume that the spectrum of H is entirely discrete. Let $E_0, E_1, \ldots E_n,$ $\ldots$ be the energy levels in increasing order. We expand the state in the eigenstates of the Hamiltonian. Thus,

$$|\psi\rangle = \sum_n C_n |\psi_n\rangle, \text{ where } H|\psi_n\rangle = E_n|\psi_n\rangle.$$

So

$$E[\psi] = \frac{\sum\limits_n |C_n|^2 E_n}{\sum\limits_n |C_n|^2}, \text{ where } \langle \psi | \psi \rangle = \sum_n |C_n|^2$$

Therefore, $\quad E[\psi] - E_0 = \dfrac{\sum\limits_n |C_n|^2 (E_n - E_0)}{\sum\limits_n |C_n|^2}$

Since $E_n \geq E_0$ for all n, where E_0 is the lowest energy eigenvalue, the right hand side is non-negative.

Hence the theorem $E[\psi] \geq E_0$ follows. The functional $E[\psi]$ gives an upper bound to the ground state energy. This inequality is used in the variational method for the approximate calculation of the lowest eigenvalue. A suitable function ψ_t known as a trial function which depends on one or more variable parameters λ_i is chosen. This is substituted into $E[\psi]$ and the parameters are varied so as to make E a minimum by solving p equations

$$\frac{\partial E}{\partial \lambda_i} = 0 \quad (i = 1, 2, \ldots, p). \tag{8}$$

The minimum value of E determined gives an upper bound to E_0 which will be near to its true value if ψ_t has a form closely resembling the actual ground state wave function ψ_0. Equality is obtained when $\psi_t = \psi_0$. Sometimes one chooses several different trial functions and the best estimate for E_0 is the lowest value obtained.

We now apply the variational method to the calculation of the ground state of the hydrogen atom. The Hamiltonian for the hydrogen atom in the center of mass frame is

$$H = -\frac{\hbar^2}{2\mu}\vec{\nabla}^2 - \frac{e^2}{r}.$$

We choose a trial function $\psi_\lambda(r) = e^{-r/\lambda}$, where λ is a variable parameter. This trial function has the spherical symmetry of the ground state of the hydrogen atom. Then

$$E[\psi_\lambda] = \frac{\langle\psi_\lambda|H|\psi_\lambda\rangle}{\langle\psi_\lambda|\psi_\lambda\rangle} = \frac{\langle T\rangle + \langle V\rangle}{\langle\psi_\lambda|\psi_\lambda\rangle},$$

where $\langle T\rangle$ and $\langle V\rangle$ are the kinetic energy and potential energy terms.

Now,
$$\langle\psi_\lambda|\psi_\lambda\rangle = \int|\psi_\lambda(r)|^2\,d^3r = 4\pi\int_0^\infty e^{-2r/\lambda}\,r^2dr = \pi\lambda^3.$$

$$\langle T\rangle = \frac{1}{2\mu}\left(\langle\psi_\lambda|\vec{p}\rangle\cdot(\vec{p}|\psi_\lambda\rangle)\right)$$

$$= \frac{\hbar^2}{2\mu}\int\left|\frac{\partial\psi_\lambda}{\partial r}\right|^2 d^3r$$

$$= \frac{\hbar^2}{2\mu}\cdot\frac{4\pi}{\lambda^2}\int_0^\infty e^{-2r/\lambda}r^2\,dr$$

$$= \frac{\hbar^2\pi\lambda}{2\mu}$$

$$\langle V\rangle = -\int|\psi_\lambda|^2\frac{e^2}{r}\,d^3r$$

$$= -e^2.4\pi\int e^{-2r/\lambda}r\,dr = -\pi e^2\lambda^2$$

$$\therefore \qquad E[\psi_\lambda] = \frac{\hbar^2}{2\mu\lambda^2} - \frac{e^2}{\lambda}.$$

The minimum value of $E|\psi_n|$ is obtained from the condition $\dfrac{\partial E}{\partial\lambda} = 0$ which is satisfied for $\lambda_0 = -\dfrac{\hbar^2}{\mu e^2}$ and corresponds to

$$E_{\mathbf{min}}(\lambda_0) = \frac{-\mu e^4}{2\hbar^2}.$$

This is the exact value of the ground state energy of the hydrogen atom. In this case, by a fortunate choice the form of the trial function is identical to the exact ground state wave function. Hence the upper bound $-\dfrac{e^2}{2a_0}$ for the ground state energy is, in fact, the exact value of the ground state energy. See Messiah (1961)for an analysis of the variational calculation of the energy for other trial functions.

Although the variational method applies to the approximate determination of the lowest eigenvalue, it can be modified to obtain upper bounds for the energy of the excited states. Let us arrange the eigenvalues in an ascending sequence $E_0, E_1, E_2, \ldots$ ($E_j > E_i$ for $j > i$)). To determine an upper bound for the eigenvalue E_m corresponding to the mth state, we choose a function ψ which is orthogonal to the eigenfunctions of all the lower states, i.e.,

$$\langle \psi_i | \psi \rangle = 0, \quad i = 0, 1, 2, \ldots, m - 1$$

Expanding ψ in the set of eigenvectors ψ_m of H, $|\psi\rangle = \sum_i C_i |\psi_i\rangle$ we have $C_i = \langle \psi_i | \psi \rangle = 0$ ($i = 0, 1, 2, \ldots, m - 1$) so that the functional

$$E[\psi] = \frac{\sum\limits_{i=m} |C_i|^2 E_i}{\sum\limits_{i=m} |C_i|^2}$$

Hence,

$$E_m \leq E[\psi]. \tag{9}$$

This method works well for low-lying excited state, as the orthogonality constraint conditions are difficult to satisfy. The Schmidt procedure can be used to satisfy the orthogonality requirements. In some cases symmetry conditions are used to ensure the constraint conditions. In many practical applications the lower eigenfunctions are not known exactly and the approximate forms of these functions may not satisfy the orthogonality conditions.

As an example, we calculate the energy of the first excited state of the harmonic oscillator. The Hamiltonian is (in the coordinate basis)

$$H = -\frac{\hbar^2}{2m} \frac{d^2}{dx^2} + \frac{1}{2} m\omega^2 x^2.$$

The ground state wave function is

$$\psi^{(0)} = e^{-\alpha x^2/2}.$$

To obtain the energy and the eigenfunction of the first excited state, we must choose a trial function $\psi_\lambda^{(1)}$ that is orthogonal to $\psi^{(0)}$ It must have one node and vanish at $x \to \pm \infty$. A simple choice for the trial wave function satisfying all the requirements is

$$\psi_\lambda^{(1)} = Nxe^{-\lambda x^2/2}$$

where

$$\langle \psi_\lambda^{(1)} | \psi_\lambda^{(0)} \rangle = 0$$

With this choice, we get

$$\langle \psi_\lambda^{(1)} | \psi_\lambda^{(1)} \rangle = N^2 \int_{-\infty}^{\infty} x^2 e^{-\lambda x^2} \, dx = 1$$

whence

$$N = \left(\frac{4\lambda^3}{\pi} \right)^{1/4}.$$

Also,

$$\langle \psi_\lambda^{(1)} | V | \psi_\lambda^{(1)} \rangle = \frac{1}{2} m\omega^2 N^2 \int_{-\infty}^{\infty} x^4 \, e^{-\lambda x^2} \, dx$$

$$= \frac{1}{2} m\omega^2 N^2 \frac{3}{4\lambda^2} \left(\frac{\pi}{\lambda} \right)^{1/2} = \frac{3}{4} \frac{m\omega^2}{\lambda}.$$

$$\langle \psi_\lambda^{(1)} | T | \psi_\lambda^{(1)} \rangle = -\frac{\hbar^2}{2m} N^2 \int_{-\infty}^{\infty} xe^{-\lambda x^2/2} \frac{d^2}{dx^2} (xe^{-\lambda x^2/2}) \, dx$$

$$= -\frac{\hbar^2}{2m} N^2 \left(-3\lambda + \lambda^2 \cdot \frac{3}{2\lambda} \right) \frac{1}{2\lambda} \left(\frac{\pi}{\lambda} \right)^{1/2}$$

$$= \frac{\hbar^2}{2m} N^2 \cdot \frac{3}{4} \left(\frac{\pi}{\lambda} \right)^{1/2}$$

$$= \frac{3}{4} \frac{\hbar^2 \lambda}{m}$$

Therefore,

$$\langle H \rangle = \frac{3}{4} \left[\frac{\lambda \hbar^2}{m} + \frac{m\omega^2}{\lambda} \right].$$

The value of the variational parameter λ for which the above expression is a minimum is obtained from the condition

$$\frac{d\langle H \rangle}{d\lambda} = 0 \quad \text{whence } \lambda_{\text{min}} = \frac{m\omega}{\hbar}.$$

The upper bound for the energy of the first excited state is

$$\langle H \rangle_{\text{min}} = E_1 = \frac{3}{2} \hbar\omega,$$

which is the exact eigenvalue because the trial function chosen happens to be the correct eigenfunction.

Using the value of N and $\lambda_{\mathbf{min}}$, the wave function

$$\psi^{(1)} = \left(\frac{2}{\pi^{1/2}}\right)^{1/2} \left(\frac{m\omega}{\hbar}\right)^{3/4} x e^{-m\omega^2 x^2/2\hbar},$$

which is the correct normalized eigenfunction for the first excited state.

The variational method provides upper bounds. There are several techniques to obtain lower bounds. Weinstein (1934) obtained an expression which gives a lower bound to the ground state energy E_0. Let

$$I = \langle\psi|(H - J_1)^2|\psi\rangle \tag{10}$$

where
$$J_m = \langle\psi|H^m|\psi\rangle, \tag{11}$$

$|\psi\rangle$ being a normalized ket. Expanding $|\psi\rangle$ in the eigenstates of H, we may write

$$I = \sum_n |C_n|^2 (E_n - J_1)^2$$

$$= (E_0 - J_1)^2 + \sum_n |C_n|^2 \{(E_n - J_1)^2 - (E_0 - J_1)^2\}$$

If J_1 is closer to E_0 than to the energy of any other state,

$$I \geq (E_0 - J_1)^2$$

Now,
$$I = J_2 - J_1^2$$

and hence
$$E_0 \geq J_1 - \sqrt{J_2 - J_1^2}. \tag{12}$$

Another simple formula (Temple's inequality) also gives a lower bound to E_0. Let

$$K = \langle\psi|(H - E_0)(H - E_1)|\psi\rangle. \tag{13}$$

Using the expansion of $|\psi\rangle$ as above, we get

$$K = \sum_m |C_n|^2 (E_n - E_0)(E_n - E_1)$$

so that
$$K \geq 0. \tag{14}$$

Hence,
$$J_2 - E_1 J_1 + E_0 (E_1 - J_1) \geq 0.$$

Thus,
$$E_0 \geq J_1 - \frac{J_2 - J_1^2}{E_1 - J_1}. \tag{15}$$

This inequality, however, requires a knowledge of E_1, the energy eigenvalue of the first excited state.

For a discussion of Kato's method of obtaining both upper and lower bounds to any eigenvalue and the calculation in the case of the screened Coulomb potential, see Ballentine (1998).

The variational method provides excellent estimates of ground state (or, higher) energy of a system. The wave function is usually not determined accurately. If one calculates the expectation values of other observables, the results are not as reliable as the calculated energy, because there may not be any minimum theorem.

13.2 THE LINEAR VARIATION METHOD

In this method, the set of trial functions is constructed by a linear combination of a finite number of linearly independent functions ϕ_1, ϕ_2, ..., ϕ_N:

$$|\psi\rangle = \sum_{n=1}^{N} C_n |\phi_n\rangle, \tag{16}$$

where the coefficients C_i are variational parameters to be determined by minimizing $E[\psi]$. The basis functions ϕ_i form a linear space (a subspace of the Hilbert space). Substituting Eq. (16) into the functional Eq. (1) we get

$$E[\psi] = \frac{\displaystyle\sum_{n=1}^{N} \sum_{n'=1}^{N} C_{n'}^* C_n H_{n'n}}{\displaystyle\sum_{n=1}^{N} \sum_{n'=1}^{N} C_{n'}^* C_n S_{n'n}}, \tag{17}$$

where

$$H_{n'n} = \langle \phi_{n'} | H | \phi_n \rangle, \tag{18}$$

and the overlap integral

$$S_{n'n} = \langle \phi_{n'} | \phi_n \rangle. \tag{19}$$

If the functions ϕ_n are orthonormal, then $S_{n'n} = \delta_{n'n}$

We rewrite Eq. (17) as

$$E[\psi] = \sum_{n=1}^{N} \sum_{n'=1}^{N} C_{n'}^* C_n S_{n'n} = \sum_{n=1}^{N} \sum_{n'=1}^{N} C_{n'}^* C_n H_{n'n} \tag{17a}$$

To obtain the values of the parameters $\{C_n\}$ for which the functional $E[\psi]$ is stationary, we differentiate Eq. (17a) with respect to each C_n or C_n^*. The set of N conditions $\dfrac{\partial E}{\partial C_n} = 0 \left(\text{or,} \dfrac{\partial E}{\partial C_{n'}^*} \right)$, yields a set of N linear and homogeneous equations in the variables C_i's

$$\sum_{n=1}^{N} C_n (H_{n'n} - S_{n'n} E) = 0; \quad n' = 1, 2, ..., N \tag{20}$$

This is an $N \times N$ matrix eigenvalue equation. For a non-trivial solution to this system of equations, the determinant of the coefficients must vanish.

So
$$\det|H_{n'n} - S_{n'n}E| = 0 \qquad (21)$$

The expansion of this determinant gives an algebraic equation of degree N in the unknown E. Let $E_0^{(N)}$, $E_1^{(N)}$, ..., $E_{N-1}^{(N)}$ be the N roots of this equation, arranged in ascending order. Here N indicates that we are dealing with an $N \times N$ determinant. The lowest root $E_0^{(N)}$ is an upper bound to the ground state energy E_0. It can be shown that each root $E_i^{(N)}$ ($i = 1, 2, ..., N - 1$) is an upper bound to the corresponding exact eigenvalue E_i. Substituting E_0 in Eq. (20) and solving this set of equations for the coefficients C_i in terms of one of them, we obtain an approximate ground state wave function. Using higher roots, one can obtain approximate wave functions for excited states. These approximate wave functions may be shown to be mutually orthogonal.

For a trial function ψ' combining an additional basis function ϕ_{N+1}, i.e.,

$$|\psi'\rangle = \sum_{n-1}^{N+1} C_n |\phi_n\rangle \qquad (22)$$

it can be shown the "new" $N + 1$ roots $E_0^{(N+1)}$, $E_1^{(N+1)}$ ..., $E_N^{(N+1)}$ of the secular equation are separated by the "old" roots $E_0^{(N)}$, $E_1^{(N)}$ ..., $E_{(N-1)}^{(N)}$. Clearly, the approximate eigenvalues for successive values of N are interleaved and in the limit $N \to \infty$ converge from above to the exact eigenvalues (see Fig. 13.1). This is known as the Hylleraas-Undheim theorem.

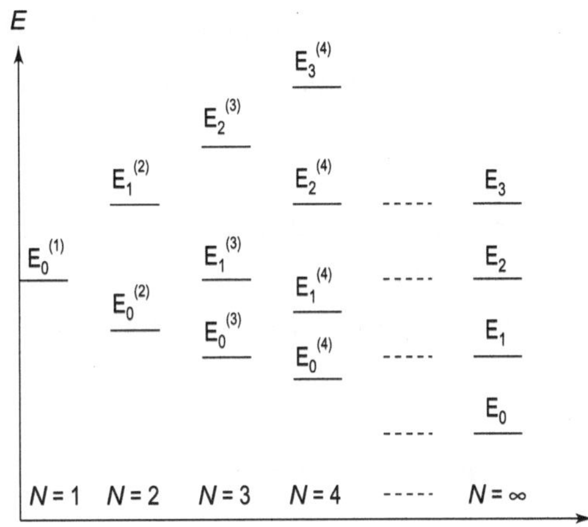

Fig. 13.1 Diagram showing the Hylleraas-Undheim theorem

13.3 THE VARIATION–PERTURBATION METHOD

The variation-perturbation method can be used to estimate the second and higher-order perturbation corrections to the energy eigenvalues and the wave functions without evaluating the infinite summations.

Let H_0 be the unperturbed Hamiltonian and H' the perturbation of the system. Let $|\psi_m\rangle$ be an eigenket of H_0 with E_m the corresponding eigenvalue: $H_0|\psi_m\rangle = E_m|\psi_m\rangle$.

The first-order correction to the energy is $E_m^{(1)} = \langle\psi_m|H'|\psi_m\rangle$. We assume that these are known for a given state m of the system. Let $\phi_m^{(1)}$ be an arbitrary trial function. We consider the functional

$$F_1[\phi_m^{(1)}] = \langle\phi_m^{(1)}|H_0 - E_m|\phi_m^{(1)}\rangle + 2\,\langle\phi_m^{(1)}|H' - E_m^{(1)}|\phi_m^{(1)}\rangle. \tag{23}$$

This functional is stationary for variations of $\phi_m^{(1)}$ about ψ_k. The variational equation

$$\delta F_1 = 0 \tag{24}$$

requires that

$$(H_0 - E_m)\,\phi_m^{(1)} + (H' - E_m^{(1)})\,\psi_m = 0. \tag{25}$$

Comparing Eq. (25) with Eq. (23), we see that the function $\phi_k^{(1)}$ which makes F_1 stationary is a solution $\psi_m^{(1)}$, the first-order perturbed wave function. Also, $F_1\lfloor\phi_k^{(1)}\rfloor$ is identical to $E_m^{(2)}$, the second-order perturbed energy, when $\phi_m^{(1)} = \psi_m^{(1)}$.

In case m refers to the lowest eigenvalue of H_0, then

$$E_m^{(2)} \leq F_1\,[\phi_m^{(1)}] \tag{26}$$

We choose a trial function $\phi_m^{(1)}$ with parameters. $E_m^{(2)}$ can be evaluated by minimizing the functional $F_1\,[\phi_m^{(1)}]$ with respect to the variational parameters. This minimum value of F_1 gives an upper bound to $E_m^{(2)}$. The corresponding function $\phi_m^{(2')}$ can be used to calculate an approximate value of $E_m^{(3)}$. Approximate expressions yielding upper bounds of higher order terms like $E_m^{(4)}$, $E_m^{(5)}$ and the wave function $\psi_m^{(2)}$ can be obtained.

There are many connections between perturbation theory and the variation method (Epstein (1974)).

13.4 THE HELLMANN-FEYNMAN AND VIRIAL THEOREMS

We consider a system with a time-independent Hamiltonian which depends on a (real) parameter λ. Given a variational approximation such that $\delta\langle\psi|H|\psi\rangle = 0$ and $\delta\langle\psi|\psi\rangle = 0$, we can prove the generalized Hellmann-Feynman theorem which states that $\dfrac{\partial E(\lambda)}{\partial\lambda}$ is equal to the expectation value of $\dfrac{\partial H(\lambda)}{\partial\lambda}$ if variations $\delta\psi$ include all variations induced by $\delta\lambda$:

$$\frac{\partial E}{\partial \lambda} = \left\langle \frac{\partial H}{\partial \lambda} \right\rangle, \tag{27}$$

where the angular brackets $\langle \text{---} \rangle$ denote the expectation value, $\dfrac{\langle \psi | ... | \psi \rangle}{\langle \psi | \psi \rangle}$, calculated using ψ.

Let each trial function depend on λ and $\delta\psi = \delta\lambda \dfrac{\partial\psi}{\partial\lambda}$ be in the Hilbert space of the set of trial functions. For variations due to $\delta\lambda$, we have

$$\delta \langle \psi | H - E | \psi \rangle = 0, \text{ i.e.,}$$

$$\langle \delta\psi | H - E | \psi \rangle \langle \psi | H - E | \delta\psi \rangle + \delta\lambda \left(\left\langle \psi \left| \frac{\partial H}{\partial\lambda} - \frac{\partial E}{\partial\lambda} \right| \psi \right\rangle \right) = 0.$$

Now, for variational wave functions and energy values,

$$\langle \delta\psi | H - E | \psi \rangle \langle \psi | H - E | \delta\psi \rangle = 0. \text{ Hence,}$$

$$\frac{\partial E}{\partial\lambda} = \left\langle \psi \left| \frac{\partial H}{\partial\lambda} \right| \psi \right\rangle \Big/ \langle \psi | \psi \rangle = \left\langle \frac{\partial H}{\partial\psi} \right\rangle,$$

which is the variational version of the generalized Hellmann-Feynman theorem.

Let us consider a simple harmonic oscillator with the Hamiltonian $H = -\dfrac{\hbar^2}{2m}\dfrac{d^2}{dx^2} + \dfrac{1}{2}kx^2$.

By taking the force constant k as the parameter λ, we have $\dfrac{\partial H}{\partial k} = \dfrac{1}{2}x^2$. Since the energy levels are given by

$$E_n = \left(n + \frac{1}{2}\right)h\nu = \left(n + \frac{1}{2}\right)\frac{h(k/m)^{1/2}}{2\pi}, \text{ we have,}$$

$$\frac{\partial E_n}{\partial k} = \frac{1}{2}\left(n + \frac{1}{2}\right)h\nu/k.$$

Equation (27) gives $\quad \langle x^2 \rangle = \left(n + \dfrac{1}{2}\right)h\nu/k.$

Using the Hellmann-Feynman theorem, we have found $\langle x^2 \rangle$ for any harmonic oscillator stationary state.

Hellmann and Feynman independently applied Eq. (27) to molecules where the molecular electronic Hamiltonian depends parametrically on the nuclear coordinate. See Levine (2000) for a discussion of the Hellmann-Feynman electrostatic theorem.

We next discuss the hypervirial theorem which derives conservation laws from invariant transformations. Let G be a hermitian operator, and let us consider a unitary transformation of the Schrödinger equation.

$U(H - E)|\psi\rangle = U(H - E)\, U^{+} U|\psi\rangle = 0$. We suppose that the set of trial functions is invariant to the continuous family of unitary transformation $U(\alpha) = e^{i\alpha G}$ parametrized by α (real). Then $iG\psi$ is a possible variation of ψ within the set and hence satisfies

$$\langle iG\psi|H - E|\psi\rangle + \langle \psi|H - E|iG\psi\rangle = 0,$$

or,
$$\langle \psi\,|[H,\,G]|\,\psi\rangle = 0,$$

where $[H,\,G]$ denotes the commutator of H and G. Since the normalization is preserved, ψ satisfies the hypervirial theorem for G:

$$\langle [H,\,G]\rangle = 0. \tag{28}$$

Incase we have a set of trial functions invariant to a family of transformations $U = e^{iG(\alpha)}$ such that $G(0) = 1$, then the hypervirial theorem assumes the form

$$\left\langle \left[H, \left(\frac{\partial G}{\partial \alpha} \right)_{\alpha \to 0} \right] \right\rangle = 0. \tag{29}$$

We now derive the virial theorem from Eq. (28). We consider a system of N-particles. We choose

$$G = \sum_i q_i p_i = -i\hbar \sum_i q_i \frac{\partial}{\partial q_i}$$

where the sum runs over the $3N$-coordinates of the N-particles. Then

$$\left[H, \sum_i q_i p_i \right] = \sum_i [H,\, q_i p_i]$$

$$= \sum_i q_i\, [H,\, p_i] + \sum_i [H,\, q_i]\, p_i$$

$$= i\hbar \sum_i q_i \frac{\partial V}{\partial q_i} - i\hbar \sum_i \frac{1}{m_i}\, p_i^2$$

$$= i\hbar \sum_i q_i \frac{\partial V}{\partial q_i} - 2i\hbar T,$$

where T and V are the kinetic – and potential-energy operators for the system. Substituting this expression into Eq. (28), we obtain for a stationary state

$$2\langle T\rangle = \left\langle \sum_i q_i \frac{\partial V}{\partial q} \right\rangle. \tag{30}$$

Equation (30) is the quantum mechanical virial theorem. If the interaction potential is spherically symmetric and V is a homogeneous function of degrees then Eq. (30) simplifies to

$$2\langle T\rangle = s\, \langle V\rangle. \tag{31}$$

Since $\langle T \rangle + \langle V \rangle = E$, we write Eq. (31) in the form

$$\left. \begin{array}{l} \langle V \rangle = \dfrac{2E}{s+2} \\[3mm] \langle T \rangle = \dfrac{sE}{s+2} \end{array} \right\} \qquad (32)$$

For the simple harmonic oscillator $V = \dfrac{1}{2} kx^2$.
Hence $s = 2$, and we get

$$\langle T \rangle = \langle V \rangle = \frac{1}{2} E = \frac{1}{2} h\nu \left(n + \frac{1}{2} \right).$$

For the hydrogen atom bound stationary state, $s = -1$, and hence $2\langle T \rangle = -\langle V \rangle$ or $\langle V \rangle = 2E$ and $\langle T \rangle = -E$.

For the application of the virial theorem to molecules and chemical bonding, see Levine (2000).

13.5 VAN DER WAALS' INTERACTION

We now apply the variation method to the calculation of the long-range interaction, or van der Waals force, between two atoms at large distances. First however, we study this problem for two hydrogen atoms in their ground states by means of the perturbation theory.

We assume that the two protons of the two hydrogen atoms are fixed in space at a distance R apart, large compared to the Bohr radius. With the notation as shown in Fig. 13.2, the Hamiltonian H of the system can be written

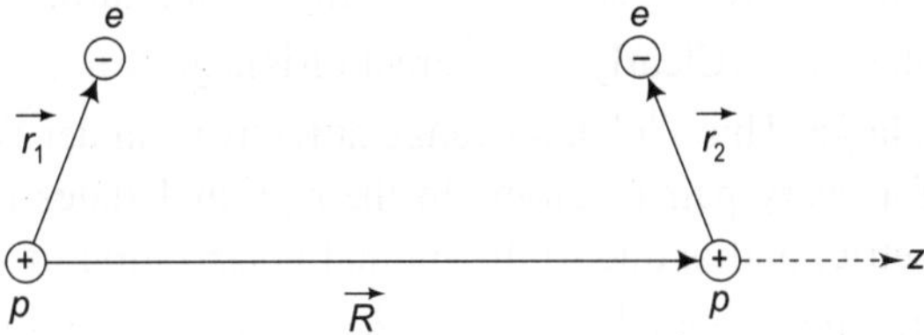

Fig. 13.2 The coordinates for calculating the van der Waals' interaction between two hydrogen atoms

$$H = H_0 + V$$

$$H_0 = -\frac{\hbar^2}{2m} (\vec{\nabla}_1^2 + \vec{\nabla}_2^2) - \frac{e^2}{r_1} - \frac{e^2}{r_2} \qquad (33)$$

$$V = \frac{e^2}{R} + \frac{e^2}{|\vec{R} + \vec{r}_2 - \vec{r}_1|} - \frac{e^2}{|\vec{R} + \vec{r}_2|} - \frac{e^2}{|\vec{R} - \vec{r}_1|}$$

The unperturbed energy and wave functions of the two non-interacting hydrogen atoms in their ground states, are given by

$$E_0^{(0)} = \frac{e^2}{a_0}, \ u_0^{(0)} (\vec{r}_1, \vec{r}_2) = u_{100}(\vec{r}_1) \ u_{100}(\vec{r}_2)$$

$$= (\pi a_0^3)^{-1} \ e^{-(r_1 + r_2)/a_0} \tag{34}$$

where a_0 is the Bohr radius.

Now, for R large ($\gg a_0$) we expand the perturbation V in powers of $\vec{r}_i/R$ (r_1, r_2 are of the order of a_0) and keep the lowest terms

$$V = \frac{e^2}{R^3} \left(\vec{r}_1 \cdot \vec{r}_2 - 3 \ \frac{(\vec{r}_1 \cdot \vec{R}) \ (\vec{r}_2 \cdot \vec{R})}{R^2} \right)$$

$$= \frac{e^2}{R^3} (x_1 x_2 + y_1 y_2 - 2 z_1 z_2). \tag{35}$$

This term represents the interaction energy of two electric dipoles $e\vec{r}$ and $e\vec{r}_2$ separated by R. The higher order terms neglected represent higher order multipole interactions. So every term in V can be expressed as spherical harmonics Y_l^m with $l_i > 0$ for each atom. Hence, each term in V will have zero expectation value for u_0, since the ground state wave function u_0 has $l_i = 0$. Thus the first-order perturbation energy matrix element is zero. The leading term in the interaction energy is the second-order perturbation of the dipole-dipole term, and is proportional to $1/R^6$. The second-order perturbation is

$$E^{(2)}(R) = \frac{e^2}{R^6} \sum_{n \neq 0} \frac{|\langle 0^{(0)}| \ x_1 x_2 + y_1 y_2 - 2 z_1 z_2 |n^{(0)} \rangle|^2}{E_0^{(0)} - E_n^{(0)}} \tag{36}$$

where the summation is over all states of the pair of unperturbed hydrogen atoms including dissociated states. Clearly this interaction is negative ($E_n^{(0)} > E_0^{(0)}$) and varies as $1/R^6$ when R is large. This $1/R^6$ long-range attractive van der Waals interaction is characteristic for every pair of atoms in their ground states (non-degenerate, spherically symmetric) and is due to the mutual polarization of the atoms.

We obtain an upper bound for $-E^{(2)}(R)$ by assuming that each energy eigenvalue E_n is replaced by E_{n*}, the energy of the lowest excited state of the two hydrogen atoms for which $\langle 0|V|n^* \rangle \neq 0$. Now, the denominator on the right hand side of Eq. (36) can be taken outside the summation. The summation is

$$\sum_{n \neq 0} |\langle 0|V|n \rangle|^2 = \sum_n \langle 0|V|n \rangle \ \langle n|V|0 \rangle - (\langle 0|V|0 \rangle)^2$$

$$= \langle 0|V^2|0 \rangle \ \text{since} \ \langle 0|V|0 \rangle = 0$$

Hence,

$$-E^{(2)}(R) \le \frac{\langle 0 | V^2 | 0 \rangle}{E_{n*} - E_0}. \tag{37}$$

The lowest excited state $n*$ of the system is the one where both the hydrogen atoms are excited to $n = 2$ states. Now, $E_0 = -2\,(e^2/2a_0)$ and $E_{n*} = -2\,(e^2/8a_0)$. From Eq. (35) we have

$$V^2 = \frac{e^4}{R^6}\,(x_1^2 x_2^2 + y_1^2 y_2^2 + 4z_1^2\,z_2^2 + 2x_1 x_2 y_1 y_2 \,...) \tag{38}$$

The expectation value of the cross terms like $x_1 x_2 y_1 y_2$ is zero. The other terms involve integrals like

$$\int x^2 |u_{100}(\vec{r})|^2\,d\vec{r} = \frac{1}{3}\int r^2 |u_{100}(\vec{r})|^2\,d^3 r$$

$$= \frac{1}{3\,\pi a_0^3}\cdot 4\pi \int_0^\infty r^2 e^{-2r/a_0} r^2\,dr = a_0^2.$$

Hence the first three terms on the right hand side (R.H.S.) of Eq. (38) contribute $a_0^2 \cdot a_0^2 + a_0^2 \cdot a_0^2 + 4a_0^2 \cdot a_0^2 = 6a_0^4$, and so $\langle 0 | V^2 | 0 \rangle = \dfrac{6e^4 a_0^4}{R^6}$.

Substituting in Eq. (37), we get

$$E^{(2)}(R) \ge -\frac{8e^2\,a_0^5}{R^6}. \tag{39}$$

Let us write the energy of both the atoms up to second-order in perturbation theory

$$E_0 = -\frac{e^2}{a_0}\left[1 + \xi\left(\frac{a_0}{R}\right)^6\right], \tag{40}$$

where ξ is a numerical positive constant. Thus perturbation theory enables us to estimate a lower bound for the ground state energy of two hydrogen atoms far apart: $\xi \le 8$.

We now obtain an upper limit on $E^{(2)}(R)$ by the variation method. We choose as trial function.

$$\psi\,(\vec{r}_1, \vec{r}_2) = u_{100}(\vec{r}_1)\,u_{100}(\vec{r}_2)\,[1 + AV]$$

$$= \frac{1}{(\pi a_0)^3}\,e^{-\frac{(r_1 + r_2)}{a_0}}\,[1 + AV]. \tag{41}$$

where A is a (real) variation parameter. The average value of the energy is

$$E\,[\psi] = \frac{\langle \psi | H_0 + V | \psi \rangle}{\langle \psi | \psi \rangle}$$

$$= \frac{\langle 0|(1 + AV)\,(H_0 + V)\,(1 + AV)|0\rangle}{\langle 0|(1 + AV)^2|0\rangle}$$

$$= \frac{E_0 + 2A\langle 0|V^2|0\rangle + A^2\,\langle 0|VH_0 V|0\rangle}{1 + A^2\,\langle 0|V^2|0\rangle}, \tag{42}$$

where $E_0 = -e^2/a_0$ and $\langle 0|V|0\rangle = \langle 0|V^3|0\rangle = 0$.

Now, $\langle 0|V[VH_0 V]|0\rangle = \langle 0|V[H_0, V]|0\rangle - \langle 0|V^2 H_0|0\rangle$

$$= \langle 0|V[H_0, V]|0\rangle - \frac{6e^6 a_0^3}{R^6}\langle 0|V\,[H_0,\,V]|0\rangle$$

$$= -\frac{a_0 e^6}{R^6}\left\langle 0\left|[(\vec{r}_1 \cdot \vec{r}_2) - 3(\vec{r}_1 \cdot \hat{R})(\vec{r}_2 \cdot \hat{R})]^2\,\frac{1}{r_1}\frac{\partial}{\partial r_1}\right|0\right\rangle$$

So $\quad \langle 0|VH_0 V|0\rangle = 0$

We expand the R.H.S. of Eq. (42) to terms of order V^2 and obtain

$$(E_0 + 2A\langle 0|V^2|0\rangle)\,(1 + A^2\,\langle 0|V^2|0\rangle)^{-1}$$

$$\simeq E_0 + (2A - E_0 A^2)\,\langle 0|V^2|0\rangle. \tag{43}$$

The minimum of Eq. (43) with respect to variation of A is obtained when $A = 1/E_0$. Hence

$$E_0 + E^{(2)}(R) \le E_0 + \frac{\langle 0|V^2|0\rangle}{E_0} = E_0 - \frac{6e^2 a_0^5}{R^6}. \tag{44}$$

Combining Eqs. (39) and (44), we have both upper and lower limits on the interaction energy.

$$-\frac{8e^2 a_0^5}{R^6} \le E^{(2)}(R) \le -\frac{6e^2 a_0^5}{R^6} \tag{45}$$

We conclude

$$6 \le \xi \le 8 \tag{46}$$

Precise calculations lead to the value $\xi = 6.499027$.

13.6 THE WKB APPROXIMATION

WKB theory is a powerful tool for obtaining approximate solutions of a linear differential equation whose highest derivative is multiplied by a small parameter. The WKB approximation order by order in powers of this parameter to the solution consists of exponentials of elementary integrals of algebraic functions. For obtaining approximate solution of the Schrödinger equation this approximation is based on an expansion of the wave function in powers of $\hbar$, which, although asymptotic, is very useful. This method is called the WKB approximation after Wentzel, Kramers and Brillouin who independently introduced it in quantum mechanics, although the general mathematical method was considered earlier by Green, Liouville, Rayleigh and Jeffreys. It is often called the JWKB method, the semi-classical approximation, the phase-integral method or the asymptotic approximation method. It is readily applicable to one dimensional problems and in higher dimensions in which the wave equation can be separated into one or more ordinary differential equations.

Let us consider the time-independent Schrödinger equation describing the one-dimensional motion of a particle of mass m in a potential $V(x)$:

$$\left[E - \frac{\hbar^2}{2m} \frac{d^2}{dx^2} - V(x) \right] \psi(x) = 0 \tag{47}$$

or,
$$\left(\frac{d^2}{dx^2} + k^2(x) \right) \psi(x) = 0, \quad k^2 > 0 \tag{47a}$$

where

$$[k(x)]^2 = \frac{1}{\hbar^2} [p(x)]^2$$

$$[p(x)]^2 = 2m \, (E - V(x)) \tag{48}$$

$\pm \, p(x)$ is the classical momentum at x. If the potential were a constant, say V_0, then solutions of Eq. (47) would be linear combination of plane waves.

$$\psi(x) = A e^{\pm \frac{1}{\hbar} p_0 x}$$

where
$$p_0 = [2m \, (E - V_0)]^{1/2}$$

and A is a constant.

We now assume that the potential is slowly varying, i.e., $V(x)$ changes only slightly over the de Broglie wavelength $\lambda = \dfrac{h}{p(x)} = \dfrac{2\pi\hbar}{p(x)}$ associated with the particle. So we may still seek the solution as linear combination of plane waves with position-dependent wavelength

$$\psi(x) = A e^{\pm \frac{i}{\hbar} \int dx' p(x')} \tag{49}$$

This is valid if the potential is slowly varying, i.e. the wavelength ($\delta\lambda$) is small compared to the characteristic distance (λ) over which the potential varies, i.e.,

$$\left|\frac{\delta\lambda}{\lambda}\right| \ll 1,$$

or,

$$\left|\frac{\frac{d\lambda}{dx} \cdot \lambda}{\lambda}\right| \ll 1,$$

or,

$$\left|\frac{d\lambda}{dx}\right| \ll 1. \tag{50}$$

To analyze this systematically, we seek solutions of Eq. (47a) in the form

$$\psi(x) = A(x)\, e^{iS(x)/\hbar} \tag{51}$$

with real amplitude $A(x)$ and real phase $S(x)$. This exponential form is not suitable for obtaining asymptotic approximations since both $A(x)$ and $S(x)$ are implicit functions of $\hbar$. As we are interested in the behavior of the solutions as $\hbar \to 0$, it is best to expand $A(x)$ and $S(x)$ as series in powers of $\hbar$ and combining them as a single exponential power series of the form

$$\psi(x) \sim \exp\left[\frac{i}{\hbar} \sum_{n=0}^{\infty} \hbar^n S_n(x)\right], \quad \hbar \to 0., \tag{51a}$$

Hence it is sufficient for our purpose that we write

$$\psi(x) = A e^{iS(x)/\hbar}, \tag{51b}$$

where A may be treated as constant.

Substituting Eq. (51b) in Eq. (47a), we have

$$\frac{i}{\hbar}\frac{d^2 S(x)}{dx^2} - \left[\frac{1}{\hbar}\frac{dS(x)}{dx}\right]^2 + \frac{p^2}{\hbar^2} = 0. \tag{52}$$

This equation is, however, non-linear. We expand $S(x)$ in powers of $\hbar$* (parameter of smallness) as in Eq. (51a)

$$S(x) = S_0(x) + \hbar S_1(x) + \hbar^2 S_2(x) + \dots . \tag{53}$$

Substituting the expansion of S into Eq. (52) and equating equal powers of $\hbar$, we get the set of equations (where primes denote differentiation with respect to x)

$$-S_0'^2 = p^2(x) = 0, \tag{54}$$

$$iS_0'' - 2S_0' S_1' = 0, \dots . \tag{54a}$$

which must be solved successively to get $S_0(x)$, $S_1(x)$, ... Integration of Eq. (54) gives

$$S_0(x) = \pm \int^x p(x')\, dx' \tag{55}$$

where the arbitrary constant of integration which can be absorbed in the coefficient A on the right side of Eq. (51) has been omitted. This provides the classical limit of $S(x)$ ($\hbar \to 0$). Using Eq. (55) in Eq. (54a), we obtain by integration.

$$S_1(x) = \frac{i}{2} \ln S_0' = i \ln \sqrt{p(x)}. \tag{56}$$

To this order of approximation, we have

$$S(x) = S_0(x) + \hbar S_1(x) = \pm \int dx'\, p(x') + i\hbar \ln (p(x))^{1/2}, \tag{57}$$

and hence we obtain the WKB (or, semiclassical) approximation to the wave function in a classically allowed region ($V(x) < E$)

$$\psi(x) = A e^{\pm \frac{i}{\hbar} \int^x dx' p(x') - \ln (p(x))^{1/2}}$$

$$= \frac{A}{(p(x))^{1/2}}\, e^{\pm \frac{i}{\hbar} \int^x p(x')\, dx'}. \tag{58}$$

Hence the general WKB solution in this region is

$$\psi(x) = \frac{1}{(p(x))^{1/2}} \left\{ A e^{\frac{i}{\hbar} \int^x p(x')\, dx'} + B e^{-\frac{i}{\hbar} \int^x p(x')\, dx'} \right\}, \; E > V \tag{58a}$$

where A and B are arbitrary constants.

The accuracy of these WKB solutions can be estimated by comparing the magnitudes of the successive terms in the expansion of S. Thus, we expect the solutions to be useful in that region of x when

$$\left| \frac{\hbar S_1}{S_0} \right| \ll 1.$$

Since $|S_0(x)|$ is a monotonic increasing function of x, unless $p(x) = 0$, $|\hbar S_1/S_0|$ is expected to be small if $|\hbar S_1'/S_0'|$ is small. Now

$$|\hbar S_1'/S_0'| = \left| \frac{\hbar p'}{2p^2} \right| = \frac{1}{4\pi} \left| \frac{d\lambda}{dx} \right| \ll 1., \tag{59}$$

$\lambda = 2\pi\hbar/p$ being the local de Broglie wavelength. This agrees with the earlier estimate Eq. (50). So the WKB solutions are useful when the potential changes so slowly that the momentum of the particle is reasonably constant over many wavelengths.

So far we have considered classically allowed region for which $V(x) < E$. In classically forbidden region when $V(x) > E$, we write the one-dimensional wave equation as

$$\left(\frac{d^2}{dx^2} - K^2(x)\right)\psi(x) = 0, \quad K^2 > 0 \tag{60}$$

where

$$[K(x)]^2 = \frac{1}{\hbar^2}[2m(V(x) - E)], \quad V(x) > E \tag{61}$$

The general WKB solution in this region is

$$\psi(x) = \frac{1}{(K(x))^{1/2}}\left\{Ce^{\int^x K(x')dx'} + De^{-\int^x K(x')dx'}\right\}, \quad E\langle V(x) \tag{62}$$

where C and D are arbitrary constants.

The WKB solutions diverge when k and K are zero, where $V(x) = E$, i.e., at the classical turning points. So near these points the basic condition Eq. (59) is violated as the wavelength becomes infinite. Thus, the two kinds of WKB solutions Eqs. (58a) and (62) are asymptotically valid, so that they can be used sufficiently far from the nearest turning point if the wavelength there is slowly varying.

13.7 THE CONNECTION FORMULAE

The wave equations Eqs. (47) and (60) are regular at a turning point and so the exact wave function is continuous and smooth for all x (including a turning point). Hence it should be possible to obtain connection formulas which allow us to connect the two types of asymptotic solutions (an oscillating solution like Eq. (58a) and an exponential one like Eq. (62)) across a turning point.

Let as consider one isolated turning point which, without loss of generality, can be taken as the origin of x (See Fig. 13.3). We also assume that $V(x) < E$ to the right of the turning point, so that $x > 0$ is a classically allowed region (I). In this region, $p(x)$ is real and the semi-classical solution is given by Eq. (58a):

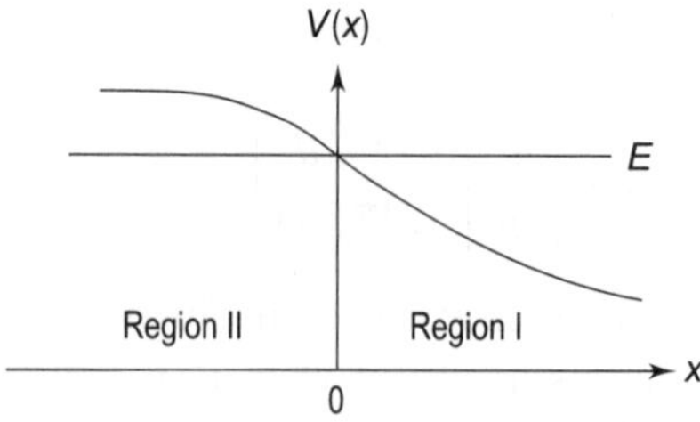

Fig. 13.3 Classical turning point at $x = 0$

$$\psi(x) = \frac{1}{(p(x))^{1/2}} \left\{ A e^{\frac{i}{\hbar}\int_0^x p(x')\,dx'} + B e^{-\frac{i}{\hbar}\int_0^x p(x')\,dx'} \right\}, \; x > 0 \tag{62a}$$

In region II $(x < 0)$, the classically inaccessible region $(V(x) > E)$ the semi-classical solution is given by (62):

$$\psi(x) = \frac{1}{(K(x)^{1/2}} \left\{ C e^{\int K(x')\,dx'} + D e^{-\int K(x')\,dx'} \right\}, \; x < 0. \tag{62b}$$

If the semi-classical solution were an analytic function of x near the turning point $x = 0$, then we could analytically continue the solution in region I to region II, and thus get the relation among the constants A, B, C and D. However, we cannot pass from region I to II along the real axis since the semi-classical approximation breaks down at $x = 0$. One can go around the point $x = 0$ in the complex plane along an arc of large radius and hope to obtain the result. This analytic continuation procedure gives the wrong results due to the Stokes' phenomenon (see Gottfried and Yan (2003), Bender and Orszag (1978), Morse and Feshbach (1953)). One must exercise caution in dealing with this subtle matter. We shall indicate later how it should be done.

Another method consists in looking for the exact solution near $x = 0$, taking the potential to be a linear function of x in this region. Near the turning point at the origin,

$$V(x) \simeq V(x = 0) + \alpha x,$$

$$= E + \alpha x, \tag{63}$$

where $\alpha = V'(x = 0) < 0$ with x measuring the distance from the turning point. The Schrödinger equation (47) becomes

$$\left(\frac{d^2}{dz^2} - z \right) \psi(z) = 0, \tag{64}$$

where $z = (2m\alpha/\hbar^2)^{1/3}\, x$ and $\psi(z(x)) = \psi(z)$. The general solution of Eq. (64) is well known and can be expressed as a linear combination of Airy functions $Ai(z)$ and $Bi(z)$. For z positive $Ai(z)$ decreases and $Bi(z)$ increases exponentally; for z negative both are oscillatory and for $|z|$ large their phase difference approaches $\pi/2$. The asymptotic expansions of Airy functions are given below

$$z > 0 \quad Ai(z)_{z\to\infty} \sim \frac{1}{2}\,\pi^{-1/2}\, z^{-1/4} \exp\left(-\frac{2}{3} z^{3/2} \right) \tag{65}$$

$$Bi(z)_{z\to\infty} \sim \pi^{-1/2}\, z^{-1/4} \exp\left(\frac{2}{3}\, z^{3/2} \right) \tag{66}$$

and

$$Ai(-z)_{z \to \infty} \sim \pi^{-1/2} \, z^{-1/4} \, \sin\left(\frac{2}{3} \, z^{3/2} + \frac{\pi}{4}\right) \qquad (67)$$

$$Bi(-z)_{z \to \infty} \sim \pi^{-1/2} \, z^{-1/4} \, \cos\left(\frac{2}{3} \, z^{3/2} + \frac{\pi}{4}\right). \qquad (68)$$

For $z > 0$

$$\hbar K(x) = \sqrt{2m\,(V(x) - E)} = \sqrt{2m\alpha x}$$

so that

$$\int_0^{-x} K(x')dx' = \frac{1}{\hbar} \, \sqrt{2m\alpha} \int_0^x dx' x'^{1/2}$$

$$= \frac{1}{\hbar} \, \sqrt{2m\alpha} \cdot \frac{2}{3} \, x^{3/2}$$

$$= \frac{2}{3} \, z^{3/2}. \qquad (69)$$

For $z < 0$

$$\hbar k(x) = \sqrt{2m\,(E - V(x))} = \sqrt{-2m\alpha x}$$

so that

$$\int_0^x k(x')dx' = \frac{1}{\hbar} \int_0^{-x} dx' = \sqrt{2m\alpha x'}$$

$$= \frac{1}{\hbar} \, \sqrt{2m\alpha} \cdot \frac{2}{3} \, (-x)^{3/2}$$

$$= \frac{2}{3} \, (-z)^{3/2}. \qquad (70)$$

Since

$$\sin\left(\phi + \frac{\pi}{4}\right) = \cos\left(\phi - \frac{\pi}{4}\right)$$

we can write

$$\frac{1}{2} \, (K(x))^{-1/2} \, e^{-\int_0^x dx' K(x')} \to (k(x))^{-1/2} \, \cos\left(\int_x^0 dx' \, k(x') - \frac{\pi}{4}\right). \qquad (71)$$

The arrow implies that the asymptotic solutions in region II that appears on the left goes into the asymptotic solution in region I that appears on the right.

Since $$\cos\left(\phi + \frac{\pi}{4}\right) = -\sin\left(\phi - \frac{\pi}{4}\right)$$

we write

$$-(K(x))^{-1/2} \, e^{-\int_0^x dx' K(x')} \leftarrow (k(x))^{-1/2} \, \sin\left(\int_x^0 dx' \, k(x') - \frac{\pi}{4}\right). \qquad (72)$$

The connection formulas can be employed in the direction indicated so that the solutions in the classically inaccessible regions are matched onto the solutions in the classically accessible regions.

The connection formulas for a right hand barrier can be obtained analogously.

We now obtain these relations without using the Airy functions. The treatment is based on Migdal and Krainov (1969). The differential equation (47) is extended from real values of x to the complex plane. In the complex plane, we write $x = \rho e^{i\phi}$, where the radius ρ around the point $x = 0$ is of the order such that quasi-classical approximation can be used and the potential can be taken to be linear in x.

Then $k(x) = \alpha' x^{1/2}$ and $\int_0^x k(x')\,dx = \left(\dfrac{2}{3}\,\alpha'\right) x^{3/2}$. In the complex plane $x = \rho e^{i\phi}$

and so $\int_0^x k(x')\,dx' = \left(\dfrac{2}{3}\,\alpha'\right) \rho^{3/2} \exp\left[\dfrac{3}{2}\,i\phi\right]$.

Hence,
$$\frac{1}{\sqrt{k}}\,\exp\left[i\int_0^x k\,dx\right]$$

$$= \frac{1}{\sqrt{\alpha'\rho^{1/2}}}\,e^{-\frac{i\phi}{4}}\,\exp\left[\frac{2}{3}\,\alpha'\rho^{3/2}\left(i\cos\frac{3}{2}\,\phi - \sin\frac{3}{2}\,\phi\right)\right] \qquad (73)$$

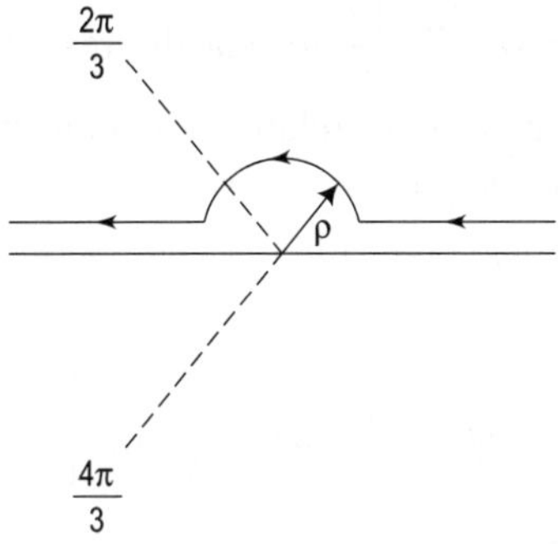

Fig. 13.4   Contour with boundaries at $\phi = 0$, $\dfrac{2\pi}{3}$, $\dfrac{4\pi}{3}$

As we go around the contour, $\sin\dfrac{3}{2}\,\phi > 0$ so long as $\phi < \dfrac{2\pi}{3}$ and Eq. (73) contains a decreasing exponential. When $\phi > \dfrac{2\pi}{3}$, the exponential is an increasing one (dominant) and then we can neglect all terms containing decreasing one.

As we go around the circle of radius ρ in the upper half plane the expression $i\left(\dfrac{2}{3}\right)\alpha' x^{3/2}$ goes at first into a decreasing exponential and then, after crossing the line $\phi = \dfrac{2\pi}{3}$ into an increasing exponential one. As soon as the increasing exponential appears the approximate nature of the semi-classical approximation introduces an exponentially small error. Hence the analytic continuation procedure gives us only the coefficient of the increasing exponential, we get

$$\frac{1}{\sqrt{k}} \exp\left[i\left(\int_0^x k\,dx - \frac{\pi}{4}\right)\right] \to \frac{1}{i\sqrt{|k|}} \exp\left[\int_0^x |k|\,dx\right], \tag{74}$$

where we have lost a term which is exponentially small compared to the right side of Eq. (74).

If we go around the point $x = 0$ in the lower half-plane then Eq. (74) goes over first into increasing exponential, and then, after crossing the line $\phi = -\dfrac{2\pi}{3}$ into a decreasing one. The exponentially small term at least in the region $\dfrac{-2\pi}{3} < \phi < 0$ goes into an exponentially large term after we cross into the region $-\pi < \phi < -\dfrac{2\pi}{3}$. So we do not get the correct value of the coefficient of the decreasing exponential.

The valid approximate solution of Eq. (47) must be single valued and thus should be stitched together from the solutions in the three sections with boundaries at arg $x = \phi = 0, \dfrac{2\pi}{3}, \dfrac{4\pi}{3}$ so that the resulting one is single-valued. The connection formulas give these relations among the approximate solutions in different regions so that the global approximate solution is single-valued.

Let us suppose that in the classically inaccessible region $(V > E)$ there is only a decreasing exponential:

$$\psi = \frac{1}{\sqrt{|k|}} \exp\left[-\int_x^0 |k|\,dx\right] \tag{75}$$

The general solution for $(V < E)$ is of the form

$$\psi = \frac{A}{\sqrt{|k|}} \exp\left[i\int_0^x k\,dx - i\phi_1\right] + \frac{B}{\sqrt{k}} \exp\left[-i\int kdx + i\phi_2\right] \tag{76}$$

To find A, B and ϕ_1, ϕ_2 we analytically continue Eq. (75) into the region $x > 0$. Analytic continuation in the upper half-plane gives

$$\frac{1}{\sqrt{|k|}} \exp\left[-\int_x^0 |k|dx\right] \to \frac{\exp\,(i\pi/4)}{\sqrt{\alpha'}\rho^{1/2} \exp\,(i\phi/2)}$$

$$\times \exp\left[\frac{2}{3}\alpha'\rho^{3/2}\left(\sin\frac{3}{2}\phi - i\cos\frac{3}{2}\phi\right)\right] \tag{77}$$

For $\phi = 0$, we get the second term in Eq. (76) with $\phi_2 = \dfrac{\pi}{4}$, $B = 1$. If we carry the analytic continuation in the lower half-plane we get the first term in Eq. (76) with $A = B = 1$, $\phi_1 = \phi_2 = \pi/4$. In each case one of the two terms is lost. Finally, we get

$$\frac{1}{\sqrt{|k|}} \exp\left[-\int_x^0 |k|\, dx\right] \rightarrow \frac{2}{\sqrt{k}} \cos\left(\int_0^x k\, dx - \frac{\pi}{4}\right) \tag{78}$$

in agreement with Eq. (71).

13.8 APPLICATIONS

The WKB approximation has been applied successfully to a wide variety of quantum mechanical problems such as the cold emission of electrons from a metal and the α-particle decay of a nucleus.

(a) Bound State Energies and the Quantization Condition: As a first application, we estimate the energies of the bound states in a one-dimensional potential $V(x)$. For each energy level E, we assume that $V(x)$ has only two classical turning points at x_1 and x_2.

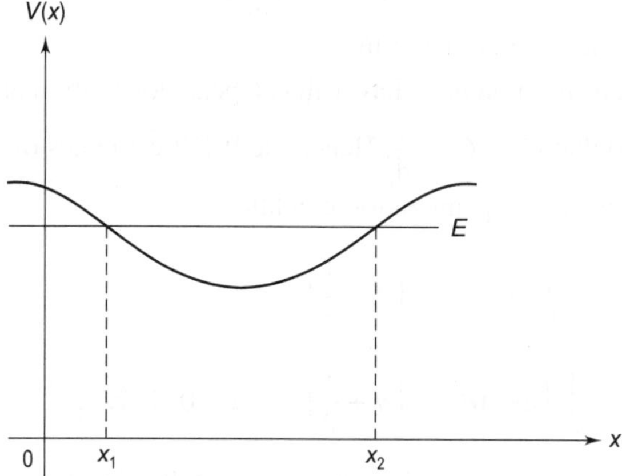

Fig. 13.5 A one-dimensional potential well

In the classically accessible region $(x_1 < x < x_2)$ there are two expressions for the semi-classical wave function. To the right of x_1, the solution is of the form

$$\psi_1 = \frac{A_1}{\sqrt{k}} \cos\left(\int_{x_1}^x k\, dx' - C_1\pi\right),$$

where $C_1\pi$ is the phase resulting from matching to the decreasing exponential for $x < x_1$. To the left of x_2 the solution has the form

$$\psi_2 = \frac{A_2}{\sqrt{k}} \cos\left(\int_x^{x_2} k\,dx' - C_2\pi\right).$$

These two solutions must agree. This requirement yields

$$\int_{x_1}^{x_2} k(x)\,dx = (n + C_1 + C_2)\pi \quad n = 0, 1, 2, \ldots. \tag{79}$$

with $A_2 = (-1)^n A_1$. As x goes from x_1 to x_2, $\cos\left(\int_{x_1}^{x} k\,dx' - C_1\pi\right)$ goes through zero

n times so that n is the number of nodes of the wave function in this interval.

Let us consider the example of a one-dimensional square well with infinitely high walls. The wave function goes to zero at both the walls. So we have $C_1 = C_2 = \frac{1}{2}$. The condition Eq. (79) gives

$$\int_{x_1}^{x_2} k_n\,dx = k_n L = (n + 1)\pi,$$

where L is the width of the well. Thus,

$$E_n = \frac{\hbar^2}{2m} k_n^2 = \frac{\pi^2 \hbar^2}{2mL^2}(n + 1)^2.$$

This is the exact energy spectrum.

In cases where the potential has a linear behavior both near x_1 and x_2, we see from Eq. (79) that $C_1 = C_2 = \frac{1}{4}$. Hence the WKB estimates of the bound state energies are given by the quantization condition

$$\int_{x_1}^{x_2} k(x)\,dx = \left(n + \frac{1}{2}\right)\pi \tag{80}$$

or,
$$\frac{1}{\hbar}\int_{x_1}^{x_2} p(x)\,dx = \left(n + \frac{1}{2}\right)\pi, \quad n = 0, 1, 2, \ldots. \tag{80a}$$

We may write the condition Eq. (80a) in the following form

$$\frac{1}{2\pi\hbar} \oint p(x)\,dx = n + \frac{1}{2}, \tag{81}$$

where the integral is over a complete cycle in phase space, i.e., from x_1 to x_2 and back to x_1. This condition is called the Bohr-Sommerfeld quantization condition although the original rule in the old quantum theory had n on the right. The number n is large in the semi-classical case, i.e., the WKB quantization condition is not expected to give reasonable estimates of energies of low lying states. In some cases, however, the WKB quantization rule gives the exact answers, e.g., the energy levels in a Coulomb field, and those of a harmonic oscillator.

In the semi-classical approximation we have dropped terms of $O(h^2)$ or $\dfrac{1}{(kl)^2}$, where l is the characteristic length over which the potential changes appreciably. Since $\int k\,dx \sim kl \sim n\pi$ the semi-classical estimates of the energies are valid to an accuracy of order $\dfrac{1}{n^2\pi^2}$. Thus, we can keep the terms C_1 and C_2 in Eq. (79) and also the extra term $\dfrac{1}{2}$ in Eq. (80), as these are of relative order $\dfrac{1}{n}$.

In normalizing the semi-classical wave functions the integration can be restricted to the range $x_1 < x < x_2$, since outside this region the wave function decreases exponentially. For almost all the region $x_1 < x < x_2$, the wave function is

$$\psi = \frac{A}{\sqrt{p}}\cos\left(\frac{1}{\hbar}\int_{x_1}^{x_2} p\,dx' - \frac{\pi}{4}\right).$$

This gives

$$\int |\psi|^2\,dx \approx \frac{1}{2}A^2\int_{x_1}^{x_2}\frac{dx}{p(x)},$$

where we have replaced the squared cosine term by its average value $\dfrac{1}{2}$ since the argument of the cosine term is a rapidly varying function. The above integral is

$$\int |\psi|^2\,dx = \frac{1}{2}A^2\cdot\frac{1}{m}\cdot\frac{T}{2} = \frac{\pi A^2}{2m\omega} = 1$$

where $T = \dfrac{2\pi}{\omega}$ is the classical period of the motion and ω is the corresponding frequency. Therefore the expression for the normalized semi-classical wave function is

$$\psi = \sqrt{\frac{2m\omega}{\pi p}}\cos\left(\frac{1}{\hbar}\int_{x_1}^{x} p\,dx' - \frac{\pi}{4}\right). \tag{82}$$

From the quantization rule Eq. (81) we can find the distribution of the energy levels. Let ΔE be the distance between two neighboring energy levels i.e., levels whose quantum numbers differ by unity. From Eq. (81) we get (for large n)

$$\Delta E \oint \left(\frac{\partial p}{\partial E}\right) dx = 2\pi\hbar,\ \text{since } \Delta E \text{ is small for the energy of the levels.}$$

Since
$$\frac{\partial E}{\partial p} = v,\ \oint \left(\frac{\partial p}{\partial E}\right) dx,$$

$$= \oint \frac{dx}{v} = T$$

Hence,
$$\frac{dE_n}{dn} \equiv \Delta E = \frac{2\pi\hbar}{T} = \hbar\omega. \tag{83}$$

This relation Eq. (83) states that at large quantum numbers the energy difference between two neighbouring levels is $\hbar\omega$, where ω is the classical frequency of motion. This is in accord with the correspondence principle. We now apply the WKB quantization condition to estimate the bound state energies in a variety of potentials.

(i) The harmonic oscillator

The potential is
$$V(x) = \frac{1}{2}\, m\omega^2 x^2$$

The classical turning points are
$$x_{\pm} = \pm \left(\frac{2E}{m\omega^2}\right)^{1/2}$$

Thus,
$$k(x) = \frac{1}{\hbar}\sqrt{2m\left(E - \frac{1}{2}m\omega^2 x^2\right)}$$

$$= \frac{m\omega}{\hbar}\sqrt{A^2 - x^2} \quad \text{where } A^2 = \frac{2E}{m\omega^2}.$$

From Eq. (80)
$$\int_{x_-}^{x_+} k(x)\,dx = \left(n + \frac{1}{2}\right)\pi$$

Now,
$$\int_{-A}^{A} k(x)\,dx = \frac{m\omega}{\hbar}\int_{-A}^{A}\sqrt{A^2 - x^2}\,dx$$

$$= \frac{\pi E}{\hbar\omega}$$

whence
$$E_n = \hbar\omega\left(n + \frac{1}{2}\right). \tag{84}$$

The quantization rule gives the exact energy eigenvalues E_n for all n.

(ii) A particle is moving in a one-dimensional potential $V(x) = a|x|$, $a > 0$.

The turning points are $x_{\pm} = \pm\, E/a$.

The quantization condition is $\displaystyle\int_{x_-}^{x_+} k(x)\,dx = \left(n + \frac{1}{2}\right)\pi$.

Now,
$$\frac{1}{\hbar}\int_{x_-}^{x_+}\sqrt{2m\,(E - a|x|)}\,dx$$

$$= \sqrt{\frac{2m}{\hbar}}\int_{-E/a}^{+E/a}(E - a|x|)^{1/2}\,dx$$

$$= 2 \cdot \frac{\sqrt{2mE}}{\hbar} \int\limits_{0}^{+E/a} \left(1 - \frac{ax}{E}\right) dx$$

$$= 2 \cdot \frac{\sqrt{2mE}}{\hbar} \cdot \frac{E}{a} \int\limits_{0}^{1} \sqrt{1 - z} \; dz, \quad z = \frac{ax}{E}$$

$$= 2 \cdot \frac{\sqrt{2mE}}{\hbar} \cdot \frac{E}{a} \cdot \frac{2}{3}$$

$$\therefore \qquad E_n^{WKB} = \alpha^{-2/3} \left[\frac{3\pi\,(2n+1)}{8}\right]^{2/3}, \; \alpha = \frac{\sqrt{2m}}{\hbar a}. \tag{85}$$

E_n exact can be expressed in terms of the Airy functions. See Galindo and Pascual (1991) for a detailed analysis.

(iii)   The family of potentials $U(x) \equiv \dfrac{2mV(x)}{\hbar^2} = |x|^\alpha$ (in appropriate units), $\alpha > 0$.

For $\alpha = 2$, this is the potential of an oscillator; for $\alpha = 1$, the potential is the example, studied in (ii). If $\alpha \to \infty$, this potential tends to a square well. If $\alpha \to 0$, $V(x) = 1$.

The turning points are $x_\pm = \pm \, \epsilon^{1/\alpha}$, where $\epsilon = \dfrac{2mE}{\hbar^2}$.

The quantization condition gives

$$\int\limits_{-\epsilon^{1/\alpha}}^{+\epsilon^{1/\alpha}} \sqrt{\epsilon - |x|^\alpha}\, dx = \left(n + \frac{1}{2}\right)\pi$$

or, $\qquad 2\int\limits_{0}^{+\epsilon^{1/\alpha}} \sqrt{\epsilon - x^\alpha}\, dx = \left(n + \frac{1}{2}\right)\pi$

Let $z = \dfrac{x^\alpha}{\epsilon}$ so that $dz = \dfrac{ax^{\alpha-1}}{\epsilon}\, dx.$

We get the left hand side as

$$2 \, \frac{\epsilon^{\frac{1}{2} + \frac{1}{\alpha}}}{\alpha} \int\limits_{0}^{1} z^{\frac{1}{\alpha} - 1} (1 - z)^{1/2}\, dz$$

$$= 2 \, \frac{\epsilon^{\frac{1}{2} + \frac{1}{\alpha}}}{\alpha} \cdot \frac{\sqrt{\pi}}{2} \cdot \frac{\Gamma\left(\dfrac{1}{\alpha}\right)}{\Gamma\left(\dfrac{1}{\alpha} + \dfrac{3}{2}\right)}.$$

Hence,

$$\epsilon_n^{WKB}(\alpha) = \left[\alpha\pi^{1/2}\left(n + \frac{1}{2}\right) \frac{\Gamma\left(\dfrac{1}{\alpha} + \dfrac{3}{2}\right)}{\Gamma\left(\dfrac{1}{\alpha}\right)}\right]^{\frac{2\alpha}{\alpha+2}}, \quad n = 0, 1, 2, \ldots \tag{86}$$

See Galindo and Pascual (1991) for a detailed analysis.

We now discuss the application of the WKB method to the solution of radial equations. For a central potential $V(r)$, the reduced radial wave function $u(r)$ satisfies the following equation

$$\frac{d^2 u(r)}{dr^2} + \left[k^2 - U(r) - \frac{l\,(l+1)}{r^2} \right] u(r) = 0, \tag{87}$$

where $k^2 = \dfrac{2m}{\hbar^2}\, E$ and $U(r) = \dfrac{2m}{\hbar^2}\, V(r),\ r \geq 0$.

It is not possible to apply the previous results directly to the radial equation because of its singularity at $r = 0$. In the one dimensional case, the range of the variable is $-\infty < x < \infty$. To bring the radial equation to the same form, we change the independent variable r in the radial equation, to x, so that $r = 0$ corresponds to $x = -\infty$. We introduce the transformation $r = e^x$.

The function u now satisfies the equation

$$\frac{d^2 u}{dx^2} - \frac{du}{dx} + e^{2x}\, [k^2 - U(e^x) - l\,(l+1)\,e^{-2x}]\, u = 0.$$

This equation is not in the form of one-dimensional Schrödinger equation. To bring this equation to the desired form, we introduce a new dependent variable $\chi(x)$ as follows

$$u = e^{x/2}\, \chi(x).$$

The function χ satisfies the equation

$$\frac{d^2 \chi(x)}{dx^2} + e^{2x} \left[k^2 - U(e^x) - \left(l + \frac{1}{2}\right)^2 e^{-2x} \right] \chi = 0. \tag{88}$$

The results of the WKB approximation can now be applied to this equation. The quantization condition is

$$\int_{x_0}^{x} e^x \left[k^2 - U(e^x) - \left(l + \frac{1}{2}\right)^2 e^{-2x} \right]^{1/2} dx = \left(n + \frac{1}{2}\right)\pi, \quad n = 0, 1, 2, \dots.$$

or, transforming back to the original independent variable r, we get

$$\int_{r_0}^{r_1} dr \left[k^2 - U(r) - \frac{\left(l + \frac{1}{2}\right)^2}{r^2} \right]^{1/2} = \left(n_r + \frac{1}{2}\right)\pi, \quad n_r = 0, 1, 2, \dots. \tag{89}$$

where r_0 and r_1 are the two roots of the integrand. A necessary condition for the validity of this formula is $n_r \gg 1$. The transition $l\,(l+1)$ to $\left(l + \frac{1}{2}\right)^2$ is known as Langer's correction.

As an illustrative example, we consider the Coulomb field $V(r) = -\dfrac{Ze^2}{r}$ $(Z > 0)$. The bound state energies in the WKB approximation are given by Eq. (89).

We make the substitutions, for bound state $(E < 0)$, $\lambda^2 = -\dfrac{2me^4}{\hbar^2 E}$, $\xi = \dfrac{2me^4}{\hbar^2}\, r$.

Hence Eq. (89) becomes

$$\int_{\xi_1}^{\xi_2} \left(-\frac{1}{\lambda^2} + \frac{Z}{\xi} - \frac{\left(l + \frac{1}{2}\right)^2}{\xi^2} \right)^{1/2} d\xi = \left(n_r + \frac{1}{2}\right)\pi, \tag{90}$$

where ξ_1 and ξ_2 are the zeros of the integrand.

$$\text{Using} \int_{\alpha}^{\beta} \frac{\sqrt{(z - \alpha)\,(\beta - z)}}{z}\, dz = \frac{\pi}{2}\,(\alpha + \beta - 2\sqrt{\alpha\beta}),$$

we get from Eq. (90),

$$\frac{1}{\lambda} \int_{\xi_1}^{\xi_2} \frac{\sqrt{(\xi - \xi_1)\,(\xi_2 - \xi)}}{\xi}\, d\xi = \left(n_r + \frac{1}{2}\right)\pi$$

or,
$$\frac{1}{\lambda} \cdot \frac{\pi}{2}\left(Z\lambda^2 - 2\lambda\left(l + \frac{1}{2}\right)\right) = \left(n_r + \frac{1}{2}\right)\pi$$

or,
$$\frac{1}{2}\,(Z\lambda - (2l + 1)) = \left(n_r + \frac{1}{2}\right)$$

so that
$$Z\lambda = 2\,(n_r + l + 1).$$

This gives

$$-E = -E_n = \frac{mZ^2 e^4}{2\hbar^2 n^2} \tag{91}$$

where $n = n_r + l + 1$.

This coincides with the exact result.

(b) Penetration of a Potential Barrier

We consider the motion of a particle in a potential barrier. This problem was solved for a rectangular barrier in Section 3.8. Here we consider a general potential barrier and apply the WKB approximation to calculate the transmission coefficient for the barrier. Classically, if the energy of the particle E is less than the height of the potential barrier it would be reflected, i.e., a potential barrier is impenetrable to a particle. In quantum mechanics, a particle can penetrate through a potential barrier. This phenomenon is called the "tunnel effect".

Let us consider a general potential $V(x)$ as shown in the Fig. 13.6. A particle with energy E is incident from the left. The energy E is insufficient to overcome the barrier classically, i.e. $E < V_0$ (the top of the barrier). There are two classical turning points at $x = x_1$ and at $x = x_2$. Regions I and III are classically accessible, while region II is classically forbidden. In region III beyond the barrier, there is a transmitted wave traveling to the right. The wave function in this region in the WKB approximation may be written as

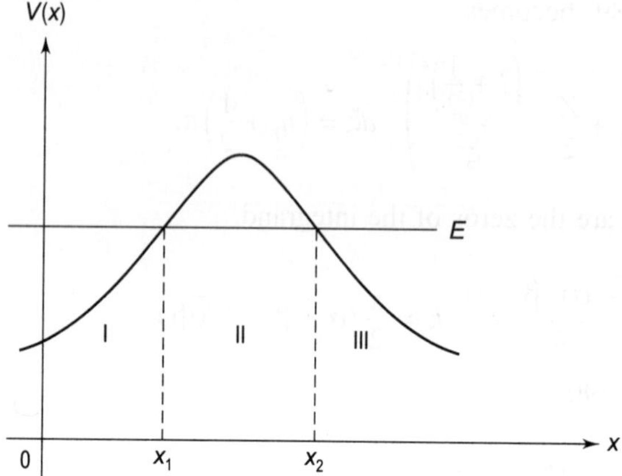

Fig. 13.6 A one-dimensional potential barrier

$$\psi_{III}(x) = \frac{A}{\sqrt{p(x)}} \exp\left(\frac{i}{\hbar}\int_{x_2}^{x} p(x')\,dx' - \frac{i\pi}{4}\right), \quad x > x_2 \tag{92}$$

where the phase factor $e^{-i\pi/4}$, which could be absorbed in the constant, is written explicitly. We rewrite

$$\psi_{III}(x) = \frac{A}{\sqrt{p(x)}}\left[\cos\left(\frac{i}{\hbar}\int_{x_2}^{x} p(x')\,dx' - \frac{\pi}{4}\right) + i\sin\left(\frac{i}{\hbar}\int_{x_2}^{x} p(x')\,dx' - \frac{\pi}{4}\right)\right], \quad x > x_2 \tag{93}$$

Using the connection formulas Eq. (71) for the case of a barrier to the left, we write the wave function in region II, within the barrier as

$$\psi_{II}(x) = \frac{A e^{-i\pi/2}}{K(x)^{1/2}} \exp\left[\int_{x_2}^{x} |K|\,dx'\right], \quad x_1 < x < x_2 \tag{94}$$

In the forbidden region II, $|\psi(x)|$ increases as x moves to the left of the turning point x_2. Hence, we write Eq. (94) in the form of a decreasing exponential for $x > x_1$ as

$$\psi_{II}(x) = \frac{-iA}{|K|^{1/2}} \exp\left[\int_{x_1}^{x_2} dx\,|K|\right] \exp\left(-\int_{x_1}^{x} |K|\,dx'\right)$$

$$= \frac{-iA}{\sqrt{|K|}}\, e^{\wedge} \exp\left(-\int_{x_1}^{x} |K|\,dx'\right), \quad x_1 \le x \le x_2 \tag{95}$$

where
$$\Lambda = \int_{x_1}^{x_2} |K(x)|\, dx.$$

Using the connection formula Eq. (72) for the case of a barrier to the right, we obtain the oscillatory wave function in region I in front of the barrier.

$$\psi_I(x) = \frac{-2iA}{\sqrt{p(x)}}\, e^{\Lambda} \cos\left[-\frac{1}{\hbar}\int_{x_1}^{x} p(x')\, dx' - \frac{\pi}{4}\right]$$

$$= \frac{-iA}{\sqrt{p(x)}}\, e^{\Lambda} \left[\exp\left\{i\left(\frac{1}{\hbar}\int_{x_1}^{x} p(x')\, dx' - \frac{\pi}{4}\right)\right\}\right.$$

$$\left. + \exp\left\{-i\left(\frac{1}{\hbar}\int_{x_1}^{x} p(x')\, dx' - \frac{\pi}{4}\right)\right\}\right], \quad x < x_1 \qquad (96)$$

The first term represents a wave incident on the barrier, and the second term a wave reflected by the barrier, moving to the left. The amptitudes of the incident and reflected waves in region I are the same. This is a consequence of the WKB approximation. The exponentially small difference between the two is lost in the quasi-classical approximation. In $\psi_{II}(x)$ a small term $\sim e^{-\Lambda} e^{\int_{x_1}^{x}|K|\, dx'}$ is dropped compared to the term in Eq. (95).

The transmission coefficient is defined as

$$T = \frac{v_{tr}|\psi_{tr}|^2}{v_{inc}|\psi_{inc}|^2} = \frac{k_{tr}|\psi_{tr}|^2}{k_{inc}|\psi_{inc}|^2}$$

$$= \frac{|A|^2}{|A|^2\, e^{2\Lambda}} = e^{-2\Lambda}$$

$$= \exp\left[-\frac{2}{\hbar}\int_{x_1}^{x_2} \{2m\,[V(x) - E]\}^{1/2}\, dx\right]. \qquad (97)$$

The reflection coefficient R, in this approximation, is unity. The semi-classical approximation is valid when $V(x)$ is slowly varying and the turning points are well separated. The above results are valid when T is very small, i.e., when Λ is large. Conservation of probability flux requires $R + T = 1$ which is approximately satisfied in this approximation since T is very small. R differs from 1 by an exponentially small (subdominant) quantity. The above result is a good approximation for $E < V_{\max}$. When the energy E is near the top of the barrier, the turning points are close and the WKB approximation is not valid. When $E = V_{\max}$, i.e., for scattering off the peak of a barrier, there is only one turning point and a careful analysis shows that half of the incident wave is reflected and half transmitted (Bender and Orszag (1978)). We note that classically one cannot define R or T in this case, because the particle can never reach the peak. If E is above the barrier but not far above the top, there are no turning points, and the method cannot be applied. When E is far above the top of the barrier there is only a transmitted wave with a phase shift relative to the incident wave. Kemble (1937), Fröman and Fröman (1965) and others have obtained the following expressions for T and R.

$$T = \frac{1}{1 + e^{2\Lambda}}$$

$$R = 1 - T = \frac{e^{2\Lambda}}{1 + e^{2\Lambda}}. \tag{98}$$

Clearly, when Λ is very large, the two expressions Eqs. (97) and (98) coincide.

We now consider two illustrative examples where exact solutions can be obtained for comparison with those found by the WKB approximation. The material draws on Jeffreys (1961) and Gottfried and Yan (2003).

(i) Let us first consider reflection from a parabolic barrier

$$V(x) = V_0 \left(1 - \left(\frac{x}{x_0} \right)^2 \right), \tag{99}$$

and take $E < V_0$.

We put
$$y = \frac{x}{x_0}, \quad \left(\frac{2m V_0 x_0^2}{\hbar^2} \right)^{1/2} = b, \, k' = \frac{V_0 - E}{V_0}.b. \tag{100}$$

Then the Schrödinger equation becomes

$$\frac{d^2\varphi}{dy^2} + b^2 \left(y^2 - \frac{k'}{b} \right) \varphi = 0 \tag{101}$$

with
$$\psi(x) = \varphi(y).$$

Substituting
$$\varphi(u) = e^{\frac{i}{2} by^2} \xi(u)$$
$$u = e^{-\frac{i}{2}\pi} by^2 \tag{102}$$

in Eq. (101), we get

$$u \frac{d^2\xi}{du^2} + \frac{d\xi}{du} \left(\frac{1}{2} - u \right) - a\xi = 0, \tag{103}$$

where
$$a = \frac{1 + ik'}{4}.$$

Equation (103) is a confluent hypergeometric equation and the two linearly independent solutions of Eq. (101) are

$$\phi_1(y) = {}_1F_1\left[a, \frac{1}{2}, by^2 \, e^{-\frac{i}{2}\pi}\right] e^{\frac{i}{2} by^2} \tag{104}$$

$$\phi_2(y) = \sqrt{b} \, e^{-\frac{i\pi}{4}} \, y \, {}_1F_1\left[a + \frac{1}{2}, \frac{3}{2}, by^2 e^{-\frac{i\pi}{2}}\right] e^{\frac{i}{2} by^2} \tag{105}$$

Using the asymptotic approximation to ${}_1F_1(\alpha, \beta, z)$

$$= 1 + \frac{\alpha}{1!\beta} z + \frac{\alpha(\alpha+1)}{2!\beta(\beta+1)} z^2 + \dots \text{ for } -\pi < \arg z < 0 \text{ and } |z| \text{ large as}$$

$$(\beta - 1)! \left[\frac{1}{(\alpha - 1)!} \, z^{\alpha-\beta} \, e^z + \frac{1}{(\beta - \alpha - 1)!} \, z^{-\alpha} e^{-i\alpha\pi}\right],$$

we find that the linear combination of ϕ_1 and ϕ_2 defined by

$$\Phi(y) = 2\left[\frac{\left(\frac{1}{2}\right)!}{\left(a - \frac{1}{2}\right)!} \, \phi_1(y) - \frac{\left(-\frac{1}{2}\right)!}{(a - 1)!} \, \phi_2(y)\right], \tag{106}$$

has the asymptotic approximation

$$\Phi(y) \sim b^{-a} \, e^{\frac{i}{2} a\pi} \, y^{-2a} \, e^{\frac{i}{2} by^2}, \; \arg y = 0. \tag{107}$$

This represents a wave going from left to right when $y > 0$ is large.

Now,

$$\varphi_1\lfloor re^{i\pi}\rfloor = \varphi_1(r)$$

$$\varphi_2\lfloor re^{i\pi}\rfloor = -\varphi_2(r)$$

where $|y| = r$, a constant of large value and $\arg y$ changes from 0 to π. Then

$$\Phi(re^{i\pi}) = 2\left[\frac{\left(\frac{1}{2}\right)!}{\left(a - \frac{1}{2}\right)!} \, \varphi_1(r) + \frac{\left(-\frac{1}{2}\right)!}{(a - 1)!} \, \varphi_2(r)\right], \tag{108}$$

and

$$\Phi(r) = 2\left[\frac{\left(\frac{1}{2}\right)!}{\left(a - \frac{1}{2}\right)!} \, \varphi_1(r) - \frac{\left(-\frac{1}{2}\right)!}{(a - 1)!} \, \varphi_2(r)\right], \tag{109}$$

so that

$$\Phi[re^{i\pi}] = \Phi(r) + \frac{4\left(-\frac{1}{2}\right)!}{(a - 1)!} \, \varphi_2(r). \tag{110}$$

The asymptotic approximation to this for large r is

$$\Phi[re^{i\pi}] \sim b^{-a} e^{\frac{i}{2}a\pi} \left[\gamma r^{-2a} e^{\frac{i}{2}br^2} + \delta r^{-2a} e^{-\frac{i}{2}br^2} \right] \tag{111}$$

where γ and δ are independent of r. Now, when y is large and negative y is $-r$ and the first term in Eq. (111) represents a wave traveling from right to left while the second a wave traveling in the opposite direction. Then

$$T = \frac{1}{|\delta|^2} = \frac{1}{1 + |\gamma|^2} \tag{112}$$

$$R = \frac{|\gamma|^2}{|\delta|^2} = \frac{|\gamma|^2}{1 + |\gamma|^2}$$

where

$$|\gamma| = \exp\left(\frac{1}{2} \pi k' \right)$$

Thus,

$$T = \frac{1}{1 + e^{\pi k'}}, \quad R = \frac{e^{\pi k'}}{1 + e^{\pi k'}} \tag{113}$$

For k' large, $T \approx e^{-\pi k'}$.

In the WKB approximation

$$\Lambda = \frac{1}{\hbar} \int_{x_1}^{x_2} \sqrt{2m[V(x) - E]}\, dx$$

where x_1, x_2 are the turning points

$$= \left(\frac{2mV_0}{\hbar^2} \right)^{1/2} \int_{x_1}^{x_2} \sqrt{\left(\frac{V_0 - E}{V_0} \right) - \frac{x^2}{x_0^2}}\, dx$$

$$\text{with } x_2 = -x_1 = x_0 \left(\frac{V_0 - E}{V_0} \right)^{1/2}$$

$$= \left(\frac{2mV_0 x_0^2}{\hbar^2} \right)^{1/2} \int_{-a}^{a} \sqrt{a^2 - x'^2}\, dx',$$

$$\text{where } a^2 = \left(\frac{V_0 - E}{V_0} \right) \text{ and } x' = x/x_0.$$

$$= \left(\frac{2mV_0 x_0^2}{\hbar^2} \right)^{1/2} \cdot \left(\frac{V_0 - E}{V_0} \right) \cdot \frac{\pi}{2}$$

so that
$$T \approx \exp(-2\Lambda)$$

$$= \exp\left\{-\pi\left(\frac{V_0 - E}{V_0}\right)\left(\frac{2mV_0 x_0^2}{\hbar^2}\right)^{1/2}\right\}$$

$$= \exp(-\pi k'). \tag{114}$$

(ii) Next we calculate the transmission coefficient for a potential barrier defined by

$$V(x) = \frac{V_0}{\cosh^2 \alpha x}, \quad V_0 > 0, \quad \alpha > 0. \tag{115}$$

The energy E of the particle is positive but less than V_0, i.e., $0 < E < V_0$. The Schrödinger equation is

$$\frac{d^2\psi(x)}{dx^2} + \left[k^2 - \frac{U_0}{\cosh^2 \alpha x}\right]\psi(x) = 0, \tag{116}$$

where
$$k^2 = \frac{2mE}{\hbar^2} \text{ and } U_0 = \frac{2mV_0}{\hbar^2}.$$

We put $\xi = \tanh \alpha x$ in Eq. (116) and obtain

$$\frac{d}{d\xi}\left[(1 - \xi^2)\frac{d\varphi(\xi)}{d\xi}\right] + \left[s(s + 1) + \frac{k^2}{\alpha^2(1 - \xi^2)}\right]\varphi(\xi) = 0, \tag{117}$$

with $s = \frac{1}{2}\left(-1 + \sqrt{1 - \frac{4U_0}{\alpha^2}}\right)$, $s(s + 1) = -\frac{U_0}{\alpha^2}$ and $\varphi(\xi) \equiv \psi(x)$. By making the

substitution $\varphi(\xi) = (1 - \xi^2)^{-ik/2\alpha} w(\xi)$, Eq. (117) becomes

$$(1 - \xi^2)\frac{d^2 w(\xi)}{d\xi^2} - 2\xi\left(1 - \frac{ik}{\alpha}\right)\frac{dw(\xi)}{d\xi} + \left[\frac{ik}{\alpha} + s(s + 1) + \frac{k^2}{\alpha^2}\right]w(\xi) = 0. \tag{118}$$

Changing the variable to $t = \frac{1}{2}(1 - \xi)$, we see that $y(t) = w(\xi)$ is a solution of a hypergeometric equation

$$t(1 - t)\frac{d^2 y(t)}{dt^2} + (\epsilon + 1)(1 - 2t)\frac{dy(t)}{dt} + (\epsilon - s)(\epsilon + s + 1)y(t) = 0, \tag{119}$$

where
$$\epsilon = -ik/\alpha.$$

The solution of Eq. (119), subject to the boundary condition that as $x \to \infty$ (i.e., as $\xi \to 1$, $(1 - \xi) \approx 2e^{-2\alpha x}$), the wave function should behave as a transmitted wave ($\sim e^{ikx}$) only, is

$$\psi(x) = C(1 - \tanh^2 \alpha x)^{\epsilon/2} \, {}_2F_1\left(\epsilon - s, \epsilon + s + 1; \epsilon + 1; \frac{1}{2}(1 - \tanh \alpha x)\right),$$

$$(120)$$

where ${}_2F_1(a, b; c; z)$ is the hypergeometric function. Using the formula

$$
{}_2F_1(\alpha, \beta; \gamma; z) = \frac{\Gamma(\gamma)\,\Gamma(\gamma - \alpha - \beta)}{\Gamma(\gamma - \alpha)\Gamma(\gamma - \beta)}
$$

$$
\times \, {}_2F_1(\alpha, \beta; \alpha + \beta + 1 - \gamma; 1 - z) + \frac{\Gamma(\gamma)\,\Gamma(\alpha + \beta - \gamma)}{\Gamma(\alpha)\Gamma(\beta)}
$$

$$
\times \, (1 - z)^{\gamma - \alpha - \beta} \, {}_2F_1(\gamma - \alpha, \gamma - \beta; \gamma + 1 - \alpha - \beta; 1 - z) \quad (121)
$$

relating hypergeometric functions of z and $(1 - z)$, we write Eq. (120) as

$$
\psi(x) = C(1 - \xi^2)^{\epsilon/2} \left\{ \frac{\Gamma(1 + \epsilon)\,\Gamma(-\epsilon)}{\Gamma(1 + s)\,\Gamma(-s)} \right.
$$

$$
\times \, {}_2F_1(\epsilon - s, \epsilon + s + 1; \epsilon + 1; (1 + \xi)/2)
$$

$$
+ \left(\frac{1 + \xi}{2}\right)^{-\epsilon} \frac{\Gamma(1 + \epsilon)\,\Gamma(\epsilon)}{\Gamma(\epsilon - s)\,\Gamma(\epsilon + s + 1)}
$$

$$
\left. \times \, {}_2F_1(1 + s, - s; - \epsilon, (1 + \xi)/2) \right\} \quad (122)
$$

As $x \to -\infty$, i.e., as $1 + \xi \to 2e^{2\alpha x}$, the asymptotic form of the wave function
is

$$
\psi(x) \sim 4^{\epsilon/2} \left[\frac{\Gamma(1 + \epsilon)\,\Gamma(-\epsilon)}{\Gamma(- s)\,\Gamma(1 + s)} \, e^{-ikx} + \frac{\Gamma(\epsilon)\,\Gamma(1 + \epsilon)}{\Gamma(\epsilon - s)\,\Gamma(\epsilon + s + 1)} \, e^{ikx} \right]. \quad (123)
$$

The exact transmission and reflection amplitudes are

$$
\tau = \frac{\Gamma(\epsilon - s)\,\Gamma(\epsilon + s + 1)}{\Gamma(\epsilon)\,\Gamma(\epsilon + 1)} \quad (124)
$$

$$
r = \frac{\Gamma(-\epsilon)\,\Gamma(\epsilon - s)\,\Gamma(\epsilon + s + 1)}{\Gamma(\epsilon)\,\Gamma(s + 1)\,\Gamma(-s)} \quad (125)
$$

r vanishes whenever s is an integer.

Using the properties of the gamma function

$$\Gamma^*(z) = \Gamma(z^*), \quad \Gamma(-iy)\,\Gamma(+iy) = \frac{\pi}{y}\sinh(\pi y), \quad y \neq 0 \text{ and}$$

$\Gamma(z)\,\Gamma(1 - z) = \pi\,\mathrm{cosec}\,\pi z \; (0 < \mathrm{Re}z < 1)$, we obtain the expression for the exact transmission coefficient T:

$$
T = \frac{\sinh^2(\pi k/\alpha)}{\sinh^2(\pi k/\alpha) + \cosh^2\left(\dfrac{\pi}{2}\sqrt{1 - \dfrac{4U_0}{\alpha^2}}\right)}, \quad \frac{4U_0}{\alpha^2} \leq 1 \quad (126)
$$

$$= \frac{\sinh^2(\pi k/\alpha)}{\sinh^2(\pi k/\alpha) + \cosh^2\left(\frac{\pi}{2}\sqrt{\frac{4U_0}{\alpha^2} - 1}\right)}, \quad \frac{4U_0}{\alpha^2} \geq 1 \tag{127}$$

The turning points $x_1(x_2)$ are given by $\cosh \alpha x_{1(2)} = \pm\sqrt{\dfrac{V_0}{E}}$.

Then

$$\Lambda = \int_{x_1}^{x_2} \sqrt{-k^2 + U_0 \cosh^{-2} \alpha x} \; dx$$

$$= \frac{1}{\alpha} \int_1^{\sqrt{U_0/k^2}} \frac{1}{z}\sqrt{\frac{U_0 - k^2 z}{z - 1}} \; dz \; \text{where } z = \cosh^2 \alpha x$$

$$= \frac{\pi}{\alpha}(U_0^{1/2} - k) \tag{128}$$

so that the transmission coefficient in the WKB approximation is

$$T = \exp\left(-\frac{2\pi}{\alpha}\left(\sqrt{U_0} - k\right)\right). \tag{129}$$

This approximation is valid for short wavelengths, i.e., $k \gg \alpha$ and when the exponent is large. For the WKB result to be a good approximation to the exact result it is required that $(\sqrt{U_0} - k) \gg \alpha$ in addition to $k \gg \alpha$.

(c) Cold Emission of Electrons from a Metal

As an important application of tunneling, we consider field emission i.e., the cold emission of electrons from a metal by an external electric field. A simple model of the potential wall (a box of finite depth) for electrons in a metal is shown below in Fig. 13.7(a). In order to extract an electron from the metal, we must impart an energy not less than the work function W:

$$W = V_0 - E_F \tag{130}$$

where V_0 is the top of the well and E_F is the Fermi energy. At any temperature above $T = 0$, some electrons are excited to higher energy levels. If some of the electrons acquire sufficient energy from thermal motion to overcome the potential barrier then thermal emission takes place. Even at room temperature the fraction of electrons in higher energy levels is small and is neglected.

A constant electric field of strength $\vec{\varepsilon}$ is now applied to the metal to extract electrons. The potential as seen by an electron is now $V(x) - e\varepsilon x$, where x is the distance from the surface of the metal. From Fig. 13.7(b) we see that the potential barrier has now a finite width and so an electron with $E > 0$ has a nonzero probability of tunneling through the barrier to come out of the metal. For an electron, at the top of the sea, the energy $E = E_F = V_0 - W$. The turning point x_2 is obtained from the relation

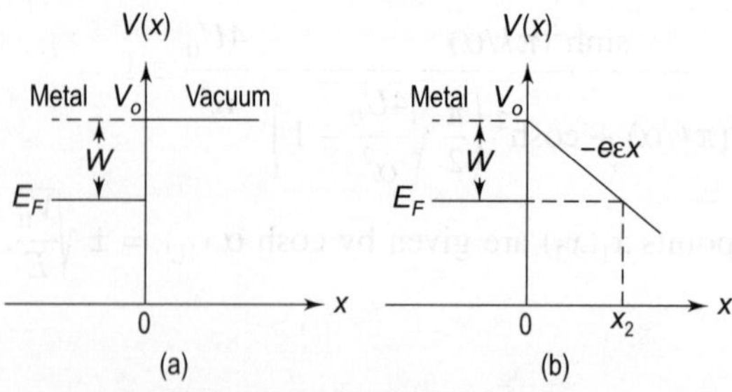

Fig. 13.7   (a) Potential binding the electrons in a metal (b) Potential modified by an external electric field

$$V_0 - e\varepsilon x_2 = V_0 - W$$

or,
$$x_2 = \frac{W}{e\varepsilon}. \tag{131}$$

The other turning point is at the surface of the metal ($x_1 = 0$). Although the potential changes abruptly near this point, we can use the WKB approximation as the distance x_2 is much larger compared to interatomic distance.

Now,

$$\Lambda = \sqrt{\frac{2m}{\hbar}} \int_0^{x_2} [V_0 - e\varepsilon x - E]^{1/2}\, dx$$

$$= \sqrt{\frac{2m}{\hbar}} \int_0^{x_2} [W - e\varepsilon x]^{1/2}\, dx,$$

$$= \frac{2}{3} \sqrt{\frac{2m}{\hbar}} \frac{W^{3/2}}{e\varepsilon}. \tag{132}$$

Hence the transmission coefficient in the WKB approximation is

$$T = e^{-2\Lambda}$$

$$= \exp\left(-\frac{4}{3} \frac{(2m)^{1/2}\, (W)^{3/2}}{\hbar e\varepsilon} \right)$$

$$= \exp\left(-\frac{\varepsilon_0}{\varepsilon} \right) \tag{133}$$

where $\varepsilon_0 = \dfrac{4}{3} \dfrac{(2m)^{1/2}\, (W)^{3/2}}{\hbar e}$ depends on W, the work function of free electrons

in a metal. Equation (133) is known as the Fowler-Nordheim formula. The cold emission current I is proportional to T:

$$I = I_0\, T = I_0 \exp\left(-\frac{\varepsilon_0}{\varepsilon} \right). \tag{134}$$

Equation (134) is in reasonable agreement with experimental data. Cold emission current is observed for $\varepsilon \gtrsim 10^6$ *V/cm*.

(d)   Alpha Decay

One of the early successes of quantum tunneling was the understanding of the phenomenon of the α-particle decay of an atomic nucleus. The theory was proposed in 1928 by Gamow and by Gurney and Condon. In this process, a heavy nucleus ("parent") decays into a lighter one ("daughter") by the spontaneous emission of an α-particle (a helium nucleus consisting of two protons and two neutrons)

$$_N^Z X^A \rightarrow {}_{(N-2)}^{(Z-2)} Y^{(A-4)} + {}_2^2 He^4 \tag{135}$$

where Z, N, A are respectively, the atomic number (the number of protons), the number of neutrons and the total number of nucleons (mass number) in a nucleus X. The simple model assumes that the α-particles exist in the nucleus and α-decay is due to the tunneling of an α-particle through a potential barrier. Inside the nucleus ($r < R$, R being the nuclear radius) the α-particles are held by strongly attractive nuclear force (extremely short ranged). We assume that the potential inside is a square well of depth V_0 (for $r < R$). Outside the nucleus ($r > R$) the α-particle experiences only mutual Coulomb repulsion with the daughter nucleus. If Ze is the charge of the parent nucleus, then the potential energy of an α-particle can be represented, to a good approximation, by

$$V(r) = - V_0, \; r < R$$

$$= \frac{zZ'e^2}{r} \, , \; r > R \tag{136}$$

where $z = 2$ for an α-particle and $Z = Z - 2$ is the atomic number of the daughter nucleus (see Fig. 13.8). We note that since the strong force inside the nucleus is extremely short ranged the α-particles inside the nucleus may be assumed to move as free particles, i.e., we may put $V_0 = 0$. The precise shape of $V(r)$ for $r < R$ is not known and is not needed for our discussion. This problem reduces to a one dimensional one and the potential barrier is expressed as a function of the radial coordinate. We may now apply the WKB approximation with

$$V_{eff} = V(r) + \frac{\hbar^2 l\,(l+1)}{2mr^2}.$$

An α-particle of energy E can penetrate a potential barrier in the region R to r_2 where r_2 is a turning point with the probability given by the transmission coefficient T.

The turning point r_2 is determined (for $l = 0$) from the relation

$$E = \frac{2\,(Z-2)\,e^2}{r^2}$$

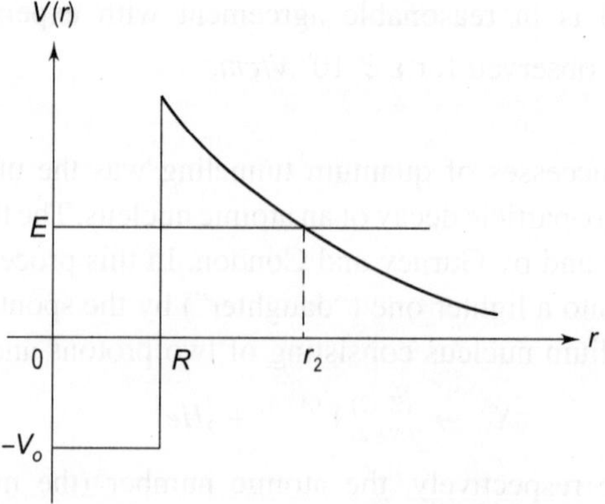

Fig. 13.8 Potential curve for alpha decay

or,
$$r_2 = \frac{2\,(Z-2)\,e^2}{E}.$$
(137)

The turning point r_1 is set at the nuclear radius R. For a nucleus of mass number A, $R \approx 1.1\,A^{1/3} \times 10^{-13}$ cm. Since $E_\alpha \sim 4 - 10\ MeV$, $r_2 - r$, is much larger than $r_1 = R$.

Now,

$$\Lambda = -\frac{\sqrt{2m_\alpha}}{\hbar} \int_{r_1}^{r_2} \left[\frac{2\,(Z-2)\,e^2}{r} - E \right]^{1/2} dr$$

$$= \left(\frac{4m_\alpha\,(Z-2)\,e^2}{\hbar^2} \right)^{1/2} \int_{R}^{r_2} \left(\frac{1}{r} - \frac{1}{r_2} \right)^{1/2} dr.$$
(138)

The integral can be evaluated in closed form. Using $\dfrac{1}{y} = \dfrac{r_2}{r}$, the integral can be written as

$$\sqrt{r_2} \int_{R/r_2}^{1} \sqrt{\frac{1}{y} - 1}\, dy = \sqrt{r_2} \left[\sqrt{y(1-y)} - \cos^{-1} \sqrt{y} \right]_{R/r_2}^{1}$$

$$= \sqrt{r_2} \left[\cos^{-1} \sqrt{\frac{R}{r_2}} - \sqrt{\frac{R}{r_2} \left(1 - \frac{R}{r_2} \right)} \right].$$

Since $R/r_2 < 1$, we expand the terms in the square bracket, and obtain, keeping terms to order $\sqrt{R/r_2}$, $\left[\dfrac{\pi}{2} - 2\sqrt{\dfrac{R}{r_2}} \right]$, using

$$\cos^{-1} x = \frac{\pi}{2} - x + O(x^3).$$

Hence,

$$\Lambda \simeq \left(\frac{4m_\alpha\,(Z-2)\,e^2 r_2}{\hbar^2} \right)^{1/2} \left[\frac{\pi}{2} - 2\left(\frac{R}{r_2} \right)^{1/2} \right].$$
(139)

The transmission coefficient T is

$$T = e^{-2\Lambda}$$

$$= \exp\left[8\left(\frac{m_\alpha(Z-2)e^2 R}{\hbar^2}\right)^{1/2} - 2\pi\left(\frac{m_\alpha(Z-2)e^2 r_2}{\hbar^2}\right)^{1/2}\right]$$

$$= \exp\left[\frac{8}{\hbar}\left((Z-2)e^2 m_\alpha R\right)^{1/2} - \frac{2\pi(Z-2)e^2}{\hbar}\left(\frac{2m_\alpha}{E}\right)^{1/2}\right]. \qquad (140)$$

We may write Eq. (140) as

$$T = T_G \exp\left(\frac{64(Z-2)e^2}{\hbar^2}\, m_\alpha R\right)^{1/2} \qquad (141)$$

where

$$T_G = \exp\left(-\frac{4\pi(Z-2)e^2}{\hbar \upsilon}\right) \qquad (142)$$

is called the Gamow factor. It inhibits the transmission of an α-particle through the potential barrier.

We make an estimate of the decay constant λ of the α-emitter. The decay constant $\lambda\left(=\frac{1}{\tau},\ \tau\ \text{being the mean lifetime}\right)$ is the product of the tunneling probability through the Coulomb barrier T and the frequency ω with which the α-particle hits the barrier

$$\lambda \equiv \frac{1}{\tau} = T\omega \qquad (143)$$

The frequency may be expressed as $v/2R$ where v is the velocity of the α-particle inside the nucleus. If we take $V_0 = 0$, then $v = \sqrt{\dfrac{2E}{m_\alpha}}$, which is the same as the velocity of the α-particle after emission. Thus

$$\lambda = \frac{v}{2R}\, T. \qquad (144)$$

Using Eq. (140), we get

$$\ln \lambda \approx \ln\frac{v}{2R} + 2.97\,(Z-2)^{1/2}\,R^{1/2} - 3.95\,(Z-2)\,E^{-1/2}, \qquad (145)$$

where E is in MeV and R in units of 10^{-13} cm. Equation (145) can be plotted, approximately, as a function of $(Z-2)E^{-1/2}$ alone, as there is little change in the other terms over the restricted range of nuclei. A small change in energy, say from 4 to 9 MeV, leads to an enormous change in λ of about a factor $e^{55} \sim 10^{25}$! This is the modern version of the empirical Geiger-Nuttall law.

The Gamow factor which also inhibits the approach of a positively charged particle to the nucleus is a crucial factor in nuclear fusion reactions.

(e) Tunnel Splitting in a One-Dimensional Double Well Potential

Let us consider a double well potential in one dimension. Such potentials are soluble examples of two-state systems. The coupled double well systems have been considered in different areas of physics. The ammonia (NH_3) molecule is a well known example (Feynman et al, Vol III (1965)). Another important example is the "strangeness oscillations" of a neutral K-meson beam. The double well has been used to study homopolar molecules like H_2 or N_2. For a review, see Leggett et al (1987).

The coupling of the two states in the quantum double well (QDW) potential leads to new eigenstates. The general QDW potential is shown below.

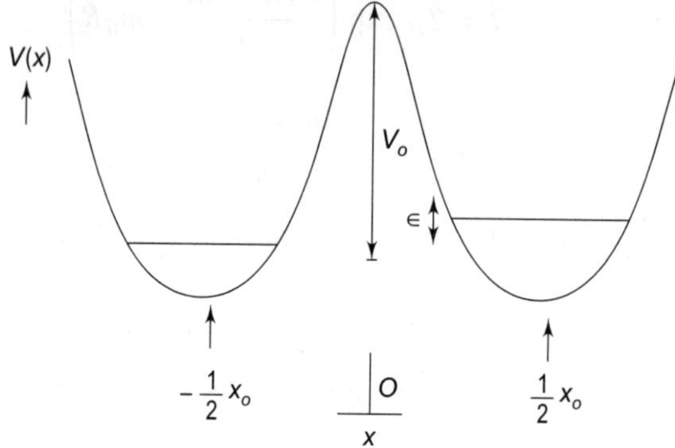

Fig. 13.9   The general QDW system

The eigenstates are a symmetric (bonding) lower state $|S>$ and an antisymmetric (antibonding) upper state $|A>$. The wave functions of these states are

$$|R> = \frac{1}{\sqrt{2}}\ (|S> + |A>), \tag{146}$$

$$|L> = \frac{1}{\sqrt{2}}\ (|S> - |A>). \tag{147}$$

The wave functions are largely localized in the right and left-sides, respectively. If we consider a symmetrical potential $V(x) = V(-x)$, then the Hamiltonian has a symmetry under parity (π). $|S>$ and $|A>$ are simultaneous eigenkets of H and π. But $R>$ and $|L>$ are not parity eigenstates. They are mapped into each other under the parity operation. There is an energy splitting $\Delta E = E_A - E_S$ due to the repulsion of the two states. The splitting is very small if the barrier between the wells is high. $|R>$ and $|L>$ are non-stationary states. A wave packet located in the right well is represented by $|R>$ at $t = 0$, i.e., the probability $P_R\ (t = 0)$. At a later time t,

$$|R,\ t_0 = 0;\ t> = \frac{1}{\sqrt{2}}\ (e^{-\frac{i}{\hbar}E_S t}|S> + e^{-\frac{i}{\hbar}E_A t}|A>)$$

$$= \frac{1}{\sqrt{2}}\, e^{-\frac{i}{\hbar} E_S t}\, (|S> + e^{-\frac{i}{\hbar}(E_A - E_S)t}|A>). \qquad (148)$$

Let $P(t) = P_R - P_L$ and $P(0) = 1$, the subsequent behavior is given by

$$P(t) = \cos \Delta E t. \qquad (149)$$

Thus there is an oscillation between $|R>$ and $|L>$ with the angular frequency

$$\omega = \frac{\Delta E}{\hbar}. \qquad (150)$$

The oscillatory behavior displays the consequences of the phase coherence between the amplitudes for being in the right and left wells, and has no classical analog. Coherent oscillations can be directly observed in the semiconductor QDW system, e.g. GaAs/AlGaAs material system.

A particle initially in the right well can tunnel through the classically forbidden region (the barrier) into the left well, and then back to the right well, and so on. Now, we consider the symmetrical QDW with an infinite barrier between the wells. The $|S>$ and $|A>$ states are now degenerate, and so $|R>$ and $|L>$ are energy eigenkets but not parity eigenkets. There is no tunneling, i.e., if there is a particle in one well, it will remain there forever. Hence there is no oscillatory behavior, i.e. the time period of oscillation $\tau = \dfrac{2\pi\hbar}{E_A - E_S}$ is infinite.

Although the Hamiltonian has a symmetry under parity, the lowest energy or ground state is not invariant under parity. So with degeneracy of the ground state the symmetry of H is broken, i.e., the energy eigenstates $|S>$ and $|A>$ are not invariant under the symmetry operation. So we say the symmetry is spontaneously broken. The breakdown is spontaneous in that the Hamiltonian exhibits the symmery. For symmetry breaking, the ground state must be degenerate, i.e., there must be more than one ground state and these states are not invariant under the symmetry operation under which the Hamiltonian is invariant. Ferromagnets provide another example of broken symmetry.

Next consider a symmetric QDW potential as typified by the potential

$$V(x) \propto (x^2 - \alpha^2)^2 \qquad (151)$$

We now analyze in the WKB approximation this one-dimensional potential with more than two turning points (Park (1974)). The potential curve with two minima shows the five regions of integration (Fig. 13.10).

Let $\psi_I(x)$ be the wave function in the WKB approximation in the region I.

$$\psi_I(x) \approx \frac{D_1}{\sqrt{K(x)}}\, \exp\left(\int_a^x K(x')\, dx'\right), \qquad (152)$$

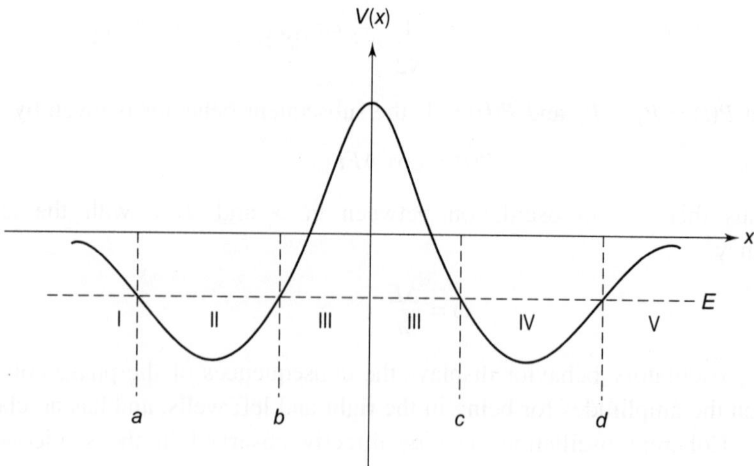

Fig. 13.10   Potential curve with two minima

$$\psi_{II}(x) \approx$$

$$\frac{A_2}{\sqrt{k(x)}}\exp\left(-i\int_a^x k(x')\,dx'\right) + \frac{B_2}{\sqrt{k(x)}}\exp\left(i\int_a^x k(x')\,dx'\right) \tag{152a}$$

$$\psi_{III}(x) \approx \frac{C_3}{\sqrt{K(x)}}\exp\left(-\int_b^x K(x')\,dx'\right)$$

$$+ \frac{D_3}{\sqrt{K(x)}}\exp\left(-i\int_b^x K(x')\,dx'\right), \tag{152b}$$

$$\psi_{IV}(x) \approx \frac{A_4}{\sqrt{k(x)}}\exp\left(-i\int_c^x k(x')\,dx'\right)$$

$$+ \frac{D_4}{\sqrt{k(x)}}\exp\left(i\int_c^x k(x')\,dx'\right), \tag{152c}$$

$$\psi_V(x) \approx \frac{C_5}{\sqrt{K(x)}}\exp\left(-\int_d^x K(x')\,dx'\right) \tag{152d}$$

Using the appropriate connection formulae at each of the four turning points

$$A_2 = e^{i\pi/4}\,D_1 B_2 = e^{-i\pi}/4\,D_1, \tag{153}$$

$$C_3 = \sin\theta_2 D_1 D_3 = 2\cos\theta_2 D_1 \tag{153a}$$

$$C_3 = 2 \cos \theta_4 \, e^{\theta_3} C_5 D_4 = \sin \theta_4 \, e^{-\theta_3} C_5, \tag{153b}$$

$$A_4 = e^{i\theta_4} \, e^{-i\pi/4} \, C_5 D_4 = e^{-i\theta_4} \, e^{i\pi/4} \, C_5, \tag{153c}$$

where the phase integrals are

$$\theta_2 = \int_a^b k(x') \, dx', \; k(x) = \left[\frac{2m}{\hbar^2} \, (V(x) - E) \right]^{1/2} \tag{154}$$

$$\theta_3 = \int_b^c K(x') \, dx', \; K(x) = \left[\frac{2m}{\hbar^2} \, (E - V(x)) \right]^{1/2} \tag{154a}$$

$$\theta_4 = \int_c^d k(x') \, dx', \tag{154b}$$

From Eqs. (153a) and (153b), we get

$$2D_1 \cos \theta_2 = \sin \theta_4 \, e^{-\theta_3} \, C_5, \tag{155}$$

$$D_1 \sin \theta_2 = 2 \cos \theta_4 \, e^{\theta_3} C_5,$$

Taken together, we get

$$\tan \theta_2 \tan \theta_4 = 4e^{2\theta_3}. \tag{156}$$

This is the eigenvalue condition.

For a symmetric QDW, $\theta_4 = \theta_2$ and we obtain

$$\tan \theta_2 = \pm \, 2e^{\theta_3}, \tag{157}$$

and so
$$C_5 = \pm \, D_1. \tag{158}$$

The signs in Eqs. (157) and (158) correspond.

The wave functions are either symmetric or antisymmetric. We shall restrict ourselves to the case when θ_3 is large. In most cases then e^{θ_3} is a large number, and θ_2 is in the neighborhood of $\frac{\pi}{2}, \frac{3\pi}{2}$, etc. To solve Eq. (157), we put

$$\theta_2 = \left(n + \frac{1}{2} \right) \pi - \theta_1, \tag{159}$$

where n is an integer and θ_1 is small and so Eq. (157)

gives
$$\theta_1 \approx \tan \theta_1 = \pm \, \frac{1}{2} \, e^{-\theta_3}. \tag{160}$$

For symmetric wave functions ($C_5 = D_1$), i.e.,

$$\tan \theta_2 = 2e \, \theta_3, \text{ we get}$$

$$\theta_2 \approx \left(n + \frac{1}{2} \right) \pi - \frac{1}{2} \, e^{-\theta_3}, \; n = 0, 1, 2, \ldots \tag{161}$$

If $\theta_2 = \left(n + \dfrac{1}{2}\right)\pi$, then $\theta_3 = \infty$, i.e., the barrier is infinite, and the energy levels become the harmonic oscillator levels, but doubly degenerate.

For antisymmetric wave functions $(C_5 = -D_1)$ we get

$$\theta_2 \approx \left(n + \frac{1}{2}\right)\pi + \frac{1}{2}\,e^{-\theta_3}. \tag{162}$$

Now, there is a splitting of energy due to tunneling.

The level E_0 is split into levels $E_1(E_S)$ and $E_2(E_A)$.

Therefore,
$$\frac{\partial\theta_2}{\partial E}\,\Delta E = -\frac{1}{2}\,e^{-\theta_3} = E_1 - E_0,$$

$$\frac{\partial\theta_2}{\partial E}\,\Delta E = +\frac{1}{2}\,e^{-\theta_3} = E_2 - E_0,$$

so that
$$E_2 - E_1 = \frac{e^{-\theta_3}}{\dfrac{\partial\theta_2}{\partial E,}}. \tag{163}$$

We have

$$\frac{\partial\theta_2}{\partial E} = \frac{\partial}{\partial E}\int_a^b \left[\frac{2m}{\hbar^z}\,(E - V(x))\right]^{1/2} dx.$$

$$= \frac{m}{\hbar^2}\int dx\left[\frac{2m}{\hbar^z}\,(E - V(x))\right]^{1/2} dx. \tag{164}$$

But

$$\int_a^b \frac{dx}{\left[\dfrac{2m}{\hbar^z}\,(E - V(x))\right]^{1/2}}$$

$$= \frac{\hbar}{m}\int_a^b \frac{dx}{\left[\dfrac{2}{m}\,(E - V(x))\right]^{1/2}}$$

$$= \frac{\hbar}{m}\int_a^b \frac{dx}{dx/dt}, \text{ where } \upsilon = \frac{dx}{dt} \text{ is the velocity of a}$$
$$\text{particle between } a \text{ and } b.$$

$$= \frac{\hbar}{m}\cdot\frac{T}{2}, \text{ where } T \text{ is the period}$$

$$= \frac{\hbar}{m}\cdot\frac{\pi}{\omega}$$

$$= \frac{\pi\hbar}{m\omega}.$$

From Eq. (164), $\dfrac{\partial\theta_2}{\partial E} = \dfrac{\pi}{\hbar\omega}$

and so

$$E_2 - E_1 = \frac{\hbar\omega}{\pi} \exp\left(-\int_{-c}^{c} |K(x)|\, dx\right), \tag{165}$$

where c is the turning point corresponding to the energy E_0, and $K(x) = \left[\dfrac{2m}{\hbar^2}(E - V(x))\right]^{1/2}$. This is the formula for splitting given by Landau and Lifshitz (1977).

This tunnel splitting has attracted considerable attention. It can also be obtained through two other methods – the semi-classical path integral and the instanton procedure (Holstein (1988)).

It is known that the WKB method is not a particularly useful approximation to the exact result for the ground state and the low-lying excited states.

A careful analysis by Garg (2000) has shown that the above formula Eq. (165) for tunnel splitting has a prefactor which is incorrect for the ground state. The prefactor is amended and the formula for the splitting which is applicable to the ground and low lying excited states is obtained below following Garg (2000).

Let us consider a smooth, symmetric double well potential in one dimension (see Fig. 13.11).

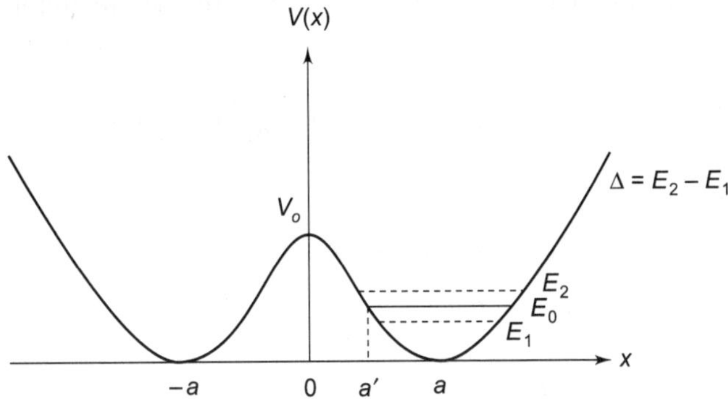

Fig. 13.11   A symmetric double well potential in one dimension

Due to tunneling the energy levels split and we are interested in calculating the tunnel splittings in such a potential. The potential is symmetric about $x = 0$, and has quadratic minima at $x = \pm a$, where the potential vanishes i.e., $V(\pm a) = 0$. Let $\psi_0(x)$ be the quasi-classical wave function with energy E_0 in one well say the right hand well, which is exponentially damped on both sides of the well. Due to tunneling, the level E_0 splits into two levels E_1 and E_2.

The ground state splitting is given by Herring's formula

$$\Delta = \frac{2\hbar^2}{m} \psi(0)\, \psi_0'(0) \tag{166}$$

where $\psi_0'(x) = \dfrac{d\psi_0(x)}{dx}$ and $\psi_0(x)$ is normalized to unit probability in the right-hand well.

In the region $(a' - x) \gg \sqrt{\dfrac{\hbar}{m\omega}}$, the WKB approximation for $\psi_0(x)$ is

$$\psi_0(x) = \frac{C_0}{\sqrt{|v|(x)}} \exp\left(\frac{1}{\hbar} \int_0^x |p(x')|\, dx'\right), \tag{167}$$

where $v(x) = \dfrac{p(x)}{m}$ and C_0 is a constant. ω is the frequency of small oscillations in the wells, about $x = \pm a$, $\pm a'$ are the classical turning points given by

$$V(a') = E_0 = \frac{1}{2}\hbar\omega + O(\hbar). \tag{168}$$

Note that $p(x) = \sqrt{2m(V(x) - E)}$ is imaginary in the classically forbidden region $-a' < x < a'$ and $V(\pm a) = 0$.

Now, $\qquad \psi_0(x) \approx \left(\dfrac{m|v(x)|}{\hbar}\right) \psi_0(x)$, so that

$$\Delta = 2\hbar C_0^2 \tag{169}$$

Now, near the minima $(x = a)$ and also in the forbidden region $\psi_0(x)$ may be represented by the ground state wave function for a harmonic oscillator

$$\psi_0(x) = \left(\frac{m\omega}{\pi\hbar}\right)^{1/4} \exp\left(-\frac{m\omega}{2\hbar}(x - a)^2\right). \tag{170}$$

We write

$$|p(x)| = m\omega\, [(a - x)^2 - (a - a')^2]^{1/2}$$

so that $\qquad |v(x)| \approx \omega\,(a - x),$

since $\qquad a - a' = \sqrt{\dfrac{\hbar}{m\omega}}. \tag{171}$

Hence, Eq. (167) may be written as

$$\psi_0(x) \approx \frac{C_0}{\sqrt{\omega(a - x)}} \exp\left(\frac{1}{\hbar} \int_0^{a'} |p(x')|\, dx' + \varphi(x)\right)$$

where

$$\varphi(x) = -\frac{m\omega}{\hbar} \int_x^{a'} [(a - x')^2 + (a - a')^2]^{1/2}\, dx'. \tag{172}$$

$$= \frac{m\omega(a-x)^2}{2\hbar} + \ln\left(\frac{2(a-x)}{a-a'}\right)^{1/2} + \frac{1}{4} + O\left(\frac{a-a'}{a-x}\right)^2. \tag{173}$$

Comparing the two expressions for $\psi_0(x)$,

$$C_0 = \left(\frac{\omega^2}{4\pi e}\right)^{1/4} \exp\left(\frac{1}{\hbar}\int_0^{a'} |p(x)|\, dx\right). \tag{174}$$

Thus the magnitude of the ground state splitting in the WKB approach is

$$\Delta = \frac{\hbar\omega}{\sqrt{e\pi}} \exp\left[-\frac{1}{\hbar}\int_{-a'}^{a'} |p|\, dx\right]. \tag{175}$$

The prefactor for Δ given by Landau and Lifshitz (Prob. 3, § 50) is corrected in the above expression. As Garg has emphasized the expression Eq. (175) is difficult to use and the simpler result for the splitting for a general potential may be written as

$$\Delta = 2\hbar\omega \left(\frac{m\omega a^2}{\pi\hbar}\right)^{1/2} e^A e^{-S_0/\hbar}, \tag{176}$$

where

$$S_0 = \int_{-a}^{a} \sqrt{2mV(x)}\, dx, \tag{177}$$

and

$$A = \int_0^a \left[\frac{m\omega}{\sqrt{2mV(x)}} - \frac{1}{a-x}\right] dx. \tag{178}$$

PROBLEMS

13.1 A particle of mass m is moving in a one dimensional quartic potential $V(x) = \lambda x^4$ ($\lambda > 0$). Use the variational method to find an approximate value for the ground state energy. Choose a trial wave function of the

from $\psi_\alpha^{(0)}(x) = \left(\frac{2\alpha}{\pi}\right)^{1/4} e^{-\alpha x^2}$ ($\alpha > 0$).

13.2 Consider a simple harmonic oscillator with the Hamiltonian

$$H = -\frac{\hbar^2}{2m}\frac{d^2}{dx^2} + \frac{1}{2}m\omega^2 x^2$$

(a) By varying the parameter α in the trial function $\psi_\alpha^{(0)}(x) = \dfrac{1}{x^2+\alpha}$

($\alpha > 0$) obtain an upper bound for the ground state energy of the oscillator.

(b) Repeat the same procedure for another one-parameter family of trial wave functions $\psi_\beta^{(0)}(x) = xe^{-\beta x^2}$ ($\beta > 0$).

13.3 Using the variational method, find an estimate of the ground state energy of hydrogen atom. Use the following expressions as trial wave functions

(a) $\psi_\alpha^{(0)} = e^{-\alpha r^2}$, ($\alpha > 0$) where α is a variational parameter.

(b) $\psi_\alpha^{(0)}(r) = \begin{cases} c\left(1 - \dfrac{r}{\alpha}\right) & \text{for } r \leq \alpha \\ 0 & \text{for } r > \alpha \end{cases}$

where c is a normalization constant and α is the variational parameter.

13.4 A particle of mass m is moving in a one-dimensional potential $V(x) = \lambda x^4$ ($\lambda > 0$). Use the WKB method to estimate the ground state energy of the particle. Compare with the result obtained by the variational method in 13.1

13.5 Consider the "bouncing ball" in Problem 3.5. Estimate the ground state energy of the particle by using the variational method, with the trial function $\psi_\alpha^{(0)} = Aze^{-\alpha z}$, where A is the normalization constant and α is the variational parameter. Compare with the result obtained in Problem 3.5.

CHAPTER 14

Methods for Time-Dependent Problems

Let us now study time-dependent problems. When the Hamiltonian is time-dependent, there are no stationary states. It is generally impossible to obtain exact solutions to the time-dependent Schödinger equation. We therefore use approximation methods. In this chapter we discuss three different approximation schemes – the time-dependent perturbation theory, the adiabatic approximation and the sudden approximation. We also analyze an exactly solvable example of time dependent problems. In Section 14.8, we discuss the geometric phases. In the next chapter we shall consider, as an application of the theory, the interaction of radiation with matter.

14.1   TIME-DEPENDENT PERTURBATION THEORY

We consider a system whose Hamiltonian may be written as

$$H = H_0 + V(t), \tag{1}$$

where H_0 is time-independent and its eigenvalues E_n and the corresponding eigenkets $|\psi_n^{(0)}\rangle$ (which form a complete, orthonormal set) are known:

$$H_0|\psi_n^{(0)}\rangle = E_0|\psi_n^{(0)}\rangle. \tag{2}$$

An arbitrary solution of the unperturbed time-independent Schrödinger equation

$$i\hbar \frac{\partial}{\partial t} |\psi^{(0)}(t)\rangle = H_0|\psi^{(0)}(t)\rangle, \tag{3}$$

can be written as

$$|\psi^{(0)}(t)\rangle = \sum_n C_n e^{-\frac{i}{\hbar}E_n t}|\psi_n^{(0)}\rangle, \tag{4}$$

where C_n are constants, the square of whose modulus $|C_n|^2$ is the probability of finding the system in the unperturbed state $|\psi_n^{(0)}\rangle$. The summation implies a sum over the discrete spectrum and the integral over the continuous spectrum.

We now seek the general solution of the perturbed equation

$$i\hbar \frac{\partial}{\partial t} |\underline{\Psi}(t)\rangle = (H_0 + V(t))|\underline{\Psi}(t)\rangle, \tag{5}$$

in the same form as in Eq. (4):

$$|\underline{\Psi}(t)\rangle = \sum_n C_n(t)\, e^{-\frac{i}{\hbar} E_n t} |\psi_n^{(0)}\rangle, \tag{6}$$

where now the expansion coefficients $C_n(t)$ are functions of time. This method corresponds to the method of variation of constants to solve linear differential equations and was applied by Dirac in 1926.

Substituting Eq. (6) into Eq. (5) and using Eq. (1) we obtain

$$i\hbar \sum_n \dot{C}_n(t)\, e^{-\frac{i}{\hbar} E_n t} |\psi_n^{(0)}\rangle + \sum_n C_n(t)\, E_n e^{-\frac{i}{\hbar} E_n t} |\psi_n^{(0)}\rangle$$

$$= \sum_n C_n(t)\, (H_0 + V(t))\, e^{-\frac{i}{\hbar} E_n t} |\psi_n^{(0)}\rangle, \tag{7}$$

where the dot indicates differentiation with respect to the time. With the help of Eq. (2), Eq. (7) becomes

$$i\hbar \sum_n \dot{C}_n(t)\, e^{-\frac{i}{\hbar} E_n t} |\psi_n^{(0)}\rangle$$

$$= \sum_n V(t)\, C_n(t)\, e^{-\frac{i}{\hbar} E_n t} |\psi_n^{(0)}\rangle. \tag{8}$$

Taking the inner product with $\langle \psi_k^{(0)}|$, and using the orthonormality property $\langle \psi_m^{(1)}|\psi_n^{(1)}\rangle = \delta_{mn}$ we have

$$i\hbar\, \dot{C}_k e^{-\frac{i}{\hbar} E_k t} = \sum_n C_n e^{-\frac{i}{\hbar} E_n t}\, V_{kn}(t), \tag{9}$$

where

$$V_{kn}(t) = \langle \psi_k^{(0)}| V(t) |\psi_n^{(0)}\rangle. \tag{10}$$

We define the Bohr (angular) frequency

$$\omega_{kn} = \frac{E_k - E_n}{\hbar}. \tag{11}$$

Thus,

$$\dot{C}_k(t) = (i\hbar)^{-1} \sum_n V_{kn}(t)\, e^{i\omega_{kn} t}\, C_n(t). \tag{12}$$

This set of Eq. (12) for all k is exactly equivalent to the Schrödinger equation (5).

In most cases of time-dependent problems, we resort to approximate solutions of the set of coupled differential Eq. (12) by perturbation expansion.

We now consider Eq. (12), replace V by λV, and use the perturbation expansion by expressing the coefficients C's as power series in λ.

$$C_n(t) = C_n^{(0)} + \lambda C_n^{(1)} + \lambda C_n^{(2)} + \dots \tag{13}$$

We assume that for $0 \leq \lambda \leq 1$, this series is a continuous analytic function of λ. Substituting Eq. (13) into Eq. (12) and equating coefficients of corresponding powers of λ, and setting $\lambda = 1$ finally, we get

$$\dot{C}_k^{(0)} = 0$$

$$\dot{C}_k^{(1)} = (i\hbar)^{-1} \sum_n V_{kn}\, e^{i\omega_{kn}t}\, C_n^{(0)},$$

$$\dots \dots \dots \dots \dots \dots \tag{14}$$

$$\dot{C}_k^{(r+1)} = (i\hbar)^{-1} \sum_n V_{kn} e^{i\omega_{kn}t}\, C_n^{(r)}, \; r = 0,\, 1,\, 2,\, \dots$$

This set of equations can, in principle, be integrated successively to any desired order in the perturbation.

The first of Eq. (14) shows that the coefficients $C_k^{(0)}$ to the zeroth approximation are independent of time. Their values are the initial conditions and characterize the state of the system before the perturbation is applied. We assume that initially the system is in a definite unperturbed energy state say, in the stationary state m. Then all except one of the $C^{(0)}$'s are zero. So we write

$$C_k^{(0)} = \delta_{km} \text{ or } \delta\,(k - m), \tag{15}$$

according as the initial state m is discrete or continuous. Integrating the first order equation we have

$$C_k^{(1)}(t) = (i\hbar)^{-1} \int_{-\infty}^{t} V_{km}(t')\, e^{i\omega_{km}t'}\, dt', \tag{16}$$

where the constant of integration is chosen to be zero so that $C_k^{(1)}$ vanish at $t = -\infty$ before the perturbation is applied. Therefore, to first order in perturbation, the probability for transition from the initial state m to a state different from m is

$$P_{m \to n}^{(1)}(t) = \hbar^{-2} \left| \int_{-\infty}^{t} V_{nm}(t')\, e^{i\omega_{nm}t'}\, dt' \right|^2 . \tag{17}$$

It is to be noted that the time-dependent perturbation causes a stationary state (an eigenstate of H_0) to evolve into a non-stationary state. The probability $P_{m \to n}$ expresses the probability of the initial state m, after the perturbation is switched on, to be found in a state with energy E_n. See Ballentine (1998).

Instead of proceeding in this manner we now discuss the time evolution operator in the interaction picture and obtain the results to all orders by a perturbation expansion of the evolution operator. This elegant and powerful method of operator expansion is applied in quantum field theory.

14.2 THE INTERACTION PICTURE

In Section 2.8, we have seen that the time evolution of a system can be described in different but equivalent pictures. From our discussion, we recall that the state vector $|\psi_I(t)\rangle$ in the interaction picture is related to the state vector $|\psi_S(t)\rangle$ in the Schrödinger picture as

$$|\psi_I(t)\rangle = e^{iH_0 t/\hbar} |\psi_S(t)\rangle \tag{18}$$

An operator $O_I(t)$ in the interaction picture is defined by the relation

$$O_I(t) = e^{iH_0 t/\hbar} \, O_S^{(t)} \, e^{-iH_0 t/\hbar}, \tag{19}$$

where $O_S^{(t)}$ is the operator in the Schrödinger picture.

The time evolution operator in the interaction picture is defined by

$$|\psi_I(t)\rangle = U_I(t, t_0)|\psi_I(t_0)\rangle, \tag{20}$$

in analogy to

$$|\psi_S(t)\rangle = U_S(t, t_0)|\psi_S(t_0)\rangle \tag{21}$$

From the Schrödinger equation (5), we see that $U_S(t, t_0)$ satisfies the differential equation

$$i\hbar \, \frac{\partial}{\partial t} U_S(t, t_0) = H \, U_S(t, t_0). \tag{22}$$

The time-dependence of the state ket in the interaction picture is given by

$$i\hbar \, \frac{\partial}{\partial t}\left|\psi_I(t)\right\rangle = - H_0 e^{iH_0 t/\hbar}|\psi_S(t)\rangle + e^{iH_0 t/\hbar} \, i\hbar \, \frac{\partial}{\partial \hbar}\left|\psi_S(t)\right\rangle, \text{ using (18).}$$

$$= - H_0 e^{iH_0 t/\hbar}|\psi_S(t)\rangle + e^{iH_0 t/\hbar} \, (H_0 + V(t))|\psi_S(t)\rangle$$

$$= e^{iH_0 t/\hbar} \, V(t) e^{iH_0 t/\hbar}|\psi_I(t)\rangle \tag{23}$$

Hence we write

$$i\hbar \, \frac{\partial}{\partial(t)}\left|\psi_I(t)\right\rangle = V_I(t)|\psi_I(t)\rangle, \tag{24}$$

where the perturbation operator in the interaction picture is

$$V_I(t) = e^{iH_0 t/\hbar} \, V(t) e^{-iH_0 t/\hbar}. \tag{25}$$

In terms of the time evolution operator we may write Eq. (24) as

$$i\hbar \frac{\partial}{\partial t} U_I(t, t_0) = V_I(t) U_I(t, t_0). \tag{26}$$

This operator differential equation with the initial condition

$$U_I(t, t_0)|_{t=t_0} = 1, \tag{27}$$

can be converted to the following integral equation

$$U_I(t, t_0) = 1 - \frac{i}{\hbar} \int_{t_0}^{t} V_I(t') U_I(t', t_0)\, dt'. \tag{28}$$

Equation (28) can be solved by iteration (the Neumann method) giving an expansion in powers of the perturbation $V_I(t)$:

$$U_I(t, t_0) = 1 - \frac{i}{\hbar} \int_{t_0}^{t} dt_1 V_I(t_1) \left[1 - \frac{i}{\hbar} \int_{t_0}^{t_1} dt_2\, V_I(t_2) U_I(t_2, t_0) \right]$$

$$= 1 - \frac{i}{\hbar} \int_{t_0}^{t} dt_1 V_I(t_1) + \left(\frac{-i}{\hbar}\right)^2 \int_{t_0}^{t} dt_1 \int_{t_0}^{t_1} dt_2 V_I(t_1) V_I(t_2)$$

$$+ \ldots + \left(-\frac{i}{\hbar}\right)^n \int_{t_0}^{t} dt_1 \int_{t_0}^{t_1} dt_2 \ldots \int_{t_0}^{t_{n-1}} dt_n V_I(t_1) V_I(t_2) \ldots V_I(t_n) + \ldots$$

$$= \sum_{n=0}^{\infty} U_I^{(n)}(t, t_0) \tag{29}$$

where
$$U_I^{(0)}(t, t_0) = 1,$$

$$U_I^{(1)}(t, t_0) = -\frac{i}{\hbar} \int_{t_0}^{t} dt_1\, V_I(t_1),$$

$$U_I^{(n)}(t, t_0) = \left(-\frac{i}{\hbar}\right)^n \int_{t_0}^{t} dt_1 \int_{t_0}^{t_1} dt_2 \ldots \int_{t_0}^{t_{n-1}} dt_n V_I(t_1) V_I(t_2) \ldots V_I(t_n). \tag{30}$$

The series Eq. (29) is known as the Dyson series. Dyson pointed out that the integration is essentially an integration over the time interval t_0 to t, where any limit t_i should be earlier than t_{i-1} ($i \leq n$). Dyson showed that this can be done by using a chronological ordering operator T which is defined for say two non-commuting operators, as

$$T\left[V_I(t_1) V_I(t_2)\right] = \begin{cases} V_I(t_1) V_I(t_2), & t_1 > t_2 \\ V_I(t_2) V_I(t_1), & t_2 > t_1 \end{cases} \tag{31}$$

Each term in the series is in time-ordered form, and since $t_1 \geq t_2 \ldots \geq t_n$, we may write $T \lfloor V_I(t_1) \ldots V_I(t_n) \rfloor$ in the series:

$$U_I(t, t_0) = 1 + \sum_{n=1}^{\infty} \left(-\frac{i}{\hbar}\right)^n \int_{t_0}^{t} dt_1 \int_{t_0}^{t_1} dt_2 \ldots \int_{t_0}^{t_{n-1}} dt_n - T\,[V_I(t_1) \ldots V_I(t_n)]. \qquad (32)$$

We rewrite Eq. (32) as

$$U_I(t, t_0) = 1 + \sum_{n=1}^{\infty} \frac{(-i/\hbar)^n}{n!} \int_{t_0}^{t} dt_1 \ldots \int_{t_0}^{t} dt_n\, T\,[V_I(t_1) \ldots V_I(t_n)],$$

$$= T\left(\exp\left[-\frac{i}{\hbar} \int_{t_0}^{t} V_I(t)\, dt \right] \right), \qquad (33)$$

where the time-ordered exponential is a short hand notation for the sum.

14.3   TRANSITION PROBABILITY

The time development of any state ket can be obtained once the time evolution operator $U_I(t, t_0)$ is known. The relationship between the time evolution operators in the two pictures is given by

$$U_I(t, t_0)|\,\psi_I(t_0)\rangle = |\,\psi_I(t)\rangle$$

$$= e^{iH_0 t/\hbar}|\,\psi_S(t)\rangle.$$

$$= e^{iH_0 t/\hbar}\, U_S(t, t_0)|\,\psi_S(t_0)\rangle$$

$$= e^{iH_0 t/\hbar}\, U_S(t, t_0)|\,\psi_I(t_0)\rangle. \qquad (34)$$

Since $|\psi_I(t_0)\rangle$ is arbitrary, we have

$$U_I(t, t_0) = e^{iH_0 t/\hbar}\, U_S(t, t_0)\, e^{-iH_0 t/\hbar}$$

or, $\qquad\qquad U_S(t, t_0) = e^{-iH_0 t/\hbar}\, U_I(t, t_0)\, e^{iH_0 t/\hbar}. \qquad (35)$

From the perturbation series for $U_I(t, t_0)$, one can obtain a corresponding perturbation series for $U_S(t, t_0)$. We can also obtain a perturbation series for the transition amplitude. The matrix element of $U_I(t, t_0)$ between an initial state $|m\rangle$ (an energy eigenstate of H_0) and a state $|n\rangle$ with energy E_n is

$$\langle n|\, U_I(t, t_0)|m\rangle = e^{i(E_n t - E_m t_0)/\hbar}\, \langle n|\, U(t, t_0)|m\rangle, \qquad (36)$$

where $U(t, t_0)$ is the time evolution operator in the Schrödinger picture (we drop the subscript S). The transition amplitude defined earlier is the quantity $\langle |U(t, t_0)|\rangle$ on the right hand side of Eq. (36). The transition probability is, of course, the same in the two pictures:

$$P_{m \to n}(t) = |\langle n| U_I(t, t_0)|m\rangle|^2 = |\langle n| U(t, t_0)|m\rangle|^2. \tag{37}$$

Let us suppose that at time $t = t_0$, the system is in a state $|i\rangle$. The state ket in the Schrödinger picture is then $|\psi_s\rangle = |i\rangle \times$ phase factor. We choose the phase factor as

$$|\psi_S(t_0)\rangle = e^{-iE_i t_0/\hbar} |i\rangle \tag{38}$$

so that in the interaction picture

$$|\psi_I(t_0)\rangle = |i\rangle. \tag{39}$$

At a later time t,

$$|\psi_I(t)\rangle = U_I(t, t_0)|i\rangle \tag{40}$$

Expanding ψ_I in a complete set $\{|k\rangle\}$,

$$|\psi_I(t)\rangle = \sum_k C_k(t)|k\rangle, \tag{41}$$

we obtain

$$C_n(t) = \langle n| U_I(t, t_0)\rangle|i\rangle. \tag{42}$$

Now, comparing the perturbation expansion of $U_I(t, t_0)$ in Eq. (29) and the expansion of $C_n(t)$ in Eq. (13), we get

$$C_n^{(0)}(t) = \delta_{ni} \ (n \neq i),$$

$$C_n^{(1)}(t) = \frac{-i}{\hbar} \int_{t_0}^{t} \langle n| V_I(t_1)|i\rangle \, dt_1$$

$$= \frac{-i}{\hbar} \int_{t_0}^{t} e^{i\omega_{ni} t_1} V_{ni}(t_1) \, dt_1,$$

$$C_n^{(2)}(t) = \left(\frac{-i}{\hbar}\right)^2 \sum_m \int_{t_0}^{t} dt_1 \int_{t_0}^{t_1} dt_2 \, e^{i\omega_{nm} t_1} V_{nm}(t_1), \times e^{i\omega_{mi} t_2} V_{mi}(t_2)$$

$$\cdots \qquad \cdots \qquad \cdots \tag{43}$$

Hence the transition probability is given by

$$P_{i \to n}(t) = |C_n^{(1)}(t) + C_n^{(2)}(t) + \ldots|^2 \tag{44}$$

It is extremely useful and convenient to represent the successive contributions to the transition amplitude by the diagrams of Fig. 14.2. We associate a line with the propagator $\exp(-iE_a(t_0 - t)/\hbar)$ and a "vertex" with $-\frac{i}{\hbar} \langle b|V|a\rangle dt_1$ (Fig. 14.1).

In Fig. 14.2, we show the lowest order diagrams contributing to the transition amplitude $i \to f$. Here time flows upward and i and f denote external lines, while an internal line joining two vertices represents an intermediate state.

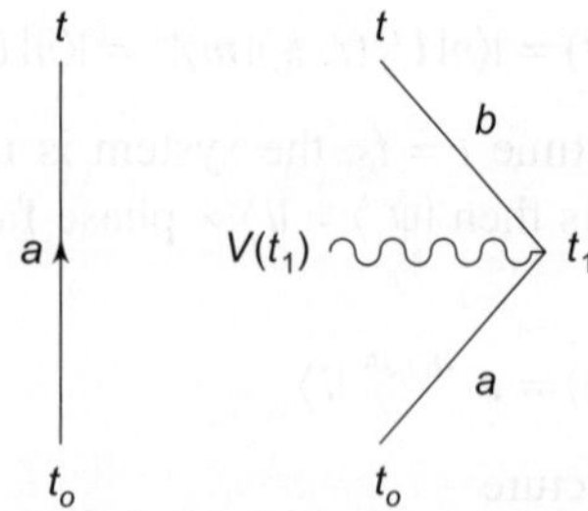

Fig. 14.1 Propagator and vertex

We associate a "Feynman diagram" with each order of perturbation and the total amplitude is the sum over all diagrams (note the similarity with the Feynman path integral).

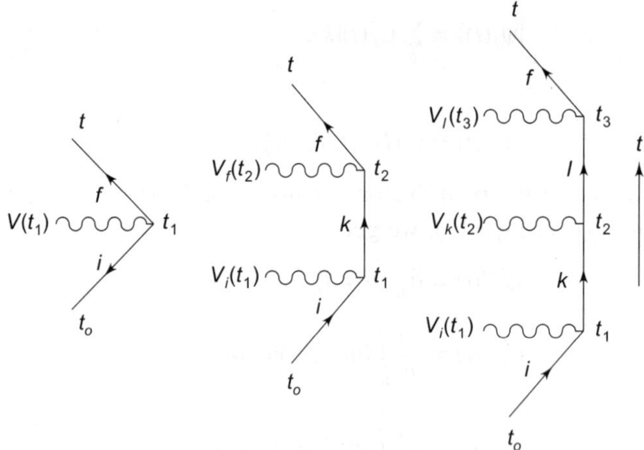

Fig. 14.2 Diagrams showing the contributions of successive order to the transition amplitude

14.4 TIME-DEPENDENT TWO-STATE PROBLEM: MAGNETIC RESONANCE

Time dependent perturbations are of various types. Before we begin our discussion of such time dependent problems in detail, we consider an exactly solvable problem with time-dependent potential. We discuss magnetic resonance—a subject of considerable practical importance.

In Section 6.3 we have seen that a magnetic moment $\vec{\mu} = \gamma \vec{S}$ where $\vec{S}$ is the spin operator (in units of $\hbar$) in a static magnetic field B_0 in the z-direction precesses at the Larmor frequency $\omega_0 = \gamma B_0$ about the axis of the magnetic field (z-axis). In magnetic resonance experiments the system is subjected to a time-varying magnetic field perpendicular to the direction of the original magnetic field. When the frequency of this magnetic field $\omega \approx \omega_0$, the system is strongly disturbed and the probability of spin-flips becomes large. This is called paramagnetic resonance (EPR or sometimes ESR). The measurement of the resonant frequency makes possible the accurate determination of the g factors.

The paramagnetic resonance principle was first applied by Rabi and his co-workers in their experiments on molecular beams to measure the magnetic moments of atoms. With known values of the magnetic moments, the resonance technique provides information about the local magnetic field at the site of the electron or nucleus. It is also possible to detect magnetic resonance in bulk samples of matter. This technique is called EPR when applied to an unpaired electron spin and is usually studied at microwave frequencies (10^9–10^{11} Hz). Nuclear magnetic resonance (NMR) is the application of paramagnetic resonance method to the magnetic moments of nuclei and is studied at radio frequency region (10^6 to 10^9 Hz). This technique is best known as a diagnostic tool for humans (magnetic resonance imaging or MRI).

The precessional motion of an angular momentum vector of a spinning particle in a static magnetic field B_0 along the z-axis is shown in Fig. 6.2.

Since the spinning particle precesses counter clockwise about the z-axis at the rate ω_0, we can make the spin appear constant (no precession) if we observe the spin in a rotating coordinate frame with angular velocity ω_0 (see Fig. 14.3).

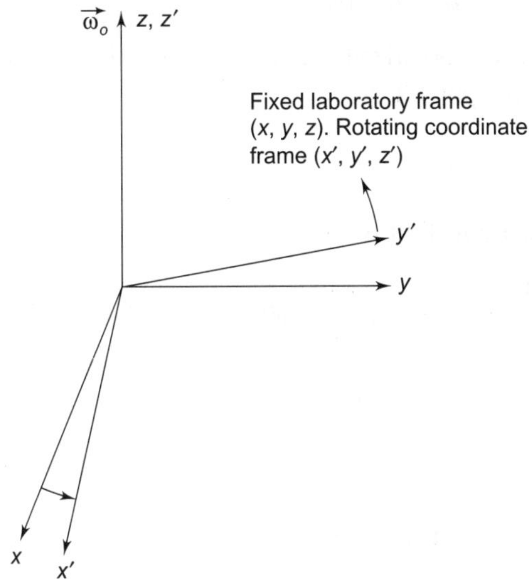

Fig. 14.3   A rotating coordinate frame

With respect to the rotating frame, we can say that there is a compensating magnetic field which cancels the applied magnetic field. The rotating axis representation was introduced by Rabi, Schwinger, and van Vleck. Therefore, if in the rotating frame we apply a constant magnetic field along x' then the spin would in this frame precess about it. In the laboratory frame this requires an oscillating magnetic field in the xy plane in addition to a constant component B_0 along the z-axis. The usual experimental arrangement has an oscillating magnetic field $B_1 \cos \omega t$ along the x-axis and a constant field B_0 along the z-axis. It is convenient to express the transverse oscillating field in terms of two rotating fields. To see this, we write

$$\vec{B}_1(t) = \hat{i} B_1 \cos \omega t = \frac{B_1}{2} (\hat{i} \cos \omega t + \hat{j} \sin \omega t)$$

$$+ \frac{B_1}{2} (\hat{i} \cos \omega t - \hat{j} \sin \omega t). \tag{45}$$

The first term is a field of constant amplitude $\dfrac{B_1}{2}$, rotating about $\vec{B}_0$ at frequency ω in the same sense as the Larmor precession. The second term is a similar field rotating in the opposite sense. This counter rotating term does not cause any magnetic resonance and produces minor modifications provided $\omega \approx \omega_0$ and $B \ll B_0$. This term can be neglected. Neglecting the far-off-resonance and the rapidly oscillating term is known as the rotating–wave approximation and is often used in magnetic resonance and in quantum optics.

We may now write the total magnetic field as

$$\vec{B} = B_1 \cos \omega t \, \hat{i} + B_0 \hat{k} \tag{46}$$

where we use a linearly polarized oscillating magnetic field. The Hamiltonian is then for a particle of spin 1/2:

$$H = - \gamma \left(\frac{\hbar}{2} \right) (B_0 \sigma_z + B_1 \cos \omega t \, \sigma_x). \tag{47}$$

The Schrödinger equation is

$$i\hbar \, \frac{\partial}{\partial t} \left| \underline{\psi}(t) \right\rangle = - \gamma \hbar \, \frac{1}{2} (B_0 \sigma_z + B_1 \cos \omega t \, \sigma_x) | \underline{\psi}(t) \rangle \tag{48}$$

Let $$|\Phi(t)\rangle = D_z(\omega t) | \underline{\psi}(t) \rangle, \tag{49}$$

where $D_z(\omega t) = e^{-i\omega t \sigma_z /2}$ is the operator that generates a rotation through the angle ωt about the z-axis.

From Eqs. (48) and (49), we get

$$i\hbar \, \frac{\partial}{\partial t} \Big| \Phi(t) \Big\rangle = - \frac{\hbar}{2} \{ (\omega_0 - \omega) \sigma_z + \omega_1 \cos \omega t \, e^{-i\omega t \sigma_z /2} \, \sigma_x \, e^{i\omega t \sigma_z /2} \} | \Phi(t) \rangle, \tag{50}$$

where $\omega_0 = \gamma B_0$ and $\omega_1 = \gamma B_1$.

Now,

$$\cos\omega t\, e^{-i\omega t\sigma_z/2}\, \sigma_x\, e^{-i\omega t\sigma_z/2}$$

$$= \cos\omega t\, \sigma_x\, e^{i\omega t\sigma_z}$$

$$= \sigma_x\,(\cos^2\omega t + i\sigma_z\cos\omega t\sin\omega t)$$

$$= \frac{1}{2}\,\sigma_x + \frac{1}{2}\,(\sigma_x\cos 2\omega t + \sigma_y\sin 2\omega t).$$

In the rotating–wave approximation, we neglect the two high frequency terms, and Eq. (50) becomes

$$i\hbar\,\frac{\partial}{\partial t}\Big|\Phi(t)\Big\rangle = \left[\frac{1}{2}\,(\omega-\omega_0)\,\sigma_z - \frac{1}{2}\,\omega_1\sigma_x\right]|\Phi(t)\rangle. \tag{51}$$

The solution to this equation in the rotating frame is.

$$|\Phi(t)\rangle = e^{-i\left[\frac{1}{2}(\omega-\omega_0)\sigma_z - \frac{1}{2}\omega_1\sigma_x\right]t}\,|\Phi(0)\rangle. \tag{52}$$

Using Eq. (49), we get the solution of the original Eq. (48) in the static frame

$$|\underline{\psi}(t)\rangle = e^{i\omega t\sigma_z/2}\, e^{-i\left[\frac{1}{2}(\omega-\omega_0)\sigma_z - \frac{1}{2}\omega_1\sigma_x\right]t}|\underline{\psi}(0)\rangle. \tag{53}$$

We write

$$(\omega-\omega_0)^2 + \omega_1^2 = \Omega^2, \tag{54}$$

$$\hat{n} = \cos\varphi\hat{k} + \sin\varphi\hat{i} \tag{55}$$

where

$$\tan\varphi = \frac{-\omega_1}{\omega-\omega_0}. \tag{56}$$

Equation (53) can be expressed as

$$|\underline{\psi}(t)\rangle = e^{i\omega t\sigma_z/2}\, e^{-i\Omega t\vec{\sigma}\cdot\hat{n}/2}|\underline{\psi}(0)\rangle \tag{56}$$

$$= D_z^{\dagger}\,(\omega t)\, D_{\hat{n}}\,(\Omega t)|\underline{\psi}(0)\rangle. \tag{57}$$

$\hat{n}$ is a vector in the xz plane and at an angle $\left(\dfrac{\pi}{2}-\phi\right)$ to the x-axis. Also

$$\sigma_n \equiv \vec{\sigma}\cdot\hat{n} = \frac{\omega-\omega_0}{\Omega}\,\sigma_z - \frac{\omega_1}{\Omega}\,\sigma_x,$$

and

$$\Omega\vec{\sigma}\cdot\hat{n} = (\omega-\omega_0)\,\sigma_z - \omega_1\sigma_x.$$

Let us now calculate the transition probability from the initial state to the final state. We suppose that initially (i.e., at $t = 0$) the spin is along z-direction: $|\underline{\psi}(0)\rangle = |\Phi(0)\rangle = |+\rangle$. The transition amplitude for the spin to flip over to the state $|-\rangle$ at time t is

$$\langle -|\overline{\Psi}(t)\rangle = e^{-i\omega t/2} - \left(\cos \frac{\Omega t}{2} - i\sigma_n \sin \frac{\Omega t}{2} \right)| + \rangle$$

$$= -ie^{-i\omega t/2} \langle -|\sigma_n|+\rangle \sin \frac{\Omega t}{2}$$

$$= -ie^{-i\omega t/2} \frac{\omega_1}{\Omega} \sin \frac{\Omega t}{2}. \tag{58}$$

Hence the probability that the spin flip has taken place at time t is

$$P(t) = |\langle -|\overline{\Psi}(t)\rangle|^2$$

$$= \frac{\omega_1^2}{\Omega^2} \sin^2 \frac{\Omega t}{2}$$

$$= \frac{\omega_1^2}{2\Omega^2} (1 - \cos \Omega t). \tag{59}$$

This is Rabi's formula. It gives the probability that a particle prepared at time $t = 0$ in the spin-up state $\left(s_z = +\frac{\hbar}{2} \right)$ and subjected to a rotating magnetic field will at a time t be found to have the spin flipped down $\left(s_z = -\frac{\hbar}{2} \right)$. The transition probability is zero when $\omega_1 = 0$. The probability is largest for times t_n where

$$t_n = \frac{(2n + 1)\pi}{\sqrt{(\omega - \omega_0)^2 + \omega_1^2}}, \, n = 0, 1, 2, \dots \tag{60}$$

The maximum value $\dfrac{\omega_1^2}{\Omega^2}$ is obtained when $\Omega t = \pi$ (180° *pulse*). At resonance, $\omega = \omega_0$, the maximum probability is one. The resonance curve has a width ω_1 at $\omega = \omega_0$. Resonance occurs when the frequency ω of the oscillating field is such that $\hbar\omega$ is equal to the energy difference ΔE of the two (Zeeman) levels of the system, i.e.,

$$\Delta E = \hbar\omega_0. \tag{61}$$

When $|(\omega - \omega_0)| \gg \omega_1$, the maximum probability is small.

If we assume that the oscillating field can be considered as a perturbation, then perturbation theory gives the transition probability to first order as

$$P^{(1)}(t) = \left(\frac{\omega_1}{\omega_0 - \omega} \right)^2 \sin^2 \left(\frac{1}{2}(\omega_0 - \omega)t \right). \tag{62}$$

This expression is a good approximation to the exact result Eq. (59) if $|(\omega_0 - \omega)| \gg \omega_1$. The perturbation result is wrong at or near the resonance, no matter how small ω_1 may be.

We have briefly discussed the basic idea of magnetic resonance. For details of EPR and NMR in actual practice, see Pake and Estle (1973), Slichter (1963).

14.5   CONSTANT AND HARMONIC PERTURBATIONS

We now turn to a detailed study of time dependent perturbation theory for different forms of time-dependence of the perturbation.

Let us suppose that a constant perturbation is turned on at $t = 0$:

$$V(t) = \begin{cases} 0, & t < 0 \\ V, & t \geq 0. \end{cases} \tag{63}$$

For times $t < 0$, the system with unperturbed Hamiltonian H_0 is in an eigenstate $|i\rangle$ with energy E_i. To first order, the transition amplitude to a state with energy E_f is

$$C_f^{(1)}(t) = (i\hbar)^{-1} V_{fi} \int_0^t e^{i\omega_{fi}t'}\, dt'$$

$$= \frac{V_{fi}}{\hbar\omega_{fi}} (1 - e^{i\omega_{fi}t}). \tag{64}$$

Thus, the transition probability to find the system in a state with energy E_f at time $t > 0$ when the initial state is $|i\rangle$ with energy E_i is

$$P_{i \to f}(t) = |C_f^{(1)}|^2$$

$$= \frac{|V_{fi}|^2}{|E_f - E_i|^2} (2 - 2\cos\omega_{fi}t)$$

$$= \frac{4|V_{fi}|^2}{|E_f - E_i|^2} \sin^2\left[\frac{(E_f - E_i)t}{2\hbar}\right]^2. \tag{65}$$

The transition probability is plotted as a function of ω_{fi} in Fig. 14.4. For very short times, the probability grows from zero as t^2 for all energies E_f. As time goes on the probability is largest for those states whose energy are in the vicinity of $\omega_{fi} = 0$ ($E_f = E_i$). The height of the central peak about $\omega_{fi} = 0$ is t^2 and it is proportional to $1/t$. Thus, as t increases, the probability is appreciable only for those final states whose energy E_f satisfy

$$|E_f - E_i| \lesssim \frac{2\pi\hbar}{t}. \tag{66}$$

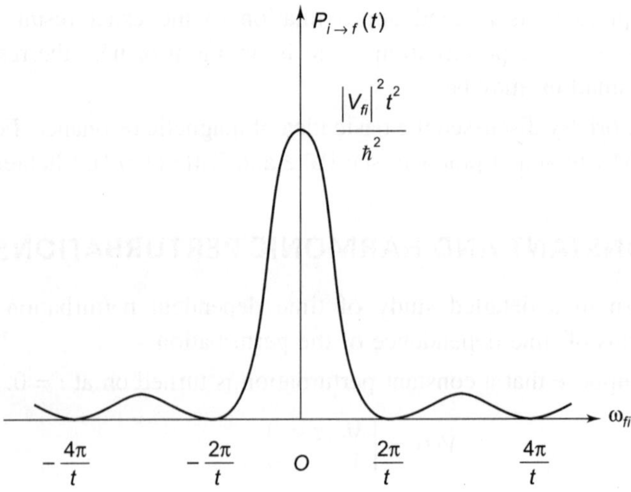

Fig. 14.4   The transition probability $P_{i \to f}(t)$ as a function of ω_{fi}

Thus, the spread in energy ΔE of the transitions of the system induced by constant perturbations with appreciable probability satisfy

$$\Delta E \cdot \Delta t \gtrsim 2\pi\hbar, \tag{67}$$

where Δt is the time interval during which the perturbation has acted. We see that if $E_f \neq E_i$, then the transition probability oscillates as a function of time with period $2\pi\hbar/|E_f - E_i|$. If $E_f = E_i$, then the transition probability to first order in perturbation grows as t^2. After a while, higher order effects become important and ensure that the probability never exceeds one. From Eq. (65), for $E_f = E_i$, perturbation theory is applicable if

$$t \ll \frac{\hbar}{|V_{fi}|}. \tag{68}$$

The time-energy uncertainty relation Eq. (67) is fundamentally different from the uncertainty relation relating x and p, since x and p are observables, while time is a parameter (see § 2.8).

We now consider the interesting case, where the final state belongs to the continuous spectrum or to a group of states of neighboring energies. All scattering processes, elastic or inelastic, and emission of radiation or decay of unstable particles are examples of this type of phenomena. To calculate the transition rate we must sum Eq. (65) over the final states about E_f.

Let $\rho(E)$ be the number of states at E per unit energy. Then $\rho(E)E_f$ is the number of states in an interval dE_f about E_f. $\rho_f(E)$ is called the density of states (levels) at energy E. The first-order transition probability $P^{(1)}$ from $|i\rangle$ to one of states in the group about the energy E_f is then

$$P^{(1)}_{\substack{i \to f \\ \text{group}}} (t) = \int_{\text{group}} P^{(1)}_{i \to f}\, \rho_f\, (E_f)\, dE_f, \tag{69}$$

where $P^{(1)}_{i \to f}$ is the first order transition probability $i \to f$. Now, we assume that a constant perturbation is switched on at time $t = 0$. Substituting Eq. (65) in Eq. (69), we get

$$P^{(1)}_{\substack{i \to f \\ \text{group}}} = \frac{1}{\hbar^2} \int_{\text{group}} dE_f |V_{fi}|\, \rho_f(E_f)\, f(\omega_{fi},\, t), \tag{70}$$

where
$$f(\omega,\, t) = \frac{2\,(1 - \cos \omega t)}{\omega^2} = \frac{4 \sin^2 (\omega t/2)}{\omega^2}. \tag{71}$$

We consider the transitions to the group of levels in the interval $\left(E_f - \frac{1}{2}\,\epsilon,\, E_f + \frac{1}{2}\,\epsilon \right)$. We assume that the width ϵ is small enough so that V_{fi} and $\rho_f\,(E_f)$ are practically constant within the group of levels. So these quantities can be taken outside the integral sign in Eq. (70). We also assume that t is sufficiently large so that the quantity ϵ is much greater than the period of oscillation of f:

$$\epsilon \gg \frac{2\pi\hbar}{t}.$$

The integral on the right hand side of Eq. (70) can now be evaluated. We consider the two cases

(i)   If the central peak of f is outside of the integration region (transitions not conserving energy), then we can replace f by its value averaged over several oscillations and obtain the time independent expression.

$$P^{(1)}_{\substack{i \to f \\ \text{group}}} \simeq 2\, \epsilon\, \rho_f\,(E_f)|\, V_{if}|^2/|E_f - E_i|^2. \tag{72}$$

Here the integral is over the interval $t.dE/2\hbar \gg 2\pi$ and the average value of $\sin^2 x$ in the interval $(0,\, 2\pi)$ is $1/2$.

(ii)   If the central peak is within the domain of integration (energy conserving transitions), then we can extend the limits of integration to infinity. Since the main contribution comes from the central peak region, only a small error is involved in extending the limits.

Now,

$$\int_{-\infty}^{\infty} f\,(\omega,\, t)\, d\omega = \int_{-\infty}^{\infty} \frac{\sin^2 (\omega t/2)}{(\omega t/2)^2}\, d(\omega t/2)\,.\,2t = 2\pi t, \tag{73}$$

since,
$$\int_{-\infty}^{\infty} \frac{\sin^2 x}{x^2}\, dx = \int_{-\infty}^{\infty} \frac{\sin x}{x} = \pi$$

Hence,

$$P^{(1)}_{\substack{i \to f \\ \text{group}}} \simeq 2\pi\hbar^{-1} |V_{if}(E_i)|^2 \rho_f(E_i)\, t.$$
(74)

From the inequality $\epsilon \gg \dfrac{2\pi\hbar}{t}$, this probability is greater than the sum of all the others. The transition rate, i.e. the probability of transition per unit time is

$$W_{\substack{i \to f \\ \text{group}}} = \frac{dP^{(1)}_{\substack{i \to f \\ \text{group}}}}{dt}.$$
(75)

From Eqs. (72) and (74), we see that only for the energy conserving transitions, the transition rate is given by

$$W_{\substack{i \to f \\ \text{group}}} = \frac{2\pi}{\hbar} |V_{fi}|^2 \rho_f(E_f),$$
(76)

where the matrix element V_{fi} and the density of final states ρ_f relate to the states about the energy E_f equal to the energy of the initial state E_i. The expression for W is independent of t, as expected. For this formula Eq. (76) to be valid the time t must be sufficiently large so that the inequality is satisfied, and also sufficiently small so that $\omega t \ll 1$ for the first order perturbation theory to be justified. The depletion of the initial state after a long time must also be considered. This equation Eq. (76) can be simply obtained if we use the asymptotic expression

$$f_{\substack{t \to \infty}}(\omega, t) \sim 2\pi t\, \delta(\omega)$$
(77)

in Eq. (70) and using
$$\delta(\omega_{fi}) = \delta\left(\frac{E_f - E_i}{\hbar}\right) = \hbar\delta(E_f - E_i)$$

we get

$$W_{\substack{i \to f \\ \text{group}}} \underset{t \to \infty}{\sim} \frac{2\pi}{\hbar} |V_{fi}|^2 \,\delta(E_f - E_i).$$
(78)

Here, Eq. (78) has to be integrated with $\int dE_f\, \rho_f(E_f)$. The expression Eq. (76) for the transition rate, due to Dirac, is of such great utility that it was called "Golden Rule No. 2" by Fermi (E. Fermi, Nuclear Physics, p. 142, Revised Ed., The University of Chicago Press, Chicago, 1950).

As a simple application of the formula Eq. (76) one may obtain the Born approximation expression for the collision cross section. Let us consider the scattering of a particle by a potential $V(\vec{r})$ We consider a particle in a momentum state $|\vec{k}\rangle$ in a box of volume L and then switch on a potential $V(\vec{r})$. The potential is treated as a perturbation. The matrix element of V between an initial state $|\vec{k}\rangle$ and a final state $|\vec{k}'\rangle$ is

$$\langle \vec{k}'|V|\vec{k}\rangle = \frac{1}{L^3} \int e^{i(\vec{k}-\vec{k}')\cdot\vec{r}}\, V(\vec{r})\, d\vec{r}. \tag{79}$$

From Eq. (78), the rate of transition from $\vec{k}$ to $\vec{k}'$ is

$$W_{\vec{k}\to\vec{k}'} = \frac{2\pi}{\hbar} \frac{\left|V_{\vec{k}'-\vec{k}}\right|^2}{L^6} \, \delta\left(E_{\vec{k}'} - E_{\vec{k}}\right), \tag{80}$$

where $V_{\vec{k}-\vec{k}'}$ is the Fourier transform of $V(\vec{r})$ and $E_{\vec{k}} = \dfrac{\hbar^2 k^2}{2m}$. The number of

states in $d\vec{k}'$ in the interval $(\vec{k}', \vec{k}'+d\vec{k}')$ with energy $(E_{\vec{k}'}, E_{\vec{k}'}+dE_{\vec{k}'})$ is $L^3 \cdot \dfrac{d\vec{k}'}{(2\pi)^3}$

$= L^3 \dfrac{m\hbar k'}{(2\pi\hbar)^3}\, d\Omega' dE_{\vec{k}'}$. So there are $L^3 \dfrac{mk'}{(2\pi)^3\,\hbar^2}$ states per unit energy per unit solid

angle. From Eq. (76), we see that the rate of transition from a state $|\vec{k}\rangle$ into one of the states about $\vec{k}'$ in the solid angle $(\Omega, \Omega + d\Omega)$ with $|\vec{k}'| = |\vec{k}| = k$ is

$$W_{\substack{\vec{k}\to\vec{k}' \\ \text{group}}} d\Omega \simeq \frac{1}{L^3} \frac{mk}{4\pi^2\hbar^3} \left|V_{\vec{k}'-\vec{k}}\right|^2 d\Omega. \tag{81}$$

In an actual scattering experiment one uses a mono-energetic beam of incident particles. Then the incident flux per particle of momentum $\hbar\vec{k}$ in a volume L^3 is $\dfrac{\hbar k}{L^3 \cdot m}$. The differential scattering cross section is given by

$$\frac{d\sigma}{d\Omega} = \frac{W}{\hbar k/mL^3} = \frac{m^2}{4\pi^2\hbar^4}\left|V_{\vec{k}'-\vec{k}}\right|^2. \tag{82}$$

This is the expression for the differential scattering cross section in the Born approximation.

Let us now look at the second order term in Eq. (43). We study this in the case of a slow-turn-on method. We assume that $V(t) = 0$ at $t = -\infty$ (remote past). Then we slowly (adiabatically) turn on the perturbation: $V(t) = Ve^{\eta t}$, where V is constant and η is a small positive quantity. Finally, we let $\eta \to 0$ and the potential then becomes constant at all times. The second-order contribution is

$$C_n^{(2)}(t) \equiv \langle n|\overline{\psi}_I(t)\rangle$$

$$= \frac{1}{(i\hbar)^2} \int_{t_0}^{t} dt_1 \int_{t_0}^{t_1} dt_2 \sum_m V_{nm}\, e^{i\omega_{nm}t_1/\hbar + \eta t_1} \times V_{mi}\, e^{i\omega_{mi}t_2/\hbar + \eta t_2} \tag{83}$$

Doing the t_1 and t_2 integrals and taking the limit $t_0 \to -\infty$, we get

$$C_n^{(2)}(t) = e^{-\frac{i}{\hbar}(E_i - E_n)t}\, \frac{e^{2\eta t}}{E_i - E_n + 2i\eta\hbar} \sum_m \frac{V_{nm}V_{mi}}{E_i - E_m + i\eta\hbar}. \tag{84}$$

This is the expression for the second-order term in the transition amplitude. It may happen that the first-order transition amplitude is zero. We can obtain the transition rate into a group of final states with energy about E_n

$$W_{\substack{i \to n \\ \text{group}}} = \frac{2\pi}{\hbar} \left| \sum_m \frac{V_{nm} V_{mi}}{E_i - E_m + i\eta\hbar} \right|^2 \delta(E_n - E_i), \qquad (85)$$

with the limit $\eta \to 0$. This is the golden rule for second-order transition. The transition in the second-order takes place in two steps. First, $|i\rangle$ goes to an intermediate state $|m\rangle$ by a virtual (energy non-conserving) transition, and then from $|m\rangle$ to the state in a group about the energy E_n by a virtual transition. Of course, there is overall energy conservation ($E_i = E_n$). Also, we sum over all intermediate states in order to calculate the rate. In case the first-order matrix element is nonzero, then the transition rate is given by the second-order term Eq. (85) (and higher order terms, if required) with the first-order term V_{ni} added inside the modulus. We note that the first-order term V_{ni} represents a direct energy-conserving real transition.

We next consider a sinusoidally varying time dependent potential with frequency ω, commonly referred to as harmonic or periodic perturbation. Since $V(t)$ is a hermitian operator, we may write.

$$V(t) = [Ae^{-i\omega t} + A^\dagger e^{i\omega t}] \qquad (86)$$

where A is a time-independent operator. This harmonic perturbation is turned on slowly (adiabatically) with an $e^{\eta t}$ factor, where $\eta > 0$. To first order in V, the probability of transition $i \to f$ is given by (taking $t_0 \to -\infty$)

$$P_{i \to f} \simeq \hbar^{-2} \left| A_{fi} \int_{-\infty}^{t} e^{i(\omega_{fi} - \omega - i\eta)t_1} \, dt_1 + A_{fi}^\dagger \int_{-\infty}^{t} e^{i(\omega_{fi} - \omega - i\eta)t_1} \, dt_1 \right|^2 \qquad (87)$$

$$= \frac{e^{2\eta t}}{\hbar^2} \left\{ \frac{|A_{fi}|^2}{(\omega_{fi} - \omega)^2 + \eta^2} + \frac{|A_{fi}^\dagger|^2}{(\omega_{fi} - \omega)^2 + \eta^2} \right.$$

$$\left. + 2A_{fi} A_{fi}^\dagger \, \mathrm{Re} \, \frac{e^{-2i\omega t}}{(\omega_{fi} - \omega - i\eta)(\omega_{fi} + \omega - i\eta)} \right\}. \qquad (88)$$

The first term in Eq. (88) is due to the positive frequency part $Ae^{-i\omega t}$ of $V(t)$, the second term is due to the negative frequency part of $V(t)$, while the third term is the interference term. The transition rate is

$$\frac{dP_{i \to f}(t)}{dt}$$

$$= \frac{e^{2\eta t}}{\hbar^2} \left\{ 2\eta \left[\frac{|A_{fi}|^2}{(\omega_{fi} - \omega)^2 + \eta^2} + \frac{|A_{fi}^\dagger|^2}{(\omega_{fi} - \omega)^2 + \eta^2} \right] (1 - \cos 2\omega t) \right.$$

$$+ (A_{fi} A_{fi}^\dagger)\, 2\sin 2\omega t \left[\frac{\omega_{fi} - \omega}{(\omega_{fi} - \omega)^2 + \eta^2} - \frac{\omega_{fi} + \omega}{(\omega_{fi} + \omega)^2 + \eta^2} \right] \bigg\}. \tag{89}$$

The sine and cosine terms arise from the interference term. As before, we assume that $|f\rangle$ is part of a group of states $[f]$, and we take the limit $\eta \to 0$. The first two terms are nonzero only if

$$\omega_{fi} - \omega \simeq 0 \quad \text{or,} \quad E_f \simeq E_i + \hbar\omega \tag{90}$$

$$\omega_{fi} + \omega \simeq 0 \quad \text{or,} \quad E_f \simeq E_i - \hbar\omega \tag{90a}$$

If we average $\dfrac{dP}{dt}$ over a few periods, then the sine and cosine terms average to zero, and we get

$$W_{i \to [f]} = \frac{2\pi}{\hbar} \left\{ \overline{|A_{fi}|^2}\, \delta(E_f - E_i + \hbar\omega) + \overline{|A_{fi}^\dagger|^2}\, \delta(E_f - E_i - \hbar\omega) \right\} \tag{91}$$

For a harmonic perturbation, the first-order transition probability is appreciable for transitions in which the unperturbed system emits or absorbs an amount of energy $\hbar\omega$ (to within $2\pi\hbar/t$). In practice, t is sufficiently large ($t \gg 2\pi/\omega$) so that the two regions about ($E_f = E_i \pm \hbar\omega$) do not overlap. We have introduced the harmonic perturbation adiabatically. The result will be the same, if one introduces it suddenly at $t = 0$.

If we write Eq. (91), as in Eq. (76).

$$W_{i \to [f]} = \frac{2\pi}{\hbar}\, \overline{|A_{fi}|^2}\, \rho(E_f) \bigg|_{E_f = E_i + \hbar\omega}$$

and

$$W_{i \to [f]} = \frac{2\pi}{\hbar}\, \overline{|A_{ft}^\dagger|^2}\, \rho(E_f) \bigg|_{E_f = E_i - \hbar\omega}, \tag{92}$$

then, since

$$\overline{|A_{fi}|^2} = \overline{|A_{fi}^\dagger|^2},$$

we obtain the detailed balance between emission and absorption

$$\frac{\text{emission rate for } i \to [f]}{\text{density of final states for } [f]} = \frac{\text{absorption rate for } f \to [i]}{\text{density of final states for } [i]}, \tag{93}$$

where we put $[i]$ for the final states for absorption. In the general case, when V is any periodic function of time, of frequency ω, one has to use the Fourier expansion of V and proceed accordingly.

14.6   IONIZATION OF A HYDROGEN ATOM

We now consider an example of the application of the first-order time-dependent perturbation theory. A hydrogen atom in its ground state is subjected to a harmonically time-varying electric field $\vec{\varepsilon}(t) = \vec{\varepsilon}_0 \sin \omega t$. We wish to calculate the probability of ionization of the hydrogen atom. Now, ω must exceed $me^4/2\hbar^3$ ($\sim 10^{16}$ cycles/sec) in order for ionization to occur.

Let $\vec{r}$ denote the radius vector of the electron. The perturbation $V(t)$ is

$$V(t) = -e\vec{r} \cdot \vec{\varepsilon}(t) = -e\vec{r} \cdot \vec{\varepsilon}_0 \sin \omega t$$

$$= Ae^{-i\omega t} + A^* e^{i\omega t} \tag{94}$$

with
$$A = \frac{e}{2i}\, \vec{r} \cdot \vec{\varepsilon}_0.$$

We calculate the matrix element

$$\langle \vec{k}|A|0\rangle = \langle \vec{k}|\frac{e}{2i}\, \vec{\varepsilon}_0 \cdot \vec{r}|0\rangle$$

where $|0\rangle$ is the initial ground state of the hydrogen atom and $|k\rangle$ represents the continuum free particle state with wave number $\vec{k}$.

Actually, the final state corresponds to the motion of an electron in the Coulomb field of the proton. We consider the much simpler approximate situation where the final state is a continuum free particle momentum eigenstate (plane waves). The initial ground state of the hydrogen atom is written as

$$u^{(\vec{r})}_{100} = (\pi a_0^3)^{-1/2}\, e^{-r/a_0}$$

and the final state as $(2\pi)^{-3/2} e^{i\vec{k} \cdot \vec{r}}$ where $\hbar \vec{k}$ is the momentum of the ejected electron. Hence,

$$\langle \vec{k}|A|0\rangle = \frac{e}{2i}\, (2\pi)^{-3/2} (\pi a_0^3)^{-1/2} \int e^{-i\vec{k} \cdot \vec{r} - r/a_0} \times (\vec{\varepsilon}_0 \cdot \vec{r})\, d\vec{r}. \tag{95}$$

To evaluate the integral we introduce spherical polar coordinates with $\vec{k}$ as the polar axis shown in Fig. 14.5.

If θ'' is the angle between $\vec{r}$ and the electric field $\vec{\varepsilon}$, then

$$\cos \theta'' = \cos \theta' \cos \theta + \sin \theta' \sin \theta \cos (\phi' - \phi), \tag{96}$$

when θ', ϕ' are the polar angles of $\vec{r}$ with respect to $\vec{k}$ as polar axis, and θ, ϕ are the polar angles of the electric field direction with respect to $\vec{k}$.

From Eq. (95), we get

$$A_{k0} = \frac{e|\vec{\varepsilon}_0|}{i \cdot 2^{5/2} \pi^2\, a_0^{3/2}} \int e^{ikr\cos\theta'} r\cos \theta''\, e^{-r/a_0} d^3 r. \tag{97}$$

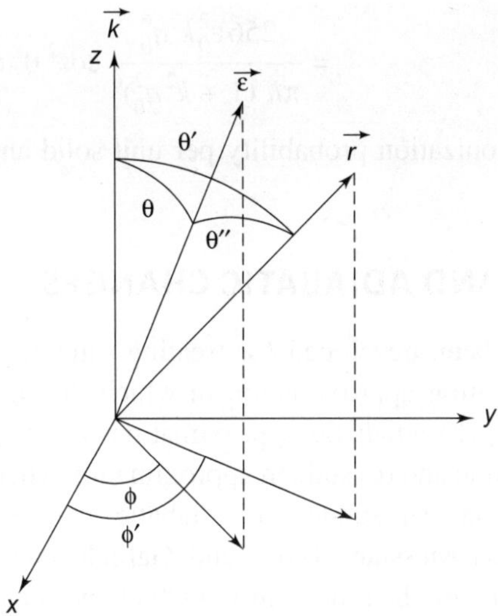

Fig. 14.5   The coordinates and the angles for the integration of the matrix element

We put $d^3r = r^2\,dr\,d(\cos\theta')\,d\phi'$ and substitute Eq. (96) in Eq. (97). The ϕ' integration causes the second term in Eq. (96) to vanish and performing the integration of the first term we get

$$A_{ko} = \frac{16\,e\varepsilon_0 k a_0^5 \cos\theta}{\pi(2a_0)^{3/2}\,(1 + k^2 \cdot a_0^2)^3}. \tag{98}$$

Energy conservation gives $E_k = E_0 + \hbar\omega$ or $\dfrac{\hbar^2 k^2}{2m} = -\dfrac{me^4}{2\hbar^2} + \hbar\omega$ so that the magnitude of k is fixed. Hence, the group of final states corresponds to an infinitesimal range of directions for the motion of the ejected electron. If $\rho_f(k)$ denotes the density of final states, then $\rho_f(k)\,dE_k\,d\Omega_k = \vec{dk} = k^2\,dk\,d\Omega_k$, where dE_k and dk are related by differentiation of the equation $E_k = \dfrac{\hbar^2 k^2}{2m}$. Thus, we obtain

$$\rho_f^{(k)} = \frac{mk}{\hbar^2}. \tag{99}$$

Substitution of the matrix element Eq. (98) and the density of final states Eq. (99), into the golden rule gives the probability per unit time of ionization of the hydrogen atom with ejection of the electron into the element of solid angle $d\Omega_k = \sin\theta\,d\theta\,d\phi$

$$dW = \frac{2\pi}{\hbar}\,|A_{ko}|^2\,\rho_f(k)$$

$$= \frac{256\varepsilon_0^2 k^3 a_0^6}{\pi\hbar \, (1 + k^2 a_0^2)^6} \, \cos^2\theta \; d\Omega. \tag{100}$$

The differential ionization probability per unit solid angle is proportional to $\cos^2\theta$.

14.7   SUDDEN AND ADIABATIC CHANGES

Three methods have been developed for treating time-dependent perturbations. These are (a) perturbation approximation in which the time dependence of the Hamiltonian is small, (b) adiabatic approximation in which the Hamiltonian is slowly varying with time and (c) sudden approximation when it is rapidly varying. In this section we consider sudden or adiabatic change of the Hamiltonian. This section draws on Messiah (1961) and Galindo and Pascual (1991). The result depends critically on the time T during which the change of the Hamiltonial takes place. The limiting cases arise when T is very small (sudden change) and very large (adiabatic change).

We suppose that at the initial time t_0 the unperturbed Hamiltonian of the system is H_0. The Hamiltonian changes in a continuous way to a value H_1 at time t_1. We parametrize time by a variable τ so that $\tau = \dfrac{t - t_0}{T}$, with $T = t_1 - t_0$.

We denote by $H(\tau)$ the Hamiltonian at time $t = t_0 + \tau T \cdot H(\tau)$ is a continuous function of τ and we have $H(0) = H_0$ and $H(1) = H_1$. The evolution of the system from t_0 to t_1 now depends only on the parameter T, measuring the time during which the change of the Hamiltonian takes place. We put $U\,(t, t_0) = U_T\,(\tau)$. The time evolution operator satisfies

$$U_T(T) = 1 - \frac{i}{\hbar}\, T \int_0^\tau d\tau' \, H(\tau') \, U_T(\tau'). \tag{101}$$

The problem is to determine $U\,(t_1, t_0)$ i.e., $U_T\,(1)$ and to find its dependence on T. In the following we derive the general results for the two limiting cases.

In the limit $T \to 0$ (sudden perturbation), the dynamical state of the system remains unchanged:

$$\lim_{T \to 0} U_T(1) = 1. \tag{102}$$

This result follows from Eq. (101). In this case, the change in the Hamiltonian is so rapid that the system has no time to adjust, and the wave function is the same at $\tau = 0$ and at $\tau = 1$.

In the limit when $T \to \infty$, i.e., in the case of an infinitely slow ("adiabatic") passage, if the system is initially in an eigenstate of $H_1 \equiv H(t_0)$ it will, at time t_1 have passed into the eigenstate of $H_0 \equiv H(t_1)$, that derives from it by continuity (i.e., will evolve with $H(t)$). The result is known as the Adiabatic Theorem.

We now return to the case of rapid passage. If T is sufficiently small, we may, in a first approximation, suppose that $U_T(1) \simeq 1$. This is known as the sudden approximation.

Let $|0\rangle$ be the initial normalized state vector of the system. In the sudden approximation, $U(t_1, t_0)|0\rangle \simeq |0\rangle$. A measure of the error involved in this approximation is obtained by calculating the probability $\overline{w}$ of finding the system in states other than the initial state:

$$\overline{w} = \langle 0| U^\dagger (t_1, t_0)\, Q_0\, U(t_1, t_0)|0\rangle$$

$$= \langle 0| U_T^\dagger(1)\, Q_0\, U_T(1)|0\rangle, \qquad (103)$$

where $Q_0 = I - |0\rangle\langle 0|$ is the projection operator onto the subspace orthogonal to $|0\rangle$. Expansion of $U_T(1)$ in powers of T gives

$$U_T(1) = I - \frac{i}{\hbar} T \int_0^1 d\tau_1 H(\tau_1) + \left(\frac{i}{\hbar}\right)^2 T^2 \int_0^1 d\tau_1 \int_0^{\tau_1} d\tau_2\, H(\tau_1)\, H(\tau_2) + \dots \qquad (104)$$

Substitution of Eq. (104) into the right-hand side of Eq. (103) gives the expansion of $\overline{w}$ in powers of T. As $Q_0|0\rangle = 0$, the lowest order term is T^2 and we get

$$\overline{w} = \frac{T^2}{\hbar^2} \langle 0|\overline{H} Q_0 \overline{H}|0\rangle + 0(T^3) \qquad (105)$$

where $\overline{H}$ is the average value of the Hamiltonian in the interval $(t_1 - t_0)$.

$$\overline{H} = \int_0^1 d\tau H(\tau) = \frac{1}{T} \int_{t_0}^{t_1} dt\, H(t). \qquad (106)$$

Since

$$\langle 0|\overline{H} Q_0 \overline{H}|0\rangle = \langle 0|\overline{H^2}|0\rangle - \langle 0|\overline{H}|0\rangle^2 = (\Delta\overline{H})^2. \qquad (107)$$

where $\Delta\overline{H}$ is the root mean square deviation of the observable $\overline{H}$ in the state $|0\rangle$, we have

$$\overline{w} = T^2 \frac{(\Delta\overline{H})^2}{\hbar^2} + O(T^3). \qquad (108)$$

Thus the condition for the validity of the sudden approximation, $\overline{w} \ll 1$, requires that

$$T \ll \frac{\hbar}{\Delta\overline{H}}. \qquad (109)$$

This condition for the validity of the sudden approximation is a form of the time-energy uncertainty relationship.

As an example of the application of the sudden approximation, let us consider what happens to an atom in a constant magnetic field when the direction of the field is suddenly reversed. We assume (i) the atom is one for which the *LS* coupling scheme holds, (ii) the magnetic field $\vec{B}$ is sufficiently strong to decouple $\vec{L}$ (the total orbital angular momentum) and $\vec{S}$ (the total spin), (iii) $\vec{B}$ is always parallel to the z-axis and passes from the value $-B_0$ to the value B_0 according to the linear law

$$\vec{B}(t) = \vec{B}_0\,(2\tau - 1) = \vec{B}_0\,\lfloor 2(t - t_0)/T - 1\rfloor.$$

The Hamiltonian of the atom may be written in the form

$$H(\tau) = H^{(0)} + \alpha\,(\vec{L}\cdot\vec{S}) - \frac{e}{2mc}\,(L_z + 2S_z)\,B(\tau), \tag{110}$$

where $H^{(0)}$ is the unperturbed Hamiltonian in the *LS* coupling scheme and $\alpha\,(\vec{L}\cdot\vec{S})$ is the spin-orbit coupling term. So

$$H_0 = H^{(0)} + \alpha\,(\vec{L}\cdot\vec{S}) + \frac{eB_0}{2mc}\,(L_z + 2S_z)$$

$$H_1 = H^{(0)} + \alpha(\vec{L}\cdot\vec{S}) - \frac{eB_0}{2mc}\,(L_z + 2S_z). \tag{111}$$

Since the time average of the magnetic field is zero, the time average of the Hamiltonian is

$$\overline{H} = H^{(0)} + \alpha(\vec{L}\cdot\vec{S}). \tag{112}$$

We assume that initially the system is in an eigenstate of H. We put $|0\rangle = |LSM_L M_S\rangle$.

From Eq. (112), we get

$$\overline{H}^2 = (H^{(0)})^2 + \alpha^2\,(\vec{L}\cdot\vec{S})^2 + H^{(0)}\,\alpha(\vec{L}\cdot\vec{S})\,H^{(0)}.$$

Then

$$(\Delta H)^2 = \langle LSM_L M_S | \overline{H}^2 | LSM_L M_S\rangle$$

$$- \langle LSM_L M_S | \overline{H} | LSM_L M_S\rangle^2$$

$$= \alpha^2\,[\langle(\vec{L}\cdot\vec{S})^2\rangle - \langle\vec{L}\cdot\vec{S}\rangle^2]$$

$$= \alpha^2\left[\left\langle\left(\frac{L_+S_- + L_-S_+}{2}\right)^2\right\rangle - \left\langle\left(\frac{L_+S_- + L_-S_+}{2}\right)\right\rangle^2\right]$$

using $\qquad \vec{L}\cdot\vec{S} = \dfrac{L_+S_- + L_-S_+}{2} + L_zS_z$

$$= \frac{1}{4}\,\alpha^2\hbar^4\,[2\,(L(L+1) - M_L^2)\,(S(S+1) - M_S^2) - 2M_L M_S].$$

Thus,

$$\Delta\overline{H} = \frac{1}{2}\,\alpha\hbar^2\,[2\,(L(L+1) - M_L^2)\,(S(S+1) - M_S^2) - 2M_L M_S]^{1/2}. \tag{113}$$

The quantity in the square bracket is a numerical factor of order unity (it vanishes for $M_L = \pm L$, $M_s = \pm S$). The deviation $\Delta \overline{H}$ is therefore of the order of $\alpha \hbar^2$, i.e. of the order of magnitude of the splitting of the LS levels by spin-orbit coupling. The condition Eq. (109) for the validity of the sudden approximation is then

$$T \ll \frac{1}{\alpha \hbar} \tag{114}$$

$(\alpha \hbar)^{-1}$ is of the order of the characteristic time associated with the spin-orbit coupling. The state vector remains practically constant if the condition Eq. (114) is met. Inequality Eq. (114) is a sufficient but not a necessary condition for the atom to remain in the same state after the sudden reversal of the magnetic field.

As another simple application of the sudden approximation we shall consider the case of a one electron atom if an abrupt change of the nuclear charge $Z \to Z \pm 1$ occurs following $\beta^{\mp}$ decay. To be specific, let us consider that a tritium atom (H^3) which has one proton and two neutrons in its nucleus suddenly changes to the nucleus He^3 containing two protons and one neutron through β^- emission:

$$H^3 \to He^3 + e^- + \overline{\nu}_e \tag{115}$$

We shall calculate the probability that the helium ion $(He^3 + e^-)$ be found in its ground state assuming that the tritium atom is in its ground state before β-decay. A large fraction of the electrons emitted in the β-decay process has energy of the order of several keV. The velocity v $\left(\dfrac{v}{c} = \dfrac{1}{4} \text{ for energy } 16 \text{ } keV\right)$ of the electron emitted is much larger than the velocity v_a of the atomic electron $\left(\dfrac{v_a}{c} \sim \dfrac{1}{137}\right)$ in the ground state of tritium. The β-electron will leave the atom in a time $\tau_0 = \dfrac{a_0}{v} \ll T_a$, where $T_a = \dfrac{2\pi a_0}{v_a}$ is the period associated with the atomic electron. The change in the Hamiltonian is thus sudden and the use of the sudden approximation is justified.

The initial and final eigenfunctions (both hydrogenic) are

$$\left. \begin{aligned} \text{(tritium atom) } \phi_{1s} &= \frac{1}{\sqrt{\pi a_0^3}} \, e^{-r/a_0} \\[2em] \text{(helium ion) } \chi_{1s} &= \frac{2\sqrt{2}}{\sqrt{\pi a_0^3}} \, e^{-2r/a_0} \end{aligned} \right\} \tag{116}$$

The transition amplitude is

$$a_{00} = \frac{2\sqrt{2}}{\pi a_0^3} \int d^3r \, e^{\frac{-3r}{a_0}}$$

$$= \frac{2\sqrt{2} \cdot 4\pi}{\pi a_0^3} \int\limits_0^\infty dr\, r^2 e^{\frac{-3r}{a_0}}$$

$$= \frac{16\sqrt{2}}{27}.$$

The probability that the helium ion be found in its ground state is

$$P_{1s,\,1s} = |a_{00}|^2 = \frac{2^9}{3^6} \simeq 0.7. \tag{117}$$

The total probability for excitation and ionization of the helium ion is ~ 0.3.

In the rest of this section we discuss the other extreme case ($T \rightarrow \infty$). In this case we have the adiabatic theorem which underlies the adiabatic approximation.

We assume that $H(\tau)$, $0 \leq \tau \leq 1$, has a purely discrete spectrum with eigenvalues $E_1(\tau), E_2(\tau) \ldots E_n(\tau), \ldots$. The projection operators onto the subspaces are respectively $P_1(\tau), P_2(\tau), \ldots P_n(\tau), \ldots$. We further assume that

(i)   The eigenvalues remain distinct (no level crossing) throughout the whole transition period:

$$E_i(\tau) \neq E_j(\tau),\ i \neq j,\ \tau \in [0, 1]. \tag{118}$$

(ii)   The derivatives $dP_i(\tau)/d\tau$ and $d^2 P_i(\tau)/d\tau^2$ are well defined and bounded and piecewise continuous in the interval $0 \leq \tau \leq 1$.

The evolution operator $U_T(\tau)$ satisfies the Schrödinger equation

$$i\hbar \frac{d}{d\tau} U_T(\tau) = H(\tau)\, U_T(\tau), \tag{119}$$

where the Hamiltonian $H(\tau)$ is given by

$$H(\tau) = \sum_i E_i(\tau)\, P_i(\tau). \tag{120}$$

The adiabatic theorem states that $U_T(\tau)$ has the asymptotic property.

$$\lim_{T \rightarrow \infty} U_T(\tau)\, P_i(0) - P_i(0) \lim_{T \rightarrow \infty} U_T(\tau) = 0\left(\frac{1}{T}\right), \quad (i = 1, 2, \ldots) \tag{121}$$

This property is equivalent to the one discussed above, for if $|i\rangle$ is an eigenvector of $H(0)$, so that $H(0)|i\rangle = E_i(0)|i\rangle$, then $P_i(0)|i\rangle = |i\rangle$ and Eq. (121) gives

$$\lim_{T \rightarrow \infty} U_T(\tau)|i\rangle = P_i(\tau) \lim_{T \rightarrow \infty} U_T(\tau)|i\rangle,\ i = 1, 2, \ldots.$$

In the limit $T \rightarrow \infty$, a vector $U_T(\tau)|i\rangle$ tends toward a vector of the subspace of $E_i(\tau)$.

For the proof of the adiabatic theorem, the reader is referred to Messiah (1961) and Galindo and Pascual (1991).

We introduce a unitary operator $A(\tau)$ such that

$$P_j(\tau) = A(\tau)\, P_j(0)\, A^\dagger(\tau), \tag{122}$$

where $A(\tau)$ satisfies the initial condition

$$A(0) = 1, \tag{123}$$

and the differential equation

$$i\hbar\, \frac{dA}{d\tau} = K(\tau)\, A(\tau), \tag{124}$$

$K(\tau)$ being an appropriate hermitian operator. The unitary transformation $A^\dagger(\tau)$ carries the vectors and operators of the Schrodinger representation into the vectors and operators in another representation (the "rotating axis representation"). The observable $H(\tau)$ transforms into

$$H^{(A)}(\tau) = A^\dagger(\tau)\, H(\tau)\, A(\tau) \tag{125}$$

and the evolution operator in the rotating axis representation is

$$U^{(A)}(\tau) = A^\dagger(\tau)\, U_T(\tau). \tag{126}$$

Let $\Phi_T(\tau)$ be the solution of the Schrödinger equation

$$i\hbar\, \frac{d\Phi_T(\tau)}{d\tau} = T H^{(A)}(\tau)\, \Phi_T(\tau). \tag{127}$$

with

$$\Phi_T(0) = 1.$$

It can be proved that

$$U_T(\tau) \underset{T \to \infty}{\sim} A(T)\, \Phi_T(\tau). \tag{128}$$

If T is sufficiently large, or, if the basis vectors of $H(t)$ rotate sufficiently slowly, we can replace $U_T(1)$ to a first approximation by its asymptotic form:

$$U(t_1, t_0) \equiv U_T(1) \simeq A(1)\, \Phi_T(1). \tag{129}$$

This is called the adiabatic approximation. If the initial normalized state vector is $|0\rangle$, then in the adiabatic approximation

$$U(t, t_0)|0\rangle \simeq A(1)\, \Phi_T(1)|0\rangle.$$

As before, we can estimate the error involved in this approximation by calculating the probability p_0 of finding the system at time t_1 in a state different from $A(1)\, \Phi_T(1)|0\rangle$. Now,

$$p_0 = \langle 0|\, U^\dagger(t_1, t_0)\, A(1)\, \Phi_T(1)\, Q_0\, \Phi_T^\dagger(1)\, A^\dagger(1)\, U(t_1, t_0)|0\rangle, \tag{130}$$

where

$$Q_0 = I - |0\rangle\langle 0|.$$

We may rewrite p_0 as

$$p_0 = \langle 0| W_T^\dagger(1)\, Q_0 W_T(1)|0\rangle, \tag{131}$$

where

$$W_T \equiv \Phi_T^\dagger\, U^{(A)} = \Phi_T^\dagger A^\dagger\, U_T.$$

$W_T(\tau)$ satisfies the integral equation

$$W_T(\tau) = I + \frac{i}{\hbar}\int_0^\tau d\tau'\overline{K}_T(\tau')\, W_T(\tau'), \tag{132}$$

where

$$\overline{K}_T(\tau) = \Phi_T^\dagger A^\dagger K(\tau) A\Phi_T.$$ We define an operator

$$F_T(\tau) = \int_0^\tau d\tau'\, \overline{K}_T(\tau'). \tag{133}$$

To first order,

$$p_0 \simeq \frac{1}{\hbar^2}\, \langle 0|F_T(1)\, Q_0 F_T(1)|0\rangle$$

$$= \frac{1}{\hbar^2}\, (\Delta F_T)^2, \tag{134}$$

where ΔF_T is the root-mean-square deviation of the self-adjoint operator F_T in the state $|0\rangle$.

The condition $p_0 \ll 1$ for the validity of the adiabatic approximation becomes

$$\Delta F_T \ll \hbar. \tag{135}$$

This condition is not useful because it is very difficult to construct the observable F_T.

Let us reexamine the condition Eq. (135) in the case when the initial state is an eigenstate of $H(0)$. Henceforth we shall use the variable t instead of τ. With appropriate definitions of various quantities in terms of t, we may write

$$F \cong F(t_1) = \int_{t_0}^{t_1} dt\, \Phi_T^\dagger(t)\, A^\dagger(t)\, K'(t)\, A(t)\, \Phi_T(t), \tag{136}$$

where

$$K'(t) = \frac{K(T)}{T}.$$

We assume that the spectrum of $H(t)$ is completely non-degenerate. We denote the set of basis vectors of $H(0)$ by $|1\rangle_0, |2\rangle_0, \ldots, \langle j\rangle_0 \ldots$ and the set of basis vector of $H(t)$ after transformation by $A(t)$ by $|1\rangle_t, |2\rangle_t, \ldots,\langle j\rangle_t,\ldots$

For any t,

$$\left.\begin{array}{l} |i\rangle_t = A(t)|i\rangle_0 \\[4pt] H(t)|i\rangle_t = E_i(t)|i\rangle_t \\[4pt] P_i(t) = |i\rangle_t\,{}_t\langle i| \end{array}\right\} \quad (i = 1, 2, \ldots) \tag{137}$$

Let $|i\rangle_0$ be the state vector of the system at time t_0. In the adiabatic approximation, the state vector at time t_1, will be equal to within a phase factor to $|i\rangle_1 \equiv A(t_1)|i\rangle_0$:

$$U(t_1, t_0)|i\rangle_0 \simeq e^{-\frac{i}{\hbar}\int_{t_0}^{t_1} dt\, E_i(t)}\,|i\rangle_1. \tag{138}$$

The probability $p_{i \to j}$ $(j \neq i)$ of finding the system in another eigenstate of $H(t_1)$, say $|j\rangle_1$ is

$$p_{i \to j} = |{}_1\langle j|\, U(t_1, t_0)|i\rangle_0|^2 \tag{139}$$

and

$$p_i = \sum_{j \neq i} p_{i \to j}$$

$$\simeq \frac{1}{\hbar^2}\sum_{j \neq i} {}_0\langle i|F|j\rangle_0\, {}_0\langle j|F|i\rangle_0. \tag{140}$$

It can be shown that

$$p_{i \to j} \simeq \left| \int_{t_0}^{t_1} \alpha_{ji}(t)\, e^{i\int_{t_0}^{t}\omega_{ji}(t')\,dt'}\, dt \right|^2, \tag{141}$$

where

$$\alpha_{ji}(t) = {}_t\!\left\langle j \left| \frac{d|i\rangle_t}{dt} \right.\right\rangle \tag{142}$$

$$\omega_{ji}(t) = [E_j(t) - E_i(t)]/\hbar \tag{143}$$

$\alpha_{ji}(t)$ is a measure of the speed of rotation of the eigenvectors of $H(t)$ and $\omega_{ji}(t)$ is the Bohr frequency of the transition $i \to j \cdot p_{i \to j}$ is of the order of $\left|\dfrac{\alpha_{ji}}{\omega_{ji}}\right|^2$.

The condition $p_i \ll 1$ is satisfied in most cases if

$$\left|\frac{\alpha_i^{\mathbf{max}}}{\omega_i^{\mathbf{min}}}\right|^2 \equiv \left|\frac{\text{maximum angular velocity of }|i\rangle_t}{\text{minimum Bohr frequency of }|i\rangle_t}\right|^2 \ll 1, \tag{144}$$

where $\alpha_i^{\mathbf{max}}$ is the maximum value of $\alpha_i^{(t)}$ given by

$$\alpha_i^2(t) = \sum_{j \neq i} |\alpha_{ji}(t)|^2$$

and $\omega_i^{\mathbf{min}}$ is the minimum value of the Bohr frequency for the transition from i to its nearest neighbor.

The adiabatic approximation is applicable when the condition Eq. (144) is satisfied. We may say that the adiabatic approximation holds when the axes which diagonalize $H(\tau)$ rotate with frequencies negligible with respect to ω, the minimum of the transition frequencies. The adiabatic approximation is useful when the Hamiltonian $H(t)$ changes slowly with time so that $\left|\dfrac{1}{H}\dfrac{\partial H}{\partial t}\right| >> \omega$.

As a simple example, we consider a charged-particle linear harmonic oscillator acted upon by a time-dependent electric field $\varepsilon(t)$. The Hamiltonian for this system is

$$H(t) = -\frac{\hbar^2}{2m}\frac{\partial^2}{\partial x^2} + \frac{1}{2}kx^2 - q\varepsilon(t)x$$

$$= -\frac{\hbar^2}{2m}\frac{\partial^2}{\partial x^2} + \frac{1}{2}k(x - a(t))^2 - \frac{1}{2}ka^2(t), \qquad (145)$$

where m and q are the mass and charge respectively of the particle and $a(t) = \dfrac{q}{k}\varepsilon(t)$. Except for the term $-\dfrac{1}{2}ka^2(t)$ which is constant at a given time, the Hamiltonian is that of a unperturbed linear harmonic oscillator with the same frequency $\omega = \sqrt{k/m}$ but with the equilibrium point shifted at $x = a(t)$. The instantaneous energy eigenfunctions are the harmonic oscillator wave functions centered at the point $a(t)$:

$$u_n(x) = N_nH_n\left[\alpha\,(x - a)\right]e^{-\frac{1}{2}\alpha^2(x - a)^2}, \qquad (146)$$

where

$$N_n = \left(\frac{\alpha}{\sqrt{\pi}\,2^n n!}\right), \ \alpha = \sqrt{\frac{m\omega}{\hbar}},$$

and the corresponding instantaneous energy eigenvalues are,

$$E_n(t) = \left(n + \frac{1}{2}\right)\hbar\omega - \frac{1}{2}ka^2(t), \ n = 0, 1, 2, \ldots. \qquad (147)$$

We assume that the oscillator is initially in its ground state ($n = 0$) and that the electric field is applied at time $t = t_0$. The electric field is supposed to vary slowly.

The time derivative of the Hamiltonian is $\dfrac{\partial H}{\partial t} = -kx\dfrac{da(t)}{dt} = -qx\dfrac{d\varepsilon(t)}{dt}$. The only non-vanishing matrix element of $\dfrac{\partial H}{\partial t}$ is the one with the first excited state. The transition probability at the time t, is

$$P_{10}(t_1) = \frac{1}{\hbar^2\omega^2}\left|e^{-i\omega t_0}\int_{t_0}^{t_1} dt\,e^{i\omega t}\left\langle 1\left|\frac{\partial H}{\partial t}\right|0\right\rangle\right|^2$$

$$= \frac{q^2}{2m\hbar\omega^3} \left| \int_{t_0}^{t_1} dt \, \frac{d\varepsilon(t)}{dt} \, e^{i\omega t} \right|^2.$$ (148)

The coefficient of the time-dependent factor in the transition amplitude is

$$\frac{q}{\omega^2 \sqrt{2m\hbar\omega}} \left| \frac{d\varepsilon}{dt} \right| = \frac{k \dfrac{da}{dt}}{\omega^2 \sqrt{2m\hbar\omega}} = \frac{\dot{a}}{\sqrt{2\hbar\omega/m}}.$$ (149)

The denominator is of the order of the maximum speed $v_c = \sqrt{\hbar\omega/m}$ of a hypothetical classical oscillator having the zero-point energy. The adiabatic approximation is good if the equilibrium point moves slowly in comparison with the classical oscillator speed. In the limit of the motion of the equilibrium point being very slow (the electric field changes very slowly with time), the transition probability is very small and hence the system would remain in the ground state, in agreement with the adiabatic theorem.

As another illustrative example, we consider the problem of the effect of adiabatic reversal of a magnetic field on an atom. We assume the same conditions as discussed in the case of the sudden approximation. From Eq. (110), we write the Hamiltonian as

$$H(t) = H^{(0)} + \alpha \, (\vec{L} \cdot \vec{S}) - \alpha\hbar\rho(t) \, (L_z + 2S_z),$$ (150)

where $\rho(t) = \mu_B B(t)/\alpha\hbar^2$ ($\mu_B \equiv e\hbar/2mc$, Bohr magneton)

$$B(t) = B_0 \left(\frac{2(t - t_0)}{T} - 1 \right).$$

The eigenvectors of $H^{(0)}$ are $|\gamma L S M_L M_S\rangle$ and the eigenvectors for $H^{(0)} + \alpha \, (\vec{L} \cdot \vec{S})$ are $|\gamma' L S J M_J\rangle$, where $|L - S| \leq J \leq (L + S)$ and $M_J = M_L + M_S$. We assume that initially the atom is in a ^2P state. There are in all 6 different ^2P states, linear combinations of the 6 basis vectors $\left| \gamma 1 \, \frac{1}{2} \, M_L M_S \right\rangle \left(M_L = 1, 0, -1; M_S = \frac{1}{2}, -\frac{1}{2} \right)$ which we denote $|M_L M_S\rangle$. We take the corresponding eigenvalue as the zero of the energy. The energies of the levels are continuous functions of the parameter ρ. We have

$$\frac{1}{\alpha\hbar^2} \, \langle M'_L M'_S | H(t) | M_L M_S \rangle$$

$$= [M_L M_S - (M_L + 2M_S) \, \rho(t)] \, \delta_{M_L M'_L} \delta_{M_S M'_S}$$

$$+ \frac{1}{2} \left[\left(M_S + \frac{1}{2} \right) \left(\frac{3}{2} - M_S \right) (2 + M_L) (1 - M_L) \right]^{1/2} \delta_{M_L + 1, M'_L} \, \delta_{M_S - 1, M'_S}$$

$$+ \frac{1}{2}\left[\left(\frac{1}{2} - M_S\right)\left(\frac{3}{2} + M_S\right)(1 + M_L)(2 - M_L)\right]^{1/2} \delta_{M_L - 1, M'_L}\, \delta_{M_S + 1, M'_S} \quad (151)$$

Notice that over the period t_0 to t_1, the eigenvalue of J_z remains constant and each level corresponds to a definite value of $M_J = M_L + M_S$. At the limits $\rho \to \pm \infty$, the corresponding eigenvector tends toward a particular $|M_L M_S\rangle$. We shall consider the two cases $M_J = \frac{3}{2}$ and $M_J = \frac{1}{2}$.

We first consider the case when $M_J = \frac{3}{2}$. There is only one state with $M_J = \frac{3}{2}$, namely $|M_L M_S\rangle = \left|1\,\frac{1}{2}\right\rangle$. This is an eigenvector of H for any t. the energy will vary with time as follows (from Eq. (151))

$$H\left|1\,\frac{1}{2}\right\rangle = \alpha\hbar^2 \left(\frac{1}{2} - 2\rho(t)\right)\left|1\,\frac{1}{2}\right\rangle. \quad (152)$$

If the atom is initially in the state $\left|1\,\frac{1}{2}\right\rangle$, then according to the adiabatic approximation, the atom remains in the same state through the time during which the reversal of the magnetic field takes place. Its state vector will be multiplied by the phase factor $e^{-i\alpha\hbar\int_{t_0}^{t}\left(\frac{1}{2} - 2\rho(t')\right)dt'}$. In the case of a linear variation of $B(t)$ as indicated in Eq. (150), at a time T the state vector is $e^{-i\alpha\hbar T}\left|1\,\frac{1}{2}\right\rangle$. If $\alpha\hbar T \ll 1$, then we get the result according to the sudden approximation, i.e., the state vector itself $\left|1\,\frac{1}{2}\right\rangle$.

Next, we consider the case $M_J = \frac{1}{2}$. Here there are two eigenstates of H, each a linear combination of $\left|0\,\frac{1}{2}\right\rangle$ and $\left|1 -\frac{1}{2}\right\rangle$. If initially the atom is in one of these states, then at any time t the state vector will be

$$\left|\frac{1}{2}\right\rangle_j = C^{(j)}_{0\frac{1}{2}}\left|0\,\frac{1}{2}\right\rangle + C^{(j)}_{1-\frac{1}{2}}\left|1 -\frac{1}{2}\right\rangle, j = 1, 2 \quad (153)$$

with

$$\left|C^{(j)}_{0\frac{1}{2}}\right|^2 + \left|C^{(j)}_{1-\frac{1}{2}}\right| = 1.$$

Using Eq. (151), we write the Hamiltonian matrix $H(t)$ for $M_J = \frac{1}{2}$ as

$$H\left(M_J = \frac{1}{2}\right) = \alpha\hbar^2 \begin{pmatrix} -\rho & \frac{1}{2}\sqrt{2} \\ \frac{1}{2}\sqrt{2} & -\frac{1}{2} \end{pmatrix} \quad (154)$$

Since any 2×2 matrix can be written as a linear combination of the 3 Pauli matrices σ_1, σ_2, σ_3 and the 2×2 unit matrix $I_2 = \begin{pmatrix} 1 & 0 \\ 0 & 1 \end{pmatrix}$, we write the Hamiltonian matrix as $H(t) = b_0 I + \vec{b} \cdot \vec{\sigma}$ whence $b_0 = -\frac{1}{4}(1 + 2\rho)$, $b_1 = \frac{1}{\sqrt{2}}$, $b_2 = 0$, $b_3 = \frac{1}{4}(1 - 2\rho)$. Thus,

$$H(t) = \alpha\hbar^2 \left[-\frac{1}{4}(1 + 2\rho)I + \vec{b}' \cdot \vec{\sigma} \right] \text{ with } \vec{b}' = 4\vec{b}. \quad (155)$$

Let $\vec{u}$ be the unit vector along b', i.e., $\vec{b}' = b'\,\vec{u}$, with $|b'| = \sqrt{8 + (1 - 2\rho)^2}$, $\vec{u} = \left(\frac{2\sqrt{2}}{b'}, 0, \frac{1 - 2\rho}{b'} \right)$. Equation (155) may be rewritten as

$$H(t) = \frac{1}{4}\,\alpha\hbar^2 \left[(-1 - 2\rho) + b'\,(\sigma \cdot \vec{u}) \right] \quad (156)$$

The eigenvalues are

$$E_\pm(t) = \frac{1}{4}\,\alpha\hbar^2 \left((-1 - 2\rho) \pm b' \right). \quad (157)$$

The corresponding eigenvectors are

$$|+\rangle \equiv \left| \frac{1}{2} \right\rangle_1 = C^{(+)}_{0\frac{1}{2}} \left| 0\,\frac{1}{2} \right\rangle + C^{(+)}_{1-\frac{1}{2}} \left| 1 - \frac{1}{2} \right\rangle,$$

$$|-\rangle \equiv \left| \frac{1}{2} \right\rangle_2 = C^{(-)}_{0\frac{1}{2}} \left| 0\,\frac{1}{2} \right\rangle + C^{(-)}_{1-\frac{1}{2}} \left| 1 - \frac{1}{2} \right\rangle, \quad (158)$$

where

$$C^{(\pm)}_{0\frac{1}{2}} = \frac{1}{\sqrt{1 + 2(\rho + E_\pm)^2}}$$

$$C^{(\pm)}_{1-\frac{1}{2}} = \frac{\sqrt{2}\,(\rho + E_\pm)}{\sqrt{1 + 2(\rho + E_\pm)^2}}. \quad (159)$$

As the magnetic field is varied from $-B_0$ to B_0 (i.e., when ρ goes from $-\infty$ to ∞), the eigenvector $|+\rangle$ goes from $\left| 0\,\frac{1}{2} \right\rangle$ to $\left| 1 - \frac{1}{2} \right\rangle$ (except for a phase factor) and the eigenvector $|-\rangle$ goes from $\left| 1 - \frac{1}{2} \right\rangle$ to $\left| 0\,\frac{1}{2} \right\rangle$. The condition for the validity of the adiabatic approximation is

$$T' \equiv \left(\frac{2\alpha\hbar^2}{\mu_B B_0} \right) T \gg (1/\alpha\hbar), \quad (160)$$

where T' is the time necessary for the magnetic coupling energy $\mu_B B$ to change from $-2\alpha\hbar^2$ to $2\alpha\hbar^2$ T is the time during which the magnetic energy changes from $-\mu_B B_0$ to $\mu_B B_0$. The adiabatic approximation is good when this period T' is large compared to $\dfrac{1}{\alpha\hbar}$, the period characteristic of the discrete transition $|+\rangle \leftrightarrow |-\rangle$ under the spin-orbit interaction.

14.8 GEOMETRIC PHASES

Considerable attention has recently been focused on the theoretical and experimental aspects of adiabatic processes. In a seminal paper Berry (1984) showed that in addition to the dynamical phase factor in the evolving wave function in an adiabatic regime there is a geometrical phase factor which has observable consequence. Since the discovery of a topological phase factor in certain applications of the quantum adiabatic theorem, an enormous amount of work has been done on geometric phases in various areas of physics, e.g., the Aharonov-Bohm effect, quantized Hall effect, the rotation of photon polarization in helical optical fibers, etc. leading to a deeper insight into the underlying physics.

Let us consider a system with a Hamiltonian

$$H = H\,(R_1, \ldots, R_N) = H(R) \tag{161}$$

which depends on a set of N time-dependent parameters. The representative point of this time-dependent Hamiltonian $H(R(t))$ traces out slowly a finite curve C on the N-dimensional parameter manifold M

$$C : R_i = R_i(t),\ \forall t \in [0,\,T],\ i = 1,\, 2\, \ldots\, N. \tag{162}$$

where T is the evolution time. For example R_i are the $3N$ coordinates of the nuclei in a molecule. In a spin-resonance experiment, R_i are the components of the magnetic field. Here, we deal with the general case. We study the time evolution of the system.

The instantaneous eigenstates and eigenvalues satisfy

$$H(R)|\,\psi_n(R) > \ = E_n\,(R)|\,\psi_n(R) >. \tag{163}$$

The time dependence of $|\psi_n(\vec{x},\,t) >$ comes from the time dependence of $R_i\,(t)$. We indicate the time dependence of ψ_n as $|\,\psi_n(t) > \ = \ |\,\psi_n(R(t) >$.

Then

$$\frac{d}{dt}\,|\,\psi_n(t) > \ = \ \vec{\nabla}_R|\,\psi_n(R(t) >. \ \dot{R}(t). \tag{164}$$

We assume that the Hamiltonian has a discrete, nondegenerate spectrum. We suppose that the rate of change of the parameters is slow compared with a typical quantum mechanical orbital frequency. The time T is large compared with any quantum mechanical oscillation period. So the adiabatic condition holds.

The general solution $|\psi(t)>$ to the Schrödinger equation

$$i\hbar \left.\frac{d}{dt}\right| \psi(t) > \; = H(R(t)|\psi(t) > \tag{165}$$

may be written as

$$\psi(t) \simeq \psi_n(t) \exp\left(-\frac{i}{\hbar} \int_0^T dt'\, E_n(t')\right) C_n(t) \tag{166}$$

Then

$$i\dot{\psi}_n(t)\, C_n \; + i\psi(t)\, \dot{C}_n = 0. \tag{167}$$

The normalization condition

$$\langle \psi_n(t)|\, \psi_n(t)\rangle \; = \langle \psi(t)|\psi(t)\rangle = 1, \tag{168}$$

gives

$$C_n(t)\, \dot{C}_n^*(t) \; + C_n^*(t)\, \dot{C}_n(t) = 0. \tag{169}$$

This implies that $C_n(t)$ is a phase factor

$$C_n(t) \; = \exp i\gamma_n(t), \tag{170}$$

where

$$\dot{\gamma}_n(t) \; = 1<\psi_n(t)|\dot{\psi}_n(t)>. \tag{171}$$

Using Eq. (164), we may write

$$\dot{\gamma}_n(t) \; = i <\psi_n\,(R(t)|\nabla_R\psi_n(R(t)) >.\, \dot{R}(t). \tag{172}$$

The existence of a phase $\gamma_n(t)$, in addition to the dynamical phase $\exp\left(-\dfrac{i}{\hbar} \int dt'\, E_n\,(R_i(t'))\right)$, has been known for a long time. If the path C in the parameter space is a closed one, i.e., when $R(T) = R(0)$, then this phase is

$$\gamma_n = i \int_0^T dt\, \dot{R}(t).<\psi_n\,(R(t))|\nabla_R\psi_n\,(R(t) >$$

$$= i \oint_C dR. <(\psi_n(R)|\,\nabla_R\psi_n(R)>.$$

This phase $\gamma_n(C)$ is known as the Berry phase and is an observable. This phase depends only on the path taken in parameter space, not on the rate at which the path is traversed, provided it is slow to satisfy the condition of adiabatic approximation. Hence the phase is called a geometric phase. We notice that in the expression for the Berry phase, $\hbar$ does not appear, in contrast to its appearance in the dynamical phase.

Let us put

$$i < \psi_n(R) | \nabla_R \psi_n(R) > = A(R). \tag{173}$$

Then we express γ_n in terms of A, *a* "vector potential":

$$\gamma_n(C) = \oint_C dR.\, A(R). \tag{174}$$

If we make a phase transformation of the eigenstates

$$|\psi_n(R)> \;\rightarrow\; |\psi_n(R) > e^{i\lambda(R)} \tag{175}$$

then

$$A(R) \;\rightarrow\; A(R) - \nabla_R \lambda(R) \tag{176}$$

which is analogous to a gauge transformation. By Stokes' theorem (suitably generalized)

$$\gamma_n = \int (\nabla_R \times A(R)).\, dS \tag{178}$$

where the integral of the curl is taken over any surface enclosed by C. Obviously, Berry's phase is independent of the choice of the gauge. Berry's phase depends on the "magnetic" field like quantity $B(R)$:

$$\gamma_n = \int B(R).\, dS \tag{179}$$

with

$$B(R) = \nabla_R \times A(R). \tag{180}$$

Berry's phase is the flux of this field through the surface.

From the normalization condition $\langle \psi_n(R)| \psi_n(R)\rangle = 1$, we get $\nabla_R \langle \psi_n(R)| \psi_n(R)\rangle = 0$

or,

$$\langle \psi_n(R)| \nabla_R \psi_n(R)\rangle + \langle \nabla_R \psi_n(R)| \psi_n(R)\rangle = 0,$$

whence $\quad \langle \psi_n(R)| \nabla_R \psi_n(R)\rangle = - \langle \nabla_R \psi_n(R)| \psi_n(R)\rangle$

$$= - \langle \psi_n(R)| \nabla_R \psi_n(R)\rangle,$$

or, $\quad Re\, \langle \psi_n(R)| \nabla_R \psi_n(R)\rangle = 0.$

We have

$$\gamma_n = i \oint dR \cdot \langle \psi_n(R)| \nabla_R \psi_n(R)\rangle$$

$$= - Im \int dS. < \nabla_R \psi_n(R)| \times \nabla_R \psi_n(R) >.$$

We insert a complete set of states

$$\sum_m |\psi_m(R) > < \psi_m(R)| = 1 \tag{181}$$

and note that the diagonal contribution vanishes, since $\langle \psi_n(R)|\nabla_R \psi_n(R)\rangle$ is purely imaginary. Differentiating the eigenvalue condition Eq. (163)

$$(\nabla_R) H(R)|\psi_n(R)\rangle + H(R)\,\nabla_R|\psi_n(R)\rangle$$

$$= E_n(R)\,\nabla_R|\psi_n(R)\rangle + (\nabla_R E_n(R))|\psi_n(R)\rangle.$$

Taking the scalar product of the above equation with $|\psi_l(R)\rangle$:

$$\langle\psi_l(R)|\nabla_R H(R)|\psi_n(R)\rangle = (E_n(R) - E_l(R)) \times \langle\psi_l(R)|\nabla_R\psi_n(R)$$

or,
$$\langle\psi_l(R)|\nabla_R\psi_n(R)\rangle = \frac{\langle\psi_l(R)|\nabla_R H(R)|\psi_n(R)\rangle}{E_n(R) - E_l(R)\rangle}. \tag{182}$$

Thus,

$$B(R) = \nabla_R \times A(R)$$

$$= -\,Im \sum_{l \neq n} \frac{\langle\psi_n(R)|\nabla_R H(R)|\psi_2(R)\rangle \times (\psi_l(R)|\nabla_R H(R)|\psi_n(R)\rangle}{\langle E_n(R) - E_l(R)\rangle^2}. \tag{183}$$

Since the states $|\nabla_R\psi_n(R) >$ do not appear in the expression of B it is independent of the way they depend on R.

$B(R)$ is infinite in case of a degeneracy. This implies the presence of a field source, and the surface integral which is the Berry phase is the flux associated with such a source.

We briefly consider the degenerate case. The eigenvalue equation is

$$H(t)|\psi_n(t, a)> = E_n(t)|\psi_n(t, a) >, a = 1, ..., f \tag{184}$$

for an f-fold degenerate eigenvalue for $0 \leq t \leq T$. Here, $H(t) = H(R(t))$, $E_n(t) = E_n(R(t))$. In the adiabatic regime, we write the linear combination of the degenerate states as

$$|\psi_n(t, a)\rangle = \sum_b C_{ab}(t)|\psi_n(t, b) > \exp\left(-\frac{i}{\hbar}\int_0^t dt' E_n(t')\right) \tag{185}$$

for a fixed t and $T \rightarrow \infty$. Substituting Eq. (185) into the Schrödinger equation yields

$$\dot{C}_{ab}(t_j) = iA_{bj}(t)\,C_{aj}(t), \tag{186}$$

where

$$A_{bj}(t_j) = i\,\langle\psi_n(t, b)|\psi_n(t, j)\rangle. \tag{187}$$

Integrating Eq. (186) as a time-ordered product, we get the expression for the $f \times f$ matrix $C(t)$:

$$[C(T)]_{ab} = \left[\left(\exp\left(i\int_0^T dt'\, A(t')\right)\right)_+\right]_{ba} \qquad (188)$$

In the adiabatic regime

$$|\psi_n(T, a)\rangle = |\psi_n(T, b)\rangle \left[\left(\exp\left[i\int_0^T dt'\, A(t')\right]\right)_+\right]_{ba} \times \exp\left(-\frac{i}{\hbar}\int_0^T dt'\, E_n(t')\right) \qquad (189)$$

with the initial condition

$$|\psi^{(a)}(0)\rangle = |\psi_n(0, a)\rangle,$$

we consider the general transformations

$$\left.\begin{array}{l} |\psi_n(t, b)\rangle \to \Omega_{bc}(t)|\psi_n(t, c)\rangle \\[4pt] \langle\psi_n(t, b)| \to \langle\psi_n(t, c)|\,\Omega_{cb}^{-1}(t) \end{array}\right\} \qquad (190)$$

Ω is a $f \times f$ matrix.

Under these transformations, $A_{bj}(t)$ transform as

$$A_{bj}(t) \to A_{cd}(t)\,\Omega_{cb}^{-1}(t)\,\Omega_{jd}(t) + i\,\dot{\Omega}_{jc}(t)\,\Omega_{cb}^{-1}(t) \qquad (191)$$

Let us consider an illustrative example of this formalism. We consider a spin $\frac{1}{2}$ particle in a time-dependent magnetic field $\vec{B}(t)$ The Hamiltonian is

$$H = -\frac{\mu}{2}\,\vec{\sigma}.\vec{B}(t). \qquad (192)$$

Here, the parameters $R_i(t)$ are the components of the magnetic field. The eigenvalues are

$$E_+(\vec{B}) = -E_-(\vec{B}) = \frac{\mu}{2}\,|\vec{B}|. \qquad (193)$$

The eigenvalues are non-degenerate. There exists a degeneracy at the origin i.e. at $B = 0$. In parameter space, this point is an isolated point at which all the parameters vanish. There are only two states $|\psi_\pm(\vec{B})\rangle$ which depend on the direction of $\vec{B}$.

Using Eq. (183), we can compute $\vec{\nabla} \times \vec{A}(\vec{B})$. $\vec{\nabla}$ refers to derivatives with respect to the parameters B_i. Then

$$\vec{\nabla}H(\vec{B}) = -\frac{\mu}{2}\,\vec{\sigma}. \qquad (194)$$

We choose $\vec{B}$ along the z-axis: $\vec{B} = B\hat{k}$. Then $\langle\psi_+|\sigma_z|\psi_-\rangle = 0$, $\langle\psi_+|\sigma_x|\psi_-\rangle = 1$, $\langle\psi_+|\sigma_y|\psi_-\rangle = -i$.

Now,

$$\vec{\nabla} \times \vec{A}\,(B\hat{k}) = -\frac{1}{4B^2(t)}\,\mathrm{Im}\,\langle\psi_+|\vec{\sigma}|\psi_-\rangle \times \langle\psi_-|\vec{\sigma}|\psi_+\rangle$$

$$= -\frac{\hat{k}}{4B^z(t)}\,\mathrm{Im}\,(\langle\psi_+|\sigma_x|\psi_-\rangle^*\langle\psi_+|\sigma_y|\psi_-\rangle - \langle\psi_+|\sigma_y|\psi_-\rangle^*\langle\psi_+|\sigma_x|\psi_-\rangle)$$

$$= -\frac{\hat{k}}{2B^z(t)}. \tag{195}$$

In general,

$$\vec{\nabla} \times \vec{A}\,(\vec{B}) = -\frac{\hat{B}}{2|\vec{B}(t)|^z}. \tag{196}$$

This is the field of a point charge or a "magnetic monopole" of strength $-\dfrac{1}{2}$ at the origin (point of degeneracy) in parameter space, where the components of the magnetic field $\vec{B}$ are the coordinates. The Berry phase is the flux of this source through the surface S bounded by the closed curve C

$$\gamma_n(c) = \int_C \vec{A}(\vec{B})\cdot d\vec{B} = \int_S (\vec{\nabla} \times \vec{A})\cdot d\vec{S} = \pm\frac{1}{2}\,\Delta\Omega, \tag{197}$$

where $\Delta\Omega$ is the solid angle subtended by C at the origin of parameter space. The sign depends on the sense in which the curve is traversed.

To verify Eq. (197), we consider the spinor along $\vec{B}$,

$$|\chi_+\rangle = \begin{pmatrix} \cos\dfrac{\theta}{2} \\[2mm] \sin\dfrac{\theta}{2}\,e^{i\phi} \end{pmatrix} \tag{198}$$

with spherical coordinates θ, ϕ. The vector potential is

$$\vec{A}_+ = i\langle\chi_+|\vec{\nabla}|\chi_+\rangle$$

$$= -\hat{a}_\phi\,\frac{1}{B\sin\theta}\,\sin^2\frac{\theta}{2}$$

$$= -\hat{a}_\phi\,\frac{1}{2B}\,\tan\frac{\theta}{2}, \tag{199}$$

we note

$$\vec{\nabla} \times \vec{A}_+ = \hat{a}_r\,\frac{1}{B\sin\theta}\,\frac{\partial}{\partial\theta}\,(\sin\theta A_\phi)$$

$$= -\hat{a}_r\,\frac{1}{2B^2}.$$

$\vec{A}_+$ is the vector potential of a magnetic monopole of strength $-\dfrac{1}{2}$ located at the origin.

The Berry phase is

$$\gamma_+ = \oint \vec{A}_+(\vec{B}).d\vec{B} = i \int_{\phi=0}^{\phi=2\pi} \left\langle \chi_+ \left| \frac{\partial}{\partial \phi} \chi_+ \right\rangle d\phi \right.$$

$$= \pm (1 - \cos \theta). \tag{200}$$

The solid angle subtended by this trajectory as seen from the origin,

$$\Delta\Omega = \int_0^\theta d\theta' \sin \theta' \int_0^{2\pi} d\phi.$$

$$= 2\pi (1 - \cos \theta). \tag{201}$$

So

$$\gamma_+ = \pm \frac{1}{2} \Delta\Omega.$$

This phase is observable and can be measured. Bitter and Dubbers (1987) carried out an experiment with a spin polarized neutron beam. The validity of the above expression for Berry's phase was confirmed.

Berry pointed out that the Aharonov-Bohm effect can be considered as an example of a geometric phase. Let us consider a particle of mass m and charge q confined to a box whose center is at $\vec{R}$ with respect to an origin fixed in space. There is a long solenoid containing a magnetic field $\vec{B}$. The box lies outside the solenoid where the magnetic field vanishes, but the vector potential $\vec{A}$ is not zero. $\vec{R}$ is an external parameter in real space (Fig. 14.6).

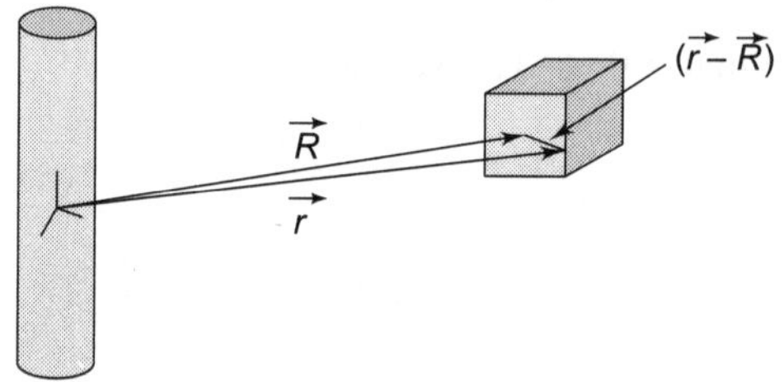

Fig. 14.6 The Aharonov-Bohm effect in a box containing a charged particle transported around a magnetic flux tube

When $\vec{A} \neq 0$ the eigenfunctions $\psi\,(\vec{r};\,\vec{R})$ depend on $\vec{A}$ and so on $\vec{R}$ as a parameter. The wave functions of the confined particle is written as

$$\psi\,(\vec{r};\,\vec{R}) = \exp\left(-\frac{iq}{\hbar c}\int_{\vec{R}}^{\vec{r}} \vec{A}\,(\vec{r}')\cdot d\vec{r}'\right) \psi_n\,(\vec{r} - \vec{R}), \tag{202}$$

where $\psi_n(\vec{r}\,')$ is the wave function of the particle at the position $\vec{r}\,'$ without the magnetic field. We now carry the box in a closed loop C enclosing the solenoid. The gradient of the wave function with respect to the external parameters is

$$\vec{\nabla}_{\vec{R}}\,\psi\,(\vec{r};\,\vec{R}) = -\frac{iq}{\hbar c}\,\vec{A}(\vec{R})\,\psi\,(\vec{r};\,\vec{R}) - \exp\left(-\frac{iq}{\hbar c}\int_{R}^{\vec{r}}\vec{A}(\vec{r}\,')\cdot d\vec{r}\,'\right)\vec{\nabla}_{\vec{r}}\,\psi_n(\vec{r}-\vec{R}),$$

$$(203)$$

and so

$$\langle\psi(\vec{R})|\vec{\nabla}\psi(\vec{R})\rangle = \int\,[\psi\,(\vec{r};\,\vec{R})]^{*}\,\vec{\nabla}_{\vec{r}}\,\psi\,(\vec{r};\,\vec{R})\,d^{3}r$$

$$= -\frac{iq}{\hbar c}\,\vec{A}(\vec{R}). \tag{204}$$

The second term under the integral vanishes because the states are normalized for any R.

From Eqs. (174) and (204), we obtain the geometrical phase.

$$\gamma(C) - \frac{q}{\hbar c}\oint_{C}\vec{A}(\vec{R})\cdot d\vec{R}$$

$$= \frac{q}{\hbar c}\int(\vec{\nabla}\times\vec{A}).\,\hat{n}\,dS$$

$$= \frac{q}{\hbar c}\,\Phi, \tag{205}$$

where Φ is the magnetic flux through the loop.

Thus, the *AB* effect may be considered as an example of geometric phase.

Let us next consider a complex system, e.g., a polyatomic molecule composed of electrons and nuclei. In a molecule, the coordinates can be divided into "fast" coordinates and "slow" coordinates. The full Hamiltonian for a diatomic molecule may be written as

$$H = H_n(\vec{R}) + H_e\,(\vec{r}_1,\,...,\,\vec{r}_N) + V(\vec{R},\,\vec{r}_1,\,...,\,\vec{r}_N). \tag{206}$$

$\vec{R}$ is the relative position coordinate between the two nuclei $\vec{r}_A$ and $\vec{r}_B$, and $\vec{r}_1,\,...,\,\vec{r}_N$ are the position coordinates of N electrons. The center of mass is taken as the origin. The first term in H denotes the kinetic energy of the nuclei, the second term the energy of the electrons, and the third term the interaction energies. In a compact notation, the full Hamiltonian of a molecular system is written as

$$H = \frac{\vec{P}^{z}}{2M} + \frac{\vec{p}^{z}}{2m} + V(\vec{R},\,\vec{r}), \tag{206a}$$

where the capital letters $(\vec{R}, \vec{P})$ denote the coordinate and the momentum of the slow moving nuclear system and the small letters $(\vec{r}, \vec{p})$ denote the fast variables of the electron system. Berry showed that non-integrable phases occur in such systems.

We use an adiabatic approximation here. We ignore the kinetic energy of the slow nuclear part and then determine the eigenstates and eigenvalues of the fast Hamiltonian of the electrons. This Hamiltonian depends on the slow nuclear variable, $\vec{R}$, as an external parameter. The eigenstates of electrons $|n: \vec{R} >$ with eigenvalues $E_n(\vec{R})$ satisfy

$$h(\vec{R})|n; \vec{R}> = E_n(\vec{R})|n; \vec{R}>, \tag{207}$$

where

$$h(\vec{R}) = \frac{\vec{p}^2}{2m} + V(\vec{R}, \vec{r}). \tag{208}$$

We approximate the state ket of the system as

$$|\psi(\vec{R}, \vec{r})> = |n, \vec{R}> \chi(\vec{R})$$

where $\chi(\vec{R})$ is the nucleus ket.

Then

$$H|\psi(\vec{R}, \vec{r})> = E_n(\vec{R})|n; \vec{R}> \chi(\vec{R})$$

$$+ n; \vec{R}> \frac{\vec{P}^2}{2M} \chi(\vec{R})$$

$$- i \vec{\nabla}_{\vec{R}}|n; \vec{R}) \cdot \frac{1}{M} \vec{P} \chi(\vec{R})$$

$$- \frac{1}{2M} \left(\nabla^2_{\vec{R}} |n; \vec{R}> \right) \chi(\vec{R}). \tag{209}$$

Projecting out the $|n; \vec{R}>$ states and neglecting the off-diagonal matrix elements– $\langle x; \vec{R}|\vec{\nabla}_{\vec{R}}\, n; \vec{R}\rangle\, \alpha\delta_{n'n}$–we obtain the Schrödinger equation for $\chi(\vec{R})$ with an effective Hamiltonian involving only the coordinates of nuclei:

$$H_{eff} = \frac{1}{2M} (\vec{P} - \vec{A}(\vec{R})^2) + U(\vec{R}), \tag{210}$$

where

$$U(\vec{R}) = E_n(\vec{R}) - \frac{1}{2M} \left(<n: \vec{R}|\nabla^2_{\vec{R}}\, |n; \vec{R} > + \vec{A}^2(R) \right) \tag{211}$$

is the effective scalar potential and

$$\vec{A}(\vec{R}) = i < n;\ \vec{R}\ |\vec{\nabla}_{\vec{R}}\ n;\ \vec{R}\rangle \tag{212}$$

is the vector potential.

The wave function for the slow variable satisfies the Schrödinger equation in presence of a geometrical gauge field $\vec{A}(\vec{R})$. $\vec{R}$ is the adiabatic parameter. In the adiabatic approximation, $\vec{R}$ changes slowly and the kinetic energy of the slow system is ignored. The time evolution of $\vec{R}(t)$ describes essentially a classical motion. This is the Born-Oppenheimer approximation for a complex system containing both fast and slow variables. In molecular physics, Mead (1983) first noted the existence of $\vec{A}$. We note that Berry's phase is the line integral of $\vec{A}$ which can be rewritten as $\gamma_n = \int\limits_{S} \vec{\nabla} \times \vec{A} \cdot d\vec{S}$.

Hannay (1985) has obtained an analogue of the geometrical phase for classical systems. A geometrical phase factor emerges when we consider the oscillation of a Foucault pendulum on the rotating earth. The direction of gravity acting on the pendulum changes adiabatically as the earth rotates. The total rotational angle θ of the pendulum after a full rotation of the earth is

$$\theta = \omega T + \Delta\theta, \tag{213}$$

where T is the period of revolution of the earth and ω is the angular velocity of the pendulum in the orbital plane. The shift $\Delta\theta$ from the dynamical angle ωT is a geometrical phase and is called Hannay's angle. $\Delta\theta$ depends on the latitude of the observation place but is independent of gravity and the length of the pendulum. The relationship of the Hannay angle $\Delta\theta$ and the quantal Berry phase angle γ_n is $\Delta\theta = -\dfrac{\partial}{\partial n}\gamma_n$. $\tag{214}$

Berry showed that a phase factor, deemed unimportant for noncyclic evolution and hence put to unity, is important for adiabatic cyclic evolution and has observable consequence. The phase angle is defined in terms of an integral over a vector-valued function

$$\vec{A}^n(\vec{R}) = i < n;\ \vec{R}|\vec{\nabla}|n;\ \vec{R}> \tag{215}$$

which is called the Berry connection or the Berry gauge potential. The Berry phase angle is

$$\gamma_n(C) = \int\limits_{C} \vec{A}^n(\vec{R}) \cdot d\vec{R}.$$

$$= \int \vec{\nabla} \times \vec{A}^n \cdot d\vec{S} \tag{216}$$

where C is a closed curve in parameter space. The Berry vector potential transforms under the phase transformations like the vector potential of electrodynamics transforms under a guage transformation.

The Berry phase angle $\gamma_n(C)$ depends only on the closed path C in parameter space (with the condition of adiabaticity). Here we have a guage theory with gauge group $U(1)$ and gauge potential $\vec{A}$. The gauge field $\vec{\nabla} \times \vec{A}$ (in parameter space) is called the Berry curvature. The Berry phase angle $\gamma_n(C)$ is analogous to the magnetic flux of electromagnetic theory.

Simon (1983) has shown that the states $|n;\ \vec{R}\rangle$ describe a Hilbert bundle with connection $\vec{A}$, having curvature $\vec{B} = \vec{\nabla} \times \vec{A}$ whose integral $\int_S \vec{B} \cdot d\vec{S}$ gives a first Chern class of the bundle.

Aharonov and Anandan (AA) (1987) generalized the Berry phase by dispensing with the adiabaticity requirement but requiring cyclic evolution. The AA geometric phase has been observed experimentally. Samuel and Bhandari (1988) showed that a geometric phase can be generated by giving up the condition of cyclicity of the parametric motion. This was done by extending the work of Pancharatnam (1956) in optics.

PROBLEMS

14.1 A simple harmonic oscillator with mass m and frequency ω is acted on by a time-dependent external force $f(t)$. The Hamiltonian of the forced harmonic oscillator is

$$H(t) = -\frac{\hbar^2}{2m}\frac{d^2}{dx^2} + \frac{1}{2}m\omega^2 x^2 - xf(t).$$

Using the creation and annihilation operators, write the Hamiltonian as

$$H(t) = H_0 + H_1(t),$$

where

$$H_0 = \hbar\omega\left(a^\dagger a + \frac{1}{2}\right)$$

and

$$H_1 = \sqrt{\frac{\hbar}{2m\omega}}\, f(t)\, (a^\dagger + a).$$

Assume that $f(t)$ acts only during $t_1 < t < t_2$.

(a) Show that the transition probability from a normalized eigenstate $|m\rangle$ at time $t' < t_1$ to $|n\rangle$ at time $t > t_2$ is

$$P_{m,n} = \begin{cases} \dfrac{m!}{n!}\, z^{n-m}\, e^{-z}\, (L_m^{n-m}(x))^2, & n \geq m \\[2mm] P_{n,m}, & n < m \end{cases}$$

where $L_m^{n-m}(x)$ is a Laguerre polynomial, and $z = \dfrac{|K|^2}{2m\omega}$ with

$$K = \frac{1}{\sqrt{2mh\omega}} \int_{t_1}^{t_2} f(t)\, e^{-i\omega t}\, dt.$$

(b) Calculate the transition probability using first- and second-order time dependent perturbation theory.

(c) If the driving force

$$f(t) = \begin{cases} 0, & t \le t_1, \\ A\cos\Omega t, & t_1 \le t \le t_2 \\ 0, & t \ge t_2 \end{cases}$$

determine the transition probability form the ground state $(t < t_1)$ to the nth excited state $(t > t_2)$.

14.2 Consider a simple harmonic oscillator with modulated frequency described by the Hamiltonain

$$H(t) = -\frac{\hbar^2}{2m}\frac{d^2}{dx^2} + \frac{1}{2}m\omega^2(t)\,x^2.$$

Assume that at time $t = 0$, the time-dependent oscillator is in the ground state. Also assume

$$\omega(t) = \omega_o + \cos(\Omega t)\,\delta\omega,$$

where $\delta\omega \ll \omega_o$ and Ω is a constant. Find the transition probability from the ground state to a final state n.

14.3 The Hamiltonian of a forced harmonic oscillator is

$$H(t) = -\frac{1}{2}\frac{d^2}{dx^2} + \frac{1}{2}x^2 - f(t)\,x.$$

(a) Show

$$U(t, t_o)|\,z(0)> = e^{i\phi(t)}\,|z(t)>,$$

where $U(t, t_o)$ is the propagator and $z(t)$ is a coherent state.

(b) Find the dependence on time of $\phi(t)$ and $z(t)$.

14.4 A hydrogen atom in its ground state is placed in a homogeneous electric field with the time dependence $E = 0$ $(t < 0)$ and $E = E_o\,e^{-t/\tau}$ $(t > 0)$. Find the probability that the atom is in the $2s$ state after a long time. What is the probability that it is in the $2p$ state?

14.5 An electron is moving in a one-dimensional δ-potential $V(x) = -\lambda\delta(x)$ $(\lambda > 0)$. Suddenly the strength of the potential is changed to $\lambda'(\lambda' > 0)$.

Using the method of sudden approximation, calculate the transition probability from the original ground state to the new ground state.

14.6 Show that for the eigenvectors $|n; \vec{R}\rangle$

(a) $\langle n; \vec{R} \,|\vec{\nabla}_{\vec{R}}|\, n; \vec{R}\rangle$ is purely imaginary and the Berry phase angle γ_n is real.

(b) $\langle m; \vec{R} \,|\vec{\nabla}|\, n; \vec{R}\rangle = \langle m; \vec{R}|\vec{\nabla} H(\vec{R})|\, n; \vec{R}\rangle \dfrac{1}{E_n(\vec{R}) - E_m(\vec{R})}, \quad m \neq n.$

CHAPTER 15

Interaction of Radiation with Matter

One of the most important applications of the time-dependent perturbation theory is the theory of the interaction of radiation with the constituents of matter. In the present chapter, we develop the theory of the interaction of an atom with radiation.

15.1 THE ELECTROMAGNETIC FIELD AND THE INTERACTION HAMILTONIAN

The classical electromagnetic field is described by the electric field $\vec{E}$ and the magnetic field $\vec{B}$ which satisfy Maxwell equations (in gaussian units)

$$\vec{\nabla} \times \vec{E} + \frac{1}{c}\frac{\partial \vec{B}}{\partial t} = 0, \quad \vec{\nabla} \times \vec{B} - \frac{1}{c}\frac{\partial \vec{E}}{\partial t} = \frac{4\pi}{c}\,\vec{j},$$

$$\vec{\nabla} \cdot \vec{E} = 4\pi\rho, \quad \vec{\nabla} \cdot \vec{B} = 0. \tag{1}$$

The equation of continuity giving the conservation of charge is

$$\vec{\nabla} \cdot \vec{j} + \frac{\partial \rho}{\partial t} = 0. \tag{2}$$

The electric and magnetic field strengths can be expressed in terms of the scalar and vector potentials Φ and $\vec{A}$ respectively by

$$\vec{E}(\vec{r}, t) = -\frac{1}{c}\frac{\partial}{\partial t}\vec{A}(\vec{r}, t) - \vec{\nabla}\Phi(\vec{r}, t),$$

$$\vec{B}(\vec{r}, t) = \vec{\nabla} \times \vec{A}(\vec{r}, t). \tag{3}$$

The potentials are not uniquely determined by the above equations, since the field strengths $\vec{E}$ and $\vec{B}$ are unaltered by the gauge transformations.

$$\vec{A} \to \vec{A}' = \vec{A} + \vec{\nabla}\chi,$$

$$\Phi \rightarrow \Phi' = \Phi - \frac{1}{c}\frac{\partial \chi}{\partial t}, \tag{4}$$

where χ is an arbitrary real differentiable function of $\vec{r}$ and t.

The gauge invariance allows us to choose the gauge which is the most convenient to work. The gauge called the radiation or the transverse gauge is one in which the scalar potential $\Phi(\vec{r}, t)$ vanishes when the field has no sources and the vector potential is divergence free:

$$\vec{\nabla} \cdot \vec{A}(\vec{r}, t) = 0. \tag{5}$$

The electric and magnetic fields are then given by

$$\vec{E}(\vec{r}, t) = -\frac{1}{c}\frac{\partial \vec{A}(\vec{r}, t)}{\partial t},$$

$$\vec{B}(\vec{r}, t) = \vec{\nabla} \times \vec{A}(\vec{r}, t). \tag{6}$$

The fields satisfy Maxwell equations provided that the vector potential $\vec{A}$ satisfies the wave equation

$$\Box^2 \vec{A}(\vec{r}, t) \equiv \left(\nabla^2 - \frac{1}{c^2}\frac{\partial^2}{\partial t^2}\right)\vec{A}(\vec{r}, t) = 0. \tag{7}$$

The energy of the electromagnetic field is given by

$$H = \frac{1}{8\pi}\int d\vec{r}\left[\vec{E}^2(\vec{r}, t) + \vec{B}^2(\vec{r}, t)\right], \tag{8}$$

and the energy-flux vector is given by the Poynting vector

$$\frac{c}{4\pi}\vec{E}(\vec{r}, t) \times \vec{B}(\vec{r}, t).$$

We now assume that the radiation field is enclosed in a large cube of volume $V (= L^3)$ and impose periodic boundary conditions on the fields at the surface of the cube. The vector potential may be expanded as a linear superposition of plane waves:

$$\vec{A}(\vec{r}, t) = \frac{1}{\sqrt{V}}\sum_{\vec{k},\alpha}\left[\hat{\epsilon}_{\vec{k}}^{\alpha}\, C_{\vec{k},\alpha}(t)\, e^{i\vec{k}.\vec{r}} + c.c\right] \tag{9}$$

where the allowed values of the wave vector $\vec{k}$ are

$$\vec{k} = \frac{2\pi}{V^{1/3}}(n_2, n_2, n_3), \quad n_i = 0, \pm 1, \pm 2, \dots \tag{10}$$
$$(i = 1, 2, 3)$$

$\hat{\epsilon}_{\vec{k}}^{\alpha}$ called a (linear) polarization vector, is a real unit vector such that $\left(\hat{\epsilon}_{\vec{k}}^{1}, \hat{\epsilon}_{\vec{k}}^{2}, \hat{k}\right)$ form a right-handed set of mutually orthogonal unit vectors. One may also use complex circular polarization vectors $\hat{e}^{\pm 1}_{\vec{k}} = \mp \dfrac{1}{\sqrt{2}}\left(\hat{e}^1_{\vec{k}} \pm i\hat{e}^2_{\vec{k}}\right)$. The phases are those for the spherical harmonics with $l = 1$, $m = \pm 1$. These satisfy

$$\hat{e}^{\lambda}_{\vec{k}} \cdot \hat{e}^{\lambda'}_{\vec{k}} = \delta_{\lambda\lambda'},$$

$$\hat{e}^{\lambda}_{\vec{k}} \times \hat{e}^{\lambda'}_{\vec{k}} = i\lambda\hat{k}\delta_{\lambda\lambda'},$$

$$(\lambda, \lambda' = \pm 1)$$

$$\vec{k} \cdot \hat{e}^{\lambda}_{\vec{k}} = 0. \tag{10a}$$

Since $\vec{A}$ is real, the complex conjugate term is added in Eq. (9). Since $\vec{\nabla} \times \vec{A} = 0$, we have

$$\hat{e}^{\alpha}_{\vec{k}}, \vec{k} = 0. \tag{11}$$

From the wave equation (7), we get

$$\ddot{C}_{\vec{k},\,\alpha}(t) + k^2 C_{\vec{k},\,\alpha}(t) = 0 \tag{12}$$

so that we may write

$$C_{\vec{k},\,\alpha}(t) = C_{\vec{k},\,\alpha}(0)\, e^{-i\omega t}, \tag{13}$$

where

$$\omega = |\vec{k}|c \tag{14}$$

Thus,

$$\vec{A}(\vec{r}, t) = \dfrac{1}{\sqrt{V}} \sum_{\vec{k}} \sum_{\alpha=1,2} \left[C_{\vec{k},\,\alpha}(0)\, \hat{e}^{\alpha}_{\vec{k}}\, e^{ik \cdot x} \right.$$

$$\left. + C^*_{\vec{k},\,\alpha}(0) \cdot \hat{e}^{\alpha}_{\vec{k}} e^{-ik \cdot x} \right] \tag{15}$$

with

$$k.x = \vec{k}.\vec{r} - \omega t. \tag{16}$$

The electric and magnetic fields corresponding to Eq. (15) are the mode expansions of the field strengths:

$$\vec{E}(x) = \dfrac{i}{c\sqrt{V}} \sum_{\vec{k}} \sum_{\alpha} \left[\omega C_{\vec{k},\alpha}(0)\hat{e}^{\alpha}_{\vec{k}} e^{ik \cdot x} + c.c \right], \tag{17}$$

$$\vec{B}(x) = \dfrac{i}{\sqrt{V}} \sum_{\vec{k}} \sum_{\alpha} \left[\vec{k} \times \hat{e}^{\alpha}_{\vec{k}}\, C_{\vec{k},\,\alpha}(0)\, e^{ik \cdot x} - c.c \right] \tag{18}$$

The energy of the electromagnetic field is given by Eq. (8). In terms of the mode amplitudes $\vec{C}_{\vec{k},\alpha}(t) = \hat{\epsilon}^{\alpha}_{\vec{k}} \, C_{\vec{k},\alpha}(t)$, we express Eqs. (17) and (18) as

$$\vec{E}(x) = \frac{i}{c\sqrt{V}} \sum_{\vec{k}} \sum_{\alpha} \omega e^{i\vec{k}.\vec{r}} \left[\vec{C}_{\vec{k},\alpha}(t) - \vec{C}^{*}_{\vec{k},\alpha}(t) \right], \tag{17a}$$

$$\vec{B}(x) = \frac{i}{\sqrt{V}} \sum_{\vec{k}} \sum_{\alpha} e^{i\vec{k}.\vec{r}} \, \vec{k} \times \left[\vec{C}_{\vec{k},\alpha}(t) + \vec{C}^{*}_{\vec{k},\alpha}(t) \right] \tag{18a}$$

Then

$$\int d\vec{r}\, \vec{E}.\vec{E}^{*} = \sum_{\vec{k}} \sum_{\alpha} k^{2} \left| \vec{C}_{\vec{k},\alpha}(t) - c.c \right|^{2} \tag{19}$$

$$\int d\vec{r}\, \vec{B}.\vec{B}^{*} = \sum_{\vec{k}} \sum_{\alpha} k^{2} \left| \vec{C}_{\vec{k},\alpha}(t) + c.c \right|^{2} \tag{20}$$

where we have used the relations

$$\frac{1}{V} \int d\vec{r}\, e^{i(\vec{k} - \vec{k}').\vec{r}} = \delta_{\vec{k}\vec{k}'}, \tag{21}$$

and

$$\frac{1}{V} \sum_{\vec{k}} e^{i\vec{k}.(\vec{r} - \vec{r}')} = \delta(\vec{r} - \vec{r}'). \tag{22}$$

Thus, the total energy of the electromagnetic field is

$$H_V = \frac{1}{2\pi} \sum_{\vec{k}} \sum_{\alpha} k^{2} \left| \vec{C}_{\vec{k},\alpha}(t) \right|^{2}. \tag{23}$$

Considering the time dependence of the amplitudes in Eq. (12), we see that the field energy is that of an infinite collection of independent and uncoupled simple harmonic oscillators, each of which is characterized by $\vec{k}$, α. Of course, t drops out in H, since it is a constant of motion. To make the analogy clear, let us define the canonical variables

$$\vec{Q}_{\vec{k},\alpha}(t) = \frac{i}{c\sqrt{4\pi}} \left[\vec{C}_{\vec{k},\alpha}(t) - \vec{C}^{*}_{\vec{k},\chi}(t) \right], \tag{24}$$

$$\vec{P}_{\vec{k},\alpha}(t) = \frac{k}{\sqrt{4\pi}} \left[\vec{C}_{\vec{k},\alpha}(t) + \vec{C}^{*}_{\vec{k},\chi}(t) \right] \tag{25}$$

Then

$$H_V = \frac{1}{2} \sum_{\vec{k}} \sum_{\alpha} \left(\vec{P}^{2}_{\vec{k},\alpha} + \omega^{2} \vec{Q}^{2}_{\vec{k},\alpha} \right) \tag{26}$$

Clearly,

$$\frac{\partial H}{\partial \vec{Q}_{\vec{k},\alpha}} = -\dot{\vec{P}}_{\vec{k},\alpha}; \quad \frac{\partial H}{\partial \vec{P}_{\vec{k},\alpha}} = \dot{\vec{Q}}_{\vec{k},\alpha} \tag{27}$$

It is to be noted that these $\vec{P}_{\vec{k},\lambda}$ and $\vec{Q}_{\vec{k},\lambda}$ are not the momenta and coordinates of any physical object.

Let us next consider the interaction of the electromagnetic field with the constituents of matter. The Hamiltonian of an electron in an electromagnetic field described by the potentials $(\Phi, \vec{A})$ is

$$H = \frac{1}{2m}\left[\vec{p} - \frac{e}{c}\vec{A}(\vec{r},\,t)\right]^2 + e\Phi(\vec{r},\,t) + V(\vec{r},\,t), \tag{28}$$

where V describes all other (non-electromagnetic) potentials. The Schrödinger equation describing the motion of the electron in an electromagnetic field is then

$$i\hbar\,\frac{\partial\underline{\psi}(\vec{r},\,t)}{\partial t} = \left[\frac{1}{2m}\left(-i\hbar\,\vec{\nabla} - \frac{e}{c}\vec{A}(\vec{r},\,t)\right)^2\right.$$
$$\left. + e\Phi(\vec{r},\,t) + V(\vec{r},\,t)\right]\underline{\psi}(\vec{r},\,t). \tag{29}$$

Under a gauge transformation Eq. (4), the state function undergoes a transformation. It is easy to show that the form of the Schrödinger Eq. (29) remains unchanged under a gauge transformation Eq. (4) provided that the wave function is replaced by

$$\underline{\psi}(\vec{r},\,t) \rightarrow \underline{\psi}'(\vec{r},\,t) = e^{i\left(\frac{e}{\hbar c}\right)\chi(\vec{r},\,t)}\,\underline{\psi}(\vec{r},\,t). \tag{30}$$

We write the Hamiltonian H in Eq. (28) as

$$H = H_0 + H_1, \tag{31}$$

where

$$H_0 = \frac{\vec{p}^2}{2m} + V(\vec{r},\,t), \tag{32}$$

and
$$H_1 = -\frac{e}{2mc}(\vec{p}.\vec{A} + \vec{A}.\vec{p}) + \frac{e^2}{2mc^2}(\vec{A}.\vec{A}) + e\Phi \tag{33}$$

describes the interaction of the system with the radiation field. Since $\vec{p}.\vec{A} + \vec{A}.\vec{p} = -i\hbar\,\vec{\nabla}\times\vec{A}$, only in the transverse gauge we may write $\vec{p}.\vec{A} = \vec{A}.\vec{p}$.

For a multielectron atom, we write the Hamiltonian as the sum of H_0, the unperturbed Hamiltonian

$$H_0 = \sum_i \frac{1}{2m}\,\vec{p}_i^2 + V, \tag{34}$$

and the interaction part H_1

$$H_1 = \sum_i\left[-\frac{e}{2mc}\left(\vec{p}_i.\vec{A}(\vec{r}_i,\,t) + \vec{A}(\vec{r}_i,\,t).\vec{p}_i\right)\right.$$
$$\left. + \frac{e^2}{2mc^2}\,\vec{A}^2(\vec{r}_i,\,t) + e\Phi(\vec{r}_i,\,t)\right]. \tag{35}$$

In terms of the particle density operator

$$\rho(\vec{r}) = \sum_i \delta(\vec{r} - \vec{r}_i) \tag{36}$$

and the current density operator

$$\vec{j}(\vec{r}) = \frac{1}{2m} \sum_i \left\{ \vec{p}_i \delta(\vec{r} - \vec{r}_i) + \delta(\vec{r} - \vec{r}_i)\vec{p}_i \right\} \tag{37}$$

we write the interaction Hamiltonian as

$$H_1 = \int d^3r \left[-\frac{e}{c} \vec{j}(\vec{r}).\vec{A}(\vec{r}, t) + \frac{e^2}{2mc^2} \rho(\vec{r})\vec{A}^2(\vec{r}, t) \right.$$
$$\left. + e\rho(\vec{r})\Phi(\vec{r}_i, t) \right] \tag{38}$$

Note that $j(\vec{r})$ is a hermitian operator. Using the gauge-invariant substitution $\vec{p} - \frac{e}{c}\vec{A}$ we see that

$$\vec{J}(\vec{r}) = \frac{1}{2m} \left[\left(\vec{p}_i - \frac{e}{c}\vec{A}(\vec{r}_i, t)\delta(\vec{r} - \vec{r}_i) \right) \right.$$
$$\left. + \delta(\vec{r} - \vec{r}_i)\left(\vec{p}_i - \frac{e}{c}\vec{A}(\vec{r}_i, t) \right) \right]$$
$$= \vec{j}(\vec{r}) - \frac{e}{mc}\vec{A}(\vec{r}, t)\rho(\vec{r}) \tag{39}$$

is the correct form of the operator for the particle current.

15.2   ABSORPTION AND INDUCED EMISSION

We now have the necessary machinery to deal with the emission and absorption of photons by nonrelativistic atomic electrons. In the semi-classical treatment the electromagnetic field is dealt with classically and the particles with which the field interacts are considered quantum mechanically. It is possible in this approximate way to get reasonably accurate account of the effects of the radiation field on a system of particles (absorption and induced emission) but not of the effect of the particles on the field (spontaneous emission). In the nonrelatitivistic regime it is advantageous to use the radiation gauge. In the transverse gauge the last term in Eq. (38) is zero. The term, quadratic in $\vec{A}$, is small compared to the linear term in $\vec{A}$. As we shall see the quadratic term leads to processes involving two photons. For very strong electromagnetic field generated by lasers, two photon emission and absorption processes are important. We shall now consider the emission and the absorption of single photons by atoms. We neglect the quadratic term in $\vec{A}$ and write the interaction Hamiltonian as

$$H_1 = -\frac{e}{c} \int d^3r \, \vec{j}(\vec{r}).\vec{A}(\vec{r}, t). \tag{40}$$

We now apply the time-dependent perturbation theory in first-order, considering H_1 as a small perturbation. With the normal mode representation of $\vec{A}(\vec{r}, t)$ in Eq. (15), we have

$$H_1 = -\frac{e}{c}\frac{1}{\sqrt{V}}\sum_{\vec{k},\alpha}\left[C_{k,\alpha}\,\vec{j}_{-\vec{k}}.\hat{\epsilon}^{\alpha}_{\vec{k}}\,e^{-i\omega t}\right.$$

$$\left. + C^{*}_{\vec{k},\alpha}\,\vec{j}_{\vec{k}}.\hat{\epsilon}^{\alpha}_{\vec{k}}e^{i\omega t}\right] \tag{41}$$

where

$$\vec{j}_{\vec{k}} = \int d^3r\,\vec{j}(\vec{r})\,e^{-i\vec{k}.\vec{r}} \tag{42}$$

is the Fourier transform of the current density.

Let us consider the absorption of a beam (incoherent) of light by an atom initially in the state $|n\rangle$. The perturbation is switched on at $t = 0$. Let the final state be $|m\rangle$. As discussed earlier, upward (downward) transitions are caused by the positive (negative) frequency component of the perturbation. There is no interference between the two terms; the first is important when $E_m \approx E_n + \hbar\omega$ and the second when $E_m \approx E_n - \hbar\omega$. The transition probability per unit time is independent of time only if the final state is part of a continuum or in a group of closely spaced states. If the incident radiation is monochromatic then we get constant rate of transition probability only if the transition occurs to any final state in the continuum or to any of a group of closely spaced states. For transitions between two discrete states, the rate of transition probability is constant in time if the radiation covers a spread of frequencies with no phase relations between the different frequency components.

The rate of absorption (upward transition) caused by $a_{\vec{k},\alpha}\left(= C_{\vec{k},\alpha}\right)$ is, from the golden rule,

$$\Gamma^{abs}_{n\to m;\,\vec{k},\alpha} = \frac{2\pi}{\hbar}\frac{e^2}{Vc^2}\,\delta(E_m - E_n - \hbar\omega)\,\left|a_{\vec{k},\alpha}\right|^2$$

$$\times\left|\left\langle m\left|\hat{\epsilon}^{\alpha}_{\vec{k}}.\vec{j}_{-\vec{k}}\right|n\right\rangle\right|^2. \tag{43}$$

Summing over all $\vec{k}$ and α, the total rate of upward transitions from $|n\rangle$ to $|m\rangle$ is

$$\Gamma^{abs}_{n\to m} = \frac{1}{V}\sum_{\vec{k},\alpha}\frac{2\pi e^2}{\hbar c^2}\,\delta(E_m - E_n - \hbar\omega)\,\left|a_{\vec{k},\alpha}\right|^2$$

$$\times\left|\left\langle m\left|\hat{\epsilon}^{\alpha}_{\vec{k}}.\vec{j}_{-\vec{k}}\right|n\right\rangle\right|^2$$

$$= \frac{2\pi e^2}{\hbar^2 c^2}\cdot\frac{\omega^2}{(2\pi c)^3}\int d\Omega\,\sum_{\alpha}\left|a_{\vec{k},\alpha}\right|^2$$

$$\times\left|\left\langle m\left|\hat{\epsilon}^{\alpha}_{\vec{k}}.\vec{j}_{-\vec{k}}\right|n\right\rangle\right|^2 \tag{44}$$

since

$$\frac{1}{V}\sum_{\vec{k}} \to \int \frac{d^3k}{(2\pi)^3} = \int k^2 \frac{dk\,d\Omega}{(2\pi)^3} = \int \omega^2 \frac{d\omega\,d\Omega}{(2\pi c)^3},$$

and

$$\hbar\omega = E_m - E_n.$$

If the incident beam of radiation is polarized and subtends a solid angle $d\Omega$, then the total rate of energy flow is

$$\frac{1}{V}\sum_{\vec{k}}\frac{\omega^2}{2\pi c}\left|a_{\vec{k},\,\alpha}\right|^2 = d\Omega \int d\omega\,\frac{\omega^4}{(2\pi c)^4}\left|a_{\vec{k},\,\alpha}\right|^2.$$

The intensity of the incident beam per unit frequency is

$$I(\omega) = \frac{\omega^4}{(2\pi c)^4}\left|a_{\vec{k},\,\alpha}\right|^2 d\Omega. \tag{45}$$

The rate of absorption may then be written as

$$\Gamma^{abs}_{n\,\to\,m} = \frac{4\pi^2 e^2}{\hbar^2 c\omega^2}\,I(\omega)\left|\left\langle m\left|\hat{\epsilon}^{\alpha}_{\vec{k}}\cdot\vec{j}_{-\vec{k}}\right|n\right\rangle\right|^2. \tag{46}$$

The rate of downward transition stimulated by the incident beam from the initial state $|m\rangle$ to the final state $|n\rangle$ is given by

$$\Gamma^{ind.em.}_{m\,\to\,n} = \frac{1}{V}\sum_{\vec{k},\,\alpha}\frac{2\pi e^2}{\hbar c^2}\,\delta(E_n - E_m + \hbar\omega)\left|a_{\vec{k},\,\alpha}\right|^2$$

$$\times\left|\left\langle n\left|\hat{\epsilon}^{\alpha*}_{\vec{k}}\cdot\vec{j}_{-\vec{k}}\right|m\right\rangle\right|^2. \tag{47}$$

$$= \frac{4\pi^2 e^2}{\hbar^2 c\omega^2}\,I(\omega)\left\langle\left|n\left|\hat{\epsilon}^{\alpha*}_{\vec{k}}\cdot\vec{j}_{-\vec{k}}\right|m\right|\right\rangle^2, \tag{48}$$

where

$$\hbar\omega = E_m - E_n.$$

Since

$$\left\langle n\left|\hat{\epsilon}^{\alpha*}_{\vec{k}}\cdot\vec{j}_{\vec{k}}\right|m\right\rangle = \left\langle m\left|\hat{\epsilon}^{\alpha}_{\vec{k}}\cdot\vec{j}_{-\vec{k}}\right|n\right\rangle^*,$$

the rate of induced emission is equal to the rate of absorption. This equality is referred to as detailed balancing.

We associate with the upward transition process the *absorption* of one photon of energy $\hbar\omega_{mn} = E_m - E_n$ from the radiation field. In similar fashion the downward transition is associated with the emission of one photon of energy $\hbar\omega = E_m - E_n$. Since the emission rate is proportional to the intensity of the radiation process, this process is called *induced (or stimulated) emission.*

We can express the absorption and emission rates in terms of the number of photons.

A photon of frequency ω has energy $\hbar\omega$. So the total energy in the beam is

$$E = \sum_{\vec{k}}\sum_{\alpha}\hbar\omega N_{\vec{k},\,\alpha} \tag{49}$$

where $N_{\vec{k}, \alpha}$ is the number of photons in the mode $\vec{k}, \alpha$.

From Eq. (23), we see that

$$\frac{k^2}{2\pi} \left| a_{\vec{k}, \alpha} \right|^2 = \hbar\omega N_{\vec{k}, \alpha}$$

or,

$$\left| a_{\vec{k}, \alpha} \right|^2 = \frac{2\pi\hbar c^2}{\omega} N_{\vec{k}, \alpha} \tag{50}$$

where $a_{\vec{k}, \alpha}$ is the amplitude of the mode $\vec{k}, \alpha$.

In terms of the number of photons, we write

$$\Gamma^{abs}_{n \to m} = \frac{1}{V} \sum_{\vec{k}, \alpha} \frac{4\pi^2 e^2}{\omega} \delta(E_m - E_n - \hbar\omega) N_{\vec{k}, \alpha} \times \left| \left\langle m \left| \hat{\epsilon}^\alpha_{\vec{k}} \cdot \vec{j}_{-\vec{k}} \right| n \right\rangle \right|^2 \tag{51}$$

$$= \Gamma^{ind.em}_{m \to n}.$$

The incident photon flux in obtained by dividing the intensity $I(\omega) = \rho(\omega)c$

$$= \frac{\hbar\omega N_{\vec{k}, \alpha}\, c}{V} \text{ by } \hbar\omega, \text{ where } \rho = N/V \text{ is the density of photons, i.e., the flux is } \frac{N_{\vec{k}, \alpha}\, c}{V}.$$

The total absorption cross section is obtained by diving Γ^{abs} by the incident flux. Then

$$\sigma_{abs} = \frac{4\pi^2 e^2}{\omega c} \sum_m \delta(E_m - E_n - \hbar\omega) \times \left\langle \left| m \left| \hat{\epsilon}^\alpha_{\vec{k}} \cdot \vec{j}_{-\vec{k}} \right| n \right| \right\rangle^2. \tag{52}$$

We can also define a stimulated emission cross section $\sigma_{ind.em}$ by dividing $\Gamma^{ind.em}$ by the incident flux. Clearly,

$$\sigma_{abs} = \sigma_{ind.em.} \tag{53}$$

15.3 QUANTIZATION OF THE ELECTROMAGNETIC FIELD

The electromagnetic field is dynamically equivalent to a collection of an infinite set of independent simple harmonic oscillators. We can quantize the theory by assuming that the classical canonical variables P and Q are now hermitian operators satisfying the canonical commutation relations

$$\left[Q_{\vec{k}, \alpha}, P_{\vec{k}', \alpha'} \right] = i\hbar\, \delta_{\vec{k}\vec{k}'}\, \delta_{\alpha\alpha'} \tag{54}$$

$$\left[Q_{\vec{k}, \alpha}, Q_{\vec{k}', \alpha'} \right] = 0, \quad \left[P_{\vec{k}, \alpha}, P_{\vec{k}', \alpha'} \right] = 0 \tag{55}$$

In terms of these operators, the Hamiltonian of the quantized electromagnetic field is

$$H_\gamma = \frac{1}{2} \sum_{\vec{k}} \sum_\alpha \left(P^2_{\vec{k}, \alpha} + \omega^2\, Q^2_{\vec{k}, \alpha} \right). \tag{56}$$

Note that we have used the linear polarization description. But we can also use the circular polarization description in which the commutation rules have the same form with the substitution $P_{\vec{k}, \alpha} \to P_{\vec{k}, \lambda}, Q_{\vec{k}, \alpha} \to Q_{\vec{k}, \lambda}$ where $\alpha = 1, 2$ and $\lambda = \pm 1$.

We next define the non-hermitian operators

$$a_{\vec{k},\lambda} = \left(\frac{1}{\sqrt{2\hbar\omega}}\right)\left(\omega Q_{\vec{k},\lambda} + iP_{\vec{k},\lambda}\right), \tag{57}$$

$$a^{\dagger}_{\vec{k},\lambda} = \left(\frac{1}{\sqrt{2\hbar\omega}}\right)\left(\omega Q_{\vec{k},\lambda} - iP_{\vec{k},\lambda}\right). \tag{58}$$

These operators satisfy the equal-time commutation relations

$$\left[a_{\vec{k},\lambda},\, a^{\dagger}_{\vec{k}',\lambda'}\right] = \delta_{\vec{k}\vec{k}'}\,\delta_{\lambda\lambda'},$$

$$\left[a_{\vec{k},\lambda},\, a_{\vec{k}',\lambda'}\right] = 0, \tag{59}$$

$$\left[a^{\dagger}_{\vec{k},\lambda},\, a^{\dagger}_{\vec{k}',\lambda'}\right] = 0.$$

With the substitutions

$$C_{\vec{k},\lambda}(t) \to -ic\,\sqrt{\frac{2\pi\hbar}{\omega}}\, a_{\vec{k},\lambda}(t)$$

$$C^{*}_{\vec{k},\lambda}(t) \to ic\,\sqrt{\frac{2\pi\hbar}{\omega}}\, a^{\dagger}_{\vec{k},\lambda}(t)$$

we have in Eq. (9),

$$\vec{A}(\vec{r},t) = \frac{1}{\sqrt{V}}\sum_{\vec{k}}\sum_{\lambda} ic\,\sqrt{\frac{2\pi\hbar}{\omega}}$$

$$\times\left[a^{\dagger}_{\vec{k},\lambda}(t)\,\hat{\epsilon}^{\lambda}_{\vec{k}}\,e^{-i\vec{k}.\vec{r}} - a_{\vec{k},\lambda}(t)\,\hat{\epsilon}^{\lambda}_{\vec{k}}\,e^{i\vec{k}.\vec{r}}\right]. \tag{60}$$

The classical vector potential function of $\vec{r}$ and t is now an hermitian operator that acts on state vectors in occupation number space. $\vec{A}$ is now a field operator parametrized by $\vec{r}$ and t. From Eq. (60), we define the field operators corresponding to the electromagnetic field strengths.

$$\vec{E}(\vec{r},t) = \frac{1}{\sqrt{V}}\sum_{\vec{k}}\sum_{\lambda}\sqrt{2\pi\hbar\omega}\left[a_{\vec{k},\lambda}(t)\,\hat{\epsilon}^{\lambda}_{\vec{k}}\,e^{i\vec{k}.\vec{r}} + h.c.\right] \tag{61}$$

$$\vec{B}(\vec{r},t) = \frac{1}{\sqrt{V}}\sum_{\vec{k}}\sum_{\lambda}\sqrt{2\pi\hbar\omega}\left[\vec{k}\times\hat{\epsilon}^{\lambda}_{\vec{k}}\,a_{\vec{k},\lambda}(t)\,e^{i\vec{k}.\vec{r}} + h.c.\right] \tag{62}$$

The electromagnetic field is now an operator which creates and destroys photons.

The Hamiltonian operator of the quantized radiation field is

$$H = \frac{1}{8\pi}\int\left(\vec{E}.\vec{E} + \vec{B}.\vec{B}\right) d^{3}r$$

$$= \frac{1}{2} \sum_{\vec{k}} \sum_{\lambda} \hbar\omega \left(a^{\dagger}_{\vec{k},\lambda} \, a_{\vec{k},\lambda} + a_{\vec{k},\lambda} \, a^{\dagger}_{\vec{k},\lambda} \right)$$

$$= \sum_{\vec{k}} \sum_{\lambda} \left(N_{\vec{k},\lambda} + \frac{1}{2} \right) \hbar\omega, \tag{63}$$

where the hermitian operator $N_{\vec{k},\lambda}$ is defined by

$$N_{\vec{k},\lambda} = a^{\dagger}_{\vec{k},\lambda} \, a_{\vec{k},\lambda}. \tag{64}$$

The number operators $N_{\vec{k},\lambda}$ have integer eigenvalues

$$n_{\vec{k},\lambda} = 0, \, 1, \, 2, \, \dots \tag{65}$$

The eigenstates are

$$\left| n_{\vec{k},\lambda} \right\rangle = \frac{\left(a^{\dagger}_{\vec{k},\lambda} \right)^{n_{\vec{k},\lambda}}}{\sqrt{n_{\vec{k},\lambda}!}} \, |0\rangle, \tag{66}$$

where $|0\rangle$ is the vacuum state, the state without photons.

An eigenvector of $N_{\vec{k},\lambda}$ is the state vector of the free field with a definite number of photons in mode $\vec{k}, \lambda$. An arbitrary state can be specified by the infinite set $\left\{ n_{\vec{k},\lambda} \right\}$. The number $n_{\vec{k},\lambda}$ is called the occupation number for the mode $(\vec{k}, \lambda)$. A state with different sets of $(\vec{k}, \lambda)$ maybe written as

$$\left| n_{\vec{k}_1,\lambda_1}, \, n_{\vec{k}_2,\lambda_2}, \, \dots, \, n_{\vec{k}_i,\lambda_i}, \, \dots \right\rangle = \left| n_{\vec{k}_1,\lambda_1} \right\rangle \otimes \left| n_{\vec{k}_2,\lambda_2} \right\rangle \otimes \dots \otimes \left| n_{\vec{k}_i,\lambda_i} \right\rangle \otimes \dots \tag{67}$$

These states can all be constructed from the vacuum state and explicitly, we write

$$\left| n_{\vec{k}_1,\lambda_1}, \, n_{\vec{k}_2,\lambda_2}, \, \dots, \, n_{\vec{k}_i,\lambda_i}, \, \dots \right\rangle = \prod_{\vec{k}_i,\lambda_i} \frac{\left(a^{\dagger}_{\vec{k},\lambda} \right)^{n_{\vec{k}_i,\lambda_i}}}{\sqrt{n_{\vec{k}_i,\lambda_i}!}} \, |0\rangle. \tag{68}$$

When acting on $\left| \left\{ n_{\vec{k},\lambda} \right\} \right\rangle$, the operator $a^{\dagger}_{\vec{k},\lambda}$ creates an additional photon of the mode $(\vec{k}, \lambda)$, leaving all other occupation numbers unchanged. Accordingly, the energy is increased by $\hbar\omega_{\vec{k}}$. $a^{\dagger}_{\vec{k},\lambda}$ is called the creation operator for a photon of mode $(\vec{k}, \lambda)$. We have

$$a^{\dagger}_{\vec{k}_i,\lambda_i} \left| n_{\vec{k}_1,\lambda_1}, \, n_{\vec{k}_2,\lambda_2}, \, \dots, \, n_{\vec{k}_i,\lambda_i}, \, \dots \right\rangle$$

$$= \sqrt{n_{\vec{k}_i,\lambda_i} + 1} \, \left| n_{\vec{k}_1,\lambda_1}, \, n_{\vec{k}_2,\lambda_2}, \, \dots, \, n_{\vec{k}_i,\lambda_i} + 1, \, \dots \right\rangle. \tag{69}$$

Correspondingly, $a_{\vec{k},\,\lambda}$ is the destruction (or annihilation) operator for a photon of mode $(\vec{k},\,\lambda)$

$$a_{\vec{k_i},\,\lambda_i} \left| n_{\vec{k_1},\,\lambda_1},\, n_{\vec{k_2},\,\lambda_2},\, ...,\, n_{\vec{k_i},\,\lambda_i},\, ... \right\rangle$$

$$= \sqrt{n_{\vec{k_i},\,\lambda_i}} \left| n_{\vec{k_1},\,\lambda_1},\, n_{\vec{k_2},\,\lambda_2},\, ...,\, n_{\vec{k_i},\,\lambda_i} - 1,\, ... \right\rangle \tag{70}$$

From the commutation relations, we see that a many-photon state is symmetric under interchange of any pairs of labels. So photons obey Bose-Einstein statistics.

Quantization of the radiation field provides us with a description of the particle aspects of the electromagnetic field. The formalism can be applied to describe the physical situation in which the number of photons of definite mode $(\vec{k},\,\lambda)$ is increased or decreased – to describe the emission and the absorption of radiation.

Let us return to the energy of the radiation field Eq. (63). The lowest bound to the energy, i.e., the energy of the vacuum state is $\Sigma \dfrac{1}{2} \hbar\omega$, which is clearly infinite. This is a difficulty, which is removed by considering the energy scale in which the energy of the vacuum state is zero, since only differences in energy matter. Thus we use the following expression for the Hamiltonian of the free radiation field.

$$H = \sum_{\vec{k},\,\lambda} \hbar\omega\, N_{\vec{k},\,\lambda} = \Sigma\, \hbar\omega\, a^\dagger_{\vec{k},\,\lambda}\, a_{\vec{k},\,\lambda}. \tag{71}$$

In classical electrodynamics, the total linear momentum of the electromagnetic field is given by the expression

$$\vec{P} = \frac{1}{4\pi c} \int d^3r\, (\vec{E} \times \vec{B}). \tag{72}$$

Using the operator expressions Eqs. (61) and (62) for $\vec{E}$ and $\vec{B}$ respectively, the momentum operator is

$$\vec{P} = \sum_{\vec{k}} \sum_{\lambda} \hbar\vec{k} \left(a^\dagger_{\vec{k},\,\lambda}\, a_{\vec{k},\,\lambda} + a_{\vec{k},\,\lambda}\, a^\dagger_{\vec{k},\,\lambda} \right)$$

$$= \sum_{\vec{k}} \sum_{\lambda} \hbar\vec{k} \left(N_{\vec{k},\,\lambda} + \frac{1}{2} \right). \tag{73}$$

Thus,

$$\vec{P} = \sum_{\vec{k},\,\lambda} \hbar\vec{k}\, N_{\vec{k},\,\lambda}, \tag{74}$$

where we have dropped the term $\sum_{\vec{k}} \hbar\vec{k}$, since for every $\vec{k}$ there is a $-\vec{k}$.

We consider the effect of H and $\vec{P}$ on a single photon state:

$$H a^{\dagger}_{\vec{k}, \lambda} |0\rangle = \hbar\omega \, a^{\dagger}_{\vec{k}, \lambda} |0\rangle,$$

$$\vec{P} a^{\dagger}_{\vec{k}, \lambda} |0\rangle = \hbar\vec{k} \, a^{\dagger}_{\vec{k}, \lambda} |0\rangle. \tag{75}$$

Thus, the photon has the energy $\hbar\omega = \hbar|\vec{k}|c$ and the momentum $\hbar\vec{k}$. The mass of the photon is obtained from the relationship $E^2 = \vec{p}^2 c^2 + m^2 c^4$. Clearly, the mass of the photon is zero and the photon moves with the speed of light.

The photon also has one unit of intrinsic angular momentum (spin). In classical electrodynamics, the total angular momentum of the electromagnetic field is

$$\vec{J} = \frac{1}{4\pi c} \int d\vec{r}\, \vec{r} \times (\vec{E} \times \vec{B}). \tag{76}$$

Now, $\left(\vec{E} \times \vec{B}\right)_i = \in_{ijk} E_j B_k$, where $\in_{ijk}$ is the Levi-Civita tensor.

$$= \in_{ijk} \in_{klm} E_j \frac{\partial A_m}{\partial x_l}$$

$$= E_m \frac{\partial}{\partial x_i} A_m - \frac{\partial}{\partial x_l} (E_l A_i)$$

since

$$\in_{ijk} \in_{klm} = \delta_{il}\delta_{jm} - \delta_{im}\delta_{jl}, \quad \frac{\partial E_i}{\partial x_i} = 0,$$

$$B_i = \in_{ijk} \frac{\partial}{\partial x_j} A_k.$$

Then

$$\left[\vec{r} \times (\vec{E} \times \vec{B})\right]_i = E_m \left(\in_{ijk} x_j \frac{\partial A_m}{\partial x_k} \right) - \frac{\partial}{\partial x_l} (\in_{ijk} x_j E_l A_k)$$
$$+ \in_{ijk} E_j A_k.$$

Integrating this expression over space, the second term vanishes for a wave of finite extent, and we get

$$\vec{J} = \vec{J}_0 + \vec{J}_S, \tag{77}$$

where

$$\vec{J}_0 = \frac{1}{4\pi} \int d\vec{r} \sum_{i=1}^{3} E_i(\vec{r} \times \vec{\nabla}) A_i, \tag{78}$$

$$\vec{J}_S = \frac{1}{4\pi} \int d\vec{r} \, (\vec{E} \times \vec{A}). \tag{79}$$

$\vec{J}_0$ has the form of an orbital angular momentum and $\vec{J}_S$ that of an intrinsic (spin) angular momentum. We now substitute the operators for $\vec{E}$ and $\vec{A}$ and the hermitian adjoints in Eq. (79).

We get
$$\vec{J}_S = \hbar \sum_{\vec{k}, \lambda} \lambda \hat{k}\, a^{\dagger}_{\vec{k}, \lambda}\, a_{\vec{k}, \lambda}$$

$$= \hbar \sum_{\vec{k}, \lambda} \lambda \hat{k}\, n_{\vec{k}, \lambda}. \tag{80}$$

A photon of momentum $\hbar \vec{k}$ has angular momentum $\pm\hbar$ along $\vec{k}$. The photon has helicity ± 1, i.e., the photon spin is either parallel or antiparallel to the direction of propagation. We note that for a particle of spin one with a well-defined momentum $\vec{k}$, the component of its spin along $\vec{k}$ can take one of the three values $\pm 1, 0$. Since the photon is massless, there is no helicity zero state, corresponding to the absence of longitudinally polarized electromagnetic wave. A massless particle of spin s has two helicity states $\lambda = \pm s$ (see Wigner (1957)).

Thus, quantization of the electromagnetic field allows us to associate the particles, photons, with the electromagnetic field. Photons can be regarded as particles of mass zero, charge zero, and spin one. Photons obey Bose-Einstein statistics.

In summary,

(a) A single photon state of definite mode $\left(\vec{k}, \lambda\right)$ is represented by $a^{\dagger}_{\vec{k}, \lambda} |0\rangle$

and has the energy $\hbar\omega_k = \hbar c k$ and the momentum $\hbar \vec{k}$.

(b) A stationary state of the free electromagnetic field is characterized by the set of occupation numbers $\left\{ n_{\vec{k}, \lambda} \right\}$. The state contains photons in different modes $\left(\vec{k}_1, \lambda_1\right), \left(\vec{k}_2, \lambda_2\right)$.

The state has energy $E\left(\left\{ n_{+\vec{k}, \lambda} \right\}\right) = \sum_{\vec{k}} \sum_{\lambda} \hbar\omega \left(n_{\vec{k}, \lambda} + \frac{1}{2}\right)$ and

momentum $\vec{P}\left(\left\{ n_{\vec{k}, \lambda_1} \right\}\right) = \sum_{\vec{k}} \sum_{\lambda} \hbar \vec{k}\, n_{\vec{k}, \lambda}.$

(c) An arbitrary state can be constructed by taking the direct product of eigenvectors

$$\left| \left\{ n_{\vec{k}, \lambda} \right\} \right\rangle = \prod_{\vec{k}} \prod_{\lambda} \frac{1}{\sqrt{n_{\vec{k}, \lambda}!}} \left(a^{\dagger}_{\vec{k}, \lambda}\right)^{n_{\vec{k}, \lambda}} |0\rangle$$

$$= \prod_{\vec{k}_i} \prod_{\lambda_i} \otimes \left| n_{\vec{k}_i, \lambda_i} \right\rangle.$$

This is symmetric under interchange of any pair of labels.

(d) Acting on $\left| \left\{ n_{\vec{k}, \lambda} \right\} \right\rangle$, the creation operator $a^{\dagger}_{\vec{k}, \lambda}$ has the property of creating an additional photon in mode $\left(\vec{k}, \lambda\right)$ leaving other occupation numbers unchanged. Similarly, the destruction operator $a_{\vec{k}, \lambda}$ removes one such photon.

15.4 EINSTEIN'S A AND B COEFFICIENTS

Before proceeding further, we discuss a statistical argument given by Einstein in 1917 to relate the rate for spontaneous emission to those for absorption and induced emission.

We consider the thermal equilibrium between atoms and a radiation field in a cavity at temperature T. Thermal equilibrium is established by the absorption and emission of photons of frequency $\omega_{fn} = (E_f - E_n)/\hbar$. We have assumed that the atoms have two levels, E_f and E_n, with $E_f > E_n$. Two processes for attaining equilibrium, namely, induced emission and absorptions, are proportional to $\rho(\omega_{fn})$, the energy density at the frequency ω_{fn}. The rate at which atoms make transition from $n \rightarrow f$ (absorption) is proportional to the number $N(n)$ of atoms in the state n and to ρ. Hence

$$\frac{dN(n \rightarrow f)}{dt} = B_{nf}\, N(n)\rho(\omega_{fn}), \tag{81}$$

where B_{nf} is called the Einstein coefficient for absorption. The third process, spontaneous emission, occurs even in the absence of external radiation and so is independent of ρ. The rate at which atoms make transition $f \rightarrow n$ (emission) is

$$\frac{dN(f \rightarrow n)}{dt} = B_{nf}\, N(f)\rho(\omega_{fn}) + A_{fn}\, N(f) \tag{82}$$

where $N(f)$ is the number of atoms in the state f. A_{fn} is called the Einstein coefficient for spontaneous emission and B_{fn} is the Einstein coefficient for induced emission. At equilibrium, the two rates are equal. The principle of detailed balance gives $B_{fn} = B_{nf}$. Thus

$$B_{fn}\, N(f)\rho(\omega_{fn}) + A_{fn}\, N(f) = B_{fn}\, N(n)\rho(\omega_{fn})$$

whence
$$\rho(\omega_{fn}) = \frac{A_{fn}/B_{fn}}{[N(n)/N(f)] - 1} \tag{83}$$

According to the Boltzmann law, at thermal equilibrium at temperature T, the ratio $N(n)/N(f)$ is given by

$$\frac{N(n)}{N(f)} = e^{\hbar\omega_{fn}/k_B T} \tag{84}$$

where k_B is the Boltzmann constant. From Eqs. (83) and (84), we find the energy density of the radiation field

$$\rho(\omega_{fn}) = \frac{A_{fn}/B_{fn}}{e^{\hbar\omega_{fn}/k_B T} - 1}.$$

(85)

Equation (85) has the form of the Planck distribution for black-body radiation. Planck's formula for the energy density of black-body radiation is

$$\rho(\omega_{fn}) = \frac{\hbar\omega^3_{fn}}{\pi^2 c^3}\frac{1}{e^{\hbar\omega_{fn}/k_B T} - 1}$$

(86)

Comparing the two expressions, we obtain

$$A_{fn} = \frac{\hbar\omega_{fn}}{\pi^2 c^3} B_{fn}$$

(87)

Einstein used the method to derive Planck's radiation formula. At that time there was no way of evaluating A_{fn}/B_{fn}.

15.5 PHOTONS AND ATOMS: EMISSION AND ABSORPTION

We now consider the emission and absorption of photons by nonrelativistic atomic electrons. The Hamiltonian of the system of matter coupled to quantized radiation field is

$$H = H_\gamma + H_m + H_1,$$

(88)

where H_m and H_γ are the Hamiltonians of matter (electrons) and the free radiation field respectively. H_1 is the interaction Hamiltonian of matter with the field. We are concerned with one photon transitions.

We write the interaction Hamiltonian (omitting the spin magnetic interaction) as

$$H_1 = -\frac{e}{\sqrt{V}} \sum_{\vec{k}, \lambda} \left[j_{\vec{k}, \lambda}\, a_{\vec{k}, \lambda} + h.c. \right]$$

(89)

where the operator $j_{\vec{k}, \lambda}$ may be written as

$$j_{\vec{k}, \lambda} = -\sum_i \sqrt{\frac{2\pi\hbar}{\omega}}\, \frac{\vec{p}_i}{m} \cdot \hat{\epsilon}^{(\lambda)}_{\vec{k}}\, e^{i\vec{k}.\vec{r}_i}.$$

(90)

Let us consider the absorption of a photon in the mode $(\vec{k}, \lambda)$. An atom in the initial state $|i\rangle$ makes a transition to the final state $|f\rangle$ by absorption of a photon of mode $(\vec{k}, \lambda)$. In the initial state of the field $\left| n_{\vec{k}, \lambda} \right\rangle$ there are $n_{\vec{k}, \lambda}$ photons present. Then in the final state $\left| n_{\vec{k}, \lambda} - 1 \right\rangle$, there are $n_{\vec{k}, \lambda} - 1$ photons. The initial and the final state of the system are written as

$$\left| i; n_{\vec{k},\lambda} \right\rangle \equiv |i\rangle \otimes \left| n_{\vec{k},\lambda} \right\rangle$$

$$\left| f; n_{\vec{k},\lambda} - 1 \right\rangle \equiv |f\rangle \otimes \left| n_{\vec{k},\lambda} - 1 \right\rangle \tag{91}$$

In the computation of the transition matrix element for the absorption process to first order only $a_{\vec{k},\lambda}$ term gives a nonvanishing contribution.

The transition amplitude is given by the matrix element

$$\left\langle f; n_{\vec{k},\lambda} - 1 \left| H_1 \right| i; n_{\vec{k},\lambda} \right\rangle = - e \sqrt{n_{\vec{k},\lambda}} \sqrt{\frac{2\pi\hbar}{V\omega}} \left\langle f \left| j_{\vec{k},\lambda} \right| i \right\rangle. \tag{92}$$

In the semi-classical theory the absorption rate is proportional to the intensity, while in the quantum theory it is proportional to $n_{\vec{k},\lambda}$, the number of incident photons. The description of the absorption process in the semi-classical theory agrees with the quantum theoretic description to lowest order.

Next we discuss the emission processes. Now, it is the creation operator $a^{\dagger}_{\vec{k},\lambda}$ which gives the nonvanishing contribution to the matrix element. First we consider the spontaneous emission. The initial and final states of the atom are $|i\rangle$ and $|f\rangle$ respectively with energies E_i and E_f. From the initial excited state the atom emits a photon with momentum $\hbar\vec{k}$ and helicity λ and makes a transition to the final state. The initial state of the radiation field is the vacuum $|0\rangle$ (no photons present) and the final state is $\left| 1_{\vec{k},\lambda} \right\rangle \equiv a^{\dagger}_{\vec{k},\lambda} |0\rangle$ with one photon in the mode $(\vec{k}, \lambda)$.

The initial and the final states of the system are

$$|i; 0\rangle \equiv |i\rangle \otimes |0\rangle$$

$$\left| f; 1_{\vec{k},\lambda} \right\rangle \equiv |f\rangle \otimes a^{\dagger}_{\vec{k},\lambda} |0\rangle \tag{93}$$

The transition matrix element is

$$\left\langle f; 1_{\vec{k},\lambda} \left| H_1 \right| i; 0 \right\rangle = - e \sqrt{\frac{2\pi\hbar}{V\omega}} \left\langle f \left| j^{\dagger}_{\vec{k},\lambda} \right| i \right\rangle, \tag{94}$$

since

$$\left\langle 1_{\vec{k},\lambda} \left| a^{\dagger}_{\vec{k}',\lambda'} \right| 0 \right\rangle = \left\langle 1_{\vec{k},\lambda} \middle| 1_{\vec{k}',\lambda'} \right\rangle$$

$$= \delta_{\vec{k}\vec{k}'} \, \delta_{\lambda\lambda'}.$$

Spontaneous emission describes the emission of a photon by an isolated exited atom when there are no electromagnetic radiation. Had we described the vector potential $\vec{A}$ classically, we would have obtained zero for the transition matrix element in the absence of radiation ($\vec{A} = 0$). Spontaneous emission may be regarded as the quantum mechanical version of the classical electromagnetic phenomenon of radiation from an accelerated charge (cf. Larmor's formula).

Induced or stimulated emission of a photon occurs when there are $n_{\vec{k},\lambda}$ photons present. Now, there are $n_{\vec{k},\lambda}$ photons in the initial state of the radiation field, and the final state contains $n_{\vec{k},\lambda} + 1$ photons. The corresponding transition matrix element is

$$\left\langle f;\, n_{\vec{k},\lambda} + 1 \middle| H_1 \middle| i;\, n_{\vec{k},\lambda} \right\rangle = -\,e\,\sqrt{\frac{2\pi\hbar}{V\omega}}\,\sqrt{n_{\vec{k},\lambda} + 1}\,\left\langle f \middle| j^{\dagger}_{\vec{k},\lambda} \middle| i \right\rangle, \tag{95}$$

since
$$\left\langle n_{\vec{k},\lambda} + 1 \middle| a^{\dagger}_{\vec{k},\lambda} \middle| n_{\vec{k},\lambda} \right\rangle = \sqrt{n_{\vec{k},\vec{k}'} + 1}\,\,\delta_{\vec{k},\vec{k}'}\,\delta_{\lambda\lambda'}.$$

We see from Eq. (95) that the matrix element for emission is nonvanishing even when there are no photons present initially $\left(n_{\vec{k},\lambda} = 0 \right)$. In the quantum field theoretic treatment spontaneous emission and induced emission are considered on equal footing.

Next, we compute the transition rates for the emission and the absorption processes using the golden rule.

Let us consider the emission process. The transition rate for spontaneous emission to $\vec{k},\lambda$ is given by

$$\Gamma^{sp.em.}_{i \to f;\, \vec{k},\lambda} = \frac{2\pi}{\hbar}\left|\left\langle f;\, 1_{\vec{k},\lambda} \middle| H_1 \middle| i;\, 0 \right\rangle\right|^2$$

$$\times\, \delta(E_f - E_i - \hbar\omega)$$

$$= \frac{4\pi^2 e^2}{\omega V}\left|\left\langle f \middle| j^{\dagger}_{\vec{k},\lambda} \middle| i \right\rangle\right|^2, \tag{96}$$

where $\hbar\omega$ satisfies

$$E_f - E_i = \hbar\omega$$

The photon states allowed become a continuum as the normalization volume $V \to \infty$. For a photon emitted into a solid angle element $d\Omega$ in the direction $\vec{k}$ the number of allowed states can be written as

$$\rho\,d\omega = \frac{V d^3 k}{(2\pi)^3} = \frac{V}{(2\pi)^3}\, k^2\, dk\, d\Omega$$

$$= \frac{V\omega^2 d\Omega}{(2\pi c)^3}\, d\omega. \tag{97}$$

The spontaneous emission rate for the transition $i \to f$ with the mode $\left(\vec{k},\lambda \right)$ of the emitted photon specified is

$$\Gamma^{sp.em.}_{\substack{i \to f;\, d\Omega \\ \vec{k},\lambda}} = \frac{\omega^2 d\omega}{(2\pi c)^3}\cdot\frac{4\pi^2 e^2}{\omega}\left|\left\langle f \middle| j^{\dagger}_{\vec{k},\lambda} \middle| i \right\rangle\right|^2\, d\Omega. \tag{98}$$

Hence the power radiated into the solid angle $d\Omega$ about the direction of $\vec{k}$ is

$$dP = \sum_{\substack{k \text{ in } d\Omega \\ \vec{k},\,\lambda}} \hbar\omega\, \Gamma^{sp.em.}_{i \to f;\, d\Omega}$$

$$= d\Omega \int \frac{\omega^2 d\omega}{(2\pi c)^3} \cdot \hbar\omega \cdot \frac{4\pi^2 e^2}{\omega} \left| \left\langle f \left| j^{\dagger}_{\vec{k},\,\lambda} \right| i \right\rangle \right|^2$$

$$\times\, \delta(E_f - E_i - \hbar\omega).$$

So

$$\frac{dP}{d\Omega} = \frac{\omega^2 e^2}{2\pi c^3} \left| \left\langle f \left| j^{\dagger}_{\vec{k},\,\lambda} \right| i \right\rangle \right|^2. \tag{99}$$

For emission rate in presence of photons, we get an extra factor of $n_{\vec{k},\,\lambda} + 1$. The stimulated emission rate is larger by a factor $n_{\vec{k},\,\lambda} + 1$ than the rate for spontaneous emission. The general expression for power radiated by the emitted photon is

$$\frac{dP}{d\Omega} = \frac{\omega^2 e^2}{2\pi c^3} \left| \left\langle f \left| j^{\dagger}_{\vec{k},\,\lambda} \right| i \right\rangle \right|^2 \left(\bar{n}_{\vec{k},\,\lambda} + 1 \right) \tag{100}$$

where $\bar{n}_{\vec{k},\,\lambda}$ is the mean number of photons. This result includes both spontaneous and induced emission.

For absorption, we note that if the final state of the atom is discrete, then we apply the golden rule with the density of initial photon states (see Eq. (97)) as these form a continuum. The absorption rate is

$$\Gamma^{abs}_{i \to f;\, d\Omega} = \frac{\omega^2 d\omega}{(2\pi c)^3} \cdot \frac{4\pi^2 e^2}{\omega}\, \bar{n}_{\vec{k},\,\lambda} \left| \left\langle f \left| j_{k,\,\lambda} \right| i \right\rangle \right|^2 d\Omega. \tag{101}$$

15.6   ELECTRIC DIPOLE TRANSITIONS

In a typical atomic transition the wavelength of the emitted radiation is of the order of several thousand angstrom units while atomic radii are of order one angstrom unit. Then

$$\vec{k}.\vec{r} \approx k a_0 = \frac{a_0}{\lambda} << 1, \tag{102}$$

and we can replace

$$e^{-i\vec{k}.\vec{r}} = 1 - i\vec{k}.\vec{r} + \frac{\left(-i\vec{k}.\vec{r}\right)^2}{2!} + \dots \tag{103}$$

in $j^{\dagger}_{\vec{k},\,\lambda}$ by its leading term. In this approximation, we put the exponential factor as unity and the rate of a radiative transition depends essentially on the absolute square of the matrix element

$$\left\langle f \left| \frac{1}{m} \, \vec{p}_i . \hat{\epsilon}_{\vec{k}}^{(\lambda)} \right| i \right\rangle = \left\langle f \left| \dot{\vec{r}} \right| i \right\rangle \cdot \hat{\epsilon}_{\vec{k}}^{(\lambda)}.$$

The equation of motion is

$$\frac{1}{m} \, \vec{p}_i \equiv \dot{\vec{r}}_i = \frac{i}{\hbar} \left[H_m, \, \vec{r}_i \right]. \tag{104}$$

The matrix element can be rewritten as

$$\left\langle f \left| \dot{\vec{r}} \right| i \right\rangle = \left\langle f \left| \frac{i}{\hbar} \left[H_m, \, \vec{r}_i \right] \right| i \right\rangle$$

$$= \frac{i}{\hbar} \, (E_f - E_i) \left\langle f \left| \vec{r}_i \right| i \right\rangle$$

$$= \pm \, i\omega \left\langle f \left| \vec{r}_i \right| i \right\rangle, \tag{105}$$

since $\hbar\omega = |E_f - E_i|$ for both emission and absorption.

Now,
$$\vec{r} = \sum_i \vec{r}_i \tag{106}$$

is the dipole moment operator and

$$\vec{D} = e\vec{r} \tag{107}$$

is the electric dipole moment operator. A radiative transition described by a nonvanishing matrix element of the electric dipole moment of the atom is called an electric dipole transition. The long wavelength approximation in which the exponential factor is put as unity and the transition matrix element involves only the $\vec{p}_i . \hat{\epsilon}_{\vec{k}}^{\lambda}$ term is called the electric dipole approximation.

From Eq. (99) the power radiated in electric dipole spontaneous emission is

$$\frac{dP}{d\Omega} = \frac{\omega^4}{2\pi c^3} \left| \left\langle f \left| \vec{D} . \hat{\epsilon}_{\vec{k}}^{(\lambda)^*} \right| i \right\rangle \right|^2$$

$$= \frac{\omega^4 e^2}{2\pi c^3} \left| \vec{r}_{fi} . \hat{\epsilon}_{\vec{k}}^{(\lambda)^*} \right|^2, \tag{108}$$

where

$$\vec{r}_{fi} = \left\langle f \left| \vec{r} \right| i \right\rangle.$$

The dipole transition rate is

$$\Gamma_{i \to f}^{dipole} = \frac{\alpha\omega^3}{2\pi c^2} \left| \vec{r}_{fi} . \hat{\epsilon}_{\vec{k}}^{(\lambda)^*} \right|^2 d\Omega, \tag{109}$$

where $\alpha = \dfrac{e^2}{\hbar c}$ is the fine structure constant.

The magnitude of the dipole transition rate is (putting $r \sim a_0$) of the order of 10^8 sec^{-1} for the optical region, 10^{11} sec^{-1} for the X-rays and 10^{14} sec^{-1} for γ-rays.

In order to have an electric dipole transition, we must have a nonvanishing matrix element. In terms of the spherical components, $\vec{r}$ can be written as

$$r^{(\pm 1)} = \mp \frac{1}{\sqrt{2}} (x \pm iy)$$

$$= r \sqrt{\frac{4\pi}{3}} \, Y_1^{\pm 1}, \tag{110}$$

$$r^{(0)} = z = r \sqrt{\frac{4\pi}{3}} \, Y_1^0.$$

$r^{(k)}$ ($k = -1, 0, 1$) are the components of a spherical tensor of rank one. We evaluate the matrix element by applying the Wigner-Eckart theorem

$$\left\langle \alpha_f J_f m_f \left| r^{(q)} \right| \alpha_i J_i m_i \right\rangle = \frac{1}{\sqrt{2J_i + 1}} \left\langle J_i 1 m_i q \,\middle|\, J_f m_f \right\rangle \times$$

$$\left\langle \alpha_f J_f \|r\| \alpha_i J_i \right\rangle. \tag{111}$$

The condition that the CG coefficient shall not vanish leads to the "selection rules" for the electric dipole transition. The selection rules are

$$J_f - J_i = 0, \pm 1, \tag{112}$$

$$m_f - m_i = -q. \tag{113}$$

Physically, the angular momentum selection rule is a consequence of the fact that the spin of the photon is one. Also, there can be no one-photon transition between 0 angular momentum states, because there are no one-photon state with zero helicity. When $q = 0$, i.e., when the radiation is polarized parallel to the z-axis, there the matrix element vanishes unless $m_f = m_i$, or $\Delta m = 0$. In case $q = \pm 1$ (radiation polarized perpendicular to the z-axis), the matrix element vanishes unless $m_f - m_i = \pm 1$, or $\Delta m = \pm 1$.

For the parity selection rule, we note that

$$\left\langle f | \vec{r} | i \right\rangle = - \left\langle f | \Pi^{-1} \vec{r} \, \Pi | i \right\rangle$$

$$= - \pi_f \pi_i \left\langle f | \vec{r} | i \right\rangle, \tag{114}$$

where $\pi_f \pi_i$ are the intrinsic parity of the states f, i respectively. Hence, $\langle f | \vec{r} | i \rangle$ vanishes unless $\pi_f = - \pi_i$.

The selection rules for the electric dipole transition are

$$\Delta J = 0, \pm 1, \quad \text{no} \quad 0 \to 0$$

$$\Delta m = 0, \pm 1 \tag{115}$$
$$\underset{\text{parity} \quad \text{change}}{}$$

This parity selection rule is known as the "Laporte rule" (1924).

The transition rate for spontaneous emission of a photon into the solid angle $d\Omega$ in the dipole approximation is given by Eq. (109). If Θ is the angle between $\hat{\epsilon}^{(\lambda)}$ and $\vec{r}_{fi}$, then Eq. (109) can be written as

$$\Gamma^{sp.em.}_{i \to f}(\Omega)\, d\Omega \;=\; \frac{\alpha\omega^3}{2\pi c^2}\, \left|\vec{r}_{fi}\right|^2 \cos^2\Theta\, d\Omega. \tag{116}$$

In this radiative transition a photon with definite $\left(\vec{k}, \lambda\right)$ is emitted. We now sum over two polarization states for given $\vec{k}$ and integrate over all propagation directions. The integrated transition rate for spontaneous emission of a photon in the dipole approximation is

$$\Gamma^{Sp.em.El}_{i \to f} \;=\; \frac{\alpha\omega^3}{2\pi c^2} \cdot \frac{8\pi}{3}\, \left|\vec{r}_{fi}\right|^2$$

$$\;=\; \frac{4}{3c^2}\, \alpha\omega^3_{fi}\, \left|\vec{r}_{fi}\right|^2. \tag{117}$$

As a concrete example, let us calculate the decay rate for the $2p \to 1s$ transition in a hydrogenic atom. Let the quantum numbers $n_i l_i m_i$ and $n_f l_f m_f$ characterize the initial and final states. Then the transition matrix element is

$$\left\langle n_f l_f m_f \left|\vec{r}. \hat{\epsilon}^{(\alpha)}\right| n_i l_i m_i \right\rangle \;=\; \sqrt{\frac{4\pi}{3}\, \frac{2l_i + 1}{2l_f + 1}}\, \sum_q \left\langle l_i 1 m_i q \middle| l_i\, 1 l_f\, m_f \right\rangle$$

$$\times \left\langle l_i\, 100 \middle| l_i\, 1 l_f\, 0 \right\rangle \left[Y_1^q\,(\hat{\epsilon}) \right]^*$$

$$\times \int_0^\infty R_{n_f l_f}(r)\, R_{n_i l_i}(r)\, r^3\, dr, \tag{118}$$

where $R_{nl}(r)$ is the real normalized radial wave function of a state characterized by n and l. Using the properties of the CG coefficients, we can evaluate the average rate of a radiative dipole transition $(n_i l_i m_i) \to (n_f l_f m_f)$, where $m_f = m_i$, $m_i \pm 1$:

$$\sum_{m_f} \Gamma^{dipole}_{n_i l_i m_i} \to n_f l_f m_f$$

$$= \frac{4\alpha\omega^3}{3c^2} \left\{ \begin{array}{c} \dfrac{l_i + 1}{2l_i + 1} \\[2mm] \dfrac{l_i}{2l_i + 1} \end{array} \right\} \left| \int_0^\infty R_{n_f l_f}(r)\, R_{n_i l_i}(r)\, r^3\, dr \right|^2$$

$$\text{for} \qquad l_f = \left\{ \begin{array}{c} l_i + 1 \\ l_i - 1 \end{array} \right. . \tag{119}$$

Let us now consider the case $2p \to 1s$. The radial wave functions are

$$R_{10}(r) \;=\; 2 \left(\frac{Z}{a_0}\right)^{3/2} e^{-Zr/a_0}$$

$$R_{21}(r) \;=\; \frac{1}{\sqrt{24}} \left(\frac{Z}{a_0}\right)^{5/2} r e^{-Zr/2a_0}, \tag{120}$$

where a_0 is the Bohr radius. The radial integral is

$$\int_0^\infty R_{10}^*(r)\, R_{21}(r) r^3\, dr = \frac{1}{\sqrt{6}} \left(\frac{Z}{a_0}\right)^4 \int_0^\infty r^4 \exp\left(-3Zr/2a_0\right) dr$$

$$= \frac{24}{\sqrt{6}} \left(\frac{2}{3}\right)^5 \left(\frac{a_0}{Z}\right). \tag{121}$$

The angular integral is

$$\int Y_{0,0}^*(\theta, \phi)\hat{\epsilon}\cdot\hat{r}\, Y_{1,m}(\theta, \phi)\, d\Omega$$

$$= \frac{1}{\sqrt{4\pi}} \int \sqrt{\frac{4\pi}{3}} \left(\epsilon_z Y_{1,0} + \frac{-\epsilon_x + i\epsilon_y}{\sqrt{2}} Y_{1,1} + \frac{\epsilon_x + i\epsilon_y}{\sqrt{2}} Y_{1,-1}\right)$$

$$\times Y_{1,m}\, d\Omega$$

$$= \frac{1}{\sqrt{3}} \left(\epsilon_z \delta_{m,0} + \frac{-\epsilon_x + i\epsilon_y}{\sqrt{2}} \delta_{m,-1} + \frac{\epsilon_x + i\epsilon_y}{\sqrt{2}} \delta_{m,1}\right) \tag{122}$$

The transition rate for a given magnetic substate of the excited atom is

$$\Gamma_{2pm \to 1s} = \sum_\alpha \int \frac{\alpha\omega^3}{2\pi c^2} \cdot \frac{2^{15}}{3^{10}} \left(\frac{a_0}{Z}\right)^2$$

$$\times \left[\delta_{m,0}\, \epsilon_z^2 + \frac{1}{2}\, (\delta_{m,1} + \delta_{m,-1})\, (\epsilon_x^2 + \epsilon_y^2)\right] d\Omega, \tag{123}$$

where ω is the frequency of the radiation emitted.

Now,

$$\omega = \frac{1}{\hbar}\, (E_f - E_i)$$

$$= \frac{3}{8}\, \frac{mc^2}{\hbar}\, (Z\alpha)^2. \tag{124}$$

The transition rate from each magnetic substate is the same and if we assume that the initial p state is unaligned, i.e. each magnetic state occurs with equal probability, then the transition rate is

$$\Gamma_{2pm \to 1s} = \frac{1}{3} \sum_{m=-1}^{1} \Gamma_{2pm \to 1s}$$

$$= \frac{2^{17}}{3^{11}}\, a_0^2\, \frac{\alpha\omega^3}{c^2}$$

$$= \left(\frac{2}{3}\right)^8 \frac{mc^2}{\hbar}\, \alpha(Z\alpha)^4$$

$$= 6.27 \times 10^8\, Z^4 s^{-1} \tag{125}$$

Electric diplole transitions satisfying the selection rules (115) are called allowed transitions. If the dipole matrix element is zero, the transition is forbidden. When the transition is forbidden, higher terms in the series (103) corresponding to higher multipoles (e.g. magnetic dipole, electric quadrupole, ...) may be nonvanishing. But the transition rates for these "forbidden" transitions are reduced by a factor $(ka)^n$, $n = 2, 3, 4 \ldots$, where the linear dimension of the wave function is $\sim a$. If the general unapproximated matrix element in Eq. (98) vanishes then the transition between these states is said to be "strictly forbidden". Even then higher order transitions involving more than one photon may occur.

15.7 MAGNETIC DIPOLE AND ELECTRIC QUADRUPOLE TRANSITIONS, HIGHER MULTIPOLE TRANSITIONS

Occasionally it may happen that because of symmetry or otherwise the electric dipole matrix element between an initial state and a final state vanishes. Then the electric dipole transition between these two states is forbidden. Transitions can still take place between these states through higher order terms in the expansion (103). The next order term beyond the dipole term gives the matrix element

$$- i \left\langle f \left| (\vec{k}.\vec{r}) \left(\hat{\epsilon}^{(\alpha)*} .\vec{p} \right) \right| i \right\rangle \tag{126}$$

We may write

$$(\vec{k}.\vec{r}) \left(\hat{\epsilon}^{(\alpha)*} .\vec{p} \right) = \frac{1}{2} \left[(\vec{k}.\vec{r}) \left(\hat{\epsilon}^{(\alpha)*} .\vec{p} \right) - \left(\hat{\epsilon}^{(\alpha)*} .\vec{r} \right) \left(\vec{k}.\vec{p} \right) \right]$$

$$+ \frac{1}{2} \left[(\vec{k}.\vec{r}) \left(\hat{\epsilon}^{(\alpha)*} .\vec{p} \right) + \left(\hat{\epsilon}^{(\alpha)*} .\vec{r} \right) \left(\vec{k}.\vec{p} \right) \right]$$

$$= \frac{-1}{2} \left(\hat{\epsilon}^{(\alpha)*} \times \vec{k} \right).\left(\vec{r} \times \vec{p} \right)$$

$$+ \frac{1}{2} \vec{k}.\left(\vec{r}\vec{p} + \vec{p}\vec{r} \right).\hat{\epsilon}^{(\alpha)*}, \tag{127}$$

where $\vec{r}\vec{p} + \vec{p}\vec{r}$ is a symmetric dyadic. The first term in the matrix element is

$$\frac{i}{2} \hat{\epsilon}^{(\alpha)*} \times \vec{k} . \vec{L} \tag{128}$$

where $\vec{L}$ is the orbital angular momentum operator. The first term in the matrix element is proportional to the matrix element of the orbital angular momentum and thus to the orbital magnetic dipole operator $\vec{M}_{orb} = \dfrac{e}{2mc} \vec{L}$. Also, $\vec{k} \times \hat{\epsilon}^{(\alpha)}$ is

the leading term in the plane wave expansion of $\vec{B}$. The radiative transition due to this term is called a magnetic dipole (M_1) transition. The contribution of the

magnetic moment due to the spin $\sim \dfrac{e\hbar}{2mc} \sigma . \left(\vec{k} \times \hat{\epsilon}^{(\alpha)} \right)$ is of the same order and

should be considered together with the orbital contribution.

The form of the matrix element is similar to EI transition with the changes

$$\vec{p} \to \vec{L}, \quad \hat{\epsilon}^{(\alpha)} \to -\frac{i}{2}\,\vec{k} \times \hat{\epsilon}^{(\alpha)*}.$$

The net matrix element for an $M1$ transition is $\sim \dfrac{\vec{k} \times \hat{\epsilon}^{(\alpha)*}}{2m}\,\langle f\,|\vec{L}|\,i\rangle \approx \dfrac{\hbar k}{m}$,

whereas $E1$ matrix element is of order ωa_0. The ratio is $\dfrac{\hbar k}{m\omega a_0} = \alpha$. The M1

transition amplitudes are $Z\alpha$ times smaller than the E1 amplitude in an atom.

Since $\vec{L}$ is a spherical tensor of rank one, we have

$$\langle \alpha_f\,J_f\,m_f\,|L^q|\,\alpha_i\,J_i\,m_i\rangle = \frac{1}{\sqrt{2J_i + 1}}\,\langle J_i 1 m_i q|\,J_f\,m_f\rangle$$

$$\times \langle \alpha_f\,J_f\,\|L\|\,\alpha_i J_i\rangle. \tag{129}$$

The selection rules for an $M1$ transition are as in the $E1$ transition.

$$|J_f - J_i| \le 1, \quad \text{no} \quad 0 \to 0 \tag{130}$$

Since $\vec{L}$ is an axial vector, we have

$$\langle f\,|\vec{L}|\,i\rangle = \langle f\,|\Pi^{-1}\,\vec{L}\,\Pi|\,i\rangle$$

$$= \pi_i\,\pi_f\,\langle f\,|\vec{L}|\,i\rangle. \tag{131}$$

In contrast to an E1 transition, the M1 selection rule is

$$\text{no parity change} \tag{132}$$

The second term in Eq. (127) is $\dfrac{1}{2}\,\vec{k}\,.\,(\vec{r}\vec{p} + \vec{p}\vec{r})\,.\,\hat{\epsilon}^{(\alpha)*}$.

Since $$\vec{r}\vec{p} + \vec{p}\vec{r} = \frac{im}{\hbar}\,[H_0,\,\vec{r}\vec{r}], \tag{133}$$

the matrix element may be written as

$$-i\,\frac{\vec{k}}{2}\,.\,\langle f\,|\vec{r}\vec{p} + \vec{p}\vec{r}|\,i\rangle\,.\,\hat{\epsilon}^{(\alpha)*}$$

$$= \frac{m}{2\hbar}\,(E_i - E_f)\,\vec{k}\,.\,\langle f\,|\vec{r}\vec{r}|\,i\rangle\,.\,\hat{\epsilon}^{(\alpha)*}.$$

$$= -\frac{m\omega}{2}\,\vec{k}\,.\,\langle f\,|\vec{r}\vec{r}|\,i\rangle\,.\,\hat{\epsilon}^{(\alpha)*} \tag{134}$$

Since $\hat{\epsilon}^{(\alpha)*}\,.\,\vec{k} = 0$ (transversality condition), we can replace $\vec{r}\vec{r}$ by its traceless part with components

$$Q_{ij} = x_i x_j - \frac{1}{3}\,\delta_{ij}\,|\vec{r}|^2 \tag{135}$$

Q_{ij} is symmetric and traceless:

$$Q_{ij} = Q_{ji}, \quad Tr\,Q_{ij} = 0 \tag{136}$$

Q_{ij} is the electric quadrupole moment operator and has five independent components. Q_{ij} is equivalent to a spherical tensor of rank two. The selection rules associated with a radiative transition due to this term, called an electric quadrupole ($E2$) transition, are

$$|J_f - J_i| \leq 2 \leq J_f + J_i \tag{137}$$

no parity change

The parity selection rule follows since Q_{ij} is even.

Higher multipole transitions have been discussed by considering higher order terms in the expansion (103). However, it is convenient to use the multipole expansion involving vector spherical harmonics in which rather than expanding in plane waves we expand in spherical waves with well defined angular momentum and parity. In electromagnetic transitions, atoms or nuclei emitting or absorbing photons are usually in states of definite total angular momentum. We therefore construct electromagnetic fields as superpositions of photon wave functions with a well-defined angular momentum and a well-defined energy and also a well-defined parity. Here we briefly describe the multipole expansion of electomagnetic fields.

Electromagnetic waves are classified according to the values of J and L components as shown in Section 10.2.

The L value determines the parity. According to their parity transformation, electromagnetic radiations are classified in the multipole language as

(a) EJ radiation or the electric 2^J pole radiation: the photons have energy $\hbar\omega$, angular momentum. J and parity $- (- 1)^{J \pm 1} = (- 1)^J$;

(b) MJ radiation or the magnetic 2^J pole radiation: the photons have energy $\hbar\omega$, angular momentum J and parity $- (- 1)^J = (- 1)^{J + 1}$.

Note that the extra minus signs are due to the fact that the photon has odd intrinsic parity.

The total angular momentum $\vec{J}$ is

$$\vec{J} = \vec{L} + \vec{S}, \tag{138}$$

and the spin associated with a vector field is unity. The three eigenvectors which correspond to its three eigenvalues 1, 0 and −1 are

$$\hat{\chi}_{+1} = - \frac{1}{\sqrt{2}} \, (\hat{e}_x + i\hat{e}_y)$$

$$\hat{\chi}_0 = \hat{e}_z$$

$$\hat{\chi}_{-1} = \frac{1}{\sqrt{2}} \, (\hat{e}_x - i\hat{e}_y) \tag{139}$$

where $\hat{e}_x$, $\hat{e}_y$, $\hat{e}_z$ are the unit vectors in the x, y, and z-directions.

We now define vector spherical harmonics $\vec{Y}(\theta, \phi)$ as vector functions of the angles only which are simultaneous eigenvectors of $\vec{J}^2$ and J_z. For each value of $J^2 = J(J+1)$ with $J \geq 1$, there are three types of vector spherical harmonics, corresponding to $L = J + 1$, $L = J$ and $L = J - 1$.

Now,
$$\vec{Y}_{Jl1}^{M}(\theta, \phi) = \sum_{m=-l}^{l} \sum_{q=-1}^{1} Y_{lm}(\theta, \phi)\, \chi_q\, (lm1q|l1JM). \tag{140}$$

For each eigenvalue J, there are three $\vec{Y}_{Jl1}^{M}$ with $l = J - 1, J, J + 1$. In the case $J = 0$, there is only one vector spherical harmonic. These satisfy

$$\vec{J}^2\, \vec{Y}_{Jl1}^{M} = J(J+1)\vec{Y}_{Jl1}^{M} \tag{141}$$

$$J_z\, \vec{Y}_{Jl1}^{M} = M\vec{Y}_{Jl1}^{M} \tag{142}$$

$$\int_0^{2\pi}\int_0^{\pi} \vec{Y}_{Jl1}^{M*}(\theta, \phi)\,.\,\vec{Y}_{J'l'1}^{M'}(\theta, \phi)\, \sin\theta\, d\theta\, d\varphi$$
$$= \delta_{JJ'}\, \delta_{ll'}\, \delta_{MM'}. \tag{143}$$

For a fixed value of l, the $(2J + 1)$ functions form an irreducible tensor of order J called an irreducible vector tensor or a vector spherical harmonic.

The vector functions $\vec{Y}_{Jl1}^{M}$ have parity $(-1)^l$. The three types of vector spherical harmonics can be divided into two classes according to their behavior under space inversion. For given values of J and M, the parity of $\vec{Y}_{JJ1}^{M}$ is $(-1)^J$, while the parity of the other two functions $\vec{Y}_{JJ\pm11}^{M}$ is $(-1)^{J+1}$. For $\vec{Y}_{JJ1}^{M}(\theta, \phi)$ we introduce the notation $\vec{X}_{JM}(\theta, \phi)$. These vector spherical harmonics may be generated from the scalar spherical harmonics:

$$\vec{X}_{JM}(\theta, \phi) \equiv \vec{Y}_{JJ1}^{M}(\theta, \phi)$$

$$= \frac{\vec{L}Y_{JM}(\theta, \phi)}{\sqrt{J(J+1)}} \tag{144}$$

where $\vec{L}$ is the vector differential operator $-i\vec{r} \times \vec{\nabla}$ and Y's are the scalar spherical harmonics.

Since the vector spherical harmonics form a complete set, any vector field $\vec{V}(\vec{r})$ can be written as a linear combination of the vector spherical harmonics with coefficients which depend on the radial coordinate r only.

A quantum with energy $\hbar\omega$ associated with a vector field represented by a vector spherical harmonic $\vec{Y}_{Jl1}^{M}(\theta, \phi)$ has total angular momentum $\hbar\sqrt{J(J+1)}$ and its component along the z-axis is $\hbar M$.

We now turn to the electromagnetic field. In a source free region of empty space, the Maxwell equations with the assumption of a time dependence $e^{-i\omega t}$ may be written as

$$c\vec{\nabla} \times \vec{E} = i\omega\vec{B}, \quad c\vec{\nabla} \times \vec{B} = -i\omega\vec{E},$$
$$\vec{\nabla} \cdot \vec{E} = 0, \quad \vec{\nabla} \cdot \vec{B} = 0. \tag{145}$$

These equations lead to the vector Helmholtz equation for $\vec{B}$

$$\left(\vec{\nabla}^2 + k^2\right)\vec{B} = 0,$$

with

$$\vec{\nabla} \cdot \vec{B} = 0, \tag{146}$$

and

$$\vec{E} = \frac{i}{k}\vec{\nabla} \times \vec{B},$$

where $k = \omega/c$. Alternatively, we may write

$$\left(\vec{\nabla}^2 + k^2\right)\vec{E} = 0,$$

with

$$\vec{\nabla} \cdot \vec{E} = 0, \tag{147}$$

and

$$\vec{B} = -\frac{i}{k}\vec{\nabla} \times \vec{E}.$$

Thus, we write

$$\left(\vec{\nabla}^2 + k^2\right)\vec{G} = 0 \tag{148}$$

where $\vec{G}$ indicates any one of $\vec{E}, \vec{B}$.

In the Coulomb gauge $\vec{\nabla} \cdot \vec{A} = 0$ and the electric and magnetic fields are

$$\vec{E} = i\omega\vec{A}, \vec{B} = \vec{\nabla} \times \vec{A} \tag{149}$$

where $\vec{A}$ is the vector potential.

We note that $\vec{A}$ and $\vec{E}$ are polar vectors while $\vec{B}$ is an axial vector.

We seek solutions to Eq. (148) in the form of outgoing spherical waves. The vector fields $\vec{E}, \vec{B}, \vec{A}$ can be expanded into multipole fields involving the vector spherical harmonics. The two types of vector spherical harmonics $l = J$ and $l = J \pm 1$ are separated because they have opposite parity under space inversion. Thus, the electromagnetic multipole fields are of two types – the electric and magnetic multipoles according to their parity transformation.

For MJ radiation the vector potential is

$$\vec{A}_{JM}^{(m)}(\vec{r}, \omega) = -\sqrt{\frac{2}{\pi}}\, j_J(\omega r)\, \vec{X}_{JM}(r), \tag{150}$$

where j_J is a spherical Bessel function.

For EJ radiation, we have

$$\vec{A}_{JM}^{(e)}(\vec{r}, \omega) = \sqrt{\frac{2}{\pi}}\left[-\sqrt{\frac{J}{2J+1}}\, j_{J+1}(\omega r)\, \vec{Y}_{JJ+11}^{(M)}(\hat{r})\right.$$

$$+ \sqrt{\frac{J}{2J+1}} \; j(\omega r) \; \vec{Y}^{(M)}_{J\,J-1\,1}(\hat{r})\Bigg]. \tag{151}$$

We have

$$\int d^3r \, \vec{A}^{(\alpha')*}_{J'M'}(\vec{r}, \omega') \cdot \vec{A}^{(\alpha)}_{JM}(\vec{r}, \omega)$$

$$= \frac{1}{\omega^2} \, \delta(\omega' - \omega) \, \delta_{\alpha'\alpha} \, \delta_{J'J} \, \delta_{M'M}, \tag{152}$$

where $\qquad\qquad \alpha = e, m.$

From Eqs. (150) and (151) we can obtain the fields. It should be noted that

$$\vec{r} \cdot \overline{X}_{JM}(r) = 0 \tag{153}$$

so that for 2^J multipole electric (magnetic) radiation the magnetic (electric) field has no radial component. Also $\vec{X}_{JM} = 0$ for $J = 0$. Hence, there is no $J = 0$ multipole radiation.

Let $a^{\dagger(\alpha)}_{JM}(\omega)$ be the multipole wave creation operator for a photon state of definite angular momentum eigenvalues (JM) and energy $\hbar\omega$, and parity $\pm(-1)^J$ indicated by $\alpha - e, m$. Then $a^{\dagger(\alpha)}_{JM}(\omega)$ creates a photon with wave function $\vec{A}^{(\alpha)}_{JM}(\vec{r}, \omega)$.

The vector potential field operator is

$$\vec{A}(\vec{r}, t) = \int_0^\infty \omega^2 \, d\omega \, \sqrt{\frac{4\pi}{2\omega}} \sum_{JM} \sum_{\alpha} \vec{A}^{(\alpha)}_{JM}(\vec{r}, \omega) \, e^{i\omega t}$$

$$\times \, a^{(\alpha)}_{JM}(\omega) + h.c. \Big]. \tag{154}$$

Let us now indicate, how the multipole expansion is used in the electromagnetic transition processes. We consider an emission process in which a physical system (atom, nucleus, ...) is in an initial state $|i\rangle = |E_i J_i M_i \pi_i\rangle$ and no photon is present. The final state contains a photon in a state $|\omega JM\alpha\rangle$ with wave function $\vec{A}^{(\alpha)}(\omega)$ and the system in a final state $|f\rangle = |E_f J_f M_f \pi_f\rangle$.

The electromagnetic interaction specified by the interaction Hamiltonian H_1 is responsible for thus transition. Since H_1 is a scalar under rotation and is invariant under space inversion the allowed multipole radiation with $(JM\pi)$ must satisfy the selection rules

$$\left.\begin{array}{c} |J_i - J_f| \le J_i + J_f, \\[4pt] M = M_i - M_f \\[4pt] \pi = \pi_f \pi_i. \end{array}\right\} \tag{155}$$

In many applications in atomic physics the electric dipole term gives the dominant contribution. Contributions from higher multipoles are rare. The Laporte rule states that the electric dipole transitions take place between states of opposite parity. Wigner showed that this rule is a consequence of the invariance of the electromagnetic interactions under space inversion. In nuclear physics, however, higher multipole transitions are very common. This is due to the different energies and dimensions involved in the two cases. The probability of the emission of a radiation of wave number k from a 2^J pole of linear dimension a is proportional to $(ka)^{2J+1}$ for EJ radiation and to $(ka)^{2J+3}$ for MJ radiation. Now, ka is of order 10^{-4} in atoms and of order 10^{-2} for nuclei. Atoms in states which cannot deexcite by an electric dipole radiation often deexite by other faster mechanisms (e.g., inelastic collisions with other atoms). Magnetic dipole transitions occur in interstellar gas (emission nebulae) containing some doubly ionized 0^{++} – ions at extremely low pressure. The frequencies of electric quadrupole lines are distinct in alkali and more complex atoms and have been measured in absorption (e.g., the transition $3s \rightarrow 3d$ in Na). The quadrupole transitions $S \rightarrow D$ in oxygen is allowed and has a long life. In the ionosphere (i.e. under low pressure) the atom has time to radiate before a collision. This is the origin of the famous red line of the aurora borealis.

15.8   LIFETIMES, LINE INTENSITIES, SUM RULES AND LINE BREADTHS

To compute the mean lifetime of an atom in an excited state n we sum over the transition rates into all possible states of lower energy, compatible with the selection rules:

$$\tau_n^{-1} = \sum_m \Gamma_{n \rightarrow m}^{sp}. \qquad (156)$$
$$E_m < E_n$$

As an example, in the $2p \rightarrow 1s$ transition in a hydrogen atom, the lifetime of the $2p$ state is

$$\tau_{2p \rightarrow 1s} = \frac{1}{6.27 \times 10^8 \text{ s}^{-1}} = 1.6 \times 10^{-9} \text{ s} \qquad (157)$$

as no other channels are available. Atomic transitions of $E1$ type have lifetimes of order 10^{-8}s while the lifetimes of typical $E2$ or $M1$ transitions are of the order of 10^{-3}s.

As another example, we consider the lifetime of a hydrogen atom in the metastable $2S_{1/2}$ state. Spontaneous transitions to the $2P_{1/2}$ state have a negligible probability. $E1$ transition to the ground state (1S) is forbidden by parity, while $M1$ matrix element vanishes in the nonrelativistic approximation to the radial wave functions. $E2$ and all other higher multipole transitions are forbidden by the angular momentum conservation. The largest probability for a $2s \rightarrow 1s$ transition is due to the simultaneous emission of two photons and the mean lifetime has been calculated to be $1/7s$.

The mean lifetime of an excited state of a hydrogenic atom increases with increasing n, the principal quantum number. For a fixed value of L,

$$\tau_{nl} \sim n^3 \tag{158}$$

whereas the average lifetime of the nth quantum state

$$\tau_n = \left(\sum_l \frac{2l+1}{n^2} \frac{1}{\tau_{nl}} \right)^{-1} \sim n^{4.5}. \tag{159}$$

The intensity of radiation can be obtained by multiplying the transition rate by $\hbar\omega$. In the dipole approximation, the intensity of a transition is proportional to the quantity $|\vec{r}_{fi}|^2$. The relative intensity of the Lyman α and the Lyman β lines of atomic hydrogen is ~ 3.

In discussion of line intensities and other applications, it is convenient to introduce a dimensionless quantity f_{ni}, called the oscillator strength.

The oscillator strength f_{ni} is defined as

$$f_{ni} = \frac{2m\omega_{ni}}{3\hbar} |\vec{r}_{ni}|^2 \tag{160}$$

where $\hbar\omega_{ni} = E_n - E_i$. For absorption, $f_{ni} > 0$, whereas for emission, $f_{ni} < 0$.

The oscillator strengths obey Thomas-Reiche-Kuhn sum rule

$$\sum_n f_{ni} = 1 \tag{161}$$

where the sum is over all energy eigenstates, that is, a summation over the discrete states and an integration over the continuum. The state is a discrete energy eigenstate. The sum rule can be derived as follows. Let

$$f_{ni}^x = \frac{2m\omega_{ni}}{3\hbar} |x_{ni}|^2 \tag{162}$$

and

$$f_{ni} = f_{ni}^x + f_{ni}^y + f_{ni}^z \tag{163}$$

Then

$$f_{ni}^x = \frac{2m}{3\hbar} \omega_{ni} |x_{ni}|^2$$

$$= \frac{2m\omega_{ni}}{3\hbar} x_{ni}^* \cdot x_{ni}$$

$$= \frac{2m\omega_{ni}}{3\hbar} \langle i|x|n \rangle \langle n|x|i \rangle$$

Since

$$\langle n|p_x|i \rangle = \left\langle n \left| \frac{im}{\hbar} [H_0, x] \right| i \right\rangle$$

$$= \frac{im}{\hbar} (E_n - E_i) \langle n|x|i \rangle,$$

$$f_{ni}^x = \frac{2i}{3\hbar} \langle i|p_x|n \rangle \langle n|x|i \rangle,$$

$$= -\frac{2i}{3\hbar} \langle i|x|n \rangle \langle n|p_x|i \rangle$$

$$= \frac{i}{3\hbar} [\langle i|p_x|n \rangle \langle n|x|i \rangle - \langle i|x|n \rangle \langle n|p_x|i \rangle]$$

Using
$$\sum_n |n\rangle\langle n| = 1,$$

we get

$$\sum_n f^x_{ni} = \frac{i}{3\hbar} \langle i|p_x x - x p_x|i \rangle = \frac{1}{3}.$$

Therefore,

$$\sum_n f^x_{ni} = \sum_n f^y_{ni} = \sum_n f^z_{ni} = \frac{1}{3}.$$

Hence, the sum rule Eq. (161) is proved. For an atom with Z electrons, the sum rule is

$$\sum_n f_{ni} = Z. \tag{164}$$

For a detailed discussion of the sum rules consult Bethe & Salpeter (1957) and Sobelman (1992).

We now consider the problem of line shapes and widths. An excited atomic level has a finite lifetime τ due to the transition probability and by the time-energy uncertainty relation, we expect the energy of the level to be uncertain by an amount

$$\Delta E \sim \frac{\hbar}{\tau}. \tag{165}$$

The energy level has a breadth ΔE. Since the ground state is stable, τ_{ground} is infinite and the width $(\Delta E)_{\text{ground}} = 0$.

Weisskopf and Wigner (1930) developed the theory of atomic decay by use of time-dependent perturbation theory. We consider the simple case of an atom with two states $|a\rangle$, $|b\rangle$ only ($E_b > E_a$). Initially, the atom is in excited state $|b\rangle$ and no photons are present. Thus,

$$|\psi(t = 0)\rangle = |b\rangle |0\rangle = |b, 0\rangle \tag{166}$$

where $|0\rangle$ is a vacuum state of the radiation field. We expand the initial state in terms of the atom photon state $|n\rangle$. Now,

$$H_0|n\rangle = E_n^{(0)}|n\rangle \tag{167}$$

with $H_0 = H_\gamma + H_{\text{atom}}$. We write

$$|\psi(t = 0)\rangle = \sum_n c_n(0)|n\rangle \tag{168}$$

Asuming H_1, the interaction Hamiltonian, to be a small perturbation, we write

$$|\psi(t)\rangle = \sum_n c_n(t)\, e^{-\frac{i}{\hbar} E_n^{(0)} t}\, |n\rangle \tag{169}$$

where $c_n(t)$ is now a slowly varying function of time.

The time dependent amplitudes satisfy the following infinite set of coupled first-order differential equations

$$i\hbar\, \frac{\partial}{\partial t}\, c_n(t) = \sum_m c_m(t)\, \langle n|H_1|m\rangle$$

$$\times\, e^{i\left(E_n^{(0)} - E_m^{(0)}\right)\frac{t}{\hbar}}, \tag{170}$$

Initially, all c_n except one, c_b, are zero. Hence,

$$i\hbar\, \frac{\partial}{\partial t}\, c_n = c_b(t)\, \langle n|H_1|b,\, 0\rangle\, e^{i\left(E_n^{(0)} - E_b^{(0)}\right)\frac{t}{\hbar}},$$

where
$$|n\rangle \neq |b,\, 0\rangle. \tag{171}$$

$$i\hbar\, \frac{\partial}{\partial t}\, c_b = \langle b,\, 0|H_1|b,\, 0\rangle\, c_b(t)$$

$$+ \sum_n c_n(t)\, \langle b,\, 0|H_1|n\rangle \times e^{i\left(E_b^{(0)} - E_n^{(0)}\right)\frac{t}{\hbar}} \tag{172}$$

Substituting,

$$c_b(t) = c_b'(t)\, e^{-i\,\wedge\, t/\hbar}, \tag{173}$$

where

$$\Lambda = \langle b,\, 0|H_1|b,\, 0\rangle. \tag{174}$$

We write the Eqs. (171) and (172) as

$$i\hbar\, \frac{\partial}{\partial t}\, c_n = c_b'\, e^{i\left(E_n^{(0)} - E_b'\right)\frac{t}{\hbar}}\, \langle n|H_1|b,\, 0\rangle, \tag{175}$$

$$i\hbar\, \frac{\partial}{\partial t}\, c_b' = \sum_n c_n\, e^{i\left(E_b' - E_n^{(0)}\right)\frac{t}{\hbar}}\, \langle b,\, 0|H_1|n\rangle, \tag{176}$$

where
$$E_b' = E_b^{(0)} + \Lambda. \tag{177}$$

Following Weisskopf and Wigner, we try to solve the Eqs. (175) and (176) by assuming.

$$c_b'(t) = e^{-\gamma t/2} \tag{178}$$

with γ a complex constant, i.e., we assume that the probability of finding the atom in the excited state decreases exponentially with a lifetime $1/\gamma$.

The differential Eq. (175) becomes

$$i\hbar\, \frac{\partial}{\partial t}\, c_n = e^{i\left(E_n^{(0)} - E_b'\right)\frac{t}{\hbar} - \frac{\gamma t}{2}}\, \langle n|H_1|b,\, 0\rangle. \tag{179}$$

The solution of Eq. (179) with the condition all $c_n(t = 0) = 0$ is

$$c_n(t) = \frac{1 - e^{i\left(E_n^{(0)} - E'_b\right)\frac{t}{\hbar} - \frac{\gamma t}{2}}}{E_n^{(0)} - E'_b + i\gamma/2}. \tag{180}$$

Equation (176) becomes

$$i\hbar\,\frac{\partial}{\partial t}\,c'_b \equiv -\,i\hbar\,\frac{\gamma}{2}\,e^{-\gamma t/2}$$

$$= \sum_n \frac{e^{-i\left(E_n^{(0)} - E'_b\right)\frac{t}{\hbar}} - e^{-\frac{\gamma t}{2}}}{E_n^{(0)} - E'_b + i\gamma/2}\,\left|\langle n|H_1|b, 0\rangle\right|^2. \tag{181}$$

The discrete sum can be replaced by an integration.

We consider an integral

$$\int_{E_a}^{\infty} f\left(E_n^{(0)}\right)\,\frac{e^{-i\left(E_n^{(0)} - E'_b\right)\frac{t}{\hbar}} - e^{-\frac{\gamma t}{2}}}{E_n^{(0)} - E'_b + i\gamma/2}\,dE_n^{(0)} \tag{182}$$

where $f\left(E_n^{(0)}\right) = \left|\langle n|H_1|b, 0\rangle\right|^2$. In most atomic decays the damping is small or the lifetime is large compared with the frequency of the atom. γ can then be neglected in the integrand. We have

$$\int_{E_a}^{\infty} f\left(E_n^{(0)}\right)\,\frac{e^{-i\left(E_n^{(0)} - E'_b\right)\frac{t}{\hbar}}}{E_n^{(0)} - E'_b + i\gamma/2}\,dE_n^{(0)}$$

$$= -\,2\pi i e^{-\gamma t/2}\,f\left(E'_b - i\gamma\right)$$

$$= -\,2\pi i e^{-\gamma t/2}\,f\left(E'_b\right),$$

since γ is small compared with E'_b.

$$= -\,2\pi i e^{-\gamma t/2}\int_{E_a}^{\infty} f\left(E_n^{(0)}\right)\,\delta(E_n^{(0)} - E'_b)\,dE_n^{(0)} \tag{183}$$

Equation (181) becomes

$$-\,i\hbar\,\frac{\gamma}{2}\,e^{-\gamma t/2} = -\,e^{-\gamma t/2}\sum_n\left|\langle n|H_1|b, 0\rangle\right|^2$$

$$\times\left[2\pi i\delta(E_n^{(0)} - E'_b) + \frac{1}{E_n^{(0)} - E'_b + i\gamma}\right]. \tag{184}$$

Using

$$\lim_{\epsilon \to 0}\frac{1}{x \pm i\epsilon} = P\left(\frac{1}{x}\right) \mp i\pi\delta(x), \tag{185}$$

we get

$$-\,i\hbar\,\frac{\gamma}{2} = -\sum_n\left|\langle n|H_1|b, 0\rangle\right|^2$$

$$\times\left(P\,\frac{1}{E_n^{(0)} - E'_b} + i\pi\delta(E'_b - E_n^{(0)})\right). \tag{186}$$

Thus γ has a real and an imaginary part:

$$\frac{\hbar}{2} \, \mathrm{Re} \, \gamma = \sum_n \pi |\langle n|H_1|b, 0\rangle|^2 \, \delta(E_b' - E_n^{(0)}). \tag{187}$$

$$\frac{\hbar}{2} \, \mathrm{Im} \, \gamma = - \sum_n |\langle n|H_1|b, 0\rangle|^2 \, P \, \frac{1}{E_n^{(0)} - E_b'}. \tag{188}$$

The real part of γ (simply denoted by γ) is

$$\gamma = \frac{2\pi}{\hbar} \sum_n \delta(E_b' - E_n^{(0)}) \, |\langle n|H_1|b, 0\rangle|^2. \tag{189}$$

γ is the decay rate of the state $|b\rangle$, i.e., it is equal to the spontaneous transition probability per unit time for emission from the state $|b\rangle$.

The imaginary part of γ

$$\mathrm{Im} \, \gamma = \frac{2}{\hbar} \sum_n |\langle n|H_1|b, 0\rangle|^2 \, P \, \frac{1}{E_b' - E_n^{(0)}} \tag{190}$$

is the second-order perturbation shift $\Delta E_b^{(2)}$.

The total energy shift is

$$\Delta E_b = \Lambda + \Delta E_b^{(2)}. \tag{191}$$

The time-dependent amplitude is

$$c_b(t) = e^{-i\left(\Delta E_b - \frac{1}{2} i\Gamma_b\right) t/\hbar} \tag{192}$$

and the probability

$$|c_b(t)|^2 = \exp\left(-\frac{\Gamma_b}{\hbar}\right) t \tag{193}$$

to remain in the state b decreases exponentially with time. The lifetime τ is

$$\gamma = \frac{\Gamma_b}{\hbar} = \frac{1}{\tau}. \tag{194}$$

The intensity distribution of the emitted line is given by the probability function for the final state. We assume that the atom in an excited state $|b\rangle$ makes a transition to the final state $\left|a; \vec{k}, \lambda\right\rangle = |a\rangle \left|1_{\vec{k}, \lambda}\right\rangle$ when the atom is in the state $|0\rangle$ with a photon in the mode $\vec{k}, \lambda$. From Eq. (180), the transition amplitude is

$$c_{a; \vec{k}, \lambda}(t) = \frac{1 - \exp i\left(E_a^{(0)} + \omega_k - E_b^{(0)} - \Delta E_b\right)\dfrac{t}{\hbar} \, e^{-\frac{\Gamma_b}{2\hbar}t} \times \left\langle a; \vec{k}, \lambda|H_1|b, 0\right\rangle}{E_a^{(0)} + \omega_k - E_b^{(0)} - \Delta E_b + i\Gamma_b/2\hbar}. \tag{195}$$

After a time $t \gg \dfrac{1}{\gamma}$, or, $\Gamma_b t/\hbar \gg 1$, when the atom has certainly made a transition, we have

$$c_{a;\,\vec{k},\,\lambda}(\infty) = \frac{\left\langle a;\vec{k},\lambda\middle|H_1\middle|b,0\right\rangle}{E_a^{(0)} + \omega_k - E_b^{(0)} - \Delta E_b + i\Gamma_b/2\hbar}. \tag{196}$$

The probability that a photon has been emitted is

$$\left|c_{a;\,\vec{k},\,\lambda}(\infty)\right|^2 = \frac{\left|\left\langle a;\vec{k},\lambda\middle|H_1\middle|b,0\right\rangle\right|^2}{\hbar^2}\,\frac{1}{(\omega_{ab} - \omega_k)^2 + \dfrac{1}{4}\,\Gamma_b^2/\hbar^2}, \tag{197}$$

where

$$\hbar\omega_{ab} = E_b^{(0)} + \Delta E_b - E_a^{(0)}. \tag{198}$$

Summing over the photon states, we get the intensity distribution as

$$I(\omega_k)d\omega_k = \hbar\omega_k\,\rho_k\,dk\,\sum_\lambda\int\left|c_{a;\,\vec{k},\,\lambda}\right|^2 d\Omega$$

$$= \frac{\Gamma_b}{2\pi\hbar}\,\frac{\hbar\omega_k\,d\omega_k}{(\omega_k - \omega_{ab})^2 + \Gamma_b^2/4\hbar^2}. \tag{199}$$

The total intensity is $\hbar\omega_k = I_0$. This formula is essentially identical with the classical formula for a damped oscillator with γ now representing the transition probability per unit time instead of $2e^2\,\omega_{ab}^2/3mc^3$.

The spectrum is a "Lorentzian" one and the maximum intensity lies at the frequency $\omega_{ab} = (E_b - E_a)/\hbar$ (corrected by a small line shift). For $\omega_k - \omega_{ab} = \pm\dfrac{\gamma}{2}$ $= \pm\dfrac{\Gamma_b}{2\hbar} = \pm\dfrac{1}{2\tau}$ the intensity is half of the maximum value. The line width at half maximum is equal to the total transition probability per unit time. The width Γ_b is called the natural width of the emitted line.

If the atom has several atomic levels $a_1, a_2, \ldots.$ in the order of their energies, then we assign to each level a_i, say, a certain width given by the sum of all transition probabilities from a_i to all lower levels:

$$\Delta E_a/\hbar \equiv \gamma_i = \sum_{j<i}\Gamma_{a_j a_i} \tag{200}$$

where $\Gamma_{a_j a_i}$ represents the probability (per unit time) for the transition $a_i \to a_j$;

The breadth of a certain line $a_i \to a_k$, say, is the sum of the breadths of the two levels a_i and a_k:

$$\gamma_{ki} = \gamma_i + \gamma_k. \tag{201}$$

The shape of the absorption line must be the same as the emission line. This follows from general equilibrium considerations (Kirchhoff's law).

In general the natural widths of spectral lines are very small. As an example, let us consider the $2p$ level of atomic hydrogen. The life time is $\tau = 1.6 \times 10^{-9}$ s, so that the width is

$$\Gamma = \frac{\hbar}{\tau} \approx 4.1 \times 10^{-7} \text{ eV}.$$

The center is at ω_k where $\hbar\omega_k = 13.6\left(1 - \frac{1}{2^2}\right) = 10.4$ eV.

Besides the damping due to the radiation itself, there are other causes which broaden the line. Among the most important are the collisional (or pressure) broadening effect and the Doppler effect.

15.9   THE PHOTOEFFECT

We now consider the transition from an atomic bound state to a continuum state. We shall mainly be concerned with the photoeffect, i.e., the absorption of radiation by an atom in a bound state accompanied by the ejection of an electron into a "free" state, (a state of positive energy in the continuum) ("bound-free" transitions).

In this photoionization process, a photon of energy $\hbar\omega$ and momentum $\hbar\vec{k}$ is incident on a (one-electron) atom in the ground state. If $\hbar\omega > I$, the ionization energy of the atom, an absorption process takes place with the ejection of an electron into a continuum state of momentum $\hbar\vec{k}$. The kinetic energy T of the electron after leaving the atom is given by Einstein's equation.

$$T = \hbar\omega - I. \tag{202}$$

The photoelectric effect plays an important part in the absorption of X-rays and γ-rays in matter. We shall carry out the calculations in the case of a hydrogenic atom. We assume that the energy of the electron in the continuum is small compared with mc^2 so that relativistic corrections are not important, i.e., the photon energies are well below mc^2:

$$\hbar\omega \ll mc^2. \tag{203}$$

When the photon energy is large compared with the ionization energy, then we can neglect the effect of the Coulomb field acting on the ejected electron. For an atom with nuclear charge Z, this condition is

$$T = \frac{p^2}{2\mu} \gg I$$

$$= -E_0 = \frac{Z^2\mu\alpha^2}{2}$$

or, $\qquad\qquad \xi \equiv \frac{Ze^2}{\hbar v} \ll 1. \tag{204}$

This condition is identical with the condition for the validity of the Born approximation. In this high energy regime, we can replace the wave function of the electron in the continuum in the matrix elements by a plane wave.

In the neighborhood of the absorption edge (threshold) when $\hbar\omega \sim I$ the velocity of the ejected electron is small so that Coulomb interaction must be taken into account. In this case the photon wavelength is large compared with the atomic dimensions and the electric dipole approximation is valid.

Electron spin is neglected in the nonrelativistic regime.

Now, we calculate the cross section for the photoelectric effect in the high energy regime in the non-relativistic case. The results are valid at large distances from absorption edge.

The initial state

$$|i\rangle = \left| K;\, 1_{\vec{k},\,\lambda} \right\rangle = |K\rangle \left| 1_{\vec{k},\,\lambda} \right\rangle \tag{205}$$

represents the electron in the K-shell and a photon of momentum $\hbar\vec{k}$ and helicity λ. The final state

$$|f\rangle = |\vec{p};\, 0\rangle = |\vec{p}\,\rangle\, |0\rangle \tag{206}$$

represents the electromagnetic vacuum (no photon) state and a one-electron state of momentum $\vec{p}$. The matrix element for the process is

$$\langle f|H_1|i\rangle = \left\langle \vec{p};\, 0 \left| H_1 \right| K;\, 1_{\vec{k},\,\lambda} \right\rangle$$

$$= \frac{e}{mc}\left(\frac{2\pi\hbar c^2}{\omega V}\right)\left(\vec{p}.\hat{\vec{\epsilon}}_{\vec{k},\,\lambda}\right)\left\langle \vec{p}\left| e^{i\vec{k}.\vec{r}}\right| K\right\rangle, \tag{207}$$

where we have used $\left\langle 0 \left| a_{\vec{k},\,\lambda},\, a^{\dagger}\vec{k}\lambda \right| 0\right\rangle = \delta_{\vec{k}\vec{k}'}\,\delta_{\lambda\lambda'}$.

In the high energy regime the photon wave length is not large compared to atomic dimensions ($|kr| \gg 1$) and the phase factor $e^{i\vec{k}.\vec{r}}$ must be retained.

Since $|\vec{p}\,\rangle$ is a momentum eigenstate of the electron we have

$$\left\langle \vec{p}\left| e^{i\vec{k}.\vec{r}}\right| K\right\rangle = \frac{1}{\sqrt{V}}\int d^3r\, e^{i\left(\vec{k}-\vec{p}/\hbar\right).\vec{r}}\,\psi_K(\vec{r}). \tag{208}$$

Here $\psi_K(\vec{r})$ is the wave function of the electron in the K-shell:

$$\psi_K(\vec{r}) = \frac{1}{\sqrt{\pi a^3}}\, e^{-r/a} \tag{209}$$

where $a = \dfrac{a_0}{Z}$ with $a_0 = \dfrac{\hbar^2}{me^2}$, the Bohr radius. Also, the wave function of the

electron in the continuous spectrum with the momentum $\vec{p}$ is

$$\psi_e(\vec{r}) = \langle \vec{p} | \vec{r} \rangle$$

$$= \frac{e^{i\vec{p}\cdot\vec{r}/\hbar}}{\sqrt{V}}. \tag{210}$$

Expressing

$$\vec{q} = \vec{k} - \vec{p}/\hbar \tag{211}$$

as the momentum transferred from photon to electron, the Fourier transform of $\psi_K(\vec{r})$ is

$$\phi_K(\vec{q}) = \frac{1}{\sqrt{\pi a^3}} \int d^3r \, e^{-r/a} \, e^{-i\vec{q}\cdot\vec{r}}$$

$$= \frac{1}{\sqrt{\pi a^3}} \cdot \frac{8\pi\left(\frac{1}{a}\right)}{\left[\left(\frac{1}{a}\right)^2 + q^2\right]^2}$$

$$= \frac{8\sqrt{\pi}\, a^{3/2}}{(1 + q^2 a^2)^2}. \tag{212}$$

The density of final electron states is

$$\rho_f = \frac{V d^3 p}{(2\pi\hbar)^3 \, dE_p}$$

$$= \frac{V}{(2\pi\hbar)^3} \frac{p^2 \, dp d\Omega}{d\left(\frac{p^2}{2m}\right)}$$

$$= \frac{V}{(2\pi\hbar)^3} \cdot mp d\Omega. \tag{213}$$

Since there is one photon in the volume V, the incident photon flux is c/V.

The differential cross section for photoelectron production (the electron being ejected into an element of solid angle $d\Omega$) is

$$d\sigma_\lambda = \frac{2\pi}{\hbar} |\langle f|H_1|i\rangle|^2 \, \rho_f \cdot \frac{V}{c}$$

$$= \frac{2\pi}{\hbar} \frac{mp}{(2\pi\hbar)^3} \left(\frac{e}{mc}\right)^2 \frac{2\pi\hbar c^2}{\omega}$$

$$\cdot \frac{1}{\pi}\left(\frac{1}{a}\right)^3 \left(\hat{e}_{k\lambda} \cdot \vec{p}\right)^2 \cdot \frac{1}{c}$$

$$\times \frac{64\pi^2 \left(\frac{1}{a}\right)^2}{\left[\left(\frac{1}{a}\right)^2 + q^2\right]^4} \, d\Omega$$

Hence,

$$\frac{d\sigma_\lambda}{d\Omega} = 32Z^5 \, a_0^2 \left(\frac{pc}{\hbar\omega}\right) \frac{1}{m^2 c^2} \left(\vec{p} \cdot \hat{\epsilon}_{k\lambda}\right)^2$$

$$\times \frac{1}{\left(Z^2 + \vec{q}^2 \, a_0^2\right)^4}. \tag{214}$$

From energy conservation,

$$\hbar\omega = E_0 + \frac{\vec{p}^2}{2m} \tag{215}$$

For energies above the threshold, we can neglect E_0 and write

$$\hbar\omega \simeq \frac{p^2}{2m}. \tag{216}$$

The angular distribution of the ejected photo-electrons is given by the factor

$$\frac{\left|\vec{p} \cdot \hat{\epsilon}_{k\lambda}\right|^2}{\left[1 + \left(\vec{k} - \dfrac{\vec{p}}{\hbar}\right)^2 a^2\right]^4}. \tag{217}$$

Now,

$$\left(\frac{pc}{\hbar\omega}\right) \frac{1}{(mc)^2} \left(\vec{p} \cdot \hat{\epsilon}_{k\lambda}\right)^2 \simeq \frac{2p}{mc} \left(\vec{p} \cdot \hat{\epsilon}_{k\lambda}\right)^2 \tag{218}$$

using Eq. (216)

$$\vec{q}^2 = \frac{\left(\hbar\vec{k} - \vec{p}\right)^2}{\hbar^2} = \frac{1}{\hbar^2}\left[\left(\frac{\hbar\omega}{c}\right)^2 - 2\frac{\hbar\omega}{c}\hat{k} \cdot \vec{p} + p^2\right]$$

$$\simeq \frac{1}{\hbar^2}\left[p^2 - \left(\frac{p^3}{mc}\right)\hat{k} \cdot \hat{p}\right],$$

since $p \ll mc^2$ for non relativistic electrons.

$$\simeq \frac{p^2}{\hbar^2}\left(1 - \beta\hat{k} \cdot \hat{p}\right). \tag{219}$$

where

$$\beta = \frac{v}{c}.$$

Thus,

$$\frac{d\sigma_\lambda}{d\Omega} = \frac{64Z^5 \, a_0^2 \, \alpha^8 \left(\dfrac{p}{mc}\right) \left(\hat{p} \cdot \hat{\epsilon}_{k\lambda}\right)^2}{\left[\alpha^2 Z^2 + \dfrac{p^2}{m^2 c^2}\left(1 - \beta\hat{k} \cdot \hat{p}\right)\right]^4}. \tag{220}$$

Denoting by θ the angle between $\hat{k}$ and $\hat{p}$ and by ϕ the angle between the planes $\left(\vec{p},\, \vec{k}\right)$ and $\left(\hat{e},\, \vec{k}\right)$, we have

$$\frac{d\sigma_\lambda}{d\Omega} = \frac{64 Z^5\, a_0^2\, \alpha^8 \left(\dfrac{p}{mc}\right) \sin^2\theta\, \cos^2\phi}{\left[\alpha^2\, Z^2 + \dfrac{p^2}{m^2 c^2}\, (1 - \beta \cos\theta)\right]^4}$$

$$= \frac{64\sqrt{2} Z^5\, a_0^2\, \alpha^8 \left(\dfrac{E}{mc^2}\right)^{1/2} \sin^2\theta\, \cos^2\phi}{\left[\alpha^2\, Z^2 + \dfrac{2E}{mc^2}\, (1 - \beta \cos\theta)\right]^4}. \tag{221}$$

Almost all of the photoelectrons are emitted in the direction of polarization of the incident photon ($\theta = \pi/2$, $\phi = 0$). In the direction of $\vec{k}$ no photoelectrons are emitted ($\theta = 0$). The maximum of the differential cross section of photoelectrons is in the direction of polarization of the photon. The denominator in Eq. (221) displaces the maximum slightly in the forward direction, which becomes more marked with increasing velocities of the electron. For unpolarized photon beam, we replace $\cos^2\phi$ by its average value $\dfrac{1}{2}$.

Thus the differential cross section for the photoelectric effect for an unpolarized photon beam incident on a hydrogenic atom target at high energies (non relativistic) is

$$\frac{d\sigma_\lambda}{d\Omega} = \frac{32\sqrt{2} Z^5\, a_0^2\, \alpha^8 \left(\dfrac{E}{mc^2}\right)^{1/2} \sin^2\theta}{\left[\alpha^2\, Z^2 + \dfrac{2E}{mc^2}\, (1 - \beta \cos\theta)\right]^4} \tag{222}$$

If, $Z\alpha \ll 1$, then we get to first order in β,

$$\frac{d\sigma}{d\Omega} = 2\sqrt{2} Z^5\, \alpha^8\, a_0^2 \left(\frac{E}{mc^2}\right)^{-7/2} (1 + 4\beta \cos\theta)\, \sin^2\theta. \tag{223}$$

The total cross section for the photoeffect of the K-shell is

$$\sigma_K = \sigma_T\, \frac{Z^5}{137^4} \cdot 4\sqrt{2} \left(\frac{\mu c^2}{k}\right)^{7/2}$$

$$= \sigma_T \cdot 64 \cdot \frac{137^3}{Z^2} \left(\frac{I}{k}\right)^{7/2}, \tag{224}$$

where $\sigma_T = \dfrac{8\pi r_0^2}{3}$ is the cross section for the Thomson scattering, $r_0 = \dfrac{e^2}{mc^2}$ being the classical electron radius. Atomic K-shell contains 2 electrons. The absorption coefficient τ_K per cm. for the K-shell is

$$\tau_K = N\sigma_T \frac{Z^5}{137^4} \cdot 4\sqrt{2} \left(\frac{\mu c^2}{k}\right)^{7/2}, \tag{225}$$

where N is the number of atoms per cm^3. In the neighborhood of threshold (absorption edge) the electric dipole approximation is valid. Here the exact wave functions of the continuous spectrum in the Coulomb field must be used instead of plane waves. The total cross section in this case is obtained by multiplying σ_K in Eq. (224) by a factor

$$f(\xi) = 2\pi \sqrt{\left(\frac{I}{k}\right)} \frac{e^{-4\xi \cot^{-1} \xi}}{1 - e^{-2\pi\xi}}, \tag{226}$$

where

$$\xi = \sqrt{\left(\frac{I}{k - I}\right)} = \frac{Ze^2}{\hbar v}. \tag{227}$$

Thus, the total cross section for photoeffect for K-shell in the neighborhood of absorption edge is

$$\sigma_K^{(abs.edge)} = 128\pi \frac{137^3}{Z^2} \sigma_T \left(\frac{I}{k}\right)^4 \frac{e^{-4\xi \cot^{-1} \xi}}{1 - e^{-2\pi\xi}}. \tag{228}$$

Consult Heitler (1954) and Bethe and Salpeter (1957) for photoeffect in the relativistic region.

15.10 RAYLEIGH SCATTERING, THOMSON SCATTERING, AND THE RAMAN EFFECT. RESONANCE FLUORESCENCE

In this section we study the scattering of photons by bound electrons, that is to say, by atoms. In a scattering process there is no net change in the number of photons because initially there is an incident photon and finally there is an outgoing photon. The linear $\left(\vec{A}.\vec{p}\right)$ term (H_1) in the interaction Hamiltonian Eq. (35) will contribute in second-order to a scattering process since a photon must be absorbed and another photon emitted. The quadratic term $\vec{A}.\vec{A}$ (H_2) will contribute in the first order.

We assume that initially the atom is in state $|a\rangle$ and the incident photon is in the mode $\left(\vec{k}, \hat{e}_{k}^{\lambda}\right)$. After scattering, the atom is in the state $|b\rangle$ and the outgoing photon is in the mode $\left(\vec{k}', \hat{e}_{k'}^{\lambda'}\right)$. The initial state of the system is

$$|i\rangle = \left|a; \vec{k}, \hat{e}^{\lambda}\right\rangle = a_{k\lambda}^{\dagger} |a, 0\rangle \tag{229}$$

and the final state is

$$|f\rangle = \left|b; \vec{k}', \hat{e}^{\lambda'}\right\rangle = a_{k'\lambda'}^{\dagger} |b, 0\rangle \tag{230}$$

with energies

$$E = E_a + \hbar \, |\vec{k}| \, c = E_b + \hbar \, |\vec{k}'| \, c. \qquad (231)$$

The matrix element of the quadratic term H_2 between these states contains four terms of the type $aa^\dagger$, $a^\dagger a$, aa and $a^\dagger a^\dagger$ of which the first two give nonzero but equal contributions. We have

$$\langle f|H_2|i\rangle = \left\langle b;\, \vec{k}',\, \hat{e}^{\lambda'} \middle| H_2 \middle| a;\, \vec{k},\, \hat{e}^\lambda \right\rangle$$

$$= \frac{e^2}{2mc^2} \left\langle b;\, \vec{k}',\, \hat{e}^{\lambda'} \middle| \vec{A}(\vec{r}) \cdot \vec{A}(\vec{r}) \middle| a;\, \vec{k},\, \hat{e}^\lambda \right\rangle$$

$$= \frac{e^2}{2mc^2} \cdot \frac{2\pi\hbar c^2}{V\sqrt{\omega\omega'}} \left\langle b;\, \vec{k}',\, \hat{e}^{\lambda'} \middle| \left(a_{\vec{k},\lambda}\, a^\dagger_{\vec{k}'}\, \hat{e}^\lambda_{\vec{k}} \cdot \hat{e}^{\lambda'*}_{\vec{k}'} + a^\dagger_{\vec{k}',\lambda'}\, a_{\vec{k}\lambda}\, \hat{e}^{\lambda'*}_{\vec{k}'} \cdot \hat{e}^\lambda_{\vec{k}} \right) \right.$$

$$\left. \times\, e^{i\,(\vec{k}-\vec{k}').\vec{r}} \middle| a;\, \vec{k},\, \hat{e}^\lambda \right\rangle.$$

$$= \frac{e^2}{2mc^2} \cdot \frac{2\pi\hbar c^2}{V\sqrt{\omega\omega'}} \cdot 2 \left(\hat{e}^{\lambda'*}_{\vec{k}'} \cdot \hat{e}^\lambda_{\vec{k}} \right) \langle b|a\rangle,$$

$$= \frac{e^2}{mc^2} \cdot \frac{2\pi\hbar c^2}{V\sqrt{\omega\omega'}} \left(\hat{e}^{\lambda'*}_{\vec{k}'} \cdot \hat{e}^\lambda_{\vec{k}} \right) \delta_{ba}, \qquad (232)$$

in the dipole approximation.

The $\vec{A}.\vec{p}\,(\sim H_1)$ term can only contribute to the matrix element for the process in second order. H_1 can only cause transitions involving two photons through intermediate atomic states I which are different from a and b. There are two kinds of such intermediate states. In the first type the atom is in the intermediate state and no photon is present. In the second type the atom is in the intermediate state and both the incident photon and the outgoing photon are present. The energy difference of the initial states and the intermediate states in the two cases are

$$E_a + \hbar\omega_a - E_I, \quad E_a - E_I - \hbar\omega'. \qquad (233)$$

We draw a space-time diagram (Feynman diagram) below in which a solid line represents the atom, and a wavy line represents a photon.

Fig 15.1(a) represents a type 1 process, while Fig. 15.1(b) represents a type 2 process. Fig. 15.1(c) represents the lowest order $\vec{A}.\vec{A}$ interaction.

The term of second order in H_1 is

$$\sum_I \frac{\langle f|H_1|I\rangle\langle I|H_1|f\rangle}{E_i - E_I}. \qquad (234)$$

The amplitude for type 1 process (Fig. 15.1(a)) in the long wavelength limit (dipole approximation) is

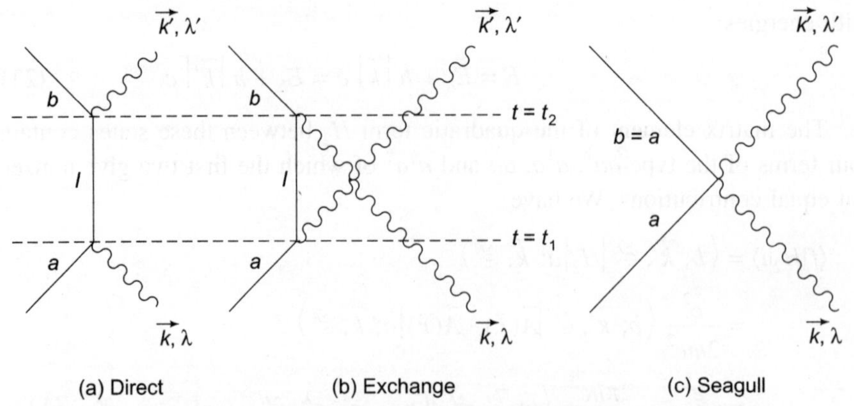

Fig. 15.1 Space-time diagrams for the scattering of light

$$\left(-\frac{e}{mc}\right)^2 \frac{2\pi\hbar c^2}{V\cdot\sqrt{\omega\omega'}} \sum_I \frac{\left\langle b\left|\vec{p}.\hat{\vec{\epsilon}}^{\,*}_{k',\lambda'}\right|I\right\rangle\left\langle I\left|\vec{p}.\hat{\vec{\epsilon}}_{k,\lambda}\right|a\right\rangle}{E_a + \hbar\omega - E_I}. \tag{235}$$

The amplitude for type 2 process (Fig. 15.1(b)) in the long wavelength limit is

$$\left(-\frac{e}{mc}\right)^2 \frac{2\pi\hbar c^2}{V\cdot\sqrt{\omega\omega'}} \sum_I \frac{\left\langle b\left|\hat{\vec{\epsilon}}_{k',\lambda'}.\vec{p}\right|I\right\rangle\left\langle I\left|\vec{p}.\hat{\vec{\epsilon}}^{\,*}_{k',\lambda'}\right|a\right\rangle}{E_a - E_I - \hbar\omega'}. \tag{236}$$

The amplitudes for the two processes add coherently. The total second order transition amplitude is

$$\left(\frac{e}{mc}\right)^2 \frac{2\pi\hbar c^2}{V\cdot\sqrt{\omega\omega'}} \sum_I \left(\frac{\left(\vec{p}.\hat{\vec{\epsilon}}^{\,*}_{k',\lambda'}\right)_{bI}\left(\vec{p}.\hat{\vec{\epsilon}}_{k,\lambda}\right)_{Ia}}{E_a + \hbar\omega - E_I} \right.$$
$$\left. + \frac{\left(\vec{p}.\hat{\vec{\epsilon}}_{k,\lambda}\right)_{bI}\left(\vec{p}.\hat{\vec{\epsilon}}^{\,*}_{k',\lambda'}\right)_{Ia}}{E_a - E_I - \hbar\omega'} \right), \tag{237}$$

where $\left(\vec{p}.\hat{\vec{\epsilon}}_{k,\lambda}\right)_{bI}$ is the matrix element of $\vec{p}.\hat{\vec{\epsilon}}_{k,\lambda}$ between b and I. To this we must add the amplitude for the quadratic interaction (Fig. 15.1(c)) in Eq. (232).

We have the matrix element

$$M = \left(\frac{e^2}{mc^2}\right)\frac{2\pi\hbar c^2}{V\sqrt{\omega\omega'}}\left[\hat{\vec{\epsilon}}^{\,*}_{k',\lambda'}\cdot\hat{\vec{\epsilon}}_{k,\lambda}\,\delta_{ab} + \frac{1}{m}\right.$$

$$\times \sum_I \frac{\left(\vec{p}.\hat{\vec{\epsilon}}^{\,*}_{k',\lambda'}\right)_{bI}\left(\vec{p}.\hat{\vec{\epsilon}}_{k,\lambda}\right)_{Ia}}{E_a + \hbar\omega - E_I}$$

$$+ \frac{\left(\vec{p}.\hat{\epsilon}_{\vec{k},\lambda}\right)_{bI} \left(\vec{p}.\hat{\epsilon}^*_{\vec{k}',\lambda'}\right)_{Ia}}{E_a - E_I - \hbar\omega'} \Bigg) \Bigg]. \tag{238}$$

The transition probability per unit time is

$$w = \frac{2\pi}{\hbar} |M|^2 \rho_f, \tag{239}$$

where

$$\rho_f = \frac{V}{(2\pi c)^3} \frac{\omega'^2}{\hbar}. \tag{240}$$

Since the incident flux is c/V, the differential cross section for the scattering of $\left(\vec{k},\lambda\right)$ photons into an element of solid angle $d\Omega$ in the mode $\left(\vec{k}',\lambda'\right)$, the atom going from $|a\rangle$ to $|b\rangle$ is

$$\frac{d\sigma}{d\Omega} = r_0^2 \left(\frac{\omega'}{\omega}\right) \Bigg| \Bigg(\hat{\epsilon}^*_{\vec{k}',\lambda'} \, \epsilon_{\vec{k},\lambda} \, \delta_{ab}$$

$$+ \frac{1}{m} \times \sum_I \frac{\left(\vec{p}.\hat{\epsilon}^*_{\vec{k}',\lambda'}\right)_{bI} \left(\vec{p}.\hat{\epsilon}'_{\vec{k},\lambda}\right)_{Ia}}{E_a + \hbar\omega - E_I}$$

$$+ \frac{\left(\vec{p}.\hat{\epsilon}_{\vec{k},\lambda}\right)_{bI} \left(\vec{p}.\hat{\epsilon}^*_{\vec{k}',\lambda'}\right)_{Ia}}{E_a - E_I - \hbar\omega'} \Bigg) \Bigg|^2, \tag{241}$$

where r_0 is the classical electron radius. Equation (241) is known as the Kramers-Heisenberg dispersion formula.

There are several special cases of Eq. (241) worth noting. Elastic or Rayleigh scattering of light occurs when $a = b$ and so $\hbar\omega = \hbar\omega'$. Let us write

$$\langle a \,|\hat{\epsilon}^{(\alpha)}.\hat{\epsilon}^{(\alpha')}|\, a\rangle = \frac{1}{i\hbar} \left\langle a \left| \left[\vec{r}.\hat{\epsilon}^{(\alpha)}, \vec{p}.\hat{\epsilon}^{(\alpha')}\right] \right| a\right\rangle$$

$$= \frac{1}{i\hbar} \sum_I \left[\langle a \,|\vec{r}.\hat{\epsilon}^{(\alpha)}|\, I\rangle\langle I \,|\vec{p}.\hat{\epsilon}^{(\alpha')}|\, a\rangle \right.$$

$$\left. - \langle a \,|\vec{p}.\hat{\epsilon}^{(\alpha')}|\, I\rangle\langle I \,|\vec{r}.\hat{\epsilon}^{(\alpha)}|\, a\rangle \right]$$

$$= \frac{1}{m\hbar} \sum_I \frac{2}{\omega_{Ia}} \left(\vec{p}.\hat{\epsilon}^{(\alpha)}\right)_{aI} \left(\vec{p}.\hat{\epsilon}^{(\alpha')}\right)_{Ia}, \tag{242}$$

using the completeness of the intermediate states I and using

$$\langle b|\vec{p}|a\rangle = \frac{im}{\hbar} (E_b - E_a)\langle b|\vec{r}|a\rangle$$

where

$$\hbar\omega_{Ia} = E_I - E_a.$$

Obviously, Eq. (242) is valid under the exchange of α and α'. We get

$$\delta_{aa}\,\hat{e}^{(\alpha)}.\hat{e}^{(\alpha')} - \frac{1}{m\hbar}\sum_{I}\left[\frac{\left(\vec{p}.\hat{e}^{(\alpha')}\right)_{aI}\left(\vec{p}.\hat{e}^{(\alpha)}\right)_{Ia}}{\omega_{Ia}-\omega}\right.$$

$$\left.+\frac{\left(\vec{p}.\hat{e}^{(\alpha)}\right)_{aI}\left(\vec{p}.\hat{e}^{(\alpha')}\right)_{Ia}}{\omega_{Ia}+\omega}\right]$$

$$=-\frac{1}{m\hbar}\sum_{I}\left[\frac{\omega\left(\vec{p}.\hat{e}^{(\alpha')}\right)_{aI}\left(\vec{p}.\hat{e}^{(\alpha)}\right)_{Ia}}{\omega_{Ia}(\omega_{Ia}-\omega)}\right.$$

$$\left.-\frac{\omega\left(\vec{p}.\hat{e}^{(\alpha)}\right)_{aI}\left(\vec{p}.\hat{e}^{(\alpha')}\right)_{Ia}}{\omega_{Ia}(\omega_{Ia}+\omega)}\right]. \tag{243}$$

As $\omega \to 0$, the cross section vanishes. For small ω, we expand the denominators as

$$\frac{1}{\omega_{Ia} \mp \omega} \approx \frac{1 \pm \dfrac{\omega}{\omega_{Ia}}}{\omega_{Ia}},$$

and we get

$$\sum_{I}\frac{1}{\omega_{Ia}^{2}}\left[\left(\vec{p}.\hat{e}^{(\alpha')}\right)_{aI}\left(\vec{p}.\hat{e}^{(\alpha)}\right)_{Ia}\right.$$

$$\left.-\left(\vec{p}.\hat{e}^{(\alpha)}\right)_{aI}\left(\vec{p}.\hat{e}^{(\alpha')}\right)_{Ia}\right]$$

$$= m^{2}\left\langle a\left|\left[\vec{r}.\hat{e}^{(\alpha')},\,\vec{r}.\hat{e}^{(\alpha)}\right]\right|a\right\rangle$$

$$= 0. \tag{244}$$

Hence, the Rayleigh or coherent scattering cross section for $\omega \ll \omega_{Ia}$ is

$$\left(\frac{d\sigma}{d\Omega}\right)_{R} = \left(\frac{r_{0}}{m\hbar}\right)^{2}\omega^{4}\left|\sum_{I}\left(\vec{p}.\hat{e}^{(\alpha')}\right)_{aI}\left(\vec{p}.\hat{e}^{(\alpha)}\right)_{Ia}\right.$$

$$\left.+\left(\vec{p}.\hat{e}^{(\alpha)}\right)_{aI}\left(\vec{p}.\hat{e}^{(\alpha')}\right)_{Ia}\right|^{2}$$

$$= \left(\frac{r_{0}m}{\hbar}\right)^{2}\omega^{4}\sum_{I}\frac{1}{\omega_{Ia}^{6}}\left|\left[\left(\vec{r}.\hat{e}^{(\alpha')}\right)_{aI}\left(\vec{r}.\hat{e}^{(\alpha)}\right)_{Ia}\right.\right.$$

$$\left.\left.+\left(\vec{r}.\hat{e}^{(\alpha)}\right)_{aI}\left(\vec{r}.\hat{e}^{(\alpha')}\right)_{Ia}\right]\right|^{2}. \tag{245}$$

Thus, we arrive at Rayleigh's law: the scattering cross section at low frequency varies as the fourth power of the frequency. This explains the blue color of the sky and the red color of the sunset (see Jackson (1975)).

Let us now consider the other extreme when the photon energy is much larger compared with the atomic binding energy. In this case, the direct and exchange contributions can be neglected, and the seagull term dominates. The form $\delta_{ab}\,\hat{\epsilon}^{(\alpha)}.\hat{\epsilon}^{(\alpha')}$ is insensitive to the binding of the electron. The problem now reduces to scattering from a free electron. The cross section becomes

$$\left(\frac{d\sigma}{d\Omega}\right)_T = r_0^2\,|\hat{\epsilon}^{(\alpha)}.\hat{\epsilon}^{(\alpha')}|^2.$$
(246)

This is the Thomson cross section and is the cross section for the scattering of light by a free electron. It is the nonrelativistic limit of the cross section for Compton scattering. If the initial photon is not polarized and the final photon polarization is not detected, then

$$\frac{d\sigma}{d\Omega} = \frac{r_0}{2}\,(1 + \cos^2\theta)$$
(247)

The total cross section for Thomson scattering is

$$\sigma_T = \frac{8\pi}{3}\,r_0^2.$$
(248)

This result is valid for photon energies much larger than the atomic binding energy, but the rest energy $mc^2 \gg \hbar\omega$. If $mc^2 \le \hbar\omega$, then we must use relativistic treatment leading to the Klein-Nishina formula.

Next, we consider the inelastic scattering of light by an atom in which the initial and final states of the atom are different $a \ne b$ and $\omega \ne \omega'$. The contribution of the Thomson term vanishes. If the initial atomic state a is the ground state, then $\omega' < \omega$, since $E_b > E_a$. The observed spectral line is called a Stokes' line.

If the initial atomic state a is an excited state, then the final state b may have higher or lower energy. In the latter case $\omega' > \omega$ and the observed spectral line is called an anti-Stokes' line.

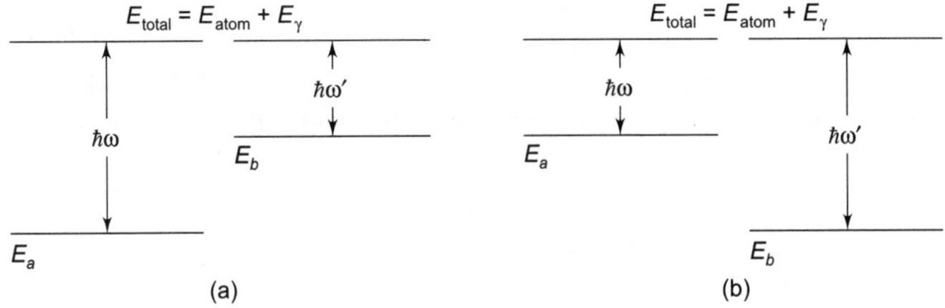

Fig. 15.2   (a) Stokes' line; (b) Anti-Stokes' line

The inelastic scattering of light is known as Raman scattering or the Raman effect. Raman in 1928 observed the shift in the frequency of a scattered radiation in liquid solutions, an effect suggested by Smekal. The effect was observed experimentally in solids by Landsberg and Mandelstamm.

The theory of the scattering of light by an atom as developed by Kramers and Heisenberg breaks down when the incident photon energy $\hbar\omega$ becomes equal to the excitation energy $E_I - E_a$ for some state I of the atom. The contribution of the direct term becomes infinite and so the cross section is then infinite. Clearly, this is unphysical but in the neighborhood of $E_I - E_a = \hbar\omega$ the intensity of the scattered radiation becomes very large compared with the ordinary scattering. This phenomenon is known as resonant scattering of light or resonance fluorescence.

The reason for this breakdown is that we did not take into account the spontaneous emission of photons from the intermediate states I. We must include a damping force due to the reaction of the emitted light on the atom. The "radiation damping" can be introduced in the Kramers-Heisenberg formula as in the case of the theory of line breadth. In the neighborhood of a nondegenerate resonance state the resonance fluorescence cross section is

$$\frac{d\sigma}{d\Omega} = r_0^2 \left(\frac{\omega'}{\omega}\right) \frac{1}{m^2} \frac{\left|\left(\vec{p}.\hat{\epsilon}^{(\alpha')}\right)_{bR}\right|^2 \left|\left(\vec{p}.\hat{\epsilon}^{(\alpha)}\right)_{Ra}\right|^2}{(E_R - E_a - \hbar\omega)^2 + \frac{1}{4}\,\Gamma_R^2} \tag{249}$$

where Γ_R is the width. This is a single level resonance formula.

If there is a degeneracy of states or if two or more excited states are very close in energy then these states participate in the resonance fluorescence process. When the excited states are coupled by external fields (e.g. a magnetic field), the angular distribution of the intensity of scattered radiation depends on the energy separation and the coupling of the excited states. In 1924 Hanle showed that in a light scattering experiment with Hg vapor the depolarization of the scattered light by the magnetic field was proportional to the atomic lifetime of the excited state. In 1959 Colegrove, Franken, Lewis and Sands (1959) performed the level-crossing experiment as a variant of the "Hanle effect". They measured the intensity of the polarized scattered radiation as a function of the applied static magnetic field. They observed a resonance when the magnetic field produced a degeneracy of the frequencies. See Fontana (1982) for a discussion of the level-crossing and anticrossing spectroscopy, optical double resonance and dynamic Stark effect.

15.11 THE CASIMIR EFFECT

In the vaccum, the electromagnetic field does not vanish but rather fluctuates. We now discuss an observable consequence of the vacuum fluctuations: the Casimir effect. The original observation of Casimir (1948) is that, in the vacuum there is a small force proportional to $\hbar$ exerted on conductors.

The introduction of conducting surfaces in the vacuum will impose macroscopic boundary conditions, there by changing the ground state energy of the electromagnetic field. The frequencies ω_i of the normal modes of the field differ in the two cases – the no conductor vacuum case (unbounded region) and the conductor present case. The zero point energies $\frac{1}{2} \Sigma_i \hbar\omega_i$ in both cases are infinite and have no direct physical meaning. The difference between these is a physically meaningful quantity that depends on the details of the boundary conditions due to the geometry of the conductor surfaces. This gives rise to pressure on the surfaces that is measurable.

Let us consider the simple configuration studied by Casimir of two large parallel square perfectly conducting plates, with sides of length L separated by a distance a with $a << L$ (Fig. 15.3). The plates are large enough so that the edge effects are negligible.

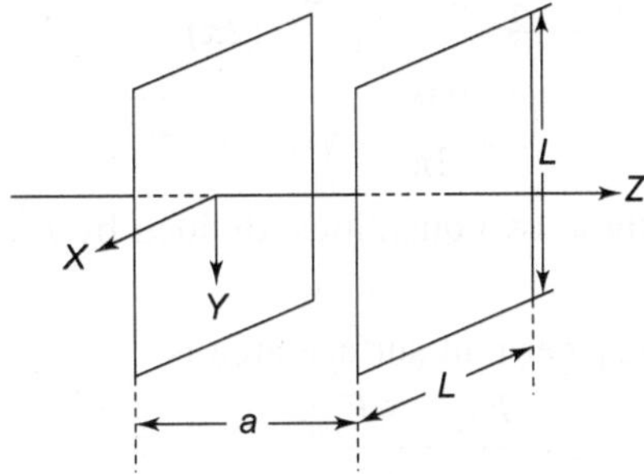

Fig. 15.3   Arrangement for the derivation of the Casimir effect

We consider the zero point energy inside the volume $L^2 a$ both with and without the plates present. The boundary conditions imply that the values of the component k_z are discrete $k_z = \dfrac{n\pi}{a}$, $n = 1, 2, \ldots$, and there are two transverse polarization states for each allowed mode. If k_z vanishes, there is only one (TE) mode.

k_x, k_y are effectively continuous since L is large. The zero point energy in this space is

$$E_{\text{walls}} = \Sigma \frac{1}{2} \hbar\omega_1$$

$$= \frac{1}{2} \hbar c \, \Sigma \sqrt{k_x^2 + k_y^2 + k_z^2}$$

$$= \frac{1}{2} \hbar c \int_{-\infty}^{\infty} \frac{L dk_x}{2\pi} \int_{-\infty}^{\infty} \frac{L dk_y}{2\pi}$$

$$\times 2 \sum_{n=0}^{\infty} \in_n k_n (a)$$

where
$$\in_n = \frac{1}{2}, \quad n = 0$$
$$= 1, \quad n \geq 1$$

and
$$k_n(a) = \sqrt{\left(\frac{n\pi}{a}\right)^2 + k_x^2 + k_y^2}, \quad n = 0, 1, \ldots$$

$$= \frac{1}{2}\,\hbar c \int_{-\infty}^{\infty} \int_{-\infty}^{\infty} L^2 \frac{dk_x dk_y}{(2\pi)^2}$$

$$\times \left[\sqrt{k_x^2 + k_y^2} + 2 \sum_{n=1}^{\infty}\left(k_x^2 + k_y^2 + \frac{n^2\pi^2}{a^2}\right)\right]. \qquad (250)$$

This expression is of course, meaningless, being infinity. To get a physically meaningful quantity we subtract from it the zero point energy if the planes were not present (E_{vac}).

Now,
$$E_{vac} = \frac{\hbar c}{2} \int_{-\infty}^{\infty} \int_{-\infty}^{\infty} L^2 \frac{dk_x dk_y}{(2\pi)^2}.$$

$$\times \int_{-\infty}^{\infty} \frac{a\,dk_z}{2\pi}\, 2\, \sqrt{k_x^2 + k_y^2 + k_z^2}. \qquad (251)$$

The difference of these two quantities, divided by L^2, is the energy per unit surface area.

Therefore, the energy per unit surface area is

$$\varepsilon = \frac{E_{walls} - E_{vac}}{L^2}$$

$$= \frac{\hbar c}{2\pi} \int_0^{\infty} k\,dk \left(\frac{k}{2} + \sum_{n=1}^{\infty} \sqrt{k^2 + \frac{n^2\pi^2}{a^2}}\right.$$

$$\left. - \int_0^{\infty} dn\, \sqrt{k^2 + \frac{n^2\pi^2}{a^2}}\right), \qquad (252)$$

where $k_x^2 + k_y^2 = k^2$, so that $dk_x dk_y = 2\pi k\,dk$. This quantity is still not defined due to the ultraviolet (large k) divergences. Also, the prefect conductor approximation is unrealistic for arbitrary short wavelengths, because wavelengths shorter than the atomic size cannot be confined. Therefore, we introduce a smooth cutoff function $f(k)$ into the integrand of the above integral, where $f(k) = 1$ for $k \ll k_m$, and $f(k) \to 0$ for $k \gg k_m$, k_m being of the order of the inverse atomic dimension.

Putting $u = \dfrac{k^2 a^2}{\pi}$, we obtain,

$$\varepsilon = \hbar c\, \frac{\pi^2}{4a^3} \int_0^{\infty} du \left[\frac{\sqrt{u}}{2} f\left(\frac{\pi}{a}\sqrt{u}\right)\right.$$

$$+ \sum_1^{\infty} \sqrt{u + n^2}\, f\left(\frac{\pi}{a}\sqrt{u + n^2}\right)$$

$$\left. - \int_0^{\infty} dn\, \sqrt{u + n^2}\, f\left(\frac{\pi}{a}\sqrt{u + n^2}\right)\right]. \qquad (253)$$

Let us put

$$F(n) = \int_0^{\infty} dn\, \sqrt{u + n^2}\, f\left(\frac{\pi}{a}\sqrt{u + n^2}\right) \qquad (254)$$

$$F(n) \to 0 \quad \text{as} \quad n \to \infty.$$

Equation (253) may be written as

$$\varepsilon = \hbar c \, \frac{\pi^2}{4a^3} \left[\frac{1}{2} F(0) + \sum_{n=1}^{\infty} F(n) - \int_0^{\infty} dn F(n) \right]. \qquad (255)$$

The interchange of sums and integrals can be justified owning to the absolute convergence in presence of the cutoff function.

The discrete sum $\sum_{n=0}^{\infty} \in_n F(n)$ in the expression Eq. (255) is the trapezoidal approximation to the integral. The difference between the sum and the integral may be estimated by an expansion in the derivatives of $F(n)$ using the Euler-Maclaurin formula.

The Bernoulli numbers are defined by the series (Arfken and Weber (1995))

$$\frac{x}{e^x - 1} = \sum_{n=0}^{\infty} B_n \frac{x^n}{n!} \qquad (256)$$

which converges for $|x| < 2\pi$. Also,

$$B_n = \left[\frac{d^n}{dx^n} \left(\frac{x}{e^x - 1} \right) \right]_{x=0}. \qquad (257)$$

The first few Bernoulli numbers are

$$B_0 = 1, \, B_1 = -\frac{1}{2}, \, B_2 = \frac{1}{6}, \, B_4 = -\frac{1}{30}, \, B_6 = \frac{1}{42}, \, \ldots \qquad (258)$$

All the odd order Bernoulli numbers but B_1 are zero.

We have

$$\frac{xe^s}{e^x - 1} = \sum_{n=0}^{\infty} B_n(s) \frac{x^n}{n!} \qquad (259)$$

defining the Bernoulli functions, $B_n(s)$. Clearly

$$B_n(0) = B_n, \quad n = 0, 1, 2, \ldots \qquad (260)$$

From the defining relation, we get a difference relation

$$B'_n(s) = nB_{n-1}(s), \quad n = 1, 2, 3, \ldots \qquad (261)$$

and a symmetry relation

$$B_n(1) = (-1)^n \, B_n(0), \quad n = 0, 1, 2, \ldots \qquad (262)$$

We now give the Euler-Maclaurin integration formula. We have the integral of a differentiable function $f(x)$,

$$\int_0^n f(x) \, dx = \frac{1}{2} f(0) + f(1) + f(2) + \ldots + f(n-1) + \frac{1}{2} f(n)$$

$$-\sum_{p=1}^{q} \frac{1}{(2p)!} B_{2p} \, [f^{(2p-1)}(n) - f^{(2p-1)}(0)]$$

$$+ \frac{1}{(2q)!} \int_0^1 B_{2q}(x) \sum_{v=0}^{n-1} f^{(2q)} (x+v) \, dx. \tag{263}$$

The terms $\frac{1}{2} f(0) + f(1) + \ldots + \frac{1}{2} f(n)$ give the trapezoidal approximation. The summation over p represents the correction to the approximation.

If the remainder term is well behaved, the limit $q \to \infty$ is allowed. If the integral converges as $n \to \infty$, then

$$\int_0^\infty f(x) \, dx = \frac{1}{2} f(0) + \sum_{n=1}^{\infty} f(n)$$

$$+ \sum_{p=1}^{\infty} \frac{1}{(2p)!} B_{2p} \, f^{(2p-1)} (0). \tag{264}$$

Therefore, the difference between the sum and the integral occurring in the square bracket in Eq. (255) may be written as

$$\frac{1}{2} F(0) + F(1) + F(3) + \ldots + \int_0^\infty dn \, F(n)$$

$$= \frac{-1}{2!} B_2 F'(0) - \frac{1}{4!} B_4 F'''(0) + \ldots$$

$$= - \frac{1}{6 \times 2!} F'(0) + \frac{1}{30 \times 4!} F'''(0) + \ldots \tag{265}$$

From Eq. (254), we have

$$F(n) = \int_{n^2}^{\infty} \sqrt{u} f\left(\frac{\pi \sqrt{u}}{a}\right) du, \tag{266}$$

and

$$F'(n) = 2n^2 f\left(\frac{\pi n}{a}\right),$$

$$F''(n) = -4nf\left(\frac{\pi n}{a}\right) - 2n^2 f'\left(\frac{\pi n}{a}\right)\left(\frac{\pi}{a}\right),$$

$$F'''(n) = -4f\left(\frac{\pi n}{a}\right) - 4nf'\left(\frac{\pi n}{a}\right)\left(\frac{\pi}{a}\right) - 2n^2 f''\left(\frac{\pi n}{a}\right)\left(\frac{\pi}{a}\right)^2, \tag{267}$$

where the prime indicates the derivative with respect to the argument. We assume that $f(0) = 1$, and all its derivatives vanish at the origin. So $F'(0) = 0$ and $F'''(0) = -4$ and higher derivatives of F are all zero. The regularized energy per unit area between the conducting plates is

$$\varepsilon = \hbar c \, \frac{\pi^2}{a^3} \frac{B_4}{4!} = - \frac{\hbar c}{720} \frac{\pi^2}{a^3}, \tag{268}$$

All reference to the cutoff has vanished. The result is valid provided $k_m a \gg 1$, or $a \gg \lambda_c$ where $\lambda_c = \dfrac{2\pi}{k_m}$ is the wavelength at the cutoff frequency.

The energy shift produces a force per unit area between the plates

$$P = -\frac{\hbar c}{720}\frac{\pi^2}{a^4} = -\frac{0.013}{(a)^4} \text{ dyne/cm}^2 \qquad (269)$$

where a is measured in units of μm.

The minus sign indicates that there is an attractive macroscopic but tiny force on the plates.

This is the Casimir effect.

In a series, of experiments, Sparnaay (1958) verified this quantum effect.

The Casimir effect has been calculated for different geometrical configurations. For example, the Casimir force between a sphere and a flat plate varies as $\frac{1}{a^3}$.

The sign of the force is geometry dependent. The Casimir force is repulsive in a spherical cavity. All these predictions have been experimentally confirmed.

One can study van der Waals forces (see Section 13.5) among neutral atoms or molecules by considering the fluctuations of the electromagnetic field. Casimir and Polder (1948) obtained an expression for the potential energy U of two atoms with static polarizabilities α_1 and α_2 and a distance r apart:

$$U = -\frac{23\hbar c}{4\pi}\frac{\alpha_1\alpha_2}{r^7}. \qquad (270)$$

The nonrelativistic treatment of van der Waals force gives $U \sim \frac{1}{r^6}$ at a large distance. The reader should consult Itzykson and Zuber (1985).

PROBLEMS

15.1   (a) What is the selection rule for allowed transitions of a linear harmonic oscillator of mass m, frequency ω_0, and charge q?

      (b) Find the spontaneous transition probability per unit time from *an* excited state $|n\rangle$ to the ground state.

15.2 A simple harmonic oscillator of mass m, frequency ω_0, and charge q is excited by a radiation field

$$\vec{A}(\vec{r},\, t) = \begin{cases} 2A_0^2\, \hat{z} \cos\,(ky - \omega t), & t > 0 \\ 0, & t \le 0 \end{cases}$$

      (a) Assuming that initially $|\psi(t = 0)\rangle = |0\rangle$, find the time-dependent state vector $|\psi(t)\rangle$, using first order perturbation theory.

      (b) Calculate the induced electric dipole moment.

15.3 For an electron in a central field, calculate the probabilities of the spontaneous transitions $(n,\, l) \to (n',\, l + 1)$ and $(n,\, l) \to (n',\, l - 1)$.

15.4 Compute the ratio of the intensities of the Balmer H_α and H_β lines of atomic hydrogen.

CHAPTER 16

Identical Particles

In this chapter we first discuss systems of identical particles and point out the consequences arising from the indistinguishability of such particles. Next the permutation group is discussed in somewhat greater detail. After this, in Section 16.6, we consider the $SU(3)$ symmetry and its application in particle physics.

16.1 PERMUTATIONS AND SYMMETRY

In quantum mechanics, identical particles are completely indistinguishable from one another. Let us consider a system of N identical particles. This means that it is not possible to distinguish any physical situation obtained by the interchange of identical particles. Therefore, any observable, and in particular, the Hamiltonian of the system is a symmetric function of the coordinates (space and spin) of the identical particles. Let the Hamiltonian of the system be H (1, 2, ..., N) and let P_{ij} be an operator that permutes the variables of particles i and j. Thus, we have

$$P_{ij} H P_{ij}^{-1} = H. \tag{1}$$

P_{ij} also commutes with the total momentum and angular momentum of the system. So P_{ij} is unitary and $P_{ij}^2 = 1$. Hence, the eigenvalues of each permutation operator is ± 1.

Let $\mathcal{H}_1$, $\mathcal{H}_2$, ..., $\mathcal{H}_N$ be the spaces of physical states of the first, second, ..., Nth one-particle system (all the spaces $\mathcal{H}$ are identical). The space of states of the N-particle system is the direct product space

$$\mathcal{H} = \mathcal{H}_1 \otimes \mathcal{H}_2 \otimes ... \otimes \mathcal{H}_N. \tag{2}$$

Let $|\xi_i\rangle_i$ denote a basis in the space $\mathcal{H}_i$.

The basis in $\mathcal{H}$ is

$$|\xi_1, \xi_2 ... \xi_N\rangle = |\xi_1\rangle_1 \otimes |\xi_2\rangle_2 \otimes ... \otimes |\xi_N\rangle_N. \tag{3}$$

Now, $P_{ij}|\xi_1\xi_2 \ldots \xi_N\rangle \equiv P_{ij}|\xi_1\xi_2 \ldots \xi_i \ldots \xi_j \ldots \xi_N\rangle.$

$$= |\xi_1\xi_2 \ldots \xi_j \ldots \xi_i \ldots \xi_N\rangle$$

$$\equiv |\xi_1\rangle_1|\xi_2\rangle_2 \ldots |\xi_j\rangle_i \ldots |\xi_i\rangle_j \ldots |\xi_N\rangle_N. \tag{4}$$

Clearly, $\qquad\qquad [P_{ij}, P_{kl}] \neq 0. \tag{5}$

We consider the case $N = 2$, that is a system of two identical particles. There is one permutation operator P_{12} and since H, P_{12} commute, they possess a complete set of common eigenvectors. Then eigenvalues of P_{12} are ± 1 and its eigenfunctions are either symmetric or antisymmetric under the interchange of the two particles. The two linearly independent and normalized sets are

$$|\xi_1, \xi_2\rangle_{\pm} = \frac{1}{\sqrt{2}}\left[|\xi_1\xi_2\rangle \pm |\xi_2\xi_1\rangle\right], \tag{6}$$

where $+$ designates the symmetric and $-$ the antisymmetric states.

For a system of more than two identical particles, the situation is complicated. We consider a system of three identical particles ($N = 3$). Since there are $3! = 6$ distinct permutations of three objects, there are 6 different permutation operators. These are the identity operator I, three interchange operators (transpositions) P_{12}, P_{23}, and P_{31}, and two cyclic permutations P_{123} and $(P_{123})^2$. The six permutation operators are not mutually commutative. We have the totally symmetric and antisymmetric states, i.e., $P_{ij}|\ldots\rangle_{\pm} = \pm|\ldots\rangle_{\pm}$ for all the permutation operators. There are 4 other states with mixed or partial symmetry.

Here, we simplify the notation by writing $|i\rangle$ for $|\xi_i\rangle$.

Symmetric (S)

$$\frac{1}{\sqrt{6}}\left[|123\rangle + |213\rangle + |132\rangle + |321\rangle + |312\rangle + |231\rangle\right] \tag{7}$$

Anti-symmetric (A)

$$\frac{1}{\sqrt{6}}\left[|123\rangle - |213\rangle - |132\rangle + |231\rangle + |312\rangle - |321\rangle\right] \tag{7a}$$

Mixed symmetric (M)

$$\frac{1}{2\sqrt{3}}\left[2|123\rangle + 2|213\rangle - |132\rangle - |321\rangle - |312\rangle - |231\rangle\right]$$

$$\frac{1}{2}\left[+|132\rangle - |321\rangle + |312\rangle - |231\rangle\right]. \tag{7b}$$

Mixed symmetric (M$'$)

$$\frac{1}{2\sqrt{3}} \left[2|123\rangle - 2|213\rangle + |132\rangle + |321\rangle - |312\rangle - |231\rangle \right]$$

$$\frac{1}{2} \left[+ |132\rangle - |321\rangle - |312\rangle + |231\rangle \right]. \tag{7c}$$

The Hilbert space for three identical particles H_3 can be decomposed into the direct sum of four invariant subspaces: $\mathcal{H}_3 = \mathcal{H}_S \oplus \mathcal{H}_A \oplus \mathcal{H}_M \oplus \mathcal{H}_{M'}$. Eqs. (7) and (7a) belong to the two distinct $1D$ IRs of the permutation group on three objects (S_3) and Eqs. (7b) and (7c) belong to two distinct $2D$ IRs. Stationary states may be classified by their symmetry under permutations. Moreover, the action of an observable O does not change the symmetry type. O has no matrix elements connecting the subspaces.

In the case of a system of N identical particles

$$\mathcal{H}_N = \mathcal{H}_S \oplus \mathcal{H}_A \oplus \mathcal{H}_{M_1} \oplus \mathcal{H}_{M_2} \oplus \ldots, \text{ where } \mathcal{H}_S \text{ and } \mathcal{H}_A$$

are the spaces of totally symmetric and anti-symmetric functions respectively. The basis functions belong to $1D$ IRs of S_N. In the ($N!$ -2) remaining subspaces, the basis functions exhibit mixed symmetry and belong to higher dimensional IRs of S_N. The symmetric and antisymmetric states may be constructed as follows:

$$|\xi_1 \ldots \xi_N\rangle_{\pm} = \frac{1}{\sqrt{N!}} \begin{vmatrix} |\xi_1\rangle_1 & |\xi_1\rangle_2 & \ldots & |\xi_1\rangle_N \\ |\xi_2\rangle_1 & |\xi_2\rangle_2 & \ldots & |\xi_2\rangle_N \\ \vdots & & & \\ |\xi_N\rangle_1 & \xi_N\rangle_2 & \ldots & |\xi_N\rangle_N \end{vmatrix}_{\substack{\text{det.} \\ \text{perm}}} \tag{8}$$

where det. stands for determinant, and perm. stands for permanent (same linear combination with all positive signs). In the antisymmetric case, Eq. (8) is called a Slater determinant. The phase ambiguity is removed by adopting a standard order for the indices.

16.2 INDISTINGUISHABILITY OF PARTICLES

The principle of indistinguishability of identical particles may be stated as follows: Dynamical states that differ only by a permutation of identical particles cannot be distinguished by any observation whatsoever. (Messiah and Greenberg (1964)).

Let θ be an operator that represents an observable and let $|\overline{\psi}\rangle$ be the state vector of a system of identical particles. Then using the principle of indistinguishability of identical particles, we have

$$\langle \overline{\Psi} | \theta | \overline{\Psi} \rangle = \langle \overline{\Psi} | (P_{ij})^\dagger \, \theta P_{ij} | \overline{\Psi} \rangle$$

Then $\qquad \theta = (P_{ij}) \, \theta P_{ij}$ since $P_{ij} = (P_{ij})^\dagger = P_{ij}^{-1}$,

we get $\qquad P_{ij}\theta = \theta P_{ij}.$

Thus, for any observable θ of a system of identical particles

$$[\theta, P] = 0 \ \forall \ P \in S_N. \tag{9}$$

The Hamiltonian of the system is obviously permutation-invariant. So, the various consequences such as the classification of the states by their symmetry properties under permutations follow from the above principle. There is also a "superselection rule" which may be stated as the following: Interference between states of different permutation symmetry is not observable.

16.3   THE SYMMETRIZATION POSTULATE

It terms out that in nature for a system of N identical particles the states are either totally symmetric or totally antisymmetric under the interchange of any pair. This property depends on the nature of the particles. The symmetrization postulate states:

The pure states of a system of identical particles must be totally symmetric or totally antisymmetric under the exchange of any two particles.

There is considerable experimental evidence that no states of mixed symmetry (belonging to higher-dimensional representations of the permutation group) exist. Messiah and Greenberg (1964) and Green have discussed the statistics ("parastatistics") obeyed by such (hypothetical) particles. There is a general connection between the spin of a particle and the possible symmetry of the states of an assembly of the particles under interchange of the particles. The spin-statistics theorem states: The pure states of a system of identical particles whose spin is an integer (a half-odd integer) multiple of $\hbar$ is totally symmetric (totally antisymmetric).

Particles that must have symmetric states under exchange of two particles obey Bose statistics and are called bosons while particles with antisymmetric states obey Fermi statistics and are called fermions. Pions, photons are bosons while electrons, neutrons, protons are fermions. Particles with integer spins ($s = 0$, 1, 2, ...) are bosons, while particles with half-odd integer spins $\left(s = \dfrac{1}{2}, \dfrac{3}{2}, \dfrac{5}{2} \dots \right)$ are fermions.

Here, particles can be composite. For example, a He^4 nucleus is a boson while a He^3 nucleus is a fermion. In the relativistic quantum theory of local fields it can be shown that half-integral spin fields obey Fermi statistics and that integral spin fields obey Bose statistics, provided that the following assumptions are made.

   (a) Lorentz invariance,

   (b) The vacuum is the state of lowest energy,

   (c) Microcausality,

   (d) Positive definite metric.

The spin-statistics theorem was first proved by Pauli (1940) for free fields. An important consequence of the fact that fermions are described by the antisymmetric state is the Pauli exclusion principle. For a system of identical fermions, the antisymmetric state $|\xi_1\xi_2 \dots \xi_N\rangle$ vanishes when any two (or more) of the quantum numbers are equal. Thus, no two identical fermions can occupy the same state. Pauli (1925) postulated the exclusion principle which states that no two electrons in a given system can occupy the same quantum state (orbital) to explain the periodic classification of elements. It forms the basis of atomic physics.

For a system of N identical fermions, the normalized ket can be written as an $N \times N$ determinant known as the Slater determinant, Eq. (8). The exchange of two coordinates will interchange two columns, which changes the sign of the determinant, demonstrating the required antisymmetry. The exchange of two states, i.e., of two orbitals, interchanges two rows, which leads to the change of sign of the determinant. Therefore, if two particles are in the same state, the determinant is zero. This is Pauli exclusion principle.

One may build a symmetric eigenfunction and an antisymmetric eigenfunction from any arbitrary eigenfunction by the use of the operators

$$S = \frac{1}{\sqrt{N!}} \sum_P P \qquad\qquad (10)$$

$$A = \frac{1}{\sqrt{N!}} \sum_P (-1)^p\, P$$

where P is any permutation and p is the parity of the permutation. S and A are known as the symmetrizer and the antisymmetrizer, respectively. $S^2 = S$ and $A^2 = A$, i.e., S and A are projection operators. For $N = 3$,

$$S = \frac{1}{\sqrt{6}}\,[I + P_{12} + P_{13} + P_{23} + P_{123} + P_{132}],$$

$$A = \frac{1}{\sqrt{6}}\,[I - P_{12} - P_{13} - P_{23} + P_{123} + P_{132}].$$

It should be noted that identical particles can be distinguished if their wave functions do not overlap, i.e., if they are in different states. Incase the wave functions overlap, there is an "exchange density" giving observable consequences. When identical particles scatter each other, their wave functions do overlap at some time, giving rise to interference terms (see the discussion on the scattering of identical particles, (Section 21.1).

16.4   THE PERMUTATION GROUP

We now discuss the symmetric or permutation group S_N. The group S_N and its representations are important for a number of reasons. The group S_N is important for physical systems consisting of identical particles. Since all finite groups of order N are subgroups of S_N (Cayley's theorem), representations of S_N are useful in the study of representations of other finite groups. The IR's of S_N are closely connected with the IR's of other important classical continuous groups – such as $GL(m)$, $U(m)$ and $SU(m)$ (m = integer).

The permutation group or the symmetric group S_N is the abstract group whose realization may be taken to be the $N!$ permutations of N distinguishable objects. We number the objects by the integers 1, 2, ..., N. An arbitrary permutation of N objects is represented by

$$P = \begin{pmatrix} 1 & 2 & 3 & \cdots & N \\ i_1 & i_2 & i_3 & \cdots & i_N \end{pmatrix} \tag{11}$$

where each entry in the first row is to be replaced by the corresponding one in the second row. The set of $N!$ permutations of N objects forms the group S_N. One may associate with every permutation Eq. (11) an inverse permutation

$$P^{-1} = \begin{pmatrix} i_1 & i_2 & i_3 & \cdots & i_N \\ 1 & 2 & 3 & \cdots & N \end{pmatrix} \tag{12}$$

The product of two permutations $P_2 P_1$ is also a permutation, the effect of which corresponds to the action of one permutation (P_1) followed by another (P_2). This defines the group multiplication. The identity element corresponds to no permutation

$$I = \begin{pmatrix} 1 & 2 & 3 & \cdots & N \\ 1 & 2 & 3 & \cdots & N \end{pmatrix} \tag{13}$$

The effect of applying a permutation P and its inverse P^{-1} is to bring the objects to their original positions, this forms the identical permutation I. The associative nature of the group multiplication can be verified. Thus, the $N!$ permutations of N objects form a group S_N.

It is convenient to decompose the elements of a permutation group into products of (commuting) cycles. A cycle is a permutation which can be written in the form

$$P_{i_1, i_2 \ldots i_k} = \begin{pmatrix} i_1 & i_2 & i_3 & \ldots & i_k & i_{k+1} & \ldots & i_N \\ i_2 & i_3 & i_4 & \ldots & i_1 & i_{k+1} & \ldots & i_N \end{pmatrix}; \qquad (14)$$

this is more easily written $(i_1 i_2 \ldots i_k)$. As an example, we consider the permutation

$$P = \begin{pmatrix} 1 & 2 & 3 & 4 & 5 & 6 \\ 3 & 5 & 4 & 1 & 2 & 6 \end{pmatrix} \qquad (15)$$

of six objects. We may write the above permutation

$$P = (134)\,(25)\,(6). \qquad (16)$$

The number of elements in a cycle is called its length. A cycle of two elements is called a transposition, i.e., a transposition is a permutation that interchanges two symbols only. We have

$$(i_1 i_2) = (i_1 i_2)^{-1}. \qquad (17)$$

Every cycle can always be written in the form of a product of transpositions so that every permutation can be expressed as a product of transpositions (in fact, of transpositions of adjacent symbols). Such a representation is not unique. But the number of transpositions which form a given permutation always has a unique parity. A permutation is called even (odd) according to whether the number of its transpositions are even (odd). The set of all even permutation of N objects forms the alternating group of order $N!/2$, which is an invariant sub-group of S_N. Two permutations of S_N are in the same class if they have the same cycle structure (i.e., the number and lengths of the cycles in the two permutations are the same). The two permutations differ only in the elements forming the cycles. For example, $P_1 = (123)(4)(56)$ and $P_2 = (456)(1)(23)$ are in the same class.

Each cycle structure specifies a class. Since the sum of the cycle lengths is N each class is characterized by a particular partition of the N elements into cycles. A partition $\lambda \equiv [\lambda_1, \lambda_2 \ldots, \lambda_r]$ of the integer N is a sequence of positive integers, arranged in descending order, whose sum is equal to N: $\lambda_i \geq \lambda_{i+1}$, $i = 1, \ldots,$ $r - 1$; and $\sum_{i=1}^{r} \lambda_i = N$. Every class is its own inverse (i.e., all classes of S_N are ambivalent).

A partition λ is represented graphically by means of a diagram ("Young diagram") which consists of N squares arranged in r rows, (or "pattern", "frame", "shapes") the ith one of which contains λ_i squares.

Example 16.1 From two squares (cells) one can form only two diagrams:

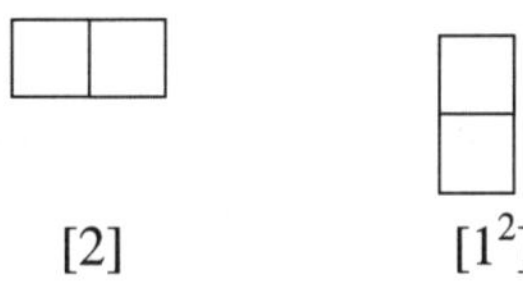

Example 16.2 For the group S_3 there are three distinct partitions: [3], [2, 1], and [1, 1, 1] = [1^3]. The corresponding Young diagrams are

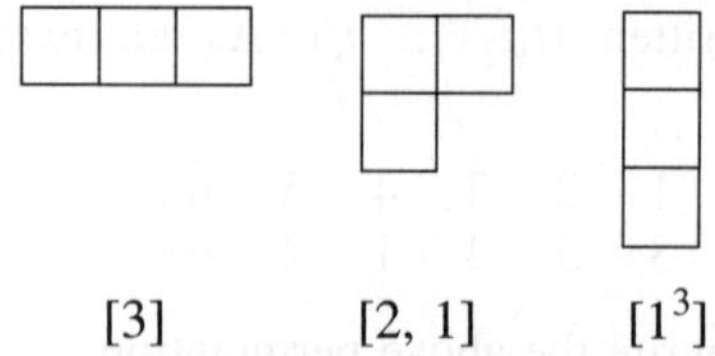

[3]   [2, 1]   [1^3]

Example 16.3 For the group S_4 there are five distinct partitions: [4], [3, 1], [2, 2], = [2^2], [2, 1, 1], = [2, 1^2] and [1, 1, 1, 1] = [1^4]

The corresponding Young diagrams are

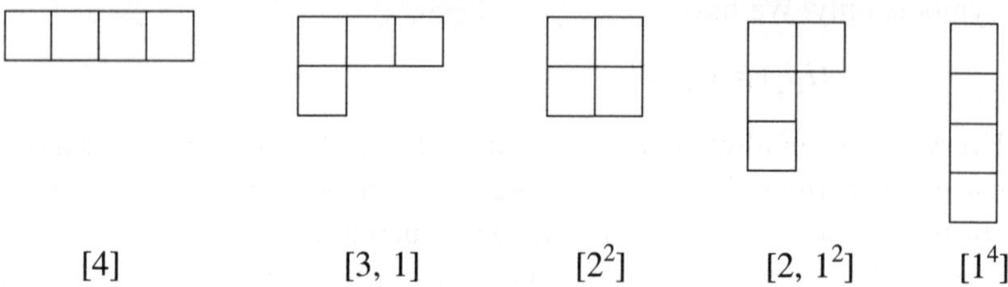

[4]   [3, 1]   [2^2]   [2, 1^2]   [1^4]

There is a one-to-one correspondence between the partitions of N and the classes of group elements in S_N.

Therefore, the number of distinct Young diagrams for any given N is equal to the number of classes of S_N which is equal to the number of inequivalent IR's of S_N. Each Young diagram uniquely corresponds to a specific. IR of the group S_N corresponding to every partition [λ] of N, there is an IR, often designated $\Gamma[\lambda]$ There are only two ID representations of S_N. One is the identity representation, designated $\Gamma_{[N]}$; the other is the alternating representation, designated $\Gamma_{[1^N]}$. The Young diagram [N] for S_N corresponds to a 1D symmetric representation, and the diagram [1^N] corresponds to an 1D antisymmetric representation. The "conjugate" (or "associated" or "dual") representation is $\Gamma[\tilde{\lambda}] = \Gamma[\lambda] \times \Gamma_{[1^N]}$. In terms of Young diagrams, the dual diagram is obtained by a reflection about the diagonal. For example,

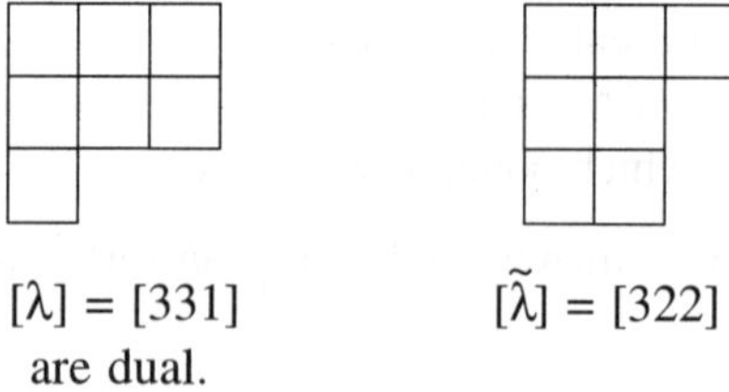

[λ] = [331]   [$\tilde{\lambda}$] = [322]
are dual.

To include the specification of the basis functions of the representation, we use the Young tableaux. A Young tableau is obtained by filling in the squares of the Young diagram [λ] with numbers 1, 2, ..., N in any order, each number being used only once. So there are $N!$ distinct Young tableaux altogether. A standard Young tableaux is one in which the numbers 1, 2, ..., N in each row appear increasing to the right and those in each column appear increasing to the bottom.

Henceforth, we shall refer to the standard Young tableaux simply as Young tableaux. For example

$$\begin{array}{|c|c|c|}\hline 1 & 2 & 4 \\\hline 3 \\\cline{1-1}\end{array} \qquad \begin{array}{|c|c|}\hline 1 & 3 \\\hline 2 & 4 \\\hline\end{array}$$

are standard tableaux of S_4.

Each basis function of an IR $\Gamma_{[\lambda]}$ can be associated with a Young tableaux. The number $g_{[\lambda]}$ of different tableaux constructed from a single Young diagram [λ] is equal to the dimension of the representation $D^{[\lambda]}$. For example, there are three tableaux for the IR [31].

$$\begin{array}{|c|c|c|}\hline 1 & 2 & 3 \\\hline 4 \\\cline{1-1}\end{array} \qquad \begin{array}{|c|c|c|}\hline 1 & 2 & 4 \\\hline 3 \\\cline{1-1}\end{array} \qquad \begin{array}{|c|c|c|}\hline 1 & 3 & 4 \\\hline 2 \\\cline{1-1}\end{array} \tag{18}$$

$$\text{(2111)} \qquad\qquad \text{(1211)} \qquad\qquad \text{(1121)}$$

The Yamanouchi notation for a Young tableau specifies the basis elements in terms of the subgroup decomposition. The Yamanouchi symbol (or, simply Y-symbol) for S_N is the N-element symbol $(r_N r_{N-1} \ldots r_1)$ where r_i is the row-number of the integer i in the tableaux. For the three tableaux shown above, we have given the corresponding Y-symbol below each tableau.

Finally, we discuss the Young–Yamanouchi treatment of S_N for the construction of irreducible representation matrices. Irreducible unitary representations (real and orthogonal) $\Gamma^{[\lambda]}$ of S_N may be constructed if we know the matrices of transpositions, since any permutation of S_N can be expressed as a product of $N - 1$ transpositions (12), (23), ..., $(N - 1, N)$.

So the representation matrix $\hat{\Gamma}^{[\lambda]}(P)$ for any permutation P can be constructed once we know the matrices $\hat{\Gamma}^{[\lambda]}(i, i + 1)$ for the transpositions $(i, i + 1)$ $(i = 1, 2, 3, \ldots, N - 1)$. The Young tableaux are the basis functions for this representation. When we specify the rows and columns of the representation matrices $\hat{\Gamma}^{[\lambda]}$ by the Young tableaux, the Young–Yamanouchi rules for constructing $\hat{\Gamma}^{[\lambda]}(i, i + 1)$ are given below.

(1) When i and $i + 1$ appear in the same row of the tableaux α, we put $\hat{\Gamma}_{\alpha\alpha}^{[\lambda]}$ $(i, i + 1) = 1$. All other elements of the row α and column α are zero.

(2) When i and $i + 1$ appear in the same column of α, we put $\hat{\Gamma}_{\alpha\alpha}^{[\lambda]}(i, i + 1) = -1$ with all the other elements of the row α and column α equal to zero.

(3) When i and $i + 1$ are located in the rth row, μth column and the r'th row, μ'th column respectively of the tableaux $\alpha(r \neq r', \mu \neq \mu')$, $\hat{\Gamma}^{[\lambda]}(i, i + 1)$ has an off–diagonal element between α and β, the latter tableaux being obtained by exchanging i and $i + 1$ in the tableaux α. The nonvanishing elements in the rows α, β and columns α, β are

$$\Gamma^{[\lambda]}_{\alpha\alpha}(i, i + 1) = -\Gamma^{[\lambda]}_{\beta\beta}(i, i + 1) = \rho_\alpha = \frac{1}{\mu' - r' - \mu + r}$$

$$\Gamma^{[\lambda]}_{\alpha\beta}(i, i + 1) = \Gamma^{[\lambda]}_{\beta\alpha}(i, i + 1) = \sqrt{1 - \rho_\alpha^2}. \tag{19}$$

As an example, we write some matrices of *IR* [31] of S_4 using the rules given above, the Young tableaux of which are given in Eq. (18)

$$\hat{\Gamma}(12) = \begin{pmatrix} 1 & 0 & 0 \\ 0 & 1 & 0 \\ 0 & 0 & -1 \end{pmatrix}$$

$$\hat{\Gamma}(23) = \begin{pmatrix} 1 & 0 & 0 \\ 0 & -\dfrac{1}{2} & \sqrt{3}/2 \\ 0 & \sqrt{3}/2 & 1/2 \end{pmatrix} \tag{20}$$

$$\hat{\Gamma}(34) = \begin{pmatrix} -\dfrac{1}{3} & \dfrac{\sqrt{8}}{3} & 0 \\ \dfrac{\sqrt{8}}{3} & \dfrac{1}{3} & 0 \\ 0 & 0 & 1 \end{pmatrix}$$

Representation matrices for other permutations of S_4 may be obtained from writing in terms of transpositions. For example, writing $(13) = (12)(23)$, we obtain

$$\hat{\Gamma}(13) = \begin{pmatrix} 1 & 0 & 0 \\ 0 & -\dfrac{1}{2} & -\sqrt{3}/2 \\ 0 & -\sqrt{3}/2 & 1/2 \end{pmatrix}$$

The representation defined by these rules is known as the Young–Yamanouchi standard orthogonal representation (or simply the standard representation). Its matrix elements satisfy the following orthogonality relations

(a) $\displaystyle\sum_r \hat{\Gamma}^{[\lambda]}_{rt}(P)\, \hat{\Gamma}^{[\lambda]}_{ru}(P) = \delta_{tu}$, \hfill (21)

$\displaystyle\sum_r \hat{\Gamma}^{[\lambda]}_{tr}(P)\, \hat{\Gamma}^{[\lambda]}_{ur}(P) = \delta_{tu}$,

(b) $\displaystyle\sum_{P} \hat{\Gamma}^{[\lambda]}_{rs}(P)\,\hat{\Gamma}^{[\lambda']}_{ut} = (N!/f_\lambda)\,\delta_{\lambda\lambda'}\delta_{ru}\delta_{st},$

$$\sum_{\lambda,r,t} (f_\lambda/N!)\,\hat{\Gamma}^{[\lambda]}_{rt}(P)\,\Gamma^{[\lambda]}_{rt}(Q) = \delta_{PQ},$$

where f_λ is the dimension of the representation. For every standard representation $\Gamma^{[\lambda]}$ we can associate a conjugate representation $\Gamma^{[\tilde\lambda]}$ of the same dimension, the matrices of which are related by

$$\Gamma^{[\tilde\lambda]}(P) = (-1)^p \Gamma^{(\lambda)}(P) \tag{23}$$

where p is the parity of the permutation.

Let us denote the general tableau as

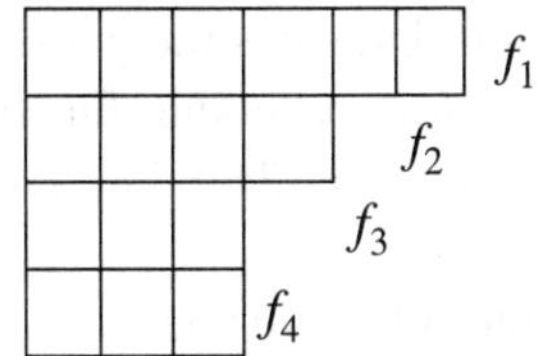

where f_i is the number of boxes in the ith row and the ordering is $f_1 \geq f_2 \geq f_3 \dots$ and written collectively as [f]. Thus, for the general Young shape, we use the notation $[f] = [f_1 f_2 f_3 \dots f_m]$. The number of standard tableaux for the shape $[f] = [f_1 f_2 f_3 \dots f_m]$ is given by

$$g_{[f]} = \frac{N!}{h_1! h_2! \dots h_m!} \prod_{i<j}^{m} (h_i - h_j), \tag{24}$$

where $h_i = f_i + m - i$. This formula gives the dimension of an *IR* of the symmetric group. For $N = 5$, the shape

$f_1 = 2,\ h_1 = 4;\ f_2 = 2,\ h_2 = 3;\ f_3 = 1,\ h_3 = 1$ so that Eq. (24) gives

$g = \dfrac{5!}{4!3!1!}\,(1\cdot 3\cdot 2) = 5$, in agreement with the number of tableaux.

The number of independent N-particle wave functions for each Young tableaux is given by

$$d_{[f]} = \prod_{i<j=1}^{s} \frac{f_i - f_j + j - i}{j - i}, \tag{25}$$

where s is the degeneracy of single particle function.

This formula Eq. (25) gives the dimension of an IR of the unitary group.

The method of counting by writing out the possible tableaux works well for N small. For N large, it is inconvenient to use this method. Then, we resort to the general formulas Eqs. (24) and (25).

16.5   SYMMETRIES OF MANY – PARTICLE WAVE FUNCTIONS AND YOUNG TABLEAUX

We now utilize the methods of the preceding section to consider the symmetries, with respect to permutations, of the wave functions of a system of N identical particles.

Let us consider as an example–states of two and three identical particles.

$N = 2$. This is a simple case since the wave function must be either totally symmetric or totally antisymmetric. We use a box to represent the wave function for a single particle. In the box we put the number of the particle. Boxes in the same row describe a symmetric state (symmetric under the exchange of all coordinates), while boxes in the same column describe an antisymmetric state. We assume that the single particle states are members of an s-fold degenerate set, $\psi_\alpha(i)$, where $\alpha = 1, 2, \ldots, s$ and i indicates the particle number. For example, for isospin function: $\alpha = m_t$; $s = 2$ for space-spin function: $\alpha = m_j$, m_t; $s = 2j + 1$; for spin-isospin function: $\alpha = m_s\, m_t, = (2s + 1)\,(2t + 1) = 4$. Thus,

Young tableaux	Nature of symmetry	Number of possible wave functions for each symmetry
$\underline{N = 2}$		
$\boxed{1\ 2}$	Symmetric	$\dfrac{s(s + 1)}{2}$
$\boxed{\begin{array}{c}1\\2\end{array}}$	Antisymmetric	$\dfrac{s(s - 1)}{2}$

Total number of
Two-particle states is
$s^2(s^{N=2})$

$\underline{N = 3}$ The situation is now more interesting and more complex since in this case there are mixed symmetric states besides the fully symmetric and fully antisymmetric states. We try to build the three-particle wave functions from the known two-particle wave functions. Let us start with $\boxed{1\ 2}$ and add $\boxed{3}$:

$\boxed{1\ 2\ 3}$ Totally symmetric $\phi_a(1)\,\phi_a(2)\,\phi_a(3)$ $\hspace{2em} s$

$$(\phi_a\phi_a\phi_b + \phi_a\phi_b\phi_a + \phi_b\phi_a\phi_a) \qquad s(s - 1)$$

$$\mathcal{S}\,(\phi_a\phi_b\phi_a) \qquad \frac{s(s - 1)\,(s - 2)}{6}$$

where $\phi_a\phi_b\phi_c = \phi_a(1)\,\phi_b(2)\,\phi_c(3)$ and $\mathcal{S}$ is a symmetrizer

$\begin{array}{|c|c|}\hline 1 & 2 \\\hline 3 \\\cline{1-1}\end{array}$ mixed symmetric $\{\lfloor\phi_a(1)\,\phi_b(2) + \phi_b(1)\,\phi_a(2)\rfloor\phi_c(3)$

$$- \phi_c(1)\,\phi_b(2)\,\phi_a(3) - \phi_c(1)\,\phi_a(2)\,\phi_b(3)\}\ \dfrac{s(s-1)(s-2)}{6}$$

$$\{\phi_a(1)\,\phi_c(2)\,\phi_b(3) - \phi_b(1)\,\phi_c(2)\,\phi_a(3)\}\qquad s(s-1)$$

Let us start with $\begin{array}{|c|}\hline 1 \\\hline 2 \\\hline\end{array}$ and add $\boxed{3}$:

$\begin{array}{|c|c|}\hline 1 & 3 \\\hline 2 \\\cline{1-1}\end{array}$ Mixed symmetric, (basically the same as $\begin{array}{|c|c|}\hline 1 & 2 \\\hline 3 \\\cline{1-1}\end{array}$ $\dfrac{s(s-1)\,(s+1)}{3}$

$\begin{array}{|c|}\hline 1 \\\hline 2 \\\hline 3 \\\hline\end{array}$ totally antisymmetric $\mathcal{A}\ (\phi_a\,\phi_b\,\phi_c)$
where $\mathcal{A}$ is an antisymmetrizer $\dfrac{s(s-1)\,(s-2)}{6}$

Total number
of functions
$= s^3(s^{N\,=\,3})$

Here, we have coupled $\boxed{1}$ and $\boxed{2}$ and then coupled in $\boxed{3}$.
We can do this coupling in six ways altogether.

$$\begin{array}{|c|c|}\hline 1 & 2 \\\hline 3 \\\cline{1-1}\end{array}\quad \left[\begin{array}{|c|c|}\hline 2 & 1 \\\hline 3 \\\cline{1-1}\end{array}\ \equiv\ \begin{array}{|c|c|}\hline 1 & 2 \\\hline 3 \\\cline{1-1}\end{array}\right]\quad \begin{array}{|c|c|}\hline 3 & 1 \\\hline 2 \\\cline{1-1}\end{array}$$

$$\begin{array}{|c|c|}\hline 1 & 3 \\\hline 2 \\\cline{1-1}\end{array}\quad \left[\begin{array}{|c|c|}\hline 2 & 3 \\\hline 1 \\\cline{1-1}\end{array}\ \equiv\ \begin{array}{|c|c|}\hline 1 \\\hline 2 & 3 \\\cline{2-2}\end{array}\right]\quad \begin{array}{|c|c|}\hline 3 & 2 \\\hline 1 \\\cline{1-1}\end{array}$$

But these can be expressed as linear combinations of the two on the left i.e.,
there are only two linearly independent symmetrics.

The fully symmetric and the fully antisymmetric states are two one-dimensional IR's of the permutation group S_3. The four mixed-symmetric states are linear combinations of two two-dimensional IR's.

We finally consider

$\underline{N = 4}$

We start with $\boxed{1}$ and $\boxed{2}$, then add $\boxed{3}$ in all possible ways and finally add $\boxed{4}$ in all possible ways. The five Young diagrams are shown below.

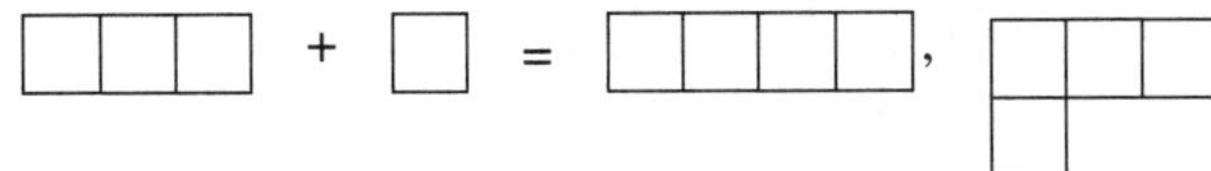

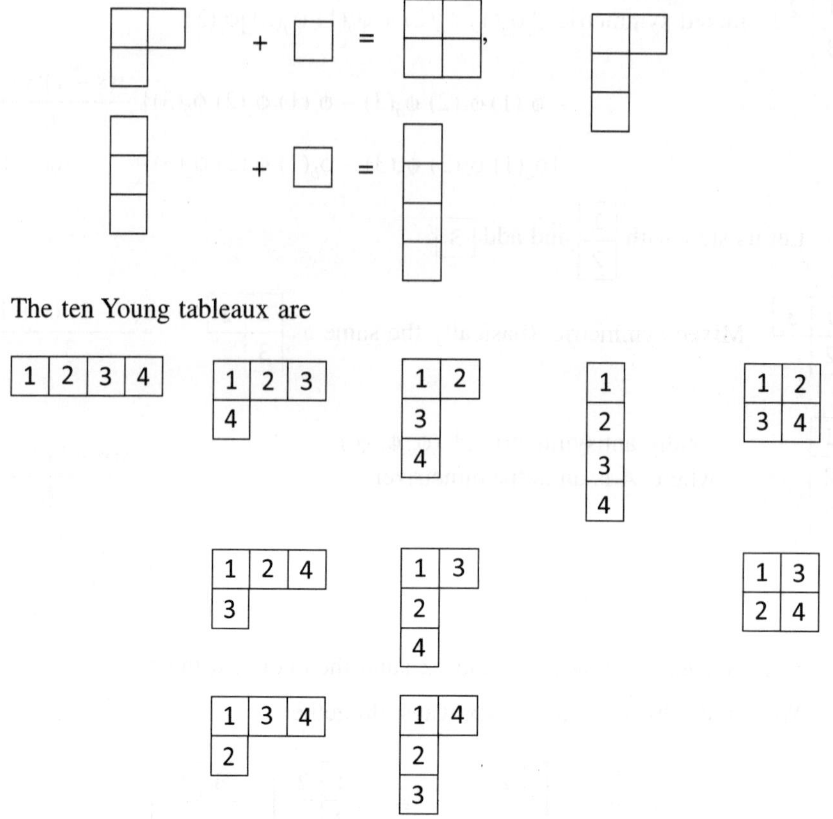

The ten Young tableaux are

We have considered here the standard Young tableaux. Each basis function of IR can be associated with a Young tableau, and the dimension of the IR is given by the number of standard Young tableau. The independent wave functions are obtained by the standard arrangement of Young tableaux.

For a system of large number of particles, we use a systematic method. The N-particle symmetries for a given Young tableau are commonly written in their normal symmetric or antisymmetric form. With a shape $[f] \equiv [f_1 f_2 \ \dots \ f_m]$ we associate a set $[k] = [k_1 k_2 \ \dots \ k_\mu]$ as shown below, where k_i = number of boxes in ith column.

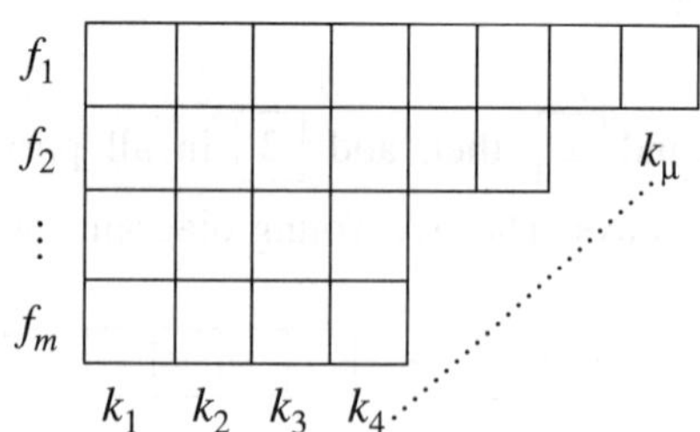

Normal symmetric form

For a given Young tableau, $[f_1 f_2 \dots f_m]$, is a function which is totally symmetric in one group of f_1 particles, and is simultaneously totally symmetric in a second group of f_2 particles, ..., and totally symmetric within a group of f_m particles, but is such that it cannot be further symmetrized.

Normal antisymmetric form

For a given Young tableau, $[k_1 k_2 \dots k_\mu]$, is a function which is totally antisymmetric in one group of k_1 particles, and is simultaneously totally antisymmetric in a second group of k_2 particles, ..., and totally antisymmetric in a μth group of k_μ particles, but is such that it cannot be further antisymmetrized.

Each Young diagram with at most n rows represents an *IR* of the group $SU(n)$. This is obtained from the following theorem. Every N-particle state that belongs to an *IR* of the group S_N and is built up from single particle states of an n-dimensional $SU(n)$ multiplet transforms under an *IR* of the group $SU(n)$.

To construct such functions we consider the symmetrizer and the antisymmetrizer defined in Eq. (10). Let the state in normal symmetric (antisymmetric) form be ψ_s (ψ_A). We take, an example, the standard tableau.

$$
\begin{array}{|c|c|c|}
\hline
1 & 2 & 3 \\
\hline
4 & 5 \\
\cline{1-2}
6 \\
\cline{1-1}
\end{array}
$$

Then

$$
\psi_S \left\{ \begin{array}{|c|c|c|}
\hline
1 & 2 & 3 \\
\hline
4 & 5 \\
\cline{1-2}
6 \\
\cline{1-1}
\end{array} \right\} = N\mathcal{S}_{(1,2,3)}\,\mathcal{S}_{(4,5)}\,\mathcal{A}_{(1,4,6)}\,\mathcal{A}_{(2,5)}\,[\phi_a(1)\phi_b(2) \dots \phi_f(6)]
$$

$$
= N'[1 + P_{12} + P_{13} + P_{23} + P_{123} + P_{132}]
$$

$$
\times [1 + P_{45}] \times [1 - P_{14} - P_{16} - P_{46} + P_{146} + P_{164}] \times \text{(where } N \text{ is a normlization}
$$

$$
\times [1 - P_{25}]\,[\phi_a(1)\phi_b(2) \dots \phi_f(6)], \qquad \text{constant)}
$$

$$\tag{26}$$

Here, we write, from right to left, the product of single particle wave functions (unsymmetrical) and then the antisymmetrizers for each column, followed by the symmetrizers for each row. ψ_S cannot be further symmetrized.

$$
\psi_A \left\{ \begin{array}{|c|c|c|}
\hline
1 & 2 & 3 \\
\hline
4 & 5 \\
\cline{1-2}
6 \\
\cline{1-1}
\end{array} \right\} = N\mathcal{A}_{(1,4,6)}\,\mathcal{A}_{(25)}\,\mathcal{S}_{(45)}\,\mathcal{S}_{(123)}\,[\phi_a(1)\phi_b(2) \dots \phi_f(6)] \tag{27}
$$

Here, we write in the same way as above with the symmetrizers on right and the antisymmetrizers on left ψ_A cannot be further antisymmetrized.

16.6 THE SU(3) SYMMETRY

We consider the special unitary group, $SU(n)$, which is the group of unitary, unimodular $n \times n$ matrices. $SU(n)$ and the orthogonal group, $O(n)$, are subgroups of the unitary group $U(n)$, which in turn, is a subgroup of $GL(n)$, the general linear group. The generators of $SU(n)$ can be chosen as n^2-1-linearly independent hermitian traceless $n \times n$ matrices. For example, the Pauli spin matrices $\sigma_1, \sigma_2, \sigma_3$ can be taken as the generators of $SU(2)$. (see § 5.7) More conveniently, we choose $S_i = \frac{1}{2}\,\sigma_i (i = 1, 2, 3)$ as generators satisfying the commutation relation

$$[S_i, S_{i'}] = i \in_{ijk} S_k. \tag{28}$$

$SU(2)$ and $SO(3)$ have the same Lie algebra. The rank of $SU(2)$ is one, and there is only one Casimir operator

$$C \equiv S^2 = \sum_{i=1}^{3} S_i^2 \tag{29}$$

IR's of $SU(2)$ and $SO(3)$ have already been discussed.

The special unitary group in three dimensions $SU(3)$ has $3^2 - 1 = 8$ generators. Eight hermitian traceless 3×3 matrices $\lambda_1, \lambda_2, ..., \lambda_8$ may be chosen as the generators. Three of these can be constructed from $\sigma_1, \sigma_2, \sigma_3$ because $SU(2)$ is a subgroup of $SU(3)$.

Following Gell-Mann, we write

$$\lambda_1 = \begin{pmatrix} 0 & 1 & 0 \\ 1 & 0 & 0 \\ 0 & 0 & 0 \end{pmatrix} \qquad \lambda_2 = \begin{pmatrix} 0 & -i & 0 \\ i & 0 & 0 \\ 0 & 0 & 0 \end{pmatrix}$$

$$\lambda_3 = \begin{pmatrix} 1 & 0 & 0 \\ 0 & -1 & 0 \\ 0 & 0 & 0 \end{pmatrix} \qquad \lambda_4 = \begin{pmatrix} 0 & 0 & 1 \\ 0 & 0 & 0 \\ 1 & 0 & 0 \end{pmatrix}$$

$$\lambda_5 = \begin{pmatrix} 0 & 0 & -i \\ 0 & 0 & 0 \\ i & 0 & 0 \end{pmatrix} \qquad \lambda_6 = \begin{pmatrix} 0 & 0 & 0 \\ 0 & 0 & 1 \\ 0 & 1 & 0 \end{pmatrix}$$

$$\lambda_7 = \begin{pmatrix} 0 & 0 & 0 \\ 0 & 0 & -i \\ 0 & i & 0 \end{pmatrix} \qquad \lambda_8 = \frac{1}{\sqrt{3}}\begin{pmatrix} 1 & 0 & 0 \\ 0 & 1 & 0 \\ 0 & 0 & -2 \end{pmatrix} \tag{30}$$

Clearly,

$$\lambda_i^\dagger = \lambda_i, \ Tr\,\lambda_i = 0. \tag{31}$$

The trace, commutator and anticommutator of two λ matrices are

$$Tr(\lambda_i\lambda_j) = 2\delta_{ij},$$

$$[\lambda_i, \lambda_j] = 2if_{ijk}\lambda_k,$$

$$\{\lambda_i, \lambda_j\} = \frac{4}{3}\delta_{ij} + 2d_{ijk}\lambda_k, \tag{32}$$

where the structure constants f_{ijk} are totally antisymmetric in its indices while the coefficients d_{ijk} are totally symmetric. For example, $f_{123} = 1, f_{367} = -\frac{1}{2}, f_{678} = \frac{\sqrt{3}}{2}$; $d_{118} = \frac{1}{\sqrt{3}}, d_{366} = -\frac{1}{2}, d_{888} = -\frac{1}{\sqrt{3}}$, etc.

For convenience, one may redefine the generators as F_i ("F-spin"):

$$F_i = \frac{1}{2}\lambda_i. \tag{33}$$

$SU(3)$ is of rank 2, and hence there are two Casimir operators:

$$C_1 = \sum_{i=1}^{8} F_i^2 = -\frac{2i}{3}\sum_{ijk} f_{ijk}\,F_iF_jF_k, \tag{34}$$

$$C_2 = \sum_{ijk} d_{ijk}F_iF_jF_k. \tag{34a}$$

From the theorem quoted in Section 16.5, we see that each Young diagram with at most n rows corresponds to an IR of $SU(n)$.

We now consider application to particle physics. All hadrons can be considered to be built from quarks. We consider only the three low-lying quark fields – three types or flavors of quark, up, down and strange.

We represent this as

$$q = \begin{pmatrix} u \\ d \\ s \end{pmatrix} \tag{35}$$

Strong interaction is approximately invariant under $SU(3)$ flavor symmetry. The $q^i s(q^1 = u, q^2 = d, q^3 = s)$ form the IR $\mathbf{3}$ (triplet) and their antiparticles $q_1 = \bar{u}, q_2 = \bar{d}$, and $q_3 = \bar{s}$ form the representation $\bar{3}$ (antitriplet). The isospin $SU(2)$ is a subgroup of $SU(3)$ with $u = \uparrow$ (*isopin up*), $d = \downarrow$ (*isopin down*) forming an isospin doublet and s an isospin singlet. The baryon states can be regarded as bound states of three quarks. The meson states ($B = 0$) can be regarded as quark-antiquark combination.

We can construct the $SU(3)$ multiplets using the Young diagrams, as discussed earlier. We obtain the following reduction of Kronecker products of IR's. Using only the $SU(3)$ flavor we get

$$\textcircled{3} \otimes \textcircled{\bar{3}} = \textcircled{8} \oplus \textcircled{1}, \tag{36}$$

which describes the eight pseudoscalar mesons and a singlet. We have

$$\textcircled{3} \otimes \textcircled{3} \otimes \textcircled{3} = \textcircled{1} \oplus \textcircled{8} \oplus \textcircled{8} \oplus \textcircled{10}, \tag{37}$$

which describes the baryons and their resonances. In Eq. (37), there is a singlet, two octets with different symmetry, and a decuplet.

Let us discuss the decuplet in detail (Lee (1988)). We write the wave functions as follows:

$$\Omega^- \sim sss$$

$$\Xi^{*0} \sim \frac{1}{\sqrt{3}}\,(uss + sus + ssu)$$

$$\Xi^* \simeq \frac{1}{\sqrt{3}}\,(dss + sds + ssd)$$

$$\Sigma^{*+} \sim \frac{1}{\sqrt{3}}\,(suu + usu + uus)$$

$$\Sigma^{*0} \sim \frac{1}{\sqrt{6}}\,(sud + sdu + usd + dsu + uds + dus)$$

$$\Sigma^* \sim \frac{1}{\sqrt{3}}\,(sdd + dsd + sdd)$$

$$\Delta^{++} \sim uuu$$

$$\Delta^+ \sim \frac{1}{\sqrt{3}}\,(duu + udu + uud)$$

$$\Delta^0 \sim \frac{1}{\sqrt{3}}\,(udd + dud + ddu)$$

$$\Delta^- \sim ddd.$$

The charges of u, d, s are $\frac{2}{3}$, $-\frac{1}{3}$, $-\frac{1}{3}$ respectively and the quarks are spin $\frac{1}{2}$ objects. The members of the decuplet are spin $\frac{3}{2}$ baryons in the lowest energy ($l = 0$) S state. The space part (orbital) of the wave function is symmetric. To form a total spin $\frac{3}{2}$, all three spin $\frac{1}{2}$ quarks are aligned. So the quarks are in a spin symmetric state. Hence the $\frac{3^+}{2}$ decuplet total wave function is symmetric in flavor, spin and space. The Fermi statistics of quarks would make it impossible for them to have total symmetry unless it is postulated that there is an additional degree of freedom, called color. We assume that each quark exists in three colors $u(c)$, $d(c)$ or $s(c)$, with $c = 1, 2, 3$ or $c =$ red, green, blue, and baryons and mesons are color singlets. For example, Δ^{++} consists of one red, one green and one blue u-quark. The statistics problem is now solved.

PROBLEMS

16.1 Show all the standard Young tableaux of the permutation group S_4. Enumerate all the inequivalent irreducible representations of S_4 and specify their dimensions. (See Section 16.5)

16.2 Repeat the analysis for the permutation group S_5.

16.3 (a) Prove Eq. (32).

(b) Show that C_1 and C_2 in (34) and (34a) are the Casimir operators of $SU(3)$.

16.4 Discuss the decomposition of the tensor product $\textcircled{8} \otimes \textcircled{8}$ for $SU(3)$.

CHAPTER 17

Many-Particle Systems

In this chapter, we give a brief introduction to the many-body techniques for studying the effects of interaction among many bodies in a many-body system. These techniques describe general methods applicable to all many-body systems e.g., in nuclear physics, condensed matter physics, etc. First, we describe the occupation number representation and introduce the creation and annihilation operators. Wick's theorem is then discussed. The Hartree-Fock method is discussed in the second quantized notation and applied to free electron gas. We discuss the correlation functions and the Hanbury Brown and Twiss effect. In the final section we consider two-electron systems.

17.1   THE OCCUPATION NUMBER REPRESENTATION

The physical world consists of many-particle systems. To deal with such systems we describe the elegant and compact representation known as the "occupation number representation". This formalism greatly simplifies, even in a nonrelativistic theory, the discussion of many identical particles interacting with each other. It enables us to deal with systems containing a variable number of particles. Even for systems with a fixed number of particles, this language is extremely useful. The occupation number formalism incorporates the statistics (symmetry properties of Fermi and Bose systems) at each step, in contrast to the complicated way of using symmetrized or antisymmetrized products of single-particle wave functions. The basic elements are the system wave function where we specify which orbitals are occupied by particle, and by how many. We specify the "occupation numbers" n_a, n_b, ... of orbitals a, b, ... If the wave function is antisymmetric, we can only have $n_i = 0$ or 1, $i = a$, b,..... (Fermi statistics). For a symmetric wave function, any occupation number is allowed (Bose statistics). This type of representation is called the "occupation number representation" and is used in modern many-body physics. The Hilbert space which is spanned by functions showing occupancies only is called the "Fock space"

Let us consider a system of N identical fermions. Such a system can be described by a Slater determinant $\overline{\Psi}_{\{\alpha\}}$:

$$\overline{\Psi}_{\{\alpha\}} = \frac{1}{\sqrt{N!}} \begin{vmatrix} \phi_{\alpha_1}(x_1) & \phi_{\alpha_1}(x_2) & \cdots & \phi_{\alpha_1}(x_N) \\ \vdots & \vdots & & \vdots \\ \phi_{\alpha_N}(x_1) & \phi_{\alpha_N}(x_2) & \cdots & \phi_{\alpha_N}(x_N) \end{vmatrix} \tag{1}$$

where $\phi_{\alpha i}(x)$ form a complete set of single-particle wave functions (orbitals) orthonormalized according to

$$\int \phi_{\alpha_i}(x)\phi_{\alpha_j}(x) = \delta_{\alpha_i \alpha_j} \tag{2}.$$

The Slater determinants $\overline{\Psi}_{\{\alpha\}}$ form a complete orthonormal set. The Slater determinants are somewhat cumbersome in taking care of the antisymmetrical nature of fermion wave functions.

We now develop an elegant and more convenient method of handling a system of identical particles. The basic idea of this formalism is as follows. For identical particles, it is not possible to say that particle 2 is in state α_1, particle 3 in state α_6, ..., i.e. we cannot say which particular particle is in which particular state. In the Slater determinant this problem is resolved by taking all possible linear combinations of all possible distributions in an appropriate manner such that the antisymmetric nature of the wave function is taken into account. But we can develop a formalism in which instead of the particle labels only the state labels occur, and in which we specify only whether a particular state is occupied by a particle or not.

We can establish a one-to-one correspondence between each Slater determinant and a ket in the Fock space $|n_1, n_2, \dots n_i \dots\rangle$ in which the occupancies of the orbitals (in the standard order) are specified. n_i is the occupation number of the ith orbital. If we assume that the number of particles is fixed, then

$$\sum_i n_i = N. \tag{3}$$

It should be noted that the change from the set of Slater determinants to the set of vectors in the occupation number space is not a unitary one.

For a system of N identical bosons, the totally symmetric and normalized basis wave functions are

$$\overline{\Psi}_{\{\alpha\}}^S = \sqrt{\frac{n_1! n_2! \dots}{N!}} \sum_P \left[P\phi_{\alpha_1}(x_1) \dots \phi_{\alpha_N}(x_N) \right]. \tag{4}$$

Each symmetrized wave function can be represented by a ket $|n_1, n_2, \dots n_i \dots\rangle$ in occupation number space.

For both bosons and fermions it is necessary to specify the occupation numbers $(n_1, n_2, \dots)$. For fermions, each occupation number is either 0 or 1 by the

Pauli principle, while for bosons these can take any integer value (0, 1, 2, ...) The set of all $|n_1, n_2, ... n_i ...\rangle$ is a complete set of orthogonal basis functions in Fock space. The set $|000...\rangle \equiv |0\rangle$ with no particles at all is called the "vacuum". For interacting particles, the eigenstates are obtained as linear combinations of these states.

17.2   CREATION AND ANNIHILATION OPERATORS

In the study of a system with N identical particles it is very useful to introduce creation and annihilation operators. We first discuss these operators for fermions and then for bosons.

(a)   Fermions

We start with the vacuum state $|0\rangle$, where all orbitals are empty. We assume that there is a "creation operator" $a_\alpha^\dagger$ which creates a fermion in the orbital α, if unoccupied (Pauli principle). We define a set of operators $a_{\{\alpha\}}^\dagger$ such that

$$a_{\alpha_1}^\dagger a_{\alpha_2}^\dagger ... a_{\alpha_N}^\dagger |0\rangle \tag{5}$$

possesses all the properties of the Slater determinant

$$\frac{1}{\sqrt{N!}} \begin{vmatrix} \phi_{\alpha_1}(1) & \cdots & \phi_{\alpha_N}(N) \\ \vdots & \cdots & \vdots \\ \phi_{\alpha_1}(1) & & \phi_{\alpha_N}(N) \end{vmatrix} \equiv \frac{1}{\sqrt{N!}} ||\alpha_1 ... \alpha_N||. \tag{6}$$

So

$$\begin{aligned} a_{\alpha_1}^\dagger a_{\alpha_2}^\dagger ... a_{\alpha_N}^\dagger |0\rangle &= \frac{1}{\sqrt{N!}} ||\alpha_1 \alpha_2 ... \alpha_N|| \\ &= \frac{1}{\sqrt{N!}} ||\alpha_2 \alpha_1 ... \alpha_N|| \\ &= - a_{\alpha_2}^\dagger a_{\alpha_1}^\dagger ... a_{\alpha_N}^\dagger |0\rangle. \end{aligned} \tag{7}$$

Hence, we see that the operators $a_\alpha^\dagger$ anticommute, i.e.,

$$a_{\alpha_1}^\dagger a_{\alpha_2}^\dagger = - a_{\alpha_2}^\dagger a_{\alpha_1}^\dagger$$

or,
$$\left\{ a_\alpha^\dagger a_\beta^\dagger \right\} = 0. \tag{8}$$

Note that $\left(a_\alpha^\dagger\right)^2 = 0$ showing the impossibility of creating two fermions in the same state.

Taking hermitian conjugate of Eq. (7), we obtain

$$\langle 0|a_{\alpha_N} ... a_{\alpha_2} a_{\alpha_1} = - \langle 0|a_{\alpha_N} ... a_{\alpha_1} a_{\alpha_2}$$

i.e.,
$$\{a_\alpha, a_\beta\} = 0. \tag{9}$$

Also,

$$\left\langle 0 \left| a_{\alpha_N} \cdots a_{\alpha_1} a_{\beta_1}^\dagger \cdots a_{\beta_N}^\dagger \right| 0 \right\rangle = \frac{1}{\sqrt{N!}} \int \left\| \alpha_1 \cdots \alpha_N \right\|^* \left\| \beta_1 \cdots \beta_N \right\| \, d\vec{r}_1 \cdots d\vec{r}_N$$

$$= \begin{cases} 0, & \text{unless the sets } \{\alpha\} \text{ and } \{\beta\} \text{ are the same,} \\ +1, & \text{if the sets are the same but differ by an even permutation,} \\ -1, & \text{if the sets are the same but differ by an odd permutation.} \end{cases} \tag{10}$$

Equation (10) is statisfied if we postulate

$$\langle 0|0 \rangle = 1, \tag{11}$$

$$a_\alpha |0 \rangle = 0, \quad \forall \alpha \tag{12}$$

$$\left\{ a_\alpha, a_\beta^\dagger \right\} = \delta_{\alpha\beta}. \tag{13}$$

To see this, suppose that the sets are not the same. Then, there is at least one α, say $\alpha = j$, which is not in the set $\{\beta\}$. This a_j may be anticommuted to the right so that it acts on $|0\rangle$ to give zero, using Eq. (12). If the sets are the same, then the anticommutation relation Eq. (8) for $a_\beta^\dagger$'s may be used until the product

$$a_{\beta_1}^\dagger \, a_{\beta_2}^\dagger \cdots a_{\beta_N}^\dagger = (\pm) a_{\alpha_1}^\dagger \cdots a_{\alpha_N}^\dagger.$$

The plus (minus) sign occurs if an even (odd) number of permutations are made to bring the set to the desired order. Now, using Eq. (13), we see that

$$\left\langle 0 \left| a_{\alpha_N} \cdots a_{\alpha_1} a_{\alpha_1}^\dagger \cdots a_{\alpha_N}^\dagger \right| 0 \right\rangle = 1.$$

Thus, the Eqs. (8), (9), (11), (12) and (13) are the basic relations for $a^\dagger$'s, a's and $|0\rangle$.

Let us now consider a state $|\overline{\psi}\rangle = |1, 1, 1, 0, 0 \dots\rangle = a_1^\dagger \, a_2^\dagger \, a_3^\dagger \, |0\rangle$. In this state, the orbitals 1, 2, 3 are occupied (one fermion in each orbital), while all other orbitals are unoccupied (empty).

Now,

$$a_1 |\overline{\psi}\rangle = a_1 a_1^\dagger \, a_2^\dagger \, a_3^\dagger \, |0\rangle$$
$$= \left(1 - a_1^\dagger a_1 \right) a_2^\dagger \, a_3^\dagger \, |0\rangle,$$

using Eq. (13)

$$= a_2^\dagger \, a_3^\dagger \, |0\rangle - a_1^\dagger \, a_1 \, a_2^\dagger \, a_3^\dagger \, |0\rangle$$
$$= a_2^\dagger \, a_3^\dagger \, |0\rangle - a_1^\dagger \, a_2^\dagger \, a_3^\dagger \, a_1 \, |0\rangle,$$

using Eq. (13)

$$= a_2^\dagger \, a_3^\dagger \, |0\rangle,$$

since $a_1|0\rangle = 0$ by Eq. (12)

$$= |0, 1, 1, 0, 0 \dots\rangle.$$

Thus, a_1 destroys a particle in orbital 1, and is known as an annihilation or destruction operator. In general, a_i destroys a particle in ith orbital, if it is occupied. If ith orbital is unoccupied, then it simply gives zero. $a_i^\dagger$ and a_1 are hermitian adjoints of each other. These are non-hermitian and not observables.

Next, we calculate $a_2|\overline{\Psi}\rangle$

$$a_2|\overline{\Psi}\rangle = a_2 a_1^\dagger\, a_2^\dagger\, a_3^\dagger\, |0\rangle.$$
$$= - a_1^\dagger\, a_2\, a_2^\dagger\, a_3^\dagger\, |0\rangle$$
$$= - a_1^\dagger\, a_3^\dagger\, |0\rangle - a_1^\dagger\, a_2^\dagger\, a_3^\dagger\, a_2\, |0\rangle$$
$$= - a_1^\dagger\, a_3^\dagger\, |0\rangle$$
$$= - |1, 0, 1, 0, ...\rangle.$$

Although a_2 destroys the particle in orbital 2, there is a change of sign. Clearly, we get a minus sign for every creation operator before the particular annihilation operator. We thus write

$$a_i|..., n_i, ...\rangle = (-1)^{v_i}\, \sqrt{n_i}|..., (1 - n_i), ...\rangle \qquad (14)$$

$$a_i^\dagger|..., n_i, ...\rangle = (-1)^{v_i}\, \sqrt{1 - n_i}\,\Big|..., (1 + n_i), ...\Big\rangle \qquad (15)$$

where
$$v_i = \sum_{j=1}^{i-1} n_j.$$

Let us construct a hermitian operator $a_i^\dagger a_i$. Let it operate on a state vector:

$$a_i^\dagger a_i|n_1, n_2, ..., 1_i, ...\rangle = (-1)^{v_i}\, a_i^\dagger|n_1, n_2, ..., 0_i, ...\rangle$$
$$= (-1)^{2v_i}\, |n_1, n_2, ..., 1_i, ...\rangle$$
$$= (+1)\, |n_1, n_2, ..., 1_i, ...\rangle$$

Similarly,

$$a_i^\dagger a_i|n_1, n_2, ..., 0_i, ...\rangle = 0|n_1, n_2, ..., 0_i, ...\rangle$$

In general,

$$a_i^\dagger a_i|n_1, n_2, ..., n_i, ...\rangle = n_i|n_1, n_2, ..., n_i, ...\rangle \qquad (16)$$

Thus, the eigenvalue of the operator $a_i^\dagger a_i$ for the state $\overline{\Psi}$ is just the occupation number for that state. We define the number operator for the orbital i as

$$N_i = a_i^\dagger a_i \qquad (17)$$

which operating on any state gives as eigenvalue n_i, the occupancy of orbital i. It is easy to verify that $N_i^2 = N_i$ so that the eigenvalues of N_i are 0 or 1 only. Also

$$[N_i, N_j] = 0. \qquad (18)$$

The total number of particles is an eigenvalue of the total number operator

$$N = \sum_{\text{all } r} a_r^\dagger a_r. \qquad (19)$$

In summary,

the creation operator $a_i^\dagger$ is defind as satisfying.

$$a_i^\dagger|n_1, n_2, ..., n_i, ...\rangle = (-1)^{v_i} \sqrt{1 - n_i}|n_1, ..., 1 + n_i, ...\rangle$$

where

$$v_i = \sum_{j=1}^{i-1} n_j;$$

the annihilation operator is defined by

$$a_i|n_1, ..., n_i, ...\rangle = (-1)^{v_i} \sqrt{n_i}|n_1, ..., 1 - n_i, ...\rangle$$

The occupancy n_i for the orbital i is 0 or 1.

These operators obey the following anti-commutation relations

$$\{a_i, a_j\} = \{a_i^\dagger, a_j^\dagger\} = 0$$

$$\{a_i, a_j^\dagger\} = \delta_{ij}$$

in which the antisymmetry properties are built in. The state $|n_1, n_2, ..., n_i, ...\rangle$ may be written as

$$|n_1, n_2, ..., n_i, ...\rangle = \left(a_1^\dagger\right)^{n_1} ... \left(a_i^\dagger\right)^{n_i} ... |0\rangle,$$

where $|0\rangle$ is the vacuum state.

The number operator $N_i = a_i^\dagger a_i$ is defined by

$$N_i|n_1, ..., n_i, ...\rangle = n_i|n_1, ..., n_i, ...\rangle.$$

(b) Bosons

Much of the formalism for bosons is the same as for fermions. The essential difference is that the creation and annihilation operators for bosons satisfy commutation relations instead of the anticommutation relations for fermions. We denote a many-boson state vector as $|n_1, n_2, ..., n_i, ...\rangle$, where the non-negative integer n_i is the occupancy of orbital i. The creation operator $b_i^\dagger$ for the orbital i satisfies

$$b_i^\dagger|n_1, ..., n_i, ...\rangle = \sqrt{n_i + 1}\ |n_1, ..., n_i + 1, ...\rangle$$

and the destruction operator b_i for the orbital i is defined by

$$b_i|n_1, ..., n_i, ...\rangle = \sqrt{n_i}\ |n_1, ..., n_i - 1, ...\rangle.$$

Also, $b_i|0\rangle = 0$ for all i. The creation and destruction operators satisfy the commutation relations

$$[b_i, b_j] = \left[b_i^\dagger, b_j^\dagger\right] = 0,$$

$$[b_i, b_j^\dagger] = \delta_{ij}.$$

The number operator $N_i = b_i^\dagger b_i$ satisfy

$$N_i |n_1, \ldots, n_i, \ldots\rangle = n_i |n_1, \ldots\rangle$$

$$|n_1, \ldots, n_i, \ldots\rangle = \frac{\left(b_1^\dagger\right)^{n_1}}{\sqrt{n_1}} \cdots \frac{\left(b_1^\dagger\right)^{n_i}}{\sqrt{n_i}} \cdots |0\rangle.$$

17.3 ONE-PARTICLE AND TWO–PARTICLE OPERATORS

We now consider how to express operators in this representation. Let us consider an additive one-body operator

$$\mathcal{O} = \sum_{i=1}^{N} O(\vec{r}_i, \vec{p}_i) \tag{20}$$

which is a sum of operators, each of which acts only on one-particle. As an example, we have the external potential

$$V\left(\vec{r}_1, \ldots, \vec{r}_N\right) = \sum_{i-1}^{N} V(\vec{r}_i).$$

The expression for operators in terms of creation and annihilation operators is essentially the same for bosons and fermions. We shall demonstrate it for fermions.

We want to put the one-body operator in a form in the occupation number formalism such that the matrix elements of the operator as computed in occupation number representation and in Slater determinant method are equal. (see Raimes (1972)). In general, any single-particle operator can be written as

$$\underset{i}{\sum} h(\vec{r}_i) \quad \Rightarrow \quad \underset{\alpha,\beta}{\sum} \left\langle \underline{\Psi}_\beta(\vec{r}) | h(\vec{r}) | \underline{\Psi}_\alpha(\vec{r}) \right\rangle a_\beta^\dagger a_\alpha \tag{21}$$

$$\text{(particle labels)} \qquad \text{(state labels)}$$

$$= \sum_{\alpha, \beta} h_{\beta\alpha}\, a_\beta^\dagger\, a_\alpha.$$

Let us consider the matrix element of h between two determinantal wave functions:

$$\frac{1}{N!} \int \left\| \beta_1 \ldots \beta_N \right\|^* h(\vec{r}_1) \left\| \alpha_1 \ldots \alpha_N \right\| d^3 r_1 \ldots d^3 r_N$$

$$= \begin{cases} 0, \text{ if two or more labels in } \{\alpha\} \text{ and } \{\beta\} \text{ are different} \\[2mm] \pm \dfrac{1}{N} \left\langle \phi_{\beta_i}(\vec{r}_1) \left| h(\vec{r}_1) \right| \phi_{\alpha_i}(\vec{r}_1) \right\rangle, \text{ if only one} \\[2mm] \text{label in each set say } \beta_i \text{ and } \alpha_i, \text{ is different} \\[2mm] \pm \dfrac{1}{N} \sum_i \left\langle \phi_i(\vec{r}_1) \left| h(\vec{r}_1) \right| \phi_i(\vec{r}_1) \right\rangle, \text{ if the sets are} \\[2mm] \text{the same but differ by an even (+) or odd (−) permutations} \end{cases} \tag{22}$$

Notice that in the second case when only one label in each set is different, then the matrix element is nonzero in those $(N-1)!$ permutations which keep r in β_i and r in α_i. This matrix element Eq. (22) is, in the occupation number representation, using Eq. (21),

$$\sum_{\alpha,\,\beta} \langle \phi_\beta(r)|h(r)|\phi_\alpha(r)\rangle \left\langle 0 \left| a_{\beta_N} \cdots a_{\beta_1} a_\beta^\dagger a_\alpha a_{\alpha_1} \cdots a_{\alpha_N} \right| 0 \right\rangle, \tag{23}$$

when summed over $h(r_i)$

If two or more labels in the set $\{\beta\}$ are different from the set $\{\alpha\}$, then one of the a_β $\left(\text{say, } a_{\beta_i}\right)$ can be commuted to the right to give zero, by Eq. (12). In the second case, if β_i is not in the set $\{\alpha\}$ and α_i is not the set $\{\beta\}$ then β must equal β_i (otherwise we could commute a_{β_i} through to the right to give zero. Using the same argument, $\alpha = \alpha_i$. So Eq. (23) is

$$\pm \left\langle \phi_{\beta_i}(\vec{r}) \left| h(\vec{r}) \right| \phi_{\alpha_i}(\vec{r}) \right\rangle$$

In the last case, if the set $\{\alpha\} = \{\beta\}$, then (23) becomes

$$\pm \sum_{\alpha,\,\beta} \langle \phi_\beta|h|\phi_\alpha\rangle \left\langle 0 \left| a_{\alpha_N} \cdots a_{\alpha_1} a_\alpha^\dagger a_\alpha a_{\alpha_1}^\dagger \cdots a_{\alpha_N}^\dagger \right| 0 \right\rangle,$$

where $+$ $(-)$ sign is taken if an even (odd) permutation is necessary to match $\beta_1 \ldots \beta_N$ to $\alpha_1 \ldots \alpha_N$. If a_α destroys the particle in α_i then $a_\beta^\dagger$ must create the particle in α_i. Hence the matrix elements Eq. (23) are

$$\pm \sum_{i=1}^{N} \left\langle \phi_{\alpha_i}(\vec{r}) \left| h(\vec{r}) \right| \phi_{\alpha_i}(\vec{r}) \right\rangle.$$

This establishes the correspondence. Therefore, the one-body operator in the occupation number formalism is

$$h^{occ} = \sum_{i,\,j} h_{ij}\, a_i^\dagger\, a_j, \tag{24}$$

where
$$h_{ij} = \langle \phi_i|h|\phi_j\rangle.$$

Next, we consider the two-body operator

$$g = \frac{1}{2} \sum_{\substack{i,\,j=1 \\ (i \neq j)}}^{N} g\left(\vec{r}_i,\, \vec{p}_i,\, \vec{r}_j,\, \vec{p}_j\right). \tag{25}$$

For example, the interaction potential is

$$V\left(\vec{r}_1,\, \ldots,\, \vec{r}_N\right) = \sum_{\substack{i,\,j=1 \\ (i \neq j)}}^{N} V\left(\vec{r}_i - \vec{r}_j\right)$$

An important quantity of this type is the Coulomb-repulsion term

$$g(i,\, j) = e^2/r_{ij}.$$

In general, any two-body operator may be written as

$$\sum_{i>j} g(r_{ij}) = \frac{1}{2} \sum_{i,j} g(r_{ij})$$

(particle labels)

$$\Rightarrow \frac{1}{2} \sum_{ijkl} \langle \phi_i(r_1)\phi_j(r_2)|V(r_1, r_2)|\phi_k(r_1)\phi_l(r_2)\rangle \, a_i^\dagger a_j^\dagger a_l a_k. \tag{26}$$

state labels

We now proceed in a similar fishion so that we require, as in the case of one-body operator the equality of the matrix elements of the two-body operator, as computed in the occupation number representation and in Slater determinant method. The details of the rules for computing matrix elements of one-body and two-body operators between determinantal functions are given in Condon & Shortley (1951). The matrix element is

$$\frac{1}{N!} \int \|\beta_1 \, ... \, \beta_N\|^* \, g(i, j) \, \|\alpha_1 \, ... \, \alpha_N\| \, d^3r_1 \, ... \, d^3r_N. \tag{27}$$

For nonvanishing value $\{\beta\}$ can differ from $\{\alpha\}$ by at most two of the individual sets. The matrix element is

(a) $\displaystyle \pm \sum_{i \neq j} \left[\int \phi_i^*(\beta_k) \, \phi_j^*(\beta_l) \, g(i, j) \, \phi_i(\alpha_k)\phi_j(\alpha_l) \right.$

$$\left. - \int \phi_i^*(\beta_k) \, \phi_j^*(\beta_l) \, g(i, j) \, \phi_i(\alpha_l)\phi_j(\alpha_k) \right], \tag{28}$$

where $\{\alpha\}$ and $\{\beta\}$ differ by two individual sets (i.e. in two one-electron states). The $\pm$ sign is determined by the parity of permutation necessary to achieve maximum matching between $\{\alpha\}$ and $\{\beta\}$.

(b) the matrix element is

$$\pm \sum_{\beta_l} \left[\int \phi_i^*(\beta_k) \, \phi_j^*(\beta_l) \, g(i, j) \, \phi_i(\alpha_k)\phi_j(\beta_l) \right.$$

$$\left. - \int \phi_i^*(\beta_k) \, \phi_j^*(\beta_l) \, g(i, j) \, \phi_i(\beta_l)\phi_j(\alpha_k) \right], \tag{29}$$

where β_l runs over the $(N - 1)$ individual sets that are common to $\{\alpha\}$ and $\{\beta\}$, and $\{\alpha\}$ differs from $\{\beta\}$ by one set.

(c) The diagonal element is

$$\pm \sum_{k > l = 1} \left| \phi_i^*(\beta_k) \, \phi_j^*(\beta_l) \, g(i, j) \, \phi_i(\beta_k)\phi_j(\beta_l) \right.$$

$$- \int \phi_i^*(\beta_k) \, \phi_j^*(\beta_l) \, g(i, j) \, \phi_i(\beta_l)\phi_j(\beta_k) \tag{30}$$

(a) Let us write the N-electron state vectors which differ by two one-electron states as

$$\left| \Phi_{\{\alpha\}}^N \right\rangle = \left| ..., 1_p, ..., 0_r, ..., 1_q, ..., 0_s, ... \right\rangle \tag{31}$$

contains ϕ_p, ϕ_q but not ϕ_r, ϕ_s

and

$$\left| \Phi^N_{\{\beta\}} \right\rangle = |..., 0_p, ..., 1_r, ..., 0_q, ..., 1_s, ...\rangle$$

contains ϕ_r, ϕ_s but not ϕ_p, ϕ_q

For a nonzero value of the matrix element $\left\langle \Phi^N_{\{\beta\}} \right| a_i^+ a_j^+ a_L a_k \left| \Phi^N_{\{\alpha\}} \right\rangle$

the annihilation operators must destroy the state ϕ_p, ϕ_q and the creation operators must create the states ϕ_r and ϕ_s.

$$\left\langle \Phi^N_{\{\beta\}} \right| a_i^+ a_j^+ a_l a_k \left| \Phi^N_{\{\alpha\}} \right\rangle \tag{32}$$

There are four possibilities (p, q, r, s are all different)

(i) $i = r, j = s, l = p, k = q$ so that

$$a_i^\dagger a_j^\dagger a_l a_k \left| \Phi_{\{\alpha\}} \right\rangle = a_r^\dagger a_s^\dagger a_p a_q \left| \Phi_{\{\alpha\}} \right\rangle$$

$$= - a_r^\dagger a_p a_s^\dagger a_q \left| \Phi_{\{\alpha\}} \right.$$

$$= - \left| \Phi_{\{\beta\}} \right\rangle.$$

The other three possibilities are

(ii) $i = r, j = s, l = q, k = p$ so that

$$a_i^\dagger a_j^\dagger a_l a_k \left| \Phi_{\{\alpha\}} \right\rangle = \left| \Phi_{\{\beta\}} \right\rangle$$

(iii) $i = s, j = r, l = p, k = q$ so that

$$a_i^+ a_j^+ a_l a_k \left| \Phi_{\{\alpha\}} \right\rangle = \left| \Phi_{\{\beta\}} \right\rangle$$

(iv) $i = s, j = r, l = q, k = p$ so that

$$a_i^\dagger a_j^\dagger a_l a_k \left| \Phi_{\{\alpha\}} \right\rangle = - \left| \Phi_{\{\beta\}} \right\rangle.$$

Thus,

$$\left\langle \Phi_{\{\beta\}} \right| g \left| \Phi_{\{\alpha\}} \right\rangle = \frac{1}{2} \sum_{i, j, k, l} \langle ij|g|kl \rangle \left\langle \Phi_{\{\beta\}} \right| a_i^\dagger a_j^\dagger a_l a_k \left| \Phi_{\{\alpha\}} \right\rangle$$

$$= -\frac{1}{2} \langle rs|g|qp \rangle + \frac{1}{2} \langle rs|g|pq \rangle$$

$$-\frac{1}{2} \langle sr|g|qp \rangle - \frac{1}{2} \langle sr|g|pq \rangle$$

$$= \langle rs|g|pq \rangle - \langle sr|g|pq \rangle. \tag{33}$$

(b) Next, we consider $\left| \Phi_{\{\alpha\}} \right\rangle$ and $\left| \Phi_{\{\beta\}} \right\rangle$ differ in only one function, say,

$$\left| \Phi_{\{\alpha\}} \right\rangle = |..., 1_p, ..., 0_q, ...\rangle \tag{34}$$

$$\left| \Phi_{\{\beta\}} \right\rangle = \left| ..., 0_p, ..., 1_q, ... \right\rangle \tag{35}$$

For a nonzero value of the matrix element, the annihilation operator must destroy ϕ_p and the creation operator creates ϕ_q. There are four possibilities

(i) $k = p, j = q, i = l$, so that

$$a_i^\dagger a_j^\dagger a_l a_k \left| \Phi_{\{\alpha\}} \right\rangle = a_i^\dagger a_q^\dagger a_i a_p \left| \Phi_{\{\alpha\}} \right\rangle$$
$$= - a_i^\dagger a_i a_q^\dagger a_p \left| \Phi_{\{\alpha\}} \right\rangle$$
$$= - n_i \left| \Phi_{\{\beta\}} \right\rangle$$

(ii) $k = p, i = q, j = l$, so that

$$a_i^\dagger a_j^\dagger a_l a_k \left| \Phi_{\{\alpha\}} \right\rangle = n_j \left| \Phi_{\{\beta\}} \right\rangle$$

(iii) $l = p, j = q, i = k$, so that

$$a_i^\dagger a_j^\dagger a_l a_k \left| \Phi_{\{\alpha\}} \right\rangle = n_j \left| \Phi_{\{\beta\}} \right\rangle$$

(iv) $l = p, i = q, j = k$, so that

$$a_i^\dagger a_j^\dagger a_l a_k \left| \Phi_{\{\alpha\}} \right\rangle = - n_j \left| \Phi_{\{\beta\}} \right\rangle$$

The matrix element becomes

$$\frac{1}{2} \sum_{i,j,k,l} \langle ij|g|kl \rangle \, \Phi_{\{\beta\}}^* \, a_i^\dagger a_j^\dagger a_l a_k \, \Phi_{\{\alpha\}} d\tau$$

$$= \frac{1}{2} \sum_j \langle qj|g|pj \rangle n_j - \frac{1}{2} \sum_i \langle iq|g|pi \rangle n_i$$

$$+ \frac{1}{2} \sum_i \langle iq|g|ip \rangle n_i - \frac{1}{2} \sum_j \langle qj|g|jp \rangle n_j$$

$$= \sum_i \left[\langle iq|g|ip \rangle - \langle qi|g|ip \rangle n_i \right]. \tag{36}$$

Now,

$$n_i = \begin{cases} 1, & \text{for } i = \alpha_1, ..., \alpha_N \text{ in this standard order} \\ 0, & \text{otherwise} \end{cases}$$

Then the matrix element is identical to the previous one.

(c) In a similar fashion, we can show that the diagonal elements are same. Therefore, any two-body operator may be written in the occupation number representation as

$$g^{ooc} = \frac{1}{2} \sum_{i,j,k,l} g_{ijkl} \, a_i^\dagger a_j^\dagger a_l a_k, \tag{37}$$

where

$$g_{ij,\,kl} = \langle ij|g|kl \rangle.$$

The expressions Eqs. (24) and (37) for the one-body and two-body operators in the occupation number representation are the same for bosons and fermions. Note the reversal of order of the last two operators relative to the order of the indices in the matrix element. This is necessary for fermions due to the anticommuting nature of the operators. In the case of bosons, this ordering is irrelevant due to the commuting nature of the operators. We can, of course, reverse the order of the first two operators, keeping the order of the last two intact.

Let us consider a system of N identical particles (bosons or fermions) with a Hamiltonian

$$H = \sum_{k=1}^{N} T(x_k) + \frac{1}{2} \sum_{k \neq l = 1}^{N} V(x_k, x_l) \tag{38}$$

where T is the kinetic energy and V is the potential energy of interaction between particles. x_i denotes the coordinates (space, spin or isospin) of the ith particle. In the occupation number representation the state vector $|\overline{\psi}(t)\rangle$ satisfies the Schridinger equation

$$i\hbar \frac{\partial}{\partial t} |\overline{\psi}(t)\rangle = H|\overline{\psi}(t)\rangle, \tag{39}$$

where H is now an operator in the abstract occupation-number space and is given by

$$H = \sum_{i,j} \langle i|T|j \rangle \, c_i^\dagger c_j^\dagger + \frac{1}{2} \sum_{i,j,k,l} \langle ij|V|kl \rangle \, c_i^\dagger c_j^\dagger c_l c_k, \tag{40}$$

where the creation and destruction operator are denoted by $c^\dagger$ and c (for generality). The creation and annihilation operators contain all of the statistics and operator properties. Here the matrix element $T_{ij} = \langle i|T|j \rangle$ of the kinetic energy and the matrix element $V_{ijkl} = \langle ij|V|kl \rangle$ of the potential energy taken between the single-particle eigenstates of the Schrödinger equation (in first quantization) are complex numbers.

17.4   PARTICLE – HOLE FORMALISM FOR FERMIONS

We can make simplification in our calculations in the occupation number language if we consider the ground state of a system of non-interacting fermions, N in number, filling the lowest N level upto the Fermi level as the "zero" ("Fermi vacuum") and recording changes from this in terms of particles and holes. We write the Fermi vacuum as

$$|0\rangle = a_{k_1}^\dagger \, a_{k_2}^\dagger \, ... \, a_{k_N = k_F}^\dagger \, |\text{vacuum}\rangle, \tag{41}$$

where k_F is the Fermi momentum. Hence, we have the two defining equations for $|0\rangle$:

$$a_k|0\rangle = 0, \quad k > k_F$$
$$a_k^\dagger|0\rangle = 0, \quad k < k_F \tag{42}$$

This is a vacuum with respect to "quasi-particles". The first equation in Eq. (42) implies that it is impossible to annihilate a particle, outside the Fermi sea (for there are none) while the second one means that it is impossible to create a particle inside the Fermi sea, for all states are occupied.

In the particle-hole formalism, we rewrite the operators for $k < k_F$ as

$$a_k^\dagger = b_k,$$
$$a_k = b_k^\dagger, \tag{43}$$

so that Eq. (42) becomes

$$a_k|0\rangle = 0, \quad k > k_F$$
$$b_k|0\rangle = 0, \quad k < k_F \tag{44}$$

Thus,

for $k_i > k_F$, we have the operators

$$a_i^\dagger \ \text{(creates particle)}$$
$$a_i \ \text{(destroys particle)}$$

while for $k_j \leq k_F$, we have the operators

$$b_j \ \text{(destroys hole)}$$
$$b_j^{\ \dagger} \ \text{(creates hole)}$$

The particle-hole transformations may be written as

$$c_i = \theta_{k_i - k_F} \, a_i + \theta_{k_F - k_i} \, b_i^\dagger \tag{45}$$
$$c_i^\dagger = \theta_{k_i - k_F} \, a_i^\dagger + \theta_{k_F - k_i} \, b_i$$

where $\theta_x = 1$ for $x > 0$ and $\theta_x = 0$ for $x < 0$.

The commutation rules for the particle and hole operators can be obtained from the fermion commutation rules. We have

$$\left\{ a_i^\dagger, a_k \right\} = \delta_{ik}, \quad \left\{ b_j^\dagger, b_l \right\} = \delta_{jl}, \tag{46}$$

and all other anticommutators vanish. This is a canonical transformation, which preserves the anticommutation relations. Obviously, the a's and b's anticommute with each other. In the particle-hole formalism, particle and hole states are distinct, so that

$$\vec{k}_{\text{hole}} \neq \vec{k}_{\text{particle}}.$$

In terms of the particle and hole operators, the many-particle Hamiltonian H_0 is easy to rewrite. We have

$$H_0 = \sum_i \in_i c_i^\dagger c_i$$

$$= \sum_{\in_j \le \in_F} \in_j c_j^\dagger c_j + \sum_{\in_j > \in_F} \in_i c_i^\dagger c_i$$

$$= \sum_{\in_j \le \in_F} \in_j b_j b_j^\dagger + \sum_{\in_j > \in_F} \in_i a_i^\dagger a_i$$

$$= \sum_{\in_j \le \in_F} \in_j - \sum_{\in_j > \in_F} \in_j b_j^\dagger b_j$$

$$+ \sum_{\in_i > \in_F} \in_i a_i^\dagger a_i,$$

using Eq. (46)

$$= \sum_{k < k_F} \in_k - \sum_{k < k_F} \in_k b_k^\dagger b_k + \sum_{k > k_F} \in_k a_k^\dagger a_k. \tag{47}$$

The first term is the energy of the Fermi vacuum, and the second term shows that the holes have negative energy, since $b_k^\dagger b_k$ is the hole number operator. We note that the hole wave function is that of the particle with energy negative of the particle energy. The absence of a particle of energy E, momentum $+\vec{k}$, from the occupied Fermi sea implies that the hole behaves with energy negative to the energy of the particle and momentum $-\vec{k}$. The hole behaves with charge negative to the charge of the particle and spin $-\vec{\sigma}$. If the total number of particle (fermions) is fixed, then particle and holes occur in pairs, with each pair having a net positive energy. For bosons, obviously, there is no particle-hole formalism.

17.5   FIELD OPERATORS

The treatment so far of the creation and annihilation operators has been based on single-particle basis functions which are eigenstates of the single-particle Hamiltonian (i.e., the energy operator). We now introduce another convenient set of single particle states – the eigenstates of the position operator – and write

$$\psi(x) = \sum_i \phi_i(x) c_i \tag{48}$$

$$\psi^\dagger(x) = \sum_i \phi_i^\dagger(x) c_i^\dagger. \tag{49}$$

Here, ϕ's are single-particle wave functions and the sum is over the complete set of single-particle quantum numbers.

These operators ψ and $\psi^\dagger$ are called "field operators" and are operators in the abstract occupation – number Hilbert space. The operator $\psi^\dagger(x)$ creates a particle at x (at position $\vec{r}$ with spins) while the operator $\psi(x)$ destroys a particle at x. The number density operator is

$$n(x) = \psi^\dagger(x)\psi(x). \tag{50}$$

Its eigenvalues are the number of particles at x. The total number operator is

$$N = \int \psi^\dagger(x)\psi(x)d^3x. \tag{51}$$

These field operators satisfy the following commutation (anticommutation) relations depending on the statistics

$$\left[\psi_\alpha(x), \psi^\dagger{}_\beta(x')\right]_\mp = \sum_i \phi_i(x)_\alpha\, \phi_i(x')^*_\beta$$

$$= \delta_{\alpha\beta}\, \delta(x - x'), \tag{52}$$

$$\left[\psi_\alpha(x), \psi_\beta(x')\right]_\mp = \left[\psi^\dagger{}_\alpha(x), \psi^\dagger{}_\beta(x')\right]_\mp = 0, \tag{53}$$

where the upper (lower) sign refers to bosons (fermions).

These relations follow from the commutation or anticommutation relations for the boson or fermion creation and destruction operators and the completeness condition for the single-particle wave functions.

Using the field operators, we may write the Hamiltonian Eq. (38) as

$$H = \int d^3x\, \psi^\dagger(x)T(x)\psi(x)$$

$$+ \frac{1}{2} \iint d^3x d^3x' \psi^\dagger(x)\psi^\dagger(x')V(x, x')\psi(x')\psi(x). \tag{54}$$

For fermions, we can split $\psi^\dagger$, ψ into particle and hole parts. First, we redefine the fermion operator as

$$c_{k\lambda} = \begin{cases} a_{k\lambda} & k > k_F \quad \text{(particle)} \\ b^\dagger{}_{-k\lambda} & k < k_F \quad \text{(hole)} \end{cases} \tag{55}$$

so that the field operator is

$$\psi(x) = \sum_{k\lambda} \phi_{k\lambda}(x)\, c_{k\lambda}. \tag{56}$$

In terms of the particle/hole operator, we write the field operator (in the Schrödinger representation)

$$\psi(x) = \sum_{k_\lambda > k_F} \phi_{k\lambda}(x)\, a_{k\lambda} + \sum_{k_\lambda < k_F} \phi_{k\lambda}(x)\, b^\dagger{}_{-k\lambda}$$

$$= \psi_{\text{part}}(x) + \psi^\dagger_{\text{hole}}(x), \tag{57}$$

and similarly,

$$\psi^\dagger(x) = \psi^\dagger_{\text{part}}(x) + \psi_{\text{hole}}(x). \tag{58}$$

17.6   WICK'S THEOREM

This theorem is very useful in evaluating matrix elements of operators in the occupation number representation. For example, the matrix element of a two-body operator between the states $|i\rangle$ and $|j\rangle$ is

$$\langle j|V|i\rangle = \frac{1}{2} \sum_{r,\,s,\,m,\,n} \langle rs|V|mn\rangle\, \langle j|c_r^\dagger c_s^\dagger c_n c_m|i\rangle$$

$$= \frac{1}{2} \sum_{r,\,s,\,m,\,n} \langle rs|V|mn\rangle\, \langle 0|c_{\alpha'} c_{\beta'} \dots c_r^\dagger c_s^\dagger c_n c_m c_\alpha^\dagger c_\beta^\dagger \dots |0\rangle. \tag{59}$$

To evaluate this by direct application of commutation or anticommutation relations is tedious and lengthy. Instead we use Wick's theorem, which is an operator identity involving the creation and the destruction operators thereby greatly simplifying the calculation. Wick's theorem provides a general procedure for evaluating these quantities in an elegant, systematic and compact manner (Schweber (1961)).

Before stating the theorem, we introduce some definitions.

(a)   Time Ordered Product (Wick T-product)

The time ordered product of N operators is that in which the times of the operators increase from right to left with a factor of -1 for each interchange of fermion operators. For example,

$$T\,[A(t_1)B(t_2)] = \begin{cases} A(t_1)B(t_2), & t_1 > t_2 \\ -B(t_2)A(t_1), & t_1 < t_2 \end{cases} \tag{60}$$

Thus,

$$T\,[A(t_2)B(t_4)C(t_1)D(t_3) \dots X(t_N)]$$

$$= (-1)^P\, X(t_N) \dots B(t_4)D(t_3)A(t_2)C(t_1),$$

$$t_N > t_{N-1} > \dots > t_2 > t_1. \tag{61}$$

where P is the number of interchanges of operators needed to get the operators in the proper time order, starting with the given order. The factor $(-1)^P$ is absent for bosons.

(b)   Normal Product (Wick N-product)

The normal order of product of operators is that in which all the creation operators are placed to the left of all the destruction operators. Thus,

$$N[ABC^\dagger DF^\dagger G^\dagger \dots] = (-1)^P\, [C^\dagger F^\dagger G^\dagger \dots ABD \dots], \tag{62}$$

where P is the number of interchanges of adjacent operators required to get from the given order to the normal order. For example,

$$N[a_k b_l^\dagger a_m^\dagger] = b_l^\dagger a_m^\dagger a_k$$

$$= -\, a_m^\dagger b_l^\dagger a_k \tag{63}$$

The order among the different destruction or creation operators is immaterial (proper sign should be put).

The normal product has the following properties:

(a) Its vacuum expectation value vanishes identically

$$\langle 0|N[ABC^\dagger D \ldots]|0\rangle = 0. \tag{64}$$

If there is at least one destruction operator in the product, then the result is true by Eq. (44). If the product consists of creation operators only then the result is true, since

$$\langle 0|a_j^\dagger = a_j|0\rangle = 0, \text{ etc.}$$

The utility of the normal product in simplifying the evaluation of matrix elements lies in this important property.

(b) Anticommutativity (for fermions):

$$N[A^\dagger B^\dagger C] = -N[B^\dagger A^\dagger C] \tag{65}$$

(c) Distributivity:

$$N[(A + B)(C + D) \ldots] = N[AC \ldots] + N[AD \ldots]$$
$$+ N[BC \ldots] + N[BD \ldots] + \ldots \tag{66}$$

The *T*-product is also distributive.

The vacuum expectation value of *T*-product for operators may be evaluated by transforming it to the corresponding *N*-product. The problem however, is the enumeration of the additional terms generated in the process of conversion of the *T*-product to the *N*-product. These additional terms are defined below.

(c) Contraction

The contracted product or the contraction of two operators A and B is denoted $\overline{AB}$ and is equal to the difference between the *T*-product and the *N*-product of A and B:

$$\overline{AB} = T[AB] - N[AB] \tag{67}$$

Evidently, the contraction is a c number in the occupation number Hilbert space, not an operator and has the magnitude of the anticommutator $\{A, B\}$, or zero, depending on whether $T[AB]$ and $N[AB]$ are the same or different. Note that

$$AB = \pm \overline{BA} \tag{68}$$

Also, $$AB = \langle 0|T[AB]|0\rangle, \tag{69}$$

using Eq. (64).

The contraction is just the unperturbed propagator.

(d)   A Convention

A normal product of operators with more than one contraction will have the contractions denoted by joining these operators by lines. Two operators that are contracted must be brought together by rearrangement of the order of the operators within the normal product, keeping track of the sign during this process. After rearrangement the values of the contractions are substituted and being numbers, can be taken outside the normal-ordered product. For example,

$$N[ABCDEFG\ldots] = N[\overline{AE}\ \overline{BD}\ CFG\ldots]$$

$$= \pm\,\overline{AE}\ \overline{BD}\ N[CFG\ldots] \tag{70}$$

(e)   For Operators A, B, …. at Equal Times

$$T[A(t)B(t)\ldots] = (-1)^P \times \text{operators, rearranged so all}$$
$$c^\dagger\text{s (or } a^\dagger_s, \text{ \& or } b\text{'s) stand to the}$$
$$\text{left of } c\text{'s (or } a\text{'s or } b^\dagger\text{s)} \tag{71}$$

The factor $(-1)^P$ is not present in the boson case. For example,

$$T\!\left[a_k a_k^\dagger\right] = -\,a_k^\dagger a_k, \quad k > k_F$$
$$T\!\left[b_k^\dagger b_k\right] = -\,b_k b_k^\dagger, \quad k \le k_F \tag{72}$$

for equal time fermion operators. The contraction of two operators at the same time vanishes.

Remark

We associate a parameter ("time") with each operator so that a time-ordered product may be defined. We shall state Wick's theorem for a time ordered product, though time dependence and time ordering are not essential. One may state Wick's theorem for a general product of operators. The contraction of two operators may then be defined as their vacuum expectation value.

We are now in a position to state Wick's theorem.

Wick's Theorem

Any time-ordered product of creation and destruction operators can be decomposed into a (unique) sum over pure normal product, normal product with one or more contractions, and completely contracted normal products as follows

$$T\,[UVW\ldots XYZ] = N[UVW\ldots XYZ] + N[\overline{U}VW\ldots XYZ]$$

$$+\, N[U\overline{VW}\ldots XYZ] + \ldots$$

$$+\, N[\overline{U}\,\overline{VW}\ldots XYZ] + N[\overline{U}V\overline{W}\ldots XYZ] + \ldots$$

$$+\, \ldots$$

$$+\, N[\overline{UVW}\ldots XYZ] + N[\overline{UVW}\ldots X\overline{YZ}] + \ldots \tag{73}$$

The theorem essentially gives all the additional terms that occur in reordering a T-product into an N-product. We see that Eq. (67) is a special case of this. If we take the vacuum expectation value of both sides, the non-fully contracted normal products vanish by Eq. (64), leaving

$$\langle 0|T[UVW....XYZ]|0\rangle = \overrightarrow{UV}\,\overrightarrow{W}\,...\,X\,\overrightarrow{YZ} + \overrightarrow{U}\,\overrightarrow{VW}\,...\,X\overrightarrow{YZ} + ... \tag{74}$$

$$= \text{sum over all possible fully contracted products.}$$

Equation (73) is of great help in evaluating matrix elements. Wick's theorem is an operator identity so that it can be used to evaluate any arbitrary matrix element. However, its important use is in the evaluation of vacuum expectation value (or, ground-state average), using Eq. (74). For example, the exact Green' function consists of all possible fully, contracted terms.

To prove Wick's theorem, we follow Wick's derivation. First we prove the following useful lemma.

Basic lemma

If $N[UV....XY]$ is a normal ordered product, Z is an operator with a time earlier than the times of U, V, ..., X, Y, then

$$N[UV ... XY]Z = N[UV ... XYZ] + N[\overrightarrow{UV ... XYZ}]$$

$$+ N[\overrightarrow{UV} ... XYZ] + ... + N[UV ... X\overrightarrow{YZ}] \tag{75}$$

$$= N[UV ... XYZ] + \underset{\text{all } U ... Y}{\sum} N[UV ... \overrightarrow{XYZ}].$$

This lemma states that if we multiply an N-product on the right by an operator at an earlier time then we get a normal product with the extra-operator included with a sum of normal products where the extra-operator is contracted with all the operators in the original product.

We note the following points.

(a) We may assume that the operator product $UV....XY$ is in normal order. If not, then the operators can be reordered on both sides with the same signature factor occurring in each term of Eq. (75) and cancels identically.

(b) If Z is a destruction operator, then $\overrightarrow{UZ} = 0$ for any U, no matter whether U is a creation or an annihilation operator. Hence the lemma is true when Z is a destruction operator.

(c) If Z is a creation operator then we may assume U, V, ..., Y are all destruction operators, since the contractions of Z with other creation operators are zero. If the lemma is proved in this form, then creation operators may be included by multiplying on the left.

The lemma Eq. (75) will be proved in general if we assume Z is a creation operator and all others U .... Y are destruction operators. The proof follows by the method of induction.

The lemma Eq. (75) is true for $n = 1$ since

$$N(Y)Z = T(YZ) = N(YZ) + \overline{YZ}. \tag{76}$$

We now assume the lemma to be true for n operators and prove it for n+1 operators.

We multiply Eq. (75) on the left by another destruction operator D with a time later than the time of Z $(t_D > t_Z)$.

$$DN[UV \ .... \ XY] \ Z = DN[UV \ .... \ XYZ]$$
$$+ DN[\overline{UV ... XYZ}] + ... + DN[UV ... X\overline{YZ}]. \tag{77}$$

The operator D can be taken inside the normal ordering in all terms wherever Z has been contracted, since the contraction of Z with any destruction operator $(U, V, ...$ are all destruction operators) is a c number. In the first term on the right D has to be kept outside the normal ordering, since here Z is an (uncontracted) operator. We write Eq. (77) as

$$[UV ... XYZ] = [UV ... XYZ] + N[D\overline{UV ... XYZ}]$$
$$+ ... + N[DUV ... X\overline{YZ}]. \tag{78}$$

Now, $\qquad N[UV ... XYZ] = (-1)^p \, ZUV ... XY$

and $\qquad DN[UV ... XYZ] = (-1)^p \, DZUV ... XYZ \tag{79}$

But, $\qquad\qquad DZ = T[DZ] = \overline{DZ} + N[DZ]$

$$= DZ + (-1)^q \, ZD,$$

since $t_D > t_Z \, (q = 1)$

From Eq. (79), we have

$$DN[UV ... XYZ] = (-1)^p \, T[DZ] \, UV... XY$$
$$= (-1)^p \, \overline{DZ}UV ... XY + (-1)^p \, N[DZ]UV ... XY$$
$$= (-1)^p \, \overline{DZ}UV ... XY + (-1)^{p+q} \, ZDUV ... XY$$
$$= (-1)^{2p} \, [\overline{DUV ... XYZ}] + (-1)^{2(p+q)} \, N[DUV ... XYZ]$$
$$= N[\overline{DUV ... XYZ}] + N[DUV ... XYZ]. \tag{80}$$

Substituting Eq. (80) into Eq. (78), we get the relationship

$$DN[UV ... XY] \, Z = N[DUV ... XYZ]$$
$$= N[DUV ... XYZ] + N[D\overline{UV ... XYZ}] + ...$$
$$... + N[DUV .... X\overline{YZ}]. \tag{81}$$

Hence, if the theorem is true for n operators it is true for $(n + 1)$ operators. Since it is true for $n = 1$, it is true for all n. The basic lemma Eq. (75) is proved. It is easy to generalize the lemma. Multiplying both sides of Eq. (75) by $\overline{RS}$, and interchanging the operators on both sides, we see that since each term has the same overall sign change which cancels identically, the lemma holds when there is any number of contractions within $UV \dots XY$.

We are now in a position to prove Wick's theorem.

The proof is again by induction.

It is true for two operators, by the definition of a contraction:

$$T[UV] = N[UV] + N[UV] = N[UV] + \overline{UV}.$$

Let us assume that the theorem is true for n factors, and multiply Eq. (73) on the right by an operator Ω with a time argument earlier than that of any other operator. We have

$$T[UVW \dots XYZ]\Omega = T[UVW \dots XYZ\Omega]$$

$$= N[UVW \dots XYZ]\Omega + N[\overline{UV}W \dots XYZ]\Omega$$

$$+ \dots + N[\overline{UV}\overline{W} \dots \overline{X}YZ]\Omega + \dots$$

$$+ \dots + \Sigma\, N[\overline{UVW \dots XYZ}]\Omega$$

The operator Ω can be included in the T-product since its time is earlier than the times of all other operators already in the T-product. Now, we apply the basic lemma Eq. (75) to each term on the right to introduce Ω into the normal ordered product. This immediately establishes Wick's theorem to be true for $n + 1$ factors provided that $t_{-\Omega} < t_U, t_V, \dots, t_Z$. Once Ω is inside the T-and N-products, we may simultaneously reorder the operators in each term. The sign conventions give the same overall sign on both sides of the equation. The equation, therefore, is true with the restriction on the time argument of the operator Ω now removed. Hence, we have proved Wick's theorem.

We now apply the theorem to evaluate the vacuum expectation value of the operator product $a_i a_j a_k^\dagger a_l^\dagger$.

We write

$$\left\langle 0 \middle| a_i a_j a_k^\dagger a_l^\dagger \middle| 0 \right\rangle = \left\langle 0 \middle| a_i(t_1) a_j(t_2) a_k^\dagger(t_3) a_l^\dagger(t_4) \middle| 0 \right\rangle,$$

with $t_1 > t_2 > t_3 > t_4$.

$$= \left\langle 0 \middle| T\!\left[a_i a_j a_k^\dagger a_l^\dagger \right] \middle| 0 \right\rangle,$$

the time arguments are suppressed.

By Wick's theorem,

$$\langle 0|T[a_i a_j a_k^\dagger a_l^\dagger]|0\rangle = \text{sum over all fully contracted products.} \tag{82}$$

Since the contraction of two creation operators or two destruction operators vanishes, the nonvanishing contractions are of the type $\overline{a_i\, a_k^\dagger}$.

Now,

$$\overline{a_i\, a_k^\dagger} = T[a_i\, a_k^\dagger] - N[a_i\, a_k^\dagger]$$
$$= a_i\, a_k^\dagger + a_k^\dagger\, a_i$$
$$= \{a_i,\, a_k^\dagger\}$$
$$= \delta_{ik}.$$

Hence,

$$\langle 0|a_i a_j a_k^\dagger a_l^\dagger|0\rangle = \overline{a_i\, a_j\, a_k^\dagger\, a_l^\dagger} + \overline{a_i\, a_j\, a_k^\dagger\, a_l^\dagger}$$
$$= \delta_{il}\,\delta_{jk} - \delta_{ik}\,\delta_{jl}. \tag{83}$$

17.7 THE HARTREE APPROXIMATION

The problems associated with many-particle systems are in general extremely complicated.

For many-fermion systems like nucleons in nuclei and electrons in atoms, molecules and solids, approximate methods have been developed to obtain the groundstate and the lowest excited states. Most of these methods are essentially based on the idea of a self-consistent field introduced by Hartree (1928) and improved upon by Fock (1930) and by Slater (1930). The Hartree approximation is a first step toward the Hartree-Fock approximation. Our discussion will focus on many-electron systems. For a detailed survey of Hartree-Fock approach to nucleons in nuclei, see Eisenberg and Greiner (1972).

The Hamiltonian for a system of N electrons is

$$H = \sum_{i=1}^{N} H(i) + \sum_{i,\,j(i \neq j)}^{N} V(ij). \tag{84}$$

The Schrödinger equation for this system is

$$H|\overline{\Psi}\rangle = E|\overline{\Psi}\rangle \tag{85}$$

The Hartree method is equivalent to a variational principle (first pointed out by Fock and Slater) in which

$$E[\Phi] = \langle \overline{\Psi}|H|\overline{\Psi}\rangle = \text{minimum} \tag{86}$$

and the many-particle wave function is chosen as the product of N single-particle wave functions

$$\overline{\Psi}(x_1, x_2, ..., x_N) = \phi_1(x_1)\,\phi_2(x_2)\,...\,\phi_N(x_1). \tag{87}$$

Since ϕ's are assumed to be normalized, $\overline{\Psi}$ is normalized. We first find the expectation value of H

$$\langle\overline{\Psi}|H|\overline{\Psi}\rangle = \sum_i \phi_i^*(x_i)H(i)\phi_i(x_i)d\tau_i$$

$$+ \frac{1}{2}\sum_{i,j}' \int \phi_i^*(x_i)\phi_j^*(x_j)V(ij)\phi_i(x_i)\phi_j(x_j)d\tau_i\,d\tau_j$$

$$= \sum_i \int \phi_i^*(x_i)H(1)\phi_i(x_i)d\tau_i$$

$$+ \frac{1}{2}\sum_{i,j}' \int \phi_i^*(x_1)\phi_j^*(x_2)V(1,\,2)\phi_i(x_1)\phi_j(x_2)d\tau_1\,d\tau_2. \tag{88}$$

Now, we need to find an extremum (hopefully, a minimum) for $\langle\overline{\Psi}|H|\overline{\Psi}\rangle$. To take into account the constraint of normalization, we use the method of Lagrange multipliers. The variational principle then gives.

$$\delta\left[\langle\overline{\Psi}|H|\overline{\Psi}\rangle - \sum \lambda_i \int \phi_i^*(x_i)\phi_i(x_i)d\tau_i\right] = 0, \tag{89}$$

where δ is an arbitrary variation of the single-particle wave functions and λ_i are Lagrange multipliers. Choosing $\delta = \delta_k$, we obtain

$$\int\delta_k\phi_k(x_1)\left\{\left[H(1)\phi_k(x_1) + \left(\sum_{j(\neq k)} \int\phi_j^*(x_2)V(1,\,2)\phi_j(x_2)d\tau_2\right)\phi_k(x_1)\right]\right.$$

$$\left. - \lambda_k\phi_k(x_1)\right\}d\tau_1 + c.c = 0. \tag{90}$$

Since $\delta_k\phi_k$ and $\delta_k\phi_k^*$ are independent and arbitrary, we obtain the Hartree equations

$$H(1)\phi_k(x_1) + \left[\sum_{j(\neq k)} \int\phi_j^*(x_2)V(1,\,2)\phi_j(x_2)d\tau_2\right]\phi_k(x_1)$$

$$= \lambda_k\phi_k(x_1) \tag{91}$$

$k = 1, 2, ... N.$

The variational principle thus leads us to N simultaneous equations for the one-particle functions ϕ's. But these are extremely complicated nonlinear integro-differential equations. Each equation describes an electron subject to an effective field due to other electrons and to the nucleus. One solves Eq. (91) by iteration. First, one guesses reasonable forms for the functions of the set $\{\phi_j\}$, and using Eq. (91) obtains a new set. This procedure is continued until one arrives at a set which does not change by further iterations. When this stage is reached, one has generated a self-consistent set. Hartree calls the set of fields as self-consistent in the sense that their own eigenfunctions are consistent with the potential field from which they are determined.

The two shortcomings of the Hartree method are

(a) the neglect of correlations between electrons (because of the use of single-particle wave functions).

(b) the neglect of the effects of antisymmetry of many-electron wave functions (because of the use of simple product N functions ϕ_i). The exclusion principle is partially satisfied by taking all ϕ's different. Despite these deficiencies the Hartree approximation is quite useful in the case of many-electron atoms. For atoms, further approximations (e.g. the central field approximation) are used.

17.8 THE HARTREE-FOCK APPROXIMATION

Fock and Slater took into account the proper symmetry by taking the trial wave function $|\overline{\psi}\rangle$ to be a Slater determinant of single-particle orbitals. This is the Hartree-Fock (HF) theory and is given below in the language of second quantitization. The correlation effects arising out of the r_{ij}^{-1} terms in the Hamiltonian are not taken into consideration in HF theory but can be evaluated by perturbation theory.

We write the Hamiltonian of the N-fermion system as

$$H = \sum_{\alpha,\,\beta} T_{\alpha\beta}\, a_\alpha^\dagger\, a_\beta + \frac{1}{2} \sum_{\alpha,\,\beta,\,\gamma,\,\delta} V_{\alpha\beta,\,\gamma\delta}\, a_\alpha^\dagger\, a_\beta^\dagger\, a_\delta\, a_\gamma, \tag{92}$$

where $T_{\alpha\beta}$'s and $V_{\alpha\beta,\,\gamma\delta}$'s are matrix elements of the one-body and two-body operators respectively, and $V_{\alpha\beta,\,\gamma\delta} = V_{\beta\alpha,\,\delta\gamma}$. We use a ground-state trial state vector as

$$|\overline{\psi}\rangle = \prod_{\mu=1}^{N} a_\mu^\dagger\, |0\rangle, \tag{93}$$

where all the μ's are different. We have $\langle \overline{\psi} | \overline{\psi} \rangle = 1$. We assume that the one-fermion states with lowest energy ϵ_α^H are filled up:

$$\sum_\alpha \epsilon_\alpha^H f_\alpha = \text{minimum} \tag{94}$$

where
$$f_\alpha = \begin{cases} 1, & \text{if } \phi_\alpha \text{ is occupied in } |\overline{\psi}\rangle \\ 0, & \text{if } \phi_\alpha \text{ is empty in } |\overline{\psi}\rangle. \end{cases} \tag{95}$$

The single-particle states form a complete orthonormal set. These states are to be varied to produce an energy extremum in a variational procedure. The variational principle states that $|\overline{\psi}\rangle$ satisfy

$$\delta\langle \overline{\psi} | H | \overline{\psi} \rangle = 0, \tag{96}$$

which gives
$$\langle \delta\overline{\psi} | H | \overline{\psi} \rangle = 0, \tag{97}$$

and the hermitian conjugate of Eq. (97). A variation in $|\overline{\psi}\rangle$ is one in which the basis functions are changed infinitesimally,

$$\phi_\lambda \rightarrow \phi_\lambda + \eta\phi_\sigma, \quad |\eta| << 1 \tag{98}$$

The variation is written as

$$|\delta\overline{\psi}\rangle = \eta a_\sigma^\dagger a_\lambda|\overline{\psi}\rangle \tag{99}$$

Obviously, the variation is nonzero only if λ corresponds to an occupied and σ an empty level (the case $\sigma = \lambda$ does not represent a true variation). $(|\overline{\psi}\rangle + |\delta\overline{\psi}\rangle)$ is still normalized to first order in η. Substituting Eq. (99) into Eq. (97), we get

$$\langle\overline{\psi}|a_\lambda^\dagger a_\sigma H|\overline{\psi}\rangle = 0, \tag{100}$$

With H given in Eq. (92), we may use Wick's theorem to evaluate Eq. (100). After making particle-hole transformation, we know that only the fully contracted terms will contribute. The nonvanishing contractions are

$$\overline{c_\lambda^\dagger\, c_\lambda} = 1 \quad \text{for } \lambda \text{ occupied}$$

$$\overline{c_\sigma\, c_\sigma} = 1 \quad \text{for } \sigma \text{ unoccupied}$$

Then Eq. (100) gives

$$T_{\sigma\lambda} + \frac{1}{2}\sum_{\mu=1}^{N}\left[V_{\mu\sigma,\,\mu\lambda} - V_{\mu\sigma,\,\lambda\mu} - V_{\sigma\mu,\,\mu\lambda} + V_{\sigma\mu,\,\lambda\mu}\right] = 0$$

or,
$$T_{\sigma\lambda} + \sum_{\mu=1}^{N}\left(V_{\sigma\mu,\,\lambda\mu} - V_{\sigma\mu,\,\mu\lambda}\right) = 0, \tag{101}$$

by using the symmetry in the V's and making dummy changes in summation variables. The hermitian conjugate form of this equation is

$$T_{\lambda\sigma} + \sum_{\mu=1}^{N}\left(V_{\lambda\mu,\,\sigma\mu} - V_{\lambda\mu,\,\mu\sigma}\right) = 0. \tag{102}$$

Equations (101) and (102) hold for any unoccupied state σ and occupied state λ in the trial state $|\overline{\psi}\rangle$.

We now define a single-particle operator called the self-consistent one-particle Hamiltonian

$$H^{S.C.} = \sum_\sigma \sum_\lambda \left[T_{\lambda\sigma} + \sum_{\mu=1}^{N} V_{\sigma\mu,\,\lambda\mu} - V_{\sigma\mu,\,\mu\lambda}\right] a_\sigma^\dagger a_\lambda. \tag{103}$$

In Eq. (103) there is no restriction on σ, μ whereas in Eqs. (101) or (102), all $\sigma > N$, $\lambda \leq N$, $H^{S.C.}$ has no matrix elements between occupied levels and unoccupied levels. To see this, let us consider

$$\langle\sigma|H^{S.C.}|\lambda\rangle = \sum_{\alpha,\,\beta} f_{\alpha\beta}\,\langle\sigma|a_\alpha^\dagger a_\beta|\lambda\rangle, \quad \begin{matrix}\sigma > N \\ \lambda \leq N\end{matrix}$$

$$= \sum_{\alpha,\,\beta} \alpha f_{\alpha\beta}\,\langle 0|a_\sigma a_\alpha^\dagger a_\beta a_\lambda^\dagger|0\rangle$$

$$= \sum_{\alpha,\,\beta} f_{\alpha\beta} \left\langle 0 \middle| \left(a_\alpha^\dagger a_\sigma - \delta_{\alpha\sigma} \right)\left(a_\lambda^\dagger a_\beta - \delta_{\beta\lambda} \right) \middle| 0 \right\rangle$$

Since $a_\beta |0\rangle = 0$, we have

$$\left\langle \sigma \middle| H^{S.C.} \middle| \lambda \right\rangle = f_{\sigma\lambda} = 0, \tag{104}$$

using Eq. (101).

It is convenient to change to the representation which diagonalizes $H^{S.C.}$ (hermitian). This implies that the new occupied levels are linear combinations of occupied levels and the new unoccupied levels. In the new representation,

$$H^{S.C.} = \sum_\alpha \in_\alpha a_\alpha^\dagger a_\alpha. \tag{105}$$

From Eqs. (103) and (105), we get the Hartree-Fock equations

$$\left\langle \alpha | T | \beta \right\rangle + \sum_{\mu=1}^{N} \left(V_{\alpha\mu,\,\beta\mu} - V_{\alpha\mu,\,\mu\beta} \right) = \in_\alpha \delta_{\alpha\beta}. \tag{106}$$

If $|\phi_\alpha\rangle$, the basis vectors are the eigenvectors of $H^{S.C.}$, then

$$H^{S.C.}|\phi_\alpha\rangle = \in_\alpha |\phi_\alpha\rangle. \tag{107}$$

The eigenvalues $\in_\alpha$ are approximate excitation energies of the levels α and are known as the "self-consistent energies" of the levels. Equation (107) is solved self-consistently.

Using the HF set of single-particle states, we may evaluate the ground-state energy of N fermion system:

$$E_0 = \left\langle \overline{\Psi} | H | \overline{\Psi} \right\rangle$$

$$= \sum_{\lambda=1}^{N} \left\langle \lambda | T | \lambda \right\rangle + \frac{1}{2} \sum_{\lambda,\,\mu=1}^{N} \left\langle \lambda\mu \, | V | \, \widetilde{\lambda\mu} \right\rangle,$$

where the antisymmetric combination $\left| \widetilde{\lambda\mu} \right\rangle = |\lambda\mu\rangle - |\mu\lambda\rangle$

$$= \sum_{\lambda=1}^{N} \in_\lambda - \frac{1}{2} \sum_{\lambda,\,\mu}^{N} \left\langle \lambda\mu \, | V | \, \widetilde{\lambda\mu} \right\rangle$$

$$= \sum_{\lambda=1}^{N} \in_\lambda - \frac{1}{2} \sum_{\lambda=1}^{N} \left\langle \lambda | U | \lambda \right\rangle, \tag{108}$$

where U is the HF self-consistent potential so that

$$\left\langle \alpha | U | \beta \right\rangle = \sum_{\mu=1}^{N} \left\langle \alpha\mu \, | V | \, \widetilde{\beta\mu} \right\rangle. \tag{109}$$

The total energy is not the sum of the single particle energies but is reduced by half the sum over the interaction energies.

We now evaluate the expectation value of H in the excited states. The lowest excited state is described by a state vector

$$\left| \overline{\Psi}_\mu^\sigma \right\rangle \equiv a_\sigma^\dagger a_\mu \left| \overline{\Psi} \right\rangle, \quad \mu \le N, \ \sigma > N \tag{110}$$

This state is called one particle-one hole ($1p - 1h$) state, as it is the configuration in which a particle originally in the level μ (in the Fermi sea) is excited to the level σ (above the Fermi sea) which is unoccupied in the groundstate. First, we consider the off-diagonal elements of a secular matrix for the Hamiltonian:

$$\left\langle \overline{\Psi}_\mu^\sigma | H | \overline{\Psi} \right\rangle = \left\langle \overline{\Psi} | a_\mu^\dagger \, a_\sigma H | \overline{\Psi} \right\rangle = 0, \tag{111}$$

using Eq. (100). This result (Brillouin's theorem) shows that there is no mixture of the ground state in a secular matrix for the excited $1p - 1h$ states. The off-diagonal matrix element between particle-hole states is

$$\left\langle \overline{\Psi}_\mu^\sigma | H | \overline{\Psi}_\nu^\tau \right\rangle = - \left\langle \sigma\nu \, |V| \, \widetilde{\tau\mu} \right\rangle, \quad \begin{array}{l} \sigma \neq \tau \text{ or} \\ \mu \neq \nu \end{array} \tag{112}$$

and the diagonal element is

$$\begin{aligned}
E_\mu^\sigma &\equiv \left\langle \overline{\Psi}_\mu^\sigma | H | \overline{\Psi}_\mu^\sigma \right\rangle \\
&= E_0 + \left\langle \sigma|T|\sigma \right\rangle - \left\langle \mu|T|\mu \right\rangle \\
&\quad + \sum_{\substack{\mu = 1 \\ \lambda \neq \mu}}^{N} \left[\left\langle \sigma\lambda \, |V| \, \widetilde{\sigma\lambda} \right\rangle - \left\langle \mu\lambda \, |V| \, \widetilde{\mu\lambda} \right\rangle \right] \\
&= E_0 + \in_\sigma - \in_\mu - \left\langle \sigma\mu \, |V| \, \widetilde{\sigma\mu} \right\rangle. \tag{113}
\end{aligned}$$

The last two terms on the R.H.S. of Eq. (113) are of order $1/N$. For N large, the excitation energy $\sim \in_\sigma - \in_\mu$.

We now discuss the physical significance of $\in_\mu$, the HF single-particle energy. Koopmans' theorem states that $\in_\mu$ is the negative of the energy required to remove an electron in the state μ. The state vector of a system of $N-1$ fermions with a hole in the single-particle state μ is

$$|\overline{\Psi}_\mu\rangle = a_\mu \, |\overline{\Psi}\rangle, \tag{114}$$

where $|\overline{\Psi}\rangle$ is the ground-state of a N-particle system and is given in Eq. (93). The energy of this system is

$$\begin{aligned}
E_\mu &\equiv \left\langle \overline{\Psi}_\mu | H | \overline{\Psi}_\mu \right\rangle \\
&= \sum_{\substack{\lambda = 1 \\ \lambda \neq \mu}}^{N} \left\langle \lambda|T|\lambda \right\rangle + \frac{1}{2} + \sum_{\substack{\lambda, \, \nu = 1 \\ \lambda, \, \nu \neq \mu}}^{N} \left\langle \lambda\nu \, |V| \widetilde{\lambda\nu} \right\rangle. \tag{115}
\end{aligned}$$

From Eqs. (105) and (115),

$$\begin{aligned}
E_\mu - E_0 &= - \left\langle \mu|T|\mu \right\rangle - \frac{1}{2} \sum_{\lambda = 1}^{N} \left\langle \lambda\mu \, |V| \widetilde{\lambda\mu} \right\rangle \\
&\quad - \frac{1}{2} \sum_{\nu = 1}^{N} \left\langle \mu\nu \, |V| \, \widetilde{\mu\nu} \right\rangle
\end{aligned}$$

$$= - \langle \mu | T | \mu \rangle - \sum_{\lambda = 1}^{N} \langle \lambda \mu \, | V | \, \widetilde{\lambda \mu} \rangle$$

$$= - \, \epsilon_\mu \tag{116}$$

Thus, it requires an energy $|\epsilon_\mu|$ to remove a fermion from the state μ. This is valid only if the other states are unaffected by the removal of an electron (or nucleon) in state μ.

It is useful to write the results in coordinate spin representation. The set of integro-differential equations are

$$\left[- \frac{\hbar^2}{2M} \vec{\nabla}^2 + W(\vec{r}) \right] \phi_{k\sigma}(\vec{r}) + \sum_{j\sigma'} \int \phi_{j\sigma'}^*(\vec{r}') \, V(1,2)\phi_{j\sigma'}(\vec{r}')d^3r' \, \phi_{k\sigma}(\vec{r})$$

$$- \sum_{j} \phi_{j\sigma}^*(\vec{r}) \int \phi_{j\sigma}^*(\vec{r}')V(1,2)\phi_{k\sigma}(\vec{r}')d^3r'$$

$$= \epsilon_k \, \phi_{k\sigma}(\vec{r}). \tag{117}$$

In the last term ("exchange term"), only states of parallel spin contribute. For Coulomb interaction, the exchange term is (taking care of the spin summations)

$$A_1\phi_i(\vec{r}_1) = - \sum_{j(\|i)} \int \frac{e^2}{r_{12}} \, \phi_j^*(r_2)\phi_i(\vec{r}_2)d^3r_2 \cdot \phi_j(\vec{r}_1)$$

$$= - \sum_{j(\|i)} \int \frac{e}{r_{12}} \left(\frac{e\phi_j^*(\vec{r}_2)\phi_i(\vec{r}_2)\phi_i(\vec{r}_2)\phi_j(\vec{r}_1)}{\phi_i(\vec{r}_1)} \right) \phi_i(\vec{r}_1)d^3r_2$$

$$= \int \frac{(-e)}{r_{12}} \rho \left(\vec{r}_1, \vec{r}_2 \right) \phi_i(\vec{r}_1)d^3r_2 \tag{118}$$

where

$$\rho \left(\vec{r}_1, \vec{r}_2 \right) = \frac{e \sum\limits_{j(\|i)} \phi_j^*(r_2)\phi_i(\vec{r}_2)\phi_j(\vec{r}_1)}{\phi_i(\vec{r}_1)}. \tag{119}$$

Thus, we can interpret the nonlocal exchange potential as the potential energy of interaction of an electron at $\vec{r}_1$ with a charge distribution of density $\rho \left(\vec{r}_1, \vec{r}_2 \right)$.

Since $\int \rho \left(\vec{r}_1, \vec{r}_2 \right)d^3r_2 = +e$, we may say that an electron (of given spin) is surrounded by a Fermi hole of charge $+e$.

The exchange term is due to the antisymmetric nature of the wave function in HF theory and is absent in the original method of Hartree.

The HF equations have been found to be useful in calculating the average potential acting on the electrons and the associated one-particle wave functions in atoms. The approximate Hamiltonian of an atom with a nucleus of change $Z|e|$ surrounded by N-electrons is of the form Eq. (92), with

$$T_{ij} = -\frac{\hbar^2}{2M} \sum_\sigma \int \phi_i^*(\vec{r}, \sigma) \vec{\nabla}^2 \phi_j(\vec{r}, \sigma)\, d^3r$$

$$- Ze^2 \sum_\sigma \int \phi_i^*(\vec{r}, \sigma)\, r^{-1} \phi_j(\vec{r}, \sigma)\, d^3r$$

$$V_{ij,\,kl} = e^2 \sum_\sigma \sum_{\sigma'} \iint \phi_i^*(\vec{r}_1, \sigma)\phi_j^*(\vec{r}_2, \sigma')\, |\vec{r}_1 - \vec{r}_2|^{-1}$$

$$\times\ \phi_k(\vec{r}_2, \sigma)\phi_l(\vec{r}_2, \sigma')\, d^3r_1 d^3r_2. \tag{120}$$

The HF equations are

$$-\frac{\hbar^2}{2M} \vec{\nabla}^2 \phi_i(\vec{r}, \sigma) - \frac{Ze^2}{r} \phi_i(\vec{r}, \sigma)$$

$$+ e^2 \sum_{j=1}^{N} \sum_{\sigma'} \int \phi_j^*(\vec{r}', \sigma')\, |\vec{r} - \vec{r}'|^{-1}\phi_j(\vec{r}', \sigma')\phi_i(\vec{r}, \sigma)d^3r'$$

$$- e^2 \sum_{j=1}^{N} \sum_{\sigma'} \int \phi_j^*(\vec{r}', \sigma')\, |\vec{r} - \vec{r}'|^{-1}\phi_j(\vec{r}, \sigma)\phi_i(\vec{r}', \sigma')d^3r'$$

$$= \in_i \phi_i(\vec{r}, \sigma). \tag{121}$$

Equation (121) gives a set of N coupled nonlinear, integrodifferential equations. These equations are solved by an iterative method. A guess is made for the self-consistent potential (using the Thomas Fermi method or otherwise) and the equation solved for that potential. The wave functions found are used to calculate a second approximation to the potential, and the process is repeated until self-consistency is achieved.

17.9   THE FREE-ELECTRON GAS

We now consider a simple model – the "electron gas"–that provides a first approximation to a metal or a plasma. In a metal there are $\sim 10^{23}$ positively charged ions arranged in a lattice with $\sim 10^{23}$ electrons moving freely (almost) among the ion cores. The ions also oscillate about their equilibrium positions. Instead of considering such a complicated model we discuss the simple electron gas model. In this system the ions are assumed motionless, and the net positive charge is smeared out to make a uniform positive back ground charge (chosen to ensure the neutrality of the system) against which the electrons move. The electrons interact by pure Coulomb force. The homogeneous electron gas is also known as the "jellium model" of a solid.

Let us assume that there are N-electrons in a large cubical box of length L so that the average density is $n = \dfrac{N}{V}$, where $V\,(= L^3)$ is the volume of the box. The positive charge of density n is smeared uniformly throughout the volume. The HF equations are

$$-\frac{\hbar^2}{2M}\,\vec{\nabla}^2\,\phi_i(\vec{r}_1) + U^{ion}(\vec{r}_1)\phi_i(\vec{r}_1) + \left[e^2 \sum_{j=1}^{N} \int d^3 r_2 \frac{|\phi_j(\vec{r}_2)|^2}{r_{12}}\right]\phi_i(\vec{r}_1)$$

$$-\,e^2 \sum_{j(\|i)}\left[\int \frac{\phi_j^{*}(\vec{r}_2)\phi_i(\vec{r}_2)\phi_j(\vec{r}_1)}{r_{12}\phi_i(\vec{r}_1)}\,d^3 r_2\right]\phi_i(\vec{r}_1)$$

$$= E_i\phi_i(\vec{r}_1), \tag{121a}$$

where U^{ion} is the potential of the ions. The eigenfunctions ϕ_i of Eq. (121a) form a set of orthonormal plane waves. Then the electronic charge distribution will be a uniform smear. Hence, the second term in Eq. (121) exactly cancels the third term ("direct term"): $U^{ion} + U^{el} = 0$. We retain the fourth term ("exchange term") and the HF equations are

$$-\frac{\hbar^2}{2M}\,\vec{\nabla}^2\,\phi_i(\vec{r}_1) - e^2 \sum\left[\int \frac{\phi_j^{*}(\vec{r}_2)\phi_i(\vec{r}_2)\phi_j(\vec{r}_1)}{r_{12}\phi_i(\vec{r}_1)}\,dr_2\right]\phi_i(\vec{r}_1)$$

$$= E_i\phi_i(\vec{r}_1). \tag{122}$$

The HF equations applied to a monovalent metal are the same as those applied to free-electron gas.

The free electron plane waves are

$$\phi_i(\vec{r}) = V^{-\frac{1}{2}}\, e^{i\vec{k}\cdot\vec{r}}\,\eta_\lambda, \tag{123}$$

where η_λ are the spin functions. The set of plane waves are self consistent solutions of Eq. (122). Let us write the Coulomb interaction in terms of its Fourier transform

$$V\left(\vec{r}_1 - \vec{r}_2\right) = \frac{e^2}{|\vec{r}_1 - \vec{r}_2|}$$

$$= \frac{1}{V}\sum_{\vec{q}}\frac{4\pi e^2}{q^2}\,e^{i\vec{q}\cdot(\vec{r}_1 - \vec{r}_2)}. \tag{124}$$

Therefore, the HF equation has the plane wave eigenfunctions with the eigenvalues

$$\in\!\left(\vec{k}\right) = \frac{\hbar^2 k^2}{2m} - \sum_{k'}{}' G(\vec{k} - \vec{k}'), \tag{125}$$

where
$$G(\vec{k}) = \int d^3\xi\, e^{-i\vec{k}.\vec{\xi}}\, V(\xi) \tag{126}$$

with $\vec{\xi} = \vec{r}_1 - \vec{r}_2$ is the Fourier transform of $V(\vec{r}_1 - \vec{r}_2)$

Then
$$G(\vec{k} - \vec{k}') = \int d^3r\, \frac{e^2}{|\vec{r}|}\, e^{i(\vec{k} - \vec{k}')\vec{r}}$$

$$= \frac{4\pi e^2}{|\vec{k} - \vec{k}'|^2}. \tag{127}$$

For the ground state,
$$\sum_{k'}' G(\vec{k} - \vec{k}') = 4\pi e^2 \sum_{k'}' \frac{1}{|\vec{k} - \vec{k}'|^2}$$

$$= \frac{4\pi e^2}{V} \cdot \frac{V}{(2\pi)^3} \int_{k' < k_F} d^3k' \frac{1}{|\vec{k} - \vec{k}'|^2},$$

where k_F is the Fermi wave vector and $V/(2\pi^3)$ is the density of points in $\vec{k}$ space.

$$= \frac{e^2}{2\pi^2} \cdot 2\pi \int_0^{k_F} k'^2\, dk' \int_{-1}^{1} d\mu\, \frac{1}{k^2 + k'^2 - 2kk'\mu}$$

$$= \frac{e^2}{\pi k} \int_0^{k_F} k'\, dk'\, \ln \frac{k' + k}{|k' - k|}$$

$$= \frac{e^2}{\pi} \left(k_F + \frac{k_F^2 - k^2}{2k} \ln \left| \frac{k_F + k}{k_F - k} \right| \right). \tag{128}$$

Combining Eqs. (125) and (128),
$$\in(\vec{k}) = \frac{\hbar^2 k^2}{2m} - \frac{e^2 k_F}{\pi} \left(1 + \frac{1 - x^2}{2x} \ln \left| \frac{1 + x}{1 - x} \right| \right), \tag{129}$$

where
$$x = \frac{k}{k_F}.$$

Thus, plane waves are solutions of the HF equations and the energy of the one-electron level with wave vector $\vec{k}$ is given by Eq. (129).

In the Hartree approximation, the energy of the electron gas is purely kinetic and hence the same for free electrons. The first term in Eq. (129) is the kinetic energy term. The direct Coulomb term is omitted as in the model of homogeneous electron gas it is exactly cancelled by the direct interaction with the uniform positive background. The next term in Eq. (129) is the exchange energy which we may write as

$$\in_{ex} = \frac{e^2 k_F}{\pi}\, F(x) \tag{130}$$

with

$$F(x) = -\left(1 + \frac{1-x^2}{2x} \ln\left|\frac{1+x}{1-x}\right|\right). \tag{131}$$

The exchange energy depends only on the wave vector of the electron and the function $F(x)$ gives this dependence. At $x = 0$, $F(x) = -2$. Near $x = 1$, it rises steeply and approaches zero at large values of x. At $x = 1$, $F(1) = -1$. The slope of the function diverges logarithmically at $x = 1$, since

$$\frac{dF}{dx} = \frac{1}{2x}\left(\frac{1+x^2}{x} \ln\left|\frac{1+x}{1-x}\right| - 2\right). \tag{132}$$

Since $\dfrac{\partial \epsilon}{\partial k}$ becomes logarithmically infinite at $k = k_F$, the velocity of the electron becomes infinite while the effective mass is zero! The electronic heat capacity at low temperatures goes as $T/\ln T$ not as T. The unphysical singularity is due to the divergence of the Fourier transform $\dfrac{4\pi e^2}{k^2}$ at $k = 0$. The reason is the infinitely long range of the Coulomb interaction. The singularity may be eliminated by using a screened Coulomb interaction. The logarithmic singularity is cancelled by dynamic correlations.

The density of the electron gas is characterized by the parameter r_S given by

$$\frac{1}{n} = \frac{4\pi}{3}(r_S a_0)^3. \tag{133}$$

where $n = \dfrac{N}{V}$ is the electron density and $a_0 =$ Bohr radius $= \dfrac{\hbar^2}{me^2}$, r_s is essentially the average distance between electrons. r_S is small for a high density electron gas (K.E. >> P.E) and large for a low-density gas (K.E. << P.E.). For real metals, r_S is in the range $2 - 6$. The Fermi momentum k_F may be found from

$$N = 2 \sum_{k < k_F} 1 = 2 \frac{V}{(2\pi)^3} \int_0^{k_F} d^3k = \frac{V}{3\pi^2} k_F^3,$$

where the factor of 2 is from the spin sum.

or,

$$n = \frac{k_F^3}{3\pi^2}. \tag{134}$$

Let us now evaluate the HF energy. The contribution of the kinetic energy term in Eq. (129) to the ground state is

$$E_0 = \sum_{\substack{k < k_F \\ \sigma}} \frac{\hbar^2 k^2}{2m} = \frac{2\hbar^2}{2m} \cdot \frac{V}{(2\pi)^3} \int_0^{k_F} d^3k\, k^2 = \frac{\hbar^2 V}{10 m \pi^2} k_F^5$$

$$= \frac{3}{10} \frac{\hbar^2 k_F^2}{m} N$$

$$= \frac{3}{5} \in_F N, \tag{135}$$

where the Fermi energy $\in_F$ is the mean kinetic energy per electron $= \dfrac{\hbar^2 k_F^2}{2m}$. In terms of the parameter r_S,

$$\langle \in_F \rangle = \frac{E_0}{N} = \frac{3}{10} \cdot \left(\frac{9\pi^2}{4} \right)^{2/3} \frac{\hbar^2}{m a_0^2} \cdot \frac{1}{r_S^2}$$

$$= \frac{2.2099}{r_S^2} \quad \text{Ry (per electron)} \tag{136}$$

where the energy is measured in rydbergs $\left(\dfrac{e^2}{2a_0} = 1 \text{ Ry} = 13.60 \text{ eV} \right)$. This is the total energy per electron in the Hartree approximation. The exchange energy contribution to the ground state energy is

$$E_{ex} = -\frac{e^2 k_F}{\pi} \sum_{\substack{k < k_F \\ \sigma}} \left[1 + \frac{k_F^2 - k^2}{2kk_F} \ln \left| \frac{k_F + k}{k_F - k} \right| \right] \tag{137}$$

We now require the value of $G\left(\vec{k} - \vec{k}' \right)$ summed over all occupied states $\vec{k}$ and $\vec{k}'$ (Kittel (1987)). Let

$$I = \int\limits_{k_1^2,\, k_2 < k_F} d^3 k_1 \, d^3 k_2 \frac{1}{\left| \vec{k}_1 - \vec{k}_2 \right|^2}.$$

Now,

$$\frac{1}{\left| \vec{k}_1 - \vec{k}_2 \right|^2} = \frac{1}{k_1^2} \cdot \frac{1}{1 + s^2 - 2s\mu},$$

where

$$\mu = \cos \theta \tag{138}$$

$$s = \frac{k_2}{k_1};$$

But for $s < 1$, $\dfrac{1}{1 + s^2 - 2s\mu} = \sum\limits_{l,\,\lambda} s^{l + \lambda} P_l(\mu) P_\lambda(\mu)$, where $P_l(\mu)$ is a Legendre polynomial of order l.

Thus,

$$I = 2 \int\limits_{k_1 < k_F} d^3 k_1 \int 2\pi k_2^2 \, dk_2 d\mu \sum_{l,\,\lambda} \left(\frac{k_2}{k_1} \right)^{l + \lambda} \cdot \frac{1}{k_1^2} P_l(\mu) P_\lambda(\mu) \tag{139}$$

Using Rodrigue's formula

$$\int_{-1}^{1} P_l(\mu)P_\lambda(\mu)d\mu = \frac{2}{2l+1}\,\delta_{l\lambda};$$

$$I = 8\pi \int_{k_1 < k_F} d^3k_1 \int_0^{k_1} dk_2 \, \Sigma \left(\frac{k_2}{k_1}\right)^{2l+2} \cdot \frac{1}{2l+1},$$

considering both $k_2 < k_1$ and $k_1 < k_2$.

$$= 8\pi \int_{k_1 < k_F} d^3k_1 k_1 \, \sum_l \frac{1}{(2l+1)(2l+3)}$$

$$= 8\pi^2 \, k_F^4 \sum_{l=0}^{\infty} \frac{1}{(2l+1)(2l+3)}$$

$$= 4\pi^2 \, k_F^4, \tag{140}$$

since

$$\sum_{l=0}^{\infty} \frac{1}{(2l+1)(2l+3)} = \frac{1}{2} \sum_{l=0}^{\infty} \frac{1}{(2l+1)} - \frac{1}{(2l+3)}$$

$$= \frac{1}{2}.$$

The average exchange energy per electron is

$$\langle \in_{ex} \rangle = -\frac{1}{2} \cdot \frac{2}{N} \, \Sigma' \, G(\vec{k} - \vec{k}')$$

$$= -\frac{4\pi e^2}{n} \sum_{ij} \frac{1}{\left|\vec{k}_i - \vec{k}_j\right|^2}$$

$$= -\frac{4\pi e^2}{(2\pi)^6 n} I$$

$$= -\frac{2e^2 k_F^4}{(2\pi^3)n}$$

$$= -\frac{3}{4} \frac{e^2 k_F}{\pi}$$

$$= -\frac{0.9163}{r_S} \quad \text{Ry (per electron)} \tag{141}$$

The HF energy is

$$\in_{HF} = \frac{2.2099}{r_s^2} - \frac{0.9163}{r_S}. \tag{142}$$

This is a better value of the energy than the Hartree energy. Gell-Mann and Brueckner (1957) obtained the ground-state energy (per particle) of the high-density electron gas as

$$E_g = \left[\frac{2.2099}{r_s^2} - \frac{0.9163}{r_S} - 0.094 + 0.0622 \ln (r_S) + ... \right]$$

Ry (per electron)   (143)

The correlation energy $\in_c$ is defined as $\in_{exact} - \in_{HF}$ so that correlation energy means the energy terms beyond HF. We write the total ground state energy (per particle) of the electron gas as

$$\in_g = \frac{2.2099}{r_s^2} - \frac{0.9163}{r_S} + \in_c$$

where

$$\in_c = -0.094 + 0.0622 \ln (r_S) + O(r_S), \qquad (144)$$

with
$$O(r_S) \to 0 \quad \text{as } r_s \to 0$$

The low-density limit is discussed in Fetter & Walecka (1971).

It is to be noted that both Hartree and HF equations have other solutions with lower energies than the plane wave solutions. These solutions, known as spin density waves, describe spatial variation of spin density.

17.10   CORRELATIONS

The correlation energy is the correction to the ground state energy beyond the HF approximation. The HF method takes into account the antisymmetry of the wave function under permutation of identical electrons. This leads to a type of correlation between the positions of two electrons with parallel spins, even if they are non-interacting. This is the "exchange" effect and is due to the anti-symmetry (Pauli principle). Dynamic correlation or simply correlation refers to the effect of the interaction among the particles. The effect of Coulomb interaction between electrons leads to a correction of the electronic motion reducing the probability of two electrons approaching each other closely.

In an electron gas, electrons of the same spin tend to avoid each other as a consequence of the Pauli principle. The "pair correlation function" $g_{ss'} (\vec{r}, \vec{r}')$ is the probability of finding an electron at $\vec{r}'$ (with spin s') if there is one at $\vec{r}$ (with spin s). We use the notation $g_{ss'} (r)$ to denote an electron (with spin s) at $\vec{r}$ if there is already an electron (with spin s') at $r = 0$. For example, $g_{\downarrow\uparrow} (r)$ is the probability that a spin down electron is at $\vec{r}$ if a spin up electron is at $r = 0$. These pair correlation functions are averages for the moving electrons. For an electron gas, there are two p.c.f. – parallel and antiparallel – so the total pair correlation function is

$$g(r) = g_{\uparrow\uparrow} (r) + g_{\uparrow\downarrow} (r) \qquad (145)$$

We now calculate the p.c.f^s. for the homogeneous electron gas in the HF approximation. The N-particle density matrix is

$$\rho_N\left(\vec{r}_1, ..., \vec{r}_N\right) = \left|\overline{\Psi}\left(\vec{r}_1, ..., \vec{r}_N\right)\right|^2, \tag{146}$$

where the N-particle wave function is a Slater determinant. The p.c.f. is

$$g\left(\vec{r}_1, \vec{r}_2\right) = V^2 \int d^3r_1 ... d^3r_N \left|\overline{\Psi}\left(\vec{r}_1, ..., \vec{r}_N\right)\right|^2 \tag{147}$$

i.e.

$$g_{ss'}\left(\vec{r}_1, \vec{r}_2\right) = \frac{V^2}{N(N-1)} \sum_{\lambda_i, \lambda_j} \left| \begin{matrix} \phi_{\lambda i}(\vec{r}_1) & \phi_{\lambda j}(\vec{r}_2) \\ \phi_{\lambda j}(\vec{r}_1) & \phi_{\lambda i}(\vec{r}_2) \end{matrix} \right|^2, \tag{148}$$

where the sum is over all occupied states. The orbitals are plane waves

$$\phi_\lambda = \chi_s \frac{e^{i\vec{k}\cdot\vec{r}}}{\sqrt{V}}. \tag{149}$$

Averaging over the spin functions χ, we obtain

$$g_{\uparrow\downarrow}\left(\vec{r}_1 - \vec{r}_2\right) = \frac{1}{n^2} \sum_{\vec{k}_1, \vec{k}_2} \left(\left| \frac{e^{i\left(\vec{k}_1\cdot\vec{r}_1 + \vec{k}_2\cdot\vec{r}_2\right)}}{V} \right|^2 \right.$$

$$\left. + \left| \frac{e^{i\left(\vec{k}_1\cdot\vec{r}_2 + \vec{k}_2\cdot\vec{r}_1\right)}}{V} \right|^2 \right). \tag{150}$$

$$= \frac{2}{N^2}\left(\frac{N}{2}\right)^2$$

$$= \frac{1}{2},$$

$$g_{\uparrow\uparrow}\left(\vec{r}_1 - \vec{r}_2\right) = \frac{1}{n^2} \sum_{\vec{k}_1, \vec{k}_2} \left| \frac{e^{i\left(\vec{k}_1\cdot\vec{r}_1 + \vec{k}_2\cdot\vec{r}_2\right)}}{V} - \frac{e^{i\left(\vec{k}_1\cdot\vec{r}_2 + \vec{k}_2\cdot\vec{r}_1\right)}}{V} \right|^2.$$

$$= \frac{2}{N^2} \sum_{\vec{k}_1, \vec{k}_2} \left(1 - e^{i\left(\vec{k}_1 - \vec{k}_2\right)\cdot\left(\vec{r}_1 - \vec{r}_2\right)}\right)$$

$$= \frac{1}{2}\left[1 - \phi(\vec{r}_1 - \vec{r}_2)^2\right] \tag{151}$$

where

$$\langle\chi_\uparrow\chi_\uparrow\rangle = 1, \quad \chi_\uparrow\chi_\downarrow = 0, ...$$

and

$$\phi(\vec{r}) = \frac{2}{N}\sum_{\vec{k}} e^{i\vec{k}\cdot\vec{r}}.$$

Thus, there is no spatial correlation for electrons of opposite spin, while for electrons with parallel spins there is spatial dependence.

Now,

$$\phi(r) = \frac{2}{n} \int \frac{d^3k}{(2\pi)^3}\, n_k\, e^{i\vec{k}\cdot\vec{r}}$$

$$= \frac{2}{n} \cdot \frac{1}{\pi^2} \int_0^{k_F} k^2\, dk \int_{-1}^{1} d\mu\, e^{ikr\mu}$$

$$= \frac{2}{n} \cdot \frac{1}{\pi^2} \cdot \frac{k_F^3}{3} \cdot \frac{3(\sin x - x \cos x)}{x^3},$$

where $x = k_F r$

$$= \frac{3}{(k_F r)^3}\, [\sin (k_F r) - (k_F r) \cos (k_F r)],$$

using Eq. (134)

$$= \frac{3}{(k_F r)^3}\, j_1\, (k_F r). \tag{152}$$

We see that $g_{\uparrow\uparrow}(r) \to 0$ at $r = 0$ nd approaches $\frac{1}{2}$ as $r \to \infty$.

Since $g(r) = \frac{1}{2} + g_{\uparrow\uparrow}(r)$, the normalization is

$$n \int d^3r\, [g(r) - 1] = -\frac{n}{2} \int d^3r\, (\phi(r))^2$$

$$= -\frac{6}{\pi} \int_0^{\infty} (j\,(x))^2\, dx, \quad x = k_F r$$

$$= -1. \tag{153}$$

The Coulomb energy is

$$E_{Coulomb} = e^2 n \int d^3r\, \frac{g(r) - 1}{r}$$

$$= \frac{e^2 n}{2} \int d^3r\, \frac{(\phi(r))^2}{r}$$

$$= -\frac{6}{\pi}\, e^2 k_F \int_0^{\infty} dx\, \frac{(j_1(x))^2}{x}$$

$$= -\frac{3}{4}\, \frac{e^2 k_F}{\pi}, \tag{154}$$

which is the same as the exchange energy (per particle), given in Eq. (141).

The electron deficiency near the origin is regarded as a hole in the electron density. This is called the Fermi hole or the exchange hole or correlation hole. The hole has positive charge density and the total charge is $+e$. Thus, each electron of charge $-e$ has its hole with charge $+e$ so that the system is neutral.

For a system of spin-zero bosons in the state $|\phi\rangle = |n_{\vec{p}_0}, n_{\vec{p}_1}, \ldots\rangle$ with density $n = \dfrac{1}{V}\sum_{\vec{p}} n_{\vec{p}}$, the p.c.f. is

$$n^2 + \left| \frac{1}{V}\sum_{\vec{p}} n_{\vec{p}}\, e^{-i\vec{p}\cdot(\vec{r}-\vec{r}')/\hbar} \right|^2$$

$$- \frac{1}{V^2}\sum_{\vec{p}} n_{\vec{p}}\,(n_{\vec{p}}+1).$$

We note that due to the symmetry of boson wave functions the second term is positive and the third term indicates that bosons tend to clump together.

17.11 INTENSITY-FLUCTUATION CORRELATIONS (THE HANBURY BROWN AND TWISS EFFECT)

During 1952-1956 Hanbury Brown and Twiss developed a new radio interferometry technique in which the signals at the aerials were detected separately and the angular diameter of the radio source was obtained from measurements of the correlation of the intensity fluctuations as a function of the aerial separation. Hanbury Brown and Twiss also demonstrated that an equivalent arrangement may be used with visible light in the laboratory. The correlation of two optical intensities expressed as the degree of second-order coherence was first measured by Hanbury Brown and Twiss.

The experimental arrangement is shown schematically in Fig. 17.1. Light from a mercury arc was filtered to allow only the 435.8 nm emission line of Hg to be split into two beams by a half silvered mirror. The intensity of each was measured by a photomultiplier detector. The signals were multiplied and averaged in a correlator. One of the photomultiplier tubes was movable across the beam. As this detector moved, the degree of coherence at the two detectors varied. The intensity correlation was measured as a function of the separation between the detectors.

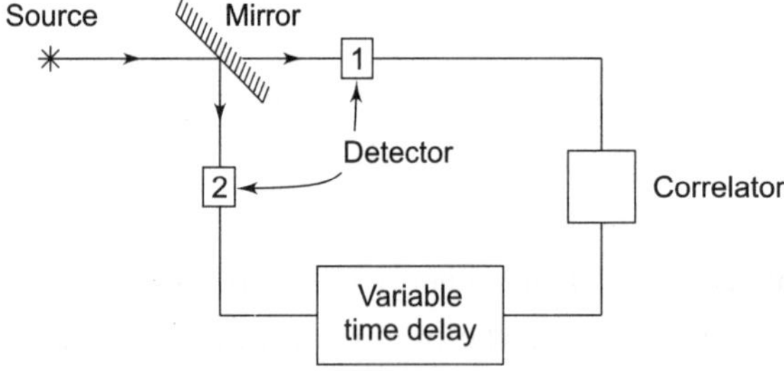

Fig. 17.1   Systematic diagram of the arrangement in an intensity interference experiment by Hanbury Brown and Twiss

Although the classical wave theory of light is able to account for many of the observed effects, we do not consider the classical theory here but refer the reader to Mandel (1963) and Loudon (1983).

We consider a photonic state $|\psi_0\rangle = \left| N_{\vec{k}}, N_{\vec{k}'} \right\rangle$ with the occupation numbers with momenta $\hbar \vec{k}$ and $\hbar \vec{k}'$. The relative probability density of detecting simultaneously one photon at detector at $\vec{r}$ and the other at detector at $\vec{r}'$ is

$$P_{12}\left(\vec{r} - \vec{r}'\right) = \frac{\langle \psi_0 | \psi^\dagger(\vec{r}) \psi^\dagger(\vec{r}') \psi(\vec{r}) \psi(\vec{r}') | \psi_0 \rangle}{\langle \psi_0 | \psi_0 \rangle} \tag{155}$$

$$= N_{\vec{k}}\left(N_{\vec{k}} - 1\right) + N_{\vec{k}'}\left(N_{\vec{k}'} - 1\right)$$
$$+ 2N_{\vec{k}}N_{\vec{k}'}\left[1 + \cos\left\{(\vec{k} - \vec{k}').(\vec{r} - \vec{r}')\right\}\right]. \tag{156}$$

The relative probability $P_1(\vec{r})$ of detecting a photon at D_1 is

$$P_1(\vec{r}) = \frac{\langle \psi_0 | \psi^\dagger(\vec{r}) \psi(\vec{r}) | \psi_0 \rangle}{\langle \psi_0 | \psi_0 \rangle}$$

$$= N_{\vec{k}} + N_{\vec{k}'}. \tag{157}$$

Also,

$$P_2(\vec{r}') = N_{\vec{k}} + N_{\vec{k}'}. \tag{158}$$

We get

$$P_{12}\left(\vec{r} - \vec{r}'\right) = P_1(\vec{r})\, P_2(\vec{r}') + 2\sqrt{P_1(\vec{r})\, P_2(\vec{r}')}$$
$$\times \cos\left\{(\vec{k} - \vec{k}').(\vec{r} - \vec{r}')\right\},$$
$$N_{\vec{k}}, N_{\vec{k}'} \gg 1. \tag{159}$$

Hanbury Brown and Twiss measured the light intensities $I(t)$ at D_1 at time t and $I_2(t + \tau)$ at D_2 at a later time $t + \tau$ and averaged the product of the intensities over t, keeping τ fixed. This is equivalent to determining the relative probability of observing two photons at two detectors separated by a distance $c\tau$ (c = speed of light). The average correlated intensities $I_1(t)\, I_2(t + \tau)$ is plotted as a function of τ. It has the same form as the pair correlation function for two (spin zero) bosons. The second term on the right side of Eq. (159) shows the boson nature of the photons. At short distances this term become positive, showing the tendency of clustering or bunching of bosons (photons). Light, in general, is described as being photon bunched if

$$g^{(2)}(0) > 1 \tag{160}$$

Chaotic light is an example ($g^{(2)}(0) = 2$). Here, $g^{(2)}(\tau)$ is the second-order correlation function (normalized). The Fig. 17.2 shows $g^{(2)}(\tau)$ as a function of τ for starlight in the original Hanbury Brown and Twiss experiment.

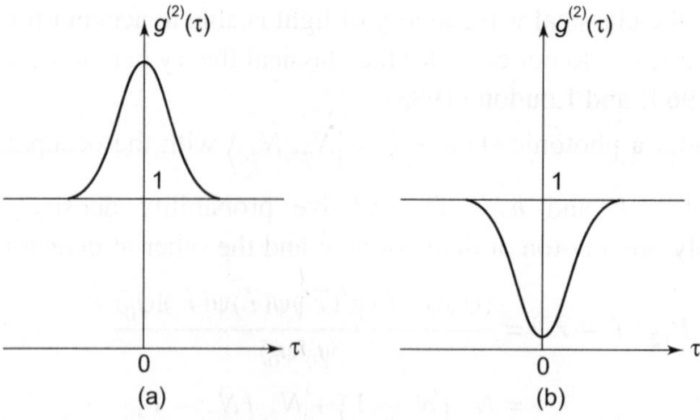

Fig. 17.2 Second-order correlation function versus time (t). (a) Photon bunching for starlight in the original Hanbury Brown and Twiss experiment (b) Photon antibunching in resonance fluorescence

Coherent light has no photon bunches ($g^{(2)}(0) = 1$).

Light for which the criterion

$$1 > g^{(2)}(0) \geq 0 \tag{161}$$

is satisfied is said to be photon-antibunched. Recent experiments in resonance fluorescence shows antibunching. (Meystre and Sargent III (1999)).

17.12   TWO ELECTRON SYSTEMS

In this section we study two-electron systems, namely atoms or ions consisting of a nucleus of charge $+Ze$ and two electrons. These include the ion $H^-(Z = 1)$, He atom ($Z = 2$), Li^+ ($Z = 3$), etc. A study of these systems is important for two reasons. Here we study the implications of the identity of the two electrons (Pauli principle) and since the two-particle Schrödinger equation cannot be solved analytically, we illustrate various approximation methods. Neglecting the motion of the nucleus, the Hamiltonian for an atom having two electrons is given in atomic units by

$$H = -\frac{1}{2}\left(\vec{\nabla}_1^2 + \vec{\nabla}_2^2\right) - \left(\frac{Z}{r_1} + \frac{Z}{r_2}\right) + \frac{1}{r_{12}}, \tag{162}$$

where r_1 and r_2 are the distances of the first and second electrons from the nucleus and r_{12} their mutual separation.

The Schrödinger equation is

$$\left[-\frac{1}{2}\vec{\nabla}_1^2 - \frac{1}{2}\vec{\nabla}_2^2 - \frac{Z}{r_1} - \frac{Z}{r_2} + \frac{1}{r_{12}}\right]\psi(\vec{r}_1, \vec{r}_2) = E\psi(\vec{r}_1, \vec{r}_2) \tag{163}$$

The spatial wave functions are either symmetric or antisymmetric on interchange of the coordinates of the two electrons. States described by space-symmetric wave functions are called "para" states, while those described by space-antisymmetric wave functions are called ortho-states.

The full eigenfunctions may be written as

$$\overline{\Psi}(x_1, x_2) = \psi\left(\vec{r}_1, \vec{r}_2\right) \chi(1, 2) \tag{164}$$

where x denotes the space and spin coordinates of electron i and $\chi(1, 2)$ denote the spin wave functions for the two electrons. We consider four normalized and mutually orthogonal spin wave functions

$$\chi_{0,\,0}(1, 2) = \frac{1}{\sqrt{2}} \, [\alpha(1)\beta(2) - \beta(1)\alpha(2)], \tag{165}$$

and

$$\chi_{1,\,1}(1, 2) = \alpha(1)\beta(2),$$

$$\chi_{1,\,0}(1, 2) = \frac{1}{\sqrt{2}} \, [\alpha(1)\beta(2) + \beta(1)\alpha(2)], \tag{166}$$

$$\chi_{1,\,-1}(1, 2) = \beta(1)\beta(2).$$

The first one Eq. (165), is antisymmetric while the last three, Eq. (166), are symmetric for an interchange of the two electrons. Let s_1 and $\vec{s}_2$ be the spin operators for the two electrons. The total spin operator is

$$\vec{S} = \vec{s}_1 + \vec{s}_2, \quad \text{whose } z\text{-component is}$$

$$S_z = s_{1z} + s_{2z}.$$

The antisymmetric spin function with quantum numbers $S = 0$, $M_s = 0$ is called a (spin) singlet, while the three symmetric spin functions with $S = 1$. $M_s = 1, 0, -1$ form a (spin)triplet.

An antisymmetric spatial (ortho) wave function must be multiplied by one of the three symmetric spin wave functions so that the ortho states are called triplet states, while the symmetric spatial (para) wave function must be multiplied by the single antisymmetric spin wave function so that the para state is called a singlet state.

Neglecting the spin-orbit interaction, the level scheme of helium and of ions with two electrons consists of two systems of levels, one containing para (singlet) levels and the other ortho (triplet) levels. Radiative transitions between a singlet and a triplet state ("intercombination line") are forbidden in the electric dipole approximation.

Let us write the Hamiltonian Eq. (162) as

$$H = H_0 + H' \tag{167}$$

with the unperturbed Hamiltonian

$$H_0 = -\frac{1}{2}\,\vec{\nabla}_1^2 - \frac{Z}{r_1} - \frac{1}{2}\,\vec{\nabla}_2^2 - \frac{Z}{r_2}$$

$$= H(1) + H(2), \tag{168}$$

where

$$H(i) = -\frac{1}{2}\vec{\nabla}_i^2 - \frac{Z}{r_i} \quad i = 1, 2 \tag{169}$$

and a perturbation

$$H' = \frac{1}{r_{12}}. \tag{170}$$

H_0 is the sum of two hydrogenic Hamiltonians and H' is the electron-electron repulsion term.

If E_{n_i} are the energy eigenvalues and $|n_i l_i m_i\rangle$ are the eigenstates of $H(i)$, then

$$H(i)|n_i l_i m_i\rangle = E_{n_i}|n_i l_i m_i\rangle \tag{171}$$

with

$$E_{n_i} = -\frac{1}{2}\frac{Z^2}{n^2} \quad \text{(in a.u.).} \tag{172}$$

The unperturbed equation

$$H_0|\psi^{(0)}\rangle = E^{(0)}|\psi^{(0)}\rangle \tag{173}$$

gives the eigenvalue

$$E^{(0)} = E_{n_1} + E_{n_2}$$

$$= -\frac{Z^2}{2}\left(\frac{1}{n_1^2} + \frac{1}{n_2^2}\right). \tag{174}$$

The eigenstate of H_0 can be written in the product form

$$|\psi^{(0)}\rangle = |n_1 l_1 m_1\rangle|n_2 l_2 m_2\rangle. \tag{175}$$

An interchange of the electron labels gives the same energy ("exchange degeneracy"). This degeneracy is removed by the interelectronic repulsion term. In the ground state, both electrons are in the 1s state ($n_1 = n_2 = 1$, $l_1 = l_2 = 0$, $m_1 = m_2 = 0$), so that the electrons must have antiparallel spin. Thus the ground state is a parastate. The normalized unperturbed (zero order) spatial wave function for the ground state of a two-electron atom is

$$\psi_0^{(0)} = \psi_{100}(\vec{r}_1)\,\psi_{100}(\vec{r}_2)$$

$$= \frac{Z^3}{\pi}\,e^{-Z(r_1 + r_2)}, \tag{176}$$

and ground state energy is

$$E_0^{(0)} = E_{n_1 = 1,\, n_2 = 1}^{(0)}$$

$$= -Z^2 \quad \text{(in a.u.).} \tag{177}$$

For helium ($Z = 2$),

$$E_0^{(0)} = -4 \text{ a.u. } (= -108.8 \text{ eV}).$$

This corresponds to an ionization potential

$$I_P^{(0)} = 2 \text{ a.u. } (= 54.4 \text{ eV}).$$

The experimental values are

$$E_0^{\text{exp}} = -2.90 \text{ a.u. } (= -79.0 \text{ eV})$$

and

$$I_P^{\text{exp}} = 0.90 \text{ a.u. } (= 24.6 \text{ eV}).$$

We now obtain a better energy value for the ground state by applying the first-order perturbation theory using the Coulomb repulsion term $1/r_{12}$ as the perturbation,

The first-order correction to the ground state energy is

$$E_0^{(1)} = \left\langle \psi_0^{(0)} \middle| H' \middle| \psi_0^{(0)} \right\rangle$$

$$= \frac{Z^6}{\pi^2} \int e^{-2Z(r_1 + r_2)} \frac{1}{r_{12}} \, d\vec{r}_1 \, d\vec{r}_2. \tag{178}$$

To evaluate the integral, we first note that

$$\frac{1}{r_{12}} = \sum_{l=0}^{\infty} \frac{r_<^l}{r_>^{l+1}} P_l (\cos \theta), \tag{179}$$

where $r_>(r_<)$ is the larger (smaller) of r_1 and r_2, and θ is the angle between $\vec{r}$ and $\vec{r}_2$. The angular integral is performed by expressing $P_l (\cos \theta)$ in terms of spherical harmonics:

$$P_l (\cos \theta) = \frac{4\pi}{2l + 1} \sum_{m = -l}^{l} Y_{lm}^*(\theta_1, \phi_1) Y_{lm}(\theta_2, \phi_2). \tag{180}$$

We obtain $E_0^{(1)} = \dfrac{Z^6}{\pi^2} \sum_{l=0}^{\infty} \sum_{m=-l}^{l} \dfrac{(4\pi)^2}{2l+1} \int_0^{\infty} dr_1 r_1^2$

$$\times \int_0^{\infty} dr_2 r_2^2 \, e^{-2Z(r_1 + r_2)} \times \frac{r_<^l}{r_>^{l+1}}$$

$$\times \int Y_{lm}^*(\theta_1, \phi_1) Y_{100} d\Omega_1 \int d\Omega_2 \, Y_{lm}(\theta_2, \phi_2) Y_{00},$$

$$= \frac{Z^6}{\pi^2} \sum_{l=0}^{\infty} \sum_{m=-l}^{l} \frac{(4\pi)^2}{2l+1} \int_0^{\infty} dr_1 r_1^2$$

$$\times \int_0^{\infty} dr_2 r_2^2 \, e^{-2Z(r_1 + r_2)} \times \frac{r_<^l}{r_>^{l+1}} \delta_{l0} \delta_{m0}$$

$$= 16Z^2 \int_0^\infty dr_1 r_1^2 \int_0^\infty dr_2 r_2^2 \, e^{-2Z(r_1 + r_2)} \frac{1}{r_>}$$

$$= 16Z^2 \int_0^\infty dr_1 r_1^2 \, e^{-2Zr_1} \left[\frac{1}{r_1} \int_0^{r_1} dr_2 r_2^2 \, e^{-2Zr_2} \right.$$

$$\left. + \int_{r_1}^\infty dr_2 r_2 e^{-2Zr_2} \right]. \tag{181}$$

Evaluating the radial integrals, we get

$$E_0^{(1)} = \frac{5}{8} Z \quad \text{a.u.} \tag{182}$$

The ground-state energy is

$$E_0 \simeq E_0^{(0)} + E_0^{(1)} = -Z^2 + \frac{5}{8} Z \quad \text{a.u.}$$

$$= -Z\left(Z - \frac{5}{8}\right). \tag{183}$$

Since $\left| E_0^{(1)} / E_0^{(0)} \right|$ decreases as Z^{-1} as Z increases we expect higher order corrections to become less important for increasing Z. For He, Eq. (183) gives the ground state energy -2.75 a.u (-74.8eV) compared to the experimental value -78.975eV.

The calculations of higher order corrections after the first order get more difficult. Instead, we now consider in place of Eq. (176) a more general unperturbed wave function for the ground state as

$$\psi_0^{(0)'} = \left(\frac{Z'^3}{\pi} \right) e^{-Z'(r_1 + r_2)} \tag{184}$$

where Z' is now arbitrary. This wave function is clearly the solution of

$$\left[-\frac{1}{2} \left(\vec{\nabla}_1^2 + \vec{\nabla}_2^2 \right) - \left(\frac{Z'}{r_1} + \frac{Z'}{r_2} \right) \right] \psi_0^{(0)'} = -Z'^2 \, \psi_0^{(0)'}. \tag{185}$$

This means that the Hamiltonian Eq. (162) is now split as the sum of an unperturbed Hamiltonian

$$H_0' = -\frac{1}{2} \left(\vec{\nabla}_1^2 + \vec{\nabla}_2^2 \right) - \left(\frac{Z'}{r_1} + \frac{Z'}{r_2} \right) \tag{186}$$

and a perturbation

$$H'' = \frac{1}{r_{12}} - (Z - Z') \left(\frac{1}{r_1} + \frac{1}{r_2} \right). \tag{187}$$

The unperturbed or zero-order energy is

$$E_0^{(0)} = -Z'^2. \tag{188}$$

The first-order correction to this energy is

$$E_0^{(1)} = \frac{(Z')^6}{\pi^2} \int e^{-2Z'(r_1 + r_2)} \frac{1}{r_{12}} \, d\vec{r}_1 \, d\vec{r}_2$$

$$- 2 \, (Z - Z') \left(\frac{Z'^3}{\pi} \right) \int \frac{e^{-2Z'r}}{r} \, d\vec{r}$$

$$= \frac{5Z'}{8} - 2\,(Z - Z')Z'. \tag{189}$$

To the first-order, the ground state energy of an atom with two electrons and charge Z is

$$E_0 = Z'^2 + \frac{5Z'}{8} - 2ZZ'. \tag{190}$$

The best choice for Z' is obtained by minimizing E_0 with respect to Z', i.e.,

$$\frac{dE_0}{dZ'} = 0$$

which gives

$$Z' = Z - \frac{5}{16}. \tag{191}$$

For this value of Z', $E_0^{(1)}$ vanishes and we get

$$E_0 = -\left(Z - \frac{5}{16}\right)^2 \tag{192}$$

This result is identical to that obtained by using the variational method with the unperturbed wave function Eq. (184) as a trial function (see later). The improvement in the result is due to a suitable choice of splitting of the Hamiltonian.

The ground state energy of two-electron systems can be calculated with extreme accuracy by means of the Ritz variational method. In order to take into account the screening of the each electron by the other electron, we choose a simple trial function of the form Eq. (176):

$$\phi(\vec{r}_1, \vec{r}_2) = \frac{Z_{\text{eff}}^2}{\pi}\, e^{-Z_{\text{eff}}\,(r_1 + r_2)}, \tag{193}$$

with the variational parameter, the effective charge Z_{eff}.

Since $\langle \phi | \phi \rangle = 1$, we evaluate the energy functional

$$E[\phi] = \langle \phi | H | \phi \rangle$$

$$= \left\langle \phi \left| -\frac{1}{2}\vec{\nabla}_1^2 - \frac{1}{2}\vec{\nabla}_2^2 - \frac{Z}{r_1} - \frac{Z}{r_2} + \frac{1}{r_{12}} \right| \phi \right\rangle. \tag{194}$$

Now,

$$\left\langle \phi \left| -\frac{1}{2}\vec{\nabla}_1^2 \right| \phi \right\rangle = \left\langle \phi \left| -\frac{1}{2}\vec{\nabla}_2^2 \right| \phi \right\rangle$$

$$= \frac{1}{2}\,Z_{\text{eff}}^2 \tag{195}$$

Using the virial theorem,

$$\left\langle \phi \left| \frac{1}{r_1} \right| \phi \right\rangle = \left\langle \phi \left| \frac{1}{r_2} \right| \phi \right\rangle$$

$$= Z_{\text{eff}} \tag{196}$$

and

$$\left\langle \phi \left| \frac{1}{r_{12}} \right| \phi \right\rangle = \frac{5}{8} Z_{eff}. \tag{197}$$

Therefore,

$$E[\phi] = Z_{eff}^2 - 2ZZ_{eff} + \frac{5}{8} Z_{eff}. \tag{198}$$

The energy is minimum when $Z_{eff} = Z - \dfrac{5}{16}$. This gives a screening constant $= 0.3125$ and for He, the effective charge is 1.6875. The least upper bound to the energy of the ground state energy is

$$E = -\left(Z - \frac{5}{16}\right)^2 \quad \text{a.u.} \tag{199}$$

For He, E = –2.848 a.u. (= –77.5 eV), quite close to the experimental value and more accurate than the first-order perturbation theory result, given in Eq. (183). For H⁻ ion the ground state energy is – 0.473 a.u., which though much better than the first-order perturbation value (– 0.375 a.u.), but still is above the ground state energy of atomic hydrogen (– 0.5 a.u.) so that a stable bound state for H⁻ is not demonstrated.

For improvement, we may choose the trial function

$$\phi(\vec{r}_1, \vec{r}_2) = \phi(\vec{r}_1)\, \phi(\vec{r}_2), \tag{200}$$

where $\phi_i(i = 1, 2)$ are obtained by solving the HF equation. For the ground state, there is no contribution of the exchange term since the spins of the electrons are antiparallel in the singlet state. Hence, Hartree and HF methods are then identical. For He ($Z = 2$), we get for the ground state energy $E_0 = -2.86168$ a.u. after numerical solutions of the HF equations. For further improvement, in variational calculations, one must introduce correlations into the trial function. Hylleraas suggested that the trial function ought to depend on r_{12}. He introduced three distance coordinates

$$s = r_1 + r_2 \quad 0 \le s \le \infty$$

$$t = r_1 - r_2 \quad -\infty \le t \le \infty$$

$$u = r_{12} \quad 0 \le u \le \infty \tag{201}$$

and trial functions of the type

$$\phi = e^{-ks} \sum_{l, m, n = 0}^{N} c_{l,\, 2m,\, n}\, s^l\, t^{2m}\, u^n, \tag{202}$$

where the c's are linear variational parameters and the scale parameter k is akin to Z_{eff}. In the ground state (para), the trial function must be an even function of t. N gives the maximum number of terms kept in the trial function. For $N =$

0, $k = Z_{eff}$, we get back the simple trial function (unnormalized). With 6 linear parameters, Hylleraas obtained $E_0 = -2.90324$ a.u. for He. Extremely accurate variational calculations with Hylleraas-type trial functions or other trial functions for the ground state eigenenergies for H^-, He, …… and other properties like diamagnetic susceptibility, etc. have been made. Pekeris obtained the values -0.52775097 a.u., -2.90372431 a.u. respectively for the ground state energies for H^- and He. The result for the ionization potential in the case of He is (including nuclear motion plus relativistic corrections and Lamb shift contributions)

$$I_P^{theory} = 198310.699 \pm 0.05 \ \text{cm}^{-1} \tag{203}$$

while the experimental result obtained by Herzberg is

$$I_P^{exp} = 198310.82 \pm 0.15 \ \text{cm}^{-1}. \tag{204}$$

This phenomenal agreement of Eq. (203) with Eq. (204) is one of the striking triumphs of quantum mechanics. We now turn our attention to the excited states of two-electron systems. We shall discuss the "genuinely discrete" excited states in which one electron is in the ground state and the other one is an excited state for two-electron atoms with $Z \geq 2$. So we consider $(1s)(n(m)$. In first-order perturbation theory, the energy shift is

$$E_\pm^{(1)} = \left\langle \psi_\pm^{(0)} \left| \frac{1}{r_{12}} \right| \psi_\pm^{(0)} \right\rangle, \tag{205}$$

where the subscript $+(-)$ refers to para (ortho) states. Since the energy shift is independent of m (due to the vanishing of the commutator $\left[\vec{L}, \frac{1}{r_{12}} \right]$, where $\vec{L} = \vec{L}_1 + \vec{L}_2$ is the orbital angular momentum), we can evaluate the energy shift for $m = 0$. We write

$$E_{nl,\,\pm}^{(1)} = J_{nl} \pm K_{nl}, \tag{206}$$

where the direct or Coulomb integral is

$$J_{nl} = \int \left| \psi_{100}(\vec{r}_1) \right|^2 \frac{1}{r_{12}} \left| \psi_{nlo}(\vec{r}_2) \right|^2 d\vec{r}_1 \, d\vec{r}_2$$

$$= \int_0^\infty dr_2 \, r_2^2 \, R_{nl}^2 (r_2) \int_0^\infty dr_1 \, r_1^2 \, R_{10}^2 (r_1) \frac{1}{r_>} \tag{207}$$

and the exchange integral

$$K_{nl} = \int \psi_{100}^*(\vec{r}_1) \, \psi_{nlo}^*(\vec{r}_2) \frac{1}{r_{12}} \psi_{100}(\vec{r}_2) \, \psi_{nlo}(\vec{r}_1) \, d\vec{r}_1 \, d\vec{r}_2$$

$$= \frac{1}{2l + 1} \int_0^\infty dr_2 \, r_2^2 \, R_{10}(r_2) R_{nl}(r_2)$$

$$\times \int_0^\infty dr_1 \, r_1^2 \, R_{10}(r_1) R_{nl}(r_1) \times \frac{r_<^l}{r_>^{l+1}}, \tag{208}$$

using the hydrogenic functions $\psi_{nlm}(\vec{r}) = R_{nl}(r)Y_{lm}(\theta, \phi)$ and Eqs. (179) and (180).

The energy of the state is

$$E_{nl,\,\pm} = E^{(0)}_{1,\,n} + E^{(1)}_{nl,\,\pm}$$

$$= -\frac{Z^2}{2}\left(1 + \frac{1}{n^2}\right) + J_{nl} \pm K_{nl}. \tag{209}$$

The direct integral J is the Coulomb interaction between the charge distributions of the two electrons and is obviously positive. The exchange integral. K is due to the identity (antisymmetry of the wave function) and is also positive. Therefore, an ortho state has an energy lower than the corresponding para state for the same configuration. According to Hund's rule: States of highest spin multiplicity lie lowest. The energy level splitting of (1s) (nl) for He is shown below (Fig. 17.3).

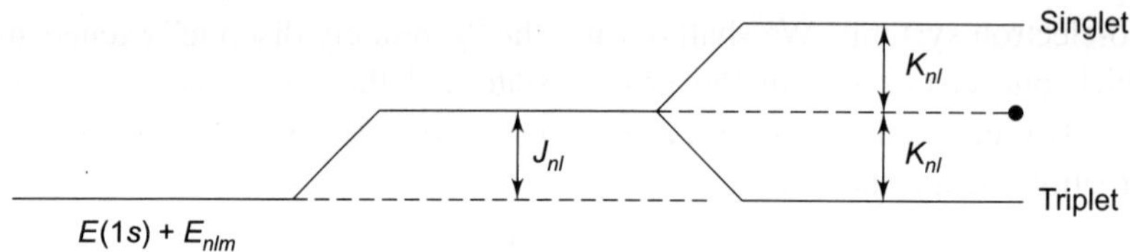

Fig. 17.3   Schematic diagram for the energy level splittings of (1s) (nl) for the He atom

The ground state $(1s)^2$ is a para state and non-degenerate. Other configurations split into the para and the ortho state with the para state higher in energy. A few low lying configurations of He are shown below (Fig. 17.4).

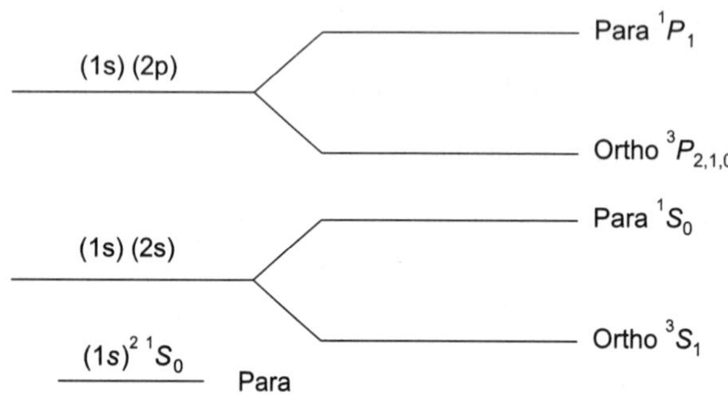

Fig. 17.4   Schematic diagram showing a few low-lying configurations of the He atom

Since

$$\vec{S}_1 \cdot \vec{S}_2 = \frac{1}{2}\vec{S}^2 - \frac{3}{4}$$

$$= \begin{cases} -\dfrac{3}{4}, & \text{singlet } (S = 0) \\[2mm] \dfrac{1}{4}, & \text{triplet } (S = 1) \end{cases}$$

the first-order energy shift Eq. (206) may be written in the form

$$E^{(1)}_{nl,\,\pm} = J_{nl} - \frac{1}{2}\,(1 + 4\vec{S}_1 \cdot \vec{S}_2)\,K_{nl}$$

$$= J_{nl} - \frac{1}{2}\,(1 + \vec{\sigma}_1 \cdot \vec{\sigma}_2)\,K_{nl} \tag{210}$$

with
$$\vec{S}_i = \frac{1}{2}\,\vec{\sigma}_i \ \text{(in a.u.).}$$

Thus, the first-order energy value depends on the spin state – the electrons with parallel spins have a lower energy – though the Hamiltonian is spin-independent and is of purely electrostatic origin. This apparent spin-dependent exchange force arises from the symmetry requirement on the wave functions and is of the same order of magnitude as the electrostatic force. It is much stronger by a factor of α^{-2} than the spin-orbit interaction. This concept is due to Heisenberg who later pointed out that this might provide an explanation of the very large spin-spin interaction in ferromagnets.

The calculation of excited states by the variation method is in general, more difficult than for the ground state because of the subsidiary condition that the trial function of an excited state must be orthogonal to the eigenfunctions of all lower states. In general, the eigenfunctions belonging to two states are orthogonal with different value of $\vec{L}$ or $\vec{S}$ or both. Hence the $2\,^3S$, $2\,^1P$, $2\,^3P$, $3\,^1D$, ... states of He can be treated by the variation method directly. However for the 2^1S term, one must make the trial wave function orthogonal to the wave function of the ground state 1^1S.

We now discuss briefly Eckart's (1930) calculation for the configuration 1s2p of He. The normalized 1s and 2p functions are

$$1s : u_{1s}(r) = \frac{\sqrt{\alpha^3}}{\pi}\,e^{-\alpha r},$$

$$2p : v_{2pm}(r) = \frac{\sqrt{6}}{12}\,\beta^{3/2}\,re^{-\beta r/2}\,Y_{1m}(\theta,\,\phi), \tag{211}$$

where α and β are the effective nuclear charges. The trial wave functions for the para and ortho states ($2\,^1P$ and $2\,^3P$ states respectively) are

$$\phi(2\,^1P,\,m) = 2^{-\frac{1}{2}}\,[u_{1s}(r_1)v_{2pm}(r_2) + u_{2pm}(r_1)u_{1s}(r_2)]|\chi^0\rangle \tag{212}$$

$$\phi(2\,^3P,\,mm_s) = 2^{-\frac{1}{2}}\,[u_{1s}(r_1)v_{2pm}(r_2) - v_{2pm}(r_1)u_{1s}(r_2)]|\chi^1_{m_s}\rangle, \tag{213}$$

where $|\chi^0\rangle$ and $|\chi^1_{ms}\rangle$ are the singlet and triplet spin states respectively. Since the energies do not depend on m and m_s we put $m = m_s = 0$. We put $v_0 = R(r)Y^{(\theta)}_{10}$. Note that the functions $\phi(2\,^3P)$ have correlation built in. The energy functional is then

$$\langle E \rangle_\pm = \int d^3r_1\,[u^*(1)H^{(1)}_0\,u(1) + v^*_0(1)H^{(2)}_0\,v^{(1)}_0]$$

$$+ \int d^3r_1 \, d^3r_2 \, \frac{1}{r_{12}} \, |u(1)v_0(2)|^2$$

$$\pm \int d^3r_1 \, d^3r_2 \, \frac{1}{r_{12}} \, [u^*(1)v_0^*(2)u(2)v_0(1)]$$

$$= \left\langle H_0^{(1)} + H_0^{(2)} \right\rangle + J(\alpha, \beta) \pm K(\alpha, \beta). \tag{214}$$

Using the expansion of $\dfrac{1}{r_{12}}$, the direct integral

$$J = 4\pi \int_0^\infty r_1^2 \, dr_1 \int_0^\infty r_2^2 \, dr_2 \, r_>^{-1} \, |u(1)R(2)|^2 \tag{215}$$

and

$$K = \frac{4\pi}{3} \int_0^\infty r_1^2 \, dr_1 \int_0^\infty r_2^2 \, dr_2 \, \frac{r_<}{r_>^2} \, u^*(1)u(2)R(1)R^*(2). \tag{216}$$

Since $J, K > 0$, the triplet state has lower energy. After evaluation of J, K, one obtains a minimum for the energy by varying the parameters α and β. The results obtained by Eckart are given below.

State	α	β	E_{Eckart} (a.u.)	E_{Pekeris} (a.u)
$2\,^3P$	1.99	1.09	-2.131	-2.133
$2\,^1P$	2.003	0.965	-2.123	-2.124

In view of the simplicity of the trial functions the agreement of Eckart's results with the very accurate calculations by Pekeris 1958)) is excellent.

So far we have considered singly excited states of two-electron atoms. However, there also exist doubly excited states in which both electrons are excited, e.g., $2s^2$, 2s2p, 3p4d, etc. All such states are discrete states embedded in the continuum, since these states lie above the ionization threshold. Any doubly excited state in He is higher than the ground state energy of a He^+-ion (H-like ion with $Z = 2$) and a free electron. For a doubly excited state in He, a radiationless transition to an ionized state plus an electron is much more probable than a radiative transition to a bound state of He. Such a transition is called autoionization (Auger effect) and the doubly excited states unstable against ionization are called autoionizing states. These states of two-electron atoms manifest in the scattering of electrons by the atom (or ion). For example consider the two-step process.

$e^- + He^+ \rightarrow He$ (autoionizing state) $\rightarrow e^- + He^+$, where a resonance is obtained in the scattering cross section.

PROBLEMS

17.1 (a) Find the possible values of the total angular momentum J for the states

$$^1S, \, ^3S, \, ^3P, \, ^2D, \, ^4D.$$

(b) Which spectral terms $^{2s\,+\,1}L_J$ are possible for the following two-electron configurations: $nsn's$; $nsn'p$; $nsn'd$; $npn'p$?

(c) List the possible terms for the following configurations:

$$(np)^3; \quad (nd)^2; \quad ns(n'p)^4.$$

(d) Using Hund's rule, find the lowest terms for the following elements:

O, Cl, Fe, Co, As.

For the electron configurations of these atoms, see Condon and Shortley (1951).

17.2 (a) Calculate the energy levels, including electrostatic and spin-orbit energy, for a p^2 configuration. Draw the energy level scheme.

(b) Repeat the calculation in the jj-coupling scheme.

17.3 The different term energies $\langle LSJM|H_{so}|LSJM\rangle = E_{so}(J)$ of a given multiplet $2S+1_L$ satisfy

$$\sum_{J\,=\,|L\,-\,S|}^{J\,=\,L\,+\,S} (2J + 1)\, E_{so}(J) = X.$$

Find X by evaluating the above sum. E_{so} is the spin-orbit interaction energy.

17.4 Atomic energy levels of the same J value cannot in general cross when one parameter is varied. In a diatomic molecule the intersection of terms of like symmetry is impossible. von Neumann and Wigner gave the non-crossing theorem in 1929.

Consider a system (e.g., a multielectron atom) with the Hamiltonian H depending on a parameter λ. The strength of the symmetry breaking perturbation due to an external field may be considered as the parameter. If the eigenenergies as functions of λ ate plotted against λ, the eigenenergies bifurcate at some value of λ ("level repulsion"). Given a Hamiltonian matrix.

$$H(\lambda) = \begin{vmatrix} H_{11} & H_{12} \\ H_{21} & H_{22} \end{vmatrix},$$

where $H_{12}(\lambda) = H_{21}(\lambda)$, show that there exists no value λ_0 of λ for which the two eigenvalues are equal.

17.5 Consider a gas of spin $\dfrac{1}{2}$ particles in their ground states at $T = 0$. Show that the pair correlation function is

$$G^{(0)}_{\sigma\sigma'}(\vec{r} - \vec{r}) = \frac{n^2}{4}\left[1 - \delta_{\sigma\sigma'}\left(\frac{\sin x - x\cos x}{x^2}\right)^2\right]$$

where n is the number of particles per unit volume, $x = p_F|\vec{r} - \vec{r}'|/\hbar$, with p_F the Fermi momentum.

17.6 Consider a gas of spin zero bosons. Obtain the general expression for the pair correlation function. The particles are subject to an attractive exchange interaction. The explicit form of the occupation number is

$$N_p = \frac{(2\pi\hbar)^3 \, \alpha^{3/2}}{(2\pi)^{3/2}} \, n \, \exp\left(-\alpha \, (\vec{p} - \vec{p}_0)^2/2\right)$$

where n is the number of particles per unit volume. Find the correlation in this case.

17.7 Two "electrons" in the presence of an isotropic harmonic oscillator potential $V(r) = \frac{1}{2}\,\mu\omega^2 r^2$ interact with each other through a harmonic potential which depends on their relative distance $C(\vec{r}' - \vec{r}'')^2$.

   (a) Solve the problem exactly and obtain the ground state energy and the orbital wave function.

   (b) Solve the HF equations to find the ground state energy and the wavefunction. Compare with the exact result.

17.8 Write the HF equation for the ground state of a ferromagnetic system in which the spins of all the electrons are parallel. Apply this to a free electron gas and show that the average HF energy per electron is

$$\frac{3.51}{r_s^2} - \frac{1.154}{r_s}\text{(in Ry)}.$$

CHAPTER 18

The Description of Scattering Processes

In this and the following three chapters we discuss the quantum theory of scattering and more generally, collision processes. Since the days of Rutherford, the analysis of collision phenomena yielded much information about the structure of matter. Indeed, most of our present-day knowledge about the interaction between particles is obtained from scattering experiments.

In the first section, we give a wave packet description of a simple scattering process. Then we discuss the scattering cross section and in the last section we provide a general description of collision processes.

18.1   THE WAVE PACKET DESCRIPTION OF SCATTERING PROCESSES

In this section we develop the description of the scattering process in terms of wave packets. We consider the scattering of particles by a fixed, finite ranged potential. In a typical scattering experiment, a collimated and controlled beam of particles is emitted by a source. Thereafter, the projectile interacts with the target. Finally, the systems produced in the interaction, reaction products, travel to the experimental measuring apparatus. We suppose that a particle is emitted by the source at the time t_0 and is represented by a wave packet

$$\psi(\bar{r}, t_0) = \int \frac{d^3k}{(2\pi)^3} \, e^{i\vec{k}\cdot\vec{r}} \, a_{\vec{k}}. \tag{1}$$

We assume that the wave packet is spatially extended, so that it does not spread appreciably during the course of the experiment. It must be large compared to the size of the target but small compared to typical dimensions of the laboratory. The packet is localized well out of the range of the potential. $a_{\vec{k}}$ is a smooth function of narrow width, centered about a wave vector $\vec{k}_0$; the wave packet moves with the velocity $\vec{v} = \hbar\vec{k}_0/m$ toward the scatterer. We seek the wave function $\psi(\vec{r}, t)$ at a time after the particle has interacted with the scattering center.

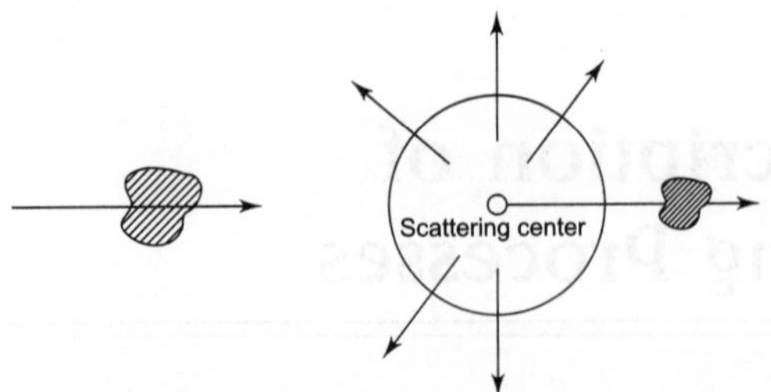

Fig. 18.1 The scattering of a wave packet

Let $\psi_{\vec{k}}(\vec{r})$ be the exact eigenstate of the Hamiltonian of the potential problem:

$$\left[-\frac{\hbar^2}{2m}\vec{\nabla}^2 + V(\vec{r})\right]\psi_{\vec{k}}(\vec{r}) = E_k\psi_{\vec{k}}(\vec{r}),\tag{2}$$

or,
$$(\vec{\nabla}^2 + \vec{k}^2)\psi_{\vec{k}}(\vec{r}) = U(\vec{r})\,\psi_{\vec{k}}(\vec{r}),$$

where $\qquad E_{\vec{k}} = \hbar^2\vec{k}^2/2m.$ and $\quad U(\vec{r}) = \dfrac{2m}{\hbar^2}\,V(\vec{r})$. All values of $E_{\vec{k}} > 0$ are

possible eigenvalues. We next expand $\psi(\vec{r}, t_0)$ in these eigenstates

$$\psi(\vec{r}, t_0) = \int \frac{d^3k}{(2\pi)^3}\,\psi_{\vec{k}}(\vec{r})A_{\vec{k}}.\tag{3}$$

In expansion Eq. (3), only states with energy $E > 0$ enter, since the wave functions of the bound states fall off rapidly to zero (exponentially) at large distances from the potential. The wave function $\psi(\vec{r}, t)$ consists of a part incident on the target and a part that has been scattered by the target. Therefore, in expansion Eq. (3), only those eigenfunctions corresponding to an incident wave (from the left) and an outgoing scattered wave appear. Once we have determined the expansion Eq. (3), the wave function at any later time is given by

$$\psi(\vec{r}, t) = \int \frac{d^3k}{(2\pi)^3}\,\psi_{\vec{k}}(\vec{r})A_{\vec{k}}e^{-iE_{\vec{k}}(t-t_0)/\hbar}.\tag{4}$$

We now determine the stationary states $\psi_{\vec{k}}(\vec{r})$. We introduce the Green function, $G_{\vec{k}}(\vec{r}, \vec{r}')$, for the Helmholtz equation

$$(\vec{\nabla}^2 + \vec{k}^2)\,G_{\vec{k}}(\vec{r}, \vec{r}') = \delta(\vec{r} - \vec{r}').\tag{5}$$

Using this Green function, Eq. (2) becomes an integral equation

$$\psi_{\vec{k}}(\vec{r}) = \phi(\vec{r}) + \int d^3 r' G_{\vec{k}}(\vec{r}, \vec{r}') \, U(\vec{r}') \, \psi_{\vec{k}}(\vec{r}'), \quad (6)$$

where $\phi(\vec{r})$ is a solution of the homogeneous equation

$$(\vec{\nabla}^2 + \vec{k}^2) \, \phi(\vec{r}) = 0 \tag{7}$$

Thus, $\psi_{\vec{k}}(\vec{r})$ is the sum of an incident plane wave and a scattered wave. We now construct G. G is a function of $|\vec{r} - \vec{r}|$ only, so we write

$$G_{\vec{k}}(\vec{r}, \vec{r}') = \int \frac{d^3 q}{(2\pi)^3} \, e^{i\vec{q}(\vec{r} - \vec{r}')} g_{\vec{k}}(\vec{q}). \tag{8}$$

Putting Eq. (8) into Eq. (5), we have

$$(k^2 - q^2) g_{\vec{k}}(\vec{q}) = 1. \tag{9}$$

In Eq. (8), we have to integrate over all positive q^2 including the singular point $k^2 = q^2$. To invert Eq. (9), we generalize Eq. (5) to complex k. We shall see that as Im $k \to 0_{\pm}$ the solutions are not equal, and the two different limits correspond to different boundary conditions. With Im $k \neq 0$, we invert Eq. (9) and obtain

$$G_k(r) = \frac{1}{8\pi^2 \, ri} \int_{-\infty}^{+\infty} qdq \, \frac{e^{iqr} - e^{-iqr}}{(k - q) \, (k + q)} \quad (r = |\vec{r} - \vec{r}'|)$$

after integrating over the orientation of $\vec{q}$. The poles of G lie in the complex q plane at $q = \pm k$. Since $r > 0$, the contour can be closed in the upper half-plane in the term with e^{iqr}, and vice versa in the e^{-iqr} term. We must now specify the contour at the poles; this corresponds to choosing the boundary condition. The contour chosen for the first term is shown in Fig. 18.2. The residue theorem then gives

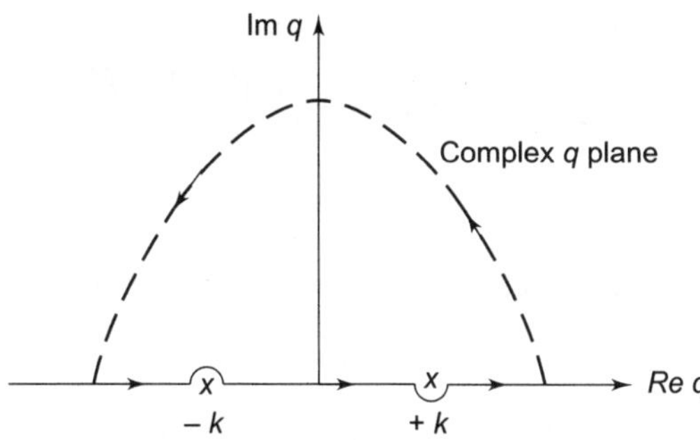

Fig. 18.2   The contour for calculating the retarded Green function

$$G_k(r) = -\frac{1}{4\pi} \frac{e^{ikr}}{r} \; (\text{Im } k > 0) \tag{10}$$

$$G_k(r) = -\frac{1}{4\pi}\frac{e^{-ikr}}{r} \quad (\text{Im } k < 0) \tag{11}$$

G_k is a multivalued function of complex k with a branch cut along the real axis. The boundary value of this function above the cut gives an outgoing spherical wave. The other value e^{-ikr}/r represents an incoming wave, while the linear combination

$$\frac{1}{2}\lim_{\epsilon \to 0}[G_{k+i\epsilon} + G_{k-i\epsilon}] = -\frac{1}{4\pi}\frac{\cos kr}{r}$$

describes a standing wave. Therefore, with our choice of an outgoing scattered wave (*Im* $k = 0_+$) we write the Green function as

$$G_k(r) = -\frac{1}{4\pi}\frac{e^{ik|\vec{r}-\vec{r}'|}}{|\vec{r}-\vec{r}'|} = \int \frac{1}{(2\pi)^3}\frac{d^3_q\, e^{i\vec{q}\cdot\vec{r}}}{k^2 - q^2 + i\epsilon} \tag{12}$$

where the limit $\epsilon \to 0_+$ after the evaluation of the integral.

Substituting Eq. (12) in Eq. (6), we have the basic integral equation of scattering theory

$$\psi_{\vec{k}}(\vec{r}) = \phi_{\vec{k}}(\vec{r}) - \frac{1}{4\pi}\int \frac{e^{ik|\vec{r}-\vec{r}'|}}{|\vec{r}-\vec{r}'|}\, U(\vec{r}')\, \psi_{\vec{k}}(\vec{r}')\, d\vec{r}', \tag{13}$$

where we have chosen for ϕ the momentum eigenfunction $\phi_{\vec{k}}(\vec{r})$ as we are interested in the solution in which we have an incident beam in addition to scattered waves. On multiplying Eq. (13) by $e^{-iE_k t/\hbar}$ we see that the wave fronts in the second term move outward from the potential.

The detectors of the scattered particles are located far from the scatterer, i.e. $|\vec{r}| \gg |\vec{r}'|$, so that

$$k|\vec{r} - \vec{r}'| = k\sqrt{\vec{r}^2 - 2\vec{r}\cdot\vec{r}' + \vec{r}'^2}$$

$$= kr\sqrt{1 + \left(\frac{r'}{r}\right)^2 - \frac{2\vec{r}\cdot\vec{r}'}{r^2}}$$

$$\approx kr - \vec{k}'\cdot\vec{r}',$$

where $\vec{k}' = k\hat{r}$ is the wave vector as observed in the far field. Note that

$$|\vec{k}| = |\vec{k}'|.$$

Thus, the wave function has the asymptotic form

$$\psi_{\vec{k}}(\vec{r}) \xrightarrow[r\to\infty]{} \phi_{\vec{k}}(\vec{r}) - \frac{1}{4\pi}\frac{e^{ikr}}{r}\int e^{-i\vec{k}'\cdot\vec{r}'}\, U(\vec{r}') \times \psi_{\vec{k}}(\vec{r}')\, d^3r'. \tag{14}$$

We may interpret the asymptotic form in Eq. (14) as the incident plane wave plus an outgoing spherical wave emanating from the target in the far away region. This is simply another form of Huygens' principle. This asymptotic requirement is Sommerfeld's "Ausstrahlungs bedingung" (Radiation condition). We may write Eq. (14) as

$$\psi_{\vec{k}}(\vec{r}) \xrightarrow[r \to \infty]{} e^{i\vec{k}\cdot\vec{r}} + \frac{e^{ikr}}{r} f(\vec{k}, \vec{k}'),\tag{15}$$

where

$$f(\vec{k}', \vec{k}) = -\frac{m}{2\pi\hbar^2} \int e^{-i\vec{k}'\cdot\vec{r}'}\, V(\vec{r}')\, \psi_{\vec{k}}(\vec{r}')\, d^3r'$$

$$= \frac{1}{4\pi}\,(\phi_{\vec{k}'},\, U\psi_{\vec{k}}),\tag{16}$$

is called the "scattering amplitude", and has the dimensions of length.

We now continue the calculation of the time evolution of the wave packet. We determine the coefficients $A_{\vec{k}}$ in Eq. (3) in the expansion of the wave packet in terms of the exact eigenstates $\psi_{\vec{k}}(\vec{r})$. Substituting $e^{i\vec{k}\cdot\vec{r}}$ from Eq. (13) into Eq. (1), we get

$$\psi(\vec{r}, t_0) = \int \frac{d^3k}{(2\pi)^3}\, a_{\vec{k}}\, \left[\psi_{\vec{k}}(\vec{r}) + \frac{m}{2\pi\hbar^2} \int d^3r'\, \frac{e^{ik|\vec{r}-\vec{r}'|}}{|\vec{r}-\vec{r}'|} \times V(\vec{r}')\, \psi_{\vec{k}}(\vec{r}').\tag{17}$$

The second term on R.H.S. of Eq. (17) is proportional to

$$\int \frac{d^3k}{(2\pi)^3}\, a_{\vec{k}}\, e^{ik|\vec{r}-\vec{r}'|}\, \psi_{\vec{k}}(\vec{r}).\tag{18}$$

Since $a_{\vec{k}}$ is peaked about $\vec{k}_0$, we can replace $\psi_{\vec{k}}(\vec{r}')$ by $\psi_{\vec{k}_0}(\vec{r}')$. This is not possible at a sharp scattering resonance where $\psi_{\vec{k}}$ varies rapidly with $\vec{k}$. Since $k_0 \gg |\vec{k} - \vec{k}_0|$.

we can write

$$k = \sqrt{(\vec{k}_0 + \vec{k} - \vec{k}_0)^2}$$

$$k = \sqrt{\vec{k}_0^2 + 2\vec{k}_0\cdot(\vec{k} - \vec{k}_0) + (\vec{k} - \vec{k}_0)^2}$$

$$\approx \sqrt{k_0^2 + 2\vec{k}_0\cdot(\vec{k} - \vec{k}_0)}$$

$$\approx \hat{k}_0 \cdot \vec{k}, \text{ where } \hat{k}_0 = \vec{k}_0/k_0.$$

Equation (18) becomes

$$\int \frac{d^3k}{(2\pi)^3}\, a_{\vec{k}}\, e^{i\vec{k}\cdot\hat{k}_0|\vec{r}\,-\,\vec{r}\,'|}\psi_{\vec{k}_0}(\vec{r}\,')$$

$$= \psi(\hat{k}_0|\vec{r}-\vec{r}\,'|,\, t_0)\, \psi_{\vec{k}_0}(\vec{r}\,'). \tag{19}$$

The first factor vanishes since $\hat{k}_0|\vec{r}-\vec{r}\,'|$ is a vector to the right of the potential, and the initial wave packet vanishes there (it is localized at the source at the time t_0). Thus, Eq. (19) equals zero, and Eq. (17) becomes

$$\psi(\vec{r},\, t_0) = \int \frac{d^3k}{(2\pi)^3}\, a_{\vec{k}}\psi_{\vec{k}}(\vec{r}) \tag{20}$$

giving $\qquad\qquad A_{\vec{k}} = a_{\vec{k}}.$

At a later time the wave packet is given by Eq. (4):

$$\psi(\vec{r},\, t) = \int \frac{d^3k}{(2\pi)^3}\, a_{\vec{k}}\psi_{\vec{k}}(\vec{r})e^{-iE_k(t-t_0)/\hbar}. \tag{21}$$

For $\vec{r}$ for from the potential we can put the asymptotic form Eq. (15) for $\psi_{\vec{k}}$ in Eq. (17) to get

$$\psi(\vec{r},\, t) = \psi_0(\vec{r},\, t) + \int \frac{d^3k}{(2\pi)^3}\, a_{\vec{k}}\, \frac{e^{\{i(kr-E_k(t-t_0)/\hbar\}}}{r} f(\vec{k},\, \vec{k}'), \tag{22}$$

where

$$\psi_0(\vec{r},\, t) = \int \frac{d^3k}{(2\pi)^3}\, a_{\vec{k}}\, e^{i\vec{k}\cdot\vec{r}-iE_k(t-t_0)/\hbar} \tag{23}$$

represents the wave packet at the time t in the absence of the scatterer (potential). In the second term we replace k by $\vec{k}\cdot\hat{k}_0$ and put $f(\vec{k},\,\hat{k}') \approx f(\vec{k}_0,\,\vec{k}')$. Then

$$\psi(\vec{r},\, t) = \psi_0(\vec{r},\, t) + \frac{f(\vec{k}_0,\,\vec{k}')}{r}\, \psi_0\,(\hat{k}_0 r,\, t). \tag{24}$$

Equation (24) is interpreted as follows. The wave function after scattering is given by a wave packet at the later time t if there were no scattering and a scattered wave packet. In the radial direction $\psi_0(\hat{k}_0 r,\, t)$ has the same shape as the incident wave packet.

Equation (24) does not hold in two situations (a) if there is a sharp scattering resonance at energy $\hbar^2 k^2/2m$ causing a strong distortion of the wave packet, (b) for long-range potentials, e.g., Coulomb potential.

18.2   THE SCATTERING CROSS SECTION

The results of scattering experiments are usually expressed in terms of cross sections.

The differential scattering cross section is defined for a beam of particles as

$$\frac{d\sigma}{d\Omega} = \frac{dN_{sc}/d\Omega}{dN_{inc}/dA} \tag{25}$$

where dN_{inc} is the number of particles traversing the element of area dA normal to $\vec{k}$ in the incident packet, and dN_{sc} is the number of particles scattered into the cone subtended by the solid angle $d\Omega$. We may write $d\sigma/d\Omega$ in terms of the total probability that a particle is scattered as

$$\frac{d\sigma}{d\Omega} = \frac{\text{Probability of being scattered into solid angle } d\Omega/d\Omega}{\text{Probability for crossing unit area (in front of the target)}}. \tag{26}$$

The total probability of being scattered into an infinitesimal solid angle $d\Omega$ at $\vec{r}$ is the rate at which a particle strikes an area $r^2 d\Omega$ integrated over all time. The rate is the velocity $v = \dfrac{\hbar k_0}{m}$ times $r^2 d\Omega$ times the square of the scattered amplitude, $\left(\dfrac{\left| f_{\vec{k}_0} \right|^2}{r^2} \right) |\psi_0(\hat{k}_0 r, t)|^2$.

Therefore, the total probability of being scattered into $d\Omega$ is

$$\left| f_{\vec{k}_0} \right|^2 d\Omega \, \frac{\hbar k_0}{m} \int_{-\infty}^{\infty} dt |\psi_0 (\hat{k}_0 r, t)|^2. \tag{27}$$

The total probability that crosses a unit area at a point $\vec{r}_0$ in front of the target in the incident beam is the flux (probability) integrated over all time :

$$\frac{\hbar k_0}{m} \int_{-\infty}^{\infty} dt |\psi_0(\vec{r}_0, t)|^2. \tag{28}$$

Neglecting the spreading of the wave packet between $\vec{r}_0$ and $\hat{k}_0 r$, the two integrals in Eqs. (27) and (28) are equal, and we have

$$\frac{d\sigma}{d\Omega} = |f_{\vec{k}0}(\theta, \phi)|^2. \tag{29}$$

This is a general result independent of the shape of the incident wave packet. Equation (29) relates the experimental quantity, $\dfrac{d\sigma}{d\Omega}$, the differential scattering cross section, to f, the scattering amplitude, which characterizes the wave function at large distances from the target. It is the fundamental relation between scattering theory and scattering experiments.

The total scattering cross section is defined as the integral of $\dfrac{d\sigma}{d\Omega}$ over all angles.

$$\sigma = \int \frac{d\sigma}{d\Omega} \, (\theta, \phi) \, d\Omega. \tag{30}$$

σ gives the cross sectional area of the incident beam that is intercepted and the particles therein deflected by the target. Cross sections are usually measured in barns ($1b = 10^{-24}$ cm^2) or millibarns or microbarns.

18.3   THE GENERAL DESCRIPTION OF SCATTERING PROCESSES

We shall consider a general interaction between two particles. The most characteristic feature of collisions of two particles is that many alternative results exist. A neutron, for example, may be elastically scattered, or inelastically scattered, or captured with subsequent emission of γ-ray, charged particles or more neutrons.

In scattering experiments, a beam of particles ("projectiles") are incident upon a beam of target particles. The "particles" may be elementary particles (neutrons, protons, electrons), but they can be atoms, molecules, nuclei.

In general, we shall be interested in scattering processes of several types symbolized by

$$a + X \rightarrow \begin{cases} X + a \\ X' + a' \\ Y + b \\ Z + b_1 + b_2 + \ldots \end{cases} \tag{31}$$

The notation means that particle a strikes a particle X to produce outgoing particles.

Collisions are classified into the following types.

(i) Elastic scattering symbolized as $a + X \rightarrow X + a$, in which the projectile a leaves with the same energy (in the c.m. system) and X is left in the initial state. There is no change in internal states of the particle.

(ii) Inelastic scattering, $a + X \rightarrow X' + a'$ in which the projectile as usually re-emerges with an energy lower than its initial one by the amount of the excitation energy given to X. We include in the inelastic scattering those events which involve a change of internal states with no energy transfer between a and X.

(iii) Rearrangement collisions, $a + X \rightarrow Y + b$ in which matter is exchanged and the outgoing particles may be of different species from the incoming particles. Nuclear reactions are examples.

(iv) Particle production, $a + X \rightarrow Z + b_1 + b_2 + \ldots$ in which three or more particles are in the final state.

Each of these possibilities, characterized by the nature of the outgoing particles and their internal states is a channel. So we must define differential cross sections for each channel. In general, the scattering process is multi-channel, but often many of these channels are closed due to energy considerations.

PROBLEM

18.1 Use the wave packet description of the scattering process to obtain the optical theorem as a consequence of probability conservation.

CHAPTER 19

Formal Scattering Theory

We describe the formal scattering theory in this chapter. The methods with which we shall deal are formal and consequently, applicable to general collision phenomena. The central concept is the knowledge of the scattering operator or S-matrix.

In 1945, Møller gave the first systematic discussion of the formal theory of scattering. The S-matrix was introduced by Heisenberg in 1943 although Wheeler (1937) used the concept of a collision matrix earlier. Important contributions were made by Jost (1947), Lippmann and Schwinger (1950), Gell-Mann and Goldberger (1953), Brenig and Haag (1959), Newton (1960), Weinberg (1964), and others.

19.1   THE LIPPMANN-SCHWINGER EQUATION

We begin with a time-independent formulation of scattering processes. We consider the scattering solutions of the Schrödinger equation

$$H|\psi^{\pm}\rangle = E|\psi^{\pm}\rangle \tag{1}$$

where $\pm$ respectively indicate "in" and "out" boundary conditions. We consider H to be of the form

$$H = H_0 + V, \tag{2}$$

with
$$H_0|\phi\rangle = E|\phi\rangle. \tag{3}$$

Both H_0 and H exhibit continuous energy spectra. From Eqs. (1), (2) and (3), we get the inhomogeneous differential equation

$$(E - H_0)|\psi^{\pm}\rangle = V|\psi^{\pm}\rangle. \tag{4}$$

The corresponding homogeneous equation is given in Eq. (3), where $|\psi^{\pm}\rangle \to |\phi\rangle$ as $V \to 0$. These differential equations can be combined to a Fredholm integral equation of the second kind:

$$|\psi^{\pm}\rangle = |\phi\rangle + (E - H^0)^{-1} V |\psi^{\pm}\rangle. \tag{5}$$

We introduce two resolvent operators

$$G(z) = (z - H)^{-1}, \ G_0(z) = (z - H_0)^{-1}. \tag{6}$$

To deal with the singular nature of these operators, the energy parameter z is chosen to be complex. The inverse operator G (or G_0.) does not exist when z is equal to an eigenvalue of H (or H_0). These operators have different limits when z approaches the positive real axis in the complex plane from above and from below.

We now define an operator $T(z)$, called the "transition matrix" (T-matrix) by

$$G(z) = G_0(z) + G_0(z) \, T(z) \, G_0(z). \tag{7}$$

From Eq. (7), we have

$$G_0 T G_0 = G - G_0$$

or,

$$T = G_0^{-1} G G_0^{-1} - G_0^{-1}$$

$$= (z - H_0) \, (G G_0^{-1} - 1)$$

$$= (z - H_0) \, (G G_0^{-1} - G G^{-1})$$

$$= (z - H_0) \, GV.$$

The above result may also be written as

$$T = VG \, (z - H_0).$$

These two results give

$$G_0 T = GV, \quad T G_0 = VG. \tag{8}$$

Combining Eqs. (7) and (8), we have

$$G(z) = G_0(z) + G(z)VG_0(z) = G_0(z) + G_0(z) \, VG(z) \tag{9}$$

Also,

$$T - V = G_0^{-1} GV - V = (G_0^{-1} G - 1)V$$

$$= (G_0^{-1} - G^{-1}) GV$$

$$= VGV$$

or,

$$T = V + VGV. \tag{10}$$

Equation (9) can be solved iteratively to obtain

$$G = G_0 + G_0VG_0 + G_0VG_0VG_0 + \dots \tag{11}$$

and hence Eq. (10) gives

$$T = V + VG_0V + VG_0VG_0V + \dots \tag{12}$$

These relations hold for z not on the real axis and a limiting process, either from above or from below the real axis, depending on the boundary condition must be used for $z = E$ (real).

The Hilbert space integral equation (5) with the prescribed boundary condition is an example of a Lippmann-Schwinger integral equation. The Lippmann-Schwinger equation for the scattering wave functions is

$$|\psi^{\pm}\rangle = |\phi\rangle + G_0^{\pm}(E)V|\psi^{\pm}\rangle. \tag{13}$$

Combining Eqs. (6) and (10), we obtain the operator Lippmann-Schwinger equation for the T-matrix

$$T(E) = V + VG_0^+(E)T(E). \tag{14}$$

The operator Lippmann-Schwinger equation for the Green function is given by Eq. (9):

$$\begin{aligned} G^{\pm}(E) &= G_0^{\pm}(E) + G_0^{\pm}(E)\,VG^{\pm}(E) \\ &= G_0^{\pm}(E) + G^{\pm}(E)\,VG_0^{\pm}(E). \end{aligned} \tag{15}$$

The notation $G_0^{\pm}(E)$ indicates that we have to take the limit of $G_0(E \pm i\,\epsilon)$ as $\epsilon \to 0$ through positive values. The prescription corresponds to the outgoing and incoming wave boundary conditions. The free (bare) Green function is

$$G_0^{\pm}(E) = (E - H_0 \pm i\epsilon)^{-\mathbf{1}}, \tag{16}$$

and the full Green function is

$$G^{\pm}(E) = (E - H \pm i\epsilon)^{-1}. \tag{16a}$$

Using the differential operators $D_0(E) = E - H_0$ and $D(E) = E - H$, we have

$$D_0(E)\,G_0^+(E) = 1, \quad D(E)G^+(E) = 1,$$

and so, in coordinate space with D diagonal

$$\langle \vec{r}|D(E)|\vec{r}'\rangle = [E - V(r) + \vec{\nabla}^2/2m]\,\delta^{(3)}(\vec{r} - \vec{r}')$$

$$= D_E(\vec{r})\,\delta^{(3)}(\vec{r} - \vec{r}')$$

whence

$$D_E(\vec{r})\,G^{\pm}(\vec{r}, \vec{r}'; E) = \delta^{(3)}(\vec{r} - \vec{r}'). \tag{17}$$

The $L - S$ Eq. (13) in coordinate representation is

$$\psi^+ (\vec{r}) = \phi(\vec{r}) + \int G_0^+ (\vec{r}, \vec{r}'; E) V (\vec{r}') \, \psi^+ (\vec{r}') \, d^3 r', \tag{18}$$

where $\quad G_0^+ (\vec{r}, \vec{r}'; E) = \langle \vec{r} | (E - H_0 + i \, \epsilon)^{-1} | \vec{r}' \rangle.$

Now,

$$G_0^+ (\vec{r}, \vec{r}'; E) = -\frac{2m}{\hbar^2} \frac{\exp(ik|\vec{r} - \vec{r}'|)}{4\pi|\vec{r} - \vec{r}'|} \tag{19}$$

Choosing ϕ to be a plane wave $e^{i\vec{k} \cdot \vec{r}}$, Eq. (18) becomes

$$\psi^+ (\vec{r}) = e^{i\vec{k} \cdot \vec{r}} - \frac{m}{2\pi\hbar^2} \int \frac{\exp(ik|\vec{r} - \vec{r}'|)}{|\vec{r} - \vec{r}'|} V(\vec{r}') \times \psi^+(\vec{r}') d^3 r' \tag{20}$$

The asymptotic limit of Eq. (20) is

$$\psi^+ (\vec{r}) \xrightarrow[r \to \infty]{} e^{i\vec{k} \cdot \vec{r}} - \frac{m}{2\pi\hbar^2} \int e^{-i\vec{k}' \cdot \vec{r}'} V (\vec{r}') \, \psi^+ (\vec{r}') \, d^3 r' \times \frac{e^{ikr}}{r}. \tag{21}$$

This means that at a large distance from the target, the wave function consists of the incident wave plus an outgoing (spherical) scattered wave. Then the $E + i\epsilon$ prescription yields outgoing scattered wave. The coefficient of the spherical outgoing wave e^{ikr}/r is the scattering amplitude. The coefficient is

$$-\frac{m}{2\pi\hbar^2} \int e^{-i\vec{k}' \cdot \vec{r}'} \, V(\vec{r}') \psi^+(\vec{r}') d^3 r'$$

$$= -\frac{m}{2\pi\hbar^2} \langle \phi | V | \psi^+ \rangle.$$

Thus, the outgoing and the incoming scattering states, ψ^+ and ψ^-, can be calculated from the $L - S$ equation using the $i\epsilon$ prescription.

These states are defined by

$$H | \psi_a^+ \rangle = E_a | \psi_a^+ \rangle, \tag{22}$$

$$H | \psi_b^- \rangle = E_b | \psi_b^- \rangle. \tag{22a}$$

We now demonstrate the orthogonality and linear dependence of the scattering functions.

We rewrite the $L - S$ equation (13) as

$$(1 - G_0^+ (E)V) | \psi_a^+ \rangle = | \phi_a \rangle.$$

The formal solution of the above equation is

$$| \psi_a^+ \rangle = (1 + G^+ (E)V) | \phi_a \rangle, \tag{23}$$

where we have used $[1 + G(z)V][1 - G_0(z)V]$

$$= [1 - G_0(z)V][1 + G(z)V] = 1. \text{ Also,}$$

$$|\psi_a^-\rangle = (1 + G^-(E)V|\phi_a\rangle. \tag{23a}$$

From Eq. (23), we see the

$$V|\psi_a^+\rangle = [V + VG^+(E)V]|\phi_a\rangle = T^+(E)|\phi_a\rangle. \tag{24}$$

The conjugate equation of Eq. (23) is

$$\langle\psi_a^+| = \langle\phi_a|(1 + G^+(E)V)^\dagger = \langle\phi_a|(1 + VG^-(E)). \tag{25}$$

Thus,

$$\langle\psi_a^+|\psi_b^+\rangle = \langle\phi_a|\psi_b^+\rangle + \langle\phi_a|V(E_a - i\epsilon - H)^{-1}|\psi_b^+\rangle$$

$$= \langle\phi_a^+|\psi_b^+\rangle + (E_a - i\epsilon - E_b)^{-1}\langle\phi_a|V|\psi_b^+\rangle$$

$$= \langle\phi_a|\phi_b\rangle + \langle\phi_a|(E_b + i\epsilon - H_0)^{-1}V|\psi_b^+\rangle$$

$$+ (E_a - i\epsilon - E_b)^{-1}\langle\phi_a|V|\psi_b^+\rangle$$

$$= \langle\phi_a|\phi_b\rangle + [(E_b + i\epsilon - E_a)^{-1} + (E_a - i\epsilon - E_b)^{-1}] \times \langle\phi_a|V|\psi_b^+\rangle$$

$$= \langle\phi_a|\phi_b\rangle$$

$$= (2\pi)^3\delta(\vec{k}_a - \vec{k}_b)\,\delta_{ab}. \tag{26}$$

Therefore, the outgoing scattering functions $\{\psi_b^+\}$ are also mutually orthogonal. Similarly, one can show that the incoming scattering functions $\{\psi_b^-\}$ are also mutually orthogonal

$$\langle\psi_a^-|\psi_b^-\rangle = (2\pi)^3\,\delta(\vec{k}_a - \vec{k}_b)\,\delta_{ab} \tag{27}$$

Now,

$$\langle\psi_b^-|\psi_a^+\rangle = \langle\phi_b|\psi_a^+\rangle + \langle\phi_b|V(E_b + i\epsilon - H)^{-1}|\psi_a^+\rangle.$$

$$= \langle\phi_b|\psi_a^+\rangle + (E_b + i\epsilon - E_a)^{-1}\langle\phi_b|V|\psi_a^+\rangle$$

$$= \langle\phi_b|\phi_a\rangle + \langle\phi_b|(E_a + i\epsilon - H_0)^{-1}V|\psi_a^+\rangle$$

$$+ (E_b + i\epsilon - E_a)^{-1}\langle\phi_b|V|\psi_a^+\rangle.$$

$$= \langle\phi_b|\phi_a\rangle + [(E_a + i\epsilon - E_b)^{-1} + (E_b + i\epsilon - E_a)^{-1}] \times \langle\phi_b|V|\psi_a^+\rangle$$

$$= \langle\phi_b|\phi_a\rangle - 2i\epsilon[(E_a - E_b)^2 + \epsilon^2]^{-1}\langle\phi_b|V|\psi_a^+\rangle.$$

Taking the limit $\epsilon \to 0$, we get

$$\langle\psi_b^-|\psi_a^+\rangle = \langle\phi_b|\phi_a\rangle - 2\pi i\,\delta(E_a^- - E_b)\langle\phi_b|V|\psi_a^+\rangle$$

$$= (2\pi)^3 \, \delta(\vec{k}_a - \vec{k}_b) \, \delta_{ab} - 2\pi i \, \delta(E_a - E_b) \, \langle \phi_b | T^+(E_a) | \phi_a \rangle, \tag{28}$$

using (24).

Thus, the two sets of scattering functions $\{\psi_a^+\}$ and $\{\psi_b^-\}$ are linearly dependent on each other. Neither set is complete, since they span only the subspace of positive energy eigenfunctions of H, but both can be completed by including the bound states of H, which span the negative energy subspace.

We define the S-matrix by the relation

$$S_{ba} = \langle \psi_b^- | \psi_a^+ \rangle. \tag{29}$$

From Eqs. (28) and (29), we obtain the relationship of the S-matrix and the T matrix:

$$S_{ba} = (2\pi)^3 \, \delta(\vec{k}_a - \vec{k}_b) \, \delta_{ab} - 2\pi i \, \delta(E_a - E_b) \, \langle \phi_b | T^+(E_a) | \phi_a \rangle \tag{30}$$

The S-matrix is unitary

$$SS^\dagger = S^\dagger S = 1 \tag{31}$$

The properties of the S-matrix are discussed later.

19.2   IN-STATES AND OUT-STATES

In this section we develop the time-dependent approach to scattering in the Schrödinger picture. The time-dependent method provides the bridge between nonrelativistic and relativistic scattering theories. We consider the scattering process as a change in the state ket from a free particle ket to a state ket influenced by the interaction. The bare and full time-dependent kets satisfy the Schrödinger equation

$$(i\hbar\partial_t - H_0) | \phi(t) \rangle = 0, \tag{32}$$

$$(i\hbar\partial_t - H_0) | \psi(t) \rangle = 0, \tag{33}$$

with $H = H_0 + V$.

We define four Green functions by the equations

$$(i\hbar\partial_t - H_0) \, G_0^\pm \, (t, t') = 1\delta(t' - t), \tag{34}$$

$$(i\hbar\partial_t - H) \, G^\pm \, (t, t') = 1\delta(t' - t), \tag{35}$$

where 1 stands for the unit operator. The time-dependent integral equations (analogous to the $L - S$ equations) are

$$| \psi^+(t') \rangle = | \phi(t') \rangle + \int G_0^+ \, (t, t') \, V(t) | \psi^+ (t) \rangle \, dt, \tag{36}$$

$$G^\pm(t', t) = G_0^\pm \, (t', t) + \int G_0^\pm \, (t', t'') V(t'') \, G^\pm \, (t'', t) \, dt''. \tag{37}$$

The physical requirements of causality impose the following boundary conditions on G and G_0:

$$G^+(t' - t) = G_0^+(t' - t) = 0 \quad \text{for } t' < t$$

$$G^-(t' - t) = G_0^-(t' - t) = 0 \quad \text{for } t' > t \tag{38}$$

Thus G^+ and G_0^+ are the retarded Green functions and G^- and G_0^- the advanced ones. For time-independent H_0 and H, the solutions to Eqs. (34) and (35) and (38) are

$$G^+(t' - t) = -\frac{i}{\hbar}\,\theta(t' - t)e^{-iH(t' - t)/\hbar},$$

$$G_0^+(t' - t) = -\frac{i}{\hbar}\,\theta(t' - t)\,e^{-iH_0(t' - t)/\hbar}, \tag{39}$$

where the step-function $\theta(t' - t)$ is defined by

$$\theta(t' - t) = \begin{cases} 1 & \text{if } t' > t, \\ 0 & \text{if } t' < t, \end{cases} \tag{40}$$

with
$$\partial_t\theta(t) = \delta(t).$$

The relationship of the causal Green function $G^+\,(t' - t)$ to the time-independent Green function $G^+(E)$ is given by

$$G^+(t' - t) = \frac{1}{2\pi} \int\limits_{-\infty}^{\infty} dE\, e^{-iE(t' - t)/\hbar}\, G^+(E), \tag{41}$$

where $G^+(E) = (E - H + i\epsilon)^{-1}$. A similar relation holds for $G^+ \cdot G^+(t)$ is the Fourier transform of $G^+(E)$ and therefore the causal boundary condition in time, given in Eq. (38), corresponds to the outgoing spherical wave prescription in coordinate space.

The functions G_0 and G describe the propagation of the state kets in time – they are called "propagators". For example, the operator G_0^+ allows us to express the ket $|\phi(t')\rangle$ for any time t' later than t $(t' > t)$ in terms of its value at $t' = t$,

$$|\phi(t')\rangle = i\hbar\, G_0^+\,(t' - t)|\phi(t)\rangle \tag{42}$$

Similarly, we have for $t' > t$

$$|\psi(t')\rangle = i\hbar\, G^+(t' - t)|\psi(t)\rangle \tag{42a}$$

and for $t' < t$,

$$|\phi(t')\rangle = -i\hbar\, G_0^-\,(t' - t)|\phi(t)\rangle \tag{42b}$$

$$|\psi(t')\rangle = -i\hbar\, G^-\,(t' - t)|\psi(t)\rangle \tag{42c}$$

Thus, the operators G_0^+ and G^+ describe the propagation due to H_0 and H respectively, in the future and G_0^- and G^- in the past.

Let us define

$$|\psi(t)\rangle = i\hbar\, G_0^+\,(t - t')\,|\psi(t')\rangle \tag{43}$$

This is a state ket whose time development for $t > t'$ is governed by the free Hamiltonian H_0 but which at the time t_o was equal to $|\psi(t_0)\rangle$ We now let $t' \to -\infty$. This defines the "in-state"

$$|\psi_{in}(t)\rangle = \lim_{t' \to -\infty}\, i\hbar\, G_0^+\,(t - t')\,|\psi(t')\rangle. \tag{44}$$

$|\psi_{in}\rangle$ is a free state ket which develops at all times according to the free Hamiltonian, H_0, but which in the infinite past was equal to the exact state ket of the system.

Next, we define the "out-state"

$$|\psi_{out}(t)\rangle = \lim_{t' \to \infty}\, - i\hbar\, G_0^-\,(t - t')\,|\psi(t')\rangle \tag{45}$$

This is a state ket which develops at all times according to the free Hamiltonian H_0. but which will be equal to the exact state ket $|\psi\rangle$ in the infinite future. We have

$$|\psi(t)\rangle = |\psi_{in}(t)\rangle + \int_{-\infty}^{\infty} dt'\, G^+(t - t')\, V\, |\psi_{in}(t')\rangle, \tag{46}$$

$$|\psi(t)\rangle = |\psi_{out}(t)\rangle + \int_{-\infty}^{\infty} dt'\, G^-(t - t')\, V\, |\psi_{out}(t')\rangle, \tag{46a}$$

or, symbolically,

$$|\psi(t)\rangle = |\psi_{in}(t)\rangle - i\hbar \int_{-\infty}^{t} dt'\, e^{-iH(t - t')/\hbar}\, V\, |\psi_{in}(t')\rangle, \tag{46b}$$

$$= |\psi_{out}(t)\rangle + i\hbar \int_{t}^{\infty} dt'\, e^{-iH(t - t')/\hbar}\, V\, |\psi_{out}(t')\rangle. \tag{46c}$$

19.3   THE SOPERATOR AND THE WAVE OPERATORS

Inserting Eq. (42a) into Eq. (46), we have

$$|\psi(t)\rangle = \Omega^{(+)}|\psi_{in}(t)\rangle \tag{47}$$

with
$$\Omega^{(+)} = 1 - i\hbar \int_{-\infty}^{\infty} dt'\, G^+(t - t')\, V G_0^-\,(t' - t)$$

$$= 1 - i\hbar \int_{-\infty}^{\infty} dt \, G^+ (- t) \, VG_0^-(t). \tag{48}$$

The time-independent operator Ω^+, called the *Møller wave operator*, converts the "in-state" to the complete state $|\psi(t)\rangle$ which corresponds to it in the sense that it was essentially equal to it in the remote past.

Similarly, we define the *Møller wave operator* $\Omega^{(-)}$ by the relation

$$|\psi(t)\rangle = \Omega^{(-)}|\psi_{out}(t)\rangle \tag{49}$$

with

$$\Omega^{(-)} = 1 + i\hbar \int_{-\infty}^{\infty} dt G^-(- t') \, VG_0^+ (t). \tag{50}$$

We may write

$$\Omega^{(\pm)} = \lim_{t \to {-\infty \atop +\infty}} G^\pm(-t) \, G_0^\mp (t)$$

$$= \lim_{t \to {-\infty \atop +\infty}} e^{-iHt/\hbar} \, e^{-iH_0t/\hbar}.$$

Inverting Eqs. (47) and (49), we obtain.

$$|\psi_{in}(t)\rangle = \Omega^{(+)\dagger}|\psi(t)\rangle, \tag{51}$$

$$|\psi_{out}(t)\rangle = \Omega^{(-)\dagger}|\psi(t)\rangle. \tag{52}$$

Combining Eqs. (47) and (51), and (49) and (52), we get

$$|\psi_{in}(t)\rangle = \Omega^{(+)\dagger} \, \Omega^{(+)}|\psi_{in}(t)\rangle. \tag{53}$$

$$|\psi_{out}(t)\rangle = \Omega^{(-)\dagger} \, \Omega^{(-)}|\psi_{out}(t)\rangle. \tag{54}$$

On the basis of the assumption that the free in-and out-states are wave packets which span the entire Hilbert space, we see that $\Omega^{(+)}$ and $\Omega^{(-)}$ are isometric

$$\Omega^{(+)\dagger} \, \Omega^{(+)} = \Omega^{(-)\dagger} \, \Omega^{(-)} = 1. \tag{55}$$

This means that the operators $\Omega^{(\pm)}$ do not change the length of a vector,

$$|\Omega\psi|^2 = (\Omega\psi, \Omega\psi)$$

$$= (\psi, \Omega^\dagger \, \Omega\psi)$$

$$= (\psi, \psi)$$

$$= |\psi|^2.$$

It does not imply that $\Omega^{(+)}$ or $\Omega^{(-)}$ is necessarily unitary. From Eqs. (53) and (54), we cannot conclude that Eq. (55) also holds with the order of the factors reversed. The scattering states do not necessarily span the Hilbert space. If H has bound states, these cannot be represented by packets which evolve in time from free packets sent in from far away. All the scattering states are orthogonal to the bound states.

Inserting Eq. (47) or Eq. (49) into Eqs. (32) and (33), we get

$$H\Omega^{(\pm)} = \Omega^{(\pm)} H_0^{(\pm)}. \tag{56}$$

Hence, $\Omega^{(\pm)}$ intertwines the full and bare Hamiltonians. Taking adjoints,

$$\Omega^{(\pm)\dagger} H = H_0 \Omega^{(\pm)\dagger} \tag{56a}$$

It can be shown (that) for Ω being either $\Omega^{(+)}$ or $\Omega^{(-)}$

$$\Omega\,\Omega^{\dagger} = 1 - \Lambda, \tag{57}$$

where Λ, called the unitary deficiency of Ω, is the projection onto the space spanned by the bound states of H,

$$\Lambda = \sum_n |\psi_{bound}^{(n)}\rangle\langle\psi_{bound}^{(n)}|. \tag{58}$$

Clearly, if H has no bound state, then $\Omega^{(+)}$ and $\Omega^{(-)}$ are unitary.

The relationship of the out-state of a given state vector to its in-state is obtained from Eqs. (52) and (47),

$$|\psi_{out}(t)\rangle = \Omega^{(-)\dagger}\,\Omega^{(+)}|\psi_{in}(t)\rangle. \tag{59}$$

We define the scattering operator S by the relation

$$S \equiv \Omega^{(-)\dagger}\,\Omega^{(+)} \tag{60}$$

i.e.,
$$|\psi_{out}(t)\rangle = S|\psi_{in}(t)\rangle. \tag{61}$$

The scattering operator converts an in-state to an out-state.

Let us consider that in the remote past the system was in the controlled state $|\psi_{in}(\alpha, t)\rangle = |\phi(\alpha, t)\rangle$ and now it is in the state $|\psi^{(+)}(\alpha, t)\rangle$. The amplitude for the probability of finding the system in the state $|\phi(\beta, t)\rangle$ in the distant future is

$$\lim_{t\to\infty} \langle\phi(\beta, t)|\psi^{(+)}(\alpha, t)\rangle$$

$$= \langle\phi(\beta, t)|\psi_{out}(t)\rangle$$

$$= \langle\phi(\beta, t)|\Omega^{(-)\dagger}\,\Omega^{(+)}\psi_{in}(t)\rangle$$

$$= \langle\phi(\beta, t)|S|\phi(\alpha, t)\rangle. \tag{62}$$

The matrix of the operator S on the basis of free states is called the S-matrix.

From the intertwining relations Eqs. (56) and (56a), we get

$$\Omega^{(-)\dagger}\,\Omega^{(+)}H_0 = \Omega^{(-)\dagger}H\Omega^{(+)} = H_0\Omega^{(-)\dagger}\,\Omega^{(+)}.$$

Hence, the scattering operator commutes with H_0

$$[S, H_0] = 0, \tag{63}$$

and so the matrix $S_{\beta\alpha}$ is independent of time. From Eq. (61), taking adjoints, we obtain

$$|\psi_{in}(t)\rangle = S^\dagger\,|\psi_{out}(t)\rangle. \tag{64}$$

Since the sets $\{\psi_{in}\}$ and $\{\psi_{out}\}$ are assumed complete we conclude that S is unitary (provided that the Hamiltonian is hermitian)

$$SS^\dagger = S^\dagger S = 1. \tag{65}$$

We next define the reactance operator K by the relation (Cayley transform)

$$K = i(1 - S)\,(1 + S)^{-1}$$

or,
$$S = (1 + iK)\,(1 - iK)^{-1}. \tag{66}$$

The unitarity of S implies that K is hermitian Equation (66) can be written in the following form

$$S - 1 = 2iK + iK(S - 1). \tag{66a}$$

Equation (66a) is known as Heitler's integral equation.

From Section 2.8 we know that there are three pictures of describing the time development of quantum states and operators – the Schrödinger, the Heisenberg and the Dirac interaction pictures. Time-independent formulation of scattering is a special case of the Heisenberg picture, while time-dependent scattering theory can be formulated in the Schrödinger picture or in the interaction picture. It is obvious that the final results obtained in any problem are independent of the pictures used. The physical content of the three pictures for scattering processes is embodied in the same S operator, whose matrix elements are identical in all three cases. In this section we have formulated time-dependent scattering in the Schrödinger picture.

We now discuss time-dependent scattering in the interaction picture. The Schrödinger equation in Eq. (33) has the solution

$$|\psi(t)\rangle = \exp\left[-(i/\hbar]\,H\,(t - t_0)\right]|\psi(t_0)\rangle, \tag{67}$$

where we have introduced an exponential operator by the relation

$$\exp\left[-(i/\hbar)Ht\right] \equiv \sum_{n=0}^{\infty} \frac{1}{n!}\left(-\frac{i}{\hbar}\,Ht\right)^n. \tag{68}$$

If we define the time-translation operator as

$$U(t, t_0) = \exp\left[- (i/\hbar)\, H(t - t_0)\right] \tag{69}$$

then we can write

$$|\psi(t)\rangle = U(t, t_0)\,|\psi(t_0)\rangle. \tag{70}$$

If H is hermitian, the time-translation operator is unitary.

Integrating Eq. (33) we get

$$|\psi(t)\rangle = |\psi(t_0)\rangle - (i/\hbar) \int_{t_0}^{t} dt'\, H\,|\psi(t')\rangle, \tag{71}$$

whence we get an integral equation for $U(t, t_0)$,

$$U(t, t_0) = 1 - \left(\frac{i}{\hbar}\right) \int_{t_0}^{t} dt'\, H U\,|\psi\,(t', t_0)\rangle. \tag{72}$$

If the equation is iterated in powers of H, we obtain

$$U(t, t_0) = 1 - \frac{i}{\hbar} \int_{t_0}^{t} dt'\, H + \left(-\frac{i}{\hbar}\right)^2 \int_{t_0}^{t} dt'\, H \int_{t_0}^{t'} dt''\, H + \ldots$$

$$= 1 - \frac{i}{\hbar}\, H(t - t_0) + \frac{1}{2} \left(-\frac{i}{\hbar}\right)^2 H^2 (t - t_0)^2 + \ldots$$

$$= \exp\left[-\left(\frac{i}{\hbar}\right) H(t - t_0)\right].$$

The integral Eq. (72) is thus equivalent to the exponential expression, provided the power series is meaningful.

A state ket in the interaction picture is related to one in the Schrödinger picture by the relation

$$|\psi_I(t)\rangle = \exp\left[(i/\hbar)H_0 t\right] |\psi(t)\rangle. \tag{73}$$

Inserting Eq. (73) in Eq. (33), we have

$$i\hbar\partial_t |\psi_I(t)\rangle = V_I(t)\,|\psi_I(t)\rangle \tag{74}$$

with
$$V_I(t) = \exp\left[(i/\hbar)\, H_0 t\right] V \exp\left[-(i/\hbar)\, H_0 t\right]. \tag{75}$$

As expected, the time-dependence of $|\psi_I(t)\rangle$ is due to the interaction. A free-particle state is a constant one. $|\psi_I(t)\rangle$ satisfies the relation

$$|\psi_I(t)\rangle = U_I(t, t_0)\,|\psi_I(t_0)\rangle, \tag{76}$$

where the time-translation operator in the interaction picture is

$$U_I(t,\, t_0) = \exp\,[(i/\hbar)H_0 t]\, \exp\,[-\,(i/\hbar)H\,(t - t_0)]\, \exp\,[-\,(i/\hbar)\,H_0 t_0]. \qquad (77)$$

Differentiating with respect to t, we get

$$\frac{\partial U_I(t,\, t_0)}{\partial t} = -\,\frac{i}{\hbar}\, V_I(t)\, U_I(t,\, t_0). \qquad (77a)$$

Integrating Eq. (74), we obtain

$$|\psi_I(t)\rangle = |\psi_I(t_0)\rangle - \frac{i}{\hbar}\int_{t_0}^{t} dt' V_I(t')\, U_I(t',\, t_0)|\psi_I(t_0)\rangle \qquad (78)$$

or,

$$|U_I(t,\, t_0) = 1 - \frac{i}{\hbar}\int_{t_0}^{t} dt' V_I(t')\, U_I(t'\, t_0), \qquad (79)$$

Iterating Eq. (79) we get

$$U_I(t,\, t_0) = 1 - \frac{i}{\hbar}\int_{t_0}^{t} dt_1 V_I(t_1) + \left(\frac{i}{\hbar}\right)^2 \int_{t_0}^{t} dt_1 \int_{t_0}^{t_1} dt_2 V_I(t_1)\, V_I(t_2) + \,\ldots \qquad (80)$$

Note that Eq. (80) should be in time-ordered form. For example, the second-order term with time-ordered integration variables may be written as

$$\int_{t_0}^{t} dt_1 \int_{t_0}^{t_1} dt_2 V_I(t_1)\, V_I(t_2) = \frac{1}{2}\int_{t_0}^{t} dt_1 \int_{t_0}^{t_1} dt_2\, T\{V_I(t_1)V_I(t_2)\}, \qquad (81)$$

where T is the Dyson time-ordering operator defined by

$$T\{V_I(t_1)\, V_I(t_2)\} = V_I(t_1)\, V_I(t_2)\, \theta\,(t_1 - t_2) + V_I(t_2)\, V_I(t_1)\, \theta\,(t_2 - t_1). \qquad (82)$$

All the expressions discussed above have no limits for very early times ($t \to -\infty$) or for very late times ($t \to +\infty$). In using this formalism to scattering experiments, we must be able to provide prescriptions for passing on to the limits ($t \to \pm\infty$).

This problem has been considered by several workers. Some authors have introduced the technique known as "adiabatic switching". To see how the prescription is used, we first note that the plane-wave limit for Schrödinger states far from the scattering is $|\psi_E(t)\rangle_S \to e^{-iEt/\hbar}|\phi_E\rangle$ as $t \to \pm\,\infty$, where $|\phi_E\rangle$ is the plane-wave state in the time-independent picture. In the interaction picture, we must also remove an additional $e^{iH_0 t/\hbar}$ dependence, giving

$$|\psi_E(t)\rangle_I \xrightarrow[t \to \pm\infty]{} e^{-iEt/\hbar}\, e^{-iHt_0/\hbar}|\phi_E\rangle = |\phi_E\rangle.$$

To insure that this limit is achieved, we introduce the adiabatic switching in which the interaction $V_I(t)$ is adiabatically switched on and off at infinity. We damp down the large oscillations of the exponentials at $t = \pm\infty$ by writing $V_I(t)$ as

$$V_{I,\epsilon}(t) = e^{-\epsilon|t|}V_I(t), \ \epsilon \to 0+ \tag{83}$$

where the symbol $0+$ indicates that ϵ is an infinitesimally small, positive quantity. $V_{I,\epsilon}(t) \to 0$ for $t \to \pm\infty$ as required. Similarly, $U_I(t, t_0) \to U_{I,\epsilon}(t, t_0)$.

Next, we apply the damped $U_{I,\epsilon}(t, t_0)$ to a full energy eigenstate at $t = -\infty$, transforming it to $t = 0$:

$$|\psi_E(0)\rangle_I = U_{I,\epsilon}(0, -\infty)|\psi_E(-\infty)\rangle_I. \tag{84}$$

Iterating $U_{I,\epsilon}(0, -\infty)$ to first order for V time independent,

$$U_{I,\epsilon}(0, -\infty) = 1 - (i/\hbar)\int_{-\infty}^{0} dt'' e^{\epsilon t''} e^{iH_0 t''/\hbar}Ve^{-iH_0 t''/\hbar} + \dots \tag{85}$$

From Eq. (84), we get

$$|\psi_E(0)\rangle_I = |\phi_E\rangle - (i/\hbar)\int_{-\infty}^{0} dt'' \ e^{\epsilon t''} \ e^{iH_0 t''/\hbar}Ve^{-iH_0 t''/\hbar} \ |\phi_E\rangle + \dots$$

$$= |\phi_E\rangle + (E - H_0 + i\epsilon)^{-1}V|\phi_E\rangle + \dots$$

$$= |\phi E\rangle + G_0^+(E)V|\psi_E(0)\rangle_I,$$

by iterating Eq. (85) to all orders in V.

This is just the $L - S$ equation for in states (outgoing spherical waves) in the time-independent formalism. Thus,

$$|\psi(0)\rangle = |\psi^+\rangle \tag{86}$$

or,
$$|\psi^+\rangle = U_{I,\epsilon}(0, -\infty)|\phi\rangle \tag{87}$$

valid for time-independent states of any energy. One can show that

$$|\psi^-\rangle = U_{I,\epsilon}(0, \infty)|\phi\rangle$$

and
$$\langle\psi^-| = \langle\phi|U_{I,\epsilon}(\infty, 0) \tag{88}$$

Equations (87) and (88) are called the "dressing relations" for a time-dependent V because these relations represent the "clothing" of a bare state at $t = -\infty$ converting it to a full state at $t = 0$.

We define the $M\phi$ller operators

$$\Omega^{(\pm)} = U_{I,\epsilon}(0, \mp\infty). \tag{89}$$

Equations (87) and (88) may be written as

$$|\psi^\pm\rangle = \Omega^{(\pm)}|\phi\rangle. \tag{90}$$

The development may be described as follows. From $t = -\infty$ to $t = 0$, the bare state is dressed. Then, it undergoes interaction from $t = 0$ to $t = \tau$ with "in" states $|\psi^+\rangle$ converted to "out" states $|\psi^-\rangle$. From $t = \tau$ to $t = \infty$, the out state is undressed, becoming another bare state (Scadron (1991)). The scattering process is

$$|\phi(\infty)\rangle_I = U_{I,\epsilon}(\infty, -\infty)|\phi(-\infty)\rangle_I \tag{91}$$

$$S_I = U_{I,\epsilon}(\infty,-\infty) = U_{I,\epsilon}(\infty, 0)U_{I,\epsilon}(0, -\infty) = \Omega^{(-)\dagger}\,\Omega^{(+)} \tag{92}$$

The S operator describes the dressing, scattering and undressing process. It is difficult to rigorously justify the "adiabatic switching" technique. For a wave-packet approach, see Rodberg and Thaler (1967).

We now consider the overlap S-matrix elements in the three pictures

$$S_{fi}^{H\cdot P} = \langle \psi_f^- | \psi_i^+ \rangle, \tag{93}$$

$$S_{fi}^{S\cdot P} = \lim_{t\to\infty} \langle \psi_f(t)| \psi_i(t)\rangle_S, \tag{93a}$$

$$S_{fi}^{I\cdot P} = \lim_{t\to\infty} \langle \psi_f^{(t)}| \psi_i(t)\rangle_I \tag{93b}$$

where the superscripts H.P., S.P. and I.P. denote the Heisenberg picture, the Schrödinger picture and the interaction picture respectively. All the three matrix elements denote the probability amplitude for the transition $i \to f$ and hence

$$S_{fi}^{H\cdot P\cdot} = S_{fi}^{S\cdot P\cdot} = S_{fi}^{I\cdot P\cdot}. \tag{94}$$

We now demonstrate the unitarity of the S-matrix. We show first that $S^\dagger S = 1$. The adjoint matrix element is

$$S_{fg}^\dagger = S_{gf}^* = \langle \psi_g^- | \psi_f^+ \rangle^* = \langle \psi_f^+ | \psi_g^- \rangle$$

so that

$$(S^\dagger S)_{fi} = \sum_g S_{fg}^\dagger S_{gi} = \sum_g \langle \psi_f^+ | \psi_g^- \rangle \langle \psi_g^- | \psi_f^+ \rangle$$

where the sum is over all scattering states. This sum can be extended to include the bound states since the S-matrix elements joining scattering and bound states are zero. Using the completeness relation.

$$\sum_g |\psi_g^-\rangle \langle \psi_g^-| + \sum_b |\psi_b\rangle\langle\psi_b| = 1$$

and the orthogonality of the "in" states $|\psi_i^+\rangle$, we get

$$(S^\dagger S)_{fi} = \langle \psi_f^+ | \psi_i^+ \rangle = \delta_{fi}$$

i.e., $\qquad\qquad\qquad\qquad S^\dagger S = 1. \tag{95}$

Next, we prove that $SS^\dagger = 1$. We have

$$(SS^\dagger)_{fi} = \sum_g S_{fg} S^\dagger_{gi} = \sum_g \langle \psi_f^- | \psi_g^+ \rangle \langle \psi_g^+ | \psi_i^- \rangle = \langle \psi_f^- | \psi_i^- \rangle = \delta_{fi},$$

where the sum over scattering states is extended to a sum over all states.

i.e. $$SS^\dagger = 1. \tag{96}$$

Thus, the scattering operator S is unitary. The unitarity implies the conservation of probabilities.

Note that in the I.P., $S_I = U(\infty, -\infty)$ is manifestly unitary.

The S-matrix embodies all possible scattering configurations, so it has a "no scattering" part. Hence, we write

$$S = 1 + R, \tag{97}$$

where the identity operator represents no scattering and the reaction matrix R represents only the definite scattering part of S.

We consider time-independent scattering. From Eqs. (23) and (23a),

$$|\psi_E^+\rangle - |\psi_E^-\rangle = [G^+(E) - G^-(E)]V|\phi_E\rangle. \tag{98}$$

Now,

$$G^+(E) - G^-(E) = \frac{1}{E - H + i\epsilon} - \frac{1}{E - H - i\epsilon}, \tag{99}$$

from Eq. (16b). But the inverse operators can be written as

$$\frac{1}{E - H \pm i\epsilon} = P\,\frac{1}{E - H} \mp i\pi\delta(E - H) \tag{100}$$

where P indicates the Cauchy principal value. From Eqs. (99) and (100), we get

$$G^+(E) - G^-(E) = -2\pi i\delta(E - H). \tag{101}$$

Combining Eqs. (98) and (101), we write

$$|\psi_{E_i}^+\rangle - |\psi_{E_i}^-\rangle = -2\pi i\,\delta(E_i - H)V|\phi_{E_i}\rangle. \tag{102}$$

Multiplying Eq. (102) on the left by $\langle \psi_{E_f}^-|$, we have

$$\langle \psi_{E_f}^- | \psi_{E_i}^+ \rangle - \delta_{fi} = -2\pi i \sum_n \delta(E_i - E_n)\delta_{fn} \langle \psi_{E_n}^- |V|\phi_{E_i}\rangle, \tag{103}$$

where we have inserted a complete set of out-states $\sum_n |\psi_n^-\rangle\langle\psi_n^-| = 1$ between $\delta(E_i - H)$ and V and used $H|\psi_n^-\rangle = E_n|\psi_n^-\rangle$ and the orthogonality of scattering states:

$$\langle \psi_f^- | \psi_i^- \rangle = \langle \psi_f^+ | \psi_i^+ \rangle = \langle \phi_f | \phi_i \rangle = \delta_{fi}. \tag{104}$$

But,

$$\langle \psi_{E_n}^- | V | \phi_{E_i} \rangle = \langle \phi_n | T(E_n) | \phi_i \rangle = T_{ni}(E_n),$$

using $\langle \psi_E^- | V = \langle \phi_E | T \langle E)$.

Hence, from Eqs. (103) and (104) we obtain,

$$R_{fi} = S_{fi} - \delta_{fi} = - i\delta(E_f - E_i) \, T_{fi}. \tag{105}$$

Note that while the T-matrix can be "off the energy shell" (there is the energy delta function multiplying it), the S-matrix must be on the energy shell.

We now investigate the consequences of the unitarity of S for the transition operator T. In term of the reaction matrix R, the unitarity of S, Eqs. (95) and (96) give

$$R + R^\dagger = - R^\dagger R. \tag{106}$$

The matrix element between two arbitrary states yields

$$R_{fi} + R_{if}^* = - \sum_n R_{nf}^* R_{ni}. \tag{107}$$

Using Eqs. (105), and (107) becomes

$$- i\delta(E_{fi}) \, (T_{fi} - T_{if}^*) = - \sum_n \delta(E_{fn}) \, \delta(E_{ni}) \, T_{nf}^* T_{ni}. \tag{108}$$

$\delta(E_{fi})$ can be cancelled from both sides, since

$$\delta(E_{fn}) \, \delta(E_{ni}) = \delta(E_{fi}) \, \delta(E_{ni})$$

provided we remain on the energy shell so that $E_f = E_i$ in all matrix elements T_{fi}. Therefore, the unitarity relation for the T-matrix on the energy shell is

$$i(T_{fi} - T_{if}^*) = \sum_n \delta(E_{ni}) \, T_{nf}^* \, T_{ni} \tag{108a}$$

The L.H.S. of Eq. (108a) is equal to $- 2 \operatorname{Im} T_{fi}$ if the T-matrix is symmetric, i.e., if $T_{fi} = T_{if}$. Time-reversal invariance implies that the T-matrix is symmetric in the angular-momentum representation. However, for the diagonal element of Eq. (108), even without this assumption,

$$- 2 \operatorname{Im} T_{ii} = \sum_n |T_{ni}|^2 \, \delta(E_{ni}). \tag{109}$$

The summation on the right becomes

$$\sum_n \rightarrow \int d\alpha_n \, dE_n \rho_n(E_n)$$

where $\rho_n(E_n)$ is the density of states and the integral over α_n includes a sum over all possible configurations or rearrangements. Equation (109) is then

$$- 2 \operatorname{Im} T_{ii} = \int d\alpha_n |T_{ni}|^2_{E_n = E_i} \, \rho_n(Ei) \tag{109a}$$

Using Eq. (108), the R.H.S. of Eq. (109a) becomes

$$\sum_n \Gamma_{ni} = \Gamma_i,$$

where Γ_i is the total transition rate. We may write $\sigma_{tot} = \dfrac{\Gamma_i}{F_i}$, where σ_{tot} is the total cross section for scattering from the initial system in the state ϕ_i and F_i the incident flux. Then, Eq. (109a) becomes

$$\sigma_{tot} = - 2 \, \frac{\operatorname{Im} T_{ii}}{F_i}. \tag{110}$$

This is the optical theorem relating the imaginary part of the forward-scattering element of the T-matrix to the total cross section. For elastic scattering in the forward direction, the scattering amplitude $f(\theta = 0) = - \dfrac{2m}{4\pi h^2} \, T_{ii}$, so that Eq. (110) becomes the familiar form of the optical theorem

$$\sigma_{tot} = \frac{4\pi}{k} \operatorname{Im} f(\theta = 0). \tag{111}$$

We note that the matrix element $T_{fi} = <f|T| \, i>$ is related to the matrix element S_{fi} by the relation

$$S_{fi} = \delta_{fi} - 2\pi i \; \delta \, (E_i - E_f) \; T_{fi}. \tag{112}$$

We introduce

$$S^{\tau}_{fi} = - T_{fi} \frac{i}{\hbar} \int_{-\frac{\tau}{2}}^{+\frac{\tau}{2}} e^{\frac{i}{\hbar}(E_i - E_f)t} \, dt \; (i \neq f), \tag{113}$$

so that

$$S_{fi} = \lim_{\tau \to \infty} S^{\tau}_{fi}. \tag{114}$$

The transition probability per unit time is

$$\omega_{i \to f} = \lim_{\tau \to \infty} \frac{|S^{\tau}_{fi}|^2}{\tau}. \tag{115}$$

$$\omega_{i \to f} = \frac{2\pi}{\hbar} \, \delta \, (E_i - E_f)|T_{fi}|^2. \tag{116}$$

The apparent singularity in Eq. (116) arises because $\omega_{i \to f}$ denotes the transition into a sharp state in the continuum. Usually, one is interested in a transition into a group of states centered about one at $E = E_f$. The transition probability into the group is

$$\omega_{i \to f}^{\text{group}} = \frac{2\pi}{\hbar} \int_{E_f - dE/2}^{E_f + dE/2} \delta(E_i - E_f) \, |T_{fi}|^2 \, \rho(E) \, dE$$

$$= \frac{2\pi}{\hbar} \, \rho(E) \, |T_{fi}|^2, \tag{117}$$

where $\rho(E)$ denote the density of states in the neighborhood of the energy E_f. The matrix element and $\rho(E)$ are taken at $E = E_i = E_f$ Eq. (117) has the same from as the Golden rule. However, while the latter and even its generalization are only approximate, Eq. (117) is rigorous The cross section is obtained by dividing $\omega_{i \to f}^{\text{group}}$ by the incident flux.

19.4 SYMMETRY CONSIDERATIONS

In this section, we study the consequences of symmetries in scattering.

The S-matrix is a function of the Hamiltonian H, and therefore the S-matrix is invariant under all the transformations that leave H invariant. The free Hamiltonian H_0 is invariant under parity, time reversal and rotations. Let U be the unitary symmetry operator implementing one of these symmetries (for example, rotation and parity). We have

$$UH_0 = H_0 U \text{ and } UV = VU$$

i.e., $$UH = HU, \tag{118}$$

where H is the total Hamiltonian. From Eqs. (89) and (92) we obtain

$$U\Omega^{\pm} = \Omega^{\pm}U$$

$$US = SU. \tag{119}$$

The form of T in Eq. (12) shows that

$$UT = TU. \tag{120}$$

We define

$$|\tilde{\vec{k}}\rangle \equiv U|\vec{k}\rangle, \ |\tilde{\vec{k}}'\rangle \equiv U|\vec{k}'\rangle. \tag{121}$$

Then

$$\langle \tilde{\vec{k}}'|T|\tilde{\vec{k}}\rangle = \langle \vec{k}'|U^{\dagger}UTU^{\dagger}U|\vec{k}\rangle = \langle \vec{k}'|T|\vec{k}\rangle. \tag{122}$$

If Π stands for space inversion (parity),

$$\Pi|\vec{k}\rangle = |-\vec{k}\rangle, \ \Pi|-\vec{k}\rangle = |\vec{k}\rangle.$$

Thus, if $H \ (= H_0 + V)$ conserves parity,

$$\langle -\vec{k}'|T|-\vec{k}\rangle = \langle \vec{k}'|T|\vec{k}'\rangle, \tag{123}$$

i.e., $$f(\vec{k} \to \vec{k}') = f(-\vec{k} \to -\vec{k}').$$

This is pictorially shown in Fig. 19.1 The effect of space inversion on the collision event (Fig. 19.1(a)) is to reverse the momenta, the spins remaining unchanged. The space-inversed collision event is shown in Fig. 19.1(b). The cross sections for these two events are equal.

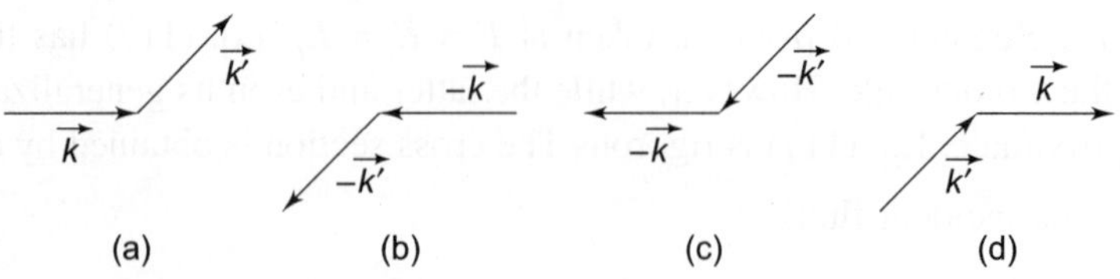

Fig. 19.1   Symmetry in collision events. (a) direct; (b) space-inverted; (c) time-reversed; (d) inverse

To see what happens when the symmetry operation is antiunitary (e.g. time reversal), we note that

$$T^{+\dagger} = V + V \frac{1}{E - i\epsilon - H_0 - V} V,$$

since E, H_0, V are hermitian and $(A^{-1})^{\dagger} = (A^{\dagger})^{-1}$. If however, H_0 and V are invariant under time reversal invariance, then

$$\Theta T^{+} \Theta^{-1} = V + V \frac{1}{E - i\epsilon - H_0 - V} V = T^{-},$$

i.e.,
$$\Theta\, T\Theta^{-1} = T^{\dagger} \tag{124}$$

From Eq. (118), one gets the most general form of the reciprocity relation

$$\langle \psi_f | T | \psi_i \rangle = \langle \Theta T \Theta^{-1} \Theta \psi_i | \Theta \psi_f \rangle = \langle \Theta \psi_i | T | \Theta \psi_f \rangle. \tag{125}$$

It is clear that

$$\Theta \Omega^{\pm} = \Omega^{\pm} \Theta$$
$$\Theta S = S^{\dagger} \Theta \tag{126}$$

Let

$$|\tilde{\alpha}\rangle = \Theta|\alpha\rangle \quad \text{and} \quad |\tilde{\beta}\rangle = \Theta\,|\beta\rangle$$

with
$$|\alpha\rangle = T|\vec{k}\rangle, \quad \langle\beta| = \langle\vec{k}'|.$$

Then

$$|\tilde{\alpha}\rangle = \Theta T|\vec{k}\rangle = \Theta T \Theta^{-1} \Theta|\vec{k}\rangle = T^{+}|-\vec{k}\rangle$$

$$|\tilde{\beta}\rangle = \Theta|\tilde{k}'\rangle = |-\vec{k}'\rangle.$$

Therefore,

$$\langle \beta | \alpha \rangle = \langle \tilde{\alpha} | \tilde{\beta} \rangle \text{ yields}$$

$$\langle \vec{k}' | T | \vec{k} \rangle = \langle -\vec{k} | T | -\vec{k}' \rangle. \tag{127}$$

$\therefore$ If H is invariant under time reversal, then

$$f(\vec{k} \to \vec{k}') = f(-\vec{k}' \to -\vec{k}). \tag{128}$$

Equation (128) is known as the principle of microreversibility or reciprocity theorem.

The time-reversed collision event (Fig. 19.1(c)) is obtained from event (a) by interchanging the initial and final momenta, and reversing the directions of momenta and spins. Time-reversal invariance implies that these two events have equal scattering amplitudes. Notice that invariance under the combined operations of time reversal and parity,

$$f(\vec{k} \to \vec{k}') = f(\vec{k}' \to \vec{k}). \tag{128a}$$

For spinless particles, Eq. (128) yields

$$\frac{d\sigma}{d\Omega}(\vec{k} \to \vec{k}') = \frac{d\sigma}{d\Omega}(\vec{k}' \to \vec{k}) \tag{129}$$

Equation (129), known as the "principle of detailed balance", shows that the cross sections for a collision event (Fig. 19.1(a)) and the inverse collision obtained by interchanging the initial and final states (Fig. 19.1(d)), are equal provided that the particles are spinless. For central potential that does not affect spin the cross sections for the direct and inverse collisions will be equal. On the other hand, when the particles have spin, then we characterize the initial and final free-particle kets by $|k, m_s\rangle$ and $|k, m_s'\rangle$ respectively, where m_s and m_s' refer to the z-components of the initial and final spin. Since

$$\Theta | \vec{k}, m_s \rangle = i^{2m_s} | -\vec{k}, -m_s \rangle,$$

Eq. (125) gives

$$\langle \vec{k}', m_s' | T | \vec{k}, m_s \rangle = i^{2m_s - 2m_s'} \langle -\vec{k}, -m_s | T | -\vec{k}', -m_s' \rangle. \tag{130}$$

The factor $i^{2m_s - 2m_s'} = \pm 1$ since there can be no transitions between half-integral and integral J systems. If parity is conserved, Eq. (125) becomes.

$$\langle \vec{k}', m_s' | T | \vec{k}, m_s \rangle = \pm \langle \vec{k}, -m_s | T | \vec{k}', -m_s' \rangle. \tag{131}$$

This is not the same as detailed balance since the spins are reversed. For unpolarized initial states, we sum over the initial spin states and devide by $(2s + 1)$; if the final polarization is not observed then we sum over the final states. We have

$$\frac{\overline{d\sigma}}{d\Omega}(\vec{k} \to \vec{k}') = \frac{\overline{d\sigma}}{d\Omega}(\vec{k}' \to \vec{k}), \tag{132}$$

where the bar over the differential cross section $d\sigma/d\Omega$ indicates that we average over the initial spin states and sum over the final spin states. Heitler calls this "semi-detailed balance". For an excellent discussion of the reciprocity theorem and detailed balance see Blatt and Weisskopf (1952).

Let us consider the reaction

$$a + A \underset{\rightharpoonup}{\leftharpoondown} b + B. \tag{133}$$

Then the principle of detailed balance is (in c.m. system)

$$\frac{d\sigma(\to)}{d\sigma} \left/ \frac{d\sigma(\leftarrow)}{d\Omega} \right. = \left(\frac{\vec{k}'}{\vec{k}} \right)^2 \frac{(2s_b + 1)\,(2s_B + 1)}{(2s_a + 1)\,(2s_A + 1)}, \tag{134}$$

where the relative momenta in the initial and final states are $\vec{k}'$ and $\vec{k}$ respectively and the incoming particles are unpolarized. (see Nishijima, (1963)). This formula has many applications in nuclear physics, and has also been applied for determining the π^+ spin to the reactions (see Sakurai (1994), Nishijima (1963), Muirhead (1965)).

$$\pi^+ + d \underset{\rightharpoonup}{\leftharpoondown} p + p.$$

If H is spherically symmetric,

$$f(\vec{k} \to \vec{k}') = f(R\vec{k} \to R\vec{k}'), \tag{135}$$

i.e., the scattering amplitude depends only on the magnitude of the incoming momentum and on the scattering angle:

$$f(\vec{k} \to \vec{k}') = f(k, \cos\theta).$$

19.5 SCATTERING FROM TWO POTENTIALS

There is a class of interesting scattering problems in which the potential may be written as the sum of two parts, one of which must be taken exactly, whereas the other part can be considered as a small perturbation. Examples are (i) scattering of particles under the combined influence of Coulomb and nuclear forces; (ii) bremsstrahlung, (iii) meson production in a nucleon-nucleon collision, (iv) the photoelectric effect, the ejection of a bound electron from an atom (or of a bound nucleon from a nucleus by a photon).

Let us write the Hamiltonian of the system as

$$H = H_0 + U + V \tag{136}$$

where U is the potential to be taken into account exactly. The initial and final configurations are described by the eigenfunctions of H.

Now, the S-matrix element is

$$S_{\gamma\beta} = \langle \psi_\gamma^- | \psi_\beta^+ \rangle = \delta_{\gamma\beta} - 2\pi i\, \delta(E_\beta - E_\gamma) T_{\gamma\beta} \tag{137}$$

with

$$T_{\gamma\beta} = \langle \phi_\gamma^- | U | \chi_\beta \rangle + \langle \phi_\gamma^- | V | \psi_\beta^+ \rangle$$

$$= \langle \phi_\gamma^- | U | \chi_\beta \rangle + \left\langle \phi_\gamma^- \left| \left(V + V \frac{1}{E_\beta - H + i\epsilon} V \right) \right| \phi_\beta^+ \right\rangle$$

$$\simeq \langle \phi_\gamma^- | U | \chi_\beta \rangle + \langle \phi_\gamma^- | V | \phi_\beta^+ \rangle, \tag{138}$$

to first order in V.

Here, $| \psi_\beta^+ \rangle$ is an eigenket of H with eigenvalue E_β, and $| \phi_\gamma^- \rangle$ is an eigenket of $(H_0 + U)$. $| \chi_\beta \rangle$ is the plane wave state. The first term in Eq. (138) is the matrix element that would have been obtained if the potential V had not been present. The second term is given to first order in V, assuming that V is small. In many problems the first term would vanish. For example, in bremesstrahlung ϕ_γ^- contains a photon, while χ_β does not, and U is a Coulomb or a nuclear potential. For the atomic or nuclear photoeffect the first term vanishes.

Let us consider another example. We consider U to be the parity-conserving internucleon potential due to strong interactions, and V the weak PNC internucleon potential. Then, $\langle \phi_\gamma^- | U | \chi_\beta \rangle = 0$, where $|\gamma\rangle$ and $|\beta\rangle$ are two kets with opposite parities. The S-matrix element is

$$S_{\gamma\beta} = \delta_{\gamma\beta} - 2\pi i\delta(E_\beta - E_\gamma)T_{\gamma\beta} \tag{139}$$

with $\qquad\qquad T_{\gamma\beta} \simeq \langle \phi_\gamma^- | V | \phi_\beta^+ \rangle.$

The two-potential approach allows us to take into account much of the elastic scattering occurring during any collision event. It also leads to the so-called "distorted-wave Born approximation" and in other circumstances, to the method of the "final-state interaction".

PROBLEMS

19.1 (a) Show that the Møller wave operators satisfy the following relations

$$\Omega^{(+)} \, e^{\frac{i}{\hbar}tH_0} = e^{\frac{i}{\hbar}tH} \, \Omega^{(+)},$$

$$\Omega^{(-)} \, e^{\frac{i}{\hbar}tH_0} = e^{\frac{i}{\hbar}tH} \, \Omega^{(-)},$$

where H_0 is the kinetic energy operator.

(b) The scattering operator is defined as
$$S = \Omega^{(-)\dagger}\Omega^{(+)}.$$
Show that the unitarity of S follows from this definition and the properties of the wave operators.

(c) S commutes with H_0.

Show that this implies conservation of energy.

19.2 Show that the time-reversal operator transforms an out-state into an in-state and vice versa.

19.3 Use the Lippmann – Schwinger equation to obtain the transmission and reflection amplitudes in the case of an attractive delta function potential
$$V(x) = - g \left(\frac{\hbar^2}{2m}\right) \delta(x) \ (g > 0).$$
Comment on the poles.

CHAPTER 20

Potential Scattering: Methods of Solution

In the previous chapter we have discussed the formal methods for the treatment of scattering process. In the current chapter we present a detailed discussion of the radial Schrödinger equation with appropriate boundary conditions. Though the differential equation is a powerful tool for the study of potential scattering the integral equation (incorporating the boundary conditions) approach to scattering theory constitutes an elegant, compact, and general treatment. Here, we deal with the different methods of solution of potential scattering problem.

20.1 THE INTEGRAL FORM OF THE SCHRÖDINGER EQUATION

We start with the Lippmann-Schwinger equation

$$|\psi^{\pm}\rangle = |\phi\rangle + \frac{1}{E - H_0 \pm i\epsilon} \, V|\psi^{\pm}\rangle, \tag{1}$$

where $H = H_0 + V$. We now go to the coordinate-space representation and so take the coordinate space projection of (1):

$$\langle \vec{x}|\psi^{\pm}\rangle = \langle \vec{x}|\phi\rangle + \int d^3x' \, \langle \vec{x}|(E - H_0 \pm i\epsilon)^{-1}|\vec{x}\rangle \, \langle \vec{x}'|V. \tag{2}$$

Equation (2) is an integral equation for scattering. The kernel' of the integral equation

$$G^{\pm}(\vec{x}, \vec{x}') \equiv \frac{\hbar^2}{2m} \, \langle \vec{x}|(E - H_0 \pm i\epsilon)^{-1}|\vec{x}'\rangle \tag{3}$$

is the Green function discussed earlier.

The explicit form is

$$G^{\pm}(\vec{x}, \vec{x}') = -\frac{1}{4\pi} \, \frac{e^{\pm ik|\vec{x} - \vec{x}'|}}{|\vec{x} - \vec{x}'|}. \tag{4}$$

Equation (2) becomes

$$\langle \vec{x}|\psi^{\pm}\rangle = \langle \vec{x}|\phi\rangle - \frac{2m}{\hbar^2} \int d^3x' \, \frac{e^{\pm ik|\vec{x} - \vec{x}'|}}{|\vec{x} - \vec{x}'|} \, \langle \vec{x}'|V|\psi^{\pm}\rangle. \tag{5}$$

For a local potential V defined by

$$\langle \vec{x}'|V|\vec{x}''\rangle = V(\vec{x}') \, \delta^{(3)} (\vec{x}' - \vec{x}'') \tag{6}$$

the Eq. (5) becomes

$$\langle \vec{x}|\psi^{\pm}\rangle = \langle \vec{x}|\phi\rangle - \frac{2m}{\hbar^2} \int d^3x' \, \frac{e^{\pm ik|\vec{x} - \vec{x}'|}}{4\pi|\vec{x} - \vec{x}'|} \, V(\vec{x}') \, \langle \vec{x}'|\psi^{\pm}\rangle. \tag{7}$$

Equation (7) is the basic integral equation of scattering theory. We write this equation in the form

$$\psi_{\vec{k}}(\vec{r}) = \phi_{\vec{k}}(\vec{r}) - \frac{1}{4\pi} \int \frac{e^{\pm ik|\vec{r} - \vec{r}'|}}{|\vec{r} - \vec{r}'|} \, U(\vec{r}') \, \psi_{\vec{k}}(\vec{r}') \, d^3r' \tag{8}$$

where $\qquad U(\vec{r}) = \dfrac{2m}{\hbar^2} \, V(\vec{r}).$

We now verify that Eq. (8) has the correct asymptotic form. Let $r \to \infty$, and V be a finite-range potential. We assume that U decreases rapidly so that $r \gg r'$ everywhere. Then

$$k|\vec{r} - \vec{r}'| = kr \sqrt{1 + \left(\frac{r'}{r}\right)^2 - \frac{2\vec{r}.\vec{r}'}{r^2}} \approx kr - \vec{k}'.\vec{r}',$$

where $\vec{k}' = k\hat{r}$. Note that we have only elastic scattering

$$\left|\vec{k}\right| = \left|\vec{k}'\right|.$$

Thus, the wave function has the correct asymptotic form

$$\psi_{\vec{k}}(\vec{r}) \xrightarrow[r \to \infty]{} \phi_{\vec{k}}(\vec{r}) - \frac{1}{4\pi} \frac{e^{ikr}}{r} \int e^{-i\vec{k}'.\vec{r}'} U(\vec{r}') \, \psi_{\vec{k}}(\vec{r}') \, d^3r' \tag{9}$$

where $\phi_{\vec{k}}(\vec{r})$ is a plane wave and is a solution of the homogeneous equation

$$\left(\vec{\nabla}^2 + \vec{k}^2\right)\phi = 0. \tag{10}$$

Thus, the wave function asymptotically is the sum of a plane wave in propagation direction $\vec{k}$ and an outgoing spherical wave with amplitude

$$f\left(\vec{k}', \vec{k}\right) = -2\pi^2 \int e^{-i\vec{k}'.\vec{r}'} U(\vec{r}') \, \psi_{\vec{k}}(\vec{r}') \, d^3r'$$

$$= -\frac{4\pi^2 m}{\hbar^2} \langle \vec{k}'|V|\psi^+\rangle. \tag{11}$$

Note that $\langle \vec{r}|\psi^-\rangle$ corresponds asymptotically to the sum of the original plane wave in propagation direction $\vec{k}$ and an incoming spherical wave with spatial dependence $\dfrac{e^{-ikr}}{r}$ and amplitude $-\dfrac{4\pi^2 m}{\hbar^2}\langle -\vec{k}'|V|\psi^-\rangle$. We may write the scattering amplitude in terms of the transition operator T as

$$f\left(\vec{k}', \vec{k}\right) = -\frac{4\pi^2 m}{\hbar^2}\langle \vec{k}'|T|\vec{k}\rangle. \tag{12}$$

The methods of solution of the integral equation will be discussed in Section 20.4.

20.2 THE OPTICAL THEOREM

In this section, we discuss the optical theorem which is a relation between the total scattering cross section and the imaginary part of the forward scattering amplitude. It is a consequence of the conservation of particles or probability. To see clearly the physics of this theorem we examine the radial component of the current, $j_r\,(\vec{r}, t)$, associated with the wave function $\psi\,(\vec{r}, t)$ Eq. (18.24). We have

$$j_r\,(\vec{r}, t) = \frac{\hbar}{m}\,\mathrm{Im}\left(\psi^*\,(\vec{r}, t)\,\frac{\partial\psi(\vec{r}, t)}{\partial r}\right). \tag{13}$$

We compute j_r in the asymptotic region (far field). We have

$$\frac{\partial\psi_0(\hat{k}_0 r, t)}{\partial r} \approx ik_0\psi_0(\hat{k}_0 r, t)$$

$$\frac{\partial\psi_0(\vec{r}, t)}{\partial r} \approx i\vec{k}_0.\hat{r}\psi_0(\vec{r}, t)$$

and drop the derivative of the $1/r$ factor, which is small in the far field. The asymptotic form of j_r consists of the three terms: the incident current $j_{r,\,inc} = j_{r,\,0}(\vec{r}, t) = \dfrac{\hbar\vec{k}_0.\hat{r}}{m}\,|\psi_0(\vec{r}, t)|^2$, the scattered current density $j_{r,\,sc} = \dfrac{\hbar\vec{k}_0}{m}\dfrac{1}{r^2}$

$\times\,|\psi_0(\hat{k}_0 r, t)|^2|f|^2$ and the current due to the interference between the scattered and incident beam $j_{r,\,int}(\vec{r}, t)$

$$= \frac{\hbar\vec{k}_0.\hat{r}}{mr}\,2\,\mathrm{Im}\left[if\,(\Omega)\psi_0^*(\vec{r}, t)\,\psi_0(\hat{k}_0 r, t)\right].$$

We have in the far region

$$j_r = j_{r,\,inc} + j_{r,\,sc} + j_{r,\,int}. \tag{14}$$

$\psi_0(\hat{k}_0 r, t)$ is nonzero only after the particle reaches the target; then $\psi_0^*(\vec{r}, t)$ and hence the interference current are nonzero only in the forward direction where $\vec{k}_0.\hat{r} = k_0$ and $f_{\vec{k}_0}\,(\Omega_r) = f_{\vec{k}_0}\,(0)$. The interference term contributes a current flowing back toward the target along the incident direction thereby reducing the net current in the forward direction. This just means that the target produce a shadow in the forward direction.

Since there are no sources or sinks of particles in the potential, the total scattered current must exactly equal that taken out of the incident beam. Therefore, the integral of Eq. (14) over all angles and times must be equal to the negative of the integral of equation (1) over all angles and times.

The total scattered current is

$$\int d\Omega \, \left| f_{\vec{k}_0} (\Omega_r) \right|^2 \frac{\hbar k_0}{m} \int_{-\infty}^{\infty} dt \, \left| \psi_0(\hat{k}_0 r, \, t) \right|^2. \tag{15}$$

Now,

$$\int d\Omega \, j_{r, \, int}(\vec{r}, \, t) = \frac{\hbar k_0}{r} \, 2 \, \mathrm{Im} \left[i f_{\vec{k}_0}(0) \, \psi_0^*(\hat{k}_0 r, \, t) \times \int d\Omega \, \psi_0(\vec{r}, \, t) \right].$$

For evaluating $\int d\Omega \, \psi_0(\vec{r}, \, t)$, where $\psi_0(\vec{r}, \, t)$ is given by Eq. (18.23), we need the following integral

$$\int d\Omega \, e^{i\vec{k}.\vec{r}} = \frac{2\pi \, (e^{ikr} - e^{-ikr})}{ikr}$$

$$= \frac{2\pi \left(e^{i\vec{k}.(\hat{k}_0 r)} - e^{-i\vec{k}.(\hat{k}_0 r)} \right)}{ik_0.r}$$

for $\vec{k}$ near $\vec{k}_0$.

Thus,

$$\int d\Omega \, \psi_0(\vec{r}, \, t) \simeq \frac{2\pi}{ik_0.r} \left[\psi_0(\hat{k}_0 r, \, t) - \psi_0(-\hat{k}_0 r, \, t) \right],$$

and so

$$r^2 \int d\Omega \, j_{r, \, int} = -\frac{4\pi\hbar}{m} \, \mathrm{Im} \, f_{\vec{k}_0}(0) \left| \psi_0(\hat{k}_0 r, \, t) \right|^2, \tag{16}$$

since $\psi_0(\hat{k}_0 r, \, t) \psi_0(-\hat{k}_0 r, \, t^*) = 0$ in the far field. Equating the negative of the integral of Eq. (16) over all t with Eq. (15) we get

$$\sigma_{\mathrm{tot}} = \frac{4\pi}{k_0} \, \mathrm{Im} \, f(0). \tag{17}$$

This important relationship is known as the optical theorem or the Bohr-Peierls-Placzek relation, although this result was first introduced in quantum mechanics by Feenberg in 1932.

20.3   PARTIAL WAVES AND PHASE SHIFTS

If the potential producing the scattering is spherically symmetric, then the angular momentum is a constant of the motion. Since H is spherically symmetric,

$$RS = SR, \quad \forall \text{ rotation } R \tag{18}$$

and therefore,

$$\vec{L}S = S\vec{L}. \tag{19}$$

The transition operator T commutes with $\vec{L}^2$ and $\vec{L}$ and clearly, T is a scalar operator. States corresponding to different values of the angular momentum independently take part in the scattering.

In Eq. (18.1) we represented the incident wave packet as a linear superposition of plane waves. We now represent a plane wave as a linear combination of angular momentum eigenfunctions (partial waves). We consider a plane wave $e^{i\vec{k}.\vec{r}}$ with the positive z-direction along $\vec{k}$:

$$e^{i\vec{k}.\vec{r}} = \sum_{l=0}^{\infty} i^l \, (2l+1) \, j_l(kr) \, P_l(\cos\theta) \tag{20}$$

$$= \frac{1}{2} \sum_{l=0}^{\infty} i^l \, (2l+1) \, P_l(\cos\theta) \, (h_l(kr) + h_l^*(kr)) \tag{21}$$

where $j_l(kr)$ is the spherical Bessel function of order l, $h_l(kr)$ is the Hankel function of order l and $P_l(\cos\theta)$ is the Legendre function of order l. Note that only $m = 0$ eigenfunction enters in the expansion. Each angular momentum component of Eq. (21) can be considered separately. Next, we expand the wave function $\psi_{\vec{k}}(\vec{r})$ in angular momentum eigenstates

$$\psi_{\vec{k}}(\vec{r}) = \sum_{l=0}^{\infty} i^l \, (2l+1) \, R_l(r) \, P_l(\cos\theta). \tag{22}$$

Because of the rotational symmetry (φ-independence), only $Y_{l0} \sim P_l(\cos\theta)$ occurs. Here, the radial wave function R_l satisfies

$$\left[\frac{d^2}{dr^2} + k^2 - \frac{l(l+1)}{r^2} \right] rR_l(r) = U(r) rR_l(r), \tag{23}$$

where $E = \dfrac{\hbar^2 k^2}{2m}$ and $U(r) = \dfrac{2m}{\hbar^2} \, V(r)$.

We write $R_l(r) = \dfrac{u_l(r)}{r}$ and so the radial equation satisfied by $u_l(r)$ is

$$\frac{d^2 u_l(r)}{dr^2} - \left[k^2 - U(r) - \frac{l(l+1)}{r^2} \right] u_l(r) = 0. \tag{23a}$$

Equation (22) is called the partial-wave expansion. To see the behavior at large distances beyond the range of the potential, for which $V(r) = 0$ or, at least $V(r) < \dfrac{1}{r^2}$, we first examine Eq. (21) and then Eq. (22).

For large x,

$$j_l(x) \xrightarrow[x \to \infty]{} \frac{1}{x} \sin\left(x - \frac{l\pi}{2}\right)$$

so that

$$e^{ikz} \xrightarrow[r \to \infty]{} \frac{1}{2ikr} \sum_{l=0}^{\infty} (2l+1) i^l \left(e^{i\left(kr - \frac{l\pi}{2}\right)}\right.$$

$$\left. - e^{-i\left(kr - \frac{l\pi}{2}\right)}\right) P_l(\cos\theta)$$

$$= \frac{1}{2ikr} \sum_{l=0}^{\infty} (2l+1)\left(e^{ikr} - e^{-i(kr - l\pi)}\right) P_l(\cos\theta) \qquad (24)$$

since $\quad i^l = e^{i\left(\frac{\pi}{2}\right)l}.$

This representation is in terms of an incoming spherical wave with spatial dependence $\left(\dfrac{e^{-ikr}}{r}\right)$ and an outgoing spherical wave $\left(\dfrac{e^{ikr}}{r}\right)$ with its phase shifted by $l\pi$ with respect to the former. This phase shift is due to the presence of the repulsive centrifugal potential.

At large distances (beyond the range of the potential), the equation reduces to a Bessel equation, as the R.H.S. becomes zero. In the far region, the radial function $R_l(r)$ must be a linear combination of $h_l(kr)$ and $h_l^*(kr)$:

$$R_l(r) = B_l\,[h_l^*(kr) + S_l(E)\,h_l(kr)], \qquad (25)$$

where B_l and S_l are functions of k. In the absence of the potential, i.e. for no scattering, we have

$$R_l(r) = j_l(kr) + \frac{1}{2}(h_l + h_l^*),$$

so $B_l = \dfrac{1}{2}$ and $S_l = 1$ in this case. h_l^* corresponds to an incoming spherical wave, and h_l to an outgoing spherical wave. In the presence of an interaction potential, the relative normalization between the incident and the outgoing spherical wave remains the same, i.e. $B_l = \dfrac{1}{2}$ and so the scattering only modifies the coefficient of the outgoing part containing h_l.

Since the potential is not a source nor a sink of particles, the radial component of the probability current for each partial wave must vanish for elastic scattering. We have

$$j_r = \frac{\hbar}{m}\, \mathrm{Im}\left(R_l^* \frac{\partial}{\partial r} R_l\right)$$

$$= \frac{\hbar k}{m}\, \mathrm{Im}\,[h_l\, h_l^* + |S_l(E)|^2\, h_l^* h_l' + 2\,\mathrm{Re}\,(h_l S_l h_l')],$$

where
$$h'_l = \frac{\partial}{\partial r} h_l$$

$$\sim -\frac{h}{mkr^2}\left(1 - |S_l(E)|^2\right), \tag{26}$$

using the asymptotic forms

$$h_l \equiv h_l^{(1)} \sim -i\frac{e^{i(kr - l\pi/2)}}{kr},$$

$$h_l^* = h_l^{(2)},$$

and omitting the factor

$$|P_l(\theta)|^2 \geq 0.$$

Therefore, for elastic scattering $|S_l(E)| = 1$ and we can write

$$S_l(E) = e^{2i\delta_l(E)} \tag{27}$$

with δ_l real (the factor 2 here is conventional). Equation (27) is known as the unitarity relation for the lth partial wave. We shall see later that $S_l(E)$ can be regarded as the lth diagonal element of the S operater which is unitary as a consequence of probability conservation. The angle $\delta_l(k)$ is known as the phase shift for the lth partial wave. The only modification due to scattering by an interaction potential in the wave function at a large distance is to change the phase of the outgoing part. $2\delta_l$ is the difference in the phase of the outgoing part of the wave function $\psi_{\vec{k}}(\vec{r})$ and the plane wave $e^{i\vec{k}\cdot\vec{r}}$. From Eqs. (22), (25) and (27), we can write the wave function asymptotically as

$$\psi_{\vec{k}}(\vec{r}) = \frac{1}{2}\sum_l i^l\,(2l + 1)\,P_l(\cos\theta)\left[h_l^*(kr) + e^{2i\delta_l}\,h_l(kr)\right] \tag{28}$$

$$= e^{i\vec{k}\cdot\vec{r}} + \frac{1}{2}\sum_l i^l\,(2l + 1)\,P_l(\cos\theta)\left(e^{2i\delta_l} - 1\right)h_l(kr). \tag{29}$$

using Eq. (21).

Since $h_l(kr) \xrightarrow[r\to\infty]{} e^{i(kr - l\pi/2)}/ikr$, taking the asymptotic form of the wave function, we obtain an expression for the scattering amplitude in terms of δ_l:

$$f(\theta) = \frac{1}{2ik}\sum_{l=0}^{\infty}(2l + 1)\left(e^{2i\delta_l} - 1\right)P_l(\cos\theta)$$

$$= \frac{1}{k}\sum_{l=0}^{\infty}(2l + 1)\,e^{i\delta_l}\sin\delta_l\,P_l(\cos\theta). \tag{30}$$

Note that $f(\theta)$ is a function of E (or k), although this is not indicated. We define the lth partial wave scattering amplitude as

$$f_l(k) = \left(e^{2i\delta_l} - 1\right)/2ik$$

$$= (S_l - 1)/2ik, \tag{31}$$

so that Eq. (30) becomes

$$f(\theta) = \sum_{l=0}^{\infty} (2l + 1)\, f_l(k)\, P_l(\cos\theta). \tag{32}$$

The differential scattering cross section is given by

$$\frac{d\sigma}{d\Omega} = \frac{1}{k^2} \sum_{l,\,l'} (2l + 1)(2l' + 1)\, e^{i(\delta_l - \delta_{l'})} \sin\delta_i \sin\delta_{l'}$$

$$\times\, P_l(\cos\theta)\, P_{l'}(\cos\theta). \tag{33}$$

The total scattering cross section is

$$\sigma_{\text{tot}} = \int \frac{d\sigma}{d\Omega}\, d\Omega$$

$$= \frac{4\pi}{k^2} \sum_{l} (2l + 1)\sin^2\delta_l$$

$$= \sum_{l=0}^{\infty} \sigma_l, \tag{34}$$

using the orthogonality property

$$\int_{-1}^{+1} P_l(x)\, P_{l'}(x) = \frac{2\delta_{l'l}}{2l + 1},$$

where σ_l is the lth partial scattering cross section. Though there are interference of the partial waves in the expression for $d\sigma/d\Omega$, there is no interference among the partial waves in the total cross section, due to the orthogonality of the P_l's. The scattering in each angular momentum channel is limited by

$$\sigma_l < \frac{4\pi}{k^2} (2l + 1). \tag{35}$$

σ_l is equal to $\dfrac{4\pi}{k^2} (2l + 1)$ when $\delta_l = \left(n + \dfrac{1}{2}\right)\pi$ $(n = 0, 1, 2, \ldots)$. In the forward direction,

$$\operatorname{Im} f(\theta = 0) = \operatorname{Im} \frac{1}{k} \sum_{l=0}^{\infty} (2l + 1)\, e^{i\delta_l} \sin\delta_l\, P_l(1)$$

$$= \frac{1}{k} \sum_{l=0}^{\infty} (2l + 1)\sin^2\delta_l$$

$$= \frac{k}{4\pi}\, \sigma_{\text{tot}} \tag{36}$$

which is the optical theorem.

We now consider the relation of the scattering amplitude with the T operator. Since T is a scalar operator, the little Wigner-Eckart theorem gives

$$\langle E', l', m' | T | E, l, m \rangle = T_l(E)\, \delta_{l'l}\, \delta_{m'm} \tag{37}$$

i.e., T is diagonal both in l and in m. Also, the diagonal element does not depend on m. Now, the scattering amplitude

$$f\left(\vec{k}', \vec{k}\right) = -\frac{1}{4\pi}\frac{2m}{\hbar^2}(2\pi)^3 \left\langle \vec{k}'|T|\vec{k}\right\rangle$$

$$= -\frac{1}{4\pi}\frac{2m}{\hbar^2}(2\pi)^3 \sum_l \sum_m \sum_{l'} \sum_{m'} \int dE \int dE' \left\langle \vec{k}'|E', l', m'\right\rangle$$

$$\times \left\langle E'l'm'|T|Elm\right\rangle \left\langle Elm|\vec{k}'\right\rangle$$

$$= -\frac{1}{4\pi}\frac{2m}{\hbar^2}(2\pi)^3 \cdot \frac{\hbar^2}{mk} \sum_l \sum_m T_l(E)\Big|_{E=\frac{\hbar^2 k^2}{2m}} Y_{lm}\left(\vec{k}'\right) Y_{lm}^*\left(\vec{k}'\right)$$

$$= -\frac{4\pi^2}{k}\sum_l \sum_m T_l(E)\Big|_{E=\frac{\hbar^2 k^2}{2m}} Y_{lm}\left(\vec{k}'\right) Y_{lm}^*\left(\vec{k}'\right) \tag{38}$$

As usual, $\vec{k}'$ is in the positive z-direction. Then

$$Y_{lm}\left(\vec{k}'\right) = \sqrt{\frac{2l+1}{4\pi}}\,\delta_{m0},$$

and so only $m = 0$ terms contribute. Also,

$$Y_{lm}\left(\vec{k}'\right) = \sqrt{\frac{2l+1}{4\pi}}\,P_l(\cos\theta)$$

where θ is the angle between $\vec{k}$ and $\vec{k}'$. Thus, from Eq. (38), we have

$$f\left(\vec{k}', \vec{k}\right) = f(\theta) = \sum_{l=0}^{\infty}(2l+1)\,f_l(k)\,P_l(\cos\theta) \tag{39}$$

where

$$f_l(k) = -\frac{\pi T_l(E)}{k} \tag{40}$$

is the lth partial wave scattering amplitude.

We consider the S-matrix in the angular momentum representation. We represent normalized in and out-states for a central force as

$$\psi_{k,l}^{\pm} = \frac{k}{\pi}\cdot\frac{2l+1}{\sqrt{2(2l+1)}}\,i^l\,e^{\pm i\delta_l(k)}\,P_l(\cos\theta)\,\frac{u_{l,k}(r)}{r}.$$

The matrix elements of S in the angular momentum representation are

$$S_{kl,k'l'} \equiv (\psi_{k,l}^{-}, \psi_{k',l'}^{+}) = e^{2i\delta_l(k)}\,\delta_{ll'}\,\delta(k-k')$$

S is diagonal in this representation. The eigenvalues of S are $\exp(2i\delta_l(k))$. We write

$$S_l(k) = e^{2i\delta_l(k)},\ \text{which is Eq. (27)}$$

The results hold, in this from, for elastic scattering only. S cannot be diagonalized by mean of stationary $|k, l> -$ states if inelastic processes are present.

We rewrite the general form of the S-matrix as

$$S = e^{2iD}, \tag{27a}$$

where D is a hermitian operator, called the "phase matrix", which has as its eigenvalues the (real) phase shifts $\delta_l(k)$. The unitarity of S is clearly exhibited. The T-matrix element is given by

$$T_l = e_{i\delta_l} \sin \delta_l. \tag{27b}$$

Let us now see how to compute the phase shift. We assume a finite-range potential $V(r)$ which vanishes for $r > R$. The phase shift δ_l is determined by fitting the radial wave function $R_l(r)$ for $r < R$, which may have an analytic form and can always be found numerically to the solution $R_l(r)$ for $r > R$. The boundary condition at $r = R$ is that both the wave function and its derivative must be continuous. The wave function for $r > R$ is

$$R_l(r) = \frac{1}{2}\left[h_l^*(kr) + e^{2i\delta_l} h_l(kr) \right], \quad h_l \equiv h_l^{(1)} \tag{41}$$

while for $r < R$ it is necessary to solve Eq. (23) to find $R_l(r)$.

Thus,

$$\left[\frac{\dfrac{\partial}{\partial r}\left[h_l^*(kr) + e^{2i\delta_l} h_l(kr) \right]}{h_l^*(kr) + e^{2i\delta_l} h_l(kr)} \right]_{r=R} = \gamma_l \tag{42}$$

where

$$\gamma_l \equiv \frac{1}{R_l(r)}\left[\frac{dR_l(r)}{dr} \right]_{r=R} \tag{43}$$

is the logarithmic derivative of R_l at $r = R$. From Eq. (42) we get

$$e^{2i\delta_l} - 1 = \left[\frac{\left(2\dfrac{d}{dr} j_l - \gamma_l j_l \right)}{\gamma_l h_l - \dfrac{d}{dr} h_l} \right]_{r=R} \qquad \left(\begin{array}{l} h_l^{(1)} = j_l + i n_l, \\ h_l^{(2)} = j_l - i n_l \end{array} \right)$$

or,

$$\cot \delta_t = \left[\frac{\dfrac{d}{dr} n_l - \gamma_l n_l}{\dfrac{d}{dr} j_l - \gamma_l j_l} \right]_{r=R}, \tag{44}$$

since

$$e^{2i\delta_l} - 1 = \frac{2i}{\cot \delta_l - i}.$$

The problem of computing the phase shift is thus reduced to that of determining γ_l.

Let us now consider an example. We consider the scattering by a perfectly rigid (hard) sphere, which is represented by the potential

$$V(r) = \begin{cases} +\infty & \text{for} \quad r < a \\ 0 & \text{for} \quad r > a \end{cases} \tag{45}$$

The radial wave function $R_l(r)$ for $r > a$ is given by, Eq. (41):

$$R_l(r) = e^{i\delta_l} [\cos \delta_l \, j_l(kr) - \sin \delta_l \, n_l(kr)]. \tag{46}$$

The sphere is impenetrable, and so the wave function must vanish at $r = a$, i.e. $R_l(r) = 0$ at $r = a$. Then γ_l is ∞. The phase shifts are obtained by setting either $R_l(r = a)$ given by, Eq. (46) equal to zero, or γ_l in, Eq. (44) equal to infinity:

$$\cot \delta_l = \frac{n_l(ka)}{j_l(ka)}. \tag{47}$$

The phase shift for s-waves ($l = 0$) is

$$\delta_0 = - ka. \tag{48}$$

We see that a repulsive potential yields a negative phase shift.

The calculation of the scattering is simple in the low-energy limit:

$$ka = 2\pi a/\lambda \ll 1.$$

Since

$$j_l(\rho) \xrightarrow[\rho \to 0]{} \frac{\rho^l}{(2l + 1)!!},$$

$$n_l(\rho) \xrightarrow[\rho \to 0]{} \frac{(2l + 1)!!}{\rho^{l+1}},$$

$$(2l + 1)!! \equiv 1.3.5...(2l + 1),$$

So

$$\tan \delta_l \approx - \frac{(ka)^{2l+1}}{(2l + 1)[(2l - 1)!!]^2}$$

for

$$ka \ll 1. \tag{49}$$

Thus, δ_l falls off very rapidly as l increases. At sufficiently low incident energy, we have s-wave scattering only, (isotropic scattering). Because $\delta_0 = - ka$, irrespective of whether k is large or small, we have

$$\frac{d\sigma}{d\Omega} = \frac{\sin^2 \delta_0}{k^2} = a^2$$

for

$$ka \ll 1. \tag{50}$$

The total cross section

$$\sigma_{\text{tot}} = \int \frac{d\sigma}{d\Omega} \, d\Omega = 4\pi a^2 \tag{51}$$

is four times the geometric cross section (classical value). Low-energy scattering means large wavelength scattering for which we do not expect a classical result.

In the high-energy limit ($ka \gg 1$), we might expect to get the classical result when the de Broglie wavelength is small. The result $\sigma_{\text{tot}} = 2\pi a^2$, twice the classical value, is a surprise.

At high energies, a large number of l values contribute, up to $l_{max} = ka$. Therefore,

$$\sigma_{tot} = \frac{4\pi}{k^2} \sum_{l=0}^{l=ka} (2l + 1) \sin^2\delta_l. \tag{52}$$

From Eq. (47)

$$\sin^2\delta_l = \frac{1}{1 + \cot^2\delta_l} = \frac{[j_l(ka)]^2}{[j_l(ka)]^2 + [n_l(ka)]^2}. \tag{53}$$

Using,

$$j_l(x) \sim \frac{1}{x} \sin\left(x - \frac{l\pi}{2}\right)$$

$$n_l(x) \sim -\frac{1}{x} \cos\left(x - \frac{l\pi}{2}\right)$$

for $x \gg 1,\ x \gg l$

Eq. (53) becomes

$$\sin^2\delta_l \simeq \sin^2\left(ka - \frac{l\pi}{2}\right). \tag{54}$$

Inserting this into Eq. (52), we get

$$\sigma_{tot} = \frac{4\pi}{k^2} \left\{ \sin^2 ka + 3 \sin^2\left(ka - \frac{l\pi}{2}\right) \right.$$
$$\left. + 5\sin^2(ka - \pi) + ... \right\}$$
$$= \frac{4\pi}{k^2} \left\{ \left[\sin^2 ka + \sin^2\left(ka - \frac{\pi}{2}\right)\right] \right.$$
$$\left. + 2\left[\sin^2\left(ka - \frac{\pi}{2}\right) + \sin^2(ka - \pi)\right] + ... \right\}$$
$$= \frac{4\pi}{k^2} \sum_{l=0}^{ka} l,$$

since for an adjacent pair of partial waves

$$\sin^2\delta_l + \sin^2\delta_{l+1} = \sin^2\delta_l + \sin^2\left(\delta_l - \frac{\pi}{2}\right)$$
$$= \sin^2\delta_l + \cos^2\delta_l = 1$$

$$\therefore \qquad \sigma_{tot} = \frac{4\pi}{k^2} \cdot \frac{ka}{2}\,(ka + 1)$$
$$= 2\pi a^2 \quad \text{for } ka \gg 1. \tag{55}$$

The scattering cross section is twice the geometric cross section πa^2. The reason is the diffraction of the wave. From Eq. (30), we write the scattering amplitude $f(\theta)$ in this case as follows

$$f(\theta) = \frac{1}{2ik} \sum_{l=0}^{ka} (2l + 1)\, e^{2i\delta_l}\, P_l(\cos\theta)$$

$$+ \frac{i}{2k} \sum_{l=0}^{ka} (2l + 1)\, P_l(\cos\theta)$$

$$= f_{\text{reflection}} + f_{\text{shadow}}. \tag{56}$$

$$\therefore \quad \int |f_{\text{refl}}|^2 \, d\Omega = \frac{2\pi}{4k^2} \sum_{l=0}^{ka} \int_{-1}^{+1} (2l+1)^2 \, [P_l(\cos\theta)]^2 \, d(\cos\theta)$$

$$= \frac{\pi}{k^2} (ka)^2$$

$$= \pi a^2. \tag{57}$$

The first term is isotropic and corresponds to the classical scattering. The second term f_{shad} is pure imaginary and is sharply concentrated in the forward direction since $P_l(\cos\theta) = 1$ for $\theta = 0$. So

$$f_{\text{shad}} = \frac{i}{2k} \sum_l (2l + 1)\, J_0(l\theta),$$

using
$$P_l(\cos\theta) \xrightarrow[\substack{\text{large } l \\ \text{small } \theta}]{} J_0(l\theta)$$

$$= ik \int_0^a b\,db\, J_0(kb\theta)$$

$$= \frac{ia J_1(ka\theta)}{\theta}, \text{ where } J_1 \text{ is a Bessel function of order 1.} \tag{58}$$

This is the expression for Fraunhofer diffraction with a strong peak near $\theta = 0$ (see Jackson (1975)). Then

$$\int |f_{\text{shad}}|^2 \, d\Omega = 2\pi \int_{-1}^{+1} \frac{a^2\, [J_1(ka\theta)]^2}{\theta^2} \, d(\cos\theta)$$

$$= 2\pi a^2 \int_0^\infty \frac{[J_1(x)]^2}{x^2} \, dx, \text{ where } x = ka\theta, \tag{59}$$

$$= \pi a^2.$$

The second term is due to diffraction and contribute πa^2 to the total scattering cross section. Since the phase of f_{refl} oscillates $(2\delta_{l+1} - 2\delta_l = \pi)$, approximately averaging to zero,

$$\text{Re}(f_{\text{shad}}^* f_{\text{refl}}) = 0, \tag{60}$$

i.e., the interference between f_{shad} and f_{refl} vanishes. The second term (coherent contribution in the forward direction) is called a *shadow*. The wave function $\psi = \psi_{in} + \psi_{sc}$. Just behind the scatterer, in the shadow there is zero probability of finding the particle, so there ψ_{sc} must be equal to ψ_{in} in magnitude and opposite in sign. The shadow results from the destructive interference of the newly scattered wave in the forward direction with the incident wave. This interference removes as much intensity from the ray propagating in the forward direction as is reflected of into finite angles. At large distances this additional contribution to the scattering amplitude is concentrated in the forward direction. Classically, however, one regards only the part reflected off the scatterer as scattering. As an exercise one may check that the optical theorem is satisfied.

The application of the partial wave method is particularly convenient when the interaction potential $V(r)$ has a finite range R. In these cases, partial waves with small values of l will participate in the scattering of particles with low energy. In the sum Eq. (34), only $l \lesssim kR$ contribute. One may see this using qualitative considerations. At distance $r > R$, only the centrifugal, repulsive force with a potential energy $\dfrac{\hbar^2 l(l + 1)}{2mr^2}$ will act upon a particle with orbital angular momentum l. The particle will thus mainly move at distances satisfying the inequality $\dfrac{\hbar^2 l(l + 1)}{2mr^2} \leq \dfrac{\hbar^2 k^2}{2m} = E$, the energy of the relative motion. For the energy E, the distance of closest approach (classical turning radius) is $r_{0l} = \sqrt{l(l + 1)}/k$. If $r < r_{0l}$, the wave function falls of exponentially and so the probability of observing the particle is exponentially small. If the range $R < r_{0l}$, the corresponding partial wave will practically not reach the region where $V(r)$ acts and will not take part in the scattering. Hence, partial waves with a quantum number l will contribute to the scattering if $r_{0l} \leq R$, i.e. $\sqrt{l(l + 1)} \approx l \leq kR$. In this case, all the partial waves $l = 0, 1, 2, \ldots, l_{\max} = kR$ will contribute.

We now discuss scattering by a complex system in which the target is not structureless as before but is a complicated system with excited states. If the projectile is sufficiently energetic, these states can be excited. It may even happen that the scatterer can absorb the projectile and emit other particles.

For the lth partial wave, the total wave function describing the elastic scattering may be written as

$$\psi_{el}^{(l)} = C\chi_0 \left[h_l^*(kr) + S_l(E)\, h_l(kr) \right] P_l(\cos \theta),$$

for $\qquad\qquad r > R \qquad\qquad$ (61)

where C is a normalization constant, χ_0 is the ground state wave function of the scatterer. For elastic scattering, $|S_l| = 1$ due to the conservation of probability. In this case, $S_l(E) = e^{2i\delta_l(E)}$. When inelastic processes are also energetically possible, then the elastically scattered wave suffers a decrease of amplitude and so $|S_l(E)| < 1$. We write $S_l(E) = \rho_l(E)e^{2i\delta_l(E)}$ with $0 \leq \rho_l(E) \leq 1$. The amplitude of the lth partial wave is

$$f_l = \frac{S_l - 1}{2ik} = \frac{\rho_l e^{2i\delta_l} - 1}{2ik}. \qquad\qquad (62)$$

The elastic scattering amplitude is

$$f_{el}(k, \theta) = \sum_{l=0}^{\infty} (2l + 1) \, f_l P_l(\cos \theta).$$ (63)

$\therefore$ The total elastic scattering cross section is

$$\sigma_{el}(k) = \int d\Omega \, |f_{el}(k, \theta)|^2$$

$$= \frac{\pi}{k^2} \sum_{l=0}^{\infty} (2l + 1) \, |S_l - 1|^2$$

$$= \frac{\pi}{k^2} \sum_{l=0}^{\infty} (2l + 1) \, (1 + \rho_l^2 - 2\rho_l \cos 2\delta_l).$$ (64)

The total cross section for all inelastic processes ("reaction cross section") can also be expressed in terms of S_l.

The total elastic flux is

$$-\int r^2 j_r \, d\Omega = -\int d\Omega \, r^2 \sum_{l, \, l'} i^{-l + l'} \, (2l + 1) \, (2l' + 1)$$

$$\times P_l P_{l'} \frac{\hbar}{m} \times \operatorname{Im} R_l^* \frac{d}{dr} R_{l'}$$

$$= \operatorname{Re} \frac{i\pi\hbar r^2}{m} \sum_{l=0}^{\infty} (2l + 1) \, R_l^* \frac{d}{dr} R_l$$

$$= \operatorname{Re} \frac{4\pi\hbar}{km} \sum_{l=0}^{\infty} (2l + 1) \, [e^{ikr} - (-1)^l \, S_l^* \, e^{-ikr}]$$

$$\times [e^{-ikr} + (-1)^l \, S_l e^{ikr}]$$

$$= + \frac{\hbar k}{m} \frac{\pi}{k^2} \sum_{l=0}^{\infty} (2l + 1) \left(1 - |S_l|^2\right).$$ (65)

This is just the flux lost due to inelastic processes. The inelastic cross section is obtained by dividing this quantity by the incident flux ($= \hbar k/m$):

$$\sigma_{inel}(k) = \frac{\pi}{k^2} \sum_{l=0}^{\infty} (2l + 1) \left(1 - |S_l|^2\right)$$

$$= \frac{\pi}{k^2} \sum_{l=0}^{\infty} (2l + 1) \left(1 - \rho_l^2\right).$$ (66)

The inelastic cross section is a measure of the depletion of the outgoing elastic wave and is therefore determined by $\left(1 - |S_l|^2\right)$.

The sum of the elastic and reaction (inelastic) cross sections is the total scattering cross section

$$\sigma_{tot}(k) = \sigma_{el}(k) + \sigma_{inel}(k)$$

$$= \frac{2\pi}{k^2} \sum_{l=0}^{\infty} (2l + 1) \, (1 - \operatorname{Re}(S_l))$$

$$= \frac{2\pi}{k^2} \sum_{l=0}^{\infty} (2l + 1) \, (1 - \rho \cos 2\delta_l),$$ (67)

where Re stands for "real part of".

From Eqs. (62) and (63),

$$\operatorname{Im} f_{el}(k, 0) = \frac{1}{2k} \sum_{l=0}^{\infty} (2l + 1)\,(1 - \operatorname{Re}(S_l)) \tag{68}$$

so that

$$\sigma_{\text{tot}}(k) = \frac{4\pi}{k} \operatorname{Im} f_{el}(k, 0). \tag{69}$$

This is the generalization of the optical theorem in the presence of inelastic processes.

The condition

$$|S_l|^2 \leq 1$$

insures that the cross section Eq. (66) does not become negative.

Equations (64) and (66) lead to interesting relations for the parameters. The maximum value of the scattering cross section for a given l value is obtained for $S_l = -1$, and σ_{inel} then vanishes for this value. The maximum of σ_{inel} for a particular value of l occurs for $S_l = 0$; this implies

$$\sigma_{el}^{(l)}(k) = \sigma_{inel}^{(l)}(k) = \frac{\pi}{k^2}(2l + 1).$$

Thus, for $\rho_l = 1$, there is no inelastic scattering, but pure elastic scattering occurs. For $\rho_l = 0$, $\sigma_{el}(k) = \sigma_{inel}(k)$. Elastic scattering always accompanies inelastic scattering, i.e., when $|S_l| < 1$, $|S_l - 1| > 0$.

"There cannot be any reaction without some scattering although scattering without reaction is possible."

The permitted range of values of σ_{el} and σ_{inel} for a given l value is shown in Fig. 20.1 (see Blatt & Weisskopf, (1952))

The scattering process, in general, changes both the amplitude and the phase of the outgoing waves. Scattering without absorption occurs if the outgoing part of the incident plane wave is not diminished in intensity but only shifted in phase. We see from Eq. (64) that the phase of S_l affects the magnitude of σ_{el}, whereas only $|S_l|^2$ enters into the expression for σ_{inel} (Eq. (66)). The incoming and outgoing waves are coherent in elastic scattering. They can interfere constructively or destructively. The maximum value of σ_{el} is four times the maximum value of σ_{inel}. The factor 4 is due to the possibility of constructive interference. For the inelastic processes outgoing waves are all incoherent with the incident wave and with each other.

The effect of the other processes on the elastic scattering is often simulated by a complex potential usually called the optical-model potential. A real potential gives real phase shifts while a complex potential generally yields complex phase shifts.

The parametrization for a general scattering event including inelastic channels is commonly expressed in one of two ways, either by using a complex phase shift δ_l.

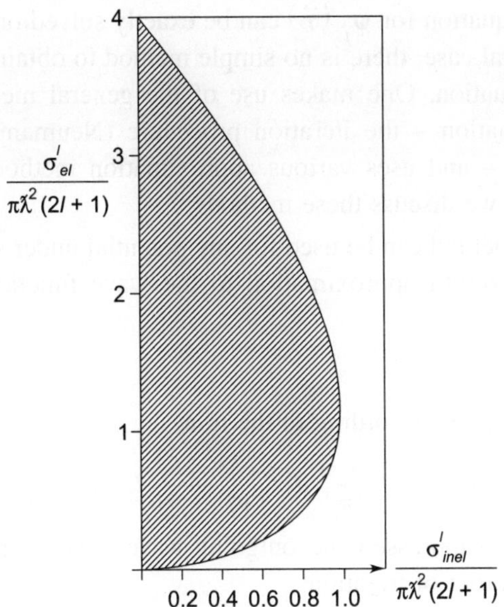

Fig. 20.1   The allowed (shaded) region in elastic and inelastic scattering cross sections for any partial wave

$$S_l = e^{2i\delta_l}, \quad \delta_l = \alpha_l + i\beta_l \tag{70}$$

or by using a real phase shift δ_l and an inelasticity parameter ρ_l:

$$S_l = \rho_l e^{2i\delta_l}; \quad \rho_l,\ \delta_l \ \text{real} \ (0 \le \rho_l \le 1). \tag{71}$$

The two approaches are obviously equivalent, and we have used the latter one.

20.4   THE BORN APPROXIMATION AND THE FREDHOLM METHOD

The basic integral equation of scattering theory is given by Eq. (8):

$$\psi_{\vec{k}}^{(+)}(\vec{r}) = \phi_{\vec{k}}(\vec{r}) + \int d^3r'\, G^{(+)}(\vec{r},\vec{r}')\, V(\vec{r}')\, \psi_{\vec{k}}^{(+)}(\vec{r}')$$

$$= e^{i\vec{k}.\vec{r}} - \frac{1}{4\pi} \int d^3r'\, \frac{\exp(ik|\vec{r}-\vec{r}'|)}{|\vec{r}-\vec{r}'|}\, U(\vec{r}')\, \psi_{\vec{k}}^{(+)}(\vec{r}')$$

where $U(\vec{r}') = \dfrac{2m}{\hbar^2} V(\vec{r}')$ and the superscript $(+)$ indicates the outgoing wave

boundary condition. The integral expression for the scattering amplitude is

$$f\left(\vec{k}',\vec{k}\right) = -\frac{1}{4\pi} \int d^3r'\, e^{i\vec{k}.\vec{r}}\, U(\vec{r}')\, \psi_{\vec{k}}^{(+)}(\vec{r}')$$

The integral equation for $\psi_{\vec{k}}(\vec{r})$ can be exactly solved only in a few special cases. In the general case, there is no simple method to obtain the exact solution of the integral equation. One makes use of the general methods of solving a linear integral equation – the iteration procedure (Neumann method) and the Fredholm method – and uses various approximation methods. In this and the following sections we discuss these methods.

The iteration method can be used for any potential under suitable conditions. One takes as the zeroth approximation to the wave function the unperturbed plane wave

$$\psi_{\vec{k}}^{(0)} = e^{i\vec{k}.\vec{r}} \tag{72}$$

and then iterates Eq. (8) according to the rule

$$\psi_{\vec{k}}^{(n)}(\vec{r}) = \psi_{\vec{k}}^{(0)}(\vec{r}) - \frac{1}{4\pi} \int d^3r' \, \frac{e^{ik|\vec{r}-\vec{r}'|}}{|\vec{r}-\vec{r}'|} \, U(\vec{r}') \, \psi_{\vec{k}}^{(n-1)}(\vec{r}') \tag{73}$$

Here, we have suppressed the outgoing-wave superscript. The superscript now indicates the order of iteration.

The first iterate is

$$\psi_{\vec{k}}^{(1)}(\vec{r}) = e^{i\vec{k}.\vec{r}} - \frac{1}{4\pi} \int d^3r' \, \frac{e^{ik|\vec{r}-\vec{r}'|}}{|\vec{r}-\vec{r}'|} \, U(\vec{r}') \, e^{i\vec{k}.\vec{r}'}. \tag{74}$$

The second iterate is

$$\psi_{\vec{k}}^{(2)}(\vec{r}) = \psi_{\vec{k}}^{(2)}(\vec{r}) + \left(\frac{1}{4\pi}\right)^2 \int d^3r' \int d^3r'' \, \frac{e^{ik|\vec{r}-\vec{r}'|}}{|\vec{r}-\vec{r}'|} \, U(\vec{r}') \, \frac{e^{ik|\vec{r}'-\vec{r}''|}}{|\vec{r}'-\vec{r}''|} \, U(\vec{r}'') \, e^{i\vec{k}.\vec{r}''}. \tag{75}$$

This process can be continued, generating the "Born expansion". The approximation in which the exact wave function is replaced by the plane wave $e^{i\vec{k}.\vec{r}}$ is called the first Born approximation, or simply the "Born approximation".

Putting the Born expansion for the wave function into the expression for the scattering amplitude Eq. (11), we get the Born expansion for the scattering amplitude

$$f\left(\vec{k}', \vec{k}\right) = -\frac{1}{4\pi} \int d^3r \, e^{-i\vec{k}'.\vec{r}} \, U(\vec{r}) \, e^{i\vec{k}.\vec{r}} + \left(\frac{1}{4\pi}\right)^2 \int d^3r' \int d^3r'$$

$$\times e^{-i\vec{k}'.\vec{r}} \, U(\vec{r}) \, \frac{e^{ik|\vec{r}-\vec{r}'|}}{|\vec{r}-\vec{r}'|} \, U(\vec{r}') \, e^{i\vec{k}.\vec{r}'} + \dots \tag{76}$$

If $f^{(n)}(\theta)$ is the nth order approximation to the scattering amplitude, then

$$f^n\left(\vec{k}', \vec{k}\right) = -\frac{1}{4\pi} \int d^3r \, e^{-i\vec{k}'.\vec{r}} \, U(\vec{r}) \psi_{\vec{k}}^{(n-1)}(\vec{r}). \tag{77}$$

The Born approximation to the scattering amplitude is

$$f^{(1)}\left(\vec{k}', \vec{k}\right) = -\frac{1}{4\pi} \int d^3r \, e^{-i\vec{k}'.\vec{r}} \, U(\vec{r}) \, e^{i\vec{k}.\vec{r}}$$

$$= -\frac{1}{4\pi} \int d^3r \, e^{-i(\vec{k} - \vec{k}').\vec{r}} \, U(\vec{r}) \tag{78}$$

Thus, apart from the factor $-\dfrac{m}{2\pi\hbar^2}$, the first Born scattering amplitude is the Fourier transform of the potential $V(\vec{r})$ with respect to the momentum transfer $\vec{q} = \vec{k} - \vec{k}'$. Note that $\vec{k}$ is the incident wave vector, $\vec{k}'$ is the wave vector of the scattered particle.

For a spherically symmetric potential, $f^{(1)}\left(\vec{k}', \vec{k}'\right)$ is a function of $|\vec{k} - \vec{k}'|$ and

$$(\vec{k} - \vec{k}')^2 = \vec{k}^2 + \vec{k}'^2 - 2\vec{k}.\vec{k}'$$

$$= 2k^2 \, (1 - \cos \theta),$$

since $|\vec{k}| = |\vec{k}'| = k$, by energy conservation; θ is the scattering angle.

$$= \left(2k \sin \frac{\theta}{2}\right)^2,$$

so
$$|\vec{k} - \vec{k}'| = q = 2k \sin \frac{\theta}{2}$$

$$\therefore \qquad f^{(1)}(\theta) = -\frac{1}{2}\frac{1}{iq} \int_0^\infty \frac{r^2}{r} \, U(r) \, (e^{iqr} - e^{-iqr}) \, dr$$

$$= -\frac{2m}{\hbar^2}\frac{1}{q} \int rV(r) \, \sin qr \, dr. \tag{79}$$

We have

$$f^{(1)}(\theta) = -\frac{2m}{\hbar^2} \int_0^\infty \frac{\sin qr}{qr} \, V(r) r^2 \, dr.$$

If $V(r)$ is an atomic field, it is convenient to transform Eq. (79) into an integral involving the charge density in the atom. We have

$$V(r) = -\frac{Ze^2}{r} + e^2 \int \rho(r') \frac{d^3r'}{|\vec{r} - \vec{r}'|}$$

where $-e\rho(r)$ denotes the charge density at any point. Substituting $V(r)$ in Eq. (78), we get

$$f(\theta) = \frac{2me^2}{\hbar^2} \frac{Z - F(\theta)}{q},$$

using
$$\int \frac{\exp\left(i\vec{k}.\vec{r}'\right)}{|\vec{r} - \vec{r}'|} \, d^3r' = \frac{4\pi}{k^2} \, e^{i\vec{k}.\vec{r}}$$

$$= \frac{e^2}{2mv^2} \, [Z - F(\theta)] \, \mathrm{cosec}^2 \frac{1}{2}\, \theta$$

where

$$F(\theta) = 4\pi \int_0^\infty \rho(r) \, \frac{\sin qr}{qr} \, r^2 \, dr$$

is known as the atomic scattering or form factor.

As an example, we consider scattering by a screened Coulomb potential

$$V(r) = V_0 \, \frac{e^{-\mu r}}{\mu r} \tag{80}$$

which was originally suggested by Yukawa as a possible nucleon-nucleon potential. Here, V_0 is independent of r and $1/\mu$ corresponds to the range of the potential. From Eq. (79), we obtain

$$f_B^{(1)}(\theta) = -\left(\frac{2mV_0}{\mu\hbar^2}\right) \frac{1}{q^2 + \mu^2}. \tag{81}$$

Therefore,

$$\frac{d\sigma}{d\Omega} = \left(\frac{2mV_0}{\mu\hbar^2}\right)^2 \frac{1}{[2k^2 \, (1 - \cos\theta) + \mu^2]^2}. \tag{82}$$

In the limit as $\mu \to 0$ (the range $\to \infty$), the Yukawa potential becomes the Coulomb potential, provided the ratio V_0/μ is fixed, say $ZZ'e^2$, Then the differential scattering cross section is given by

$$\left(\frac{d\sigma}{d\Omega}\right) = \frac{(2m)^2 \, (ZZ'e^2)^2}{\hbar^4} \frac{1}{16k^4 \, \sin^4(\theta/2)}$$

$$= \frac{1}{16} \left(\frac{ZZ'e^2}{E_{ke}}\right)^2 \operatorname{cosec}^4 \frac{\theta}{2}$$

$$= \frac{1}{4} \left(\frac{ZZ'e^2}{mv^2}\right)^2 \operatorname{cosec}^4 \frac{\theta}{2} \tag{83}$$

where $E_{ke} = \dfrac{1}{2} \, mv^2 = \dfrac{|\vec{p}|^2}{2m}$ with $|\vec{p}| = \hbar\vec{k}$. This is the classical Rutherford scattering cross section. This happens to be the result (non-relativistic) obtained by an exact quantum mechanical treatment. But its derivation from the Born approximation is accidental, since Eq. (78) is invalid for the long-ranged Coulomb potential.

The total cross section for the screened Coulomb field is

$$\sigma_{\text{tot}} = 2\pi \left(\frac{2mV_0}{\mu\hbar^2}\right)^2 \frac{1}{k^2} \int_0^{2k} \frac{K dK}{(K^2 + \mu^2)^2},$$

where $K = 2k \sin \dfrac{\theta}{2}$ and $\sin\theta \, d\theta = \dfrac{K dK}{k^2}$

$$= \frac{4\pi \left(\frac{2mV_0}{\mu^2\hbar^2}\right)^2}{4k^2 + \mu^2}. \tag{84}$$

As $\mu \to 0$, the total cross section becomes infinite, as expected.

The radial wave function $u_l(r)$ satisfies the integral equation

$$u_l(r) = F_l(kr) + \int_0^\infty dr' \, g_l(r, r') \, U(r') \, u_l(r') \tag{85}$$

The iterative solution of Eq. (85) is

$$u_l(r) = F_l(kr) + \int_0^\infty dr' \, g_l(r, r') \, U(r') \, F_l(kr') + \dots \tag{86}$$

The solution converges under proper conditions. The exact phase shift is given by

$$\tan \delta_l = - k^{-1} \int_0^\infty dr \, F_l(kr) \, U(r) \, u_l(r) \tag{87}$$

where

$$g_l(r, r') = - k^{-1} \, F_l(kr_<) \, G_l(kr_>).$$

The Born approximation to the phase shift is obtained by putting $F_l(kr)$ for $u_l(r)$ in Eq. (87):

$$\tan \delta_l^{(B)} = - k^{-1} \int_0^\infty dr \, F_l^2(kr) \, U(r)$$
$$= - k \int_0^\infty r^2 \, [j_l(kr)]^2 \, U(r) \, dr. \tag{88}$$

The phase shift is small in most cases, so

$$\tan \delta_l^{(B)} \approx \delta_l^{(B)}.$$

From Eq. (88), we see that a (weak) repulsive potential produces negative phase shifts ($\delta_l < 0$) while a (weak) attractive potential produces positive phase shifts ($\delta_l > 0$).

For $(kr)^2 \ll 4l + 6$ we have (Moiseiwitsch (1961))

$$j_l(kr) \cong \frac{l!}{(2l + 1)!} \, (2kr)^l.$$

From Eq. (88)

$$\sin \delta_l = - k^{2l + 1} \left\{ \frac{2^l \, l!}{(2l + 1)!} \right\}^2 \int_0^\infty r^{2(l + 1)} \, U(r) \, dr \tag{89}$$

If R is the range of the potential and if $\left| r^2 U(r) \right| \leq C$ for $0 < r < R$, we get

$$|\sin \delta_l| \langle \frac{C(kR)^{2l + 1}}{2l + 1} \left\{ \frac{2^l \, l!}{(2l + 1)!} \right\}^2. \tag{90}$$

$\therefore$ For large l,

$$\ln |\sin \delta_l| < 2l(\ln kR - \ln 2 - \ln l), \tag{91}$$

using Stirling's formula $\ln l! = l \ln l$.

So, δ_l is a rapidly decreasing function of l.

We now discuss the validity of the first-order Born approximation. The Born approximation is applicable if the first first-order correction to the wave function is small compared to the incident wave in the range of the potential. Taking the origin as the center of this interaction region, this condition is

$$\left| \frac{1}{4\pi} \int d^3r \left(\frac{e^{ikr}}{r} \right) U(\vec{r}) \, e^{i\vec{k}\cdot\vec{r}} \right| \ll 1. \tag{92}$$

For a square well of radius a and depth V_0, this implies

$$\left| \left(\frac{mV_0}{\hbar^2 k^2} \right) \left(e^{ika} \sin ka - ka \right) \right| \ll 1. \tag{93}$$

For low energies,

$$\left| \frac{mV_0 a^2}{\hbar^2} \right| \ll 1 \tag{94}$$

and for high energies,

$$\left| \frac{mV_0 a}{\hbar^2 k} \right| \ll 1. \tag{95}$$

A bound state exists if $\left| \dfrac{mV_0 a^2}{\hbar^2} \right| \gtrsim 1$. The Born approximation will not be valid if the potential is strong enough to develop a bound state. On the other hand the criterion can be satisfied for any potential by going to a sufficiently high energy. The Born approximation tends to get better at higher energies.

The Born approximation to the phase shift is expected to be valid when the correction to the wave function is small near the origin. The condition for this is

$$\left| k^{-1} \int_0^\infty dr \, G_l(kr) \, F_l(kr) \, U(r) \right| \ll 1. \tag{96}$$

For the square well of radius a and depth V_0, this implies

$$(2l + 1)^{-1} \left| \frac{mV_0 a^2}{\hbar^2} \right| \ll 1 \quad \text{if} \quad l \gg ka$$

and

$$\left| \frac{mV_0 a^2}{\hbar^2 k^2} \right| \ll 1 \quad \text{if} \quad l \ll ka. \tag{97}$$

Thus, the Born approximation for the phase shift is expected to be valid for large values of l. For small l, the approximation is valid for weak potentials and high energies. In both cases, the phase shift will be small.

Let us now discuss the higher-order Born approximation. We consider the operator form of the Lippmann-Schwinger equation (equation (19.14)).

$$T(z) = V + VG_0(z)\,T(z). \tag{98}$$

The formal solution is

$$T(z) = (1 - VG_0(z))^{-1}\, V. \tag{99}$$

The power series expansion of $(1 - VG_0)^{-1}$ leads to the iterative solution for T, the Born expansion ("Born series");

$$T(z) = V + VG_0(z)V + VG_0(z)VG_0(z)V + \ldots \tag{100}$$

Since the scattering amplitude can be written as

$$f\!\left(\vec{k}',\, \vec{k}\right) = -\frac{1}{4\pi}\,\frac{2m}{\hbar^2}\left\langle \vec{k}' | T | \vec{k} \right\rangle$$

we can expand f as follows

$$f\!\left(\vec{k}',\, \vec{k}\right) = \sum_{n=1}^{\infty} f^{(n)}\!\left(\vec{k}',\, \vec{k}\right). \tag{101}$$

We have

$$f^{(1)}\!\left(\vec{k}',\, \vec{k}\right) = -\frac{1}{4\pi}\,\frac{2m}{\hbar^2}\left\langle \vec{k}' | v | \vec{k} \right\rangle$$

$$f^{(2)}\!\left(\vec{k}',\, \vec{k}\right) = -\frac{1}{4\pi}\,\frac{2m}{\hbar^2}\left\langle \vec{k}' \left| V\, \frac{1}{E - H_0 + i\epsilon}\, V \right| \vec{k} \right\rangle. \tag{102}$$

$$\ldots\ldots\ldots\quad\ldots\ldots\ldots$$

Here, $f^{(2)}$, $f^{(3)}$, … correspond to scattering as a two-step, three-step, ….. processes (not multiple-scattering!). Let us write $U(\vec{r})$ as $gU(\vec{r})$, where g is a dimensionless strength parameter. The Born expansion is a power series expansion in g. The convergence of the series can be discussed in terms of the radius of convergence of the power series in g. We may write from Eq. (101)

$$f(k,\,\theta,\,g) = \sum_{n=1}^{\infty} g^n f^{(n)}(k,\,\theta) \tag{103}$$

The Born approximation to the scattering amplitude does not satisfy the optical theorem. In fact, $f^{(1)}(k,\,\theta)$ is real. The reason is that the Born approximation uses an unscattered plane wave function and so the removal of the scattered particles from the incident flux is not taken into account. If f is computed to order n in the potential, the total cross section will be of order 2n, and the optical theorem is not satisfied in general. Inserting the series Eq. (103) into the optical theorem we obtain

$$\int d\Omega \left| \sum_{n=1}^{\infty} g^n f^{(n)}(k,\,\theta) \right|^2 = \frac{4\pi}{k}\, \mathrm{Im} \sum_{n=1}^{\infty} g^n f^{(n)}(k,\,0)$$

Equating the coefficients of equal powers of g, we get

$$\operatorname{Im} f^{(1)}(k, 0) = 0, \tag{104}$$

$$\int d\Omega \, |f_1^{(1)}(k, \theta)|^2 = \frac{4\pi}{k} \operatorname{Im} f^{(2)}(k, 0). \tag{105}$$

Of course, the optical theorem holds to each order in g.

The second Born approximation to the scattering amplitude is given by

$$f^{(2)}(\theta) = f^{(1)}(\theta) + \frac{1}{4\pi} \iint e^{i\vec{k}'\cdot\vec{r}} \, U(r) \, \frac{e^{ik\,|\vec{r} - \vec{r}'|}}{4\pi|\vec{r} - \vec{r}'|}$$

$$\times \, U(\vec{r}') \, e^{i\vec{k}\cdot\vec{r}'} \cdot d^3r\, d^3r' \tag{106}$$

For the Yukawa potential we now consider the second Born approximation term $f^{(2)}$. We obtain (Moiseiwitsch (1961))

$$f^{(2)}(\theta) = \left(\frac{2mV_0}{\mu\hbar^2}\right) \frac{1}{2kA \, \sin\left(\dfrac{\theta}{2}\right)} \left[\tan^{-1}\left(\frac{\mu k \, \sin\left(\dfrac{\theta}{2}\right)}{A}\right) \right.$$

$$\left. + \frac{i}{2} \ln\left(\frac{A + 2k^2 \, \sin\left(\dfrac{\theta}{2}\right)}{A - 2k^2 \, \sin\left(\dfrac{\theta}{2}\right)}\right) \right] \tag{107}$$

where
$$A = \mu^4 + 4\mu^2 k^2 + 4k^4 \, \sin^2\left(\frac{\theta}{2}\right).$$

As $\mu \to 0$, we see that the real part of $f^{(2)}$ vanishes and so there is no contribution to the differential scattering cross section for a Coulomb field from terms of the third order in V_0. Since the differential scattering cross section given by the first Born approximation agrees with the exact result, all terms higher than the second order in V_0 must vanish. The imaginary part of the forward scattering amplitude given by the second Born approximation is

$$\operatorname{Im} f^{(2)}(0) = \left(\frac{2mV_0}{\mu^2\hbar^2}\right)^2 \frac{k}{4k^2 + \mu^2}. \tag{108}$$

From Eqs. (84) and (108), we see that Eq. (105) is satisfied.

We discuss now the application of variational methods for the determination of phase shifts and scattering amplitudes. Two types of variational methods are most commonly used. These were developed by (a) Hulthen and Kohn and (b) Schwinger. Here we consider Schwinger's variational method for the scattering amplitudes.

The solution of the Schrödinger equation

$$\left[\vec{\nabla}^2 + k^2 - U(r)\right] \psi(\vec{r}) = 0 \tag{109}$$

with the asymptotic boundary condition

$$\psi(\vec{r}) \sim e^{i\vec{k}.\vec{r}} + \frac{e^{ikr}}{r} f\left(\vec{k}', \vec{k}\right) \tag{110}$$

is

$$\psi(\vec{r}) = e^{i\vec{k}.\vec{r}} - \frac{1}{4\pi} \int \frac{e^{ik|\vec{r} - \vec{r}'|}}{|\vec{r} - \vec{r}'|} U(r')\psi(\vec{r}')d^3r' \tag{111}$$

with the scattering amplitude given by

$$f\left(\vec{k}', \vec{k}\right) = -\frac{1}{4\pi} \int e^{-i\vec{k}'.\vec{r}'} U(r')\psi(\vec{r}')d^3r'. \tag{112}$$

Let $\tilde{\psi}(\vec{r})$ be the solution of Eq. (109) corresponding to a wave incident in the direction $-\vec{k}'$ so that

$$\tilde{\psi}(\vec{r}) = e^{-i\vec{k}'.\vec{r}'} - \frac{1}{4\pi} \int \frac{e^{ik|\vec{r} - \vec{r}'|}}{|\vec{r} - \vec{r}'|} U(r')\tilde{\psi}(\vec{r}')d^3r'. \tag{113}$$

Multiplying both sides of Eq. (111) by $\tilde{\psi}(\vec{r})U(r)$ and integrating, we obtain

$$\int \tilde{\psi}(\vec{r})U(r)\, \psi(\vec{r})d^3r = \int \tilde{\psi}(\vec{r})U(r)e^{i\vec{k}.\vec{r}}d^3r$$

$$-\frac{1}{4\pi} \iint \tilde{\psi}(\vec{r})U(r) \frac{e^{ik|\vec{r} - \vec{r}'|}}{|\vec{r} - \vec{r}'|} U(r')\tilde{\psi}(\vec{r}')d^3rd^3r'. \tag{114}$$

The scattering amplitude is given by

$$\left[f\left(\vec{k}', \vec{k}\right)\right] = \frac{-\dfrac{1}{4\pi} \int \tilde{\psi}(\vec{r})U(r)e^{i\vec{k}.\vec{r}}d^3r \int e^{-i\vec{k}'.\vec{r}'} U(r')\psi(\vec{r}')d^3r'}{\int \tilde{\psi}(\vec{r})U(r)\psi(\vec{r})d^3r + \dfrac{1}{4\pi} \iint \tilde{\psi}(\vec{r})U(r) \times}$$

$$\times \frac{e^{ik|\vec{r} - \vec{r}'|}}{|\vec{r} - \vec{r}'|} U(r')\psi(\vec{r}')d^3rd^3r'. \tag{115}$$

If $\psi(\vec{r})$ and $\tilde{\psi}(\vec{r})$ are exact solutions of Eq. (109), then $[f]$ is stationary with respect to infinitesimal changes in $\psi(\vec{r})$ and $\tilde{\psi}(\vec{r})$.

To apply this method, one introduces a set of trial functions ϕ and $\tilde{\phi}$ for ψ and $\tilde{\psi}$ involving a set of unknown parameters C_n. By solving the set of simultaneous equations.

$$\frac{\partial f\left(\vec{k}', \vec{k}\right)}{\partial C_n} = 0, \tag{116}$$

one obtains the "best" values of the parameters C_n. We note here that the Born approximation is obtained if we set $\psi(\vec{r}) = e^{i\vec{k}.\vec{r}}$ and $\tilde{\psi}(\vec{r}) = e^{-i\vec{k}'.\vec{r}'}$ and the second term in the denominator of $[f]$, in Eq. (115) is omitted. The reciprocity theorem

$$f\left(\vec{k} \to \vec{k}\right) = f\left(-\vec{k}' \to -\vec{k}\right) \text{ immediately follows from Eq. (115).}$$

This method has been applied to nucleon scattering, electron-atom scattering, etc.

The following expression for the phase shift δ_l

$$[\cot \delta_l] = \frac{\int_0^\infty u_l^2(r)U(r)dr + \frac{1}{4\pi}\int_0^\infty\int_0^\infty u_l(r)U(r)G_k^{(l)}(r,\,r')U(r')u_l(r')drdr'}{\left[\int_0^\infty krj_l(kr)U(r)u_l(r)dr\right]^2}$$

$$= \frac{k\int_0^\infty u_l^2(r)U(r)dr - 2\int_0^\infty krn_l(kr)U(r)u_l(r)\int_0^r kr'j_l(kr')U(r)u_l(r')\,drdr'}{\left[\int_0^\infty krj_l(kr)U(r)u_l(r)dr\right]^2},$$

$$\tag{117}$$

is the variational integral. Blatt and Jackson (1950) used this approach to obtain an expression for $k \cot \delta_l$ as a power series in k^2 for a short-ranged potential. This is of great importance in low energy nuclear physics.

The Fredholm method provide a way of finding the solution to the integral equation for scattering to an arbitrary accuracy for all potentials that go to zero faster than r^{-1} as $r \to 0$ and are less singular than $\dfrac{1}{r^2}$ at $r = 0$.

We first consider a central potential and write the radial integral equation as

$$u_l(r) = F_l(kr) + \int_0^\infty dr'g_l(r,\,r')V(r')u_l(r').\tag{118}$$

The solution of the Fredholm integral equation of the second kind is given by (Whittaker and Watson (1965))

$$u_l(r) = F_l(kr) + \int_0^\infty dr'\,\{D_l(r,\,r';\,\lambda)/D_l(\lambda)\}\,F_l(kr')\tag{119}$$

$D_l(\lambda)$ is called the Fredholm determinant belonging to the integral equation for the lth partial wave and the function $D_l(r,\,r';\,\lambda)$ is called the Fredholm minor. Here $D_l\,(r,\,r';\,\lambda)$ and $D_l(\lambda)$ are power series in λ, the strength of the potential:

$$D_l(\lambda) = 1 - \lambda \int_0^\infty drg_l(r,\,r)V(r) + \frac{\lambda^2}{2!}\iint drdr'$$

$$\times \begin{vmatrix} g_l(r,\,r)V(r) & g_l(r,\,r')V(r') \\ g_l(r',\,r)V(r) & g_l(r',\,r')V(r') \end{vmatrix}$$

$$- \frac{\lambda^3}{3!}\iiint drdr'dr'' \begin{vmatrix} g_l(r,\,r)V(r) & g_l(r,\,r')V(r') & g_l(r,\,r'')V(r'') \\ g_l(r',\,r)V(r) & g_l(r',\,r')V(r') & g_l(r',\,r'')V(r'') \\ g_l(r'',\,r)V(r) & g_l(r'',\,r')V(r') & g_l(r'',\,r'')V(r'') \end{vmatrix}$$

$$+ \dots\tag{120}$$

$$D_l(r,\,r';\,\lambda) = g_l(r,\,r')V(r') - \lambda \int_0^\infty dr'' \begin{vmatrix} g_l(r,\,r')V(r') & g_l(r,\,r'')V(r'') \\ g_l(r'',\,r')V(r') & g_l(r'',\,r'')V(r'') \end{vmatrix}$$

$$+ \frac{\lambda^2}{2!} \int_0^\infty \int_0^\infty dr'' dr'''$$

$$\times \begin{vmatrix} g_l(r, r')V(r') & g_l(r, r'')V(r'') & g_l(r, r''')V(r''') \\ g_l(r'', r')V(r') & g_l(r'', r'')V(r'') & g_l(r'', r''')V(r''') \\ g_l(r''', r')V(r') & g_l(r''', r'')V(r'') & g_l(r''', r''')V(r''') \end{vmatrix}.$$

$$+ \ldots \tag{121}$$

The series for $D_l(\lambda)$ and $D_l(r, r'; \lambda)$ converge for all values of λ, and hence are entire analytic function of λ. The resolvent kernel

$$R_l(r, r'; \lambda) = \frac{D_l(r, r'; \lambda)}{D_l(\lambda)} \tag{122}$$

is an analytic function of λ, except for those values of λ which are zeros of the function $D_l(\lambda)$. The latter are the poles of the resolvent kernel.

As an example, let us consider the separable potential (Rodberg and Thaler (1967))

$$V(r, r') = v_l(r)v_l(r'). \tag{123}$$

Then,

$$D_l(\lambda) = 1 - \lambda \int_0^\infty dr dr' v_l(r)g_l(r, r')v_l(r') \tag{124}$$

and

$$D_l(r, r') = \int_0^\infty dr'' g_l(r, r'')v_l(r'')v_l(r').$$

Inserting Eq. (124) into Eq. (119) we get the solution

$$u_l(r) = F_l(kr) + \left\{ \int_0^\infty dr' g_l(r, r')v_l(r') \right\}$$

$$\times \left\{ \int_0^\infty dr'' v_l(r'')F_l(kr'') \right\} \Big/ \left\{ 1 - \int_0^\infty dr \int_0^\infty dr' v_l(r)g_l(r, r')v_l(r') \right\}. \tag{125}$$

We now give the Fredholm expression for the phase shift

$$-\frac{1}{k} D_l(k) \cot \delta_l = 1 - \frac{2}{\pi} P \int_0^\infty \frac{dk'}{k^2 - k'^2} D_l(k') \tag{126}$$

where

$$D_l(k) = V_l(k, k) - \lambda \frac{2}{\pi} P \int_0^\infty \frac{dk'}{k^2 - k'^2} \begin{vmatrix} V_l(k, k) & V_l(k, k') \\ V_l(k', k) & V_l(k', k') \end{vmatrix}$$

$$+ \frac{\lambda^2}{2!} \left(\frac{2}{\pi} \right)^2 P \int_0^\infty \int_0^\infty \frac{dk' dk''}{(k^2 - k'^2)(k^2 - k''^2)}$$

$$\times \begin{vmatrix} V_l(k, k) & V_l(k, k') & V_l(k, k'') \\ V_l(k', k) & V_l(k', k') & V_l(k', k'') \\ V_l(k'', k) & V_l(k'', k') & V_l(k'', k'') \end{vmatrix} \tag{127}$$

with

$$V_l(k', k) = \int_0^\infty dr F_l(k'r) V(r) F_l(kr).$$
(128)

The Fredholm determinant $D_l(\lambda)$ is related to $D_l(k)$ by the expression

$$D_l(\lambda) = 1 - \lambda \frac{2}{\pi} P \int_0^\infty \frac{dk'}{k^2 - k'^2} D_l(k').$$
(129)

The power series of $D_l(k)$ is convergent, regardless of the strength of the potential. The right-hand side of Eq. (126) can be approximated by a polynomial in k^2 (or E) at low energy ($E \to 0$). The effective range parameters can be computed for $D_l(k)$. One can determine the positions of resonances or bound states from the zeros of D_l i.e., from the R.H.S. of Eq. (126).

Finally, we obtain the Fredholm solution for the three dimensional equation. One cannot proceed in a direct manner because of the singularity in the Green function $G(\vec{r}, \vec{r}')$ at $\vec{r} = \vec{r}'$. To bypass this difficulty we iterate the integral equation once. Inserting the second term in the Born expansion into the integral equation we have

$$\begin{aligned}
\psi_{\vec{k}}(\vec{r}) &= e^{i\vec{k}.\vec{r}} + \int d^3r' G(\vec{r}, \vec{r}') V(\vec{r}') \psi_{\vec{k}}(\vec{r}') \\
&= e^{i\vec{k}.\vec{r}} + \int d^3r' G(\vec{r}, \vec{r}') V(\vec{r}') e^{i\vec{k}.\vec{r}} \\
&\quad + \int d^3r' G(\vec{r}, \vec{r}') V(\vec{r}') \left(\psi_{\vec{k}}(\vec{r}') - e^{i\vec{k}.\vec{r}} \right) \\
&= \psi_{\vec{k}}^{(1)}(\vec{r}) + \int d^3r' G_2(\vec{r}, \vec{r}') V(\vec{r}') \psi_{\vec{k}}(\vec{r}'),
\end{aligned}$$
(130)

where, $\psi_{\vec{k}}^{(1)}(\vec{r}')$, the first iterate of the wave function, is given by Eq. (74) and

$$G_2(\vec{r}, \vec{r}') = \int d^3r'' G(\vec{r}, \vec{r}'') V(\vec{r}'') G(\vec{r}'', \vec{r}').$$
(131)

Since $G_2(\vec{r}, \vec{r}')$ is not singular at $\vec{r} = \vec{r}'$, the Fredholm solution is now obtained. The solution is

$$\psi_{\vec{k}}(\vec{r}) = \psi_{\vec{k}}^{(1)}(\vec{r}) + \int d^3r' \frac{D_l(r, r'; \lambda)}{D(\lambda)} \psi_{\vec{k}}^{(1)}(\vec{r}'),$$
(132)

where

$$\begin{aligned}
D(\vec{r}, \vec{r}') &= G_2(\vec{r}, \vec{r}') V(\vec{r}') \\
&\quad - \frac{\lambda}{1!} \int d^3r'' \begin{vmatrix} G_2(\vec{r}, \vec{r}') V(\vec{r}') & G_2(\vec{r}, \vec{r}'') V(\vec{r}') \\ G_2(\vec{r}'', \vec{r}') V(\vec{r}') & G_2(\vec{r}'', \vec{r}'') V(\vec{r}'') \end{vmatrix}
\end{aligned}$$
(133)

and the Fredholm determinant

$$D(\lambda) = 1 - \lambda \int d^3r\, G_2(\vec{r}, \vec{r}) V(\vec{r})$$

$$+ \frac{\lambda}{2!} \int d^3r\, d^3r' \begin{vmatrix} G_2(\vec{r}, \vec{r})V(\vec{r}) & G_2(\vec{r}, \vec{r}')V(\vec{r}') \\ G_2(\vec{r}', \vec{r})V(\vec{r}) & G_2(\vec{r}', \vec{r}')V(\vec{r}') \end{vmatrix} \qquad (134)$$

These power series converge for all potentials provided

$$\lim_{r \to 0} r^2\, V(r) = 0$$

$$\lim_{r \to \infty} r^2\, V(r) = 0. \qquad (135)$$

20.5 THE EIKONAL APPROXIMATION

We now consider a semi-classical approximation scheme which is applicable if the potential $V(r)$ changes so slowly that the local momentum $\hbar k(\vec{r})$ is reasonably constant over many wavelengths λ. Under these conditions the exact wave function for the scattering amplitude in Eq. (11) can be replaced by the semi-classical wave function (see § 3.7).

$$\psi_{\vec{k}}^{+}(\vec{r}) = e^{iS_{\vec{k}}(\vec{r})/\hbar}, \qquad (136)$$

where $S_{\vec{k}}(\vec{r})$ is the solution of the Hamilton–Jacobi equation

$$\left| \vec{\nabla} S_{\vec{k}}(\vec{r}) \right| = \sqrt{2m\,[E - V(\vec{r})]} \qquad (137)$$

It is a formidable task in general to be able to integrate Eq. (137) exactly to determine the classical trajectory. So we make a further approximation. We replace the exact classical trajectory by a straight line in evaluating S. Clearly, this approximation is satisfactory at high energy and for small scattering angle. We thus have a high energy approximation in which $V(r)$ does not change rapidly. Note that V need not be small so long as $E \gg |V|$.

We decompose the radial vector $\vec{r}$ as $\vec{r} = \vec{b} + \hat{k}z$, where $\vec{b}$ is perpendicular to $\vec{k}$ with length equal to the impact parameter b (see Fig. 20.2)

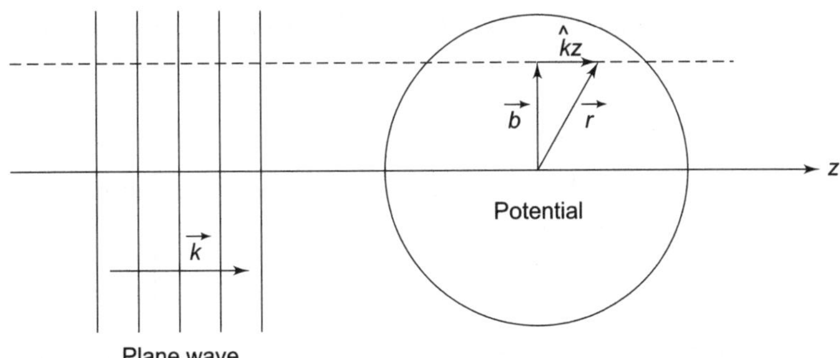

Fig. 20.2 Schematic diagram showing certain coordinate vectors of eikonal approximation in scattering

Equation (137) becomes

$$\frac{dS_{\vec{k}}}{dz} = \left\{ 2m \left[E - V(\vec{b} + \hat{k}z) \right] \right\}^{1/2} \tag{138}$$

assuming that the trajectory is a straight line. Integrating Eq. (138), we have

$$\frac{S_{\vec{k}}}{\hbar} = kz + \int_{-\infty}^{z} \left\{ \left[k^2 - U(\vec{b} + \hat{k}z') \right]^{1/2} - k \right\} dz', \tag{139}$$

where the constant of integration is so chosen that $S_{\vec{k}}/\hbar \to kz$ as $U \to 0$. Since $E \gg V$, i.e., $k^2 \gg |U(\vec{r})|$, Eq. (139) becomes

$$\frac{S_{\vec{k}}}{\hbar} = kz - \frac{1}{2k} \int_{-\infty}^{z} U(\vec{b} + \hat{k}z') \, dz'. \tag{140}$$

The wave function is

$$\psi_{\vec{k}}^{+}(\vec{b} + \hat{k}z') \cong \phi_{\vec{k}}(\vec{b} + \hat{k}z') \exp\left[-\frac{i}{2k} \int_{-\infty}^{z} U(\vec{b} + \hat{k}z') \, dz' \right]. \tag{141}$$

Therefore, the exact wave function is approximated by the undistorted wave function everywhere except in a cylinder of radius R (range of the potential) to the right of the scattering center. This wave function, however, does not have the correct asymptotic form Eq. (9). But we can use it in the expression for the scattering amplitude and obtain an approximate expression for it. We have,

$$f(\vec{k}', \vec{k}) \cong -\frac{1}{4\pi} \int e^{i q . (\vec{b} + \hat{k}z)} \, U(\vec{b} + \hat{k}z) \, dz d^2 b$$

$$\times \exp\left\{ -\frac{i}{2k} \int_{-\infty}^{z} U(\vec{b} + \hat{k}z') \, dz' \right\} \tag{142}$$

where $\vec{q} = \vec{k} - \vec{k}'$ and d^2b denotes integration over the plane of impact vectors. For elastic scattering. $|\vec{k}'| = |k|$, so that for small scattering angles, the vector $\vec{q} = \vec{k} - \vec{k}'$ is nearly perpendicular to $\vec{k}$. The error in approximating $\exp[i\vec{q}.\hat{k}z]$ by 1 is of order $(1 - \cos\theta)kz \sim \frac{1}{2}\theta^2 kz$. So for small scattering angles, θ satisfies $\theta^2 \ll \frac{1}{kR}$.

We replace $e^{i\vec{q}.\hat{k}z}$ by 1. With this simplification, the z-integration is that of an exact differential and hence we obtain

$$f(\vec{k}', \vec{k}) = \frac{k}{2\pi i} \int \left[e^{i\vec{q}.\vec{b}} \, e^{-\frac{i}{2k} \int_{-\infty}^{\infty} U(\vec{b} + \hat{k}z')dz'} - 1 \right] d^2 b. \tag{143}$$

This is the basic result for the sacattering amplitude for elastic scattering in the eikonal approximation (Glauber (1959)). It corresponds to a picture in which each part of the incident wave passes through the potential along a straight line and suffers a phase shift.

For potentials with azimuthal symmetry, the integration can be carried out one step further. Using the small angle expression

$$\vec{q}.\vec{b} = -\vec{k}', \ b \simeq kb\theta \cos \phi,$$

where ϕ is the azimuthal angle specifying the orientation of $\vec{b}$. Noting that

$$\frac{1}{2\pi} \int_0^{2\pi} e^{i\lambda \cos \phi \, d\phi} = J_0(\lambda) \tag{144}$$

where $J_0(\lambda)$ is the Bessel function of order zero, we obtain the final expression for the scattering amplitude in the eikonal approximation for axially symmetric potentials.

$$f\left(\vec{k}', \vec{k}\right) \simeq -ik \int_0^{\infty} J_0(kb\theta) \ [e^{2i\Delta(b)} - 1] \ bdb, \tag{145}$$

where

$$\Delta(b) = -\frac{1}{4\pi} \int_{-\infty}^{\infty} U\left(\sqrt{b^2 + z^2}\right) dz. \tag{146}$$

In Eq. (145), the factor $e^{2i\Delta} - 1$ tends to zero if b is greater than R, the range of the potential. The integral over b covers the classical impact parameters. The expression Eq. (145) is the type of formula one uses in the Fraunhofer diffraction by a spherically symmetric object.

The eikonal approximation scheme is justified when the finite-ranged potential is a smooth function of r and the following conditions are satisfied

$$kR \gg 1 \ (\bar{\lambda} \ll R),$$

$$V/E \ll 1,$$

and the small scattering angle requirement

$$\theta^2 \, kR \ll 1.$$

If we replace θ in the argument of the Bessel function by $2 \sin \dfrac{\theta}{2}$ in Eq. (145), the angular range of the approximation may be increased. The expression for the scattering amplitude then becomes

$$f\left(\vec{k}', \vec{k}\right) = -ik \int_0^{\infty} J_0\left(2kb \sin \frac{\theta}{2}\right)[e^{2i\Delta(b)} - 1] \ bdb. \tag{147}$$

Note that when $|\Delta| \ll 1$, $[e^{2i\Delta} - 1] = 2i\Delta$, and from Eq. (145) we get back the Born approximation for the scattering amplitude:

$$f_B\left(\vec{k}', \vec{k}\right) \simeq -\frac{1}{2} \int_0^{\infty} bdb \int_{-\infty}^{\infty} J_0(kb\theta) U\left(\sqrt{b^2 + z^2}\right) dz. \tag{148}$$

The eikonal approximation satisfies the optical theorem. The total cross section is

$$\sigma_{tot} = 2\pi k^2 \int_0^\infty b\,db\,b'\,db' \int_0^\pi \sin\theta \; d\theta J_0(kb\theta)J_0(kb'\theta)$$

$$\times \left[e^{2i\Delta(b)} - 1\right] \left[e^{2i\Delta(b')} - 1\right]. \tag{149}$$

Since the scattering is concentrated near the forward direction, we replace $\sin\theta$ by θ and extend the upper limit in the θ-integral to infinity. Using the completeness relation for the Bessel functions

$$\int_0^\infty J_0(x\theta)\, J_0(x'\theta)\theta\, d\theta = \frac{1}{x}\, \delta(x - x'),$$

we get

$$\sigma_{tot} = 2\pi \int_0^\infty b\,db \; |e^{2i\Delta(b)} - 1|^2$$

$$= 8\pi \int_0^\infty b\,db \; \sin^2\Delta(b). \tag{150}$$

Now, the imaginary part of the forward scattering amplitude is

$$\mathrm{Im}\, f(0) = - k\int_0^\infty b\,db \; \mathrm{Re}\, [e^{2i\Delta(b)} - 1]. \tag{151}$$

The optical theorem is satisfied.

If there is absorption, then Eq. (149) must be modified as

$$\sigma_{tot} = 4\pi \int_0^\infty b\,db \; [1 - \mathrm{Re}\, e^{2i\Delta(b)}]. \tag{152}$$

The relationship of the eikonal approximation with the partial wave analysis is considered below. Since $kb = l + \frac{1}{2}$ in the semi classical correspondence and using the large l, small-angle approximation of the Legendre polynomial of order l

$$P_l(\cos\theta) \sim J_0\left(2\left(l + \frac{1}{2}\right)\sin\frac{\theta}{2}\right) + \frac{1}{4}\sin^2\frac{\theta}{2} + \ldots \tag{153}$$

we see that equation Eqs. (145) or (146) is the replacement of

$$f\left(\vec{k}', \vec{k}\right) \sim \frac{1}{2ik} \sum_l (2l + 1)\, (e^{2i\Delta_l} - 1)\, P_l(\cos\theta) \tag{154}$$

by an integral. The function $\Delta(b)$ corresponds to the lth angular-momentum phase shift $\delta_l(k)$ in the following way

$$\Delta\left(\frac{\left(l + \frac{1}{2}\right)}{k}\right) \leftrightarrow \delta_l(k) \tag{155}$$

when all the conditions are satisfied.

We now discuss some illustrative examples of the high energy approximation. We consider the scattering from a "black" sphere of radius R. We assume that the region inside R is completely absorptive ("black"). Here $kR \gg 1$.

We have

$$e^{2i\Delta(b)} = \begin{cases} 0, & b < R \\ 1, & b > R \end{cases} \tag{156}$$

so that the scattering amplitude becomes

$$f(\theta) = +ik \int_0^R J_0\left(2kb \sin \frac{\theta}{2}\right) b\,db$$

$$= iR \, \frac{J_1\left(2kR \sin \frac{\theta}{2}\right)}{2 \sin \frac{\theta}{2}}. \tag{157}$$

The differential elastic scattering cross section is

$$\frac{d\sigma_{sc}}{d\Omega} = |f(\theta)|^2 = (kR^2)^2 \left[\frac{J_1\left(2kR \sin \frac{\theta}{2}\right)}{2kR \sin \frac{\theta}{2}}\right]^2. \tag{158}$$

This is the formula for diffraction from a perfectly absorbing sphere of radius R. The cross section is sharply peaked in the forward direction and is concentrated within the region having $\theta < (kR)^{-1}$ (Fig. 20.3).

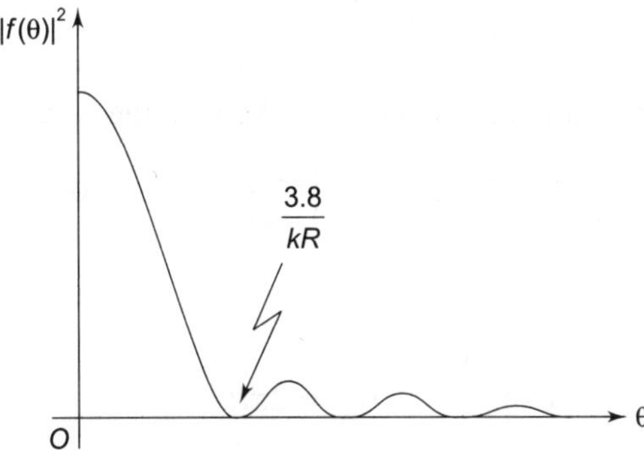

Fig. 20.3 The differential cross section for scattering by a "black" sphere in the eikonal approximation

The diffraction formula is useful for high-energy scattering.

The elastic scattering cross section is

$$\sigma_{el} = \int |e^{2i\Delta(b)} - 1|^2 \, b\,db = \pi R^2, \tag{159}$$

while the absorption cross section is

$$\sigma_{abs} = \int (1 - |e^{2i\Delta(b)}|^2) \, b\,db = \pi R^2, \tag{160}$$

So,

$$\sigma_{tot} = 2\pi R^2 \tag{161}$$

If the scattering at high energy is not totally absorptive, we may introduce an opacity factor a as

$$a = |1 - S_l| \qquad (l < kR)$$
$$S_l = 1 \qquad (l > kR)$$

Assuming that the (real) phase shifts are small, one can show that

$$\sigma_{el} = \pi(aR)^2, \quad \sigma_{tot} = 2\pi aR^2,$$

$$\frac{d\sigma_{el}}{d\Omega} = (aR)^2 \left[\frac{J_1\left(2kR \sin\frac{\theta}{2}\right)}{2 \sin\frac{\theta}{2}} \right]^2. \tag{162}$$

Next, we consider the case of a square potential well of radius R:

$$V(r) = \begin{cases} V_0, & r < R \\ 0, & r > R \end{cases}. \tag{163}$$

We have

$$2\Delta(b) = \begin{cases} -\dfrac{2V_0}{\hbar v}\sqrt{R^2 - b^2}, & b < R \\ 0, & b > R \end{cases}. \tag{164}$$

Here the integration is not elementary. We give the result (Glauber (1959))

$$\frac{\sigma_{tot}}{\pi R^2} = 2 + \frac{1}{\alpha^2} + \frac{2}{\alpha}\left(\frac{\cos 2\alpha}{2\alpha} + \sin 2\alpha\right) \tag{165}$$

where

$$\alpha = \frac{V_0 R}{\hbar v}$$

For

$$\alpha \to \infty, \qquad \sigma_{tot} = 2\pi R^2,$$

which is the result for a "black" sphere of radius R.

For $\alpha \ll 1$, we get the Born approximation

$$\frac{\sigma_{tot}}{\pi R^2} = 2\alpha^2 \quad (\alpha \ll 1)$$

A plot of $\dfrac{\sigma_{tot}}{\pi R^2}$ vs α is shown in Fig. 20.4.

20.6 THE DISTORTED-WAVE BORN APPROXIMATION (DWBA)

The discussion in Section 19.5 on scattering from two potential gives rise in a natural way to DWBA. This approximation called the distorted wave Born

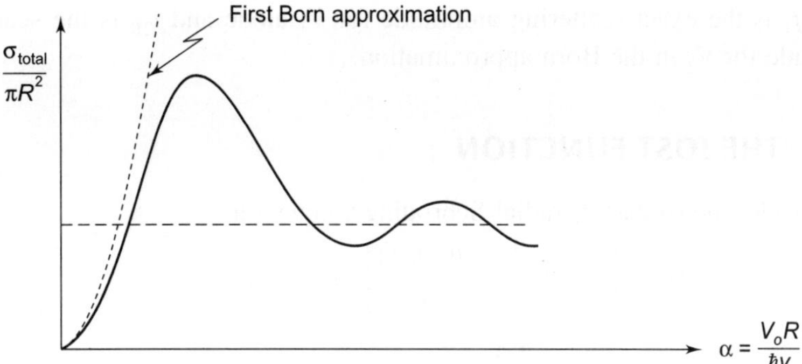

Fig. 20.4   A plot of the ratio of the total cross section to the geometrical cross section versus α

approximation, is useful when the scattering potential V can be written as the sum of two parts $V = V_1 + V_2$ and V_2 is small. If V_2 is sufficiently weak then it can be treated in a first-order approximation. Equation (19.137) shows that the second term in the T-matrix element is given to first-order in V_2 by approximating $\psi_i^{(+)}$ by $\chi_i^{(+)}$. The T-matrix element is now

$$T_{fi} \simeq T_{1fi} + \left\langle \chi_f^{(-)} \middle| V_2 \middle| \chi_i^{(+)} \right\rangle \tag{166}$$

where T_{1fi} is the exact T-matrix for the potential V_1. The second term is a generalization of the Born approximation in which distorted-wave functions are used to compute the scattering due to V_2 in presence of the distorting potential V_1. This is called the "distorted-wave Born approximation" or simply the "distorted-wave approximation". It is to be noted that for the calculation of the matrix element one needs an outgoing wave state for the initial state and for the final state, an incoming state.

As an example, in the scattering of protons from nuclei, one may take V_1 to be the Coulomb potential and V_2 the short-range nuclear potential. The resulting T-matrix is then given by the Coulomb T-matrix plus another term which describes the scattering due to the nuclear force, taking account of the Coulomb repulsion.

In the case in which V_1 and V_2 are spherically symmetric, the DWBA can be expressed in terms of the phase shifts δ_l associated with V_1 alone. Schiff (1968) gives the following expression for the scattering amplitude in the DWBA for spherically symmetric scattering:

$$f_{DWBA}(\theta) = f_1(\theta) + f_{2B}(\theta)$$

$$-\frac{2m}{\hbar} \sum_{l=0}^{\infty} (2l + 1) P_l(\cos\theta) \int_0^\infty V_2(r)$$
$$\times \, [R_l^2(r) - j_l^2(kr)] r^2 \, dr, \tag{167}$$

where f_1 is the exact scattering amplitude for V_1 alone and f_{2B} is the scattering amplitude for V_2 in the Born approximation.

20.7 THE JOST FUNCTION

We consider the (reduced) radial Schrodinger equation

$$u_l''(r) + \left[k^2 - U(r) - \frac{l(l+1)}{r^2} \right] u_l(r) = 0, \tag{168}$$

where

$$k^2 = \frac{2mE}{\hbar^2}$$

and

$$U(r) = \frac{2mV(r)}{\hbar^2}.$$

The regular solution of this equation is specified by the boundary condition near the origin:

$$\phi_l(k, r) \xrightarrow[r \to 0]{} \frac{r^{l+1}}{(2l+1)!!}. \tag{169}$$

For fixed r, $\phi_l(k, r)$ is an entire function of k.

We define another solution $f_l(k, r)$ by an asymptotic boundary condition:

$$f_l(k, r) \xrightarrow[r \to \infty]{} i^l e^{-ikr}. \tag{170}$$

Clearly, this is irregular at the origin where it behaves like r^{-l}. Here, we assume $\int_0^\infty r \, |V(r)| dr < \infty$ and $\int_0^\infty r^2 \, |V(r)| dr < \infty$. A solution $f_l(-k, r)$, except at $k = 0$, which is linearly independent of $f_l(k, r)$ has the asymptotic form

$$f_l(-k, r) \xrightarrow[r \to \infty]{} i^l e^{ikr}. \tag{171}$$

Thus, $\phi_l(k, r)$ can be expressed as a linear combination of the Jost solutions $f_l(\pm k, r)$. We write

$$\phi_l(k, r) = \frac{1}{2} ik^{-l-1} [f_l(-k)f_l(k, r) - (-1)^l f_l(k)f_l(-k, r)]. \tag{172}$$

The coefficient $f_l(k) = f_l(k, 0)$ is called the Jost function.

The asymptotic form of $\phi_l(k, r)$ is

$$\phi_l(k, r) \xrightarrow[r \to \infty]{} \frac{1}{2} \left(\frac{i}{k} \right)^{l+1} [f_l(-k)e^{-ikr} - (-1)^l f_l(k)e^{ikr}] \tag{173}$$

Comparison with Eq. (25) gives the S-matrix element for the lth pertial wave

$$S_l(k) = \frac{f_l(k)}{f_l(-k)} \tag{174}$$

From the asymptotic boundary conditions Eqs. (170), (117) and the reality of the differential Eq. (168) we see that

$$[f_l(-k, r)]^* = (-1)^l f_l(k, r) \tag{175}$$

for real k. For complex k,

$$[f_l(-k^*, r)]^* = (-1)^l f_l(k, r) \tag{176}$$

while

$$[\phi_l(k^*, r)]^* = \phi_l(k, r) \tag{177}$$

for all finite complex k.

The complex conjugate of Eq. (172) is

$$[\phi_l(k, r)]^* = -\frac{1}{2} i(k^*)^{-l-1} [f_l(-k)f_l(k, r) - (-1)^l f_l(k)f_l(-k, r)]^*.$$

Putting k^* for k and using Eqs. (176) and (177) we get

$$[f_l(-k^*)]^* = f_l(k). \tag{178}$$

It follows that for real k,

$$f_l(k) = |f_l(k)|\, e^{i\delta_l(k)}. \tag{179}$$

For real k, $|S_l(k)| = 1$, so that S is unitary and $\delta_l(k)$ is real.

Let us now find the Wronskian of the two solutions $f_l(k, r)$ and $\phi_l(k, r)$ of the differential equation (168). We have

$$W[f_l(k, r), \phi_l(k, r)] \equiv f_l(k, r) \frac{\partial}{\partial r} \phi_l(k, r) - \phi_l(k, r) \frac{\partial}{\partial r} f_l(k, r)$$

$$= \frac{1}{2} i(-k)^{-l-1} f_l(k) \cdot W[f_l(k, r), f_l(-k, r)].$$

Now, $W[f_l(k, r), f_l(-k, r)] = (-1)^l\, 2ik$, using the asymptotic forms of $f_l(k, r)$ and $f_l(-k, r)$.

The above expression then gives

$$f_l(k) = k^l\, W[f_l(k, r), \phi_l(k, r)] \tag{180}$$

From Eqs. (171) and (179)

$$u_l(r) \equiv rR_l(r) = \frac{k^l}{f_l(-k)}\, \phi(k, r). \tag{181}$$

Thus, as $r \to 0$, $rR_l(r) \to k^l\, r^{l+1}/f_l(-k)(2l+1)!$. while the free solution $j_l(kr) \to k^l\, r^{l+1}/(2l+1)!!$. Then

$$\mathfrak{S}_l = \frac{e^{i\delta_l(k)}}{|f_l(k)|}. \tag{182}$$

The quantity $|\mathfrak{S}_l|^2$, the enhancement factor, is the ratio of the probability of finding the two particles at $r = 0$ when their interaction is $V(r)$, to that for no interaction between the particles. This is of importance in computing the final state interaction or in describing production and absorption of particles, e.g. pions. Therefore, the scattering phase shift $\delta_l(k)$ is the phase of the complex function $f_l(k)$ and its magnitude $|f_l(k)|$ has a simple physical interpretation.

The zeros of the Jost function in the lower half-plane correspond to bound states. A bound state is described by a regular solution $\phi_l(k, r)$ if k^2 is real and negative and a decreasing exponential behavior in r, for large r. Then one of the two terms in Eq. (172) must be zero. With $k = \pm iK$, a bound state is given by

$$\phi_l(\pm iK, r) \xrightarrow[r \to \infty]{} \frac{1}{2}(-K)^{-l-1} f_l(iK)e^{-Kr}.$$

$$f_l(-iK) = 0, \quad K > 0$$

provided that the point $k = iK$ lies within the region of analyticity of $f_l(k)$, connected with the real axis.

Then $f_l(iK) \neq 0$. The energy of the bound state is $-\dfrac{\hbar^2}{2m}K^2$. From Eq. (174), we see that the S-matrix element has a zero at $k = -iK$ and a pole at $k = iK$. Each such zero and pole are simple.

Although all negative imaginary zeros of the Jost function give bound states, it is not true of all negative imaginary zeros or positive imaginary poles of $S_l(k)$.

Bargmann has shown that the number of bound states $n_{b,\,l}$ of angular momentum l satisfies the inequality

$$(2l + 1)n_{b,\,l} < \int_0^\infty dr\, r\, |V(r)|,$$

where the conditions on $V(r)$, given earlier, are satisfied. The Jost function is identical with the Fredholm determinant of the radial integral equation.

20.8 LOW ENERGY SCATTERING AND BOUND STATES

In this section we discuss the low energy behavior of scattering phase shifts with special emphasis on s-wave scattering. For a finite-range potential, low-energy scattering excites only a few low-lying partial waves. We know that the phase shift $\delta_l(k) \propto k^{2l+1}$ as $k \to 0$ (threshold behavior). Near threshold, $l = 0$ s-waves and perhaps, $l = 1$ p-waves dominate elastic scattering.

A very useful analysis of s-wave scattering is based on the expression for the zero order phase shift δ which is valid in the limit of small energies. Let $u(k, r)$ be the solution of the radial Schrödinger equation for $l = 0$ partial wave. We have

$$\frac{d^2u(k,r)}{dr^2} + [k^2 - U(r)]u(k, r) = 0 \tag{183}$$

with $u(k, 0) = 0$ and $u(k, r) \xrightarrow[r \to \infty]{} w(k, r) = \dfrac{\sin (kr + \delta)}{\sin \delta}$. Then $u(0, r)$ satisfies

$$\frac{d^2 u(0, r)}{dr^2} - U(r) u(0, r) = 0. \tag{184}$$

Multiplying Eq. (183) by $u(0, r)$ and Eq. (184) by $u(k, r)$ and subtracting we get

$$u(0, r) \frac{d^2 u(k, r)}{dr^2} - u(k, r) \frac{d^2 u(0, r)}{dr^2} + k^2 u(0, r) u(k, r) = 0. \tag{185}$$

Since $w(k, r)$ satisfies the radial Schrödinger equation for $V = 0$, we have

$$\frac{d^2 w(k, r)}{dr^2} + k^2 w(k, r) = 0. \tag{186}$$

Hence, we also have

$$w(0, r) \frac{d^2 w(k, r)}{dr^2} - w(k, r) \frac{d^2 w(0, r)}{dr^2} + k^2 w(0, r) w(k, r) = 0. \tag{187}$$

Subtracting Eq. (187) from Eq. (185) and integrating the result from zero to infinity, we get

$$\left[w(0, r) \frac{dw(k, r)}{dr} - w(k, r) \frac{dw(0, r)}{dr} \right]_{r = 0}$$

$$= k^2 \int_0^\infty dr\, [w(k, r) w(0, r) - u(k, r) u(0, r)] \tag{188}$$

Since the Wronskian $W[w(0, r), w(k, r)]_{r = 0}$

$$= \frac{1}{a} + k \cot \delta$$

where

$$\frac{1}{a} = - \lim_{k \to 0} k \cot \delta. \tag{189}$$

Therefore, we obtain,

$$k \cot \delta(k) = - \frac{1}{a} + k^2 \int_0^\infty dr\, [w(k, r) w(0, r) - u(k, r) u(0, r)]. \tag{190}$$

Putting $k = 0$ in the integral, we get the low-energy approximation

$$k \cot \delta(k) = - \frac{1}{a} + \frac{1}{2} r_0 k^2 \tag{191}$$

where

$$r_0 = 2 \int_0^\infty dr\, [w^2(0, r) - u^2(0, r)]. \tag{192}$$

The parameter a is called the "scattering length" and r_0 the "effective range". Expression Eq. (191), called the "effective-range approximation", is one of the most important parametrizations for low energy scattering. For a finite-range potential which vanishes for $r = R$, $u = w$ for all $r > R$ and the integral in Eq. (192) extends to R. The function $w(k, 0) = w(0, 0) = 1$ and $w(0, r) = 1 - r/a$. So the function $w(0, r)$ is a straight line from 1 at $r = 0$ and it cuts the abscissa at $r = a$. The function u is equal to w for $r > R$ but goes down to zero at $r = 0$. Hence the integral in r_0 is expected to have a value about half the range of the potential. r_0 is therefore called the effective-range of the potential.

The effective-range approximation can be generalized to higher angular momenta. For the lth order partial wave, we get

$$k^{2l+1} \cot \delta_l = -\frac{1}{a_l} + \frac{1}{2} r_{0,\,l} \, k^2 \tag{193}$$

where a_l and $r_{0,\,l}$ are independent of k. This formula is, however, less useful.

For s-wave scattering, the scattering length a has the dimension of length. In terms of a, the zero energy cross section is $\sigma_{tot} = 4\pi a^2$. Thus, at very low energy the scattering by a finite-range potential is the same as if the scatterer were a hard sphere of radius a.

Let us now consider s-wave scattering by an attractive square-well potential

$$V(r) = \begin{cases} -V_0 & \text{for} \quad r \leq R \\ 0 & \text{for} \quad r > R \end{cases} \tag{194}$$

The Schrödinger equation can now be easily solved. The outside wave function behaves like $\dfrac{e^{i\delta_0} \sin(kr + \delta_0)}{kr}$ while the inside wave function is $\sim \sin$ Kr, where $K^2 = k + K_0^2$, $K_0^2 = \dfrac{2\mu V_0}{\hbar^2}$. By matching the logarithmic derivatives of the functions at $r = R$, we get

$$\tan \delta_0 = \frac{k \tan KR - K \tan KR}{K + k \tan KR \tan KR} \tag{195}$$

or,

$$\delta_0 = \tan^{-1}\left(\frac{k}{K} \tan(KR)\right) - KR. \tag{195a}$$

For low energies,

$$\delta_0 = KR\left(\frac{\tan(KR)}{KR} - 1\right) \tag{196}$$

unless $\tan KR$ is infinite. The total scattering cross section is then

$$\sigma_{\text{tot}} = \frac{4\pi}{k^2} \sin^2 \delta_0 = 4\pi R^2 \left[\frac{\tan(KR)}{KR} - 1\right]^2. \tag{197}$$

For $\tan KR = KR$, the cross section vanishes; the phase shift δ_0 is now π. This is the Ramsauer – Townsend effect. In the scattering of electrons by rare gas atoms (Ar, Kr, Xe) the cross section is very small at about 0.7eV bombarding energy.

Let us now return to the Eqs. (195) and (195a). For a weak potential ($|K_0|R \ll 1$) the phase shift is small everywhere, and $\delta_0 \to 0$ as $k \to 0$. If we now increase the strength of the attractive potential, the phase shift will increase. At low energies ($kR \ll 1$), it is approximated by the expression Eq. (195). When KR lies just below $\frac{\pi}{2}$, the function $\left(\frac{\tan(KR)}{KR} - 1\right)$ is large and positive until KR goes through $\frac{\pi}{2}$. Then the approximation Eq. (196) does not hold; the phase shift turns over and decreases. When $K_0R = \frac{\pi}{2}$, the phase shift is $\frac{\pi}{2}$ at $k = 0$. When K_0R is slightly greater than $\frac{\pi}{2}$, then the phase shift goes through $\frac{\pi}{2}$ and goes on increasing to π as the energy decreases. We then replace Eq. (196) by

$$\delta_0 \simeq \pi - kR\left(1 - \frac{\tan(KR)}{KR}\right). \tag{197a}$$

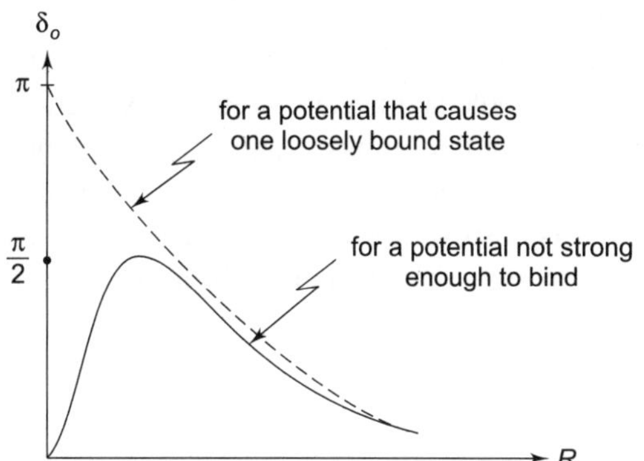

Fig. 20.5 The behavior of *s*-wave phase shift in the low-energy region

Now, $K_0R = \frac{\pi}{2}$ is the minimum value necessary for a bound state to exist. We recall the equation determining the bound states $K \cot KR = -\frac{|k|}{\hbar} = -\frac{(2\mu|E|)^{1/2}}{\hbar}$ so that for a bound state to exist the potential must have a minimum value of $\frac{\pi^2\hbar^2}{8\mu R^2}$. Thus, the *s*-wave phase shift at zero energy has the value zero if there is no bound state and the value π if there is one. At the transitional value of the potential strength, $\delta_0 = \frac{\pi}{2}$ at zero energy, and the cross section is infinite (sometimes this is called a "zero-energy resonance"). One finds that every introduction of an additional bound state raises the phase shift at zero energy by an additional π. So

$$\delta_0(0) = n_B \pi, \tag{198}$$

where n_B is the number of s-wave bound states, except at the "transitional" strengths when

$$\delta_0(0) = \left(n_B + \frac{1}{2}\right)\pi. \tag{198a}$$

Equations (198) and (198a) are known as Levinson's theorem. For a detailed discussion, consult Newton (1966).

We know

$$e^{2i\delta_0} - 1 = \frac{2i}{\cot\delta_0 - i} = \frac{2ka}{i - ka} \quad \text{(for } |kR| \ll 1) \tag{199}$$

where a is the scattering length. So the scattering amplitude has a pole at $k = \dfrac{i}{a}$ or at $K = -ik = -\dfrac{1}{a}$ (provided $R \ll |a|$). This corresponds to a bound state of energy $E_b = -\dfrac{\hbar^2 K^2}{2\mu} = -\dfrac{\hbar^2}{2\mu a^2}$ if K is positive and small. Since the bound state wave function $\sim e^{-Kr/a}$ at large distances, its extension is K^{-1} so that a is essentially the size of the bound state. A pole at a negative value of K corresponds to a solution of the Schrödinger equation that grows exponentially at large distances, and so is not a physical state. Such a pole is on the second Riemann sheet of $e^{2i\delta_0(E)}$. So the scattering length a is positive and large if there is an s-wave bound state near zero energy. The low energy cross section is given by

$$\sigma_{tot} = \frac{4\pi}{k^2}\frac{1}{\cot^2\delta_0 + 1} = \frac{\dfrac{2\pi\hbar^2}{\mu}}{-E_b + E}. \tag{200}$$

This is a remarkable result: if there is an s-wave bound state near zero energy it completely determines the low-energy cross section.

If there is no bound state, then δ_0 increases at $k = 0$ and a is negative. At the potential strength for which $\delta_0 = \dfrac{\pi}{2}$, a is infinite. If there is a shallow bound state, then a is positive. Before the potential reaches the strength to introduce a second bound state, a goes through zero and changes sign. Near each bound state a becomes large and goes through infinity when a new bound state appears. When the potential is repulsive a is positive while the phase shift is negative.

These considerations have been applied to the case of the low-energy scattering of neutrons from a proton target. Experimentally it is found that the incoherent low-energy cross section $\sigma = \pi\,(|a_s|^2 + 3|a_t|^2) \approx 20$ barns, where a_s and a_t are the singlet and triplet np scattering lengths, is very large. A detailed analysis of the coherent NN scattering gives $a_s \approx 23.7\ fm$ and $a_t \approx -5.4\ fm$. These values are much larger than the size of the nucleons ($\sim 1\ fm$). But the np system has a shallow bound state – the deuteron is a spin 1 ($J = 1$), triplet ($S = 1$) and isoscalar singlet ($I = 0$) bound state of n and p with binding energy $E_B = -2.22$ MeV. This predicts a scattering length $a_t \sim -4.3$ *fm*. The large positive *np* scattering length $a_s \approx 24$ *fm* and also the *pp* scattering length ≈ 8 *fm* indicates almost bound ("virtual") states, consistent with the absence of any nucleon bound state except the deuteron. In contrast, the s-wave scattering lengths for pion-nucleon scattering, $a \approx \mp 0.4$ *fm* for $\pi^\pm p \rightarrow \pi^\pm p$ at low energy. These are much smaller than *NN* scattering lengths, indicating the absence of any real or virtual πN bound states. For the shallow bound state, the effective range $r_0 \approx 1.74$ *fm*, using $a_t \approx -5.4$ *fm*.

20.9 RESONANCE SCATTERING

Resonance phenomena constitute some of the most striking features of scattering. The scattering cross section exhibits interesting behavior as a function of energy. A striking feature is the appearance of a sharp peak superimposed on a back ground. Such resonances are frequently encountered in atomic and nuclear physics. Its characteristic is, not only the peak in the cross section but the fact that the phase shift increases with energy as it goes through $\dfrac{\pi}{2}$ (modulo π). Interactions among "elementary" particles have also shown resonances. Resonances are quasistationary states. For an excellent discussion of resonance phenomena, consult Bohm (1993).

There are two main types of experiments in which resonances are obtained: formation process and production process. A formation process is schematically shown in Fig. 20.6 and can be represented as

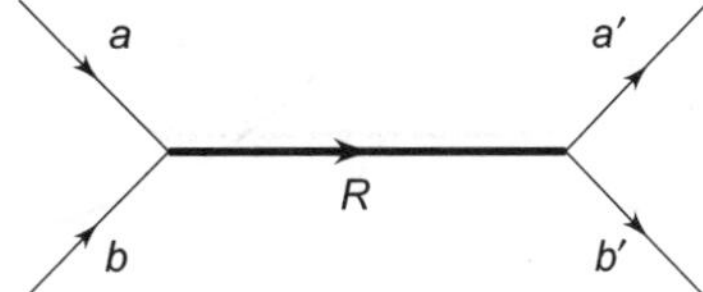

Fig. 20.6 Schematic diagram of a formation process

$$a + b \rightarrow R \rightarrow a' + b' \tag{201}$$

A production process, schematically depicted in Fig. 20.7 can be represented as

$$a + b \rightarrow c + R \rightarrow c + d + f. \tag{202}$$

As an example of the production process consider the helium energy-loss experiment in which a long-lived intermediate state He^* is produced. This subsequently decays either into a helium ion and an electron or into a helium atom and a photon:

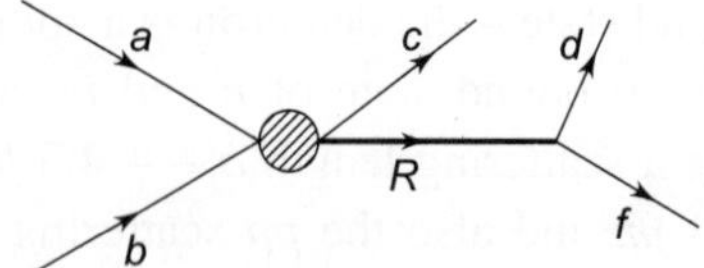

Fig. 20.7 Schematic diagram of a production process

$$e^- + \text{He} \rightarrow e^- + \text{He}^*$$
$$\rightarrow \text{He}^+ + e^-$$
$$\rightarrow \text{He} + \gamma \tag{203}$$

An example of the formation process is provided by the elastic scattering process

$$e^- + \text{He} \rightarrow \text{He}^- \rightarrow \text{He} + e^- \tag{204}$$

The scattering cross section changes sharply at an energy $E = E_R = 19.31\,\text{eV}$.

For p-wave $\pi^+ p$ scattering the cross section in the $\Delta(1232)$ resonance region is shown schematically in the Fig. 20.8

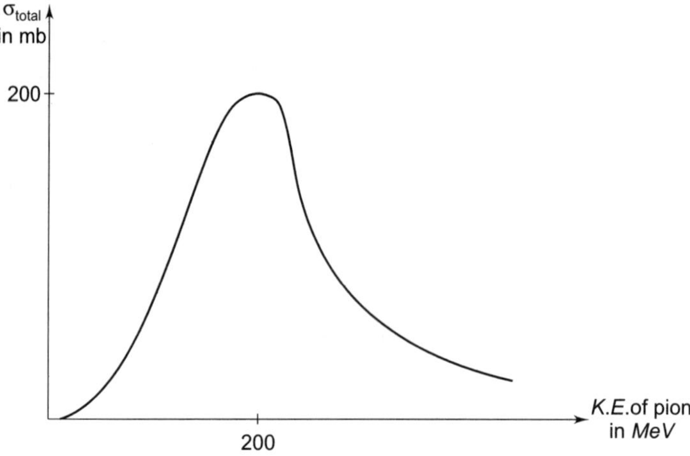

Fig. 20.8 First resonance $\Delta(1232)$ region of $\pi^+ p$ scattering. Total resonance cross section is plotted versus the kinetic energy of the pion

The resonance is approximately non-relativistic, since the Δ mass, $m_\Delta \approx 1232$ MeV, is only slightly greater than the sum of the constituent masses $m_\pi + m_N \approx 1080$ MeV. The resonant pion kinetic energy in the lab (nucleon at rest) frame is found from the invariant form

$$p_\Delta^2 = (p_\pi + p_N)^2 = m_\Delta^2 = m_\pi^2 + m_N^2 + 2m_N\,(m_\pi + K.E_\pi)$$

whence $\qquad K.E_{\pi} \approx 195$ MeV (Scadron (1991)).

(See Muirhead (1965) for strong-interaction resonancs, e.g., ρ-system, ω-system, etc.)

The contribution to the cross section from the lth partial wave is maximum when $\sin^2\delta_l = 1$, i.e. when $\cot \delta_l$ passes through zero from above. Then $\sigma_l \sim k^{-2}$ and is quite large at low energy. Such a bump is a manifestation of a scattering resonance. Then the numerator of Eq. (44) must vanish at the resonance energy $E = E_r$. The logarithmic derivative of the lth partial wave function in the neighborhood of $E = E_r$ may be written as $\alpha_l \approx c - b(E - E_r)$, where $c = kn'_l(kR)/n_l(kR)$. This approximation is valid only if $n_l(kR) \neq 0$. Since α_l is a decreasing function of energy we have $b > 0$. In the neighborhood of $E = E_r$, we have from Eq. (44),

$$\cot \delta \approx \frac{n_l(kR)b(E - E_r)}{kj'_l(kR) - \alpha_l \, j_l(kR)}$$

$$= \frac{n_l b(E - E_r)}{kj'_l - cj_l}$$

$$= \frac{n_l^2 b(E - E_r)}{k(j'_l \, n_l - n'_l j_l)}$$

$$= \frac{E_r - E}{\dfrac{1}{2}\,\Gamma}, \tag{205}$$

using $\qquad W[j_l(x), n_l(x)] \equiv j'_l(x)n_l(x) - j_l(x)n_l(x) = -x^2.$

Here $\Gamma = 2\,[kR^2 b(n_l(kR))^2]^{-1}$. From Eq. (204), we obtain

$$\sin \delta_l \, e^{i\delta_l} = \frac{\Gamma}{2(E_r - E) - i\Gamma}. \tag{206}$$

An expression for the scattering amplitude may be obtained from Eq. (30), using Eq. (206). Then near the resonance, the lth partial cross section is given by

$$\sigma_l^{res}(E) = \frac{\left(\dfrac{\pi}{k^2}\right)(2l + 1)\Gamma^2}{(E - E_r)^2 + \left(\dfrac{1}{2}\,\Gamma\right)^2}$$

$$= \frac{\sigma_l^{max}\,\dfrac{1}{4}\,\Gamma^2}{(E - E_r)^2 + \dfrac{1}{4}\,\Gamma^2} \tag{207}$$

where $\sigma_l^{max} = \dfrac{4\pi}{k^2}\,(2l + 1)$ is the maximum unitary limit at resonance, i.e.

$\sigma_l^{res}(E_r) = \sigma_l^{max}$. The resonance cross section has a maximum at E_r, the resonance energy. The resonance width Γ is the energy spread of the peak at half maximum:

$\sigma_l^{res}\left(E_r \pm \dfrac{1}{2}\Gamma\right) = \dfrac{1}{2}\,\sigma_l^{max}$. Also for $E_r \gg \dfrac{1}{2}\Gamma$, the bump becomes a delta-function spike.

Equation (207) is a specific example of the Breit-Wigner one-level resonance formula.

For s-waves, σ_0 does not have a maximum when $1 + R\alpha_0(E) = 0$, unless $\dfrac{\partial \alpha_0}{\partial k^2} > \dfrac{1}{\sqrt{2k}}$ at E_r. Even then the maximum of σ_0 is much less pronounced than it is at a higher angular momentum resonance.

A resonance occurs for $l \geq 1$, whenever $\cot \delta_l$ vanishes, i.e. whenever

$$\delta_l = \left(n + \frac{1}{2}\,\pi\right). \tag{208}$$

Near the resonance

$$\delta_l = \frac{\pi}{2} + \tan^{-1}\left(\frac{E - E_r}{\dfrac{\Gamma}{2}}\right). \tag{209}$$

Thus, as E passes through E_r, the phase shift increases suddenly from a value near zero (plus $n\pi$) to a value near π (plus $n\pi$). It jumps by π in a small energy interval $\sim \Gamma$. At E_r, the phase shift has the value $\pi/2$ and the slope of δ_l is $\dfrac{2}{\Gamma}$. Therefore, the sharper the rise, the sharper is the resonance.

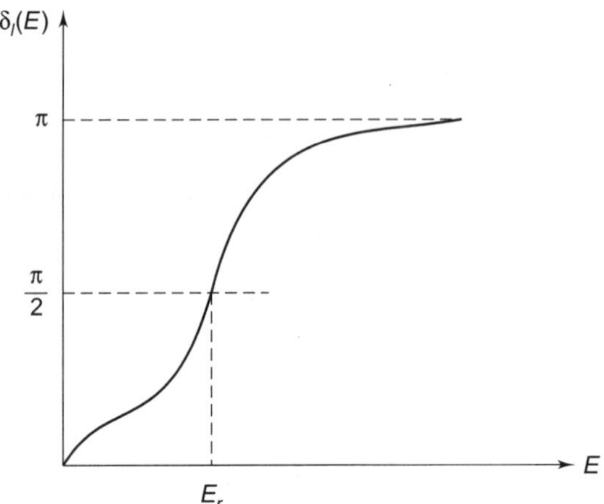

Fig. 20.9 The behavior of the phase shift near a resonance

To understand the physical nature of a resonance in scattering, we consider its behavior in time. The lifetime of a quasistationary state formed in a scattering state is roughly the time by which the projectile is delayed in the scattering region due to the interaction. We consider a time dependent state

$$\psi(\vec{r}, t) = \int a(\vec{k}) \, \psi_{\vec{k}}^{(+)}(\vec{r}) \, e^{-iEt/\hbar} \, d^3k, \tag{210}$$

where $\psi_{\vec{k}}^{(+)}(\vec{r})$ is a stationary scattering state. This state function can be split as usual into an incident wave and a scattered wave where the latter is given by

$$\psi_{sc}(\vec{r}, t)_{\text{larger } r} \int a(\vec{k}) f_k(\theta, \phi) \frac{e^{ikr}}{r} \, e^{-iEt/\hbar} \, d^3k. \tag{211}$$

We now assume that the lth partial wave is resonant so that all phase shifts are small except δ_l. Using Eq. (206) we get

$$\psi_{sc}(\vec{r}, t) \sim (2l + 1) P_l(\cos\theta) \int a(\vec{k}) \frac{e^{ikr}}{kr} \left[\frac{\dfrac{\Gamma}{2}}{(E_r - E) - i\dfrac{\Gamma}{2}} \right]$$

$$\times e^{-iEt/\hbar} \, d^3k. \tag{212}$$

$$\sim \frac{\mu}{\hbar^2} (2l + 1) P_l(\cos\theta) \frac{F(r, t)}{r}, \text{ writing } d^3k = k^2 \, d\Omega_k \, dk \tag{213}$$

$$= \frac{\mu}{\hbar^2} \, d\Omega_k \, k dE$$

where

$$F(r, t) = \int_0^\infty \alpha(E) \, \Gamma \, \frac{\exp\left(i(kr - Et/\hbar)\right)}{2(E_r - E) - i\Gamma} \, dE, \tag{214}$$

with

$$\alpha(E) = \int a(\vec{k}) \, d\Omega_k.$$

In the resonance condition, we replace $\alpha(E)$ by $\alpha(E_r)$ in the integral, and write $k \approx k_r + (E - E_r)/\hbar v_r$, where $E_r = \hbar^2 k_r^2/2\mu$ and $v_r = \hbar k_r/\mu$. Using the dimensionless variable $z = (E - E_r)/\Gamma$, we rewrite Eq. (213) as

$$F(r, t) = -\alpha(E_r) \, \Gamma \exp i \left(k_r r - \frac{Et}{\hbar} \right)$$

$$\times \int_{-E_r/\Gamma}^\infty \frac{\exp\left(-\dfrac{i\tau\Gamma z}{\hbar}\right)}{2z + i} \, dz, \tag{214a}$$

where $\tau = t - \dfrac{r}{v_r}$ is a retarded time. If $\Gamma \ll E_r$, the lower limit can be extended to $-\infty$. This integral can be evaluated for positive τ by closing the contour with an infinite semicircle in the lower half of the complex z-plane. From the residue of the pole at $z = -\dfrac{i}{2}$, we get the time dependence $e^{-\tau\Gamma/2\hbar}$. For negative τ the contour must be closed in the upper half plane, where there are no poles, and so the integral vanishes. Hence, the time dependence of the scattered wave for large r is

$$\overline{\Psi}_{sc}(\vec{r},\, t) \sim e^{-t\Gamma/2\hbar} \quad \text{for} \quad t > \frac{r}{v_r}$$

$$\overline{\Psi}_{sc}(\vec{r},\, t) = 0 \quad \text{for} \quad t < \frac{r}{v_r}. \tag{215}$$

It is zero before $t = \dfrac{r}{v_r}$, because it is the time needed for propagation from the scatterer to the detector. For times greater than this, the time-dependent probability $|\overline{\Psi}_{sc}|^2 \propto e^{-\Gamma t/2\hbar}$. The average lifetime of the resonant state is

$$\tau = \frac{\hbar}{\Gamma}. \tag{216}$$

Thus, the lifetime of a resonant state considered as an unstable system with exponential decay, is equal to the inverse width. The width Γ is the transition probability rate for the decay of the resonance into its constituents. By extending the energy variable E into the complex plane $E \to E_r - \dfrac{i\Gamma}{2}$, we see that the time dependent phase $\overline{\Psi}(t) \sim e^{-iEt/\hbar} \to e^{-iE_r t/\hbar}\, e^{-\frac{1}{2}\frac{\Gamma t}{\hbar}}$ so that the probability $\sim e^{-\Gamma t/\hbar}$. Hence, the resonant width Γ is the same as the Γ in Eq. (216).

The average time delay is

$$t^l_D(E) = 2\, \frac{d\delta_l(E)}{dE}, \tag{217}$$

and the resonant time delay is well approximated by

$$\frac{1}{2}\, t^l_D(E) = \frac{\dfrac{\Gamma}{2}}{(E_r - E)^2 + \dfrac{\Gamma^2}{4}} \tag{218}$$

The time delay for a monoenergetic beam with energy $E = E_r$ is

$$t^l_D(E)\big|_{E\,=\,E_r} = \frac{4\hbar}{\Gamma}. \tag{219}$$

The l th partial S-matrix element may be written as

$$S_l(E) = e^{2i\delta_l(E)}$$

$$= e^{2ir}\, \tan^{-1}\left[\frac{\left(\dfrac{\Gamma}{2}\right)}{(E_r - E)}\right] e^{2i\gamma_l} \tag{220}$$

where $r = 0,\, 1,\, 2\, \dots$ and γ_l is a constant. The phase shift in the neighborhood of a quasistationary state (or, slowly varying term) is written as

$$\delta_l(E) = \delta_l^{(R)}(E) + \gamma_l, \tag{221}$$

where $\delta_l^{(R)}(E)$ is a rapidly varying function ($\Gamma << 2E_r$)

$$\delta_l^{(R)}(E) = r \tan^{-1} \frac{\left(\dfrac{\Gamma}{2}\right)}{(E_r - E)}. \tag{222}$$

The value of $\delta_l^{(R)}$ changes by almost $r\pi$ when E passes through E_r. The quantity $\gamma_l(E) = \delta_l^{bg}(E)$ is called the background phase shift, or the potential part of the phase shift, while $\delta_l^{(R)}(E)$ is called the resonant part of the phase shift. The lth partial scattering cross section is

$$\sigma_l = \frac{4\pi}{k^2}(2l + 1)\sin^2\left(\delta_l^{bg} + \delta_l^{(R)}\right). \tag{223}$$

The parameter r is usually taken to be 1. Neglecting any background, we get the pure resonance case discussed earlier.

We now consider the scattering amplitude near resonance

$$f_l(k) = \frac{1}{k} \frac{1}{\cot \delta_l(k) - i}$$

$$\xrightarrow[E \to E_r]{} \frac{1}{k} \frac{\dfrac{\Gamma}{2}}{E_r - E - \dfrac{i\Gamma}{2}}. \tag{224}$$

Near resonance δ_l must increase i.e., $\cot \delta_l = \dfrac{E_r - E}{\dfrac{\Gamma}{2}}$. This corresponds to a counter clockwise circle in the complex f_l plane with a resonance occurring when f_l becomes pure imaginary at $E = E_r$. As the scattering becomes inelastic, the complex vector f_l must shrink in length off the "unitarity circle". Also,

$$S_l^{resonance}(E) = \frac{E - E_r - i\Gamma/2}{E - (E_r - i\Gamma/2)}, \tag{225}$$

so that $\qquad\qquad S_l(E_r) = -1.$

We note that a true bound state appears as a pole in the scattering amplitude at negative energy and a sharp resonance is related to a pole at the slightly complex energy $E = E_r - i\Gamma/2$ in the lower half-plane (on the second Riemann sheet). For angular momenta other than zero, an "almost" bound state causes a resonance, but for $l = 0$, it does not. If the potential $V(r)$ is an attractive well, then $V(r) + \hbar^2 l(l + 1)/\mu r^2$ is an effective potential of the form in Fig. 20.10. Resonances are the "bound" states of the well at positive energy, indicated by dotted lines in the Fig. 20.10. A particle in such a state is held by the centrifugal barrier produced by the angular momentum. Eventually, however, the particle tunnels through the barrier with a time constant $\sim \dfrac{\hbar}{\Gamma}$, and leaves.

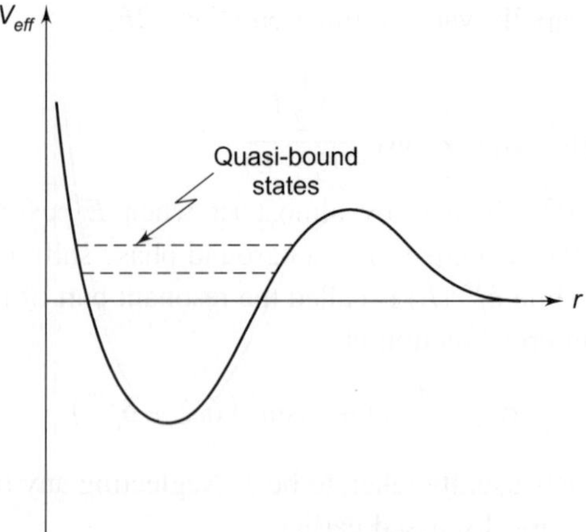

Fig. 20.10   The effective potential versus *r*

At resonance, the probability of the particle to "tarry awhile" in such a quasibound state becomes large giving the increase in cross section. As we deepen the potential well, the resonance energies move downward, where the centrifugal barrier is thicker. As a result, their lifetimes increase, and the widths decrease. Finally, at a specific potential depth a resonance may move to zero, and thereafter, to negative energy giving a true bound state where of course, no leakage occurs. The absence of the centrifugal barrier explains the non-occurrence of low-energy resonances for *s*-waves. If the potential contains its own barrier, it can cause low-energy resonances also in the *s*-wave.

20.10   VARIABLE PHASE METHOD

Calogero (1967) obtained a first-order nonlinear differential equation which on integration from the origin to the asymptotic region gives the value of the scattering phase shift. This approach is called the "variable phase method" or simply the "phase method". This method provides a useful tool for deriving qualitative and quantitative properties of the phase shifts and is very suitable for numerical computation. We follow closely the exposition given by Calogero.

We introduce the phase function $\delta_l(r)$ and the amplitude function $\alpha(r)$ through the equations

$$u_l(r) = \alpha_l(r) \, [\cos \delta_l(r) \tilde{j}_l(kr) - \sin \delta_l(r) \tilde{n}_l(kr)], \qquad (226)$$

$$u_l'(r) = k\alpha_l(r) \, [\cos \delta_l(r) \tilde{j}_l'(kr) - \sin \delta_l(r) \tilde{n}_l'(kr)], \qquad (226a)$$

where $u_l(r)$ is the radial wave function and the prime indicates differentiation with respect to the argument. Here we have introduced the notation $\tilde{f}(z) = zf(z)$, where $f(z)$ is any of the spherical Bessel, Neumann and Hankel functions. These functions $\tilde{f}(z)$ are called Riccati-Bessel, Riccati-Neumann and Riccati-Hankel functions, depending on the function $f(z)$. Since, $u_l(r)$ near the origin $\sim \tilde{j}_l(kr)$, the boundary conditions are

$$\delta_l(0) = 0, \tag{227}$$

$$\alpha_l(0) = 1. \tag{227a}$$

Differentiating Eq. (226) and comparing the result with Eq. (226a), we get

$$\alpha'_l(r) \, [\cos \delta_l(r)\tilde{j}_l(kr) - \sin \delta_l(r)\tilde{n}_l(kr)]$$

$$= \delta'_l(r)\alpha_l(r) \, [\sin \delta_l(r)\tilde{j}_l(kr) + \cos \delta_l(r)\tilde{n}_l(kr)] \tag{228}$$

Also, differentiating Eq. (227) and using the radial Schrödinger equation $u_l''(r) + [k^2 - l(l + 1)/r^2 - V(r)]u_l(r) = 0$ and the Riccati-Bessel equation, we find

$$V(r)u_l(r) = k\alpha'_l(r) \, [\cos \delta_l(r)\tilde{j}'(kr) - \sin \delta_l(r)\tilde{n}'(kr)]$$

$$- k\alpha_l(r)\delta_l(r) \, [\sin \delta_l(r)\tilde{j}'(kr) + \cos \delta_l(r)\tilde{n}'_l(kr)] \tag{229}$$

Multiplying both sides of Eq. (229) by
$$\frac{u_l(r)}{\alpha_l^2(r)} = \frac{\cos \delta_l(r)\tilde{j}'_l(kr) - \sin \delta_l(r)\tilde{n}'_l(kr)}{\alpha_l(r)}$$

and using Eq. (228) and the Wronskian relation

$$\tilde{j}_l(z)\tilde{n}'_l(z) - \tilde{j}'_l(z)\tilde{n}_l(z) = 1,$$

we obtain

$$\delta'_l(r) = - k^{-1} \, V(r) \left[\frac{u_l(r)}{\alpha_l(r)} \right]^2$$

i.e.,

$$\delta'_l(r) = - k^{-1} \, V(r) \, [\cos \delta_l(r)\tilde{j}_l(kr) - \sin \delta_l(r)\tilde{n}_l(kr)]^2. \tag{230}$$

The Eq. (230) is known as the "phase equation". The phase function $\delta_l(r)$ satisfies a first-order nonlinear differential equation of the Riccati type. Note that

$$\delta_l(r) \xrightarrow[r \to 0]{} 0$$

and
$$\lim_{r \to \infty} \delta_l(r) \equiv \delta_l(\infty) = \delta_l$$

(we assume $V(r) \xrightarrow[r \to 0]{} V_0 r^{-m}$, $m < 2$).

The phase function yields asymptotically directly the value of the scattering phase shift. Once $\delta_l(r)$ is known, $\alpha_l(r)$ may be obtained from the following equation

$$\alpha_l(r) = \exp \left\{ (2k)^{-1} \int_0^r dr' \, \tilde{D}_l^2(kr') \sin 2 \left[\tilde{\delta}_l(kr') + \delta_l(r') \right] \right\}, \tag{231}$$

where
$$\tilde{D}_l^2(z) = \tilde{j}_l^2(z) + \tilde{n}_l^2(z)$$

and

$$\tilde{\delta}_l(z) = -\tan^{-1}\left[\frac{\tilde{j}_l(z)}{\tilde{n}_l(z)}\right].$$

If both $\delta_l(r)$ and $\alpha_l(r)$ are determined, one can obtain $u_l(r)$ and its first derivative $u_l'(r)$ from Eqs. (226) and (226a). It is to be noted that $\alpha_l(r)$ is the modulus of the Jost function produced by the potential truncated at r.

20.11　ANALYTIC PROPERTIES OF THE S-MATRIX

We turn now to a brief description of the analyticity properties of the S-matrix. We shall state the results with a few remarks. Detailed discussions including derivations are given in the references.

We have

$$S_l(k) = e^{2i\delta_l(k)} = \frac{f_l(k)}{f_l(-k)}. \tag{174$'$}$$

The analytic properties of the S-matrix element are rather more complicated than those of the Jost functions. $S_l(k)$ generally has poles both at the poles of $f_l(k)$ and at the zeros of $f_l(-k)$. On the real axis $S_l(k) \to 1$ as $k \to \pm \infty$. From our discussion in Section 20.7 we see that the zeros of $f_l(k)$ have the following interpretation (for central potentials with certain conditions).

(i)　bound state: $\mathrm{Im}\ k > 0$, $\mathrm{Re}\ k = 0$

(ii)　zero energy bound state: $k = 0$, $l > 0$

　　　zero energy resonance: $k = 0$, $l = 0$

(iii)　virtual state: $\mathrm{Im}\ k < 0$, $\mathrm{Re}\ k = 0$

(iv)　resonance: $\mathrm{Im}\ k < 0$, $\mathrm{Re}\ k > 0$

$S_l(k)$ has a zero at $k = -iK$ and a pole at $k - iK$. Each such zero and pole are simple. With $k = \pm iK$, there is a bound state with energy $-\hbar^2 K^2/2\mu$. There are redundant zeros which do not correspond to bound states. This does not happen for finite-range potential. We can eliminate such zeros when evaluating the bound states by replacing the actual potential by a potential with a cut-off at a large distance.

The lth partial matrix element is usually considered as a function of the energy E, i.e., $S_l(E)$. Then the complex k-plane is mapped onto a two-sheeted Riemann surface with the branch cut from 0 to ∞. The upper half k-plane, $\mathrm{Im}\ k > 0$ corresponds to the first sheet; the first quadrant of the k-plane corresponds to the upper half of the first sheet. The first sheet is called the "physical" sheet, as the physically meaningful values of k in scattering states, k real, $k \geq 0$, correspond to the upper rim of the first sheet. If one continues through the cut one comes

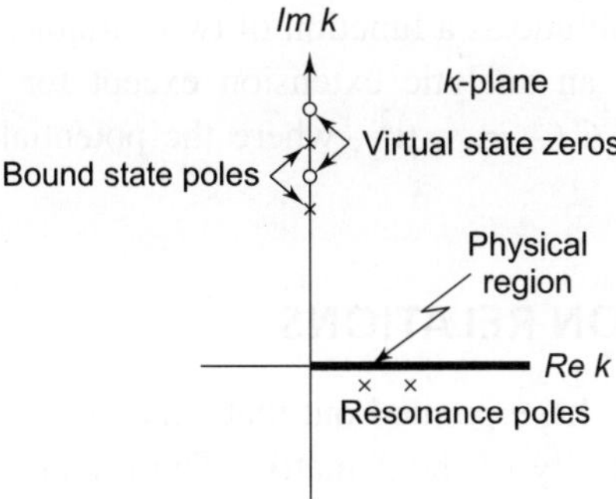

Fig. 20.11 The complex *k*-plane

to the second sheet ("unphysical" sheet) which corresponds to the lower half of the *k*-plane. $S_l(E)$ is a function on the two-sheeted Riemann surface. The bound-state poles of $S_l(k)$ on the positive imaginary *k*-axis correspond to poles on the negative real *E*-axis of the "physical" sheet.

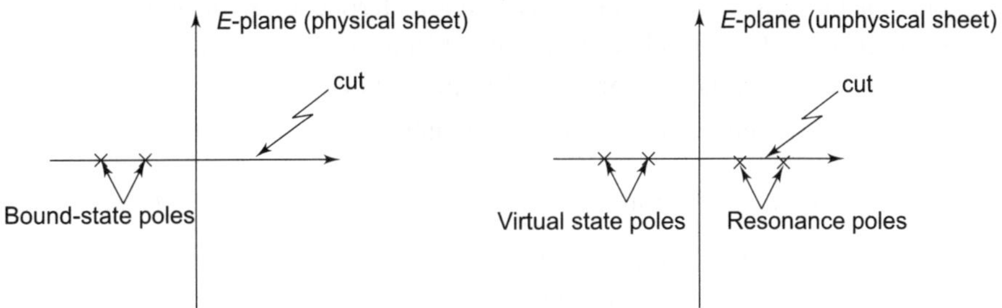

Fig. 20.12 The complex *E*-plane

$S_l(E)$ is a meromorphic function on the two-sheeted Riemann surface with a branch point at $E = 0$ and a cut from 0 to ∞. The physical values of E in scattering processes lie on the upper edge of the cut on the physical sheet. Bound-state poles lie on the negative real axis of the physical sheet, and except for these bound-state poles $S_l(E)$ is an analytic function on the physical sheet. Further poles (of any order) may lie on the second, "unphysical" sheet, coming from zeros on the first sheet. Poles on the negative real axis of the unphysical sheet, coming from zeros of $S_l(k)$ on the positive imaginary axis are called virtual-state poles. Poles of $S_l(E)$ on the unphysical sheet, close to the positive real axis, are called resonance poles or Siegert poles. A pole in the second sheet below the real axis describes a decaying state. Its symmetric counterpart above the real axis describes a forming (capture) state. The scattering amplitude is given by

$$f(k, \theta) = \sum_{l=0}^{\infty} (2l + 1) \, P_l(\cos \theta) f_l(k) \tag{39$'$}$$

The scattering amplitude as a function of two variables E and q^2 (momentum transfer/$\hbar$)^2, admits of an analytic extension except for "poles" to the region $E \in \mathcal{C} - [0, \infty]$, $q^2 \in \mathcal{C}\,(-\infty, -\mu^2)$, where the potential is Yukawa type with range μ^{-1}.

20.12   DISPERSION RELATIONS

In scattering theory, we have pointed out that general results can be obtain from the symmetry and unitarity of the S-matrix. The principle of microcausality is another general concept which leads to other general results. The original use of dispersion relations was by Kramers and Kronig in 1926/27. They established integral relations between the real and imaginary parts of the dielectric constant of a substance. The physical basis of the analytic behavior underlying the dispersion relations is the principle of microcausality: a signal in a medium cannot propagate at a speed exceeding the speed of light. Causality connects events occurring at different times giving the relationship between Fourier components of the electromagnetic field corresponding to different frequencies. In non-relativistic scattering theory the causality requirement has to be reformulated for the dispersion relations in the absence of a limiting velocity.

Let $f(z)$ be a function of the complex variable z and is analytic in the upper half-plane. We take an integral $\oint \dfrac{f(z)}{z - a}\, dz$ along the path shown in Fig. 20.13.

If a is in the upper half-plane, then by Cauchy's integral theorem the value of the integral is $2\pi i f(a)$. If $f(z)$ vanishes rapidly enough at infinity such that the contribution of the infinite semicircle (see Fig. 20.13) vanishes, we get

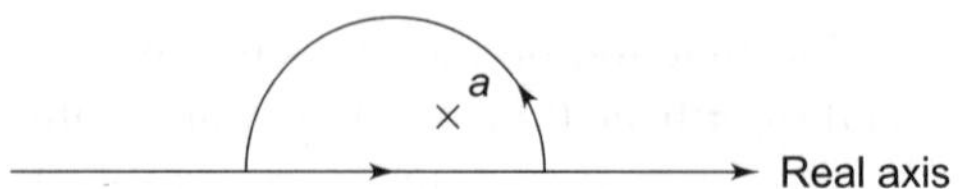

Fig. 20.13   The contour of integration

$$2\pi i f(a) = \int_{-\infty}^{\infty} \frac{f(x)}{x - a}\, dx.$$

Replacing a by z, we get

$$2\pi i f(a) = \int_{-\infty}^{\infty} \frac{f(x')}{x' - z}\, dx'.$$

Thus, we obtain by putting $\mathrm{Im}\, z = \epsilon$, infinitesimal,

$$\lim_{\epsilon \to 0} f(x + i\epsilon) = \frac{1}{2\pi i} \lim_{\epsilon \to 0} \int_{-\infty}^{\infty} \frac{f(x')dx'}{x' - x - i\epsilon}$$

$$= \frac{1}{2\pi i}\left[P \int_{-\infty}^{\infty} \frac{f(x')}{x' - x}\, dx' + i\pi f(x) \right]$$

or,
$$f(x) = \frac{1}{i\pi} P \int_{-\infty}^{\infty} \frac{f(x')}{x' - x} \, dx', \tag{232}$$

where P indicates that we take the Cauchy principal value at $x' = x$.

It is convenient to write the relations between the real (dispersive) and imaginary (absorptive) parts of $f(x)$. From Eq. (232) we find

$$\text{Re } f(x) = \frac{1}{\pi} P \int_{-\infty}^{\infty} dx' \, \frac{\text{Im } f(x')}{x' - x}, \tag{233}$$

$$\text{Im } f(x) = -\frac{1}{\pi} P \int_{-\infty}^{\infty} dx' \, \frac{\text{Re } f(x')}{x' - x}. \tag{233a}$$

The pair of relations, Eqs. (233) amd (233a) are known as Hilbert transforms. This discussion can be made rigorous by the use of the Titchmarsh theorem.

We suppose that an input signal $I(t_1)$ is sent on a system which then gives an output signal $O(t_2)$. Assuming that I and O are related linearly, we write

$$O(t_2) = \int_{-\infty}^{\infty} dt_1 \, g(t_1, t_2) I(t_1). \tag{234}$$

If these are causally connected, then

$$g(t_1, t_2) = 0 \quad \text{for} \quad t_2 - t_1 < 0. \tag{235}$$

Asuming that the internal properties of the system remain constant in time, we write

$$O(t_2) = \int_{-\infty}^{\infty} dt_1 \, g(t_1, t_2) I(t_1), \tag{236}$$

and
$$g(t_2 - t_1) = 0 \quad \text{for} \quad t_2 - t_1 < 0. \tag{237}$$

Putting $t = t_2 - t_1$, this becomes

$$g(t) = 0 \quad \text{for} \quad t < 0. \tag{238}$$

The relationship between causality and dispersion relations is brought out by the theorem of Titchmarsh which is stated below.

Let $g(t)$ be a bounded and square integrable function of the real variable t which vanishes identically for $t < 0$. If $f(\omega)$ is the Fourier transform.

$$f(\omega) = \int_0^{\infty} g(t) \, e^{i\omega t} \, dt$$

$$= \int_{-\infty}^{\infty} g(t) \, e^{i\omega t} \, dt \tag{239}$$

and $\quad \int_{-\infty}^{\infty} d\omega \, |f(\omega)|^2 < \infty,$

then $f(\omega)$ is the boundary value of a function $f(w + i\gamma)$, $\gamma \to 0$ [the boundary is $\gamma \to 0_+$], which is analytic in the upper half-plane. Furthermore, $\int_{-\infty}^{\infty} d\omega |f(w + i\gamma)|^2 < \infty$ (all $\gamma > 0$).

These are necessary and sufficient conditions for the Hilbert transform relation

$$f(\omega) = \frac{P}{i\pi} \int_{-\infty}^{\infty} d\omega' \, \frac{f(\omega')}{\omega' - \omega}, \tag{240}$$

or,
$$f(\omega) = \frac{1}{2\pi i} \int_{-\infty}^{\infty} d\omega' \, \frac{f(\omega')}{\omega' - \omega - i\epsilon} . \tag{241}$$

The real and the imaginary parts of the Eq. (240) are the pair of Eqs. (233) and (233a).

We now state the causality requirement in scattering theory. If an incident particle (or wave packet) arrives at a scatterer at a certain time, then the scattered particle should not appear at a distance r from the scatterer until at least a time $\frac{r}{c}$ has elapsed, where c is the speed of light.

Following Hagedorn (1961, 1966) the theorem can be reformulated in a manner suitable for dispersion relations. A function $f(\omega)$ with any one of the following properties

(i)   it satisfies Hilbert transforms,

(ii)   it has a Fourier transform which vanishes for $t < 0$,

(iii)   it is analytic in the upper half plane,

automatically has the other two properties.

We give only a part of the proof. We assume that $f(\omega)$ satisfies Eq. (240). Then taking the inverse Fourier transform

$$g(t) = \int_{-\infty}^{\infty} d\omega \, e^{-i\omega t} f(\omega)$$

$$= \int_{-\infty}^{\infty} d\omega \, e^{-i\omega t} \frac{1}{i\pi} P \int_{-\infty}^{\infty} d\omega' \, \frac{f(\omega')}{\omega' - \omega} . \tag{242}$$

Interchanging the order of the integrations, we get

$$g(t) = \int_{-\infty}^{\infty} d\omega' \, f(\omega') \left(-\frac{1}{i\pi} \right) P \int_{-\infty}^{\infty} d\omega \, \frac{e^{-i\omega t}}{\omega - \omega'} \tag{243}$$

But,

$$-\frac{1}{i\pi} P \int_{-\infty}^{\infty} d\omega \, \frac{e^{-i\omega t}}{\omega - \omega'} = \begin{cases} e^{-i\omega' t} & \text{for} \quad t > 0 \\ -e^{-i\omega' t} & \text{for} \quad t < 0 \end{cases} . \tag{244}$$

Hence,

$$g(t) = \int_{-\infty}^{\infty} d\omega' \, e^{-i\omega' t} f(\omega')$$

$$= g(t) \quad \text{for} \quad t > 0$$

$$g(t) = -\int_{-\infty}^{\infty} d\omega' \, e^{-i\omega' t} f(\omega')$$

$$= -g(t) \quad \text{for} \quad t < 0. \tag{245}$$

So we get the causality condition

$$g(t) = 0 \quad \text{for} \quad t < 0. \tag{246}$$

In dispersion relations, there are crossing relations connecting physical and unphysical regions. A function $f(w)$ can, in general assume the following values

$$f^*(w) = f(-w)$$

$$f^*(w) = -f(-w). \tag{247}$$

These are crossing relations, analogues of the so-called crossing symmetry. From Eq. (247), we find that

$$\text{Re } f(\omega) = \text{Re } f(-\omega); \ \text{Im } f(\omega) = -\text{Im } f(-\omega) \text{ if } f^*(\omega) = f(-\omega),$$

and

$$\text{Re } f(\omega) = -\text{Re } f(-\omega); \ \text{Im } f(\omega) = \text{Im } f(-\omega) \text{ if } f^*(\omega) = -f(-\omega), \tag{248}$$

Therefore,

$$\text{Re } f(\omega) = \frac{P}{\pi} \int_{-\infty}^{\infty} d\omega' \ \frac{\text{Im } f(\omega')}{\omega' - \omega}$$

$$= \frac{2}{\pi} P \int_{0}^{\infty} d\omega' \ \omega' \ \frac{\text{Im } f(\omega')}{\omega'^2 - \omega^2} \text{ for } f^*(\omega) = f(-\omega)$$

$$\text{Re } f(\omega) = -\frac{2}{\pi} P \int_{0}^{\infty} d\omega' \ \omega \ \frac{\text{Im } f(\omega')}{\omega'^2 - \omega^2} \text{ for } f^*(\omega) = -f(-\omega)$$

It is to be noted that the second relation in Eq. (248) is more convergent than the first one. If $f(\omega)$ does not satisfy the required convergence behavior, then greater convergence may be achieved by using a subtraction technique.

The function $f_l(R) - 1$ is analytic for $\text{Im } k \leq 0$, provided that the potential $V(r)$ satisfies $\int_{0}^{\infty} r|V(r)| \ dr < \infty$ and $\int_{0}^{\infty} r^2|V(r)|dr < \infty$. The dispersion relations for the Jost function $f_l(k)$ are written as

$$\text{Re } [f_l(k) - 1] = \frac{1}{\pi} P \int_{-\infty}^{\infty} dk' \ \frac{\text{Im } [f_l(k') - 1]}{k' - k},$$

$$\text{Im } [f_l(k) - 1] = \frac{1}{\pi} P \int_{-\infty}^{\infty} dk' \ \frac{\text{Re } [f_l(k') - 1]}{k' - k}. \tag{250}$$

One may obtain a dispersion relation for the forward scattering amplitude.

20.13   COULOMB SCATTERING

The most familiar problem of scattering by the Coulomb potential is outside the bounds of our discussion so far. The difficulty arises from the infinite range of the Coulomb potential. It does not belong to the class of well-behaved potentials discussed previously.

We consider the scattering of a particle of charge Z_1e by a particle of charge Z_2e. With $V_c(r) = Z_1Z_2e^2/r$ the Schrödinger equation is

$$-\frac{\hbar^2}{2\mu} \vec{\nabla}^2 \psi(\vec{r}) + \frac{Z_1Z_2e^2}{r} \psi(r) = E\psi(\vec{r}). \tag{251}$$

The equation for the Coulomb potential may be separated in parabolic and in prolate spheroidal coordinates as well as in spherical coordinates. We introduce the parabolic coordinates

$$\xi = r - z = r(1 - \cos\theta)$$

$$\zeta = r + z = r(1 + \cos\theta)$$

and the azimuthal angle ϕ about the z-axis. We try a solution of the form

$$\psi = e^{ikz} f(\xi). \tag{252}$$

The Schrödinger equation reads

$$\left[-\frac{4}{\xi + \zeta} \left(\frac{\partial}{\partial\xi} \xi \frac{\partial}{\partial\xi} + \frac{\partial}{\partial\zeta} \zeta \frac{\partial}{\partial\zeta} \right) - \frac{1}{\xi\zeta} \frac{\partial^2}{\partial\phi^2} + \frac{nk}{\xi + \zeta} \right] \psi = k^2\psi \tag{253}$$

where
$$n = \frac{Z_1Z_2e^2}{\hbar^2 k} = \frac{Z_1Z_2e^2}{\hbar v},$$

with v the relative velocity. Putting Eq. (252) in Eq. (253) we obtain the confluent hypergeometric equation

$$\xi f'' + (1 - ik\xi)f' - nkf = 0, \tag{254}$$

with the regular solution

$$f(\xi) = C_1 F_1(-in, 1, ik\xi). \tag{255}$$

$_1F_1(a, b, z)$ is the confluent hypergeometric function

$$_1F_1(a, b; z) = \sum_{s=0}^{\infty} \frac{\Gamma(a + s)\Gamma(b)z^s}{\Gamma(b + s)\Gamma(a)s!}. \tag{256}$$

We set

$$F(a, b; z) = W_1(a, b; z) + W_2(a, b; z). \tag{257}$$

The asymptotic expansions of W_1 and W_2 may be written as

$$W_1 \sim \frac{\Gamma(b)}{\Gamma(b - a)} (-z)^{-a} G(a, a - b + 1; -z)$$

$$W_2 \sim \frac{\Gamma(b)}{\Gamma(a)} e^z z^{a-b} G(1 - a, b - a; z) \tag{258}$$

where G denotes the semi-convergent series

$$G(a, b; z) = 1 + \frac{ab}{z.1!} + \frac{a(a + 1)b(b + 1)}{z^2.2!} + \dots$$

From Eq. (257) we obtain the asymptotic expansion of $F(a, b; z)$ for $|z|$ large, a and b remaining constant.

The solution of the confluent hypergeometric equation that is irregular at the origin may be written as

$$i(W_1 - W_2).$$

The asymptotic form of the Coulomb wave function is

$$\psi \xrightarrow[r \to \infty]{} \frac{Ce^{\frac{1}{2}n\pi}}{\Gamma(1 + in)} \left\{ e^{i[kz + n \ln k(r - z)]} \left[1 - \frac{n^2}{ik(r - z)} + \dots \right] \right.$$

$$\left. + f_c(\theta) r^{-1}\, e^{i(kr - n \ln 2kr)} \left[1 - \frac{(1 + in)^2}{ik(r - z)} + \dots \right] \right\}, \tag{259}$$

where the Coulomb scattering amplitude

$$f_c(\theta) = -\frac{n}{2k\, \sin^2 \frac{1}{2}\theta}\, e^{-in \ln \sin^2 \frac{1}{2}\theta + 2i\eta_0}, \tag{260}$$

$$\eta_0 = \arg \Gamma(1 + in). \tag{261}$$

Both the incident "plane" wave and the "scattered" wave are modified at infinite distances by logarithmic phase factors. This is a consequence of the infinite range of the Coulomb field. For the leading terms in both the square brackets in Eq. (259) we must have

$$\left| \frac{n^2}{k(r - z)} \right| \ll 1. \tag{262}$$

This implies that r is large but also that we are not too close to the forward direction. Thus, if we avoid the forward direction, then

$$\frac{d\sigma}{d\Omega} = |f_c(\theta)|^2 = \frac{n^2}{4k^2 \sin^4 \frac{1}{2}\theta}$$

$$= \left(\frac{Z_1 Z_2 e^2}{4E \sin^2 \frac{1}{2}\theta} \right)^2. \tag{263}$$

This is the Rutherford cross-section. (See Eq. (83)). The cross section is independent of whether the force is attractive or repulsive, although the phase of $f_c(\theta)$ depends on this.

The Coulomb wave function is normalized to unit incident flux by setting

$$C = e^{-\frac{1}{2}n\pi}\, \Gamma(1 + in)\, v^{-\frac{1}{2}}. \tag{264}$$

It is then given by

$$\psi_c(\vec{k}, \vec{r}) = v^{-\frac{1}{2}} e^{-\frac{1}{2}n\pi} \Gamma(1 + in) \exp\left(i\vec{k}, \vec{r}\right)$$

$$\times {}_1F_1\left(- in, 1; i\left(kr - \vec{k}.\vec{r}\right)\right). \tag{265}$$

$\therefore$ The particle density at $r = 0$ is given by

$$\left|\psi_c(\vec{k}, 0)\right|^2 = v^{-1} \frac{2n\pi}{e^{2n\pi} - 1}, \tag{266}$$

since

$$\Gamma(1 + x) = x\Gamma(x), \quad \Gamma(ix)\Gamma(1 - ix) = \frac{\pi}{\sin i\pi x}$$

so $$|\Gamma(1 + in)|^2 = |n\Gamma(in)\Gamma(1 - ix)| = \frac{2\pi|n|e^{n\pi}}{e^{2n\pi} - 1}.$$

For small velocities ($|n| \gg 1$),

$$|\psi_c(0)|^2 \simeq \frac{2\pi|n|}{v} \quad \text{attractive force } (n < 0).$$

$$|\psi_c(0)|^2 \simeq \frac{2\pi n}{v} e^{- 2n\pi} \quad \text{repulsive force } (n > 0) \tag{267}$$

The factor by which the particle density at $r = 0$ differs in the two cases, namely, $\exp\left(\dfrac{- 2\pi Z_1 Z_2 e^2}{\hbar v}\right)$, is called the Gamow factor.

Next we make a partial-wave analysis. We put

$$\psi_c(\vec{r}) = (kr)^{-1} \sum_{l = 0}^{\infty} (2l + 1)i^l u_l(r) P_l(\cos\theta) \tag{268}$$

in Eq. (251).

The radial wave function $u_l(r)$ satisfies the equation

$$\frac{d^2 u_l(r)}{dr^2} + \left[k^2 - \frac{2\mu}{\hbar^2} \frac{Z_1 Z_2 e^2}{r} - \frac{l(l + 1)}{r^2}\right] u_l(r) = 0, \tag{269}$$

or, using the dimensionless variable

$$\rho = kr,$$

$$\frac{d^2 u_l(\rho)}{d\rho^2} + \left[1 - \frac{2n}{\rho} - \frac{l(l + 1)}{\rho^2}\right] u_l(\rho) = 0. \tag{270}$$

On substitution

$$u_l(\rho) = \rho^{l + 1} e^{i\rho} v_l(\rho) \tag{271}$$

in Eq. (270), we obtain the confluent hypergeometric equation

$$\rho v_l''(\rho) + (2l + 2 + 2i\rho)v_l'(\rho) + 2i(l + 1 + in)v_l(\rho) = 0, \tag{272}$$

whose regular solution is

$$v_l(\rho) = C_l \, {}_1F_1(l + 1 + in, 2l + 2; - 2i\rho). \tag{273}$$

The asymptotic form of Eq. (273) can be found from Eq. (258). This gives the radial wave function at large distances

$$u_l(r) \xrightarrow[r \to \infty]{} C_l(2l + 1)! \; \frac{e^{\frac{1}{2} n\pi}}{|\Gamma(l + 1 + in)|2^l}$$

$$\times \sin\left(kr - \frac{1}{2}\, l\pi - n \ln 2kr + \eta_l\right), \tag{274}$$

where

$$\eta_l = \arg \Gamma(l + 1 + in), \tag{275}$$

is called the Coulomb phase shift. Again, there appears the ubiquitous logarithmic phase. Notice that an irregular solution may be defined.

The normalization constants C_l is determined so that the partial wave expansion Eq. (268) is identical with the solution Eq. (265) in parabolic coordinates. Then

$$C_l = v^{-\frac{1}{2}} e^{\frac{1}{2} n\pi} \frac{\Gamma(l + 1 + in)2^l}{(2l + 1)!}$$

so that

$$\psi_c = v^{-\frac{1}{2}} e^{-\frac{1}{2} n\pi} \sum_{l=0}^{\infty} \frac{\Gamma(l + 1 + in)}{(2l)!} (2ikr)^l \, e^{ikr}$$

$$\times {}_1F_1(l + 1 + in, 2l + 2; - 2ikr) \, P_l(\cos\theta). \tag{276}$$

The nearest analog of the Jost function is (Newton (1966))

$$\mathfrak{J}_{l\pm}^{(c)}(k) = \frac{(2k)^{-l} \, e^{\frac{1}{2} n\pi \pm i \frac{1}{2} l\pi} \, \Gamma(2l + 2)}{\Gamma(l + 1 \pm in)} \tag{277}$$

and the Coulomb S-matrix is given by

$$S_l^{(c)} \equiv e^{2i\eta_l} = \frac{\Gamma(l + 1 + in)}{\Gamma(l + 1 - in)}. \tag{278}$$

The poles of $S_l^{(c)}$ are the poles of the Γ function

$$\frac{1}{\Gamma(l + 1 + in)} = 0. \tag{279}$$

If $Z_1 Z_2 < 0$, the bound states are given by

$$k_m = i \frac{Z_1 Z_2 e^2 \mu}{m + l + 1}, \quad m = 0, 1, 2, \ldots \tag{280}$$

Due to the infinite range of the Coulomb potential, there are infinitely many bound states for each angular momentum. The ground state has the binding energy

$$E_0 = \frac{1}{2}\,(Z_1 Z_2 e^2)^2 \mu = \frac{1}{2}\,(Z_1 Z_2)^2\,\alpha^2\,c^2 \tag{281}$$

where $\alpha = \dfrac{e^2}{\hbar c}$ is the fine-structure constant.

In actual experiments the Coulomb potential does not extend to infinity. The Coulomb interaction is almost always "shielded" at large distances by the physical environment. Let us assume a "screened potential" $V_c(r)g_c(r)$, where

$$g_c(r) = 1 \quad \text{for} \quad r < R$$
$$= 0 \quad \text{for} \quad r \gg R, \tag{282}$$

where R is the "screening radius", or the distance at which the two charges are effectively shielded from each other. For some applications, such as experiments in nuclear and elementary particle physics the simple expression Eq. (282) is adequate. Then the relevant collision distances are $\sim 1f$ while $R \sim 1\text{Å}$, since screening is due to the orbital electrons. For other applications, such as the scattering of low-energy electrons by atoms or molecules the function $g_c^{(r)}$ has to be determined.

Next, we consider scattering by Coulomb plus a short-range force. Classic examples include the scattering of protons by protons and by atomic nuclei. Let $V(r)$ denote the potential associated with the short-range nuclear interaction. The radial wave equation is

$$\left\{ \frac{1}{r^2}\frac{d}{dr}\left(r^2\frac{d}{dr}\right) - \frac{l(l+1)}{r^2} + k^2 + \frac{2nk}{r} + U(r) \right\} \chi_l\,(k;\,r) = 0, \tag{283}$$

where $U(r) = \dfrac{2\mu}{\hbar^2}\,V(r)$. At large distances ($r \to \infty$), the Coulomb field dominates both $U(r)$ and $\dfrac{l(l+1)}{r^2}$. Thus,

$$\chi_l(r) \xrightarrow[r\to\infty]{} -\frac{1}{2ikr}\left[e^{-i\left(kr - \frac{1}{2}l\pi + \ln \ln 2kr\right)} \right.$$
$$\left. - e^{2i\Delta_l(k)}\,e^{i\left(kr - \frac{1}{2}l\pi + n\ln 2kr\right)} \right] \tag{284}$$

where Δ_l is real. When $V = 0$, $\Delta_l = \eta_l$. We thus write Δ_l in the form

$$\Delta_l = \eta_l + \delta_l',$$

where δ_l' is the additional phase shift due to V (commonly called the "nuclear phase shift"). It is to be noted that δ_l' is not identical with the phase shift which governs the purely nuclear scattering by the potential $V(r)$ in the absence of the Coulomb interaction. Equation (284) may be written as

$$\chi_l \sim -\frac{1}{2ikr}\left[e^{-i\left(kr - \frac{1}{2}l\pi + n\ln 2kr\right)} \right.$$

$$- e^{2i(\eta_l + \delta'_l)} \, e^{i\left(kr - \frac{1}{2} l\pi + \ln \ln 2kr\right)} \Big]. \tag{285}$$

The total wave function is

$$\overline{\Psi}_{\vec{k}}(\vec{r}) = \sum_{l=0}^{\infty} \left(\frac{2l+1}{2\pi^2}\right)^{\frac{1}{2}} i^l \, \chi_l(k;\, r) \, Y_{l0}(\theta)$$

$$\underset{r \to \infty}{\sim} \; \Psi_{\vec{k}}(\vec{r}) + \frac{1}{2ikr} \, e^{i(kr + n \ln 2kr)}$$

$$\times \sum_{l=0}^{\infty} \left(\frac{2l+1}{2\pi^2}\right)^{\frac{1}{2}} e^{2i\eta_l} \left(e^{2i\delta'_l} - 1\right) Y_{l0}(\theta) \tag{286}$$

Using Eq. (259) we obtain

$$\overline{\Psi}_{\vec{k}}(\vec{r}) \sim \frac{1}{\sqrt{2\pi}} \left[e^{i\left(\vec{k}.\vec{r} - n \ln (kr - \vec{k}.\vec{r})\right)} \right.$$

$$+ \frac{e^{i(kr + n \ln 2kr)}}{r} \, \{f_c(\theta) + f'(\theta)\}, \tag{287}$$

where

$$f'(\theta) = \frac{1}{k} \sum_{l=0}^{\infty} \sqrt{4\pi(2l+1)} \; e^{2i\eta_l} \, e^{i\delta'_l} \sin \delta'_l \, Y_{l0}^{(\theta)}. \tag{288}$$

The differential scattering cross section is

$$\frac{d\sigma}{d\Omega} = \left(\frac{d\sigma}{d\Omega}\right)_c + |f'(\theta)|^2 + 2 \text{ Re } |f_c^*(\theta)f'(\theta)|. \tag{289}$$

Equation (289) is applicable when the two interacting particles are not identical.

This analysis has been applied to proton-proton scttering, using the appropriate formulas for identical particle scattering. Since the proton has a spin 1/2 units, the total spin for two protons may be $S = 0$ or $S = 1$. At low energies (< 25 MeV), we can describe the scattering by spin-dependent central forces. So we indicate this by giving a spin index to $f'(\theta)$ and to δ'_l. At 10 MeV, $n = \frac{1}{20}$ and the Coulomb phase shifts are small and only nuclear s-wave scattering is important. So we may put $e^{2i\eta_0} \sim 1$ and $\delta'_0 \sim \delta_0$. The cross section is then

$$\frac{d\sigma}{d\Omega} = \left(\frac{d\sigma}{d\Omega}\right)_c + \lambdabar^2 \sin^2\delta_0 + \frac{n\lambdabar \sin \delta_0 \cos \left(\delta_0 - 2n \ln \sin \frac{\theta}{2}\right)}{k \sin^2 \frac{\theta}{2}}. \tag{290}$$

The interference term enables one to determine $\text{sgn}\delta_0$ which tells us the sign of the nuclear potential.

Dollard (1964) has given a rigorous treatment of Coulomb scattering. A modified "free evolution" operator $\exp\left[\dfrac{-iH_{0,\,c}(t)}{\hbar}\right]$ is defined where

$$H_{0,\,c}(t) = H_0 t + (\text{sgn } t) \frac{\mu e_1 e_2}{|\vec{p}|} \ln\left(\frac{4H_0|t|}{\hbar}\right) \tag{291}$$

The Møller operators are defined as

$$\Omega_c^{(\pm)} = \lim_{t \to \pm \infty} e^{\frac{iHt}{\hbar}}\, e^{\frac{-iH_{0,\,c}(t)}{\hbar}}, \quad H = H_0 + V \tag{292}$$

They are isometric and asymptotically complete. The S operator

$$S_c = \Omega_c^{(-)}\, \Omega_c^{(+)} \tag{293}$$

is unitary.

20.14 INELASTIC ELECTRON-ATOM SCATTERING

We consider the scattering of electrons by atoms. The incident electron may be scattered elastically when the internal state of the atom remains the same after the collision and the kinetic energy of the electron does not change (the atom is assumed infinitely heavy). In inelastic scattering, the kinetic energy of the final outgoing electron is less than that of the initial incident electron. The change in energy is absorbed by the Z-electron atom, leading to excitation or ionization. The Hamiltonian for the system is

$$H = H_a + H_p + H_{int}, \tag{294}$$

where H_a is the Hamiltonian of the atom, H_p the free-particle (electron) Hamiltonian and H_{int} is the interaction Hamiltonian describing the interaction of the incident electron with the electrons and the nucleus of the atom. We confine ourselves to high energy scattering where the Born approximation is valid.

The transition amplitude in the Born approximation from state m to state n is

$$T_{mn} = \langle \psi_n | H_{int} | \psi_m \rangle \tag{295}$$

where $|\psi\rangle$ is an eigenket of $H_a + H_p$. So we write the wave function

$$\psi = e^{i\vec{k}\cdot\vec{r}}\, \phi(A) \tag{296}$$

where $\vec{k}$ is the wave vector of the incident electron and $\phi(A)$ is an atomic eigenstate, A representing the Z spatial coordinates of the atomic electrons (we neglect electron spin). We write

$$\psi_m = e^{i\vec{k}_0\cdot\vec{r}}\, \phi_0$$
$$\psi_n = e^{i\vec{k}_n\cdot\vec{r}}\, \phi_n \tag{297}$$

where ϕ_0 is the ground state of the atom and ϕ_n is the nth excited state of the atom after the collision. The initial and final momenta of the electron are $\hbar \vec{k}_0$ and $\hbar \vec{k}_n$ respectively so that the momentum transfer suffered by the incident electron is $\hbar \vec{q} = \hbar \left(\vec{k}_0 - \vec{k}_n \right)$. If the intial and final energies of the atom are E_0 and E_n respectively, then by energy conservation

$$\hbar^2 \left(\vec{k}_0'^2 - k_n^2 \right) = 2m \left(E_n - E_0 \right)$$

$$= - 2m \left(W_n - W_0 \right), \tag{298}$$

where W_n is the total energy (including the rest energy) of the incident particle. Thus the Born approximation to the transition amplitude is

$$T_{0n} = \int e^{i\vec{q}.\vec{r}} \, \phi_n^* (A) \, H_{int} \left(\vec{r}, A \right) \phi_0(A) \, d\tau dA. \tag{299}$$

The interaction Hamiltonian is

$$H_{int} = - \frac{Ze^2}{r} + \sum_{i=1}^{Z} \frac{e^2}{\left| \vec{r} - \vec{r}_i \right|}. \tag{300}$$

Here the first term represents the interaction of the incident electron with the nucleus, which is considered to be a point charged particle situated at the origin. The second term describes the interaction of the incident electron with the atomic electrons. $\vec{r}$ is the position vector of the incident electron.

The matrix element is

$$\int d^3 r e^{i\vec{q}.\vec{r}} \, \phi_n^* (A) \left[- \frac{Ze^2}{r} + \sum_{i=1}^{Z} \frac{e^2}{\left| \vec{r} - \vec{r}_i \right|} \right] \phi_0(A) dA.$$

To evaluate the first term we note that

$$\int \frac{\exp \left(i\vec{q}.\vec{X} \right) d^3 X}{X} = \lim_{\mu \to 0} \int d^3 X \, \frac{e^{i\vec{q}.\vec{X} - \mu X}}{X}$$

$$= \lim_{\mu \to 0} \frac{4\pi}{q} \int_0^\infty \sin qX \, e^{-\mu X} \, dX$$

$$= \frac{4\pi}{q^2}.$$

To evaluate the second term, we find the Fourier transform of $1/\left| \vec{x} - \vec{x}_i \right|$. We have

$$\sum_i \int \frac{d^3 x \, e^{i\vec{q}.\vec{x}}}{\left| \vec{x} - \vec{x}_i \right|} = \sum_i \int \frac{d^3 x \, e^{i\vec{q}.(\vec{x} + \vec{x}_i)}}{\left| \vec{x} \right|}, \quad \vec{x} \to \vec{x} + \vec{x}_i$$

$$= \sum_i e^{i\vec{q}.\vec{x}_i} \int \frac{e^{i\vec{q}.\vec{x}}}{x} \, d^3 x$$

$$= \frac{4\pi}{q^2} \sum_i e^{i\vec{q}.\vec{x}_i}.$$

Thus, Eq. (299) becomes

$$T_{0n} = \frac{4\pi Z e^2}{q^2} \left[F_n(\vec{q}) - \delta_{n0} \right], \qquad (301)$$

where δ_{n0} arises from the orthonormality of ϕ_n and ϕ_0. We define the electronic form factor

$$ZF_n(\vec{q}) = \int \phi_n^*(A) \sum_{j=1}^{Z} e^{i\vec{q}.\vec{r}_j} \phi_0(A) \, dA$$

$$= \left\langle n \left| \sum_{j=1}^{Z} e^{i\vec{q}.\vec{r}_j} \right| 0 \right\rangle. \qquad (302)$$

Note that as $q \to 0$,

$$\frac{1}{Z} \left\langle n \left| \sum_{j=1}^{Z} e^{i\vec{q}.\vec{r}_j} \right| 0 \right\rangle \to 1 \quad \text{for} \quad n = 0.$$

Hence the form factor tends to unity for elastic stattering as $q \to 0$. For inelastic scattering $(n \neq 0)$, $F_n(\vec{q}) \to 0$ as $q \to 0$ by orthogonality of $|n\rangle$ and $|0\rangle$.

In general the form factor arises in a situation where the incident particle is scattered by a composite system consisting of identical scatterers distributed in space. Measurement of the form factor in high energy scattering experiments gives information about the distribution of the scatterers.

The differential scattering cross section for the inelastic (or elastic) scattering of electrons by atoms is

$$\frac{d\sigma}{d\Omega}(0 \to n) = \left(\frac{k_n}{k_0} \right) \left| \frac{1}{4\pi} \frac{2m_e}{\hbar^2} \frac{4\pi Z e^2}{q^2} \left[F_n(\vec{q}) - \delta_{n0} \right] \right|^2$$

$$= \frac{4m_e^2}{\hbar^4} \left(\frac{Z e^2}{q^2} \right)^2 \left(\frac{k_n}{k_0} \right) \left| F_n(\vec{q}) - \delta_{n0} \right|^2. \qquad (303)$$

The total cross section is obtained by integrating the differential cross section

$$\sigma_{tot}(0 \to n) = \int d\Omega \, \frac{d\sigma}{d\Omega}(0 \to n).$$

The condition for Born approximation to be applicable is that the velocity of the incident electron should be large compared with those of the atomic electrons. The above treatment ignores the spin of the electron and also we have not taken into account the exchange effect arising from the identity of the colliding particles. For the treatment of the exchange effect and other inelastic collision processes, e.g., inelastic scattering of slow particles, inelastic collisions between heavy particles and atoms, the reader is referred to the literature.

At large values of q $\left(q \gtrsim \dfrac{1}{R_{nucleus}} \gtrsim 10^{12}\ \text{cm}^{-1}\right)$ the structure of the nucleus becomes important. Instead of a point nucleus with charge Ze we now consider the nucleus as an extended charge distribution. The Coulomb potential $-\dfrac{Ze}{r}$ is now replaced by

$$-\int \frac{Ze^2 N(\vec{r}')}{|\vec{r} - \vec{r}'|}\, d^3r',$$

where $N(\vec{r})$ the nuclear charge distribution is normalized so that $\int N(\vec{r})d^3r = 1$. The Fourier transform of this quantity can be evaluated as follows

$$Ze^2 \int d^3x \int \frac{d^3x'\ e^{i\vec{q}.\vec{x}}\ N(\vec{x})}{|\vec{x} - \vec{x}'|}$$

$$= Ze^2 \int dx'\ e^{i\vec{q}.\vec{x}'} N(\vec{x}') \int \frac{d^3x\, e^{i\vec{q}.\vec{x}}}{x},$$

by shifting $\qquad\qquad \vec{x} \to \vec{x} + \vec{x}'$

$$= Ze^2 \frac{4\pi}{q^2} F_N(\vec{q}),$$

where

$$F_N(\vec{q}) = \int d^3r\ e^{i\vec{q}.\vec{r}}\ N(\vec{r}) \tag{304}$$

is the nuclear form factor. If $N(\vec{r})$ is spherically symmetric $N(\vec{r}) = N(r)$, for small q we can expand

$$F_N(\vec{q}) = \int d^3r\ e^{i\vec{q}.\vec{r}}\ N(r)$$

$$= \int d^3r \left(1 + i\vec{q}.\vec{r} - \frac{1}{2}(\vec{q}.\vec{r})^2 + \ldots\right) N(r)$$

$$= 1 - \frac{1}{6}\, q^2 R_N^2, \quad \left(q \ll R_N^{-1}\right), \tag{305}$$

where $R_N = \left[\langle r^2 \rangle_{Nucleus}\right]^{1/2}$ is the r.m.s. radius of the nucleus $\left(\sim Z^{1/3}\ fm\right)$. The second term $\vec{q}.\vec{r}$ vanishes because of spherical symmetry and the angular average of $\cos^2\theta$ (θ is the angle between $\vec{q}$ and $\vec{r}$) is 1/3.

Thus, the resulting differential cross section can be written as

$$\frac{d\sigma}{d\Omega} = \left(\frac{d\sigma}{d\Omega}\right)_{point} \times |F_N(q)|^2, \tag{306}$$

where $\left(\dfrac{d\sigma}{d\Omega}\right)_{point}$ is the differential cross section for the scattering of an electron by a point nucleus of charge $Z|e|$ (Rutherford formula). This provides a measure of the "size" of the extended object. The pioneering experimental work on the scattering of electrons from protons and deuterous was done at Stanford by Hofstadter and coworkers.

If we include the effect of the extended nucleus, the transition matrix element is

$$T_{0n} = \frac{4\pi Z e^2}{q^2} \left[F_N(\vec{q}) - \delta_{n0} F_N(q) \right]. \tag{307}$$

The quantity $F_N(q) - F_0(q)$ is called the elastic form factor of the atom. For small q, the form factor of the electronic charge distribution for elastic scattering may be written as

$$F_0(q) = 1 - \frac{1}{6} q^2 R_e^2, \quad \left(q \ll R_e^{-1} \right), \tag{308}$$

where $R_e^2 = \frac{1}{Z} \sum_{j=1}^{Z} \langle 0 | \vec{r}_j^2 | 0 \rangle$ is the mean square radius of the electronic charge

distribution $\left(\sim Z^{-1/2} a_0 \text{ where } a_0 \text{ is the Bohr radius} \right)$. Near the forward direction the amplitude for elastic scattering at small q values will be a constant.

$$T_{0n} \xrightarrow[q=0]{} \frac{4\pi Z e^2}{6} R_e^{2}.$$

When a charged particle passes through matter, it makes many collisions and loses energy by such collisions and eventually is brought to rest. The charged particle loses energy by exciting the atoms it passes. So the range and energy loss (per unit length) in matter is determined by the inelastic atom cross sections. An extremely relativistic electron may lose energy by bremsstrahlung. A fast electron can produce a sharp track in a bubble or cloud chamber or a photographic emulsion. A large number of collisions involve small momenta transfers. So all the inelastic cross sections are sharply peaked toward the forward direction at high energy, and the mean free path between large angle collisions is long. So we are interested in the energy loss per unit length due to all the inelastic processes.

We define the stopping power as the energy loss of the incident charged particle per unit length traversed in the material:

$$-\frac{dE}{dx} = N \sum_n (E_n - E_0)\, \sigma_n$$

$$= N \sum_n (E_n - E_0) \int \frac{d\sigma}{dq} (0 \to n)\, dq \tag{309}$$

where N is the number of atoms per unit volume, E_n the energy of the bound atomic state $|n\rangle$ and σ_n is the total cross section for exciting this state by collisions. Here we have used, $\dfrac{d\sigma}{dq} = \dfrac{2\pi q}{k_n k_0} \dfrac{d\sigma}{d\Omega}$, using $q^2 = \left| \vec{k}_n - \vec{k}_0 \right|^2 = k_n^2 - 2 k_n k_0 \cos\theta + k_0^2$.

From Eq. (303), we may write Eq. (308) as

$$-\frac{dE}{dx} = N \sum_n (E_n - E_0) \frac{4Z^2}{a_0^2} \int_{q_{min}}^{q_{max}} \frac{k_n}{k_0} \frac{1}{q^4} \frac{2\pi q}{k_n k_0}$$

$$\times \left| F_n(\vec{q}) \right|^2 dq.$$

$$= \frac{8\pi N}{k_0^2 a_0^2} \sum_n (E_n - E_0) \int_{q_{min}}^{q_{max}} \frac{1}{q^3}$$

$$\times \left| \left\langle n \left| \sum_{j=1}^{Z} e^{i\vec{q}\cdot\vec{r}_j} \right| 0 \right\rangle \right|^2 dq. \tag{310}$$

Bethe, Bloch and others have obtained the following expression.

$$-\frac{dE}{dx} = \frac{4\pi N Z e^4}{m_e v^2} \ln\left(\frac{2m_e v^2}{I}\right) \tag{311}$$

where v is the incident velocity and I is an average excitation energy. The stopping power is expressed in terms of one free parameter I, which depends on the atom. If the charged particle carries charge $\pm ze$, then we put $z^2 Z e^4$ for $Z e^4$.

The classical theory of stopping power was first given by Bohr in 1915. His result is essentially Eq. (311) with a slightly different parameter I.

20.15 REGGE POLES

We discuss the Regge pole theory for the scattering of spinless particles. Regge (1959) studied the analytic properties of the partial wave scattering amplitude as a function of energy and angular momentum l as a continuous, complex variable. Although Regge worked in the non-relativistic potential theory, his ideas were applied to high-energy particle physics by Chew, Frautschi, and Mandelstam (1962). Regge theory has become an extremely useful model in particle physics (Perkins (1982)).

We consider the scattering of two spinless particles producing two other spinless particles

$$a + b \rightarrow c + d. \tag{312}$$

The scattering amplitude depends, in the center of mass (c.m.) frame, on two scalar quantities the energy E and the scattering angle θ only. The particles have four momenta p_a, p_b, p_c, and p_d respectively (Fig. 20.14). The four momentum $p_a = \left(\vec{p}_a, iE_a\right)$ and $p_i^2 = -m_i^2$ where m_i. $(i = a, b, c, d)$ denotes the (rest) mass of the particle i. Now, we can form only two independent Lorentz invariants from these four-vectors. The scattering amplitude, being an invariant function, must be a function of the invariants. Instead of using the energy and the scattering angle

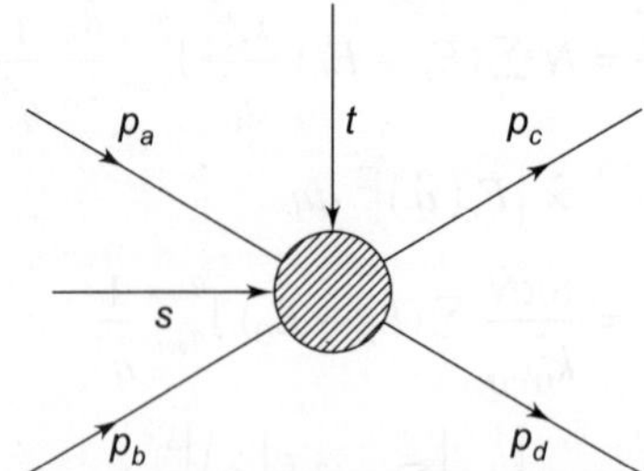

Fig. 20.14   The Mandelstam invariants for a two-body scattering process

in the c.m. frame, we shall use energy and four momentum transfer squared, which is a Lorentz invariant quantity.

We introduce the Mandelstam variables

$$s = -(p_a + p_b)^2 = -(p_c + p_d)^2 = E^2, \tag{313}$$

$$t = -(p_a - p_c)^2 = -(p_b - p_d)^2 = -q^2, \tag{313a}$$

$$u = -(p_a - p_d)^2 = -(p_b - p_c)^2 \tag{313b}$$

with

$$s + t + u = m_a^2 + m_b^2 + m_c^2 + m_d^2 = \sum_i m_i^2. \tag{313c}$$

Only two of the variables in the triplet are independent. In the c.m. system, a and b have equal and opposite three momentum, so that $s = E^2$, the square of the total c.m. energy (Fig. 20.15) t denotes the square of the four-momentum transfer between a and c or b and d. u is the square of the crossed four-momentum transfer. For an elastic collision, there is no transfer of energy, and $|\vec{k}| = |\vec{k}'|$,

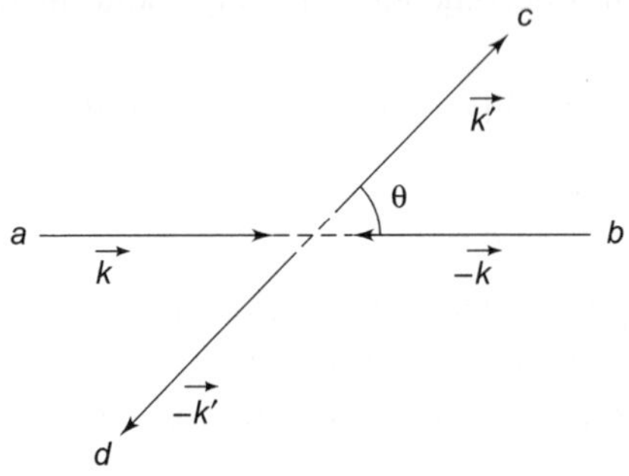

Fig. 20.15   A collision in the CM frame

$$t = -\left(\vec{k} - \vec{k}'\right)^2 = -2k^2 \left(1 - \cos\theta\right), \tag{314}$$

where θ is the scattering angle in the c.m. system.

The Schrödinger equation for the radial function $u = rR(r)$ in a central field $V(r)$ is

$$u_l'' + \left(k^2 - \frac{2m}{\hbar^2}V(r) - \frac{l(l+1)}{r^2}\right)u_l = 0. \tag{315}$$

For small values of r, the general form of the solution is

$$u_l \approx C_1 r^{l+1} + C_2 r^{-l}, \quad \text{for} \quad r \to 0. \tag{316}$$

The increasing solution as $r \to 0$ is eliminated by putting $C_2 = 0$. So

$$\frac{u_l}{r}\xrightarrow{} 0 \sim r^{l+1}. \tag{317}$$

In order for the second solution to be eliminated and distinguished from the first solution, the term in r^{-l} must exceed that in r^{l+1} as $r \to 0$. For complex values of l, we have the condition

$$\text{Re}\,(l+1) > \text{Re}\,(-l)$$

or,
$$\text{Re}\left(l + \frac{1}{2}\right) > 0. \tag{318}$$

Thus, the radial part of the Schrödinger equation – a second-order differential equation – has the asymptotic behavior

$$\frac{u_l}{r}\xrightarrow{} 0 \sim r^{l+1} \quad \left(l \geq -\frac{1}{2}\right)$$

$$\frac{u_l}{r}\xrightarrow{} \infty \sim e^{-ikr} + S(l, E)e^{ikr} \quad \text{(scattering)}$$

$$\frac{u_l}{r}\xrightarrow{} \infty \sim e^{kr} + S(l, E)e^{-kr} \quad \text{(bound-state)}$$

At a resonance or bound-state $S(l, E)$ has a dominant pole so that the terms e^{-ikr} and e^{kr} of the asymptotic boundary conditions may be ignored. Solutions of the entire Schrödinger equation for non-integral l do not correspond to physical states because the singular solution then has singularities in the unit sphere. However, the radial Schrödinger equation can be used to define solutions for other values of l. The poles of $S(l, E)$ in the complex angular momentum plane are called Regge poles.

We start from the partial-wave series which relates the full and partial scattering amplitudes. In the t channel

$$f(t, z_t) = \sum_{J=0}^{\infty} (2J + 1)\, f_J(t) P_J(z_t), \tag{319}$$

where $z_t = \cos\theta_t$. Using a method due to Sommerfeld and Watson, we transform the sum into an integral in the complex J plane along the contour C shown in Fig.: 20.16(a)

$$f(t, z_t) = -\frac{1}{2\pi i}\int_C dJ H[f(J, t)] \tag{320}$$

where

$$H[f(J, t)] = \frac{\pi(2J + 1)\, f(J,t)P_J(- z_t)}{\sin \pi J}.$$

Because $P_J(- z) = (- 1)^J P_J(z)$, J integer, Eq. (320) known as the Sommerfeld-Watson representation, is equivalent to Eq. (319) when both the sum and the series exist. The contour C surrounds the real J axis from 0 to ∞. C encircles all the integers. In this integral $f(J, t)$ is the partial amplitude analytically continued to complex $J.P_J(z)$ is the solution of the Legendre differential equation for complex J:

$$P_J(z) = \frac{1}{\pi} \int_0^\pi d\phi \left[z - \sqrt{z^2 - 1}\, \cos \phi \right]^J \tag{321}$$

$P_J(z)$ is entire in J and analytic in z except on a cut ($- \infty < z \le - 1$). $\sin \pi J$ has poles at all integers and the argument of P_J has been changed to $- z$ to compensate for the alternating sign of $\sin \pi J$.

If $f(J, t)$ has no poles on the real positive J axis, the poles of the integrand occur at the integral values of $J = 0, 1, 2, 3, \ldots$ If there are any poles of $f(J, t)$ inside C we modify C in such a way as to exclude these poles. Application of the theory of residues to the S-W representation gives the partial wave sum.

Regge investigated the analytic continuation of the scattering amplitude for a class of potentials. For $ReJ \ge -\dfrac{1}{2}$, it was found that

   (i) $f(J, t)$ is meromorphic (its only singularities are simple poles) in J and continuous on $ReJ = -\dfrac{1}{2}$.

   (ii) Below the threshold energy, the poles of $f(J, t)$ are on the real J axis.

   (iii) Above the threshold, the poles of the amplitude are in the upper half J plane.

   (iv) For $ReJ \ge -\dfrac{1}{2}$, the scattering amplitude has a finite number of poles. Consequently, there is always a pole furthest to the right of $ReJ = -\dfrac{1}{2}$.

   (v) The positions of all the poles vary with energy (t).

   (vi) As $|J| \to \infty$, the amplitude $f(J, t)$ tends to zero faster than $P_J/\sin \pi J$ diverges. Here, we note

$$\lim_{J \to \infty} P_J(\cos \theta) \sim \frac{1}{\sqrt{J}} \left(C_1 e^{iJ\theta} + C_2 e^{- iJ\theta} \right)$$

$$\lim_{J \to \infty} \sin \pi J \sim e^{\pi J}$$

The series Eq. (319) and the integral Eq. (320) are both convergent in $\cos \theta$ within the small Lehmann ellipse with foci at ± 1.

We now deform the contour of integration C into a closed loop C' in the complex J-plane (Fig. 20.16(b))

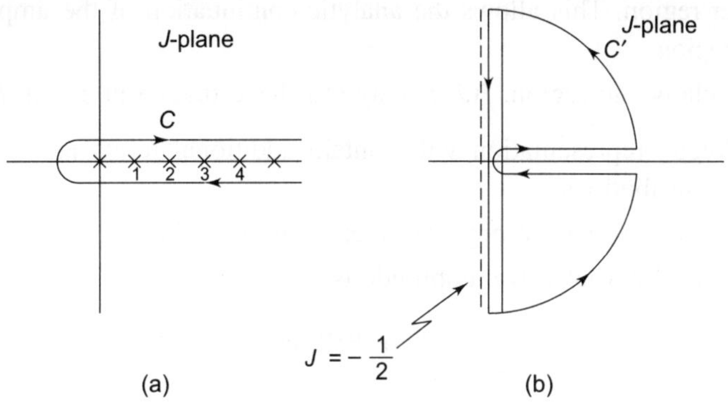

Fig. 20.16   The two contours used

From property (vi), the large semicircle does not contribute to the integral about C'. Hence,

$$-\frac{1}{2\pi i} \int_{C'} dJ H[f(J,\, t)] = -\sum_{i=1}^{N} ResH \, [f'(J = \alpha_i,\, t)]$$

$$= -\frac{1}{2\pi i} \left[\int_C dJ - \int_{-\frac{1}{2} - i\infty}^{-\frac{1}{2} + i\infty} dJ \right] H[f(J,\, t)]$$

$$= f(t,\, z_t) + \frac{1}{2\pi i} \int_{-\frac{1}{2} - i\infty}^{-\frac{1}{2} + i\infty} dJ \, H[f(J,\, t)]$$

and

$$f(t,\, z_t) = -\frac{1}{2\pi i} \int_{-\frac{1}{2} - i\infty}^{-\frac{1}{2} + i\infty} dJ \, H[f(J,\, t)]$$

$$- \sum_{i=1}^{N} Res \, H[f(J = \alpha_i,\, t)]. \tag{322}$$

The integral along the line $ReJ = -\frac{1}{2}$ is called the background integral (B.I.).

The final result is

$$f(t,\, z_t) = \text{B.I.} - \sum_{i=1}^{N} \frac{\pi(2\alpha_i + 1)}{\sin \pi\alpha_i} \, \beta_i(t) P_{\alpha_i(t)} (-z_t), \tag{323}$$

where $\alpha_i = \alpha_i(t)$ denotes a Regge pole in $ReJ > -\frac{1}{2}$. The locus of $\alpha(t)$ as t varies is called a Regge trajectory. The function $\beta_i(t)$ is the residue of $f(J,\, t)$ at the ith Regge pole:

$$\beta_i(t) = \lim_{J \to \alpha_i} (J - \alpha_i) f(J, t) \tag{324}$$

The passage from the integral Eq. (320) to that of is Eq. (323) called the Sommerfeld-Watson transform. Although the two expressions agree inside the small Lehmann ellipse, the Regge representation Eq. (323) is convergent in a much larger region. This allows the analytic continuation of the amplitude into the larger region.

In the relativistic region, $f(J, t)$ may also have branch points in $ReJ > -\dfrac{1}{2}$. Here the Regge representation will contain additional terms arising from the branch cut contributions.

Let us first consider the physical interpretation of the Regge poles.

The physical partial wave amplitude is

$$f_J(t) = \frac{1}{2} \int_{-1}^{+1} d(\cos\theta) P_J(\cos\theta) f(t, z_t) \tag{325}$$

using orthogonality.

The identity

$$\frac{1}{2} \int_{-1}^{1} dz P_J(z)\, P_\alpha(-z) = \frac{\sin \pi\alpha}{\pi(\alpha - J)(\alpha + J + 1)}, \tag{326}$$

yields for the pole contribution

$$f_J(t) = \frac{\beta(t)}{J - \alpha}\left(\frac{2\alpha + 1}{\alpha + J + 1}\right). \tag{327}$$

Near $\alpha = J$, a Taylor expansion gives

$$\alpha(t) = \left(J + (t - t_J) \frac{d}{dt} (Re\alpha(t))\Big|_{t = t_J} + \ldots\right) + iIm\alpha(t).$$

Then

$$f_J(E) \approx \frac{\beta(t)}{(t_J - t)\, Re\alpha' - iIm\alpha}$$

or,

$$f_J(E) \approx \frac{B(t)/Re\alpha'}{t_J - t - i\dfrac{Im\alpha}{Re\alpha'}} \quad \text{(Breit-Wigner form)}. \tag{328}$$

If the particles a and b possess a bound state it occurs at an energy below the threshold ($E_{thr} = m_a + m_b$). Below threshold, if $Im\alpha(E) = 0$, $f_J(E)$ has a bound state or a pole at $Re\alpha(E) = n$, an integer. Increasing the energy still further through E_{thr}, the trajectory acquires a positive imaginary part ($Im\alpha(E) > 0$). In this case, there is a resonance whenever the trajectory passes $Re\alpha(E) = 0 =$ an integer. The term $\dfrac{Im\alpha}{Re\alpha'}$ is identified with the product of the mass and width of the resonance, provided $Re\alpha' = \dfrac{d}{dt}[Re\alpha(t)]_{t=m_R^2} > 0$. A typical Regge trajectory (for two opposite Yukawa potentials) is indicated in Fig. 20.17 below.

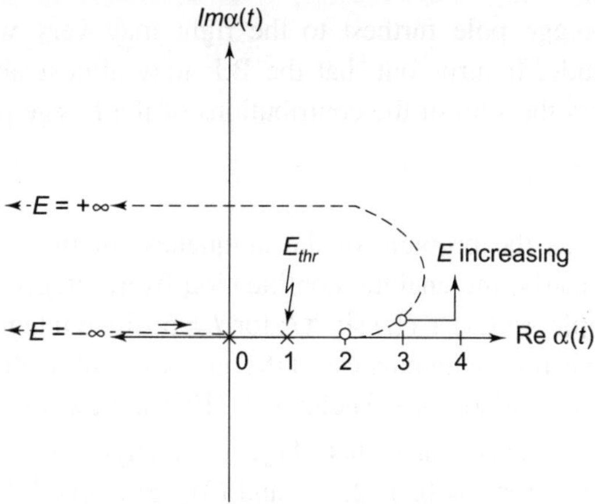

Fig. 20.17   Typical behavior of a Regge trajectory

Here, below E_{thr}, $Im\alpha = 0$, bound states at $J = 0, 1$.

above E_{thr}, $Im\alpha > 0$, $Re\alpha' > 0$, resonance at $J = 2, 3$.

We note that this trajectory turns over (has a maximum value of $Re\alpha$) and heads to $E \to +\infty$, and there are no further resonances as the potential is not strong enough to produce more resonant states. The number of bound states or resonances on a particular trajectory depends on the strength of the interaction potential.

The Regge representation is an exact functional expression for resonance scattering in the t channel. The crossing symmetry hypothesis states that this same amplitude $f(t, z_t)$ must describe a direct channel exchange in an entirely different region of the variables (high energy (s) and negative momentum transfer (t)). The Regge representation converts the infinite partial wave sum into a finite sum plus an integral.

We note the asymptotic limits

$$\lim_{z \to \infty} P_\alpha(z_t) = \frac{2^\alpha}{\sqrt{\pi}} \frac{\Gamma\left(\alpha + \dfrac{1}{2}\right)}{\Gamma(\alpha + 1)} z_t^\alpha \left[1 + O\left(\frac{1}{z^2}\right)\right]$$

$$+ \frac{2^{-a-1}}{\sqrt{\pi}} \frac{\Gamma\left(-\alpha - \dfrac{1}{2}\right)}{\Gamma(-\alpha)} z_t^{-\alpha-1} \left[1 + O\left(\frac{1}{z^2}\right)\right] \tag{329}$$

$$\lim_{z \to \infty} P_{-\frac{1}{2}}(z_t) \sim \frac{1}{\sqrt{z_t}} \sim \frac{1}{\sqrt{s}}. \tag{330}$$

Consequently, in the high energy limit, the Regge pole furthest to the right – with the largest real part – will completely overshadow the B.I.:

$$z_t^{Re\alpha_N(t)} \sim s^{Re\alpha_N(t)} >> \text{B.I.} \sim z_t^{-\frac{1}{2}} \sim s^{-\frac{1}{2}}. \tag{331}$$

Indeed the Regge pole furthest to the right may very well dominate the scattering amplitude. It turns out that the B.I. may almost always be ignored. The B.I. represents the sum of the contributions of the Regge poles to the left of

$$J = -\frac{1}{2}.$$

We now discuss the problem of the uniqueness of the definition of $f(J, t)$ for complex J. Clearly, the analytic continuation from integer values of J is not unique. For example, $[f(J, t) + g(t) \sin \pi J]$ for $J = $ positive integer, is also equal to $f_J(t)$. The conditions for the uniqueness of the analytic continuation are established in a theorem due to Carlson (see Titchmarsh (1939), Newton (1966)):

If a function $f(z)$ exists such that (1) $f(z)$ is analytic in the right half plane $Re z \geq 0$, (2) $f(z) = 0$ for $z = 0, 1, 2, \ldots$, and (3) $f(z)$ is $O(e^{\lambda|z|})$ there as $|z| \to \infty$, with $\lambda < \pi$, then $f(z)$ is identically zero.

As an example, let us consider the motion in an attractive Coulomb field. The Coulomb scattering amplitude is

$$f_l = \frac{e^{i\gamma \ln \sin^2 \frac{\theta}{2}}}{2lk \sin^2 \frac{\theta}{2}} \frac{\Gamma(1 + i\gamma)}{\Gamma(1 - i\gamma)}, \tag{332}$$

where $\gamma = -\dfrac{me^2}{\hbar^2 k}$. The partial wave amplitudes are

$$f_l = \frac{1}{2ik} \frac{\Gamma(l + 1 + i\gamma)}{\Gamma(l + 1 - i\gamma)} \tag{333}$$

$$= \frac{1}{2ik} e^{i\delta_l}. \tag{334}$$

In the complex l-plane f_l has poles whenever $(l + 1 + i\gamma)$ is a negative integer or zero. Regge poles occur whenever the above condition is satisfied, i.e., at

$$l = -p - i\gamma = -p + i\frac{me^2}{\hbar^2 k}, \quad p = 1, 2, 3, \ldots \tag{335}$$

Note that for $E < 0$, $k = i\sqrt{-E}$.

For an attractive Coulomb interaction, the Regge trajectory is along the positive real axis, since there are infinitely many bound states for each angular momentum due to the long range of the Coulomb field.

As discussed earlier, a short range potential supports only a finite number of bound states upto some maximum value of l, due to the centrifugal barrier. So, for a short range potential the Regge trajectory goes off the positive real axis at some finite value of Rel. In the repulsive case, there are only virtual states so the Regge trajectory never passes through a non-negative value of l.

PROBLEMS

20.1 To describe the nucleon–nucleon interaction, Yukawa introduced the potential

$$V(r) = \frac{V_0}{r} \exp\left(-\mu r\right)$$

The potential also serves as a model for screened Coulomb field of an atom.

Calculate in the first Born approximation the s-wave phase shift for scattering, the scattering amplitude, the differential and the total cross sections for scattering.

20.2 Find in the first Born approximation the scattering amplitude and the differential and the total cross sections for scattering for the following potentials:

(a) Exponential potential

$$V(r) = V_0 \exp\left(-\mu r\right)$$

(b) Gaussian potential

$$V(r) = V_0 \exp\left(-\mu^2 r^2\right)$$

(c) Square-well potential

$$V(r) = V_0, \quad r < r_0$$
$$= 0, \quad r > r_0$$

(d) Polarization potential

$$V(r) = \frac{V_0}{(r^2 + a^2)^2}$$

20.3 Investigate the scattering from a delta shell potential

$$V(r) = - g\, \frac{\hbar^2}{2m}\, \delta(r - a).$$

(a) Find the partial wave scattering amplitude and the phases $\delta_l(k)$.
(b) Find the cross section for s waves.
(c) Determine the poles of $e^{2i\delta_l} - 1$ on the negative real E-axis.

20.4 (a) Consider the scattering of particles by a spherical potential well

$$V(r) = -V_0, \quad r < r_0$$

$$= 0, \quad r > r_0$$

Find the Jost functions $F_l(k)$ and $S_l(k)$ for s-wave.

(b) Find the Jost function for $l = 0$ for scattering by an exponential potential $V(r) = -V_0 \exp(-r/a)$. Comment on the poles of S_0.

20.5 Prove that if the Jost function has a zero in the upper half plane, this zero is simple.

20.6 Particles are scattered by the potential $V(r) = C/r^2$. Find the phase shifts and the differential cross section.

Calculate the differential cross section in the Born approximation and compare with the above result.

20.7 (a) Compute the total Born cross section for the $1s$-$2s$ excitation of atomic hydrogen by fast positrons. Show that the cross section falls of as E^{-1} for large incident energies.

(b) Repeat the same for the $1s - 2p$ excitation and show that this cross section at high energies behaves like $E^{-1} \log E$.

CHAPTER 21

Scattering of Identical Particles and Polarized Particles

In this chapter, we discuss the effect of the identity of particles on scattering. Then we consider the scattering of polarized particles.

21.1 SCATTERING OF IDENTICAL PARTICLES

The scattering of two identical particles exhibits striking effects due to the quantum mechanical principle of indistinguishability. The effect depends on whether the particles have integral spin or whether their spins are half integrals. The differential cross section in classical theory of collisions of identical (and distinguishable) particles is given by

$$\left(\frac{d\sigma}{d\Omega}\right)_{cl} = |f(k, \theta)|^2 + |f(k, \pi - \theta)|^2 \tag{1}$$

in terms of scattering amplitudes (Note that in a classical theory we add probabilities).

(a) Boson-Boson Scattering

We consider the scattering of two identical spin 0 bosons viewed in the CM frame. The wave function has asymptotically the form (ignoring the symmetry)

$$\psi(\vec{r}) \underset{r \to \infty}{\sim} e^{i\vec{k} \cdot \vec{r}} + f(k, \theta) \frac{e^{ikr}}{r} \tag{2}$$

where $\vec{r} = \vec{r}_1 - \vec{r}_2$ is the relative coordinate of the two particles. For identical particles, the spatial wave function is symmetric (if total spin is even) under the interchange of the particles ($\vec{r} \to - \vec{r}$).

The correct asymptotic wave function is

$$\psi(\vec{r}) = e^{i\vec{k}\cdot\vec{r}} + e^{-i\vec{k}\cdot\vec{r}} + [f(k, \theta) + f(k, \pi - \theta)] \times \frac{e^{ikr}}{r}. \tag{3}$$

In the CM frame, the first two terms represent two equal incident plane waves, propagated in opposite directions, since we cannot distinguish the incident particle and the target particle. The coefficient of the outgoing spherical wave takes into account the scattering of both particles and is the scattering amplitude $f(k, \theta) + f(k, \pi - \theta)$. Under interchange of the two particles, $\vec{r} \to -\vec{r}$, we have $\theta \to \pi - \theta$ and $|\vec{r}| \to |\vec{r}|$. The two scattering events shown in the Figure 21.1 cannot be distinguished for identical particles. In both cases one particle emerges at angle θ and an identical one emerges at angle $\pi - \theta$. The two terms $f(k, \theta)$ and $f(k, \pi - \theta)$ are called the "direct" and "exchange"–amplitudes respectively. The total amplitude is the sum of the direct and exchange terms. (Note that in quantum mechanics we add amplitudes).

The differential scattering cross section is

$$\frac{d\sigma}{d\Omega} = |f(k, \theta) + f(k, \pi - \theta)|^2$$

$$= |f(k, \theta)|^2 + |f(k, \pi - \theta)|^2 + 2\,\mathrm{Re}\,f(k, \theta)\,f^*(k, \pi - \theta). \tag{4}$$

The last term in Eq. (4) is the interference term due to the exchange symmetry. For scattering at 90°, $f(\theta) = f(\pi - \theta) = f(\pi/2)$. For distinguishable particles, the differential cross section at 90° is $2|f(\pi/2)|^2$ whereas for indistinguishable particles, $\left(\frac{d\sigma}{d\Omega}\right)_{90°} = 4\left|f\left(\frac{\pi}{2}\right)\right|^2$. This enhancement has been observed in the elastic scattering of C^{12}-nuclei by C^{12}.

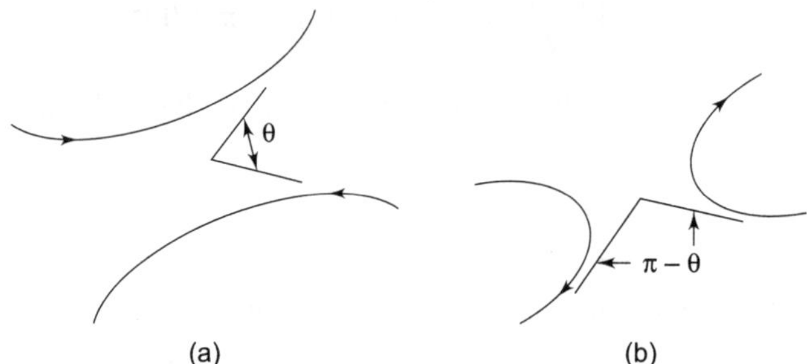

Fig. 21.1   Two indistinguishable processes in the scattering of identical particles

The partial wave decomposition of the scattering amplitude is

$$f(k, \theta) = \frac{1}{k} \sum_l (2l + 1)\, e^{i\delta_l} \sin \delta_l P_l(\cos \theta) \tag{5}$$

But, $\quad P_1 (\cos (\pi - \theta)) = P_l (- \cos \theta) = (-1)^l P_l (\cos \theta).$

Hence,

$$\frac{d\sigma}{d\Omega} = \frac{4}{k^2} \sum_{l=0,2,4,\ldots} (2l + 1) \, e^{i\delta_l} \sin \delta_l P_l (\cos \theta). \tag{6}$$

Therefore, partial waves of only even angular momenta are allowed in identical boson-boson scattering. Although we have considered elastic scattering, this applies also to reactions involving identical bosons in the incident state.

In Coulomb scattering of identical particles, the effect of the interference term can be seen as follows. Classically, the cross section is the sum of the direct and recoil cross sections,

$$\left(\frac{d\sigma}{d\Omega}\right) = \frac{d\sigma(\theta)}{d\Omega} + \frac{d\sigma \, (\pi - \theta)}{d\Omega}$$

$$= \left(\frac{Z^2 e^2}{4E}\right)^2 \left(\mathrm{cosec}^4 \frac{\theta}{2} + \sec^4 \frac{\theta}{2}\right). \tag{7}$$

Quantum mechanically,

$$\frac{d\sigma}{d\Omega} = |f_c (k, \theta) + f_c (k, \pi - \theta)|^2$$

$$= \left(\frac{Z^2 e^2}{4E}\right)^2 \left[\mathrm{cosec}^4 \frac{\theta}{2} + \sec^4 \frac{\theta}{2} + 8 \, \mathrm{cosec}^2\theta \times \cos\left(n \ln \tan^2 \frac{\theta}{2}\right)\right], \tag{8}$$

where $n = \dfrac{Z^2 e^2}{\hbar v}$. This formula Eq. (8) is referred to as the Mott scattering cross section. The interference term depends on $\hbar$, unlike the first two terms in Eq. (8).

(b) Fermion-Fermion Scattering

We now discuss the scattering of identical half-integral spin particles. We consider the scattering of two spin $\frac{1}{2}$ particles, e.g. electrons. Assuming that the interaction is spin-independent, we write the wave function as

$$\psi(1, 2) = e^{i\vec{p} \cdot (\vec{r}_1 + \vec{r}_2)} \, \psi(\vec{r}_1 - \vec{r}_2) \, \chi \, (s_1, s_2), \tag{9}$$

where χ is the spin part of the wave function. For the singlet state, the relative wave function $\psi(\vec{r})$ must be symmetric. So the scattering in the singlet channel is the same as in the case of two zero spin bosons. Thus

$$\left(\frac{d\sigma}{d\Omega}\right)_{\text{singlet}} = |f (k, \theta) + f (k, \pi - \theta)|^2. \tag{10}$$

For the triplet state, the total wave function is anti-symmetric only if the relative wave function $\psi(\vec{r})$ is antisymmetric, i.e., $\psi(-\vec{r}) = -\psi(\vec{r})$. In terms of partial waves, only odd l terms are allowed for triplet spin states. The asymptotic form of $\psi(\vec{r})$ is obtained by antisymmetrizing the wave function

$$\psi(\vec{r}) \underset{r \to \infty}{\sim} e^{i\vec{k}\cdot\vec{r}} - e^{-i\vec{k}\cdot\vec{r}} + [f(k,\,\theta) - f(k,\,\pi-\theta)] \times \frac{e^{ikr}}{r} \qquad (11)$$

so that the scattering amplitude is obtained in the triplet case by subtracting the exchange term from the direct amplitude. Clearly, at $\theta = \dfrac{\pi}{2}$ fermions in the triplet state cannot scatter. The differential scattering cross section is

$$\left(\frac{d\sigma}{d\Omega}\right)_{\text{triplet}} = |f(k,\,\theta) - f(k,\,\pi-\theta)|^2$$

$$= |f(k,\,\theta)|^2 + |f(k,\,\pi-\theta)|^2 - 2\,\text{Re}\,f^*(k,\,\theta)\,f(k,\,\pi-\theta). \qquad (12)$$

For unpolarized scattering, the observed cross section is an average of three parts triplet and one part singlet:

$$\left(\frac{d\sigma}{d\Omega}\right)_{\text{unpol}} = \frac{1}{4}\left[3\left(\frac{d\sigma}{d\Omega}\right)_{\text{triplet}} + \left(\frac{d\sigma}{d\Omega}\right)_{\text{singlet}}\right]$$

$$= |f(k,\,\theta)|^2 + |f(k,\,\pi-\theta)|^2 - \text{Re}\,f^*(k,\,\theta)\,f(k,\,\pi-\theta). \qquad (13)$$

In general, the differential cross section for scattering into the direction θ is

$$\frac{d\sigma}{d\Omega} = |f(k,\,\theta) \pm f(k,\,\pi-\theta)|^2, \qquad (14)$$

where the $\pm$ is taken according as the total spin S is even or odd. But in a scattering experiment, the total spin of the system is usually not specified. If the spin of either particle is s the total spin S assumes $(2s+1)$ different values ranging from 0 to $2s$. For each state of total spin S, there are $(2s+1)$ states (with different S_z). So the total number of states is

$$\sum_{S=0}^{2s} (2S+1) = (2s+1)^2.$$

S even

Total number of states

$$\sum_{S=0,2,\ldots 2s-1} (2S+1) = (2s+1)\,(s+1), \; s\text{-integral}$$

$$\sum_{S=0,2,\ldots,2s-1} (2S+1) = s\,(2s+1), \; s \text{ half-integral}$$

S odd

Total number of states

$$\sum_{S=1,3,\dots,2s} (2S + 1) = s\,(2s + 1), \quad s\text{-integral}$$

$$\sum_{S=1,3,\dots,2s} (2S + 1) = (2s + 1)\,(s + 1), \quad s \text{ half-integral.}$$

For boson-boson scattering with integral spin s of either particle,

$$\frac{d\sigma}{d\Omega} = \frac{s+1}{2s+1}\left| f(k, \theta) + f(k, \pi - \theta)\right|^2 + \frac{s}{2s+1}\left| f(k, \theta) - f(k, \pi - \theta)\right|^2,$$

since the fraction of states with even total spin S is $\dfrac{s+1}{2s+1}$ and with odd total

spin $|(S)|$ is $\dfrac{s}{2s+1}$

$$= |f(k, \theta)^2| + |f(k, \pi - \theta)|^2 + \frac{2}{2s+1}\,\operatorname{Re} f(k, \theta) \times f^*(k, \pi - \theta). \tag{15}$$

Note that for $s = 0$, $S = 0$, and

$$\frac{d\sigma}{d\Omega} = |f(k, \theta)|^2 + |f(k, \pi - \theta)|^2 + 2\operatorname{Re} f(k, \theta) \times f^*(k, \pi - \theta)$$

which is Eq. (4).

For fermion-fermion scattering with half-integral spin s of either particle,

$$\frac{d\sigma}{d\Omega} = \frac{s}{2s+1}\left| f(k, \theta) + f(k, \pi - \theta)\right|^2 + \frac{s+1}{2s+1}\left| f(k, \theta) - f(k, \pi - \theta)\right|^2,$$

since the fraction of states with even total spin (S) is $\dfrac{s}{2s+1}$ and with odd total

spin (S) is $\dfrac{s+1}{2s+1} = |f(k, \theta)|^2 + |f(k, \pi - \theta)|^2 - \dfrac{2}{2s+1}\operatorname{Re} f(k, \theta)\, f^*(k, \pi - \theta).$
$$\tag{16}$$

The final expressions refer to unpolarized scattering which can be considered as a statistical mixture of even and odd total spin states.

21.2 SCATTERING OF POLARIZED PARTICLES

We now consider in detail the scattering of particles with spin. For simplicity, we confine ourselves to elastic scattering. Let us consider a beam of particles that have just two states of polarization, e.g., a beam particles of spin $\dfrac{1}{2}$. The spin part of the wave function can be written as a linear combination of two orthogonal components, each giving an amplitude and phase of a possible spin orientation (or polarization)

$$\chi = a_1 \chi_{1/2}^{1/2} + a_2 \chi_{1/2}^{-1/2}, \tag{17}$$

where the superscripts indicate that the spins are directed in either positive or negative sense along a certain direction in space. The two possible spin states are referred to as the two states of polarization of the beam. The intensity of the beam is the sum of the probabilities for the two states of polarization.

$$I = |a_1|^2 + |a_2|^2. \tag{18}$$

The degree of polarization along the axis is defined as the net fraction of the beam aligned parallel to the axis (our chosen direction say positive z-direction)

$$P = (|a_1|^2 - |a_2|^2)/I \tag{19}$$

The beam can be specified by four parameters $-I$, and a polarization vector $\vec{P} = \hat{i} P_x + \hat{j} P_y + \hat{k} P_z \cdot P_z$ gives the difference between "up" and "down" spins, i.e., the difference between the number of particles with z-component of spin $\dfrac{1}{2}$ and the number with z-component of spin $-\dfrac{1}{2}$. P_z is given in Eq. (19). The four quantities I, P_x, P_y, P_z are called the Stokes' parameters – first introduced by Stokes to describe the polarization of light Photons have intrinsic spin of 1 unit, but have only two transverse directions of polarization. This is a consequence of the fact that the rest mass of photon is zero. In practice, one seldom has fully polarized beams, but rather partially polarized beams. The description of polarization is most conveniently done by density matrix technique.

The density matrix $\vec{\rho}$ is an operator (2×2 matrix) in the spin space. Since any 2×2 matrix can be written as a linear combination of the three Pauli spin matrices σ_i ($i = 1, 2, 3$) and the unit 2×2 matrix 1, we can write

$$\rho = a1 + \sum_{i=1}^{3} b_i \sigma_i. \tag{20}$$

The total beam intensity I is given by

$$I = Tr\rho = 2a, \tag{21}$$

since $Tr\, 1 = 2$, $Tr\, [\sigma_i] = 0$. Let us calculate the average value of σ_i i.e., $\langle \sigma_i \rangle = Tr\rho\sigma_i/Tr\rho$. Since $Tr\sigma_i\sigma = 2\delta_{ij}$, we have

$$\langle \sigma_i \rangle = \frac{2b_i}{2a}. \tag{22}$$

Therefore, the density matrix for an arbitrary beam may be written as

$$\rho = \frac{1}{2} I \,(1 + \vec{\sigma} \cdot \langle \vec{\sigma} \rangle) = \frac{1}{2} I \,(1 + \vec{\sigma} \cdot \vec{P}), \tag{23}$$

where $\vec{P} = \langle \vec{\sigma} \rangle$ is called the polarization of the beam.

An unpolarized beam whose polarization vector $\vec{P} = 0$ is described by

$$\rho = \frac{1}{2} I \begin{pmatrix} 1 & 0 \\ 0 & 1 \end{pmatrix} = \frac{1}{2} I1. \tag{24}$$

If the beam is in a pure spin state, then $\rho^2 = \rho$ and so $|\vec{P}| = 1$. Hence, a completely polarized beam polarized in the z-direction, say, is represented by

$$\rho = \frac{1}{2} I (1 + \sigma_z) = \begin{pmatrix} I & 0 \\ 0 & 0 \end{pmatrix}. \tag{25}$$

Note that for the system of spin 1/2, described by Eq. (17), the density matrix may be written as

$$\rho = \begin{pmatrix} a_1 a_1^* & a_1 a_2^* \\ a_2 a_1^* & a_2 a_2^* \end{pmatrix} \tag{26}$$

so that $Tr\rho\, I = |a_1|^2 + |a_2|^2$. Usually one writes $Tr\rho = 1$ by appropriate normalization. Let $|\chi\uparrow\rangle$ and $|\chi\downarrow\rangle$ be the spin "up" and "down" kets with the direction of $\vec{P}$ as the z axis. Then

$$\vec{\sigma}\cdot\vec{P}|\chi\uparrow\rangle = P|\chi\uparrow\rangle;\ \vec{\sigma}\cdot\vec{P}|\chi\downarrow\rangle = - P|\chi\downarrow\rangle. \tag{27}$$

In this diagonal representation

$$\rho = \begin{pmatrix} \dfrac{1}{2}(1 + P) & 0 \\ 0 & \dfrac{1}{2}(1 - P) \end{pmatrix} \tag{27a}$$

so that $|\vec{P}|$ is the probability of being in $|\chi\uparrow\rangle$ minus that of being in $|\chi\downarrow\rangle$.

Let us now consider the elastic scattering of two particles with spin. The analysis can be conveniently made by introducing a spin scattering matrix M. (Wolfenstein and Ashkin, 1952, Wolfenstein 1956). To begin, we consider a spin $\dfrac{1}{2}$ particle with momentum $\hbar \vec{k}_i$ incident on a spin 0 target. For a pure incident state $|\vec{k}_i\chi_i\rangle$, the probability of finding the scattered particle in state $|\vec{k}_f\chi_f\rangle$ depends on four amplitudes corresponding to the four possible processes $\left|\pm\dfrac{1}{2}\right\rangle \rightarrow \left|\pm\dfrac{1}{2}\right\rangle$ and $\left|\pm\dfrac{1}{2}\right\rangle \rightarrow \left|\mp\dfrac{1}{2}\right\rangle$. We introduce as the scattering amplitude a 2×2 matrix $M(\hat{k}_i, \hat{k}_f)$ in the composite spin space.

The asymptotic form of the spinor is

$$\psi(\vec{r}) \underset{r\to\infty}{\sim} \left[e^{i\vec{k}_i\cdot\vec{r}} + \frac{e^{ikr}}{r} M\,(\vec{k}_f,\, \vec{k}_i) \right] \times |\chi_i\rangle, \tag{28}$$

where $e^{i\vec{k}_i\cdot\vec{r}}|\chi_i\rangle$ is the incident spinor.

If the incident beam is represented by a density matrix ρ_i, then the beam scattered in the direction $\vec{k}_f$ is described by the density matrix ρ_f:

$$\begin{aligned}
\rho_f &= |\chi_f\rangle \langle\chi_f| \\
&= [M|\chi_i\rangle][M|\chi_i\rangle]^\dagger \\
&= M|\chi_i\rangle \langle\chi_i|M^\dagger \\
&= M\rho_i M^\dagger.
\end{aligned} \qquad (29)$$

The differential scattering cross section is given by

$$\frac{d\sigma}{d\Omega} = \frac{I_{sc}}{I_i} = \frac{Tr\rho_f}{Tr\rho_i} = \frac{TrM\rho_i M^\dagger}{Tr\rho_i}, \qquad (30)$$

where I_{sc} and I_i are the intensities of the scattered and incident beams respectively. Usually $Tr\rho_i$ is normalized to unity. The expectation value of the polarization of the scattered beam is given by

$$\vec{P}_f = Tr\,(\vec{\sigma}\rho_f)/Tr\rho_f = Tr\,(\sigma M\rho_i M^\dagger)/Tr\rho_f. \qquad (31)$$

Equations (29), (30) and (31) apply to scattering of particles of arbitrary spin.

The 2×2 M-matrix can be written in the form Eq. (20). M is a scalar for rotational invariant interaction. With space and time reversal invariances, we write

$$M = g1 + (\vec{\sigma} \times \hat{n})h, \qquad (32)$$

where g and h are complex functions of the angle θ and k, and $\hat{n} = (\vec{k}_i \times \vec{k}_f)/|\vec{k}_i \times \vec{k}_f|$ is a unit vector normal to the scattering plane. We ordinarily suppress the unit matrix 1 in Eq. (32). We assume that the incident beam of particles is unpolarized. Then

$$\frac{d\sigma}{d\Omega} = \frac{TrM\rho_i M^\dagger}{Tr\rho_i} = \frac{1}{2}\frac{I\,TrMM^\dagger}{I} = \frac{1}{2}MM^\dagger.$$

$$= Tr\frac{1}{2}(g + \vec{\sigma}\cdot\hat{n}h)(g^* + \sigma\cdot\hat{n}h^*), \qquad (33)$$

using Eq. (32), since $\vec{\sigma}$ is hermitian.

If we use the general relation

$$(\vec{\sigma}\cdot\vec{A})(\vec{\sigma}\cdot\vec{B}) = \vec{A}\cdot\vec{B} + i\vec{\sigma}\cdot(\vec{A}\times\vec{B})$$

we obtain

$$\frac{d\sigma}{d\Omega} = |g|^2 + |h|^2. \qquad (34)$$

The polarization of the scattered beam is

$$\vec{P}_f = \frac{1}{2} I_i Tr \, \vec{\sigma} \, (g + \vec{\sigma}\cdot\hat{n}h)(g^* + \vec{\sigma}\cdot\hat{n}h^*) \, / Tr\rho_f$$

$$= \frac{1}{2} Tr\vec{\sigma} \, (\vec{\sigma}\cdot\hat{n})(gh^* + g^*h)/(I_{sc}/I_i)$$

$$= \frac{\hat{n}(gh^* + g^*h)}{|g|^2 + |h|^2}$$

$$= \frac{2\hat{n}\,\mathrm{Re}gh^*}{|g|^2 + |h|^2}. \tag{35}$$

Hence, the polarization vector lies perpendicular to the scattering plane. This is a general result and one can show that the polarization of the scattered beam is perpendicular to the scattering plane for the scattering of an unpolarized incident beam on an unpolarized target, provided the interaction is invariant under space reflection.

From Eq. (34), we see that the differential cross section does not depend on the azimuthal angle for the scattering of an unpolarized incident beam. If, however, a polarized beam is scattered, there is an azimuthal dependence in the cross section. We calculate the differential cross section for the scattering of an incident beam with polarization $\vec{P}_i \neq 0$. We have

$$\frac{d\sigma}{d\Omega} = \frac{1}{2} Tr \, (1 + \vec{P}_i\cdot\vec{\sigma})(g^* + \vec{\sigma}\cdot\hat{n}h^*)(g + \vec{\sigma}\cdot\hat{n}h)$$

$$= \frac{1}{2} Tr \, (1 + \vec{P}_i\cdot\vec{\sigma})(|g|^2 + |h|^2 + 2\vec{\sigma}\cdot\hat{n}\,\mathrm{Re}gh^*)$$

$$= |g|^2 + |h|^2 + 2\vec{P}_i\cdot\hat{n}\,\mathrm{Re}gh^*. \tag{36}$$

If the scattering plane is perpendicular to $\vec{P}_i$, then $\left(\dfrac{d\sigma}{d\Omega}\right)_{\mathrm{right}}$ and $\left(\dfrac{d\sigma}{d\Omega}\right)_{\mathrm{left}}$, depending on the scattering through $\phi = \pm \, \pi/2$, are not equal (ϕ is measured from $\vec{P}_i$). We get

$$\frac{\left(\dfrac{d\sigma}{d\Omega}\right)_{\mathrm{right}} - \left(\dfrac{d\sigma}{d\Omega}\right)_{\mathrm{left}}}{\left(\dfrac{d\sigma}{d\Omega}\right)_{\mathrm{right}} + \left(\dfrac{d\sigma}{d\Omega}\right)_{\mathrm{left}}} = \frac{2\,|\mathrm{Re}gh^*|}{|g|^2 + |h|^2} P_i. \tag{37}$$

If $|\vec{P}_i| = 1$, this is just the polarization of the scattered beam produced by the scattering of an unpolarized incident beam, Eq. (35). This equality of left-right asymmetry and polarization is a general result which depends on the invariance of the interaction under space inversion and time reversal, provided that the target is unpolarized.

Let us now discuss the physical significance of the terms g and h in the scattering matrix M. The vector $\hat{n}$ is perpendicular to the scattering plane defined by the incident and scattered particles. Assuming that the incident particle is along the z-direction, we have

$$k'_x = k' \sin \theta \cos \phi, \quad k' = |\vec{k'}|$$

$$k'_y = k' \sin \theta \sin \phi,$$

$$k'_z = k' \cos \theta.$$

Then $\hat{n}$ has the components $- \sin \phi, \cos \phi, 0$. Let the spin functions be

$$\chi_{1/2}^{1/2} = \begin{pmatrix} 1 \\ 0 \end{pmatrix} \text{ and } \chi_{1/2}^{-1/2} = \begin{pmatrix} 0 \\ 1 \end{pmatrix}$$

Then

$$M\chi_{1/2}^{1/2} = (g1 + h\vec{\sigma}\cdot\hat{n}) \begin{pmatrix} 1 \\ 0 \end{pmatrix}$$

$$= g \begin{pmatrix} 1 \\ 0 \end{pmatrix} + ihe^{i\phi} \begin{pmatrix} 0 \\ 1 \end{pmatrix}$$

$$- g \, \chi_{1/2}^{1/2} + ihe^{i\phi} \chi_{1/2}^{-1/2}. \tag{38}$$

Hence, g may be considered as the spin non-flip amplitude and h as the spin-flip amplitude.

The measurement of polarization of the scattered beam can be achieved by a double scattering experiment. (Fig. 21.2). An unpolarized beam of spin $\frac{1}{2}$ particles say nucleons, of well-defined energy and direction is scattered through an angle θ by an unpolarized target. The scattered beam is partially polarized and so the target acts as a polarizer. A polarization-insensitive detector measures the differential cross section, i.e., $|g|^2 + |h|^2$. The once-scattered beam is now allowed to scatter from an identical target through θ in the same plane, to the right and to the left. The second taget acts as an analyzer. The left-right asymmetry of the doubly scattered beam, i.e., $4 \left(\dfrac{\mathrm{Re} g h^*}{|g|^2 + |h|^2} \right)^2$ is measured. Thus, $|\mathrm{Re} g h^*|$ is determined. Thus, we determine the polarization of the scattered beam in a double scattering experiment. The sign of $\mathrm{Re} g h^*$ giving the absolute phase can be determined by interference with some other process (arising from different interaction). A triple-scattering experiment serves to determine the scattering matrix completely, upto a phase factor.

If we allow spin-dependent interaction, then the Hamiltonian. of a spin $\frac{1}{2}$ particle interacting with a spin 0 target is

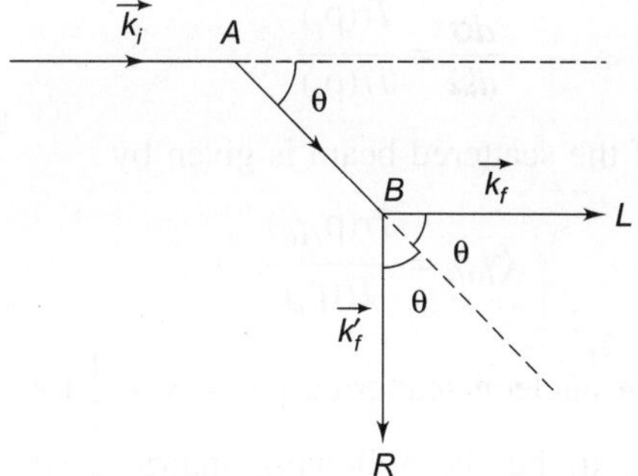

Fig. 21.2 Schematic diagram of a double scattering experiment. *A* and *B* are the first and the second target, both unpolarized

$$H = \frac{\vec{p}^2}{2m} + V(r) + U(r)\, \vec{\sigma}\cdot\vec{L}, \tag{39}$$

where we have assumed rotational and reflection invariance, and V and U are spherically symmetric. The spin-orbit coupling term $U(r)\,\vec{\sigma}\cdot\vec{L}$ causes a small correction in atomic spectroscopy. In nuclear physics such spin-orbit interactions are of great importance.

Next, we give a brief discussion of the density matrix formalism for the general case. This can be applied to the interesting case of nucleon-nucleon scattering. Let us consider the scattering of a beam of particles with spin S_p by a target of particles with spin S_t. The scattering matrix M is then a $(2S_p + 1)\,(2S_t + 1) \times (2S_p + 1)\,(2S_t + 1)$ matrix in the combined spin space. The density matrix ρ_i of the system before scattering has $(2S_p + 1)^2 (2S_t + 1)^2$ independent components. We define a complete, orthogonal set of $(2S + 1)^2 (2S_t + 1)^2$ hermitian operators γ_α in the combined spin space. Then we can express ρ_i as

$$\rho_i = \frac{Tr\,(\rho_i)}{(2S_p + 1)\,(2S_t + 1)} \sum_{\alpha=1}^{(2S_p + 1)^2 (2S_t + 1)^2} \langle\gamma_\alpha\rangle_i\, \gamma_\alpha, \tag{40}$$

with the normalization condition

$$Tr\,(\gamma_\alpha\gamma_\beta) = (2S_p + 1)\,(2S_t + 1)\,\delta_{\alpha\beta} \tag{41}$$

Here,
$$\langle\gamma\alpha\rangle_i = \frac{Tr\,(\rho_i\gamma_\alpha)}{Tr(\rho_i)} \tag{42}$$

denotes the initial orientation of the system. After scattering, the density matrix of the system is given by

$$\rho_f = M\rho_i M^\dagger \tag{43}$$

and the differential scattering cross section is

$$\frac{d\sigma}{d\Omega} = \frac{Tr(\rho_f)}{Tr(\rho_i)}.$$ (44)

The orientation of the scattered beam is given by

$$\overline{\langle\gamma_\alpha\rangle_f} = \frac{Tr(\rho_f\gamma_\alpha)}{Tr(\rho_f)}.$$ (45)

For elastic nucleon-nucleon scattering $\left(S_p = S_t = \frac{1}{2}\right)$ we can take the sixteen hermitian matrices γ_α to be the following matrices (in the combined four-dimensional spin space)

$$\sigma_i^{(1)}\, \sigma_j^{(2)},\ \sigma_i^{(1)}\, 1^{(2)},\ 1^{(1)}\, \sigma_j^{(2)},\ 1^{(1)}1^{(2)}\ (i, j = 1, 2, 3),$$

where σ_i and 1 are the Pauli spin matrices and the unit matrix respectively, and the superscripts refer to the nucleons. These matrices satisfy Eq. (41). The density matrix ρ_i for a polarized beam incident on the unpolarized target is

$$\rho_i = \frac{1}{4}\,(1 + \vec{\sigma}_1\cdot\vec{P}_{\text{beam}}),$$ (46)

and for an unpolarized beam incident on a polarized target is

$$\rho_i = \frac{1}{4}\,(1 + \vec{\sigma}_2\cdot\vec{P}_{target})$$ (47)

where

$$\vec{P} = \overline{\langle\sigma\rangle} = Tr(\vec{\sigma}\rho).$$ (48)

The differential scattering cross section for the elastic scattering of a nucleon beam with initial polarization $\vec{P}_i$ by unpolarized nucleons is

$$\frac{d\sigma}{d\Omega} = \frac{1}{4}\,TrM\,(1 + \vec{\sigma}\cdot\vec{P}_i)\,M^\dagger$$

$$= \left(\frac{d\sigma}{d\Omega}\right)_0 + \frac{1}{4}\,Tr\,(M\vec{\sigma}M^\dagger)\,\vec{P}_i$$ (49)

where $\left(\dfrac{d\sigma}{d\Omega}\right)_0 = \dfrac{1}{4}\,Tr\,(MM^\dagger)$ is the differential cross section of the unpolarized beam. Particles 1 and 2 refer to the beam and target particles respectively.

PROBLEMS

21.1 Using the Coulomb scattering amplitude, obtain the Mott formula for the Coulomb scattering of two identical spin 1/2 fermions.

21.2 The scattering of neutrons on protons depends on the resultant spin of the neutron and proton. We assume that the interaction potential energy between a neutron and a proton is a square well of depth V_0 and a range a. Find the scattering length and the effective range in the triplet case and in the singlet case. Use the following data:

triplet: $\qquad\qquad V_0 = 35.5$ Mev

$\qquad\qquad\qquad a = 2.03 \times 10^{-13}$ cm

singlet: $\qquad\qquad V_o = 14.2$ Mev

$\qquad\qquad\qquad a = 2.57 \times 10^{-13}$ cm.

What is the cross section for an unpolarized beam and target?

21.3 Slow neutrons are scattered by the hydrogen molecule. In the low energy regime when the wavelength of the incident neutrons is much larger than proton-proton distance in the molecule (0.74×10^{-8} cm), the scattering amplitude is equal to the sum of the scattering amplitudes of the two protons. Calculate the cross section for the scattering of neutrons on parahydrogen and orthohydrogen.

21.4 Consider the scattering of two spin $\dfrac{1}{2}$ particles whose interaction is $V = V_1(r) + V_2(r)\,\vec{\sigma}_1 \cdot \vec{\sigma}_2$. Find the elastic scattering amplitude and the inelastic (spin flip) scattering amplitude in the Born approximation.

21.5 Consider the scattering of a spin $\dfrac{1}{2}$ particle by a spin 0 target. The Hamiltonian is $H = \dfrac{\vec{p}^2}{2m} + V_1(r) + V_2(r)\,\vec{\sigma}\cdot\vec{L}$, assuming rotation and reflection invariance. The interaction includes a spin-orbit term:

(a) Find the scattering amplitude due to the spin-orbit interaction, using DWBA.

(b) Solve the problem by using the projection operators onto the states $j = l \pm \dfrac{1}{2}$ and the radial functions and obtain the total cross section.

21.6 Give a geometric argument to establish the following. For the scattering of a beam of projectiles of any spin from a target of particles of any spin, the final spin polarizations of the projectiles are normal to the scattering plane, assuming parity invariance, The projectiles and the target particles are both unpolarized.

21.7 Consider identical spin $\dfrac{1}{2}$ – spin $\dfrac{1}{2}$ particle scattering with spin-dependent interaction. Even if the initial state is unpolarized there will be correlations among the spins of the particles after scattering. The scattering matrix is four dimensional in spin space. Find the spin correlation coefficients in terms of the singlet-and triplet-scattering amplitudes.

CHAPTER 22

Interpretation

> On Rupnarayan's bank
> I awoke
> and knew the world
> was no dream.
>
> ....
>
> Truth is hard,
> and I loved the hard;
> it never deceives.
>
> — Rabindranath Tagore (1941)

Quantum mechanics is a most successful physical theory as illustrated in many examples covering the atomic, nuclear, and sub-nuclear regimes described in this book. It lies at the heart of our understanding of physics and chemistry. We owe to quantum mechanics many modern developments in technology. To date, none of its predictions has been proved wrong experimentally. Inspite of this success, from its inception many conceptual issues arise in quantum theory, some of which lie at the boundaries of physics and philosophy. Quantum mechanics introduces many radically new concepts which seem strange to the established classical concepts about nature. We address ourselves to many of these conceptual difficulties in this chapter and discuss the interpretation of quantum mechanics.

We consider the following issues. We first discuss the extraordinary features of quantum entanglement. We then discuss the role of hidden variables in quantum theory. We consider the Einstein-Podolsky-Rosen criticism and the incompleteness of quantum mechanics. Bell's inequalities and the quantum measurement problem are also considered. These issues are linked together and the overall interpretative scheme still remains controversial in spite of the enormous growth in the literature in this field.

22.1 QUANTUM ENTANGLEMENT

Recently the physics of entanglement has been the focus of considerable attention. The phenomenon of quantum entanglement is an important feature of quantum mechanics and is central to our understanding of several weird features of quantum theory. A state of a composite quantum system (or, suitable ensemble of such systems) is called "entangled" if it cannot be represented as a mixture of factorizable pure states. Otherwise, the state is called "separable". State vectors of entangled systems are not tensor products of state vectors. Entangled states have peculiar non-classical correlations. These are of great importance in the theories of the He spectrum, and magnetism, and in Bell's theorem. Quantum entanglement is the basis of quantum information processes, e.g., quantum cryptography, quantum computation.

Quantum entanglement stems from the tensor product structure of the Hilbert space of a composite quantum system (or an ensemble of such systems). Let us consider a system S composed of two subsystems S_1 and S_2 with Hilbert spaces $\mathcal{H}_1$ and $\mathcal{H}_2$ respectively. The state vectors of separable states of S belong to the tensor product $\mathcal{H}_1 \otimes \mathcal{H}_2$. This corresponds to the classical situation. However, quantum mechanics allows for entangled states which have the state vectors that are not tensor products of state vectors.

Let $|\psi>$ be a pure state of S with wave function $\psi\,(q_1 q_2)$ Its density matrix is

$$\langle q_1 q_2 | \rho | q_1' q_2' \rangle = \psi^*(q_1 q_2)\,\psi^*(q'q_2'). \tag{1}$$

The expectation value of O_1, an observable of S_1 in ψ, is

$$\langle O_1 \rangle_\psi = \int dq_1\, dq_1' dq_2\, \psi^*(q_1 q_2) < q_1|O_1|q_1' > \psi(q_1' q_2)$$

$$= \int dq_1 dq_1' <q_1|O_1|q_1'> <q_1'|\rho_1|q_1> = Tr\,O_1\rho_1, \tag{2}$$

where ρ_1 is the density matrix of S_1:

$$<q_1|\rho_1|q_1'> = \int dq_2 <q_1 q_2|\rho|q_1' q_2'>. \tag{3}$$

We consider

$$\psi(q_1 q_2) = a_1 u_1(q_1) v_1(q_2) + a_2 u_2(q_1)\, v_2(q_2),$$

$$|a_1|^2 + |a_2|^2 = 1, \tag{4}$$

with u_i and v_j orthonormal,

$$\int dq_1 u_i^*(q_1)\, u_j(q_1) = \delta_{ij} \cdot \int dq_2 v_i^*(q_2)\, v_j(q_2) = \delta_{ij}. \tag{5}$$

When the coefficients a_1 and a_2 are non-vanishing, ψ describes an entangled state, i.e., a state that cannot be written as a product

$$\psi(q_1 q_2) = \phi(q_1)\, \chi(q_2). \tag{6}$$

Using Eq. (5), we obtain from Eq. (4),

$$<q_1|\rho_1|q_1'> = |a_1|^2 u_1(q_1) u_1^*(q_1') + |a_2|^2 u_2(q_1) u_2^*(q_1') \tag{7}$$

Now,

$$<q_1|(\rho_1)_1^2|q_1'> = |a_1|^4 u_1(q_1) u_1^*(q_1') + |a_2|^4 u_1(q_1) u_1^*(q_1'), \tag{8}$$

whence

$$Tr(\rho_1)^2 = |a_1|^4 + |a_2|^4 < 1 \tag{9}$$

Inequality Eq. (9) shows that ρ_1 is not pure and cannot be represented by a state in $\mathcal{H}_1$, unless one of a_i vanishes, in which case ψ is in the product form and is not entangled. The result can be generalized. So, even if a composite system is in a pure state ψ, its sub-systems are in general in mixed states. Only when ψ is in a product form, a subsystem is in a pure state.

For the state given by Eq. (4), the two-body probability distribution is

$$P(q_1, q_2) = |a_1|^2 |u_1(q_1)|^2 |v_1(q_2)|^2 + |a_2|^2 |u_2(q_1)|^2 |v_2(q_2)|^2$$
$$+ 2\,\mathrm{Re}\,((a_1 a_2^* u_1(q_1) u_2^*(q_1) v_1(q_2) v_2^*(q_2)). \tag{10}$$

The two-body interference term describes strange non-classical correlations for quantum systems that are far apart (Gottfried and Yan (2003)).

Entangled states cannot be prepared locally by acting on each sub-system separately. The amount of entanglement E of the state $|\psi>$ of the composite system S is given by the von Neumann entropy

$$S_{vN} = S(\rho_1) = S(\rho_2). \tag{11}$$

E is invariant under the action of local unitary operators.

A simple example is given by a pair of spin $\frac{1}{2}$ particles (electrons) prepared in a singlet state. The spin part of the state of the two particles can be written as

$$|\psi> = \frac{1}{\sqrt{2}}\, (|+>_1|->_2 - |->_1|+>_2),$$

where $|+>_1$, $|->_2$ indicate the state in which particles 1 and 2 have spin $1/2\,\hbar$ and $-1/2\,\hbar$ respectively along the z-axis. This state is not factorable as the product of two states and is referred to as an entangled state. The entangled nature of the state is due to the Pauli principle.

22.2 HIDDEN VARIABLES

Lack of determinism is a fundamental quantum doctrine. Quantum mechanics introduces probabilities (or in some cases expectation values) as a postulate without the formulation of rules from an underlying sample space of predictable individual events. We note here the contrasting passage to classical statistical physics. The aim of a hidden-variables (HV) theory is to consider the possibility that although quantum mechanics is a very accurate description of nature there is a deeper level of description in which properties of individual systems have preexisting values revealed by the act of measurement. In this approach one creates a unobservable sample space of nature by introducing one, or more parameters (hidden variables). This is a statistical theory in which the wave function describes the macrostate of the system obtained by averaging over the microstates which are characterized by the hidden variables.

There are various HV theories which are broadly classified as (Belinfante (1973)) (i) HV theories of the first kind-the motivation is to maintain the stochastic predictions of quantum mechanics; (ii) HV theories of the second kind or local hidden variables (LHV) theories–the motivation is to save the local action principle. The theories of the first kind are compatible with quantum mechanics while those of the second kind have the objective of eliminating the nonlocal feature of quantum mechanics.

We now specify the general features of a HV theory. In any given state ψ, any observable A possesses an objectively existing value A $(\psi, \lambda_1 \lambda_2, ..., \lambda_n)$ determined by ψ and by the values of a set of hidden variables $(\lambda_1, \lambda_2, ..., \lambda_n)$ belonging to some space Λ. It is assumed that there exists a probability density μ_ψ on Λ such that the expectation value of A in the state ψ is

$$\langle A \rangle_\psi = \int_\Lambda \mu_\psi (\lambda_1, \lambda_2, ..., \lambda_n) A (\psi, \lambda_1, \lambda_2, ..., \lambda_n) d\lambda_1, d\lambda_2 ... d\lambda_n. \qquad (12)$$

The basic problem is to find the space Λ, the probability density μ_ψ and the value function A $(\lambda_1, \lambda_2, ..., \lambda_n)$ such that the expression Eq. (12) reproduces the predictions of standard quantum theory.

The progress of HV theories was hindered for three decades by von Neumann's "proof" on the mathematical impossibility of such variables in quantum theory, until Bell established that von Neumann's proof is irrelevant for realistic HV theories von Neumann's theorem states. "It is therefore not, as is often assumed, a question of reinterpretation of quantum mechanics–the present system of quantum mechanics would have to be objectively false in order that another description of the elementary process than the statistical one be possible." Bell showed that von Neumann's demonstration that dispersion free states, and so hidden variables, are impossible was based on an arbitrary assumption (and impossible) relation between the results of incompatible measurements. Hence, the proof does not apply to realistic HV theories.

Two barriers to the HV extension of quantum theory are the no-hidden-variables theorems (or, "no-go" theorems) proved by Kochen and Specker (KS) (also by Bell, see Mermin (1993)) and by Bell. We first discuss the KS argument. We describe the three-dimensional state space in terms of observables built of the angular momentum components of spin 1 operator.

The observables we consider are the squares of the spin along various directions. Since these operators have eigenvalues 0 or 1, and since $S_1^2 + S_2^2 + S_3^2$ $= s\,(s + 1) = 2$ where 1, 2, 3 are any three mutually orthogonal directions, any measurement of two of the three operators must take the value 1, and the other one the value 0. The operators $S_1^2 + S_2^2 + S_3^2$ constitute a mutually commuting set and so they can be simultaneously measured. Next, we assume that there are hidden variables in one-to-one correspondence with the operators S_i^2 which take preassigned values such that the measurement outcome is 0 or 1, consistent with the constraint. KS showed that this is impossible, by identifying a state of three-dimensional vectors (i.e., directions) on the unit sphere that have mutually commuting spin-squared operators, but which admit of no consistent assignment of the values 0 and 1 subject to the constraint that every three mutually orthogonal vectors contain just two 1's and one 0. So the quantum mechanics of a spin 1 system cannot be reproduced by a set of HV. The theorem of KS now reads:

There is no value function V_ψ for a state ψ subject to the set of assumptions if the Hilbert space $\mathcal{H}$ of the system is such that dim $\mathcal{H} > 2$.

Gleason's work (see Section 2.4) was not explicitly addressed to the HV problem, although there are a number of links with KS theorem. Gleason's work shows that the additivity requirement for expectation values of commuting operators cannot be statisfied by HV (dispersion free) states. So this work puts strong constraints on the possible extention of the standard quantum theory by HV.

Bell analyzed the Einstein-Podolsky–Rosen argument from the viewpoint of HV. A crucial element in Bell's reasoning is the locality hypothesis which states that there is no direct, non-local interaction between two measuring apparatuses far apart!. Here, the problematic assumption of non-contextuality is replaced by the assumption of reality. To establish the incompatibility of quantum mechanics and the LHV theories, Bell showed that the correlations predicted by any LHV theory are limited by inequalities ("Bell inequalities") that are violated by certain quantum predictions.

22.3   NONLOCALITY AND THE BELL INEQUALITIES

(a)   The Critique of Einstein, Podolsky, and Rosen and the Incompleteness of Quantum Theory

Conceptual issues confronting hidden variables and possessed properties were posed by Einstein, Podolsky, and Rosen (EPR) in their seminal contribution in 1935 – "Can Quantum – Mechanical Description of Physical Reality be Considered Complete?". The argument – sometimes referred to as a paradox – advanced by

EPR was that quantum mechanical description could not be complete but should be supplemented by additional variables required to restore to the theory causality and locality.

EPR began by giving a definition of "complete" in the form of two propositions

(i) Every element of physical reality must have a counterpart in the physical theory.

(ii) If, without in any way disturbing a system, we can predict with certainty (i.e., with probability equal to unity) the value of a physical quantity, then there exists an element of physical reality corresponding to this physical quantity.

EPR then discussed a class of Gedanken experiment by considering a system composed of two parts which interacted in the past but which separated subsequently such that a measurement on one part did not influence the other one.

The original analysis of EPR referred to a particle that disintegrated into two identical particles moving in opposite directions with equal speeds, the incompatible variables being the position and the momentum. We now consider a simpler version of the EPR formulation introduced by Bohm – a pair of spin $\frac{1}{2}$ particles in a spin singlet state |0> and moving freely in opposite directions. The EPR argument for this version is as follows. Measurements are made on selected components of the spin $\vec{S}_1$ and $\vec{S}_2$ of particles 1 and 2 respectively. A measurement of spin $\vec{S}_1$ along a particular direction $\vec{n}$, i.e., $\vec{S}_1 \cdot \vec{n}$ will yield the value $\pm \hbar/2$. The constraint on the total spin means that the second particle will have the value $\mp \hbar/2$ of the spin component along $\vec{n}$ ($\vec{S}_2 \cdot \vec{n}$) with certainty without ever disturbing this particle which is spatially separated from the first particle. Therefore, the component of $\vec{S}_2$ along $\vec{n}$ is an element of physical reality that exists independently for the particle 2 alone. This value of $S_2 \cdot \vec{n}$ for particle 2 must have existed prior to a measurement done on the particle 1.

Since $\vec{n}$ was orbitrary, we may conclude that all the components of spin of the particle 2 were known to begin with, i.e., definite values must exist for all the components of a particle. But this is in contradiction with quantum mechanics which asserts that "when two operators corresponding to two physical quantities do not commute the two quantities cannot have simultaneous physical reality". EPR argued that quantum mechanics fails to satisfy the proposition (i) and hence concluded that "the quantum mechanical description of physical reality given by the wave function is not complete". The answer to the question posed in the title of the EPR paper is negative.

Bohr emphasized that the non-commuting observables of particle 1 (say $\vec{S}_1 \cdot \hat{x}$ and $\vec{S}_1 \cdot \hat{z}$) and hence also corresponding ones (say $\vec{S}_2 \cdot \hat{x}$ and $S_2 \cdot \hat{z}$) of particle 2 are not simultaneously elements of reality but are complementary attributes of the particle. Bohr argued "that the procedure of measurement has an essential influence on which the very definition of the physical quantities rests".

Einstein concluded that the EPR argument forced us to relinquish one of the following two assertions

(1) the description by means of the ψ-function is complete;

(2) the real states of spatially separated objects are independent of each other (locality).

Thus EPR established that quantum mechanics is incompatible with local realism. EPR came to the conclusion that quantum mechanics is indeed incomplete, as they rejected a non-local theory with "spooky action at a distance". EPR nevertheless did not suggest how quantum mechanics could be extended to a local realist theory keeping all the established consequences intact. One natural path would be to introduce additional hidden variables.

(b) The Bell Inequalities

In 1964 Bell analyzed the EPR experiment by introducing the HV and showed that one cannot understand EPR-type correlations by extending quantum mechanics along the lines envisaged by Einstein. This result, known as Bells' theorem, established that no LHV theory can reproduce all the results of quantum theory. To establish the incompatibility between quantum mechanics and LHV theories, Bell showed that one can derive inequalities for correlation functions – called the Bell inequalities (BI) – which are different in quantum mechanics as opposed to any LHV model. Furthermore, such inequalities are amenable to experimental tests. Several tests have been proposed by other authors. These are referred to as Bell-like or Bell-type tests. We shall discuss one such test in this section. In contrast to the EPR consideration about two observers making measurements along the same axis, Bell considered what happens if two observers make measurements along different axes. For directions $\vec{a}$ and $\vec{b}$, the correlation function between measurements made by the two observers along these directions is

$$C(\vec{a}, \vec{b}) = \lim_{N \to \infty} \frac{1}{N} \sum_{n=1}^{N} a_n b_n, \tag{13}$$

where $a_n(b_n)$ is (2/$\hbar$ times) the value of $\vec{a} \cdot \vec{S} \, (\vec{b} \cdot \vec{S})$ possessed by particle 1 (2) in the nth element of the collection. If $C(\vec{a}, \vec{b})$ is $+1$ (-1) then the results are totally correlated (anticorrelated).

Now, consider the quantity (Isham (1995))

$$g_n = a_n b_n + a_n b'_n + a'_n b_n - a'_n b'_n. \tag{14}$$

Clearly,

$$a_n = \pm 1 \; (b_n = \pm 1) \text{ as } \vec{a} \cdot \vec{S} = \pm \frac{1}{2} \hbar \left(\vec{b} \cdot \vec{S} = \pm \frac{1}{2} \hbar \right).$$

Each term in the sum g_n takes the value $+1$ or -1. If $a_n = b_n = \pm 1 = a'_n = b'_n$, then $g_n = 2$. If $a_n = -b_n = \pm 1 = a'_n = -b'_n$, then $g_n = -2$. Considering the various possibilities, we see that g_n can take only the values ± 2. The average value of g_n

$$\left| \frac{1}{N} \sum_{n=1}^{N} g_n \right| = \left| \frac{1}{N} \sum_{n=1}^{N} a_n b_n + \frac{1}{N} \sum_{n=1}^{N} a_n b'_n + \frac{1}{N} \sum_{n=1}^{N} a'_n b_n - \frac{1}{N} \sum_{n=1}^{N} a'_n b'_n \right|, \tag{15}$$

must be less than or equal to 2.

We obtain, in the limit $N \to \infty$, the Bell inequality

$$|C(\vec{a}, \vec{b}) + C(\vec{a}, \vec{b}') + C(\vec{a}', \vec{b}) - C(\vec{a}', \vec{b}')| \leq 2. \tag{16}$$

In proving Eq. (16) two assumptions are made.

(1) Each particle has a definite value of the projection of spin $\vec{S}$ along any direction at all times.

(2) Locality holds in the sense that the value of any quantity is not affected by altering the position of a distant measuring apparatus. This implies that a_n has the same value independently of the direction ($\vec{b}$ or $\vec{b}'$) along which the other observer is measuring the spin projection of particle 2. So context dependent values as in the KS theorem are ruled out.

According to "orthodox" quantum mechanics, the correlation between the spin measurements along axes $\vec{a}$ and $\vec{b}$ is the expectation value

$$C(\vec{a}, \vec{b}) = \frac{\langle \psi | \vec{a} \cdot \vec{S}_1 \otimes \vec{b} \cdot \vec{S}_2 | \psi \rangle}{(\hbar/2)^2}, \tag{17}$$

where $\vec{S}_1$ and $\vec{S}_2$ are the spin operators for particle 1 and 2 respectively. Spherical symmetry requires that the correlation is a function of the angle between the two directions. So $C(\vec{a}, \vec{b})$ is a function of $\cos \theta_{ab} = \vec{a} \cdot \vec{b}$ only and hence without loss of generality we can assume that $\vec{a}$ points along the z-axis and that $\vec{b}$ lies in the x-z plane. Equation (17) becomes

$$C(\vec{a}, \vec{b}) = \langle \psi | \sigma_{1z} \otimes (\sigma_{2z} \cos \theta_{ab} + \sigma_{2x} \sin \theta_{ab} | \psi \rangle \tag{18}$$

where

$$\sigma_{iz} = \begin{pmatrix} 1 & 0 \\ 0 & -1 \end{pmatrix}, \quad i = 1, 2$$

$$\sigma_{ix} = \begin{pmatrix} 0 & 1 \\ 1 & 0 \end{pmatrix}, \quad i = 1, 2 \tag{19}$$

are the Pauli spin matrices along the z-axis and the x-axis respectively for the ith particle.

Here ψ is the spin-singlet part of the state vector of the pair of spin $\frac{1}{2}$ particles 1 and 2 (Eq. (11)).

Now,

$$C(\vec{a}, \vec{b}) = \frac{1}{2} \cos \theta_{ab} \{(|+>_1|->_2 - |->_1|+>_2)$$

$$(\sigma_{1z}|+>_1 \sigma_{2z}|->_2 - \sigma_{1z}|->_1 \sigma_{2z}|+>_2)\},$$

$$= -\frac{1}{2} \cos \theta_{ab} \{(|+>_1|->_2 - |->_1|+>_2)(|+>_1|->_2 - |->_1|+>_2)\}$$

$$= -\cos \theta_{ab}, \tag{20}$$

where we have used

$$\sigma_{iz}|+>_i = |+>_i,$$

$$\sigma_{iz}|->_i = -|->_i, \tag{21}$$

According to quantum mechanics, the decay of a spin zero particle into a pair of spin $\frac{1}{2}$ particles must satisfy this relation. Here, $C(0) = -1$ and $C(\pi) = 1$, as required.

Let us consider the particular case in which the unit vectors $\vec{a}$, $\vec{b}$, $\vec{a}'$, $\vec{b}'$, are coplanar with $\vec{a}$ and $\vec{b}$ parallel to each other. If $\theta_{ab'} = \theta_{a'b} = \phi$, then the Bell inequality will be satisfied provided

$$|1 + 2\cos \phi - \cos 2\phi| \leq 2. \tag{22}$$

If ϕ lies between 0 and $\frac{\pi}{2}$, the inequality is violated. This shows that the idea of systems possessing individual values for observables is not tenable unless one accepts an essential non-locality. The LHV model is untenable.

(c) The Clauser-Horne Inequality

Bell's inequality is not amenable to experimental tests. Inequalities of the Bell-type have been proposed by a number of authors. We now consider one such inequality, called the Clauser-Horne inequality, originating from the works of Clauser, Horne, Shimony, and Holt (1969) and Clauser and Horne (1974). [See also Clauser and Shimony (1978)]. This inequality refers only to detected results, and is amenalde to experimental tests. Also, the proof of Bell's theorem is stronger in the sense that in contrast to Bell's derivation, it eliminates the assumption that the elaboration of quantum mechanics is deterministic.

We consider EPR type processes and let a, a', ... be the values of the observable A of subsystem S_1 and also let b, b', ... be the values of the observable B of subsystem S_2. Here a, b, etc. may denote spin directions or photon polarization directions, etc. We are interested in determining single counts and coincidence counts. Let $N_1(a)$ and $N_2(b)$ denote the number of single counts measured at detectors for systems S_1 and S_2 respectively. Let $N_{12}(a, b)$ be the number of coincidence counts when measurements are made on both and

N the total number of EPR pairs. We introduce the probabilities (Gottfried and Yan (2003))

$$p_1(a) = \frac{N_1(a)}{N}, \ p_2(b) = \frac{N_2(b)}{N},$$

variables
$$p_{12}(a, b) = \frac{N_{12}(a, b)}{N}. \tag{23}$$

Hidden variables (uncontrollable parameters) are now introduced. So the single count probabilities are denoted by $p_1(a; \lambda)$ and $p_2(b; \lambda)$. The values of these functions lie between 0 and 1. In contrast to Bell's analysis, we assume that the outcome (measured value) of a, say, is not uniquely determined by a and λ, but only that the probability of the outcome is determined by a and λ. The probability of a coincidence or simultaneous detection is denoted by $p_{12}(a, b; \lambda)$. The set of parameters λ have the probability distribution $w(\lambda)$ normalized as

$$\int_\Lambda dw(\lambda) = 1, \tag{24}$$

summed over the set Λ of all values of λ. $w(\lambda)$. is independent of the instrumental settings.

Averaged over this distribution, we have the single-count probabilities

$$p_1(a) = \int_\Lambda p_1(a; \lambda) \, w(\lambda) \, d\lambda, \tag{25}$$

$$p_2(b) = \int_\Lambda p_2(b; \lambda) \, w(\lambda) \, d\lambda, \tag{26}$$

and the probability of detecting a coincidence

$$p_{12}(a, b) = \int_\Lambda p_{12}(a, b; \lambda) \, w(\lambda) \, d\lambda. \tag{27}$$

We now make the key assumption of locality: if the system is in a state specified by λ, the joint or coincidence probability distribution for specific values of a and b does not have correlations between the distant subsystems S_1 and S_2

$$p_{12}(a, b; \lambda) = p_1(a; \lambda) \, p_2(b; \lambda). \tag{28}$$

This factorization implies the probability count of the measurement on one particle is independent of the probability count of the measurement performed on the other particle after both particles emerged.

We next consider the lemma. For four real numbers

$0 \leq x, x', y, y' \leq 1$ the following inequality holds

$$-1 \leq xy - xy' + x'y + x'y' - x' - y \leq 0. \tag{29}$$

Let $U = xy - xy' + x'y + x'y' - x' - y.$

To prove the upper bound $U \leq 0$ we rewrite

$$U = (x - 1)y + x'(y - 1) + y'(x' - x) \leq 0$$

for $x \geq x'$ since every term is non-positive

For $x \leq x'$ we rewrite U as

$$U = x(y - y') + (x' - 1)y - x'(1 - y')$$

$$\leq x(y - y') + (x' - 1)y - x(1 - y')$$

$$= xy + (x' - 1)y - x$$

$$= x(y - 1) + (x' - 1)y \leq 0.$$

Hence,
$$U \leq 0.$$

To prove the lower bound $-1 \leq U$, we consider for $x' \geq x$,

$$U + 1 = (1 - x')(1 - y) + xy + y'(x' - x) \geq 0,$$

since every term is non-negative.

For $x > x'$,

$$U + 1 = (1 - x')(1 - y) - (x - x')(y' - y) + x'y,$$

$$\geq (x - x')(1 - y) - (x - x')(y' - y) + x'y$$

$$= (x - x')(1 - y') + x'y \geq 0.$$

Hence,
$$-1 \leq U.$$

We have proved the lemma $-1 \leq U \leq 0$. The lower bound is not used in the experiment.

Now, let $x = p_1(a; \lambda)$, $x' = p_1(a'; \lambda)$, $y = p_2(b; \lambda)$ $y' = p_2(b'; \lambda)$. Multiplying Eq. (29) by $w(\lambda)$ and integrating over λ, we obtain

$$-1 \leq p_{12}(a, b) - p_{12}(a, b') + p_{12}(a', b) + p_{12}(a', b') - p_1(a') - p_2(b) \leq 0, \tag{30}$$

or,
$$\frac{p_{12}(a, b) - p_{12}(a, b') + p_{12}(a', b) + p_{12}(a', b')}{p_1(a') + p_2(b)} \leq 1. \tag{31}$$

This is the Clauser-Horne (CH) inequality. In terms of Eq. (23), we express the left side as

$$\frac{N_{12}(a, b) - N_{12}(a, b') + N_{12}(a', b) + N_{12}(a', b')}{N_1(a') + N_2(b)}, \tag{32}$$

which is independent of the unknown number N, the total number of EPR pairs. The CH inequality, in contrast to the Bell inequality, is, therefore amenable to an experimental test.

A considerable number of experiments have been performed to test whether EPR states have correlations that violate Bell-type inequalities. Some experiments were done with γ-ray photons emitted in positronium annihilation or with protons, but the most convincing ones were based on visible light photon pairs. A series of experiments by Aspect and collaborators (Aspect, Grangier and Roger (1981, 1982), Aspect, Dalibard and Roger (1982)) in the early 1980s in which two photons traveling in opposite directions and having specified linear polarizations were detected obtained results in agreement with quantum theory that violated Bell-type inequality. Recent experiments by Tapster et al (1994), Weihs et al (1999), Tittel et al (1999) have confirmed these results.

The overwhelming conclusion is that the predictions of quantum theory are correct and no LHV theory underlies quantum theory.

22.4 THE PROBLEM OF MEASUREMENT

The concept of measurement plays a fundamental role in quantum theory. Quantum theory underlies the detailed understanding of microscopic systems. The application of the formalism to macroscopic systems lead to conceptual issues. The "quantum measurement problem" arises when one applies the rules of quantum theory simultaneously to a microscopic object and to a macroscopic apparatus which is measuring the state of the object. The problem is that although the Hilbert space of possible states is vast owing to the principle of linear superposition, there are only a few classical macroscopic states. The question arises then how to reconcile the "reality" of the world appearing classical to us with the underlying quantum nature.

The quantum measurement process differs from that in classical physics where isolated systems were studied. A measurement is an operation on a system which probes the quantum state just before the measurement and yields a definite recordable number which determines an observable entity of the system. State preparation is an operation in which a system (or an ensemble of systems) is prepared in a reproducible state just after the operation. This is closely connected with the state reduction. The typical process of measurement follows the state preparation.

(a) Quantum measurement scheme

The ingredients of a measurement process are an object, a typically microscopic system S, and an apparatus A, and an interaction that produces a correlation between some dynamical variable of the object and an appropriate indicator variable of the apparatus A. So the apparatus A includes its readout. We now investigate an ideal measurement (von Neumann (1932)). The object S is represented by the basis set of vectors $\{ |s_n > \}$ in a Hilbert space $\mathcal{H}_s$. The measuring apparatus

A is represented by the basis set of vectors $\{|a_n>\}$ in a Hilbert space $\mathcal{H}_A. |a_n>$. corresponds to macroscopically distinguishable "pointer" position recording the outcome of a measurement when S is in the state $|s_n>$. Before measurement, at time $t = 0$, the whole system (SA) is represented by the state vector

$$|\psi(0)> = \sum_n c_n|s_n> |\phi_0>, \tag{33}$$

where $|\phi_0>$ is the initial eigenvector describing A and S is in a superposition. At a later time t, after interaction of S with A, the state is

$$|\psi(t)> = \sum_n c_n|s_n> |a_n> \tag{34}$$

in the Hilbert product space $\mathcal{H}_s \otimes \mathcal{H}_A$.

The density matrix for the pure state is

$$\rho(t) = |\psi(t)> < \psi(t)|. \tag{35}$$

The density matrix for observables relevant to S is

$$\rho = Tr_A \rho(t) = \sum_n |c_n|^2 s_n > < s_n|, \tag{36}$$

if we do not read the pointer position In this case, we see that after the measurement we have a mixed state.

Suppose we read off a particular value, say $|a_m>$, then the appropriate density matrix is

$$|s_m > < s_m| \tag{37}$$

and the probability of measuring the value a_m on the apparatus is $|c_m|^2$.

The change in a measurement from a pure state Eq. (35) to a mixed state described by the density matrix $\rho_{nm} = |c_n|^2 \delta_{nm}$ is known as the reduction of the wave packet.

(b) Reduction of the State Vector

The postulate of reduction of the state vector introduces great difficulty in interpretation of the theory of measurement. The problems are how to account for the mechanism of the reduction and the question of when it takes place. The basic issue is how to reconcile the entangled, Schrödinger time-evolution of the object and the apparatus with the concept of the reduction of the state vector.

Schrödinger (1935) pointed out the difficulties of measurement theory by a thought experiment. He considered a chamber containing a cat, a radioactive atom and a device which releases lethal poison gas on the decay of the atomic nucleus. The state of the system after one half-life is

$$\frac{1}{\sqrt{2}} (|u >_{atom}| \text{cat} >_{live} + |d >_{atom}| \text{cat} >_{dead}), \tag{38}$$

where $|u>_{atom}$ and $|d>_{atom}$ denote the undecayed and decayed states. Such a state is an entangled state. The cat is a macroscopic measurement device, and in a measurement process, it is a superposition of macroscopically distinct states (live cat and dead cat). On opening the chamber, the observer finds the cat in only one of the states – the cat is either alive or dead. This is an act of measurement which forces the system into one of the eigenstates of being alive or being dead. This famous, picturesque thought experiment ("Schrödinger's cat") explains and illustrates the problems of measurement theory. The cat is a very complex thermodynamical system and there are no pure alive or dead cat states.

In the standard ("orthodox") interpretation of quantum mechanics, the unitary time-evolution of the state vector does not apply to a measurement process. The postulate of the reduction of state vector is invoked in which a collapse mechanism transforms a pure-state density matrix into a proper mixture.

(c)   The Watched Pot

There is a long history of the role of measurements on quantum decay transitions of a (metastable) system during the course of its time evolution. Repeated observations on a metastable atom make the decay go to zero. The suppression of the decay of a metastable quantum system due to measurements on it in the course of its evolution is known as the "quantum Zeno effect" (Misra and Sudarshan (1977)).

Let us prepare a system in a state $|\psi_u>$ at time $t = 0$. After a finite time t the state vector evolves to be $e^{-\frac{i}{\hbar}Ht}|\psi_u>$, where H is the Hamiltonian of the system. The probability that the system has not decayed that is the probability of survival in a time interval t is

$$P_u(t) = |A(t)|^2, \tag{39}$$

where the amplitude is

$$A(t) = <\psi_u|e^{-\frac{i}{\hbar}Ht}|\psi_u>. \tag{40}$$

If t is small, then

$$A(t) = 1 - \frac{i}{\hbar} <H> t - \frac{<H>^2 t^2}{\hbar^2} + \dots \text{ and the survival probability is}$$

$$P_u(t) \approx 1 - \frac{< (H - <H>)^2 > t^2}{\hbar^2}, \tag{41}$$

where the expectation values are taken in the state $|\psi_u>$. Now, suppose we divide the time interval t into n equal subintervals and make measurements on or observe the system initially in the state $|\psi_u>$ at the time $\frac{t}{n}, \frac{2t}{n}, \frac{3t}{n}, \dots, t$. If n is large so that $\frac{t}{n}$ is small, the probability that the system has not decayed at the time of the first observation is

$$P_u(t/n) = 1 - \left(\frac{\sigma t}{n\hbar}\right)^2, \text{ where } \sigma^2 = \langle(H - <H>)^2\rangle.$$

The probability of survival in state $|\psi_u>$ at the time t after a sequence of n independent observations is

$$P_u(t) = \left[P_u\left(\frac{t}{n}\right)\right]^n$$

$$= \left[1 - \frac{t^2}{n^2}\,\alpha\right]^n, \text{ where } \alpha = \left(\frac{\sigma}{\hbar}\right)^2$$

$$= 1 - \alpha\,\frac{t^2}{n} + \dots \tag{42}$$

In the limit $n \to \infty$, the system is continuously observed and $P_u(t)$ tends to unity. Hence we conclude like the old adage "a watched pot never boils" that a continuously observed system never changes its state. Continuous or extremely rapid repeated observations on a metastable state make the decay go to zero. This is called the quantum Zeno effect.

Itano et al (1990) performed an experiment using induced transitions with the result in agreement with the theory. However, we note that the result can also be interpreted differently (see Ballentine (1998)). Hence, this experiment cannot be considered a confirmation of the notion of "reduction of the state vector".

(d)   The No-Cloning Theorem

An attractive possibility of state preparation is to make exact replicas or clones of the state of a system and make measurements on these copies. One cannot make a duplicate of a quantum state. One cannot thus build a Xerox machine for a quantum state. This is the content of the no – cloning theorem of Wootters and Zurek (1982). Of course, classical states can, in principle, be copied perfectly.

Let $|\psi^0>$ be an arbitrary quantum state of a system. We try to make a copy ("clone"), of $|\psi^0>$ by a copying device which is in a "known" blank state $|\psi_0^c>$ (white paper in a copying machine). The copying procedure is the following transition

$$|\psi^0> \otimes |\psi_0^c> \to |\psi^0> \otimes |\psi^0>. \tag{43}$$

The transition is effected by a unitary time development operator U.

$$U|\psi^0> \otimes |\psi_0^c> = |\psi^0> \otimes |\psi^0> |\psi_f^c>, \tag{44}$$

where $|\psi_f^c>$ is the final state vector of the copying device. To prove the impossibility of cloning, we assume to the contrary that there are two states $|\psi_1^0>$ and $|\psi_2^0>$ for which Eq. (44) holds:

$$U|\psi_1^0> \otimes |\psi_0^c> = |\psi_1^0> \otimes |\psi_1^0> \otimes |\psi_f^c>, \tag{45}$$

$$U|\psi_2^0> \otimes |\psi_0^c> = |\psi_2^0> \otimes |\psi_2^0> \otimes |\psi_f^c> \tag{45a}$$

We consider a third state

$$|\psi_3^0> = \frac{1}{\sqrt{2}}\,(c_1|\psi_1^0> + c_2|\psi_2^0>) \tag{46}$$

and try to make a copy of it. We have

$$U|\psi_3^0> \otimes |\psi_0^c> = \frac{1}{\sqrt{2}}\,(c_1|\psi_1^0> \otimes |\psi_1^0> \otimes |\psi_f^{c'}>$$

$$+ c_2|\psi_2^0> \otimes |\psi_2^0> \otimes |\psi_f^{c''}> \tag{47}$$

But this is an entangled state and is definitely different from

$$|\psi_3^0> \otimes |\psi_3^0> \otimes |\psi_f^{c'''}>.$$

Therefore, no cloning is possible. It is impossible to build a copying device to copy an arbitrary, unknown quantum state.

22.5 OTHER VIEWS

A number of approaches for the foundations of quantum mechanics have been explored in recent times. In this section we present a brief overview of some of these approaches regarding the interpretation of the foundations of quantum theory.

(a) Consistent–Histories Interpretation

The consistent– (or decoherent) histories approach presents a language for a generalized version of the Copenhagen interpretation. The basic idea of this approach is to study quantum histories which make pricise the notions of the state preparation, an apparatus, the outcome of a measurement. A history is defined as a sequence of properties (predicates) represented by a set of projectors at well-defined successive instants of time, and describing the events occurring in an experiment. A probability is assigned for each history in a family. Griffiths showed that consistency conditions exist due to the consistency of additive probabilities with additive amplitudes yielding meaningful histories. It was shown that formal logic holds in a family of consistent histories. Gell-Mann and Hartle (1990) related the consistency conditions to decoherence.

Let us consider a physical system S. At the initial time t_0, S is described by the density matrix ρ_0. We consider a set of mutually orthogonal hermitian projectors $\{E_{\alpha_i}^{(i)}(t_i)\}$,

$\alpha_i = 1, \ldots, m_i$, $1 \le i \le n$ at an instant of time t_i in a sequence of times $t_1 < t_2 < \ldots < t_n$ with $t_1 > t_0$. We have

$$\sum_{\alpha_i} E^{(i)}_{\alpha_i}(t_i) = 1; \quad E^{(i)}_{\alpha_i}(t_i)\, E^{(i)}_{\beta_i}(t_i) = E^{(i)}_{\alpha_i}(t_i)\, \delta_{\alpha_i \beta_i} \tag{48}$$

Also,

$$E^{(i)}_{\alpha_i}(t) = U^\dagger(t_0, t)\, E^{(i)}_{\alpha_i}(t_0)\, U(t_0, t). \tag{49}$$

A "maximally fine-grained" history is defined by the set

$$H_{(\alpha)} = \{E^{(1)}_{\alpha_1}(t_1),\ E^{(2)}_{\alpha_2}(t_2),\ \ldots,\ E^{(n)}_{\alpha_n}(t_n)\}. \tag{50}$$

This set of projectors acting on the Hilbert space of S is associated with the series of events (defined as properties at given times) or a sequence of sentences of propositions.

This probability p of a history is given by

$$p(H_{(\alpha)}) = D\,(\alpha, \alpha) \tag{51}$$

where the decoherence functional $D(\alpha, \beta)$ is defined by

$$D\,(\alpha, \beta) = Tr\,[E^{(n)}_{\alpha_n}(t_n)\ \ldots\ E^{(1)}_{\alpha_1}(t_1)\,\rho_0\, E^{(1)}_{\beta_1}(t_1)\ \ldots\ E^{(n)}_{\beta_n}(t_1)] \tag{52}$$

The necessary and sufficient consistency condition for two histories $H_{(\alpha)}$ and $H_{(\beta)}$ is

$$\mathrm{Re}\, D(\alpha, \beta) = \delta_{\alpha, \beta}\, D(\alpha, \alpha) \text{ (weak decoherence)} \tag{53}$$

or,
$$D(\alpha, \beta) = \delta_{\alpha\beta}\, D(\alpha, \alpha) \text{ (medium decoherence)} \tag{53a}$$

From Eqs. (51) with (53) or (53a) we obtain all properties of quantum probabilities. In this approach, the reduction of the state vector follows from a Bayesian rule relating the density matrix before the measurement to that after measurement.

Detailed expositions of this approach are given by Griffiths (2002), Omnès (1992, 1994, 1999) and Schlosshauer (2004).

(b) The "Many Worlds" Interpretation

The approach, proposed by Everett (1957) and elaborated by Wheeler, Dewitt, Dewitt and Graham (1973) was to consider a unitarily evolving state vector representing the state of the entire universe. The basic idea is to abandon the system-observer duality and the notion of the state vector reduction of the orthodox quantum mechanics. It is postulated that all terms in the superposition of the total state consistent with the initial conditions correspond to physical states. So physical reality becomes multiple. In the "many worlds" interpretation each physical state corresponds to a particular branch of a constantly splitting universe. This is the basis for obtaining the Born rule as the probability for the occurrence of a particular outcome.

(c) Decoherence

Considerable attention has been devoted in recent years to decoherence. In this approach the standard formulation of quantum theory is assumed to be correct, but the reduction of the state vector is used as a heuristic device. One assumes that realistic quantum systems are not isolated but are coupled to the environment. This effectively leads to a reduction of the state vector of the system by a process of decoherence.

The central idea of the decoherence program is that a quantum system is immersed in its environment and continuously interacts with it. The resulting correlations destroy coherence among the states of the system. According to Zurek (2003) decoherence is the destruction of quantum coherence between preferred states associated with the observables monitored by the environment. As a consequence there is fast, local suppression of interference among the states. This process is especially effective for macroscopic systems. Another consequence is the environment-induced super-selection ("einselection") which is defined as the decoherence – imposed selection of the preferred sets of states ("pointer states") that remain stable despite being immersed in the environment. Einselected pointer sets correspond to the "classical states". Zurek recently introduced a symmetry termed "environment – assisted invariance" or "envariance" exhibited by a system immersed in and interacting with the environment. Envariance allows one to explain the emergence of probabilities and leads to Born's rule and to reduced density matrices. One can define ignorance of the state of the system as a consequence of envariance. Decoherence program claims to have found a satisfactory explanation for the manner in which states of some quantum systems become effectively classical.

Let us now reconsider the von Neumann model of ideal measurement discussed earlier with the environment ε included (Schlosshauer (2004)). Let $|e_0>$ be the initial state vector of ε in the Hilbert space $\mathcal{H}_\varepsilon$. The Hilbert space of the entire system $SA\varepsilon$ is the tensor product space $\mathcal{H}_s \otimes \mathcal{H}_A \otimes \mathcal{H}_\varepsilon$. The evolution of the system $SA\varepsilon$ is indicated below

$$\left(\sum_n c_n|s_n>\right)|a_r>|e_0> \xrightarrow{\ (1)\ } \left(\sum_n c_n|s_n > |a_n >\right)|e_0 >$$

$$\xrightarrow{\ (2)\ } \sum_n c_n|s_n >|a_n|e_n > \tag{54}$$

where the $|e_n >$ are the states of the environment associated with the different pointer states $|a_n >$ of A.

Let O_{SA} be an observable of SA only. Its expectation value is

$$\langle O_{SA} \rangle = Tr \left(\rho_{SA\varepsilon} [O_{SA} \otimes I_\varepsilon]\right).$$

$$= Tr_{SA}(\rho_{SA} O_{SA}). \tag{55}$$

The density matrix of the entire system $SA\varepsilon$ is given by

$$\rho_{SA\varepsilon} = \sum_{mn} c_m c_n^* |s_m>|a_m>|e_m><s_n|<a_n|<e_n|. \tag{56}$$

In a measurement, we do not read off ε, i.e., $|e_n>$ since it is not possible to keep track of all the large number of macroscopic consequences. So in Eq. (56) we have replaced $\rho_{SA\varepsilon}$ by ρ_{SA} obtained from $\rho_{SA\varepsilon}$ by taking the trace over all the unobserved degrees of the environment:

$$\rho_{SA} = Tr_\varepsilon \, (\rho_{SA\varepsilon})$$

$$= \sum_{mn} c_m c_n^* |s_m>|a_m><s_n|<a_n|<e_n||e_m>. \tag{57}$$

ρ_{SA} is the local or reduced density matrix. Clearly, ρ_{SA} contains interference terms, e.g., $|s_m>|a_m><s_n|<a_n|$, $m \neq n$. By considering explicit models for the environment and its interaction with the system, it has been shown that due to the large number of degrees of freedom of the environment, the pointer states $|e_n>$ of the environment extremely rapidly approach orthogonality, $\langle e_n | e_m \rangle \, (t) \rightarrow \delta_{nm}$ such that ρ_{SA} becomes orthogonal in the pointer basis $\{|a_n>\}$. We have

$$\rho_{SA} \xrightarrow{t} \rho_{SA}^d = \sum_n |c_n|^2 |s_n>|a_n><s_n|<a_n|$$

$$= \sum_n |c_n|^2 \, P_n^s \otimes P_n^A, \tag{58}$$

where P_n^S and P_n^A are the projectors onto the eigenstates of S and A respectively. Hence the interference terms have dropped out, i.e., phase coherence has been locally lost. This is called the environment-induced decoherence.

Detailed studies have shown how rapidly the decoherence takes place for a macroscopic system. Even microscopic systems are very rapidly decohered. The 3K cosmic microwave background radiation produces decoherence for mesoscopic systems on a decoherence time scale $t_D \sim 10^{-23}$s.

Raimond, Brune, Haroche, and their colleagues at the Ecole Normale Superieure (Brune et al (1996), Raimond, Brune, and Haroche (2001)) carried out an experimental study of quantum decoherence. The system was the field in a microwave cavity interacting with an environment consisting of the cavity walls. The conclusion was that quantum decoherence exists and the experimental results agree with theory.

22.6 EPILOGUE

En toutes choses il faut considérer le fin (J. de La Fontaine)

In the course of this book we have discussed the standard or orthodox formulation of quantum theory. The development of quantum mechanics led to a radical change in the concepts used in classical physics to describe the world. From its inception, there have been conceptual issues which confront our attempts to build a coherent, complete interpretation of the foundations of quantum mechanics. The present chapter discusses the interpretation.

After intense efforts, Bohr and his followers (Heisenberg, Born, and Pauli) carved out the interpretation of quantum theory known as the Copenhagen interpretation (CI). According to Bohr, quantum theory is a "completely rational description of physical phenomena". From the beginning, there were major difficulties in the interpretation of quantum theory to which the CI tries to provide the answers.

In the standard or orthodox formalism of quantum mechanics it is postulated that the most complete possible description of the state of a system is given by a state vector. The principle of linear superposition admits of linear combination of states as possible quantum states. This concept, well-established for microscopic domain, gives rise to serious problem in case of extension to macroscopic systems. The problem of macroscopic interferences is illustrated by the famous Schrödinger's cat. The probabilistic nature of quantum mechanics and the measurement process give rise to a variety of conceptual problems. In the standard formalism probabilities are postulated, but do not follow from the sample space of predictable individual events. The unitary evolution of the state vector has to be supplemented by the additional postulate of the state vector reduction which purports to give the state vector after measurement. The incisive critique of Einstein and his epic debates with Bohr clarified the nature of quantum entanglement. Among the founding fathers, Einstein and also de Broglie and Schrödinger believed that quantum theory, although a highly successful physical theory, was incomplete. To Einstein, reality meant physical reality. EPR gave an operational definition of physical reality and finally concluded "We are thus forced to conclude that the quantum-mechanical description of physical reality given by wave functions (i.e., state functions) is not complete".

The quaternity of fundamental problems in quantum theory is (Isham (1995))

  (i) The meaning of probability;

 (ii) The role of measurement;

(iii) The problem of actualizing potentialities (the reduction of the state vector);

(iv) Quantum entanglement.

These issues are linked together.

Bohr believed that quantum mechanics was complete and physical reality was Janus-faced. There was a macroscopic causal classical domain (includes measuring apparatuses, observers) and the quantum world of atoms and particles. Bohr emphasized that the boundary between the domains of classical and quantum worlds must be movable. Bohr formulated the principle of complementarity to interpret the wave-particle duality and the uncertainty relations. It is postulated that a measurement induces a break in the unitary state vector evolution through the state vector reduction or collapse of the wave function. The principle of superposition does not apply in the classical domain. In his reply to the argument of EPR, Bohr maintained that the definition of physical reality proposed by EPR contained an element of ambiguity regarding the meaning of the expression "without in any way disturbing a system". Bohr argued that "the procedure of measurement has an essential influence on which the very definition of the physical quantities rests". Bohr refuted the conclusion of EPR that quantum-mechanical description is essentially incomplete.

The profound analysis of Bell of the EPR problem (i.e., entanglement, hidden variables leading to Bell's theorems, inequalities and their significance, locality condition, the nature of the measurement problem, the quantum jumps, the separation between the classical and the quantum worlds) provides a source and inspiration for the intense research activity in the field of interpretational foundation of quantum mechanics.

For dealing with these interpretive problems, a number of different approaches have been explored in recent years, e.g., the consistent-histories approach, the many-worlds interpretation, the decoherence program, etc.

Contemporary research both theoretical and experimental on the foundations of quantum theory, has yielded answers to some but many others remain undecided.

The following has been established

(i) Local hidden variable theories have been experimentally ruled out. Thus, no LHV theory underlies quantum theory. Non-local HV theories are still viable options but the prospects of such a theory are not good.

(ii) The phenomenon of quantum entanglement is an objective feature of systems.

(iii) Although CI claimed that quantum theory is applicable to individual objects, all experimental evidence were obtained for large ensembles. Recent experiments opened the possibility of continuous observation of single objects. Many experiments have been performed on single ion or a neutral atom inside a trap. A direct observation of quantum jumps in the fluorescence of a single trapped ion was made in 1986. The conceptual issue of the ensemble approach versus the individual system approach was discussed by Bohr and Einstein.

(iv) Recent experiments confirm the existence of quantum decoherence.

Many leading adherents of the decoherence program for the foundations of quantum mechanics claim that decoherence through einselection helps solve the measurement problem. Others assert "Decoherence by itself does not yet solve the measurement problem."

Though much progress has been made, the decoherence approach is yet to provide consistent answers to many questions about the foundations of quantum mechanics.

Intensive efforts in recent years clarified many conceptual issues in quantum theory although we are far from resolving many other conceptual issues. Future progress in the interpretation of quantum mechanics will depend on asking searching questions about quantum mechanics and on performing new experiments. We are still far from answering Pilate's famous question : "What is truth?"

Bibliography

Abers, E.S. (2004), Quantum Mechanics (Pearson Education, New Jersey).

Abramowitz, M., and Stegun, I.A. (1964), Handbook of Mathematical Functions (National Bureau of Standards).

Adler, S.L. (2004), Quantum Theory As An Emergent Phenomenon (Cambridge University Press, Cambridge).

Aharonov, Y., and Bohm, D. (1959), Significance of Electromagnetic Potentials in the Quantum Theory, Phys. Rev. **115**, 485-491.

Aharonov, Y., and Anandan, J. (1987), Phase Change during a Cyclic Quantum Evolution, Phys. Rev. **58**, 1593-1596.

Albeverio, S., Gesztesy, F., Høegh-Krohn, R. and Holden, H. (1988), Solvable Models in Quantum Mechanics, (Springer Verlag, New York).

Arfken, G.B., and Weber, H.J. (1995), Mathematical Methods for Physicists, (Fourth Edn. Academic, San Diego).

Ashcroft, N.W., and Mermin, N.D. (1976), Solid State Physics (Saunders, Philadelphia).

Aspect, A., Grangier, P., and Roger, G. (1981), Experimental Tests of Realistic Local Theories via Bell's Theorem, Phys. Rev. Lett. **47**, 460-467.

— (1982) Experimental Realization of Einstein-Podolsky-Rosen-Bohm Gedanken experiment: A New Violation of Bell's Inequality, Phys. Rev. Lett. **49**, 91-94.

Aspect, A., Dalibard, J. and Roger, G. (1982), Experimental Test of Bell's Inequalities Using Time-Varying Analyzers, Phys. Rev. Lett. **49**, 1804-1807.

Auletta, G. (2000), Foundations and Interpretation of Quantum Mechanics, (World Scientific, Singapore).

Ballentine, L.E. (1998), Quantum Mechanics: A Modern Development (World Scientific, Singapore).

Bander, M., and Itzykson, C. (1966), Group Theory and the Hydrogen Atom (I), (II), Rev. Mod. Phys. **38**, 330, 346-358.

Barclay, D., and Maxwell, C. (1991) Phys. Lett. A157, 357-360.

Barut, A.O., and Raczka, R. (1986). Theory of Group Representations and Applications (Second Revised Edn., World Scientific, Singapore).

Bates, D.R. (Ed.) (1961-62), Quantum Theory in Three Volumes (Academic, New York).

Bayen, F., Flato, M., Fronsdal, C., Lichnerowicz, A., and Sternheimer, D. (1978), Deformation Theory and Quantization I, II, Ann. Phys. (N.Y.) **111**, 61-110, 111-151.

Baym, G. (1969), Lectures on Quantum Mechanics (Benjamin, New York).

Belinfante, F.J. (1973), A Survey of Hidden Variable Theories (Pergamon, Oxford).

Bell, J.S. (1987), Speakable and Unspeakable in Quantum Mechanics (Cambridge University Press, Cambridge) (Second Edn. (2004)).

Bell, M. Gottfried, K., and Veltman, M. (Eds.) (2001), John S.Bell on the Foundations of Quantum Mechanics (World Scientific, Singapore).

Bender, C.M., and Wu, T.T.(1969), Anharmonic Oscillator, Phys. Rev. **184**, 1231-1260; (1971), Large Order Behavior of Perturbation Theory, Phys. Rev. Lett. **27**, 461-465; (1973), Anharmonic Oscillator II. A Study of Perturbation Theory in Large Order, Phys. Rev. **D7**, 1620-1636.

Bender, C.M., and Orszag, S.A. (1978), Advanced Mathematical Methods for Scientists and Engineers (McGraw-Hill, New York).

Berry, M.V., (1984), Quantal Phase Factor Accompanying Adiabatic Changes Proc. R. Soc. Lond. **A392**, 45-57; (1988), The Geometric Phase, Sci. Am. **259(6)**, 46-52.

Berry, M.V., and Mount, K.E. (1972), Semiclassical Wave Mechanics, Rep. Prog. Phys. **35**, 315-377.

Bethe, H.A., and Salpeter, E.E. (1957), Quantum Mechanics of One- and Two-Electron Atoms, (Springer-Verlag, Berlin).

Bethe, H.A., and Jackiw, R.W. (1968), Intermediate Quantum Mechanics, (Second Edn., Benjamin, New York).

Biedenharn, L.C. (1963), Group Theoretical Approaches to Nuclear Spectroscopy in Lectures in Theoretical Physics, Boulder, Vol. V, 258-421, Eds. W.E. Brittin, B.W. Downs and J. Downs (Interscience, New York).

Biedenharn, L.C., and Van Dam, H. (Eds.) (1965), Quantum Theory of Angular Momentum (Academic, New York).

Biedenharn, L.C., and Louck, J.D. (1981), Angular Momentum in Quantum Physics, Encycl. of Maths and its Applications (Academic, New York).

Bjorken, J.D., and Drell, S.D. (1964), Relativistic Quantum Mechanics (McGraw-Hill, New York).

Blatt, J.M., and Weisskopf, V.F. (1952), Theoretical Nuclear Physics (Wiley, New York).

Blockley, C.A., and Stedman, G.E. (1985) Simple Supersymmetry: I. Basic Examples, Eur. J. Phys. **6**,218-224.

Bohm, A., and Gadella, M. (1989), Dirac Kets, Gamow Vectors, and Gelfand Triplets (Springer, Berlin).

Bohm, A. (1993), Quantum Mechanics: Foundations and Applications (Third Edn., Springer-Verlag, New York).

Bohm, D. (1951), Quantum Theory (Prentice-Hall, New York) (1989), (Reprinted by Dover, New York).

Bohm, D., (1952), A Suggested Interpretation of the Quantum Theory in Terms of "Hidden Variables", I and II (1952), Phys. Rev. **85**,166-193.

Bohm, D., and Aharonov, Y. (1957), Discussion of Experimental Proofs for the Paradox of Einstein, Rosen and Podolsky, Phys. Rev. **108**, 1070-1075.

Bohr, A., and Mottelson, B.R. (1969, 1975), Nuclear Structure, Vols. I, II (Benjamin, New York).

Bohr, N. (1935), Can Quantum Mechanical Description of Physical Reality be Considered Complete?, Phys. Rev. **48**, 696-699.

Bohr, N. (1949), Discussion with Einstein on Epistemological Problems in Atomic Physics in Albert Einstein: Philosopher-Scientist, P.A. Schilpp (Ed.) (Library of Living Philosphers, Inc.).

Bransden, B.H., and Joachain, C.J. (1983), Physics of Atoms and Molecules, (Longman, London).

Bransden, B.H., and Joachain, C.J. (2000), Quantum Mechanics (Second Edn., Pearson Education).

Brezin, E., Parisi, G. and Zinn-Justin, J. (1977), Perturbation Theory at Large Orders for a Potential with Degenerate Minima, Phys. Rev. **D16**, 408-412.

Brink, D.M., and Satchler, G.R. (1968), Angular Momentum (Second Edn. Clarendon, Oxford).

Brune, M., Hagley, E., Dreyer, J., Maitre, X., Wunderlich, C., Raimond, J.-M. and Haroche, S. (1996), Phys. Rev. Lett. **77**, 4887-4890.

Calogero, F. (1967), Variable Phase Approach to Potential Scattering (Academic, New York).

Carruthers, P., and Nieto, M.M. (1965), Coherent States and the Forced Quantum Oscillator, Am. J. Phys. **33**, 537-544.

Carruthers, P., and Nieto, M.M. (1968), Phase and Angle Variables in Quantum Mechanics, Rev. Mod. Phys. **40**, 411-440.

Casimir, H.B.G. (1948), On the Attraction Between Two Perfectly Conducting Plates, Proc. Kon. Ned. Akad. Wet. **51**, 793-795.

Casimir, H.B.G., and Polder, D. (1948), The Influence of Retardation on the London – van der Waals Forces, Phys. Rev. **73**, 360-372.

Chaichian, M., Sheikh-Jabbari, M.M. and Tureanu, A. (2001), Hydrogen Atom Spectrum and the Lamb Shift in Non-commutative Quantum Electrodynamics, Phys. Rev. Lett. **86**, 2716-2719.

Clauser, J.F., Horne, M.A., Shimony, A. and Holt, R.A. (1969), Proposed Experiment to Test Local Hidden-Variable Theories, Phys. Rev. Lett. **23**, 880-884.

Clauser, J.F., and Horne, M.A. (1974), Experimental Consequences of Objective Local Theories, Phys. Rev. **D10**, 526-535.

Clauser, J.F., and Shimony, A. (1978), Bell's Theorem: Experimental Tests and Implications, Rep. Prog. Phys. **41**, 1881-1927.

Cohen-Tannoudji, C., Diu, B. and Laloë, F. (1977), Quantum Mechanics, (Wiley, New York).

Colegrove, F.D., Franken, P.A., Lewis, R.R. and Sands, R.H. (1959), Novel Method of Spectroscopy with Applications to Precision Fine Structure Measurements, Phys. Rev. Lett. **3**, 420-422.

Commins, E.D., and Bucksbaum, P.H. (1983), Weak Interactions of Leptons and Quarks (Cambridge University Press, Cambridge).

Condon, E.U., and Shortley, G.H. (1951), The Theory of Atomic Spectra, (Cambridge University Press, Cambridge).

Cooper, F., and Freedman, B. (1983), Aspects of Supersymmetric Quantum Mechanics, Ann. Phys. (N.Y.), **146**, 262-288.

Cooper, F., Khare, A. and Sukhatme, U.P. (1995), Supersymmetry and Quantum Mechanics, Phys. Rep. **251**, 267-.

Das, A., and Melissinos, A.C. (1986), Quantum Mechanics: A Modern Introduction (Gordon and Breach, New York).

Das, A. (2003), Lectures on Quantum Mechanics, (Hindustan Book Agency, New Delhi).

Davydov, A.S. (1965), Quantum Mechanics (Second Edn., Pergamon, Oxford).

De Alfaro, V., and Regge, T. (1965), Potential Scattering (North-Holland, Amsterdam).

Delabaere, E., Dillinger, H. and Pham, F. (1997), Exact Semiclassical Expansions for One-Dimensional Quantum Oscillators, J. Math. Phys. **38**, 6126-6184.

de la Madrid, R. (2005), The Role of the Rigged Hilbert Space in Quantum Mechanics, Eur. J. Phys. **26**, 287-312.

DeLange, O.L., and Raab, R.E. (1991), Operator Methods in Quantum Mechanics, (Clarendon, Oxford).

Dennery, P., and Krzywicki, A. (1967), Mathematics for Physicists, (Harper and Row, New York).

de Shalit, A., and Talmi, I. (1963), Nuclear Shell Theory (Academic, New York).

d'Espagnat, B. (1988), Conceptual Foundations of Quantum Mechanics (Second Edn., Persens, Reading).

d'Espagnat, B. (1995), Veiled Reality (Addison-Wesley, Reading).

Dewitt, B.S., and Graham, N. (1973), The Many-Worlds Interpretation of Quantum Mechanics. (Princeton University Press, Princeton).

Dirac, P.A.M. (1958), The Principles of Quantum Mechanics. (Fourth Edn., Charendon, Oxford).

Dodonov, V.V., Man'ko, O.V. and Man'ko, V.I. (1993), Time-Dependent Oscillator with Kronig-Penney Excitation, Phys. Lett. **A175**, 1-4.

Douglas, M.R., and Nekrasov, N.A. (2001), Non-commutative Field Theory, Rev. Mod. Phys. **73**, 977-1029.

Dutt, R., Khare, A., and Sukhatme, U.P. (1988), Supersymmetry, Shape Invariance and Exactly Solvable Potentials (1988), Am. J. Phys. **56**, 163-168.

Edmonds, A.R. (1957), Angular Momentum in Quantum Mechanics (Princeton University Press, Princeton).

Einstein, A. (1909), Zum gegenwärtigen Stande des Strahlung Problems, Phys. Zeits, **10**, 185-193.

Einstein, A., Podolsky, B. and Rosen, N. (1935), Can Quantum Mechanical Description of Physical Reality be Considered Complete? Phys. Rev. **47**, 777-780.

Einstein, A. (1949), Autobiographical Notes; Remarks to the Essays Appearing in this Collective Volume in Albert Einstein: Philosopher Scientist, Ed. P.A. Schilpp. (Library of Living Philosophers, Inc.)

Eisenberg, J.M., and Greiner, W. (1972), Microscopic Theory of the Nucleus (North-Holland Amsterdam)

Elizaide, E. and Romeo, A. (1991), Essentials of the Casimir Effect and Its Computation, Am. J. Phys. **59**, 711-719.

El'yashevich, M.A. (1977), From the Origin of Quantum Concepts to the Establishment of Quantum Mechanics, Sov. Phys. Usp. **20**, 656-682.

Emmerson, J. McL., (1972), Symmetry Principles in Particle Physics (Clarendon, Oxford).

Englert, B.-G., (2006), Lectures in Quantum Mechanics, (3 Vols) (World Scientific, Singapore).

Epstein, S.T. (1974), The Variation Method in Quantum Chemistry, (Academic New York).

Erdös, P. and Herndon, R.C. (1982), Theories of Electrons in One-Dimensional Disordered Systems, Adv. in Phys. **31**, 65-163.

Esposito, G., Marmo, G. and Sudarshan, G. (2004), From Classical to Quantum Mechanics (Cambridge University Press, Cambridge).

Everett III, H., (1957), "Relative State", Formulation of Quantum Mechanics, Rev. Mod. Phys. **29**, 454-462.

Falicov, L.M., (1966), Group Theory and Its Physical Applications, (The University of Chicago Press, Chicago).

Fano, U. (1957), Description of States in Quantum Mechanics by Density Matrix and Operator Techniques, Rev. Mod. Phys. **29**, 74-93.

Fano, U., and Racah, G. (1959), Irreducible Tensorial Sets (Academic, New York).

Felsager, B. (1987), Geometry, Particles and Fields (Fourth Edn., Odense University Press).

Fermi, E. (1950), Nuclear Physics (The University of Chicago Press, Chicago).

Fetter, A.L., and Walecka, J.D. (1971), Quantum Theory of Many-Particle Systems, (McGraw-Hill, New York).

Feynman, R.P., Leighton, R.B., and Sands, M. (1965), The Feynman Lectures on Physics, Vol. III, Quantum Mechanics (Addison-Wesley, Reading).

Feynman, R.P., and Hibbs, A.R. (1965), Quantum Mechanics and Path Integrals (McGraw-Hill, New York).

Flügge, S. (1971), Practical Quantum Mechanics I and II (Springer-Verlag, Berlin).

Folland, N.O. (1983), Energy Bands and Forbidden Gaps in the Krong-Penney Model, Phys. Rev. **28**, 6068-6070.

Fonda, L., and Ghirardi, G.C. (1970), Symmetry Principles in Quantum Physics. (Dekker, New York).

Fontana, P.R. (1982), Atomic Radiative Processes. (Academic, New York).

Fortson, E.N., Sandars, P.G.H. and Barr, S.M. (2003), The Search for a Permanent Electric Dipole Moment, Phys. Today **56(6)**, 33-39.

Frautschi, S. (1963), Regge Poles and S-Matrix Theory (Benjamin, New York).

Friedrich, H., and Wintgen, D. (1989), The Hydrogen Atom in a Uniform Magnetic Field — An Example of Chaos, Phys. Rep. **183**, 37-79.

Friedrich, B. and Herschbach, D. (2003), Stern and Gerlach; How a Bad Cigar Helped Reorient Atomic Physics, Physics Today, **56(12)**, 53-59.

Fröman, N. and Fröman, P.O. (1965), JWKB Approximation. Contribution to the Theory (North Holland, Amsterdam)

Furry, W.H. (1963), Behavior of de Broglie Wave and Wave Packets in Lectures in Theoretical Physics, Boulder, Vol. V, 1-112.

Galindo, A., and Pascual, P. (1991), Quantum Mechanics I, II (Springer-Verlag, Berlin).

Gamboa, J., Loewe, M. and Rojas, J.C. (2001), Noncommutative Quantum Mechanics, Phys. Rev. **D64**, 067901.

Garg, A. (2000), Tunnel Splittings for One-Dimensional Potentials Revisited, Am. J. Phys. **68**, 430-437.

Garstang, R.H. (1977), Atoms in High Magnetic Fields, Rep. Prog. Phys. **40**, 105-154.

Gasiorowicz, S. (1996), Quantum Physics, (Second Edn. Wiley, New York).

Gasparian, V.M., Alltshuler, B.L., Aronov, A.G. and Kasamanian, Z.A. (1988), Resistance of One-Dimensional Chains in Kronig-Penney Like Models, Phys. Lett. **A132**, 201-305.

Gel'fand, I.M., and Shilov, G.E. (1964), Generalized Functions, Vol. I, Properties and Operations. (Academic, New York).

Gel'fand, I.M., and Vilenkin, N. Ya. (1964), Generalized Functions, Vol. IV. Applications of Harmonic Analysis (Academic, New York).

Gendenstein, L.E. (1983), Derivation of Exact Spectra of the Schrödinger Equation by means of Supersymmetry, JETP Lett. **38**, 356-359.

Georgi, H. (1982), Lie Algebras in Particle Physics (Benjamin, Reading).

Gieres, F. (2000), Mathematical Surprises and Dirac's Formalism in Quantum Mechanics, Rep. Prog. Phys. **63**, 1893-1931.

Gibson, W.M., and Pollard, B.R. (1976), Symmetry Principles in Elementary Particle Physics (Cambridge University Press, Cambridge).

Glauber, R.J. (1959), High Energy Collision Theory in Lectures in Theoretical Physics, Boulder, Vol. I, Eds. Brittin, W.E. and Dunham, L.G., 315-414 (Interscience, New York).

Glauber, R.J. (1963), Photon Correlations, Phys. Rev, Lett. **10**, 84-86. The Quantum Theory of Optical Coherence, Phys. Rev. **130**, 2529-2539; Coherent and Incoherent States of the Radiation Field, Phys. Rev. **131**, 2766-2788.

Gleason, A.M. (1957), Measures on the Closed Subspaces of Hilbert Space, J. Math. Mech. **6**, 885-893.

Goldberger, M.L., and Watson, K.M. (1964), Collision Theory, (Wiley, New York).

Goldstein, J., Lebiedzik, C., and Robinett, R.W., (1994), Supersymmetric Quantum Mechanics: Examples with Dirac- δ-Function, Am. J. Phys. **62**, 612-618.

Gottfried, K. (1966), Quantum Mechanics I. Fundamentals (Benjamin, New York).

Gottfried, K., and Yan, T-M. (2003), Quantum Mechanics: Fundamentals, (Second Edn. Springer, New York).

Gradshteyn, I.S., and Ryzhik, I.M. (1980), Tables of Integrals, Series and Products (Acdemic, New York).

Greenberger, D.M., Horne, M.A., Shimony, A. and Zeilinger, A. (1990), Bell's Theorem Without Inequalities, Am. J. Phys. **58**, 1131-1143.

Greiner, W. (1989), Quantum Mechanics: An Introduction (Springer-Verlag, Berlin).

Greiner, W., and Müller, B. (1989), Quantum Mechanics: Symmetries, (Springer-Verlag, Berlin).

Griffiths, D.J., and Steinke (2001), Waves in Locally Periodic Media, Am. J. Phys. **69**, 137-154.

Griffiths, D.J. (2005), Introduction to Quantum Mechanics (Pearson Education, N.J.).

Griffiths, R.B. (2002), Consistent Quantum Theory (Cambridge University Press, Cambridge).

Groeneweld, H.J. (1946), On the Principles of Elementary Quantum Mechanics, Physica, **12**, 405-460.

Gutzwillar, M.C. (1990), Chaos in Classical and Quantum Mechanics (Springer Verlag, New York).

Hamermesh, M. (1962), Group Theory and Its Application to Physical Problems (Addison-Wesley, Reading).

Hagedorn, R. (1966), Causality and Dispersion Relations in Preludes in Theoretical Physics. Eds. de-Shalit, A., Feshback, H. and van Hovel, (North-Holland, Amsterdam).

Hanbury Brown, R., and Twiss, R.Q. (1954), Phil. Mag. **45**,663: (1956) Nature **177**, 27-, 178-1447-; (1957) Proc. Roy. Soc. (London) **242A**,300-, **243A**, 291-.

Hancock, J., Walton, M.A., and Wynder, B. (2004), Quantum Mechanics Another Way, Eur. J. Phys. **25**, 525-534.

Hassani, S. (1999), Mathematical Physics: A Modern Introduction to its Foundations (Springer, New York).

Haymaker, R.W., and Rau, A.R.P. (1986), Supersymmetry in Quantum Mechanics, Am. J. Phys. **54**, 928-936.

Heine, V. (1970), The Pseudopotential Concept, in Solid State Physics: Advances in Research and Applications, Eds. Ehrenreich, H., Seitz, F. and Turnbull, D. Vol. **24**, 1-36.

Heitler, W. (1954), The Quantum Theory of Radiation (Third Edn., Clarendon, Oxford).

Henley, E.M. (1969), Parity and Time-Reversal Invariance in Nuclear Physics in Ann. Rev. Nucl. Sci. **19**, 367-427.

Hilgevoord, J. (1996), The Uncertainty Principle for Energy and Time, Am. J. Phys. **64**,1451-1456; (2002) Time in Quantum Mechanics, Am. J. Phys. **70**, 301-306.

Hillery, M., O'Connel, R.F., Scully, M.O. and Wigner, E.P. (1984), Distribution Functions in Physics: Fundamentals, Phys. Rep. **106**, 121-167.

Hirschfeld, A.C., and Henselder, P. (2002), Deformation Quantization in the Teaching of Quantum Mechanics, Am. J. Phys. **70**, 537-547.

Ho, P-M., and Kao, H-C. (2002), Non-commutative Quantum Mechanics from Non-commutative Quantum Field Theory, Phys. Rev. Lett. **88**, 151602.

Holstein, B.R., and Swift, A.R. (1982), Path Integrals and the WKB Approximation, Am. J. Phys. **50**, 829-832.

Holstein, B.R. (1984), Semi-classical Treatment of Above Barrier Scattering, Am. J. Phys. **52**, 321-325; (1988), Semi-classical Treatment of the Double Well, Am. J. Phys. **56**, 338-348; (1989), The Adiabatic Theorem and Berry's Phase, Am. J. Phys. **57**, 1079-1084; (1992), Topics in Advanced Quantum Mechanics. (Addison-Wesley, Calif.) (2001), The van der Waals Interaction, Am. J. Phys. **69**, 441-449.

Hori, J. (1968), Spectral Properties of Disordered Chains and Lattices, (Pergamon, Oxford).

Infeld, L., and Hull, T.E. (1951), The Factorization Method, Rev. Mod. Phys. **23**, 21-68.

Inui, T., Tanabe, Y., and Onodera, Y. (1990), Group Theory and Its Applications in Physics (Springer-Verlag, Berlin).

Isham, C.J. (1995), Lectures on Quantum Theory. Mathematical and Structural Foundations (Imperial College Press, London).

Itano, W.M., Heinzen, D.J., Bollinger, J.J. and Wineland, D.J., (1990), Quantum Zeno Effect, Phys. Rev. **A41**, 2295-2300.

Itzykson, C., and Zuber, J.B. (1985), Quantum Field Theory, (McGraw-Hill, New York).

Jackson, J.D. (1975), Classical Electrodynamics (Second Edn. Wiley, New York).

Jacob, M., and Wick, G.C. (1959), On the General Theory of Collisions for Particles with Spin, Ann. Phys. (N.Y.) **7**, 404-428.

Jammer, M. (1974), The Philosophy of Quantum Mechanics (Wiley, New York); (1966), The Conceptual Development of Quantum Mechanics (McGraw-Hill, New York) (1989) (Second Edn., American Institute of Physics, New York).

Jauch, J.M. (1968), Foundations of Quantum Mechanics. (Addison-Wesley, Reading).

Joachain, C.J. (1983), Quantum Collision Theory, (Third Edn. North-Holland, Amsterdam).

Jordan, T.F. (1969), Linear Operators for Quantum Mechanics (Wiley, New York).

Junker, G. (1996), Supersymmetric Methods in Quantum and Statistical Physics (Springer, Berlin).

Kaplan, I.G. (1975), Symmetry of Many-Electron Systems (Academic, New York).

Kato, T. (1980), Perturbation Theory for Linear Operators (Second Edn. Springer-Verlag, Berlin).

Kemble, E.C. (1958), The Fundamental Principles of Quantum Mechanics (Dover, New York).

Kittel, C. (1987), Quantum Theory of Solids (Second Revised Edn., Wiley, New York).

Klauder, J.R., and B.S. Skagerstam (1985), Coherent States (World Scientific, Singapore).

Kleber, M. (1994), Exact Solutions for Time-Dependent Phenomena in Quantum Mechanics, Phys. Rep. **236**, 331-395.

Kochen, S., and Specker, E.P. (1967), The Problem of Hidden Variables in Quantum Mechanics, J. Math. Mech. **17**, 59-87.

Kramers, H.A. (1957), Quantum Mechanics (Dover, New York)

Kronig, R. de L., and Penney, W.G. (1931), Quantum Mechanics of Electrons in Crystal Lattices, Proc. R. Soc. (Lond.) **A130**, 499-513.

Lai, D. (2001), Matter in strong magnetic fields, Rev. Mod. Phys. **73**, 629-661.

Landau, L.D., and Lifshitz (1977), Quantum Mechanics (Non-relativistic Theory) (Third Edn. Pergamon, Oxford).

Landau, R.H. (1996), Quantum Mechanics II (Wiley, New York).

Landauer, R. (1970), Transmission Coefficient, Phil. Mag. **21**, 863-.

Lee, H.W. (1995), Theory and Applications of the Quantum Phase-Space Distribution Functions, Phys. Rep. **259**, 147-211.

Lee, T.D. (1988), Particle Physics and Introduction to Field Theory, (Harwood, Chur).

Lee, T.D., and Yang, C.N. (1957), Elementary Particles and Weak Interactions, BNL 443(T-91), Bookhaven National Laboratory Report.

Leggett, A.J., Chakravarty, S., Dorsey, A.T., Fisher, M.P.A., Garg, A., and Zwerger, W. (1987) Dynamics of the Dissipative Two-State System, Rev. Mod. Phys, **59**, 1-85.

Le Guillou, J.C., and Zinn-Justin, J. (Eds.) (1990), Large-Order Behavior of Perturbation Theory (North-Holland, Amsterdam).

Levine, I.N. (2000), Quantum Chemistry, (Fifth Edn., Pearson Education, N.J.).

Lévy-Leblond, J.M. (1971), Galilei Group and Galilean Invariance in Group Theory and Its Applications, Vol. II, 222-299 Ed. Loebl, E.M. (Academic, London).

Liboff, R.L. (1998), Introductory Quantum Mechanics (Third Edn. Addison-Wesley, Reading).

Lipkin, H.J. (1973), Quantum Mechanics: New Approaches to Selected Topics (North-Holland, Amsterdam).

Lindgren, I., and Morrison, J. (1986), Atomic Many-Body Theory (Second Edn. Springer-Verlag, Berlin).

Loebl, E.M. (1971), Group Theory and Its Applications, Vols. I and II (Academic, New York).

Loudon, R. (1983), The Quantum Theory of Light (Second Edn. Clarendon, Oxford).

Luo, S. (2005), Heisenberg Uncertainty Relations for Mixed States, Phys. Rev. **A72**, 042110.

Mahan, G.D. (2000), Many-Particle Physics (Third Edn. Plenum, New York).

Mandel, L. (1963), Fluctuations of Light Beams in Prog. in Optics, Vol. II (North-Holland, Amsterdam).

Marshak, R.E. (1993), Conceptual Foundations of Modern Particle Physics, (World Scientific, Singapore).

Manoukian, E.B. (2006), Quantum Theory: A Wide Spectrum, (Springer, New York).

Mattis, D.C. (Ed.) (1993), The Many-Body Problem: An Encyclopedia of Exactly Solved Models in One Dimension, (World Scientific, Singapore).

Mattuck, R.D. (1976), A Guide to Feynman Diagrams in the Many-Body Problem (Second Edn. McGraw-Hill, New York).

McIntosh, H.V. (1971), Symmetry and Degeneracy in Group Theory and Its Applications, Vol. II, 75-144.

Mermin, N.D. (1990), Boojums All the Way Through (Cambridge University Press, New York) (1993), Hidden Variables and the Two Theorems of John Bell, Rev. Mod. Phys. **65**, 803-815.

Merzbacher, E. (1998), Quantum Mechanics (Third Edn. Wiley, New York).

Messiah, A. (1961), Quantum Mechanics (Interscience, New York).

Meystre, P., and Sargent III, M. (1999), Elements of Quantum Optics (Third Edn. Springer, New York).

Migdal, A.B., and Krainov, V. (1969), Approximation Methods in Quantum Mechanics (Benjamin Reading).

Milonni, P.W. (1994), The Quantum Vacuum (Academic, Boston).

Misra, B., and Sudarshan, E.C.G. (1977), The Zeno Paradox in Quantum Theory, J. Math. Phys. **18**, 756-763.

Moiseiwitsch, B.L. (1961), The Scattering of Electrons by Atoms in Lectures in Theoretical Physics, Boulder, Vol. III, 142-194, Ed. Brittin, W.E., Downs, B.W. and Downs, J.

Morse, P.M. and Feshbach, H. (1953), Methods of Theoretical Physics, Part I and Part II (McGraw-Hill, New York).

Mostepanenko, V.M. and Trunov, N.N. (1997), The Casimir Effect and Its Applications (Clarendon, Oxford).

Mott, N.F. and Massey, H.S.W. (1965), Theory of Atomic Collisions (Third Edn. Clarendon, Oxford).

Moyal, J.E. (1949), Quantum Mechanics as a Statistical Theory, Proc. Camb. Phil, Soc. **45**, 99-124.

Muirhead, H. (1965), The Physics of Elementary Particles (Pergamon, Oxford).

Nauenberg, M. (2000), Wave Packets: Past and Present in the Physics and Chemistry of Wave Packets, Yeazell, J.A. and Uzer, T. (Eds.) (Wiley, New York).

Negele, J.W., and Orland, H. (1988), Quantum Many-Particle Systems, (Addison-Wesley, Reading).

Newton, R.G. (1966), Scattering Theory of Waves and Particles (McGraw-Hill, New York) (1982) (Second Edn. Springer, New York).

Nishijima, K. (1964), Fundamental Particles (Benjamin, New York).

Normand, J.M. (1980), A Lie Group: Rotations in Quantum Mechanics (North-Holland, Amsterdam).

Omnès, R. (1994), The Interpretation of Quantum Mechanics (Princeton University Press, Princeton); (1999), Understanding Quantum Mechanics (Princeton University Press, Princeton).

O'Raifeartaigh, L., Straumann, N., and Wipf, A. (1991), On the Origin of the Aharonov-Bohm Effect, Comm. Nucl. Part. Phys. **20**, 15-22.

Pake, G.E., and Estle, T.L. (1973), Electron Paramagnetic Resonance (Second Edn., Benjamin, New York).

Park, D. (1974), Introduction to the Quantum Theory (Second Edn. McGraw-Hill, New York).

Patterson, J.D. (1971), Introduction to The Theory of Solid State Physics (Addison-Wesley, Reading).

Pauli, W. (1973), Lectures on Physics: Vol. 5, Wave Mechanics, Ed. Enz., C.P. (The MIT Press, Cambridge) (1958), Die Allgemeine Prinzipien der Wellenmechanik, Handbuch der Physik, Band V, Teil 1 (Springer-Verlag, Berlin); (1980) Eng. Edn. General Principles of Quantum Mechanics (Allied Publ., New Delhi).

Peebles, P.J.E. (1992), Quantum Mechanics (Princeton University Press, Princeton).

Perelomov, A.M. (1986), Generalized Coherent States and Their Applications, (Springer, Berlin).

Peres, A. (1983), Transfer Matrices for One-Dimensional Potentials, J. Math. Phys. **24**, 1110-19. (1993), Quantum Theory: Concepts and Methods, (Kluwer, Dordrecht).

Perkins, D.H. (2000), Introduction to High Energy Physics (Fourth Edn., Cambridge University Press, Cambridge).

Pfeifer, P., and Frohlich, J. (1995), Generalized Time-Energy Uncertainty Relations and Bounds on Lifetimes of Resonances, Rev. Mod. Phys. **67**, 759-779.

Platt, D.E. (1992), A Modern Analysis of the Stern-Gerlach Experiment, Am. J. Phys. **60**, 306-309.

Prange, R.E., and Girvin, S.M. (1990), The Quantum Hall Effect (Springer-Verlag, Berlin).

Prugovecki, E. (1981), Quantum Mechanics in Hilbert Space (Second Edn., Academic, New York).

Quigg, C., and Rosner, J.L. (1979), Quantum Mechanics with Applications to Quarkonia, Phys. Rep. **56**, 167.

Raimes, S. (1972), Many-Electron Theory (North-Holland, Amsterdam).

Raimond, J.M., Brune, M., and Haroche, S. (2001), Manipulating Quantum Entanglement with Atoms and Photons in a Cavity, Rev. Mod. Phys. **73**, 565.

Redhead, M. (1989), Incompleteness, Non-locality and Realism (Clarendon, Oxford).

Reed, M., and Simon, B. (1975). Methods of Modern Mathematical Physics II: Fourier Analysis and Self-Adjointness.

(1978). IV. Analysis of Operators.

(1979). III. Scattering Theory.

(!980). I. Functional Analysis (Revised and Enlarged Edn.) (Academic, New York).

Robertson, H.P. (1929), The Uncertainty Principle, Phys. Rev. **34**, 163-164.

(1930), A General Formulation of the Uncertainty Principle and Its Classical Interpretation, Phys. Rev. **35**, 667-.

(1934), An Indeterminacy Relation for Several Observables and Its Classical Interpretation, Phys. Rev. **46**, 794-801.

Robinett, R.W. (1997), Quantum Mechanics: Classical Results, Modern Systems and Visualized Examples (Oxford University Press, New York).

Rodberg, L.S. and Thaler, R.M. (1967), Introduction to the Quantum Theory of Scattering (Academic, New York).

Rogalsky, M.S., and Palmer, S.B. (1999), Quantum Physics (Gordon and Breach, Amsterdam).

Roman, P. (1965), Advanced Quantum Theory: An Outline of the Fundamental Ideas, (Addison-Wesley, Reading).

Rose, M.E. (1957), Elementary Theory of Angular Momentum, (Wiley, New York).

Rotenberg, M., Bivins, R., Metropolis, N. and Wooten, J.K. (1959), The 3-j and 6-j Symbols (Technology Press, M.I.T., Cambridge).

Sachs, R.G. (1987), The Physics of Time Reversal (The University of Chicago Press, Chicago).

Sakurai, J.J. (1964), Invariance Principles and Elementary Particles (Princeton University Press, Princeton).

(1967) Advanced Quantum Mechanics (Addison-Wesley, Reading).

(1994) Modern Quantum Mechanics (Revised Edition) (Ed. Tuan, S.F.) (Addison-Wesley, Reading).

Santhanam, T.S. (2000), Higher-order Uncertainty Relations, J. Phys. **A33**, L83-L85.

Scadron, M.D. (1991), Advanced Quantum Theory and Its Application Through Feynman Diagrams (Second Edn., Springer-Verlag, Berlin)

Scheck, F. (2007), Quantum Physics (Springer-Verlag, Berlin).

Schiff, L.I. (1968), Quantum Mechanics (Third Edn. McGraw-Hill, New York).

Schlosshauer, M. (2004), Decoherence, the Measurement Problem and Interpretation of Quantum Mechanics, Rev. Mod. Phys. **76**, 1267-1305.

Schrödinger, E. (1930), Zum Heisenbergschen Unschaerfeprinzip, Ber. Kgl. Akad. Wiss. **29**, 296-303. (1935), Discussion of Probability Relations between Separated Systems, Proc. Cam. Phil. Soc. **31**, 555; (1936), ibid, **32**, 446-; (1941), Further Studies on Solving Eigenvalue Problems by Factorization, Proc. Roy Irish. Acad. **A46**, 183-206.

Schulman, L.S.(1981), Techniques and Applications of Path Integration (Wiley, New York)

Schwabl, F. (1992), Quantum Mechanics (Springer-Verlag, Berlin).

Schweber, S.S. (1961), An Introduction to Relativistic Quantum Field Theory (Row, Peterson, Evanston).

Schwinger, J. (1952), On Angular Momentum, U.S. Atomic Energy Commission, NYO-3071, Reprinted in Biedenharn and van Dam.

(2001), Quantum Mechanics: Symbolism of Atomic Measurements, Ed. Englert, B-G. (Springer-Verlag, Berlin).

Selleri, F. (1990), Quantum Paradoxes and Physical Reality (Kluwer, Dordrecht).

Shankar, R. (1994), Principles of Quantum Mechanics (Second Edn. Plenum, New York).

Shapere, A., and Wilczek, F. (Eds.) (1989), Geometric Phases in Physics (World Scientific, Singapore).

Silverman, M.P. (1995), More than One Mystery: Explorations in Quantum Interference (Springer-Verlag, New York).

Sobelman, I.I. (1992), Atomic Spectra and Radiative Transitions (Second Edn., Springer-Verlag, Berlin).

Sparnaay, M.J. (1958), Measurement of Attractive Forces Between Flat Plates, Physica, **24**, 751-764.

Stedman, G.S. (1985), Simple Supersymmetry: II. Factorisation Method in Quantum Mechanics, Eur. J. Phys. **6**,225-231.

Sternberg, S. (1994), Group Theory and Physics, (Cambridge University Press, Cambridge).

Streater, R.F., and Wightman, A.S. (1964), PCT, Spin, Statistics, and All that (Benjamin, New York).

Styer, D.F. et al. (2002), Nine Formulations of Quantum Mechanical, Am. J. Phys. **70**, 288-297.

Sudarshan, E.C.G. (1963), Equivalence of Semiclassical and Quantum Mechanics Descriptions of Statistical Light Beams, Phys. Rev. Lett. **10**, 277-279.

Sudarshan, E.C.G., Chin, C.B., and Bhamathi, G. (1995), Generalized Uncertainty Relations and Characteristic Invariants for the Multimode States, Phys. Rev. **A52**, 43-54.

Sukumar, C.V. (1985), Supersymmetric Quantum Mechanics of One-Dimensional Systems, J. Phys. **A18**, 2917-2936.

Swanson, M.S. (1992), Path Integrals and Quantum Processes (Academic, New York).

Tapster, P.R., Rarity, J.G., and Owens, P.C.M. (1994), Violation of Bell's Inequality Over 4 km of Optical Fiber, Phys. Rev. Lett. **73**, 1923-.

Tanner, G., Richter, K., and Rost, J-M. (2000), The Theory of Two-Electron Atoms: Between Ground State and Complete Fragmentation, Rev. Mod. Phys. **72**, 497-544.

Taylor, J.R. (1972), Scattering Theory (Wiley, New York).

Thirring, W. (1981), A Course in Mathematical Physics, Vol. III: Quantum Mechanics of Atoms and Molecules (Springer-Verlag, New York).

Tinkham, M. (1964), Group Theory and Quantum Mechanics (McGraw-Hill, New York).

Tittel, W., Brendel, J., Zbinden, H. and Gisin, N. (1999), Violation of Bell Inequalities by Photons More Than 10 km. Apart, Phys. Rev. Lett. **81**, 3563-.

Townsend, J.S. (2000), A Modern Approach to Quantum Mechanics (University Science Books, Sausalito).

Tung, W-K. (1985), Group Theory in Physics (World Scientific, Singapore).

Uffink, J. (1993), The Rate of Evolution of a Quantum State, Am. J. Physics **61**, 935-936.

von Neumann, J., and Wigner, E.P. (1929), Phys. Zeits. **30**, 467.

von Neumann, J. (1931), Mathmatische Grundlagen der Quantentheorie (Springer, Berlin); (1955) Mathematical Foundations of Quantum Mechanics. (Princeton University Press, Princeton) (Engl. Translation).

Walker, J.S., and Gathright, J. (1994), Exploring One-Dimensional Quantum Mechanics with Transfer Matrices, Am. J. Phys. **62**, 408-422.

Weihs, G., Jennewein, T., Simon, C., Weinfurter, H., and Zeilinger, A. (1999), Violation of Bell's Inequality under Strict Einstein Locality Condition, Phys. Rev. Lett. **81**, 5039-5043.

Weinberg, S. (1996). The Quantum Theory of Fields, Vol. I (Cambridge University Press, Cambridge).

Weyl, H. (1927), Quantenmechanik and Gruppentheorie, Z. Phys. **46**, 1-46.

(1931), The Theory of Groups and Quantum Mechanics (Dover, New York).

Wheeler, J.A., and Zurek, W.H. (Eds.) (1983), Quantum Theory and Measurement (Princeton University Press, Princeton).

Whittaker, E.T., and Watson, G.N. (1969), A Course of Modern Analysis (Fourth Edn., Cambridge University Press, Cambridge).

Wick, G.C. (1958), Invariance Principles of Nuclear Physics, in Ann. Rev. Nucl. Sci. **8**,1-48, Ed. Segre, E. (Ann. Rev. Inc, PaloAlto).

Wigner, E.P. (1932), On the Quantum Correction for Thermodynamic Equilibrium, Phys. Rev. **40**, 749-759.

(1931) Gruppentheorie und Ihre Anwendung anf die Quantenmechanik der Atomospektren (Vieweg, Braunschweig) (Engl. Trans.) (1959). Group Theory and Its Application to the Quantum Mechanics of Atomic Spectra (Academic, New York).

Witten, E. (1981), Dynamical Breaking of Supersymmetry, Nucl. Phys. **B188**, 513-554.

Wolfe, J.C. (1978), Summary of the Kronig-Penney Electron, Am. J. Phys. **46**, 1012-1014.

Wolfenstein, L. (1956), Ann. Rev. Nucl. Sci. **6**, 43.

Wu, T.Y., and Ohmura, T. (1962), Quantum Theory of Scattering (Prentice Hall, Englewood Cliffs).

Yeazell, J.A., and Uzer, T. (Eds.) (2000). The Physics and Chemistry of Wave Packets (Wiley, New York).

Zachos, C. (2002), Deformation Quantization: Quantum Mechanics Lives and Works in Phase Space, Int. J. Mod. Phys. **A17**, 297-316.

Zettili, N. (2001), Quantum Mechanics Concepts and Applications (Wiley, New York).

Zhang, W.M., Feng, D.H. and Gilmore, R. (1990), Coherent States: Theory and Some Applications, Rev. Mod. Phys. **62**, 867-927.

Ziman, J.M. (1979), Models of Disorder (Cambridge University Press, Cambridge).

Zurek, W.H. (2003), Decoherence, Einselection and the Quantum Origins of the Classical, Rev. Mod. Phys. **75**, 715-775.

Index